THEORETICAL ACOUSTICS

Page	Location	Printed form	Corrected form
181	Eq.(5.1.11)	$y = \ldots$	$Y(x) = \ldots$
290	line 14	$\sqrt{\dfrac{K\omega}{2\rho c C_p}}$	$\sqrt{\dfrac{K\omega}{2\rho c^2 C_p}}$
290	line 15	$\sqrt{\dfrac{\mu\omega}{2\rho c}}$	$\sqrt{\dfrac{\mu\omega}{2\rho c^2}}$
303	Problem 8	$\left[\dfrac{1+i(\Phi/\rho_p\omega)}{1-(c_p/c)^2\sin^2\delta}\right]^2$	$\left[\dfrac{1+i(\Phi/\rho_p\omega)}{1-(c_p/c)^2\sin^2\delta}\right]^{1/2}$
304	line 2	$\sqrt{\tfrac{1}{2}kl_v}$	$\sqrt{2kl_v}$
304	line 2	$\sqrt{\tfrac{1}{2}kl_h}$	$\sqrt{2kl_h}$
304	line 6	$p_h \simeq -i\gamma\alpha kl_h\tau_h$	$p_h \simeq -i\gamma\alpha k(l_h - l_v')\tau_h$
304	line 7	$\tau_h \simeq \dfrac{\gamma-1}{\gamma\alpha}$	$\tau_h \simeq -\dfrac{\gamma-1}{\gamma\alpha}$
304	line 11	$u_{ty} \simeq -(1-i)\dfrac{\mu^2 d_v}{\rho ck}$	$u_{ty} = -(1-i)\dfrac{\mu^2 d_v}{2\rho ck}$
304	line 13	$u_x = 0$	$u_y = 0$
416	line -2	$= \dfrac{VT_c^3 l}{(2\pi)^6\omega_c}$	$= \dfrac{VTl_c^3}{(2\pi)^6\omega_c}$
520	line -2	$\left[(1+i)\dfrac{r}{d_v}\right]^{e^{ik_t z}}$	$\left[(1+i)\dfrac{r}{d_v}\right]e^{ik_t z}$
520	line -1	$\left[(1+i)\dfrac{r}{d_v}\right]^{e^{ik_t z}}$	$\left[(1+i)\dfrac{r}{d_v}\right]e^{ik_t z}$
580	line -14	$(\xi \simeq 1.55)$	$(1/\xi \simeq 1.55)$
702	line 10	..for irrotational fluid motion	..for isentropic gas flow
703	line 2	$p[(\gamma-1)/\gamma]M_0\cos\varphi$	$pM_0\cos\varphi$
709	line 18	$= \dfrac{c_1}{c_1+V}$	$= \dfrac{c_1}{c_1+\Delta V}$
723	Eq.(11.2.12)	$(1-\beta^2)$	$(1-M^2)$
740	Fig. 11.19	$\tau = \dfrac{2\pi}{\Omega}$	$\tau = \dfrac{2\pi}{B\Omega}$
746	Eq.(11.3.23)	$\beta_{nB-m}, \delta_{nB-m}$	$\beta_{nB-l}, \delta_{nB-l}$
799	Eq.(12.2.18)	$\dfrac{\beta}{\omega}$	$-\dfrac{\beta}{\omega}$
803	Eq.(12.3.14)	$\rho\dfrac{\partial b_y}{\partial t} =$	$\rho\dfrac{\partial u_y}{\partial t} =$

THEORETICAL ACOUSTICS

PHILIP M. MORSE

K. UNO INGARD

PRINCETON UNIVERSITY PRESS
PRINCETON, NEW JERSEY

Published by Princeton University Press,
41 William Street,
Princeton, New Jersey 08540
Copyright © 1968 by McGraw-Hill, Inc.
All rights reserved
First Princeton University Press edition, with errata page, 1986
LCC 86-42860
ISBN 0-691-08425-4
ISBN 0-691-02401-4 (pbk.)
Reprinted by arrangement with McGraw-Hill, Inc.

Princeton University Press books are printed on acid-free
paper and meet the guidelines for permanence and durability of the
Committee on Production Guidelines for Book Longevity of the
Council on Library Resources

Printed in the United States of America

9 8 7 6 5

PREFACE

In the thirty years since the textbook "Vibration and Sound" first appeared, the science of acoustics has expanded in many directions. Products of the jet age have added economic incentive to the solution of problems related to the generation and transmission of noise; the behavior of underwater sound now is of commercial as well as of scientific and military importance; new mathematical techniques for the solution of problems involving coupled acoustic systems have been developed and the correspondingly great improvement in acoustical measurement techniques has made the theoretical solutions of more than academic interest. Acoustical measurements are being increasingly used in exploring the properties of matter; the interaction between sound fields and electromagnetic waves is an important part of plasma physics; and magneto-hydrodynamic wave motion is a phenomenon of growing importance in the sciences of meteorology and of astrophysics. The phenomena of acoustics have taken on new importance and significance, both scientifically and technologically.

With these developments in mind, the decisions involved in a revision of "Vibration and Sound" were difficult indeed. The second edition of "Vibration and Sound" is still useful as a fairly elementary text; simple addition of much new material, necessarily more advanced, would destroy its compactness and probably would reduce its utility to beginning students. What has been decided is to write a new book, with a new title. Some of the material in the earlier book has been retained, as introduction to the more advanced portions and to keep the book as self-contained as possible. The result is a graduate, rather than an undergraduate text, and one which we hope will be useful to physical scientists and engineers, who wish to learn about new developments in acoustical theory. As with "Vibration and Sound," each chapter begins with a discussion of the basic physical concepts and the more elementary theory; the later sections can be returned to later for detailed study or for reference.

The choice of new material has had to be limited, of course, by the constraints of size and unity. Experimental equipment and techniques are not discussed; other books cover this field. Specialized aspects of ultrasonics, of underwater sound, and of wave motion in elastic solids are not treated. Coverage is restricted primarily to the subjects of the generation, propagation, absorption, reflection, and scattering of compressional waves in fluid media, in the distortion of such waves by viscous and thermal effects as well as by solid boundaries, and in their coupling through the induced vibration of walls and transmission panels.   New material has been included on the acoustics of moving media, on plasma acoustics, on nonlinear effects, and on the interaction between light and sound.   Much of this material has not previously been presented in unified form.

Theoretical techniques, developed for quantum mechanics, field theory, and communications theory, to mention a few, have been adapted to solve these acoustical problems.   Autocorrelation functions are introduced to deal with the radiation from randomly vibrating surfaces and with the scattering of sound from turbulence, for example.   The Green's function, which enables differential equation and boundary conditions to be combined into an integral equation for the wave motion, is introduced early and is used extensively throughout the rest of the book.   Variational techniques are discussed and are used to obtain improved approximations in many complex problems.   In all of these subjects we have tried to explain and motivate the theoretical methodology, as well as to use it to elucidate the physics.   Problems, illustrating other applications, are given at the end of each chapter.   While questions of rigor could not be deeply explored, it is hoped the book will be its own introduction to the more advanced mathematical techniques which have been used.   Anyone familiar with calculus and vector analysis should be able to understand the mathematical techniques used here; if he is willing also to spend some time and mental effort, he should be able to master them.

The harmonic oscillator and coupled oscillators are reviewed primarily as a simple means of introducing the concepts of resonance, impedance, transient and steady-state motion, and the techniques of the Fourier transform.   Transverse waves on a string are examined in some detail, again as a means of introducing the concepts of wave motion, of the flow of energy and momentum, and as a simple model with which to demonstrate the use of Green's functions and variational methods.   There is a short chapter on the vibratory properties of those solid bodies, bars, membranes, and thin plates, which are of importance in later discussions.

The rest of the book is concerned with acoustic wave motion in fluid media.   The equations of mass, momentum, and energy flow are discussed in some detail and the interrelations between pressure, density, velocity, temperature, and entropy are examined.   The range of validity of the linear wave equation, the origin and consequences of the additional terms present in the Stokes-Navier equation, and the significance of the elements of the stress-energy tensor are developed in a unified manner.   Based on these equations, the chapters on radiation, scattering, transmission, and acoustical coupling follow.   The final chapters are concerned with more recent ramifications of the theory and with an introductory discussion of nonlinear oscillations and the nonlinear behavior of high intensity sound.

The authors are indebted to Professors Herman Feshbach and Wallace Dean for many suggestions regarding the organization and contents of the textual material. The authors must bear the responsibility for any lapses in clarity and the errors that may remain. We would welcome notice from our readers of such errors they may find, so later printings can be corrected.

Philip M. Morse
K. Uno Ingard

CONTENTS

Chapter 5   Bars, Membranes, and Plates

Chapter 6   Acoustic Wave Motion

Chapter 8   The Scattering of Sound

Chapter 9   Sound Waves in Ducts and Rooms

THEORETICAL ACOUSTICS

CHAPTER

I

INTRODUCTORY

I.I DEFINITIONS AND METHODS

The subject of this book is the application of the methods of classical dynamics to the description of acoustical phenomena. Its object is twofold: to present the salient facts regarding acoustical phenomena, in both descriptive and mathematical form, to students and specialists in the field; and to provide an extended series of illustrations of the way mathematical physics can aid in the understanding of a small facet of nature.

The discussion of any problem in science or engineering has two aspects: the physical side, the statement of the facts of the case in everyday language and of the results in a manner that can be checked by experiment; and the mathematical side, the working out of the intermediate steps by means of the symbolized logic of calculus. These two aspects are equally important and are used side by side in every problem, one checking the other.

The solution of the problems that we shall meet in this book will, in general, involve three steps: the posing of the problem, the intermediate symbolic calculations, and the statement of the answer. The stating of the problem to be solved is not always the easiest part of an investigation. One must decide which properties of the system to be studied are important and which can be neglected, what facts must be given in a quantitative manner and what others need only a qualitative statement. When all these decisions are made for problems of the sort discussed in this book, we can write down a statement somewhat as follows: Such and such a system of bodies is acted on by such and such a set of forces.

We next translate this statement in words into a set of equations and solve the equations (if we can).

The mathematical solution must then be translated back into the physical statement of the answer: If we do such and such to the system in question, it will behave in such and such a manner. It is important to realize that the mathematical solution of a set of equations is not the answer to a physical problem; we must translate the solution into physical statements before the problem is finished.

Units

The physical concept that force causes a change in the motion of a body has its mathematical counterpart in the equation

$$F = \frac{d}{dt}(mv) \qquad (1.1.1)$$

In order to link the two aspects of this fact, we must define the physical quantities concerned in a quantitative manner; we must tell how each physical quantity is to be measured and what standard units of measure are to be used. The fundamental quantities, distance, mass, and time, can be measured in any arbitrary units, but for convenience we use those arbitrary units which most of the scientific world is using: the meter, the kilogram, and the second (occasionally, the centimeter, gram, and second). The units of the few other quantities needed, electrical, thermal, etc., will be given when we encounter them.

The units of measure of the other mechanical quantities are defined in terms of these fundamental ones. The equation $F = d(mv)/dt$ is not only the mathematical statement of a physical law, it is also the definition of the unit of measure of a force. It states that the amount of force, measured in *newtons*, equals the rate of change of momentum, in kilogram meters per second per second. If force were measured in other units than newtons, this equation would not be true; an extra numerical factor would have to be placed on one side or the other of the equality sign.

Energy

Another physical concept which we shall often use is that of work, or energy. The wound-up clock spring can exert a force on a gear train for an indefinite length of time if the gears do not move. It is only by motion that the energy inherent in the spring can be expended. The work done by a force on a body equals the distance through which the body is moved by the force times the component of the force in the direction of the motion; and if the force is in newtons and the distance is in meters, the work is given in *joules*. The mathematical statement of this is

$$W = \int \mathbf{F} \cdot \mathbf{ds} \qquad (1.1.2)$$

where both the force $\mathbf{F}$ and the element of distance traveled $\mathbf{ds}$ are vectors and their scalar product is integrated.

If the force is used to increase the velocity of the body on which it acts, the work that it does is stored up in energy of motion of the body and can be given up later when the body slows down again. This energy of motion is called *kinetic energy*, and when measured in joules, it is equal to $mv^2/2$.

If the force is used to overcome the forces inherent in the system—the "springiness" of a spring, the weight of a body, the pull between two unlike charges, etc.—the work can be done without increasing the body's velocity. If we call the inherent force $\mathbf{F}$ (a vector), the work done opposing it is $W = -\int \mathbf{F} \cdot d\mathbf{s}$. In general, this amount depends not only on the initial and final position of the body, but also on the particular way in which the body travels between these positions. For instance, because of friction, it takes more work to run an elevator from the basement to the third floor, back to the first, and then to the fourth than it does to run it directly to the fourth floor.

In certain ideal cases where we can neglect friction, the integral W depends only on the position of the end points of the motion, and in this case it is called the *potential energy* V of the body at its final position with respect to its starting point. If the body moves only in one dimension, its position being given by the coordinate x, then

$$V = -\int F \, dx \qquad \text{or} \qquad F = -\frac{dV}{dx} \qquad (1.1.3)$$

In such a case we can utilize Newton's equation $F = m(dv/dt)$ to obtain a relation between the body's position and its velocity.

$$V = -\int F \, dx = -m \int \frac{dv}{dt} \, dx = -m \int v \, dv = -\tfrac{1}{2}mv^2 + \text{const}$$

or

$$\tfrac{1}{2}mv^2 + V = \text{const} \qquad (1.1.4)$$

This is a mathematical statement of the physical fact that when friction can be neglected, the sum of the kinetic and potential energies of an isolated system is a constant. This proof can be generalized for motion in three dimensions. In case V depends on the three rectangular coordinates x, y, z, the corresponding force $\mathbf{F} = -\text{grad } V$ is a vector, with components

$$F_x = -\frac{\partial V}{\partial x} \qquad F_y = -\frac{\partial V}{\partial y} \qquad \text{and} \qquad F_z = -\frac{\partial V}{\partial z} \qquad (1.1.5)$$

1.2 A LITTLE MATHEMATICS

With these definitions given, we can go back to our discussion of the twofold aspect of a physical problem. Let us take an example.

Suppose we have a mass m on the end of a spring. We find that to keep this mass displaced a distance x from its equilibrium position requires a force of Kx newtons. The farther we push it away from its equilibrium position, the harder the spring tries to bring it back. These are the physical facts about the system. The problem is to discuss its motion.

We see physically that the motion must be of an oscillatory nature, but to obtain any quantitative predictions about the motion we must have recourse to the second aspect, the mathematical method. We set up our equation of motion $F = m(d^2x/dt^2)$, using our physical knowledge of the force of the spring $F = -Kx$.

$$\frac{d^2x}{dt^2} + \omega^2 x = 0 \tag{1.2.1}$$

where ω^2 stands for K/m.

The physical statement corresponding to this equation is that the body's acceleration is always opposite in sign and proportional to its displacement, i.e., is always toward the equilibrium point $x = 0$. As soon as the body goes past this point in one of its swings, it begins to slow down; eventually it stops and then returns to the origin again. It cannot stop at the origin, however, for it cannot begin to slow down until it gets *past* the origin on each swing.

The trigonometric functions

Equation (1.2.1) is a differential equation. Its solution is well known, but since we shall meet more difficult ones later, it is well to examine our method of obtaining its solution, to see what we should do in other cases. We usually state that the solution to Eq. (1.2.1) is

$$x = a_0 \cos(\omega t) + a_1 \sin(\omega t) \tag{1.2.2}$$

but if we had no table of cos or sin, this statement would be of very little help. In fact, the statement that Eq. (1.2.2) is a solution of Eq. (1.2.1) is really only a definition of the symbols cos and sin. We must have more than just symbols in order to compute x for any value of t.

What we do—what is done in solving any differential equation—is to guess the answer and then see if it checks. In most cases the guess was made long ago and the solution is familiar to us, but the guess had nevertheless to be made.

So we shall guess the solution to Eq. (1.2.1). We shall make a pretty general sort of guess, that x can be expanded in a power series in t, and then see if some choice of the coefficients of the series will satisfy the equation. We set

$$x = a_0 + a_1 t + a_2 t^2 + a_3 t^3 + a_4 t^4 + \cdots = \sum_{n=0}^{\infty} a_n t^n$$

where the symbol Σ is shorthand for a sum of terms of the sort shown after the Σ sign, the limits of the sum being indicated below and above the Σ sign. For example,

$$\sum_{n=M}^{N} a_n f[nx] = a_M f[Mx] + a_{M+1} f[(M+1)x] + \cdots$$
$$+ a_{N-1} f[(N-1)x] + a_N f[Nx]$$

For this series, for x to be a solution of Eq. (1.2.1) we must have that d^2x/dt^2, which equals $2a_2 + 6a_3t + 12a_4t^2 + 20a_5t^3 + \cdots$, plus $\omega^2 x$, which equals $\omega^2a_0 + \omega^2a_1t + \omega^2a_2t^2 + \omega^2a_3t^3 + \cdots$, equals zero for every value of t. The only way this can be done is to have the sums of the coefficients of each power of t each separately equal to zero, i.e., to have $a_2 = -\omega^2a_0/2$, $a_3 = -\omega^2a_1/6$, $a_4 = -\omega^2a_2/12 = \omega^4a_0/24$, etc. Therefore the series that satisfies Eq. (1.2.1) is

$$x = a_0\left(1 - \frac{\omega^2t^2}{2} + \frac{\omega^4t^4}{24} - \frac{\omega^6t^6}{720} + \cdots\right)$$
$$+ a_1\left(t - \frac{\omega^2t^3}{6} + \frac{\omega^4t^5}{120} - \frac{\omega^6t^7}{5,040} + \cdots\right)$$

By comparing this with (1.2.2) we see that what we actually mean by the symbols cos, sin is

$$\cos z = 1 - \frac{z^2}{2} + \frac{z^4}{24} - \frac{z^6}{720} + \cdots$$

$$\sin z = z - \frac{z^3}{6} + \frac{z^5}{120} - \frac{z^7}{5,040} + \cdots$$

(1.2.3)

(in the problem $z = \omega t$) and that when we wish to compute values of cos or sin, we use the series expansion to obtain them. For instance,

$$\cos 0 = 1 - 0 + 0 - \cdots = 1$$
$$\cos \tfrac{1}{2} = 1 - 0.125 + 0.003 - \cdots = 0.878$$
$$\cos 1 = 1 - 0.5 + 0.042 - 0.002 + \cdots = 0.540$$

We see that the mathematical solution involves certain arbitrary constants a_0 and a_1. These must be fixed by the physical conditions of the particular experiment we have made, and will be discussed later (Sec. 1.3).

Of course, we usually mean by the symbols cos z, sin z certain ratios between the sides of a right triangle whose oblique angle is z. To make our discussion complete, we must show that the trigonometric definitions correspond to the series given above. In any book on elementary calculus it is shown that the trigonometric functions obey the following relations:

$$\frac{d}{dz}\cos z = -\sin z \qquad \frac{d}{dz}\sin z = \cos z$$

$$\cos 0 = 1 \qquad \sin 0 = 0$$

By combining the first two of these relations, we see that both the sine and cosine of trigonometry obey the equation $d^2y/dz^2 = -y$, which is equivalent to (1.2.1). Taking into account the third and fourth relations, we see that the series solutions of this equation which correspond to the two trigonometric functions must be the ones given in (1.2.3).

Once this is known, we can utilize the trigonometric properties of the sine and cosine to simplify our discussion of the solution (1.2.2) [although, when it comes to computing values, the series (1.2.3) are always used]. For instance, we can say that x is a periodic function of time, repeating its motion every time that ωt increases by 2π or every time t increases by $(2\pi/\omega)$. The value of $2\pi/\omega$ is called the *period* of oscillation of the mass and is denoted by the symbol T_p. The number of periods per second, $\omega/2\pi$, is called the *frequency* of oscillation and is denoted by ν. The quantity $\omega = 2\pi\nu$, which equals the rate of increase of the angle ωt in the trigonometric functions $\cos(\omega t)$ and $\sin(\omega t)$, is called the *angular velocity* of the motion. Remembering the definition of ω, we see that

$$T_p = 2\pi\sqrt{\frac{m}{K}} \qquad \nu = \frac{1}{2\pi}\sqrt{\frac{K}{m}}$$

Bessel functions

More complicated differential equations can be solved by means of the same sort of guess that we used above. For instance, we can solve the equation

$$\frac{d^2y}{dx^2} + \frac{1}{x}\frac{dy}{dx} + \left(1 - \frac{1}{x^2}\right)y = 0$$

by again guessing that $y = a_0 + a_1x + a_2x^2 + a_3x^3 + \cdots$ and setting this guess in the equation. Thus

$$y = a_0 + a_1x + a_2x^2 + a_3x^3 + \cdots$$

minus
$$\frac{1}{x^2}y = -\frac{a_0}{x^2} - \frac{a_1}{x} - a_2 - a_3x - a_4x^2 - a_5x^3 - \cdots$$

plus
$$\frac{1}{x}\frac{dy}{dx} = \frac{a_1}{x} + 2a_2 + 3a_3x + 4a_4x^2 + 5a_5x^3 + \cdots$$

plus
$$\frac{d^2y}{dx^2} = 2a_2 + 6a_3x + 12a_4x^2 + 20a_5x^3 + \cdots$$

must equal zero. Equating coefficients of powers of x to zero as before, we have

$$a_0 = 0 \qquad a_2 = -\frac{a_0}{3} = 0 \qquad a_3 = -\frac{a_1}{8}$$

$$a_4 = 0 \qquad a_5 = -\frac{a_3}{24} = \frac{a_1}{8 \cdot 24} \cdots$$

Therefore

$$y = a_1\left(x - \frac{x^3}{2 \cdot 4} + \frac{x^5}{2 \cdot 4 \cdot 4 \cdot 6} - \frac{x^7}{2 \cdot 4 \cdot 6 \cdot 4 \cdot 6 \cdot 8} \cdots\right)$$

We shall call this series a *Bessel function*, just as we call each of the series in (1.2.3) a trigonometric function. To save having to write the series out every time, we shall give it a symbol. We let

$$J_1(x) = \frac{1}{2}\left(x - \frac{x^3}{2 \cdot 4} + \frac{x^5}{2 \cdot 4 \cdot 4 \cdot 6} - \cdots\right)$$

just as we represent series (1.2.3) by the symbols cos and sin.

Essentially, this series is no more complicated than series (1.2.3). We can compute values of $J_1(x)$ in the same manner as we computed values of cos z.

$$J_1(0) = 0 - \cdots = 0$$

$$J_1(\tfrac{1}{2}) = 0.250 - 0.008 + \cdots = 0.242$$

$$J_1(1) = 0.5 - 0.063 + 0.003 - \cdots = 0.440$$

Once someone has obtained this series, given us the symbol $J_1(x)$, and computed its values, we can use the symbol with as great freedom as we use the symbol cos z. We can say that the solution of (1.2.4) is $y = AJ_1(x)$, where the arbitrary constant A is to be determined by the physical conditions of the problem, as we shall discuss later.

There are other Bessel functions, in addition to the one we have been discussing. The general equation they satisfy,

$$\frac{d^2y}{dx^2} + \frac{1}{x}\frac{dy}{dx} + \left(1 - \frac{m^2}{x^2}\right)y = 0 \qquad (1.2.4)$$

is called the Bessel equation. The solution which is finite at $x = 0$ can be represented in terms of the series

$$J_n(x) = \frac{(\tfrac{1}{2}x)^n}{n!} - \frac{(\tfrac{1}{2}x)^{n+2}}{1!\,(n+1)!} + \frac{(\tfrac{1}{2}x)^{n+4}}{2!\,(n+2)!} - \cdots \qquad (1.2.5)$$

which is called the nth-order Bessel function of x. The solution $J_1(x)$, worked out in detail, is thus the first-order Bessel function.

Addition and differentiation of the series solution will show that there are two fairly simple relationships between Bessel functions of consecutive order,

$$J_{n-1}(x) + J_{n+1}(x) = \frac{2n}{x}J_n(x) \qquad J_{n-1}(x) - J_{n+1}(x) = 2\frac{d}{dx}J_n(x)$$

or

$$(1.2.6)$$

$$J_{n-1}(x) = \frac{d}{dx}J_n(x) + \frac{n}{x}J_n(x) \qquad J_{n+1}(x) = -\frac{d}{dx}J_n(x) + \frac{n}{x}J_n(x)$$

These are called *recursion formulas*. Together they correspond to the differential equation (1.2.4) for $J_n(x)$; successive use of Eqs. (1.2.6) results in

$$\frac{d^2}{dx^2} J_n = \frac{1}{2} \frac{d}{dx} J_{n-1} - \frac{1}{2} \frac{d}{dx} J_{n+1} = -J_n + \frac{n-1}{2x} J_{n-1} + \frac{n+1}{2x} J_{n+1}$$

$$= -J_n + \left[\frac{n-1}{2x} - \frac{n+1}{2x} \right] \frac{d}{dx} J_n + \left[\frac{n(n-1)}{x^2} + \frac{n(n+1)}{x^2} \right] J_n$$

which reduces to the equation for J_n. The recursion formulas also reflect the choice of the coefficient multiplying the x^n term in the series representation of Eq. (1.2.5).

To have these formulas apply for negative values of n as well as for positive values, we must define the functions for negative order:

$$J_{-n}(x) = (-1)^n J_n(x) \qquad \text{for } n \text{ an integer}$$

The validity of the following useful formulas can be demonstrated by differentiation of the series and application of Eqs. (1.2.6).

$$\int J_1(x) \, dx = J_0(x) \qquad \int J_n(x) \, dx = 2 \sum_{m=0}^{\infty} J_{n+2m+1}(x)$$

$$1 = \sum_{m=-\infty}^{\infty} J_{2m}(x) = J_0(x) + 2 \sum_{m=1}^{\infty} J_{2m}(x)$$

$$x = 2 \sum_{m=0}^{\infty} (2m + 1) J_{2m+1}(x)$$

So far we have been talking about one solution of Eq. (1.2.4); since it is a second-order equation, there must be a second solution [just as Eq. (1.2.1) has two solutions]. In the case of the Bessel equation, however, the second solution, which is usually denoted by $N_n(x)$, becomes infinite at $x = 0$. Its properties will be taken up later. We can say here, however, that the complete solution of Eq. (1.2.4) is $AJ_m(x) + BN_m(x)$, where A and B are arbitrary constants.

Complex quantities

Another very useful way of dealing with the solution of (1.2.1) can be obtained by the following chain of reasoning: We utilize the series method to show that a solution of the equation $dy/dz = y$ is the series called the *exponential*:

$$e^z = 1 + z + \frac{z^2}{2} + \frac{z^3}{6} + \frac{z^4}{24} + \frac{z^5}{120} + \cdots$$

sometimes written exp z. By repeated differentiation we see that $x = Ce^{at}$ is a solution of the equation $(d^2x/dt^2) - a^2x = 0$. If a^2 were to equal $-\omega^2$, this equation would be the same as (1.2.1). Therefore we can say that a solution of (1.2.1) is $Ce^{-i\omega t}$, where $i = \sqrt{-1}$ (we could also use $e^{+i\omega t}$).

The function e^{-iz} is a complex number, with real and imaginary parts, and can be represented in the usual manner by a point on the "complex plane" whose abscissa is the real part of the function and whose ordinate is the imaginary part.   It is also represented by the vector drawn from the origin to this point in the complex plane.

If we expand e^{-iz} in its series form,

$$e^{-iz} = \left(1 - \frac{z^2}{2} + \frac{z^4}{24} - \cdots\right) - i\left(z - \frac{z^3}{6} + \frac{z^5}{120} - \cdots\right)$$

we see immediately that the relation between the imaginary exponential and the trigonometric functions is

$$e^{-iz} = \cos z - i \sin z$$
$$\cos z = \tfrac{1}{2}(e^{iz} + e^{-iz}) \tag{1.2.7}$$
$$\sin z = -\tfrac{1}{2}i(e^{iz} - e^{-iz})$$

This shows that the number e^{-iz} can be represented on the complex plane (Fig. 1.1) by a vector of unit length inclined at an angle $-z$ with respect to the real axis.   It also shows that the complex number $a + ib$ can be expressed as $Ae^{i\Phi}$, where $A = |a + ib| = \sqrt{a^2 + b^2}$ is called the *magnitude* of $a + ib$, and Φ is called its *phase*.   For $Ae^{i\Phi} = A \cos \Phi + iA \sin \Phi$, and since $\cos \Phi = a/A$ and $\sin \Phi = b/A$, we have $Ae^{i\Phi} = a + ib$.   The number $a - ib = Ae^{-i\Phi}$ is called the *complex conjugate* of $a + ib$ and is denoted by an asterisk, $(a + ib)^*$.   Thus, if $C = a + ib$ is a complex number, its magnitude is the square root of $C \cdot C^* = |C|^2 = A^2 = a^2 + b^2$.   The factor $e^{i\Phi}$ signifies a rotation of the vector A through an angle Φ in the complex plane.

Thus we see that $Ce^{-i\omega t}$ is a solution of Eq. (1.2.1) and can be expressed as a combination of $\cos(\omega t)$ and $\sin(\omega t)$, as is demanded by Eq. (1.2.2). But this new solution is a complex number, and the results of physics are, in general, real numbers; so the new solution would seem to be of little value to us.   The compactness and ease of manipulation of the complex notation, however, impel us to find a convention which will enable us to express physical quantities in terms of complex numbers.   Two conventions are possible.

One convention is to write down the solution as being $Ce^{-i\omega t}$ and agree that we use only the *real part* of this solution as a representation of the physical quantity.   The other convention is somewhat more sophisticated, but in the end somewhat more convenient; it is to consider energy and other quadratic functions of the variables as the basic physical quantities and require that they be real quantities in the convention.   Since the square of the magnitude of a complex number $|C|^2 = A^2 = a^2 + b^2$ is always real and nonnegative, we can make the convention that when a complex number C is to be squared, we multiply by its complex conjugate $C^* = a - ib$.   This latter convention is particularly convenient for quantities which depend on time through the factor $e^{-i\omega t}$, for then the square of the magnitude of the quantity does not contain the exponential factor, and is thus independent of time. We postpone details as to how this second convention works until Chap. 6 [Eq. (6.2.18)]; by that time sufficient background material will have been presented so that we can see how to use it.

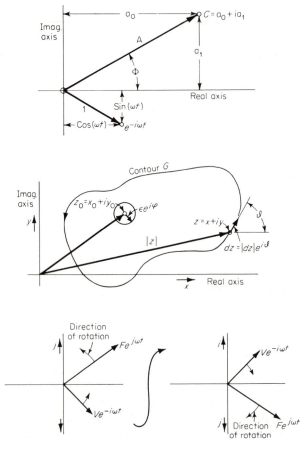

FIGURE 1.1
Representation of complex numbers on the complex plane.
Integration around a contour G in the complex plane.
Relation between i and j.

Until Chap. 6, therefore, we shall use the former convention, that the real part of the complex quantity has physical significance. It is possible to do this with the solution of any *linear* differential equation (i.e., equations containing only the first power of the unknown function and its derivatives); for if a complex function is a solution of a linear differential equation, both its real and imaginary parts by themselves are also solutions. (Why is this not true for nonlinear differential equations?) We could, of course, make a convention that we use only the imaginary part of the function, for the imaginary part is also a solution of the equation, but the usual convention is to take the real part.

We can therefore express solution (1.2.2) as the real part of

$$x = Ce^{-i\omega t} \qquad C = a_0 + ia_1$$

This can be checked by the use of (1.2.7), for

$$Ce^{-i\omega t} = a_0 \cos(\omega t) + ia_1 \cos(\omega t) - ia_0 \sin(\omega t) + a_1 \sin(\omega t)$$

and the real part of this corresponds to (1.2.2). The advantages of this method will become apparent as we use it. For instance, since $C = Ae^{i\Phi}$, $A^2 = a_0{}^2 + a_1{}^2$, $\tan(\Phi) = a_1/a_0$, we see that the real part of $Ce^{-i\omega t} = Ae^{-i(\omega t - \Phi)}$ is $A\cos(\omega t + \Phi)$. Therefore we can express the solution of (1.2.1) in any of four ways:

$$x = a_0 \cos(\omega t) + a_1 \sin(\omega t) = A \cos(\omega t - \Phi)$$
or
$$x = Ce^{-i\omega t} = Ae^{-i(\omega t - \Phi)} \tag{1.2.8}$$

where our present convention requires that we take only the real part of the last expressions. The constant A is called the *amplitude* of oscillation of the mass, since it is the value of the maximum displacement of the mass from equilibrium.

This convention as to the use of complex quantities must be applied with discretion when we come to compute powers and energies, where squares of quantities enter. For instance, in a given problem the power radiated may turn out to be Rv^2, where v is the velocity of a diaphragm and R is a real number. The velocity may be represented by the expression $v = Ce^{-2\pi i\nu t}$, $C = a_0 + ia_1$, just as x is in Eq. (1.2.8). But to compute the average power radiated, we must take the real part of the expression before squaring and averaging,

$$Rv^2 = RA^2 \cos^2(2\pi\nu t - \Phi)$$

according to Eq. (1.2.8). Since the average value of $\cos^2$ is $\frac{1}{2}$, the average power radiated will be

$$\tfrac{1}{2}RA^2 = \tfrac{1}{2}R(a_0{}^2 + a_1{}^2) = \tfrac{1}{2}R\,|C|^2 = \tfrac{1}{2}R\,|v|^2$$
$$|v|^2 = (\text{real part of } v)^2 + (\text{imaginary part of } v)^2$$

If $z = x + iy$ is any complex number, the quantity $|z|$ is called the *magnitude* of z. It is the distance from the origin to the point in the complex plane corresponding to z, and it equals the square root of the sum of the squares of the real and imaginary parts of z. Therefore another important rule concerning the use of complex quantities in physical problems is the following: The *average value of the square* of a quantity represented by the complex function $Ce^{-i\omega t}$ is equal to *one-half of the square of the magnitude of C.*

The angle Φ in Eq. (1.2.8) is called the *phase angle* of the complex quantity C, since it measures the angle of lag of the quantity $Ce^{-i\omega t}$ behind the simple exponential $e^{-i\omega t}$.   We notice that the phase angle of i is 90°, that of -1 is 180°, and that of $-i$ is 270°, or 90° lead.   If $x = Ce^{-i\omega t}$, the velocity $v = dx/dt = -i\omega Ce^{-i\omega t}$ leads the displacement x by 90°, as is indicated by the fact that $v = (-i)\omega x$.

In the present volume we use the letter i to stand for $\sqrt{-1}$ and the symbol $e^{-i\omega t}$ to express simple-harmonic dependence on time.   Many books on electrical engineering use the symbol j instead of i and $e^{j\omega t}$ with the positive sign in the exponential.   Since we intend to take the real part of the result, the choice of symbol and sign is but a convention; either convention is satisfactory if we are consistent about it.

As long as we are studying simple systems, with displacements a function only of time, there is little to choose between the two conventions; as a matter of fact, the positive exponential $e^{j\omega t}$ would be slightly preferable. However, as soon as we come to study wave motion, with displacement a function of position as well as of time, it turns out that the form involving the negative exponential $e^{-i\omega t}$ is rather more satisfactory than the positive. This will become apparent later.

Since we are going to deal with problems of wave motion in this book, we shall consistently use the negative exponential $e^{-i\omega t}$.   And since this convention differs from the positive exponential used in most electrical engineering books, we shall use the symbol i instead of j, to emphasize the difference.   It will, however, always be possible to *transform any of the formulas developed in this book into the usual electrical engineering notation by replacing every i in the formulas by $-j$.*   We might, if we wish, consider i and j as the two roots of -1, so that

$$i = -j \qquad (i)^2 = (j)^2 = -1$$

In this notation, the impedance of a circuit with resistance and inductance in series will be $R - i\omega L$.

Another example of the interconnections between the various functions we have been describing, and how they can be extended by the use of the complex plane, is the relationship between the exponential function and the Bessel functions of Eq. (1.2.5).   By expanding the exponential in a power series in z and t, and then by equating coefficients of the various powers of t, we can show that

$$\exp\left[\tfrac{1}{2}z\left(t - \frac{1}{t}\right)\right] = \sum_{n=-\infty}^{\infty} t^n J_n(z) = J_0(z) + \sum_{n=1}^{\infty}\left[t^n + \left(\frac{-1}{t}\right)^n\right] J_n(z)$$

Next, by inserting $ie^{i\theta}$ for t on both sides of this equation, and by using Eq.

(1.2.7), we obtain a relationship between the exponential, the trigonometric, and the Bessel functions,

$$e^{iz \cos \theta} = J_0(z) + 2 \sum_{n=1}^{n} i^n \cos (n\theta) J_n(z) \qquad (1.2.9)$$

which will be of considerable use later.

Incidentally, for future reference, we should point out that the *hyperbolic* functions are related to the real exponential as the trigonometric functions are to the complex exponential.

$$\cosh x = \tfrac{1}{2}(e^x + e^{-x}) = \cos (ix) \qquad \sinh x = \tfrac{1}{2}(e^x - e^{-x}) = -i \sin (ix)$$
$$\qquad (1.2.10)$$

$$\cosh (ix) = \cos x \qquad \sinh (ix) = i \sin x \qquad \tanh x = \frac{\sinh x}{\cosh x} = -i \tan (ix)$$

Sometimes it is more convenient to use hyperbolic functions than trigonometric ones; as Eq. (1.2.10) has just demonstrated, one can easily transform back and forth from one to the other.

Other solutions

Power series are not always the best guesses for the solutions of differential equations. Now that we have defined the exponential function, we can sometimes express solutions of other equations in terms of exponentials. Consider the equation

$$\frac{d^2x}{dt^2} + n^2x = ae^{-i\omega t}$$

where, to represent a physical problem, we must use our convention on the right-hand side of the equation. This could be solved by expanding the exponential in a series and solving for the series for x as before. But since an exponential is in the equation, and since we know that Ce^{-int} is the solution of this equation when a is zero, it will be simpler to guess that

$$x = Ce^{-int} + Be^{-i\omega t}$$

Setting this in the equation, we have that $n^2x = n^2Ce^{-int} + n^2Be^{-i\omega t}$ plus $d^2x/dt^2 = -n^2Ce^{-int} - \omega^2Be^{-i\omega t}$ must equal $ae^{-i\omega t}$. This means that $B = a/(n^2 - \omega^2)$. If we have used our convention on the term $ae^{-i\omega t}$, we must use it on the answer. The real part of

$$x = Ce^{-int} + \frac{a}{n^2 - \omega^2} e^{-i\omega t}$$

is

$$x = a_0 \cos (nt) + a_1 \sin (nt) + \frac{a}{n^2 - \omega^2} \cos (\omega t) \qquad (1.2.11)$$

if a is a real quantity.   If a is complex and equals $De^{i\psi}$, we can write

$$x = A \cos (nt - \Phi) + \frac{D}{n^2 - \omega^2} \cos (\omega t - \psi)$$

We notice again that in (1.2.11) our solution has two arbitrary constants, a_0 and a_1.   The arbitrariness corresponds to the fact that this solution must represent all the possible sorts of motion which the system can have when it is acted on by the forces implied by the equation.   A mass on a spring can have different motions, depending on how it is started into motion at time $t = 0$. Therefore the particular values of the arbitrary constants a_0 and a_1 in (1.2.2) are determined entirely by the physical statements as to how the system was started into motion.   These physical statements are called *initial conditions* and are usually stated by giving the position and velocity of the system at $t = 0$.   More will be said on this point in the next section.

Contour integrals

In a number of cases discussed in this volume it will be necessary to use integrals of complex quantities of the general form $\int F(z)\, dz$, where $F = U + iV$ is a function of the complex variable $z = x + iy$.   Since z can vary over the xy plane, it will in general be necessary to specify the path over which the integration is carried out.   A natural extension of our usual definition of integration indicates that the integral is the limit of a series of terms, a typical term being the product of the value of $F(z)$ for a value of z on the path, and a quantity $dz = |dz|e^{i\vartheta}$, with length $|dz|$ and with phase angle ϑ determined by the direction of the tangent to the path at the point z.   The integral is the limit of such a sum as $|dz|$ goes to zero.   This is shown in Fig. 1.1.   Another definition is that $\int F\, dz = \int (U + iV)(dx + i\, dy)$, where x and y are related in such a manner that the integral follows the chosen path.

When the path chosen is a closed one, the integral is called a *contour integral* (labeled $\oint$), and the chosen path is called a *contour*.   Contour integrals have certain remarkable properties, proved in standard texts on theory of functions of a complex variable.   We have space here to outline only some of these properties.   For simplicity it will be assumed that the function $F(z)$ to be dealt with is a smoothly varying, reasonable sort of function over most of the z plane.

The value of a contour integral depends largely on whether the integrand $F(z)$ becomes zero or infinity for some value or values of z *inside* the contour. Near such points, $F(z)$ would take on the form $(z - z_0)^n R(z)$, where $R(z_0)$ is neither zero nor infinity and $R(z)$ is not discontinuous near z_0.   When n is *not* an integer, positive or negative, the point z_0 is called a *branch point*, and the problem becomes more complicated than is necessary to discuss here. When n is a *negative integer*, the point z_0 is called a *pole* of $F(z)$, the pole for $n = -1$ being called a *simple pole* and that for $n = -2, -3, \ldots$ being called a *pole* of second, third, etc., *order*.   The statement at the beginning of this paragraph can now be made more specific: The value of the contour integral $\oint F(z)\,dz$ depends on the behavior of F at the poles and branch points that happen to be inside the contour.

When F has *no* branch points or poles inside the contour, the value of the contour integral is *zero*. This can be verified by a tedious bit of algebraic manipulation for those cases where F can be expressed as a convergent power series in z.

When F has a simple pole at $z = z_0$, but has no other pole or branch point inside the contour (Fig. 1.1), the contour integral can be shown to be equal to that around a vanishingly small circle drawn about z_0. Since z_0 is specified as being a simple pole, $F(z)$ has the form $R(z)/(z - z_0)$ near z_0, where $R(z)$ is continuous and finite in value near z_0. We can write the equation for the resulting circle as $z - z_0 = \epsilon e^{i\varphi}$ and $dz = i\epsilon e^{i\varphi}\,d\varphi$, where ϵ is vanishingly small. The contour integral then reduces to

$$\oint F(z)\,dz = R(z_0)\int_0^{2\pi} \frac{i\epsilon e^{i\varphi}\,d\varphi}{\epsilon e^{i\varphi}} = iR(z_0)\int_0^{2\pi} d\varphi = 2\pi i R(z_0)$$

where $R(z_0) = \lim_{z \to z_0}[(z - z_0)F(z)]$ is called the *residue* of $F(z)$ at its simple pole z_0.

When $F(z)$ has N simple poles, at $z_0, z_1, \ldots, z_N$, and no other poles or branch points, the contour integral $\oint F(z)\,dz$ equals $2\pi i$ times the *sum of the residues of F at the poles that are inside* the chosen contour. We note that the direction of integration around the contour is counterclockwise; changing the direction will change the sign of the result.

This is a remarkably simple result; in fact, it seems at first too simple to be true, until we realize that the function $F(z)$ we are dealing with is a very specialized form of function. F is not just any complex function of x and y; its dependence on these variables is severely limited by the requirement that it be a function of $z = x + iy$. It can have the form $bz^2 + (C/z)$ or $z \sin z$. but *not* the form $x \sin(z) + iy$ or $|z|$. For such a specialized function $F(z)$, which is called an *analytic* function, the positions of its branch points and poles, and its behavior near these branch points and poles, completely determine its behavior *everywhere else on the complex plane.* When this is understood, the result we have quoted for the value of a contour integral does not seem quite so surprising.

As an example of the rule of residues, we can take the integral

$$\oint \sin(z)\,dz/(z^2 - a^2)$$

where the contour is a circle of radius greater than a, with center at the origin $z = 0$. The integrand has two simple poles, at $z = a$ and $z = -a$, with residues $(1/2a)\sin a$ and $-(1/2a)\sin(-a)$. The result is therefore

$$\oint \frac{\sin z}{z^2 - a^2}\,dz = 2\pi i\left[\frac{\sin a}{a}\right]$$

A special case of the residue theorem can be stated in equation form:

$$\oint \frac{R(z)}{z - z_0}\, dz = 2\pi i R(z_0) \tag{1.2.12}$$

where the contour of integration does *not* enclose a branch point or pole of $R(z)$. By differentiating once with respect to z_0 on both sides of the equation, we obtain the equation

$$\oint \frac{R(z)}{(z - z_0)^2}\, dz = 2\pi i R'(z_0) \tag{1.2.13}$$

where $R'(z) = dR(z)/dz$. This indicates how we can evaluate contour integrals around poles of second order. Cases for poles of higher order may be dealt with by further differentiation of Eq. (1.2.13).

Infinite integrals

A class of integral often encountered may be evaluated in terms of contour integrals. Consider the integral

$$\int_{-\infty}^{\infty} F(z)e^{-izt}\, dz$$

where the integral is taken along the real axis. If $F(z)$ has no branch points or poles of higher order, if the simple poles of F are not on the real axis, and if $F(z)$ goes to zero as $|z|$ goes to infinity, the integral can be made into a contour integral by returning from $+\infty$ to $-\infty$ along a semicircle of infinite radius. When t is positive (to make it more general we should say, when the *real part* of t is positive, but let us assume that t is real), the exponential factor ensures that the integral along this semicircle is zero if the semicircle encloses the *lower* half of the z plane. For along this path, $z = \rho e^{i\varphi} = \rho \cos\varphi + i\rho \sin\varphi$, with ρ infinitely large and with φ going from 0 to $-\pi$, so that $\sin\varphi$ is negative. Therefore, along the semicircle,

$$e^{-izt} = e^{-i\rho t \cos\varphi + \rho t \sin\varphi}$$

As long as $\sin\varphi$ is negative this vanishes, because of the extremely large value of ρ. Therefore the addition of the integral around the semicircle turns it into a contour integral but does not change its value.

The final result is that the integral equals $-2\pi i$ times the sum of the residues of $F(z)e^{-izt}$ at every one of its poles located *below* the real axis. The negative sign is due to the fact that the contour in this case is in the *clockwise* direction.

When t is negative, we complete the contour along a semicircle enclosing the upper half of the z plane, which now has the vanishing integral, and the resulting value is $2\pi i$ times the sum of the residues at each of the poles *above* the real axis. The limitations on F have already been given.

As an example, we can write down the result

$$\frac{a^2}{2\pi} \int_{-\infty}^{\infty} \frac{e^{-izt}}{z^2 + a^2} \, dz = \begin{cases} (a/2)e^{at} & t < 0 \\ (a/2)e^{-at} & t > 0 \end{cases}$$

practically by inspection. Another result of interest later is

$$\frac{ia}{2\pi} \int_{-\infty}^{\infty} \frac{e^{-izt}}{z + ia} \, dz = \begin{cases} 0 & t < 0 \\ ae^{-at} & t > 0 \end{cases} \tag{1.2.14}$$

In some cases of interest, F has poles on the real axis. These cases can also be treated by making the convention that the integration is not exactly along the real axis but along a line an infinitesimal amount *above* the real axis. With this convention, and for functions $F(z)$ that have *only simple poles* and that vanish at $|z| \to \infty$, we have the general formula

$$\int_{-\infty}^{+\infty} F(z)e^{-izt} \, dz = \begin{cases} -2\pi i \begin{pmatrix} \text{sum of residues of} \\ Fe^{-izt} \text{ at all its} \\ \text{poles } on \text{ and } below \\ \text{the real axis} \end{pmatrix} & t > 0 \\ \\ +2\pi i \begin{pmatrix} \text{sum of residues of} \\ Fe^{-izt} \text{ at all its} \\ \text{poles } above \text{ the} \\ \text{real axis} \end{pmatrix} & t < 0 \end{cases} \tag{1.2.15}$$

As an example, we can write down the very useful formulas

$$u(t) = \frac{-1}{2\pi i} \int_{-\infty}^{\infty} e^{-izt} \frac{dz}{z} = \begin{cases} 0 & t < 0 \\ 1 & t > 0 \end{cases} \tag{1.2.16}$$

$$\frac{-a}{2\pi} \int_{-\infty}^{\infty} \frac{e^{-izt}}{z^2 - a^2} \, dz = \begin{cases} 0 & t < 0 \\ \sin(at) & t > 0 \end{cases} \tag{1.2.17}$$

$$-\frac{1}{2\pi i} \int_{-\infty}^{\infty} \frac{e^{-izt}}{z - a} \, dz = \begin{cases} 0 & t < 0 \\ e^{-iat} & t > 0 \end{cases} \tag{1.2.18}$$

1.3 VIBRATORY MOTION

A final introductory topic, appropriate for this chapter, has to do with how we are to describe, mathematically, the motion of an acoustic system, and how such a description corresponds to the various measurements we can make on the system. Most systems of interest in acoustics vibrate; they move to and fro rather than just in one direction. With a few exceptions, such as the single pulse produced by an explosion, their mean-square velocity is much larger than the square of their mean velocity. In many cases we can ignore the mean velocity entirely and concentrate on analyzing their vibrations about their average position.

The energy equation

The oscillator of Eq. (1.2.1) is the simplest example; here the force on the mass point is proportional to its displacement and the motion is simple-harmonic, i.e., proportional to a sine or cosine of ωt. The displacement, as well as the velocity, averages to zero over a period of the oscillation. Next in order of complication would be a mass point moving in one dimension, acted on by any sort of force. Such a force may be of several different types. One type, like the force of the spring of Eq. (1.2.1), depends only on the value of the displacement x of the mass at the given instant; we label it $F(x)$ and call it the *internal* force since it depends only on the state of the oscillator. A force imposed from outside, on the other hand, would depend explicitly on time t, though it may also depend on x; we call this external force $G(x,t)$. In addition, there may be a force, or combination of forces, which depends on the velocity $v = dx/dt$ of the mass (the viscous drag of the air is such a force); we label this frictional force $D(v,t)$.

The differential equation of motion

$$m \frac{d^2x}{dt^2} - F(x) = G(x,t) + D(v,t) \qquad (1.3.1)$$

then states that the acceleration of the mass is proportional to the sum of all the forces acting. A somewhat more enlightening form of this relationship can be obtained by integrating the equation with respect to x. As indicated in Eq. (1.1.4), the left-hand side can be integrated to give the sum $H(v,x)$ of the kinetic energy, $\frac{1}{2}mv^2$, and the potential energy, $-\int F\,dx = V(x)$; thus the energy equation is

$$H(v,x) = \tfrac{1}{2}mv^2 + V(x) = \int_0^t [G(x,t) + D(v,t)]\,v\,dt + E \qquad (1.3.2)$$

where E is the value of H at $t = 0$. This states that x and v so adjust themselves at each succeeding instant that the change in H is equal to the change in the right-hand integral. The sum H of the kinetic and potential energy of the mass point, when explicitly written as a function of x and v, as was done here, is called the *total energy* of the system.

The effect of the frictional-force term D is such as to diminish H continually; the external force G may increase or decrease H. When *both* G and D are zero, when the mass moves under the influence of forces which depend *only* on the position of the mass, then H is equal to the constant value E. Therefore force $F(x)$, responsible for the potential energy $V(x)$, is called a *conservative* force, because energy is conserved when it alone acts; and a system acted on solely by conservative forces is called a conservative system. For the time being, we shall concentrate on the one-dimensional conservative system.

Periodic motion

The nature of the motion of such a system depends, of course, on the shape of the potential energy curve, V against x. If it has a minimum, where $dV/dx \equiv -F = 0$, the mass can stay at rest at this point of equilibrium (which can be placed for convenience at the origin $x = 0$). If it really is a minimum, then d^2V/dx^2 would be positive there, so that if the mass is displaced a small amount dx, the force on it, $(dF/dx)\, dx = -(d^2V/dx^2)\, dx$, would be in the

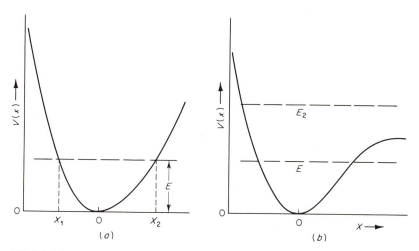

FIGURE 1.2
Potential-energy curves for bounded (curves a and b, E) and unbounded (curve b, E_2) motion of a one-dimensional oscillator.

opposite direction of dx, forcing the mass to return to equilibrium. Thus a minimum of the potential energy is a point of *stable* equilibrium.

In a conservative system, when D and G are zero, the energy equation, $\frac{1}{2}mv^2 + V(x) = E$ (a constant), indicates that at every instant the kinetic energy must adjust itself so that it equals $E - V(x)$. For a stable system the kinetic energy (and therefore the speed of the mass point) is a maximum at the minimum point of V, at $x = 0$; it is zero whenever $V = E$, for example, at x_1 and x_2 in Fig. 1.2a. We can make the relationship between v and x in a conservative system more specific by writing

$$v = \frac{dx}{dt} = \pm \sqrt{\frac{2}{m}[E - V(x)]} \qquad G \text{ and } D \text{ zero}$$

and more graphic by plotting v as a function of x, as in Fig. 1.3. The vx plane is called the *phase plane* (no relation to the phase angle Φ), and the curve of v against x, for a given value of E, is called a phase contour. The position and velocity of the system at any given instant are represented by a point on the phase plane, called the system point; as time goes on this system point traces out a curve on the phase plane, the phase contour.   The speed of traverse of the contour is determined by the ordinate of the point, for v is

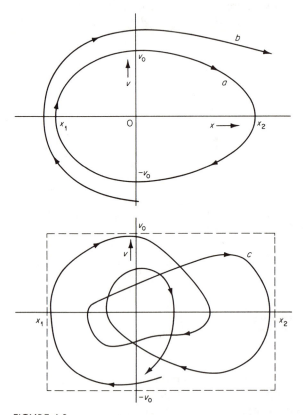

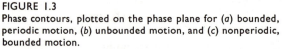

FIGURE 1.3
Phase contours, plotted on the phase plane for (a) bounded, periodic motion, (b) unbounded motion, and (c) nonperiodic, bounded motion.

the rate of change of x with time.   The contour crosses the v axis ($x = 0$) at $v = \pm v_0$, where the speed is maximum; it crosses the x axis ($v = 0$) at $x = x_1$ and x_2, where $V(x) = E$ and the mass is farthest from the origin.

   If the potential energy has only the one minimum at $x = 0$ and rises monotonically with increasing $|x|$, approaching infinity as $|x| \to \infty$ (as in Fig. 1.2a), the contour is a closed loop, and neither $|v|$ nor $|x|$ will ever become infinite.   In this case the motion is *bounded* and, as we shall see shortly, is periodic.   On the other hand, if $V(x)$ does not go to infinity as $x \to \infty$ (as shown in Fig. 1.2b), then whenever E is larger than the asymptotic value of V (as with $E = E_2$), the contour is an open curve (as shown in contour b of Fig. 1.3), and the motion is unbounded.

To see that the bounded motion is periodic, we note that we can integrate for t. Since

$$v = \frac{dx}{dt} = \sqrt{\frac{2}{m}[E - V(x)]}$$

then

$$t - t_1 = \sqrt{\frac{m}{2}} \int_{x_1}^{x} \frac{dx}{\sqrt{E - V(x)}}$$

(1.3.3)

gives t as a function of x; by inversion we can obtain x as a function of t. If the path of the system point on the phase plane is a closed loop, as with Fig. 1.3a, the point retraces its path periodically, once a cycle. The mean value of v over the cycle is zero, the mean values of v^2, of x, and of x^2 are finite and are the same for each cycle. Each traverse of the contour is exactly like all the others, which is what we mean when we call the motion *periodic*.

Thus, when the one-dimensional system is conservative, when G and D are zero, the total energy $H(v,x)$ of the system is constant, and the motion, if it is bounded, is periodic. From the symmetry of the phase contour with respect to the x axis, we see that the time required to go from x_1 to x_2 is equal to the time to go from x_2 back to x_1 [from Eq. (1.3.3)], so that the *period* T_p of the oscillation, the length of time it takes the mass point to go through a full cycle, is

$$T_p = 2(t_2 - t_1) = \sqrt{2m} \int_{x_1}^{x_2} \frac{dx}{\sqrt{E - V(x)}}$$

(1.3.4)

where $E = V(x_1) = V(x_2)$.

If the restoring force is linear, $F(x) = -Kx$, the potential energy $V = \frac{1}{2}m\omega_0^2 x^2$, where $\omega_0^2 = K/m$, is quadratic in x and Eq. (1.3.3) becomes

$$t - t_1 = \frac{1}{\omega_0} \int_{x_1}^{x} \frac{dx}{\sqrt{x_1^2 - x^2}} = \frac{1}{\omega_0} \cos^{-1} \frac{x}{x_1}$$

(1.3.5)

where we have set $E = \frac{1}{2}m\omega_0^2 x_1^2$, relating E and the maximum displacement x_1. Inversion of this result produces

$$x = x_1 \cos \omega_0(t - t_1)$$

(1.3.6)

which is equivalent to Eq. (1.2.2). An oscillation of this sort, with displacement which is a sinusoidal function of time (and thus with velocity and acceleration which also are sinusoidal), is called *simple-harmonic* oscillation. As was shown in Sec. 1.2, and was shown again in the preceding paragraphs, the motion is periodic, with period $T_p = 2\pi/\omega_0 = 2\pi\sqrt{m/K}$.

Simple-harmonic motion is a particularly simple sort of periodic motion; the phase contour is an ellipse, or if the scale of the v coordinate of the phase plane is adjusted properly, it is a circle. If we plot v/ω_0 against x, we obtain

$$\frac{v}{\omega_0} = \frac{1}{\omega_0}\sqrt{\frac{2}{m}(E - \tfrac{1}{2}m\omega_0^2 x^2)} = \sqrt{x_1^2 - x^2} \tag{1.3.7}$$

which is the equation of a circle of radius $x_1 = \sqrt{2E/m\omega_0^2}$, with center at the origin. In this simple case the system point moves with uniform speed around a circle of radius proportional to the square root of the system's energy. The angular velocity ω_0 of the radius vector to the system point is, in this case, independent both of time and of the radius of the circle (and thus of E). This is not true of other periodic oscillations, when $V(x)$ is not a quadratic function of x.

Fourier series representation

When V is not proportional to x^2, the free oscillations of the mass point are still periodic as long as the motion is bounded, but x is not a simple sinusoidal function of t. However, the motion can be expressed in terms of a series of trigonometric functions, a generalization of (1.2.8).

$$x(t) = \sum_{n=0}^{\infty} A_n \cos(n\omega t - \Phi_n) = \sum_{n=0}^{\infty}[a_n \cos(n\omega t) + b_n \sin(n\omega t)]$$

$$= \sum_{n=-\infty}^{\infty} C_n e^{-in\omega t} \tag{1.3.8}$$

where $A_n^2 = a_n^2 + b_n^2$, $a_n = A_n \cos\Phi_n$, $b_n = A_n \sin\Phi_n$
$C_0 = a_0$, $C_n = \tfrac{1}{2}(a_n + ib_n)$, $C_{-n} = \tfrac{1}{2}(a_n - ib_n)$ for $n \neq 0$

If all the C's are real, then all the b's are zero, $x(t)$ is real and an even function of t, that is, $x(-t) = x(t)$. If $C_{-n} = C_n^*$ (i.e., if all the a's and b's are real), then $x(t)$ is real but not necessarily symmetric about $t = 0$.

If we take $\Phi_0 = 0$ in Eq. (1.3.8), then $A_0 = a_0$ is the mean value of x, averaged over the cycle. The values of the amplitudes A_n and of the phase angles Φ_n (or of the C's) are, of course, determined by the value of E and the form of $V(x)$. How they are computed will be taken up shortly. Here we wish only to discuss the general properties of the series, assuming that the values of the A's and Φ's are known. As long as only integer values of n are included in the sums of Eq. (1.3.8), each term in the sum will return to its original value when t increases by $T_p = 2\pi/\omega$. The nth term will meanwhile have executed n cycles, but at the end of time T_p it will be back where it was originally, ready to start the $(n + 1)$st cycle. Therefore any series of the general form of Eq. (1.3.8) is periodic in time, with period $T_p = 2\pi/\omega$ (provided that the series converges, of course). Fourier first showed that the inverse is true; that any periodic motion which is physically feasible (i.e., continuous and bounded) can be expressed in terms of such a series; therefore the series is called a *Fourier series*.

The quantity $1/T_p = \omega/2\pi$ is called the *fundamental* frequency of the oscillation; the higher terms, for $n > 1$, are called *harmonics*. Motion of the sort which can be represented by a Fourier series is the basis of all musical sounds. We shall see later in this volume that the vibrations of string and wind instruments can be (approximately) so represented and that the "tone quality" of the sounds produced by these instruments is determined to a great extent by the relative amplitudes of the various harmonics present in the series. Therefore the representation of periodic motion in terms of a Fourier series is more than a mathematical dodge to express a complicated function in terms of a series of simple functions. It corresponds somehow to the way we hear and distinguish musical sounds; specifying the relative amplitudes and phases of the different harmonics is an appropriate way to specify the tone quality of a musical sound (or a nonmusical one, as long as it is periodic). Experimentally, we use electric filters to analyze periodic sounds into their component harmonics, though the analyzers usually measure only the amplitudes A_n, and not the phase angles Φ_n.

The mathematical counterpart to the filter analyzer is based on the integral properties of the trigonometric functions,

$$\int_0^{T_p} \cos(n\omega t) \cos(m\omega t)\, dt = \tfrac{1}{2} T_p \delta_{nm} = \int_0^{T_p} \sin(n\omega t) \sin(m\omega t)\, dt$$

$$\int_0^{T_p} \cos(n\omega t) \sin(m\omega t)\, dt = 0 \qquad n \geqslant 0;\, m > 0 \qquad (1.3.9)$$

where $T_p = 2\pi/\omega$, and the *Kronecker delta symbol* δ_{nm} is zero if $n \neq m$ and unity if $n = m$. Thus, if we multiply Eq. (1.3.8) by $\cos(m\omega t)\, dt$ and integrate over a cycle, the only term in the second form of the series which does not integrate to zero is the term with $n = m$, which integrates to $\tfrac{1}{2} T_p a_m$. Therefore, if we know the form of the curve of $x(t)$ as periodic function of t for one cycle, the coefficients a_n, b_n in the Fourier series expansion of $x(t)$ can be obtained by computing the integrals.

$$a_m = \frac{2}{T_p} \int_0^{T_p} x(t) \cos(m\omega t)\, dt \qquad b_m = \frac{2}{T_p} \int_0^{T_p} x(t) \sin(m\omega t)\, dt$$

$$a_0 = \frac{1}{T_p} \int_0^{T_p} x(t)\, dt \qquad\qquad\qquad (1.3.10)$$

where $T_p = 2\pi/\omega$. Alternatively,

$$C_m = \frac{1}{T_p} \int_0^{T_p} x(t) e^{+in\omega t}\, dt$$

The electric filter analyzer essentially carries out these integrals when it measures the amplitude of the mth harmonic in a periodic sound.

To return to the one-dimensional, conservative system of Eq. (1.3.3), we have shown that its motion is periodic and that its period can be computed in terms of its potential energy and its total energy E. Its motion can be expressed in terms of a Fourier series of the sort shown in Eq. (1.3.8). How the coefficients a_n, b_n are computed from $V(x)$ and E will be taken up in Chap. 2.

Mean and mean-square displacement and velocity may be given in terms of the Fourier series coefficients.

$$\langle x \rangle = a_0 \qquad \langle x^2 \rangle = a_0{}^2 + \tfrac{1}{2} \sum_{n=1}^{\infty} A_n{}^2$$

$$\langle v \rangle = 0 \qquad \langle v^2 \rangle = \tfrac{1}{2} \sum_{n=1}^{\infty} n^2 \omega^2 A_n{}^2$$

We note that these averages depend on the amplitudes A_n, but not on the phases Φ_n. We note also that the ratio between $\langle v^2 \rangle$ and $\langle x^2 \rangle$ provides a rough measure of the degree of presence of the higher harmonics. The ratio $\langle v^2 \rangle / \langle x^2 \rangle$ is equal to ω^2 for simple-harmonic oscillations (for which $A_n = 0$ for $n > 1$). If a large number of harmonics are present in the motion, $\langle v^2 \rangle / \langle x^2 \rangle$ would be considerably large than ω^2.

We can now return to Eq. (1.2.9) to point out that the right-hand side is the Fourier series expansion of the periodic function $e^{iz \cos \theta}$. This, together with Eqs. (1.3.10) (setting $\theta = \omega t$), provides us with *integral representations* of the Bessel functions

$$J_n(z) = \tfrac{1}{2} \pi i^n \int_0^{2\pi} \exp{(iz \cos \theta)} \cos{(n\theta)}\, d\theta \qquad (1.3.11)$$

giving the function $J_n(z)$ in terms of an integral of the exponential of a trigonometric function. These expressions will be of use in later calculations.

Nonperiodic vibration; autocorrelation

When the vibrating system has several degrees of freedom, such as the systems to be discussed in Chap. 3, the equations of motion are less simple than Eq. (1.3.1), and the motion is correspondingly more complicated. The complete description of the configuration of the system requires the specification of more than one displacement (let it be $N > 1$), some of the coordinates being components of linear displacements in several dimensions, and some possibly being angles of rotation of solid portions of the system. The nth displacement component, whether it be angle or distance, will have a corresponding moment of inertia or effective mass, which we can call m_n, and a corresponding set of torques or forces, F_n, G_n, and D_n, corresponding to Eq. (1.3.1). The displacements can be considered to be the N components of a vector displacement in N-dimensional space, which vector can be symbolized by $\mathbf{r}$. The corresponding N velocities can also be treated as components of a vector $\mathbf{v} = d\mathbf{r}/dt$.

The equations of motion for each degree of freedom, the generalization of Eqs. (1.3.1), are

$$m_n \frac{d^2 x_n}{dt^2} - F_n(\mathbf{r}) = G_n(\mathbf{r},t) + D_n(\mathbf{v},t) \qquad n = 1, 2, \ldots, N$$

where the forces depend, in general, upon all the x's or v's, and perhaps on t for G_n and D_n. To obtain the energy equation we multiply the nth equation by $dx_n = v_n \, dt$, integrate, and sum over n, obtaining

$$H(\mathbf{r},\mathbf{v}) = \sum_{n=1}^{N} \tfrac{1}{2} m_n v_n{}^2 + V(\mathbf{r}) = \sum_{n=1}^{N} \int_0^t [G_n(\mathbf{r},t) + D_n(\mathbf{v},t)] v_n \, dt + E \qquad (1.3.12)$$

where $V(\mathbf{r})$ is the N-dimensional potential energy, such that $\partial V / \partial x_n = -F_n(\mathbf{r})$. When all the G's and D's are zero, the system is conservative and the function H has the constant value E.

With most acoustical systems the potential function V has a minimum at the configuration of stable equilibrium of the system. If E is larger than this minimal value of V, the system will move in accord with the equations of motion. As long as E is not too large, the equation $E = V(\mathbf{r})$ defines a surface of finite extent, within which the N-dimensional vector $\mathbf{r}$ moves about. Because the kinetic energy of the system is a sum of the terms $\tfrac{1}{2} m_n (dx_n/dt)^2$, however, we cannot solve directly for the nth component v_n of velocity, as we were able to for one dimension in Eq. (1.3.3). At a given configuration, represented by vector $\mathbf{r}$, and for a total energy E, the magnitude of vector $\mathbf{v}$ is determined, but its direction is undetermined by the energy equation. The configuration point $\mathbf{r}$ will bounce around inside the surface $V(\mathbf{r}) = E$, but will not usually do it periodically, returning always along the same path; each bounce will usually start it off in a different direction.

Thus, if we plot the behavior of one component, the x_n and related v_n on the nth phase plane, for example, the contour will not be closed-loop as in Fig. 1.3a, but usually will be a complicated nonperiodic curve such as the one shown in Fig. 1.3c, confined within definite limits determined by E, but never repeating itself. See, for example, the curves of Fig. 3.3 for a system of two degrees of freedom. We shall see in Chap. 3 that the motion of a system with N degrees of freedom involves N different frequencies, usually mutually incommensurate, so that the combination is not periodic. In later chapters we shall see that an extended system has an infinite number of frequencies in its motion, so that the motion of any part may be quite complicated indeed.

Thus, for more complex systems, the displacement component x_n will not usually be a periodic function of time; so it cannot be represented by a Fourier series. In special cases, of course, it may be periodic, though seldom simple-harmonic; it may be as shown in Fig. 1.4a. In other cases the motion will be nearly periodic, as in Fig. 1.4b, but if the intercoupling is large, the motion will be completely aperiodic, though bounded, as shown in Fig. 1.4c.

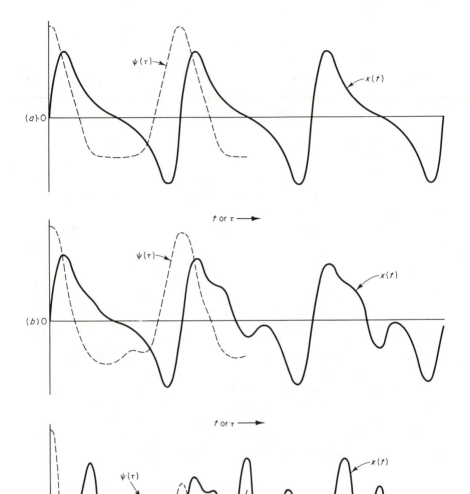

FIGURE 1.4
Solid lines are curves of displacement $x(t)$ versus time t;
dashed curves are for corresponding autocorrelation index
$\psi(\tau)$ against τ. Curves a are for periodic motion; curves b
and c for bounded, nonperiodic motion.

The pressure fluctuation in the sound from a turbulent jet of air is also of this type; in fact, curves of this type are characteristic of steady nonperiodic sound. We should learn how to describe mathematically these motions, too.

Oscillations like those pictured in Fig. 1.4b and c are bounded, nonperiodic, and stationary. The average values of x or x^2 vary with time, but we can take their averages over a sufficiently long interval of time to even out these fluctuations.

$$\langle x \rangle = \frac{1}{T} \int_{t-\frac{1}{2}T}^{t+\frac{1}{2}T} x(t)\, dt \qquad \langle x^2 \rangle = \frac{1}{T} \int_{t-\frac{1}{2}T}^{t+\frac{1}{2}T} x^2\, dt$$

In the limit of very large T, these averages are independent of T. If the vibrations are stationary, they are independent of t also. There are two ways of describing such motion, one corresponding to the action of a frequency analyzer, the other amounting to a comparison of the motion at one instant with the motion a time τ later. Periodic motion repeats itself exactly, every successive cycle. Some nonperiodic motion tends to repeat itself, but not exactly, the degree of duplication departing further and further from the original as time goes on (an example is the curve of Fig. 1.4b).

This tendency toward duplication, or lack of it, can be measured by the *autocorrelation function* Υ, which is defined as follows:

$$\Upsilon(\tau) = \lim_{T \to \infty} \left[\frac{1}{T} \int_{-\frac{1}{2}T}^{\frac{1}{2}T} x(t)x(t+\tau)\, dt \right] \tag{1.3.13}$$

We see that $\Upsilon(0) = \langle x^2 \rangle$; thus it is sometimes useful to express correlation in terms of the dimensionless *autocorrelation index* $\psi = (\Upsilon / \langle x^2 \rangle)$.

$$\psi(\tau) = \lim_{T \to \infty} \frac{\int_{-\frac{1}{2}T}^{\frac{1}{2}T} x(t)x(t+\tau)\, dt}{\int_{-\frac{1}{2}T}^{\frac{1}{2}T} x^2(t)\, dt} \qquad \psi(0) = 1 \tag{1.3.14}$$

We can find the autocorrelation function for either periodic or nonperiodic functions of time. If x is periodic and can be represented by the Fourier series (1.3.8) (curve of Fig. 1.4a), its autocorrelation index is

$$\psi(\tau) = \frac{a_0^2 + \frac{1}{2} \sum_{n=1}^{\infty} A_n^2 \cos(n\omega\tau)}{a_0^2 + \frac{1}{2} \sum_{n=1}^{\infty} A_n^2} \tag{1.3.15}$$

which is itself a Fourier series, with its maximum values, unity, occurring at $\tau = 0, T_p, 2T_p, \ldots$ ($T_p = 2\pi/\omega$). If x is periodic, then $x(t)x(t+\tau)$ is exactly equal to $x^2(t)$ whenever $\tau = nT_p$; so $\Upsilon(\tau)$ will equal $\langle x^2 \rangle$ whenever $\tau = nT_p$, and thus $\Upsilon(\tau)$ will also be periodic in τ with period T_p. Note that Υ and ψ depend only on the amplitudes A_n of the Fourier series for x, not on the phase angles Φ_n; Υ and ψ always have their maxima at $\tau = nT_p$ if x is periodic.

Now suppose $x(t)$ is "not quite periodic," as with the curve of Fig. 1.4b. It almost repeats itself in the next cycle, but not quite, and it differs more and more from the original cycle as time goes on.   In this case $x(t)x(t + \tau)$ will almost equal $x^2(t)$ if τ is a small integer times T_p, but will differ considerably from $x^2(t)$ if τ is large.   Instead of being positive everywhere (as x^2 is), it will be as often negative as positive when τ is large.   In this case the autocorrelation index will start at unity when $\tau = 0$ and will have a succession of maxima, spaced T_p apart, of decreasing magnitude until, for very large τ, $\psi(\tau) \to 0$. If there is very little correlation between x at one instant and x at the next instant as in the curve of Fig. 1.4c, then $\psi(\tau)$ will drop rapidly to zero as τ is increased.   In fact, the rapidity with which $\Upsilon(\tau)$ drops to zero as τ is increased is a measure of the degree of randomness of the oscillations of x.

Therefore the autocorrelation function (or index) is one way of measuring the nature of bounded vibratory motion.   It is not a complete measure of the vibrations, however, for it neglects the phase angles Φ.   Thus a knowledge of $\Upsilon(\tau)$ is not sufficient to allow us to compute $x(t)$.

The Fourier integral

Another way of representing bounded vibrations is to analyze the motion into its various component frequencies.   We did this already for periodic motion, with the Fourier series of Eq. (1.3.8).   Nonperiodic motion cannot be represented in terms of a fundamental component plus a sequence of harmonics; it must be considered as a superposition of motions of all frequencies.   Thus the *series* of harmonic terms must be generalized into an *integral* over all values of ω.   Using the complex exponential form for the component simple-harmonic terms, the basic equations for the Fourier integral representation, corresponding to Eq. (1.3.12) for the series, are that

$$\text{If } f(t) = \int_{-\infty}^{\infty} F(\omega)e^{-i\omega t} \, d\omega \qquad \text{then} \qquad F(\omega) = \frac{1}{2\pi} \int_{-\infty}^{\infty} f(t)e^{i\omega t} \, dt \quad (1.3.16)$$

Function $F(\omega)$ is called the *Fourier transform* of $f(t)$, and $f(t)$ is the *inverse* Fourier transform of $F(\omega)$.   The first equation shows that any function of t (within certain limits, which we shall discuss shortly) can be expressed in terms of a superposition of simple-harmonic motion of all possible frequencies $\omega/2\pi$, each frequency having for its amplitude the magnitude $|F(\omega)|$ of the complex quantity $F(\omega)$ and having for its phase angle the phase angle of $F(\omega)$.   A little thought will show that if $f(t)$ is to be real (when t is real), then $F(-\omega)$ must equal the complex conjugate of $F(\omega)$ [that is, $F(-\omega) = F^*(\omega) = |F|e^{-i\Phi}$, so that $F(\omega)F(-\omega) = |F|^2$ is real].   Quantity $|F(\omega)|^2$ is called the *spectrum density* of $f(t)$ at the frequency $\omega/2\pi$.   An electric filter which cuts out all the frequencies in $f(t)$ except those between $\omega/2\pi$ and $(\omega + d\omega)/2\pi$ would deliver a measurable power proportional to $|F(\omega)|^2$ and also proportional to $d\omega/2\pi$, the width of the frequency band passed by the filter.

The second equation of (1.3.16) shows how the Fourier transform $F(\omega)$ can be computed if we know the mathematical form of $f(t)$. It is analogous to Eqs. (1.3.10) or (1.3.8) for the Fourier series coefficients. It is, of course, the reason we discussed the infinite integrals of Eqs. (1.2.14) to (1.2.18). The reciprocal relationship displayed in Eqs. (1.3.16) has meaning only if the infinite integrals converge, and are of practical interest only if they converge for real values of t and ω. Books on the theory of the Fourier transform prove that convergence will certainly occur if the integral of $|f(t)|^2$ over t from $-\infty$ to ∞ is finite, in which case

$$\int_{-\infty}^{\infty} |f(t)|^2\, dt = 2\pi \int_{-\infty}^{\infty} |F(\omega)|^2\, d\omega \qquad (1.3.17)$$

and thus the integral of $|F(\omega)|^2$ is also finite.

This last relationship makes it look as though we should have difficulty in using the Fourier integral to represent stationary nonperiodic motion, of the sort we discussed in the previous subsection and portrayed in Fig. 1.4. If the vibration is stationary, $|x(t)|^2$ does not diminish as $t \to \infty$ and the integral of $|x|^2$ will not converge. We can get around the difficulty by a trick, however. Suppose we shut off $x(t)$ except for the time interval $-\tfrac{1}{2}T < t < \tfrac{1}{2}T$, and study the function

$$f(t) = \begin{cases} 0 & -\infty < t < -\tfrac{1}{2}T \\ x(t) & -\tfrac{1}{2}T < t < \tfrac{1}{2}T \\ 0 & \tfrac{1}{2}T < t < \infty \end{cases} \qquad (1.3.18)$$

The integral of the square of this function will converge; in fact,

$$\int_{-\infty}^{\infty} |f(t)|^2\, dt = T\langle x^2 \rangle = 2\pi \int_{-\infty}^{\infty} |F(\omega)|^2\, d\omega \qquad T \text{ large} \qquad (1.3.19)$$

where now

$$F(\omega) = \frac{1}{2\pi} \int_{-\tfrac{1}{2}T}^{\tfrac{1}{2}T} x(t) e^{i\omega t}\, dt$$

is a function which falls off sufficiently rapidly as ω goes to $\pm\infty$, so that the integral of $|F|^2$ converges as long as T is not infinite. Thus $|F(\omega)|^2/T$, which we can call the spectrum density per unit time, is a quantity whose integral over ω is equal to $(1/2\pi)\langle x^2 \rangle$ for T sufficiently large. This trick of analyzing only an interval T worth of $x(t)$ has its counterpart in measurement. The length of time required for the filter to separate out the components within a band $d\omega$ is longer, the narrower the bandwidth $d\omega$; however, we certainly cannot afford to wait forever, although the only way we could obtain a minutely detailed analysis would be to analyze it over an infinite length of time.

Thus the Fourier transform, with the use of the tricks of Eqs. (1.3.18) and (1.3.19), is another way of representing the bounded vibrations $x(t)$, which emphasizes its frequency analysis rather than its temporal correlation, as $\Upsilon(\tau)$ did. It is a more complete representation than is Υ, for Eqs. (1.3.16) show that if $F(\omega)$ is known completely, then $x(t)$ can be reproduced in its entirety. This is usually more detailed than is useful, however; usually we are interested only in the spectrum density $|F(\omega)|^2$ and are willing to remain ignorant of the phase of F.

Actually, the two methods of representation are closely related. If we consider $\Upsilon(\tau)$, the autocorrelation function of $x(t)$, defined in Eq. (1.3.13) as a function which also can have a Fourier transform, by using the trick of Eq. (1.3.18), then this transform would be

$$\frac{1}{2\pi} \int_{-\infty}^{\infty} e^{i\omega\tau} \, \Upsilon(\tau) \, d\tau = \frac{1}{2\pi T} \int_{-\frac{1}{2}T}^{\frac{1}{2}T} e^{i\omega\tau} \, d\tau \int_{-\frac{1}{2}T}^{\frac{1}{2}T} x(t)x(t+\tau) \, dt$$

$$= \frac{1}{2\pi T} \int_{-\frac{1}{2}T}^{\frac{1}{2}T} e^{-i\omega t} \, x(t) \, dt \int_{t-\frac{1}{2}T}^{t+\frac{1}{2}T} e^{i\omega u} x(u) \, du \xrightarrow[T \text{ large}]{} \frac{2\pi \, |F(\omega)|^2}{T} \quad (1.3.20)$$

where $F(\omega)$ is defined in Eq. (1.3.19) as the transform of $x(t)$ over the interval T. We thus reach the interesting and rather unexpected result that the Fourier transform of the autocorrelation function of $x(t)$ over the interval T is $2\pi/T$ times the spectrum density $|F(\omega)|^2$ of x, the square of the magnitude of the Fourier transform of x. Vice versa, if we know the spectrum density of x for a time interval T, its correlation function for the same interval is the inverse transform

$$\Upsilon(\tau) = \frac{2\pi}{T} \int_{-\infty}^{\infty} |F(\omega)|^2 \, e^{-i\omega\tau} \, d\omega \quad (1.3.21)$$

Thus $\Upsilon(0)$, which equals $\langle x^2 \rangle$, is proportional to the frequency integral of $|F|^2$, which also is apparent from Eq. (1.3.19).

Properties of Fourier transforms

A few general properties of Fourier transforms will be useful when we come to use them later. In the first place, mathematically speaking, both t and ω can be considered to be complex quantities, and thus the integrals of Eqs. (1.3.16) are contour integrals, of the sort discussed in connection with Eqs. (1.2.15). From Eqs. (1.2.16) to (1.2.18) we can obtain Fourier transforms for several important functions of t.

$$\text{If } f(t) = u(t) = \begin{cases} 0 & t < 0 \\ 1 & t > 0 \end{cases} \quad \text{then} \quad 2\pi F(\omega) = \frac{i}{\omega}$$

$$\text{If } f(t) = u(t) \sin(at) \quad \text{then} \quad 2\pi F(\omega) = \frac{a}{a^2 - \omega^2} \quad (1.3.22)$$

$$\text{If } f(t) = u(t)e^{-iat} \quad \text{then} \quad 2\pi F(\omega) = \frac{i}{\omega - a}$$

where the imaginary part of ω must be positive in order that the integral for

F converge. In each of these cases $F(\omega)$ has a pole or poles on the real ω axis. Setting $\omega = w + i\epsilon$, the integrand $e^{i\omega t}f(t)$ of Eqs. (1.3.16) for F has an exponential factor $e^{iwt-\epsilon t}$, which ensures convergence of the integral as long as ϵ is positive. We noted, in the discussion of Eqs. (1.2.15), that the contour over which ω is integrated should go somewhat above the real axis of ω, which is the same thing as saying that ϵ should be positive.

This arrangement for bringing convergence to the integrals is most useful in representing functions $f(t)$, which are zero for negative values of t, for then we need only worry about convergence at $t \to \infty$. In fact, if we are only interested in representing functions $f(t)$ which are zero for $t < 0$ (as would be the case if the system were at rest till $t = 0$, when it is started into motion), it may be more convenient to rotate the real-imaginary axes of ω, setting $\omega = is$ $(s > 0)$, so that

$$F_L(s) = 2\pi F(is) = \int_0^\infty f(t)e^{-st}\, dt \qquad f(t) = \frac{1}{2\pi i}\int_{-i\infty+\epsilon}^{i\infty+\epsilon} F_L(s)e^{st}\, ds \quad (1.3.23)$$

Function $F_L(s)$ is called the *Laplace transform* of $f(t)$. As we can see, it is just a different way of writing the Fourier transform, which happens to be useful for those functions which are zero for $t < 0$. We use the Fourier transform in this book, since it is more immediately related to the physical process of frequency analysis. We also shall use the Laplace transform when it is more convenient. Tables of the Fourier and Laplace transforms of various functions have been published; we mention only a few here. A short table of Laplace transforms is given in Chap. 2 (Table 2.1).

Differentiation of the integral for $f(t)$ or integration by parts of the integral for $F(\omega)$ shows that if $F(\omega)$ is the Fourier transform of $f(t)$, then $-i\omega F(\omega)$ is the Fourier transform of df/dt, and $-\omega^2 F(\omega)$ is the transform of d^2f/dt^2. From this we can obtain the transform of a differential equation; the transform of

$$\frac{d^2x}{dt^2} + \omega_0^2 x = f(t) \qquad \text{is} \qquad (\omega_0^2 - \omega^2)X(\omega) = F(\omega)$$

where $X(\omega)$ is the Fourier transform of $x(t)$, and $F(\omega)$ is the transform of $f(t)$. If $f(t)$ is known, we can compute $X(\omega) = F(\omega)/(\omega_0^2 - \omega^2)$ and then compute the inverse transform $x(t)$ (or else look it up in a table of Fourier transforms).

The pathological, but extremely useful, *Dirac delta function* is defined as follows:

$$\delta(t) = \frac{d}{dt}u(t) \qquad \int_{-\infty}^\infty \delta(t - a)g(t)\, dt = g(a) \qquad (1.3.24)$$

where $u(t)$ is defined in Eqs. (1.3.22) and where $g(t)$ is any function which is finite and continuous at $t = a$. The curve for $\delta(t)$ is the limiting form for a very high, very narrow peak, centered at $t = 0$, having unit area under the peak. Direct insertion in Eqs. (1.3.16) shows that

$$\text{The Fourier transform of } \delta(t - t_0) \text{ is } \frac{1}{2\pi} e^{i\omega t_0}$$

(1.3.25)

$$\text{The inverse transform of } e^{-i\omega_0 t} \text{ is } \delta(\omega - \omega_0)$$

This shows us that the Fourier transform of the Fourier series

$$x(t) = \sum_{n=0}^{\infty} A_n e^{-i(n\omega_0 t - \Phi_n)}$$

is

$$X(\omega) = \sum_{n=0}^{\infty} A_n e^{i\Phi_n} \delta(\omega - n\omega_0)$$

(1.3.26)

Therefore the frequency analysis of a periodic motion results in a set of equally spaced isolated peaks, one for the fundamental frequency ω_0 and one for each harmonic ($\omega = n\omega_0$). The magnitude of these peaks is proportional to A_n, the respective Fourier amplitude.

We shall use the Fourier transform often in our subsequent discussions, because it is natural, in acoustics, to analyze sounds into their simple-harmonic motions. We shall use the Laplace transform rather less often, when we are interested in studying the transient effects caused by the sudden initiation of some driving force. But there are still other integral transforms, which we shall have occasional recourse to. For example, there is the so-called Hankel transform,

$$F_m(\mu) = \int_0^{\infty} f(w) J_m(\mu w) w \, dw \qquad f(w) = \int_0^{\infty} F_m(\mu) J_m(\mu w) \mu \, d\mu \quad (1.3.27)$$

which we shall use in connection with cylindrical waves.

Problems

1 A mass m slides without friction on a horizontal table. It is attached to a light string which runs, without friction, through a hole in the table. The other end of the string is pulled downward by a constant force F. The mass (which is too large to drop through the hole) is held at rest a distance D from the hole and then let go. Set up the equation of motion of the mass, and solve it. Is the motion periodic? If so, what is its frequency, and how does the frequency depend on D?

2 A bead of mass m slides without friction on a straight wire that is whirling in a horizontal plane, about one end, with a constant angular velocity ω radians per sec. It is found that the centrifugal "force" on the bead is $m\omega^2 r$, where r is the distance of the bead from the center of rotation. The direction of the force is away from the center. Set up the equation of motion of the bead, and solve it. Is the motion periodic, and if so, what is the frequency?

3 Show that a solution of the equation

$$\frac{d^2y}{dx^2} + \frac{1}{x}\frac{dy}{dx} + n^2y = 0$$

can be represented by the series

$$y = C\left(1 - \frac{n^2x^2}{2^2} + \frac{n^4x^4}{2^2 \cdot 4^2} - \frac{n^6x^6}{2^2 \cdot 4^2 \cdot 6^2} \cdots\right)$$

The series inside the parentheses is called the Bessel function $J_0(nx)$, where $J_0(z) = 1 - (z^2/2^2) + \cdots$. Compute values of $J_0(0)$, $J_0(\tfrac{1}{2})$, and $J_0(1)$ to three places of decimals.

4 Show that the solution of the equation

$$\frac{d^2y}{dx^2} + (1 - kx^2)y = 0$$

is $y = a_0 D_e(k,x) + a_1 D_o(k,x)$

where D_e and D_o are symbols for the following series:

$$D_e(k,x) = \cos x + \frac{k}{12}x^4 - \frac{7k}{360}x^6 + \frac{11k + 15k^2}{10,080}x^8 \cdots$$

$$D_o(k,x) = \sin x + \frac{k}{20}x^5 - \frac{13k}{2,520}x^7 + \frac{17k + 63k^2}{90,720}x^9 \cdots$$

Compute values of D_e and D_o to three places of decimals for $k = 1$ and for $x = 0, \tfrac{1}{2}$, and 1.

5 Show that the solution of the equation

$$\frac{d}{dx}\left[(1 - x^2)\frac{dy}{dx}\right] + \lambda y = 0 \qquad \text{is} \qquad y = a_0 P_e(\lambda,x) + a_1 P_o(\lambda,x)$$

where P_e and P_o are symbols for the following series:

$$P_e(\lambda,x) = 1 - \frac{\lambda}{2!}x^2 - \frac{\lambda(6 - \lambda)}{4!}x^4 - \frac{\lambda(6 - \lambda)(20 - \lambda)}{6!}x^6 - \cdots$$

$$P_o(\lambda,x) = x + \frac{2 - \lambda}{3!}x^3 + \frac{(2 - \lambda)(12 - \lambda)}{5!}x^5$$

$$+ \frac{(2 - \lambda)(12 - \lambda)(30 - \lambda)}{7!}x^7 + \cdots$$

Compute values of P_e and P_o, to three places of decimals, for $\lambda = 0, 1$, and 2, for $x = 0, \tfrac{1}{2}$, and 1.

6 Show that the solution of the equation

$$\frac{1}{x^2}\frac{d}{dx}\left(x^2\frac{dy}{dx}\right) + \left(1 - \frac{2}{x^2}\right)y = 0 \qquad \text{is} \qquad y = a_0 j_1(x) + a_1 n_1(x)$$

where j_1 and n_1 are symbols for the following functions:

$$j_1(x) = \frac{\sin x}{x^2} - \frac{\cos x}{x} \qquad n_1(x) = -\frac{\cos x}{x^2} - \frac{\sin x}{x}$$

What are the values of j_1 and n_1 at $x = 0, \frac{1}{2}$, and 1? What is the solution of the equation

$$\frac{1}{x^2}\frac{d}{dx}\left(x^2\frac{dy}{dx}\right) + y = 0$$

7 What is the length of the line drawn from the origin to the point on the complex plane represented by the quantity $(a + ib)^{-1}$? What is the angle that this line makes with the real axis? What are the amplitudes and phase angles of the following quantities:

$$(a - ib)^{-\frac{1}{2}}; \qquad (a + ib)^{-1} + (c + id)^{-1}; \qquad (a + ib)^n e^{-i\omega t}$$

8 What are the real and imaginary parts, amplitudes, and phase angles of

$$\sqrt{a + ib}, \qquad \log(a + ib), \qquad e^{-2\pi i(v+i\mu)t}, \qquad e^{-ix}(1 + e^{a+ib})^{-1}$$

9 The *hyperbolic* functions are defined in a manner analogous to the trigonometric functions [Eq. (1.2.10)].

$$\cosh z = \tfrac{1}{2}(e^z + e^{-z}) \qquad \sinh z = \tfrac{1}{2}(e^z - e^{-z})$$

$$\tanh(z) = \frac{\sinh z}{\cosh z} = \frac{1}{\coth(z)} \qquad \text{(Tables I and II)}$$

Show that $\cosh(iy) = \cos y$; $\cos(iy) = \cosh y$; $\sinh(iy) = i\sin y$; $\sin(iy) = i\sin y$. Find the real and imaginary parts of $\cosh(x + iy)$, $\cos(x + iy)$, $\sinh(x + iy)$, $\sin(x + iy)$. What are the magnitudes and phase angles of these quantities?

10 Find the real part R and imaginary part X of $\tanh(x + iy)$. Show that

$$R^2 + [X + \cot(2y)]^2 = \frac{1}{\sin^2(2y)}$$

$$[R - \coth(2x)]^2 + X^2 = \frac{1}{\sinh^2(2x)}$$

Plot the curves on the complex plane for $\tanh(x + iy)$ when y is allowed to vary but $x = 0, \pi/10, \infty$; when x is allowed to vary but $y = 0, \pi/4, \pi/2$.

11 Where on the complex plane are the poles of the functions $1/\cosh z$; $\tan z$; $e^{iz}/z^2(z^2 - a^2)$; $\tan z/z(z^2 + a^2)$? Are they all simple poles? Calculate the residues at all the simple poles of these functions.

12 Compute the values of the integrals

$$\int_{-\infty}^{\infty} \frac{e^{-izt}}{z(z^2 + a^2)}\,dz \qquad \int_{-\infty}^{\infty} \tan(z)e^{-izt}\frac{dz}{z}$$

13 Transform the ordinary integral $\int_0^{2\pi}(1 - 2p\cos\theta + p^2)^{-1}\,d\theta$ to a contour integral over z by setting $e^{i\theta} = z$. What is the shape of the contour? Where are the poles of the integrand when p is real and smaller than unity? Calculate the value of the integral when $0 < p < 1$.

14 Calculate the value of the integral

$$\int_{-\infty}^{\infty} \frac{x^2\,dx}{(1 + x^2)(1 - 2x\cos\theta + x^2)}$$

15 Show that

$$-\frac{\omega}{2\pi} \int_{-\infty}^{\infty} e^{-izt} \left[\frac{1 - (-1)^n e^{iz(n\pi/\omega)}}{z^2 - \omega^2} \right] dz = \begin{cases} 0 & t < 0 \\ \sin(\omega t) & 0 < t < \dfrac{n\pi}{\omega} \\ 0 & \dfrac{n\pi}{\omega} < t \end{cases}$$

16 A mass m moves in the x direction under the influence of a potential energy $V(x) = Ax^2/(x + d)^2$. Plot the potential energy and use the plot to discuss the behavior of the system for different total energies E. Plot the phase contour for $E = \frac{3}{4}A$ and for $E = \frac{5}{4}A$. Show that, for $E < A$, the motion is periodic, with period $T_p = \sqrt{2m} \, Ad\pi/(A - E)^{\frac{3}{2}}$.

17 Plot the Fourier series $x = 10 \cos(\omega t) + 5 \cos(3\omega t) + \cos(5\omega t)$. Compute and plot the autocorrelation function $\Psi(\tau)$ for this x.

18 Compute the coefficients and phase angles for the Fourier series expansion of the following periodic function of t:

$$x(t) = \begin{cases} t & -\frac{1}{2}\pi < t < \frac{1}{2}\pi \\ \pi - t & \frac{1}{2}\pi < t < \frac{3}{2}\pi \\ t - 2\pi & \frac{3}{2}\pi < t < \frac{5}{2}\pi \\ \cdots\cdots \end{cases}$$

19 A particular function $x(t)$ can be represented by the sum

$$x(t) = \sum_{n=1}^{N} \cos(\omega_n t - \Phi_n)$$

where no pair of the ω_n's are commensurate, and the Φ_n's are arbitrary functions of n. Show that x is nonperiodic but bounded, that the mean value of x^2 is $\frac{1}{2}N$, and that the maximum possible value of x^2 is N^2. What is the expression for the autocorrelation function for x?

20 Compute the Fourier transforms of the following functions: e^{-at^2}, $a^2/(a^2 + t^2)$. What are the inverse transforms of $-1/2\pi\omega^2$ and $A/(\omega^4 - a^4)$?

21 Frequency analysis of a certain noise shows that its spectrum density $|F(\omega)|^2$ is independent of ω over the range $|\omega| < \omega_m$ and is zero for $|\omega| > \omega_m$. What is the autocorrelation function?

22 The autocorrelation function for pressure in a turbulent gas is $\Psi(\tau) = P_0^2 e^{-\omega_m |\tau|}$. What is the spectrum density?

23 If

$$F(\omega) = \frac{1}{2\pi} \int_{-\infty}^{\infty} f(t)e^{i\omega t} \, dt \quad \text{and} \quad G(\omega) = \frac{1}{2\pi} \int_{-\infty}^{\infty} g(t)e^{i\omega t} \, dt$$

show that the inverse transform of the product FG, $\displaystyle\int_{-\infty}^{\infty} F(\omega)G(\omega)e^{-i\omega t} \, d\omega$,

is equal to $\displaystyle\frac{1}{2\pi} \int_{-\infty}^{\infty} f(\tau)g(t - \tau) \, d\tau$, which is called the *convolution integral* of f and g. What about the Fourier transform of $f(t)$ times $g(t)$? Demonstrate both these results in the special case where $F(\omega) = (\omega + ia)^{-1}$, $G(\omega) = (\omega + ib)^{-1}$, by performing all the integrals, using Eqs. (1.2.15) when necessary.

CHAPTER

2

THE LINEAR OSCILLATOR

2.1 FREE OSCILLATIONS

Now that we have discussed, to some extent, the mathematical methods that we shall need in our work, we shall come back to the physics.

The whole study of sound is a study of vibrations. Some part of a system has *stiffness*; when it is pulled away from its position of equilibrium and then released, the system oscillates. We shall first study the simplest possible sort of vibrations for the simplest sort of system, a mass m fastened to some sort of spring, so that it can oscillate back and forth in just one direction. A very large number of vibrators with which we deal in physics and engineering are of this type or are approximately like it. All pendulums (the "spring" here is the force of gravity) are like this, and all watch balance wheels. Loudspeaker diaphragms which are loaded so that their mass is concentrated near their center are approximately like this (at least at low frequencies), as are loaded tuning forks, etc. Even when an oscillating system is more complex than the simple oscillator, many of its properties are like it. Later, when we study these more complicated systems, we can simplify our discussion considerably by pointing out first the properties wherein the system behaves like the simple case and then showing where it differs.

In this chapter we shall simplify the problem still further; we shall investigate the motion of the one-dimensional oscillator with an internal stiffness force which is *linearly proportional* to x, $F_i(x) = -Kx$. As mentioned in Sec. 1.3, if the system is to perform bounded vibrations, F_i must be zero at some equilibrium point (which can be set at $x = 0$) and must be negative when x is positive and positive when x is negative, so as to push the mass back to $x = 0$. This usually means that F_i can be expanded in a Taylor series about the equilibrium point,

$$F_i(x) = -Kx + K_2x^2 + K_3x^3 + \cdots$$

If x is sufficiently small, the terms in x^2, x^3, etc., can be neglected compared with the linear term $-Kx$. Many acoustical vibrations are sufficiently small in amplitude for this linear approximation to be quite good. For instance, an air molecule needs to vibrate with an amplitude of motion of only about a tenth of a millimeter to do its full part in transmitting away the racket generated at Times Square in New York on New Year's Eve.   Seldom does the amplitude of oscillation of a loudspeaker diaphragm exceed a millimeter.

The general solution

We shall assume, then, that the "springiness" force acting on the simple oscillator can be expressed by the equation

$$F = -Kx \tag{2.1.1}$$

where x is the displacement of the mass from equilibrium, K is called the *stiffness constant* (its value depends on the sort of spring we use), and the minus sign indicates that the force opposes the displacement.   This will be a very good assumption to make for most of the cases dealt with in this book. The reciprocal of the stiffness constant, $C_m = 1/K$, is called the *compliance* of the spring.

To start with, we shall consider that no other forces act on the oscillator. This is, of course, not a good assumption in many cases; usually there are frictional forces acting, and sometimes external forces come in.   There are many cases, however, where the frictional force is negligible compared with the springiness force, and we shall treat these first, bringing in friction and external forces later.

This brings us to the equation of motion discussed in Sec. 1.2:

$$\frac{d^2x}{dt^2} = -\omega^2 x \qquad \omega^2 = \frac{K}{m} = \frac{1}{mC_m} \tag{2.1.2}$$

We have already seen that the solution of this equation can be expressed, in terms of our convention, as

$$x = Ce^{-i\omega t} \qquad C = a_0 + ia_1$$

or as $\qquad\qquad x = a_0 \cos(\omega t) + a_1 \sin(\omega t) \tag{2.1.3}$

We must now discuss the physical implications of this solution.

It has been shown in Sec. 1.2 that the motion of the body is periodic, having a frequency $\nu_0 = \omega/2\pi = (1/2\pi)\sqrt{K/m}$.   This frequency is larger for a stiffer spring (for a larger K) and is smaller for a heavier mass.

Initial conditions

We have seen that the specific values of the constants a_0 and a_1 are determined by the way the mass is started into motion.   Ordinarily, we start an oscillator into motion by giving it a push or by pulling it aside and letting it go, i.e., by giving it, at $t = 0$, some specified initial displacement and initial velocity.   Once we have fixed these two *initial conditions*, the motion is completely determined from then on, unless we choose to interfere with it again.   It is not hard to see that a_0 is the value of the initial displacement and ωa_1 that of the initial velocity.   Solution (2.1.3) can then be rewritten in the following forms:

$$x = x_0 \cos (2\pi v_0 t) + \frac{v_0}{2\pi v_0} \sin (2\pi v_0 t)$$

$$= A \cos (2\pi v_0 t - \Phi) \qquad v_0 = \frac{1}{2\pi} \sqrt{\frac{K}{m}} \qquad (2.1.4)$$

$$A^2 = x_0^2 + \left(\frac{v_0}{2\pi v_0}\right)^2 \qquad \tan \Phi = \frac{v_0}{2\pi v_0 x_0}$$

where x_0 is the initial displacement, and v_0 the initial velocity of the mass.

These equations reemphasize the fact that *only* the initial value of x and of dx/dt need be given to determine the subsequent motion of an oscillator completely.   Once x_0 and v_0 are specified, then, even if we did not have a completely worked-out solution of (2.1.2), we could find the initial value of the body's acceleration by inserting x_0 on the right-hand side of (2.1.2).   The initial value of the third derivative can be obtained by differentiating (2.1.2) with respect to t and placing the value of v_0 on the right-hand side in place of dx/dt; and so on.   If it is recalled that the behavior of a function is completely specified by its Taylor's-series expansion

$$f(t) = f(0) + t\left(\frac{df}{dt}\right)_{t=0} + \frac{t^2}{2}\left(\frac{d^2f}{dt^2}\right)_{t=0} + \frac{t^3}{6}\left(\frac{d^3f}{dt^3}\right)_{t=0} + \cdots$$

(within a certain range of t whose limits are of interest in specific problems, but which need not bother us here), we see that once x_0 and v_0 are given, thus fixing the values of all the higher derivatives at $t = 0$, the future motion is determined.

It is not hard to generalize this reasoning, to see that a body, acted on by any sort of force that depends on x and v, will have its motion specified completely just by assigning definite values to its initial position and velocity. The mathematical counterpart to this statement is the rule that the solution of any second-order differential equation (one having a second derivative term in it but no higher derivative) has two arbitrary constants in it.

Another very important physical fact which can be deduced from (2.1.4) is that the frequency of the oscillation depends *only* on K and m and *not at all* on x_0 or v_0.   This means that for a given mass and a given spring, as long as the law of force of the spring is $F = -Kx$, the frequency of oscillation will be the same no matter how we start the system to oscillate, whether it oscillates with an amplitude of motion of 1 cm or 0.001 mm.   This is a very important fact in its practical applications, for if the law of force of actual springs were not nearly $F = -Kx$, or if this property did not hold for solutions of Eq. (2.1.2), no musical instrument could be played in tune.   Imagine trying to play a piano when the frequency of each note depended on how hard one struck the keys!   Compare this with Probs. 1 and 16 of Chap. 1.

We might have found this fact for a number of cases by a long series of experimental observations, but our mathematical analysis tells us immediately that *every* mass acted on by a force $F = -Kx$ has this property. Oscillations of this type are called *simple-harmonic oscillations*, as noted in the discussion of Eq. (1.3.6).

Energy of vibration

We shall need an expression for the energy of a mass oscillating with simple-harmonic motion of amplitude A and frequency v_0. The energy is the sum of the potential and kinetic energies

$$E = \tfrac{1}{2}mv^2 + \int_0^x Kx\, dx = \tfrac{1}{2}mv^2 + \tfrac{1}{2}Kx^2$$

$$= 2\pi^2 mv_0^2 A^2 \sin^2(2\pi v_0 t - \Phi) + \tfrac{1}{2}KA^2 \cos^2(2\pi v_0 t - \Phi)$$

But $4\pi^2 v_0^2 = K/m$, so that

$$E = \tfrac{1}{2}KA^2[\sin^2(2\pi v_0 t - \Phi) + \cos^2(2\pi v_0 t - \Phi)]$$

or
$$E = \tfrac{1}{2}KA^2 = 2\pi^2 mv_0^2 A^2 = \tfrac{1}{2}mU^2 \tag{2.1.5}$$

where $U = 2\pi v_0 A$ is the *velocity amplitude* of the motion.

The total energy is thus equal to the potential energy at the body's greatest displacement, $\tfrac{1}{2}KA^2$, or is equal to the kinetic energy at the body's greatest speed, $\tfrac{1}{2}mU^2$. Expressed in terms of v_0 and A, we see that E depends on the square of these two quantities.

2.2 DAMPED OSCILLATIONS

So far, we have not considered the effects of friction on oscillating systems. In general, friction does not play a very important role in the problems we shall consider in the first part of the book. If we show its effects on the simplest system with which we deal, we can deduce its effect, by analogy, on more complicated systems.

One variety of friction that is important in vibrational problems is the resistance to motion which the air surrounding the body manifests. Energy in the form of sound waves is sent out into the air. From the point of view of the vibrating system, this can be looked on as friction, for the energy of the system diminishes, being drained away in the form of sound. This resisting force depends on the velocity of the vibrator, and unless the velocity is large (much larger than those with which we shall usually deal), it is proportional to the velocity. It can be expressed mathematically as $-R_m(dx/dt)$, where the constant R_m is called the *resistance constant*. The total force on a simple oscillator acted on by both friction and springiness is therefore $-R_m(dx/dt) - Kx$, and the equation of motion becomes

$$\frac{d^2x}{dt^2} + 2k\frac{dx}{dt} + 4\pi^2v_0^2x = 0 \qquad\qquad \begin{aligned} k &= \frac{R_m}{2m} \\[2mm] 4\pi^2v_0^2 &= \frac{K}{m} \end{aligned} \qquad (2.2.1)$$

The value of v_0 is the frequency that the oscillator would have if the friction were removed (R_m were zero), and is called the *natural frequency* of the oscillator. Other resistive forces, representing the transformation of acoustic energy into heat, also are proportional to the velocity for small-amplitude motion; so Eq. (2.2.1) again holds. It is interesting to notice that the equation for the free oscillation of charge in a circuit containing inductance, resistance, and capacitance has the same form as (2.2.1). The inductance is analogous to the mass m, the resistance to the resistance factor R_m, and the inverse of the capacitance to the stiffness constant K.

The general solution

To solve Eq. (2.2.1) we make use of the exponential function again. We guess that the solution is Ce^{bt} and solve for b. Substitution in (2.2.1) shows that $(b^2 + 2kb + 4\pi^2v_0^2)e^{bt}$ must be zero for all values of the time. Therefore

$$b^2 + 2kb + 4\pi^2v_0^2 = 0 \qquad \text{or} \qquad b = -k \pm \sqrt{k^2 - 4\pi^2v_0^2}$$

In all the problems that we shall consider, the stiffness constant K is much larger than the resistance constant R_m (i.e., friction will not be big enough to make the motion much different from that discussed in the last section). Since $4\pi^2v_0^2$ is supposed to be larger than k^2, the square root in the expression for b is an imaginary quantity, and we had better write

$$b = -k \pm 2\pi iv_f, \qquad v_f = v_0\sqrt{1 - \left(\frac{k}{2\pi v_0}\right)^2} \qquad (2.2.2)$$

This means that, following our convention, we can write

$$x = Ce^{-kt-2\pi iv_f t}$$

or

$$x = e^{-kt}[a_0 \cos(2\pi v_f t) + a_1 \sin(2\pi v_f t)] = A_0 e^{-kt} \cos(2\pi v_f t - \Phi) \quad (2.2.3)$$

The values of a_0 and a_1 are again fixed by the initial conditions for the oscillator; a_0 must equal the initial displacement x_0, and the initial velocity in this case is equal to $2\pi v_f a_1 - ka_0$, so that

$$a_1 = \frac{v_0 + kx_0}{2\pi v_f}$$

The solution is *not* periodic, since the motion never repeats itself, each swing being of somewhat smaller amplitude than the one before it. However, if k is quite small compared with ν_f, we can say that it is very nearly periodic. In any case, the frequency of the oscillations is ν_f, which is very nearly equal to ν_0 if k is small. It again turns out that the frequency is independent of the amplitude of the motion. Of course, strictly speaking, we should not use the word frequency in connection with nonperiodic motion. But when the damping is small, the motion is almost periodic, and the word will have some meaning, although a rather vague one.

There are several respects in which the motion of the damped oscillator differs from that of the simple oscillator. The most important difference is that the amplitude of motion of the damped oscillator decreases exponentially with time; it is $A_0 e^{-kt}$, instead of being just A (A_0 is the initial value of the amplitude). The amplitude decreases by a factor $1/e$ in a time $1/k$ sec ($e = 2.718$). This length of time is a measure of how rapidly the motion is damped out by the friction, and is called the *modulus of decay* τ of the oscillations. The fraction of this decrease in amplitude which occurs in one cycle, i.e., the ratio between the period of vibration and the modulus of decay, is called the *decrement* δ of the oscillations. Another method of expressing this is in terms of the "Q of the system," where $Q = \omega_0 m / R_m$ is the number of cycles required for the amplitude of motion to reduce to $1/e^\pi$ of its original value. If these quantities are expressed in terms of the constants of the system (the small difference between ν_f and ν_0 being neglected), it turns out that

$$Q = \frac{\omega_0 m}{R_m} = \frac{\pi \nu_0}{k} \quad , \quad \omega_0 = 2\pi\nu_0 = \sqrt{\frac{K}{m}} \quad k = \frac{R_m}{2m}$$

$$\tau = \frac{1}{k} = \frac{Q}{\pi\nu_0} \qquad\qquad \delta = \frac{\pi}{Q} = \frac{k}{\nu_0} = \frac{1}{\tau\nu_0} \tag{2.2.4}$$

The smaller R_m is, the larger Q and τ are, indicating that it takes a longer time for the oscillations to damp out, and the smaller δ is, indicating that the reduction in amplitude per cycle is smaller. These properties, of course, are independent of the way the oscillator is started into motion.

Another difference between the damped and the undamped oscillator is the difference in frequency. When $k/2\pi\nu_0$ is small, the expression for ν_f can be expanded by means of the binomial theorem, and all but the first two terms can be neglected.

$$\nu_f = \nu_0 \sqrt{1 - \left(\frac{k}{2\pi k_0}\right)^2} = \nu_0 - \frac{k^2}{8\pi^2\nu_0} + \cdots \tag{2.2.5}$$

In most of the cases with which we shall be dealing, k and ν_0 have such values that even the second term in the series is exceedingly small, so that the change in frequency is usually too small to notice.

Energy relations

The subject of damped oscillations can be considered from a quite different point of view—that of energy loss. We must first develop an expression for the average energy of the system at any instant. We cannot use the formulas (2.2.5) because the amplitude of oscillation in the present case is not constant. The sum of the kinetic and potential energies of a body of mass m acted on by a springiness force Kx, whose displacement is given by the formula $A(t) \cos (2\pi vt - \Phi)$, is

$$E(t) = \tfrac{1}{2}mv^2 + \tfrac{1}{2}Kx^2$$
$$= 2\pi^2mv^2A^2 - 2\pi mv \frac{dA}{dt} A \sin (2\pi vt - \Phi) \cos (2\pi vt - \Phi)$$
$$+ \tfrac{1}{2}m \left(\frac{dA}{dt}\right)^2 \cos^2 (2\pi vt - \Phi)$$

When averaging this value over a single oscillation, the second term on the right drops out. If A is very slowly varying, so that dA/dt is small compared with vA, then the third term can be neglected, and we have the approximate formula for the energy of motion and position (i.e., the energy that can be recovered, that is not yet irrevocably lost in heat):

$$E(t) \simeq \tfrac{1}{2}K[A(t)]^2 = 2\pi^2mv^2[A(t)]^2 = \tfrac{1}{2}m[U(t)]^2 \qquad (2.2.6)$$

where the symbol $\simeq$ means "is approximately equal to." In the case of the damped oscillator, this "free energy" is $2\pi^2mv_0^2A_0^2e^{-2kt}$, which diminishes exponentially with time.

The rate of loss of energy due to friction is equal to the frictional force opposing the motion R_mv multiplied by the velocity v (since force times distance is energy, rate of change of energy is force times velocity). The rate of loss of energy is

$$P = R_mv^2 = [4\pi^2R_mv_f^2 \sin^2 (2\pi v_ft - \Phi)$$
$$+ 4\pi R_mv_fk \sin (2\pi v_ft - \Phi) \cos (2\pi v_ft - \Phi)$$
$$+ R_mk^2 \cos^2 (2\pi v_ft - \Phi)]A_0^2e^{-2kt}$$

Using the same approximations as before, we have for the average loss of energy per second

$$P = -\frac{dE}{dt} \simeq 2\pi^2v_0^2R_mA_0^2e^{-2kt} = \tfrac{1}{2}R_m[U(t)]^2 \qquad (2.2.7)$$

If we had started out without formula (2.2.3) for the details of the motion but had simply said that the free energy at any instant was given by (2.2.6) and that the energy loss was given by (2.2.7), where R_m is small, we could have found the dependence of E on the time by means of (2.2.7). Eliminating U from Eqs. (2.2.6) and (2.2.7) results in $P \simeq (R_m/m)E = 2kE$, so that we have

$$P = -\frac{dE}{dt} \simeq 2kE \qquad \text{or} \qquad \frac{dE}{dt} \simeq -2kE \qquad (2.2.8)$$

The solution of this is

$$E \simeq E_0 e^{-2kt} = \tfrac{1}{2}mU_0{}^2 e^{-2kt} = 2\pi^2\nu_0{}^2 m A_0{}^2 e^{-2kt}$$

which checks with Eq. (2.2.6). We see from this that the damping out of the motion is required by the fact that the energy is being lost by friction. We might point out here that the fraction of free energy lost per cycle is just $2k/\nu_f \simeq 2k/\nu_0 = 2\pi/Q$, where Q is given in Eq. (2.2.4).

In nearly every more complicated case of vibrations, the effect of friction is the same as in this simple case. The amplitude of vibration slowly decreases, and the frequencies of natural oscillation are very slightly diminished. Usually, the change of frequency is too small to be of interest.

2.3 FORCED OSCILLATIONS

It often happens that a system is set into vibration because it is linked in some way with another oscillating system (which we shall call the *driving system*). For instance, the diaphragm of a microphone vibrates because it is linked, by means of sound waves, to the vibrations of a violin string; and a loudspeaker diaphragm vibrates because it is linked to the current oscillations in the output circuit of an amplifier. The system picks up energy from the driving system and oscillates. In the two instances mentioned, and in many others, the driven system does not feed back any appreciable amount of energy to the driving system, either because the linkage between the two is very weak (as is the case with the violin and microphone) or else because the driving system has so much reserve energy that the amount fed back is comparatively negligible (as is the case with the amplifier and loudspeaker). In these cases the only property of the driving system that we need to know is that it supplies a periodic force which acts on the driven system. The more complicated case, where the feedback of energy cannot be neglected, will be considered later.

The general solution

For the present, we ask what happens to a simple oscillator when it is acted on by a periodic force $F\cos(2\pi\nu t)$, or $Fe^{-2\pi i\nu t}$, according to our convention. We wish to know what its motion is just after the force has been applied, and more important, what its motion is after the force has been acting for a long time. We are also interested in how this behavior depends on the frequency ν of the driving force (which does not have to be the same as the natural frequency ν_0 of the oscillator).

The total force on the oscillator is a combination of the springiness and the frictional and the driving force $-R_m v - Kx + Fe^{-2\pi i v t}$. The equation of motion is

$$\frac{d^2 x}{dt^2} + 2k \frac{dx}{dt} + \omega_0^2 x = ae^{-2\pi i v t}$$

(2.3.1)

$$a = \frac{F}{m} \qquad k = \frac{R_m}{2m} \qquad \omega = 2\pi v \qquad \omega_0^2 = \frac{K}{m}$$

We discussed a similar equation in Sec. 1.2, where we showed that a choice of two exponentials, one corresponding to the free vibration of the oscillator and the other to the forced motion, was a good guess for the solution. We try $x = Ce^{-kt - i\omega_f t} + De^{-i\omega t}$ in Eq. (2.3.1), where k and $\omega_f = 2\pi v_f$ are the damping and frequency parameters of free oscillation, as defined in Eqs. (2.2.2) and (2.2.5). The differential operator on the left side of Eq. (2.3.1) cancels out the free-oscillation term in x. For the second term, however, the equation is not satisfied unless

$$(-m\omega^2 - iR_m\omega + K)De^{-i\omega t} = Fe^{-i\omega t} \qquad \text{or} \qquad D = \frac{F}{-i\omega Z_m} \qquad (2.3.2)$$

where

$$Z_m = -i\omega m + R_m + i\frac{K}{\omega} = R_m - iX_m = |Z_m| e^{-i\varphi}$$

$$X_m = \omega m - \frac{K}{\omega} \qquad \tan \varphi = \frac{X_m}{R_m} = \frac{m(\omega^2 - \omega_0^2)}{\omega R_m}$$

Some time after the motion has started, the free-oscillation term $Ce^{-kt - i\omega_f t}$ has died out, and the second term remains.

$$x \to De^{-i\omega t} = \frac{F}{-i\omega Z_m} e^{-i\omega t} \qquad t \gg \frac{1}{k}$$

$$v \to -i\omega De^{-i\omega t} = \frac{F}{Z_m} e^{-i\omega t} = \frac{F}{|Z_m|} e^{-i(\omega t - \varphi)}$$

The ratio between the driving force and the velocity of the steady-state driven motion, in amplitude and phase angle, is thus equal to the quantity $Z_m = R_m - iX_m$. The steady-state amplitude of motion $|D|$ is F/K for a static ($\omega = 0$) force. As the driving frequency is increased from zero, the amplitude $|D|$ increases (if R_m is not too large) from the static value $D_{st} = F/K$ to a peak (Fig. 2.2) at *resonance*, when $\omega^2 = \omega_0^2 - 2k^2 \simeq \omega_0^2$. The resonance amplitude is $F/2km\omega_f \simeq QD_{st}$, where $Q = \omega_0 m/R_m$ is discussed in connection with (2.2.4). For higher frequencies the amplitude drops back toward zero. Thus the Q of the oscillator is the ratio between the displacement amplitude at resonance to displacement for an equal-amplitude static force (of course, if $Q < 1$, there is no peak at resonance, $|D|$ is maximum at $\omega = 0$).

We recall that the usual electrical-engineering notation is obtained by substituting $-j$ for i; so we can see that the equation for Z_m is exactly analogous to the equation for the complex electric impedance of a series circuit, with the *mechanical resistance* R_m analogous to the electric resistance, the mass m analogous to the electric inductance, and the mechanical compliance $C_m = 1/K$ analogous to the electric capacitance. The quantity $X_m = \omega m - (K/\omega)$ can be called the *mechanical reactance* of the system. The units in which this mechanical impedance is expressed are *not* ohms, for the quantity is a ratio between force and velocity rather than between voltage and current. The symbolic analogy is close enough, however, to warrant the use of the same symbol Z with the subscript m to indicate "mechanical." The units of mechanical impedance are dyne-seconds per centimeter, or grams per second.

The solution of Eq. (2.3.1) can therefore be written in either of the two alternative forms

$$x = Ce^{-kt-2\pi i\nu_f t} - \frac{F}{2\pi i\nu Z_m} e^{-2\pi i\nu t}$$

or

$$x = e^{-kt}[a_0 \cos (2\pi \nu_f t) + a_1 \sin (2\pi \nu_f t)] + \frac{F}{2\pi \nu |Z_m|} \cos (2\pi \nu t - \vartheta)$$

where $\vartheta = \varphi + \tfrac{1}{2}\pi$. The constants a_0 and a_1 are determined, as before, by the initial position and velocity of the mass.

Transient and steady state

When the force is first applied, the motion is very complicated, being a combination of two harmonic motions of (in general) different frequencies. But even if the friction is small, the first term, representing the free, or "transient," vibrations, damps out soon, leaving only the second term, which represents simple-harmonic motion of frequency equal to that of the driving force (see examples of this in Fig. 2.1).

$$x \to -\frac{F}{2\pi i\nu Z_m} e^{-i\omega t} \quad \text{or} \quad x \to \frac{F}{2\pi \nu |Z_m|} \sin (2\pi \nu t - \varphi)$$

$$v \to \frac{F}{Z_m} e^{-i\omega t} \quad \text{or} \quad v \to \frac{F}{|Z_m|} \cos (2\pi \nu t - \varphi) \tag{2.3.3}$$

where
$$\tan \varphi = -\cot \vartheta = \frac{\omega m - (K/\omega)}{R_m} \qquad \omega = 2\pi \nu$$

This part of the motion is called the *steady state*. We see that it is completely independent of the way in which the oscillator is started into motion, its amplitude, phase, and frequency depending only on the constants F and ν

of the force and on the oscillator constants m, R_m, and K.   No matter how we start the oscillator, its motion will eventually settle down into that represented by (2.3.3).

Steady-state motion is motion of a system that has forgotten how it started.

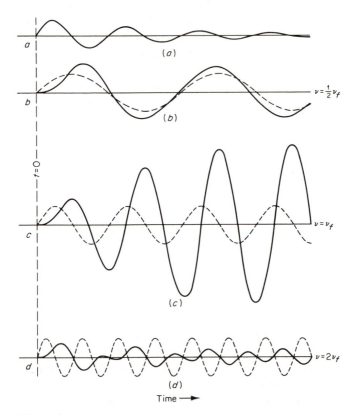

FIGURE 2.1
Forced motion of a damped-harmonic oscillator ($k = \omega_f/10$). Curve a shows free oscillations, and curves b, c, and d show forced oscillations due to sudden application of force at $t = 0$. Dotted curves give force as function of time; solid curves give displacement.   Effect of transient is apparent at the left side of the curves; at the right side the steady state is nearly reached.

Impedance and phase angle

The amplitude and velocity amplitude are proportional to the amplitude of the driving force and are inversely proportional to the magnitude of the mechanical impedance Z_m.   The analogy with electric circuits is therefore complete.   The velocity corresponds to the current, the mass to the inductance, the frictional constant to the resistance, and the stiffness constant to the reciprocal of the capacitance.

The impedance is large except when $v = v_0$, but at this frequency, if the friction is small, it has a sharply defined minimum. Therefore the amplitude of motion in the steady state is small except when $v = v_0$, where it has a sharp maximum. The case where the frequency v of the driving force equals the natural frequency v_0 of the oscillator, when the response is

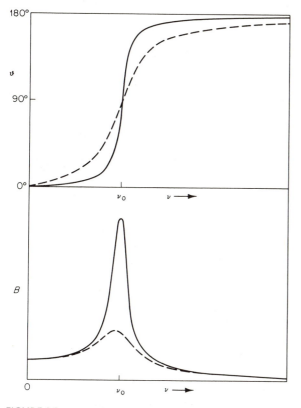

FIGURE 2.2
Phase ϑ and amplitude $B = |F/\omega Z_m|$ of forced motion as functions of the frequency v of the driving force. The frictional constant R_m for the dotted curve is eleven times that for the solid curve, all other constants being equal.

large, is called the condition of *resonance*. The peak in the curve of amplitude against v is sharp if R_m is small and is broad and low if R_m is large, as is shown in Fig. 2.2. This figure also shows that the steady-state motion of the oscillator is not very sensitive to the value of the frictional constant except in the range of frequencies near resonance. The dotted curve for the amplitude of motion is for a value of R_m eleven times that for the solid curve, yet the two are practically equal except in this frequency range.

The motion is not usually in phase with the force, the angle of lag of the displacement behind the force being given by the angle ϑ, which is zero when $v = 0$, is $\pi/2$ when $v = v_0$, and approaches π as v approaches infinity (indicating that the displacement is opposite in direction to the force).   The angle of lag of the velocity behind the force, $\varphi = -(\pi/2) + \vartheta$, is analogous to the phase angle in ac theory.   It is $-\pi/2$ when $v = 0$, zero when $v = v_0$, and $+\pi/2$ when v is very large.

In other words, when the frequency v of the driving force is much smaller than the natural frequency v_0 of the oscillator, the amplitude is small and the displacement is in phase with the force.   As v is increased, the amplitude increases and gets more and more out of phase with the force, until at resonance the amplitude is very large (if R_m is small) and the velocity is in phase with the force.   As v is still further increased, the amplitude drops down and eventually becomes very small.   For these large values of v the displacement is opposed to the force.   Figures 2.1 and 2.2 illustrate this behavior.

We use systems driven by periodic forces in two very different ways. One type of use requires the system to respond strongly only to particular frequencies (examples of this are the resonators below the bars of a xylophone, the strings of a violin, and the human mouth when shaped for a sung vowel). In this case we must make the friction as small as possible, for then the only driving frequency that produces a large response is that equal to the natural frequency of the driven system.   The other type of use requires the system to respond more or less equally well to all frequencies (examples are the diaphragms of microphones and loudspeakers and the sounding board of a violin).   In some cases we wish the steady-state amplitude to be independent of the frequency; in others we wish the amplitude of the velocity to be constant; and in still other cases we should like the acceleration to have a constant amplitude.   One or another of these requirements can be met *within a certain range of frequencies* by making the stiffness, the friction, or the mass large enough so that its effects outweigh those of the other two in the desired range of frequency.

These three limiting types of driven oscillators are called *stiffness-controlled*, *resistance-controlled*, and *mass-controlled* oscillators, respectively. Their properties and useful ranges of frequency are

Stiffness-controlled: K large $Z_m \simeq i \dfrac{K}{2\pi v}$ $x \simeq \dfrac{F}{K} e^{-i\omega t}$

$\qquad v$ considerably less than both $\dfrac{1}{2\pi} \sqrt{\dfrac{K}{m}}$ and $\dfrac{K}{2\pi R_m}$

Resistance-controlled: R_m large $Z_m \simeq R_m$ $\dfrac{dx}{dt} \simeq \dfrac{F}{R_m} e^{-i\omega t}$

$$(2.3.4)$$

$\qquad v$ considerably less than $\dfrac{R_m}{2\pi m}$, larger than $\dfrac{K}{2\pi R_m}$

Mass-controlled: m large $Z_m \simeq -2\pi i \nu m$ $\dfrac{d^2x}{dt^2} \simeq \dfrac{F}{m} e^{-i\omega t}$

ν considerably larger than both $\dfrac{1}{2\pi}\sqrt{\dfrac{K}{m}}$ and $\dfrac{R_m}{2\pi m}$

It is to be noticed that every driven oscillator is mass-controlled in the frequency range well above its natural frequency ν_0, is resistance-controlled near ν_0 (though this range may be very small), and is stiffness-controlled for frequencies much smaller than ν_0. It simply requires the proper choice of mechanical constants to place one or another of these ranges in the desired place in the frequency scale. It is also to be noted that there is *always* an upper limit to the frequency range over which an oscillator is stiffness-controlled, a lower limit to the range over which one is mass-controlled, and both an upper and a lower limit to the range over which an oscillator can be resistance-controlled. We can move these limits about by changing the mechanical constants, but we never remove the limits entirely.

Energy relations

The average energy lost per second by the oscillator due to' friction, when in the steady state, is $P = \frac{1}{2}R_m U^2 = \frac{1}{2}R_m(F/|Z_m|)^2$ [Eq. (2.2.7) of last section]. The rate of supply of energy from the driver to the driven oscillator is

$$vF \cos(2\pi\nu t) = \frac{F^2}{|Z_m|}\cos(2\pi\nu t)\cos(2\pi\nu t - \varphi)$$

$$= \frac{F^2}{|Z_m|}\cos(2\pi\nu t)[\cos\varphi\cos(2\pi\nu t) + \sin\varphi\sin(2\pi\nu t)]$$

The average value of this energy supplied per second by the driver is $\frac{1}{2}(F^2/|Z_m|)\cos\varphi = \frac{1}{2}R_m(F/|Z_m|)^2$ (since $\cos\varphi = R_m/|Z_m|$), which equals the loss of energy to friction P. One can say that the amplitude and phase of the driven oscillator so arrange themselves that the energy delivered by the driver just equals the energy lost by friction. Note the correspondence with the discussion following Eq. (1.3.2).

Response to transient forces

A simple-harmonic driving force is, of course, a very specialized way of setting the simple oscillator in motion. Driving forces are much more likely to be nonperiodic. For example, the force setting the piano string into motion is the blow of a hammer, and the oscillatory forces acting on an airplane wing are the nonperiodic reactions of turbulent air. As long as the internal restoring force $F_i(x)$ is linear in x, however, this complication adds no basically new complication. For if all the terms on the left-hand side of Eq. (2.3.1) are linear in x or its derivatives, a superposition of solutions is a solution. For example, if the driving force has two components, $F_e = F_1 e^{-i\omega_1 t} + F_2 e^{-i\omega_2 t}$, then the steady-state solution of (2.3.1) is the sum $(-F_1/i\omega_1 Z_1)e^{-i\omega_1 t} + (-F_2/i\omega_2 Z_2)e^{-i\omega_2 t}$ of the responses for each component separately, Z_1 being the impedance of the system, given in Eq. (2.3.2), for $\omega = \omega_1$, and Z_2 is the impedance appropriate for the driving frequency $\omega_2/2\pi$. Since, as we saw in Sec. 1.3, any function of t can be built up as a sum or integral of simple-harmonic terms, the corresponding motion of the linear oscillator is the corresponding sum or integral of the responses to each of the components separately.

This additive property of the solution works only for the linear oscillator; in Chap. 14 we shall see that it does not work if either or both the frictional and elastic restoring forces are nonlinear. In the linear case, the equation for a nonperiodic driving force $f(t)$,

$$m\frac{d^2x}{dt^2} + R_m\frac{dx}{dt} + Kx = f(t)$$

has a Fourier transform which can be expressed in terms of the transforms X and F of x and f, respectively. From page 31 we see that we obtain

$$(-m\omega^2 - iR_m\omega + K)X(\omega) \equiv -i\omega Z(\omega)X(\omega) = F(\omega) \qquad (2.3.5)$$

where $X(\omega) = \dfrac{1}{2\pi}\displaystyle\int_{-\infty}^{\infty} x(t)e^{i\omega t}\,dt \qquad F(\omega) = \dfrac{1}{2\pi}\displaystyle\int_{-\infty}^{\infty} f(t)e^{i\omega t}\,dt$

[Eqs. (1.3.16)] and where $Z(\omega)$ is the impedance defined in Eq. (2.3.2) for a simple-harmonic driving force of frequency $\omega/2\pi$. Therefore, for the linear oscillator, the response to any driving force can be computed by first analyzing the force into its frequency components $F(\omega)$, finding the response $X(\omega)$ for each component separately, and finally combining these by the Fourier integral

$$x(t) = \int_{-\infty}^{\infty} X(\omega)e^{-i\omega t}\,d\omega = \int_{-\infty}^{\infty} \frac{F(\omega)}{-i\omega Z(\omega)} e^{-i\omega t}\,d\omega \qquad (2.3.6)$$

to make up the complete response $x(t)$.

Examples of the Fourier transform method

Suppose, for example, the mass is given a push centered at $t = 0$, represented by the formula

$$f(t) = \begin{cases} \dfrac{a}{2}\,e^{at} & t < 0 \\[2mm] \dfrac{a}{2}\,e^{-at} & t > 0 \end{cases}$$

where the mean duration of the push is proportional to $1/a$, and the magnitude of the push has been adjusted so that its total impulse $\int f\,dt$ is unity.

The Fourier transform of this function is

$$F(\omega) = \frac{1}{2\pi} \int_{-\infty}^{\infty} f(t)e^{i\omega t}\, dt = \frac{a/4\pi}{a + i\omega} + \frac{a/4\pi}{a - i\omega} = \frac{a^2/2\pi}{a^2 + \omega^2}$$

Thus, according to Eq. (2.3.5), the Fourier transform of the displacement is

$$X(\omega) = \frac{F(\omega)}{-m\omega^2 - iR_m\omega + K}$$

$$= \frac{-(a^2/2\pi m)}{(\omega + ia)(\omega - ia)(\omega - \omega_f + ik)(\omega + \omega_f + ik)}$$

where $k = R_m/2m$, $\omega_f{}^2 = \omega_0{}^2 - k^2$, and $\omega_0{}^2 = K/m$ are the damping constant, the squares of the resonance and natural frequencies (times 2π) discussed earlier in this section.

We are now in a position to use Eq. (1.2.15) to compute $x(t)$, the inverse transform of $X(\omega)$, from Eq. (2.3.6). The integrand of the integral for x, as a function of the complex variable ω, has four poles, at $\omega = \pm ia$ and $\omega = \pm\omega_f - ik$, one above the real axis and the other three below. The residue of $X(\omega)e^{-i\omega t}$ at $\omega = ia$ is $ae^{at}/4\pi im[\omega_f{}^2 + (k + a)^2]$, that at $\omega = -ia$ is $-ae^{-at}/4\pi im[\omega_f{}^2 + (k - a)^2]$, and those at the other two poles are $-a^2e^{-kt-i\omega_f t+i\Phi}/4\pi m\omega_f A$ and $a^2e^{-kt+i\omega_f t-i\Phi}/4\pi m\omega_f A$, respectively, where $a^2 + (\omega_f + ik)^2 = Ae^{i\Phi}$ (A being real). Consequently, the instructions of Eq. (1.2.15) produce

$$x(t) = \begin{cases} \dfrac{ae^{at}}{2m[\omega_f{}^2 + (k + a)^2]} & t < 0 \\[3mm] \dfrac{ae^{-at}}{2m[\omega_f{}^2 + (k - a)^2]} + \dfrac{a^2}{m\omega_f A}\, e^{-kt}\sin(\omega_f t - \Phi) & t > 0 \end{cases}$$

Since $A^2 = (a^2 + \omega_f{}^2 + k^2)^2 - 4k^2a^2$ and $\sin\Phi = 2k\omega_f/A$, it is not difficult to verify that $x(t)$ and dx/dt are both continuous at $t = 0$, though df/dt is discontinuous at $t = 0$. The displacement x increases exponentially during the time $t < 0$, when the applied force is smoothly increasing. At $t = 0$ there is a discontinuity in the rate of change of the force, which generates a transient free vibration in x, exhibiting the damped oscillations of Eq. (2.2.3).

When $a \to \infty$, the force $f(t)$ becomes the Dirac delta function of Eq. (1.3.24), corresponding to an instantaneous unit impulse applied at $t = 0$. In this case $F(\omega) \to 1/2\pi$, and the solution becomes

$$x(t) = \begin{cases} 0 & t < 0 \\[3mm] \dfrac{1}{m\omega_f}\, e^{-kt}\sin(\omega_f t) & t > 0 \end{cases} \tag{2.3.7}$$

which could, of course, have been obtained from Eqs. (2.2.3) (for initial displacement zero, initial velocity $1/m$).

When the driving force starts at $t = 0$ and the oscillator is at rest before this, the specialized techniques of the Laplace transform [see Eq. (1.3.22)] are useful. For example, suppose the driving force is

$$f(t) = \begin{cases} 0 & t < 0 \\ \sin{(ut)} & t > 0 \end{cases}$$

which has a transform

$$F(\omega) = F(is) = \frac{1}{2\pi} \int_0^\infty e^{-st} \sin{(ut)}\, dt = \frac{u/2\pi}{u^2 + s^2} = \frac{-(u/2\pi)}{(\omega + u)(\omega - u)}$$

The Fourier transform of the displacement is then

$$X(\omega) = \frac{F(\omega)}{-i\omega Z(\omega)} = \frac{u/2\pi m}{(\omega + u)(\omega - u)(\omega - \omega_f + ik)(\omega + \omega_f + ik)}$$

with all its poles on or below the real axis of ω. Thus, according to Eqs. (1.2.15), $x(t) = 0$ for $t < 0$ (as, of course, it should, since no force acts on the system until $t = 0$). The inverse transform for $t > 0$ is

$$x(t) = \frac{1}{mA}\left[\sin{(ut - \vartheta)} - \frac{u}{\omega_f} e^{-kt} \sin{(\omega_f t - \Phi)}\right] \tag{2.3.8}$$

where $\quad \omega_f{}^2 - (u - ik)^2 = Ae^{i\vartheta} \qquad \tan\vartheta = \dfrac{2uk}{(\omega_0{}^2 - u^2)} = \dfrac{uR_m}{K - mu^2}$

$(\omega_f + ik)^2 - u^2 = Ae^{i\Phi} \qquad \tan\Phi = \dfrac{2k\omega_f}{\omega_f{}^2 - k^2 - u^2}$

$A^2 = (\omega_f{}^2 + k^2 - u^2)^2 + 4u^2k^2 = \left(\dfrac{u}{m}\right)^2 |Z(u)|^2$

Curves for x against t, for different values of $u/2\pi$, the driving frequency, are shown in Fig. 2.1. At $t = 0$ the oscillator starts from zero displacement and velocity, with linearly increasing acceleration at first; its motion for a while is a combination of forced and transient oscillations; eventually, the transient term [the second term in (2.3.8)] dies out and leaves the steady-state motion (the first term). The angles ϑ and Φ are those discussed in Eqs. (2.3.2) and (2.3.3). This time we have used $u/2\pi$ for the driving frequency, instead of ν or $\omega/2\pi$, since we are now using ω as the variable in the Fourier transform.

There is still another way of looking at Eq. (2.3.6), which will have some application later. We insert the definition of $F(\omega)$ into Eq. (2.3.6) and invert the order of integration.

$$x(t) = \int_{-\infty}^\infty \frac{e^{-i\omega t}\, d\omega}{-i\omega Z(\omega)} \frac{1}{2\pi} \int_{-\infty}^\infty f(\tau)e^{i\omega\tau}d\tau = \int_{-\infty}^\infty f(\tau)\, d\tau \left[\frac{1}{2\pi} \int_{-\infty}^\infty \frac{e^{-i\omega(t-\tau)}}{-i\omega Z(\omega)}d\omega\right]$$

To elucidate the implications of this expression, we return to Eq. (2.3.7) for the response of an oscillator to a unit impulsive driving force. Formally speaking, this can be represented by the inverse Fourier transform of the product of $1/-i\omega Z(\omega)$ and the Fourier transform of a delta function, which is unity. If an impulsive force $f(\tau)\,\delta(t - \tau)$ is applied at $t = \tau$, the Fourier transform of the force is

$$F(\omega) = \frac{1}{2\pi} \int_{-\infty}^{\infty} f(\tau)e^{i\omega t}\,\delta(t - \tau)\,dt = \frac{1}{2\pi}f(\tau)e^{i\omega\tau}$$

and the response of the oscillator is $f(\tau)g(t - \tau)$, where

$$g(t - \tau) = \frac{1}{2\pi} \int_{-\infty}^{\infty} \frac{e^{-i\omega(t-\tau)}}{-i\omega Z(\omega)}\,d\omega = \begin{cases} 0 & t < \tau \\ \dfrac{1}{m\omega_f}\,e^{-k(t-\tau)}\sin\left[\omega_f(t - \tau)\right] & t > \tau \end{cases}$$

$$m\frac{d^2g}{dt^2} + R_m\frac{dg}{dt} + Kg = \delta(t - \tau)$$

is just the quantity in brackets in the double integral for $x(t)$.

In other words, the effect of each portion $f(\tau)$ of the force, for each separate time interval $d\tau$, can be computed separately, and the separate responses can then be combined,

$$x(t) = \int_{-\infty}^{\infty} f(\tau)g(t - \tau)\,d\tau \qquad (2.3.9)$$

to obtain the complete response. Instead of analyzing the force into its frequency components, we can analyze it into its *time components* $f(\tau)\,\delta(t - \tau)$, solve the equation of motion for each impulsive force separately, and then combine the solutions to obtain $x(t)$. Again, this is possible only because the equation of motion is linear in x.

To finish this discussion, we write down the general rule for finding the response of the simple oscillator to a driving force $f(t)$, which starts at $t = 0$ and has a Fourier transform $F(\omega) = \dfrac{1}{2\pi} \int_0^{\infty} e^{i\omega t}f(t)\,dt$, as long as $F(\omega)$ has only simple poles (none of them will be above the real axis if $f = 0$ for $t < 0$). Using Eqs. (1.2.15), we have

$$x(t) = 0 \qquad\qquad\qquad\qquad\qquad \text{for } t < 0$$

$$x(t) = \frac{\pi}{m}\frac{1}{i\omega_f}\,e^{-kt}[e^{i\omega_f t}F(-\omega_f - ik) - e^{-i\omega_f t}F(\omega_f - ik)]$$

$$-\left(\begin{array}{l} 2\pi i \text{ times the sum of the residues of} \\ Fe^{-i\omega t}/(-i\omega Z) \text{ at all simple poles of} \\ F \text{ on and below the real axis of } \omega \end{array}\right) \qquad \text{for } t > 0 \qquad (2.3.10)$$

$$k = \frac{R_m}{2m} \qquad \omega_0{}^2 = \frac{K}{m} \qquad \omega_f{}^2 = \omega_0{}^2 - k^2$$

The first term is the transient motion of the system due to the sudden onset of $F(t)$ at $t = 0$. The terms due to the poles of $F(\omega)$ are a sort of generalized steady-state motion and depend on the specific form of $f(t)$.

The corresponding expression for $v(t)$ is

$$v(t) = 0 \qquad\qquad\qquad\text{for } t < 0$$

$$v(t) = \frac{\pi}{m\omega_f} e^{-kt}[(\omega_f + ik)e^{i\omega_f t}F(-\omega_f - ik)$$
$$+ (\omega_f - ik)e^{-i\omega_f t}F(\omega_f - ik)]$$
$$- \left(\begin{matrix} 2\pi i \text{ times the sum of the residues of} \\ Fe^{-i\omega t}/Z \text{ at all poles of } F \text{ on or below} \\ \text{the real axis of } \omega \end{matrix}\right) \qquad \text{for } t > 0$$

Further discussion requires choosing specific forms for f and F.

The two examples we have worked out have displayed both transient and steady-state terms.

Autocorrelation of the response

In addition to impulses and other suddenly applied forces, acoustic systems are often subjected to stationary, bounded, nonperiodic driving forces. The pressure fluctuations caused by turbulent flow and the random impulses imparted to automobile springs, while moving over a rough road, are examples. The spectrum density $|F(\omega)|^2$ and autocorrelation function $\Upsilon(\tau)$, discussed in Sec. 1.3, are appropriate means of describing such forces; we can also use them to describe the response of the acoustic system.

A typical driving force of this kind is the random fluctuation called "white noise," which is a superposition of all frequencies below some natural limit $\omega_n/2\pi$, the amplitude of each frequency component below this limit being roughly equal. A white-noise driving force will thus have a spectrum density $|F(\omega)|^2$, that is independent of ω for $|\omega| < \omega_n$ and drops to zero for $|\omega| > \omega_n$. In such a random noise, the phase of each frequency component, and thus the phase angle of $F(\omega)$, is a random function of ω, any value from 0 to 2π being equally likely. Thus the magnitude of $F(\omega)$ is a rather simple function of ω, but its phase cannot be expressed in terms of any simple, continuous function. Because of this irregular dependence of phase angle Φ on ω, it is impossible to carry out the inverse Fourier transform and express a white-noise force $f(t)$ as an analytic function of t.

We can approximate white-noise behavior by taking the sum (instead of the integral) of a number of sinusoidal terms, with randomly chosen frequencies and phase angles. The curve for $f(t)$ shown in Fig. 2.3, for example, has been computed from a series of 10 unit-amplitude terms,

$$f(t) = \sum_{m=1}^{10} \cos(\omega_m t - \Phi_m)$$

where the Φ's were chosen at random from the range 0 to 2π and the ω's were random choices from the range 0 to 10π. Thus the plotted curve for $f(t)$ approximates white noise with a "natural limit frequency" of 5. No periodicity is apparent; once in a while the components nearly all reinforce each other and a peak occurs, but these peaks are not periodically related. Most of the time the curve meanders aimlessly above and below the t axis.

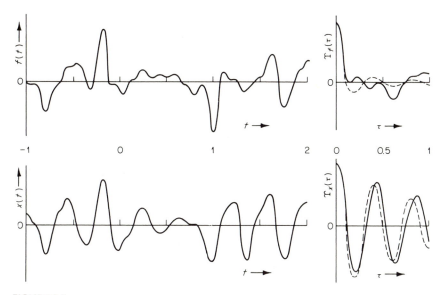

FIGURE 2.3
White-noise force $f(t)$, its autocorrelation function $\Upsilon_f(\tau)$; corresponding response $x(t)$ of simple oscillator and its autocorrelation function $\Upsilon_x(\tau)$. Solid curves are for the sample curve shown; dotted curves are the mean autocorrelation function for white noise.

As we pointed out in Sec. 1.3, the appropriate way to describe such a random force is in terms of its autocorrelation function

$$\Upsilon_f(\tau) = \lim_{T \to \infty} \left[\frac{1}{T} \int_{-\frac{1}{2}T}^{\frac{1}{2}T} f(t) f(t + \tau) \, dt \right] \qquad (2.3.11)$$

and its related Fourier transform [Eq. (1.3.20)]

$$\frac{1}{2\pi} \int_{-\infty}^{\infty} e^{i\omega\tau} \Upsilon_f(\tau) \, d\tau = \frac{2\pi |F(\omega)|^2}{T}$$

which is proportional to the spectrum density $|F|^2$ of $f(t)$. Although the phase angle of the Fourier transform $F(\omega)$ of $f(t)$ is a random function of ω, the spectrum density $|F|^2$ is a fairly smooth function of ω, which can be measured and approximated in terms of simple analytic functions. Similarly with the autocorrelation function: dealing with squares of amplitudes avoids difficulties with irregularities of phase. For example, the $f(t)$ of Fig. 2.3 corresponds roughly to a density $|F|^2$ which is constant $(F = F_0)$ for the frequency range $-\omega_n < \omega < \omega_n$ $(\omega_n = 10\pi$ in the figure) and is zero for $|\omega| > \omega_n$. The autocorrelation function of $f(t)$ is therefore

$$\Upsilon_f(\tau) = \frac{2\pi F_0^2}{T} \int_{-\omega_n}^{\omega_n} e^{-i\omega\tau} \, d\omega = \frac{4\pi\omega_n F_0^2}{T} \frac{\sin(\omega_n\tau)}{\omega_n\tau} \qquad (2.3.12)$$

which is the dotted line in the right-hand top part of Fig. 2.3. The solid line was obtained by integrating the product $f(t)f(t + \tau)$ as read from the plot, over t from $t = -1$ to $t = +1$ [i.e., for $T = 2$ in Eq. (2.3.11)]. This is, of course, only approximately equal to the Υ_f for a function which has uniform spectrum density from $-\omega_n$ to $+\omega_n$, given by the dotted line, but the correspondence is fairly good.

The white-noise forces encountered in practice do not usually have quite as sharp a cutoff in their spectrum density at $\omega = \omega_n$. All frequencies are represented in the pressure fluctuations of turbulence, for example, though the density diminishes in magnitude for ω larger than some ω_n. A better approximation for the spectrum density would be

$$|F(\omega)|^2 = F_0^2 \exp \frac{-\omega^2}{2\omega_n^2}$$

which has an autocorrelation function

$$\Upsilon_f(\tau) = \frac{2\pi F_0^2}{T} \int_{-\infty}^{\infty} e^{-i\omega\tau - (\omega^2/2\omega_n^2)} \, d\omega = (2\pi)^{\frac{3}{2}} \frac{\omega_n F_0^2}{T} e^{-\frac{1}{2}\omega_n^2\tau^2} \qquad (2.3.13)$$

having a "peak" at $\tau = 0$, of "width" $2/\omega_n$, dropping exponentially to zero on both sides of the peak, without any of the oscillations of the Υ_f of Eq. (2.3.12). The small amount of high frequencies present in this variety of white noise makes the plot of $f(t)$ more jagged than that shown in Fig. 2.3; this high frequency cancels out the last trace of oscillations in Υ_f.

Having now seen how a randomly fluctuating force can be described in terms of its spectrum density and its autocorrelation function, we must next determine the response of a simple oscillator to such a driving force. Reference to Eq. (2.3.5) shows that the spectrum density for the motion of the system is

$$|X(\omega)|^2 = \frac{|F(\omega)|^2}{\omega^2 |Z|^2}$$

$$= \frac{|F(\omega)|^2/m^2}{(\omega + \omega_f + ik)(\omega - \omega_f + ik)(\omega + \omega_f - ik)(\omega - \omega_f - ik)} \qquad (2.3.14)$$

where ω_f and k are defined in Eqs. (2.2.1) and (2.2.2) and after Eq. (2.3.10).

In most cases of practical interest the "cutoff" frequency $\omega_n/2\pi$ for the random force is considerably larger than the resonance frequency $\omega_f/2\pi$ of the oscillator, so that, over the range of ω, where $|X(\omega)|^2$ is nonnegligible, $|F(\omega)|^2$ is practically equal to F_0^2, and we may substitute F_0^2 for $|F(\omega)|^2$ in Eq. (2.3.14).

Several interesting results are immediately apparent. In the first place, as long as the driving force is essentially white noise over the range covering ω_f, then the spectrum density $|X(\omega)|^2$ of the oscillator displacement is determined primarily by the properties of the oscillator (its ω_f and k or Q), and not by the presence or absence of frequency components in $f(t)$, for frequencies much larger than $\omega_f/2\pi$. In fact, the only effect of the white noise on the spectrum density of the displacement is on its magnitude, which is proportional to F_0^2. In the second place, unless the frictional term $k = R_m/2m$ is larger than $\omega_0 = \sqrt{K/m}$, the resonance peak in $1/|Z|^2$ ensures that the spectrum density $|X|^2$ of the displacement has a peak at $\omega = \omega_f$. If k is much smaller than ω_f ($Q \gg 1$), the frequencies near $\omega_f/2\pi$ predominate in the motion of the oscillator; the system has picked out those frequencies in the driving force which are near its own resonance frequency. This, of course, has an effect on the form of the autocorrelation function $\Upsilon_x(\tau)$ of the oscillator's displacement, which can be seen as soon as we compute it.

$$\Upsilon_x(\tau) = \frac{2\pi F_0^2}{T} \int_{-\infty}^{\infty} \frac{e^{-i\omega\tau}\, d\omega}{\omega^2\, |Z(\omega)|^2} = \left(\frac{\pi^2 F_0^2}{m^2 k \omega_0 \omega_f T}\right) e^{-k\tau} \cos\left(\omega_f \tau - \theta\right) \quad (2.3.15)$$

where $\omega_f + ik = \omega_0 e^{i\theta}$, or $\sin\theta = k/\omega_0 = \frac{1}{2}Q$, and T is very large compared with $2\pi/\omega_f$. The expression given is for positive values of τ; for negative values, change the sign of τ and of θ in the exponential and the cosine. This function Υ_x has zero slope at $\tau = 0$, drops down to zero value at $\tau = \pm[(\frac{1}{2}\pi + \theta)/\omega_f]$, exhibiting damped oscillations beyond this, showing that, when k is small, there is considerable long-range correlation in $x(t)$. Indeed, the behavior of Υ_x is similar to the free vibrations of x, as a mental analysis of the system's response will demonstrate it should be. The factor in parentheses, multiplying the damped cosine, is of course the mean-square amplitude of motion of the oscillator.

The curve for $x(t)$ in Fig. 2.3 is a typical response curve for a simple oscillator with $\omega_f = 5\pi$ and $k = 0.5$ to a random driving force of the sort shown above it. The increased emphasis of the frequencies near $\nu_f = 2.5$ (period = 0.4) is apparent, but it is also apparent that $x(t)$ is far from being periodic in time; the nonperiodic driving force continually interferes with whatever resonance oscillations it may start. When one calculates the autocorrelation function, the random disturbances average out, and Υ_x resembles the undisturbed vibration of the oscillator. The solid curve for Υ_x is obtained by integrating the curve for $x(t)$ for $T = 2$; it checks fairly well with the dotted curve, obtained from Eq. (2.3.15), considering the smallness of the value of T.

Use of the Laplace transform

In many cases we do not know, or do not wish to know, what has happened to the system in the past. We know the position and velocity at $t = 0$ and the nature of the forces acting after $t = 0$, and we wish to compute the motion thereafter. In such a case it is somewhat easier to use the modification of the Fourier transform called the Laplace transform. It was defined in Eqs. (1.3.23). The basic properties may be obtained from those of the Fourier transform. The following equations will be of use later: $F_L(s)$ is the Laplace transform of $f(t)$ if

$$F_L(s) = \int_0^\infty f(t)e^{-st}\, dt \qquad f(t) = \frac{1}{2\pi i}\int_{-i\infty+\epsilon}^{i\infty+\epsilon} F_L(s)e^{st}\, ds$$

where $\omega = is$ is the corresponding variable for the Fourier transform.

To solve the equation for the forced motion of a simple oscillator, for $t > 0$, when the force is suddenly applied at $t = 0$,

$$m\frac{d^2x}{dt^2} + R\frac{dx}{dt} + Kx = f(t)u(t)$$

we take its Laplace transform,

$$(ms^2 + Rs + K)X_L(s) - mx'(0) - (ms + R)x(0) = F_L(s)$$

where $x(0)$ is the initial displacement of the oscillator and $x'(0)$ is its initial velocity. We can then solve for $X_L(s)$, the Laplace transform of $x(t)$.

$$X_L(s) = \frac{(1/m)F_L(s) + x'(0) + (s + 2k)x(0)}{(s + k)^2 + \omega_f^2} \qquad (2.3.16)$$

where $k = R/2m$, and $\omega_f^2 = (K/m) - k^2$ as before [Eq. (2.3.10) et seq.]. Since $x(0)$ and $x'(0)$ are present, this formula already includes initial conditions.

When we take the inverse transform of this expression, the initial values $x(0)$ and $x'(0)$ are numerical constants, independent of s, so that the transient part of the solution can be obtained from (9), (15), and (16) of Table 2.1.

$$x_{tr} = \text{inverse transform of } \frac{x'(0) + kx(0) + (s + k)x(0)}{(s + k)^2 + \omega_f^2}$$

$$= \frac{x'(0) + kx(0)}{\omega_f} e^{-kt}\sin(\omega_f t) + x(0)e^{-kt}\cos(\omega_f t) \qquad (2.3.17)$$

which holds for the transient part for $t > 0$. To this must be added the inverse transform of $F_L(s)/[(s + k)^2 + \omega_f^2]$ to obtain the complete expression for x for $t > 0$. It is not difficult to see that x_{tr} reduces to $x(0)$ at $t = 0$, and that its time derivative reduces to $x'(0)$.

In most cases of interest the inverse transform of the F_L term (the steady-state motion) can be obtained by appropriate manipulation of Table 2.1. For example, if the mass is hit by a hammer just after $t = 0$, being given an impulse I_0, then the driving force is $f(t) = I_0 \delta(t - \epsilon)$ $(\epsilon \rightarrow 0)$, and $F_L(s) = I_0 e^{-\epsilon s}$ $(\epsilon \rightarrow 0)$. The initial values $x(0)$ and $x'(0)$ are the displacement and velocity of the oscillator just before the hammer stroke. Formula

TABLE 2.I

FUNCTION	LAPLACE TRANSFORM
(1) $fa(t)$	$aF_L(s)$
(2) $f(at)$	$(1/a)F_L(s/a)$
(3) $f'(t) = df/dt$	$sF_L(s) - f(0)$
(4) $f''(t) = d^2f/dt^2$	$s^2F_L(s) - f'(0) - sf(0)$
(5) $\int_0^t f(x)dx$	$(1/s)F_L(s)$
(6) $t^nf(t)$	$(-1)^n(d^nF_L/ds^n)$
(7) $(1/t)f(t)$	$\int_s^\infty F_L(q)dq$
(8) $f(t - a) \quad t > a$ $0 \qquad\quad t \leqslant a$	$e^{-as}F_L(s)$
(9) $e^{bt}f(t)$	$F_L(s - b)$
(10) $\int_0^t f(x)g(t - x)\,dx$	$G_L(s)F_L(s)$
(11) $\delta(t - a)$ [Eq. (1.3.25)]	e^{-as}
(12) $u(t - a)$ [Eq. (1.3.22)]	$(1/s)e^{-as}$
(13) $e^{bt}u(t)$	$1/(s - b)$
(14) $t^ne^{bt}u(t)$	$n!/(s - b)^{n+1}$
(15) $\sin(bt)u(t)$	$b/(s^2 + b^2)$
(16) $\cos(bt)u(t)$	$s/(s^2 + b^2)$
(17) $t^n \sin(bt)u(t)$	$\dfrac{n!}{2i}[(s - ib)^{-n-1} - (s + ib)^{-n-1}]$
(18) $t^n \cos(bt)u(t)$	$\dfrac{n!}{2}[(s - ib)^{-n-1} + (s + ib)^{-n-1}]$
(19) $2\sum_{n=0}^{\infty} f(t - 2na - a)u(t - 2na - a)$	$F_L(s)/\sinh(as)$
(20) $2\sum_{n=0}^{\infty} (-1)^nf(t - 2na - a)u(t - 2na - a)$	$F_L(s)/\cosh(as)$
(21) $2\sum_{n=0}^{\infty} e^{-(2n+1)b} \delta(t - 2na - a)$	$1/\sinh(as + b)$

(15) of Table 2.1 can then be used to complete the formula for the displacement of the oscillator after the hammer blow,

$$x(t) = \frac{I_0 + mx'(0) + mkx(0)}{m\omega_f} e^{-kt} \sin(\omega_f t) + x(0)e^{-kt} \cos(\omega_f t) \quad (2.3.18)$$

the velocity just after the hammer blow being greater than that just before by the amount I_0/m.

In some cases the F_L term must be expanded in partial fractions before the inverse transforms can be found in Table 2.1. The following partial fraction expansions will be useful later:

$$\frac{1}{(s-a)(s-b)} = \frac{1}{(s-a)(a-b)} + \frac{1}{(s-b)(b-a)}$$

$$\frac{1}{(s-a)(s-b)(s-c)} = \frac{1}{(s-a)(a-b)(a-c)} + \frac{1}{(s-b)(b-a)(b-c)}$$

$$+ \frac{1}{(s-c)(c-a)(c-b)} \qquad (2.3.19)$$

$$\frac{1}{(s-a)^2(s-b)} = \frac{1}{(s-a)^2(a-b)} - \frac{1}{(s-a)(a-b)^2} + \frac{1}{(s-b)(b-a)^2}$$

$\cdots \quad \cdots \quad \cdots \quad \cdots \quad \cdots \quad \cdots \quad \cdots \quad \cdots$

The last expression has a pole of second order at $s = a$; its inverse transform would have a term proportional to te^{-at}.

For example, if $f(t) = \sin(ut)u(t)$, so that $F_L = u/(s^2 + u^2)$, the steady-state term in X_L can be expanded into two terms, one inversely proportional to $(s + k)^2 + \omega_f^2$, the other to $s^2 + u^2$; the resulting inverse transform is, of course, equal to Eq. (2.3.8).

Problems

1   A vibrator consists of a 100-g weight on the end of a spring. The spring's restoring force is proportional to the weight's displacement from equilibrium; if the weight is displaced 1 cm, this force is 10,000 dynes. The frictional force opposing its motion is proportional to its velocity and is 100 dynes when its velocity is 1 cm per sec. What is the modulus of decay of the oscillator? What is its decrement? What "frequency" do the vibrations have? What frequency would they have if there were no friction? If the weight were originally at rest and then were struck so that its initial velocity was 1 cm per sec, what would be its subsequent motion? What would be its maximum displacement from equilibrium?

2   The diaphragm of a loudspeaker weighs 1 g, and the displacement of its driving rod 1 mm from equilibrium requires a force of 1 million dynes. The frictional force opposing motion is proportional to the diaphragm's velocity and is 300 dynes when the velocity is 1 cm per sec. If it is assumed that the diaphragm moves like a simple oscillator, what will be its natural frequency, and what its modulus of decay? The driving rod is driven by a force of 100,000 $\cos(2\pi\nu t)$ dynes. Plot curves of real and imaginary parts of the mechanical impedance of the diaphragm as function of frequency, from $\nu = 0$ to $\nu = 1,000$ cps.

3   The diaphragm of Prob. 2 is driven by a force of 100,000 $\cos(2\pi\nu t)$ dynes. Plot a curve of the amplitude of motion of the diaphragm as function of frequency, from $\nu = 0$ to $\nu = 1,000$ cps. Over what frequency range is this loudspeaker mass-controlled?

4 What is the mechanical impedance of a mass m without spring or friction? What is the impedance of a spring without mass? What will be the angle by which the displacement of the mass lags behind an oscillating force? What will be the angle of lag of the velocity of the mass behind the force? What are the corresponding angles for the spring?

5 The driving rod of the loudspeaker of Prob. 2 is driven by a force of $10,000[15 \sin (200\pi t) - 10 \sin (600\pi t) + 3 \sin (1000\pi t)]$ dynes. Plot the displacement of the diaphragm during one cycle, and compare it with the curve for the force.

6 The sharpness of resonance of a forced damped oscillator is given by the "half-breadth of the resonance peak," the difference between the two frequencies for which the amplitude of oscillation is half that at the resonance frequency ν_0. Prove that if the natural period of oscillation is negligibly small compared with 2π times the modulus of decay (i.e., if $k/4\pi$ is small compared with ν_0), then this half breadth equals $\sqrt{3}/\pi$ times the reciprocal of the modulus of decay of the oscillator. What is the half breadth for the diaphragm of Prob. 2? What would the frictional force have to be in order that the half breadth may be 20 cycles?

7 A mass m is attached to the lower end of a spring of stiffness constant K. The upper end of the spring is moved up and down with an amplitude $Be^{-i\omega t}$, and the frictional force on the mass is proportional to the relative velocity of the mass and the upper end of the spring (dr/dt), where $r = x - Be^{-i\omega t}$. Show that the equation of motion of the mass is

$$m \frac{d^2x}{dt^2} + R_m \frac{dx}{dt} + Kx = (K - i\omega R_m)Be^{-i\omega t}$$

that the steady-state motion of the mass is

$$x = \frac{R_m + i(K/\omega)}{R_m - i[\omega m - (K/\omega)]} Be^{-i\omega t}$$

and that the phase lag of x behind the displacement of the top of the spring is $\tan^{-1}\{[\omega m - (K/\omega)]/R_m\} + \tan^{-1}(K/\omega R_m)$. What is the amplitude of motion of x? What are the phase and amplitude of x at very low frequencies? At very high frequencies?

8 A simple mechanical system of impedance $Z_m = R_m - i[\omega m - (K/\omega)]$ has applied to it a force

$$F(t) = \begin{cases} 0 & t < 0 \\ \sin (at) & 0 < t < \dfrac{n\pi}{a} \\ 0 & \dfrac{n\pi}{a} < t \end{cases}$$

Use the results of Prob. 15 of Chap. 1 and of Eq. (2.3.10) to compute the displacement of the system.

9 Using the integral formula

$$\int_{-\infty}^{\infty} \exp\left[-i\omega t - \frac{t^2}{2T^2}\right] dt = \sqrt{\frac{\pi}{2}} \, T \exp\left(-\tfrac{1}{2}\omega^2 T^2\right)$$

calculate the response of a simple linear oscillator with mass, resistance, and stiffness to a transient driving force

$$f(t) = F_0 \exp\left[-\tfrac{1}{2}\left(\frac{t}{T}\right)^2\right]$$

10   A simple oscillator of mass m, stiffness constant K, and resistance R is acted on by a stationary, nonperiodic driving force $f(t)$ having an autocorrelation function

$$\Psi_f(\tau) = \langle F^2\rangle \exp\left(-a\,|\tau|\right)$$

where $\langle F^2\rangle$ is the mean-square amplitude of the force.   Use Eq. (1.3.20) to compute the spectrum density $|F(\omega)|^2$ of the driving force over a long time period T.   Show that the spectrum density of the amplitude of the steady-state motion of the simple oscillator, under the influence of the force, is

$$|X(\omega)|^2 = \frac{(aT/2\pi^2m^2)\langle F^2\rangle}{(\omega^2 + a^2)[(\omega^2 - \omega_f^2)^2 + 2(\omega^2 + \omega_f^2)k^2 + k^4]}$$

where k and ω_f are given in Eqs. (2.2.1) and (2.3.10).   Discuss (and sketch) the shape of this frequency-response curve as a function of ω, relating the heights and widths of its peaks when $k \ll \omega_0 \ll a$, $k \simeq \omega_0 \ll a$, and $k \ll \omega_0 \ll a$.

11   A force $f(t) = Qte^{-bt}u(t)$ is applied to a simple oscillator of mass m, stiffness K, and resistance R, initially at rest.   Use the Laplace transform to calculate the displacement of the mass as a function of time for $t > 0$.

12   A periodically pulsating force

$$f(t) = F_0 \sum_{n=0}^{\infty} \delta\left[t - \frac{\pi}{\omega_f}(2n + 1)\right] \qquad \text{[see Eq. (1.3.24) for } \delta\text{]}$$

is applied to a simple oscillator, where the ω_f of the force is adjusted to be equal to the ω_f of the oscillator, as defined in the discussion preceding Eqs. (2.3.7).   Show that, between the Nth and the $(N + 1)$st pulse, when $N \gg \omega_f/k$, so that steady state has been reached, the displacement of the oscillator is

$$x(t) \simeq \frac{F_0/m\omega_f}{1 - \exp\left(-2\pi k/\omega_f\right)}\, e^{-ktN} \sin\left(\omega_f t_N\right)$$

$$\text{for } 0 < t_N = t - \frac{\pi}{\omega_f}(2N + 1) < \frac{2\pi}{\omega_f}$$

CHAPTER 3

COUPLED LINEAR OSCILLATORS

3.1 TWO DEGREES OF FREEDOM

Perhaps the next step in a purely logical presentation of the subject would be to discuss the motion of a one-dimensional oscillator under the influence of a nonlinear stiffness force. However, the effects of a nonlinear force are complicated enough, and the techniques used in solving nonlinear equations differ sufficiently from those developed in Chap. 2, so that it will be preferable to defer nonlinear problems until a later chapter. Instead, we shall turn to the study of the behavior of two or more simple linear oscillators coupled together by forces which also are linear.

Examples of coupled oscillators

The simplest coupled system is that shown in Fig. 3.1a, two masses constrained to move in one direction and coupled by three springs to each other and to rigid supports. If the motion is longitudinal and of small

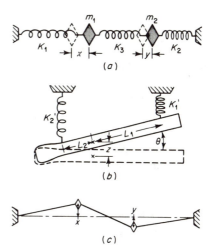

FIGURE 3.1
Examples of coupled oscillators.

enough amplitude so that the springs' restoring forces are linear functions of their extension, the equations of motion for the two masses are

$$m_1 \frac{d^2x}{dt^2} = -K_1 x - K_3(x - y) = -K_{11}x + K_{12}y$$

$$m_2 \frac{d^2y}{dt^2} = -K_2 y - K_3(y - x) = -K_{22}y + K_{21}x$$

(3.1.1)

where K_1, K_3, K_2 are the stiffness constants of the three springs, and $K_{11} = K_1 + K_3$, $K_{22} = K_2 + K_3$, $K_{12} = K_{21} = K_3$. If we hold m_2 at $y = 0$, then the restoring force on m_1 is $-K_{11}x$, and thus m_1 will oscillate with isolated simple-harmonic motion of frequency $\nu_1 = \omega_1/2\pi$, where $\omega_1^2 = K_{11}/m_1$. Similarly, the frequency of isolated vibration of m_2 is $\omega_2/2\pi$, where $\omega_2^2 = K_{22}/m_2$. To simplify the wording of our discussion, we shall assume that m_1 has the higher frequency, $\omega_1 \geqslant \omega_2$.

The transverse oscillations of two masses on a string under tension, as in Fig. 3.1c, correspond to equations of motion like (3.1.1), as long as the displacements of the masses are small compared with the length of the string. Similarly, the system shown in Fig. 3.1b, where a bar of mass M and moment of inertia I about its center of mass moves up and down and rotates in a vertical plane about its center of mass. Here again there are two degrees of freedom, the vertical displacement of the center of mass, z, and the angle θ of rotation of the bar. The equations of motion are

$$M \frac{d^2z}{dt^2} = -K_1'(z + L_1\theta) - K_2'(z - L_2\theta)$$

$$I \frac{d^2\theta}{dt^2} = -L_1 K_1'(z + L_1\theta) + L_2 K_2'(z - L_2\theta)$$

if the displacements are small enough. These equations are similar to Eqs. (3.1.1), M and I taking the place of m_1 and m_2 and $K_1' + K_2'$, $L_1^2 K_1' + L_2^2 K_2'$, and $L_2 K_2' - L_1 K_1'$ taking the place of K_{11}, K_{22}, and $K_{21} = K_{12}$, respectively.

Analysis of the motion

Both of the equations of motion (3.1.1) involve the displacements of both oscillators; the equation for x_1 has a term $K_{12}x_2$, and the equation for x_2 has a similar one, $K_{21}x_1$, where $K_{12} = K_{21}$. The coupling spring, which these terms represent, enables momentum and energy to pass from one oscillator to the other and back again. Instead of each oscillator performing simple-harmonic motion with its own frequency of isolated motion, ν_1 or ν_2, as the case may be, both oscillators exhibit a combination of two frequencies in their motion.

But if the general motion of the linear system is a combination of periodic motions of different frequencies, it should be possible, by starting the system properly, to suppress all but one of these frequencies and have the system perform simple-harmonic motion. To see whether this is possible and, if it is, to find the possible frequencies of such motion, we try inserting $x = Ae^{-i\omega t}$ and $y = Be^{-i\omega t}$ into Eqs. (3.1.1). Performing the differentiation and then dividing out the exponential, we obtain

$$(K_{11} - m_1\omega^2)A = K_{12}B \qquad (K_{22} - m_2\omega^2)B = K_{21}A$$

These two equations can be solved for A/B and ω; therefore it is possible for the system to perform simple-harmonic motion if it is started just right (i.e., if A/B has the right value). For the two equations to be mutually consistent, ω^2 must satisfy the quadratic equation

$$(m_1\omega^2 - K_{11})(m_2\omega^2 - K_{22}) = K_{12}^2 \qquad \text{or} \qquad (\omega^2 - \omega_1^2)(\omega^2 - \omega_2^2) = \mu^4$$

(3.1.2)

where $\omega_1^2 = (2\pi\nu_1)^2 = K_{11}/m_1$ and $\omega_2^2 = (2\pi\nu_2)^2 = K_{22}/m_2$ define the frequencies of isolated oscillation, ν_1 and ν_2, of the two masses (we have assumed, to simplify the discussion, that $\nu_1 > \nu_2$), and $\mu^4 = K_{12}^2/m_1m_2$ is a measure of the coupling between the two oscillators.

The two roots of Eq. (3.1.2)

$$\omega_+^2 = \tfrac{1}{2}(\omega_1^2 + \omega_2^2) + \tfrac{1}{2}\sqrt{(\omega_1^2 - \omega_2^2)^2 + 4\mu^4}$$

$$\omega_-^2 = \tfrac{1}{2}(\omega_1^2 + \omega_2^2) - \tfrac{1}{2}\sqrt{(\omega_1^2 - \omega_2^2)^2 + 4\mu^4}$$

(3.1.3)

determine the frequencies $\nu_+ = \omega_+/2\pi$ and $\nu_- = \omega_-/2\pi$ at which the system can oscillate with simple-harmonic motion. We note that, if there is no coupling ($\mu = 0$), ω_+ equals the larger of the isolated angular velocities, ω_1, and ω_- equals the smaller, ω_2. When $\mu \neq 0$, the oscillators are no longer isolated, and ω_+ is a bit larger than ω_1 and ω_- a bit smaller than ω_2. Since $K_{12} = K_{21}$ is always smaller than K_{11} or K_{22}, and thus μ is always smaller than ω_1 or ω_2, ω_-^2 can never be negative.

We have thus shown that if the system is started into motion with just the right relationship between the displacements of m_1 and m_2 (the right value of A/B), it can oscillate with simple-harmonic motion of frequency $\omega_-/2\pi$. The appropriate amplitude ratio is given by the equation $(\omega_1^2 - \omega_-^2)A = \mu^2\sqrt{m_2/m_1}\,B$, or what is the same thing, $\mu^2\sqrt{m_1/m_2}\,A = (\omega_2^2 - \omega_-^2)B$, since ω_- is a root of Eq. (3.1.2). In other words, a possible motion of the system is represented by the following equations:

$$x = \sqrt{\frac{1}{m_1}}\,\mu^2\,C_-e^{-i\omega_- t} \qquad y = \sqrt{\frac{1}{m_2}}\,(\omega_1^2 - \omega_-^2)C_-e^{-i\omega_- t}$$

Another possible simple-harmonic motion has frequency $\omega_+/2\pi$, for which the displacements are

$$x = \sqrt{\frac{1}{m_1}}\,(\omega_+^2 - \omega_2^2)C_+e^{-i\omega_+t} \qquad y = -\sqrt{\frac{1}{m_2}}\,\mu^2 C_+e^{-i\omega_+t}$$

where the complex constants C_+ and C_- may have any magnitude or phase angle.

These are very specialized kinds of motion of the system; in each case a particular relationship between the amplitudes and phases of the two masses is required. For example, if the system is to oscillate with frequency $\omega_-/2\pi$, the two masses must move in phase (we have assumed that $\omega_1 \gg \omega_2$) and the amplitude of motion of m_1 must be $\sqrt{m_2/m_1}[\mu^2/(\omega_1^2 - \omega_-^2)]$ times that of mass m_2. When $\mu \to 0$, this ratio tends to zero; in the limit of small coupling, m_2 carries most of the energy of the system, and m_1 moves only a relatively small amount (as though m_2 were the driver and m_1 a driven oscillator). When $\mu = 0$, this mode of oscillation corresponds to m_2 vibrating with its isolated frequency $\omega_2/2\pi$ and m_1 being at rest.

Correspondingly, the mode of oscillation with frequency $\omega_+/2\pi$ has the two masses moving with opposing displacements, with

$$y = -\sqrt{m_1/m_2}[\mu^2/(\omega_+^2 - \omega_2^2)]x$$

and if μ is small, m_1 has the majority of the energy and m_2 acts like the driven oscillator.

Neither of these vibrations is the general sort of motion which the system would exhibit if it were started into motion in an arbitrary manner. The more general motion, however, is just a sum of both modes. Since Eqs. (3.1.1) are linear in both x and y, a sum of solutions is also a solution. Consequently, the general motion of the system is given by the real part of

$$x = \sqrt{1/m_1}[(\omega_+^2 - \omega_2^2)C_+e^{-i\omega_+t} + \mu^2 C_-e^{-i\omega_-t}]$$

$$= A_+ \cos(\omega_+t - \Phi_+) + \frac{\mu^2\sqrt{m_2/m_1}}{\omega_1^2 - \omega_-^2}\,A_- \cos(\omega_-t - \Phi_-)$$

$$y = \sqrt{1/m_2}[-\mu^2 C_+e^{-i\omega_+t} + (\omega_1^2 - \omega_-^2)C_-e^{-i\omega_-t}] \qquad (3.1.4)$$

$$= -\frac{\mu^2\sqrt{m_1/m_2}}{\omega_+^2 - \omega_2^2}\,A_+ \cos(\omega_+t - \Phi_+) + A_- \cos(\omega_-t - \Phi_-)$$

where the phases and amplitudes of C_+ and C_-, or the values of A_+, A_-, and Φ_+, Φ_-, are adjusted to fit the initial conditions for both x and y.

Transfer of energy

For example, if m_1 is held motionless a distance A from equilibrium and m_2 is held at its equilibrium position, and both are released at $t = 0$, the subsequent motion is

$$x = \frac{A}{\omega_+^2 - \omega_-^2}[(\omega_+^2 - \omega_2^2)\cos(\omega_+t) + (\omega_2^2 - \omega_-^2)\cos(\omega_-t)]$$

$$= A\cos[\tfrac{1}{2}(\omega_+ - \omega_-)t]\cos[\tfrac{1}{2}(\omega_1 + \omega_2)t]$$

$$- A\frac{\omega_1^2 - \omega_2^2}{\omega_+^2 - \omega_-^2}\sin[\tfrac{1}{2}(\omega_+ - \omega_-)t]\sin[\tfrac{1}{2}(\omega_1 + \omega_2)t] \qquad (3.1.5)$$

$$y = \frac{\mu^2\sqrt{m_1/m_2}}{\omega_+^2 - \omega_-^2}A[-\cos(\omega_+t) + \cos(\omega_-t)]$$

$$= 2A\frac{\mu^2\sqrt{m_1/m_2}}{\omega_+^2 - \omega_-^2}\sin[\tfrac{1}{2}(\omega_+ - \omega_-)t]\sin[\tfrac{1}{2}(\omega_1 + \omega_2)t]$$

where we have used Eqs. (3.1.2) and (3.1.3) several times to simplify the formulas. Displacement y is zero at $t = 0$ (as required by the initial conditions), but it does not stay zero long. If ω_+ does not differ much from ω_-, the two cosine terms in y take several cycles to get out of phase. The second formula for y shows that when $\omega_+ - \omega_- \ll \omega_+$, the motion of m_2 may be described, roughly, as that of an oscillator with frequency equal to the average of $\omega_+/2\pi$ and $\omega_-/2\pi$, having an amplitude of motion proportional to the magnitude of $\sin[\tfrac{1}{2}(\omega_+ - \omega_-)t]$, a relatively slowly varying, periodic function. At $t = 0$, m_1 has all the energy and m_2 is at rest. As time goes on, energy is fed to m_2, and its amplitude of motion increases to its maximum value, $2\mu^2A\sqrt{m_1/m_2}/(\omega_+^2 - \omega_-^2)$, after which the energy flows back again to m_1.

An approximate expression for the energy possessed by each of the masses can be obtained by using the second forms of Eqs. (3.1.5). It is, of course, not strictly correct to speak of the energy of only part of a system, nor is it correct to use Eq. (2.2.5), appropriate for the energy of a simple oscillator, for an oscillator with sinusoidally varying amplitude. But if the variation in amplitude is slow (i.e., if $\omega_+ - \omega_- \ll \omega_+$), the following formulas are a fairly good approximation:

$$E_1 \simeq \tfrac{1}{2}m_1A^2[\tfrac{1}{2}(\omega_+ + \omega_-)]^2\left\{1 - \frac{4\mu^4}{(\omega_1^2 - \omega_2^2)^2 + 4\mu^4}\sin^2[\tfrac{1}{2}(\omega_+ - \omega_-)t]\right\}$$

$$\qquad (3.1.6)$$

$$E_2 \simeq \tfrac{1}{2}m_1A^2\frac{\mu^4(\omega_+ + \omega_-)^2}{(\omega_1^2 - \omega_2^2)^2 + 4\mu^4}\sin^2[\tfrac{1}{2}(\omega_+ - \omega_-)t]$$

for the mean "energies of vibration" of m_1 and m_2, respectively. We see that the energy surges back and forth from m_1 to m_2 with a period $4\pi/(\omega_+ - \omega_-)$, part of m_1's energy going to m_2 and then back again. If $\omega_1 = \omega_2$, then m_2 will drain all the energy from m_1 before returning it.

The case of weak coupling

We have already mentioned that when the coupling spring is weak, so that $\mu^2 \ll \omega_1^2$ or ω_2^2, the system behaves like two separate simple oscillators, each slightly influencing the motion of the other. If m_1 has the majority of the energy, m_2 acts like a driven oscillator, and vice versa. To show this more clearly, we work out the first approximations to Eqs. (3.1.3) when μ^2 is small.

$$\omega_+^2 \simeq \omega_1^2 + \frac{\mu^4}{\omega_1^2 - \omega_2^2} \quad \text{or} \quad \omega_+ \simeq \omega_1 + \frac{\mu^4}{2\omega_1(\omega_1^2 - \omega_2^2)}$$

$$\omega_-^2 \simeq \omega_2^2 - \frac{\mu^4}{\omega_1^2 - \omega_2^2} \quad \text{or} \quad \omega_- \simeq \omega_2 - \frac{\mu^4}{2\omega_2(\omega_1^2 - \omega_2^2)}$$

(3.1.7)

where $\omega_1/2\pi$ is, as we said, the larger of the two isolated frequencies of oscillation and $\omega_2/2\pi$ is the smaller.

Now suppose we set the system into oscillation with frequency $\omega_+/2\pi$. The displacements will be, to the first approximation,

$$x \simeq A_+ \cos(\omega_+ t - \Phi_+) \qquad y \simeq - \frac{K_{12}A_+}{m_2(\omega_+^2 - \omega_2^2)} \cos(\omega_+ t - \Phi_+)$$

and since $K_{12}/m_2 \ll \omega_+^2 - \omega_2^2$, the amplitude of motion of m_2 is considerably smaller than that of m_1. If coupling constant K_{12} were zero, m_1 would be oscillating by itself, with its own frequency, $\omega_1/2\pi$. If K_{12} is small but not zero, m_1 still has most of the energy, but its frequency of oscillation is slightly increased by the coupling, becoming $\omega_+/2\pi$. Mass m_2 behaves like a driven oscillator, vibrating with the frequency $\omega_+/2\pi$ of the driver and having an amplitude of motion equal to $K_{12}A_+$ divided by $m_2(\omega_2^2 - \omega_+^2)$. However, $K_{12}A_+ = F$ is the amplitude of the force applied to m_2, caused by the motion of m_1, and $m_2(\omega_2^2 - \omega_+^2) = -i\omega_+ Z_2$, where $Z_2 = i(K_{22}/\omega_+) - i\omega_+ m_2$ is m_2's mechanical impedance [Eqs. (2.3.2)]. Therefore m_2's motion is just the steady-state motion of a simple undamped oscillator, driven by m_1.

The other mode of oscillation corresponds to the slower oscillator, m_2, being the driver. Its natural frequency is decreased slightly by the coupling; the fact that m_1 is moving changes the effective restoring force on m_2. As before, the amplitude of motion of m_1 is equal to $K_{12}A_-$, the amplitude of the driving force on m_1, divided by $m_1(\omega_1^2 - \omega_-^2)$, which is $-i\omega_-$ times the mechanical impedance of m_1-plus-spring at frequency $\omega_-/2\pi$. In each case the driver, the oscillator having the majority of the energy, has its frequency of motion slightly modified by the presence of the second oscillator, coupled to it.

Of course, when the coupling is large, the interchange of energy between the oscillators is more rapid, and it no longer is appropriate to call one the driver and the other the driven.

So far we have been assuming that ω_1 is larger than ω_2 and, recently, have been assuming that $\mu^2 \ll \omega_1{}^2 - \omega_2{}^2$. What happens when $\omega_1 = \omega_2$, when the two oscillators are identical? Clearly, Eqs. (3.1.7) no longer hold. In this case, when $\mu^2 \ll \omega_1{}^2 = \omega_2{}^2$, Eqs. (3.1.3) become, to the first approximation,

$$\omega_+{}^2 = \omega_1{}^2 + \mu^2 \qquad \omega_+ \simeq \omega_1 + \frac{\mu^2}{2\omega_1} \qquad \omega_- \simeq \omega_1 - \frac{\mu^2}{2\omega_1} \qquad (3.1.8)$$

Thus the coupled system still cannot oscillate with the isolated frequencies $\omega_1/2\pi$ of either oscillator by itself; its frequency of oscillation is either a bit larger or a bit smaller than this.

The reason for the continued difference between the natural frequencies is not hard to find. If m_1 were the driver and oscillated at a frequency $\omega_1/2\pi$, equal to the resonance frequency of m_2, the impedance of m_2 at this frequency would be zero and m_2 could only reach steady state if it could absorb an infinite amount of energy from m_1 (since we neglect friction in this example). But m_1 does not have an infinite amount of energy to donate to m_2; so it has to change its frequency enough so that m_2 is no longer in exact resonance and the system can reach steady state with a reasonable amount of energy given to m_2.

The motion with frequency $\omega_+/2\pi$ is

$$x = A_+ \cos\left(\omega_+ t - \Phi_+\right) \qquad y \simeq -\sqrt{\frac{m_1}{m_2}}\, A_+ \cos\left(\omega_+ t - \Phi_+\right)$$

corresponding to an equal sharing of the energy of the system (energy of $m_1 = \frac{1}{2}m_1 A^2 \omega_+{}^2 =$ energy of m_2). The same would be true if the system were oscillating with frequency $\omega_-/2\pi$, except that the two masses would move in phase instead of having opposed motions.

On the other hand, if we try to start the system with frequency $\omega_1/2\pi = \omega_2/2\pi$ by, for example, setting m_1 into motion with m_2 initially at rest, the subsequent motions are

$$x = \tfrac{1}{2}A[\cos\left(\omega_- t\right) + \cos\left(\omega_+ t\right)] = A \cos \frac{\mu^2 t}{2\omega_1} \cos\left(\omega_1 t\right)$$

$$y = \sqrt{\frac{m_1}{4m_2}}\, A[\cos\left(\omega_- t\right) - \cos\left(\omega_+ t\right)] \qquad (3.1.9)$$

$$= \sqrt{\frac{m_1}{m_2}}\, A \sin \frac{\mu^2 t}{2\omega_1} \sin\left(\omega_1 t\right)$$

This is really an oscillation consisting of equal components of the two frequencies $\omega_+/2\pi$ and $\omega_-/2\pi$, but when $\mu \ll \omega_1$, we can consider both masses as oscillating with frequency $\omega_1/2\pi$, each having a periodically varying amplitude, with the comparatively long period $2\pi\omega_1/\mu^2$. Although m_1 starts off with all the energy, in a time $\pi\omega_1/\mu^2$ it will be at rest and m_2 will have all the energy, and so on, energy being fed back and forth from one mass to the other, ad infinitum.

3.2   NORMAL MODES OF OSCILLATION

Two noncoupled linear oscillators have two natural frequencies of oscillation, one exhibited when one mass is in motion, the other when the other mass is moving.   We have just seen that two coupled oscillators also have two frequencies of oscillation, but to perform motion exhibiting just one of these frequencies, *both* masses must be moving, with a particular relationship between the motions.   This relationship can be given a geometrical interpretation, which will help to visualize the behavior of more complex systems, to be considered later.   But let us start with a simple example.

Normal coordinates

Figure 3.2 is a vertical view of a mass m, suspended at the lower end of a long stiff bar of rectangular cross section, represented in the figure by the shaded rectangle.   The mass can thus move in two directions, each parallel to the plane of the paper.   The displacement of the mass from equilibrium can be expressed either in terms of the coordinates X, Y, in line with the bar's cross section, or else in terms of x, y, also with origin at the equilibrium point, but inclined at an angle α with respect to X, Y.

Suppose we find that a displacement X, in the direction of the greater cross-sectional dimension of the bar, is resisted by a force $-K_+X$, pointed in the negative X direction, and that a displacement Y, perpendicular to this, is opposed by a force $-K_-Y$, antiparallel to the Y axis.   Therefore, if the mass m is set in motion in the X direction, it will keep swinging in the X direction, with a frequency $\omega_+/2\pi = (\tfrac{1}{2}\pi)\sqrt{K_+/m}$; likewise, oscillations in the Y direction will remain in the Y direction and will have frequency $\omega_-/2\pi = (1/2\pi)\sqrt{K_-/m}$.   Coordinates X and Y are the "normal" coordinates to describe the motion.

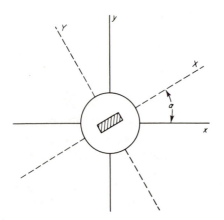

FIGURE 3.2
Vertical view of a two-dimensional oscillator, a mass on the lower end of stiff bar of rectangular cross section.

But suppose we wish to set the mass into oscillation in the x direction, at an angle α with respect to the X axis. What will its subsequent motion be? First we have to find out what force acts on m when it is displaced a distance x in this direction. This force can be calculated by transformation of axes. We find that the restoring force caused by the displacement x is not in the direction of the displacement; it has a component $-(K_+ \cos^2 \alpha + K_- \sin^2 \alpha)x$ in the x direction, but it also has a component $-(K_+ - K_-)(\sin \alpha \cos \alpha)x$ in the y direction, perpendicular to x (assuming still that the displacement is small enough so that the force is linearly proportional to the displacement). The angle between the displacement and the force is small if the bar is nearly rectangular, so that K_+ is nearly equal to K_-, but the two are not parallel unless $\alpha = 0, 90, 180$, or $270°$ if $K_+ \neq K_-$.

This is also true for a displacement along the y axis; the restoring force will then have a component $-(K_+ \sin^2 \alpha + K_- \cos^2 \alpha)y$ along y and a component $-(K_+ - K_-)(\sin \alpha \cos \alpha)y$ along x. The equations of motion of the mass, expressed in terms of the coordinates x, y, are thus

$$m \frac{d^2x}{dt^2} = -(K_+ \cos^2 \alpha + K_- \sin^2 \alpha)x - (K_+ - K_-)(\sin \alpha \cos \alpha)y$$

$$m \frac{d^2y}{dt^2} = -(K_+ \sin^2 \alpha + K_- \cos^2 \alpha)y - (K_+ - K_-)(\sin \alpha \cos \alpha)x$$

(3.2.1)

which are just Eqs. (3.1.1) for coupled oscillators, if we set

$$K_{11} = K_+ \cos^2 \alpha + K_- \sin^2 \alpha \qquad K_{22} = K_+ \sin^2 \alpha + K_- \cos^2 \alpha$$

$$K_{12} = K_{21} = -(K_+ - K_-) \sin \alpha \cos \alpha$$

Therefore, for the system of Fig. 3.2, the use of axes at an angle α to the normal coordinates converts a problem equivalent to two uncoupled oscillators into one equivalent to two coupled oscillators. The change occurs because a displacement in any direction other than "normal" produces a restoring force not in line with the displacement, and therefore gives rise to motion which is not in a straight line through the origin. Figure 3.3 shows the motion, both as plots of x and y as functions of time, and also as the path of m in the xy plane.

Transformation of coordinates

It would be well to work all this out in vector notation. The potential energy for a small displacement away from equilibrium, in terms of the normal coordinates X, Y, is [Eq. (1.1.3)]

$$V(X,Y) = \tfrac{1}{2}K_+X^2 + \tfrac{1}{2}K_-Y^2 \tag{3.2.2}$$

The corresponding force vector $\mathbf{F} = -\text{grad } V$ has component $-(\partial V/\partial X) = -K_+X$ along X and component $-(\partial V/\partial Y) = -K_-Y$ in the Y direction.

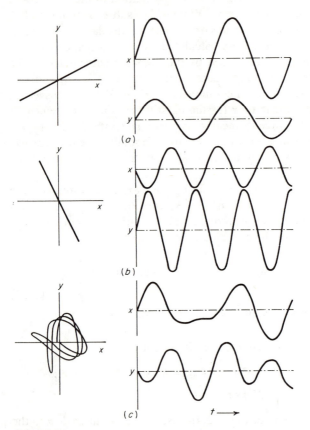

FIGURE 3.3
Motion of two coupled oscillators. Curves to the right show the displacements x and y as functions of time; those to the left show the path of the point representing the system in the xy plane (configuration space). Cases a and b show the two normal modes of vibration when the system point travels along a normal coordinate. Case c shows the general type of motion.

If we change to new axes, at an angle α to the normal coordinates, so that

$$x = X \cos \alpha - Y \sin \alpha \qquad y = X \sin \alpha + Y \cos \alpha$$
$$X = x \cos \alpha + y \sin \alpha \qquad Y = -x \sin \alpha + y \cos \alpha \qquad (3.2.3)$$

then simple substitution shows that the potential energy is

$$V(x, y) = \tfrac{1}{2}(K_+ \cos^2 \alpha + K_- \sin^2 \alpha)x^2 + (K_+ - K_-)(\sin \alpha \cos \alpha)xy$$
$$+ \tfrac{1}{2}(K_+ \sin^2 \alpha + K_- \cos^2 \alpha)y^2 \qquad (3.2.4)$$

The components of force $\mathbf{F} = -\text{grad } V$ in the x and y directions are thus

$$F_x = -\frac{\partial V}{\partial x} = -(K_+ \cos^2 \alpha + K_- \sin^2 \alpha)x - (K_+ - K_-)(\sin \alpha \cos \alpha)y$$

$$F_y = -\frac{\partial V}{\partial Y} = -(K_+ \sin^2 \alpha + K_- \cos^2 \alpha)y - (K_+ - K_-)(\sin \alpha \cos \alpha)x$$

The vector equation of motion of the mass m is thus

$$m\frac{d^2\mathbf{R}}{dt^2} = -\text{grad } V \tag{3.2.5}$$

where $\mathbf{R}$ is the vector displacement of the mass from equilibrium.

If we express this equation in terms of the normal coordinates X, Y, the component equations are

$$m\frac{d^2X}{dt^2} + m\omega_+^2 X = 0 \qquad m\frac{d^2Y}{dt^2} + m\omega_-^2 Y = 0$$

where $\omega_+^2 = K_+/m$ and $\omega_-^2 = K_-/m$, which are the equations for two uncoupled harmonic oscillators. But if we insist on using the x, y coordinates to describe the motion, the component equations (3.2.1) are those for a pair of coupled oscillators. Evidently, we can change from coupled to uncoupled oscillators by rotating the axes into line with the normal axes of the bar. Use of Eqs. (3.1.3), with the special values

$$\omega_1^2 = \omega_+^2 \cos^2 \alpha + \omega_-^2 \sin^2 \alpha \qquad \omega_2^2 = \omega_+^2 \sin^2 \alpha + \omega_-^2 \cos^2 \alpha$$

$$\mu^2 = (\omega_-^2 - \omega_+^2) \sin \alpha \cos \alpha \tag{3.2.6}$$

appropriate for the system of Fig. 3.2, results in the following values of natural frequencies for the system:

$$\omega_+^2 \equiv \tfrac{1}{2}(\omega_+^2 + \omega_-^2) + \tfrac{1}{2}[(\omega_+^2 - \omega_-^2)^2(\cos^2 \alpha - \sin^2 \alpha)^2$$
$$+ 4(\omega_+^2 - \omega_-^2) \sin^2 \alpha \cos^2 \alpha]^{\frac{1}{2}} = \omega_+^2$$

and similarly for ω_-^2, which always come out equal to the frequencies of the normal coordinates, independent of the angle α of rotation of the axes. Also use of Eqs. (3.1.4), resulting in

$$x = \sqrt{\frac{1}{m}}(\omega_+^2 - \omega_-^2)(\cos^2 \alpha C_+ e^{-i\omega_+ t} - \sin \alpha \cos \alpha\, C_- e^{-i\omega_- t})$$

$$y = \sqrt{\frac{1}{m}}(\omega_+^2 - \omega_-^2)(\sin \alpha \cos \alpha\, C_+ e^{-i\omega_+ t} + \cos^2 \alpha C_- e^{-i\omega_- t})$$

reduces these equations to those of transformation of coordinates [Eqs.

(3.2.3)] when we set

$$X = \sqrt{\frac{1}{m}} (\omega_+{}^2 - \omega_-{}^2) \cos \alpha C_+ e^{-i\omega_+ t} = A_+ e^{-i\omega_+ t}$$

$$Y = \sqrt{\frac{1}{m}} (\omega_+{}^2 - \omega_-{}^2) \cos \alpha C_- e^{-i\omega_- t} = A_- e^{-i\omega_- t}$$

where C_+ and C_- are arbitrary constants. Thus each of the components of the motion, along each of the normal coordinates, has just one frequency, $\omega_+/2\pi$ or $\omega_-/2\pi$, respectively.

In other words, it is possible to describe the motion most simply in terms of a pair of normal coordinate axes, the component of motion along either of which exhibits just one of the natural frequencies of the system. Components of the motion along any other axes exhibit both frequencies. The motion is, of course, the same, no matter which axes are used; we have only changed our way of describing it by transforming axes.

More complex systems

The methods discussed for analyzing the motion of two coupled oscillators can be extended to systems of more than two oscillators. One determines the ways in which the system can oscillate with simple-harmonic motion and then builds up the free oscillations of the system as combinations of these various normal modes of oscillation. The complexities of the mathematics of this direct approach make it preferable to use the machinery of the Laplace transform to work out the general solution, even though this machinery obscures the physical picture to some extent.

We consider a system of N degrees of freedom, held together by linear forces which tend to bring each of the N displacements back to equilibrium. Thus each of the N coordinates ξ_i ($i = 1, 2, \ldots, N$), which together specify the configuration of the system, can be given as displacements from equilibrium. Some of the coordinates may be displacements, in some specified direction, of some portion of the system, measured in units of distance; others could be angular displacements of some portion from its equilibrium direction; and so on. To the ith coordinate corresponds an effective mass or moment of inertia, which can be labeled m_i. The coordinates themselves, the units in which they are measured, and the units of the corresponding mass factors are chosen so that the total kinetic energy of the system is the sum of terms of the form $\frac{1}{2}m_i(d\xi_i/dt)^2$.

The internal forces and torques which hold the system together are linear functions of the displacements as long as these displacements are all small enough. Of course, the displacement of one degree of freedom may tend to make other degrees of freedom move away from their equilibrium positions; this is the usual situation with coupled systems. In mathematical language, if $x, y, z, \ldots$ are the displacements from equilibrium of the various degrees of freedom, expressed as distances or angles or any other units, the force (or torque) acting on the x coordinate, tending to make it return to equilibrium, is

$$F_{ex} = -A_{xx}x - A_{xy}y - A_{xz}z - \cdots$$

The coefficients A_{xx}, etc., can be called stiffness coefficients. If the forces are conservative, they can be obtained by partial differentiation of a potential energy function

$$V = \tfrac{1}{2}A_{xx}x^2 + \tfrac{1}{2}A_{xy}xy + \cdots + \tfrac{1}{2}A_{yx}yx + \tfrac{1}{2}A_{yy}y^2 + \cdots$$

where $A_{yx} = A_{xy}$ and $F_{ex} = -(\partial V/\partial x)$, etc. As mentioned, the kinetic energy of the system is

$$\tfrac{1}{2}m_x\left(\frac{dx}{dt}\right)^2 + \tfrac{1}{2}m_y\left(\frac{dy}{dt}\right)^2 + \cdots$$

If there are frictional forces, they will be proportional to the rates of change of the various displacements and, for small displacements, should be linear functions

$$D_x = -R_x\frac{dx}{dt} - \cdots$$

Therefore, for small displacements, the equations of motion are

$$m_x\frac{d^2x}{dt^2} = -R_x\frac{dx}{dt} - A_{xx}x - A_{xy}y - \cdots$$

$$m_y\frac{d^2y}{dt^2} = -R_y\frac{dy}{dt} - A_{yx}x - A_{yy}y - \cdots \qquad (3.2.7)$$

$$\cdots\cdots\cdots\cdots\cdots\cdots\cdots\cdots\cdots\cdots\cdots$$

$$A_{xy} = A_{yx} \quad \cdots$$

We now simplify these equations by reducing all coordinates, whether distances or angles, to the dimensionality of distance multiplied by the square root of mass. We set $x_1 = \sqrt{m_x}\,x$, $x_2 = \sqrt{m_y}\,y$, etc. We also reduce the stiffness coefficients to dimensions of frequency squared $(1/t^2)$ by setting $U_{11} = A_{xx}/m_x$, $U_{12} = A_{xy}/\sqrt{m_xm_y} = U_{21}$, etc. The frictional resistances also are changed, by defining $k_1 = R_x/2m_x$, etc. In terms of these reduced coordinates and coefficients, the equations of motion of the system take on the simplified form

$$\frac{d^2x_i}{dt^2} + 2k_i\frac{dx_i}{dt} = -\sum_{j=1}^{N}U_{ij}x_j \qquad i = 1, 2, \ldots, N \qquad (3.2.8)$$

To begin with, we shall neglect the frictional forces, setting the k's equal to zero.   In this case the total energy of the system,

$$E = \tfrac{1}{2}m_x \left(\frac{dx}{dt}\right)^2 + \cdots + V$$

$$= \tfrac{1}{2}\sum_{i=1}^{N}\left(\frac{dx_i}{dt}\right)^2 + \tfrac{1}{2}\sum_{i,j=1}^{N} x_i U_{ij} x_j \tag{3.2.9}$$

is constant.   As before, we ask whether it is possible for the system to perform simple-harmonic motion.   We set $x_i = C_i e^{-i\omega t}$ and try to find solutions for the N simultaneous equations,

$$\sum_{j=1}^{N} U_{ij}C_j - \omega^2 C_i = 0 \qquad i = 1, 2, \ldots, N \tag{3.2.10}$$

The frequencies for which such solutions are possible are given by the roots of the equation $\Delta_0(-i\omega) = 0$, called the *secular equation*, where $\Delta_0(-i\omega)$ is the determinant of the coefficients of the C's in Eqs. (3.2.10).

$$\Delta_0(-i\omega) \equiv \begin{vmatrix} U_{11} - \omega^2 & U_{12} & U_{13} & \cdot & \cdot & \cdot \\ U_{21} & U_{22} - \omega^2 & U_{23} & \cdot & \cdot & \cdot \\ \cdot & \cdot & \cdot & \cdot & \cdot & \cdot \end{vmatrix} = 0 \tag{3.2.11}$$

This is an Nth-order equation for the quantity ω^2, having N roots, $\omega_{01}^2, \ldots, \omega_{0N}^2$, usually all distinct.   It can be shown that, since all the coefficients U_{ii} are positive and since the nondiagonal coefficients U_{ij} are symmetric ($U_{ij} = U_{ji}$), none of the roots ω_{0n}^2 can be negative.

Once the roots of the secular equation have been found, we could substitute them back into Eqs. (3.2.10) to find the particular combination of amplitudes C_j which allows the system to vibrate in a single normal mode, with one frequency $\omega_{0n}/2\pi$, and then find the combination of normal modes which satisfies specified initial conditions.   This procedure is often tedious and does not lead to better understanding; so we shall turn to a discussion of the forced motion of the system, using the Laplace transform as the means of solution.

Forced motion

Suppose the system defined by Eqs. (3.2.8) is acted on by applied forces, the ith degree of freedom (distance or angle) being acted on by the applied force (or torque, as the case may be) F_{ai}.   In reduced form the force would become $f_i = F_{ai}/\sqrt{m_i}$, with dimensions $l\sqrt{m}/t^2$, and the equations of motion are

$$\frac{d^2x_i}{dt^2} + 2k_i\frac{dx_i}{dt} + \sum_{j=1}^{N}U_{ij}x_j = f_i(t) \tag{3.2.12}$$

If the applied forces are simple-harmonic, so that $f_i = F_i e^{-i\omega t}$, the steady-state motion of the system can be obtained in terms of the determinant

$$\Delta(s) = \begin{vmatrix} s^2 + 2k_1 s + U_{11} & U_{12} & \cdot & \cdot & \cdot \\ U_{21} & s^2 + 2k_2 s + U_{22} & \cdot & \cdot & \cdot \\ \cdot & \cdot \cdot \cdot \cdot \cdot \cdot \cdot \cdot \cdot \cdot \cdot \cdot \end{vmatrix} \quad (3.2.13)$$

and the cofactors $\Delta_{ij}^{(c)}(s) = \partial\Delta/\partial U_{ij}$,

$$x_i = \sum_{j=1}^{N} F_j \left[\frac{\Delta_{ji}^{(c)}(-i\omega)}{\Delta(-i\omega)} \right] e^{-i\omega t} \quad (3.2.14)$$

for steady-state motion.

The equation $\Delta(s) = 0$ has $2N$ roots for s. Because s occurs in the diagonal terms only and since $U_{ii} \geqslant 0$ and $U_{ij} = U_{ji}$, these roots occur in pairs,

$$\Delta(s) = 0 \quad \text{for } s = -\kappa_n \pm i\omega_n; \; n = 1, 2, \ldots, N \quad (3.2.15)$$

where, if all the frictional coefficients k_i are zero, all the κ_n's are zero and the ω_n's are equal to the roots ω_{0n} of the determinant of Eq. (3.2.11). Expansion of determinant $\Delta(s)$,

$$\Delta(s) = s^{2N} + 2(\textstyle\sum k_i)s^{2N-1} + (\textstyle\sum U_{ii})s^{2N-2} + \cdots$$

indicates that the real and imaginary parts of the roots of $\Delta(s)$ are related to the frictional and elastic coefficients by

$$\sum_n \kappa_n = \sum_i k_i \qquad \sum_n (\kappa_n^2 + \omega_n^2) = \sum_i U_{ii} \quad (3.2.16)$$

and also that we can express $\Delta(s)$ as a product of N factors,

$$\Delta(s) = [(s + \kappa_1)^2 + \omega_1^2][(s + \kappa_2)^2 + \omega_2^2] \cdots [(s + \kappa_N)^2 + \omega_N^2] \quad (3.2.17)$$

Thus we can write Eqs. (3.2.14), for the steady-state motion of a system driven by simple-harmonic forces of frequency $\omega/2\pi$, as

$$x_i = \sum_{j=1}^{N} \frac{F_j \Delta_{ji}^{(c)}(-i\omega)e^{-i\omega t}}{[\omega_1^2 - (\omega + i\kappa_1)^2] \cdots [\omega_N^2 - (\omega + i\kappa_N)^2]}$$

If the frictional coefficients k_i are small, the real parts κ_n of the roots are small, and the steady-state motion will exhibit resonance. Whenever ω nearly equals one of the values ω_N of free vibration of the system, one of the factors in the denominator of the expressions for x_i will become small and the x_i's will become large, as illustrated in Fig. 2.2 for the simple oscillator.

To find the response of the system to a more general driving force, we use the Laplace transform. If the system is at rest and the driving forces are all zero when $t < 0$, then the transform of Eq. (3.2.12) is

$$(s^2 + 2k_i s)X_i + \sum_j U_{ij}X_j = F_i(s) \quad (3.2.18)$$

where F_i is the transform of the known function f_i, and X_i is the transform of the unknown function x_i.

$$F_i = \int_0^\infty e^{-st} f_i(t)\, dt \qquad X_i(s) = \int_0^\infty e^{-st} x_i(t)\, dt$$

The simultaneous equations (3.2.18) can be solved in terms of the determinant $\Delta(s)$ already defined, and its cofactors $\Delta_{ij}^{(c)}(s)$,

$$X_i(s) = \sum_{j=1}^N F_j(s) \left[\frac{\Delta_{ji}^{(c)}(s)}{\Delta(s)} \right] \qquad (3.2.19)$$

Cofactor $\Delta_{ji}^{(c)}(s)$ is $(-1)^{j-i}$ times the $(N-1)$st-order determinant formed by removing the jth row and the ith column from the determinant $\Delta(s)$ of Eq. (3.2.13). Use of the factored form of $\Delta(s)$, given in Eq. (3.2.17), shows that the transforms $X_i(s)$ have poles when s equals any of the roots $-\kappa_n \pm i\omega_n$ of the secular equation $\Delta(s) = 0$. Usually, all $2N$ roots are distinct, in which case all these poles are simple poles. There may, of course, be additional poles if any of the transforms F_i have poles. The cofactors $\Delta_{ji}^{(c)}(s)$ have zeros but no poles.

To calculate the inverse transforms of the X_i's, we use an extension of Eqs. (2.3.19). If $G(s)$ is a function which is finite and continuous for all finite values of s, and if $H(s)$ is the simple product $(s - a_1)(s - a_2) \cdots (s - a_M)$, with roots a_n all differing from each other, the ratio $G(s)/H(s)$, which has M simple poles at the locations $s = a_n$, may be expanded in terms of its residue at each pole.

$$\frac{G(s)}{H(s)} = \sum_{n=1}^M \frac{G(a_n)}{s - a_n} \lim_{s \to a_n} \frac{s - a_n}{H(s)} \qquad (3.2.20)$$

This partial fraction expansion enables us to express the transforms $X_i(s)$ in terms of quantities contained in Table 2.1 so that an inverse transform can be found.

For example, if the jth degree of freedom of the system is given a sudden blow, at $t = 0$, of an impulse equal to $1/\sqrt{m_j}$, so that $f_j(t) = \delta(t)$ and all the other f's are zero, the transforms of the applied forces are

$$F_j(s) = 1 \qquad F_i(s) = 0 \qquad i \neq j$$

and the expansion of the transforms of the displacements become

$$X_i(s) = \frac{\Delta_{ji}^{(c)}(s)}{(s + \kappa_1 + i\omega_1)(s + \kappa_1 - i\omega_1) \cdots (s + \kappa_N - i\omega_N)}$$

$$= 2 \sum_{n=1}^N [(s + \kappa_n)(A_{ji}^{(n)} + \omega_n B_{ji}^{(n)})][(s + \kappa_n)^2 + \omega_n^2]^{-1} \qquad (3.2.21)$$

where

$$A_{ji}^{(n)} + iB_{ji}^{(n)} = \lim \left[\Delta_{ji}^{(c)}(s) \frac{s + \kappa_n + i\omega_n}{\Delta(s)} \right] \qquad s \to -\kappa_n - i\omega_n$$

is a complex constant, independent of s. Applying formulas (9), (15), and (16) of Table 2.1 to this expansion, we have, for the deflection of the ith degree of freedom, in response to an impulse $1/\sqrt{m_j}$, delivered to the jth degree of freedom,

$$x_i(t) = \xi_{ji}^{(v)} \equiv \begin{cases} 0 & t < 0 \\ \displaystyle\sum_{n=1}^{N} e^{-\kappa_n t}[A_{ji}^{(n)} \cos(\omega_n t) + B_{ji}^{(n)} \sin(\omega_n t)] & t > 0 \end{cases} \quad (3.2.22)$$

which is a combination of the N different normal modes of oscillation of the system. We see that the nth mode oscillates with damped harmonic motion of frequency $\omega_n/2\pi$ and with damping constant κ_n.

If friction is neglected, all the κ's become zero and Δ becomes Δ_0, with roots $\pm i\omega_{0n}$ of Eq. (3.2.11). It can be shown that the constants $A_{ji}^{(n)}$ all go to zero, and the $B_{ji}^{(n)}$ becomes

$$B_{0ji}^{(n)} = \frac{1}{\omega_{0n}} \lim_{s \to -i\omega_{0n}} \left[\frac{(s^2 + \omega_{0n}{}^2)\Delta_{0ji}^{(c)}(s)}{\Delta_0(s)} \right] = \frac{1}{\omega_{0n}} E_j(n)E_i(n) \quad (3.2.23)$$

where

$$[E_i(n)]^2 = \lim_{s \to -i\omega_{0n}} \left[\frac{(s^2 + \omega_{0n}{}^2)\Delta_{0ii}^{(c)}(s)}{\Delta_0(s)} \right]$$

and

$$\sum_{i=1}^{N} E_i(n)E_i(m) = \begin{cases} 1 & n = m \\ 0 & n \neq m \end{cases}$$

so that

$$x_i(t) = \xi_{ji}^{(v)} \equiv \begin{cases} 0 & t < 0 \quad \text{zero-friction case} \\ \displaystyle\sum_{n=1}^{N} E_j(n)E_i(n)\left(\frac{1}{\omega_{0n}}\right) \sin(\omega_{0n}t) & t > 0 \end{cases} \quad (3.2.24)$$

The factor $E_i(n)$ is a measure of the participation of the ith degree of freedom in the nth normal mode of oscillation; $C_i = CE_i(n)$ is the solution of Eqs. (3.2.10) when $\omega = \omega_{0n}$. The factor $E_j(n)$ also measures the relative effectiveness of striking the jth degree of freedom, in exciting the nth normal mode.

It should also be apparent that when the initial conditions are that $x_j = 1$ and $x_i = 0$ $(i \neq j)$ at $t = 0$ (initial velocities all zero), the subsequent motion of the system is

$$x_i(t) = \xi_{ji}^{(d)} \equiv \sum_{n=1}^{N} E_j(n)E_i(n) \cos(\omega_{0n}t) \quad t > 0 \quad (3.2.25)$$

for the zero-friction case. Combining Eqs. (3.2.24) and (3.2.25), we can write down the motion of the system when it is started into motion at $t = 0$

with $x_j = x_{0j}$ and $dx_j/dt = v_{0j}$ and allowed to oscillate freely thereafter.

$$x_i(t) = \sum_{j=1}^{N} [x_{0j}\xi_{ji}^{(d)} + v_{0j}\xi_{ji}^{(v)}]$$

$$= \sum_{n=1}^{N} E_i(n) \sum_{j=1}^{N} E_j(n) \left[x_{0j} \cos (\omega_{0n}t) + \frac{v_{0j}}{\omega_{0n}} \sin (\omega_{0n}t) \right] \quad (3.2.26)$$

Once the limiting values of the ratios defined in Eqs. (3.2.23) are worked out, the degree of participation of the various normal modes is given by the factors $E_i(n)$, and the transient motion of the zero-friction case can be written down. The forced motion, in the zero-friction case, can be obtained by integrating $\xi_{ji}^{(v)}$, using a generalization of Eqs. (2.3.9).

The form of Eqs. (3.2.23) to (3.2.26) can be visualized in terms of a geometrical model, analogous to the normal modes of Eqs. (3.2.2) to (3.2.5). The quantities $E_i(n)$ can be considered the components of a unit vector in N-dimensional space, which defines the "direction" of the nth normal mode, the components measuring the configuration assumed by the system when it is oscillating with frequency $\omega_{0n}/2\pi$. The unit vectors are mutually perpendicular, since $\sum_i E_i(n)E_i(m) = 0$ unless $m = n$. The contribution of the initial displacement x_{0j} to the initial value of the nth normal mode is thus x_{0j} times the direction cosine $E_j(n)$, and thus the amplitude of vibration of the nth normal mode, because of initial displacements, is $\sum_j E_j(n)x_{0j}$, as indicated in Eq. (3.2.26). The contribution of the nth normal mode to the motion of the ith degree of freedom of the system is also measured by the direction cosine $E_i(n)$, so that, finally, the displacement of the ith degree of freedom, in response to initial displacements, is the double sum

$$x_i = \sum_{n=1}^{N} E_i(n) \sum_{j=1}^{N} E_j(n)x_{0j} \cos (\omega_{0n}t)$$

as in Eq. (3.2.26).

The introduction of friction complicates this simple model of mutually perpendicular normal-mode axes. The Laplace transform technique, however, still works; the inverse transform of Eq. (3.2.19) is the solution. Examples of the procedure will be given in the problems.

3.3 LINEAR ARRAY OF OSCILLATORS

A system of some interest is the linear array of equal masses m, free to move longitudinally, coupled together by springs, all of stiffness constant $K = \frac{1}{4}m\omega_0^2$, with equilibrium spacing l, as shown in Fig. 3.4. Here the coupling is between neighboring masses only; mass 1 can affect mass n only through the intermediacy of the ones between. Hence this system exhibits wave motion; displacement of mass 1 moves mass 2 after a delay, and this in turn moves mass 3, and so on.   The delay is longer, the larger the mass m, and is shorter, the stiffer the spring; so we shall not be surprised to find that the velocity of waves along the array is proportional to $l\sqrt{K/m}$ (the square root is the power of K/m which has the dimensions of t^{-1}).

Propagation of momentum and energy

To demonstrate the wavelike nature of the motion, let us suppose that, up to $t = 0$, all the masses are at rest at their respective positions of equilibrium, each a distance l from its neighbor.   And further, suppose that, at $t = 0$, the left-hand end of the spring between masses 0 and 1 is started into

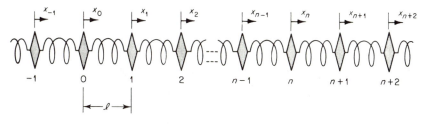

FIGURE 3.4
Infinite linear array of masses m, coupled by springs of stiffness $K = \frac{1}{4}m\omega_0^2$.   Displacement of nth mass from equilibrium is x_n.   Equilibrium spacing is l.

motion with uniform velocity u (we shall concentrate for the time being on what happens to the right of mass 0, so we can imagine that mass 0 is detached from the spring to its right).   The motion of the spring end will compress the 0–1 spring, thus producing a force f_1 on mass 1, so that it will start moving; this in turn will start mass 2 into motion, and so on.   Eventually, mass 1 and, somewhat later, mass 2 will be moving with velocity u, as is the spring end at 0; the new distance of separation will be l' instead of l.   One after the other the successive masses will be started from rest and brought to velocity u.

The motion is plotted in Fig. 3.5 as curves of displacements of the successive masses as functions of time as the ordinate.   The curve for point 0 is a straight line, of slope $1/u$, corresponding to its specified uniform motion with velocity u.   The curve for the nth mass is nearly vertical (corresponding to no motion) for some time after $t = 0$, because it takes a while for the disturbance to reach the nth mass.   The exact motion of this nth mass is rather complicated, as we shall see later, but, roughly speaking, it is at rest until shortly before a time T (where T is proportional to n), and shortly after time T it is moving with velocity u, along with the $n - 1$ masses to the left of it.   A *wave* of starting up has passed along the array, with one mass after the other starting into motion, the starting of each successive mass being delayed a time $\tau = T/n$ behind the previous one.

A time delay of motion proportional to distance is the essential characteristic of wave motion.   The velocity c of the wave motion exhibited here is
equal to the distance nl, from the origin, of the mass just starting into motion
at time T, divided by T.

$$c = \frac{nl}{T} = \frac{l}{\tau}$$

The value of the wave velocity c is determined by the balance between the
force transmitted by the springs and the momentum gained by the array per

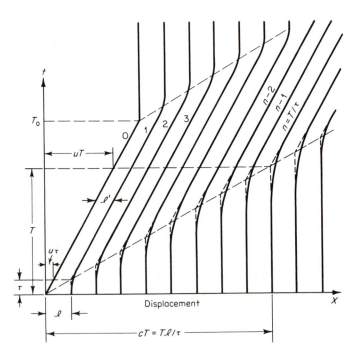

FIGURE 3.5
Wave motion in a linear array.   Mass 0 is started into motion
with velocity u at $t = 0$; effects of this disturbance are propagated with velocity c.   Solid lines represent the exact motion
of the masses; dotted lines are a simplified version of the
motion.

second, or else (what is the same thing) the balance between the work done by
the force pushing the spring end at 0 and the gain in energy of the system per
second.

Setting the masses into motion requires a force f_1 on the left-hand end of
the spring 0–1; this force is transmitted along the array by a compression of
the springs (the moving masses are closer together than the stationary ones).
If the stiffness constant of each spring is K, then the force f_1, transmitted along

the array in order to accelerate the next mass, is equal to $K(l - l')$. A glance at Fig. 3.5 shows that f_1 also can be written as $K(u\tau)$, where $\tau = T/n = l/c$ is the time delay of the motion of one mass behind the preceding one in the wave motion. The impulse $f_1 T$ which the force imparts to the system in time T must be equal to the momentum nmu which the force has generated in the n masses set into motion in time T. Thus

$$nmu = K(u\tau)T = Ku\,\frac{l}{c}\,\frac{nl}{c}$$

or

$$m = K\left(\frac{l}{c}\right)^2 \quad \text{or} \quad c = l\sqrt{\frac{K}{m}} \qquad (3.3.1)$$

which corresponds to the formula we arrived at earlier, by dimensional reasoning.

The same equation can be obtained from energy considerations. The work done by force f_1 on the left-hand end of the spring, from $t = 0$ to $t = T = n\tau = nl/c$, is f_1 times the distance uT the spring end has moved in time T. The energy gained by the system in time T is n times the kinetic energy $\frac{1}{2}mu^2$ gained by the n masses set into motion, plus n times the potential energy $\frac{1}{2}K(l - l')^2 = \frac{1}{2}K(ul/c)^2$ stored when each of the n springs is compressed by an amount $l - l'$. The equation is

$$f_1 uT \equiv \frac{Kul}{c}\,u\,\frac{nl}{c} = \tfrac{1}{2}mu^2 + \tfrac{1}{2}nK\left(\frac{ul}{c}\right)^2$$

which again reduces to Eq. (3.3.1). Thus a force f_1 applied at one end of the array feeds both momentum and energy into the array, both of which are transmitted along the array with a velocity $c = l\sqrt{K/m}$, to accelerate and donate energy to the masses and springs at the front of the wave (or to be withdrawn from the array at its far end).

If, now, we stop moving the left-hand spring end at $t = T_0$, mass 1 will begin to decelerate as spring 0–1 is restretched; after mass 1 comes to rest, mass 2 will decelerate and come to rest a time $\tau = l/c$ later; and so on. Reapplication of the previous arguments shows that the velocity of the wave of stopping is equal to $c = l\sqrt{K/m}$, that of the wave of starting.

If, after $t = T_0$, no force is applied to the left-hand spring end, it will remain at rest, and no further momentum or energy will enter the system. The array still possesses internal momentum and energy, however. A certain number, $n' \simeq T_0 c/l = T_0\sqrt{m/K}$, of the masses are moving with velocity u, though all the rest of the masses are at rest. Which of them are moving changes as time progresses. During a time $\tau = c/l$, the rearmost moving mass comes to rest, and the mass just ahead of the frontmost moving mass gets pushed into moving with velocity u. Each successive mass, as the wave reaches it and passes beyond, starts into motion, moves with velocity u

for a time $T_0 = n'\tau$, and then comes to rest after having moved a distance $uT_0 = n'l(u/c)$. Thus the wave carries $n'mu$ momentum and $n'mu^2$ energy (kinetic and potential) along with it.

The expression for the wave velocity c can be expressed in terms of properties per unit length of the array. For example, its mass per unit length is $\epsilon = m/l$. The stiffness per unit length is the reciprocal of the longitudinal compressibility κ, which is defined as the ratio between the fractional shortening $dl/l = (l - l')/l$ of any compressed portion of the spring system and the force $K(l - l')$ applied; $\kappa = (l - l')/lK(l - l') = 1/Kl$. Thus the wave velocity can be written

$$c = \sqrt{\frac{1}{\kappa\epsilon}} \tag{3.3.2}$$

We have repeated several times that this elementary discussion has not been exact, that the motion of the individual masses is actually more complicated than the simple stop-start-stop assumed here and displayed by the dotted lines of Fig. 3.5, though the conclusions we have reached are generally true and the formulas are approximately correct. We must now analyze the motion in detail to find out exactly what happens.

Simple-harmonic wave motion

Returning to the system of Fig. 3.4, we denote by x_n the longitudinal displacement from equilibrium of mass n, and we denote by f_n the force exerted on both mass $n - 1$ and mass n by the spring between them. This force is zero if the spring has its equilibrium length l, but if it is longer or shorter than l (i.e., if $x_n - x_{n-1} \neq 0$), the force will be proportional to the change of length, the proportionality constant being the stiffness constant K of the spring. Therefore the equation for f_n in terms of the displacements x_n and x_{n-1} and the related equation for the acceleration of mass n in terms of forces f_n and f_{n+1} of the springs attached to it are

$$f_n = K(x_{n-1} - x_n) = \tfrac{1}{4}m\omega_0^2(x_{n-1} - x_n)$$

$$m\frac{d^2x_n}{dt^2} = f_n - f_{n+1} \tag{3.3.3}$$

where we have introduced the constant $\omega_0^2 = 4K/m$ to simplify some of the later formulas. If all the x's are kept zero except x_n, mass n will be a simple oscillator with natural frequency $(1/2\pi)\sqrt{2K/m} = \omega_0/\pi\sqrt{8}$.

These equations of motion can be put into a more symmetric form if we use the momenta $p_n = m(dx_n/dt)$ of the masses rather than their displacements.

$$\frac{df_{n+1}}{dt} = \tfrac{1}{4}\omega_0^2(p_n - p_{n+1})$$

$$\frac{dp_n}{dt} = f_n - f_{n+1} \qquad \omega_0 = 2\sqrt{\frac{K}{m}} \tag{3.3.4}$$

We can now ask, as we have done before with other systems, whether this system can oscillate with simple-harmonic motion. To find out, we assume that both the f's and the p's have factors $e^{-i\omega t}$, so that the time derivative of either function is equal to $-i\omega$ times the function itself. In this case Eqs. (3.3.4) become

$$f_{n+1} = \frac{i\omega_0^2}{4\omega}(p_n - p_{n+1}) \qquad -i\omega p_n = f_n - f_{n+1}$$

or

$$f_{n+1} = f_n + i\omega p_n \qquad p_{n+1} = p_n\left(1 - \frac{4\omega^2}{\omega_0^2}\right) + \frac{4i\omega}{\omega_0^2}f_n$$

which expresses the force f_{n+1} and momentum p_{n+1} of one unit of the chain in terms of the force f_n and momentum p_n of the preceding unit. Evidently, these equations admit several solutions; we now ask whether one of them can represent wave motion in the direction of increasing n.

If the system can exhibit wave motion at frequency $\omega/2\pi$, the motion of p_{n+1} should lag behind that of p_n by a time delay which was called τ in Fig. 3.5. In other words, if $p_n = P_n e^{-i\omega t}$, then p_{n+1} should equal $P_n e^{-i\omega(t-\tau)}$, or p_{n+1}/p_n should equal $e^{i\omega\tau}$ (and similarly for f_{n+1}/f_n, where τ must be real if there is to be true wave motion at frequency $\omega/2\pi$ (if τ were a complex quantity, p_{n+1} would not only be delayed, but would also differ in magnitude from p_n, which does not occur in true wave motion). For the moment, setting $e^{i\omega\tau} = \gamma$ to simplify typesetting, if wave motion is to be present, Eqs. (3.3.4) must become

$$-i\omega\gamma f_n = \tfrac{1}{4}\omega_0^2(1 - \gamma)p_n \qquad \text{and} \qquad -i\omega p_n = (1 - \gamma)f_n$$

For this to be true, γ must satisfy the equation

$$\omega^2\gamma f_n = \tfrac{1}{4}\omega_0^2(1 - \gamma)^2 f_n \qquad \text{or} \qquad \gamma^2 - \left(2 - \frac{4\omega^2}{\omega_0^2}\right)\gamma + 1 = 0$$

so that

$$\gamma = 1 - \frac{2\omega^2}{\omega_0^2} \pm i\sqrt{1 - \left(1 - \frac{2\omega^2}{\omega_0^2}\right)^2} = \cos(\omega\tau) \pm i\sin(\omega\tau) = e^{\pm i\omega\tau} \quad (3.3.5)$$

where

$$\cos(\omega\tau) = 1 - \frac{2\omega^2}{\omega_0^2} \qquad \sin(\tfrac{1}{2}\omega\tau) = \frac{\omega}{\omega_0}$$

as long as $\omega < \omega_0 = 2\sqrt{K/m}$.

The frequency $\omega_0/2\pi$ is $\sqrt{2}$ times the frequency with which any one of the masses in the array would oscillate if all the other masses were held fixed at their equilibrium positions. We see that simple-harmonic wave motion is possible only as long as its frequency is less than $\omega_0/2\pi$. Referring to Fig. 3.5, we see that the velocity of propagation of such waves,

$$c = \frac{l}{\tau} = \frac{\omega l}{2 \sin^{-1}(\omega/\omega_0)} \to l \sqrt{\frac{K}{m}} \qquad \omega \ll \omega_0 \qquad (3.3.6)$$

depends on the value of ω, except in the limit of ω small. Only in this limit is the wave velocity equal to that of Eq. (3.3.1), the formula obtained by approximately balancing force and momentum. Therefore our previous discussion of Fig. 3.5 is valid only as long as the wave motion traversing the array can be analyzed into simple-harmonic components, all of which have frequencies considerably less than ω_0. Any sudden changes in velocity (such as the sudden start-up shown by the dotted lines) would contain high-frequency components which would not propagate.

A system which transmits waves with a wave velocity which depends on the frequency of the wave is called a *dispersive* medium for waves, by analogy with the dispersion of light in glass (see Eq. 9.1.17). The linear array of masses and springs is thus a dispersive medium.

When the driving frequency $\omega/2\pi$ is larger than the basic resonance frequency $\omega_0/2\pi$, true wave motion cannot exist. Instead of successive time delay, the motions of the successive masses are all in phase (alternating in sign), but successively reduced in *amplitude*, by the factor

$$\gamma = -\left(\frac{2\omega^2}{\omega_0^2} - 1\right) + \sqrt{\left(\frac{2\omega^2}{\omega_0^2} - 1\right)^2 - 1} \to -\frac{\omega_0^2}{4\omega^2} \qquad \omega \gg \omega_0$$

In other words, if the motion of mass 0 is given by $P_0 e^{-i\omega t}$, then, when $\omega > \omega_0$, the motion of mass n is $\gamma^n P_0 e^{-i\omega t}$ (γ real, negative, and of magnitude less than 1), which diminishes with n the more rapidly the larger ω is. As a transmission line, the array is a low-pass filter with a cutoff frequency at $\omega_0/2\pi$. Of course, the wave can go in either direction, as is implied by the $\pm$ sign in Eq. (3.3.5). The various forms for the expression for the displacement of the nth mass, for simple-harmonic motion, are thus

$$x_n = \begin{cases} A \exp\left(-i\omega t \pm 2in \sin^{-1}\dfrac{\omega}{\omega_0}\right) & \text{when } \omega \leqslant \omega_0 \\[4mm] A \exp\left(-i\omega t \mp i\pi n \mp 2n \cosh^{-1}\dfrac{\omega}{\omega_0}\right) & \text{when } \omega \geqslant \omega_0 \end{cases} \qquad (3.3.7)$$

the first form representing true wave motion, the second representing the vibrations above the transmission cutoff frequency.

Sinusoidal wave motion

The formulas come out somewhat simpler when we ask under what circumstances the array can propagate sinusoidal waves, rather than asking for the configuration of the array when it is performing simple-harmonic motion. A sinusoidal configuration means that, at any instant, the displacement of the nth mass would depend on n, according to the real part of the exponential $e^{i\alpha n}$, so that $x_n = x_0(t)e^{i\alpha n}$. Every time that n increases by the amount $2\pi/\alpha$, the motion is again in phase; so we can call l times this amount the *wavelength* $\lambda = l(2\pi/\alpha)$ of the sinusoidal wave. For this case Eqs. (3.3.3) become

$$m\frac{d^2x_n}{dt^2} = f_n - f_{n+1} = \tfrac{1}{4}m\omega_0^2 (x_{n+1} - 2x_n + x_{n-1})$$

or   (3.3.8)

$$\frac{d^2x_n}{dt^2} = \tfrac{1}{4}\omega_0^2(e^{i\alpha} - 2 + e^{-i\alpha})x_n = -\omega_0^2 \sin^2 (\tfrac{1}{2}\alpha)x_n$$

where we have used the last of Eqs. (1.2.7) to take the last step.

But this is the equation for simple-harmonic oscillation, of frequency $\omega/2\pi = (\omega_0/2\pi) \sin (\tfrac{1}{2}\alpha)$, which is independent of n. Thus, when a sinusoidal wave of wavelength $\lambda = 2\pi l/\alpha$ traverses the periodic array, all parts of the array move with simple-harmonic motion, all with the same frequency $(\omega_0/2\pi) \sin (\tfrac{1}{2}\alpha)$.

$$x_n = Ae^{i(\alpha n - \omega t)} \qquad \omega = \omega_0 \sin \frac{\alpha}{2}$$

                                                                        (3.3.9)

$$c = \lambda \cdot \text{frequency} = \omega_0 \frac{l}{\alpha} \sin \frac{\alpha}{2}$$

which is not an unexpected result.

For small values of α (long wavelength), the frequency of the simple-harmonic motion is approximately proportional to α and inversely proportional to λ $(\omega/2\pi \simeq \alpha\omega_0/4\pi = l\omega_0/2\lambda)$. Therefore long-wavelength sinusoidal waves travel with a velocity $c \simeq \tfrac{1}{2}l\omega_0 = l\sqrt{K/m}$, which is independent of λ. As α is increased (or λ decreased), the frequency of the wave increases, but not proportionally to α (so c is no longer independent of λ), until finally, when $\alpha = \pi$ (or $\lambda = 2l$), $\sin (\alpha/2)$ and thus $\omega/2\pi$, reaches a maximum value $(\omega/2\pi \to \omega_0/2\pi)$ which is $\sqrt{2}$ times the natural frequency of isolated motion of each mass. For this limiting wavelength, $\lambda = 2l$, we have $x_n = e^{i\pi}x_{n-1} = -x_{n-1} = -x_{n+1}$; each mass is moving in opposition to its neighbors. No shorter wavelength is possible for the system of Fig. 3.4, for if we make α larger than π (λ shorter than $2l$), the phase factor $e^{i\alpha}$ can be written $e^{-i(2\pi-\alpha)}$, which corresponds to an $\alpha' = -2\pi + \alpha$ of magnitude less than π but of negative sign. This can be interpreted as the phase factor for a wave traveling in the negative direction, of wavelength $2\pi l/\alpha' = 2\pi l/(2\pi - \alpha)$, longer than $2l$. Because the array is a discrete sequence of masses, spaced a finite distance l apart, wavelengths smaller than $2l$ have no meaning, and thus motion with frequency greater than the cutoff value $\omega_0/2\pi$ cannot be propagated as wave motion.

The only way we can make $(\omega_0/2\pi) \sin(\alpha/2)$ larger than $\omega_0/2\pi$ is to make α a complex quantity, $\alpha = \pi + 2i\beta$, so that

$$\omega = \omega_0 \sin(\tfrac{1}{2}\pi + i\beta) = \omega_0[\sin(\tfrac{1}{2}\pi)\cos(i\beta) + \sin(i\beta)\cos(\tfrac{1}{2}\pi)]$$
$$= \omega_0 \cosh\beta > \omega_0$$

and

$$x_n = (-1)^n A \exp(-2n\beta - i\omega_0 t \cosh\beta)$$

Such motion is not wave motion; there is no phase lag, proportional to n, and the amplitude of motion decreases as n increases. This result is identical with Eqs. (3.3.7); only α (or β) is now the independent variable, not ω.

Wave impedance

The driving force necessary to send a wave along the linear array can be measured by removing the spring between masses 0 and 1 and determining the force f_1 which must be applied to mass 1 to produce the specified wave form in the array to the right of 1. If the applied force is simple-harmonic, of frequency $\omega/2\pi$, then Eqs. (3.3.7) or (3.3.9) show that the time delay τ or the phase lag α between successive masses is related to ω according to either Eq. (3.3.5) or Eqs. (3.3.9). The impedance with which the right-hand half of the array will resist a simple-harmonic driving force can be computed from Eq. (3.3.4) [according to Eq. (2.3.2), impedance is the ratio between force and velocity, for a driving-force frequency of $\omega/2\pi$].

$$Z = \frac{f_1}{p_1/m} = m\frac{f_{n+1}e^{-in\alpha}}{p_{n+1}e^{-in\alpha}} = -im\omega\frac{f_{n+1}}{f_{n+1} - f_{n+2}}$$

$$= im\omega\frac{f_{n+1}}{f_{n+1}(e^{i\alpha} - 1)} = \tfrac{1}{2}m\omega\frac{e^{-i\alpha/2}}{\sin(\alpha/2)} = \tfrac{1}{2}m\omega_0 e^{-i\alpha/2}$$

$$= \tfrac{1}{2}m\omega_0\cos\frac{\alpha}{2} - \tfrac{1}{2}im\omega = \tfrac{1}{2}m\sqrt{\omega_0{}^2 - \omega^2} - \tfrac{1}{2}im\omega \qquad (3.3.10)$$

At very low frequencies, $\omega \ll \omega_0$, this impedance is real, indicating that the load is purely resistive. The force is in phase with the velocity of mass 1; power is generated by the force, which is transmitted along the array with velocity $l\sqrt{K/m}$. As ω is increased, the resistive part of the impedance diminishes, and its imaginary part increases. As long as $\omega < \omega_0$, this imaginary part is equivalent to the reactance of a mass $\tfrac{1}{2}m$; the mass-spring chain acts as though it were a mass load equal to half of one of the masses, plus a frictional load of $\tfrac{1}{2}m\sqrt{\omega_0{}^2 - \omega^2}$. When $\omega = \omega_0$, the impedance is entirely the reactive load of the mass $\tfrac{1}{2}m$. For $\omega > \omega_0$, the impedance is entirely reactive, being $-\tfrac{1}{2}im(\omega + \sqrt{\omega^2 - \omega_0{}^2})$, the magnitude of which increases until, when $\omega \gg \omega_0$, it is just the impedance $-im\omega$ of mass 1. At high frequencies the springs are not stiff enough to transmit the motion, so that mass 1 is isolated.

Transient motion

The preceding pages have dealt with the steady-state motion of sinusoidal waves in the linear array of Fig. 3.4. The transient behavior, and the motion for other forms of waves, can be obtained by using the methods of the Laplace transform, as we have done several times before (examples of this will be given in the problems). Or it can be explored by taking advantage of an interesting parallel between the equations of motion (3.3.3) of the array and the set of recursion formulas (1.2.6), relating successive Bessel functions. The second set of recursion formulas, for the Bessel functions $J_m(\omega_0 t)$, is

$$\frac{d}{dt} J_m(\omega_0 t) = \tfrac{1}{2}\omega_0 J_{m-1}(\omega_0 t) - \tfrac{1}{2}\omega_0 J_{m+1}(\omega_0 t)$$

where m is any integer, positive or negative. Referring to Eqs. (3.3.3), we see that if we set

$$y_{2n} = \frac{dx_n}{dt} \quad \text{and} \quad y_{2n+1} = \frac{2}{m\omega_0} f_{n+1} = \tfrac{1}{2}\omega_0(x_n - x_{n+1})$$

then the first of Eqs. (3.3.3) (differentiated once) corresponds to

$$\frac{d}{dt} f_{n+1} = \tfrac{1}{2}m\omega_0 \frac{d}{dt} y_{2n+1} = \tfrac{1}{4}m\omega_0{}^2(y_{2n} - y_{2n+2})$$

and the second equation becomes

$$m \frac{d}{dt} y_{2n} = \tfrac{1}{2}m\omega_0(y_{2n-1} - y_{2n+1})$$

Consequently, the quantities y_m must be proportional to the Bessel functions $J_m(\omega_0 t)$, where $m = 2n + $ any integer. For example, if we set $x_n = A_p J_{2n-2p}(\omega_0 t)$, then

$$y_{2n} = \frac{dx_n}{dt} = \tfrac{1}{2}\omega_0 A_p[J_{2n-2p-1}(\omega_0 t) - J_{2n-2p+1}(\omega_0 t)]$$

and

$$y_{2n-1} = \frac{2f_n}{m\omega_0} = \tfrac{1}{2}\omega_0 A_p[J_{2n-2p-2}(\omega_0 t) - J_{2n-2p}(\omega_0 t)]$$

is a self-consistent solution of the equations of motion (3.3.3) for the array of Fig. 3.4.

Since these equations of motion are linear, a sum of solutions is also a solution. Therefore we can arrange a sum to satisfy initial conditions, based on the fact that $J_n(0) = \delta_{0n}$ [Eq. (1.2.5)]. If, at $t = 0$, $x_n = a_n$ and $dx_n/dt = v_n$, then the subsequent motion of the system is given by the following equations:

$$x_n = \sum_{p=-\infty}^{\infty} \left[a_p J_{2n-2p}(\omega_0 t) + \left(\frac{2}{\omega_0} \sum_{m=p+1}^{\infty} v_m \right) J_{2n-2p-1}(\omega_0 t) \right]$$

$$\frac{d}{dt} x_n = \sum_{p=-\infty}^{\infty} [v_p J_{2n-2p}(\omega_0 t) + \tfrac{1}{2}\omega_0(a_p - a_{p+1}) J_{2n-2p-1}(\omega_0 t)] \qquad (3.3.11)$$

$$f_n = \tfrac{1}{2}m\omega_0 \sum_{p=-\infty}^{\infty} [\tfrac{1}{2}\omega_0(a_{p-1} - a_p) J_{2n-2p}(\omega_0 t) + v_p J_{2n-2p-1}(\omega_0 t)]$$

Curves for x_n ($n = 0, 1, 2, 3, 4$) versus time are plotted in Fig. 3.6a for the case where mass 0 is given an initial impulse ($a_n = 0$, $v_n = \delta_{n0}u_0$), so that $x_n = \sum_{m=0}^{\infty} (2u_0/\omega_0) J_{2n+2m+1}(\omega_0 t)$. These curves exhibit the difference between the actual motion of a linear array and the simple approximation discussed earlier in this section. For a nondispersive wave medium, the curve for x_1 would be the same as for x_0, except that it would be delayed a time $\tau = l/c = 2/\omega_0$ after that for x_0, and so on; so $x_n(t)$ would equal $x_0(t - n\tau)$. The dashed lines of Fig. 3.6a correspond to the velocity $c = \tfrac{1}{2}l\omega_0 = l\sqrt{K/m}$. We see that in a general way each x_n curve duplicates the preceding one; it

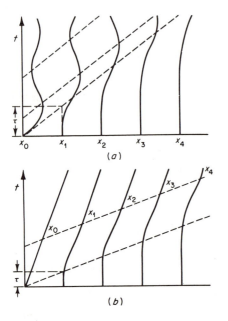

FIGURE 3.6
Wave motion in a linear array. Curves *a* are displacements of successive masses when mass 0 is given an initial impulse; curves *b* are for the situation of Fig. 3.5.

starts from rest at about $t = n\tau$, goes through a maximum, and then oscillates about a new equilibrium position, a distance u_0/ω_0 from its original position. Actually, each curve differs somewhat from the others; each successive x_n starts from zero a little earlier than $t = n\tau$, and its first maximum is broader and not so high, though the subsequent oscillations are quite similar to those for x_0.

Such differences in the motion of successive parts of the linear array are, of course, due to the fact that the velocity of simple-harmonic waves is not the same for all frequencies; according to Eq. (3.3.6) it decreases as ω increases, going from $l\omega_0/2$, for the low-frequency components of the wave, down to $l\omega_0/\pi$ when $\omega = \omega_0$; for $\omega > \omega_0$ no wave motion is possible. The relatively low-frequency oscillations of the masses, after their first large displacement, propagate without much change from mass to mass. But the sharp change of velocity of x_0 at $t = 0$ (caused by the impulsive force applied then) does not propagate; the onset of the motion of each successive mass is more and more gradual (i.e., has less and less of the high frequencies). These are typical characteristics of wave transmission in dispersive media.

We can also use the Bessel function series to obtain the exact solution for the motion pictured in Fig. 3.5, where x_0 is moved with constant velocity after $t = 0$. Application of the equations following Eqs. (1.2.6) shows that the appropriate solution is

$$x_n = \frac{2u}{\omega_0} \sum_{m=0}^{\infty} (2m + 1)J_{2n+2m+1}(\omega_0 t)$$

This solution is plotted in Fig. 3.6b. We see that the sharp change in velocity, at $t = 0$ for x_0, does not propagate; each successive mass accelerates more smoothly, slightly overshoots the velocity u, and subsequently oscillates about the velocity u with diminishing amplitude. Here again the dispersive character of the array for wave motion is apparent.

Problems

1 Two oscillators, each of mass m and natural frequency ν_0 (i.e., if one oscillator is held at equilibrium, the other will oscillate with a frequency ν_0), are coupled so that moving one mass 1 cm from equilibrium produces a force on the other of C dynes. Show that if C is small compared with $4\pi^2 m\nu_0^2$ and if one oscillator is held 1 cm from equilibrium and the other at equilibrium and both are released at $t = 0$, the subsequent displacements of the masses will be

$$y_1 = \cos\left(\frac{Ct}{4\pi m\nu_0}\right)\cos(2\pi\nu_0 t) \qquad y_2 = \sin\left(\frac{Ct}{4\pi m\nu_0}\right)\sin(2\pi\nu_0 t)$$

2 Suppose that each oscillator of Prob. 1 is acted on by a frictional force equal to R times its velocity. Show that the modulus of decay of the oscillations equals $2m/R$.

3 Discuss the forced vibrations of the coupled oscillators described in Prob. 2.

4 Three masses, each of m g, are equally spaced along a string of length $4a$. The string is under a tension T dynes. Show that the three allowed frequencies and the corresponding relations between the displacements for the normal modes are

$$\nu = \nu_0 \sqrt{1 - \sqrt{\tfrac{1}{2}}} \quad \text{if} \quad y_1 = y_3 = \frac{y_2}{\sqrt{2}}$$

$$\nu = \nu_0 \quad \text{if} \quad y_1 = -y_2 \text{ and } y_2 = 0$$

$$\nu = \nu_0 \sqrt{1 + \sqrt{\tfrac{1}{2}}} \quad \text{if} \quad y_1 = y_2 = \frac{-y_2}{\sqrt{2}}$$

where $T/a = 2\pi^2 \nu_0^2 m$.

5 An arrangement sometimes used for vibration damping is shown schematically in the figure. A large mass M (a machine of some sort) is supported by a spring with stiffness constant K. A linear driving force $f(t) = F_0 e^{-i\omega t}$ is acting on M. By adding a second oscillator with mass m and stiffness constant κ as shown, the vibration amplitude of M can be eliminated. How are m and κ to be chosen to make this possible?

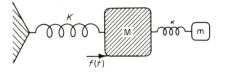

6 Four coupled harmonic oscillators have coupling constants U_{ij}, as defined in Eq. (3.2.8), related as follows: $U_{11} = U_{22} = U_{33} = U_{44} = \omega_0^2$, $U_{12} = U_{21} = U_{23} = U_{32} = U_{34} = U_{43} = U_{41} = U_{14} = \alpha\omega_0^2$, $U_{13} = U_{31} = U_{24} = U_{42} = 2\alpha\omega_0^2$. Show that the allowed natural frequencies of the system are $\omega_1^2 = \omega_0^2(1 - 4\alpha)$, $\omega_1^2 = \omega_0^2$, $\omega_3^2 = \omega_4^2 = \omega_0^2(1 + 2\alpha)$. At which frequency will all the masses be moving in phase with equal amplitude? What are the relative phases and amplitudes of motion of the masses when vibrating with the other frequencies? Compute the 16 unit components $E_i(n)$ of Eq. (3.2.23) for this system. Then calculate the motion of each mass if the first mass is struck a unit impulse at $t = 0$, the system being previously at rest.

7 Suppose the masses in the system of Fig. 3.4 move at right angles to the system axis, instead of longitudinally. The lattice is stretched to a tension $S = K(l - l_0)$, where l is the equilibrium spacing and l_0 the relaxed length of the spring, and the transverse displacement of each mass is much smaller than l. The average mass per unit length of the system being $\epsilon = m/l$, show that the speed of simple-harmonic transverse wave motion is $\sqrt{S/\epsilon}$, at sufficiently low frequencies. Prove that the wave speed of transverse waves is always smaller than the wave speed of longitudinal waves for this system.

8 With the system of Fig. 3.4, suppose there is a frictional force $R(dx_n/dt)$ acting on each mass. Consider a wave motion of the lattice in which the displacement of the nth mass is proportional to $\exp(ikx_n - i\omega t)$. Determine k as function of ω, and show that for small values of R the amplitude of oscillation decreases by a factor $\exp(4R/\omega_0 m)$ per lattice spacing [see Eq. (3.3.4) for definition of ω_0].

9 Two one-dimensional mass-spring systems are connected as shown in the figure, with $m_2 > m_1$ and $K_2 \neq K_1$. A simple-harmonic longitudinal wave of frequency $\omega/2\pi$ is traveling on the first lattice, from the left, and is partially reflected at the junction, so that the motion of the first lattice is a superposition of incident and reflected waves. In the low-frequency limit, determine the amplitudes of the transmitted and reflected waves, in terms of the amplitude of the incident wave. What is the reflected wave amplitude when the frequency of the incident wave exceeds the cutoff frequency of the second lattice?

10 A mass-spring lattice of the sort shown in Fig. 3.4 has constants $m = 1$, $K = 10^4$, and $l = 1$. (a) Determine the frequency range in which a harmonic wave can be transmitted without attenuation. (b) What is the phase velocity when the wavelength of the wave is much greater than l? When the wavelength is $4l$? (c) If the lattice is driven at frequency $\sqrt{8K/m}$, determine the spatial attenuation of the amplitude of oscillation, and also the phase shift, per lattice spacing l.

11 Two longitudinal waves with slightly different frequencies, $\omega/2\pi$ and $(\omega + d\omega)/2\pi$, and corresponding wavenumbers, k and $k + dk$, are transmitted to the right along the mass-spring lattice of Fig. 3.4. Show that the resulting wave can be described as a wave of spatially varying amplitude, and that a point of constant displacement amplitude travels forward with a velocity $d\omega/dk$ (called the *group velocity* of the wave). What is the group velocity of the system, and what is its phase velocity ω/k, when the wavelength of the wave motion is four times the lattice spacing l? Express your answers in terms of the value of the phase velocity in the limit of low frequencies.

12 A one-dimensional mass-spring lattice is driven at one end by a simple-harmonic force $F = F_0 \exp(-i\omega t)$ as shown in the figure. (a) Determine the velocity u of the first mass element (with mass $m/2$) in the lattice; show that, for $\omega \ll \sqrt{K/m}$, this is given by the simple formula $F = \epsilon c u$, where c is the phase velocity of the wave and $\epsilon = m/l$ is the mass per unit length of the lattice. Is this result consistent with Eq. (3.3.10)? (b) What is the average power transmitted to the lattice by the force when $\omega \ll \sqrt{K/m}$? When $\omega > 2\sqrt{K/m}$? (c) Suppose the first mass element is removed and the force is applied to the end of the spring as shown in the lower part of the figure. What, then, is the power delivered to the lattice when $\omega \ll \sqrt{K/m}$?

13 The linear array of Fig. 3.4 is made finite by fixing masses $n = 0$ and $n = N + 1$ at rest, so that the potential energy is

$$V = \tfrac{1}{8}m\omega_0^2[x_1^2 + (x_2 - x_1)^2 + (x_3 - x_2)^2 + \cdots + (x_N - x_{N-1})^2 + x_N^2]$$

Show that the N allowed frequencies of longitudinal vibration of the system are given by

$$\omega_\nu = \omega_0 \sin \frac{\pi\nu}{2(N+1)} \qquad \nu = 1, 2, 3, \ldots, N$$

and that, corresponding to the νth frequency, the unit solutions for the displacement amplitudes [Eq. (3.2.23)] are

$$E_n(\nu) = \sqrt{\frac{2}{N}} \sin \frac{\pi\nu n}{N+1}$$

Show that

$$\sum_{n=1}^{N} E_n(\nu)E_n(\mu) = \delta_{\nu\mu} \qquad \sum_{\nu=1}^{N} E_n(\nu)E_m(\nu) = \delta_{nm}$$

and therefore that if at $t = 0$ the nth mass is at rest but displaced a distance x_n^0 from equilibrium, the subsequent motion of the system is given by the equations

$$x_n(t) = \sum_{\nu=1}^{N} E_n(\nu) \left[\sum_{m=1}^{N} x_m^0 E_m(\nu) \right] \cos(\omega_\nu t)$$

14 A finite circular array (with mass 1 connected to mass N) has a potential energy

$$V = \tfrac{1}{8}m\,\omega_0^2[(x_2 - x_1)^2 + (x_3 - x_2)^2 + \cdots + (x_N - x_1)^2]$$

Show that the possible motions of the array are given by

$$x_n = C_\nu \exp\left(\frac{2\pi i\nu n}{N} - i\omega_\nu t\right)$$

where

$$\omega_\nu = \omega_0 \sin \frac{\pi\nu}{N} \qquad 0 \leqslant \nu < N$$

What is the motion corresponding to $\nu = 0$? What is the relation between motions for ν and for $N - \nu$? Show that if the system is originally in equilibrium and, at $t = 0$, the mth mass is given a velocity v_m^0, the subsequent motion is given by the equations

$$x_n(t) = \frac{1}{N}\left\{ v_m^0 t + \left[\sum_{\nu=0}^{N-1} \frac{v_m^0}{\omega_\nu} \exp\left(2\pi i\nu \frac{n-m}{N}\right) \right] \sin(\omega_\nu t) \right\}$$

15 For the circular array of Prob. 14, show that if mass j is held displaced an amount x_j^0 and the rest held at equilibrium, the whole being released at $t = 0$, the subsequent motion will be

$$x_n(t) = \frac{1}{N}\left[\sum_{\nu=0}^{N-1} x_j^0 \exp\left(2\pi i\nu \frac{n-j}{N}\right) \right] \cos(\omega_\nu t)$$

Utilize the equation [Eq. (1.2.10)]

$$\cos(z \sin \theta) = \sum_{m=-\infty}^{+\infty} J_{2m}(z)\, e^{-2im\theta}$$

to show that the motion of the system can also be expressed in terms of the series

$$x_n(t) = \sum_{k=-\infty}^{\infty} x_j^0 J_{2j-2n+2kN}(\omega_0 t)$$

For $N = 6$, $j = 6$, plot the displacements of masses 3 and 6 from $\omega_0 t = 0$ to $\omega_0 t = 10$.

CHAPTER

4

THE FLEXIBLE STRING

4.1 WAVES ON A STRING

Up to now we have been discussing systems which behave as though they were constructed of point masses and massless springs. A few systems which approximate this simplicity are encountered in acoustic practice. The mathematical models we have developed to describe their motion, however, are more important in practice than the systems they represent, for they exhibit in elementary form many of the properties displayed by more complex systems. Thus they have permitted us to introduce concepts of energy and impedance and resonance, for example, which will turn out to be useful aids in analyzing situations of greater intricacy.

Neglecting atomicity, systems have their mass distributed over their whole extent; every part has both density and elasticity. The vibrating string of a violin, for example, cannot be considered as having all its mass concentrated at the center of the string, or even concentrated at a finite number of points along the string; an essential property of the string is that its mass is spread *uniformly* along its length. Similarly, a loudspeaker diaphragm has a good portion of its mass spread out uniformly over its extent. In these cases, each portion of the system will vibrate with a somewhat different motion from that of any other portion. The position of just a few parts of the system will not suffice to describe its motion; the position of *every* point must be specified.

Of course, a rough approximation to the continuous string can be achieved by carrying the analysis of Sec. 3.3 to its limit. It might, for example, be considered to be a limiting case of the system portrayed in Fig. 3.1c, when the number of point masses is increased indefinitely and their mutual distance is correspondingly reduced to zero. The techniques developed in the preceding chapter allow this asymptotic solution to be worked out (indeed, it has been the subject of a problem), and this solution turns out to predict the right answers. But such a technique is really an awkward way of solving a problem which is essentially simple. What is needed is a new point of view, a new method of attack.

95

The new point of view can be summarized as follows: We must not concern ourselves with the motion of each of the infinite number of points on the string, considered as separate points; we must consider the *shape* of the string *as a whole*.   At any instant the string will have a definite shape, which can be expressed mathematically by saying that y, the displacement from equilibrium of that part of the string a distance x cm from one end, is a function of x.   The motion of the string at any instant will depend on the shape of the string, and the subsequent shape will depend on the motion; what we must do is to find the relation between the shape and the motion. In other words, the string's displacement y is a function of both x and t, and we must discover the relation between y's dependence on x and its dependence on t.   The same point of view will be necessary in studying the vibrations of diaphragms and of air, as we shall see later.

Wave motion

In this chapter we shall use one of the simplest continuous systems, the flexible string under tension, to illustrate this new point of view and how it can be used.   We assume that the string has negligible transverse dimensions and that its mass is distributed uniformly, ϵ per unit length.   We shall not be concerned here with longitudinal waves, analogous to the limiting form of the motions of the system of Fig. 3.4; we shall study only the transverse motion of the string.   If the string is stretched under a tension T newtons, its equilibrium position (in the absence of gravity or other forces) will be straight-line.   We can designate a point on the string by giving its distance x from some origin (perhaps one end of the string) while in its equilibrium position.   The motions we shall study involve the displacement of the point x an amount y at right angles to the equilibrium line.   We need only concern ourselves with displacements in one plane through the equilibrium line, for we can show that the general transverse motion is a linear combination of motions in two perpendicular planes (at least, when the displacements are small enough).

Transverse displacement y is thus a function of both x and t.   Holding t constant and plotting y against x gives us a picture of the shape of the whole string at a given instant.   Holding x constant and plotting y against t gives us a plot of the motion of a given point on the string as it moves in time.   The basic question is, how are these two plots related?   What is the relationship between the dependence of y on x and its dependence on t?   To answer this is to determine the nature of wave motion.

Even a casual observation of the behavior of ropes and strings indicates that transverse waves are propagated along them; a disturbance moves from one part to another with a velocity c.   This, of course, has a bearing on the relationship between the dependence of y on x and its dependence on t.   At $t = 0$ a string may have a shape given by $y = F(x)$ in Fig. 4.1.   If this shape is to represent a wave traveling to the right, then a time τ later the displaced

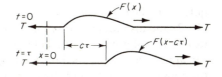

FIGURE 4.1
Wave motion along a flexible string.

region should have moved to the right by a distance $c\tau$, where c is the wave velocity. If c is the same for all frequencies and shapes of wave (this was *not* the case for the waves of Sec. 3.3), the shape of the wave should be unchanged during its motion; at time $t = \tau$ the displacement should be $F(x - c\tau)$.

Thus, for "perfect" wave motion to the right, the relationship between dependence on x and on t should be especially simple. If the displacement at $t = 0$, as function of x, is $F(x)$, then the displacement at time t should be $F(x - ct)$, where c is the wave velocity, independent of the shape of F. In other words, t and x enter only in the combination $x - ct$, as, for instance, $\sin [k(x - ct)]$, or $A(x - ct)^3 \exp [-\beta^2(x - ct)^2]$, etc. For "perfect" wave motion to the left, the displacement should be $f(x + ct)$, independent of the shape of function f. If the equation of motion is linear, so that the sum of solutions is a solution, the general expression for "perfect" wave motion along the string will be

$$y(x,t) = F(x - ct) + f(x + ct) \tag{4.1.1}$$

where $F(z)$ and $f(z)$ are any continuous functions of variable z. Therefore many of our basic questions regarding wave motion on the string would be answered if we could show under what circumstances the displacement $y(x,t)$ of Eq. (4.1.1) is a solution of the equation of motion of the string.

The wave equation

The partial derivative $\partial y/\partial x = y'(x)$ represents the slope of the displacement curve, at point x on the string, at time t. The derivative $\partial y/\partial t = \dot{y}(x)$ represents the transverse velocity of point x at time t. If the shape of the string can be described by the function $F(x - ct)$, we shall have a particularly simple relationship between these partial derivatives, $\partial y/\partial t = -cF'(x - ct) = -c(\partial y/\partial x)$, where $F'(z)$ is the derivative of $F(z)$ with respect to its argument z. This says that, as the wave moves to the right, a portion having negative slope ($\partial y/\partial x < 0$) will move upward ($\partial y/\partial t > 0$), and vice versa.

For wave motion to the left, the relationship between the derivatives is $\partial y/\partial t = +c(\partial y/\partial x)$. For *both* wave motions, and thus for the motion represented by Eq. (4.1.1), the second-order equation

$$\frac{\partial^2 y}{\partial t^2} = c^2 \frac{\partial^2 y}{\partial x^2} \tag{4.1.2}$$

holds.   In fact, we can say that a solution of the form (4.1.1) implies Eq.
(4.1.2) and also that an equation of motion of the form (4.1.2) has (4.1.1) as
its general solution.   If (4.1.1) represents "perfect" wave motion, Eq. (4.1.2)
can appropriately be called the *wave equation*.   We now must show when it is
the equation of motion of the flexible string.

The left-hand term in Eq. (4.1.2) is the transverse acceleration of point x
on the string at time t.   This should be proportional to the net transverse
force acting on the portion of the string near point x, if (4.1.2) is its equation
of motion.   Figure 4.2 shows the forces acting on the portion of string
between the points which were at x and $x + dx$ when the string was in
equilibrium.   At equilibrium the string was under uniform tension T and had
a mass ϵ per unit length.   At time t, point x is displaced transversely an
amount $y(x,t)$, and $x + dx$ is displaced an amount $y(x + dx, t)$.   Figure 4.2
shows some of the complications which limit the validity of Eq. (4.1.2).

If y has a nonzero slope at x, either the length of the string between x
and $x + dx$ is greater than dx or else the displacements $y(x)$ and $y(x + dx)$
are not exactly perpendicular to the equilibrium line.   But if the string is
stretched by the displacement, the tension in that part of the string must be
greater than the value T, which it had in equilibrium.   However, if $\partial y/\partial x$ is
negligibly small compared with unity over the whole extent of the string,
these complications can be disregarded; we can then neglect all quantities of
second and higher order in $\partial y/\partial x$.   To within the first order in this small
quantity the displacement of the element of string will be perpendicular to the
equilibrium line, and the tension in the element is equal to its equilibrium
value T (see Chap. 14).

Also, to first order in $\partial y/\partial x$, the net horizontal force on the element of
string is zero.   The net vertical force is not zero to this order, however; it is
equal to the difference in the vertical components of the tensions at the two
ends, as shown in Fig. 4.2.

$$F_\perp = Ty'(x + dx) - Ty'(x) = T\frac{\partial^2 y}{dx^2}\,dx$$

If the string is perfectly flexible, so that no force is required to bend it into
curved form, then $F_\perp$ is the only transverse force acting and the element's
transverse acceleration must be proportional to $F_\perp$.   In fact, since the mass

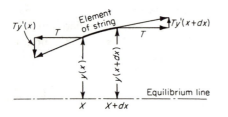

FIGURE 4.2
Forces acting on an elementary length of
flexible string.

of the element is $\epsilon\,dx$, the equation of motion must be

$$\epsilon\frac{\partial^2 y}{\partial t^2}=T\frac{\partial^2 y}{\partial x^2}\qquad\text{or}\qquad\frac{\partial^2 y}{\partial t^2}=c^2\frac{\partial^2 y}{\partial x^2}\qquad(4.1.3)$$

where $c^2=T/\epsilon$ is independent of x and t to first order in $\partial y/\partial x$.

This is, of course, the wave equation (4.1.2). We have thus shown that if the string is perfectly flexible, is uniform in mass density and in tension along its length, and is nowhere much displaced from its equilibrium configuration, so that $(\partial y/\partial x)^2$ can be neglected compared with unity, the string's motion obeys the wave equation, and wave motion of the sort represented by (4.1.1) is possible, with a wave velocity $c=\sqrt{T/\epsilon}$ which is independent of waveshape and size. Under these circumstances "perfect" wave motion along the string is possible.

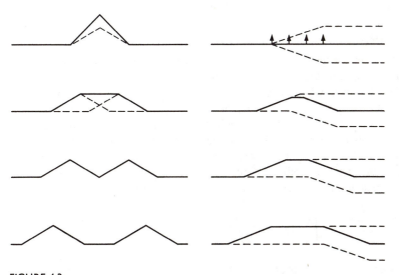

FIGURE 4.3
Motions of plucked and struck strings. The solid lines give the shapes of the strings at successive times; the dotted lines give the shapes of the two "partial waves" traveling in opposite directions whose sum is the actual shape of the string.

The forms of functions F and f in Eq. (4.1.1) can be determined from the initial transverse displacement and velocity of the string at $t=0$. Since the equation of motion is a second-order equation, only the initial displacement and velocity are needed to determine the motion [see discussion of Eqs. (2.1.4)]. Suppose, at $t=0$, the shape of the string is given by the function $y_0(x)$, and the initial transverse velocity of point x of the string is $v_0(x)$. Then we can express F and f in terms of y_0 and the function $S(x)=\displaystyle\int_0^x v_0(u)\,du$.

$$y(x,t) = \tfrac{1}{2}y_0(x - ct) + \tfrac{1}{2}y_0(x + ct) - \frac{1}{2c}S(x - ct) + \frac{1}{2c}S(x + ct) \quad (4.1.4)$$

as can be proved by computing y and $\partial y/\partial t$ for $t = 0$. In other words, the solution satisfying the initial conditions $y = y_0(x)$ and $\partial y/\partial t = v_0(x) = dS(x)/dx$ is that given in (4.1.1), with

$$F(z) = \tfrac{1}{2}y_0(z) - \frac{1}{2c}S(z) \qquad f(z) = \tfrac{1}{2}y_0(z) + \frac{1}{2c}S(z)$$

Figure 4.3 shows the behavior of an infinite string subjected to different initial conditions. A string of infinite extent, a short length of which is pulled aside and let go at $t = 0$, returns to its initial equilibrium line; one given an impulse at $t = 0$ will come to rest along a line parallel to the original line.

Wave energy

When we come to compute the energy and momentum of wave motion, we must take into account the second-order terms we neglected in deriving Eq. (4.1.3), even when they are small. For example, in computing the potential energy gained by displacing a string from equilibrium, we must calculate the work required to move it from the shape $y(x) = 0$ to the shape $y(x)$, as shown in Fig. 4.4. The distance along such a shape, from $x = 0$ to $x = l$, would inevitably be greater than the distance l measured along the equilibrium line. Therefore, if the two end supports are kept a distance l apart, the string must increase its length by an amount

$$\int_0^l \sqrt{1 + \left(\frac{\partial y}{\partial x^2}\right)^2}\,dx - l \simeq \frac{1}{2}\int_0^l \left(\frac{\partial y}{\partial x}\right)^2 dx$$

when we neglect all terms of higher than second order in the expansion of the square root.

If the string is quite elastic, it has already been stretched quite a bit when

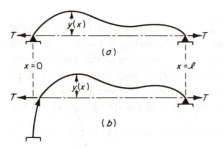

FIGURE 4.4
Displacing the string stretches the string or distorts the support.

it was put under tension T; so the small increase in length involved in the displacement would not increase the tension by an appreciable amount. Consequently, the work required to stretch the string by displacing it from its equilibrium line to the shape $y(x)$ is equal to T times the change in length of the string, and this, to second order in $\partial y/\partial x$, is

$$V = \int_0^l T\sqrt{1 + \left(\frac{\partial y}{\partial x}\right)^2}\,dx - lT \simeq \tfrac{1}{2}T\int_0^l \left(\frac{\partial y}{\partial x}\right)^2 dx \qquad (4.1.5)$$

If, on the other hand, the string does not elongate when placed under tension, one or both of the supports will be pulled inward when the string is displaced (assume it is the one at $x = 0$, as shown in Fig. 4.4b). Here again, if the tension is unchanged to the first order during the displacement, the potential energy is T times the displacement of the support, the identical formula for the potential energy of displacement given in Eq. (4.1.5), as long as $\partial y/\partial x \ll 1$. In the intermediate case, where the string stretches a bit and the supports move a bit, the expression for the potential energy of displacement is still given by Eq. (4.1.5), correct to the second order in the small quantity $\partial y/\partial x$.

One might be tempted to localize the potential energy and say that the element of string between x and $x + dx$ has the potential energy $\tfrac{1}{2}(\partial y/\partial x)^2 T\,dx$. But if the string does the stretching, this stretching is uniform along the string, not proportional to $(\partial y/\partial x)^2$ at each point; and if the support does the "stretching," the work is all done moving the end support. In any case it is more correct to say that the potential energy of the string as a whole is given by the integral of Eq. (4.1.4). If, later in this chapter, we find it convenient to talk about a "potential energy density" at x as being $V(x) = \tfrac{1}{2}T(\partial y/\partial x)^2$, we must keep in mind that the density has a certain degree of arbitrariness about it. We could just as well have said that $V(x)$ is equal to $\tfrac{1}{2}T(\partial y/\partial x)^2 + (\partial\Phi/\partial x)$, where $\Phi(x)$ is any continuous function of x which goes to zero at the two ends of the string. For really, the only measurable quantity is the integral of $V(x)$ over the whole length of the string, as given in (4.1.5), and $\Phi(x)$ has been chosen so that the integral of $\partial\Phi/\partial x$ over the string is zero.

To cast further light on the amount of energy and momentum carried by the wave, let us parallel the discussion given earlier, in connection with Fig. 3.5, for wave motion in the sequence of springs and masses. Suppose we start wave motion in a very long string, with one end at $x = 0$, by moving this support sideward with a velocity u. After a time τ the transverse displacement of the support is $u\tau$, the wave started by the displacement has reached a distance $c\tau$ along the string, and the shape of the string is that shown by the solid line of Fig. 4.5. The part of the string from $x = 0$ to $x = c\tau$ is moving upward with velocity u; the part near $x = 0$ has been moving since $t = 0$; the part near $x = c\tau$ has just started to move; the part beyond $x = c\tau$ is still at rest.

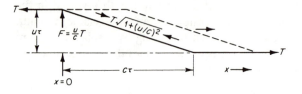

FIGURE 4.5
Initiation of wave motion
by transverse displacement
of support at $t = 0$ with
velocity u.

The transverse force necessary to keep the support moving sideward, with velocity u, is nearly equal to $T(u/c) = -Ty'(0)$ [where $y'(x) = \partial y(x)/\partial x$] as long as $y' \ll 1$. The work done on the string during time τ, in generating the wave motion, is the transverse force F times the distance $u\tau$, which equals $T(u/c)u\tau$. This must equal the kinetic energy $\frac{1}{2}\epsilon c\tau u^2$ of the length $c\tau$ of the string which is in motion at time τ, plus the potential energy of this part, which from Eq. (4.1.5) equals $\frac{1}{2}T\int(y')^2\,dx = \frac{1}{2}T(u/c)^2 c\tau$. The resulting equation for the energy balance leads directly to a relation between tension, mass density, and wave velocity c.

$$T\left(\frac{u}{c}\right)u\tau = \tfrac{1}{2}\epsilon c\tau u^2 + \tfrac{1}{2}T\left(\frac{u}{c}\right)^2 c\tau$$

or

$$\frac{Tu^2\tau}{2c} = \tfrac{1}{2}\epsilon u^2 c\tau \quad \text{or} \quad c^2 = \frac{T}{\epsilon} \tag{4.1.6}$$

The velocity c required to balance energy input with energy flow along the string is of course the same as that given in Eq. (4.1.3).

The total energy density of the wave motion can thus be written as the sum of a kinetic energy density and a potential energy density.

$$H(x,t) = \tfrac{1}{2}\epsilon\left(\frac{\partial y}{\partial t}\right)^2 + \tfrac{1}{2}T\left(\frac{\partial y}{\partial x}\right)^2 = W_{tt} \tag{4.1.7}$$

where we must keep in mind the reservations mentioned three paragraphs above. In addition, Eq. (4.1.6) shows that the energy flow along the string, across point x, during the wave motion, is $T(u/c)u = -T(\partial y/\partial x)(\partial y/\partial t)$, in watts. Finally, if the transverse force ceases at $t = \tau$, the sideward motion of the support stops, and the string, a short time later, will have the shape of the dotted line of Fig. 4.5. Energy no longer is fed into the string, but the energy it collected between $t = 0$ and $t = \tau$ continues to be propagated to the right.

Wave momentum

When we consider the momentum imparted to the string, we must be still more careful to keep track of quantities of the second order in $\partial y/\partial x$. For example, if the string elongates enough so the motion of the end at $x = 0$

is vertical, as shown in Fig. 4.5, then to second order in $\partial y/\partial x = u/c$, the tension in the moving part of the string must be $T\sqrt{1 + (u/c)^2}$. In this case the horizontal components of force at $x = c\tau$ exactly cancel and the vertical component, $T(u/c)$ to second order, serves to bring the next portion of the string into motion. On the other hand, if the string does not change length, the end $x = 0$ must move upward at an angle u/c to the vertical, as shown in Fig. 4.6. In this case, as can be seen from Fig. 4.6, the tension in the moving part of the string is $T\sqrt{1 - (u/c)^2}$, to second order, so that the unbalanced force at $t = c\tau$, with vertical component $T(u/c)$ and horizontal component $T(u/c)^2$ (to second order), starts the next part of the string up at the required angle u/c to the vertical.

These results do not invalidate our earlier derivation of Eq. (4.1.3) or (4.1.7); we there used values of the tension only to the first order in u/c, and the tension, in either case, is T throughout the string to first order.

The balance of forces at the end $x = 0$ also is affected by the degree of elasticity of the string. We first deal with the limiting case where the string does not stretch and the end support moves inward, as well as transversally, as shown in Fig. 4.6. In this case the motion of the end support is perpendicular to the string, at an angle u/c radians to the vertical, when u/c is small. The applied force therefore has a horizontal component $F(u/c) = T(u/c)^2$ to the second order in the small quantity $u/c = -(\partial y/\partial x)$. Consequently, the driving force produces both a vertical and a horizontal component of momentum in the string. The momentum transverse to the equilibrium line is the one usually considered; it amounts to ϵu times the length $c\tau$ of the string in motion at $t = \tau$ (to the second order in u/c). This must equal the transverse impulse, $F\tau = T(u\tau/c)$, of the driving force. Again equating the two gives rise to the equation $c^2 = T/\epsilon$ for the wave velocity.

The momentum in the longitudinal direction is not quite as obvious a matter. The horizontal impulse, for the inelastic string of Fig. 4.6, is $F(u/c)\tau = T\tau(u/c)^2$. This must impart a longitudinal momentum to the string. For a nonstretchable string the x component of momentum is $\epsilon(u/c)u$ per unit length, over the length $c\tau$ which is in motion. Again equating impulse to momentum leads to the equation $c^2 = T/\epsilon$; this time it indicates a *longitudinal momentum density* of $-\epsilon(\partial y/\partial t)(\partial y/\partial x)$ per unit length of string. This momentum would be "felt" by the support at the other end

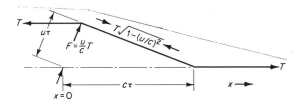

FIGURE 4.6
Balance of forces during wave motion along an inelastic string.

and would produce a "radiation pressure" equal to c times the longitudinal momentum density, $-\epsilon c(\partial y/\partial t)(\partial y/\partial x) = \epsilon u^2 = T(u/c)^2 = F(u/c)$ for the case of Fig. 4.6.

If the string is elastic, as in Fig. 4.5, so that the motion of the support at $x = 0$ is exactly transverse, it is altered longitudinal stress which is carried, not longitudinal momentum; but the effect on the support at the far end is just the same (at least to second order in u/c). The tension in the moving part of the string is $T\sqrt{1 + (u/c)^2}$, greater than T. This stretches the string and moves it vertically upward and acts like a longitudinal momentum in its effects on the support at the other end [see discussion following Eq. (4.1.12)]. A longitudinal force equal to c times $-\epsilon(\partial y/\partial t)(\partial y/\partial x)$ will be produced. Thus, in either limit, or in the intermediate case of some string elasticity and some horizontal motion of the support at $x = 0$, the string acts as though it carries a longitudinal wave momentum of density $-\epsilon(\partial y/\partial t)(\partial y/\partial x)$ per unit length of string (as long as $\partial y/\partial x$ is small), which is carried along the string with the velocity c of the wave. We note that if the direction of the transverse force F and the motion of the string had been downward, this longitudinal momentum would still be to the right. The motion of both energy and longitudinal momentum is always in the direction of the wave.

Lagrange equation and stress-energy tensor

We can now recapitulate our findings for the stress, energy, and momentum of transverse waves in a perfectly flexible string, when $\partial y/\partial x$ is small throughout its length, and we can recast its equation of motion in the form of the Lagrange equation of classical dynamics. The kinetic energy of transverse motion is $\frac{1}{2}\epsilon(\partial y/\partial t)^2$ per unit length, and its potential energy density is $\frac{1}{2}T(\partial y/\partial x)^2$, to second order in $\partial y/\partial x$. If the system were a point mass m acted on by forces having a potential energy $V(x)$, we could define a Lagrange function $L = \frac{1}{2}m(dx/dt)^2 - V(x)$, and write the equation of motion as

$$\frac{d}{dt}\frac{\partial L}{\partial \dot{x}} = \frac{\partial L}{\partial x} \qquad \dot{x} = \frac{dx}{dt}$$

In the case of the string, with its mass distributed along its length and with its dependent variable y a function of two independent variables, x and t, we must define a *Lagrange density* $L = \frac{1}{2}\epsilon(\partial y/\partial t)^2 - \frac{1}{2}T(\partial y/\partial x)^2$ and write the equation of motion, to be satisfied at every point of the string,

$$\frac{\partial}{\partial t}\frac{\partial L}{\partial \dot{y}} + \frac{\partial}{\partial x}\frac{\partial L}{\partial y'} = \frac{\partial L}{\partial y} \qquad \dot{y} = \frac{\partial y}{\partial t} \qquad y' = \frac{\partial y}{\partial x} \qquad (4.1.8)$$

or

$$\epsilon\frac{\partial^2 y}{\partial t^2} - T\frac{\partial^2 y}{\partial x^2} = 0$$

which is the wave equation. For the derivation of the equation of motion from the Lagrange density, see the discussion preceding Eq. (5.2.6).

The transverse momentum density is $\partial L/\partial \dot{y} = \epsilon(\partial y/\partial t) = p(x,t)$. The transverse force by which the string to the left of point x acts on the part to its right is $\partial L/\partial y' = -T(\partial y/\partial x) = s(x,t)$. By analogy with particle dynamics we can define an energy density $H(x,t) = p\dot{y} - L$, but whereas H is stationary for the point mass, it may flow along the string. Thus we must define an energy flux W_{tx} as well as an energy density $W_{tt} = H$.

$$W_{tt}(x,t) = \dot{y}\frac{\partial L}{\partial \dot{y}} - L = \tfrac{1}{2}\epsilon\left(\frac{\partial y}{\partial t}\right)^2 + \tfrac{1}{2}T\left(\frac{\partial y}{\partial x}\right)^2$$

$$W_{tx}(x,t) = \dot{y}\frac{\partial L}{\partial y'} = -T\frac{\partial y}{\partial t}\frac{\partial y}{\partial x}$$

(4.1.9)

From the first expression, again by analogy with point-mass dynamics, we can obtain a pair of Hamilton's equations, still another way of writing the equation of motion, which will be useful later.

$$H(p,s,y) = \frac{p^2}{2\epsilon} + \frac{s^2}{2T}$$

$$\frac{\partial y}{\partial t} = \frac{\partial H}{\partial p} \qquad \frac{\partial p}{\partial t} = \frac{\partial}{\partial x}\frac{\partial H}{\partial y'} - \frac{\partial H}{\partial y}$$

(4.1.10)

which will again produce the wave equation.

The energy density W_{tt} and the energy flux W_{tx} must be related. If, for example, more energy flows across the point $x + dx$ than flows across x, then the energy contained in length dx of the string must diminish.

$$W_{tx}(x + dx) - W_{tx}(x) = -dx\frac{\partial W_{tt}}{\partial t} \quad \text{or} \quad \frac{\partial}{\partial x}W_{tx} + \frac{\partial}{\partial t}W_{tt} = 0$$

(4.1.11)

which is the *equation of continuity* for energy. That the quantities given in Eqs. (4.1.9) indeed satisfy the equation of continuity can be quickly demonstrated, for

$$\frac{\partial}{\partial x}W_{tx} + \frac{\partial}{\partial t}W_{tt} = -T\frac{\partial^2 y}{\partial x\,\partial t}\frac{\partial y}{\partial x} - T\frac{\partial y}{\partial t}\frac{\partial^2 y}{\partial x^2} + \epsilon\frac{\partial y}{\partial t}\frac{\partial^2 y}{\partial t^2} + T\frac{\partial y}{\partial x}\frac{\partial^2 y}{\partial t\,\partial x}$$

$$= \frac{\partial y}{\partial t}\left(-T\frac{\partial^2 y}{\partial x^2} + \epsilon\frac{\partial^2 y}{\partial t^2}\right) = 0 \qquad \text{[from Eq. (4.1.8)]}$$

The quantities W_{tt} and W_{tx} are two elements in a tensor; its other components are

$$W_{xt} = y' \frac{\partial L}{\partial \dot{y}} = \epsilon \frac{\partial y}{\partial t} \frac{\partial y}{\partial x}$$

$$W_{xx} = y' \frac{\partial L}{\partial y'} - L = -\tfrac{1}{2}\epsilon \left(\frac{\partial y}{\partial t}\right)^2 - \tfrac{1}{2}T \left(\frac{\partial y}{\partial x}\right)^2 \qquad (4.1.12)$$

Reference to the special case of Fig. 4.6 indicates that $W_{xt} = -\epsilon u(u/c)$ is minus the longitudinal momentum density of the wave (distinguished from the transverse momentum $\epsilon\, \partial y/\partial t$, which is a first-order quantity). Another application of the equation of continuity would indicate that a time rate of change of momentum density should equal minus the space rate of change of stress in the string; net force must equal time rate of change of momentum. In the special case of Fig. 4.6, W_{xx} for the moving part of the string equals $-\tfrac{1}{2}\epsilon u^2 - \tfrac{1}{2}T(u/c)^2 = -T(u/c)^2$ and is zero for the rest of the string. However, Fig. 4.6 also shows that the net change in horizontal component of the stress (stress is minus the tension) caused by the string's displacement and motion is $-T[1 - (u/c)^2] + T = T(u/c)^2$ to the second order in u/c. Therefore W_{xt} is minus the longitudinal momentum density, and W_{xx} the net longitudinal stress in the string caused by the wave motion. The two satisfy the equation of continuity $\partial W_{xx}/\partial x + \partial W_{xt}/\partial t = 0$, as substitution from Eqs. (4.1.12) will show.

Thus we have obtained an expression, correct to the second order in $\partial y/\partial x$, for the Lagrange density $L(\dot{y}, y', y)$, from which we can obtain, by partial differentiation, the transverse momentum $p(x,t)$ and stress $s(x,t)$, the longitudinal momentum density W_{xt} and stress W_{xx}, and the energy density $W_{tt} = H$ and energy flux W_{tx}. The Lagrange equation (4.1.8) is the wave equation. Each of the pairs p and s, W_{xt} and W_{xx}, W_{tt} and W_{tx} satisfies an equation of continuity. We note again that the elements of the stress-energy tensor are not localized quantities which can be measured at each point of the string, but that the expressions given in Eqs. (4.1.9) and (4.1.12), when integrated over the whole length of the string, give correct values for the energy of the whole string and the energy and momentum input and output at the ends of the string, which are measurable.

4.2   FREE OSCILLATIONS

We have now demonstrated that the transverse motion of a flexible string obeys the wave equation (4.1.3) if the slope of its shape, $|\partial y/\partial x|$, is everywhere very small compared with unity and if the string is perfectly flexible. Neither limitation is exactly met in practice, but in a large number of cases of interest acoustically, they are conformed to so nearly that the actual motion corresponds fairly well to the solutions of Eq. (4.1.3). These solutions, having the general form of (4.1.1), are noninterfering waves, traveling in either direction with a velocity $c = \sqrt{T/\epsilon}$, independent of the shape or size of the wave (within the limits mentioned). Of course, it is just the shape of the wave which moves along the string; the individual parts of the string just move up and down.

Initial conditions

For a string of infinite length, its initial transverse displacement and velocity at $t = 0$ being $y_0(x)$ and $v_0(x)$, respectively, the subsequent displacement of the string is given by Eqs. (4.1.4).

$$y(x,t) = \frac{1}{2c}[cy_0(x - ct) - S(x - ct)] + \frac{1}{2c}[cy_0(x + ct) + S(x + ct)]$$

$$(4.2.1)$$

where

$$S(z) = \int_0^z v_0(x)\, dx$$

The first bracket represents a wave going in the positive x direction; the second, a wave in the negative x direction. For this wave the elements of the stress-energy tensor are

$$W_{tt} = H = \tfrac{1}{4}\{[J(x - ct)]^2 + [K(x + ct)]^2\} = -W_{xx}$$

$$W_{tx} = \frac{c}{4}\{[J(x - ct)]^2 - [K(x + ct)]^2\} = -c^2 W_{xt} \qquad (4.2.2)$$

$$J(z) = \sqrt{T}\, y_0'(z) - \sqrt{\epsilon}\, v_0(z) \qquad K(z) = \sqrt{T}\, y_0'(z) + \sqrt{\epsilon}\, v_0(z)$$

The waves in the two directions each have their own energy, momentum, and stress, there being no cross-product terms. The energy and stress terms are both positive; these quantities are independent of wave direction. On the other hand, the momentum density and the energy flux are positive for the wave in the positive x direction, and negative for the wave in the negative x direction.

Boundary conditions

So far we have been treating the string as though it had an infinite extent; actually, it is fastened down somewhere, and this fastening affects the motion of the string. The fact that the string is fastened to a support is an example of a *boundary condition*. It is a requirement on the string at a given point in space which must be true *for all time*, as opposed to initial conditions, which fix the dependence of y and v on x *at a given time*. Boundary conditions are more important in determining the general behavior of the string, its allowed frequencies, etc., than are initial conditions.

If the support is rigid and the distance along the string is measured from it, the boundary condition is that y must be zero when $x = 0$, for all values of the time. If the support is springy, so that it is displaced sideward a distance

CF cm for a sideward force F dynes, the boundary condition is that y must always be equal to C times the component of the string's tension perpendicular to the equilibrium line: $y = CT(\partial y/\partial x)$ at $x = 0$.   Many other sorts of boundary conditions are possible.

Reflection at a boundary

Let us take the simple case of the rigid support, requiring that $y = 0$ when $x = 0$, and see what effect this has on the motion of the string.   The solutions $y = F(x - ct)$ or $y = f(x + ct)$ cannot be used in this case, for they will not always be zero when $x = 0$.   However, the solution

$$y = -f(x - ct) + f(-x - ct)$$

will satisfy the boundary condition.   At $x = 0$,

$$y = -f(-ct) + f(-ct) = 0$$

for all values of t.   To see what motion of the string this expression corresponds to, let us suppose that $f(z)$ is a function that is large only when z is zero and drops off to zero on both sides of this maximum.   Then, when $t = -10$, the function $f(-x - ct)$ will have a peak at $x = 10c$ representing a single pulse traveling leftward along the string.   The function $f(x - ct)$ at this time would have its peak at $x = -10c$ if there were any string to the left

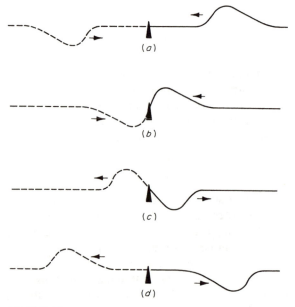

FIGURE 4.7
Reflection of a wave from the end support of a string.   The solid lines show the shape of the string at successive instants of time; the dotted lines, the imaginary extension of the wave-form beyond the end of the string.

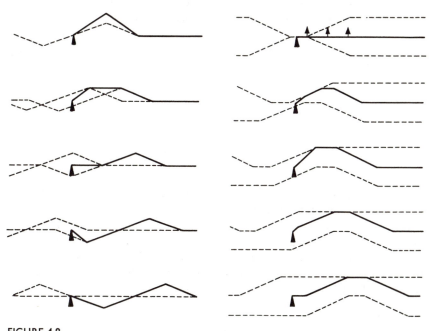

FIGURE 4.8
Motions of plucked and struck strings fixed at one end. Dotted lines show the traveling partial waves; their sum is the solid line, the actual shape of the string.

of the support, but since no string is there, the term $-f(x - ct)$ is not apparent in the shape of the string at $t = -10$. It has been represented in Fig. 4.7a by a dotted line to the left of $x = 0$.

As t increases, the wave in the actual string moves to the left, and the wave in the imagined extension of the string moves to the right, until they begin to coalesce at $x = 0$. During the coalescing the displacement of the point at $x = 0$ is always zero, for the effects of the two waves just cancel each other here. A little later the waves have passed by each other, the wave that had been on the imaginary part of the string now being on the actual string, and vice versa. The succession of events is pictured in Fig. 4.7. What has happened is that the pulse which had originally been traveling leftward is *reflected* at the point of support $x = 0$ and comes back headed toward the right, as a pulse of similar form but of opposite sign. The boundary condition at $x = 0$ has required this reflection, and the particularly simple sort of condition that we have imposed has required this very symmetric sort of reflection. Most other boundary conditions would require a greater difference between the original and the reflected wave. When a wave strikes it, a rigid support must pull up or down on the string by just the right amount to keep y zero, and in doing so it "generates" a reflected wave.

The expression for the motion of a string satisfying the boundary condition $y = 0$ at $x = 0$ and the initial conditions $y = y_0(x)$, $v = v_0(x)$ at $t = 0$ is

$$y = \frac{1}{2}\left[Y(x - ct) + Y(x + ct) - \frac{1}{c} H(x - ct) + \frac{1}{c} H(x + ct) \right] \quad (4.2.3)$$

where

$$Y(z) = \begin{cases} y_0(z) & z > 0 \\ -y_0(-z) & z < 0 \end{cases}$$

$$H(z) = \begin{cases} S(z) & z > 0 \\ S(-z) & z < 0 \end{cases} \qquad S(z) = \int_0^z v_0(x)\, dx$$

These definitions of Y and H are necessary because y_0 and v_0 are defined only for positive values of x (where the string actually is), whereas the form of the partial waves used to build up the subsequent forms of the string must be given for all values of $z = x \pm ct$. The particular forms of Y and H are chosen so that they automatically satisfy the boundary conditions at $x = 0$ for all values of t. Two examples of the way in which the motion of the string can be built up by the use of these partial waves are given in Fig. 4.8.

Motion of the end support

When the end support at $x = 0$ is not perfectly rigid but moves transversally in response to transverse motion of the string, the problem becomes somewhat more complicated, though it can be worked out by means of the Laplace transform. If the end support yields to a force, it presumably yielded when the tension T was applied. But this has already been taken into account, and the support is, by definition, at $x = 0$ when the string is in equilibrium under tension T.

If the string is not in equilibrium, if it has a nonzero slope $y'(0)$ at the origin, the support will be subject to a transverse force and may move transversally. Suppose the support responds to this transverse force like a simple mass on a spring, like the systems discussed in Sec. 2.1. The transverse force is $Ty'(0)$, as long as $y'(0)$ is small, and the reaction of the support to it prescribes a relationship between the slope $y'(0)$, the displacement $y(0)$ of both string and support, its time derivatives and the effective mass m, frictional resistance R, and transverse stiffness K of the support.

$$Ty'(0,t) = m\ddot{y}(0,t) + R\dot{y}(0,t) + Ky(0,t) \qquad (4.2.4)$$

where $y'(x,t) = \partial y / \partial x$, $\dot{y} = \partial y / \partial t$, and $\ddot{y} = \partial^2 y / \partial t^2$. This formula exhibits the stiffness, frictional, and inertive reaction of the support to transverse motion of the string end.

In accordance with Eq. (4.1.1), we assume that the displacement of the string is

$$y(x,t) = f(ct + x) + g(ct - x) \qquad x \geqslant 0$$

where $f(x)$ is the shape of the left-bound wave, which is to be reflected from the support. Thus f is known; we must use Eq. (4.2.4) to determine the shape of $g(ct - x)$, the reflected wave. We assume that, at $t = 0$, the start of the calculation, the incoming wave has not yet reached the support, so that $f(x) = 0$ for $x < b$, where $b > 0$. Then, until $t = b/c$, the support will be at rest, and we can be sure that g is zero until that time.

Inserting the expression for $y(x,t)$ into Eq. (4.2.4), we get

$$mc^2 \frac{d^2g}{dz^2} + (Rc + T)\frac{dg}{dz} + Kg = -\left[mc^2 \frac{d^2f}{dz^2} + (Rc - T)\frac{df}{dz} + Kf \right] \quad (4.2.5)$$

for $g(z)$ and $f(z)$, where $z = ct$ in this equation [since Eq. (4.2.4) holds only for $x = 0$], but will be $ct - x$ when we come to compute the shape of the reflected wave $g(ct - x)$. Next we take Laplace transforms of both sides. The formulas of Table 2.1 show that if

$$F_L(s) = \int_0^\infty e^{-sz}f(z)\, dz$$

is the Laplace transform of the known function $f(z)$ (the variable is now z instead of t, but the computations are the same), the corresponding transform of the unknown function $g(z)$ is given by

$$[mc^2s^2 + (Rc + T)s + K]G_L(s) = -F_L(s)[mc^2s^2 + (Rc - T)s + K]$$

or

$$\begin{aligned}
G_L(s) &= -F_L(s) + \frac{2TsF_L(s)}{mc^2s^2 + (Rc + T)s + K} = -F_L(s) + \frac{2TF_L(s)}{cZ(ics)} \\
&= -F_L(s) + \frac{2\epsilon}{m} \frac{sF_L(s)}{[s + (k_r/c) - i(\omega_f/c)][s + (k_r/c) + i(\omega_f/c)]}
\end{aligned} \quad (4.2.6)$$

where $k_r = (Rc + T)/2mc = (R + \epsilon c)/2m$ is a resistance parameter, like the k of Eq. (2.2.1) for the simple oscillator. In fact, k_r is the sum of the $k = R/2m$ of Eq. (2.2.1) for the transverse motion of the support, plus an additional term, $\epsilon c/2m$, produced by the attached string. We shall see later that ϵc is the "wave resistance" of the string; any transverse motion of the support produces a wave in the string, which carries energy away. Frequency $\omega_f/2\pi$ is related to the ν_f of Eq. (2.2.2); $\omega_f^2 = \omega_0^2 - k_r^2$, where $\omega_0^2 = K/m$; $\omega_0/2\pi$ is the resonance frequency of the support for transverse motion. Function $Z(\omega) = -i\omega m + (R + \epsilon c) + i(K/\omega)$ is the mechanical impedance of the support for transverse motion. Factor c in $\omega = ics$ is the ratio between wavelength and period.

The inverse transform of $G_L(s)$ thus has two terms. The first term, $-f(ct - x)$, is the wave which would be generated by a rigid boundary at $x = 0$, as shown in Fig. 4.7. The second term, the inverse transform of $2TF_L(s)/cZ(ics)$, is the wave motion generated by the motion of the support. It reflects the frequency response of the support, because of the impedance term in the denominator; this response continues after the originating wave has been reflected.

An example may be in order.   Suppose the original shape of the incident wave, going in the negative x direction, is $f(ct + x)$, where

$$f(z) = \begin{cases} 0 & 0 \leqslant z < l - a \\ \dfrac{b}{a}(z - l + a) & l - a < z < l \\ \dfrac{b}{a}(l + a - z) & l < z < l + a \\ 0 & l + a < z \end{cases}$$

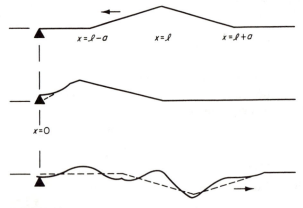

FIGURE 4.9
Reflection of wave from a nonrigid support.   Dotted line is the shape if the support is rigid.

as shown in Fig. 4.9.   The Laplace transform of $f(z)$ is then

$$F_L(s) = \frac{b}{s^2 a}(e^{s(a-l)} + e^{-s(a+l)} - 2e^{-sl})$$

and the difference between $G_L(s)$ and $-F_L(s)$ is

$$\frac{2TF_L(s)}{cZ(ics)} = \frac{2Tb}{mc^2 a}\frac{e^{-s(l-a)} + e^{-s(l+a)} - 2e^{-sl}}{s[s + (k_r/c) - (i\omega_f/c)][s + (k_r/c) + (i\omega_f/c)]}$$

$$= \frac{2Tb}{Ka}(e^{-s(l-a)} + e^{-s(l+a)} - 2e^{-sl})$$

$$\times \left\{ \frac{1}{s} - \frac{\omega_0^2}{2i\omega_f(k_r - i\omega_f)[s + (k_r/c) - i(\omega_f/c)]} \right.$$

$$\left. + \frac{\omega_0^2}{2i\omega_f(k_r + i\omega_f)[s + (k_r/c) + i(\omega_f/c)]} \right\}$$

Utilization of Table 2.1 shows that the inverse transform of this produces the solution

$$y(x,t) = f(ct + x) + g(ct - x)$$

where

$$g(z) = -f(z) + \frac{2Tb}{Ka} [Q(z - l + a) + Q(z - l - a) - 2Q(z - l)]$$

$$Q(w) = u(w) \left\{ 1 - e^{-k_r w/c} \frac{\cos [(\omega_f/c)w - \theta]}{\cos \theta} \right\} \rightarrow \begin{cases} -u(w)\left(\dfrac{\omega_f w}{c}\right)^2 & w \rightarrow 0 \\ 1 & w \rightarrow \infty \end{cases}$$

and $\cos \theta = \omega_f/\omega_0$, $\sin \theta = k_r/\omega_0$, $\omega_0{}^2 = K/m$, and $\omega_f{}^2 = \omega_0{}^2 - k_r{}^2$.

Figure 4.9 shows the shape of the string at $t = 0$ just after the incident wave strikes the support and after the wave has been reflected. The dotted lines show the shape the string would have if the support were rigid. When the incident wave strikes the support, it first is pulled upward and then oscillates with damped motion. This motion would be damped even if the inherent-friction term R were zero, because of the wave resistance ϵc of the attached string. Each of the three changes of slope of the wave generates a transient oscillation of the support, represented by the function Q of the formula. After the incident wave has gone, the support oscillation damps out rapidly. If k_r is larger than ω_0, ω_f is imaginary, and the support returns exponentially to equilibrium without oscillation.

Strings of finite length

Actual strings are fastened at *both* ends, so that, really, two boundary conditions are imposed. For instance, the string can be fastened to rigid supports a distance l cm apart, so that y must always be zero both at $x = 0$ and at $x = l$. The most important effect of a second boundary condition of this sort is to require that the motion of the string be *periodic*. A pulse started at $x = 0$ travels to the other support at $x = l$ in a time l/c, is reflected, travels back to $x = 0$, and is again reflected. If the supports are rigid, the shape of the pulse after its second reflection is just the same as that of the original pulse, and the motion is periodic with a period equal to $2l/c$. The motion in this case is not, in general, harmonic, as we shall see, but it is always periodic. This periodicity of all motion of the string depends entirely on the fact that we have imposed a particular sort of boundary condition; if other conditions are imposed at $x = 0$ and $x = l$ (i.e., if the supports are not perfectly rigid), then it may not be true that every motion is periodic; in fact, it may never be periodic.

The quantitative manner of dealing with the two boundary conditions is by means of the partial waves f. When the string is only l cm long, we are free to give any shape to $f(z)$ for z larger than l or smaller than zero. "Free" is not the correct word, however, for we must choose that form of f which satisfies both boundary conditions. If we start out at $t = 0$ with a pulse of

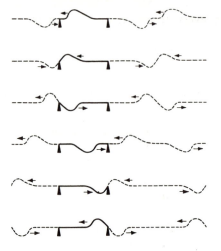

FIGURE 4.10
Periodic motion of a string fixed at both ends. Solid lines give the shape of the actual string at successive instants; dotted lines show the imaginary extension of the waveform beyond the ends of the string. The motion is made up of two partial waves going in opposite directions, each being periodic in x with period $2l$.

the form f traveling to the left, then, as before, we can satisfy the condition $y = 0$ at $x = 0$ by setting $y = -f(x - ct) + f(-x - ct)$. To have $y = 0$ at $x = l$, we must arrange the rest of the function f, beyond the limits of the actual string, so that $f(l - ct) = f(-l - ct)$ or, setting $z = -l - ct$, so that $f(z) = f(z + 2l)$ for all values of z. This means that the function $f(z)$, which must be defined for all values of z, must be *periodic* in z, repeating itself at intervals of $2l$ all along its length. An illustration of how this sort of partial wave can be used to determine the motion of the string is given in Fig. 4.10.

To satisfy the boundary conditions $y = 0$ at $x = 0$ and at $x = l$ and the initial conditions $y = y_0(x)$, $v = v_0(x)$ at $t = 0$, we build up a combination similar to that given in Eq. (4.2.3):

$$y = \frac{1}{2}\left[Y(x - ct) + Y(x + ct) - \frac{1}{c} H(x - ct) + \frac{1}{c} H(x + ct)\right] \quad (4.2.7)$$

where

$$Y(z) = \begin{cases} -y_0(-z) & -l < z < 0 \\ y_0(z) & 0 < z < l \\ -y_0(2l - z) & l < z < 2l \\ y_0(z - 2l) & 2l < z < 3l \\ \quad \cdots\cdots \end{cases}$$

$$H(z) = \begin{cases} S(-z) & -l < z < 0 \\ S(z) & 0 < z < l \\ S(2l - z) & l < z < 2l \\ S(z - 2l) & 2l < z < 3l \\ \quad \cdots\cdots \end{cases} \qquad S(z) = \int_0^z v_0(x)\, dx$$

Two examples of the motion of such strings are given in Fig. 4.11. In Fig. 4.12 the displacement of a point on the string is plotted as a function of time, showing that the motion is periodic but not simple-harmonic.

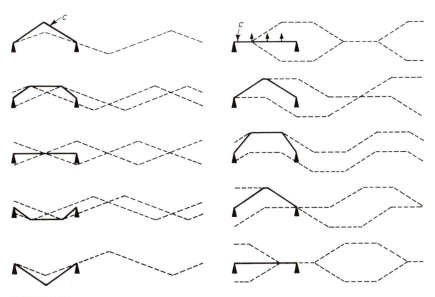

FIGURE 4.11
Motions of plucked and struck strings fixed at both ends. The solid lines show the successive shapes of the string during one half cycle. Shapes for the other half cycle are obtained by reversing the sign of the curves.

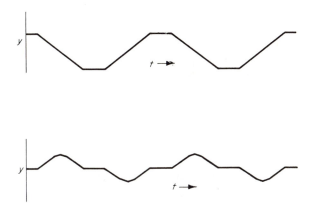

FIGURE 4.12
Displacements of the points marked c on the strings shown in Fig. 4.11, plotted as functions of the time.

4.3 SIMPLE-HARMONIC OSCILLATIONS

It has been seen in the last section that imposing boundary conditions limits the sorts of motion that a string can have, and that if the boundary conditions correspond to the fixing of both ends of the string to rigid supports, the motion is limited to *periodic* motion. The latter result is an unusual one, for we found in the last chapter that even as simple a system as a pair of coupled oscillators does not, in general, move with periodic motion. It is not unusual for a system to oscillate with simple-harmonic motion (which is a special type of periodic motion) when it is started off properly (we shall see that practically every vibrating system can do this); what is unusual in the string between rigid supports is that *every* motion is periodic, no matter how it is started.

Our problem in this section is to find the possible simple-harmonic oscillations of the string (the normal modes of vibration) and to see what the relation is between the frequencies of these vibrations that makes the resulting combined motion always periodic. The problem of determining the normal modes of vibration of a system is not just an academic exercise. For systems more complicated than that of the string between rigid supports, we have no method of graphical analysis similar to that of the last section, and the only feasible method of discussing the motion is to "take it apart" into its constituent simple-harmonic components. There is also a physiological reason for studying the problem, for the ear itself analyzes a sound into its simple-harmonic parts (if there are any). We distinguish between a note from a violin and a note from a bell, for instance, because of this analysis. If the frequencies present in a sound are all integral multiples of a fundamental frequency, as they are in a violin, the sound seems more musical than when the frequencies are not so simply related, as in the note from a bell.

Traveling and standing waves

We start our discussion with the wave equation (4.1.2), which, we showed, determines the motion of a perfectly flexible string as long as it is not displaced too far from equilibrium (as long as $|\partial y/\partial x| \ll 1$).

$$\frac{\partial^2 y}{\partial t^2} = c^2 \frac{\partial^2 y}{\partial x^2} \qquad c^2 = \frac{T}{\epsilon} \qquad (4.3.1)$$

The wave equation corresponds to a number of statements concerning the motion of a string. We saw in the last section that it implies that the wave motion travels with its shape unchanged, at a velocity c, independent of this shape. Since the derivative $\partial^2 y/\partial x^2$ is proportional to the curvature of the shape of the string at a given instant, Eq. (4.3.1) states that the *acceleration* of any portion of the string is *directly proportional to the curvature* of that portion.   If the curvature is downward, the acceleration is downward, and vice versa; and the greater the curvature, the faster the velocity changes.

If the string is infinite in extent, it can carry waves which travel exclusively in one direction.   In that case, as was pointed out at the beginning of this chapter, if the time dependence of the wave is to be sinusoidal, its space dependence must also be sinusoidal.   All simple-harmonic waves traveling in the positive x direction must have the form

$$y(x,t) = A \cos \left[\frac{\omega}{c} (x - ct) - \Phi \right]$$

or

$$y(x,t) = C \exp \left[\frac{i\omega}{c} (x - ct) \right] \tag{4.3.2}$$

if $C = Ae^{-i\Phi}$ and if physical meaning is attached only to the real part of the second expression.   For a simple-harmonic wave in the negative x direction we substitute $-(x + ct)$ for $(x - ct)$ in these expressions.   Incidentally, the reason we have chosen the time factor to be $e^{-i\omega t}$ rather than $e^{i\omega t}$ is that then the sign of the x part of the exponent, $e^{\pm i\omega x/c}$, indicates the direction of the wave.

For the wave of Eq. (4.3.2), the energy and momentum densities are [Eqs. (4.1.9) and (4.1.12)]

Kinetic energy density $= U = \frac{1}{2}\epsilon\omega^2 A^2 \sin^2 \left[\frac{\omega}{c} (x - ct) - \Phi \right]$

Potential energy density $= V = \frac{1}{2}T \left(\frac{\omega}{c} \right)^2 A^2 \sin^2 \left[\frac{\omega}{c} (x - ct) - \Phi \right]$

Total energy density $= W_{tt} = \epsilon\omega^2 A^2 \sin^2 \left[\frac{\omega}{c} (x - ct) - \Phi \right] = H \qquad$ (4.3.3)

Energy flux $= W_{tx} = cH$

Longitudinal momentum density $= W_{xt} = \dfrac{H}{c}$

Longitudinal stress $= W_{xx} = \dfrac{H}{c^2}$

The energy density is greatest where the string's slope and transverse velocity are greatest, each packet of energy spaced a half wavelength from its neighbor, each traveling with a velocity c.   Consequently, the energy flux W_{tx} is equal to c times the energy density W_{tt}.   The wavelength of these waves is, of course, the distance between one wave peak and the next, a distance such that an increase of x by λ will increase $(\omega/c)(x - ct)$ by 2π, so that $(\omega/c)\lambda = 2\pi$, or $\lambda = 2\pi c/\omega$.

We could reach the same conclusions by asking what sort of shape the string will have when it vibrates with simple-harmonic motion, i.e., when its time dependence is through the factor $e^{-i\omega t}$. Setting $y(x,t) = Y(x)e^{-i\omega t}$ into the equation of motion (4.1.3) or (4.1.8), we obtain a familiar equation for $Y(x)$.

$$\frac{d^2 Y}{dx^2} + \left(\frac{\omega}{c}\right)^2 Y = 0 \qquad c^2 = \frac{T}{\epsilon} \qquad (4.3.4)$$

which is identical with Eq. (1.2.1). This is the equation for simple-harmonic dependence on x, with "angular frequency" $k = \omega/c$ and "period" $\lambda = 2\pi/k = 2\pi c/\omega$. The quantity k is called the *wavenumber* of the wave; its dimensions are inverse length. The quantity λ is the wavelength of the wave, the distance from crest to crest of a sinusoidal wave traveling in one direction.

The general solution of Eq. (4.3.4) can be written

$$Y(x) = C_+ e^{i\omega x/c} + C_- e^{-i\omega x/c}$$

so that

$$y(x,t) = C_+ e^{i\omega(x - ct)/c} + C_- e^{-i\omega(x + ct)/c}$$

$$= A_+ \cos\left[\frac{\omega}{c}(x - ct) - \Phi_+\right] + A_- \cos\left[\frac{\omega}{c}(x + ct) + \Phi_-\right] \quad (4.3.5)$$

representing two waves, of the same frequency and wavelength, traveling in opposite directions along the string. Since the wave equation is linear, neither wave has any effect on the other.

This mutual independence of the waves extends to expressions for their energy-momentum-stress terms, such as the total energy,

$$H = U + V = \omega^2(\theta_+^2 + 2\theta_+\theta_- + \theta_-^2)$$

where

$$\theta_+ = A_+ \sin\left[\frac{\omega}{c}(x - ct) - \Phi_+\right] \qquad \theta_- = A_- \sin\left[\frac{\omega}{c}(x + ct) + \Phi_-\right]$$

The mean value of the square terms is $\langle\theta_+^2\rangle = \tfrac{1}{2}A_+^2$ and $\langle\theta_-^2\rangle = \tfrac{1}{2}A_-^2$, neither of which is zero. But the cross terms can be written

$$\theta_+\theta_- = \tfrac{1}{2}A_+ A_- \left[\cos(2\omega t + \Phi_+ + \Phi_-) - \cos\left(2\frac{\omega}{c}x - \Phi_+ + \Phi_-\right)\right]$$

When averaged over space and time, the average is zero. The energy flux has no cross term;

$$W_{tx} = -T\frac{\partial y}{\partial t}\frac{\partial y}{\partial x} = \epsilon c \omega^2(\theta_+^2 - \theta_-^2)$$

so that even the instantaneous values of the flux are simply the differences between the two individual fluxes.   Thus the average values of the stress-energy tensor are

$$H = \langle W_{tt} \rangle = \tfrac{1}{2}\epsilon\omega^2(A_+{}^2 + A_-{}^2) = \frac{1}{c^2}\langle W_{xx} \rangle$$

$$\Upsilon = \langle W_{tx} \rangle = \tfrac{1}{2}\epsilon\omega^2 c(A_+{}^2 - A_-{}^2) = \frac{1}{c}\langle W_{xt} \rangle$$

$$(4.3.6)$$

The energy and stress terms are the sum of the terms arising from each wave. The energy and momentum fluxes are the difference of the terms, since the two waves are flowing in opposite directions.

If the amplitudes of the two simple-harmonic waves are equal, there is no net flow of energy or momentum, and the combination is called a *standing wave*.

$$y(x,t) = C_+ e^{i\omega(x-ct)/c} + C_- e^{-i\omega(x+ct)/c}$$

$$= 2A \cos\left(\frac{\omega}{c}x + \tfrac{1}{2}\Phi_+ - \tfrac{1}{2}\Phi_-\right) \exp\left(-i\omega t + \tfrac{1}{2}\Phi_+ + \tfrac{1}{2}\Phi_-\right) \quad \text{(real part)}$$

$$= 2A \cos\left(\frac{\omega}{c}x + \tfrac{1}{2}\Phi_+ - \tfrac{1}{2}\Phi_-\right) \cos\left(\omega t - \tfrac{1}{2}\Phi_+ - \tfrac{1}{2}\Phi_-\right) \quad (4.3.7)$$

where $C_+ = Ae^{i\Phi_+}$ and $C_- = Ae^{i\Phi_-}$ have the same amplitude but different phases.   In this case the shape of the wave does not move along the string; it simply oscillates in amplitude with simple-harmonic motion.   At points where $\cos[(\omega/c)x + \tfrac{1}{2}\Phi_+ - \tfrac{1}{2}\Phi_-] = 0$, the two traveling waves always cancel each other and the string never moves.   These points are called the *nodal points* of the wave motion.   In the case that we are considering, where the density and tension are uniform, the nodal points are equally spaced along the string a distance $c/2\nu$ apart, two for each wavelength.   Halfway between each pair of nodal points is the part of the string having the largest amplitude of motion, where the two traveling waves always add their effects.   This portion of the wave is called a *loop*, or *antinode*.

We should ask how a standing wave gets established and is maintained, for if there is no motion at each node, there can be no flow of energy from one loop to its neighbors.   The answer is that a standing wave is a steady-state situation.   During the transient state, when energy is being distributed along the string, the nodes are not perfect (that is, y is not exactly zero there) and energy does pass from one loop to the next.   Also, even for the steady-state situation, the nodes are only perfect when there is no friction.   With zero friction, once a loop has acquired its energy, it can oscillate forever.   If friction is present, the "nodes" are simply places of minimal (but not zero) amplitude of vibration; some energy flows from loop to loop.

Normal modes

So far, we have neglected boundary conditions. If we require that $y = 0$ when $x = 0$, the general form of (4.3.5) can no longer be used; the number of possible harmonic motions is limited. The expression for y that must be used is the standing-wave form (4.3.7) with the angles Φ so chosen that *a nodal point coincides with the point of support $x = 0$*:

$$y = A \sin \left(\frac{2\pi v}{c} x\right) \cos (2\pi v t - \Phi) \qquad (4.3.8)$$

This agrees with the discussion in the previous section. For the simple boundary condition that we have used, the reflected wave has the same amplitude as the incident wave; and when the incident one is sinusoidal, the result is a set of standing waves. Any frequency is allowed, however.

When the second boundary condition $y = 0$ at $x = l$ is added, the number of possible simple-harmonic motions is still more severely limited. For now, of all the possible standing waves indicated in (4.3.8), *only those which have a nodal point at $x = l$ can be used*. Since the distance between nodal points depends on the frequency, the string fixed at both ends cannot vibrate with simple-harmonic motion of any frequency; only a discrete set of frequencies is allowed, the set that makes $\sin[(2\pi v/c)l]$ zero. The distance between nodal points must be l, or it must be $l/2$ or $l/3$, etc. The allowed frequencies are therefore $c/2l$, $2c/2l$, $3c/2l$, etc., and the different allowed simple-harmonic motions are all given by the expression

$$y = A_n \sin \left(\frac{\pi n x}{l}\right) \cos \left(\frac{\pi n c}{l} t - \Phi_n\right) \qquad n = 1, 2, 3, 4, \ldots$$

$$\qquad (4.3.9)$$

$$v_n = \frac{nc}{2l} = \frac{n}{2l} \sqrt{\frac{T}{\epsilon}}$$

The lowest allowed frequency $v_1 = c/2l$ is called the *fundamental frequency* of vibration of the string. It is the frequency of the general periodic motion of the string, as we showed in the last section. The higher frequencies are called *overtones*, the first overtone being v_2, the second v_3, and so on.

The equation for the allowed frequencies given in Eq. (4.3.9) expresses an extremely important property of the uniform flexible string stretched between rigid supports. It states that the frequencies of all the overtones of such a string are *integral multiples of the fundamental frequency*. Overtones bearing this simple relation to the fundamental are called *harmonics*, the fundamental frequency being called the first harmonic, the first overtone (twice the fundamental) being the second harmonic, and so on.

Very few vibrating systems have harmonic overtones, but these few are the bases of nearly all musical instruments. For when the overtones are harmonic, the sound seems particularly satisfying, or musical, to the ear.

Fourier series

To recapitulate: The string has an infinite number of possible frequencies of vibration; and if the supports are rigid, these frequencies have a particularly simple interrelation. If such a string is started in just the proper manner, it will vibrate with just one of these frequencies, but its general motion will be a combination of all of them:

$$y = A_1 \sin \frac{\pi x}{l} \cos \left(\frac{\pi c}{l} t - \Phi_1 \right) + A_2 \sin \frac{2\pi x}{l} \cos \left(\frac{2\pi c}{l} t - \Phi_2 \right) + \cdots$$

or symbolically,

$$y = \sum_{n=1}^{\infty} A_n \sin \frac{\pi n x}{l} \cos \left(\frac{\pi n c}{l} t - \Phi_n \right)$$

$$= \sum_{n=1}^{\infty} \sin \frac{\pi n x}{l} \left(B_n \cos \frac{\pi n c t}{l} + C_n \sin \frac{\pi n c t}{l} \right)$$

(4.3.10)

where the symbol Σ indicates the summation over the number n, going from $n = 1$ to $n = \infty$. The value of A_n is called the *amplitude of the nth harmonic.*

Equation (4.3.10) is just another way of writing Eq. (4.2.7). The present form, however, shows clearly why all motion of the string must be periodic in character. Since all the overtones are harmonic, by the time the fundamental has finished one cycle, the second harmonic has finished just two cycles, the third harmonic just three cycles, and so on, so that during the second cycle of the fundamental the motion is an exact repetition of the first cycle. This is, of course, what we mean by periodic motion.

Equation (4.3.10) is in many ways more useful for writing the dependence of y on t and x than is Eq. (4.2.7). For it gives us a means of finding the relative intensities of the different harmonics of the sound given out by the string (corresponding to the analysis that the ear makes of the sound) and thus gives us a method of correlating the motion of the string with the tone quality of the resulting sound. We shall have to wait until farther along in the book to discuss the quantitative relations between the vibrations of bodies and the intensity of the resulting sounds, but it is obvious that the intensity of the nth harmonic in the sound depends on the value of the amplitude A_n. Once the values of all the A_n's are determined, the future motion of the string and the quality of the sound which it will emit will both be determined.

The energy of vibration of this string is the integral of the energy density,

$$W_{tt} = H = \tfrac{1}{2}\epsilon \left(\frac{\partial y}{\partial t} \right)^2 + \tfrac{1}{2} T \left(\frac{\partial y}{\partial x} \right)^2$$

$$= \sum_{n=1}^{\infty} \left[\tfrac{1}{2}\epsilon \left(U_n{}^2 + \sum_{m \neq n} U_n U_m \right) + \tfrac{1}{2} T \left(S_n{}^2 + \sum_{m \neq n} S_n S_m \right) \right]$$

over the length of the string, where

$$U_n = -\frac{\pi n c}{l} A_n \sin\frac{\pi n x}{l} \sin\left(\frac{\pi n c t}{l} - \Phi_n\right)$$

$$S_n = \frac{\pi n}{l} A_n \cos\frac{\pi n x}{l} \cos\left(\frac{\pi n c t}{l} - \Phi_n\right)$$

In the double sum which is the square of the derivative, we have written the squared terms ($m = n$) separately and left the other terms ($m \neq n$) together. The reason for this is that the terms for which $m \neq n$ all disappear when integrated over x [Eq. (1.3.9)],

$$\int_0^l \sin\frac{\pi n x}{l} \sin\frac{\pi m x}{l}\, dx = \begin{cases} \tfrac{1}{2}l & n = m \\ 0 & n \neq m \end{cases}$$

$$\int_0^l \cos\frac{\pi n x}{l} \cos\frac{\pi m x}{l}\, dx = \begin{cases} \tfrac{1}{2}l & n = m \\ 0 & n \neq m \end{cases} \tag{4.3.11}$$

leaving only the sum of the squared terms to constitute the energy.

Thus the total energy of vibration of a string oscillating transversally between two rigid supports a distance l apart is

$$E = \int_0^l H\, dx$$

$$= \sum_{n=1}^{\infty} \tfrac{1}{4}\epsilon\left(\frac{\pi c}{l}\right)^2 n^2 A_n{}^2 l \sin^2\left(\frac{\pi n c t}{l} - \Phi_n\right) + \sum_{n=1}^{\infty} \tfrac{1}{4}T\left(\frac{\pi}{l}\right)^2 n^2 A_n{}^2 l \cos^2\left(\frac{\pi n c t}{l} - \Phi_n\right)$$

$$= \frac{\pi^2 T}{4l} \sum_{n=1}^{\infty} n^2 A_n{}^2 \tag{4.3.12}$$

since

$$c^2 = \frac{T}{\epsilon}$$

There are no cross terms in this expression; each mode of oscillation of the string contributes its separate part to the total energy, without any interaction with other modes. We remind ourselves that these formulas represent the string's motion only when $|\partial y/\partial x| < \sum (\pi n/l)A_n \ll 1$ [which, incidentally, is also sufficient to ensure that series (4.3.10) and (4.3.12) converge].

Determining the series coefficients

The A_n's and Φ_n's, or the B_n's and C_n's, are an infinite number of arbitrary constants whose values are fixed by the initial conditions, corresponding to the infinite number of points along the string whose positions and velocities must all be specified at $t = 0$. Our analysis will not be complete until we devise a method for determining their values when the initial shape and velocity shape of the string are given.

The initial conditions must satisfy the equations, obtained from (4.3.10) by setting $t = 0$,

$$y(x,0) = y_0(x) = \sum_{n=1}^{\infty} B_n \sin \frac{\pi n x}{l}$$

$$\left(\frac{\partial y}{\partial t}\right)_{t=0} = v_0(x) = \frac{\pi c}{l} \sum_{n=1}^{\infty} n C_n \sin \frac{\pi n x}{l}$$

(4.3.13)

Series like the right-hand sides of these equations are called *Fourier series* [Eq. (1.3.8)].

The initial shape and velocity shape of the string, the functions $y_0(x)$ and $v_0(x)$, are any continuous functions of x which go to zero at $x = 0$ and $x = l$ and are finite in value and slope in between.   It is not immediately apparent that the Fourier series on the right-hand sides of the equations for y_0 and v_0 can represent any function of this sort.   A proof is possible, however; it is given in any text on Fourier series.   As physicists, of course, we know that the motion of the string is uniquely determined by its initial configuration and velocity, and since Eq. (4.3.10) represents the most general solution for the free vibration of a string of length l stretched between rigid supports, we perhaps are justified in assuming that the series can be made to fit any physically possible shape, as closely as one pleases, by adjusting the coefficients B_n, C_n appropriately.   What is needed is a technique to determine their appropriate values.

The desired trick is embodied in Eqs. (4.3.11).   To show this, we multiply both sides of the equations for y_0 and v_0 by $\sin(\pi m x/l)$ and integrate over x from 0 to l.   The integrals on the right-hand side are all zero, except for the one for $n = m$, which is $(l/2)B_m$ or $(l/2)C_m$, respectively.   For instance,

$$\int_0^l \sin\left(\frac{\pi m x}{l}\right) y_0(x)\, dx = \sum_{n=1}^{\infty} B_n \int_0^l \sin \frac{\pi n x}{l} \sin \frac{\pi m x}{l}\, dx$$

$$= \frac{l}{2} B_m$$

so that

$$B_m = \frac{2}{l} \int_0^l \sin\left(\frac{\pi m x}{l}\right) y_0(x)\, dx$$

and similarly,   (4.3.14)

$$C_m = \frac{2}{\pi c m} \int_0^l \sin\left(\frac{\pi m x}{l}\right) v_0(x)\, dx \qquad m = 1, 2, 3, 4, \ldots$$

These equations provide a means of determining the values of the B_n's and C_n's in terms of the initial conditions.

A reference to Eq. (3.2.26) indicates that each of the standing waves in the series (4.3.13) is a normal mode of oscillation.   When the string is vibrating with frequency $\omega_v = \pi v c/l$ (v an integer), its shape is given by

$y = A_\nu \sin(\pi\nu x/l)$; thus $\sin(\pi\nu x/l)$ is equivalent to the components $E_m(\nu)$ of Eq. (3.2.26), with the continuous variable x taking the place of the discrete index m (or n). The orthogonality equations following Eq. (3.2.23), which state that the eigenvectors $\mathbf{E}(\nu)$ are mutually perpendicular (orthogonal), correspond to the orthogonality relations (4.3.11) for the continuous variable x. Series (4.3.10) is analogous to the series

$$x_m = \sum_{\nu=1}^{N} E_m(\nu)B_\nu e^{-i\omega_\nu t}$$

for the discrete set of N oscillators, and Eqs. (4.3.14), determining the normal-mode amplitudes B_m, C_m, are equivalent to the equations $B_\nu = \sum_{m=1}^{N} C_m E_m(\nu)$, which determine the normal-mode amplitudes B_ν for the discrete set of oscillators. The string is equivalent to a system of an infinite number of coupled oscillators, with the particular oscillator specified by the continuous variable x rather than the discrete index m. The energy of the νth mode is, by Eq. (4.3.12), equal to $\pi^2 T\nu^2 A_\nu^2/4l$, independent of the amplitudes of the other modes.

Plucked string; struck string

A few examples will indicate how the method works. For instance, if we pull the center of the string out h cm and then let it go at $t = 0$, all the C_m's will be zero, and

$$B_m = \frac{2}{l}\left[\int_0^{l/2} \frac{2hx}{l} \sin\frac{\pi mx}{l}\,dx + \int_{l/2}^{l} \frac{2h}{l}(l-x)\sin\frac{\pi mx}{l}\,dx\right]$$

$$= \frac{8h}{\pi^2 m^2}\sin\frac{\pi m}{2}$$

$$= \begin{cases} 0 & \text{if } m \text{ is an even integer} \\ (-1)^{(m-1)/2}\dfrac{8h}{\pi^2 m^2} & \text{if } m \text{ is an odd integer} \end{cases}$$

Therefore

$$y = \frac{8h}{\pi^2}\left[\sin\frac{\pi x}{l}\cos\frac{\pi ct}{l} - \frac{1}{9}\sin\frac{3\pi x}{l}\cos\frac{3\pi ct}{l}\cdots\right] \qquad (4.3.15)$$

Computing this series for y as a function of x and t gives the same values for the shape of the string at successive instants as are shown in the first sequence of Fig. 4.11. Figure 4.13 shows how the correct form is approached closer and closer, the more terms of the series are used.

At first sight the foregoing series appears to be simply a more awkward way of finding the shape of the string than the method used in the previous section. However, the series can tell us more about the string's motion than the results of the last section can. It tells us, for instance, that the second, fourth, sixth, etc., harmonics will be absent from the sound given out by the string, for they are not present in the motion. It tells us that if, for example, the intensity of the sound emitted is proportional to the square of the amplitude of motion of the string, the fundamental frequency will be 81 times more intense than the third harmonic, 625 times more intense than the fifth harmonic, etc.

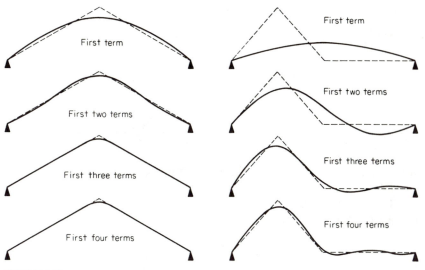

FIGURE 4.13
Fourier series representations of the initial form of the string given in Eq. (4.3.15) and the initial-velocity form given in Eq. (4.3.16). Successive solid curves show the effect of adding successive terms of the series; dotted curves show the actual form, given by the entire series.

The absent harmonics correspond to standing waves that have a nodal point at the center, the point pulled aside. This is an example of the general rule (which can be proved by computing the required integrals for B_m) that in the motion of any plucked string all those harmonics are absent which have a node at the point pulled aside.

If the string is struck, so that $y_0 = 0$ and

$$v_0(x) = \begin{cases} \dfrac{4\mu x}{l} & 0 < x < \dfrac{l}{4} \\[2ex] \dfrac{4\mu}{l}\left(\dfrac{l}{2} - x\right) & \dfrac{l}{4} < x < \dfrac{l}{2} \\[2ex] 0 & \dfrac{l}{2} < x < l \end{cases}$$

then all the B_m's are zero, and

$$\frac{\pi cm}{l} C_m = \frac{2}{l} \left(\int_0^{l/4} \frac{4\mu x}{l} \sin \frac{\pi m x}{l} \, dx + \int_{l/4}^{l/2} \frac{4\mu}{l} \left(\frac{l}{2} - x \right) \sin \frac{\pi m x}{l} \, dx \right)$$

$$= \frac{8\mu}{\pi^2 m^2} \left(2 \sin \frac{\pi m}{4} - \sin \frac{\pi m}{2} \right)$$

so that

$$y = \frac{8\mu l}{\pi^3 c} \left[(\sqrt{2} - 1) \sin \frac{\pi x}{l} \sin \frac{\pi c t}{l} + \tfrac{1}{4} \sin \frac{2\pi x}{l} \sin \frac{2\pi c t}{l} \right.$$

$$+ \frac{1 + \sqrt{2}}{27} \sin \frac{3\pi x}{l} \sin \frac{3\pi c t}{l}$$

$$\left. - \frac{1 + \sqrt{2}}{125} \sin \frac{5\pi x}{l} \sin \frac{5\pi c t}{l} \cdots \right] \quad (4.3.16)$$

We note that the fourth, eighth, etc., harmonics, those having nodes at $x = l/4$, are absent in this case.

The energy of these free vibrations may be obtained from Eq. (4.3.12). For the plucked string, it is

$$E = \frac{16h^2 T}{\pi^2 l} \left(1 + \frac{1}{9} + \frac{1}{25} + \cdots \right) = \frac{2h^2 T}{l}$$

since the sum in parentheses equals $\pi^2/8$. This is, of course, the same as the potential energy of the plucked string just before it has been released, $V_0 = \tfrac{1}{2} T \int (\partial y / \partial x)^2 \, dx = \tfrac{1}{2} T (2h/l)^2 l$. It is similarly possible to show that the energy of vibration of the struck string is equal to its initial kinetic energy,

$$E = \tfrac{1}{2} \epsilon \int_0^l \left(\frac{\partial y}{\partial t} \right)^2 dx = \epsilon \int_0^{l/4} \left(\frac{4\mu x}{l} \right)^2 dx = \frac{\epsilon l \mu^2}{12}$$

one-twelfth of the kinetic energy if all the mass of the string were concentrated at the point of greatest motion, $x = \tfrac{1}{4} l$. The total energy E remains independent of time because we have here neglected frictional forces.

Effect of friction

In the foregoing analysis we have neglected friction, although it is present in every vibrating string. To complete our discussion we should show, as with the simple oscillator, that the effect of friction is to damp out the free vibrations and to change slightly the allowed frequencies. To show that this is so is not difficult by the use of operational calculus, although it would be difficult by any other method.

The difficulty lies with the nature of the frictional term. The resistive force per unit length opposing the string's motion is due to the medium surrounding the string, the medium gaining the energy that the string loses. Part of the energy goes into heating the medium, the amount depending on the *viscosity* of the medium; and part goes into outgoing sound waves in the medium, the amount depending on the *radiation resistance* of the medium. The medium also adds an effective mass per unit length, which may not be negligible if the medium is a liquid.   The important point, however, the one that is responsible for our difficulties, is that the effective resistance due to the medium (and also its added effective mass) *depends on the frequency* of the string's motion.

The equation of motion for the string when friction is included is

$$\epsilon \frac{\partial^2 y}{\partial t^2} = T \frac{\partial^2 y}{\partial x^2} - R(\omega) \frac{\partial y}{\partial t} \qquad (4.3.17)$$

where R is the effective frictional resistance per unit length of string.   To find the "normal modes" of the string involves a sort of circular process, since we cannot solve for the natural frequencies until we know the value of R, and we cannot know R unless we know the frequency.   However, we can determine R as a function of the variable ω and note that, even if ω turns out to be a complex quantity, R is still computable.   Usually, we can assume that the added mass of the medium is negligible compared with ϵ.   For further discussion, see Eqs. (7.3.6) and (10.1.10).

A solution of Eq. (4.3.17) is the exponential $e^{i\beta x - i\omega t}$, with β and ω adjusted so that $\epsilon \omega^2 + i\omega R(\omega) = T\beta^2$.   If the string is driven by a simple-harmonic force of frequency $\omega/2\pi$, then ω is real and known, and thus R is known, and the constant β can be computed (it turns out to be complex). For free vibrations of a string between rigid supports, however, β must be real and ω will then be complex.   We set

$$y(x,t) = \sin \left(\frac{\pi n x}{l} \right) e^{-i\alpha_n t}$$

and solve the implicit equation $\alpha^2 + 2i\alpha k(\alpha) - (\pi n c/l)^2 = 0$ for α, where $k(\alpha) = (1/2\epsilon)R(\alpha)$.   If $k(\alpha)$ is a slowly varying function of α, or if it is very small compared with $\pi n c/l$, we can solve this equation by guessing a value of α, using this to find a value of $k(\alpha)$, solving the equation to find a better value of α, and so on.   In any case the value of α we finally find is related to n, c, l, etc., by the relationship

$$\alpha_n = -ik_n + \omega_n \qquad (4.3.18)$$

where

$$\omega_n^2 = \left(\frac{\pi n c}{l} \right)^2 - k_n^2$$

and where

$$k_n = k(\alpha_n) = \frac{1}{2\epsilon} R(\alpha_n)$$

Therefore the possible motions of free vibration are given by the series

$$y(x,t) = \sum_{n=1}^{\infty} \sin\left(\frac{\pi n x}{l}\right) e^{-k_n t}(B_n \cos \omega_n t + C_n \sin \omega_n t) \qquad (4.3.19)$$

which differs from Eq. (4.3.10) in the presence of the exponential damping factor, and in the fact that the successive frequencies $\omega_n/2\pi$ are not integral multiples of the fundamental. The series coefficients B_n and C_n can be determined from the initial conditions by use of Eqs. (4.3.11), as were Eqs. (4.3.14) for the frictionless case.

4.4 FORCED MOTION

The forced motion of a flexible string exhibits some of the properties exhibited by the simple oscillator; it also displays new properties, arising from the system's spatial extension. The applied transverse force may be viewed as a combination of simple-harmonic components, in which case we can analyze the force into its frequency components by means of the Fourier transform, or else we can express it as a sequence of pulses and represent the motion as a sequence of responses to the impulses, as was done for the simple oscillator in Eq. (2.3.9).

Likewise, the transverse driving force can be concentrated at a single point x, or it can be spread out over a length of the string. Again we can analyze the spread-out force in terms of a Fourier integral (or series) in x, or we can consider the distribution to be a sum of point forces, with effects to be added. This latter method, the use of an analog of Eq. (2.3.9), will turn out to have a quite wide range of applicability.

String driven from one end

First let us examine the very simple case of a very long string, driven by the transverse motion of its support at the end $x = 0$. This support is supposed to maintain the longitudinal tension in the string at T dynes, but it can be moved transversely, as shown in Fig. 4.5, by an amount $y_0(t)$. This motion will generate a transverse wave along the string, and if the string is long enough, the waves reflected from the far end will not have time to return, so that, for a while, the wave motion will be all in the positive x direction. In this case the displacement of the string which fits the boundary condition $y(0,t) = y_0(t)$ is $y(x,t) = y_0[t - (x/c)]$, a solution of the wave equation (4.1.2), valid as long as $(1/c)y_0'[t - (x/c)]$ is everywhere small. The transverse velocity is $y_0'[t - (x/c)]$, where the prime denotes differentiation with respect to the argument.

Reference to Fig. 4.5 shows that the transverse force necessary to displace the end of the string is $-T(\partial y/\partial x)$ at $x = 0$, which is

$$\left(\frac{T}{c}\right)y_0'(t) = \frac{T}{c}\frac{\partial y}{\partial t} \qquad \text{at} \qquad x = 0$$

for waves entirely in the positive direction. There is thus a constant ratio between the transverse force $F_0(t)$ necessary to move the end of the string and the transverse velocity of the end,

$$\frac{F_0(t)}{y_0'(t)} = \frac{T}{c} = \epsilon c = \sqrt{\epsilon T} \tag{4.4.1}$$

for waves in the positive direction. In other words, if the end of the string is driven transversally by a force $F_0(t)$, then the wave produced in the string has the shape $y_0[t - (x/c)]$, where $(d/dt)y_0(t) = (c/T)F_0(t)$, or

$$\epsilon c y_0(t) = \int_0^t F_0(u)\, du.$$

A ratio between a force and a velocity is a mechanical impedance; in this case, being real and independent of frequency, it is a resistance. This ratio, ϵc, is called the *wave resistance* of the string; it measures the rate at which work must be done by the force to send a wave out along the string. The power transferred from driving force to string is

$$P = \frac{|F_0|^2}{\epsilon c} = \epsilon c \left|\frac{\partial y}{\partial t}\right|^2 \tag{4.4.2}$$

These simple results occur only when there is no wave present which is going in the negative x direction.

String driven along its length

Sometimes the transverse driving force is applied at points other than the end of the string. If it is applied along a length of the string, we must talk about the *force density* $f(x,t)$ in newtons per meter at point x at time t, tending to move the string sideward. Thus the element dx of string in Fig. 4.2 would have an additional force $f(x,t)\, dx$ acting on it, and the equation of motion (4.1.3) then becomes

$$\epsilon \frac{\partial^2 y}{\partial t^2} - T\frac{\partial^2 y}{\partial x^2} = f(x,t) \tag{4.4.3}$$

if the string has mass ϵ per unit length and is under tension T.

Let us first consider the consequent motion when the string is so long that no reflected waves return, from either end, to the region of application of the force, during the time period under consideration. If the force is concentrated on an infinitesimal length of the string, at $x = x_0$, so that

$f(x,t)$ can be represented as $F(x_0,t)\delta(x - x_0)$ (with F in newtons) in terms of the delta function of Eq. (1.3.24), we can consider the string as two separate strings, stretching in opposite directions, each ending at x_0 and each being driven transversely by half the force $F(x_0,t)$, since both strings are equivalent and both share the force. Thus, from our discussion of Eq. (4.4.1), there would be a wave going to the right, in the half-string $x > x_0$, of the form $Y\{t - [(x - x_0)/c]\}$, where $dY(t)/dt = (c/2T)F(x_0,t)$, and a wave going to the left, in the region $x < x_0$, of the form $Y\{t + [(x - x_0)/c]\}$.

Put another way, the displacement of the string acted on by a force $F(x_0,t)$ concentrated at point x_0, in the absence of reflected waves, is given by the function

$$G(x,x_0;t) = \begin{cases} Y\left(x_0,t - \dfrac{x - x_0}{c}\right) & x > x_0 \\[3mm] Y\left(x_0,t + \dfrac{x - x_0}{c}\right) & x < x_0 \end{cases} \tag{4.4.4}$$

$$Y(x_0,z) = \frac{c}{2T}\int_0^z F(x_0,t)\, dt$$

To show that function $G(x,x_0;t)$ is a solution of

$$\epsilon\frac{\partial^2 G}{\partial t^2} - T\frac{\partial^2 G}{\partial x^2} = F(x_0,t)\,\delta(x - x_0) \tag{4.4.5}$$

we integrate both sides of this equation over x from $x_0 - \alpha$ to $x_0 + \alpha$ and then let α go to zero. The first term on the left integrates to $2\epsilon\alpha(\partial^2 G/\partial t^2) \to 0$; the second term integrates to

$$-T\left[\frac{\partial G}{\partial x}\right]_{x_0-\alpha}^{x_0+\alpha} = -\frac{c}{2}\left[-\frac{1}{c}F\left(x_0,t - \frac{\alpha}{c}\right) - \frac{1}{c}F\left(x_0,t + \frac{\alpha}{c}\right)\right] \to F(x_0,t)$$

and the term on the right, because of the properties of the delta function, also integrates to $F(x_0,t)$, thus proving that (4.4.4) is a solution of (4.4.5).

However, most forces are distributed along a length of string, not concentrated at a point. What is the solution of Eq. (4.4.3) if $f(x,t)$ is nonzero over the region between $x = a$ and $x = b$? Here the discussion preceding and following Eq. (2.3.9) may give us an idea. A distributed force function, nonzero between $x = a$ and $x = b$, can be represented as a sum of a whole series of point forces $f(x_0,t)\,\delta(x - x_0)$, distributed along the string from a to b,

$$f(x,t) = \int_a^b f(x_0,t)\,\delta(x - x_0)\, dx_0 \tag{4.4.6}$$

according to Eq. (1.3.24). We should then expect that the solution of Eq. (4.4.3) for a distributed force would be the sum of the solutions of Eq. (4.4.5) for all the component point forces along the string. In other words, we should expect that the solution of (4.4.3) is

$$y(x,t) = \int_a^b G(x,x_0;t)\, dx_0 \qquad (4.4.7)$$

where G is the solution of Eq. (4.4.5) for a point force $f(x_0,t)\, \delta(x - x_0)$ at x_0. We prove this by inserting (4.4.7) into Eq. (4.4.3)

$$\epsilon \frac{\partial^2 y}{\partial t^2} - T\frac{\partial^2 y}{\partial x^2} = \int_a^b \left(\epsilon \frac{\partial^2 G}{\partial t^2} - T\frac{\partial^2 G}{\partial x^2} \right) dx_0 = \int_a^b f(x_0,t)\, \delta(x - x_0)\, dx_0 = f(x,t)$$

and using (4.4.5) and (4.4.6).

Function $G(x,x_0;t)$ is called a *Green's function*. What we have shown is that a solution for a more general distributed force can be built up out of a sum of Green's functions.

An example may be desirable. Suppose that the transverse force is distributed uniformly along the length of the string from $x = 0$ to $x = b$ and that it is applied from $t = 0$ to $t = \tau$, so that

$$f(x,t) = \begin{cases} 0 & x < 0 \quad\text{or}\quad t < 0 \\ A & 0 < x < b \quad\text{and}\quad 0 < t < \tau \qquad (4.4.8) \\ 0 & x > b \quad\text{or}\quad t > \tau \end{cases}$$

The point force to use in Eq. (4.4.5), to obtain the required Green's function, is thus one which is $A\, \delta(x - x_0)$ when $0 < t < \tau$ and which is zero for $t < 0$ and $t > \tau$. The Green's function is then

$$G(x,x_0;t) = \frac{A}{2T} \begin{cases} 0 & ct < |x - x_0| \\ ct - |x - x_0| & |x - x_0| < ct < |x - x_0| + c\tau \\ c\tau & |x - x_0| + c\tau < ct \end{cases}$$

which is zero until a time $|x - x_0|/c$, when the wave first reaches the point x from the source point x_0, then rises linearly until the end of the wave produced by the force arrives at time $t = \tau + (|x - x_0|/c)$. This wave is pictured in the left-hand sequence of Fig. 4.14.

The solution for the force function of Eq. (4.4.8) is then obtained by integrating G over x_0 from 0 to b. The results are simple enough in principle, though rather complicated to write out because of the sequence of limits involved. If $c\tau > b$, then

$$y(x,t) = \begin{cases} 0 & x > ct + b; \quad ct < x - b \\ A(ct + b - x)^2 & ct < x < ct + b; \quad x - b < ct < x \\ A[b^2 + 2b(ct - x)] & \\ \qquad ct - c\tau + b < x < ct; \quad x < ct < x - b + c\tau \\ A[2bc\tau - (x - ct + c\tau)^2] & \\ \qquad ct - c\tau < x < ct - c\tau + b; \quad x - b + c\tau < ct < x + c\tau \\ A2bc\tau & ct - c\tau > x; \quad x + c\tau < ct \end{cases}$$

for the range $x > b$, and

$$y(x,t) = \begin{cases} 0 & ct < 0 \\ A(ct)^2 & 0 < ct < b - x \\ A[(ct + b - x)^2 - 2(b - x)^2] & b - x < ct < x \\ A[2b(ct + x) - b^2 - 2x^2] & x < ct < c\tau \\ A[2b(ct + x) - b^2 - 2x^2 - 2(ct - c\tau)^2] & \\ & c\tau < ct < c\tau + b - x \\ A[2bc\tau - (x - ct + c\tau)^2] & c\tau + b - x < ct < c\tau + x \\ A \cdot 2bc\tau & c\tau + x < ct \end{cases}$$

for $\frac{1}{2}b < x < b$, the shape for $x < \frac{1}{2}b$ being the mirror image of the shape for $x > \frac{1}{2}b$. This wave motion is pictured in the right-hand sequence of Fig. 4.14.

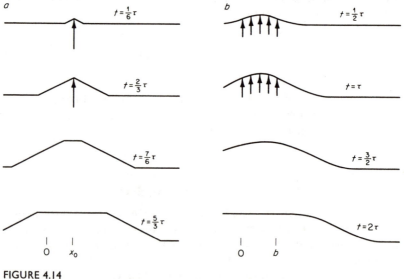

FIGURE 4.14
Infinite string driven (a) by a force concentrated at $x = x_0$ and (b) by a force distributed from 0 to b along the string.

A step-function force, with discontinuous beginning and end in both time and space, thus produces a rounded wave, made up of pieces of parabolas and straight lines, because of the double integration, once to go from f to G and again to integrate G over x_0.

Simple-harmonic driving force

If the driving force is simple-harmonic, of frequency $\omega/2\pi$, so that $f(x,t) = F(x)e^{-i\omega t}$, we can represent the steady-state shape of the string with the same time factor, $y(x,t) = Y(x)e^{-i\omega t}$. Each point on the string has its amplitude of motion, $Y(x)$, which does not change with time; if $Y(x)$ is real for all x, then all points on the string vibrate in phase, reaching maximum or going through zero simultaneously.

Again we start with the Green's function, the solution for a driving force concentrated at point x_0 on the string. In fact, we start with the solution for a concentrated force of magnitude equal to the tension in the string, $f(x,t) = T \delta(x - x_0)e^{-i\omega t}$, with corresponding steady-state shape $y(x,t) = g(x \mid x_0)e^{-i\omega t}$, where g is a solution of the Green's function equation,

$$\frac{d^2 g}{dx^2} + \left(\frac{\omega}{c}\right)^2 g = -\delta(x - x_0) \qquad (4.4.9)$$

analogous to Eq. (4.4.5), but with the time dependence removed and the inhomogeneous term normalized to unit amplitude.

If the string is very long, so that no reflected waves return to complicate the motion, the wave on the positive side of x_0 must be a constant times $e^{i(\omega x/c) - i\omega t}$, corresponding to a wave of frequency $\omega/2\pi$ and wavelength $2\pi c/\omega$, traveling in the positive x direction. The wave in the region $x < x_0$ must be represented by a constant times $e^{-i(\omega x/c) - i\omega t}$, a wave in the negative x direction. Since the string must be continuous in value across $x = x_0$, the constants must be adjusted so that we can write $g(x \mid x_0) = Ae^{i(\omega/c) \mid x - x_0 \mid}$, where $|x - x_0| = x - x_0$ when $x > x_0$ and equals $x_0 - x$ when $x < x_0$. Thus g has a discontinuity in *slope* at $x = x_0$. The magnitude of this discontinuity, and therefore the value of A, is adjusted to satisfy Eq. (4.4.9). According to the discussion following Eq. (4.3.1), a distributed force corresponds to a curvature of the string; therefore a concentrated force would require an infinite curvature at the point of application—a discontinuity of slope there.

To find the magnitude of the discontinuity in slope at x_0 we proceed as before to integrate Eq. (4.4.9) over x from $x_0 - \alpha$ to $x_0 + \alpha$, and then let α go to zero.

$$\int_{x_0 - \alpha}^{x_0 + \alpha} \frac{d^2 g}{dx^2} \, dx + \left(\frac{\omega}{c}\right)^2 \int_{x_0 - \alpha}^{x_0 + \alpha} g \, dx = -\int_{x_0 - \alpha}^{x_0 + \alpha} \delta(x - x_0) \, dx$$

or

$$\left[\frac{dg}{dx}\right]_{x_0 - \alpha}^{x_0 + \alpha} + \left(\frac{\omega}{c}\right)^2 2\alpha g = -1$$

or

$$\lim_{\alpha \to 0} \left[\left(\frac{dg}{dx}\right)_{x_0 + \alpha} - \left(\frac{dg}{dx}\right)_{x_0 - \alpha}\right] = -1 \qquad (4.4.10)$$

The discontinuity in slope at $x = x_0$, for a solution of the Green's function equation, must be minus unity.

For an infinite string, with no reflected waves, the slope of the solution $g = Ae^{i(\omega/c) \mid x - x_0 \mid}$ at $x_0 + \alpha$ is $i(\omega/c)Ae^{i(\omega/c)\alpha}$, and that at $x_0 - \alpha$ is

$-i(\omega/c)e^{i(\omega/c)x}$; so Eq. (4.4.10) becomes $-2i(\omega/c)A = 1$, or

$$g(x \mid x_0) = i\frac{c}{2\omega} e^{i(\omega/c)\,|x-x_0|} \qquad (4.4.11)$$

This is the Green's function for the infinite string (outgoing waves only). Point x is the point at which g is measured; so it can be called the *observation point*. Point x_0 is the point of application of the driving force which generates the wave; so it can be called the *source point*. We note that g is symmetric in x and x_0; its value is not changed if x and x_0 are interchanged. This general property of Green's functions, that their value is not changed when source and observation points are interchanged, is called the *principle of reciprocity*. We shall discuss its range of validity and its utility in wave calculations several times in this book.

By extension of Eq. (4.4.7) we should expect that a solution of the inhomogeneous equation for a general distributed force on a string,

$$\epsilon\frac{\partial^2 y}{\partial t^2} - T\frac{\partial^2 y}{\partial x^2} = F(x)e^{-i\omega t} \qquad (4.4.12)$$

would, in the case when no reflected waves are present, be

$$y(x,t) = \frac{1}{T}\int_{-\infty}^{\infty} F(x_0)g(x \mid x_0)\,dx_0 e^{-i\omega t}$$

$$= \frac{i}{2\epsilon c\omega}\int_{-\infty}^{\infty} F(x_0)e^{i(\omega/c)\,|x-x_0|}\,dx_0 e^{-i\omega t} \qquad (4.4.13)$$

This can be proved by inserting the integral into Eq. (4.4.12) and using (4.4.9), as was done with Eq. (4.4.7).

As an example, suppose the string were acted on by a force $f(x,t) = A\cos(\pi x/a)e^{-i\omega t}$ between $x = \frac{1}{2}a$ and $x = -\frac{1}{2}a$ and $f(x,t) = 0$ for $x < -\frac{1}{2}a$ and $x > \frac{1}{2}a$. The integral solution works out to be

$$y(x,t) = i\frac{cA}{2\omega T}\int_{-\frac{1}{2}a}^{\frac{1}{2}a} \cos\left(\frac{\pi x_0}{a}\right) e^{i(\omega/c)\,|x-x_0|}\,dx_0$$

$$= \begin{cases} \dfrac{iaA}{\pi\omega\epsilon}\dfrac{\cos(\omega a/2c)}{1-(\omega a/\pi c)^2}e^{i(\omega/c)x-i\omega t} & x > \frac{1}{2}a \\[2ex] \dfrac{iaA/\pi\omega\epsilon}{1-(\omega a/\pi c)^2}\left[e^{i\omega a/2c}\cos\dfrac{\omega x}{c}\right. \\[1ex] \qquad\qquad \left.- i\dfrac{\omega a}{\pi c}\cos\dfrac{\pi x}{a}\right]e^{-i\omega t} & -\frac{1}{2}a < x < \frac{1}{2}a \\[2ex] \dfrac{iaA}{\pi\omega\epsilon}\dfrac{\cos(\omega a/2c)}{1-(\omega a/\pi c)^2}e^{-i(\omega/c)x-i\omega t} & x < -\frac{1}{2}a \end{cases}$$

representing simple-harmonic waves going outward from the region $-\frac{1}{2}a <$ $x < \frac{1}{2}a$, where the force is applied, with a mixed oscillation in the region of application of the force. The formulas look as though they become infinite at $\omega = \pi c/a$, but in each range of x the numerator of the function is also zero at $\omega = \pi c/a$; so $y(x,t)$ is finite for all values of ω except for $\omega = 0$. For a steady force $\omega = 0$, the region of application is pushed steadily outward with a constant velocity $aA/\pi \epsilon c$, so that the steady-state amplitude is infinite. Since the string must be infinite in length to have no reflected waves and for (4.4.11) to be valid, it is possible to have an arbitrarily large displacement of the region near $x = 0$.

Effect of wave reflections

Strings held under tension, however, have to terminate in supports which maintain the tension, and usually these supports produce reflected waves. If the driving force acts for a long enough time, these reflected waves return to the region of application of the force, and the simple outgoing-wave solutions of Eqs. (4.4.11) and (4.4.13) will not suffice. The Green's function for a wave of finite length must satisfy Eq. (4.4.9) if the point driving force is simple-harmonic. It must also satisfy the boundary conditions at the two ends of the string.

In other words, the Green's function for a string must be a solution of the homogeneous equation $(d^2g/dx^2) + (\omega/c)^2 g = 0$ for all appropriate values of x, *except* at the point $x = x_0$, where it must have a discontinuity of slope of -1. It must also satisfy the boundary conditions at the ends of the string, imposed by the nature of the supports.

If the string supports are rigid and a distance l apart, $g(x \mid x_0)$ must be zero at $x = 0$ and $x = l$. A solution of the homogeneous equation which is zero at $x = 0$ is $D \sin(\omega x/c)$, and one which is zero at $x = l$ is

$$E \sin [\omega(l - x)/c].$$

For g to be continuous in value (though not in slope) at $x = x_0$, we must have $D \sin(\omega x_0/c) = E \sin [\omega(l - x_0)/c]$. This can be done by setting

$$G(x \mid x_0) = \begin{cases} A \sin \dfrac{\omega(l - x_0)}{c} \sin \dfrac{\omega x}{c} & 0 < x < x_0 < l \\[2ex] A \sin \dfrac{\omega(l - x)}{c} \sin \dfrac{\omega x_0}{c} & 0 < x_0 < x < l \end{cases}$$

where A must be adjusted so that the discontinuity in slope at $x = x_0$ is -1. The slope, for $x > x_0$, is $-(A\omega/c) \cos [\omega(l - x)/c] \sin(\omega x_0/c)$, and that for $x < x_0$ is $(A\omega/c) \sin [\omega(l - x_0)/c] \cos(\omega x/c)$; so the difference between them at $x = x_0$ is

$$-\left(\frac{A\omega}{c}\right) \sin \left[\frac{\omega}{c}(l - x_0) + \frac{\omega}{c}x_0\right] = -\frac{A\omega}{c} \cdot \sin \frac{\omega l}{c}$$

which must equal -1. Therefore the Green's function for the flexible string of length l between rigid supports, representing the steady-state response to a simple-harmonic force of amplitude T concentrated at the point x_0, is

$$G(x \mid x_0) = \frac{c/\omega}{\sin(\omega l/c)} \begin{cases} \sin\dfrac{\omega(l-x_0)}{c} \sin\dfrac{\omega x}{c} & 0 < x < x_0 < l \\[2ex] \sin\dfrac{\omega x_0}{c} \sin\dfrac{\omega(l-x)}{c} & 0 < x_0 < x < l \end{cases}$$

$$\xrightarrow[(\omega/c)\to 0]{} \begin{cases} (l-x_0)\dfrac{x}{l} & x < x_0 \\[2ex] \dfrac{x_0}{l}(l-x) & x_0 < x \end{cases} \tag{4.4.14}$$

which is the solution of Eq. (4.4.9), satisfying the boundary conditions $g = 0$ at $x = 0$ and $x = l$.

A comparison of this Green's function with that of Eq. (4.4.11), for the infinite string, shows that the presence of wave reflections from the rigid-end supports introduces a factor $\sin(\omega l/c)$ in the denominator of the amplitude factor. Since this term goes to zero whenever $\omega l/c = n\pi$ $(n = 0, 1, 2, \ldots)$ or whenever $n\lambda \equiv n(2\pi c/\omega) = 2l$, the Green's function (and therefore any solution built up out of Green's functions for different x_0's) exhibits a sequence of resonance peaks at the natural frequencies of free vibration $\nu_n = n(c/2l)$ of the string, as given in Eq. (4.3.9). Just as with the simple oscillator, whenever the driving force has frequency equal to a frequency of free vibration of the system, resonance occurs, and in the absence of frictional forces, the amplitude of motion becomes infinite. Since the string has an infinity of such frequencies, there are an infinite number of different resonant frequencies, the fundamental $(n = 1)$ and all the harmonics $(n > 1)$. At resonance the reflections from the string ends return just in time to reinforce the driving force, and the amplitude builds up without limit, unless friction and nonlinear effects are taken into account.

The factors $\sin(\omega x_0/c)$ and $\sin[\omega(l-x_0)/c]$ express the dependence of the solution on the location of the source point. If the distance x_0 of this point of application of the force from the left-hand end of the string is an integral multiple of half wavelengths $\pi c/\omega$, then a node in the standing wave, set up to the left of the source point, occurs exactly at x_0. In this case the portion of the string to the right of the force, which is not in resonance, is motionless. Conversely, if $l - x_0$ is an integral number of half wavelengths, the region between 0 and x_0 is at rest, as is shown in Fig. 4.15 for $\gamma = 1.5$.

In either case the point of application of the force does not move. This is strictly true only when friction is zero; for the case of small friction, the motion of the point of application would be small compared with the amplitude of the resonating region, and it would take a long time for the steady-state condition to be built up.   In the cases where the whole string resonates, on the other hand, the source point x_0 is not motionless, but oscillates with the string, with infinite amplitude (if there is no friction).

The one exception to both rules is when there is an integral number of half wavelengths between 0 and x_0 and also between x_0 and l (for example,

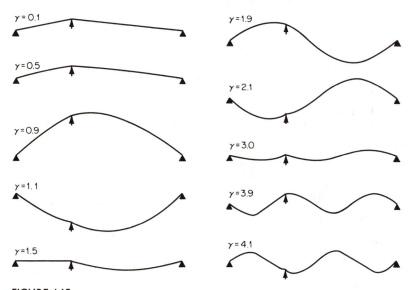

FIGURE 4.15
Forced vibration of a string driven at the point $x_0 = \frac{1}{3}l$.
Curves give amplitude of vibration of the string for various
values of $\gamma = \omega l/\pi c = 2\nu l/c = 2l/\lambda$.

when $nx_0 = ml$, n and m integers, and when $n\pi c/\omega = l$).   In this case the limiting expression is

$$G(x \mid x_0) = -\frac{l}{\pi n}\sin\left(\frac{\pi n}{l}|x - x_0|\right) \qquad \text{when } l = \frac{\pi n c}{\omega} \qquad x_0 = \frac{\pi m c}{\omega}$$

where again the source point is at rest (see Fig. 4.15 for $\gamma = 3$).   Figure 4.15 shows the shapes of function G, for $x_0 = \frac{1}{3}l$, for different driving frequencies $\omega/2\pi = \gamma c/2l$.   The discontinuity in slope at x_0 is apparent at all frequencies.

Examination of Eq. (4.4.14) indicates that this Green's function, like that of Eq. (4.4.11), is symmetric in x and x_0 and is thus subject to the principle of reciprocity.

As before, the steady-state response, to a distributed force $F(x)e^{-i\omega t}$, is the combination

$$y(x,t) = \frac{1}{T}\int_0^l F(x_0)G(x \mid x_0)\,dx_0 e^{-i\omega t} \tag{4.4.15}$$

This is the solution of Eq. (4.4.12), which goes to zero at $x = 0$ and $x = l$.

As an example, we compute the motion of the string when the driving force is distributed uniformly between $x = a$ and $x = b$.

$$F(x) = \begin{cases} 0 & 0 < x < a \\ F_0 & a < x < b \qquad 0 < a < b < l \\ 0 & b < x < l \end{cases}$$

Inserting this equation and expression (4.4.14) for g into Eq. (4.4.15), we find

$$y(x,t) = \begin{cases} \dfrac{F_0}{T} \dfrac{\sin(\omega x/c)}{\sin(\omega l/c)} \left\{ \cos\left[\dfrac{\omega}{c}(l-b)\right] - \cos\left[\dfrac{\omega}{c}(l-a)\right] \right\} & 0 < x < a \\[4mm] \dfrac{F_0/T}{\sin(\omega l/c)} \left\{ \sin\dfrac{\omega x}{c} \cos\left[\dfrac{\omega}{c}(l-b)\right] \right. & \\[2mm] \qquad \left. + \sin\left[\dfrac{\omega}{c}(l-x)\right] \cos\dfrac{\omega a}{c} - \sin\dfrac{\omega l}{c} \right\} & a < x < b \\[4mm] \dfrac{F_0}{T} \dfrac{\sin[\omega(l-x)/c]}{\sin(\omega l/c)} \left\{ \cos\dfrac{\omega a}{c} - \cos\dfrac{\omega b}{c} \right\} & b < x < l \end{cases} \qquad (4.4.16)$$

times the factor $e^{-i\omega t}$, which of course exhibits the resonance behavior discussed earlier.

Wave impedance

When the string is part of a larger mechanical system, the contribution of the string to the system's behavior can usefully be measured in terms of the mechanical impedance of the string at the point of contact of the coupling force. Early in this section we mentioned that for a very long string, driven from one end, the ratio between transverse driving force and transverse velocity of the end of the string is $(T/c) = \epsilon c = \sqrt{\epsilon T}$, the wave resistance of an infinite string [Eq. (4.4.1)]. For an infinite string driven at point x_0 by transverse force $F(t)$, Eq. (4.4.4) indicates that the transverse velocity of point x on the string at time t is $(c/2T)F[t - (1/c)|x - x_0|]$. Consequently, the impedance at x_0, the ratio between applied force and the velocity of the point of contact, is $2T/c = 2\epsilon c$, twice the impedance of an infinite string, driven at one end. Since the application of the force to a point x_0 on an infinite string is equivalent to driving two infinite strings, end to end, from the ends of each, the factor 2 is not surprising. We note that these impedances are real and are independent of the driving frequency. The energy supplied to the string by the force continually radiates away, the power lost by the force being $F(t)v(x_0,t) = F^2(t)/2\epsilon c$.

For a string of length l, driven at the end $x = 0$ and held rigidly at the end $x = l$, the situation is quite different. Here the reflected wave does return to alter the nature of the string reaction at $x = 0$. Since no energy is lost in reflection from the rigid support at $x = l$, when steady-state conditions are reached, the force is no longer feeding energy into the string; so the impedance must be reactive rather than resistive. Furthermore, since the phase relation of the reflected wave at $x = 0$ depends on the frequency of the driving force, the impedance must depend on the frequency.

A string of length l, with its end at $x = l$ held rigidly, must have a displacement $y(x,t) = A \sin [(\omega/c)(l - x)]e^{-i\omega t}$ for it to oscillate with frequency $\omega/2\pi$. Its slope at $x = 0$ would be $-(\omega/c) A \cos (\omega l/c)e^{-i\omega t}$, and its velocity there would be $v(0,t) = -i\omega A \sin (\omega l/c)e^{-i\omega t}$. The driving force would be $-T$ times the slope, $f(t) = (T\omega/c)A \cos (\omega l/c)e^{-i\omega t}$. The ratio between this force and the velocity at $x = 0$ is the impedance,

$$Z_m(\omega) = \frac{f(t)}{v(0,t)} = i\frac{T}{c} \cot \frac{\omega l}{c} = \frac{i\epsilon c}{\tan (\omega l/c)} \qquad (4.4.17)$$

which is imaginary (reactive) and a function of ω, as predicted.

For frequencies very much smaller than $c/2\pi l$, the function $\tan (\omega l/c)$ is approximately equal to $\omega l/c$; so the impedance is $i(T/\omega l)$. Comparison with Eq. (2.3.4) shows that at these low frequencies the string end is stiffness-controlled for transverse oscillation, with a stiffness constant T/l, which checks.

As the driving frequency is increased, the impedance $Z(\omega)$ decreases in magnitude, from $i\infty$ to 0 when $\omega l/c = \frac{1}{2}\pi$ or $\lambda = 2\pi c/\omega = 4l$. This is the condition of resonance; at this frequency the shape of the string is such that its slope is zero at $x = 0$; so it cannot oppose any reaction to the driving force. At a still higher frequency, when $\omega = \pi c/l$, the cotangent is again infinity; no amount of transverse force of this frequency will move the end $x = 0$. A node of the standing wave occurs at $x = 0$; so motion of $x = 0$ at this frequency is impossible (as long as there is no friction or the support at $x = l$ does not move). When the driving frequency is further increased, the impedance alternates between positive-imaginary and negative-imaginary values, being zero (resonance) whenever $\omega = (\pi c/l)(n + \frac{1}{2})$ and $\pm i\infty$ (driven end a standing-wave node) whenever $\omega = (\pi c/l)m$, where n and m are integers.

Use of the Laplace transform

Finally, we can utilize the Laplace-transform technique of Table 2.1 and Eq. (2.3.16) to calculate the behavior of the string when driven by a transient force applied at the end $x = 0$. If both force $f(t)$ and velocity $v(0,t)$ are zero for $t < 0$, we can separate the dependence of both force and velocity into their frequency components by taking Laplace transforms.

$$P_L(s) = \int_0^\infty e^{-st}f(t) \, dt$$

$$V_L(s) = \int_0^\infty e^{-st}v(0,t) \, dt$$

$$= \int_0^\infty e^{-st}\frac{d}{dt}y(0,t) \, dt = s Y_L(s)$$

where the usual relationship, $\omega = is$, holds between s and the frequency parameter ω of a Fourier transform. The relationship between the components P_L and V_L at a given frequency ω is, by definition, $P_L = Z(\omega)V_L$, where $Z(\omega) = Z(is)$ is the impedance. Therefore we find the velocity $v(0,t)$ or the displacement $y(0,t)$ of the string end by taking the inverse Laplace transforms of

$$V_L(s) = \frac{P_L(s)}{Z(is)} \qquad Y_L(s) = \frac{P_L(s)}{sZ(is)} \qquad (4.4.18)$$

For a string of length l, driven at one end and held rigidly at the other, the transforms of the velocity and displacement of the end $x = 0$, when acted on by a force having Laplace transform $P_L(s)$, are

$$V_L(s) = \frac{cP_L(s)}{iT} \frac{\sin (isl/c)}{\cos (isl/c)} = \frac{cP_L(s) \sinh (sl/c)}{T \cosh (sl/c)} = sY_L(s) \qquad (4.4.19)$$

where f, v, and y were all zero before $t = 0$. If we know the transform of $f(t)$, Table (2.1) (or a more complete table) will enable us to compute $v(0,t)$ or $y(0,t)$, the corresponding velocity and displacement of the end of the string.

For example, if the transient force $f(t)$ is a unit impulse $\delta(t)$ at $t = 0$, its Laplace transform is $P_L(s) = 1$, and we can use formula (20) in Table 2.1 to find $y(0,t)$. The transform $F_L(s)$ of this formula is $(c/sT) \sinh (sl/c) = (c/2sT)(e^{sl/c} - e^{-sl/c})$, which has an inverse transform $(c/2T)\{u[t + (l/c)] - u[t - (l/c)]\}$, where u is the step function of Eq. (1.3.22). Therefore

$$y(0,t) = \frac{c}{T} \sum_{n=0}^{\infty} (-1)^n \left\{ u\left[t - 2n\frac{l}{c}\right] - u\left[t - 2(n+1)\frac{l}{c}\right] \right\}$$

$$= \frac{c}{T}u(t) + \frac{2c}{T} \sum_{n=1}^{\infty} (-1)^n u\left(t - 2n\frac{l}{c}\right) \qquad (4.4.20)$$

which is periodic function with period $4l/c$. From $t = 0$ to $t = 2l/c$, y is equal to c/T; from $t = 2l/c$ to $t = 4l/c$, y is equal to $-(c/T)$, after which it repeats the square-wave motion indefinitely. The unit impulsive force pushes the end at $x = 0$ an amount T/c transversally; this motion is transmitted down the string and reflects back, to arrive at $x = 0$ after a time $2l/c$. By that time there is no transverse force acting on that end; so the slope $\partial y/\partial x$ must remain zero there. This requires a reflected wave of negative sign, which goes down and back, after which it changes back to positive and starts a new cycle.

The solution (4.4.20) is for an impulsive force $\delta(t)$. If the end $x = 0$ is acted on by a force $f(t)$, which differs from zero only between $t = 0$ and $t = b$, then an extension of Eq. (2.3.9) (or a further application of Table 2.1) produces the following solution for the transverse displacement of the end $x = 0$:

$$y(0,t) = \frac{c}{T} Y(t) + \frac{2c}{T} \sum_{n=1}^{\infty} (-1)^n Y\left(t - 2n\frac{l}{c}\right) \qquad (4.4.21)$$

where

$$Y(t) = \int_0^t f(u)\, du$$

For example, the lower curve of Fig. 4.16 displays the motion of the driven end of a string acted on by a force which occurs in the interval between $t = 0$

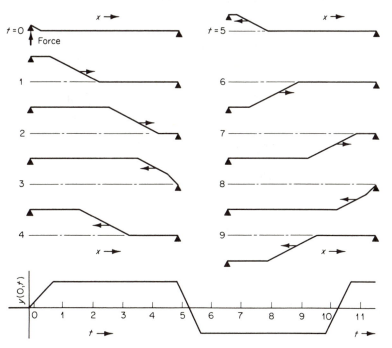

FIGURE 4.16
Motion of string fixed at $x = l$, driven by transverse force applied at $x = 0$, between $t = -\frac{1}{6}$ and $t = \frac{2}{3}$. Upper curves are successive shapes of string; lower curve is displacement of point of application $x = 0$. Period of free vibration is 10; c is thus $l/5$.

and $t = l/3c$. The upper curves show the successive waveforms at the times indicated on the lower curve. The wave reflects from the fixed end at $x = l$ without change of sign; from the "open" end $x = 0$ with change of sign. The motion is periodic after the force has ceased acting.

Uniform friction

When the frictional resistance of the medium surrounding the string is appreciable, the solutions of this section must be modified. In accordance with Eq. (4.3.17), the equation of motion of a string, driven by a force

$F(x)e^{-i\omega t}$ and acted on by a frictional force, proportional to its transverse velocity, along its length, is

$$\frac{\partial^2 y}{\partial t^2} + 2k\,\frac{\partial y}{\partial t} - c^2\,\frac{\partial^2 y}{\partial x^2} = \frac{F}{\epsilon}\,e^{-i\omega t}$$

where $k = R(\omega)/2\epsilon$. If the force has amplitude T and is concentrated at the point x_0, the solution will be a Green's function $g(x\,|\,x_0)e^{-i\omega t}$, where g satisfies an inhomogeneous equation like (4.4.9).

$$\frac{d^2 g}{dx^2} + \left(\frac{\omega}{c}\right)^2\left(1 + \frac{2ik}{\omega}\right)g = -\delta(x - x_0) \tag{4.4.22}$$

In contrast to the free-vibration solutions of Eq. (4.3.16), where the frequency turns out to be complex, for forced motion (where ω must be real), the space dependence is complex.

For a string of length l, between rigid supports, the Green's function is an obvious modification of Eq. (4.4.14).

$$G(x\,|\,x_0) = \begin{cases} \dfrac{1}{\beta}\,\dfrac{\sin(\beta x)}{\sin(\beta l)}\,\sin[\beta(l - x_0)] & 0 < x < x_0 < l \\[4mm] \dfrac{1}{\beta}\,\dfrac{\sin[\beta(l - x)]}{\sin(\beta l)}\,\sin(\beta x_0) & 0 < x_0 < x < l \end{cases} \tag{4.4.23}$$

where

$$\beta = \frac{\omega}{c}\sqrt{1 + \frac{2ik}{\omega}} \simeq \frac{\omega}{c} + i\frac{k}{c} \qquad k \ll \omega$$

When $k \ll \omega$, the resonance denominator becomes

$$\sin(\beta l) \simeq \sin\frac{\omega l}{c}\cosh\frac{kl}{c} + i\cos\frac{\omega l}{c}\sinh\frac{kl}{c}$$

$$\simeq \sin\frac{\omega l}{c} + i\frac{kl}{c}\cos\frac{\omega l}{c}$$

Consequently, at the resonance frequencies $\nu_n = \omega_n/2\pi = nc/2l$, this denominator does not go to zero; it just becomes quite small and imaginary. The corresponding amplitude of motion becomes large but not infinite, and the motion is out of phase with the force.

The fact that the x dependence of g is also complex means that the motion of the string is not all in phase; the parts farther away from x_0 lag farther behind than the parts nearby. Figure 4.17 shows sequences for three different driving frequencies. They should be compared with the amplitudes given in Fig. 4.15 for $\gamma = 1.5$, 1.9, and 3.0, for the case of no friction, when the motion is all in phase.

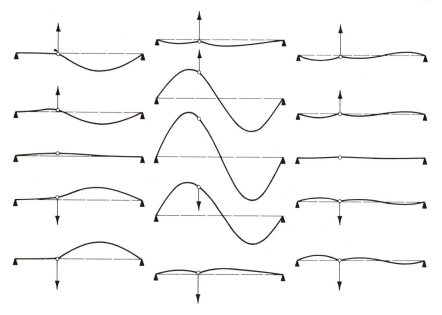

FIGURE 4.17
Sequences showing the successive shapes of a damped string, driven by a force of frequency ν, during a half cycle of the steady-state motion.   The force is concentrated at the point marked by the circle, and the arrows show its successive magnitudes and directions.   The three sequences are for the driving frequency ν, equal to $\frac{3}{2}$, two and three times the fundamental frequency of the corresponding undamped string.

4.5 THE EFFECT OF INHOMOGENEITIES

A wave in a uniform flexible string travels with constant velocity and unchanged shape, at least as long as the wave amplitude is not too great. Any irregularity in the mass density of the string, however, will distort this perfect wave motion.   If the wave is traveling along a very long string, the irregularity will reflect part of the wave and will distort the rest as it is transmitted beyond the irregularity.   If the motion is a standing wave, the irregularity will change the values of the frequencies of free vibration of the string. In this section we shall discuss various methods for computing these effects.

It is often useful to consider the behavior of an inhomogeneous region of a string as a simple example of the interaction between two vibrating systems, the uniform string being one system and the inhomogeneity another. As such, we should expect the combination to have some of the properties of coupled oscillators.   Later we shall deal with the coupled motions of a wide variety of acoustic systems; in fact, most acoustic problems of interest involve the interaction of two or more systems.   A string vibrating in vacuum between two rigid supports is a very simple—but a quite idealized—subject for study.

When we add the interaction between the string's motion and the air, and when we also consider the effects of the motion on supports which are not completely rigid, the problem becomes both very much more difficult to solve and also of much greater practical interest.

Thus the effect of inhomogeneities on the wave motion of a string deserves study because it is one of the simplest examples of the interaction between an extended system and a concentrated one.

Wave reflection from a point inhomogeneity

We start by considering an infinite string, under tension T and with uniform mass ϵ per unit length, having attached to it a simple oscillator, which can move transversally with the string, at point $x = 0$. The string's displacement $y(0,t)$ equals the displacement of the oscillator, and the shape of the string, near $x = 0$, must be such as to move the oscillator along with the string. As we saw in the previous section, the force required to produce the motion must be caused by a discontinuity in the string's slope at $x = 0$.

$$F = \lim_{\Delta \to 0} \left\{ T \left[\left(\frac{\partial y}{\partial x} \right)_{x=\Delta} - \left(\frac{\partial y}{\partial x} \right)_{x=-\Delta} \right] \right\} \tag{4.5.1}$$

The relationship between this coupling force and the displacement $y(0,t)$ of both string and oscillator at $x = 0$ will, of course, depend on the nature of the oscillator. The discontinuous change in slope required by Eq. (4.5.1) will distort the waveform as it passes through $x = 0$.

To see what happens we consider the simple case of a steplike wave $u(ct - x)$ approaching the origin from the left. Of course, a perfect step-function waveform would not satisfy the requirement that $|\partial y/\partial x|$ be small for the wave equation to be valid; but the step function can be considered to be the limiting form of an actual wave with steep wavefront. Such a wave, when it reaches the origin, would act like an impulsive force, of total impulse T. What this impulse does to the attached oscillator depends on the nature of the oscillator.

For example, if the "oscillator" at $x = 0$ is simply an extra mass m, attached to the string at $x = 0$, the impulse will give the mass an initial upward velocity T/m; but at $t = 0$ the initial displacement $y(0,0)$ will still be zero. Thus, to the steep-fronted wave, the extra mass will behave as though there were an initially rigid support at $x = 0$. As was shown in Sec. 4.2 (Fig. 4.7), a wave is reflected from a rigid support with change of sign, so the initial effect of the mass will be to start a wave of cancellation $-u(ct + x)$ back toward $-\infty$, canceling the $+1$ of the incident wave $u(ct - x)$. But the mass at $x = 0$ does not stay at rest; immediately after $t = 0$ it begins to rise, with initial velocity T/m, bringing the string along with it. The "delayed reaction" of the mass produces a wave in both directions from $x = 0$; the generation of such a wave requires energy, which must come from the kinetic energy of the mass. Thus the mass will slow down, eventually coming to rest a unit distance from its initial position. The sequence is shown in the left-hand sequence of Fig. 4.18. The reflected wave has a steep wavefront, followed by a "wake" produced by the motion of the mass m. The transmitted wave only registers the motion of the mass.

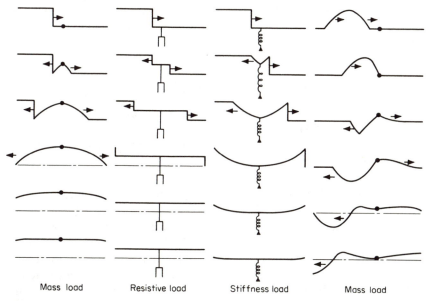

Mass load   Resistive load   Stiffness load   Mass load

FIGURE 4.18
Reflection and transmission of a wave at a point load on a string. First three time sequences are for an incident step-function wave. Last sequence is for half a sine wave.

If the load at $x = 0$ is purely resistive, with no effective mass, the impulsive force of the wavefront will jerk the load upward by an amount T/R, where R is the load's mechanical resistance. As soon as the wavefront has passed, the frictional load will stop moving. Therefore both reflected and transmitted waves are steep-fronted, as shown in the second sequence of Fig. 4.18.

Finally, if the load is a massless spring attached to the string at $x = 0$, the impulsive force will jerk the spring end up so that $y(0,t)$, just after $t = 0$, is unity and the wavefront will pass unchanged into the region $x > 0$. However, now that the spring is displaced from equilibrium, it will exert a downward force on the string, creating a wave which spreads out in both directions from the origin. The sequence of events is shown in the third sequence of Fig. 4.18; in this case the steep wavefront stays with the transmitted wave, not the reflected one.

To work out the events quantitatively, we assume that the wave on the string for $x < 0$ has both an incident and a reflected wave, whereas the wave in the region $x > 0$ has only a transmitted part.

$$y(x,t) = \begin{cases} I(ct - x) + Q(ct + x) & x < 0 \\ P(ct - x) & x > 0 \end{cases} \qquad (4.5.2)$$

where we can assume that $I(z)$, $Q(z)$, and $P(z)$ are all zero for $z < 0$, without serious loss of generality. In order that the string be continuous at $x = 0$ we must have

$$I(z) + Q(z) = P(z) \qquad z = ct$$

and also

$$F(t) = T[I'(ct) - Q'(ct) - P'(ct)] \qquad (4.5.3)$$

where $I'(z) = (d/dz)I(z)$ represents the force on the extra load at $x = 0$, as specified by Eq. (4.5.1).

If the load is purely inertial, a mass m, then the force F will cause an acceleration

$$m\frac{d^2}{dt^2}y(0,t) = mc^2 P''(ct) = F = \epsilon c^2[I'(ct) - Q'(ct) - P'(ct)]$$

where $P''(z)$ is the second derivative of $P(z)$ with respect to its argument. This equation and the first of Eqs. (4.5.3), between them, enable us to eliminate the unknown function Q and obtain an equation for the transmitted wave P in terms of the known incident wave $I(z)$. We integrate the last equation, obtaining $I(z) - Q(z) - P(z) = (m/\epsilon)P'(z)$, and thence obtain

$$\frac{m}{2\epsilon}P'(z) + P(z) = I(z)$$

which has for solution

$$P(z) = \frac{2\epsilon}{m} e^{-(2\epsilon/m)z} \int_0^z I(u)e^{(2\epsilon/m)u}\, du \qquad z > 0 \qquad (4.5.4)$$

and the reflected wave is $Q(z) = P(z) - I(z)$ from Eq. (4.5.3).

The transmitted wave $P(ct - x)$ is thus not $I(ct - x)$, the incident wave, but a delayed response of the mass to the incident wave, expressible in terms of the integral involving exponentials, of Eq. (4.5.4). This is illustrated by the first sequence of Fig. 4.18; the reflected and transmitted waves are exponentials. The last sequence of Fig. 4.18 is the reaction of the string-mass combination to an incident wave of the form

$$I(z) = \begin{cases} 0 & z < 0 \\ \sin\dfrac{\pi z}{a} & 0 < z < a \\ 0 & a < z \end{cases}$$

The mass is initially impelled upward, but after the incident wave has gone by, the reflected and transmitted waves pull the mass back down again, then slow it up, so that it eventually comes to rest at $y = 0$.

The waves produced by resistive or springy loads can be worked out in a very similar manner; they will be left for a problem.

Simple-harmonic waves

When a sinusoidal wave is incident on the point $x = 0$, to which is attached a simple oscillator, the Green's function of Eq. (4.4.11) can be used to calculate the reflected and transmitted waves. The force which the string must exert on the oscillator to make it move with the string is, by Newton's third law, equal and opposite to the reactive force exerted by the oscillator on the string at $x = 0$. If the string's motion is sinusoidal, $y(0,t) = Y(0)e^{-i\omega t}$, then the reactive force, which is equal to the oscillator's velocity times its impedance $Z(\omega)$ [Eq. (2.3.3)],

$$F(t) = -Z(\omega)\frac{d}{dt}y(0,t) = i\omega Z(\omega)Y(0)e^{-i\omega t} \qquad (4.5.5)$$

must also vary sinusoidally with time. Thus the equation of motion of both string and oscillator is given by a modification of Eq. (4.4.12).

$$\epsilon\frac{\partial^2 y}{\partial t^2} - T\frac{\partial^2 y}{\partial x^2} = i\omega Z(\omega)\,\delta(x)\,y(x,t)$$

or, since y depends on time only through the factor $e^{-i\omega t}$ so that $y(x,t) = Y(x)e^{-i\omega t}$, the equation for $Y(x)$ is

$$\frac{d^2 Y}{dx^2} + \omega^2\frac{\epsilon}{T}\,Y(x) = -i\omega\,\frac{Z}{T}\,Y(x)\,\delta(x) \qquad (4.5.6)$$

The displacement of the oscillator is $Y(0)e^{-i\omega t}$, equal to the displacement of the string at $x = 0$. The reactive force is concentrated at $x = 0$ (which accounts for the delta function) and is proportional to the displacement $Y(0)$ itself. Comparison with Eqs. (4.4.9) and (4.4.11) indicates that this reactive force must generate a wave

$$Y_g(x) = i\,\frac{c}{2\omega}\,i\omega\,\frac{Z}{T}\,Y(0)e^{i\omega\,|x|/c}$$

or

$$y_g(x,t) = -\frac{Z}{2\epsilon c}\,Y(0)\exp\left[\frac{i\omega}{c}\,(|x| - ct)\right]$$

which spreads out equally in both directions from the point of application $x = 0$ of the force. In the present case the force is produced by the motion, so that if y_g were the whole solution and $Y(0)$ on the right-hand side were equal to $Y_g(0)$, then $Y(0)$ would have to be zero. With no external driving force, no motion could result.

The driving force results from the incident wave

$$y_i(x,t) = Ae^{i\omega(x-ct)/c} \qquad Y_i(x) = Ae^{i\omega x/c}$$

which is produced at the far end of the string, $x = -\infty$, by some external force. We note that $Y_g(x)$ is only a particular solution to the equation of motion (4.5.6); to it we can add any amount of a solution of the homogeneous equation $(d^2Y/dx^2) + (\omega/c)^2 Y = 0$ [see also the discussion preceding Eq. (4.4.14)]. For example, the function $Y(x) = Y_i(x) + Y_g(x)$, or

$$Y(x) = Ae^{i\omega x/c} - \frac{Z}{2\epsilon c} Y(0)e^{i\omega|x|/c}$$

$$= \begin{cases} Ae^{i\omega x/c} - \dfrac{Z}{2\epsilon c} Y(0)e^{-i\omega x/c} & x < 0 \\[2mm] \left[A - \dfrac{Z}{2\epsilon c} Y(0)\right]e^{i\omega x/c} & x > 0 \end{cases} \qquad (4.5.7)$$

is a solution of Eq. (4.5.6), which corresponds to the physical situation we have described. The first term in the expression for $Y(x)$ is the incident wave, generated at $x = -\infty$ and going clear to $x = +\infty$ without change of amplitude or phase. The second term is the wave going out from $x = 0$, caused by the reaction of the oscillator attached to the string there. Thus, in the region $x < 0$, there is an incident and a reflected wave, each going in opposite directions; in region $x > 0$ the incident and generated waves combine to produce a transmitted wave, of amplitude $A - (Z/2\epsilon c)Y(0)$, going in the positive direction.

But we still have not solved the problem, for the unknown Y appears on both sides of Eq. (4.5.7). The reason is that the equation of motion (4.5.6) is a homogeneous equation, not an inhomogeneous one like Eq. (4.4.5). The force term contains the unknown Y, since the reaction force is proportional to the displacement itself, though we assumed it to be known when we wrote down the particular solution Y_g. An inhomogeneous term emerges only when we introduce the incident wave Y_i into the equation for the solution. This term enables us to solve for the magnitude of Y by setting $x = 0$ in Eq. (4.5.7).

$$Y(0) = A - \frac{Z}{2\epsilon c} Y(0) \qquad \text{or} \qquad Y(0) = \frac{2\epsilon c}{Z + 2\epsilon c} A$$

or

$$Y(x) = Ae^{i\omega x/c} - \frac{AZe^{i\omega|x|/c}}{Z + 2\epsilon c} = \begin{cases} Ae^{i\omega x/c} - \dfrac{AZe^{-i\omega x/c}}{Z + 2\epsilon c} & x < 0 \\[2mm] \dfrac{2\epsilon c Ae^{i\omega x/c}}{Z + 2\epsilon c} & x > 0 \end{cases} \qquad (4.5.8)$$

$$y(x,t) = Y(x)e^{-i\omega t}$$

This gives the amplitudes of the reflected and transmitted waves in terms of the amplitude A of the incident wave. The proportionality factors involve the impedance Z with which the oscillator opposes the motion of the string, and the impedance $2\epsilon c$ with which the string opposes the motion of the oscillator [see discussion of Eq. (4.4.1)]. If the impedance of the load Z is much larger than $2\epsilon c$, the reflected wave is large and the transmitted wave is small compared with the incident wave. The reverse is true if $|Z|$ is small compared with $2\epsilon c$.

Energy flow and resonance effects

The mechanical impedance of a simple oscillator, $-i\omega m + R + i(K/\omega)$, has a real part, the mechanical resistance term R, and an imaginary part, the reactance $-iX = -i[\omega m - (K/\omega)]$, caused by the oscillator's mass and stiffness. Unless X is zero, both reflected and transmitted waves are out of phase with the incident wave. Unless R is zero, the energy carried away by the reflected and transmitted waves is less than the energy introduced by the incident wave.

According to Eq. (4.4.2), the mean incident power Υ_i transmitted by the incident wave from $-\infty$ to 0, the mean power Υ_r reflected back to $-\infty$, and the mean power Υ_t transmitted from $x = 0$ on to $+\infty$ are

$$\Upsilon_i = \tfrac{1}{2}\epsilon c\omega^2 |A|^2 \qquad \Upsilon_r = \tfrac{1}{2}\epsilon c\omega^2 |A|^2 \frac{R^2 + X^2}{(R + 2\epsilon c)^2 + X^2}$$

$$\Upsilon_t = \tfrac{1}{2}\epsilon c\omega^2 |A|^2 \frac{4(\epsilon c)^2}{(R + 2\epsilon c)^2 + X^2} \tag{4.5.9}$$

The sum of Υ_t and Υ_r is the power which leaves the region $x = 0$, whereas Υ_i is the power fed into the region by the incident wave. The difference,

$$\Upsilon_i - \Upsilon_t - \Upsilon_r = \frac{\tfrac{1}{2}\epsilon c\omega^2 |A|^2}{(R + 2\epsilon c)^2 + X^2} [(R + 2\epsilon c)^2 + X^2 - R^2 - X^2 - 4(\epsilon c)^2]$$

$$= \tfrac{1}{2}R\omega^2 |Y(0)|^2 = \tfrac{1}{2}R \left| \left(\frac{\partial y}{\partial t}\right)_{x=0} \right|^2 \tag{4.5.10}$$

is the amount absorbed in moving the oscillator against its resistance R.

For a simple oscillator the reactance $iX = -i\omega m + i(K/\omega)$ is zero at the resonance frequency $\omega_0/2\pi$, where $\omega_0^2 = K/m$. Near this frequency the transmitted wave carries away most of the incident energy, if $R < 2\epsilon c$; far from this frequency the reflected wave carries more. The effect is more pronounced the larger the Q of the oscillator, $\sqrt{Km}/R$ [Eq. (2.2.4)]. If this quantity is large, the incident wave usually finds the region $x = 0$ hard to penetrate; for all frequencies except those very near $\omega_0/2\pi$, the transmitted power Υ_t is much smaller than the incident power Υ_i or the reflected power Υ_r. At resonance, however, X is zero, and (if $R < \sqrt{Km}$ and $R < 2\epsilon m$)

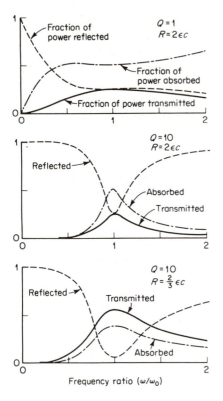

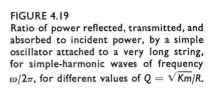

FIGURE 4.19
Ratio of power reflected, transmitted, and absorbed to incident power, by a simple oscillator attached to a very long string, for simple-harmonic waves of frequency $\omega/2\pi$, for different values of $Q = \sqrt{Km}/R$.

the point $x = 0$ is nearly transparent to the incident wave; Υ_r is small compared with Υ_i and Υ_t.   Figure 4.19 shows examples of this; the transmitted and reflected fractions of the incident power are there plotted as a function of ω/ω_0.   For the cases of $Q = 10$, the effects of the resonance are quite noticeable; for $Q = 1$ ($R = \sqrt{Km}$), the resistance has broadened out the resonance peak and damped out the string reflection at high frequencies. In all cases, however, for frequencies sufficiently above resonance so that $|X| \rightarrow \omega m$ is larger than $R + 2\epsilon c$, the transmitted fraction falls off inversely with ω^2.   In this frequency range the inertia of the point load effectively opposes motion of the string and reflects most of the incident energy.

Extended range of inhomogeneity

Now suppose that the oscillator is attached to an extended range of length of the string, rather than at the point $x = 0$.   In other words, suppose that over the length of string from $x = -\frac{1}{2}a$ to $x = \frac{1}{2}a$, there is an additional load on the string, which opposes its displacement by a reaction force which is proportional to the string's displacement at x.   For a simple-harmonic wave of frequency $\omega/2\pi$, the ratio between reaction-force density and displacement could be written as $-i\omega z(x)$, where $z(x)$ is the *impedance density* of the load in mechanical ohms per unit length, at x. Setting $y(x,t) = Y(x)e^{-i\omega t}$ as before, the equation for the amplitude-phase function $Y(x)$ is

$$\frac{d^2 Y}{dx^2} + \left(\frac{\omega}{c}\right)^2 Y = -i\frac{\omega}{T} z(x) Y(x) \qquad (4.5.11)$$

which is to be compared with Eq. (4.5.6). Here the load spreads out over a range of x (z differs from zero only in the range $-\frac{1}{2}a < x < +\frac{1}{2}a$) rather than being concentrated at $x = 0$. In a few cases, where z is a simple mathematical function (but not very physical), this equation can be solved exactly; in most cases of practical interest we must be satisfied with an approximate solution.

To solve the equation we use the Green's function technique of Eqs. (4.4.13) and (4.4.15), assuming for the moment that the right-hand term in (4.5.11) is a known force density. For example, if the string is long enough so that reflections from the ends may be neglected, we can use the Green's function of Eqs. (4.4.11) and (4.4.13) and transform Eq. (4.5.11) into an integral equation, similar to Eq. (4.5.7) for a load concentrated at $x = 0$. We multiply Eq. (4.5.11) (expressed in terms of x_0 instead of x) by $g(x_0 \mid x)$ of Eq. (4.4.11) and subtract from this $Y(x_0)$ times Eq. (4.4.9) for $g(x_0 \mid x)$, with x_0 and x interchanged, giving

$$g(x_0 \mid x) \frac{d^2}{dx_0^2} Y(x_0) - Y(x_0) \frac{d^2}{dx_0^2} g(x_0 \mid x)$$

$$= \delta(x_0 - x) Y(x_0) - \frac{i\omega}{T} \frac{ic}{2\omega} e^{i\omega |x_0 - x|/c} z(x_0) Y(x_0) \qquad (4.5.12)$$

Now integrate both sides of this equation with respect to x_0 from $-d$ to $+b$, where $-\frac{1}{2}a > -d < x < +b > \frac{1}{2}a$, so that the range of integration includes the point $x_0 = x$ and also the range $-\frac{1}{2}a < x_0 < \frac{1}{2}a$, for which the load impedance $z(x_0)$ differs from zero. On the right-hand side, the integral of the first term is equal to $Y(x)$; the second term becomes an integral of $[z(x_0)/2\epsilon c]e^{i\omega |x_0 - x|/c} Y(x_0)$ over x_0 from $-\frac{1}{2}a$ to $\frac{1}{2}a$, the range for which $z(x_0)$ differs from zero. On the left-hand side the terms can be integrated directly; so the integral becomes the difference between an expression evaluated at b and one evaluated at $-d$.

$$\left[g(x_0 \mid x) \frac{d}{dx_0} Y(x_0) - Y(x_0) \frac{d}{dx_0} g(x_0 \mid x) \right]_{-d}^{b}$$

$$= \frac{ic}{2\omega} \left[e^{i\omega |x - x_0|/c} \frac{d}{dx_0} Y(x_0) - Y(x_0) \frac{d}{dx_0} e^{i\omega |x - x_0|/c} \right]_{-d}^{b}$$

At this point we have to specify something about the general properties of the solution $Y(x)$, in order to evaluate these terms. Points b and $-d$

are outside both the point x and the range of x_0 upon which the load falls. If we are investigating the effect of the load on an incident wave, we are assuming that the solution $Y(x_0)$, to the left of the load, has an incident wave $Ae^{i\omega x_0/c}$ plus a reflected wave $Qe^{-i\omega x_0/c}$ (where A is known but Q has, as yet, to be determined); and we are assuming that $Y(x_0)$, to the right of the load, is a transmitted wave $Pe^{i\omega x_0/c}$ (where P is as yet unknown).

This degree of specification of the solution is sufficient to enable us to evaluate the integral of the left-hand side of (4.5.12). Since $b > x$ and $-d < x$, at $x_0 = b$ the quantity $|x - x_0|$ equals $x_0 - x$ and at $x_0 = -d$ it equals $x - x_0$. Therefore the integral becomes

$$\frac{ic}{2\omega}\left\{\left[e^{i\omega(x_0-x)/c}\frac{d}{dx_0}(Pe^{i\omega x_0/c}) - Pe^{i\omega x_0/c}\frac{d}{dx_0}e^{i\omega(x_0-x)/c}\right]_{x_0=b}\right.$$

$$- \left[e^{i\omega(x-x_0)/c}\frac{d}{dx_0}(Ae^{i\omega x_0/c} + Qe^{-i\omega x_0/c})\right.$$

$$\left.\left. - (Ae^{i\omega x_0/c} + Qe^{-i\omega x_0/c})\frac{d}{dx_0}e^{i\omega(x-x_0)/c}\right]_{x_0=-d}\right\} = Ae^{i\omega x/c}$$

which is just the expression for the incident wave. Therefore the integral of Eq. (4.5.12), for a long string on which an incident wave $Ae^{i\omega x/c}$ impinges on a load between $-\frac{1}{2}a$ and $\frac{1}{2}a$, becomes the integral equation

$$Y(x) = Ae^{i\omega x/c} - \frac{1}{2}\int_{-\frac{1}{2}a}^{\frac{1}{2}a} e^{i\omega|x-x_0|/c}\zeta(x_0)Y(x_0)\,dx_0 \qquad (4.5.13)$$

where $\zeta = z/\epsilon c$, the ratio between the impedance density of the load and the radiation resistance of the string, may be called the *impedance ratio* of the load. It has the dimensions of inverse length.

This is an inhomogeneous integral equation for the waveform $Y(x)$. It contains more specifications about the solution than does the equation of motion (4.5.11), for it specifies the nature of the solution at large distances from the load; in particular, it includes the fact that an incident wave $Ae^{i\omega x/c}$ is coming in from $-\infty$. A solution of Eq. (4.5.13) is a solution of Eq. (4.5.11), but not all solutions of (4.5.11) are solutions of (4.5.13). The integral equation is a natural extension of Eq. (4.5.7) for a point load; it states that the solution $Y(x)$ is a sum of an incident wave plus a term representing the distortion of the wave by the added load. Each loaded point x_0 generates a wave $e^{i\omega|x-x_0|/c}$, radiating in both directions, of amplitude proportional to the amplitude $Y(x_0)$ of the string at x_0 and also proportional to the ratio $\zeta(x_0) = z(x_0)/\epsilon c$ between the transverse mechanical impedance density at x_0 and the wave impedance of the string.

Unfortunately, exact solutions of integral equations like (4.5.13) are no easier to obtain than are exact solutions of differential equations, about which we talked in Sec. 1.2. There are several advantages to the integral-equation formulation, however. In the first place, it contains in it boundary conditions, as well as dynamic specifications, as discussed in the preceding paragraph. Because of this, one can derive general properties of the solution without exact knowledge of its form. And finally, the integral equation provides a more natural starting point for approximate solutions than does the equivalent combination of differential equation and boundary conditions.

For example, if we know the true form of $Y(x)$ in the range $-\frac{1}{2}a < x < \frac{1}{2}a$, where the load is attached, we can determine its form everywhere else in terms of the function

$$\varphi(\eta) = \begin{cases} 0 & \eta < -\dfrac{a}{2c} \\[2mm] \dfrac{\pi c \zeta(c\eta)\,Y(c\eta)}{A} = \pi\,\dfrac{z(c\eta)}{\epsilon}\,\dfrac{Y(c\eta)}{A} & -\dfrac{a}{2c} < \eta < \dfrac{a}{2c} \\[2mm] 0 & \dfrac{a}{2c} < \eta \end{cases}$$

The Fourier transform of this quantity with respect to the variable η, as defined in Eqs. (1.3.16), is

$$\Phi(\omega) = \frac{1}{2\pi} \int_{-a/2c}^{a/2c} e^{i\omega\eta}\varphi(\eta)\,d\eta = \frac{1}{2A} \int_{-a/2}^{a/2} e^{i\omega x_0/c}\zeta(x_0)\,Y(x_0)\,dx_0 \quad \textbf{(4.5.14)}$$

Conversely, if $\Phi(\omega)$, as a function of ω, can be determined, its inverse transform will determine the value of $Y(x)$ in the range $-\frac{1}{2}a < x < \frac{1}{2}a$.

The function $\Phi(\omega)$ is intimately related to the amplitude and phase of the waves reflected and transmitted from the loaded portion of the string, for an incident wave of frequency $\omega/2\pi$. This can be seen if we make $x < -\frac{1}{2}a$ or $x > \frac{1}{2}a$ in Eq. (4.5.13).

$$Y(x) = Ae^{i\omega x/c} - \tfrac{1}{2}e^{-i\omega x/c}\int_{-a/2}^{a/2} e^{i\omega x_0/c}\zeta(x_0)\,Y(x_0)\,dx_0$$

$$= Ae^{i\omega x/c} - A\Phi(\omega)e^{-i\omega x/c} \qquad x < -\tfrac{1}{2}a \qquad \textbf{(4.5.15)}$$

Also $\qquad Y(x) = A[1 - \Phi(-\omega)]e^{i\omega x/c} \qquad x > \tfrac{1}{2}a$

expresses the transmitted wave in terms of the transform of φ for the value $-\omega$. Thus a measurement of the ratios (amplitude and phase angle) between the incident and reflected waves and between the incident and transmitted waves, for the complete range of frequencies of the incident wave, will serve to measure the transform $\Phi(\omega)$ for ω from $-\infty$ to $+\infty$. The inverse transform will then give us $Y(x)$ in the range $-\frac{1}{2}a < x < \frac{1}{2}a$, and Eq. (4.5.15) provides the form of $Y(x)$ in the unloaded part of the string. Contrariwise, if we can compute the form of $Y(x)$ in the range $-\frac{1}{2}a < x < \frac{1}{2}a$, we can compute it for all other ranges of x and can thus calculate the reflection and transmission of the wave.

Successive approximations

If a, the length of string over which the load is applied, is very small compared with the wavelength $2\pi c/\omega$, then the term $e^{i\omega|x-x_0|/c}Y(x_0)$ in the integrand of (4.5.13) can be replaced by $e^{i\omega|x|/c}Y(0)$, and Eq. (4.5.13) reduces to Eq. (4.5.7) for a load applied at point $x = 0$, with Z equal to the integral of $z(x)$ over x from $-\frac{1}{2}a$ to $\frac{1}{2}a$. For larger values of a the form of $Y(x)$ in the range $-\frac{1}{2}a < x < \frac{1}{2}a$ is usually computed by means of successive approximations.

We can, for example, assume that the load does not change the form of the string very much, and insert $Y = Y_0(x) = Ae^{i\omega x/c}$ into the integral of Eq. (4.5.13), giving a first approximation.

$$Y_1(x) = A\left[e^{i\omega x/c} - \frac{1}{2}\int_{-a/2}^{a/2} e^{i\omega|x-x_0|/c}\zeta(x_0)e^{i\omega x_0/c}\,dx_0\right]$$

$$= A\left\{e^{i\omega x/c}\left[1 - \frac{1}{2}\int_{-a/2}^{x}\zeta(x_0)\,dx_0\right]\right.$$

$$\left. - \frac{1}{2}e^{-i\omega x/c}\int_{x}^{a/2}e^{2i\omega x_0/c}\zeta(x_0)\,dx_0\right\} \qquad -\frac{1}{2}a < x < \frac{1}{2}a \quad (4.5.16)$$

For x close to $-\frac{1}{2}a$, the positive exponential wave is nearly that of the incident wave, $Ae^{i\omega x/c}$; the negative exponential part is nearly that of the reflected wave. As x increases from $-\frac{1}{2}a$ to $\frac{1}{2}a$, the wave in the positive direction differs more and more from the incident wave, and the negatively directed wave decreases in amplitude, until, at $x = \frac{1}{2}a$, the wave is all in the positive direction. The reflected and transmitted amplitude functions of Eqs. (4.5.15) are, to this approximation,

$$\Phi_1(\omega) = \frac{1}{2}\int_{-a/2}^{a/2}e^{2i\omega x_0/c}\zeta(x_0)\,dx_0 = \pi\Xi\left(\frac{2\omega}{c}\right)$$

$$\Phi_1(-\omega) = \frac{1}{2}\int_{-a/2}^{a/2}\zeta(x_0)\,dx_0 = \pi\Xi(0)$$

$$(4.5.17)$$

where

$$\Xi(w) = \left(\frac{1}{2\pi}\right)\int_{-a/2}^{a/2}\zeta(x_0)e^{iwx_0}\,dx_0$$

is the Fourier transform of the impedance ratio $\zeta(x)$. This approximation is usually called the *Born approximation*.

Better approximations can be obtained by reinserting $Y_1(x)$ into the integral of Eq. (4.5.13). In fact, one can compute a sequence of approximations, $Y_n(x)$, where

$$Y_{n+1}(x) = Ae^{i\omega x/c} - \frac{1}{2}\int_{-a/2}^{a/2}e^{i\omega|x-x_0|/c}\zeta(x_0)Y_n(x_0)\,dx_0 \qquad (4.5.18)$$

which converges on the correct answer as n increases. The computations become progressively more tedious to carry out analytically, but can often be carried out numerically on a digital computer.

A variational technique

If we are more interested in calculating the transmission and reflection factors $\Phi(-\omega)$ and $\Phi(\omega)$ than in determining the exact shape of the wave in the loaded region $-\frac{1}{2}a < x < \frac{1}{2}a$, we can be satisfied with a process which leads directly to improved approximations for these factors. To explain the process, which will be of considerable use in solving many difficult problems, we consider an integral equation which is in some ways more general than Eq. (4.5.13) and in some ways simpler:

$$M(x) = \int_a^b K(x \mid x_0) Y(x_0) \, dx_0 \qquad \text{(equality holds for } a < x < b) \quad (4.5.19)$$

where $M(x)$ is a known function of x; $K(x \mid x_0)$ is symmetric [that is, $K(x_0 \mid x) = K(x \mid x_0)$]; and both M and K are real functions. The unknown function is $Y(x)$.

The first step is to multiply both sides of the equation by $Y(x)$ and then integrate over x from a to b, obtaining

$$0 = \int_a^b M(x) Y(x) \, dx - \int_a^b Y(x) \, dx \int_a^b K(x \mid x_0) Y(x_0) \, dx_0 \quad (4.5.20)$$

It usually happens that the integral $\int M Y \, dx$ has some physical importance in its own right. For example, in Eq. (4.5.13), $(1/2A)A\zeta(x)e^{i\omega x/c}$ is equivalent to $M(x)$; so $\int M Y \, dx$ is the transmission factor $\Phi(\omega)$, the quantity of interest in the present case. Thus we give this quantity a special symbol,

$$V(Y) = \int_a^b M(x) Y(x) \, dx \qquad (4.5.21)$$

Adding the two equations and substituting for the solution $Y(x)$ the general function $u(x)$, we define the functional

$$V(u) = 2 \int_a^b M(x) u(x) \, dx - \int_a^b u(x) \, dx \int_a^b K(x \mid x_0) u(x_0) \, dx_0 \quad (4.5.22)$$

where $u(x)$ is any function integrable between a and b. Thus $V(u)$ is defined so that, when $u = Y$, the solution of Eq. (4.5.19), then $V(u)$ will have the correct value of the integral $\int M Y \, dx$.

This does not seem to be a very productive equation; a functional $V(u)$ is equal to a set of integrals involving the undetermined function u. Of course, if we knew the exact form of $Y(x)$ between a and b, we could use the equation to calculate $V(Y)$, but we know that already; also, Eq. (4.5.21) is an easier formula to use to compute $V(Y)$ (if we know Y). However, the right-hand side of Eq. (4.5.22) has several very important properties. Its first term is linear in the function u, and the second term, quadratic in u, is also symmetric in the u's. In other words, if we had multiplied Eq. (4.5.19) by $v(x)$ and integrated, the integral $\int v \, dx \int K(x \mid x_0) u \, dx_0$ would remain the same if u and v were interchanged (this is because K is symmetric in x and x_0). Because of this symmetry, Eq. (4.5.22) can be made the basis of a variational principle, similar to Hamilton's principle.

To see what is meant by this statement, we note that the value of $V(u)$ is changed if the shape of $u(x)$ is changed. A small change, say, changing u to $u + \lambda v(x)$, where λ is a small quantity and v an integrable function, will usually produce a change in V proportional to λ. What we propose to show is that if, and only if, $u(x)$ is equal to $Y(x)$, the solution of Eq. (4.5.19), then the change in V is not proportional to λ, but is proportional to λ^2. Variation in V produced by small changes in shape of $u + \lambda v$ around the function $Y(x)$ is *stationary* to this variation, to first order in λ, *if* u is Y. We cannot say that V is necessarily maximum or minimum for this choice of shape of u; all we can say is that, if V is considered as a function of the shape-varying parameter λ, then $dV/d\lambda$ will be zero when $\lambda = 0$ if $u = Y$.

To prove these statements, suppose we set $u(x) = y(x) + \lambda v(x)$. Then

$$V(u) = V(y) + 2\lambda \int_a^b v(x)\, dx \left[M(x) - \int_a^b K(x \mid x_0) y(x_0)\, dx_0 \right]$$

$$- \lambda^2 \int_a^b v(x)\, dx \int_a^b K(x \mid x_0) v(x_0)\, dx_0$$

In order that $dV/d\lambda$ be zero at $\lambda = 0$, the quantity in square brackets must be zero; in other words, y must be the solution of Eq. (4.5.19). In many cases, of course, it is difficult to find the exact form of solution $Y(x)$; indeed, if we could find it easily, we should not need the assistance of functional $V(u)$. Often we are reduced to devising a form for u, containing parameters, which is too simple to be the exact solution for any set of values of the parameters; we should then like to know for what values of the parameters u comes as close as possible to Y. And more important, we want to know how close to $V(Y)$ is the corresponding value of $V(u)$ for these "best values" of the parameters. Equation (4.5.22) is eminently successful in attaining these more modest aims.

Suppose we choose a form for the trial function u, equipped with parameter λ, which we fear will never exactly equal Y for any value of λ but which we hope will not differ much from Y for some value of λ. In other words, we hope that u will have the form $u(\lambda,x) = Y(x) + \epsilon w(x) + \lambda v(x)$, where we do not know the exact form of function w but we hope that constant ϵ is small. Inserting this form into the integrals for $V(u)$, we have

$$V(u) = V(Y) - \epsilon^2 \int w \int Kw - 2\lambda\epsilon \int v \int Kw - \lambda^2 \int v \int Kv$$

where the meaning of the shorthand symbols for the integrals should be obvious. If we knew the shape of w and the magnitude of ϵ, constituting the difference between $u(0,x)$ and Y, we could obtain the exact solution. But we do not know $\epsilon w(x)$; so all we can do is to vary the explicit parameter λ until $dV/d\lambda = 0$, to obtain a "best" value for λ and a corresponding "best" shape for u, as well as the stationary value for V.

$$\lambda_s = -\epsilon \frac{\int w \int Kv}{\int v \int Kv} \qquad u_s = Y(x) + \epsilon \left[w(x) - v(x) \frac{\int w \int Kv}{\int v \int Kv} \right]$$

$$V(u_s) = V(Y) - \epsilon^2 \left[\int w \int Kw - \frac{\left(\int w \int Kv \right)^2}{\int v \int Kv} \right]$$

Note that $V(u_s)$ may be maximum or minimum for V (or neither), depending on the relationship between w, v, and K.

We see that, although u_s differs from the correct solution Y by a term of the first order in ϵ, it actually is closer to Y than is $Y + \epsilon w$, and it may be quite close if v happens to have a shape like w in the region where K is large. We also see that the stationary value of the variational expression $V(u_s)$ is very much closer to the correct value $V(Y)$ if ϵ is small than is the value $V(Y + \epsilon w)$, which one would obtain by arbitrarily setting $\lambda = 0$. It thus is obvious that the success of the variational procedure is critically dependent on the user's physical intuition regarding the shape of the unknown function $Y(x)$ in the range of x from a to b, and particularly in the range where K and M are large. If one can devise a parametrized form $u(\lambda,x)$ which is fairly close in form to Y, then setting $dV/d\lambda$ equal to zero will produce a good approximation for u and a very good (second-order) value for V.

Finally, we should note that the variational calculation has concerned itself entirely with the shape of Y in the range $a < x < b$; once this is known, Eq. (4.5.19) can usually be extended to enable us to compute Y elsewhere [as will be shown for Eq. (4.5.13)].

Solutions and adjoint solutions

The procedure just described must be generalized somewhat to handle Eq. (4.5.13), since both terms on the right are complex quantities. To symmetrize the physical situation we should also consider the case when a wave is incident on the inhomogeneity, from $-\frac{1}{2}a$ to $+\frac{1}{2}a$, from the right, with a reflected wave going back to $x \to \infty$ and a transmitted wave going to $-\infty$. The solution corresponding to this inverse situation will be written $Y^-(x)$ and will be called the *adjoint* solution. The equation it must satisfy is like Eq. (4.5.13) but with a different inhomogeneous term,

$$Y^-(x) = Ae^{-i\omega x/c} - \frac{1}{2} \int_{-\frac{1}{2}a}^{\frac{1}{2}a} e^{i\omega |x-x_0|/c} \zeta(x_0) Y^-(x_0) \, dx_0 \qquad (4.5.23)$$

with reflection and transmission factors $\Phi^-(\pm\omega)$ similar in form to those of Eq. (4.5.14).

The variational form for both these functions can be found by multiplying Eq. (4.5.13) by $\zeta(x) Y^-(x)$ and integrating; by adding the equation for $2A^2\Phi(-\omega)$ to this, we obtain the analogue to Eq. (4.5.22).

$$V(u,u^-) = A\int_{-\frac{1}{2}a}^{\frac{1}{2}a} \zeta(x)u(x)e^{-i\omega x/c}\,dx + A\int_{-\frac{1}{2}a}^{\frac{1}{2}a} \zeta(x)u^-(x)e^{i\omega x/c}\,dx$$

$$-\tfrac{1}{2}\int_{-\frac{1}{2}a}^{\frac{1}{2}a} \zeta(x)u^-(x)\,dx\int_{-\frac{1}{2}a}^{\frac{1}{2}a} e^{i\omega\,|x-x_0|/c}\zeta(x_0)u(x_0)\,dx_0$$

$$-\int_{-\frac{1}{2}a}^{\frac{1}{2}a} u^-(x)\zeta(x)u(x)\,dx \quad (4.5.24)$$

If we regard u and u^- as separate but related functions, we see that a requirement that the change in V, because of a variation in u^-, be stationary (i.e., be second order in the variation) is equivalent to requiring that u satisfy Eq. (4.5.13); whereas the requirement that V be stationary to variation of the form of u is equivalent to stating that u^- be a solution of Eq. (4.5.23). If we set $u(\lambda,x)$ as a trial function with a shape we hope is close to that of Y, and also devise an adjoint $u^-(\lambda,x)$ for every value of λ, then the value λ_s, for which $dV/d\lambda = 0$, is the "best" value of λ, the corresponding value of V should be quite close to the exact value $2A^2\Phi(-\omega)$, and the corresponding functions $u(\lambda_s,x)$ and $u^-(\lambda_s,x)$ should be fairly close to the correct forms $Y(x)$ and $Y^-(x)$.

For example, we could guess that the general form of the solution in the region $-\frac{1}{2}a < x < \frac{1}{2}a$ resembles the function $\chi(x)$, but not be sure about its amplitude. In this case we could assume the trial function $u(x)$ to be $\lambda\chi(x)$, and work out that the adjoint function is $\lambda\chi^*(x)$ (because of the symmetry of the problem, u^- should be the complex conjugate of u in the range $-\frac{1}{2}a < x < +\frac{1}{2}a$), λ, the amplitude, being the variational parameter. Then

$$V = 2\lambda AC - \lambda^2(D + G)$$

where

$$D = \int_{-\frac{1}{2}a}^{\frac{1}{2}a} \chi(x)\zeta(x)\chi^*(x)\,dx$$

$$C = \tfrac{1}{2}\int_{-\frac{1}{2}a}^{\frac{1}{2}a} \zeta(x)\chi(x)e^{-i\omega x/c}\,dx + \tfrac{1}{2}\int_{-\frac{1}{2}a}^{\frac{1}{2}a} \zeta(x)\chi^*(x)e^{i\omega x/c}\,dx$$

and

$$G = \tfrac{1}{2}\int_{-\frac{1}{2}a}^{\frac{1}{2}a} \zeta\chi\,dx\int_{-\frac{1}{2}a}^{\frac{1}{2}a} e^{i\omega\,|x-x_0|/c}\zeta\chi^*\,dx_0$$

$$= \tfrac{1}{2}\int_{-\frac{1}{2}a}^{\frac{1}{2}a} \zeta\chi^*\,dx\int_{-\frac{1}{2}a}^{\frac{1}{2}a} e^{i\omega\,|x-x_0|/c}\zeta\chi\,dx_0$$

Setting $dV/d\lambda = 0$ results in a value for λ and a "best" approximation for Y,

Y^-, $\Phi(\omega)$, and $\Phi(-\omega)$ for this trial function.

$$\lambda_s = \frac{AC}{D+G} \qquad Y \simeq \lambda_s\chi \qquad Y^- \simeq \lambda_s\chi^*$$

$$\Phi(-\omega) \simeq \frac{1}{2A^2} V_s = \frac{C^2}{2(D+G)} \tag{4.5.25}$$

$$\Phi(\omega) = \frac{1}{2A}\int_{-\frac{1}{2}a}^{\frac{1}{2}a} e^{i\omega x/c}\zeta Y\,dx \simeq \frac{CE}{2(D+G)} \qquad E = \int_{-\frac{1}{2}a}^{\frac{1}{2}a}\zeta\chi e^{i\omega x/c}\,dx$$

where we have inserted the "best" form for Y in Eq. (4.5.14) to obtain $\Phi(\omega)$. Since $\lambda_s\chi$ differs by first order in ϵ from Y (see page 157), this expression for $\Phi(\omega)$ is not as accurate as the one for $\Phi(-\omega)$, which is the stationary value for V, and thus is good to second order in ϵ. A value of $\Phi(\omega)$ of the same accuracy as that for $\Phi(-\omega)$ can be obtained by multiplying Eq. (4.5.13) by $(1/2A^2)\zeta Y$ and integrating.

$$\Phi(\omega) = \frac{1}{2A^2}\int_{-\frac{1}{2}a}^{\frac{1}{2}a} Y^2\zeta\,dx + \frac{1}{4A^2}\int_{-\frac{1}{2}a}^{\frac{1}{2}a} Y\zeta\,dx\int_{-\frac{1}{2}a}^{\frac{1}{2}a} e^{i\omega|x-x_0|/c}\zeta Y\,dx_0$$

Inserting $\lambda_s\chi$ for Y in this, we obtain

$$\Phi(\omega) \simeq \frac{C^2}{2(D+G)^2}\left(\int\chi^2\zeta\,dx + \tfrac{1}{2}\int\zeta\chi\,dx\int e^{i\omega|x-x_0|/c}\chi\zeta\,dx_0\right) \tag{4.5.26}$$

which is of the same order of approximation as the formula for V.

More complex trial functions, with more than one parameter, can be used, of course. One could try $u = \lambda\chi(x) + \mu\psi(x)$, or have another parameter implicit in χ. The "best" values are then determined by requiring that $\partial V/\partial\lambda = 0$ and $\partial V/\partial\mu = 0$ and solving the resulting simultaneous equations.

An example

To indicate the range of validity of the various approximate formulas we have presented here, we should consider a case which can be solved exactly, so as to compare exact with various approximate results. Suppose the string, which extends from $-\infty$ to $+\infty$, is covered with a uniform load having mechanical impedance $z(x) = \epsilon c\zeta$ per unit length in the range $-\frac{1}{2}a < x < +\frac{1}{2}a$, and zero elsewhere. The equation of motion (4.5.11) of the string is thus

$$\frac{d^2Y}{dx^2} + \left(\frac{\omega}{c}\right)^2 Y = \begin{cases} 0 & x < -\frac{1}{2}a \\[2mm] -i\dfrac{\omega}{c}\,\zeta Y & -\frac{1}{2}a < x < +\frac{1}{2}a \\[2mm] 0 & x > +\frac{1}{2}a \end{cases}$$

which can be solved exactly by joining solutions, valid for each region, at the boundaries $x = -\frac{1}{2}a$ and $x = +\frac{1}{2}a$. For a solution representing an incident wave coming from $-\infty$, being reflected in part and in part being transmitted through the load, on to $+\infty$, the three solutions must have the following form:

$$y(x) = \begin{cases} Ae^{i\omega x/c} + Re^{-i\omega x/c} & x < -\tfrac{1}{2}a \\ Be^{i\beta\omega x/c} + Ce^{-i\beta\omega x/c} & -\tfrac{1}{2}a < x < +\tfrac{1}{2}a \\ Pe^{i\omega x/c} & x > +\tfrac{1}{2}a \end{cases}$$

where $\beta^2 = 1 + i(c\zeta/\omega)$. Since there are no concentrated loads anywhere, both the string displacement and its slope must be continuous everywhere. Thus the three solutions must match values and slopes at their joining points, which suffices to determine R, B, C, and P in terms of A, the incident amplitude.

$$R = -2\zeta a \frac{A}{Q} \frac{\sin(\beta\omega a/c)}{\omega a/c} e^{-i\omega a/c} = -A\Phi(\omega)$$

$$P = 4\beta \frac{A}{Q} e^{-i\omega a/c} = A[1 - \Phi(-\omega)] \qquad \beta^2 = 1 + i\frac{c\zeta}{\omega} \qquad (4.5.27)$$

$$B = 2(\beta + 1)\frac{A}{Q} \qquad C = 2(\beta - 1)\frac{A}{Q}$$

$$Q = (\beta + 1)^2 e^{-i\beta\omega a/c} - (\beta - 1)^2 e^{i\beta\omega a/c}$$

For purposes of comparison with various approximate solutions, we write the expansion, in powers of ζ, to second order, of the expressions for $\Phi(\omega)$ and $\Phi(-\omega)$ and their limiting values, as $\omega a/c$ goes to zero.

$$\Phi(\omega) = -\frac{R}{A} \simeq \frac{c}{2\omega} \frac{\sin(\omega a/c) + \tfrac{1}{2}i\zeta a \cos(\omega a/c)}{1 + \tfrac{1}{2}\zeta a + \tfrac{1}{2}i(\zeta c/\omega)} \rightarrow \frac{\zeta a}{2 + \zeta a} \qquad \frac{\omega a}{c} \rightarrow 0$$

$$(4.5.28)$$

$$\Phi(-\omega) = 1 - \frac{P}{A} \simeq \tfrac{1}{2}\zeta a \frac{1 + \tfrac{1}{4}\zeta a + \tfrac{1}{4}i(\zeta c/\omega) + \tfrac{1}{8}(\zeta c^2/a\omega^2)(e^{2i\omega a/c} - 1)}{1 + \tfrac{1}{2}\zeta a + \tfrac{1}{2}i(\zeta c/\omega)}$$

$$\rightarrow \frac{\zeta a}{2 + \zeta a} \qquad\qquad \frac{\omega a}{c} \rightarrow 0$$

The ratios $|R/A|^2 = |\Phi(\omega)|^2$ and $|P/A|^2 = |1 - \Phi(-\omega)|^2$, the ratios between the power reflected and transmitted by the load to the incident power, may be called the reflection and transmission ratios, respectively. If $z = \epsilon c\zeta$ is purely reactive, the sum of the ratios will equal unity; no energy is lost to the load in that case.

Both ratios oscillate as $\beta\omega a/c$ increases, because of interference between reflections at the discontinuities at $-\tfrac{1}{2}a$ and $+\tfrac{1}{2}a$. If z has a resistive part, then β is complex and the sum of the two ratios will be less than unity. The solid lines of Fig. 4.20 are plots of the reflection ratio for the simple case of pure mass loading, $z = -i\omega\mu\epsilon$, $\zeta = -i\mu(\omega/c)$, the mass density of the string being $(1 + \mu)\epsilon$ in the range $-\tfrac{1}{2}a < x < +\tfrac{1}{2}a$ and ϵ elsewhere. One curve shows the dependence of the reflection ratio on μ, the mass parameter; the other curve displays its dependence on the frequency parameter $\eta = \omega a/c = 2\pi a/\lambda$, where λ is the wavelength of the waves in the unloaded part of the string. This curve shows the effect of interference by dropping to zero at $\beta \omega a/c = \pi$, or $\eta\sqrt{1 + \mu} = \pi$.

This exact solution is now to be compared with the results of various approximate solutions. The Born approximation, for example, consists in inserting the incident wave $Ae^{i\omega x/c}$ into the integral of Eq. (4.5.13) and thus

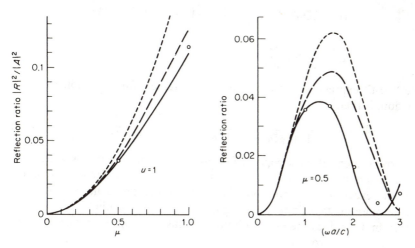

FIGURE 4.20
Ratio of power reflected to incident power for wave incident on an added mass load of density $\mu\epsilon$ over length a of string, as function of μ for $\eta = 1$ and of $\eta = 2\pi a/\lambda$ for $\mu = 0.5$. Solid lines are exact solution; dashed lines are first-order variational solution of Eq. (4.5.31); dotted lines are Born approximation; and circles are values from Eq. (4.5.32).

into the integral of Eq. (4.5.14) for the reflection factor Φ. For the problem under discussion, the approximations for these factors are

$$\Phi(\omega) \simeq \tfrac{1}{2}\zeta \int_{-\frac{1}{2}a}^{\frac{1}{2}a} e^{2i\omega x/c}\, dx = \tfrac{1}{2}\zeta\, \frac{c}{\omega}\sin\frac{\omega a}{c}$$

$$\Phi(-\omega) \simeq \tfrac{1}{2}\zeta \int_{-\frac{1}{2}a}^{\frac{1}{2}a} dx = \tfrac{1}{2}\zeta a \qquad \text{(Born approximation)}$$

(4.5.29)

Comparison with Eqs. (4.5.27) indicates that the Born approximation is a first-order approximation in the quantities ζa and $\zeta c/\omega$. The dotted lines in Fig. 4.20 plot this approximation to $|\Phi(\omega)|^2$ for $\zeta = -i\mu(\omega/c)$, as function of μ and of $\eta = \omega a/c$. It turns out to be a fairly good approximation as long as the mass ratio μ is less than about 0.4 and the frequency parameter u is less than about 0.6. The interference minimum comes at $\eta = \pi$ instead of $\beta\eta = \pi$ because the approximation neglects the change of wavelength in the loaded portion of the string.

A better approximation can be obtained by using the variational technique just outlined. If we assume a trial function $\chi = e^{i\omega x/c}$, which is again the unperturbed incident wave, but with an amplitude λ adjusted for stationarity of V, we obtain, from Eqs. (4.5.25),

$$\chi = e^{i\omega x/c} \qquad \chi^* = e^{-i\omega x/c} \qquad C = D = \zeta a$$

$$G = \tfrac{1}{2}\zeta^2 a^2 \left[1 + i\left(\frac{c}{\omega a}\right) - \frac{1}{2}\left(\frac{c}{\omega a}\right)^2 (e^{2i\omega x/c} - 1) \right]$$

$$E = \frac{\zeta c}{\omega} \sin \frac{\omega a}{c}$$

The "best" expression for the transmission factor, for this choice of trial function, is then

$$\Phi(-\omega) \simeq \frac{\tfrac{1}{2}\zeta a}{1 + \tfrac{1}{4}\zeta a + \tfrac{1}{4}i(\zeta c/\omega) - \tfrac{1}{8}(\zeta c^2/a\omega^2)(e^{2i\omega a/c} - 1)} \qquad (4.5.30)$$

A rearrangement of the second of Eqs. (4.5.28) so that the power series is all in the denominator shows that the variational expression for $\Phi(-\omega)$ agrees with the exact solution through the second-order terms in ζa and $\zeta c/\omega$. Furthermore, it stays finite as $\omega a/c$ goes to zero, approaching $\tfrac{1}{2}\zeta a/(1 + \tfrac{1}{2}\zeta a)$ in this limit, as does the exact solution. Thus this approximation is better than the Born approximation.

Formula (4.5.25) for the reflection factor $\Phi(\omega)$ gives

$$\Phi(\omega) \simeq \frac{(\zeta c/2\omega) \sin (\omega a/c)}{1 + \tfrac{1}{4}\zeta a + \tfrac{1}{4}i(\zeta c/\omega) - \tfrac{1}{8}(\zeta c^2/a\omega^2)(e^{2i\omega a/c} - 1)} \qquad (4.5.31)$$

The square of this is plotted as a dashed line in Fig. 4.20. As mentioned earlier, it does not check with the exact formula to second order; in particular, the interference minimum comes at $\eta = \pi$ instead of $\beta\eta = \pi$. Nevertheless, this formula fits the exact solution better than the Born approximation, and is satisfactory for η less than about 1 and μ less than about 0.8.

The more appropriate expression for the reflection factor is, of course, Eq. (4.5.26), which uses the integral equation to improve the shape of the strong displacement in the loaded region, before inserting it into the integral for $\Phi(\omega)$. In the present case we have

$$\Phi(\omega) \simeq \frac{\zeta c}{2\omega} \frac{\sin (\omega a/c) + \tfrac{1}{2}i\zeta a e^{-i\omega a/c}}{[1 + \tfrac{1}{4}\zeta a + \tfrac{1}{4}i(\zeta c/\omega) - \tfrac{1}{8}(\zeta c^2/a\omega^2)(e^{2i\omega a/c} - 1)]^2} \qquad (4.5.32)$$

which does check with Eq. (4.5.28) to second order. The circles in Fig. 4.20 correspond to this formula. The fit is excellent for μ and η less than 2. The formula greatly improves the interference effect; it has its minimum near $\beta\eta = \pi$, though the minimum is not zero.

Of course, if we had chosen better trial functions χ and χ^* to insert in Eqs. (4.5.25) and (4.5.26), for example, if we had let $\chi = e^{i\beta\omega x/c}$ and $\chi^* = e^{-i\beta\omega x/c}$, to allow for the difference in wavelength between the loaded and unloaded parts of the string, the resulting formulas would have produced curves which would be impossible to distinguish from the solid lines of Fig. 4.20, over the range there plotted. And of course, if we set $u = \lambda e^{i\beta\omega x/c} + \sigma e^{-i\beta\omega x/c}$ and $u^- = \lambda e^{-i\beta\omega x/c} + \sigma e^{i\beta\omega x/c}$ in Eq. (4.5.24) and had varied both λ and σ to find the stationary value of V, the resulting "best" form will be the exact solution.

Distortion of standing waves

When an extra load is attached to a string of finite length, stretched between rigid supports, the load changes the frequencies of free vibration of the string and also changes the shape of the standing waves of free vibration. If the string supports are at $x = 0$ and $x = l$ and the load is an impedance Z, attached at the point x_0, then the string, when it is oscillating at frequency $\omega/2\pi$, experiences a reaction force

$$-Z(\partial y/\partial t)_{x=x_0}\,\delta(x - x_0) = i\omega Z Y(x)e^{-i\omega t}\,\delta(x - x_0)$$

where $y(x,t) = Y(x)e^{-i\omega t}$ for simple-harmonic motion. Reference to Eq· (5.1.14) indicates that the shape of the string will be

$$Y(x) = \frac{iZ}{\epsilon c}\frac{Y(x_0)}{\sin(\omega l/c)}G(x \mid x_0)$$

where

$$G(x \mid x_0) = \begin{cases} \sin\dfrac{\omega x}{c}\sin\left[\dfrac{\omega}{c}(l - x_0)\right] & 0 < x < x_0 < l \\[3mm] \sin\dfrac{\omega x_0}{c}\sin\left[\dfrac{\omega}{c}(l - x)\right] & 0 < x_0 < x < l \end{cases} \tag{4.5.33}$$

This is, of course, an equation from which both shape and allowed frequencies can be determined. Setting $x = x_0$ provides the equation for the resonance frequencies.

$$\sin\frac{\omega l}{c} = \frac{iZ}{\epsilon c}\sin\frac{\omega}{c}x_0\sin\frac{\omega}{c}(l - x_0)$$

or

$$\tan\frac{\omega l}{c} = \frac{-(iZ/\epsilon c)\sin^2(\omega x_0/c)}{1 - (iZ/2\epsilon c)\sin(2\omega x_0/c)} \tag{4.5.34}$$

This can usually be solved by successive approximations, inserting a trial value of ω in the right-hand side and computing a better value by taking the inverse tangent of the result. If the load is a mass m, so that $Z = -i\omega m$, and if $m > \epsilon l$, then the lowest mode of oscillation has a natural frequency considerably lower than that of the fundamental, $\omega_1 = \pi c/l$, of the unloaded string.   Here we can assume that $\omega l/c \ll 1$, and the equation becomes, to the first approximation in the small quantity $\omega l/c$,

$$\frac{\omega l}{c} \simeq \frac{-(\omega m/\epsilon c)(\omega x_0/c)^2}{1 - (\omega m/2\epsilon c)(2\omega x_0/c)} \qquad \text{or} \qquad 1 \simeq \frac{m}{\epsilon}\left(\frac{\omega}{c}\right)^2 x_0\left(1 - \frac{x_0}{l}\right)$$

or

$$\omega_0 \simeq \sqrt{\frac{T}{mx_0}\left(1 - \frac{x_0}{l}\right)^{-1}} \qquad \text{where} \qquad T = \epsilon c^2$$

This would be an exact expression for a mass load on a string of negligible mass, as a simple analysis will demonstrate.   To this approximation the shape of the string is

$$Y(x) \simeq \begin{cases} A\dfrac{x}{x_0} & 0 < x_0 < x < l \\[2mm] A\dfrac{l-x}{l-x_0} & 0 < x < x_0 < l \end{cases}$$

A better approximation can be found by inserting the expression for ω_0 into the right-hand side of (4.5.34) and then finding the value of ω such that $\tan(\omega l/c)$ equals this right-hand side.

As a second example, we consider the case where the load is resistive, $Z = R = \epsilon c \rho$, where $\rho \ll 1$.   Since the right-hand side of Eq. (4.5.34) is small in this case, the quantity $\omega l/c$ must be close to an integral multiple of π for $\tan(\omega l/c)$ to be small.   We let $\omega l/c = \pi n + u_n$, where $u_n \ll 1$.   Then, to the second order in the small quantities u_n and ρ, we have

$$u_n \simeq \frac{-i\rho[\sin^2(\pi n x_0/l) + 2(u_n x_0/l)\cos(\pi n x_0/l)\sin(\pi n x_0/l)]}{1 - i\rho\sin(\pi n x_0/l)\cos(\pi n x_0/l)}$$

Setting the first-order solution $u_n \simeq -i\rho\sin^2(\pi n x_0/l)$ into the second term in the numerator, we eventually obtain, to second order,

$$\omega_n \simeq \frac{\pi n c}{l} - \rho^2\left(\frac{c}{l}\right)\left(2\frac{x_0}{l} - 1\right)\sin^3\frac{\pi n x_0}{l}\cos\left(\frac{\pi n x_0}{l}\right) - i\rho\left(\frac{c}{l}\right)\sin^2\frac{\pi n x_0}{l}$$

$$(4.5.35)$$

for the allowed values of angular frequency ω_n.   Once ω_n is determined, the corresponding shape of the string is obtained from Eq. (4.5.33), with the value of $Y(x_0)$ taking the part of an amplitude factor.   We note that the real part of ω is affected by the load (though only to the second order in ρ) as well as the imaginary part, the damping constant of the free vibration.   Whenever the load is attached at a node of the standing wave (i.e., whenever $\pi n x_0/l$ is an integer times π), the load has no effect on the motion.

Effect of a distributed load

If the inhomogeneity is distributed along a length of the finite string so that the equation of motion is given by (4.5.11), this equation can be changed into an integral equation, as was done for the infinite string. The integral equation for the motion of the finite string differs from Eq. (4.5.13) for the infinite string, however. Here there is no incident wave; so the integral equation is homogeneous; also, the Green's function (4.4.14) for a finite string between rigid supports differs from the Green's function for an infinite string.

We use the method outlined between Eqs. (4.5.11) and (4.5.13) to obtain the equation. The shape of the string, which is to be determined, is a solution of Eq. (4.5.11), going to zero at $x = 0$ and $x = l$. The solution of the auxiliary equation (4.4.9) which satisfies the same boundary conditions is the Green's function of Eq. (4.4.14). As before, we multiply Eq. (4.5.11) by g and subtract from it Eq. (4.4.9) multiplied by Y, interchange x and x_0, and integrate over x_0 from 0 to l. In this case the terms $Y(dg/dx_0) - g(dY/dx_0)$ go to zero at 0 and l; so the result is simply

$$Y(x) = \frac{i\omega}{c} \int_0^l \zeta(x_0)g(x \mid x_0) Y(x_0)\, dx_0$$

$$= \frac{i}{\sin(\omega l/c)} \int_0^l \zeta(x_0) Y(x_0) G(x \mid x_0)\, dx_0 \tag{4.5.36}$$

where

$$G(x \mid x_0) = \begin{cases} \sin\dfrac{\omega x}{c} \sin\left[\dfrac{\omega}{c}(l - x_0)\right] & 0 < x < x_0 < l \\[2ex] \sin\dfrac{\omega x_0}{c} \sin\left[\dfrac{\omega}{c}(l - x)\right] & 0 < x_0 < x < l \end{cases}$$

and where $\zeta(x)$ is the impedance density ratio $z(x)/\epsilon c$.

This equation is quite different in form from Eq. (4.5.13), yet both equations are for a solution of the same differential equation (4.5.11); the difference in the two lies in the boundary conditions, which are included in the integral equation. Solutions of (4.5.36) occur only for a discrete set of values of $\omega l/c$, which are $2\pi l/c$ times the allowed frequencies of oscillation of string-plus-load. Solutions may be obtained by successive approximations or a variational procedure, as before.

For example, we can multiply the second form of Eq. (4.5.36) by $Y(x)\zeta(x)$ and integrate over x, obtaining a symmetric expression for the resonance factor $\sin(\omega l/c)$.

$$\sin\frac{\omega l}{c} = i\frac{\displaystyle\int_0^l Y(x)\zeta(x)\, dx \int_0^l G(x,x_0)\zeta(x_0) Y(x_0)\, dx_0}{\displaystyle\int_0^l Y(x)\zeta(x) Y(x)\, dx} \tag{4.5.37}$$

which is appropriate for a variational calculation of ω. Just as before, we can show that, if an approximate form is used in the integrals for $Y(x)$, the computed value of the fraction is a better approximation to the correct value of $\sin(\omega l/c)$ than is the approximation for Y. Also, if the form for Y has parameters $\alpha, \ldots$ which change its form, then the form, for which the partials of the fraction with respect to $\alpha, \ldots$ are all zero, comes closest to the correct form of Y, and the corresponding value of the fraction is closest to the correct value of $\sin(\omega l/c)$.

To show how the technique works, we take the case where

$$\zeta(x) = \begin{cases} 0 & 0 < x < a \\ \dfrac{Z}{\epsilon c(b - a)} & a < x < b \\ 0 & b < x < l \end{cases}$$

and use the crudest approximation possible for the shape of Y between a and b, namely, $Y = A$. The denominator of the fraction is then, simply, $A^2(Z/\epsilon c)$, and Eq. (4.5.37) becomes

$$\sin\frac{\omega l}{c} \simeq -\frac{iZ}{\epsilon c}\frac{c}{\omega(b-a)}\left\{\sin\left(\frac{\omega l}{c}\right)\left[1 - \frac{c}{\omega(b-a)}\left(2\cos\frac{\omega a}{c}\sin\frac{\omega b}{c}\right.\right.\right.$$
$$\left.\left.- \cos\frac{\omega a}{c}\sin\frac{\omega a}{c} - \cos\frac{\omega b}{c}\sin\frac{\omega b}{c}\right)\right]$$
$$\left.- \cos\left(\frac{\omega l}{c}\right)\frac{c}{\omega(b-a)}\left(\cos\frac{\omega a}{c} - \cos\frac{\omega b}{c}\right)^2\right\}$$

or

$$\tan\frac{\omega l}{c}$$
$$\simeq \frac{-\left(\dfrac{iZ}{\epsilon c}\right)\left[\dfrac{c}{\omega(b-a)}\right]^2\left[2\sin\dfrac{\omega}{2c}(a+b)\sin\dfrac{\omega}{2c}(b-a)\right]^2}{1 - \left(\dfrac{iZ}{\epsilon c}\right)\dfrac{c}{\omega(b-a)}\left\{\sin\dfrac{\omega}{c}(b-a) + \sin\dfrac{\omega}{c}(b+a)\left[1 - \cos\dfrac{\omega}{c}(b-a)\right]\right\}}$$

$$(4.5.38)$$

When $b - a \to 0$ and $b + a \to 2x_0$, this formula reduces to formula (4.5.34), the exact expression for a concentrated load at x_0. When $b - a$ is not small, the formula is only an approximate one. A better approximation for the nth natural frequency could be obtained by setting $Y = A\sin(\pi n x/l)$ in Eq. (4.5.37).

Fourier series expansions

An alternative way of calculating the effect of an impedance load $z(x) = \epsilon c\zeta(x)$ on the free vibrations of a finite string is to expand Eq. (4.5.36) in a Fourier series. To do this we must find an expansion of the Green's function $g(x \mid x_0)$ of Eq. (4.4.14) in a Fourier series. But g satisfies the equation of motion (4.4.9),

$$\frac{d^2g}{dx^2} + \left(\frac{\omega}{c}\right)^2 g = -\delta(x - x_0)$$

The boundary conditions on g (and on Y) are that they go to zero at $x = 0$ and $x = l$. According to the arguments leading up to Eq. (4.3.13), it must be expressible in terms of a Fourier series,

$$g(x \mid x_0) = \sum_{n=0}^{\infty} B_n \sin \frac{\pi n x}{l}$$

Inserting this in the equation for g, we have

$$\sum_{m=0}^{\infty} B_m \left[\left(\frac{\omega}{c}\right)^2 - \left(\frac{\pi m}{l}\right)^2 \right] \sin \frac{\pi m x}{l} = -\delta(x - x_0)$$

Multiplying both sides by $\sin(\pi n x/l)$ and integrating, utilizing Eqs. (4.3.11) and (1.3.24), we find that

$$\frac{l}{2} \left[\left(\frac{\pi n}{l}\right)^2 - \left(\frac{\omega}{c}\right)^2 \right] B_n = \sin \frac{\pi n x_0}{l}$$

so that the Green's function of Eq. (4.4.14) equals the Fourier series.

$$G(x \mid x_0) = 2l \sum_{n=0}^{\infty} \frac{\sin(\pi n x_0/l) \sin(\pi n x/l)}{(\pi n)^2 - (\omega l/c)^2}$$

This form again demonstrates the fact that G is symmetric to interchange of x and x_0, thus satisfying the principle of reciprocity.

The series can be inserted in the integral equation (4.5.36), giving

$$Y(x) = 2i \frac{\omega l}{c} \sum_{n=0}^{\infty} \frac{\sin(\pi n x/l)}{(\pi n)^2 - (\omega l/c)^2} \int_0^l \zeta(x) Y(x) \sin \frac{\pi n x}{l} \, dx \qquad (4.5.39)$$

This equation can be modified to obtain the allowed frequencies of free vibration, and also to obtain the series expansion for the shapes of the corresponding standing waves of the loaded string.

For example, if $\zeta l \ll 1$, the load cannot distort the shape very much; so the allowed values of ω must be close to the values $\pi n c/l$ which hold for an unloaded string. Therefore, for the mth allowed frequency, $\omega l/c$ must be close to πm in value, and thus the mth term in the series for Y must be very much larger than any of the others. Consequently, $Y_m(x)$ must be approximately equal to $A \sin(\pi m x/l)$, and Eq. (4.5.39) is approximately equivalent to

$$A \sin \frac{\pi m x}{l} \simeq 2i \frac{\omega l}{c} \frac{\sin(\pi m x/l)}{(\pi m)^2 - (\omega l/c)^2} \int_0^l \zeta(x) A \sin^2 \frac{\pi m x}{l} \, dx$$

which can be rearranged into an equation for the value $\omega_m/2\pi$ of the mth natural frequency, with

$$\omega_m{}^2 \simeq \left(\frac{\pi mc}{l}\right)^2 - 2i\frac{\omega_m c}{l}\,D_{mm}$$

or

$$\frac{\omega_m l}{c} \simeq \pi m - iD_{mm} - \frac{(D_{mm})^2}{2\pi m} - \cdots$$

where

$$D_{nm} = \int_0^l \sin\left(\frac{\pi nx}{l}\right) \zeta(x) \sin\frac{\pi mx}{l}\,dx \tag{4.5.40}$$

With these first approximations, we can go on to higher approximations. Inserting $Y \simeq A \sin(\pi mx/l)$ and (4.5.40) into the right-hand side of (4.5.39), we obtain

$$Y_m(x) \simeq A\left[\sin\frac{\pi mx}{l} + 2i\pi m \sum_{n\neq m} \frac{\sin(\pi nx/l)}{(\pi n)^2 - (\pi m)^2}\,D_{nm}\right]$$

Inserting this for $Y_m(x)$ on the left-hand side of (4.5.39) and $Y_m(x) = A \sin(\pi mx/l)$ in all but the $n = m$ term in the right-hand series, we obtain

$$A \sin\frac{\pi mx}{l} \simeq 2i\frac{\omega l}{c}\,\frac{\sin(\pi mx/l)}{(\pi m)^2 - (\omega l/c)^2}\int_0^l \sin\frac{\pi mx}{l}\,\zeta(x)\,Y_m(x)\,dx$$

which allows us to calculate the next approximation for ω_n.

$$\left(\frac{\omega_n l}{c}\right)^2 \simeq (\pi n)^2 - 2i\frac{\omega_n l}{c}\left[D_{nn} + 2i(\pi n)\sum_{m\neq n}\frac{D_{nm}D_{mn}}{(\pi m)^2 - (\pi n)^2}\right]$$

or

$$\frac{\omega_n l}{c} \simeq \pi n - iD_{nn} - \frac{1}{2\pi n}(D_{nn})^2 - 2\pi n\sum_{m\neq n}\frac{D_{mn}D_{nm}}{(\pi n)^2 - (\pi m)^2} \tag{4.5.41}$$

Returning again to the load $\zeta(x)$, which is equal to $Z/\epsilon c(b - a)$ between $x = a$ and $x = b$ and is zero elsewhere, the matrix components are

$$D_{nn} = \frac{Z}{2\epsilon c}\left[1 - \frac{l}{b - a}\frac{\sin(2n\beta) - \sin(2n\alpha)}{2\pi n}\right] \qquad \alpha = \frac{\pi a}{l};\quad \beta = \frac{\pi b}{l}$$

$$D_{mn} = \frac{Z}{\epsilon c}\frac{l}{b - a}\left[\frac{\sin\beta(m - n) - \sin\alpha(m - n)}{2\pi(m - n)}\right.$$
$$\left. - \frac{\sin\beta(m + n) - \sin\alpha(m + n)}{2\pi(m + n)}\right]$$

so that, to the first approximation,

$$\frac{\omega_m l}{c} \simeq \pi m - \frac{iZ}{2\epsilon c}\left[1 - \frac{l}{b - a}\frac{\sin(2m\beta) - \sin(2m\alpha)}{2\pi m}\right]$$

$$\rightarrow \pi m - \frac{iZ}{\epsilon c}\sin^2\frac{\pi ma}{l} \qquad b - a \ll l$$

The limiting form for $b - a \ll l$ agrees with the first-order-approximation solution of Eq. (4.5.34).   If the load is masslike ($Z = -i\omega m$), each natural frequency is reduced, and the higher allowed frequencies are no longer integral multiples of the fundamental.   Introduction of a resistance brings an imaginary part to ω_n, which corresponds to an exponential damping of the vibrations.

Problems

1 A string clamped at one end is struck at a point a distance D from the clamp by a hammer of width $D/4$.   The head of the hammer is shaped so that the initial velocity given the string is maximum at the center of the head and is zero at the edge, the initial "velocity shape" of this portion being like an inverted V.   Plot the shapes of the string at the times $t = 0$, $D/2c$, D/c, $3D/2c$, $2D/c$.   Draw a curve showing the vertical component of the force on the clamp as a function of time.

2 A harp string is plucked so that its initial velocity is zero and its initial shape is

$$
y_0(x) = \begin{cases} \dfrac{20h}{9l} x & 0 < x < \dfrac{9l}{20} \\[2mm] \dfrac{20h}{l}\left(\dfrac{l}{2} - x\right) & \dfrac{9l}{20} < x < \dfrac{11l}{20} \\[2mm] \dfrac{20h}{9l}(x - l) & \dfrac{11l}{20} < x < l \end{cases}
$$

Plot the successive shapes of the string during one cycle of the motion.   Draw a curve showing the vertical component of the force on one of the supports as a function of time.

3 What are the total energies of the two strings shown in Fig. 4.11?

4 An infinite string with tension T and wave velocity c is subjected to a transient transverse force of the form

$$
f(x,t) = \begin{cases} A \sin \dfrac{\pi t}{\tau} \cos \dfrac{\pi x}{L} & 0 < t < \tau; \; -\tfrac{1}{2}L < x < \tfrac{1}{2}L \\[2mm] 0 & \text{otherwise} \end{cases}
$$

Show that the resulting motion of the string is given by

$$
y(x,t) = A \frac{c\tau}{\pi T} \int_{-\frac{1}{2}L}^{\frac{1}{2}L} \cos \frac{\pi x_0}{L} \, Y\!\left(t - \frac{x - x_0}{c}\right) dx_0
$$

where

$$
Y(z) = \begin{cases} 0 & z < 0 \\[2mm] \sin^2 \dfrac{\pi z}{2T} & 0 < z < \tau \\[2mm] 1 & z > \tau \end{cases}
$$

5 A uniform string with no friction is stretched between rigid supports a distance l apart.   It is driven by a force $F_0 e^{-i\omega t}$ concentrated at its midpoint.   Show that the amplitude of motion of the midpoint is $(F_0/2\epsilon c\omega) \tan(\omega l/2c)$.   What is the amplitude of motion of the point $x = l/4$?

6 A uniform string of small electric resistance is stretched between rigid supports a distance l apart, in a uniform magnetic field of B webers per sq m perpendicular to the string. A current $I_0 e^{-i\omega t}$ amp is sent through the string; what is the force on the string per unit length? Show that the velocity shape of the string (assume zero friction) is

$$v = \frac{B I_0 e^{-i\omega t}}{i\omega\epsilon} \left\{ \frac{\cos\,[(\omega/c)(x - \frac{1}{2}l)]}{\cos\,(\omega l/2c)} - 1 \right\} \text{ m per sec}$$

Use the formula $E = \int Bv\,dx$ volts to compute the motional emf induced in the string by the motion.

7 A condenser is discharged through the string of Prob. 6, producing a current of $\delta(t)$ amp. Compute the subsequent shape of the string.

8 The string of Prob. 5 is acted on by a force $F(\xi)u(t)$ concentrated at the point $x = \xi$ [see Eq. (1.2.16) for a definition of u, and Eq. (2.3.10) for its use]. Show that the shape of the string after $t = 0$ is

$$y_u(\xi,x) = \frac{lF(\xi)}{\pi^2\epsilon c^2} \sum_{n=1}^{\infty} \frac{1}{n^2} \sin\frac{\pi n\xi}{l} \sin\left(\frac{\pi nx}{l}\right)\left(1 - \cos\frac{\pi nct}{l}\right)$$

From this formula compute the shape of the string when it has been subjected to the constant (independent of time) force $F(\xi)$ up to $t = 0$ and then released to vibrate freely.

9 A length D of an infinite string is in a magnetic field B, perpendicular to the string; the rest of the string is not in a field. A current $Ie^{-i\omega t}$ is sent through the string, which is conductive. What is the displacement of the string? Suppose the string is finite in length and is stretched between rigid supports a distance L apart, with the magnetic field centered along its length ($D < L$). What, then, is the displacement of the string if a current $Ie^{-i\omega t}$ is passed through it?

.0 An electrically conducting string of length l and total electric resistance R is clamped at both ends. The ends are connected electrically by a wire of negligible resistance. The string is set into oscillation transverse to a magnetic field $B(x)$, thus generating a current in the string, which in turn gives rise to a force on the string. Show that the equation of motion of the string is

$$\frac{\partial^2 y}{\partial t^2} - c^2\frac{\partial^2 y}{\partial x^2} = -\frac{B(x)}{\epsilon R}\int_0^l \frac{\partial y}{\partial t} B(x_0)\,dx_0 \qquad c^2 = \frac{T}{\epsilon}$$

When both y and B are expanded in Fourier series,

$$y = \sum_m y_m(t)\sin\frac{\pi mx}{l} \qquad B(x) = \sum_n B_n \sin\frac{\pi nx}{l}$$

show that the equations for the time-dependent amplitudes are given by the set of coupled equations

(a)
$$\frac{d^2 y_m}{dt^2} = -\left(\frac{\pi mc}{l}\right)^2 y_m - \sum_n a_{mn}\frac{dy_n}{dt}$$

where the mode-coupling coefficients are given by

$$a_{mn} = \frac{l}{2\epsilon R} B_m B_n$$

For the case where $y = Y(x)e^{-i\omega t}$, show that a variational expression

(b)
$$V(Y) = \frac{\displaystyle\int_0^l (dY/dx)^2\, dx - (i\omega/TR)\left(\int_0^l BY\, dx\right)^2}{\displaystyle\int_0^l Y^2\, dx}$$

has a stationary value equal to $(\omega/c)^2$, the shape of Y for which V is stationary being the solution of the equation of motion. For $B = B_1 \sin(\pi x/l)$, compute the frequency and damping factor for the lowest resonance, first by using equation (a), then by using equation (b) and setting $Y = \sin(\pi x/l)$; compare the results. Carry out the same calculations and comparisons, to second order in B_n, for the case when $B(x)$ is constant from $x = 0$ to $x = l$.

11 A string of mass ϵ per unit length and length L is whirled about one end, with angular velocity ω_0, so that the motion is in a plane (neglect gravity). Show that the tension in the string is $\frac{1}{2}\epsilon\omega_0^2(L^2 - x^2)$, where x is the distance from the stationary end. Show that the equation of motion of a string with variable tension $T(x)$ is

$$\frac{\partial}{\partial x}\left(T\frac{\partial y}{\partial x}\right) = \epsilon\frac{\partial^2 y}{\partial t^2}$$

and thus that the equation for simple-harmonic motion of the string perpendicular to the plane of rotation is

$$\frac{d}{dz}\left[(1 - z^2)\frac{\partial\Psi}{\partial z}\right] + 2\left(\frac{\omega}{\omega_0}\right)^2\Psi = 0 \qquad \begin{aligned} x &= lz \\ y &= \Psi(z)e^{-i\omega t} \end{aligned}$$

Show that when $\omega = \omega_0\sqrt{n(2n - 1)}$ $(n = 1, 2, 3, \ldots)$ the displacement is proportional to the Legendre function $P_{2n-1}(z)$ [see Eq. (7.2.4)] and that no other frequencies are possible. Is the general motion periodic?

12 A string of length l and mass ϵ g per cm is hung from one end, so that gravitational forces are the only ones acting. Show that if the free end of the string is taken as origin, the normal modes of vibration have the form

$$y = J_0\left(\pi\beta_n\sqrt{\frac{x}{l}}\right)e^{-2\pi i\nu_n t}$$

where $\nu_n = (\beta_n/4)\sqrt{g/l}$; the function $J_0(z)$ is given in Eq. (1.2.5) and further in Eqs. (5.2.21); and the constants β_n are the solutions of the equation $J_0(\pi\beta_n) = 0$ [Eq. (5.2.23)]. Is the motion periodic in general? What are the ratios of the lowest three allowed frequencies to the frequency of oscillation that the string would have if all its mass were concentrated at its lower end?

13 Utilizing the general formulas of Eqs. (4.3.13) and (5.2.26), obtain formulas for the shape of the string of Prob. 12 when driven by a periodic force and when struck by an impulsive force.

14 Choosing suitable values for the properties of the string of Prob. 12, plot the shape of the string when it is vibrating at its lowest three allowed frequencies. Plot its shape when driven at its free end at a frequency 1.5. times its fundamental.

15 The tip of the string of Prob. 11 is struck an impulsive blow $\delta(t)$ in a direction perpendicular to the equilibrium plane of motion. Calculate the series representing the subsequent displacement. What is the amplitude of motion of the lowest mode?

16 A string carrying an electric current J is stretched along the lines of a homogeneous magnetic field B (in the x direction).   As a result, the z and y motions of the string will be coupled.   (*a*) Obtain the equations of motion describing this coupled motion, for small amplitudes.   (*b*) Show that the solution predicts two characteristic modes of motion, representing a right- and left-hand "circularly polarized" motion of the string, with displacements y and z related either as $y = iz$ or $y = -iz$.   Calculate the phase velocity for these two modes and show that for small B they are related as $\sqrt{T/\epsilon} - \delta$ and $\sqrt{T/\epsilon} + \delta$. Determine δ.

17 A uniform string of length l and tension T is fastened at $x = l$ to a support having transverse mechanical resistance $R = \epsilon c \tanh(\pi\alpha - \frac{1}{2}i\pi)$ and zero reactance.   The string is originally at equilibrium, and the end at $x = 0$ is suddenly acted on by a transverse impulsive force $\delta(t)$.   Utilize the Laplace transform to obtain the formula for the shape of the string.

$$y(x,t) = \frac{1}{\pi\epsilon c} \sum_{n=0}^{\infty} \frac{e^{-(\pi\alpha/l)(x+ct)}}{(n+\frac{1}{2})^2 + \alpha^2} \left\{ (n+\tfrac{1}{2}) \sin\left[\frac{\pi}{l}(n+\tfrac{1}{2})(x+ct)\right]\right.$$

$$\left. + \alpha \cos\left[\frac{\pi}{l}(n+\tfrac{1}{2})(x+ct)\right]\right\}$$

$$- \frac{1}{\pi\epsilon c} \sum_{n=0}^{\infty} \frac{e^{-(\pi\alpha/l)(x-ct)}}{(n+\frac{1}{2})^2 + \alpha^2} \left\{ (n+\tfrac{1}{2}) \sin\left[\frac{\pi}{l}(n+\tfrac{1}{2})(x-ct)\right]\right.$$

$$\left. + \alpha \cos\left[\frac{\pi}{l}(n+\tfrac{1}{2})(x-ct)\right]\right\} + \frac{1}{\epsilon c} \tanh(\pi\alpha)$$

18 A string of density 0.1 g per cm is stretched with a tension of 10^5 dynes from a support at one end to a device for producing transverse periodic oscillations at the other end.   When the driving frequency has a given value, it is noted that the points of minimum amplitude are 10 cm apart, that the amplitude of motion of the minimum is 0.557 times the amplitude at the maximum, and that the nearest maximum is 6 cm from the support.   What is the driving frequency and what is the value of the transverse impedance of the support?

19 A string is stretched between two supports having transverse mechanical resistance R large compared with ϵc, and zero reactance.   The string is driven by a periodic force concentrated at $x = \xi$.   What are the amplitudes of motion of the end supports, and by how much do they lag behind the driving force?

20 The string of Prob. 19 is struck an impulsive blow at the point $x = \xi$.   Compute the subsequent motion of the two end supports.

21 An infinite string has a distribution of additional mass corresponding to the impedance ratio

$$\xi(x) = \begin{cases} \dfrac{Z(x)}{\epsilon c} = i\mu \dfrac{\omega}{c} \cos\dfrac{\pi x}{a} & -\tfrac{1}{2}a < x < \tfrac{1}{2}a \\ 0 & \text{otherwise} \end{cases}$$

By use of Eq. (4.5.17), obtain the Born approximation for the ratio between the amplitude of the wave reflected from the load to the amplitude of the incident wave.   By use of Eqs. (4.5.25) and (4.5.26), obtain a variational value of this ratio, using the trial functions $\chi = e^{i\omega x/c}$, $\chi^* = e^{-i\omega x/c}$.   Compare the results.

22   Show that an alternative variational expression to Eq. (4.5.37), equivalent to Eq. (4.5.11) plus the boundary conditions $Y = 0$ at $x = 0$ and $x = l$, is

$$V(Y) = \frac{\displaystyle\int_0^l [(dY/dx)^2 - i(\omega/c)\zeta Y^2]\,dx}{\displaystyle\int_0^l Y^2\,dx}$$

where the shape of $Y(x)$ is varied, and the stationary values of V are the allowed values of $(\omega/c)^2$.

23   A string of length l is loaded at its center by a point mass equal to $\tfrac{1}{2}\epsilon l$, half the total mass of the string, so that $Z = -\tfrac{1}{2}i\omega\epsilon l$ and $\zeta = -i(\omega l/2c)\,\delta(x - \tfrac{1}{2}l)$. Use Eq. (4.5.34) to show that the lowest frequency of the system is $\omega_1/2\pi$, where $\omega_1 l/c = 2.1418\ldots$. Plot the shape of the string. Show that the variational formula of Eq. (4.5.37), using $Y = \sin(\pi x/l)$, yields the same value of ω_1/c. Show that the variational formula of Prob. 22, again using $Y = \sin(\pi x/l)$ and the same ζ, gives $\omega_1 l/c \simeq 2.22$. Show that the first-approximation formula (4.5.40) also yields $\omega_1 l/c \simeq 2.22$.

24   Integral equation (4.5.39) is homogeneous, so that the amplitude of the displacement $Y(x)$ is undetermined. Suppose we "normalize" this amplitude so that

$$\frac{2i(\omega l/c)}{\pi^2 - (\omega l/c)^2}\int_0^l \zeta Y(x)\sin\frac{\pi x}{l}\,dx = 1$$

in which case there results an inhomogeneous integral equation

$$Y(x) = \sin\left(\frac{\pi x}{l}\right) + 2\left(\frac{\omega l}{c}\right)\sum_{n=2}^{\infty}\frac{U_n(Y)}{(\pi n)^2 - (\omega l/c)^2}\sin\frac{\pi n x}{l}$$

where

$$U_n(Y) = i\int_0^l \zeta Y(x)\sin\frac{\pi n x}{l}\,dx$$

which is particularly useful in computing the standing wave of lowest frequency. (Can a similar equation be obtained for the nth wave?) Show that this equation can be manipulated to yield still another variational form,

$$V(Y) = 2U_1(Y) - i\int_0^l \zeta Y^2(x)\,dx + 2\frac{\omega l}{c}\sum_{n=1}^{\infty}\frac{U_n{}^2(Y)}{n^2\pi^2 - (\omega l/c)^2}$$

Show that, when Y is varied to make V stationary, the resulting value of V is the correct value of U_1. This can then be inserted in the normalizing equation, which becomes

$$\left(\frac{\omega l}{c}\right)^2 + 2\frac{\omega l}{c}U_1 = \pi^2$$

and which thus will result in the exact value of ω/c for the lowest frequency of the system. Show that if a trial function $Y \simeq u(x)$, with one or more parameters, is inserted in the variational expression, setting the derivatives of V with respect to these parameters will produce a "best" form for u and a "best" value for U_1 from which to compute a value of ω/c which will be good to second order in the difference between u and Y.

25 For the point load of Prob. 23 ($Z = -\frac{1}{2}i\omega\epsilon l$), use the variational technique of Prob. 24 to determine the lowest eigenvalue $\omega_1 l/c$. Use $u = A \sin(\pi x/l)$ as trial function, and use the relation

$$\sum_{n=1}^{\infty} \frac{1}{(2n+1)^2 - 1} = \tfrac{1}{4}$$

to show that the "best" value of A is $[1 - (\omega l/2\pi c)^2]^{-1}$. Thence, by use of the normalizing equation, show that the "best" value of $\omega_1 l/c$ is 2.145. Compare this with the other solutions discussed in Prob. 23.

26 A string of length l, stretched between rigid supports with tension T, has mass density ϵ between $x = 0$ and $x = \alpha l$, mass density $\beta^2\epsilon$ between $x = \alpha l$ and $x = l$. By joining appropriate sine functions in value and slope at $x = \alpha l$, obtain an implicit equation for the natural frequencies of oscillation. For $\alpha = \frac{1}{2}$ and $\beta = 2$, compute the lowest frequency. Use Eqs. (4.5.37) and (4.5.41) to obtain approximate expressions for this lowest frequency. How close do they come to the exact value?

27 A coil spring has a relaxed length l_0; when it is stretched to length l, the tension in the spring is $K(l - l_0)$. Accounting for the fact that the mass per unit length $\epsilon = m/l$ depends on the stretched length of the spring, show that the ratio between the speed of transverse and longitudinal waves on the spring is

$$\frac{c_t}{c_l} = \sqrt{1 - \frac{l_0}{l}}$$

and hence show that the transverse wave velocity is always less than the longitudinal wave velocity.

28 A string of mass density ϵ, under tension T, extends at equilibrium along the x axis from $x = 0$ to $x = \infty$. The $x = 0$ end is attached to a mass M, which slides on a frictionless guide bar in the y direction, the mass also being attached to a spring with stiffness constant $K = M\omega_0^2$, tending to keep it at $y = 0$.

(*a*) With the system originally in equilibrium, show that, if the mass is given an initial velocity v_0 at $t = 0$, the transverse displacement of the string thereafter is

$$y(x,t) = \frac{v_0}{k_f} \exp\left[-\frac{\epsilon}{2M}(ct - x)\right]\sin[k_f(ct - x)]$$

where

$$k_f^2 = \left(\frac{\omega_0}{c}\right)^2\left[1 - \left(\frac{\epsilon c}{2M\omega_0}\right)^2\right]$$

(*b*) If an incident step wave $Au(ct + x)$ comes from $+\infty$ along the string, impinging on the mass originally at rest, show that the string displacement is

$$y(x,t) = Au(ct + x) - Au(ct - x)$$
$$+ A\frac{2\epsilon}{Mk_t}\exp\left[-\frac{\epsilon}{2M}(ct - x)\right]\sin[k_f(ct - x)]$$

(*c*) In case *a*, what is the energy in the string after the mass has come to rest?

CHAPTER

5

BARS, MEMBRANES, AND PLATES

5.1 THE FLEXIBLE BAR

Although the primary subject of study of this book is acoustic wave motion in fluids, the first five chapters are concerned with the vibration of solid objects. The reasons for the roundabout approach are twofold. Vibrating systems produce sound waves in the medium surrounding them, or they are affected by sound waves striking them, and any analysis of the waves must take this into account. But also, many computational techniques are more easily explained in connection with vibrating systems than with acoustic waves. Although our discussion of vibration has been lengthy, it has been limited to examples which conform to one or both of these reasons.

This restraint in choice of material will be particularly noticeable in this chapter. A detailed discussion of wave motion in elastic solids would be a lengthier task than the one set for this book. All we can do here is to touch on those aspects of elasticity which will be needed later in discussing sound waves in fluids.

It must have been rather obvious in the preceding chapter that we were analyzing the motions of a somewhat idealized string. In the first place, we assumed that the string was perfectly flexible; that the only restoring force was due to the tension. Second, we made no mention of the possibility of longitudinal motion of alternate compression and tension, which can be set up in any actual string, as well as in any other piece of solid material. This longitudinal wave motion will be disregarded for a while longer; we shall spend the rest of this book discussing it. However, we can no longer put off studying the effect of stiffness on the string's motion. And we shall begin the study by discussing the transverse vibrations of bars.

There is no sharp distinction between what we mean by a bar and what we mean by a string. In general, tension is more important as a restoring force than stiffness for a string, and stiffness is more important for a bar; but there is a complete sequence of intermediate cases from stiff strings to bars under tension. The perfectly flexible string is one limiting case, where the restoring force due to stiffness is negligible compared with that due to the tension. The rod or bar under no tension is the other limiting case, the restoring force being entirely due to stiffness. The first limiting case was studied in the preceding chapter. The second case, the bar under no tension, will be studied in the first part of this chapter, and the intermediate cases will be dealt with in a later part.

Stresses in a bar

To start with, we shall study the bending of a straight bar, with uniform cross section, symmetrical about a central plane. The motion of the bar is supposed to be perpendicular to this plane, and we shall call the displacement from equilibrium of the plane y. (See Fig. 5.1.)

When the bar is bent, its lower half is compressed and its upper half stretched (or vice versa). This bending requires a moment M, whose relation to the amount of bending we must find. To compress a rod of cross-sectional area S and length l by an amount dl requires a force $QS(dl/l)$, where Q is a constant, called *Young's modulus*. Now, imagine the bar to be a bundle of fibers of cross-sectional area dS, all running parallel to the center plane of the bar. If the bar is bent by an angle ϕ in a length dx, then the fibers which are a distance z down from the center surface (it is no longer a plane now that the bar is bent) will be compressed by a length $z\phi$; the force required to compress each fiber will be $Q\,dS(z\phi/dx)$; and the moment of this force about the center line of the bar's cross section will be $(Q\phi/dx)z^2\,dS$. The total moment of these forces required to compress and to stretch all the fibers in the bar will be

$$M = \frac{Q\phi}{dx} \int z^2 \, dS \qquad (5.1.1)$$

where the integration is over the whole area of the cross section.

We define a constant κ, such that $\kappa^2 = (1/S) \int z^2 \, dS$, where S is the area of the cross section. This constant is called the *radius of gyration* of the

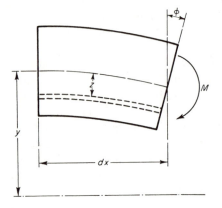

FIGURE 5.1
Moment acting on a bent element of a bar.

cross section, by analogy with the radius of gyration of solids.   Its values for some of the simpler cross-sectional shapes are as follows.

Rectangle, length parallel to center line b, width perpendicular to center line a:

$$\kappa = \frac{a}{\sqrt{12}}$$

Circle, of radius a:

$$\kappa = \frac{a}{2}$$

Circular ring, outer radius a, inner radius b:

$$\kappa = \tfrac{1}{2}\sqrt{a^2 + b^2}$$

Bending moments and shearing forces

Equation (5.1.1), giving the moment required to bend a length dx of rod by an angle ϕ, is then

$$M = \frac{Q\phi S\kappa^2}{dx} \tag{5.1.2}$$

If the rod does not bend much, we can say that ϕ is practically equal to the difference between the slopes of the axial line of the rod at the two ends of the element dx:

$$\phi = -\left(\frac{\partial y}{\partial x}\right)_{x+dx} + \left(\frac{\partial y}{\partial x}\right)_{x} = -dx\left(\frac{\partial^2 y}{\partial x^2}\right)$$

Therefore the bending moment is

$$M = -QS\kappa^2\frac{\partial^2 y}{\partial x^2} \tag{5.1.3}$$

This bending moment is not the same for every part of the rod; it is a function of x, the distance from one end of the rod.   In order to keep the element of bar in equilibrium, we must have the difference in the moments acting on the two ends of the element balanced by a shearing force represented by F in Fig. 5.2 (moment and shear are, of course, two different aspects of the single stress which is acting on the bar).   The moment of the shearing force is $F\,dx$, and this must equal dM for equilibrium, which means that

$$F = \frac{\partial M}{\partial x} = -QS\kappa^2\frac{\partial^3 y}{\partial x^3} \tag{5.1.4}$$

This equation is not exactly true when the bar is vibrating (since a certain part of the moment must be used in getting the element of bar to turn as it bends), but it is very nearly correct when the amplitude of vibration is not large compared with the length of bar.

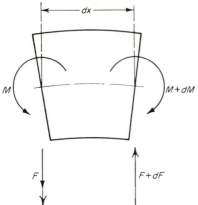

FIGURE 5.2
Bending moments and shearing forces to balance.

Properties of the motion of the bar

The shearing force F is also a function of x and may be different for different ends of the element of bar. This leaves a net force $dF = (\partial F/\partial x)\,dx$ acting on the element, perpendicular to the bar's axis; and this force must equal the element's acceleration times its mass $\rho S\,dx$, where ρ is the density of the material of the rod. Therefore the equation of motion of the bar is $dx\,(\partial F/\partial x) = \rho S\,dx\,(\partial^2 y/\partial t^2)$, or

$$\frac{\partial^4 y}{\partial x^4} = -\frac{\rho}{Q\kappa^2}\frac{\partial^2 y}{\partial t^2} \qquad (5.1.5)$$

This equation differs from the wave equation in that it has a fourth derivative with respect to x instead of a second derivative. The general function $F(x + ct)$ is not a solution, so that a bar, satisfying Eq. (5.1.5), cannot have waves traveling along it with constant velocity and unchanged shape. In fact, the term "wave velocity" has no general meaning in this case, although it can be given certain special meanings. For instance, a simple-harmonic solution of Eq. (5.1.5) is

$$y = Ce^{2\pi i(\mu x - \nu t)} = A\cos\left[2\pi(\mu x - \nu t) - \Phi\right] \qquad (5.1.6)$$

where
$$\mu = \left(\frac{\nu^2 \rho}{4\pi^2 Q\kappa^2}\right)^{\frac{1}{4}} \qquad \nu = 2\pi\mu^2\kappa\sqrt{\frac{Q}{\rho}}$$

This represents a sinusoidal wave traveling in the positive direction, of just the sort one finds on strings. There is an important difference between the two waves, however: in the case of the bar, the velocity of the wave $u = \nu/\mu = (4\pi^2 Q\kappa^2/\rho)^{\frac{1}{4}}\sqrt{\nu}$ *depends on the frequency of oscillation of the wave,* whereas in the case of the string it does not. The velocity of propagation of a simple-harmonic wave is called the *phase velocity,* and is one of the special kinds of "velocity" that have meaning for a bar. In the case of the string, the phase velocity is independent of ν and is equal to the velocity of all waves c. For the bar the phase velocity depends on ν, and there is no general velocity for all waves. This is analogous to the case of the transmission of light through glass, where light of different frequencies (i.e., of different colors) travels with different velocities and dispersion results. A bar is sometimes said to be a *dispersing* medium for waves of bending.

Wave motion of an infinite bar

Although the phase velocity is not a constant, nevertheless we should be able to build up any sort of wave out of a suitably chosen combination of sinusoidal waves of different frequencies, in a Fourier integral. At present we are considering the bar to be infinite in length, so that all frequencies are allowed, and the sum is an integral

$$y = \int_{-\infty}^{\infty} e^{2\pi i\mu x}[B(\mu) \cos (\gamma^2\mu^2 t) + C(\mu) \sin (\gamma^2\mu^2 t)] \, d\mu \qquad (5.1.7)$$

where $\gamma^2 = 4\pi^2\kappa\sqrt{Q/\rho}$, and the functions $B(\mu)$ and $C(\mu)$, corresponding to the coefficients B_n and C_n of a Fourier series, are determined by the initial conditions. This integral is analogous to the Fourier integral of Eq. (1.3.16). If the initial shape and velocity shape of the bar are $y_0(x)$ and $v_0(x)$, Eq. (1.3.16) shows that

$$B(\mu) = \frac{1}{4\pi^2} \int_{-\infty}^{\infty} y_0(x)e^{-2\pi i\mu x} \, dx$$

$$C(\mu) = \frac{1}{4\pi^2\gamma^2\mu^2} \int_{-\infty}^{\infty} v_0(x)e^{-2\pi i\mu x} \, dx \qquad (5.1.8)$$

This may seem a rather roundabout way of obtaining solutions of Eq. (5.1.5) which satisfy given initial conditions; but since the useful functions $F(x - ct)$ and $f(x + ct)$ are not solutions of (5.1.5), there is no other feasible method.

The case $y_0 = e^{-x^2/4a^2}$, $v_0 = 0$, is one that can be integrated, and can be used to show the change in shape of the wave as it traverses the bar. Solving Eqs. (5.1.8) and (5.1.7), we obtain

$$y = \left(1 + \frac{\gamma^4 t^2}{a^4}\right)^{-\frac{1}{4}} e^{-\frac{a^2 x^2}{4(a^4+\gamma^4 t^2)}} \cos \left[\frac{\gamma^2 t x^2}{4(a^4 + \gamma^4 t^2)} - \tfrac{1}{2}\tan^{-1}\frac{\gamma^2 t}{a^2}\right] \qquad (5.1.9)$$

We can now see the utility of Eq. (5.1.7); we can easily see that (5.1.9) is a solution of the equation of motion (5.1.5) by direct substitution; we can also see that it would be exceedingly difficult to obtain (5.1.9) from (5.1.5) directly.

The shape of this function, at successive instants of time, is shown by the solid lines in Fig. 5.3. For comparison, the sequence of dotted lines shows the variation in shape of a flexible string with equal initial conditions. In the case of the string, the two partial waves move outward with unchanged shape. The shape of each "partial wave" for the bar, however, changes

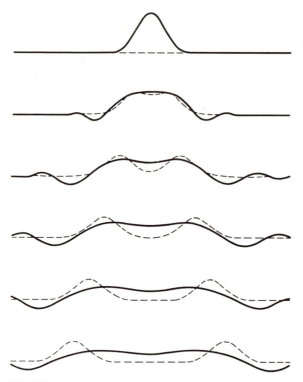

FIGURE 5.3

Comparison between the motion of waves on a bar (solid line) and on a string (dotted line), both of infinite length. Each sequence shows the shapes at successive instants, after starting from rest in the form given at the top of the sequence. The constant γ was chosen to make the average "velocity" of the waves on the bar approximately equal to c, the velocity of the waves on the string.

continuously as the wave travels outward.   In particular, notice the formation of subsidiary "ripples" ahead of the principal "crest" of each wave. This is due to the fact, which we have already noted, that the high-frequency short-wavelength parts of the wave spread outward faster than the rest.

Simple-harmonic motion

   We cannot pursue our study of the motions of the bar much further without examining its normal modes of vibration.   As with the string, we must ask if there are any ways in which the bar can vibrate with simple-harmonic motion.   We try setting $y = Y(x)e^{-2\pi i v t}$ and find that Y must satisfy the equation

$$\frac{d^4 Y}{dx^4} = 16\pi^4 \mu^4 Y \qquad \mu^4 = \frac{\rho v^2}{4\pi^2 Q \kappa^2} = 4\pi^2 \frac{v^2}{\gamma^2} \qquad (5.1.10)$$

The general solution of this is

$$y = C_1 e^{2\pi\mu x} + C_2 e^{-2\pi\mu x} + C_3 e^{2\pi i \mu x} + C_4 e^{-2\pi i \mu x}$$
$$= a \cosh(2\pi\mu x) + b \sinh(2\pi\mu x) + c \cos(2\pi\mu x) + d \sin(2\pi\mu x) \quad (5.1.11)$$

where $\cosh u = \cos(iu)$ and $\sinh u = -i \sin(iu)$. See Eq. (1.2.10) and Tables I and II.

This general solution satisfies Eq. (5.1.10) for any value of the frequency ν. It is, of course, the boundary conditions that pick out the set of allowed frequencies.

Bar clamped at one end

For example, if we have a bar of length l clamped at one end $x = 0$, the boundary conditions at this end are that *both* y and its slope $\partial y/\partial x$ must be zero at $x = 0$. The particular combination of the general solution (5.1.11) that satisfies these two conditions is the one with $c = -a$ and $d = -b$.

$$Y = a[\cosh(2\pi\mu x) - \cos(2\pi\mu x)] + b[\sinh(2\pi\mu x) - \sin(2\pi\mu x)] \quad (5.1.12)$$

If the other end is free, y and its slope will not be zero, but the bending moment $M = QS\kappa^2(d^2Y/dx^2)$ and the shearing force $F = -QS\kappa^2(d^3Y/dx^3)$ must both be zero, since there is no bar beyond $x = l$ to cause a moment or a shearing stress. We see that *two* conditions must be specified for each end instead of just one, as in the string. This is due to the fact that the equation for Y is a fourth-order differential equation, and its solution involves four arbitrary constants whose relations must be fixed, instead of two for the string. It corresponds to the physical fact that whereas the only internal stress in the string is tension, the bar has two, bending moment and shearing force, each depending in a different way on the deformation of the bar.

The two boundary conditions at $x = l$ can be rewritten as $\dfrac{1}{4\pi^2\mu^2}\dfrac{d^2Y}{dx^2} = 0$ and $\dfrac{1}{8\pi^3\mu^3}\dfrac{d^3Y}{dx^3} = 0$ at $x = l$. Substituting expression (5.1.12) in these, we obtain two equations that fix the relationship between a and b and between μ and l:

$$a[\cosh(2\pi\mu l) + \cos(2\pi\mu l)] + b[\sinh(2\pi\mu l) + \sin(2\pi\mu l)] = 0$$

$$a[\sinh(2\pi\mu l) - \sin(2\pi\mu l)] + b[\cosh(2\pi\mu l) + \cos(2\pi\mu l)] = 0$$

or

$$b = a\frac{\sin(2\pi\mu l) - \sinh(2\pi\mu l)}{\cos(2\pi\mu l) + \cosh(2\pi\mu l)} = -a\frac{\cos(2\pi\mu l) + \cosh(2\pi\mu l)}{\sin(2\pi\mu l) + \sinh(2\pi\mu l)} \quad (5.1.13)$$

By dividing out a and multiplying across, we obtain an equation for μ:

$$[\cosh(2\pi\mu l) + \cos(2\pi\mu l)]^2 = \sinh^2(2\pi\mu l) - \sin^2(2\pi\mu l)$$

Utilizing some trigonometric relationships, this last equation can be reduced to two simpler forms:

$$\cosh(2\pi\mu l)\cos(2\pi\mu l) = -1 \qquad \text{or} \qquad \coth^2(\pi\mu l) = \tan^2(\pi\mu l) \quad (5.1.14)$$

where $\coth z = \cosh z / \sinh z$.

The allowed frequencies

We shall label the solutions of this equation in order of increasing value. They are $2\pi\mu_1 l = 1.8751$, $2\pi\mu_2 l = 4.6941$, $2\pi\mu_3 l = 7.8548$, etc. To simplify the notation, we let $1/\pi$ times the numbers given above have the labels β_n, so that

$$\mu_n = \frac{\beta_n}{2l} \qquad (5.1.15)$$

where $\beta_1 = 0.597$, $\beta_2 = 1.494$, $\beta_3 = 2.500$, etc. It turns out that β_n is practically equal to $n - \frac{1}{2}$ when n is larger than 2.

By fixing μ, we fix the allowed values of the frequency. Using Eq. (5.1.10), we have

$$\nu_n = \frac{\gamma^2 \mu_n^2}{2\pi} = \frac{\pi}{2l^2}\sqrt{\frac{Q\kappa^2}{\rho}}\,\beta_n^2 \qquad (5.1.16)$$

or

$$\nu_1 = \frac{0.55966}{l^2}\sqrt{\frac{Q\kappa^2}{\rho}} \qquad \begin{array}{l} \nu_2 = 6.267\nu_1 \\ \nu_3 = 17.548\nu_1 \\ \nu_4 = 34.387\nu_1 \\ \cdots\cdots\cdots \end{array}$$

Notice that the allowed frequencies depend on the inverse *square* of the length of the bar, whereas the allowed frequencies of the string depend on the inverse first power.

Equation (5.1.16) shows how far from harmonics are the overtones for a vibrating bar. The first overtone has a higher frequency than the sixth harmonic of a string of equal fundamental. If the bar were struck so that its motion contained a number of overtones with appreciable amplitude, it would give out a shrill and nonmusical sound. But since these high-frequency overtones are damped out rapidly, the harsh initial sound will quickly change to a pure tone, almost entirely due to the fundamental. A tuning fork can be considered to be two vibrating bars, both clamped at their lower ends. The fork exhibits the preceding behavior, the initial metallic "ping" rapidly dying out and leaving an almost pure tone.

The characteristic functions

The characteristic function corresponding to the allowed frequency ν_n is given by the equation

$$\psi_n = a_n\left(\cosh\frac{\pi\beta_n x}{l} - \cos\frac{\pi\beta_n x}{l}\right) + b_n\left(\sinh\frac{\pi\beta_n x}{l} - \sin\frac{\pi\beta_n x}{l}\right) \qquad (5.1.17)$$

where

$$-b_n = a_n \frac{\cosh(\pi\beta_n) + \cos(\pi\beta_n)}{\sinh(\pi\beta_n) + \sin(\pi\beta_n)} = a_n \frac{\sinh(\pi\beta_n) - \sin(\pi\beta_n)}{\cosh(\pi\beta_n) + \cos(\pi\beta_n)}$$

We shall choose the value of a_n so that $\int_0^l \psi_n^2 \, dx = l/2$, by analogy with the sine functions for the string. The resulting values for a_n and b_n are $a_1 = 0.707$, $b_1 = -0.518$, $a_2 = 0.707$, $b_2 = -0.721$, $a_3 = 0.707$, $b_3 = -0.707$, etc. For n larger than 2, both a_n and b_n are practically equal to $1/\sqrt{2}$. Some of the properties of these functions that will be of use are

$$\int_0^l \psi_m(x)\psi_n(x) \, dx = \begin{cases} 0 & m \neq n \\ \dfrac{l}{2} & m = n \end{cases} \qquad \psi_n(l) = (-1)^{n-1}\sqrt{2}$$

$$\left(\frac{d\psi_1}{dx}\right)_{x=l} = 1.040\,\frac{\pi\beta_1}{l} \qquad \left(\frac{d\psi_2}{dx}\right)_{x=l} = -1.440\,\frac{\pi\beta_2}{l}$$

$$\tag{5.1.18}$$

$$\left(\frac{d\psi_n}{dx}\right)_{x=l} \simeq (-1)^{n-1}\sqrt{2}\,\frac{\pi\beta_n}{l} \qquad \text{and} \qquad \beta_n \simeq n - \tfrac{1}{2} \qquad n > 2$$

$$\psi_n \simeq \frac{1}{\sqrt{2}} \left[e^{-\pi\beta_n x/l} + (-1)^{n-1} e^{\pi\beta_n(x-l)/l} \right] + \sin\left(\frac{\pi\beta_n x}{l} - \frac{\pi}{4}\right) \qquad n > 2$$

The shapes of the first five characteristic functions are shown in Fig. 5.4. Note that for the higher overtones most of the length of the bar has the sinusoidal shape of the corresponding normal mode of the string, with the nodes displaced toward the free end. In terms of the approximate form given above for ψ_n, the sine function is symmetrical about the center of the bar; the first exponential alters the sinusoidal shape near $x = 0$ enough to make ψ_n have zero value and slope at this point; and the second exponential adds enough near $x = l$ to make the second and third derivatives vanish. Note also that the number of nodal points in ψ_n is equal to $n - 1$, as it is for the string.

In accordance with the earlier discussion of series of characteristic functions, we can now show that a bar started with the initial conditions, at $t = 0$, of $y = y_0(x)$ and $\partial y/\partial t = v_0(x)$ will have a subsequent shape given by the series

$$y = \sum_{n=1}^{\infty} \psi_n(x)[B_n \cos(2\pi\nu_n t) + C_n \sin(2\pi\nu_n t)] \tag{5.1.19}$$

where

$$B_n = \frac{2}{l} \int_0^l y_0(x)\psi_n(x) \, dx$$

$$C_n = \frac{1}{\pi\nu_n l} \int_0^l v_0(x)\psi_n(x) \, dx$$

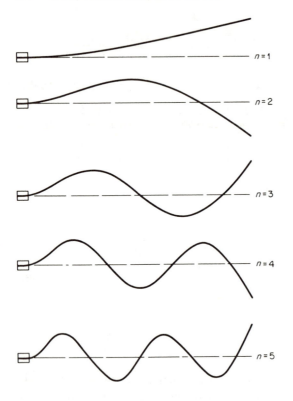

FIGURE 5.4
Shapes of the first five characteristic functions for a vibrating bar clamped at one
end and free at the other.

Plucked and struck bar

Two examples of such calculations will be given.   The first example is
that of a rod suddenly released from an initial shape $y_0 = hx/l$, an undesirable
case in actual practice (for it bends the bar quite severely at $x = 0$) but one
that can be easily solved.   The values of the coefficients B_n turn out to be

$$B_n = \frac{2h}{l^2} \int_0^l x\psi_n(x)\, dx$$

$$= \frac{2h}{\pi^2\beta_n{}^2} \{a_n[2 + \pi\beta_n(\sinh \pi\beta_n - \sin \pi\beta_n) - \cosh \pi\beta_n - \cos \pi\beta_n]$$

$$+ b_n[\pi\beta_n(\cosh \pi\beta_n + \cos \pi\beta_n) - \sinh \pi\beta_n - \sin \pi\beta_n]\}$$

$$= \frac{4h}{\pi^2\beta_n{}^2} a_n$$

when the ratios between a_n and b_n, given in Eq. (5.1.17), are used to simplify
the expression.   Since the overtones are not harmonics, the motion is not
periodic.

The other case to be considered is that of a rod struck at its free end in
such a manner that its initial velocity is zero everywhere except at $x = l$,

where it is very large, large enough so that $\int v_0 \, dx = U$. This case corresponds to that of a tuning fork struck at one end. If the total impulse given to the end of the fork is P, then $U = P/\rho S$. The coefficients B_n are zero and

$$C_n = \frac{1}{\pi v_n l} \int_0^l v_0 \psi_n \, dx = \frac{U}{\pi v_n l} \psi_n(l) = (-1)^{n-1} \frac{lU}{\pi^2 \beta_n^2} \sqrt{\frac{8\rho}{Q\kappa^2}}$$

when the value given for $\psi_n(l)$ in Eqs. (5.1.18) is used. This motion is also nonperiodic, as are all motions of the bar that correspond to more than one normal mode of vibration.

Clamped-clamped and free-free bars

Other boundary conditions will give rise to other characteristic functions and frequencies. The bar may be clamped at both ends, in which case expression (5.1.12) must again be used for Y, but instead of Eq. (5.1.13) we must have

$$b = a \frac{\sin (2\pi\mu l) + \sinh (2\pi\mu l)}{\cos (2\pi\mu l) - \cosh (2\pi\mu l)} = -a \frac{\cos (2\pi\mu l) - \cosh (2\pi\mu l)}{\sin (2\pi\mu l) - \sinh (2\pi\mu l)}$$

and instead of Eq. (5.1.14) we have

$$\cosh (2\pi\mu l) \cos (2\pi\mu l) = 1 \qquad \text{or} \qquad \tanh^2 (\pi\mu l) = \tan^2 (\pi\mu l)$$

The allowed frequencies can be obtained from the formula

$$v_n = \frac{\gamma^2 \mu_n^2}{2\pi} = \frac{\pi}{2l^2} \sqrt{\frac{Q\kappa^2}{\rho}} \beta_n^2 \qquad \begin{aligned} \beta_1 &= 1.5056 \\ \beta_2 &= 2.4997 \\ \beta_n &\simeq n + \tfrac{1}{2} \quad (n > 2) \end{aligned} \qquad (5.1.20)$$

Still another set of boundary conditions is the one for a completely free bar, where the second and third derivatives must be zero at both ends. The characteristic functions (which we can call ψ_n'') for this case can be obtained from the ones for the bar clamped at both ends (which we can call ψ_n, to distinguish them from ψ_n'') by simply differentiating twice with respect to x, $\psi_n'' = \partial^2 \psi_n / \partial x^2$. By using Eq. (5.1.10) we can show that $\partial^2 \psi_n'' / \partial x^2 = (4\pi^2 v_n^2 \rho / Q\kappa^2) \psi_n$, so that if ψ_n and $\partial \psi_n / \partial x$ are both zero at $x = 0$ and $x = l$, then $\partial^2 \psi_n'' / \partial x^2$ and $\partial^3 \psi_n'' / \partial x^3$ are both zero at the end points. The functions ψ_n'' therefore satisfy the boundary conditions for a completely free bar. It is not difficult to see that ψ_n'' is a solution of the equation of motion, corresponding to the same allowed frequency as does the function ψ_n. The allowed values are given in Eq. (5.1.20).

We thus obtain the rather surprising result that the allowed frequencies for a free bar are the same as those for a similar bar clamped at both ends and that the corresponding characteristic functions are related by simple differentiation, although their shapes are quite different.

Energy of vibration

It is sometimes useful to know the energy of vibration of a bar. This energy can be computed by using the expressions obtained in the previous section. We saw there [Eqs. (5.1.2) and (5.1.3)] that the moment required to bend an element dx of the bar by an angle ϕ is $M = QS\kappa^2(\phi/dx)$. The amount of work required to bend it from equilibrium to this angle is

$$\int_0^\phi M\, d\phi = \frac{\frac{1}{2}QS\kappa^2\phi^2}{dx} = \frac{1}{2}QS\kappa^2\left(\frac{\partial^2 y}{\partial x^2}\right)^2 dx$$

Therefore the potential energy of the bar, the work required to bend the bar into its final instantaneous shape $y(x,t)$, is

$$V = \frac{1}{2}QS\kappa^2\int_0^l \left(\frac{\partial^2 y}{\partial x^2}\right)^2 dx$$

The kinetic energy is, of course,

$$U = \frac{1}{2}\rho S\int_0^l \left(\frac{\partial y}{\partial t}\right)^2 dx$$

so that the total energy turns out to be

$$E = \frac{1}{2}\rho S\int_0^l \left[\left(\frac{\partial y}{\partial t}\right)^2 + \frac{Q\kappa^2}{\rho}\left(\frac{\partial^2 y}{\partial x^2}\right)^2\right] dx \qquad (5.1.21)$$

The energy of a bar, subject to one or another of the various boundary conditions discussed above, is obtained by substituting $y = \Sigma A_n\psi_n \cos(2\pi\nu_n t - \Omega_n)$ in Eq. (5.1.21). This results in a superfluity of terms, each containing a product of two trigonometric functions of time and an integral of a product of two characteristic functions, or two second derivatives of these functions. The kinetic-energy terms, containing integrals of the products of two functions, can be integrated by means of the first of Eqs. (5.1.18) and reduced to the single sum

$$\Sigma \pi^2\nu_n^2\rho l S A_n^2 \sin^2(2\pi\nu_n t - \Omega_n)$$

The integrals in the potential-energy terms can be integrated by parts twice:

$$\int_0^l \frac{\partial^2\psi_n}{\partial x^2}\frac{\partial^2\psi_m}{\partial x^2}\, dx = \left[\frac{\partial\psi_n}{\partial x}\frac{\partial^2\psi_m}{\partial x^2}\right]_0^l - \left[\psi_n\frac{\partial^3\psi_m}{\partial x^3}\right]_0^l + \int_0^l \psi_n\frac{\partial^4\psi_m}{\partial x^4}\, dx$$

$$= \frac{4\pi^2\nu_m^2\rho}{Q\kappa^2}\int_0^l \psi_n\psi_m\, dx$$

by using the equation of motion. The terms in the square brackets are zero no matter which of the boundary conditions discussed above are used. The resulting potential-energy series is therefore $\Sigma \pi^2\nu_n^2\rho l S A_n^2 \cos^2(2\pi\nu_n t - \Omega_n)$.

The $\sin^2$ of the kinetic-energy terms combines with the $\cos^2$ in the potential-energy terms and gives for the total energy the simple series

$$E = \sum_{n=1}^{\infty} 2\pi^2 \frac{\rho S l}{2} v_n{}^2 A_n{}^2 \qquad (5.1.22)$$

Note the similarity with the corresponding expression for the energy of the string, given in Eq. (4.3.12).

Forced motion

We can discuss the forced motion of a uniform bar, now that we know the characteristic functions. If the bar is driven by a force $F(x)e^{-2\pi i v t}$ dynes per cm length, the equation for shape of the bar during the steady state is

$$\frac{Q\kappa^2}{\rho}\frac{d^4 Y}{dx^4} - 4\pi^2 v^2 Y = f(x) \qquad f = \frac{F(x)}{\rho S} \qquad y = Y(x)e^{-2\pi i v t} \quad (5.1.23)$$

If we expand Y and f in series of characteristic functions,

$$Y = \sum h_n \psi_n(x) \qquad f = \sum g_n \psi_n(x) \qquad g_n = \frac{2}{\rho S l}\int_0^l F(x)\psi_n(x)\,dx$$

we obtain for the steady-state motion

$$y = \frac{1}{4\pi^2}\left[\sum_{n=1}^{\infty}\frac{g_n}{v_n{}^2 - v^2}\psi_n(x)\right]e^{-2\pi i v t} \qquad (5.1.24)$$

Resonance occurs whenever the driving frequency v equals one of the natural frequencies v_n, unless the corresponding g_n is zero.

These results are analogous to those discussed at the beginning of Sec. 4.4. The effects of friction can be handled as they were handled for the string on page 142 and in Sec. 4.3. The results are so similar that we do not need to go into detail about them here. Some cases will be taken up in the problems.

The stiff string

When a string is under a tension of T dynes, and also has stiffness, its equation of motion is

$$T\frac{\partial^2 y}{\partial x^2} - QS\kappa^2\frac{\partial^4 y}{\partial x^4} = \rho S\frac{\partial^2 y}{\partial t^2} \qquad (5.1.25)$$

This equation can be obtained by combining the derivations in Secs. 4.1 and 5.1. The constant S is the area of cross section, κ its radius of gyration, and ρ and Q are the density and modulus of elasticity of the material.

Sinusoidal waves can travel along such a wire, for if we set $y = Ce^{2\pi i(\mu x - v t)}$, we obtain an equation relating μ and v:

$$T\mu^2 + 4\pi^2 QS\kappa^2\mu^4 = \rho S v^2$$

or

$$\mu^2 = -\frac{T}{8\pi^2 QS\kappa^2} + \sqrt{\left(\frac{T}{8\pi^2 QS\kappa^2}\right)^2 + \frac{\nu^2 \rho}{4\pi^2 Q\kappa^2}}$$

If the frequency is very small (i.e., if ν^2 is very much smaller than $T^2/16\pi^2\rho QS^2\kappa^2$), we expand the radical by the binomial theorem and keep only the first two terms, and μ^2 turns out to be approximately equal to $\rho S\nu^2/T$. The phase velocity ν/μ is practically equal to the constant value $\sqrt{T/\rho S}$, which would be the velocity of every wave if the wire had no stiffness (if $QS\kappa^2$ were zero). The phase velocity for the wire is not constant, however. It increases with increasing ν until for very high frequencies it is $\sqrt{2\pi\nu}\,(Q\kappa^2/\rho)^{\frac{1}{4}}$, the phase velocity given in Eq. (5.1.6) for a stiff bar without tension. In other words, the wire acts like a flexible string for long wavelengths and like a stiff bar for short ones. This is not surprising.

The boundary conditions

The usual boundary conditions correspond to clamping the wire at both ends, making $y = 0$ and also $\partial y/\partial x = 0$ at $x = 0$ and $x = l$. Setting $y = Y(x)e^{-2\pi i\nu t}$, we have for Y

$$\frac{d^4 Y}{dx^4} - 8\pi^2\beta^2 \frac{d^2 Y}{dx^2} - 16\pi^4\gamma^4 Y = 0 \qquad \beta^2 = \frac{T}{8\pi^2 QS\kappa^2}$$

$$\gamma^2 = \frac{\nu}{2\pi}\sqrt{\frac{\rho}{Q\kappa^2}}$$

(5.1.26)

Setting $Y = Ae^{2\pi\mu x}$, we obtain an equation for the allowed values of μ:

$$\mu^4 - 2\beta^2\mu^2 - \gamma^4 = 0$$

This equation has two roots for μ^2 and therefore four roots for μ:

$$\mu = \pm\mu_1 \qquad\qquad \mu_1^2 = \sqrt{\beta^4 + \gamma^4} + \beta^2$$

or $\qquad\quad \mu = \pm i\mu_2 \qquad\quad \mu_2^2 = \sqrt{\beta^4 + \gamma^4} - \beta^2 \qquad$ (5.1.27)

$$\mu_1^2 = 2\beta^2 + \mu_2^2 \qquad \mu_1\mu_2 = \gamma^2$$

The general solution of Eq. (5.1.26) can then be written

$$Y = \begin{cases} ae^{2\pi\mu_1 x} + be^{-2\pi\mu_1 x} + ce^{2\pi i\mu_2 x} + de^{-2\pi i\mu_2 x} \\ A\cosh(2\pi\mu_1 x) + B\sinh(2\pi\mu_1 x) \\ \qquad\qquad + C\cos(2\pi\mu_2 x) + D\sin(2\pi\mu_2 x) \end{cases}$$

(5.1.28)

If the boundary conditions are symmetrical, it will be useful to place the point $x = 0$ midway between the supports. The normal functions will then be even functions, $\psi(-x) = \psi(x)$; or they will be odd ones, $\psi(-x) = -\psi(x)$. In either case, if we fit the boundary conditions at one end, $x = l/2$, they will also fit at the other end, $x = -(l/2)$. The even functions are built up out of the combination

$$Y = A \cosh (2\pi\mu_1 x) + C \cos (2\pi\mu_2 x)$$

and the odd functions from the combination

$$Y = B \sinh (2\pi\mu_1 x) + D \sin (2\pi\mu_2 x)$$

The boundary conditions $Y = 0$ and $dY/dx = 0$ at $x = l/2$ correspond to the following equations for the even functions:

$$A \cosh (\pi\mu_1 l) = -C \cos (\pi\mu_2 l) \qquad \mu_1 A \sinh (\pi\mu_1 l) = \mu_2 C \sin (\pi\mu_2 l)$$

The allowed frequencies

By the use of Eq. (5.1.27), these reduce to

$$\tan (\pi l/\mu_2) = -\sqrt{1 + \frac{2\beta^2}{\mu_2{}^2}} \tanh (\pi l \sqrt{\mu_2{}^2 + 2\beta^2}) \qquad (5.1.29)$$

which can be solved for the allowed values of μ_2. The allowed values of the frequency are obtained from the equation

$$\nu = 2\pi\gamma^2 \sqrt{\frac{Q\kappa^2}{\rho}} = 2\pi\mu_2 \sqrt{(\mu_2{}^2 + 2\beta^2) \frac{Q\kappa^2}{\rho}}$$

These allowed values of ν can be labeled $\nu_1, \nu_3, \nu_5, \ldots$, in order of increasing size.

The corresponding equation for μ_2 for the odd functions is

$$\sqrt{1 + \frac{2\beta^2}{\mu_2{}^2}} \tan (\pi l/\mu_2) = \tanh (\pi l \sqrt{\mu_2{}^2 + 2\beta^2}) \qquad (5.1.30)$$

The allowed values of ν obtained from this equation can be labeled $\nu_2, \nu_4, \nu_6, \ldots$; and then the whole sequence of allowed values will be in order of increasing size, ν_1 being the smallest (the fundamental), ν_2 the next (the first overtone), and so on.

When $\beta (= \sqrt{T/8\pi^2 Q S \kappa^2})$ goes to zero (i.e., when the tension is zero), the equations reduce to those of a bar clamped at both ends, with the following allowed values of ν,

$$\nu_1 = \frac{3.5608}{l^2} \sqrt{\frac{Q\kappa^2}{\rho}} \qquad \begin{matrix} \nu_2 = 2.7565\nu_1 \\ \nu_3 = 5.4039\nu_1 \\ \nu_4 = 8.9330\nu_1 \\ \cdots\cdots\cdots\cdots \end{matrix} \qquad (5.1.31)$$

obtained from Eq. (5.1.20). When β is infinity (i.e., when the stiffness is zero), the equations reduce to those of a flexible string, and the allowed values of ν are

$$\nu_1 = \frac{0.5000}{l} \sqrt{\frac{T}{\rho S}} \qquad \begin{matrix} \nu_2 = 2\nu_1 \\ \nu_3 = 3\nu_1 \\ \nu_4 = 4\nu_1 \\ \cdots\cdots\cdots \end{matrix} \qquad (5.1.32)$$

If β is large but not infinite (i.e., if the wire is stiff but the tension is the more important restoring force), it is possible to obtain an approximate expression for the allowed frequencies. When β is large, $\tanh(\pi l\sqrt{2\beta^2 + \mu_2{}^2})$ is very nearly unity, and $\sqrt{1 + (2\beta^2/\mu_2{}^2)}$ is a large quantity for the lower overtones (i.e., as long as $\mu_2{}^2$ is small compared with β). Equation (5.1.29) for the even functions then states that $\tan(\pi l\mu_2)$ is a very large negative number, which means that $\pi l\mu_2$ is a small amount larger than $\pi/2$ or a small amount larger than $3\pi/2$. In general, $\pi l\mu_2 = (m + \tfrac{1}{2})\pi + \delta$, where δ is a small quantity and m is some integer. Expanding both sides of Eq. (5.1.29) and retaining only the first terms, we obtain

$$\frac{1}{\delta} = \sqrt{2\beta^2}\,\frac{l}{(m + \tfrac{1}{2})}$$

or
$$\delta = \pi(2m + 1)\sqrt{\frac{QS\kappa^2}{Tl^2}}$$

where $2m + 1$ is any odd integer.

Equation (5.1.30) for the odd functions states that $\tan(\pi\mu_2 l)$ must be very small, which means that for these cases $\pi l\mu_2$ is a small amount larger than π or a small amount larger than 2π, or in general, $\pi l\mu_2 = k\pi + \delta$, where δ is small and k is some integer. The corresponding equation for δ is $\delta = \pi(2k)\sqrt{QS\kappa^2/Tl^2}$, where $2k$ is any even integer.

We have shown that for either even or odd functions we must have $\pi l\mu_2 = (n\pi/2) + \delta$, where n is any integer and where $\delta = \pi n\sqrt{QS\kappa^2/Tl^2}$. The even solutions correspond to the odd values of n ($n = 2m + 1$), and the odd functions correspond to the even values of n ($n = 2k$). The allowed values of μ_2 are given approximately by

$$\mu_2 \simeq \frac{n}{2l}\left(1 + \frac{2}{l}\sqrt{\frac{QS\kappa^2}{T}}\right)$$

and the allowed values of the frequency are

$$v_n \simeq \frac{n}{2l}\sqrt{\frac{T}{\rho S}}\left(1 + \frac{2}{l}\sqrt{\frac{QS\kappa^2}{T}}\right)$$

A more accurate formula can be obtained by retaining the next terms in the series expansions that we have made:

$$v_n \simeq \frac{n}{2l}\sqrt{\frac{T}{\rho S}}\left[1 + \frac{2}{l}\sqrt{\frac{QS\kappa^2}{T}} + \left(4 + \frac{n^2\pi^2}{2}\right)\frac{QS\kappa^2}{Tl^2}\right]$$

$$n = 1, 2, 3, \ldots \quad (5.1.33)$$

This formula is valid as long as n^2 is smaller than $l^2 T/\pi^2 QS\kappa^2$.

When the "stiffness constant" $QS\kappa^2$ becomes negligible compared with l^2T, this formula reduces to (5.1.32), the equation for the allowed frequencies of a flexible string. As the stiffness is increased, the allowed frequencies increase, the frequencies of the overtones increasing relatively somewhat more rapidly than the fundamental, so that they are no longer harmonics.

To obtain values for the higher overtones or for the cases where $QS\kappa^2$ is the same size as, or larger than, l^2T, we must solve Eqs. (5.1.29) and (5.1.30) numerically.

5.2 WAVE MOTION IN MEMBRANES

Before going on to study wave motion in air, it also is useful to consider some examples of two-dimensional wave motion, as illustrated by the vibrations of a flexible membrane under tension or of an elastic plate. We shall find that the addition of another dimension introduces a number of additional phenomena, which are most easily exhibited in the case of the membrane.

For example, the tension in the membrane, which tends to return the membrane to its equilibrium configuration, cannot be expressed in terms of a single magnitude, as was the case with the string. Here the tension at a point must be given in terms of the pull across an elementary length of a line drawn through the point, equaling the force tending to split the membrane along this line. The pull $T\,ds$ will be proportional to the length ds of the line element; the proportionality factor T, the force per unit length of line, is called a *stress*. It is a vector, which may or may not be perpendicular to the line element. Thus the stress in the membrane, in general, is a function of the position of the line element on the membrane, but also of its orientation.

We shall not enter into any discussion of the general case, but confine our attention to membranes in which the tensile stress is the same at every point on the membrane and for every orientation of the line element, being always perpendicular to this line element. In this special case the stress can be specified by a single quantity T, which we can call the tension. For this to be the case, the membrane must be elastic and deformable, like a sheet of rubber rather than a sheet of paper. Also, the outward pull across each unit length of the membrane's perimeter must be T.

It is possible for a membrane of this sort to behave like an assemblage of parallel strings, having waves whose crests are in parallel lines, perpendicular to their direction of propagation (like waves on the ocean). The behavior of such waves is exactly like the behavior of waves on a flexible string: the waves travel with unchanged shape (when friction is neglected), and every such wave travels with equal speed. But this is only the simplest form of wave motion that the membrane can have. It can have circular waves, radiating out from a point or in toward it; it can have elliptical ones, going out from or in to a line segment; and so on. This more complicated sort of wave motion does not conserve its form as it travels, nor does it have a speed independent of its form.   Much of it is so complicated in behavior that we are unable to analyze it at all.   In this book we shall content ourselves with a treatment of parallel and circular waves.

If the sheet of material is not perfectly flexible (i.e., is a plate), the motion is still more complicated than that for a membrane.   We shall deal only with the simplest possible case of such motion later in this chapter.

Forces on a membrane

Our first task is to set up the equation of motion for the membrane. The procedure is similar to that for the string.   We shall find that any part of the membrane having a bulge facing away from the equilibrium plane will be accelerated toward the plane, and vice versa.   In general, the acceleration of any portion is proportional to its "bulginess" and opposite in direction. We must find a quantitative measure of this bulginess.

Suppose that the membrane has a density of σ g per sq cm and that it is pulled evenly around its edge with a tension T dynes per cm length of edge. If it is perfectly flexible, this tension will be distributed evenly throughout its area, the material on opposite sides of any line segment of length dx being pulled apart with a force of $T\,dx$ dynes.   The displacement of the membrane from its equilibrium position will be called η.   It is a function of time and of the position on the membrane of the point in question.   If we use rectangular coordinates to locate the point, η will be a function of x, y, and t.

Referring to the analogous argument on page 98 for the string and to the first drawing of Fig. 5.5, we see that the net force on an element $dx\,dy$ of the

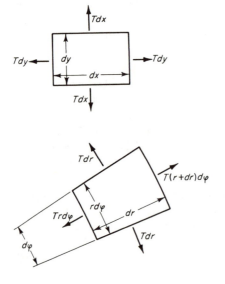

FIGURE 5.5
Forces on an element of a membrane in rectangular and polar coordinates.

membrane due to the pair of tensions $T\,dy$ will be

$$T\,dy\left[\left(\frac{\partial\eta}{\partial x}\right)_{x+dx}-\left(\frac{\partial\eta}{\partial x}\right)_{x}\right]=T\frac{\partial^2\eta}{\partial x^2}\,dx\,dy$$

and that due to the pair $T\,dx$ will be $T(\partial^2\eta/\partial y^2)\,dx\,dy$. The sum of these is the net force on the element and must equal the element's mass $\sigma\,dx\,dy$ times its acceleration. The wave equation for the membrane is therefore

$$\frac{\partial^2\eta}{\partial x^2}+\frac{\partial^2\eta}{\partial y^2}=\frac{1}{c^2}\frac{\partial^2\eta}{\partial t^2}\qquad c=\sqrt{\frac{T}{\sigma}}\qquad(5.2.1)$$

The Laplacian operator

The left-hand side of this equation is the expression giving the measure of the bulginess (or rather, the negative bulginess) of the portion of the membrane under consideration.

We now encounter one of the extra difficulties of the two-dimensional analysis; for we find that if we had picked polar instead of rectangular coordinates, so that η was a function of r and ϕ, then the resulting wave equation would have a different aspect. The net force due to the tensions perpendicular to the radius (shown in the second drawing of Fig. 5.5) is

$$T\,dr\left[\left(\frac{\partial\eta}{r\,\partial\phi}\right)_{\phi+d\phi}-\left(\frac{\partial\eta}{r\,\partial\phi}\right)_{\phi}\right]=\frac{T}{r^2}\frac{\partial^2\eta}{\partial\phi^2}\,r\,dr\,d\phi$$

and that due to the tensions parallel to the radius is

$$T\,d\phi\left[\left(r\frac{\partial\eta}{\partial r}\right)_{r+dr}-\left(r\frac{\partial\eta}{\partial r}\right)_{r}\right]=\frac{T}{r}\frac{\partial}{\partial r}\left(r\frac{\partial\eta}{\partial r}\right)r\,dr\,d\phi$$

The resulting equation of motion in polar coordinates is

$$\frac{1}{r}\frac{\partial}{\partial r}\left(r\frac{\partial\eta}{\partial r}\right)+\frac{1}{r^2}\frac{\partial^2\eta}{\partial\phi^2}=\frac{1}{c^2}\frac{\partial^2\eta}{\partial t^2}\qquad(5.2.2)$$

The left-hand side of this equation has a different form from the left-hand side of (5.2.1). This does not mean that it represents a different property of the membrane; for the value of the left side of (5.2.2) for some point on a given membrane will have the same value as the left side of (5.2.1) for the same point on the same membrane. It simply means that when we wish to measure the bulginess of a portion of a membrane by using polar coordinates, we must go at it differently from the way we should have gone at it if we had used rectangular coordinates.

We often emphasize the fact that the left side of both equations represents the same property of the surface η, by writing the wave equation as

$$\nabla^2\eta=\frac{1}{c^2}\frac{\partial^2\eta}{\partial t^2}\qquad(5.2.3)$$

where the symbol ∇^2 is called the *Laplacian operator*, or simply the *Laplacian*. It stands for the operation of finding the bulginess of the surface at some point. In different coordinates the operator takes on different forms:

$$\nabla^2 = \begin{cases} \dfrac{\partial^2}{\partial x^2} + \dfrac{\partial^2}{\partial y^2} & \text{(rectangular coordinates)} \\[2ex] \dfrac{1}{r}\dfrac{\partial}{\partial r}\left(r\dfrac{\partial}{\partial r}\right) + \dfrac{1}{r^2}\dfrac{\partial^2}{\partial \phi^2} & \text{(polar coordinates)} \end{cases}$$

The fact that we have different forms for the Laplacian operator corresponds to the fact that the membrane can have different sorts of waves. The form for rectangular coordinates is the natural one for parallel waves; that for polar coordinates is best for circular waves; and so on. Although the Laplacian is a measure of the same property of the membrane, no matter what coordinate system we use, nevertheless there is a great variation in the facility with which we can solve the wave equation in different coordinate systems. In fact, the known methods for its solution are successful only in the case of a few of the simpler coordinate systems.

Tension and shear in the membrane

As with the string, the equation of motion (5.2.3) is valid only when the displacement is small enough so that $|\partial\eta/\partial x|$ and $|\partial\eta/\partial y|$ are everywhere small compared with unity, and even then only when the membrane is flexible enough so that resistance to bending and shear is negligible compared with the transverse force produced by the tension acting on a bulge in the membrane. The membrane must be flexible in bending (as when a sheet of paper is rolled up); it also must not resist the sort of shear distortion produced by a bulge (such as that which causes a sheet of paper to wrinkle or tear when it is forced to conform to a sphere). A small circle drawn on the membrane at equilibrium is no longer a circle when the membrane is displaced if the curvatures $\partial^2\eta/\partial x^2$ and $\partial^2\eta/\partial y^2$ differ in value. In addition, if a circle is drawn on the displaced surface $\eta(x,y)$, the ratio between its length of perimeter l and its radius s, measured along the curved membrane surface, is

$$\frac{l}{s} = 2\pi\left[1 - \frac{s^2}{6}\frac{\partial^2\eta}{\partial x^2}\frac{\partial^2\eta}{\partial y^2} - \frac{s^2}{6}\left(\frac{\partial^2\eta}{\partial x\,\partial y}\right)^2 + \text{higher-order terms}\right]$$

This usually differs from the ratio 2π for a plane surface, implying a distortion of the surface, which material like paper, for example, would resist with internal forces, independent of any tension. For a membrane to move in accordance with Eq. (5.2.3), these forces must be negligible compared with the force of tension.

This limitation is not trivial. Although the shape distortion is second order in the derivatives of η, compared with the first-order terms produced by the tension, the deflection of the membrane produced by a transverse force can be quite large if the force is applied to a small area of the membrane.   To show this we work out the static deflection of a circular membrane, subject to a transverse force F, distributed evenly over a circular area of radius s at the center of the membrane, as shown in Fig. 5.6.   The equation for static deflection, under the influence of a transverse applied force of density $f(x,y)$ per unit area, is, according to Eq. (5.2.3),

$$T \nabla^2 \eta = -f(x,y)$$

For the membrane of Fig. 5.6, the force density is $F/\pi s^2$ for $0 < r < s$ and zero for $s < r < R$.   Therefore the equation for η is

$$\frac{1}{r} \frac{\partial}{\partial r} \left(r \frac{\partial \eta}{\partial r} \right) + \frac{1}{r^2} \frac{\partial^2 \eta}{\partial \phi^2} = \begin{cases} -\dfrac{F}{\pi s^2 T} & 0 \leqslant r < s \\ 0 & s < r \leqslant R \end{cases}$$

with the boundary condition that $\eta = 0$ at $r = R$.

Symmetry requires that η be independent of ϕ.   The solution in the region of application of the force is $\eta(r) = \Delta + (F/4\pi s^2 T)(s^2 - r^2)$, and in the outer region it is $A \ln (R/r)$ (which goes to zero at $r = R$), where Δ and A must be adjusted so that η and $d\eta/dr$ are continuous at $r = s$.   The result is

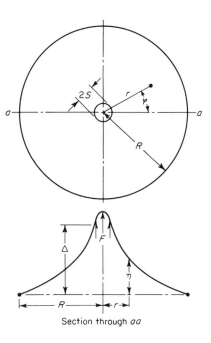

FIGURE 5.6
Top and sectional view of circular membrane displaced by force F concentrated on
a small circular area at its center.

Section through aa

$A = F/2\pi T$ and $\Delta = (F/2\pi T) \ln (R/s)$, so that

$$
\eta =
\begin{cases}
\dfrac{F}{2\pi T}\left[\ln \dfrac{R}{s} + \dfrac{1}{2s^2}(s^2 - r^2)\right] & 0 \leqslant r < s \\[1.5em]
\dfrac{F}{2\pi T}\ln \dfrac{R}{r} & s < r \leqslant R
\end{cases}
\tag{5.2.4}
$$

The displacement Δ of the region of application of the force goes to infinity as $s \to 0$, as the area of application of the force is made small enough. No matter how small the applied force F is, compared with the tension T, the deflection caused by F can be made as large as we wish by concentrating its push into a small enough area of the membrane.

This is in contrast with the flexible string. A transverse force F concentrated at a point x of the string is balanced by the discontinuity in slope of the string.

$$
F = T \lim_{\delta \to 0}\left[\left(\frac{dy}{dx}\right)_{x-\delta} - \left(\frac{dy}{dx}\right)_{x+\delta}\right]
$$

If F is small compared with T, the discontinuity is small and the slopes themselves are small. But to balance a force F concentrated within a circle of radius s requires a downward slope at the edge of the circle of

$$
F = -T \int_0^{2\pi}\left(\frac{\partial \eta}{\partial r}\right)_s s\, d\phi \quad \text{or} \quad \left\langle\frac{\partial \eta}{\partial r}\right\rangle_s \equiv \frac{1}{2\pi}\int_0^{2\pi}\left(\frac{\partial \eta}{\partial r}\right)_s d\phi = -\frac{F}{2\pi s T}
$$

If radius s is made small enough, the mean downward slope $-\langle\partial\eta/\partial r\rangle_s$ at the edge of the circle can be large even if F is much smaller than T. The reason is that the tension is applied over a length of perimeter, and if this length $2\pi s$ is small, the mean slope must be large to balance the force F.

As a result, Eq. (5.2.3) is not adequate to describe the actual motion of a membrane if the forces involved are concentrated or if the membrane is not flexible to shear as well as to bending. The formulas which follow are all subject to this limitation.

Energy and momentum density

The energy and stress densities in the membrane are given by expressions similar to Eqs. (4.1.9) and (4.1.12) for the string. In continuation of this earlier discussion, and in preparation for the similar discussion for sound waves, we shall work out in some detail the formalism of the Lagrange equation for the membrane. If the mass of the membrane is σ per unit area, its kinetic energy of transverse motion of an element $dx\, dy$ of its area can be written $\frac{1}{2}\sigma\eta_t^2 dx\, dy$, where we have written η_t instead of $\partial\eta/\partial t$, to save space. To place a membrane which is elastic enough to satisfy the requirements mentioned in the previous pages under a tension T will stretch its dimensions laterally by an amount proportional to T, even when the membrane is not displaced transversally (i.e., when $\eta = 0$). Transverse displacement from equilibrium ($\eta \neq 0$) will increase its area still further, and thus will require further work, equal to the increase in area times the tension per unit length. If the displacement is not large, the increase in area caused by the displacement η is the integral of

$$dx\, dy(\sqrt{1 + \mathrm{grad}^2\, \eta} - 1) \simeq \tfrac{1}{2}(\eta_x{}^2 + \eta_y{}^2)\, dx\, dy$$

where $\eta_x = \partial\eta/\partial x$ and $\eta_y = \partial\eta/\partial y$. Because the additional increase in area is small (if η_x and η_y are small), the tension is not changed, to the first order, so that the work required to displace the membrane is the integral of $\tfrac{1}{2}T(\eta_x{}^2 + \eta_y{}^2)\, dx\, dy$, which is the potential energy of displacement.

The Lagrange density L is, as before, the difference between the kinetic and the potential energy densities,

$$L(\eta,\eta_t,\eta_x,\eta_y) = \tfrac{1}{2}\sigma\eta_t{}^2 - \tfrac{1}{2}T(\eta_x{}^2 + \eta_y{}^2) \tag{5.2.5}$$

and the total Lagrange function for the membrane is the integral of L over the whole area of the membrane. Hamilton's principle of dynamics states that the displacement of the system (in this case η) adjusts itself in shape and velocity so that the integral of this Lagrange function over time, from some initial time t_0 to a specified final time t_1, is minimal.

For example, if $\eta_c(x,y,t)$ is the correct solution, any slight deviation away from this form, such as that given by $\eta = \eta_c + \delta\zeta(x,y,t)$, which does not change the membrane's boundary conditions or its shape at t_0 or t_1 (i.e., where ζ is zero at t_0 and t_1 and along the membrane's boundary) will produce *no first-order change* in the integral of L over t and over area. Any deviation in the value of η from its correct value η_c will increase the kinetic energy over certain parts of the region and increase the potential energy over some other parts, and if the departure from correct shape is not large, the two kinds of changes will cancel to the first order. Thus Hamilton's principle provides us with a mathematical test, which we can apply to some assumed η, to tell whether the assumed η does satisfy the equations of motion.

In fact, Hamilton's principle is equivalent to the equation of motion of the membrane, plus the boundary and initial conditions, as the following argument will show. We substitute $\eta = \eta_c + \delta\zeta$ in the form for $L(\eta,\eta_t,\eta_x,\eta_y)$ and expand L in a power series in δ; the change in L caused by the change in η is then

$$\delta L \equiv L(\eta_c + \delta\zeta,\, \eta_{ct} + \delta\zeta_t,\, \eta_{cx} + \delta\zeta_x,\, \eta_{cy} + \delta\zeta_y) - L(\eta_c,\eta_{ct},\eta_{cx},\eta_{cy})$$

$$= \delta\left[\zeta\left(\frac{\partial L}{\partial \eta}\right)_{\eta=\eta_c} + \frac{\partial\zeta}{\partial t}\left(\frac{\partial L}{\partial \eta_t}\right)_{\eta=\eta_c} + \frac{\partial\zeta}{\partial x}\left(\frac{\partial L}{\partial \eta_x}\right)_{\eta=\eta_c}\right.$$

$$\left. + \frac{\partial\zeta}{\partial y}\left(\frac{\partial L}{\partial \eta_y}\right)_{\eta=\eta_c} + \text{higher-order terms}\right]$$

where $\eta_t = \partial\eta/\partial t$, etc.   The integral of this difference can be simplified by integration by parts.   For example,

$$\int_{t_0}^{t_1} \frac{\partial \zeta}{\partial t} \left(\frac{\partial L}{\partial \eta_t}\right) dt = \left[\zeta\left(\frac{\partial L}{\partial \eta_t}\right)\right]_{t_0}^{t_1} - \int_{t_0}^{t_1} \zeta \frac{\partial}{\partial t} \left(\frac{\partial L}{\partial \eta_t}\right) dt$$

and the first term on the right-hand side is zero because ζ goes to zero at t_0 and t_1.   Therefore

$$\iiint \delta L \, dx \, dy \, dt = \delta \iiint \zeta \left[\frac{\partial L}{\partial \eta} - \frac{\partial}{\partial t}\left(\frac{\partial L}{\partial \eta_t}\right) - \frac{\partial}{\partial x}\left(\frac{\partial L}{\partial \eta_x}\right) - \frac{\partial}{\partial y}\left(\frac{\partial L}{\partial \eta_y}\right)\right] dx \, dy \, dt$$

plus terms in higher powers of δ.   For this to be zero to the first order in δ, η_c must satisfy the equation

$$\frac{\partial}{\partial t}\left(\frac{\partial L}{\partial \eta_t}\right) + \frac{\partial}{\partial x}\left(\frac{\partial L}{\partial \eta_x}\right) + \frac{\partial}{\partial y}\left(\frac{\partial L}{\partial \eta_y}\right) - \frac{\partial L}{\partial \eta} = 0 \qquad (5.2.6)$$

which is the *Lagrange-Euler equation* for the membrane.

If, in addition, a transverse driving force $f(x,y,t) \, dx \, dy$ is applied to the element of membrane at x, y, the Lagrange density has a term $f\eta$ added to it and the Lagrange-Euler equation is still Eq. (5.2.6).   For a Lagrange density of the form of Eq. (5.2.5), plus $f\eta$, with the first part quadratic in the partials of η and only the $f\eta$ term dependent on η itself, this becomes

$$\frac{\partial}{\partial t}(\sigma\eta_t) - \frac{\partial}{\partial x}(T\eta_x) - \frac{\partial}{\partial y}(T\eta_y) \equiv \sigma \frac{\partial^2 \eta}{\partial t^2} - T\nabla^2\eta = f \qquad (5.2.7)$$

which is, of course, the wave equation.

We note that if we had used polar coordinates r, φ instead of x, y, the integral would have been $\iiint \delta L r \, dr \, d\varphi \, dt$, and the integration by parts over the space derivative part of δL is

$$\delta \iint \left[\zeta_r\left(\frac{\partial L}{\partial \eta_r}\right) + \zeta_\varphi r\left(\frac{\partial L}{\partial \eta_\varphi}\right)\right] dr \, d\varphi = -\delta \iint \zeta \left[\frac{1}{r}\frac{\partial}{\partial r} r\left(\frac{\partial L}{\partial \eta_r}\right)\right.$$
$$\left. + \frac{\partial}{\partial \varphi}\left(\frac{\partial L}{\partial \eta_\varphi}\right)\right] r \, dr \, d\varphi$$

which leads, when $L = \frac{1}{2}\sigma\eta_t^2 - \frac{1}{2}T\eta_r^2 - \frac{1}{2}(T/r^2)\eta_\varphi^2$ is used, to

$$\frac{\partial}{\partial t}(\sigma\eta_t) - \frac{1}{r}\frac{\partial}{\partial r}(Tr\eta_r) - \frac{\partial}{\partial \varphi}\left(\frac{T\eta_\varphi}{r^2}\right) \equiv \sigma\frac{\partial^2\eta}{\partial t^2} - T\nabla^2\eta = f(r,\varphi,t)$$

precisely the modified form for the differential operator ∇^2 which was given following Eq. (5.2.3).

The stress-energy tensor

We now can extend the symbolism of Eqs. (4.1.8) to (4.1.12) to obtain the energy density and energy flux for wave motion. We define

$$W_{tt} = \eta_t \left(\frac{\partial L}{\partial \eta_t} \right) - L = \tfrac{1}{2}\sigma \left(\frac{\partial \eta}{\partial t} \right)^2 + \tfrac{1}{2} T \left[\left(\frac{\partial \eta}{\partial x} \right)^2 + \left(\frac{\partial \eta}{\partial y} \right)^2 \right] = H \quad (5.2.8)$$

which is called the *Hamiltonian density*. It is the sum of the kinetic and potential energy densities of transverse motion of the element $dx\,dy$ of membrane at x, y. Related to this is the two-dimensional vector $\mathbf{W}_t$, with components

$$W_{tx} = \eta_t \left(\frac{\partial L}{\partial \eta_x} \right) = -T \left(\frac{\partial \eta}{\partial t} \right) \left(\frac{\partial \eta}{\partial x} \right) \qquad W_{ty} = \eta_t \left(\frac{\partial L}{\partial \eta_y} \right) = -T \left(\frac{\partial \eta}{\partial t} \right) \left(\frac{\partial \eta}{\partial y} \right)$$

$$(5.2.9)$$

which is the generalization of the energy flux W_{tx} of Eq. (4.1.9) for the string. The direction of $\mathbf{W}_t$ is the direction of the energy flow; the magnitude is the rate of energy flow, along the membrane, per unit length perpendicular to the flow.

That $\mathbf{W}_t$ is actual energy flux is shown in several ways. In the first place, the dimensions of $\mathbf{W}_t$ are force per unit length times velocity, in other words, power per unit length. More importantly, however, is the relationship between $\mathbf{W}_t$ and the energy density W_{tt}. If $\mathbf{W}_t$ is really the energy flux vector, the divergence of $\mathbf{W}_t$,

$$\text{div}\,(\mathbf{W}_t) = \frac{\partial W_{tx}}{\partial x} + \frac{\partial W_{ty}}{\partial y} \qquad \text{(for rectangular coordinates)}$$

$$= \frac{1}{r}\frac{\partial r W_{tr}}{\partial r} + \frac{1}{r}\frac{\partial W_{t\varphi}}{\partial \varphi} \qquad \text{(for polar coordinates)}$$

$$= \cdots\cdots\cdots\cdots\cdots\cdots\cdots\cdots\cdots\cdots\cdots\cdots$$

represents the net outflow of energy from a unit area of membrane, and thus must equal the rate of loss $-(\partial W_{tt}/\partial t)$ of energy from this unit area. The resulting equation is $(\partial W_{tt}/\partial t) + \text{div}\,(\mathbf{W}_t) = Q$, where Q, the rate of introduction of energy into the area, is called the equation of continuity. It has been used before [Eq. (4.1.11)] and will be used many times again [see, for example, Eq. (6.1.12)]. That W_{tt} and $\mathbf{W}_t$ do satisfy the equation of continuity, and thus, that if W_{tt} is energy density, $\mathbf{W}_t$ must be energy flux, is demonstrated by inserting Eqs. (5.2.8) and (5.2.9) into

$$\frac{\partial W_{tt}}{\partial t} + \text{div}\,\mathbf{W}_t = \partial \eta_t \eta_{tt} + T(\eta_x \eta_{tx} + \eta_y \eta_{ty})$$

$$- T(\eta_t \eta_{xx} + \eta_x \eta_{tx} + \eta_t \eta_{yy} + \eta_y \eta_{ty})$$

$$= \eta_t [\partial \eta_{tt} - T(\eta_{xx} + \eta_{yy})] = \eta_t f(x,y,t)$$

where we have used the wave equation (5.2.7) to obtain the last expression. This is, of course, the rate of introduction of energy into the membrane by the applied force density f. If $f = 0$, then div $\mathbf{W}_t$ is just equal to $-(\partial W_{tt}/\partial t)$, which again demonstrates that $\mathbf{W}_t$ is energy flux.

The quantities

$$W_{xx} = \eta_x\left(\frac{\partial L}{\partial \eta_x}\right) - L = \tfrac{1}{2}\sigma\eta_t^2 - \tfrac{1}{2}T(\eta_x^2 - \eta_y^2)$$

$$W_{yy} = \eta_y\left(\frac{\partial L}{\partial \eta_y}\right) - L = \tfrac{1}{2}\sigma\eta_t^2 + \tfrac{1}{2}T(\eta_x^2 - \eta_y^2)$$

$$W_{xy} = \eta_x\left(\frac{\partial L}{\partial \eta_y}\right) = -T\eta_x\eta_y = W_{yx} = \eta_y\left(\frac{\partial L}{\partial \eta_x}\right)$$

are components of the tensor which represents the stress in the membrane surface produced by the deflection. For example, the force $\mathbf{W}_x\,dy = (W_{xx}\,dy,\ W_{xy}\,dy)$ is the stress across a line of length dy, perpendicular to the x axis, at point x, y. Its x component, $W_{xx}\,dy$, is the increment of tension in the x direction produced by the displacement; its y component, $W_{xy}\,dy$, is the component of shear, discussed on page 194. The force $\mathbf{W}_y\,dx = (W_{yx}\,dx,\ W_{yy}\,dx)$ is the related force across a line element perpendicular to y.

Corresponding to these stress components there is an along-the-membrane momentum vector, with components

$$W_{xt} = \eta_x\left(\frac{\partial L}{\partial \eta_t}\right) = \partial\eta_x\eta_t \qquad W_{yt} = \eta_y\left(\frac{\partial L}{\partial \eta_t}\right) = \sigma\eta_y\eta_t$$

analogous to the longitudinal momentum of the string, discussed on page 103. These quantities also satisfy equations of continuity

$$\left(\frac{\partial W_{xt}}{\partial t}\right) + \text{div } \mathbf{W}_x = \eta_x[\sigma\eta_{tt} - T(\eta_{xx} + \eta_{yy})] = \eta_x f$$

where $\eta_x f = (\partial\eta/\partial x)f(x,y,t)$ is the effective applied force density in the x direction at point x, y which produces momentum flow in the x direction (Fig. 4.6).

Beyond setting up the foregoing equations, to be compared with those for sound, it is not advantageous to investigate the motion of the membrane in much detail here. Motion of an actual membrane is strongly affected by its reaction with the air, and calculations of such coupled motion must be discussed later. Here we treat a few simple motions of membranes *in vacuo*.

Wave motion on an unlimited membrane

First let us neglect even the effects of the membrane boundaries, which support its perimeter and maintain its tension, and discuss traveling waves, as we did at the beginning of the discussion of the string.

As was mentioned earlier, a membrane can act like an assemblage of strings, each transmitting an identical wave, in parallel. For example, the function

$$\eta(x,y,t) = F_\theta(x \cos \theta + y \sin \theta - ct)$$

is a solution of the wave equation (5.2.1). This represents a wave traveling in a direction at angle θ with respect to the x axis, with velocity $c = \sqrt{T/\sigma}$. The wavefronts [for example, the line $x \cos \theta + y \sin \theta - ct = 0$ corresponding to $F(0)$] are all lines perpendicular to this direction; so it can be called a *linear wave*. The membrane bends like a sheet of paper, and there is no surface shear; the wave continues, without change of form until it strikes a boundary, just as does a wave on a flexible string.

If this were the only possible kind of wave, there would be little to say beyond what has been said for the string. But since the wave equation is linear in η, a superposition of waves, going in different directions, is also a solution. This superposition is also possible in the case of the string, of course, but in the case of the membrane, the waves may be in different directions, producing a motion which cannot be duplicated by waves in parallel strings. A general sort of wave can be built up by superposing a sequence of linear waves, each going in a different direction θ, and each, if we wish, having a different form, the combination represented by the integral

$$\eta(x,y,t) = \int_0^{2\pi} F_\theta(x \cos \theta + y \sin \theta - ct) \, d\theta \qquad (5.2.10)$$

Since each component of the integral is a solution of the wave equation, the integral is a solution. However, this solution is not a linear wave; it will exhibit properties differing from waves on a string.

We can see the difference more clearly if we express the waveform in terms of its initial conditions, as we did in Eq. (4.2.1), by specifically putting in the formula and limits for S and by integrating and differentiating y_0. The displacement of the string, in terms of its initial displacement $y_0(x)$ and its initial velocity $v_0(x)$, is

$$y(x,t) = \frac{1}{2c} \frac{\partial}{\partial t} \int_{x-ct}^{x+ct} y_0(X) \, dX + \frac{1}{2c} \int_{x-ct}^{x+ct} v_0(X) \, dX \qquad (5.2.11)$$

Point x is the *point of observation*, where the displacement of the string is observed at time t. Point X is an initial point used to specify the displacement and velocity of the string at $t = 0$. Equation (5.2.11) states that the displacement of the observation point at time t is a sum of the effects of all initial points within a distance ct of the observation point. When t is small, only the initial values for nearby parts of the string play a part; when t is large, the initial values of a long segment of the string contribute to the displacement at x.

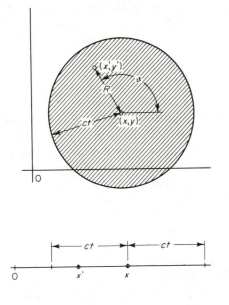

FIGURE 5.7
Dependence of the wave motion of a
string and membrane on initial conditions.
The displacement of the point (x,y) on the
membrane at the time t depends on the
initial conditions for that part of the membrane enclosed by a circle of radius ct
with center (x,y). The displacement of
point x on the string depends on the
initial conditions for the portion within
a distance ct on either side of x.

For the string, all initial values between $x - ct$ and $x + ct$ contribute
with equal weight, which ensures that the waveform is unchanged as it
progresses. For example, if y_0 and v_0 differ from zero only in the range
$-\Delta < X < 0$, then at $x > 0$ the string will be at rest until time $t = x/c$,
since neither of the integrals of Eq. (5.2.11) would include the active region
of X before that time. Moreover, after $t = (1/c)(x + \Delta)$, the string at point
x is again at rest. A simple example is shown in Fig. 5.8; we see that waves
on the string propagate without change of form, leaving no "wake" behind.

The analogous expression for the displacement of a point x, y on a
membrane at time t, in terms of the initial displacement η_0 and velocity v_0
of the point X, Y, is

$$\eta(x,y,t) = \frac{1}{2\pi c} \frac{\partial}{\partial t} \int_0^{2\pi} d\theta \int_0^{ct} \frac{y_0(X,Y)}{\sqrt{c^2 t^2 - R^2}} R \, dR$$

$$+ \frac{1}{2\pi c} \int_0^{2\pi} d\theta \int_0^{ct} \frac{v_0(X,Y)}{\sqrt{c^2 t^2 - R^2}} R \, dR \quad (5.2.12)$$

where R is the magnitude of the vector joining the observation point x, y to
the initial point X, Y, and θ is its direction, as shown in Fig. 5.7. Thus all
initial points within a range ct of the observation point contribute to the
displacement η at time t, but in contrast to the string, different points contribute different amounts, in accord with the factor $(c^2 t^2 - R^2)^{-\frac{1}{2}}$. The
initial points on the perimeter of the circle of radius ct, around the observation
point, contribute the most; the points inside this circle contribute progressively

less, the closer they are to the observation point.   If the nearest point on the membrane for which η_0 or v_0 is nonzero is a distance R_0 from the observation point, the observation point will be first displaced at time (R_0/c).   Thus the initial rise of the circular wave generated by a localized disturbance travels

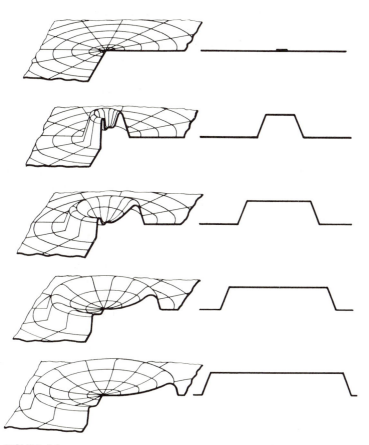

FIGURE 5.8
Comparison between the behavior of a struck string and a struck membrane. The initial displacements are zero, and the center portions are started upward; the lower sketches show the shapes at successive instants later. One-quarter of the membrane has been cut away to show the shape of the cross section.

with velocity c, as for the string.   But for the membrane, the shape of the wave changes, its amplitude decreases as it spreads out, and the wave leaves a "wake" of residual motion behind it.   Figure 5.8 illustrates this effect and compares it with the simpler behavior of waves on the string.

Rectangular membrane; standing waves

As with the string, when the motion has gone on long enough so that reflections from the membrane boundaries contribute to the motion, standing waves are set up, and the motion resembles that of a series of simple oscillators. For example, suppose the membrane is stretched between a rectangular frame of sides a and b, each unit length of boundary exerting a pull T on the membrane. The motion can most easily be expressed in terms of the rectangular coordinates, x being the axis parallel to the side of length a, and y being in the direction of the sides of length b, with the origin at one corner of the boundary.

The wave equation for the displacement $\eta(x,y,t)$ of the point x, y of the membrane, away from its equilibrium position, at time t is

$$\frac{\partial^2 \eta}{\partial x^2} + \frac{\partial^2 \eta}{\partial y^2} = \frac{1}{c^2}\frac{\partial^2 \eta}{\partial t^2} \qquad c^2 = \frac{T}{\sigma} \qquad (5.2.13)$$

if we can assume that there is no force on the membrane other than that caused by the tension T, and if the magnitude of grad η is everywhere small compared with unity. To analyze the motion we can first ask whether the membrane can oscillate with simple-harmonic motion.

If we wish to fit rectangular boundaries, we must use a standing wave having a set of nodal lines (lines along which the displacement is zero) parallel to the x axis and another set parallel to the y axis. Now, the only way that we can have a nodal line parallel to the y axis (i.e., for η to be zero for a given value of x for all values of y) is to have a factor of η which is a function of x only and which goes to zero at the value of x corresponding to the nodal line. The nodal lines parallel to the x axis would require a factor depending on y only. Finally, if the motion is simple harmonic, the time dependence must come in the factor $e^{-2\pi i v t}$. Therefore an expression for the form of the membrane satisfying our requirements must be

$$\eta = X(x)\,Y(y)e^{-2\pi i v t}$$

where substitution in Eq. (5.2.13) shows that

$$Y\frac{\partial^2 X}{\partial x^2} = -\frac{4\pi^2 v^2}{c^2}\,XY - X\frac{\partial^2 Y}{\partial y^2}$$

or

$$\frac{1}{X}\frac{\partial^2 X}{\partial x^2} = -\frac{1}{Y}\left(\frac{4\pi^2 v^2}{c^2}\,Y + \frac{\partial^2 Y}{\partial y^2}\right)$$

The left-hand side of this equation is a function of x only, whereas the right-hand side is a function of y only. Now, a function of y cannot equal a function of x for all values of x and y if both functions really vary with x and y, respectively; so that the only possible way for the equation to be true is for both sides to be *independent of both* x and y, that is, to be a constant. Suppose that we call the constant $-(4\pi^2\zeta^2/c^2)$. Then the equation reduces to two simpler equations:

$$\frac{\partial^2 X}{\partial x^2} = -\frac{4\pi^2 \zeta^2}{c^2} X \qquad \frac{\partial^2 Y}{\partial y^2} = -\frac{4\pi^2}{c^2}(\nu^2 - \zeta^2)Y$$

The solution of this pair of equations is

$$\eta = A \cos\left(\frac{2\pi\zeta}{c}x - \Omega_x\right)\cos\left(\frac{2\pi\tau}{c}y - \Omega_y\right)\cos\left(2\pi t \sqrt{\zeta^2 + \tau^2} - \Phi\right)$$

$$(5.2.14)$$

where we have set $\tau = \sqrt{\nu^2 - \zeta^2}$.

The normal modes

If the boundary conditions are that η must be zero along the edges of a rectangle composed of the x and y axes, the line $x = a$ and the line $y = b$, it is not difficult to see that $\Omega_x = \Omega_y = \pi/2$, that $2\zeta a/c$ must equal an integer, and that $2\tau b/c$ must equal an integer (not necessarily the same integer). The characteristic functions, giving the possible shapes of the membrane as it vibrates with simple-harmonic motion, are

$$\eta = A\psi_{mn}(x,y)\cos(2\pi\nu_{mn}t - \Phi)$$

$$\psi_{mn}(x,y) = \sin\frac{\pi m x}{a}\sin\frac{\pi n y}{b} \qquad (5.2.15)$$

where $\qquad \nu_{mn} = \frac{1}{2}\sqrt{\frac{T}{\sigma}}\sqrt{\left(\frac{m}{a}\right)^2 + \left(\frac{n}{b}\right)^2} \qquad \begin{matrix} m = 1, 2, 3, \ldots \\ n = 1, 2, 3, \ldots \end{matrix}$

The shapes of the first normal modes are shown in Fig. 5.9. We notice that the number of nodal lines parallel to the y axis is $m - 1$ and the number parallel to the x axis, $n - 1$.

The fundamental frequency is ν_{11}, depending on T and σ in a manner quite analogous to the case of the string. Among the allowed frequencies are all the harmonics of the fundamental, $\nu_{22} = 2\nu_{11}$, $\nu_{33} = 3\nu_{11}$, etc. But there are many more allowed frequencies which are not harmonics. When a is nearly equal to b, 2 extra overtones, ν_{12} and ν_{21}, come in between the first and second harmonic; 6 extra, $\nu_{13}, \nu_{31}, \nu_{23}, \nu_{32}, \nu_{14}, \nu_{41}$, all come between the second and third; 10 extra come between the third and fourth; and so on. No matter what the ratio between a and b is, it is possible to show that the average number of overtones between the nth and the $(n + 1)$st harmonics is $\frac{1}{2}\pi n(a^2 + b^2)/ab$, or that the average number of allowed frequencies in the frequency range between ν and $\nu + \Delta\nu$ is $(2\pi\nu ab/c^2)\Delta\nu$.

In the case of the string, the allowed frequencies are equally spaced along the frequency scale, but in the case of the membrane, the allowed frequencies get closer and closer together, the higher the pitch. The higher the pitch, the more overtones there are in a range of frequency of a given size. This property is true of all membranes, no matter what shape their boundary has. It can be shown for any membrane that the average number of allowed frequencies between ν and $\nu + \Delta\nu$ is $(2\pi\nu/c^2)\Delta\nu$ times the area of the membrane. We shall show later how this can be proved.

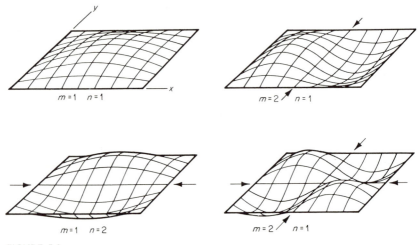

FIGURE 5.9
Shapes of the first four normal modes of a rectangular membrane. Arrows point to the nodal lines.

The degenerate case

An interesting phenomenon occurs when the rectangular membrane is a square one (that is, $a = b$), for then the allowed frequencies become equal in pairs, ν_{mn} being equal to ν_{nm}. There are fewer different allowed frequencies, but there are just as many characteristic functions as when a is not equal to b. This is called a condition of *degeneracy*. There are two different functions, ψ_{mn} and ψ_{nm}, each corresponding to the same frequency (except for the cases $n = m$, which are not degenerate). In such cases the membrane can vibrate with simple-harmonic motion of frequency ν_{mn} with any one of an infinite number of different shapes, corresponding to the different values of γ in the combination

$$\psi_{mn} \cos \gamma + \psi_{nm} \sin \gamma$$

Figure 5.10 shows the shapes of the normal modes of vibration of the square membrane corresponding to ν_{12} and ν_{13}, for different values of γ. The vibrations can be standing waves, corresponding to

$$\eta = (\psi_{mn} \cos \gamma + \psi_{nm} \sin \gamma) \cos (2\pi\nu_{mn}t)$$

The nodal lines have a different shape for each different value of γ.   It is also possible to have traveling waves, corresponding to the expression

$$\eta = \psi_{mn} \cos(2\pi\nu_{mn}t) + \psi_{nm} \sin(2\pi\nu_{mn}t)$$

In this case the nodal lines go through the whole range of possible shapes during each cycle.   It is only in degenerate cases that it is possible to have traveling waves of simple-harmonic motion in a membrane of finite size.

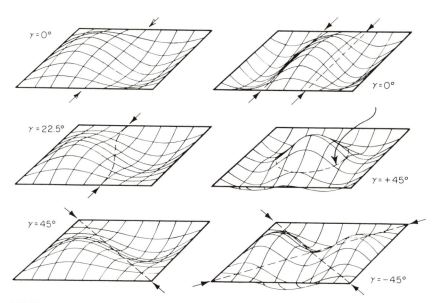

FIGURE 5.10
Various modes of simple-harmonic vibration of a square membrane, for the degenerate cases $\nu_{12} = \nu_{21}$ and $\nu_{13} = \nu_{31}$. Arrows point to the nodal lines.

The characteristic functions

The characteristic functions for the rectangular membrane have the following integral properties:

$$\int_0^a \int_0^b \psi_{mn}\psi_{m'n'} \, dx \, dy = \begin{cases} 0 & m' \neq m \quad \text{or} \quad n' \neq n \\ \dfrac{ab}{4} & m' = m \quad \text{and} \quad n' = n \end{cases} \quad (5.2.16)$$

The behavior of a membrane having an initial shape $\eta_0(x,y)$ and an initial velocity $v_0(x,y)$ is therefore

$$\eta = \sum_{m=1}^{\infty} \sum_{n=1}^{\infty} \psi_{mn}(x,y)[B_{mn} \cos(2\pi\nu_{mn}t) + C_{mn} \sin(2\pi\nu_{mn}t)] \quad (5.2.17)$$

where

$$B_{mn} = \frac{4}{ab} \iint \eta_0 \psi_{mn} \, dx \, dy$$

$$C_{mn} = \frac{2}{\pi \nu_{mn} ab} \iint v_0 \psi_{mn} \, dx \, dy$$

This expression is the correct one for both the degenerate and the non-degenerate cases.

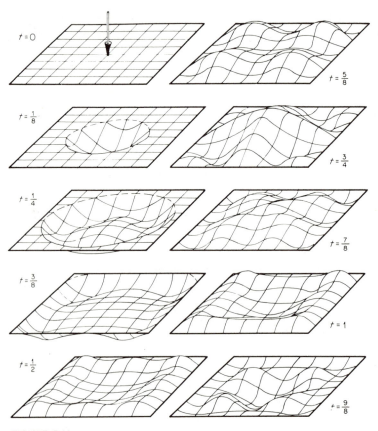

FIGURE 5.11
Successive shapes of a rectangular membrane struck at its center. Times are given in terms of fractions of the fundamental period of vibration.

Figure 5.11 shows the motion of a membrane that has been struck so that a small area around its center is started downward at $t = 0$. In this case $\eta_0 = 0$, and v_0 is zero except near the point $x = a/2$, $y = b/2$, where it has a large enough value so that $\iint v_0 \, dx \, dy = U$. The integrations for the coefficients C_{mn} become simple, and the resulting expression for η,

$$\eta = \frac{2U}{\pi ab} \sum_{m,n} \frac{1}{\nu_{mn}} \, \psi_{mn}\!\left(\frac{a}{2}, \frac{b}{2}\right) \psi_{mn}(x,y) \sin{(2\pi\nu_{mn}t)}$$

The series does not converge if m and n both run to infinity, a corollary of the fact that a concentrated force produces an infinite displacement of the membrane. However, if the series does not continue for m larger than some number M or for n larger than N, there will be only a finite number of terms to be added together, and the result will never be infinite. Such a finite series corresponds approximately to a starting area of dimensions a/M by b/N.

The case shown in Fig. 5.11 is for $M = N = 10$. The initial shape and the shapes of successive eighths of the fundamental cycle are shown. We see the initial pulse spread out and then reflect back to pile up in the center in a "splash." Since not all the overtones are harmonics, the motion is not periodic. We notice that the shape of the pulse changes as it spreads out to the edge, as has been mentioned in connection with Fig. 5.8.

The circular membrane

Rectangular coordinates are useful to describe parallel waves and normal modes for a rectangular boundary, but for the discussion of circular waves and for the study of the normal modes for a circular boundary, we shall find it easiest to use polar coordinates. In this study we shall encounter more forcibly than before the essential differences between wave motion on a string and on a membrane. We have already seen, in Fig. 5.8, that circular waves, spreading out from a concentrated source, do not maintain shape or amplitude as they progress. We also have noted that the Laplace operator ∇^2, the analogue of the curvature operator for the string, has a form, for polar coordinates, quite different from that of a sum of second derivatives. For this reason even the standing waves in a circular membrane differ markedly from stringlike behavior.

Simple-harmonic waves

To show this we turn to the circular waves that vibrate with simple-harmonic motion. As in the rectangular case, we separate the wave equation. In polar coordinates the equation is

$$\frac{1}{r}\frac{\partial}{\partial r}\!\left(r\frac{\partial \eta}{\partial r}\right) + \frac{1}{r^2}\frac{\partial^2 \eta}{\partial \phi^2} = \frac{1}{c^2}\frac{\partial^2 \eta}{\partial t^2}$$

and by setting $\eta = R(r)\Phi(\phi)e^{-2\pi i\nu t}$, we obtain the equations for R and Φ:

$$\frac{d^2\Phi}{d\phi^2} = -\mu^2\Phi \qquad \Phi = \cos{(\mu\phi)} \quad \text{or} \quad \sin{(\mu\phi)} \qquad (5.2.18)$$

$$\frac{d^2R}{dr_2} + \frac{1}{r}\frac{dR}{dr} + \left(\frac{4\pi^2\nu^2}{c^2} - \frac{\mu^2}{r^2}\right)R = 0 \qquad (5.2.19)$$

For these circular waves one "boundary condition" is required even before the shape of the boundary line is decided upon. The requirement is simply that the displacement η be a single-valued function of position; for the coordinate ϕ is a periodic one, repeating itself after an angle 2π, and we must have $\eta(r,\varphi)$ equal to $\eta(r, \varphi + 2\pi)$. This restricts the allowed values of μ to integers:

$$\Phi_{em} = \cos(m\phi) \qquad \Phi_{om} = \sin(m\phi) \qquad m = 0, 1, 2, 3, \ldots$$

This is not true for a membrane whose boundary is shaped like a sector of a circle, where ϕ cannot go from zero clear around to 2π, but such cases are of no great practical importance.

The foregoing requirement is the third different type of boundary condition that we have encountered, the first type being the fixing of the displacement or its slope (or both) at some point or along some line (as with the string, the bar, and the rectangular membrane), the second being simply that the displacement have no infinite values in the range of interest. The third, the condition of periodicity, will be used whenever any of the coordinates are angles that repeat themselves.

Bessel functions

Equation (5.2.19) for the radial factor is Bessel's equation (1.2.4). The general solution $J_m(z)$, where $z = 2\pi\nu r/c$ and m is an integer, has the following properties:

$$\frac{1}{z}\frac{d}{dz}\left(z\frac{d}{dz}J_m\right) + \left(1 - \frac{m^2}{z^2}\right)J_m = 0$$

$$J_m(z) = \frac{1}{m!}\left(\frac{z}{2}\right)^m - \frac{1}{(m+1)!}\left(\frac{z}{2}\right)^{m+2} + \frac{1}{2!(m+2)!}\left(\frac{z}{2}\right)^{m+4} \cdots$$

$$J_m(z) \xrightarrow[z\to\infty]{} \sqrt{\frac{2}{\pi z}}\cos\left(z - \frac{2m+1}{4}\pi\right)$$

$$J_m(z) = \frac{1}{2\pi i^m}\int_0^{2\pi} e^{iz\cos w}\cos(mw)\, dw \qquad (5.2.20)$$

$$J_{m-1}(z) + J_{m+1}(z) = \frac{2m}{z}J_m(z)$$

$$\frac{d}{dz}J_m(z) = \tfrac{1}{2}[J_{m-1}(z) - J_{m+1}(z)]$$

$$\frac{d}{dz}[z^m J_m(z)] = z^m J_{m-1}(z) \qquad \frac{d}{dz}z^{-m}J_m(z) = -z^{-m}J_{m+1}(z)$$

$$\int J_1(z)\, dz = -J_0(z) \qquad \int z J_0(z)\, dz = z J_1(z)$$

$$\int J_0^2(z)z\, dz = \frac{z^2}{2}[J_0^2(z) + J_1^2(z)] \qquad (5.2.21)$$

$$\int J_m^2(z)z\, dz = \frac{z^2}{2}[J_m^2(z) - J_{m-1}(z)J_{m+1}(z)]$$

$$\int J_m(\alpha z)J_m(\beta z)z\, dz = \frac{z}{\alpha^2 - \beta^2}[\beta J_m(\alpha z)J_{m-1}(\beta z) - \alpha J_m(\beta z)J_{m-1}(\alpha z)]$$

All these properties are proved in books on Bessel functions. Values of J_0, J_1, and J_2 are given in Table V.

The function $J_m(2\pi\nu r/c)$ is not the only solution of Eq. (5.2.19), for it is a second-order equation, and there must be a second solution. This other solution, however, becomes infinite at $r = 0$, and so is of no interest to us at present. It will be discussed in Sec. 7.3. We should note, however, that the use of J_m is in accord with the boundary condition of the second kind, mentioned above.

The allowed frequencies

Coming back to our problem, we can now say that a simple-harmonic solution of the wave equation which is finite over the range from $r = 0$ to $r = \infty$, which is single-valued over the range from $\phi = 0$ to $\phi = 2\pi$, is

$$\eta = \cos m(\phi - \alpha)\, J_m\left(\frac{2\pi\nu r}{c}\right) \cos(2\pi\nu t - \Omega) \qquad (5.2.22)$$

If the membrane is fastened along a boundary circle of radius a, the allowed frequencies must be those that make $J_m(2\pi\nu a/c) = 0$. For each value of m there will be a whole sequence of solutions. We shall label the allowed values of the frequency ν_{mn}, so that $\nu_{01}, \nu_{02}, \nu_{03}, \ldots$ are the solutions of $J_0(2\pi\nu a/c) = 0$; $\nu_{11}, \nu_{12}, \nu_{13}, \ldots$ are the solutions of $J_1(2\pi\nu a/c) = 0$; and so on. The values of ν_{mn} are given by the equations

$$\beta_{01} = 0.7655, \quad \beta_{02} = 1.7571, \quad \beta_{03} = 2.7546, \ldots$$
$$\beta_{11} = 1.2197, \quad \beta_{12} = 2.2330, \quad \beta_{13} = 3.2383, \ldots$$
$$\nu_{mn} = \frac{c}{2a}\beta_{mn} \qquad \beta_{21} = 1.6347, \quad \beta_{22} = 2.6793, \quad \beta_{23} = 3.6987, \ldots$$
$$\cdots\cdots\cdots\cdots\cdots\cdots\cdots\cdots\cdots\cdots\cdots\cdots\cdots\cdots \qquad (5.2.23)$$
$$\beta_{mn} \simeq n + \frac{m}{2} - \frac{1}{4} \qquad \text{if } n \text{ is large}$$

The frequency ν_{01} is the fundamental. Another way of writing these results is

$$\nu_{01} = 0.38274\,\frac{1}{a}\sqrt{\frac{T}{\sigma}}, \quad \nu_{11} = 1.5933\nu_{01}, \quad \nu_{21} = 2.1355\nu_{01}$$
$$\nu_{02} = 2.2954\nu_{01}, \qquad \nu_{31} = 2.6531\nu_{01}, \quad \nu_{12} = 2.9173\nu_{01}, \ldots \qquad (5.2.24)$$

It is to be noticed that none of these overtones is harmonic.

The characteristic functions

Corresponding to the frequency ν_{0n} is the characteristic function

$$\psi_{e0n}(r,\phi) = J_0\left(\frac{\pi\beta_{0n}r}{a}\right)$$

and corresponding to the frequency $v_{mn}(m > 0)$ are the two characteristic functions

$$\psi_{emn} = \cos{(m\phi)} J_m\left(\frac{\pi\beta_{mn}r}{a}\right) \qquad \psi_{0mn} = \sin{(m\phi)} J_m\left(\frac{\pi\beta_{mn}r}{a}\right) \quad (5.2.25)$$

Except for $m = 0$, the normal modes are degenerate, there being two characteristic functions for each frequency. The shapes of a few of the normal modes are shown in Fig. 5.12. We notice that the (m,n)th characteristic function has m diametrical nodal lines and $n - 1$ circular nodes.

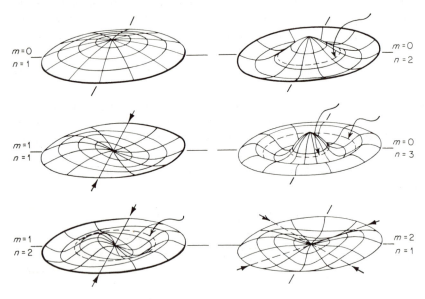

FIGURE 5.12
Shapes of some of the normal modes of vibration of the circular membrane. Arrows point to the nodal lines.

The integral properties of these functions are obtained from Eqs. (5.2.21):

$$\int_0^a \int_0^{2\pi} \psi_{emn}\psi_{em'n'}r \, dr \, d\phi = \int_0^a \int_0^{2\pi} \psi_{0mn}\psi_{0m'n'}r \, dr \, d\phi$$

$$= \begin{cases} 0 & m \neq m' \quad \text{or} \quad n \neq n' \\ \pi a^2 \Lambda_{mn} & m = m' \quad \text{and} \quad n = n' \end{cases} \quad (5.2.26$$

where $\quad \Lambda_{0n} = [J_1(\pi\beta_{0n})]^2 \qquad \Lambda_{mn} = \tfrac{1}{2}[J_{m-1}(\pi\beta_{mn})]^2 \qquad m > 0$

The values of the constants Λ_{mn} can be computed from the following value of J_m:

$$J_1(\pi\beta_{01}) = +0.5191 \qquad J_1(\pi\beta_{02}) = -0.3403$$

$$J_1(\pi\beta_{03}) = +0.2715 \qquad J_1(\pi\beta_{04}) = -0.2325$$

$$J_0(\pi\beta_{11}) = -0.4028 \qquad J_0(\pi\beta_{12}) = +0.3001$$

$$J_0(\pi\beta_{13}) = -0.2497 \qquad J_0(\pi\beta_{14}) = +0.2184$$

$$J_1(\pi\beta_{21}) = -0.3397 \qquad J_1(\pi\beta_{22}) = +0.2714$$

$$J_1(\pi\beta_{23}) = -0.2324 \qquad J_1(\pi\beta_{24}) = +0.2066$$

With the values of the constants Λ known, it is possible to compute the behavior of a circular membrane when started with the initial shape $\eta_0(r,\varphi)$ and the initial velocity $v_0(r,\varphi)$. By methods that we have used many times before, we can show that this behavior is governed by the series

$$\eta = \sum_{n=1}^{\infty} \left\{ \sum_{m=0}^{\infty} \psi_{emn}[B_{emn} \cos(2\pi\nu_{mn}t) + C_{emn} \sin(2\pi\nu_{mn}t)] \right.$$

$$\left. + \sum_{m=1}^{\infty} \psi_{0mn}[B_{0mn} \cos(2\pi\nu_{mn}t) + C_{0mn} \sin(2\pi\nu_{mn}t)] \right\} \quad (5.2.27)$$

where
$$B_{emn} = \frac{1}{\pi a^2 \Lambda_{mn}} \int_0^a \int_0^{2\pi} \eta_0 \psi_{emn} r \, dr \, d\phi$$

$$C_{emn} = \frac{1}{2\pi^2 \nu_{mn} a^2 \Lambda_{mn}} \int_0^a \int_0^{2\pi} v_0 \psi_{emn} r \, dr \, d\phi$$

A similar set of equations holds for B_{0mn} and C_{0mn}.

This completes our preliminary discussion of wave motion on a membrane. We shall return to the subject in Chaps. 9 and 10, when we have learned something of wave motion in air, and can thus take into account the interactions between membrane and surrounding air, which are so important in actual membrane behavior.

5.3   THE VIBRATION OF PLATES

The study of the vibrations of plates bears the same relation to the study of the membrane as the study of the vibrations of bars does to the study of the flexible string. The effect of stiffness in both cases increases the frequencies of the higher overtones more than it does those of the lower overtones, and so makes the fundamental frequency very much lower than all the overtones. However, the motions of a plate are very much more complicated than those of a bar; so much more complicated that we shall have to be satisfied with the study of one example, that of the circular plate, clamped at its edge and under no tension. The diaphragm of an ordinary telephone receiver is a plate of this type; so the study will have some practical applications.

The equation of motion

The increased complications encountered in the study of plates come partly from the increased complexity of wave motions in two dimensions over those of one, but also come about because of the complex sorts of stresses that are set up when a plate is bent.   The bending of a plate compresses the material on the inside of the bend and stretches it on the outside.   But when a material is compressed it tries to spread out in a direction perpendicular to the compressional force, so that when a plate is bent downward in one direction, there will be a tendency for it to curl up in a direction at right angles to the bend.   The ratio of the sidewise spreading to the compression is called *Poisson's ratio*, and will be labeled by the letter s.   It has a value about equal to 0.3 for most materials.   This complication was not considered when we studied the vibration of bars, for we tacitly assumed that the bar was thin enough compared with its length so that the effects of a sidewise curl would be negligible.

The derivation of the wave equation for the plate involves more discussion than is worthwhile here (it is given in books on theory of elasticity). The equation is

$$\nabla^4 \eta + \frac{3\rho(1 - s^2)}{Qh^2} \frac{\partial^2 \eta}{\partial t^2} = 0 \qquad (5.3.1)$$

where ρ is the density of the material, s its Poisson's ratio, Q its modulus of elasticity, and h the half-thickness of the plate.

We shall not spend any time discussing the general behavior of waves on a plate of infinite extent, but shall simply remark that, like the bar, the plate is a dispersive medium; waves of different wavelength travel with different velocities.

Simple-harmonic vibrations

To study the simple-harmonic motion of a plate, we insert the exponential dependence on time and separate the factors depending on the individual coordinates.   The differential operator ∇^4 is difficult to separate in most coordinate systems, but for polar coordinates the results turn out to be sufficiently simple to justify our analyzing them in detail.   Here, if we set $\eta = Y(r,\phi)e^{-2\pi i v t}$, where Y's dependence on ϕ is by the factor cos or sin $(m\phi)$, then the differential equation for Y can be written

$$(\nabla^2 - \gamma^2)(\nabla^2 + \gamma^2) Y = 0 \qquad \gamma^4 = \frac{12\pi^2 v^2 \rho(1 - s^2)}{Qh^2} \qquad (5.3.2)$$

Therefore Y can be a solution either of $\nabla^2 Y + \gamma^2 Y = 0$ or of $\nabla^2 Y - \gamma^2 Y = 0$.

Since ∇^2 and Y are to be expressed in polar coordinates, the solution of the first equation, which is finite at $r = 0$, is

$$Y = \frac{\cos}{\sin} (m\phi)J_m(\gamma r) \qquad \text{where } m \text{ is an integer}$$

This is the usual solution for the membrane, with γ instead of $2\pi\nu/c$. The solution of the second equation is obtained from the first by changing γ into $i\gamma$, and necessitates a little discussion of Bessel functions of imaginary values of the independent variable. Let us call these *hyperbolic Bessel functions* and define them by the equation $I_m(z) = i^{-m}J_m(iz)$. The properties of the function $I_m(z)$ can be obtained from Eqs. (5.2.20) and (5.2.21) for $J_m(z)$. The more useful formulas are

$$I_{m-1}(z) - I_{m+1}(z) = \frac{2m}{z} I_m(z) \qquad \frac{d}{dz} I_m(z) = \tfrac{1}{2}[I_{m-1}(z) + I_{m+1}(z)]$$

$$\int I_0(z)z\, dz = zI_1(z) \qquad \int I_1(z)\, dz = I_0(z) \qquad (5.3.3)$$

Values of the functions I_s, I_1, and I_2 are given in Table IV.

The normal modes

Possible solutions for the simple-harmonic oscillations of a plate are therefore given by the expressions

$$Y(r,\phi) = \frac{\cos}{\sin} (m\phi)[AJ_m(\gamma r) + BI_m(\gamma r)]$$

The boundary conditions corresponding to a circular plate of radius a, clamped at its edges, are that $Y(a,\phi) = 0$ and $(\partial Y/\partial r)_{r=a} = 0$. The first condition is satisfied by making

$$B = -A\frac{J_m(\gamma a)}{I_m(\gamma a)}$$

and the second condition is satisfied by requiring that γ have those values that make

$$I_m(\gamma a)\frac{d}{dr} J_m(\gamma r) - J_m(\gamma a)\frac{d}{dr} I_m(\gamma r) = 0 \qquad \text{at } r = a \qquad (5.3.4)$$

This equation fixes the allowed values of the frequency, for γ depends on ν. We shall label the solutions of Eq. (5.3.4) by the symbols γ_{mn}, where $\gamma_{mn} = (\pi/a)\beta_{mn}$, and where

$$\beta_{01} = 1.015, \qquad \beta_{02} = 2.007, \qquad \beta_{03} = 3.000$$
$$\beta_{11} = 1.468, \qquad \beta_{12} = 2.483, \qquad \beta_{13} = 3.490 \qquad (5.3.5)$$
$$\beta_{21} = 1.879, \qquad \beta_{22} = 2.992, \qquad \beta_{23} = 4.000$$

$$\beta_{mn} \xrightarrow[n\to\infty]{} n + \frac{m}{2}$$

The allowed values of the frequency are therefore

$$\nu_{mn} = \frac{\pi h}{2a^2} \sqrt{\frac{Q}{3\rho(1-s^2)}} (\beta_{mn})^2$$

$$\nu_{01} = 0.9342 \frac{h}{a^2} \sqrt{\frac{Q}{\rho(1-s^2)}}$$

$$\nu_{11} = 2.091\nu_{01}, \qquad \nu_{21} = 3.426\nu_{01}, \qquad \nu_{02} = 3.909\nu_{01}$$

$$\nu_{12} = 5.983\nu_{01}, \ldots$$

The allowed frequencies are spread apart much farther than those for the membrane, given in Eq. (5.2.24).   The overtones are not harmonic.

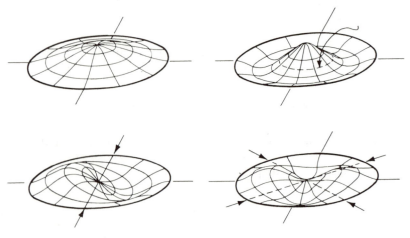

FIGURE 5.13
Shapes of a few of the normal modes of vibration of a circular
plate clamped at its edge.

The characteristic functions are

$$\psi_{emn} = \cos{(m\phi)}\left[J_m\left(\frac{\pi\beta_{mn}r}{a}\right) - \frac{J_m(\pi\beta_{mn})}{I_m(\pi\beta_{mn})} I_m\left(\frac{\pi\beta_{mn}r}{a}\right)\right] \qquad (5.3.6)$$

and a similar expression for ψ_{omn} (for $m > 0$), where $\sin{(m\phi)}$ is used instead of $\cos{(m\phi)}$.   Some of these functions are shown in Fig. 5.13.

The free vibrations of the plate corresponding to arbitrary initial conditions can be expressed in terms of a series of these characteristic functions.

To compute the coefficients of the series, we need the expressions for the integrals, over the area of the plate, of the squares of the characteristic functions.   These can be obtained by using Eqs. (5.2.21) and (5.3.4).

$$\int_0^{2\pi} d\phi \int_0^a [\psi_{\sigma mn}(\phi,r)]^2 r \, dr = \pi a^2 \Lambda_{mn} \qquad \sigma = e \text{ or } o; m, n \text{ integers}$$

$$\Lambda_{mn} = \frac{2}{\epsilon_m} \{[J_m(\pi\beta_{mn})]^2 + [J'_m(\pi\beta_{mn})]^2\}$$

(5.3.7)

where $J'_m(u) = (d/du)J_m(u)$.

Forced motion

As an example of the calculation of forced motion, let us consider the case of a diaphragm subject to a uniform simple-harmonic transverse driving force $F_\omega e^{-i\omega t}$ per unit area of its surface.   The equation of motion is

$$2h\rho \frac{\partial^2 \eta}{\partial t^2} = -\frac{2Qh^3}{3(1 - s^2)} \nabla^4 \eta + F_\omega e^{-i\omega t}$$

(5.3.8)

where $2h\rho$ is the mass of the plate per unit area, h being its half-thickness. Setting $\eta = Y_\omega(r,\phi)e^{-i\omega t}$ for the steady-state solution, we have

$$\nabla^4 Y_\omega - \gamma^4 Y_\omega = \frac{3(1 - s^2)F_\omega}{2Qh^3} \qquad \gamma^4 = \frac{3\omega^2\rho_p(1 - s^2)}{Qh^2}$$

(5.3.9)

For a circular plate, the solutions of the homogeneous equation $\nabla^4 Y - \gamma^4 Y = 0$, which are finite at $r = 0$ and independent of ϕ, are the two Bessel functions $J_0(\gamma r)$ and $I_0(\gamma r)$.   A combination of these would not satisfy the boundary conditions $Y = 0$ and $\partial Y/\partial r = 0$ at $r = a$, appropriate for a plate clamped at its edge, for an arbitrarily chosen value of ω. In fact, we have just shown that this is possible only for those frequencies given in Eq. (5.3.5). Only by adding a constant term can it satisfy these conditions for any frequency.   Specifically, the combination

$$Y = A[I_1(\gamma a)J_0(\gamma r) + J_1(\gamma a)I_0(\gamma r) - I_1(\gamma a)J_0(\gamma a) - J_1(\gamma a)I_0(\gamma a)]$$

does satisfy the two boundary conditions at $r = a$, since

$$\gamma J_1(\gamma r) = -[dJ_0(\gamma r)/dr] \quad \text{and} \quad \gamma I_1(\gamma r) = dI_0(\gamma r)/dr$$

Of course, the combination does not satisfy the homogeneous equation; in fact, it satisfies the inhomogeneous equation

$$\nabla^4 Y - \gamma^4 Y = A\gamma^4[I_1(\gamma a)J_0(\gamma a) + J_1(\gamma a)I_0(\gamma a)]$$

But the equation we are to solve is not homogeneous; in fact, we need only adjust the value of A to satisfy Eq. (5.3.9).   The final result is

$$Y_\omega = \frac{F_\omega}{2h\rho\omega^2} \frac{I_1(\gamma a)[J_0(\gamma r) - J_0(\gamma a)] + J_1(\gamma a)[I_0(\gamma r) - I_0(\gamma a)]}{I_1(\gamma a)J_0(\gamma a) + J_1(\gamma a)I_0(\gamma a)}$$

$$\to \frac{3F_\omega(1 - s^2)}{32h^3 Q}(a^2 - r^2)^2 \qquad \gamma a \to 0$$

(5.3.10)

At certain frequencies, those for which the denominator $I_1(\gamma a)J_0(\gamma a) + J_1(\gamma a)I_0(\gamma a)$ is zero, the diaphragm resonates; the formula predicts an infinite amplitude because the frictional term has been omitted. These values of γ (and thus of frequency) for resonance are, of course, those specified by Eq. (5.3.4) for $m = 0$ for the standing waves of free vibration of the plate. We note that, as the driving frequency goes to zero, the static displacement of the diaphragm, given by the limiting form, is small.

The Green's function

If instead of being spread over the plate surface, the driving force is concentrated at the point r_0, ϕ_0, the steady-state motion is given in terms of the Green's function G, the solution of

$$\nabla^4 G - \gamma^4 G = \frac{1}{r}\,\delta(r - r_0)\,\delta(\phi - \phi_0) \qquad \gamma^4 = \omega^2 \frac{3\rho(1 - s^2)}{Qh^2} \quad (5.3.11)$$

which is the analogue of Eq. (4.4.9). This function can be expressed as a series of the characteristic functions $\psi_{\sigma mn}(\phi,r)$ of Eq. (5.3.6). Inserting $G = \sum A_{\sigma mn}\psi_{\sigma mn}$ in Eq. (5.3.11), we obtain

$$\sum_{skl} A_{skl}(\gamma_{kl}{}^4 - \gamma^4)\psi_{skl}(\phi,r) = \frac{1}{r}\,\delta(r - r_0)\,\delta(\phi - \phi_0) \qquad \gamma_{mn} = \frac{\pi\beta_{mn}}{a}$$

Multiplying through by $\psi_{\sigma mn}(\phi,r)r\,d\phi\,dr$ and integrating over the diaphragm area, using Eq. (5.3.7) and the fact that the integral of $\psi_{\sigma mn}\psi_{skl}$ is zero unless $s = \sigma, k = m$, and $l = n$, and considering the property of the delta functions, we see that

$$G_\omega(\phi r \mid \phi_0 r_0) = \frac{1}{\pi a^2} \sum_{\sigma,m,n} \frac{\psi_{\sigma mn}(\phi,r)\psi_{\sigma mn}(\phi_0,r_0)}{\Lambda_{mn}(\gamma_{mn}{}^4 - \gamma^4)} \qquad (5.3.12)$$

This has properties similar to the Green's function of Eq. (4.4.14), for the string between rigid supports. For example, it obeys the principle of reciprocity since its value is unchanged when r, ϕ and r_0, ϕ_0 are interchanged. It also has resonance poles whenever $\omega = \pm 2\pi\nu_{mn}$, with ν_{mn} defined below Eq. (5.3.5). The series converges fairly rapidly, except at resonance, since $\gamma_{mn}{}^4 \simeq (\pi/a)^4(n + \tfrac{1}{2}m)^4$ for n large. From Eq. (5.3.9) the steady-state displacement of the plate, when driven by a force $F_\omega e^{-i\omega t}$ concentrated at the point ϕ_0, r_0, is $[3(1 - s^2)F_\omega/2Qh^3]G_\omega(\phi r \mid \phi_0 r_0)e^{-i\omega t}$, and the transverse velocity of the plate is $-i\omega$ times this. Therefore the mechanical impedance which the plate presents to a driving force of frequency $\omega/2\pi$, applied at r_0, ϕ_0, is

$$Z_m = \frac{F_\omega}{u} = \frac{i}{\omega}(2\pi a^2 h\rho)\left\{\sum_{\sigma mn} \frac{[\psi_{\sigma mn}(\phi_0,r_0)]^2}{\Lambda_{mn}(\omega_{mn}{}^2 - \omega^2)}\right\}^{-1} \qquad (5.3.13)$$

where $\omega_{mn}{}^2 = (2\pi\nu_{mn})^2 = [Qh^2/3\rho(1 - s^2)](\pi\beta_{mn}/a)^4$. At very low frequencies this behaves like a stiffness reactance, being inversely proportional to ω. At a resonance ($\omega = \omega_{mn}$), Z_m is zero; just above a resonance, Z_m is negative-imaginary, acting like a mass reactance. As long as we neglect internal friction of the plate material and assume that the plate is moving in a vacuum, Z_m is purely reactive at all frequencies.

Green's function for an infinite plate

Later we shall be studying the reaction between a plate of very large extent and the acoustic waves in a medium in contact with the plate. At that time we shall need to know the expression for the deflection of a plate of infinite extent, driven by a simple-harmonic force concentrated at a single point. This is, of course, proportional to a Green's function, the solution of Eq. (5.3.11) for an infinite plate. Because there are no boundaries, we can place the origin anywhere; we might as well place it at the point of application of the force. Thus we rewrite Eq. (5.3.11) as

$$\nabla^4 g - \gamma^4 g = \delta(x)\,\delta(y)$$

the function on the right-hand side being zero everywhere but at the origin, where its integral is unity. (We habitually use the lowercase symbol g for the Green's function in an infinite medium.)

With the force symmetrically placed, the radiating waves must be symmetrical, so that g should be independent of ϕ. Thus, referring to the definition of ∇^2 in polar coordinates, given in Eq. (5.2.2), we have

$$\frac{1}{r}\frac{d}{dr}\left\{ r\frac{d}{dr}\left[\frac{1}{r}\frac{d}{dr}\left(r\frac{dg}{dr} \right) \right] \right\} - \gamma^4 g = \delta(x)\,\delta(y) \qquad (5.3.14)$$

Except at the origin, function g must be a combination of Bessel functions of order zero, so arranged that the solution represents an outgoing wave, $\sqrt{1/r}\,e^{i\gamma r}$, when r is large. Part of the solution will be the two functions $J_0(\gamma r)$ and $I_0(\gamma r) = J_0(i\gamma r)$ we have already discussed. But these two functions by themselves cannot produce a traveling wave; to do this we must also use the second solution of the Bessel equation, mentioned on page 8. This solution, called the *Neumann function* of zero order, is defined by the following series and asymptotic formula:

$$N_0(z) = \frac{2}{\pi}[\ln(z) - 0.11593\ldots]J_0(z) - \frac{2}{\pi}\sum_{m=1}^{\infty}\frac{(-1)^m(z/2)^{2m}}{(m!)^2}$$

$$\times\left(1 + \tfrac{1}{2} + \tfrac{1}{3} + \cdots + \frac{1}{m} \right)$$

$$\to \sqrt{\frac{2}{\pi z}}\sin(z - \tfrac{1}{4}\pi) \qquad z \to \infty$$

$$N_0(iz) = iJ_0(iz) + \frac{2}{\pi}[\ln(z) - 0.11593\ldots]J_0(iz) \qquad (5.3.15)$$

$$-\frac{2}{\pi}\sum_{m=1}^{\infty}\frac{(z/2)^{2m}}{(m!)^2}\left(1 + \tfrac{1}{2} + \tfrac{1}{3} + \cdots + \frac{1}{m} \right)$$

$$\to \sqrt{\frac{2}{\pi z}}\sinh(z + \tfrac{1}{4}i\pi) \qquad z \to \infty$$

The function N_0 has a logarithmic infinity at $z = 0$; so we were not able to use it to describe the free vibrations of a circular plate or membrane. It must be used for the forced motion of a circular plate or membrane, however, in order to be able to express an outgoing wave. Referring to Eq. (5.2.20), we see that this combination must be

$$J_0(z) + iN_0(z) \to \sqrt{\frac{2}{\pi z}} \left[\cos\left(z - \tfrac{1}{4}\pi\right) + i \sin\left(z - \tfrac{1}{4}\pi\right)\right] = \sqrt{\frac{2i}{\pi z}}\, e^{iz}$$

and thus $\qquad J_0(iz) + iN_0(iz) \to \sqrt{\frac{2}{\pi z}}\, e^{-z} \qquad z \to \infty$

Therefore the solution of Eq. (5.3.14) must be a combination of these two combinations. The first one must be used because it is the only combination of the Bessel functions which represents purely outgoing waves, with no term containing e^{-iz} ($z = \gamma r$). The second one must be used because it is the only combination of $J_0(i\gamma r)$ and $N_0(i\gamma r)$ which does not contain a term with $e^{+\gamma r}$ (which would go to infinity at $r \to \infty$). Both of these combinations have a logarithmic infinity at $r = 0$, but these can be canceled against each other. Therefore the only combination of the four which stays finite at the origin and at infinity and which represents a purely outgoing wave is

$$g = A[J_0(\gamma r) + iN_0(\gamma r) - J_0(i\gamma r) - iN_0(i\gamma r)]$$

$$\to A\sqrt{\frac{2i}{\pi \gamma r}} \cdot e^{i\gamma r} \qquad r \to \infty$$

$$\to A\left\{1 - \tfrac{1}{4}(\gamma r)^2 + \frac{i}{\pi}(\gamma r)^2 + \frac{i}{\pi}(\gamma r)^2[\ln(\gamma r) - 0.1159]\right\} \qquad r \to 0 \quad (5.3.16)$$

where we must determine the value of A to satisfy Eq. (5.3.14).

If we integrate both sides of Eq. (5.3.14) over the area of a circle of small radius α centered at the origin, the integral on the right will be 1, no matter how small α is. The integral of $\gamma^4 g$ will go to zero as α goes to zero, but the integral of the first term will be

$$\int_0^{2\pi} d\phi \int_0^\alpha dr\, \frac{d}{dr}\left\{r\frac{d}{dr}\left[\frac{1}{r}\frac{d}{dr}\left(r\frac{dg}{dr}\right)\right]\right\} = 2\pi\left\{r\frac{d}{dr}\left[\frac{1}{r}\frac{d}{dr}\left(r\frac{dg}{dr}\right)\right]\right\}_{r=\alpha}$$

From Eq. (5.2.20), however, if g has the form given in Eq. (5.3.16),

$$\frac{1}{r}\frac{d}{dr}\left(r\frac{dg}{dr}\right) = -A\gamma^2[J_0(\gamma r) + iN_0(\gamma r) + J_0(i\gamma r) + iN_0(i\gamma r)]$$

$$\to -A\gamma^2\left\{1 - \tfrac{1}{4}(\gamma r)^2 + \frac{4i}{\pi}[\ln(\gamma r) - 0.1159]\right\} \qquad r \to 0$$

so that

$$2\pi r \frac{d}{dr}\left[\frac{1}{r}\frac{d}{dr}\left(r\frac{dg}{dr}\right)\right] \to -8i\gamma^2 A \qquad r = \alpha \to 0$$

Since this must be equal to the integral of the delta functions, which is 1, the value of A is determined, and the Green's function for a plate of infinite extent is

$$g_\omega(r \mid r_0) = \frac{i}{8\gamma^2}[J_0(\gamma R) + iN_0(\gamma R) - J_0(i\gamma R) - iN_0(i\gamma R)]$$

$$\to \frac{i}{8\gamma^2} \qquad R \to 0; \qquad \gamma^4 = \omega^2\frac{3\rho(1-s^2)}{Qh^2}$$

$$\to -\frac{e^{i\gamma R}}{\sqrt{32\pi i\gamma^5 R}} \qquad R \to \infty \qquad (5.3.17)$$

where R is the distance between the source point at r_0, ϕ_0 and the observation point at r, ϕ.

We first notice that the plate displacement is finite, even at the point of application of the force (the source point). The stiffness of the plate prevents the yielding to a point force which characterizes the elastic membrane, as illustrated in Fig. 5.6. Mathematically, the functions available to represent the membrane displacement are but two, $J_0(\omega r/c)$ and $N_0(\omega r/c)$, and these must be arranged, for an infinite membrane, in the combination $J_0 + iN_0$ in order to have outgoing waves. Thus the logarithmic infinity in the N_0 solution requires that any material which satisfies the wave equation in two dimensions exactly must exhibit a logarithmic singularity at the point of application of a point force. In contrast, the equation $\nabla^4 Y - \gamma^4 Y = 0$ allows both the combinations $J_0(\gamma r) + iN_0(\gamma r)$ and $J_0(i\gamma r) + iN_0(i\gamma r)$; between them it is possible to devise a solution which is finite at $r = 0$ and represents outgoing wave motion at $r \to \infty$.

If the driving force at the origin is $F_\omega e^{-i\omega t}$, the steady-state plate displacement is

$$\eta_\omega(r,t) = \frac{iF_\omega}{16\omega h^2\rho}\sqrt{\frac{3\rho(1-s^2)}{Q}}[J_0(\gamma r) + iN_0(\gamma r)$$
$$- J_0(i\gamma r) - iN_0(i\gamma r)]e^{-i\omega t} \quad (5.3.18)$$

The transverse velocity of the plate is $-i\omega\eta$, and the mechanical impedance of the plate to a concentrated simple-harmonic force of frequency $\omega/2\pi$ is

$$Z_m = 16h^2\rho\sqrt{\frac{Q}{3\rho(1-s^2)}} \qquad (5.3.19)$$

a pure resistance, in contrast to the impedance, given in Eq. (5.3.13) for the plate of finite size, clamped at the edges, which is purely reactive (if the plate is in a vacuum). The difference, of course, lies in the fact that energy can escape from the source if the plate is infinite in extent, but cannot escape (after steady state has set in) from the plate of finite size in a vacuum.   These formulas will be useful later, when we investigate the effect of the plate vibrations on the medium surrounding it, and the reaction, back on the plate, of the induced wave motion in the medium.

Problems

I   An annealed steel bar 20 cm long is clamped at one end.   If its cross section is a square 1 cm on a side, what will be the lowest four frequencies of vibration?   If the cross section is a circle 0.5 cm in radius, what will they be?   If the cross section is a rectangle of sides a and $2a$, what must be the value of a to have the bar's lowest frequency be 250 cps?   For steel, $\rho = 7.7$, $Q = 2 \times 10^{11}$ dynes per sq cm.

2   A bar of steel of length 10 cm, whose cross section is a rectangle of sides 1 cm and 0.5 cm, is clamped at one end.   It is struck at the midpoint of one of its wider sides, so that $v_0(x) = 0$ except at $x = l/2$, and $\int v_0\, dx = 100$.   What is the shape of the bar a time $T_1/4$ after the blow?   (T_1 is the period of the bar's fundamental.)   What is its shape at a time $T_1/2$?   What is the motion of the end point of the bar?   What is the amplitude of that part of the motion of the end point corresponding to the fundamental?   To the first overtone?   To the second overtone?

3   Plot the shapes of the first three normal modes of vibration of a bar clamped at both ends.   Of a completely free bar.

4   What is the energy of vibration of the bar, struck at one end, which is discussed on page 185?   What is the ratio between the energy of the fundamental and that of the first overtone?   Of the second overtone?

5   A cylindrical bar of radius a, clamped at one end, is damped by the air by a force equal to $1.25 \times 10^{-9}\, a^4 v^3$ ($\partial y/\partial t$) dynes per cm length of bar, where v is the frequency of vibration and $\partial y/\partial t$ the bar's velocity.   Show that the modulus of decay of the nth allowed frequency is 16×10^8 $(\pi \rho/a^2 v_n^3)$, where v_n is given by Eq. (5.1.16).

6   The Green's function $g(x\,|\,x_0)$ for transverse oscillations of a bar satisfies the equation

$$\frac{d^4g}{dx^4} - \gamma^4 g = \delta(x - x_0) \qquad \gamma^4 = (2\pi\mu)^4 = \frac{\rho\omega^2}{Q\kappa^2}$$

with requirements that g, dg/dx, and d^2g/dx^2 be continuous at $x = x_0$.
(a) Show that, for a bar of infinite extent,

$$g(x\,|\,x_0) = \frac{i}{4\gamma^3}\left(e^{i\gamma|x-x_0|} + ie^{-\gamma|x-x_0|}\right)$$

(b) When the bar is acted on by a transverse force $F(x)e^{-i\omega t}$ per unit length of bar, so that

$$\frac{d^4Y}{dx^4} - \gamma^4 Y = \frac{F}{Q\kappa^2 S} \equiv f(x)$$

show that [Eq. (4.5.13)]

$$Y(x) = \int_a^b f(x_0)g(x\,|\,x_0)\, dx_0$$

$$+ \left[Y(x_0)\frac{d^3g}{dx_0{}^3} - \frac{dY}{dx_0}\frac{d^2g}{dx_0{}^2} + \frac{dg}{dx_0}\frac{d^2Y}{dx_0{}^2} - g(x\,|\,x_0)\frac{d^3Y}{dx_0{}^3}\right]_{x_0=a}^{x_0=b}$$

where x is inside the region $a < x < b$, and a is considerably to the left of, and b considerably to the right of, the region where f differs from zero. Show that, if all the transverse motion of the bar is caused by force F, the quantity in brackets is zero at $x_0 = a$ and $x_0 = b$.

7 A distributed load of transverse mechanical impedance $z(x)$ $(a \ll x \ll b)$ per unit length is attached to the infinite bar. A transverse wave $A \exp(i\gamma x - i\omega t)$ is incident from the left. (a) Use the results of Prob. 6 to show that the displacement of the bar satisfies the integral equation

$$Y(x) = Ae^{i\gamma x} - \tfrac{1}{4} \int_a^b \zeta(x_0) \, Y(x_0) \, (e^{i\gamma|x-x_0|} + ie^{-\gamma|x-x_0|}) \, dx_0$$

where $\zeta(x) = \gamma z(x)/\rho\omega S$ is the ratio between the impedance density $z(x)$ of the load and the characteristic impedance $\rho S u = \rho S \omega/\gamma$ of the bar for the frequency $\omega/2\pi$. (b) What integral equals the ratio $-\Phi(\omega)$ between the amplitude of the reflected wave and that of the incident wave? Compare with Eq. (5.3.15). (c) The impedance load is concentrated at the origin, so that $z(x) = [-i\omega M + R + i(K/\omega)]\delta(x)$. Show that the ratio between the wave power reflected from the point load to the incident power is

$$|\Phi|^2 = \frac{R^2 + [\omega M - (K/\omega)]^2}{[(4\omega\rho S/\gamma) + R + \omega M - (K/\omega)]^2 + [R - \omega M + (K/\omega)]^2}$$

Compare this with Eq. (4.5.8). (d) For $Q = \sqrt{MK}/R = 10$ and $\omega\rho S/\gamma = R\sqrt{\omega/\omega_0}$ and $\omega_0 = \sqrt{K/M}$, plot $|\Phi|^2$ as a function of ω/ω_0 from 0 to 2. Compare this with the dashed curves of Fig. 4.19. Why, as $\omega \to 0$, does the reflection ratio go to $\tfrac{1}{2}$, whereas for the string it goes to 1?

8 Show that the Green's function for the clamped-free bar is

$$G(x,x_0) = \frac{2Q\kappa^2}{l\rho} \sum_n \frac{\psi_n(x)\psi_n(x_0)}{\omega_n^2 - \omega^2}$$

where $\omega_n^2 = \pi^4 Q \kappa^2 \beta_n^4/l^4 \rho = (2\pi\nu_n)^2$, and the other quantities are defined in Eqs. (5.1.16) and (5.1.17). Show that when the bar is pushed sideward by a force $F(x)e^{-i\omega t}$, the steady-state displacement of the bar is $Y(x)e^{-i\omega t}$, where

$$Y(x) = \frac{1}{Q\kappa^2 S} \int_0^l G(x,x_0) F(x_0) \, dx_0$$

Compare this with Eq. (5.1.24). What is the corresponding formula for a clamped-clamped bar?

9 A clamped-free bar is in a homogeneous magnetic field F perpendicular to the bar axis. A wire of negligible mass and stiffness is connected to the free end, so that a current $Ie^{-i\omega t}$ is passed along the bar. (a) Obtain the series expansion for the steady-state displacement. (b) Do the same for the clamped-clamped bar. Plot the shape of the bar when $\omega = \tfrac{1}{2}\omega_1$; when $\omega = 2\omega_1$ (to do this you will have to compute the integral of the square of ψ_n for the clamped-clamped bar for $n = 1, 2, 3$).

10 A point mass α times the total mass $\rho l S$ of the bar is fastened to the free end of a clamped-free bar. Use the result of Prob. 8 to show that the integral equation for the bar's displacement becomes

$$Y(x) = 2^{\frac{3}{2}}\omega^2\alpha \, Y(l) \sum_{n=1}^{\infty} \frac{(-1)^{n-1}\psi_n(x)}{\omega_n^2 - \omega^2}$$

remembering that $\psi_n(l) = (-1)^{n-1}\sqrt{2}$ for this bar. (a) Play the game of Chap. 4, Prob. 24, by "normalizing" Y so that $\omega^2[1 + 2\frac{3}{2}\alpha Y(l)] = \omega_1{}^2$, and thus obtain the implicit equation for $Y_1(l)$ (and thus for the lowest natural frequency of the loaded bar),

$$Y(l) = \sqrt{2}\left[1 - 4\alpha \sum_{n=2}^{\infty} \frac{(\omega_1/\omega_n)^2}{1 + 2\frac{3}{2}\alpha Y(l) - (\omega_1/\omega_n)^2}\right]^{-1}$$

where ω_n is given in Prob. 8 and Eq. (5.1.16). (b) For $\alpha = \frac{1}{2}$ and $\alpha = 1$ solve this for $Y(l)$, then compute the ratio (ω/ω_1) between the loaded and unloaded natural frequency for the lowest mode; finally, plot the corresponding $Y(x)$ for the two values of α. Are these exact solutions if the equation for $Y(l)$ is solved exactly?

II   An oscillating driving force of frequency ν is applied to the free end of a clamped-free bar. Show that the mechanical impedance of the bar is

$$Z = \tfrac{1}{4}il\rho S\left(\sum_n \frac{\nu}{\nu_n{}^2 - \nu^2}\right)^{-1}$$

Plot $|4Z/\rho lS\nu_1|$ against ν/ν_1 from 0 to 4.

12   A steel wire ($\rho = 7.7$ g per cu cm, $Q = 2 \times 10^{11}$ dynes per sq cm) 2 mm in diameter is stretched between two rigid supports 100 cm apart, with a tension of 2×10^6 dynes. Calculate the ratios between the first three overtone frequencies and the fundamental.

13   A membrane is made of material of density 0.1 g per sq cm and is under a tension of 100,000 dynes per cm. It is wished to have the membrane respond best to sound of frequency 250 cps. If the membrane is square, what will be the length of one side? What will be the frequencies of the two lowest overtones?

14   A square membrane 20 cm on a side, of mass 1 g per sq cm, is under a tension of 10^8 dynes per cm. Its motion is opposed by a frictional force of $42(\partial\eta/\partial t)$ dynes per sq cm. Find the modulus of decay of the oscillations. What are the first four "frequencies" of the damped motion?

15   A square membrane b cm on a side, of density σ and under a tension T, is loaded at its center with a mass of M g. Show that the allowed frequencies are, approximately,

$$\nu = \left\{\tfrac{1}{4}\frac{T}{\sigma}\left(\frac{m^2}{b^2} + \frac{n^2}{b^2}\right)\left[1 - \frac{4M}{\sigma b^2}\psi_{mn}{}^2\left(\frac{b}{2},\frac{b}{2}\right)\right]\right\}^{\frac{1}{2}}$$

as long as $4M/\sigma b^2$ is small. What will be the expression for the frequency if the membrane is rectangular but not square?

16   Show that the energy of vibration of a rectangular membrane of sides a and b is

$$W = \tfrac{1}{2}\sigma c^2 \int_0^a \int_0^b \left[\left(\frac{\partial\eta}{\partial x}\right)^2 + \left(\frac{\partial\eta}{\partial y}\right)^2 + \frac{1}{c^2}\left(\frac{\partial\eta}{\partial t}\right)^2\right] dx\,dy$$

and that when the motion is given in terms of the series

$$\eta = \sum_{m,n} A_{mn} \sin\frac{\pi m x}{a} \sin\frac{\pi n y}{b} \cos(2\pi\nu_{mn}t + \Phi_{mn})$$

the series for the energy is

$$W = \frac{\pi^2}{2}\sigma ab \sum_{m,n} \nu_{mn}{}^2 A_{mn}{}^2$$

17 A square membrane 20 cm on a side, with $\sigma = 1$ and $T = 10^6$, is started from rest at $t = 0$ with an initial shape $\eta = 10^{-5}x(20 - x)y(20 - y)$. What are the energies of vibration corresponding to the fundamental and the lowest three overtones, and what is the total energy of vibration of the membrane?

18 A rectangular membrane is pushed aside at the point (x_0,y_0) by a steady force F_0 and is then suddenly released at $t = 0$. What is the expression for the subsequent motion of the membrane? Neglect the reaction of the air.

19 The tensile strength of aluminum is 2.5×10^9 dynes per sq cm, and its density is 2.7 g per cc. What is the highest value of fundamental frequency that can be attained with an aluminum membrane stretched over a circular frame 3 cm in radius? If the aluminum is 0.005 cm thick, what will be the maximum tension attainable?

20 If the natural frequencies and characteristic functions for a membrane are $\omega_{mn}/2\pi = \nu_{mn}$ and $\psi_{mn}(\mathbf{r})$, where $\mathbf{r}$ stands for x, y or r, ϕ, etc., and where the mean square of ψ_{mn} is Λ_{mn}, show that the Green's function for the membrane is

$$G_\omega(\mathbf{r},\mathbf{r}_0) = \frac{c^2}{S} \sum_{m,n} \frac{\psi_{mn}(\mathbf{r})\psi_{mn}(\mathbf{r}_0)}{\Lambda_{mn}(\omega_{mn}^2 - \omega^2)}$$

If the membrane is acted on by a transverse force $F(\mathbf{r})e^{-i\omega t}$, show that the steady-state displacement of the membrane is

$$\eta = \frac{1}{\sigma c^2} \iint G_\omega(\mathbf{r},\mathbf{r}_0)\, F(\mathbf{r}_0)\, dS_0\, e^{-i\omega t}$$

where the integration is over the whole area S of the membrane.

21 A square membrane of side a is driven by a uniform force $Fe^{-i\omega t}$ per unit area. Show that the displacement of the center of the membrane is

$$\frac{16a^2F}{\pi^4\sigma c^2} \sum_{m,n} \left\{ (2m + 1)(2n + 1)\left[(2m + 1)^2 + (2n + 1)^2 - 2\left(\frac{\omega}{\omega_{11}}\right)^2 \right] \right\}^{-1}$$

times $e^{-i\omega t}$, where $\omega_{11}^2 = 2(\pi c/a)^2$ is the lowest natural frequency. Plot $\pi^4\sigma c_2/16a^2F$ times this as a function of ω/ω_{11} from 0 to 10.

22 A circular membrane is driven by a force per unit area equal to

$$F_0 \cos (m\phi)J_m(qr/a)e^{-i\omega t}$$

Since $\nabla^2 \cos (m\phi)J_m(qr/a) = -(q/a)^2 \cos (m\phi)J_m(qr/a)$, show that the solution of the equation

$$\nabla^2\eta + \left(\frac{\omega}{c}\right)^2 \eta = -\frac{F_0}{\sigma c^2} \cos (m\phi)J_m\left(\frac{qr}{a}\right)$$

for steady-state motion of the membrane is

$$\frac{A}{\sigma[(qc/a)^2 - \omega^2]} \left[J_m\left(\frac{qr}{a}\right) - \frac{J_m(q)}{J_m(\omega a/c)}J_m\left(\frac{\omega r}{c}\right) \right] \cos (m\phi)$$

which goes to zero at $r = a$ and goes to infinity at the resonant frequencies, when $J_m(\omega a/c) = 0$, but does *not* become infinite when $q = \omega a/c$. Thus, by comparison with the results of Prob. 20, show that

$$\int_0^{2\pi} d\phi \int_0^a \cos (m\phi_0)G_\omega(\mathbf{r},\mathbf{r}_0)J_m\left(\frac{qr}{a}\right) r\, dr$$

$$= \frac{a^2}{q^2 - (\omega a/c)^2} \left[J_m\left(q\frac{r}{a}\right) - \frac{J_m(q)}{J_m(\omega a/c)} J_m\left(\omega\frac{r}{c}\right) \right] \cos (m\phi)$$

Show that the Green's function for this equation is

$$G_\omega(\mathbf{r},\mathbf{r}_0) = \frac{1}{\pi} \sum_{m,n} \frac{\epsilon_m}{J_{m-1}^2(\pi\beta_{mn})} \cos m(\phi - \phi_0) \frac{J_m(\pi\beta_{mn}r/a)J_m(\pi\beta_{mn}r_0/a)}{(\pi\beta_{mn})^2 - (\omega a/c)^2}$$

where $\epsilon_0 = 1$, $\epsilon_m = 2$ $(m > 0)$.

23 By use of Eq. (1.2.9) show that a plane acoustic wave of velocity c_a, incident on the xy plane at an angle θ to the z axis, produces a pressure distribution on the xy plane of

$$P_0 \exp\left[i\left(\frac{\omega}{c_a}\right)x \sin\theta - i\omega t\right] = P_0 \sum_{m=0}^{\infty} \epsilon_m i^m \cos(m\phi)J_m\left(\frac{\omega r}{c_a}\sin\theta\right)e^{-i\omega t}$$

If this is the driving force on a circular membrane in the xy plane, show that its displacement is [see Eq. (6.3.7)]

$$\eta = \frac{P_0 a^2/\sigma c^2}{(\omega a/c_a)^2[\sin^2\theta - (c_a/c)^2]}\sum_m \epsilon_m i^m \cos(m\phi)\left[J_m\left(\frac{\omega r}{c_a}\sin\theta\right)\right.$$
$$\left. - \frac{J_m[(\omega a/c_a)\sin\theta]}{J_m(\omega a/c)}J_m\left(\frac{\omega r}{c}\right)\right]e^{-i\omega t}$$

where $c = \sqrt{T}/\sigma$ is the speed of transverse waves on the membrane.

24 A membrane is fastened over the top of a vessel of volume V_0 so that the integral of the membrane's displacement is equal to the net increase in volume δV occupied by the air in the vessel.   If the oscillations are not too rapid, this will produce a change in pressure $-(\delta V/V_0\kappa_s)$, where $\kappa_s = 1/\rho_a c_a^2$ is the compressibility of the air in the vessel, and c_a the speed of sound.   Show that the equation of motion for transverse motion of the membrane is

$$\frac{1}{c^2}\frac{\partial^2\eta}{\partial t^2} = \nabla^2\eta - \frac{\rho_a c_a^2}{V_0 T}\iint \eta\, dS \qquad T = \sigma c^2$$

where the integration is over the whole membrane area.   If it is circular of radius a, show that the vibrations of the form $A\cos(m\phi)J_m(\pi\beta_{mn}r/a)$ $(m > 0)$ will have frequencies given by Eqs. (5.2.24).   Show that, on the other hand, the radially symmetric vibrations $(m = 0)$ have different frequencies and that the form of the displacement is $\eta = A[J_0(\omega r/c) - J_0(\omega a/c)]e^{-i\omega t}$, the allowed frequencies being obtained by substituting this form in the equation of motion. Show that the resulting equation reduces to

$$(\omega a/c)^2 J_0(\omega a/c) = -(\pi\rho_a c_a^2 a^4/V_0\sigma c^2)J_2(\omega a/c)$$

Solve this equation for the lowest allowed value of $\omega a/c$ when the fraction $\pi\rho_a c_a^2 a^4/V_0\sigma c^2 = 1$.

25 Carry through Prob. 22 when the circular diaphragm over the vessel is a plate, as given in Eq. (5.3.10) and preceding.

26 The circular plate of Eq. (5.3.6) is loaded with a point mass $M = 2\alpha\pi a^2 h\rho$ (α times the total mass of the plate) placed at its center.   Use the methods of Prob. 10 [see also Eq. (5.3.12)] to compute the allowed frequencies and corresponding characteristic functions for the loaded plate.

CHAPTER

ACOUSTIC WAVE MOTION

6.1 THE DYNAMICS OF FLUID MOTION

We now come to the study of wave motion in air, the most important type of wave motion studied in the science of acoustics. Sound waves differ from the waves that we have discussed heretofore in several important respects. They are waves in three dimensions and as such can be more complicated in behavior than waves in two dimensions or in one. Sound waves also differ from waves on a string or on a membrane by being *longitudinal* waves. So far, we have been studying *transverse* waves, where the material transmitting the wave moves in a direction perpendicular to the direction of propagation of the wave. Each part of the string, for instance, moves in a direction at right angles to the equilibrium shape of the string, whereas the wave travels along the string. The molecules of air, however, move in the direction of propagation of the wave, so that there are no alternate crests and troughs, as with waves on the surface of water, but alternate compressions and rarefactions. The restoring force responsible for keeping the wave going is simply the opposition that the fluid exhibits against being compressed.

Since there are so many points of difference between the waves discussed earlier and the more complicated forms of sound waves, it is well not to introduce all the complications at once. Accordingly, we shall first study the motion of *plane waves* of sound, waves having the same direction of propagation everywhere in space, whose "crests" are in planes perpendicular to the direction of propagation. They correspond to the parallel waves on a membrane. Waves traveling along the inside of tubes of uniform cross section will usually be plane waves. Waves that have traveled unimpeded a long distance from their source will be, very nearly, plane waves.

The detailed properties of the acoustic wave motion in a fluid depend on the ratios between the amplitude and frequency of the acoustic motion and the molecular mean-free-path and the collision frequency, on whether the fluid is in thermodynamic equilibrium or not, and on the shape and thermal properties of the boundaries enclosing the fluid. All these effects will be discussed in this or in later chapters. At present we shall make a start by looking at the simplest case, that of an idealized fluid, uniform and continuous in its properties, at rest in thermodynamic equilibrium, except for the motion caused by the sound waves themselves, and with this acoustic motion small enough in magnitude so that many nonlinear effects are negligible and the equation of motion takes on its simplest form. We shall find this simplified model of acoustic wave motion an adequate representation in the majority of the cases of practical interest.

General characteristics of sound

Fluids, such as air or water, have mass density and volume elasticity; thus they have many of the characteristics of the chain of masses and springs of Fig. 3.4. The elasticity causes the fluid to resist being compressed, tending to return to its original state, and the inertia of the mass density causes the motion to "overshoot," thus providing the two requisites of wave motion. At equilibrium the fluid has density ρ, in units of kilograms per cubic meter; is under a uniform pressure P in units of newtons per square meter (or in atmospheres, where 1 atm $= 10^5$ newtons per sq m); and is at a uniform temperature $T°$K. These three quantities are connected by an equation of state, which can be given either as an explicit equation, relating ρ, P, and T, or in terms of the partial derivatives, assuming density as a function of T and P.

The equation of state is usually expressed in terms of the volume V occupied by n moles of the fluid (a mole is M kg of a substance which has molecular weight M). If nM kg of the fluid occupies volume V, the density of the fluid is obviously $\rho = nM/V$, or $V = nM/\rho$. The two partial derivatives, which are usually employed to measure the equation of state of the fluid near the equilibrium state, are

$$\kappa_T = -\frac{1}{V}\left(\frac{\partial V}{\partial P}\right)_T = \frac{1}{\rho}\left(\frac{\partial \rho}{\partial P}\right)_T \qquad \beta = \frac{1}{V}\left(\frac{\partial V}{\partial T}\right)_P = -\frac{1}{\rho}\left(\frac{\partial \rho}{\partial T}\right)_P \qquad (6.1.1)$$

Quantity κ_T, the fractional rate of change of volume (or density) with pressure at constant temperature, is called the *isothermal compressibility* of the fluid. As was indicated in Eq. (3.3.2), the compressibility of a spring is $\kappa = \frac{1}{l}\frac{\Delta l}{F} = \frac{1}{lK}$, where $F = K \Delta l$ is the force applied to shorten the spring by an amount Δl, and K is the stiffness constant. Quantity β, the fractional change in volume (or density) with temperature at constant pressure, is called the *coefficient of thermal expansion* of the fluid. Both κ_T and β are, of course, functions of P and T, though usually their rates of change are not very large. From κ_T and β we can compute the other partial derivatives of ρ, P, and T near the equilibrium state of the fluid; for example, if P, β, and κ_T are expressed as functions of ρ and T, then $(\partial P/\partial T)_\rho = \beta/\kappa_T$.

For a perfect gas the equation of state can be given as an explicit equation, $PV = nRT$, or $MP = RT\rho$, where $R = 8.3 \times 10^3$ joules per mole degree absolute. Therefore, for perfect gases, κ_T and β take on particularly simple values:

$$\kappa_T = \frac{1}{P} \quad \text{and} \quad \beta = \frac{1}{T}$$

For liquids, κ_T and β have much smaller values.

Kinetic theory indicates that the pressure of a perfect gas is caused by its molecular motion and that $P = \frac{1}{3}\rho\langle v^2\rangle$, where $\langle v^2\rangle$ is the mean-square molecular velocity and ρ is the gas density. The compressibility of a perfect gas is thus inversely proportional to the mean kinetic energy of the gas molecules. We noted in Eq. (3.3.2) that the velocity of propagation of waves along a spring-mass system is $\sqrt{1/\kappa\epsilon}$, where κ is the compressibility of the spring, and ϵ is the mass per unit length of the chain. We shall show that the velocity of propagation of sound waves in a fluid is $\sqrt{1/\kappa\rho}$, where κ is the compressibility and ρ is the density. For perfect gases, where $\kappa_T = 1/P = 3/\rho\langle v^2\rangle$, we see that the speed of sound

$$c = \sqrt{\frac{1}{\kappa\rho}} \simeq \sqrt{\frac{\langle v^2\rangle}{3}}$$

is approximately equal to the root-mean-square of the molecular speed. In liquids and solids, where intermolecular forces are large, the speed of sound is considerably larger than the root-mean-square speed of vibration of an atom about its equilibrium position. It can be shown, in fact, that c in solids is roughly equal to the mean speed a molecule would have if its amplitude of vibration were equal to the mean distance between molecules. The actual amplitudes are much smaller than this; otherwise the solid would melt or vaporize.

Thus the presence of a sound wave produces changes in density, pressure, and temperature in the fluid, each change being proportional to the amplitude of the wave. The pressure changes are usually the most easily measurable, though some sound detectors measure the motion of the fluid, caused by the sound. Direct measurement of temperature or density change is difficult, and the readings are not particularly accurate; so they are usually computed from the measured pressure change.

Pressure measurements are usually expressed in dynes per square centimeter ($= 10^{-1}$ newton per sq m $= 10^{-6}$ atm) above or below equilibrium pressure. Sometimes the pressure unit is 0.00022 dyne per sq cm, the amplitude of pressure change in a barely audible sound wave of 1,000 cps frequency [this is in terms of rms amplitudes; see Eq. (6.2.18)]. For simple-harmonic waves the pressure amplitude can be given in terms of the *pressure level* in *decibels*, 20 times the logarithm to the base 10 of the pressure amplitude in dynes per square centimeter, above or below the level of 1 dyne per sq cm. Or it can be given as pressure level in decibels above or below 0.00022 dyne per sq cm; on this scale a pressure amplitude of 1 dyne per sq cm rms amplitude is 74 db.

In fluids with very large thermal conductivity the temperature of the fluid is practically unchanged by the passage of the sound wave. In this case the isothermal compressibility κ_T is directly applicable. If the pressure change caused by the sound is p (the *acoustic pressure*) and the equilibrium pressure is P, so that the total pressure in the presence of sound is $P + p$, then the density of the fluid is

$$\rho + \delta = \rho + \left(\frac{\partial \rho}{\partial P}\right)_T p = \rho + \rho \kappa_T p \quad \text{or} \quad \delta = (\kappa_T \rho)p \quad \text{(6.1.2)}$$

where ρ is the equilibrium density, and δ is the small increase in density produced by the sound.

For sound of frequencies below about 10^9 cps or so, in gases, however, it is a better approximation to assume that the compression is *adiabatic*, i.e., that the entropy content of the gas remains nearly constant during the compression [see the discussion following Eqs. (6.4.23)]. In this case the differential relations between density and pressure and between temperature and pressure are

$$\left(\frac{\partial \rho}{\partial P}\right)_s = \rho \kappa_s = \frac{1}{\gamma}\left(\frac{\partial \rho}{\partial P}\right)_T \quad \text{or} \quad \kappa_s = \frac{\kappa_T}{\gamma}$$

$$\left(\frac{\partial T}{\partial P}\right)_s = \frac{\gamma - 1}{\gamma}\left(\frac{\partial T}{\partial P}\right)_\rho = (\gamma - 1)\frac{\kappa_s}{\beta}$$

$$\text{(6.1.3)}$$

so that $\qquad \rho + \delta = \rho + \rho \kappa_s p \qquad T + \tau = T + (\gamma - 1)\frac{\kappa_s}{\beta}p$

for adiabatic compression, where $\gamma = (C_p/C_v)$ is the ratio of the specific heat at constant pressure of the gas to the specific heat at constant volume, and τ is the small change in temperature caused by the sound wave. For a perfect monatomic gas (such as helium), $\gamma = \frac{5}{3}$; for a perfect diatomic gas (such as hydrogen or air), $\gamma = \frac{7}{5}$; for polyatomic gases, $\gamma = \frac{4}{3}$. Therefore, for perfect gases, where $MP = RT\rho$,

$$\kappa_s = \frac{1}{\gamma P} \quad \text{and} \quad \left(\frac{\partial \rho}{\partial P}\right)_s = \frac{\rho}{\gamma P} = \frac{M}{\gamma RT}$$

$$\left(\frac{\partial T}{\partial P}\right)_s = \frac{\gamma - 1}{\gamma}\frac{T}{P} = (\gamma - 1)\frac{M}{\gamma R\rho}$$

$$\text{(6.1.4)}$$

and if the pressure in the presence of the sound is $P + p$, the density and temperature are

$$\rho + \delta = \rho + \left(\frac{\rho}{\gamma P}\right)p \quad \text{or} \quad \delta = (\rho\kappa_s)p = \left(\frac{M}{\gamma RT}\right)p$$

$$T + \tau = T + \frac{(\gamma - 1)T}{\gamma P}\, p \quad \text{or} \quad \tau = \frac{(\gamma - 1)M}{\gamma R\rho}\, p$$

(6.1.5)

for adiabatic compression, for a perfect gas.

We see that, in this simple case, the ratio between the change in density δ and the acoustic pressure p is independent of the equilibrium pressure but inversely proportional to the equilibrium temperature of the gas. Also, the ratio between the change in temperature τ and the sound pressure p is independent of temperature but inversely proportional to the equilibrium density ρ. All these relationships are valid only to the first order in the small quantity p/P.

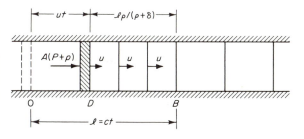

FIGURE 6.1
Pressure wave generated by motion of piston D in pipe of area A.

Acoustic energy and momentum

The discussion of the energy and momentum carried in a sound wave can be carried through in a manner quite parallel to that given on wave motion along a linear array of masses and springs, pictured in Fig. 3.5. Suppose the fluid is confined to the interior of a long pipe, of uniform cross section A, having rigid walls. At one end, as shown in Fig. 6.1, is a piston of negligible mass, which constrains the fluid to stay to its right. Originally, the piston was at rest at 0, held stationary by a force AP, counteracting the equilibrium pressure P of the fluid (this may, of course, be supplied by the presence of similar fluid to the left of the piston).

Now suppose, at time $t = 0$, the piston is started into motion to the right with constant velocity u. It will push the fluid ahead of it, compressing it somewhat and thus increasing its pressure to $P + p$. Thus, in order to get the piston into motion, the force on the piston must be increased from AP to $A(P + p)$, where p is related to the change in density δ of the fluid just ahead of the piston by the equation $\delta = \kappa \rho p$, with κ either κ_T or κ_s, whichever is appropriate.

But δ is related to the change in volume of the compressed fluid, and this is related to the amount of fluid which has been set in motion at any specified time t, as shown in Fig. 6.1. Because of the fluid's compressibility and inertia, it takes time to get it into motion; a wavefront of starting will travel away from the piston, reaching point B at time t. Suppose at time t the wavefront has traveled a distance ct, where c is the velocity of the wave. Then all the fluid to the right of this front, at B, is in its original equilibrium condition, at rest with density ρ. The fluid in the region D to B, in motion with the velocity u of the piston, is compressed to a density $\rho + \delta$, and thus occupies a volume $\rho/(\rho + \delta)$ times its original volume $Al = Act$. The difference between this volume and the original volume Act is, of course, the volume Aut, swept out by the piston in time t, so that

$$Act - A\,\frac{\rho ct}{\rho + \delta} = Aut$$

or
$$\frac{u}{c} = 1 - \frac{\rho}{\rho + \delta} = \frac{\delta}{\rho + \delta} \simeq \frac{\delta}{\rho} \qquad (6.1.6)$$

where the last formula is valid only as long as the density change δ, caused by the motion, is small compared with the equilibrium density ρ.

In traveling the distance ut, the piston has done an amount of work $A(P + p)ut$, which must have gone into additional energy of the fluid. Part of this energy will be kinetic energy, one-half of the mass of the fluid in motion, $\frac{1}{2}A\rho ct$, times the square of its velocity, u^2. The other part is potential energy of compression, the work done in compressing the fluid from volume Act to volume $Act[\rho/(\rho + \delta)]$. This work is the integral of the pressure $P + p$ times the change in volume dV, integrated between these limits. But from Eq. (6.1.1), or its equivalent for adiabatic compression, $dV = V\kappa\,dp$, where, as long as p is small compared with P, V can be set equal to Act and κ can be considered constant. Thus the potential energy is

$$\int (P + p)V\kappa\,dp = PV\kappa p + \tfrac{1}{2}V\kappa p^2 \simeq Aut\left(P + \frac{u}{2\kappa c}\right)$$

where, to obtain the last form, we have set $p = \delta/\rho\kappa = u/\kappa c$, from Eqs. (6.1.5) and (6.1.6). Thus the equation for energy balance becomes

Force $\times$ distance = potential plus kinetic energy

$$Aut(P + p) = AutP + Aut\,\frac{u}{2\kappa c} + Aut\,\frac{2}{\rho cu}$$

The first terms on both sides cancel. In fact, the work $APut$ should not be charged to energy of the sound wave, since it is the work done pushing against the equilibrium pressure; if there had been fluid pressure P on the left-hand surface of the piston, it would have provided the AP part of the force, and only force Ap would have been required to move the piston.   In fact, from now on, we shall use the expressions

$$\tfrac{1}{2}\kappa p^2 = \text{potential energy density}$$
$$\tfrac{1}{2}\rho u^2 = \text{kinetic energy density} \tag{6.1.7}$$

for the energies per unit volume of the sound, correct to second order in the small quantities p and u.

Returning to the energy-balance equation, which can be reduced to $p = (u/2\kappa c) + (\rho cu/2)$, we can again set $p = u/\kappa c$ and obtain $u/\kappa c = (u/2\kappa c) + (\rho cu/2)$, which in turn becomes

$$c^2 = \frac{1}{\rho \kappa} \tag{6.1.8}$$

What we have shown by this manipulation is that the mass of the fluid, represented by the equilibrium density ρ, and its elasticity, represented by κ, so combine that it takes time for the motion of one part of the fluid to be transmitted to another part, the speed of transmission being equal to $\sqrt{1/\rho\kappa}$ as long as the fluid motion is small enough so that the excess pressure p is small compared with the equilibrium pressure P.

The impulse-momentum relationship could also be used to compute c. The excess force Ap contributes an impulse Apt to the fluid in time t, which must equal the momentum contributed to the fluid newly set into motion in time t.   The acoustic momentum density of the fluid is $u\rho$, so that

$$Apt = (u\rho)(Act) \simeq Ac^2\,\delta t$$

from Eq. (6.1.6).   The resulting equation, $p \simeq c^2\delta$, is identical with either Eq. (6.1.2) or (6.1.3), depending on whether the compression is isothermal or adiabatic, if c^2 is given by Eq. (6.1.8).

It is obvious that the wave velocity c will be independent of the amplitude and shape of the sound wave only as long as the sound pressure p is small compared with the equilibrium pressure P.   This must be true in order for Eq. (6.1.6) to hold, and also in order to consider κ to be independent of p. For fluids like water, where κ is small enough so that δ/ρ is small even though p/P is not small, another limitation enters.   For example, if the piston were pushed to the left, a wave of rarefaction would be sent along the tube, and if u were large enough, the p computed from the formulas might be equal to or larger than P; the piston would be withdrawn rapidly enough to produce a vacuum behind it.   When this happens the fluid loses contact with the piston, leaving a large bubble of vapor in the interspace, and it is obvious that our equations no longer hold.   Long before this happens the value of κ begins to change, so that κ cannot be considered independent of p unless $|p|$ is small compared with P.

Fluid displacement and velocity

Our next task is to work out an approximate equation of motion for small-amplitude sound waves in a fluid. But before we do this, we must develop a notation which will adequately describe the motion. As a matter of fact, we shall develop two notations, one of which will be generally employed; but the other will be occasionally useful.

The most straightforward way, the extension of the notation we have used for the chain of masses and springs, is to label each portion of the fluid (each molecule in it, so to speak) by giving its position at $t = 0$, and then to give the subsequent position of each portion as a function of time and of its initial position. In this notation, called the *Lagrange* description, we follow the fluid in its motion. In one-dimensional motion, the portion of fluid initially at x_0 has position X at time t, where X is a function of t and of x_0, as shown in Fig. 6.2. All other quantities similarly follow the motion of the fluid. The density $\rho(x_0,t)$ is the density of the portion of the fluid which was initially at x_0 and which is at X at time t.

This notation gives us a good insight into what is going on inside the fluid as it moves about. For example, if the time rate of change $d\rho/dt$ of this density is less than zero, we can be sure the fluid is expanding. We shall reserve the *total derivative* notation, d/dt, for the time rate of change in the Lagrange description, the change in a particular portion of the fluid as it moves along. The derivative du/dt, where $u = dX/dt$ is the fluid velocity, is the acceleration of the element of fluid which was originally at x_0, and is thus equal to the force acting on the element, divided by its mass. However, the notation also has its disadvantages, for the coordinate system moves with the fluid, and it is not easy to determine what the fluid is doing at some specified point in space at a given time.

So the alternative notation, the *Euler description*, sets up a coordinate system (x, in one dimension) fixed in space and describes the properties of the fluid which happens to be at a given point at a given time. Therefore, in this notation, the various quantities are functions of x and of t, for one dimension, as shown in the lower half of Fig. 6.2. These are quantities, such as density, velocity, etc., measured at a *fixed point in space* at a given

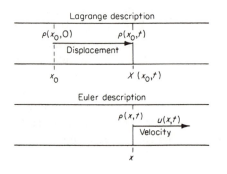

FIGURE 6.2
Alternative ways of describing the motion of a fluid.

time, and therefore corresponding to *different portions* of the fluid, as the fluid streams past the point. This is the description we shall be using most of the time, with ρ and u given as explicit functions of x and t; by it we can tell what is happening where, without having to go back to find which part of the fluid happens to be there at time t. In particular, the Euler description enables us to compute easily the spatial variation of a given quantity at time t, the partial derivative $\partial\rho/\partial x$ being the rate of change of density at x, as might be measured from an instantaneous photograph of the density at time t.

Comparison of Lagrange and Euler descriptions

We should note that the displacement of a given portion of the fluid at time t ($X - x_0$ in the Lagrange description) is not usually specified in the Euler notation; in fact, there is no simple way to write this quantity. The Euler description starts with the fluid velocity $u(x,t)$, at point x at time t, which equals the quantity dX/dt for that portion of the fluid for which $X(x_0,t) = x$ at time t, in the Lagrange description, and also deals with the other properties of the fluid, such as density $\rho(x,t)$, temperature $T(x,t)$, and so on. To show how the two descriptions differ, consider the situation pictured in Fig. 6.3. We have assumed that a one-dimensional sinusoidal wave, of frequency $\omega/2\pi$, is traveling in the positive x direction with velocity c, so that the wavelength is $\lambda = 2\pi/k = 2\pi c/\omega$. We also assume that the amplitude of motion of the fluid is *not* small compared with the wavelength, so that the effects will be easily seen.

By the Lagrange description the displacement from its equilibrium position of that portion of the fluid which was at x_0 before the wave came along is a traveling sinusoidal wave,

$$X - x_0 = A \sin(\omega t - kx_0) = -A \sin[k(x_0 - ct)]$$

In accordance with the Lagrange description, X is a function of t and of the parameter x_0, which describes where the fluid was before the wave started, not where it is while the wave is present. The velocity of the portion of the fluid, which is at X at time t, is

$$u(x_0,t) = \frac{dX}{dt} = A\omega \cos(\omega t - kx_0)$$

The various curves in Fig. 6.3a are plots of X as a function of t for the various values of x_0 which are marked on the right-hand end of each curve. We see that the fluid, in the region of $x = 0$, is compressed at $t = 0$, when the fluid is moving in the direction of the wave motion, and is correspondingly rarefied at $t = 10$, when the motion is in the opposite direction. We have chosen the scales of x and t so that the dashed lines, which indicate the motion of the wavefronts, are inclined at 45° to the t axis. Parenthetically, if the amplitude of motion in an actual sound wave were as large as this, the displacement would not be sinusoidal. This does not invalidate the point of our discussion, however, as will be appreciated later.

To see the difference between the Lagrange and the Euler descriptions, let us try to answer the following question: What is the displacement $X - x_0$ of the portion of fluid which happens to be at point x at time t?

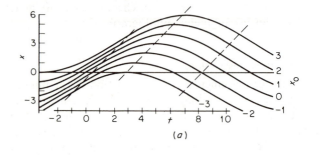

(a)

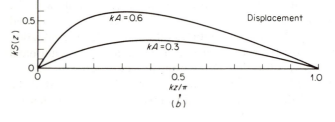

(b)

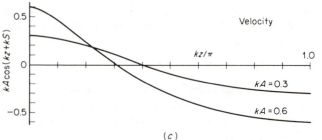

(c)

FIGURE 6.3
Comparison between Lagrange and Euler descriptions of one-dimensional fluid motion.

To answer this we must change our description from initial position x_0 to the final position x at time t. For example, at $t = 0$ the fluid which was at $x_0 = 0$ at equilibrium (i.e., the fluid following curve $x_0 = 0$) is back at $x = 0$. But at $t = 1$ the fluid following curve $x_0 = -2$ (i.e., the fluid which was at $x_0 = -2$ before the wave motion) is at $x = 0$, and at $t = 3$ the fluid of $x_0 = -3$ is at $x = 0$. To translate what happens to the fluid following each curve into what fluid happens to be at x, we must change from x_0, t to x, t.

Since
$$x = (X - x_0) + x_0 = x_0 - A \sin [k(x_0 - ct)]$$
we shall write $x_0 - ct = -z_0$ and $x - ct = -z$, so that $z = z_0 - A \sin (kz_0)$.
To find z_0 as a function of z, we write $A \sin (kz_0) = S(z)$ and see that
the function S must satisfy the implicit equation $z_0 = z + S(z)$

$$S(z) = A \sin \{k[z + S(z)]\} \quad A \sin(k z_0) = A \sin(kz + kS(z))$$

which can be solved with little difficulty by successive approximations.
Some solutions are plotted in Fig. 6.3b for $kA = 0.3$ and $kA = 0.6$. Function $S(ct - x) = -S(x - ct) = -A \sin [k(x_0 - ct)] = A \sin (kz_0)$ gives the
displacement $X - x_0$ from equilibrium of that portion of the fluid which
happens to be at point x at time t. In particular, $S(ct)$ is the displacement
which has occurred to that portion of the fluid which is at the origin at time
t. It may be obtained from curves a by plotting, against t, the negative of
x_0 (that is, $x - x_0$ for $x = 0$) for the curve which crosses the $x = 0$ axis
at time t. We see that this displacement is far from being a sine wave.
During the first part of the cycle, when the fluid is compressed, S rises rapidly
to its maximum, then goes more slowly back to zero during the part of the
cycle when the motion is opposed to the direction of propagation and the
fluid is rarefied.

As pointed out, the velocity of the portion of the fluid which was
originally at x_0 is given by a simple cosine function in terms of x_0 and t.
But the Eulerian velocity $u(x,t)$, the velocity of the part of the fluid which is
at x at time t, is

$$dX/dt = kcA \cos \{k[z + S(z)]\} = \omega A \cos \{k[ct - x + S(ct - x)]\}$$

which is plotted as a function of kz in Fig. 6.3c, for $kA = 0.3$ and 0.6. We
see that it is far from being a simple cosine function of t. But also, it is *not*
the time derivative $\partial S/\partial t$ of the implicit function S. If we were to forget
the difference between the Lagrange and the Euler descriptions, and try to
find the displacement X of some part of the fluid by integrating $u(x,t)$ with
respect to t, Fig. 6.3c shows that we should become erroneously convinced
that the fluid had a negative mean displacement (since u is negative for a
longer time than it is positive). Of course, the curves of Fig. 6.3a show
that there is no such average drift; the mistake would come from the assumption that $u(x,t)$ is the velocity of a given part of the fluid, instead of the velocity
of whichever fluid happens to be at x at time t.

To be quantitative about the difference between fluid displacement and
the integral of $u(x,t)$ and to generalize the equations for S, we assume that
the displacement $X - x_0$ of the fluid originally at x_0 is some function
$F(ct - x_0)$. Then the displacement $D(ct - x)$ of that portion of the fluid
which happens to be at point x at time t must satisfy the equation

$$D(ct - x) = F[ct - x + D(ct - x)]$$

The actual velocity $u(x,t)$ of that portion of the fluid which is at x at time t is, of course, *not* $\partial D/\partial t$ (holding x constant), but is $\partial F/\partial t$ (holding x_0 constant) $= cF'(ct - x + D)$, where F' is the derivative of F with respect to its argument. Since

$$\frac{\partial D}{\partial t} = cD'(ct - x) = c[1 + D'(ct - x)]F'(ct - x + D)$$

the relationship between u and D' is

$$u(x,t) = cF' = \frac{cD'(ct - x)}{1 + D'(ct - x)} = c[D' - (D')^2 + (D')^3 \cdots]$$

the series expansion converging if D' is smaller than unity (i.e., if the fluid velocity is everywhere smaller than c).

If F is a periodic function [as, for example, $F(u) = A \sin(ku)$], function D also is periodic, and the time average of all odd powers of D' will be zero. In this case the time integral of the Eulerian velocity, over a long time interval,

$$\int_0^T u(x,t)\, dt \to -cT \left\langle \frac{(D')^2}{1 - (D')^2} \right\rangle_{\text{avg}} \qquad \text{for } T \to \infty$$

is *negative*, roughly equal to $-cT$ times the time average of u^2/c^2 when $u \ll c$, even though the actual fluid displacement

$$\int_0^T D'c\, dt = c \int_0^T [1 + D'(u)]F'[u + D(u)]\, du$$

is zero when F and D are periodic functions and T is an integral number of periods. Thus *only* when the second-order term $(u/c)^2$ can be neglected compared with u/c can we assume that the integral of $u(x,t)$ is equal to the fluid displacement.

The main conclusion to be drawn from this rather long digression is that the time derivative of the displacement of some part of the fluid differs from $u(x,t)$ by an amount which is small only for small amplitudes. The difference between the Euler and the Lagrange descriptions can be neglected only when we can neglect the squares of u, p, and δ; even in this case there is a difference in inhomogeneous fluids.

The time derivatives

The crucial difference between the two notations comes in the time rates of change. In the Lagrange notation the *total derivative df/dt* of some property f of the fluid is the time rate of change of f in a portion of fluid, *as it moves through* point x at time t. In contrast, in the Euler notation, the *partial derivative $\partial f/\partial t$* (we shall always use the partial notation for the Eulerian time derivative) is the change of f at *the fixed point* x, as the fluid streams past. Thus $\partial f/\partial t$ includes not only the time rate of change of f in the moving fluid, but also the change in f at x because the fluid is moving past, if f differs from point to point in the fluid.

The chief advantage of the Lagrange description lies in the time derivatives; the total derivative $d\rho/dt$ measures whether the fluid is expanding or not as it moves; du/dt is the true acceleration of a portion of the fluid, and thus is proportional to the net force acting on the fluid at x, t. But the Euler description has many other advantages; so it will be useful to find a measure of the difference between the two time derivatives.

To find the total derivative df/dt of some property of the fluid, in terms of the partial derivatives at t, x, we compare the value of f at x, t with its value at time $t + dt$ *at the point $x + u\,dt$ to which the fluid has moved in time dt.* In accord with the definitions of derivatives, this is

$$\frac{df}{dt}\,dt = f(x + u\,dt, t + dt) - f(x,t)$$

$$= f(x,t) + \frac{\partial f}{\partial x}\,u\,dt + \frac{\partial f}{\partial t}\,dt - f(x,t)$$

or

$$\frac{df}{dt} = \frac{\partial f}{\partial t} + u\,\frac{\partial f}{\partial x} \tag{6.1.9}$$

The actual change in f, as the fluid moves past x with velocity u, is the change in f with time at the fixed point x, plus the spatial change in f times its velocity u of flow past x. Thus the true acceleration of the fluid at x at time t is given, in terms of the Eulerian velocity function $u(x,t)$, by

$$\frac{du}{dt} = \frac{\partial u}{\partial t} + u\,\frac{\partial u}{\partial x} \tag{6.1.10}$$

We see that the correction term $u(\partial u/\partial x)$ is second order in the fluid velocity, and therefore, if u is small (as it often is for sound waves), $du/dt \simeq \partial u/\partial t$ to the first order of approximation. The distinction cannot be forgotten, however, because the sound energy and intensity are second-order expressions, and the correction term may have to be included. Also, when the fluid is moving as a whole with respect to the observer, its velocity may not be small; so the second terms in Eqs. (6.1.9) and (6.1.10) may be as large as the first terms. We should note that Eq. (6.1.10) is really just a restatement of the discussion on page 238.

The equation of continuity

In addition to these questions of notation, we must also take into account the fact that what happens in one part of a fluid affects what happens in other parts. For example, we must express in equation form the fact that the net flow across a closed surface in the fluid produces a change in the properties of the fluid inside the surface. Suppose f is some property like entropy density or electric charge density that can be created in the fluid and, once created, can be carried along with the fluid as it moves. For one-dimensional situations, where both f and u depend only on x and t, the *flux* of f is the total amount of f that passes per second in the positive x direction across a unit cross-sectional area perpendicular to x. If f travels with the fluid, this flux $J(x,t)$ must be equal to the density f times the fluid velocity u at x and t.

Now consider the slice of fluid between a plane surface of unit area, normal to the x axis, at $x + dx$, and a parallel surface at x. The flux $J(x + dx)$ represents a loss of f from the region, and the flux $J(x)$ into the region represents a gain. The difference must equal the rate of loss by outflow of f from the region. If, in addition, f is created at a rate $Q(x,t)$ per unit volume of fluid, the net rate of change of f within the region is

$$\frac{\partial f}{\partial t}\,dx = Q\,dx - J(x + dx) + J(x) = Q\,dx - \frac{\partial J}{\partial x}\,dx$$

where partial derivatives are used because we are talking about a region fixed in space. Therefore

$$\frac{\partial f}{\partial t} = Q - \frac{\partial fu}{\partial x} = Q - f\frac{\partial u}{\partial x} - u\frac{\partial f}{\partial x} \qquad (6.1.11)$$

This equation, which states that any increase in f in a region must have been brought there by fluid flow or else by specific creation there, as represented by Q, is called the *equation of continuity*. If we wish to find the net gain in f in a region which moves with the fluid, we substitute from Eq. (6.1.9),

$$\frac{df}{dt} = Q - u\frac{\partial f}{\partial x} - f\frac{\partial u}{\partial x} + u\frac{\partial f}{\partial x}$$

$$= Q(x,t) - f\frac{\partial u}{\partial x} \qquad (6.1.12)$$

If the fluid flow is the same throughout the fluid, $\partial u/\partial x = 0$ and Q equals the total derivative df/dt, as might be expected.

In particular, if f is the mass density of fluid ρ and Q is the rate of creation of fluid (sometimes this is possible), then $(\partial\rho/\partial t) + (\partial\rho u/\partial x) = Q$, or if we write the density as $\rho + \delta(x,t)$, where ρ is the constant equilibrium density and δ is the variable part,

$$Q = \frac{\partial\delta}{\partial t} + u\frac{\partial\delta}{\partial x} + (\rho + \delta)\frac{\partial u}{\partial x} = \frac{d\delta}{dt} + (\rho + \delta)\frac{\partial u}{\partial x} \qquad (6.1.13)$$

Even if $Q = 0$, the density of a given portion of fluid can change with time if the fluid velocity changes with x. This equation is valid only for homogeneous fluids.

It is not difficult to generalize these equations to three dimensions if we use vector notation. Property f in the Euler description is a function of position, denoted by vector $\mathbf{r}$, with components x, y, and z, and fluid velocity at $\mathbf{r}$, t is a vector $\mathbf{u}$, with components u_x, u_y, and u_z. The flux of f is vector $\mathbf{J} = f\mathbf{u}$. Then Eq. (6.1.9), relating the total time derivative of $f(\mathbf{r},t)$ and its partial time derivative, is

$$\frac{df}{dt}\, dt = f(\mathbf{r} + \mathbf{u}\, dt, t + dt) - f(\mathbf{r},t)$$

$$= \left(\frac{\partial f}{\partial t} + u_x \frac{\partial f}{\partial x} + u_y \frac{\partial f}{\partial y} + u_z \frac{\partial f}{\partial z}\right) dt$$

or
$$\frac{df}{dt} = \frac{\partial f}{\partial t} + (\mathbf{u} \cdot \text{grad})f \tag{6.1.14}$$

where the operator

$$\mathbf{u} \cdot \text{grad} \equiv \mathbf{u} \cdot \mathbf{\nabla} = u_x \frac{\partial}{\partial x} + u_y \frac{\partial}{\partial y} + u_z \frac{\partial}{\partial z}$$

measures the rate of change of f at $\mathbf{r}$, t, caused by the fluid flow; it can be used on either a scalar or a vector.

To obtain the three-dimensional equation of continuity we first show that the net flux out of volume element $dx\, dy\, dz$ is

$$dy\, dz\left(\frac{\partial J_x}{\partial x}\right) dx + dx\, dz\left(\frac{\partial J_y}{\partial y}\right) dy + dx\, dy\left(\frac{\partial J_z}{\partial z}\right) dz = \text{div } \mathbf{J}\, dx\, dy\, dz$$

where
$$\text{div } \mathbf{J} = \mathbf{\nabla} \cdot \mathbf{J} = \frac{\partial J_x}{\partial x} + \frac{\partial J_y}{\partial y} + \frac{\partial J_z}{\partial z}$$

The equation of continuity is therefore

$$\frac{\partial f}{\partial t} \equiv Q(r,t) - \text{div } \mathbf{J} = Q - f\, \text{div } \mathbf{u} - \mathbf{u} \cdot \text{grad } f \tag{6.1.15}$$

or
$$\frac{df}{dt} = Q - f\, \text{div } \mathbf{u}$$

where grad $f \equiv \mathbf{\nabla} f$ is a vector with components $\partial f/\partial x$, $\partial f/\partial y$, and $\partial f/\partial z$.

6.2 WAVE MOTION, ENERGY, AND MOMENTUM

We now are ready to work out the equation of acoustic wave motion which is valid when the wave amplitude is small. The more accurate form of the equation will be considered later, when we discuss large-amplitude waves and waves in moving media. We start here with a fluid that, in the absence of sound, has uniform density ρ, pressure P, and temperature T and is everywhere at rest. We assume in this section that it is nonviscous and has zero heat conductivity, so that the only energy involved in the acoustic motion is mechanical, and the only forces are those of compressive elasticity, measured by the compressibility κ (the adiabatic compressibility is the appropriate one to use in the case of zero heat conduction, but we need not continue to write the subscript s).

In the presence of a sound wave the pressure becomes $P + p(x,t)$ if the wave is one-dimensional, the wavefronts being planes parallel to the yz plane. The density becomes $\rho + \delta(x,t)$, and the temperature becomes $T + \tau(x,t)$ over the whole of the plane a distance x from the yz plane. Waves of this sort are appropriately called *plane waves*. As mentioned previously, the change in pressure $p(x,t)$ is called the acoustic, or sound pressure. It is responsible for the motion of the fluid, and the motion is in turn responsible for the change in density, $\delta(x,t) \simeq \rho\kappa p(x,t)$.

The one-dimensional equation

The net force on the mass $(\rho + \delta)\,dx$ of fluid in the volume enclosed between two plane surfaces, each of unit area, one at x and the other at $x + dx$, is $p(x) - p(x + dx)$, which must equal the mass of the fluid times its acceleration,

$$p(x) - p(x + dx) = -\frac{\partial p}{\partial x}\,dx = \frac{du}{dt}(\rho + \delta)\,dx$$

or

$$\frac{\partial p}{\partial x} = -(\rho + \delta)\left(\frac{\partial u}{\partial t} + \frac{1}{2}\frac{\partial u^2}{\partial x}\right) \qquad (6.2.1)$$

from Eq. (6.1.10), changing from the total to the partial time derivative of u, when u depends only on x and t.

Also, from Eq. (6.1.13), if the source function Q is zero (if no new fluid is introduced at x), the equation of continuity for the density $\rho + \delta$ is

$$\frac{\partial \delta}{\partial t} = -(\rho + \delta)\frac{\partial u}{\partial x} - u\frac{\partial \delta}{\partial x} \quad using\ 6.1.9 \atop \delta \to \delta$$

which becomes

$$\rho\kappa\frac{\partial p}{\partial t} = -\rho(1 + \kappa p)\frac{\partial u}{\partial x} - \rho\kappa u\frac{\partial p}{\partial x} \qquad (6.2.2)$$

if we can use the approximate formula $\delta = \rho\kappa p$ relating density and pressure change. Eliminating u between these two equations will produce the equation of wave motion.

We have been assuming that δ and p are small compared with ρ and P, and also that u is small; thus there are several hierarchies of smallness in these two equations. In both equations the term on the left is first order in the small quantities u, δ, and p. In Eq. (6.2.1), $\rho(\partial u/\partial t)$ is first order, $\delta(\partial u/\partial t)$ and $\rho(\partial u^2/\partial x)$ are second order, and $\delta(\partial u^2/\partial x)$ is third order. In Eq. (6.2.2) $\rho(\partial u/\partial x)$ is the first-order term on the right and $-\rho\kappa p(\partial u/\partial x)$ and $-\rho\kappa u(\partial p/\partial x)$ are second-order terms. [Actually, there is a second-order term on the left of Eq. (6.2.2) involving $\partial\kappa/\partial P$, but this term is usually much smaller than the other second-order terms and is usually neglected.]

The first-order equation of motion is obtained by neglecting higher-order terms in both equations, leaving

$$\frac{\partial p}{\partial x} = -\rho \frac{\partial u}{\partial t} \qquad \kappa \frac{\partial p}{\partial t} = -\frac{\partial u}{\partial x} \qquad (6.2.3)$$

The first of these states that a pressure gradient produces an acceleration of the fluid; the second states that a velocity gradient produces a compression of the fluid.

Differentiating the first by x and the second by t enables us to eliminate u, thus obtaining the first-order equation of acoustic motion,

$$\frac{\partial^2 p}{\partial x^2} = \frac{1}{c^2} \frac{\partial^2 p}{\partial t^2} \qquad c^2 = \frac{1}{\kappa \rho} \qquad (6.2.4)$$

which is, of course, the familiar wave equation, with wave velocity inversely proportional to the square root of the density ρ times the compressibility κ of the fluid in equilibrium, as predicted by Eq. (6.1.8).

The extension of these equations to three-dimensional motion is not difficult. The equation relating fluid acceleration and pressure gradient becomes

$$(\rho + \delta) \frac{d\mathbf{u}}{dt} \equiv (\rho + \delta)\left[\frac{\partial \mathbf{u}}{\partial t} + (\mathbf{u} \cdot \mathrm{grad})\mathbf{u}\right] = -\mathrm{grad}\, p \qquad (6.2.5)$$

where the velocity $\mathbf{u}$ is now a vector, and we have used Eq. (6.1.4) to change from total to partial time derivative. The equation of continuity (6.1.15) for mass flow becomes

$$\frac{\partial \delta}{\partial t} \equiv \kappa \rho \frac{\partial p}{\partial t} = -\mathrm{div}\,[(\rho + \delta)\mathbf{u}] \equiv -(\rho + \delta)\,\mathrm{div}\,\mathbf{u} - \mathbf{u} \cdot \mathrm{grad}\, \delta$$

$$= -\rho(1 + \kappa p)\,\mathrm{div}\,\mathbf{u} - \kappa \rho \mathbf{u} \cdot \mathrm{grad}\, p \qquad (6.2.6)$$

where we have again assumed that κ is independent of p.

To first order in the small quantities p, $\mathbf{u}$, we then have

$$\rho \frac{\partial \mathbf{u}}{\partial t} = -\mathrm{grad}\, p \qquad \kappa \frac{\partial p}{\partial t} = -\mathrm{div}\,\mathbf{u} \qquad (6.2.7)$$

Taking the divergence of the first equation and using the second to eliminate $\mathbf{u}$ results in

$$\mathrm{div}\,\mathrm{grad}\, p \equiv \left(\frac{\partial^2}{\partial x^2} + \frac{\partial^2}{\partial y^2} + \frac{\partial^2}{\partial z^2}\right)p \equiv \nabla^2 p = \kappa \rho \frac{\partial^2 p}{\partial t^2} = \frac{1}{c^2} \frac{\partial^2 p}{\partial t^2} \qquad (6.2.8)$$

which is the three-dimensional form of the wave equation. The symbol ∇^2, the three-dimensional counterpart to the operator of Eq. (5.2.3), is again called the Laplacian. We shall discuss its various forms, for different coordinate systems, in a later chapter.

For future reference, though we shall not use the equations in this chapter, we write down the equation of motion to the second order in the small quantities **u**, p (though still not including the effects of viscosity, thermal conduction, or change of compressibility with pressure). To this order it is more convenient to express both p and u in terms of a scalar function Ψ, which we call the *velocity potential*, since **u** is obtained by taking the gradient of Ψ, $\mathbf{u} = -\operatorname{grad} \Psi$. Then, if we set, to second order,

$$p = \rho \frac{\partial \Psi}{\partial t} - \tfrac{1}{2}\rho \left[|\operatorname{grad} \Psi|^2 - \kappa\rho \left(\frac{\partial \Psi}{\partial t} \right)^2 \right]$$

the equation of motion for Ψ, to second order, is

$$\frac{\partial^2 \Psi}{\partial t^2} - \frac{\partial}{\partial t} \left[|\operatorname{grad} \Psi|^2 - \kappa\rho \left(\frac{\partial \Psi}{\partial t} \right)^2 \right] = \frac{1}{\kappa\rho} \nabla^2 \Psi \qquad (6.2.9)$$

assuming that κ is a constant. Thus, even when viscosity and thermal conduction are neglected, second-order terms modify the equation for simple wave motion. We shall see later what effect this correction term has on the solutions.

For small motions of the fluid, however, such that the acoustic pressure p is small compared with the equilibrium pressure P, the motion, with its accompanying small changes of pressure, propagates as simple waves with velocity $c = \sqrt{1/\rho\kappa}$. From Eq. (6.1.5), if the fluid is a perfect gas, $c = \sqrt{\gamma RT/M}$, where T is the absolute temperature of the gas at equilibrium, M its molecular weight, γ the ratio between its specific heats, and R the universal gas constant. In this case the sound velocity is proportional to the square root of T, being independent of the gas pressure or density (as long as the first-order equation is valid). Also, in the case of the perfect gas, the temperature change τ accompanying the wave pressure p is $\tau = [(\gamma - 1)M/\gamma R\rho]p = (\gamma - 1)(T/\rho c^2)p = (\gamma - 1)(u/c)T$, a quantity much smaller than T, since p is much smaller than ρc^2 if Eq. (6.2.8) is valid.

Fluid momentum and energy

It is interesting, and will also be useful later, to demonstrate that the equation of motion (6.2.5) can be written as an equation of continuity for fluid momentum. The quantity $\rho(\mathbf{r},t)\mathbf{u}(\mathbf{r},t)$ is the momentum density of the fluid at point **r** and time t. According to Eq. (6.1.15), any change in this density must be produced by the divergence of the flux of momentum (analogous to the div **J** term) or else by the introduction of "new" momentum at **r**, t (analogous to the Q term). Since momentum density is a vector, there must be three equations of continuity, one for each component of $\rho\mathbf{u}$, and therefore three flux vectors, J_x, J_y, J_z, and three "source" terms, Q_x, Q_y, Q_z.

A "source" of momentum must be a net force on a unit volume of the fluid; a force causes a change in momentum. Considering the amount of fluid inside an element of volume $dx\,dy\,dz$ at the point $\mathbf{r} = (x,y,z)$, the force on this amount, transmitted from outside across the face $dy\,dz$ at x, y, z, can be written $\mathbf{P}_x(x,y,z,t)\,dy\,dz$, where the vector $\mathbf{P}_x$ is called the x component of the fluid *stress*. In a nonviscous fluid, of the sort considered in this section, $\mathbf{P}_x$ is simply the pressure, a vector in the x direction, of magnitude equal to the pressure $P(x,y,z,t)$ at $\mathbf{r}$, t. Later we shall show that when viscosity is taken into account, $\mathbf{P}_x$ is not pointed in the x direction, but has components P_{xy} and P_{xz} as well.

The force across the $dy\,dz$ face of the element farthest from the origin, pushing back on the interior of $dx\,dy\,dz$, is then $-\mathbf{P}_x(x + dx, y, z, t)\,dy\,dz = -\mathbf{P}_x(x,y,z,t)\,dy\,dz - (\partial\mathbf{P}_x/\partial x)\,dx\,dy\,dz$; so the net force on the fluid in the element, transmitted across the faces perpendicular to the x axis, is $-(\partial\mathbf{P}_x/\partial x)\,dx\,dy\,dz$. For a nonviscous fluid this force is $-(\partial P/\partial x)\,dx\,dy\,dz$, pointed in the x direction, where P is the pressure. Similarly, the net stress force transmitted across the $dx\,dz$ faces, perpendicular to the y axis, is $-(\partial\mathbf{P}_y/\partial y)\,dx\,dy\,dz$, and that across the $dx\,dy$ faces is $-(\partial\mathbf{P}_z/\partial z)\,dx\,dy\,dz$. For a nonviscous fluid, these forces are $-(\partial P/\partial y)\,dx\,dy\,dz$, pointed in the y direction, and $-(\partial P/\partial z)\,dx\,dy\,dz$, pointed in the z direction, respectively. The total stress force is thus $-\operatorname{grad}(P)\,dx\,dy\,dz$ for the nonviscous case, and $-[(\partial\mathbf{P}_x/\partial x) + (\partial\mathbf{P}_y/\partial y) + (\partial\mathbf{P}_z/\partial z)]\,dx\,dy\,dz$, with x component

$$-\left(\frac{\partial P_{xx}}{\partial x} + \frac{\partial P_{yx}}{\partial y} + \frac{\partial P_{zx}}{\partial z}\right)dx\,dy\,dz \quad \cdots$$

when viscosity is included.

In addition, there may be an external force $\mathbf{F}(\mathbf{r},t)$, such as the force of gravity, acting on each unit volume of the fluid. This force on the fluid inside the elementary volume is $\mathbf{F}\,dx\,dy\,dz$, with x component $F_x(\mathbf{r},t)\,dx\,dy\,dz$, etc. Therefore the momentum "source" term for the x component of the momentum conservation equation is

$$Q_x\,dx\,dy\,dz = \left(F_x - \frac{\partial P_{xx}}{\partial x} - \frac{\partial P_{yx}}{\partial y} - \frac{\partial P_{zx}}{\partial z}\right)dx\,dy\,dz$$

The momentum flux $\mathbf{J}_x$ in the x direction is the momentum density $\rho\mathbf{u}$ times the x component of the fluid velocity, u_x. Thus $\mathbf{J}_x$ has components $\rho u_x u_x$, $\rho u_y u_x$, and $\rho u_z u_x$; similarly for the other fluxes. We thus see that the total stress and the total flux each have nine components rather than three; they constitute what are called *tensors*, which can be written in matrix form,

$$\mathfrak{P} \equiv \begin{pmatrix} P_{xx} & P_{xy} & P_{xz} \\ P_{yx} & P_{yy} & P_{yz} \\ P_{zx} & P_{zy} & P_{zz} \end{pmatrix} \qquad \mathfrak{J} \equiv \begin{pmatrix} \rho u_x u_x & \rho u_x u_y & \rho u_x u_z \\ \rho u_y u_x & \rho u_y u_y & \rho u_y u_z \\ \rho u_z u_x & \rho u_z u_y & \rho u_z u_z \end{pmatrix} \qquad (6.2.10)$$

instead of as trios of vectors, $\mathbf{P}_x$, $\mathbf{P}_y$, $\mathbf{P}_z$ and $\mathbf{J}_x$, $\mathbf{J}_y$, $\mathbf{J}_z$. We note that, for nonviscous fluids, the stress tensor has the particularly simple form

$$\mathfrak{P} = \begin{pmatrix} P & 0 & 0 \\ 0 & P & 0 \\ 0 & 0 & P \end{pmatrix}$$

The flux of x momentum $\mathbf{J}_x(x,y,z,t)$ may bring in to the volume element more momentum than it takes out over the other face. The net amount left, via the $dy\,dz$ faces, is $J_{xx}(x,y,z,t)\,dy\,dz - J_{xx}(x + dx, y, z, t)\,dy\,dz = -(\partial J_{xx}/\partial x)\,dx\,dy\,dz$; and the total amount, from all faces, of x momentum is

$$-\left(\frac{\partial J_{xx}}{\partial x} + \frac{\partial J_{xy}}{\partial y} + \frac{\partial J_{xz}}{\partial z}\right) dx\,dy\,dz = -\operatorname{div} \mathbf{J}_x\,dx\,dy\,dz$$

In addition, there is the amount $Q_x\,dx\,dy\,dz$ "created" per second by the external force and the internal stresses. Therefore the net gain in x momentum in the element per second is

$$\frac{\partial}{\partial t}(\rho u_x)\,dx\,dy\,dz = (Q_x - \operatorname{div} \mathbf{J}_x)\,dx\,dy\,dz$$

with similar equations for the y and z components. The equations of continuity for the fluid momentum are thus

$$\frac{\partial}{\partial t}(\rho u_x) = F_x - \frac{\partial}{\partial x}(P_{xx} + \rho u_x u_x) - \frac{\partial}{\partial y}(P_{yx} + \rho u_x u_y) - \frac{\partial}{\partial z}(P_{zx} + \rho u_x u_z)$$

and similarly for the y and z components. These can be combined symbolically into a single equation of the general form of Eq. (6.1.15),

$$\frac{\partial \mathbf{f}}{\partial t} + \nabla \cdot \mathfrak{T} = \mathbf{F} \tag{6.2.11}$$

where, in the present case, the vector density $\mathbf{f}$ is $\rho\mathbf{u}$, the momentum per unit volume of fluid, the tensor flux $\mathfrak{T}$ is the sum $\mathfrak{P} + \mathfrak{J}$ of the stress tensor $\mathfrak{P}$ and the momentum flux tensor $\mathfrak{J}$, and the source vector $\mathbf{F}$ is the external force per unit volume, such as that of gravity, or possible electric forces.

For nonviscous fluids the tensor $\mathfrak{P}$ is the simple one, with pressure P on the main diagonal and zeros elsewhere. Going through the components, we see that $\nabla \cdot \mathfrak{P}$ then becomes simply $\operatorname{grad} P$. To identify these equations with the equation of motion (6.2.5), we perform the differentiations indicated by $\nabla \cdot \mathfrak{J}$. For the x components,

$$\operatorname{div} \mathbf{J}_x = u_x \frac{\partial}{\partial x}\rho u_x + u_x \frac{\partial}{\partial y}\rho u_y + u_x \frac{\partial}{\partial z}\rho u_z$$

$$+ \rho u_x \frac{\partial}{\partial x}u_x + \rho u_y \frac{\partial}{\partial y}u_x + \rho u_z \frac{\partial}{\partial z}u_x$$

$$= u_x \operatorname{div}(\rho\mathbf{u}) + \rho(\mathbf{u} \cdot \operatorname{grad})u_x$$

and for the whole vector we can write

$$\mathbf{\nabla} \cdot \mathfrak{I} = \mathbf{u} \operatorname{div}(\rho\mathbf{u}) + \rho(\mathbf{u} \cdot \operatorname{grad})\mathbf{u}$$

Finally, we use the equation of continuity (6.1.15) for the scalar density, $\partial\rho/\partial t = -\operatorname{div}(\rho\mathbf{u})$, to change the term $\partial\rho\mathbf{u}/\partial t$ into $\rho(\partial\mathbf{u}/\partial t) - \mathbf{u}\operatorname{div}(\rho\mathbf{u})$ and thus obtain (for $\mathbf{F} = 0$)

$$\rho\,\frac{\partial\mathbf{u}}{\partial t} - \mathbf{u}\operatorname{div}(\rho\mathbf{u}) = -\operatorname{grad} P - \mathbf{u}\operatorname{div}(\rho\mathbf{u}) - \rho(\mathbf{u} \cdot \operatorname{grad})\mathbf{u}$$

which does indeed correspond to the equation of motion (6.2.5) if we cancel the divergence terms and substitute $\rho + \delta$ for ρ and $P + p$ for P, in accordance with our present notation.

We also could work out an equation of continuity for the energy density of the fluid which would include the kinetic energy density $\frac{1}{2}\rho\,|\mathbf{u}|^2$ as well as the internal (heat) energy of the fluid. This will be done in a later section, when we take up the effects of heat conduction and other internal energy changes. Here we go on to discuss the momentum and energy of the wave motion, which differ from the momentum and energy of the fluid, just as the longitudinal energy and momentum of wave motion on the string differ from the string's transverse energy and momentum, as noted in connection with Eq. (4.1.12). In the present case, however, both sets are longitudinal.

Bernoulli's equation

We shall have occasion, several times, to refer to Bernoulli's equation, which is a special form of the momentum equation (6.1.15) for a nonviscous fluid,

$$\rho\,\frac{d\mathbf{u}}{dt} = \rho\,\frac{\partial\mathbf{u}}{\partial t} + \rho(\mathbf{u} \cdot \mathbf{\nabla})\mathbf{u} = -\operatorname{grad} P + \mathbf{F}$$

For isentropic flow the pressure is a function of density alone; so we can express $(1/\rho)\operatorname{grad} P$ as

$$\frac{1}{\rho}\frac{dP}{d\rho}\operatorname{grad}\rho = \operatorname{grad}\int\frac{1}{\rho}\frac{dP}{d\rho}\,d\rho = \operatorname{grad}\int\frac{dP}{\rho}$$

Introducing the velocity potential Ψ ($\mathbf{u} = -\operatorname{grad}\Psi$) and the vector relationship

$$(\mathbf{u} \cdot \mathbf{\nabla})\mathbf{u} = \operatorname{grad}\frac{u^2}{2} - (\mathbf{u} \times \operatorname{curl}\mathbf{u})$$

the momentum equation becomes

$$\operatorname{grad}\left(\frac{u^2}{2} + V + \int\frac{dP}{\rho} - \frac{\partial\Psi}{\partial t}\right) = \mathbf{u} \times \operatorname{curl}\mathbf{u}$$

where we have assumed that the external force $\mathbf{F}$ is minus the gradient of a potential V.

We now multiply both sides of this equation by $\mathbf{u}$ (scalar product). Since $\mathbf{u} \times \mathbf{A}$ is perpendicular to $\mathbf{u}$ for any vector $\mathbf{A}$, we have

$$\mathbf{u} \cdot \text{grad} \left(-\frac{\partial \Psi}{\partial t} + \frac{u^2}{2} + V + \int \frac{dP}{\rho} \right) = 0$$

where $\mathbf{u} \cdot \text{grad} = u(\partial/\partial n)$ is the derivative in the direction of $\mathbf{u}$, that is, along the streamline defined by $\mathbf{u}$. Integrating the quantity along a streamline then results in Bernoulli's equation

$$-\frac{\partial \Psi}{\partial t} + \frac{u^2}{2} + V + \int \frac{dP}{\rho} = C(t) \tag{6.2.12}$$

where constant C may be a function of time.

For time-independent flow this reduces to the well-known equation $\frac{1}{2}u^2 + V + \int (dP/\rho) = \text{constant}$, related to the constancy of energy density. When $P/P_0 = (\rho/\rho_0)^\gamma$ and $c^2 = \gamma P/\rho$, this becomes

$$\frac{c^2}{\gamma - 1} + \tfrac{1}{2}u^2 + V = \text{constant} = V + \tfrac{1}{2}u^2 + \frac{\gamma}{\gamma - 1}\frac{P}{\rho}$$

The Lagrange density

To find the momentum and energy flux specifically connected with the wave motion, we first work out the Lagrange density [Eq. (4.1.8)] for sound waves in a fluid. As noted in this section, we are neglecting viscosity and heat conduction for the time being. Therefore the basic equations, to first order, are

$$\rho \frac{\partial \mathbf{u}}{\partial t} = -\text{grad } p \qquad \kappa \frac{\partial p}{\partial t} = -\text{div } \mathbf{u}$$

as before. According to Eq. (6.1.7), the Lagrange density, the difference between the kinetic and potential energy densities, is $\tfrac{1}{2}\rho |\mathbf{u}|^2 - \tfrac{1}{2}\kappa p^2$. To apply the machinery of the Hamilton principle to this, we must express $\mathbf{u}$ and p in terms of some common function of $\mathbf{r}$ and t. If we use p itself, there is no expression for $\mathbf{u}$ in terms of derivatives of p. However, if we let p equal ρ times the time derivative of some scalar function Ψ, we see, from the first of Eqs. (6.2.7), that $\mathbf{u}$ is minus the gradient of Ψ; so Ψ, the velocity potential mentioned earlier, is a possible function with which to express L.

$$p = \rho \frac{\partial \Psi}{\partial t} \qquad \mathbf{u} = -\text{grad } \Psi$$

$$L = -\tfrac{1}{2}\rho \left[\left(\frac{\partial \Psi}{\partial x}\right)^2 + \left(\frac{\partial \Psi}{\partial y}\right)^2 + \left(\frac{\partial \Psi}{\partial z}\right)^2 - \rho\kappa \left(\frac{\partial \Psi}{\partial t}\right)^2 \right] \tag{6.2.13}$$

This formalism has the somewhat anomalous property that the kinetic-energy term has the space derivatives of Ψ, and the potential-energy term has the time derivative, thus forcing us to change the sign of L. We could, of course, have taken **u** to be the time derivative of some vector **q**. It is more effort, however, to deal with the three components of a vector than with the one function Ψ'. In addition, we might be tempted to call **q** the displacement of the fluid (forgetting the difference between the Euler and the Lagrange description) and thus draw erroneous conclusions regarding actual fluid displacement.

So we shall use the L of Eq. (6.2.13). The result of using Hamilton's principle [see the discussion of Eq. (5.2.6)] is the Lagrange equation

$$\frac{\partial}{\partial t}\left(\frac{\partial L}{\partial \Psi'_t}\right) + \frac{\partial}{\partial x}\left(\frac{\partial L}{\partial \Psi'_x}\right) + \frac{\partial}{\partial y}\left(\frac{\partial L}{\partial \Psi'_y}\right) + \frac{\partial}{\partial z}\left(\frac{\partial L}{\partial \Psi'_z}\right) - \frac{\partial L}{\partial \Psi'} = 0 \quad (6.2.14)$$

where the subscripts indicate partial differentiation. Applying this to the L of Eq. (6.2.13), we obtain

$$\rho^2 \kappa \left(\frac{\partial^2 \Psi'}{\partial t^2}\right) - \rho \, \nabla^2 \Psi' = 0$$

the wave equation, to first order.

We could use the Lagrange density

$$L = \tfrac{1}{2}\rho \left[1 + \kappa\rho\left(\frac{\partial \Psi'}{\partial t}\right)\right]\left[\kappa\rho\left(\frac{\partial \Psi'}{\partial t}\right)^2 - \left(\frac{\partial \Psi'}{\partial x}\right)^2 - \left(\frac{\partial \Psi'}{\partial y}\right)^2 - \left(\frac{\partial \Psi'}{\partial z}\right)^2\right]$$

correct to third order (but omitting viscosity and heat conduction) to obtain the second-order wave equation (6.2.9). It is interesting to see that the second-order correction term in this equation is itself proportional to the first-order Lagrange density of Eq. (6.2.13).

As was shown in Eqs. (4.1.9) for the string and Eqs. (5.2.8) for the membrane, the energy density $w = W_{tt}$ and the energy flow vector $\mathbf{I} = (W_{tx}, W_{ty}, W_{tz})$ of the wave motion are (using the L of Eq. 6.2.13)

$$W_{tt} = \Psi'_t\left(\frac{\partial L}{\partial \Psi'_t}\right) - L = \tfrac{1}{2}\rho \, |\mathrm{grad}\, \Psi'|^2 + \tfrac{1}{2}\kappa\rho^2\Psi'^2_t = \tfrac{1}{2}\rho \, |\mathbf{u}|^2 + \tfrac{1}{2}\kappa p^2 = w$$

$$(6.2.15)$$

$$W_{tx} = \Psi'_t\left(\frac{\partial L}{\partial \Psi'_x}\right) = -\rho\Psi'_t\Psi'_x = pu_x \quad \cdots \quad \mathbf{I} = p\mathbf{u}$$

These quantities satisfy the equation of continuity $(\partial w/\partial t) + \mathrm{div}\, \mathbf{I} = 0$. Since, from Eq. (6.1.7), W_{tt} is the total energy density w of the wave motion, we can thus expect the vector **I** to be the flux of energy in the wave. Sound pressure times fluid velocity **u** is the power flux per unit area, generated by the wave motion; so vector **I** is called the *intensity* of the wave.

The wave momentum **M** (as contrasted with the fluid momentum $\rho\mathbf{u}$) has the three components

$$M_x = W_{xt} = \Psi'_x\frac{\partial L}{\partial \Psi'_t} = \kappa\rho^2\Psi'_x\Psi'_t = -\kappa\rho p u_x$$

or
$$\mathbf{M} = -\kappa\rho p\mathbf{u} \quad\quad\quad (6.2.16)$$

It is a second-order vector, proportional to the intensity $\mathbf{I}$, but usually much smaller since $\kappa\rho$ is usually small. It also satisfies a vector equation of continuity (6.2.11), $(\partial\mathbf{M}/\partial t) + \nabla\cdot\mathfrak{W} = 0$ (the source term is zero), with the tensor

$$W_{xx} = \Psi'_x \frac{\partial L}{\partial \Psi'_x} - L$$

$$\mathfrak{W} = \begin{pmatrix} W_{xx} & W_{xy} & W_{xz} \\ W_{yx} & W_{yy} & W_{yz} \\ W_{zx} & W_{zy} & W_{zz} \end{pmatrix} \qquad = \tfrac{1}{2}\rho(-u_x{}^2 + u_y{}^2 + u_x{}^2) - \tfrac{1}{2}\kappa p^2$$

$$W_{xy} = \Psi'_x \frac{\partial L}{\partial \Psi'_y} = -\rho u_x u_y \qquad (6.2.17)$$

$$\cdots\cdots\cdots\cdots\cdots\cdots$$

being the *wave-stress tensor*, measuring the flux of momentum which accompanies the wave motion. Vector $\mathbf{M}$ has the dimensions of momentum per unit volume (or pressure divided by velocity); $\mathbf{M}$ multiplied by wave velocity, $c\mathbf{M} = (1/c)p\mathbf{u}$, is the *radiation* pressure exerted by the wave.

Complex notation

It is apparent by now that many calculations are more easily carried out if we use the complex quantity $(a + ib)e^{-i\omega t}$ rather than the real quantity $a\cos(\omega t) + b\sin(\omega t)$ to represent velocity or displacement. If we are to use complex quantities, however, we should learn how to use them to express such second-order quantities as kinetic energy, wave momentum, and the like. The question is one of interpretation. If we say that the displacement x of a simple oscillator is $(a + ib)e^{-i\omega t}$, do we mean that the measured displacement is the real part of this quantity, its imaginary part, or what? Furthermore, we have said that the potential energy of the oscillator is $\tfrac{1}{2}Kx^2$; does x in this case mean the real part of $(a + ib)e^{-i\omega t}$? If so, we must abandon the use of complex quantities when we discuss the second-order quantities.

Suppose we turn the representation around and define the second-order quantities in terms of the complex notation; what, then, is the meaning of the first-order quantities? Suppose we insist that the basic second-order quantities, the kinetic and potential energies, are always real quantities. This can be achieved by using the square of the magnitude of the complex quantity instead of the square of the quantity. As shown on page 9, the quantity $|C|^2 = C^*C = (a - ib)(a + ib) = a^2 + b^2$ is real and nonnegative. Let us try using a notation in which the potential energy is $\tfrac{1}{2}K|x|^2$ and the kinetic energy is $\tfrac{1}{2}m|v|^2$.

In our previous discussion we said that the notation $x = Ae^{-i\omega t} = (a + ib)e^{-i\omega t}$ meant that the measured displacement is the real part, $a\cos(\omega t) + b\sin(\omega t)$, representing simple-harmonic motion of amplitude $A = \sqrt{a^2 + b^2}$ and phase $\Phi = \tan^{-1}(b/a)$. In this case the potential energy is

$$\tfrac{1}{2}K[a^2\cos^2(\omega t) + 2ab\cos(\omega t)\sin(\omega t) + b^2\sin^2(\omega t)]$$
$$= \tfrac{1}{4}K(a^2 + b^2) + \tfrac{1}{4}K(a^2 - b^2)\cos(2\omega t) + \tfrac{1}{2}Kab\sin(2\omega t)$$

which is a rather complicated combination of the real and imaginary parts of A and of $e^{-i\omega t}$. However, the *mean value* of the potential energy, averaged over time, is just $\frac{1}{4}K(a^2 + b^2) = \frac{1}{4}K|x|^2$, a much simpler expression. Likewise, the mean value of the kinetic energy is $\frac{1}{4}m(a^2\omega^2 + b^2\omega^2) = \frac{1}{4}m|dx/dt|^2$. This simplicity of the mean values of energy extends to cases where x is a combination of simple-harmonic motions. If

$$x = \sum A_n e^{-i\omega_n t} \tag{6.2.18}$$

then time average of $|x|^2$ is $\frac{1}{2}\sum |A_n|^2$.

Since we usually measure only the time average of the energy (or power) anyway, we may as well use a representation which will give us these average values directly, and interpret the velocities and displacements accordingly. Therefore we shall use the complex quantity x as the displacement, with the convention that the square of its magnitude (which enters into the energy expressions) is the time average of the square of the actual displacement. The magnitude of x, in this notation, will be called the *root-mean-square* (usually written rms) amplitude of x, and the kinetic and potential energies, averaged over time, are

$$\text{KE} = \frac{1}{2}m\left|\frac{dx}{dt}\right|^2 \qquad \text{PE} = \frac{1}{2}K|x|^2$$

Then if, later, we should wish to discuss the measured displacement or velocity, as a function of time, we can take $\sqrt{2}$ times the real part of x or $\sqrt{2}$ times the imaginary part of x, whichever is more convenient. The $\sqrt{2}$ factor enters to take care of the factor $\frac{1}{2}$ in Eq. (6.2.18), because the rms amplitude is only $1/\sqrt{2}$ times the maximum displacement in simple-harmonic motion.

The rms representation is particularly adaptable to the Lagrange-Hamilton approach to the equations of motion, because the basic integral of the Lagrangian [of Eq. (5.2.5), for example] is over time and space, so that time-average quantities can be used without changing the results in the slightest. Thus we can change over completely to the complex notation by using the square of the magnitude of the quantity in place of its square, if we remember that the resulting energies are time averages and the resulting amplitudes are rms amplitudes. In this notation, for example, the Lagrange density for small-amplitude motion of a compressible fluid is

$$L = \frac{1}{2}\kappa\rho^2 \frac{\partial\Psi}{\partial t}\frac{\partial\Psi^*}{\partial t} - \frac{1}{2}\rho\,\text{grad}\,\Psi^* \cdot \text{grad}\,\Psi \tag{6.2.19}$$

where now, in contrast to the Lagrangian of Eq. (6.2.13), Ψ is a complex quantity $\text{Re}\,\Psi + i\,\text{Im}\,\Psi$ and Ψ^* is its complex conjugate, $\text{Re}\,\Psi - i\,\text{Im}\,\Psi$. The amplitudes of Ψ and of all its derivatives are then rms amplitudes.

Since the real and imaginary parts of Ψ will be varied independently in the variational process (see page 158) to obtain the Lagrange equations of motion, we can consider Ψ and Ψ^* as being varied independently. Therefore the variation of the integral of L gives rise to two equations, one for Ψ and one for Ψ^*.

$$\frac{\partial}{\partial t}\left(\frac{\partial L}{\partial \Psi^*_t}\right) + \frac{\partial}{\partial x}\left(\frac{\partial L}{\partial \Psi^*_x}\right) + \frac{\partial}{\partial y}\left(\frac{\partial L}{\partial \Psi^*_y}\right) + \frac{\partial}{\partial z}\left(\frac{\partial L}{\partial \Psi^*_z}\right) - \frac{\partial L}{\partial \Psi^*} = 0$$

$$\frac{\partial}{\partial t}\left(\frac{\partial L}{\partial \Psi_t}\right) + \frac{\partial}{\partial x}\left(\frac{\partial L}{\partial \Psi_x}\right) + \frac{\partial}{\partial y}\left(\frac{\partial L}{\partial \Psi_y}\right) + \frac{\partial}{\partial z}\left(\frac{\partial L}{\partial \Psi_z}\right) - \frac{\partial L}{\partial \Psi} = 0$$

(6.2.20)

For the symmetric Lagrange density of Eq. (6.2.19) these two equations are identical; so in this case Ψ^* and Ψ satisfy the same equation, and thus can be complex conjugates of each other.

The forms of the energy-momentum matrix $\mathfrak{W}$, however, must be modified, since the second-order terms must be real. The modifications are fairly obvious, since W_{tt} must be

$$w = \tfrac{1}{2}\rho^2\kappa \left|\frac{\partial \Psi}{\partial t}\right|^2 + \tfrac{1}{2}\rho \,|\text{grad }\Psi|^2$$

the time average of the energy density. Therefore

$$W_{tt} = \Psi_t\left(\frac{\partial L}{\partial \Psi_t}\right) + \Psi^*_t\left(\frac{\partial L}{\partial \Psi^*_t}\right) - L = w$$

$$W_{xx} = \Psi_x\left(\frac{\partial L}{\partial \Psi_x}\right) + \Psi^*_x\left(\frac{\partial L}{\partial \Psi^*_x}\right) - L \;\cdots$$

(6.2.21)

$$W_{tx} = \Psi_t\left(\frac{\partial L}{\partial \Psi_x}\right) + \Psi^*_t\left(\frac{\partial L}{\partial \Psi^*_x}\right) = \tfrac{1}{2}(p^*u_x + u^*_x p) = I_x \;\cdots$$

Given that Ψ and Ψ^* satisfy Eqs. (6.2.20), one can show that the elements of $\mathfrak{W}$ satisfy the equations of continuity, similar to those of page 199.

$$\frac{\partial W_{tt}}{\partial t} + \text{div } \mathbf{I} = 0 \qquad \mathbf{I} = (W_{tx}, W_{ty}, W_{tz})$$

$$\frac{\partial \mathbf{M}}{\partial t} + \boldsymbol{\nabla} \cdot \mathfrak{W} = 0 \qquad \mathbf{M} = (W_{xt}, W_{yt}, W_{zt})$$

and $\mathfrak{W}$ is the wave-stress matrix of Eq. (6.2.17). Thus the interrelations between the second-order terms are the same as before; only now the quantities are time averages, not instantaneous values.

Propagation in porous media

In some cases of interest the fluid will be interpenetrating a porous solid, which resists the fluid's flow and also modifies the fluid's compressibility. Suppose that the pores in the material are interconnected in a random but isotropic way, so that the fluid can percolate through it with equal facility in any direction, and so that the fraction of volume which is not occupied by the solid, which the fluid can occupy, is Ω, called the *porosity*. Also suppose that the solid, though porous, is perfectly rigid and incompressible. This is not a bad approximation for gases flowing through porous materials, though it would not be very good for liquids, such as water, flowing through sand, which is not rigid and not much less compressible than water.

The actual velocity of the fluid, as it passes through the pores, will not be constant in direction or amount. What is of practical interest is the mean volume of fluid passing across a unit area inside the medium, normal to the flow, per second, which we can call $\mathbf{u}(x,t)$, the mean velocity. At the inter-surface between the porous material and the outside, the normal component of $\mathbf{u}$ will thus be continuous. The fluid in the pores will have to be moving faster, normal to the surface, than the air outside, so that the volume of fluid leaving the outside per second is equal to the volume of fluid entering the pores per second, both being equal to the area of the surface times the normal component of $\mathbf{u}$.

The equation of continuity for $\mathbf{u}$ still holds for fluid flow in the pores if we consider a large enough volume so that irregularities of pore size average out, and if we consider the density $\rho + \delta$ of the fluid as being the mass of the fluid occupying the fraction Ω of the space available to the fluid. The equation of continuity becomes $\Omega(\partial\delta/\partial t) + \rho \operatorname{div} \mathbf{u} = 0$ to the first order. Also, to the first order, the relation between density change δ and acoustic pressure is $\delta = \kappa_p \rho p$, where κ_p is the effective compressibility of the fluid in the pores. At very high frequencies this would be the adiabatic compressibility κ_s of the fluid in the open; at low frequencies the near presence of the solid material would tend to hold the temperature constant, and it would equal the iso-thermal compressibility κ_T, larger than κ_s. For some acoustic materials (for air) the change-over occurs in the frequency range of greatest interest, so that κ_p must often be considered to be a function of ω. Thus one of the equations for sound waves in a porous material is $\kappa_p \Omega(\partial p/\partial t) + \operatorname{div} \mathbf{u} = 0$, or, for simple-harmonic waves,

$$i\omega\kappa_p\Omega p = \operatorname{div} \mathbf{u} \qquad (6.2.22)$$

The other equation, giving the fluid acceleration in terms of the forces acting, must be modified to include the change in inertia of, and the frictional drag suffered by, the fluid as it moves through the pores. Some fibers of the material may move with the fluid, thus adding to its effective mass. Also, as will be indicated in Eq. (6.4.14), when fluid flows through a series of constrictions, its effective mass increases. Therefore, to express the inertial properties of the fluid in the pores, we use an effective density ρ_p, greater than the density ρ of the fluid in the open. In actual acoustic materials ρ_p is from 1.5 to 5 times ρ.

We express the frictional retardation to flow through the pores in terms of a *flow resistance* Φ, the pressure drop required to force a unit flow through the material. Most porous materials used in sound absorption have flow resistances which lie between 0.2 and 2 atm-sec per sq m; put another way, the quantity $\Phi/\rho c$ for air in these materials lies between 50 and 500 m^{-1}. Thus the equation of motion is $\rho_p(\partial \mathbf{u}/\partial t) + \Phi \mathbf{u} + \operatorname{grad} p = 0$, or, for simple-harmonic waves,

$$i\omega \rho_p \left[1 + i \frac{\Phi}{\rho_p \omega} \right] \mathbf{u} = \operatorname{grad} p \qquad (6.2.23)$$

for materials which are uniform and isotropic in their porous properties. We note that the effect of flow resistance is largest at low frequencies. For porous sound absorbers in air, $\Phi/\rho_p \omega$ lies between 0.5 and 10 for a frequency of 1,000 cps, in the intermediate range, neither very large nor very small.

To obtain the wave equation we define a velocity potential such that $\mathbf{u} = -\operatorname{grad} \Psi$, as before. Then $p = \rho_p(\partial \Psi/\partial t) + \Phi \Psi$ and

$$\nabla^2 \Psi = \kappa_p \rho_p \Omega \frac{\partial^2 \Psi}{\partial t^2} + \kappa_p \Omega \Phi \frac{\partial \Psi}{\partial t} \qquad (6.2.24)$$

Except for the last term, this is the usual wave equation with a wave velocity $c_p = \sqrt{1/\kappa_p \rho_p \Omega}$, instead of $c = \sqrt{1/\rho \kappa}$, the wave velocity in the open. Velocity c_p may be larger than c if Ω is small (material not very porous), but it usually is smaller than c because $\rho_p \Omega$ is usually larger than ρ. The last term in Eq. (6.2.24) represents the frictional loss, which is evidenced by either oscillations damped in time or attenuation in space of simple-harmonic waves.

Since κ_p and ρ_p are to some extent dependent on frequency, we can specialize Eq. (6.2.24) for motions having the time factor $e^{-i\omega t}$ and later utilize the Fourier transform to deal with nonharmonic motion. The basic equations (6.2.7) relating p and $\mathbf{u}$ for simple-harmonic motion can be written $(i\omega/\rho c^2)p = \operatorname{div} \mathbf{u}$ and $i\omega \rho \mathbf{u} = \operatorname{grad} p$. Equations (6.2.22) and (6.2.23) can be brought into this form by defining an effective fluid density ρ_e, compressibility κ_e, and wave velocity c_e, for the fluid in the pores, by the equations

$$\rho_e = \rho_p \left(1 + \frac{i\Phi}{\rho_p \omega} \right) \qquad \kappa_e = \Omega \kappa_p$$

$$c_e = \frac{1}{\sqrt{\rho_e \kappa_e}} = \frac{1}{\sqrt{\rho_w \kappa_p \Omega}} = c_p \left(1 + \frac{i\Phi}{\rho_p \omega} \right)^{-\frac{1}{2}} \qquad (6.2.25)$$

where $c_p = \sqrt{1/\rho_p \kappa_p \Omega}$. Therefore

$$\mathbf{u} = -\operatorname{grad} \Psi \qquad p = i\omega \rho_e \Psi \qquad \frac{i\omega}{\rho_e c_e^2} p = \operatorname{div} \mathbf{u}$$

and thus

$$\nabla^2 \Psi + \left(\frac{\omega}{c_e} \right)^2 \Psi = 0$$

which is the wave equation for simple-harmonic waves.

Both effective density and effective wave velocity are complex quantities, because of the frictional term in Φ. At very high frequencies both are real; $\rho_w \to \rho_p$ is somewhat larger than ρ, the density of the fluid in the open; and $c_w \to c_p$ is usually a little smaller than the wave velocity $c = \sqrt{1/\rho\kappa}$ in the open. Thus a plane wave of frequency $\omega/2\pi$, traveling in the positive x direction, is represented by the expression

$$\Psi = Ae^{i(\omega/c_e)x - i\omega t} = Ae^{i(k_e x - \omega t) - \gamma_e x}$$

$$k_e = \omega\sqrt{\rho_p\kappa_p\Omega}\left[\frac{1}{2}\sqrt{1 + \left(\frac{\Phi}{\rho_p\omega}\right)^2} + \frac{1}{2}\right]^{\frac{1}{2}} \to \begin{cases} \omega\sqrt{\rho_p\kappa_p\Omega} & \omega \text{ large} \\ \sqrt{\omega\kappa_p\Omega\Phi} & \omega \text{ small} \end{cases}$$

$$\gamma_e = \omega\sqrt{\rho_p\kappa_p\Omega}\left[\frac{1}{2}\sqrt{1 + \left(\frac{\Phi}{\rho_p\omega}\right)^2} - \frac{1}{2}\right]^{\frac{1}{2}} \to \begin{cases} \frac{1}{2}\Phi\sqrt{\dfrac{\kappa_p\Omega}{\rho_p}} & \omega \text{ large} \\ \sqrt{\omega\kappa_p\Omega\Phi} & \omega \text{ small} \end{cases}$$

$$(6.2.26)$$

The phase velocity ω/k_e is, roughly, equal to $1/\sqrt{\rho_p\kappa_p\Omega} = c_p$ when $\omega \gg \Phi/\rho_p$, somewhat smaller than the velocity c in the open for most porous materials. The wave amplitude in the porous medium is attenuated by a factor $1/e$ in a distance $1/\gamma_e \simeq 2\rho_p c_p/\Phi$ when Φ is small compared with $\rho_p\omega$. For sound-absorbing materials this distance is of the order of a centimeter or less, for frequencies above 2,000 cps.

Of course, this procedure of representing the effect of the pores in terms of an effective density and wave velocity is possible only for simple-harmonic waves. Its usefulness can be extended by means of the Fourier transform, however. Any waveform can be expressed, by this means, as a sum of simple-harmonic components, each of which will behave as though the fluid had density and wave velocity as given in Eqs. (6.2.26). We return to this in the next section.

Energy and momentum

In a porous medium energy is lost to frictional forces, so that the equation of continuity between energy density w and energy flow $\mathbf{I}$ must take this into account. If the amplitude of the velocity potential Ψ is given in rms terms, the time average of the energy density in the sound wave would be, according to Eq. (6.1.7),

$$w = \tfrac{1}{2}\kappa_p |p|^2 + \tfrac{1}{2}\rho_p |\mathbf{u}|^2 = \tfrac{1}{2}\kappa_p |\rho_p\Psi_t + \Phi\Psi|^2 + \tfrac{1}{2}\rho_p \text{ grad }\Psi^* \cdot \text{grad }\Psi$$

and the intensity vector would be

$$\mathbf{I} = \tfrac{1}{2}p^*\mathbf{u} + \tfrac{1}{2}\mathbf{u}^*p = -\tfrac{1}{2}[(\rho_p\Psi_t^* + \Phi\Psi^*)\text{ grad }\Psi + (\rho_p\Psi_t + \Phi\Psi)\text{ grad }\Psi^*]$$

The mean rate of loss of energy per unit volume should be equal to the flow resistance times the mean square of the velocity, $\Phi |\mathbf{u}|^2$, so that the equation of continuity for the energy should be

$$\frac{\partial w}{\partial t} + \text{div } \mathbf{I} = -\Phi \, |\mathbf{u}|^2 = -\Phi \, \text{grad } \Psi \cdot \text{grad } \Psi^* \qquad (6.2.27)$$

If we insert the expressions for w and $\mathbf{I}$ into the left-hand side of this equation, carry out the algebra, and use Eq. (6.2.23), we find that it does indeed equal the right-hand side.

The dissipative term $\kappa_p \Omega \Phi (\partial \Psi / \partial t)$ in Eq. (6.2.23) impedes our use of the Lagrange-Hamilton formalism, however. The only way in which we can use it is to combine the solution Ψ of Eq. (6.2.23) with a solution Ψ^a of an *adjoint* system, having negative flow resistance, and thus having a complementary rate of energy *gain* per unit volume. We set the combined Lagrange density to be

$$L = \tfrac{1}{2}\kappa_p \rho_p{}^2\Omega \, \frac{\partial \Psi^a}{\partial t} \frac{\partial \Psi}{\partial t} - \tfrac{1}{2}\rho_p \, \text{grad } \Psi^a \cdot \text{grad } \Psi$$

$$+ \tfrac{1}{4}\rho_p \kappa_p \Phi \Omega \left(\frac{\partial \Psi^a}{\partial t} \Psi - \Psi^a \frac{\partial \Psi}{\partial t} \right) \qquad (6.2.28)$$

The first two terms of this expression are similar to those of Eq. (6.2.19), except that, instead of the complex conjugate Ψ^*, the adjoint potential Ψ^a appears. The last term ensures that Ψ^a satisfies a different equation from Ψ. Applying the variational procedure to the integral of L [see the discussion of Eq. (5.2.5)], the variation of Ψ^a results in an equation like the first of Eqs. (6.2.20), except that Ψ^a occurs instead of Ψ^*. This, when applied to Eq. (6.2.28), results in the damped-wave equation (6.2.24).

On the other hand, variation of Ψ produces the second of Eqs. (6.2.20), which becomes

$$\nabla^2 \Psi^a = \kappa_p \rho_p \Omega \, \frac{\partial^2 \Psi^a}{\partial t^2} - \kappa_p \Phi \Omega \, \frac{\partial \Psi^a}{\partial t} \qquad (6.2.29)$$

when the L of Eq. (6.2.28) is used. This has the first-derivative time term with a sign opposite to that of Eq. (6.2.23), thus representing energy accretion rather than dissipation.

The corresponding energy-momentum tensor has elements

$$W_{tt} = \Psi_t{}^a \left(\frac{\partial L}{\partial \Psi_t{}^a} \right) + \Psi_t \left(\frac{\partial L}{\partial \Psi_t} \right) - L = w_0$$

$$W_{tx} = \Psi_t{}^a \left(\frac{\partial L}{\partial \Psi_x{}^a} \right) + \Psi_t \left(\frac{\partial L}{\partial \Psi_x} \right) = I_{0x} \qquad (6.2.30)$$

....

with the same general form as those of Eqs. (6.2.21), with Ψ^a substituted for Ψ^*. Since Ψ^a represents an adjoint system, however, which gains energy as rapidly as Ψ loses it, the quantities ϵ_0, I_0, etc., are *constant in time* and satisfy nondissipative equations of continuity $(\partial \epsilon_0 / \partial t) + \text{div } \mathbf{I}_0 = 0$, etc. Thus, only by introducing an adjoint system can we retain the Lagrange-Hamilton formalism when friction enters the picture.

6.3 WAVES IN AN INFINITE MEDIUM

We have now sketched the general properties of wave motion of small amplitude in a uniform medium. We have seen that, as long as the fluid is uniform and in equilibrium, if viscosity and thermal conduction can be neglected and if acoustic pressure p is small compared with equilibrium pressure P, then p, $\mathbf{u}$, etc., will satisfy a linear wave equation of the form of Eqs. (6.2.8) or of Eq. (6.2.24). For the next several chapters we shall work out the implications of these equations.

There are several reasons for spending so much time on so specialized a model of acoustic behavior. In the first place, it turns out to represent many acoustic phenomena surprisingly well. Sounds of ordinary intensity involve acoustic pressures of 10^{-4} atm or smaller (by definition $P = 1$ atm under standard conditions). In many situations of acoustic importance (such as the interior of an auditorium during a concert), we strive to keep the air quiescent and uniform, except for the sound. Solutions of the first-order equations (6.2.8) represent quite accurately much of the acoustic phenomena we are accustomed to. When we come to deal with the products of modern technology, however, such as jet aircraft or atom-bomb explosions, it becomes imperative that we study higher-order effects and propagation in a moving medium.

In the second place, the first-order wave equation is very much easier to handle mathematically than is the equation appropriate for high-intensity sounds or for a viscous or a moving medium. Equations (6.2.4), (6.2.8), and (6.2.24) are linear in the dependent variable, and we have commented in earlier chapters on the great advantages of linear equations. In fact, even when we come later to study the effects of motion of the fluid medium, we shall usually begin by studying linearized approximations to the correct equations, in order to obtain analytic solutions. Nonlinear equations usually must be solved numerically, a separate solution being worked out for each different situation of interest. Analytic solutions, though approximate, provide a quick survey of the general features of the phenomena.

In this chapter we shall concentrate on the behavior of plane-wave solutions of the simple wave equation. Later chapters will take up cylindrical and spherical waves, in which the peculiarities of two- and three-dimensional motion become apparent.

Traveling waves

We start by considering the propagation of a plane wave in a uniform nonviscous medium of infinite extent, at rest in the absence of sound. We choose the x axis to point in the direction of propagation, in which case the sound pressure p and all the other characteristics of the wave are functions of x and t only, and the fluid velocity $\mathbf{u}$ and all other vectors related to the wave are pointed in the x direction. The equation of motion which the sound pressure must satisfy is the simple wave equation (6.2.4), which we repeat here:

$$\frac{\partial^2 p}{\partial x^2} = \frac{1}{c^2}\frac{\partial^2 p}{\partial t^2} \qquad \frac{1}{c^2} = \rho\kappa \qquad (6.3.1)$$

where ρ is the equilibrium density of the medium, and κ is the adiabatic compressibility of the fluid, at equilibrium, as defined in Eqs. (6.1.3) and (6.1.5). As explained earlier, the change in density δ and in temperature τ of the fluid, and the fluid velocity u, caused by the wave, are in the first order linearly related to p, according to the equations

$$\delta = \rho\kappa p \qquad \tau = (\gamma - 1)\frac{\kappa}{\beta}p$$

and

$$\frac{\partial p}{\partial x} = -\rho\frac{\partial u}{\partial t} \qquad \kappa\frac{\partial p}{\partial t} = -\frac{\partial u}{\partial x}$$

(6.3.2)

Therefore δ, τ, and u also are solutions of Eq. (6.3.1).

We have already seen, in Sec. (4.1), that a general solution of the wave equation (6.3.1), traveling in the positive x direction, is given by the function $p(x - ct)$. This represents a wave of constant form $p(z)$, traveling with constant speed c (which equals 331 m per sec for air at 273°K and equals 1,500 m per sec for water at the same temperature). In the case of the string, the displacement is transverse to the direction of wave propagation; in the case of sound waves, the fluid motion is in the direction of propagation. By referring back to Eqs. (6.2.11), (6.2.13), and (6.2.15), we find we can write out all the important quantities characterizing the wave motion in terms of the velocity potential $\Psi(x - ct)$, where

$$\Psi(z) = -\frac{1}{\rho c}\int_0^z p(u)\,du \qquad \Psi'(z) = \frac{d\Psi}{dz}$$

$$u = -\Psi'(x - ct) \qquad p(x - ct) = -\rho c\Psi'(x - ct) = \rho c u(x - ct)$$

$$\text{Mass flux} = \rho u = \text{fluid momentum density} = \frac{p}{c}$$

$$\text{Fluid momentum flux} = \rho|u|^2 = \frac{|p|^2}{\rho c^2} = w$$

(6.3.3)

$$\text{Wave energy density } w = \tfrac{1}{2}\rho|u|^2 + \tfrac{1}{2}\kappa|p|^2 = \frac{|p|^2}{\rho c^2}$$

$$\text{Sound intensity } I_x = \tfrac{1}{2}(p^*u + pu^*) = \rho c|u|^2 = \frac{|p|^2}{\rho c} = cw$$

to the first order in the small quantities p, u, if the result is nonzero; otherwise, to the second order. We are here using complex quantities for Ψ, p, u, and τ; the amplitudes of the first-order quantities are rms, and the second-order quantities are time averages.   We thus are assuming that the variables can be expressed as sums of simple-harmonic terms of the type $A_n e^{-i\omega_n t}$ and that Eq. (6.2.19) can express the system's Lagrange density.

It should be pointed out again that we can neglect the difference between the Lagrange and the Euler descriptions, discussed in connection with Fig. 6.3, only to the first order.   For example, if we tried to improve the accuracy of the mass-flux term by using $(\rho + \delta)u = \rho u + (\rho/c)u^2$, we might be tempted to conclude that, even if u were purely sinusoidal, there was a net forward flow of fluid, corresponding to the u^2 term.   This would be wrong, of course, because one cannot compute displacement by integrating the Eulerian velocity or flux.   In fact, on page 238, we showed that the fluid displacement is, to third order, equal to the integral of $u + (u^2/c)$, not the integral of u.   When we come to discuss higher-order effects, we shall find it necessary to start afresh, keeping in mind the distinction between Euler and Lagrange descriptions at each step.   Here we can neglect the distinction *as long as we adhere to the degree of approximation* given in Eqs. (6.3.3) *as written*.

The relationships given in Eqs. (6.3.3) are simple enough; they are the bases of most acoustical measurements.   In a plane wave, pressure is in phase with fluid velocity; the region which is moving most rapidly in the direction of propagation is the region of greatest sound pressure (this is apparent from examination of Fig. 6.3*a*).   The ratio between acoustic pressure p and fluid velocity u, $\rho c = \sqrt{\rho/\kappa}$ (which equals 429 newton-sec per cu m for air at 0°C and 1 atm pressure and equals 1.5×10^6 newton-sec per cu m for water at 0°C), is called the *characteristic acoustic impedance* of the medium.

The analogy with an electrical transmission line is fairly obvious; pressure corresponds to voltage, and u to current; furthermore, the intensity I, the mean power transmitted in joules per second per unit area, is equal to the impedance ρc times the square of u.   Both sound pressure p and quantity $\rho c^2 = 1/\kappa$ have the dimensions of energy density.   Sound pressures p are usually smaller than 1 dyne per sq cm = 0.1 newton per sq m or joule per cu m, whereas ρc^2 for air at 0°C and 1 atm is 1.42×10^5 joules per cu m. The energy density w of the wave is thus the pressure times the ratio of the pressure to this standard energy density, usually much smaller than 10^{-5} times the value of p itself.

Reflection from a locally reacting surface

Suppose this small-amplitude plane wave is incident on a plane surface, such as a wall.   In general, some of the energy is reflected, and some is absorbed by the surface.   For convenience, we take the surface to be the xz plane, with the fluid occupying the half-space above this plane, for positive values of y.   The incident plane wave will have its direction of propagation in the xy plane, inclined at an angle ϑ_i to the y axis, as shown in Fig. 6.4

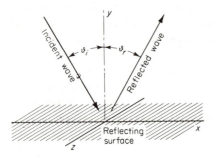

FIGURE 6.4

Reflection from a plane surface in the xz plane.

(ϑ_i is the *angle of incidence*). A rotation of axes of the expressions (6.3.3) shows that we can write them in terms of the velocity potential $\Psi_i(Z_i)$, with $Z_i = x \sin \vartheta_i - y \cos \vartheta_i - ct$.

$$p_i(x,y) = -\rho c \Psi_i'(Z_i) \qquad u_{ix} = -\sin \vartheta_i \Psi_i'(Z_i)$$

$$u_{iy}(x,y) = \cos \vartheta_i \Psi_i'(Z_i) \qquad w = \frac{|p_i|^2}{\rho c^2}$$

$$I_x = \frac{|p_i|^2}{\rho c} \sin \vartheta_i \qquad I_y = -\frac{|p_i|^2}{\rho c} \cos \vartheta_i$$

A possible reflected wave would be represented by the velocity potential $\Psi_r(x \sin \vartheta_r + y \cos \vartheta_r - ct)$ moving away from the reflecting surface at an angle ϑ_r to the y axis (the *angle of reflection*). Resulting pressures, velocities, etc., are correspondingly propagated upward at an angle ϑ_r, which is to be determined. The relationship between the functions Ψ_i and Ψ_r must also be found.

To make this determination, the acoustical nature of the reflecting surface must be specified. The acoustic pressure p acts on the surface and tends to make it move, or else tends to force more fluid into the pores of the surface. If any fluid motion normal to the surface is possible, there will be wave motion in the material forming the surface. Thus the motion of the surface at one point will be related to motion at another point of the surface, the relationship being determined by the wave motion inside the material, as well as by the incident and reflected waves.

In certain limiting cases, however, the various parts of the surface are not strongly coupled together, and we can consider that the motion, normal to the surface, of one portion of the surface is dependent only on the acoustic pressure incident on that portion and independent of the motion of any other part of the area. When this is the case, we say that the surface is one of *local reaction*; the consequent interaction of such a surface with incident and reflected waves is particularly simple. In the next subsection we shall work out the circumstances under which a surface may have local reaction; at present we assume it has.

The motion of the surface (or the motion of the fluid into the surface pores) at a given point is thus determined (we assume) by the acoustic pressure at that point. For small amplitudes the relationship is linear, and the velocity u_y normal to the surface is proportional to the pressure. Motion of the fluid tangential to the surface will not affect the surface motion; we are assuming at present that viscosity is negligible. The ratio between pressure and normal fluid velocity at a point on the surface is called the *acoustic impedance* of the surface. This impedance may depend on the frequency of the incident wave; it may be complex, in which case the normal velocity is not in phase with the pressure. But if the surface is one of local reaction, the impedance will not depend on the spatial distribution of the wave causing the pressure.

We usually symbolize this acoustic impedance by the letter z. Its natural units, as we shall soon see, are the characteristic impedance ρc of the medium. For convenience, we define the following terms:

$$\zeta = \frac{z}{\rho c} = \theta - i\chi \qquad \text{and} \qquad \beta = \frac{\rho c}{z} = \xi - i\sigma$$

$$\xi = \frac{\theta}{\theta^2 + \chi^2} \qquad \sigma = -\frac{\chi}{\theta^2 + \chi^2}$$

(6.3.4)

The ratio ζ between the acoustic impedance z of the locally reacting surface and ρc will be called the *specific acoustic impedance* of the surface. Its real part is the specific acoustic resistance θ; its reciprocal β is the specific acoustic admittance; and so on. As will be demonstrated, the interaction between sound waves and a surface of local reaction is completely defined by specifying the specific acoustic impedance (or its admittance) as a function of frequency.

For the moment let us assume that the impedance z of the surface is independent of frequency; that the ratio $(p/-u_y)$ at the surface is the same no matter what the shape of the wave. In this case we must adjust the reflected wave $\Psi_r(x \sin \vartheta_r + y \cos \vartheta_r - ct)$ so that, together with the incident wave $\Psi_i(x \sin \vartheta_i - y \cos \vartheta_i - ct)$, it has a pressure which is just z times the combined normal velocity u_y. At $y = 0$ the combined pressure is

$$-\rho c[\Psi_i''(x \sin \vartheta_i - ct) + \Psi_r''(x \sin \vartheta_r - ct)]$$

We certainly shall not be able to match $(p/-u_y)$ to z at every point unless Ψ_r'' bears the same relation to Ψ_i'' at every point. Therefore ϑ_r, the angle of reflection, must equal ϑ_i, the angle of incidence ($\vartheta_r = \vartheta_i = \vartheta$), and Ψ_r'' must be a constant C_r times Ψ_i''. We thus have the combined pressure and normal velocity at $y = 0$.

$$p(x,0) = -\rho c(1 + C_r)\Psi_i''(x \sin \vartheta - ct)$$
$$u_y(x,0) = \cos \theta(1 - C_r)\Psi_i''(x \sin \vartheta - ct) \quad = u_x - u_y$$

Note that the velocity of the wave *along the surface* is $c/\sin \vartheta$, greater than the velocity of propagation c. The ratio C_r between incident and reflected pressure waves is called the *reflection coefficient*.

We next must adjust C_r so that $(p/-u_y)$ equals z, the impedance of the surface. Going through the algebra, we find that

$$C_r = \frac{z \cos \vartheta - \rho c}{z \cos \vartheta + \rho c} \tag{6.3.5}$$

For this kind of surface, therefore, the reflection coefficient depends on the ratio between the surface impedance z and the characteristic impedance ρc of the fluid, and also on the cosine of the angle of incidence (and of reflection).

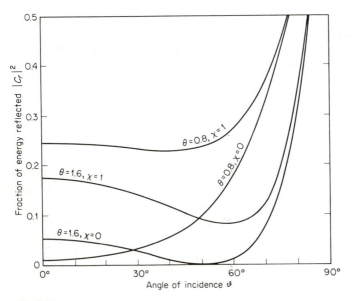

FIGURE 6.5
Fraction of energy reflected from a locally reacting surface of point impedance $z = \rho c(\theta - i\chi)$, for different values of resistance parameter θ and reactance parameter χ, as function of angle of incidence ϑ.

We note that if z is purely imaginary (reactive), then the magnitude of C_r is unity, the reflected wave differs only in phase angle from the incident wave, the reflected intensity $\rho c |C_r|^2 |\Psi_i''|^2$ is equal to the incident intensity $\rho c |\Psi_i''|^2$, and no energy is absorbed by the surface. On the other hand, if z is real (resistive), then the reflection coefficient C_r is smaller than unity, the reflected intensity is smaller than the incident intensity, and the surface absorbs the difference. Since the waves are at an angle ϑ to the normal to the surface, a unit-area portion of the wave is spread over an area $1/\cos \vartheta$ of the surface, so that the energy absorbed per unit area of surface is

$$\rho c (1 - |C_r|^2) |\Psi_i''|^2 \cos \vartheta = \rho^2 c^2 \cos^2 \vartheta \frac{4z}{(z \cos \vartheta + \rho c)^2} |\Psi_i''|^2 \tag{6.3.6}$$

when z is real. If z is also larger than ρc, there is an angle of incidence, $\vartheta = \cos^{-1}(\rho c/z)$, at which no wave is reflected; all the incident energy is absorbed. At this angle the component of the wall resistance, $z \cos \theta$, just equals the characteristic impedance ρc of the medium. Figure 6.5 shows curves of $|C_r|^2$ as functions of θ for a few values of $z/\rho c = \theta - i\chi$.

Frequency-dependent surface impedance

If the surface impedance z is a function of frequency, though still a point function, independent of the spatial distribution of the wave, we can start by calculating the reflection of a simple-harmonic wave of frequency $\omega/2\pi$. By procedures similar to those which produced Eq. (6.3.5), we find that the appropriate combination for an incident wave of rms amplitude A_i is

$$\Psi(x,y;\omega,t) = A_i e^{-i\omega T_i} + A_i C_r e^{-i\omega T_r}$$

where
$$C_r = \frac{z(\omega)\cos\vartheta - \rho c}{z(\omega)\cos\vartheta + \rho c} \qquad T_i = t - \frac{1}{c}(x\sin\vartheta - y\cos\vartheta)$$

$$T_r = t - \frac{1}{c}(x\sin\vartheta + y\cos\vartheta)$$

with the incident wave going in the direction of the unit vector

$$\mathbf{a}_i = \mathbf{a}_x \sin\vartheta - \mathbf{a}_y \cos\vartheta$$

and a reflected plane wave, of amplitude $A_i C_r$, going in the direction of the vector $\mathbf{a}_r = \mathbf{a}_x \sin\vartheta + \mathbf{a}_y \cos\vartheta$, where $\mathbf{a}_x$, $\mathbf{a}_y$ are unit vectors in the x and y directions, respectively. The pressure, fluid velocity, mean energy density, and intensity for this wave combination are

$$p = \rho\frac{\partial\Psi}{\partial t} = B(\omega)(e^{-i\omega T_i} + C_r e^{-i\omega T_r}) \qquad B = -i\omega\rho A_i$$

$$\mathbf{u} = -\text{grad}\,\Psi = \frac{B}{\rho c}(\mathbf{a}_i e^{-i\omega T_i} + C_r \mathbf{a}_r e^{-i\omega T_r})$$

$$w = \tfrac{1}{2}\rho\,|u|^2 + \tfrac{1}{2}\kappa\,|p|^2 = \frac{1}{\rho c^2}|B|^2\,(1 + |C_r|^2) \tag{6.3.7}$$

$$\mathbf{I} = \tfrac{1}{2}(p^*\mathbf{u} + p\mathbf{u}^*) = \frac{1}{\rho c}|B|^2\,(\mathbf{a}_i + |C_r|^2\,\mathbf{a}_r) \qquad c^2 = \frac{1}{\rho\kappa}$$

where $B(\omega)$ is the rms amplitude of the incident pressure wave, and BC_r that of the reflected pressure wave. The cross terms, with factors $e^{i\omega(T_i-T_r)}$, etc., in w and $\mathbf{I}$ vanish when averaged over time and space, leaving the mean energy density w as the sum of the energy densities of the two separate waves, and the mean energy flux $\mathbf{I}$ as the sum of the separate flows, each in their separate direction, as given by the unit vectors $\mathbf{a}_i$ and $\mathbf{a}_r$.

To take a typical example, let us assume that each point of the reflecting surface acts like a simple-harmonic oscillator, with effective mass m per unit area, mechanical resistance factor R, and stiffness $K = m\omega_0^2$ for motion normal to the surface ($\omega_0/2\pi$ is the resonance frequency of the wall surface). The acoustic impedance z of the wall would then be $-i\omega m + R + im(\omega_0^2/\omega)$, or $(m/i\omega)[\omega^2 + i\omega(R/m) - \omega_0^2]$, according to Eq. (2.3.2), having a simple quadratic dependence on ω. For this case the ratio C_r, the reflection coefficient, and the square of its magnitude $|C_r|^2$, the ratio between reflected and incident intensities, are

$$C_r = \frac{(\omega^2 - \omega_0^2) + i\omega[(R/m) - (\rho c/m \cos \vartheta)]}{(\omega^2 - \omega_0^2) + i\omega[(R/m) + (\rho c/m \cos \vartheta)]}$$

$$= 1 - \frac{2i\omega(\rho c/m \cos \vartheta)}{(\omega + iq - \omega_0)(\omega + iq + \omega_0)} \qquad q = \frac{R}{2m} + \frac{\rho c}{2m \cos \vartheta} \qquad (6.3.8)$$

$$|C|^2 = 1 - \frac{4\omega^2 \rho c R \cos \vartheta}{(m \cos \vartheta)^2 (\omega^2 - \omega_0^2)^2 + \omega^2(\rho c + R \cos \vartheta)^2}$$

For very low frequencies ($\omega \ll \omega_0$), the reflection coefficient C_r becomes unity for this wall, independent of θ; the incident pressure wave is reflected with no change in amplitude or phase; thus no energy is absorbed by the wall. This is also true for very high frequencies ($\omega \gg \omega_0$). At resonance ($\omega \simeq \omega_0$), however, C_r is smaller than unity and dependent on the angle of incidence θ. The reflected intensity at resonance is smallest when the angle of incidence is $\cos^{-1}(\rho c/R)$, if $R > \rho c$, or for normal incidence, if $R < \rho c$.

The power absorbed *per unit area* of the reflecting surface is

$$(|B|^2/\rho c)(1 - |C_r|^2) \cos \vartheta$$

which is the following fraction of the incident intensity ($|B|^2/\rho c$):

$$\frac{4\omega^2 \rho c R \cos^2 \vartheta}{(m \cos \vartheta)^2 (\omega^2 - \omega_0^2)^2 + \omega^2(\rho c + R \cos \vartheta)^2} \rightarrow \begin{cases} \dfrac{4\rho c R}{m^2 \omega_0^2} \dfrac{\omega^2}{\omega_0^2} & \omega \ll \omega_0 \\[2ex] \dfrac{4\rho c R \cos^2 \vartheta}{(\rho c + R \cos \vartheta)^2} & \omega = \omega_0 \\[2ex] \dfrac{4\rho c R}{m^2 \omega^2} & \omega \gg \omega_0 \end{cases}$$

for this sort of reflecting surface. This fraction goes to zero as $\omega \to 0$ and as $\omega \to \infty$, being independent of the angle of incidence at these limiting frequencies. At resonance the absorbed power is strongly dependent on ϑ, being zero at grazing incidence and greatest at normal incidence.

Reflection of a pulse-wave

Having worked out the reflection of a simple-harmonic wave from a plane surface of local reaction, we can now call in the Fourier transform to compute the reflection of a wave of any form. From Eq. (1.3.25) a unit pulse-wave, traveling in the direction of $\mathbf{a}_i = \mathbf{a}_x \sin \vartheta - \mathbf{a}_y \cos \vartheta$, can be built up out of combination of simple-harmonic waves,

$$\delta(T_i - t_0) = \frac{1}{2\pi} \int_{-\infty}^{\infty} e^{i\omega(t_0 - T_i)} \, d\omega$$

Comparing this with Eqs. (6.3.7), we see that if we set the amplitude $B(\omega)$ of the simple harmonic wave equal to $(1/2\pi)e^{i\omega t_0}$ and integrate the result over ω, we shall obtain

$$p_\delta(x,y;t,t_0) = \frac{1}{2\pi} \int_{-\infty}^{\infty} [e^{i\omega(t_0 - T_i)} + C_r(\omega)e^{i\omega(t_0 - T_r)}] \, d\omega$$

$$= \delta(T_i - t_0) + \delta(T_r - t_0) + W(T_r - t_0)$$

$$W(T_r - t_0) = -\frac{\rho c}{\pi} \int_{-\infty}^{\infty} \frac{e^{i\omega(t_0 - T_r)} \, d\omega}{z(\omega) \cos \vartheta + \rho c} \tag{6.3.9}$$

note:
$C_r = 1 - A_r$
$A_r = 1 - C_r = \frac{2\rho c}{\pi} \frac{1}{z(\omega)\cos\vartheta + \rho c}$

$$T_i = t - \frac{1}{c}(x \sin \vartheta - y \cos \vartheta) \qquad T_r = t - \frac{1}{c}(x \sin \vartheta + y \cos \vartheta)$$

representing a unit pulse-wave, incident on the reflecting plane at an angle of incidence ϑ, and a reflected wave, consisting of a unit pulse plus a "wake" W, represented by the integral, which is produced by the motion of the surface as it reacts to the pulse.

If, for example, the impedance is $z = -i\omega m + R + im(\omega_0^2/\omega)$, as in Eqs. (6.3.8), the integral becomes

$$W(T) = \frac{2\rho c}{2\pi i m \cos \vartheta} \int_{-\infty}^{\infty} \frac{\omega e^{i\omega T} \, d\omega}{(\omega + iq - \omega_0)(\omega + iq + \omega_0)} \qquad T = T_r - t_0$$

Referring to Eqs. (1.2.15) or Eq. (2.3.10), we see that

$$W(T_r - t_0) = \begin{cases} 0 & T < 0 \\ \dfrac{-2\rho c}{m \cos \vartheta} \exp\left[-\dfrac{T}{2m}\left(R + \dfrac{\rho c}{\cos \vartheta}\right)\right]\left[\cos(\omega_0 T)\right. & \\ \left. \qquad - \dfrac{1}{2m\omega_0}\left(R + \dfrac{\rho c}{\cos \vartheta}\right)\sin(\omega_0 T)\right] & T > 0 \end{cases} \tag{6.3.10}$$

which is a wave that damps out with time, the rate of damping being least for $\vartheta = 0$ and becoming infinite as $\vartheta \to \frac{1}{2}\pi$ (i.e., for grazing incidence). There is no wave when $T < 0$, that is, if t is less than $t_0 + (1/c)(x \sin \vartheta + y \cos \vartheta)$. At $T = 0$, along the reflected wavefront, $x \sin \vartheta + y \cos \vartheta = c(t - t_0)$, there is a unit pulse, immediately followed by the initially negative pressure of the "wake" W. As time goes on (T becomes larger than 0), W exhibits damped oscillations, dying out more rapidly, the smaller $\cos \vartheta$. As $\vartheta \to \frac{1}{2}\pi$ (grazing incidence), function W approaches the pulse function $-2\delta(T)$, which more than cancels the reflected pulse.

What happens is somewhat like the wave reflection illustrated in Fig. 4.9. The pressure pulse of the incident wave cannot move the mass load of the wall instantaneously; so the pulse is reflected unchanged. It has given an impulse to the surface, however; so the surface executes damped oscillation, first receding from the blow, which generates a pressure wave initially negative. The damping of the surface oscillations is proportional to the internal resistance R of the wall *plus* the radiative resistance $\rho c/\cos \vartheta$ engendered by the presence of the fluid medium.

Finally, as with Eq. (2.3.9), the complete solution for the reflection of an incident plane wave of arbitrary form $F(t)$, at angle of incidence ϑ, on a plane surface having point impedance $z(\omega)$, is

$$p = \int_{-\infty}^{\infty} F(t_0)p_\delta(x,y;t,t_0)\, dt_0 = F(T_i) + F(T_r) + \int_{-\infty}^{\infty} F(t_0)W(T_r - t_0)\, dt_0$$

(6.3.11)

where the integral involving the function W represents the modification of the reflected wave because of the wall motion.

Reflection from a surface of extended reaction

Although many surfaces react to sound waves, at least approximately, as though each portion of the surface responded to local pressure without knowledge of motion elsewhere, many surfaces do not so react. When the surface behavior at one point depends on the behavior at neighboring points, so that the reaction is different for different incident waves, the surface can be called one of *extended reaction*. Such surfaces are of many sorts: ones which behave like membranes and ones with laminated structure, in which waves are propagated parallel to the surface; ones in which waves penetrate into the material of the wall; and so on. We shall have to defer discussion of the more complex structures until later; in this chapter we can discuss only the case where the plane surface separates two media, both of which can transmit pressure waves but which differ in their acoustic properties.

The situation to be analyzed is pictured in Fig. 6.4. The boundary surface is the xz plane; the region above the surface $(y > 0)$ is filled with a fluid of uniform density ρ and wave velocity $c = \sqrt{1/\rho\kappa}$, which will be called the *medium*. The region below the surface $(y < 0)$ is filled with material of density ρ_w and wave velocity $c_w = \sqrt{1/\rho_w\kappa_w}$, which will be called the *wall material*. As before, the incident wave is pointed at an angle ϑ with respect to the y axis; its velocity potential may be written $\Psi(T_i)$, where $T_i = t - (1/c)(x \sin \vartheta - y \cos \vartheta)$, as before. The angle of reflection is also ϑ, as before, and the amplitude of the reflected wave may differ from that of the incident wave; so we can write its velocity potential as $C_r\Psi(T_r)$, where $T_r = t - (1/c)(x \sin \vartheta + y \cos \vartheta)$, as before.

Now, however, there is also a wave moving downward into the wall material, with a dependence on time like that of the incident wave but with a different amplitude and velocity and a direction of propagation at a different angle, ϑ_p, called the angle of refraction, or of penetration.   The total velocity potential, pressure, and velocity for this case are

$$\Psi = \begin{cases} \Psi(T_i) + C_r\Psi(T_r) \\ C_p\Psi(T_p) \end{cases} \qquad p = \begin{cases} \rho\Psi''(T_i) + \rho C_r\Psi''(T_r) & y > 0 \\ \rho_w C_p\Psi''(T_p) & y < 0 \end{cases}$$

$$u_y = \begin{cases} \dfrac{1}{c} \cos \vartheta[-\Psi''(T_i) + C_r\Psi''(T_r)] & y > 0 \\[2mm] -\dfrac{C_p}{c_w} \cos \vartheta_p\Psi''(T_p) & y < 0 \end{cases} \qquad (6.3.12)$$

where $T_p = t - (1/c_w)(x \sin \vartheta_p - y \cos \vartheta_p)$, and Ψ'' is the derivative of Ψ with respect to its argument.

At the surface, $y = 0$, the pressure and the normal velocity u_y are continuous, which enables us to calculate the reflection coefficient C_r and the penetration coefficient C_p, and also to determine the angle of penetration ϑ_p. Since both incident and reflected waves, at the surface, depend on t and x via the function $\Psi''[t - (x/c) \sin \vartheta]$, for the dependence of the penetrating wave to be similar, $(1/c_w) \sin \vartheta_p$ must equal $(1/c) \sin \theta$, or

$$\sin \vartheta_p = \frac{c_w}{c} \sin \vartheta \qquad \text{and} \qquad \cos \vartheta_p = \left[1 - \left(\frac{c_w}{c}\right)^2 \sin^2 \vartheta\right]^{\frac{1}{2}} \quad (6.3.13)$$

which is Snell's law for acoustic waves.   The continuity of pressure and normal velocity at $y = 0$ produces the following equations:

$$1 + C_r = \frac{\rho_w}{\rho} C_p \qquad \text{and} \qquad 1 - C_r = \frac{c \cos \vartheta_p}{c_w \cos \vartheta} C_p$$

so that

$$C_r = \frac{\rho_w c_w \cos \vartheta - \rho c \cos \vartheta_p}{\rho_w c_w \cos \vartheta + \rho c \cos \vartheta_p} \qquad C_p = \frac{2\rho c_w \cos \vartheta}{\rho_w c_w \cos \vartheta + \rho c \cos \vartheta_p} \qquad (6.3.14)$$

The formula for the reflection coefficient C_r is similar to that for a surface of local reaction, with point impedance z, as given in Eq. (6.3.5), only now the effective impedance of the surface is $z = \rho_w c_w/\cos \vartheta_p$, which depends on the angle of incidence ϑ.   Thus the degree to which the surface yields to the incident wave depends on the distribution of the incident wave; the surface is aware of the waveshape; so it is one of extended reaction.   However, if the velocity c_w of the sound waves in the wall material is considerably smaller than the speed of sound in the medium, then $\cos \vartheta_w = [1 - (c_w/c)^2 \sin^2 \vartheta]^{\frac{1}{2}}$ is practically independent of ϑ, and the wall is one of local reaction.   Comparison with Eq. (6.3.5) shows that the equivalent impedance of the wall material is $\rho_w c_w \sec \vartheta$, dependent on ϑ.

On the other hand, if c_w is larger than c, then when ϑ is larger than $\sin^{-1}(c/c_w)$, the quantity called $\cos \vartheta_p$ becomes imaginary and the reflection coefficient C_r has unit amplitude; all the incident energy is reflected unless ρ_w or c_w is complex. This is the condition of total reflection, well known in optics. As long as ρ_w, c_w, and $\cos \vartheta_p$ are real, the fraction of the incident energy which enters each unit area of the surface, thus being lost from the reflected wave, is

$$(1 - C_r^2) \cos \vartheta = \frac{4\rho c \rho_w c_w \cos^2 \vartheta \cos \vartheta_p}{(\rho_w c_w \cos \vartheta + \rho c \cos \vartheta_p)^2}$$

which, of course, equals the fraction $(\rho_w c/\rho c_w) C_p^2 \cos \vartheta_p$ of the incident energy, per unit surface area, which is carried into the wall material by the refracted wave [the factor $\rho_w c/\rho c_w$ comes about because the intensity in the wall is $(\rho_w/c_w)|\Psi''|^2$, whereas it is $(\rho/c)|\Psi''|^2$ in the medium].

If the wall material is porous, with the acoustic properties as defined in Eqs. (6.2.23) and (6.2.26), the reflection properties of the surface depend on the frequency of the incident wave, as well as on the angle of incidence. The incident, reflected, and transmitted pressure waves are obtained by inserting the ρ_e and c_e of Eq. (6.2.25) into Eqs. (6.3.12) and (6.3.14) and by specializing the waveform to a sinusoidal shape.

$$p(x,y;t,\omega) = \begin{cases} A(\omega)(e^{-i\omega T_i} + C_r e^{-i\omega T_r}) & y > 0 \\ A(\omega)C_p e^{-i\omega T_p} & y < 0 \end{cases} \qquad (6.3.15)$$

where T_i and T_r are, as before,

$$T_p = t - \frac{x}{c} \sin \vartheta + \frac{y}{c_p}\left(\cos^2 \vartheta_p + i\frac{\gamma}{\omega}\right)^{\frac{1}{2}}$$

with $c^2 = 1/\rho\kappa$, $c_p^2 = 1/\rho_p \kappa_p \Omega$, $\cos^2 \vartheta_p = 1 - (c_p/c)^2 \sin^2 \vartheta$, and $\gamma = \Phi/\rho_p$. Since c_p is usually smaller than c, $\cos \vartheta_p$ does not go to zero at grazing incidence. The reflection and penetration coefficients are

$$C_r = \frac{\rho_p c_p(\omega + i\gamma) \cos \vartheta - \rho c(\omega^2 \cos^2 \vartheta_p + i\gamma\omega)^{\frac{1}{2}}}{\rho_p c_p(\omega + i\gamma) \cos \vartheta + \rho c(\omega^2 \cos^2 \vartheta_p + i\gamma\omega)^{\frac{1}{2}}}$$

$$C_p = \frac{2\omega \rho c_p \cos \vartheta}{\rho_p c_p(\omega + i\gamma) \cos \vartheta + \rho c(\omega^2 \cos^2 \vartheta_p + i\gamma\omega)^{\frac{1}{2}}}$$

$(6.3.16)$

These quantities are complex, the imaginary parts being proportional to the flow resistance Φ of the porous material and inversely proportional to ω. At frequencies large compared with $\gamma = \Phi/\rho_p$, this imaginary part vanishes and

$$C_r \to \frac{\rho_p c_p \cos \vartheta - \rho c \cos \vartheta_p}{\rho_p c_p \cos \vartheta + \rho c \cos \vartheta_p} \qquad C_p \to \frac{2\rho c_p \cos \vartheta}{\rho_p c_p \cos \vartheta + \rho c \cos \vartheta_p}$$

similar to Eq. (6.3.14), only with the limiting values ρ_p, c_p used instead of

ρ_e, c_e. As indicated in Eq. (6.2.26), even at these high frequencies, the refracted wave is attenuated in accordance with a factor

$$\exp\left[y(\Phi/2\rho_p c_p \cos\vartheta_p)\right]$$

which diminishes as y increases in the negative direction, the attenuation being greatest for grazing incidence. At these higher frequencies there can

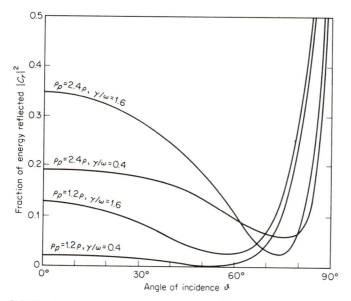

FIGURE 6.6
Fraction of energy reflected from the surface of a porous medium of effective density ρ_p, effective wave velocity $c_p = c$, and flow resistance $\Phi = \gamma\rho_p$ for frequencies $\omega/2\pi$ and angles of incidence ϑ. Compare with Fig. 6.5.

be an angle of incidence at which the reflected wave vanishes; if $c_p < c$, then $C_r = 0$ when $\cos\vartheta = (\rho c/\rho_p c_p)\cos\vartheta_p$. This is usually close to grazing incidence. At grazing incidence, of course, $|C_r|^2 \to 1$.

At very low frequencies, C_p approaches zero and nearly all the incident energy is reflected. At intermediate frequencies, the reflected wave differs from the incident wave both in phase and amplitude, depending quite strongly on frequency and angle of incidence. A few typical curves for $|C_r|^2$ as a function of angle of incidence, for different frequencies, are shown in Fig. 6.6. They are not greatly different from those of Fig. 6.5, for a surface of local reaction.

To compute the reflection of a pulse-wave from a porous surface, we parallel the derivation of Eq. (6.3.9), integrating $(e^{i\omega t_0}/2\pi)p(x,y;t,\omega)$ of Eq. (6.3.15) over ω from $-\infty$ to $+\infty$. The integrals for the reflected and penetrant waves are complicated because of the radicals in C_r and in the wave for $y < 0$. The branch points are at $\omega = 0$ and $\omega = -i(\gamma/\cos^2 \vartheta_p)$, so that the integral for the "wake" runs between these values. The formulas become

$$p = \begin{cases} \delta(T_i - t_0) + \dfrac{\rho_p c_p \cos \vartheta - \rho c \cos \vartheta_p}{\rho_p c_p \cos \vartheta + \rho c \cos \vartheta_p} \delta(T_r - t_0) \\ \qquad\qquad + W_r(T_r - t_0) \qquad\qquad y > 0 \\[2mm] \dfrac{2\rho c_p \cos \vartheta}{\rho_p c_p \cos \vartheta + \rho c \cos \vartheta_p} \delta(T_p - t_0) \exp \dfrac{y\Phi}{\rho_p c_p \cos \vartheta_p} \\ \qquad\qquad + W_p(T_p - t_0, y) \qquad y < 0 \end{cases} \quad (6.3.17)$$

where $T_i = t - (x/c) \sin \vartheta + (y/c) \cos \vartheta$, $T_r = t - (x/c) \sin \vartheta + (y/c) \cos \vartheta$, $T_p = t - (x/c) \sin \vartheta + (y/c_p) \cos \vartheta_p$, and $\cos^2 \vartheta_p = 1 - (c_p/c)^2 \sin^2 \vartheta$.

Both wake functions are zero when their arguments are negative; for positive arguments they have the general form

$$W_r(T) = -\frac{2}{\pi} \alpha g \cos \vartheta \cos \vartheta_p \int_0^1 \frac{(v - \beta^2 \cos^2 \vartheta)\sqrt{v(1-v)} \, dv \, e^{-vgT}}{\cos^2 \vartheta(v - \beta^2 \cos^2 \vartheta_p)^2 + \alpha^2 \cos^2 \vartheta_p v(1-v)}$$

$$W_p(T,y) = -\frac{2g\alpha}{\pi\beta} \cos \vartheta \int_0^1 \frac{\exp\{-vg[T - (y/c_p) \cos \vartheta_p]\}v \, dv}{\cos^2 \vartheta(v - \beta^2 \cos^2 \vartheta_p)^2 + \alpha^2 \cos^2 \vartheta_p v(1-v)}$$

$$\times \left\{ (v - \beta^2 \cos^2 \vartheta_p) \cos \vartheta \sin\left[\frac{yg}{c_p} \cos \vartheta_p \sqrt{v(1-v)}\right] \right.$$

$$\left. + \alpha \cos \vartheta_p \sqrt{v(1-v)} \cos\left[\frac{yg}{c_p} \cos \vartheta_p \sqrt{v(1-v)}\right] \right\}$$

where $\alpha = \rho c/\rho_p c_p$, $\beta = c/c_p$, and $g = \Phi/\rho_p \beta^2 \cos^2 \vartheta_p$. Both wake functions attenuate as their arguments increase, the more rapidly, the smaller $\cos \vartheta_p$. In contrast to the reactions of the point-impedance wall of Eq. (6.3.10), the wake from a porous wall has no damped oscillations; the air in the pores yields immediately to the pulse and recovers without oscillation. The integrals for W_r and W_p cannot be expressed in closed form, but they can be integrated numerically for any given situation. Expressions for T small and T large can also be worked out, though they need not be given here.

6.4   INTERNAL ENERGY LOSS

In the foregoing analysis the effects of viscosity, heat conduction, and other dissipative processes have been neglected. As long as we deal with gases at ordinary pressures and frequencies that are not too high, this omission is not very important in most problems; calculated sound pressures agree well with experimental results over a wide range of amplitudes and frequencies.

Only when we come to calculate sound transmission over long distances in the open, or through narrow ducts, will the omission of internal dissipation become apparent. Also, in the neighborhood of sharp corners in boundaries, the dissipationless analysis will predict too large velocities.

To correct these defects we must include the effects of viscosity, heat conduction, and hysteresis in internal heat absorption, as well as the effects of nonlinearities. In this section we consider only the dissipative effects; consequently, the results, along with those of the whole chapter, will be valid only for sufficiently small amplitudes. The question of what we mean by "sufficiently small" cannot be answered until we investigate the nonlinear effects, which will be postponed until Chap. 14.

Thermal conduction and viscosity

From one point of view, the phenomena of viscosity and thermal conduction in gases are a consequence of their molecular constitution. Temperature is a manifestation of random molecular motion; the higher the temperature, the faster the average molecular speed. A region at higher temperature will have more high-speed molecules than the average, and these faster particles will diffuse out into the cooler regions, tending to equalize the temperature. As with any other diffusion, the flux $\mathbf{J}$ (in this case the flux of heat) is proportional to the rate of change of the concentration (in this case to the temperature gradient)

$$\mathbf{J}_h = -K \operatorname{grad} T \qquad (6.4.1)$$

where the proportionality constant K is called the *thermal conductivity* of the gas.

Kinetic theory of gases indicates that this conductivity is simply related to other properties of the gas, such as its density ρ, its heat capacity C_v per unit mass at constant volume, the mean-free-path l of its molecules, and the speed of sound c, by the equation

$$K \simeq 1.6 \frac{l\rho c C_v}{\sqrt{\gamma}} \qquad (6.4.2)$$

where γ is the constant used in Eq. (6.1.3), and discussed later, in connection with Eqs. (6.4.19). Heat conductivity of liquids and solids is, of course, a much more complicated function of the material's properties.

Since the net gain in heat in a unit volume, because of the flux $\mathbf{J}_h$, is $-\operatorname{div} \mathbf{J}_h$, and since this influx, divided by the heat capacity at constant pressure $\rho C_p = \gamma \rho C_v$ of a unit volume, is equal to the rise in temperature of the volume of gas, we have

$$\frac{\partial T}{\partial t} = \frac{K}{\rho C_p} \operatorname{div} \operatorname{grad} T \simeq \frac{lc}{\gamma^{\frac{3}{2}}} \nabla^2 T \qquad \text{(for a gas)} \qquad (6.4.3)$$

which will be used to determine the loss of energy in the sound wave, caused by thermal diffusion.

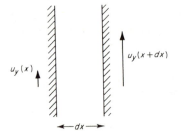

FIGURE 6.7
Rate of change of shear in a gas, producing viscous
shear stress.

Viscosity, on the other hand, is produced by momentum diffusion. If one portion of the gas slides past another so that, for example, the velocity in the y direction is a function of x, as illustrated in Fig. 6.7, the slower molecules at x will diffuse over to reduce the fluid momentum at $x + dx$, and the faster ones at $x + dx$ will tend to speed up those at x. As a result, the region between x and $x + dx$ is subject to shear stress D_{xy}, proportional to $\partial u_y/\partial x$, the proportionality factor being called the *coefficient of viscosity* μ,

$$D_{xy} = -\mu \frac{\partial u_y}{\partial x} \qquad (6.4.4)$$

As might be expected, since both depend on the diffusion of the gas molecules, there is a close relationship between the coefficient of viscosity and the thermal conductivity of a gas. Kinetic theory shows that, approximately,

$$K \simeq \tfrac{5}{3}\mu C_v \qquad \text{or} \qquad \mu \simeq \frac{l\rho c}{\sqrt{\gamma}} \qquad \text{(for a gas)} \qquad (6.4.5)$$

where c = velocity of sound

ρ = density of gas

l = mean-free-path

γ = ratio of specific heats = C_p/C_v

To obtain the complete expression for the stress tensor $\mathfrak{P}$, introduced in Eqs. (6.2.10), we might expect its nondiagonal components P_{ij} to be proportional to the differential $\partial u_i/\partial x_j$ ($u_1 = u_x$, $x_1 = x$, $x_2 = y$, etc.), as suggested by Eq. (6.4.4). But two basic considerations prevent this. In the first place, a part of the tensor $\partial u_i/\partial x_j$ represents rotation without shear. For example, the quantity $\tfrac{1}{2}$ curl $\mathbf{u}$, where

$$(\text{curl } \mathbf{u})_x = \frac{\partial u_z}{\partial y} - \frac{\partial u_y}{\partial z}$$

$$(\text{curl } \mathbf{u})_y = \frac{\partial u_x}{\partial z} - \frac{\partial u_z}{\partial x} \qquad (6.4.6)$$

$$(\text{curl } \mathbf{u})_z = \frac{\partial u_y}{\partial x} - \frac{\partial u_x}{\partial y}$$

is the vector representing the angular velocity of pure rotation of the fluid at (x,y,z). Thus we must subtract off the rotational part of tensor $\partial u_i/\partial x_j$ to obtain the true rate of strain.

$$U_{ij} = \frac{1}{2}\frac{\partial u_i}{\partial x_j} + \frac{1}{2}\frac{\partial u_j}{\partial x_i} = U_{ji} \qquad i \neq j$$

The second consideration lies in the fact that, although the nondiagonal components U_{ij} ($i \neq j$) represent pure shear strains with no change in fluid content, the elements U_{ii} in part measure a rate of change of amount of fluid inside the element of volume around point (x,y,z). For example, the combination

$$\text{div }\mathbf{u} = \frac{\partial u_x}{\partial x} + \frac{\partial u_y}{\partial y} + \frac{\partial u_z}{\partial z}$$

measures the rate of outflow of fluid from unit volume around (x,y,z) [Eq. (6.1.15)]. To ensure that the shear-strain tensor does not include this outflow part, we subtract div $\mathbf{u}$ in a symmetric way from the diagonal parts of the tensor, arriving finally at

$$\underline{\mathfrak{U}} = \begin{pmatrix} U_{xx} & U_{xy} & U_{xz} \\ U_{yx} & U_{yy} & U_{yz} \\ U_{zx} & U_{zy} & U_{zz} \end{pmatrix} \qquad \begin{array}{l} U_{ii} = \dfrac{\partial u_i}{\partial x_i} - \frac{1}{3}\text{ div }\mathbf{u} \\[2mm] U_{ij} = \dfrac{1}{2}\dfrac{\partial u_i}{\partial x_j} + \dfrac{1}{2}\dfrac{\partial u_j}{\partial x_i} \end{array} \qquad (6.4.7)$$

which is a measure of the rate at which one part of the fluid slides past another at point (x,y,z), excluding the effects of rotation and of efflux.

The corresponding *viscous-shear tensor* $\mathfrak{D}$ should be proportional to this; to have it correspond to the simple case of Eq. (6.4.4), we might expect $\mathfrak{D}$ to be equal to -2μ times $\mathfrak{U}$. But still another term should be added to this to obtain the complete viscous-stress tensor. As was mentioned in the discussion of Eq. (6.2.10), the total stress tensor for the fluid must also include the effects of the usual hydrostatic pressure P, which acts perpendicularly on each face of the element of fluid volume, as shown in Fig. 6.8. The stress on the x faces thus has components $P + D_{xx}$, D_{xy}, D_{xz}, and so on, for the other pairs of sides. As indicated in Eqs. (6.2.7), the acoustic pressure p is proportional to the change in density of the gas; its rate of change $\partial p/\partial t$ is proportional to the efflux div $\mathbf{u}$ of fluid. Thus the compressibility κ is analogous to the reciprocal $1/K$ of the stiffness constant of a spring. Related to the stiffness of the gas to compression, there may be a stiffness force proportional to the rate of change of compression, just as the frictional force on the simple oscillator is proportional to the oscillator's velocity.

This compressional resistance will be labeled η, and will be called the *coefficient of bulk viscosity;* the term to be included in the stress tensor is $-\eta$ div $\mathbf{u}$, representing a stress opposing the rate of change of compression.

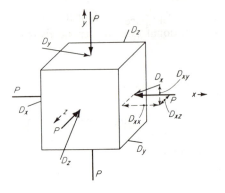

FIGURE 6.8
Total stress on an element of volume is the compressional stress P, normal to the surface, plus the various components D_{ij} of the viscous stress.

This term is too small to be measured in monatomic gases such as helium; it may be quite large in polyatomic gases at certain frequencies. The physical mechanism for most, if not all, of its effect is related to the time required to establish thermodynamic equilibrium between the translational and the rotation-vibration motions of the gas molecules. At present we simply note its presence and write down the complete expression for the components of the viscous-stress tensor $\mathfrak{D}$.

$$D_{ii} = -(\eta - \tfrac{2}{3}\mu)\,\text{div }\mathbf{u} - 2\mu\,\frac{\partial u_i}{\partial x_i}$$

$$D_{ij} = -\mu\,\frac{\partial u_i}{\partial x_j} - \mu\,\frac{\partial u_j}{\partial u_i} \qquad i \neq j$$

(6.4.8)

Thus the stress across the faces of an element of gas, perpendicular to the x axis, as shown in Fig. 6.8, is the vector sum of the normal pressure force and the viscous force $\mathbf{D}_x$, including the bulk-viscosity term. The components of this stress tensor $\mathfrak{P}$, introduced in Eqs. (6.2.10), are

$$P_{ii} = P + D_{ii} = P - (\eta - \tfrac{2}{3}\mu)\,\text{div }\mathbf{u} - 2\mu\,\frac{\partial u_i}{\partial x_i}$$

$$P_{ij} = D_{ij} = -\mu\,\frac{\partial u_i}{\partial x_j} - \mu\,\frac{\partial u_j}{\partial x_i} = P_{ji} \qquad i \neq j$$

(6.4.9)

which introduces viscous friction into the equations of motion.

Energy loss

The two processes thermal conduction and viscosity tend to transform the organized motion of the sound wave into the disorganized motion of heat. The energy content of the wave decreases, and the gas warms up. Since both the viscosity and the thermal conductivity of a gas are small, the energy transfer takes place slowly, and to first approximation, the shape of the wave is the same as that calculated neglecting the losses, except that the wave amplitude diminishes exponentially with time or space or both. Thus it is useful to work out a simple first-order expression for energy loss before turning to more accurate, though more complex, formulas.

The rate of energy loss from viscosity in a volume element $dx\,dy\,dz$ can be calculated by summing the scalar product of the stress vector $D_x\,dy\,dz$ by the difference in velocities $\mathbf{u}(x) - \mathbf{u}(x + dx)$ plus similar products for the other pairs of faces; the loss rate per unit volume of gas is thus

$$D = \sum_{ij} \frac{\partial u_i}{\partial x_j} D_{ij} = (\eta + \tfrac{4}{3}\mu)(\operatorname{div}\mathbf{u})^2 + \mu\,|\operatorname{curl}\mathbf{u}|^2$$

$$- 4\mu\left(\frac{\partial u_x}{\partial x}\frac{\partial u_y}{\partial y} + \frac{\partial u_x}{\partial x}\frac{\partial u_z}{\partial z} + \frac{\partial u_y}{\partial y}\frac{\partial u_z}{\partial z}\right.$$

$$\left. - \frac{\partial u_x}{\partial y}\frac{\partial u_y}{\partial x} - \frac{\partial u_x}{\partial z}\frac{\partial u_z}{\partial x} - \frac{\partial u_y}{\partial z}\frac{\partial u_z}{\partial y}\right) \quad (6.4.10)$$

which turns out to be everywhere positive (as it must be, since energy cannot be gained by frictional processes).

In a plane wave, where $\mathbf{u}$ is pointed along, and is a function of, the distance along the direction of propagation (call it x), this loss rate becomes, simply,

$$D = (\eta + \tfrac{4}{3}\mu)\left|\frac{du_x}{dx}\right|^2$$

If the wave is sinusoidal, D is the time average of the loss rate per volume when rms amplitude of $\mathbf{u}$ is used, in accord with the convention of Eq. (6.2.18).

The loss of energy caused by heat conduction comes about because the fluctuations of pressure in a sound wave do not represent a condition of thermodynamic equilibrium. In the regions where p is positive, the compression has not been isothermal; there has not been time for the temperature to equalize. In fact, as we have indicated in Eqs. (6.1.3) and (6.1.5), it is more nearly correct to consider the compression to be adiabatic, to assume that no heat energy moves from the region of positive p to one of negative p. If this were strictly true, of course, no energy would be lost; the compression would be reversible. But according to Eq. (6.4.1), as soon as there is a temperature differential in the gas, heat energy flows irreversibly from the hotter to the cooler regions, and the sound wave loses energy.

This would make it seem that the loss would be greater at low frequencies, when there is a longer time per cycle for the heat to leak away. But at the higher frequencies, the shorter wavelength means that there is a shorter distance between hotter and colder regions in the wave, and thus a greater temperature gradient, which more than counterbalances the shortness of the period, as we shall see.

To compute the energy lost we refer to thermodynamic equations. The sound wave produces a threefold change of internal energy U in a portion of the gas which occupies unit volume at equilibrium. In the first place, the heat flow $\mathbf{J}_h$ may concentrate heat there of an amount $-\text{div } \mathbf{J}_h$; next, heat is generated by viscous friction of an amount D, as given in Eq. (6.4.10). And finally, U is decreased by an increase in volume of the amount $-P \, dV$, and the equation of continuity indicates that this change in volume, because of expansion, is div $\mathbf{u}$. Thus, if E is the internal energy *per unit mass* of gas, we have [using Eq. (6.4.1)]

$$\frac{dE}{dt} = \frac{1}{\rho}\frac{dU}{dt} = \frac{1}{\rho}[\text{div }(K \text{ grad } T) + D - P \text{ div } \mathbf{u}] \qquad (6.4.11)$$

This is a statement of the first law of thermodynamics, which holds whether the process is reversible or not.

We know, of course, that the process is not exactly reversible, so that we cannot calculate change in entropy by its means. However, we can imagine suddenly isolating, at time t, a small portion of the gas near point (x,y,z) and letting it come to equilibrium. The entropy content of this portion can be computed by the second law; on a per-unit-volume basis it is given by the usual equation $dU = \rho T \, dS - P \, dV$ (with S the entropy content per unit mass). On a per-unit-mass basis (inserting $dV = \text{div } \mathbf{u}$ again), it is

$$\frac{dE}{dt} = T\frac{dS}{dt} - \frac{1}{\rho}P \text{ div } \mathbf{u}$$

We thus define an entropy content S of a unit mass of the gas, in spite of the fact that it is not quite in equilibrium; from it we can compute, with sufficient accuracy, heat production and other properties. For example, comparison with Eq. (6.4.11) shows that the change in entropy during the passage of the sound wave is

$$\rho\frac{dS}{dt} = \frac{D}{T} + \frac{1}{T}\text{div }(K \text{ grad } T)$$

$$= \frac{D}{T} + \frac{K}{T^2}|\text{grad } T|^2 + \text{div }\left(\frac{K}{T}\text{grad } T\right) \qquad \text{per unit volume} \quad (6.4.12)$$

If we integrate the last form over a region enclosing the whole sound wave, so that grad T is zero along the region's surface, the last term in the second line integrates to zero and the total change in S over the whole disturbed region is the integral of the first two terms, both of which are positive. Therefore S continuously increases on the whole, as it must with an irreversible process (though S may decrease at some points).

An entropy increase corresponds to a change in amount of heat energy per unit volume.

$$dQ = \rho T \frac{dS}{dt} = D + \text{div}\,(K\,\text{grad}\,T)$$

$$= D + \frac{K}{T}|\text{grad}\,T|^2 + T\,\text{div}\left(\frac{K}{T}\text{grad}\,T\right)$$

(6.4.13)

which must be equal to the loss of energy per unit volume per second from the sound wave.

For a simple-harmonic plane wave, with acoustic pressure $p = \rho c u_0 e^{ik(x-ct)}$ and fluid velocity $u_x = u_0 e^{ik(x-ct)}$, where u_0 is the rms velocity amplitude and $k = \omega/c$, the viscous loss rate is, as we have shown, $D = (\eta + \frac{4}{3}\mu)\,|ku_0|^2$. If the thermal conductivity is small, we can, to the first approximation, use Eq. (6.1.3) to compute the temperature, as though the compression were purely adiabatic, and then use Eq. (6.4.13) to calculate the power lost by heat conduction. We have

$$\tau \simeq (\gamma - 1)\frac{\rho c \kappa_s}{\beta} u_x = (\gamma - 1)\frac{1}{\beta c} u_x = \frac{c}{C_p} u_x$$

$$\frac{K}{T}|\text{grad}\,T|^2 \simeq \frac{K}{T}(\gamma - 1)^2\left(\frac{k}{\beta c}\right)^2 |u_0|^2 = (\gamma - 1)\frac{K}{C_p}|ku_0|^2$$

where the last expressions in each line are appropriate for perfect gases. The third term of Eq. (6.4.13) is zero, on the average.

Thus both loss rates increase with the square of the frequency. For a plane wave propagating through the perfect gas, with intensity $I = \rho c\,|u_0|^2$, a fraction

$$2a \equiv \frac{dQ}{I} \simeq (\eta + \frac{4}{3}\mu)\frac{k^2}{\rho c}\left[1 + \frac{K}{C_p}\frac{\gamma - 1}{\eta + \frac{4}{3}\mu}\right]$$

(6.4.14)

of its energy is withdrawn each unit distance it traverses. Hence a is called the attenuation factor for acoustic propagation. The dimensionless ratio $(C_p\mu/K)$, called the Prandtl number, is about 0.7 for air at atmospheric pressure [note Eq. (6.4.5)]. When η is much smaller than μ, we see that the second term in brackets is approximately 0.4 for air at standard conditions. Thus viscosity and heat conduction contribute roughly equally to acoustic absorption in the gas, at all frequencies. The small quantity $(4\mu/3\rho c)$ is approximately equal to 10^{-5} cm for air at standard conditions; for sound of 1,000 cps, a is approximately 3×10^{-7} cm^{-1}. Such sound must travel roughly 10 km in the open for its intensity to be reduced by a factor of 3 because of viscosity and thermal diffusion. As we shall see, viscosity and thermal diffusion produce much greater effects near a boundary. Also bulk absorption caused by relaxation effects (to be discussed shortly) in the presence of certain polyatomic impurities, such as water vapor, induces much greater absorption.   Consequently, we should work out the general equations of motion for sound, including viscosity and heat conduction.

The Stokes-Navier equation

The effects of viscosity in the equation of motion may be included by using the full expression for the stress tensor $\mathfrak{P}$ of Eq. (6.4.9).   As was done in Sec. 6.2, we write the fluid density as $\rho + \delta$, the sum of the equilibrium density ρ and the fluctuation δ caused by the sound wave; we also write the pressure as $P + p$.   If, in addition, we include a potential energy Φ caused by a possible force per unit mass of fluid, such as the force of gravity, the corrected form of the equation of motion (6.2.5) is

$$(\rho + \delta)\frac{d\mathbf{u}}{dt} \equiv (\rho + \delta)\left[\frac{\partial \mathbf{u}}{\partial t} + (\mathbf{u} \cdot \text{grad})\,\mathbf{u}\right] = -\text{grad}\,(\rho\Phi) - \boldsymbol{\nabla} \cdot \mathfrak{P}$$

$$= -\text{grad}\,[p + \rho\Phi - (\eta + \tfrac{4}{3}\mu)\,\text{div}\,\mathbf{u}] - \mu\,\text{curl}\,(\text{curl}\,\mathbf{u}) \quad (6.4.15)$$

where we have used the relationships and formulas

$$\mathfrak{P} = [P - (\eta - \tfrac{2}{3}\mu)\,\text{div}\,\mathbf{u}]\mathscr{I} + \mu(\boldsymbol{\nabla}\mathbf{u} + \mathbf{u}\boldsymbol{\nabla})$$

$$\mathscr{I}_{ij} = \delta_{ij} \qquad (\boldsymbol{\nabla}\mathbf{u})_{ij} = \frac{\partial u_j}{\partial x_i} \qquad (\mathbf{u}\boldsymbol{\nabla})_{ij} = \frac{\partial u_i}{\partial x_j}$$

$$[\boldsymbol{\nabla} \cdot (\mathbf{u}\boldsymbol{\nabla})]_j = \sum_i \frac{\partial}{\partial x_i}\frac{\partial u_i}{\partial x_j} = \frac{\partial}{\partial x_j}\sum_i \frac{\partial u_i}{\partial x_i} = [\text{grad}\,(\text{div}\,\mathbf{u})]_j$$

$$[\boldsymbol{\nabla} \cdot (\boldsymbol{\nabla}\mathbf{u})]_j = \nabla^2 u_j = [\text{grad}\,(\text{div}\,\mathbf{u})]_j - [\text{curl}\,(\text{curl}\,\mathbf{u})]_j$$

This is the Stokes-Navier equation of motion of a viscous, compressible fluid.   As was done with Eq. (6.2.11), we can show that this is equivalent to the equation of continuity for the fluid momentum density,

$$\frac{\partial}{\partial t}(\rho\mathbf{u}) + \boldsymbol{\nabla} \cdot (\mathfrak{P} + \mathfrak{J}) = -\text{grad}\,(\rho\Phi)$$

where the elements of tensor $\mathfrak{P}$ are given in Eq. (6.4.9), and of tensor $\mathfrak{J}$ in Eq. (6.2.10), their sum being the stress-momentum-flux tensor $\mathfrak{T}$ for the fluid, and $-\text{grad}\,(\rho\Phi)$ the applied force per unit volume of the fluid.

To include the effects of thermal diffusion, we must alter the relationship between p and δ to take into account the fact that the compression is not exactly adiabatic; that entropy now is produced locally because of the flow of heat.   Thus we must recognize that density, entropy, and other properties of the gas are functions of two variables, pressure and temperature, rather than just pressure.   We consider seven variables, each separated into its equilibrium value plus its small acoustic part:

Pressure: $P + p$ Temperature: $T + \tau$

Mass density: $\rho + \delta$ Entropy: $S + \sigma$ per unit mass and the fluid velocity **u** (which is a vector and which therefore is really three variables). We are, in this chapter, assuming that the equilibrium velocity is zero, so that u is small compared with c. To determine all these, we need five equations relating them. For the present we shall write out only the first-order relationships, leaving for Chap. 14 the nonlinear effects.

The first four relationships are the equation of continuity for mass flow, which, to first order, is

$$\frac{\partial \delta}{\partial t} + \rho \, \text{div } \mathbf{u} = 0 \tag{6.4.16}$$

and the first-order Stokes-Navier equation, really three equations,

$$\rho \frac{\partial \mathbf{u}}{\partial t} = -\text{grad } p + (\eta + \tfrac{4}{3}\mu) \, \text{grad (div } \mathbf{u}) - \mu \, \text{curl (curl } \mathbf{u})$$

It can be shown that any vector function of position, such as **u**, can always be uniquely separated into a longitudinal (or lamellar) part $\mathbf{u}_l$, for which curl $\mathbf{u}_l = 0$, and a transverse (or rotational) part $\mathbf{u}_t$, for which div $\mathbf{u}_t = 0$. Since the gradient of a scalar function is entirely longitudinal, i.e., since curl (grad f) = 0 for any f, the equation of motion can be split into two separate equations, one relating p to the longitudinal part of **u**, the other giving the behavior of the transverse part of **u**, unrelated to pressure waves,

$$\rho \frac{\partial \mathbf{u}_l}{\partial t} = -\text{grad } p + (\eta + \tfrac{4}{3}\mu)\nabla^2 \mathbf{u}_l$$

$$\rho \frac{\partial \mathbf{u}_t}{\partial t} = -\mu \, \text{curl (curl } \mathbf{u}_t) \tag{6.4.17}$$

where

$$(\nabla^2 \mathbf{F})_x = \nabla^2 F_x \qquad (\nabla^2 \mathbf{F})_y = \nabla^2 F_y \qquad \cdots$$

$$[\text{grad (div } \mathbf{F})]_x = \frac{\partial}{\partial x}\left(\frac{\partial F_x}{\partial x} + \frac{\partial F_y}{\partial y} + \frac{\partial F_z}{\partial z}\right)$$

$$= \nabla^2 F_x + \frac{\partial}{\partial z}\left(\frac{\partial F_z}{\partial x} - \frac{\partial F_x}{\partial z}\right) - \frac{\partial}{\partial y}\left(\frac{\partial F_x}{\partial y} - \frac{\partial F_y}{\partial x}\right)$$

$$= \nabla^2 F_x + [\text{curl (curl } \mathbf{F})]_x$$

[Eq. (6.4.6)]. Thus the two parts of the velocity solution, $\mathbf{u}_t$ and $\mathbf{u}_l$, can be solved for separately and need not be combined until we come to satisfying the boundary conditions. When we come to it we shall see that the transverse part attenuates rapidly, so that it is important only near the boundaries. The thermal-conductivity corrections and Eq. (6.4.16) have only to do with the longitudinal part of **u**.

Thermodynamic relationships

The next equation, the first-order form of Eq. (6.4.12),

$$T \frac{\partial \sigma}{\partial t} \simeq K \nabla^2 \tau \tag{6.4.18}$$

is the equation of continuity for heat flow.   We have omitted the viscosity-loss term D because it is second order in u; the first-order effects of viscosity enter in Eqs. (6.4.17).   The last two relationships between the seven variables are the equation of state, relating pressure, volume, and temperature in the gas, and the second law of thermodynamics, relating the entropy content of the gas, defined in Eq. (6.4.12), to the other variables.   These equations are most conveniently written in terms of the measurable properties of the gas, such as its heat capacity C_p per unit mass at constant pressure, its isothermal compressibility

$$\kappa_T = -\frac{1}{V}\left(\frac{\partial V}{\partial P}\right)_T$$

and coefficient of thermal expansion

$$\beta = \frac{1}{V}\left(\frac{\partial V}{\partial T}\right)_P$$

as defined in Eqs. (6.1.1).   There is also the rate of increase of pressure with temperature at constant volume, $(\partial P/\partial T)_V = \beta/\kappa_T$, which we shall call α.

In addition, there is the ratio of the specific heat at constant pressure to that at constant volume (C_P/C_V), which has been written γ.   For any material, $C_V \kappa_T = C_P \kappa_S$, or $\kappa_T = \gamma \kappa_S$; so we have a means of relating properties for an adiabatic expansion (i.e., isentropic expansion) to those for an isothermal expansion.   Since acoustic expansions are more nearly adiabatic than isothermal, we shall define the constant c appearing in the ensuing equations as the speed of sound in the limit of zero thermal conduction, when the compression is purely adiabatic.   From Eq. (6.1.8) we have $c^2 = 1/\rho\kappa_s = 1/\gamma\rho\kappa_T$.   An additional relationship between C_P and C_V, for any material, is

$$\gamma - 1 = \frac{T\beta^2}{\kappa_S C_P \rho}$$

All the quantities α, β, γ, C_P, and κ_S are functions of pressure and temperature but, if the pressure and temperature fluctuations in the sound wave are small enough for the first-order equations to be valid, we can consider these quantities to be constants, evaluated for the equilibrium values of P and T.

The pertinent quantities and relationships are then

$$\kappa_s = \frac{1}{\gamma\rho}\left(\frac{\partial\rho}{\partial P}\right)_T = \frac{1}{\rho c^2} \rightarrow \frac{1}{\gamma P} \qquad \text{(for a perfect gas)}$$

$$\beta = -\frac{1}{\rho}\left(\frac{\partial\rho}{\partial T}\right)_P \rightarrow +\frac{1}{T} \qquad \text{(for a perfect gas)}$$

$$\frac{\beta V T}{C_P} = (\gamma-1)\frac{\kappa_s}{\beta} = (\gamma-1)\frac{1}{\rho\beta c^2} \rightarrow (\gamma-1)\frac{T}{\gamma P}$$

$$\alpha = \left(\frac{\partial P}{\partial T}\right)_V = \frac{\beta}{\kappa_T} = \frac{\rho\beta c^2}{\gamma} \rightarrow \frac{P}{T} \qquad \text{(for a perfect gas)}$$

(6.4.19)

We can also express conductivity, specific heat, and viscosity in terms of characteristic lengths l_h, l_v, and l_v' by the equations

$$\frac{K}{\rho C_P c} \equiv l_h \simeq 1.6\frac{l}{\sqrt{\gamma}} \qquad \frac{\mu}{\rho c} \equiv l_v \simeq \frac{l}{\sqrt{\gamma}}$$

$$l_v' \equiv \frac{\eta + \frac{4}{3}\mu}{\rho c} = \left(\frac{4}{3}+\frac{\eta}{\mu}\right)l_v$$

(6.4.20)

The approximate formulas expressing these lengths in terms of the molecular mean-free-path l are obtained from Eqs. (6.4.2) and (6.4.5). They show that both l_h and l_v are of the order of the molecular mean-free-path, which is, roughly, 10^{-5} cm at atmospheric pressure and 20°C.

With these definitions out of the way, we return to the remaining two equations relating σ, p, τ, δ, and $\mathbf{u}$. The first is the equation of state, giving the change of density δ in terms of the acoustic pressure p and temperature τ, assumed small; the second is the equation for the change of entropy content with change of pressure and temperature:

$$\delta = \left(\frac{\partial\rho}{\partial P}\right)_T p + \left(\frac{\partial\rho}{\partial T}\right)_P \tau = \gamma\rho\kappa_s(p - \alpha\tau) \rightarrow \frac{\rho}{P}p - \frac{\rho}{T}\tau$$

$$\sigma = \left(\frac{\partial S}{\partial T}\right)_P \tau + \left(\frac{\partial S}{\partial P}\right)_T p = \frac{C_p}{T}\left(\tau - \frac{\gamma-1}{\alpha\gamma}p\right) \rightarrow C_p\left(\frac{\tau}{T} - \frac{\gamma-1}{\gamma}\frac{p}{P}\right)$$

(6.4.21)

where the expressions introduced by arrows are those valid for a perfect gas. If p is equal to $\alpha\tau$, there will be no change in density of the gas; but if $p = [\alpha\gamma/(\gamma-1)]\tau$, there will be no change in entropy; i.e., the compression will be adiabatic.

The modified wave equation

The five equations consisting of (6.4.16), the first of (6.4.17), and Eqs. (6.4.18) and (6.4.21) could serve to eliminate all but one of the five variables and determine a single equation governing the dependence of this variable on space and time.   This, plus the solution of the second of Eqs. (6.4.17), could then be adjusted to fit the appropriate boundary conditions.   It will be more convenient to relax this condition and obtain two equations, involving two variables, which then can be solved simultaneously.   The natural ones to choose are p and τ; they are more easily measured than the others, and we have already expressed δ and σ in their terms.

We first eliminate $\mathbf{u}_l$ between Eqs. (6.4.16) and (6.4.17) and then substitute for δ in terms of Eqs. (6.4.21).   The final results, in terms of the constants defined in Eqs. (6.4.19) and (6.4.20), are

$$\nabla^2 p = \frac{\gamma}{c^2}\left(\frac{\partial^2}{\partial t^2} - l'_v c \frac{\partial}{\partial t}\nabla^2\right)(p - \alpha\tau)$$

$$l_h c \nabla^2 \tau = \frac{\partial}{\partial t}\left(\tau - \frac{\gamma - 1}{\gamma\alpha}p\right) \qquad (6.4.22)$$

$$\rho\frac{\partial \mathbf{u}_l}{\partial t} = -\operatorname{grad}\left[p + \frac{\gamma l'_v}{c}\frac{\partial}{\partial t}(p - \alpha\tau)\right]$$

After the first two equations are solved to determine p and τ, the last equation and Eqs. (6.4.21) are used to compute $\mathbf{u}_l$, δ, and σ if they are needed.

The first of Eqs. (6.4.22) is the wave equation, modified for the effects of viscosity and thermal conduction.   If there were no conduction, l_h would be zero, τ would have to be equal to $(\gamma - 1)p/\gamma\alpha$, and $\gamma(p - \alpha\tau)$ would equal p.   If there were no viscosity, l'_v would be zero, and the first equation would become the simple wave equation for wave velocity c.   If the conductivity were infinite, τ would have to be zero, and the wave velocity in the first equation would be $c/\sqrt{\gamma} = \sqrt{1/\rho\kappa_T}$, the isothermal velocity.   For intermediate cases the pressure and temperature are coupled together, the fluid tending to propagate pressure waves and the heat tending to diffuse; the coupling lies in the l_h and l'_v terms.   There are thus two kinds of "waves," depending on which flow predominates.   One of them corresponds to nearly adiabatic wave motion; this velocity has a real part equal to c (the adiabatic velocity) and a small imaginary part corresponding to the energy-loss term in l_h and l'_v.   The other "wave" corresponds chiefly to heat diffusion; p is small compared with $\alpha\tau$, and the "wave velocity" is proportional to $\sqrt{i}$, indicating rapid attenuation.

To put numbers to these statements, we consider simple-harmonic separable motion, where all five acoustical quantities are proportional to p, which satisfies the equations

$$\nabla^2 p = -k^2 p \qquad \frac{\partial p}{\partial t} = -i\omega p$$

where the relationship between k and ω is obtained by substitution in Eqs. (6.4.22).   Solutions of this sort can be combined by use of the Fourier transform, to obtain transient and other solutions, as will be demonstrated later. When l_h and l_v' are small, the relation between k and ω can be obtained by using the statements of the preceding paragraph.

For example, the nearly adiabatic wave will have k^2 nearly equal to $(\omega/c)^2$. Inserting this into the second of Eqs. (6.4.22) shows that, to first order in $\omega l_h/c$,

$$-\frac{\omega^2 l_h}{c}\tau = -i\omega\tau + i\omega\frac{\gamma - 1}{\gamma\alpha}p \quad \text{or} \quad \tau \simeq \frac{\gamma - 1}{\gamma\alpha}\left(1 - i\frac{\omega}{c}l_h\right)p$$

Inserting this into the first of Eqs. (6.4.22) yields a first-order correction to k^2 and corresponding relationships between p and the other variables.

$$k^2 \simeq \left(\frac{\omega}{c}\right)^2\left[1 + i\frac{\omega l_v'}{c} + i(\gamma - 1)\frac{\omega l_h}{c}\right]$$

$$\nabla^2 p = -k^2 p \qquad \frac{\partial p}{\partial t} = -i\omega p \qquad c^2 = \frac{1}{\rho\kappa_s}$$

$$\tau \simeq \frac{\gamma - 1}{\gamma\alpha}\left(1 - i\frac{\omega}{c}l_h\right)p \qquad \delta \simeq \frac{1}{c^2}\left[1 + i(\gamma - 1)\frac{\omega}{c}l_h\right]p$$

(6.4.23)

$$\sigma \simeq -i\frac{C_p}{T}\frac{\gamma - 1}{\gamma\alpha}\frac{\omega}{c}l_h p \qquad \mathbf{u}_l \simeq \left(\frac{1}{i\omega\rho} - \frac{l_v'}{\rho c}\right)\text{grad }p$$

where $\alpha = \beta\rho c^2/\gamma$, as indicated in Eqs. (6.4.19). This kind of wave will be called the *propagational* mode. If the motion is driven, so that ω is real, then k is complex, corresponding to a wave which attenuates in space; since l_h and l_v' are so small, the attenuation is slight. If the wave is a standing wave in space, so k is real, then ω is complex, corresponding to the damping out of the standing wave in time. In either case, τ, δ, and $\mathbf{u}$ are slightly out of phase with p; the entropy fluctuation σ is small and 90° out of phase with p. As long as ω is less than either c/l_h or c/l_v' (i.e., less than about 10^9 for air at standard conditions), the speed of propagation is closely equal to $c = \sqrt{1/\rho\kappa_s}$, corresponding to adiabatic compressibility, as assumed for Eq. (6.1.3). For $\omega l_h/c$ small but not negligible, the effective compressibility is

$$\kappa_p = \frac{\delta}{\rho p} \simeq \kappa_s\left[1 + i(\gamma - 1)\frac{\omega l_h}{c}\right]$$

(6.4.24)

For the second mode of wave motion predicted by Eqs. (6.4.22), p is small compared to $\alpha\tau$, so that the second of Eqs. (6.4.22) is the primary one, the "zeroth-order" equation being $-l_h c k^2\tau \simeq -i\omega\tau$, or $k^2 \simeq i\omega/l_h c$. The first equation is then used to determine the magnitude of p, after which the first-order expressions for the other quantities can be obtained for this mode:

$$k^2 \simeq \frac{i\omega}{l_h c} \qquad \nabla^2 \tau = -k^2 \tau \qquad \frac{\partial \tau}{\partial t} = -i\omega\tau$$

$$p \simeq \frac{i\gamma\alpha\omega}{c}(l_h - l_v')\tau \qquad \delta \simeq \frac{-\alpha\gamma}{c^2}\left[1 - \frac{i\gamma\omega}{c}(l_h - l_v')\right]\tau \qquad (6.4.25)$$

$$\sigma \simeq \frac{C_p}{T}\left[1 - i(\gamma - 1)\frac{\omega}{c}(l_h - l_v')\right]\tau \qquad \mathbf{u}_l \simeq \frac{\alpha\gamma}{\rho c} l_h \operatorname{grad} \tau$$

which will be called the *thermal* mode. Constant k is large and complex, corresponding to the rapid attenuation typical of diffusion.

For liquids, or for very high frequencies in gases, we may have to solve Eqs. (6.4.22) exactly, in which case we insert $-k^2$ for ∇^2 and $-i\omega$ for $\partial/\partial t$ and solve the resulting quadratic equation for k^2. The resulting formulas are

$$k^2 = \frac{i\omega^2}{2\Omega c^2}\frac{1 - i\Upsilon - i\gamma\Omega \mp Q}{1 - i\gamma\Upsilon} \qquad \Upsilon = \frac{\omega}{c}l_v' = (\eta + \tfrac{4}{3}\mu)\frac{\omega}{\rho c^2}$$

$$\Omega = \frac{\omega}{c}l_h = \frac{K\omega}{\rho c^2 C_p} \qquad Q^2 = (1 - i\Upsilon + i\gamma\Omega)^2 - 4i(\gamma - 1)\Omega$$

$$p = \frac{\gamma\alpha}{\gamma - 1}\frac{1 - i(2\gamma - 1)\Upsilon + i\gamma\Omega \pm Q}{2(1 - i\gamma\Upsilon)}\tau \qquad (6.4.26)$$

$$\tau - \frac{\gamma - 1}{\gamma\alpha}p = \frac{1 - i\Upsilon - i\gamma\Omega \mp Q}{2(1 - i\gamma\Upsilon)}\tau$$

$$p - \alpha\tau = \frac{\alpha}{\gamma - 1}\frac{2 - \gamma - i\gamma\Upsilon + i\gamma^2\Omega \pm Q}{2(1 - i\gamma\Upsilon)}\tau$$

The propagational mode corresponds to the upper signs in the formulas; for Ω and Υ small, the formulas reduce to those of Eqs. (6.4.23). The thermal mode corresponds to the lower signs; this case reduces to Eqs. (6.4.24) when Ω and Υ are small.

Plane-wave solutions

The first example of the use of these formulas will be for plane waves in free space, though the effects of viscosity and heat diffusion are usually quite small for this case. In gases, for frequencies less than the molecular-collision frequency, we can use the first-order solutions of Eqs. (6.4.23) and (6.4.25). We set $p = f(x)e^{-i\omega t}$, where $f(x)$ is a solution of $(d^2 f/dx^2) + k^2 f = 0$. For a traveling wave, $f = Ae^{ikx}$, so that

$$p \simeq Ae^{i(\omega/c)(x - ct)}\exp\left\{-\frac{1}{2}\left(\frac{\omega}{c}\right)^2[l_v' + (\gamma - 1)l_h]x\right\} \qquad (6.4.27)$$

and

$$u_{lx} \simeq \left[1 - \tfrac{1}{2}i\frac{\omega}{c}l_v' + \tfrac{1}{2}i(\gamma - 1)\frac{\omega}{c}l_h\right]\frac{p}{\rho c} \qquad \text{(for the propagational mode)}$$

The wave attenuates as it proceeds, the attenuation factor being

$$a \simeq \frac{1}{2}\left(\frac{\omega}{c}\right)^2[l_v' + (\gamma - 1)l_h]$$

which is identical with Eq. (6.4.14). As noted before, for gases at normal pressures and temperatures and for frequencies less than 10^4 cps, this attenuation is very small [see, however, Eq. (6.4.50)].

In contrast, the thermal mode has a rapidly varying exponential factor,

$$\tau \simeq B \exp\left[\sqrt{\frac{\omega}{2l_h c}}(i - 1)x - i\omega t\right] \qquad p \simeq \frac{i\gamma a\omega}{c}(l_h - l_v')\tau \quad (6.4.28)$$

which indicates that this mode is important only near boundaries; its effects are negligible except within a layer, next to the boundary, of thickness roughly equal to the geometric mean $\sqrt{l_h \lambda}$ of the characteristic length $l_h = K/\rho c C_p$ and the wavelength $\lambda = 2\pi c/\omega$ of the sound.

The fluid motion corresponding to the rotational or transverse solution $\mathbf{u}_t$ of Eq. (6.4.17) also is important only near a boundary. Transverse flow of this sort involves no change of density or pressure (to the first approximation, at least). A possible solution would be a shear wave,

$$u_{ty} = Ae^{ikx-i\omega t} \qquad \text{div } \mathbf{u} = \frac{\partial u_{ty}}{\partial y} = 0$$

Insertion in the second of Eqs. (6.4.17) produces

$$-\rho\omega u_{ty} = -\mu \text{ curl curl } \mathbf{u} = -\mu k^2 u_{ty} \qquad \text{or} \qquad k^2 = \frac{i\omega}{l_v c} \quad (6.4.29)$$

by use of Eqs. (6.4.6) and (6.4.20). (Note that the characteristic length l_v does not involve the bulk modulus η.) Thus one possible solution is the shear wave

$$u_{ty} = A \exp\left[\sqrt{\frac{\omega}{2l_v c}}(i - 1)x - i\omega t\right] \qquad (6.4.30)$$

which dies out as rapidly as the thermal mode, since l_v is the same order of magnitude as l_h in gases. As noted earlier, $\mathbf{u}_t$ obeys a diffusion, rather than a wave equation; no wave propagation is possible.

Boundary conditions

The thermal and shear-wave modes are important near the boundaries, where they are generated because the propagational wave cannot, by itself, fit all the boundary conditions. Now that we have improved our mathematical model of wave motion to include heat diffusion and shear viscosity, we can no longer neglect to specify the values of temperature and of tangential velocity at the boundary surface. And if the propagational mode of Eqs. (6.4.23) is adjusted to fit the ratio between pressure and normal velocity specified at the surface, it cannot at the same time satisfy the boundary conditions on the acoustic temperature and tangential velocity.

The thermal and shear modes, generated by this discrepancy, are negligible outside a thin *boundary layer* just outside the boundary surface. As noted under the preceding heading, the thicknesses of these two layers are roughly equal, at least for air at standard conditions. Equations (6.4.28) and (6.4.30) show that the thickness d_v of the viscous boundary layer and d_h of the thermal layer are

$$d_v = \sqrt{\frac{2l_v c}{\omega}} = \sqrt{\frac{2\mu}{\rho\omega}} \simeq 0.21 \frac{1}{\sqrt{\nu}} \quad \text{cm}$$

$$d_h = \sqrt{\frac{2l_h c}{\omega}} = \sqrt{\frac{2K}{\rho\omega C_p}} \simeq 0.25 \frac{1}{\sqrt{\nu}} \quad \text{cm}$$

(6.4.31)

where ν is the frequency of the sound wave, and the last expressions on each line are for air at atmospheric pressure and room temperature.

Thus, except for fluids with high viscosity or conductivity or for capillary tubes of small diameter, the propagational mode of Eqs. (6.4.23) will adequately describe the acoustic behavior of the medium everywhere outside a boundary layer, usually a fraction of a millimeter thick, next to the boundaries. Indeed, in many cases of practical interest [for exceptions see Eq. (6.4.50)], the attenuation of the propagational mode can be neglected and the simple wave equation (6.2.8), with the relations of Eqs. (6.1.3) and (6.2.7), for τ, δ, and $\mathbf{u}$, can be used for the region outside the boundary layers. Since the majority of the power lost to viscosity and thermal conduction is lost within the boundary layers, and since these layers are so thin, we can consider the power to be lost *at* the boundary and can incorporate the effects of the other two modes into the boundary conditions on the propagational mode. Once this is done, we can usually forget the presence of the other two modes.

To find the effect of the thermal and shear modes on the boundary conditions for the propagational mode, we consider the situation shown in Fig. 6.4, with the boundary in the xz plane, confining the wave to the region $y \geqslant 0$. The propagational mode, the solution of Eqs. (6.4.23), is assumed to be a separable simple-harmonic wave, with the tangential motion of the fluid in the x direction. In other words, the wave for $y \geqslant 0$ is assumed to have the general form defined by the following equations:

$$p_p = \psi(x)f(y)e^{-i\omega t} \qquad \frac{d^2\psi}{dx^2} = -k_t^2\psi$$

$$\frac{d^2f}{dy^2} = (k_t^2 - k^2)f \qquad \tau_p \simeq \frac{\gamma - 1}{\alpha\gamma}\psi(x)f(y)e^{-i\omega t} \qquad (6.4.32)$$

$$u_{py} \simeq \frac{1}{i\omega\rho}\frac{df}{dy}\psi e^{-i\omega t} \qquad u_{px} \simeq \frac{1}{i\omega\rho}\frac{d\psi}{dx}fe^{-i\omega t}$$

where $k^2 = (\omega/c)^2$, or is given by Eqs. (6.4.23) if we need to include the viscous and thermal losses in the region outside the boundary layers. We also have assumed that we need not include the l_h and l_v' corrections in the formulas for τ and for $\mathbf{u}$, for this mode.

The boundary surface is usually fairly rigid; so we should expect the normal and tangential components of $\mathbf{u}$ to be quite small at $y = 0$. Also, the thermal conductivity of the boundary material is usually much greater than that of the medium; so we should expect the temperature fluctuation τ to be much smaller at $y = 0$ than it is outside the boundary layers. If the boundary surface is one of local reaction [see discussion preceding Eq. (6.3.4)] with an acoustic impedance $z = p/-u_y$, a reasonable approximation to the true boundary conditions at $y = 0$ would be

$$u_y = -\frac{p}{z} \qquad u_x = 0 \qquad \tau = 0 \qquad \text{at } y = 0 \qquad (6.4.33)$$

But the wave of Eqs. (6.4.32) does not meet these conditions. We could adjust the value of f or of df/dy (but not both, since f must satisfy its own equation) to satisfy one of these, but not all three. Thus we must add enough of the other two modes to the wave so that the combination satisfies all three conditions. In this way the two solutions, which are negligible outside the boundary layer, affect the boundary conditions on the propagational mode.

The pure plane-wave solutions of Eqs. (6.4.28) and (6.4.30) cannot be used because at $y = 0$ there must be a dependence on x in order to match the propagational solution at all points of the surface. Thus the thermal mode must have the form

$$\tau_h = \psi(x)F_h(y)e^{-i\omega t} \qquad F_h = B \exp(ik_h x)$$

with function ψ the same as in Eqs. (6.4.32). Since, according to Eqs. (6.4.25),

$$\nabla^2\tau_h \equiv \left(\frac{d^2\psi}{dx^2}F + \psi\frac{d^2F}{dy^2}\right)e^{-i\omega t} = -\frac{2i}{d_h^2}\tau_h$$

or since

$$\frac{d^2\psi}{dx^2} = -k_t^2\psi$$

the equation for F_h must be

$$\frac{d^2 F_h}{dy^2} = \left(-\frac{2i}{d_h^2} + k_t^2\right) F_h \qquad \frac{2}{d_h^2} = \frac{\omega}{l_h c}$$

However, the tangential part of the wave number k_t cannot be larger than $k = \omega/c = 2\pi/\lambda$ for the propagational wave to propagate. Usual acoustic wavelengths λ are of the order of tens of centimeters, whereas d_h, as we have seen, is usually of the order of tenths of millimeters or smaller. Therefore the quantity in parentheses multiplying F_h in the equation for F_h is nearly equal to $-(2i/d_h^2)$, and the solution which is largest at $y = 0$ and vanishes as $y \to \infty$ is $F_h = B \exp[(i - 1)(y/d_h)]$, to a good approximation.

We then adjust the value of B so that this wave will just cancel out the temperature fluctuations τ_p of the propagational wave at $y = 0$, which means that $B \simeq -[(\gamma - 1)f(0)/\alpha\gamma]$, and

$$\tau_h \simeq -\frac{\gamma - 1}{\alpha\gamma} f(0)\psi(x)e^{(i-1)(y/d_h)-i\omega t}$$

$$p_h \simeq -i\frac{\omega}{c}(\gamma - 1)(l_h - l_v')\psi(x)f(0)\exp\left(iy\sqrt{\frac{i\omega}{l_h c}} - i\omega t\right)$$

$$u_{hy} \simeq -i\frac{\gamma - 1}{\rho c}l_h\sqrt{\frac{i\omega}{l_h c}}\tau = (1 - i)(\gamma - 1)\frac{\omega d_h}{2\rho c^2}\psi(x)f(0)e^{(i-1)(y/d_h)-i\omega t}$$

When d_h is small compared with λ, $l_h = -(\omega/c)d_h^2$ is usually several orders of magnitude smaller than d_h; so, to first order in d_h, we can neglect the pressure and the tangential velocity components of this thermal mode. Not so with the normal component u_{hy}; since the normal gradient of F_h is so large, $u_{hy} \gg p_h/\rho c$. Therefore this mode can be adjusted to cancel out the temperature fluctuations at $y = 0$ at the expense of a nonnegligible contribution to the normal velocity (which will be taken care of later). But the tangential component u_{px} of the propagational wave is not yet canceled out.

To do this we must introduce a certain amount of shear wave, a solution of the second of Eqs. (6.4.17). We wish the tangential velocity component u_{tx} of this solution to equal $-u_{px}$ at $y = 0$; also, the divergence of $\mathbf{u}_t$ must be zero. The appropriate solution is

$$u_{tx} = -\frac{1}{i\omega\rho}\frac{\partial\psi}{\partial x}f(0)e^{ik_v y - i\omega t}$$

$$u_{ty} = \frac{k_t^2}{k_v\rho\omega}\psi(x)f(0)e^{ik_v y - i\omega t}$$

so that

$$\text{div }\mathbf{u}_t = \frac{\partial u_{tx}}{\partial x} + \frac{\partial u_{ty}}{\partial y} = 0 \qquad \text{and} \qquad \text{curl curl }\mathbf{u}_t = (k_v^2 + k_t^2)\mathbf{u}_t$$

or

$$k_v^2 = \frac{2i}{d_v^2} - k_t^2 \simeq \frac{2i}{d_v^2} = \frac{i\omega}{l_v c} = \frac{i\rho\omega}{\mu} \qquad \text{when } \lambda \gg l_v$$

since curl curl $\mathbf{u}_t$ must equal $i\rho\omega/\mu$ times $\mathbf{u}_t$. Therefore the shear mode, to be added to the others to fit the boundary conditions, is

$$u_{tx} \simeq -\frac{1}{i\omega\rho}\frac{\partial\psi}{\partial x}f(0)e^{(i-1)(y/d_v)-i\omega t}$$

$$u_{ty} \simeq (1-i)\frac{k_t^2 c^2}{\omega^2}\frac{\omega}{2\rho c^2}d_v\psi(x)f(0)e^{(i-1)(y/d_v)-i\omega t}$$

to the first order in l_h; no pressure or temperature changes accompany this mode.

Adding all three together, the combination of thermal, shear, and propagational waves which satisfy boundary conditions (6.4.33) is, to first order in the small quantities d_h and d_v,

$$p \simeq \psi(x)f(y)e^{-i\omega t} \qquad \frac{d^2\psi}{dx^2} = -k_t^2\psi \qquad \frac{d^2f}{dy^2} = \left[k_t^2 - \left(\frac{\omega}{c}\right)^2\right]f$$

$$\tau \simeq \frac{\gamma-1}{\alpha\gamma}\psi(x)[f(y) - f(0)e^{(i-1)(y/d_h)}]e^{-i\omega t}$$

$$\delta \simeq \frac{1}{c^2}\psi(x)[f(y) + (\gamma-1)f(0)e^{(i-1)(y/d_h)}]e^{-i\omega t}$$

$$\tag{6.4.34}$$

$$u_x \simeq \frac{1}{i\rho\omega}\frac{\partial\psi}{\partial x}[f(y) - f(0)e^{(i-1)(y/d_v)}]e^{-i\omega t}$$

$$u_y \simeq \frac{1}{i\rho\omega}\psi(x)\left\{\frac{\partial f}{\partial y} + i(1-i)\frac{\omega^2}{2c^2}\left[\frac{k_t^2 c^2}{\omega^2}d_v e^{(i-1)(y/d_v)}\right.\right.$$

$$\left.\left. + (\gamma-1)d_h e^{(i-1)(y/d_h)}\right]f(0)\right\}e^{-i\omega t}$$

Inside the boundary layer the thermal and shear waves are important, rising rapidly from nearly zero at the outer edge to a sufficient size, at $y = 0$, to cancel the temperature and tangential-velocity oscillations of the primary propagational wave. Indeed, unless the region enclosing the sound waves is so large that the minute attenuation factor of Eq. (6.4.27) becomes noticeable, we can neglect thermal and viscous effects in the main portion of the region [except perhaps for relaxation effects; see Eq. (6.4.50)]. The only place these effects show up is inside the boundary layer, and the one effect they have on the propagational wave is on its boundary conditions.

Inside the boundary layer we find an effective compressibility, given by

$$\kappa_h = \frac{\delta}{\rho p} \simeq \frac{1 + (\gamma-1)e^{(i-1)(y/d_h)}}{\rho c^2} \tag{6.4.35}$$

if we assume that d_h is so small compared with λ that $f(d_h) \simeq f(0)$. This compressibility varies from $\gamma/\rho c^2 = \kappa_T$ at $y = 0$ to $1/\rho c^2 = \kappa_s$ for $y \gg d_h$, as expected; the compression is isothermal at $y = 0$ and is nearly adiabatic outside the layer.

If the boundary surface is one of local reaction, with acoustic impedance $p/-u_y = z$, Eqs. (6.4.34) and (6.4.33) show that the ratio between p and its normal derivative at the surface must be

$$p_p \equiv \psi(x) f(0)$$

$$\simeq \frac{-z}{i\rho\omega} \psi(x) \left\{ \left(\frac{df}{dy}\right)_0 + \frac{i\omega^2}{2c^2} (1 - i) \left[\left(\frac{k_t c}{\omega}\right)^2 d_v + (\gamma - 1)d_h \right] f(0) \right\}$$

or

$$\left(\frac{\partial p_p}{\partial n}\right)_0 \equiv \psi(x) \left(\frac{df}{dy}\right)_0$$

$$\simeq -i\omega\rho \left\{ \frac{1}{z} + \frac{1 - i}{\rho c} \left[(\gamma - 1)\sqrt{\frac{\omega l_h}{2c}} + \left(\frac{k_t c}{\omega}\right)^2 \sqrt{\frac{\omega l_v}{2c}} \right] \right\} p_p \qquad y = 0$$

$$\text{(6.4.36)}$$

with the dimensionless quantities

$$\sqrt{\frac{\omega l_h}{2c}} = \frac{\omega}{2c} d_h = \sqrt{\frac{K\omega}{2\rho c C_p}} = \sqrt{\frac{\nu}{\nu_h}}$$

$$\sqrt{\frac{\omega l_v}{2c}} = \frac{\omega}{2c} d_v = \sqrt{\frac{\mu\omega}{2\rho c^2}} = \sqrt{\frac{\nu}{\nu_v}}$$

where, for air at standard conditions, $\nu_h \simeq 1.9 \times 10^9$ cps and $\nu_v \simeq 2.7 \times 10^9$ cps. These are the sole vestiges of the effects of viscosity and thermal conduction on the acoustic wave outside the boundary layer, at least to the first order in d_h and d_v. The sole boundary condition to be satisfied is this relation between the value and the normal derivative of the propagational wave pressure at the boundary.

Equation (6.4.36) establishes several conclusions of interest, perhaps the only conclusions we need to carry forward to later chapters. In the first place, if the boundary surface is moderately "soft," if the magnitude of its specific acoustic admittance $|\rho c/z|$ [Eqs. (6.3.4)] is large compared with the quantity in square brackets in (6.4.36), we need not consider the effects of thermal conduction and viscosity at all; even the surface effects are negligible compared with those of the wall's admittance. Second, we see that the boundary effect of thermal diffusion is, to the first order in d_h, completely equivalent to that of an additional specific acoustic conductance ξ_h and susceptance σ_h of the boundary surface, where $\xi_h = \sigma_h = (\gamma - 1)\sqrt{\omega l_h/2c}$. The addition of this to the admittance of the boundary surface itself will account for the effect on the propagational mode of thermal diffusion in the boundary layer.

The equivalence for the viscous losses is not so simple, because of the factor $k_t^2 c^2/\omega^2$. Thermal losses are caused by compression, and thus are related to the normal velocity of the medium, the same pair of variables which enter into the specification of acoustic impedance. Viscous losses, on the other hand, are related to the tangential velocity of the medium just outside the boundary layer; the relationship between tangential velocity and pressure is proportional to $k_t c/\omega$. This point will be discussed in more detail under the next heading.

Power lost at a surface

The power absorbed in the thermal boundary layer can most easily be calculated by using the equivalent conductance $\xi_h/\rho c$ mentioned under the previous heading. As with any surface of local reaction, the power lost per unit area of surface is the conductance times the square of the rms value of the pressure at the surface. The thermal loss per unit area at the boundary is thus

$$L_{bh} = \frac{\xi_h}{\rho c}\,|p(y=0)|^2 \simeq \left(\frac{\gamma-1}{2\rho c^2}\right)\omega d_h\,|p_p(y=0)|^2 \qquad (6.4.37)$$

This same expression can also be obtained from the formula for compressibility of Eq. (6.4.35). The acoustic energy of compression per unit volume is $-(P\,dV/V) = (P+p)(\delta/\rho)$; the rate of loss of energy per unit volume is then the real part of $-\dfrac{P+p}{\rho}\dfrac{\partial\delta^*}{\partial t}$, which equals $-\operatorname{Re}[(P+p)\kappa(\partial p^*/\partial t)]$. The term $-P\kappa(\partial p^*/\partial t)$ averages out to zero over a cycle, but $-p\kappa(\partial p^*/\partial t)$ has a real part if κ has an imaginary part. From Eq. (6.4.35) we see that κ does have an imaginary part in the boundary layer; thus the power loss in the thermal layer, per unit area of surface, is

$$-\operatorname{Re}\int_0^\infty p\kappa\,\frac{\partial p^*}{\partial t}\,dy \simeq -\operatorname{Re}\int_0^\infty \frac{i\omega}{\rho c^2}(\gamma-1)e^{(i-1)(y/d_h)}\,|p_p|^2\,dy$$

$$= -\operatorname{Re}\left[\frac{(\gamma-1)\omega}{2\rho c^2}\,d_h(i-1)\right]|p_p|^2 = L_{bh} \quad \text{of Eq. (6.4.37)}$$

The integrand, the power lost per volume in the boundary layer, is zero at $y=0$, rises to a maximum at $y=\pi d_h/4$, and drops exponentially to zero as $y\to\infty$. In the integration we have assumed that d_h is so much smaller than λ that p_p is practically independent of y within the boundary layer; so $|p|^2$ can be taken outside the integral.

The power lost by viscosity, per unit volume, is of course the function D defined in Eq. (6.4.10). In the present case, to first order, the velocity tangential to the surface, producing the shear, is

$$\frac{i}{\rho\omega}\frac{\partial\psi}{\partial x}\,f(0)e^{(i-1)(y/d_v)-i\omega t}$$

and the power lost in a unit area of boundary layer is

$$\int_0^\infty D \, dy \simeq \mu \int_0^\infty \left| \frac{\partial u_x}{\partial y} \right|^2 dy = \tfrac{1}{2}\rho\omega d_v^2 \frac{2}{\rho^2\omega^2 d_v^2} \left| \frac{\partial\psi}{\partial x} \right|^2 |f(0)|^2 \int_0^\infty e^{-2y/d_v} \, dy$$

where we have used Eq. (6.4.31) to express μ in terms of d_v. In this case the loss is greatest next to the surface, and drops off exponentially as y increases. Averaging the loss over x, we use the relation

$$\int_0^X \left| \frac{d\psi}{dx} \right|^2 dx = \int_0^X k_t^2 |\psi|^2 \, dx + \int_0^X \frac{d}{dx}\left(\psi \frac{d\psi^*}{dx} \right) dx \to X k_t^2 |\psi_{avg}|^2$$

$$X \to \infty$$

so that the mean power loss per unit area, averaged over the surface area, from viscous forces, is

$$L_{bv} \simeq \frac{k_t^2}{\rho\omega} |\psi f|^2 \int_0^\infty e^{-2y/d_v} \, dy = \left(\frac{k_t c}{\omega} \right)^2 \left(\frac{\omega}{2\rho c^2} \right) d_v |p(y=0)|^2 \quad (6.4.38)$$

which also can be obtained by considering the viscous "surface conductance" to be the term

$$\frac{1}{2} \frac{\omega}{\rho c^2} \left(\frac{k_t c}{\omega} \right)^2 d_v = \frac{1}{\rho c} \left(\frac{k_t c}{\omega} \right)^2 \sqrt{\frac{\omega l_v}{2c}}$$

as implied by Eq. (6.4.38). Multiplying this by $|p(y=0)|^2$ results in the energy loss.

If $u_{tx} = -(1/i\omega\rho)(d\psi/dx)f(0)e^{-i\omega t}$ is the component of the shear wave parallel to the surface at $y = 0$, $-u_{tx}$ is the corresponding component of the propagational wave there, since tangential $\mathbf{u}$ must be zero at the surface. When d_v is much smaller than a wavelength, the tangential component of the propagational wave, just outside the boundary layer, is thus $-u_{tx}$. Consequently, an alternative formula to Eq. (6.4.38) for the viscous surface power loss per unit area is

$$L_{bv} \simeq \tfrac{1}{2}\rho\omega d_v |\mathbf{u}_{tan}|^2 \quad (6.4.39)$$

where $\mathbf{u}_{tan}(= -u_{tx}$ in the example) is the rms amplitude of the tangential component of the propagational wave just outside the boundary layer. This formula also allows us to calculate the viscous surface losses without having to deal with the details of the shear wave. It is generally applicable for wave frequencies such that $d_v \ll \lambda$.

The power lost per unit volume in the region outside the boundary layer is another order of magnitude smaller. As already indicated in Eq. (6.4.14), for a plane wave of intensity $I = |p|^2/\rho c = \rho c |\mathbf{u}|^2$, this loss rate is

$$L_m = \left(\frac{\omega}{c} \right)^2 [l_v' + (\gamma - 1)l_h]I = \frac{1}{2}\left(\frac{\omega}{c} \right)^3 [(d_v')^2 + (\gamma - 1)(d_h)^2]I$$

where d_v and d_h are given in Eqs. (6.4.31) and $(d_v')^2 = [\tfrac{4}{3} + (\eta/\mu)](d_v)^2$.

Boundary losses in plane-wave reflection

As an example of the utility of Eq. (6.4.36), which substitutes a simple ratio between propagational pressure and normal gradient of propagational pressure at the surface for the details of viscous and thermal forces inside the boundary layer, we shall work out the reflection of a plane wave from a plane surface. The procedure is similar to that for Eq. (6.3.5); only we use Eq. (6.4.36) for the boundary conditions, instead of the formula $(\partial p/\partial n)_0 = -(i\omega\rho/z)p(\text{surface})$, which was used for the surface of local reaction when viscous and thermal losses are neglected.

For a plane wave of frequency $\omega/2\pi$ incident on the yx plane at angle of incidence ϑ, as shown in Fig. 6.4, the pressure outside the boundary layer is

$$p_p(x,y,t) = A \exp\left(\frac{i\omega x}{c} x \sin\vartheta - \frac{i\omega y}{c}\cos\vartheta - i\omega t\right)$$

$$+ AC_r \exp\left(\frac{i\omega x}{c}\sin\vartheta + \frac{i\omega y}{c}\cos\vartheta - i\omega t\right) \quad (6.4.40)$$

where C_r is, as before, the reflection coefficient for the pressure wave. If C_r is unity, the reflected pressure equals the incident pressure at the surface, the two adding to produce a pressure amplitude $2A$ there. If $C_r = -1$, the reflected wave just cancels the incident wave at the surface, and the pressure there is only the small amplitude caused by the thermal wave inside the boundary layer.

We now compare this formula with those of Eqs. (6.4.32) for the assumed form of the propagational wave and find that

$$\psi(x) = Ae^{i(\omega x/c)\sin\vartheta} \qquad f(y) = e^{-i(\omega y/c)\cos\vartheta} + C_r e^{i(\omega y/c)\cos\vartheta}$$

$$k_t^2 = \left(\frac{\omega}{c}\right)^2 \sin^2\vartheta \qquad \left(\frac{\omega}{c}\right)^2\cos^2\vartheta = \left(\frac{\omega}{c}\right)^2 - k_t^2$$

if we neglect the attenuation of the propagational wave in free space (i.e., if we neglect small quantities of the order of $l_v = \omega d_v^2/2c$ in the formula for k^2). Thus, for the reflection of a plane wave, the factor $(k_t c/\omega)^2$, which enters the viscosity term in the boundary conditions, is simply the square of the sine of the angle of incidence ϑ. Inserting these values into Eq. (6.4.36), we find, for the reflection coefficient,

$$C_r \simeq \frac{\cos\vartheta - (\rho c/z) - (\xi_b - i\sigma_b)}{\cos\vartheta + (\rho c/z) + (\xi_b - i\sigma_b)} \quad (6.4.41)$$

where

$$\xi_b = \sigma_b = \frac{\omega d_v}{2c}\sin^2\vartheta + (\gamma - 1)\frac{\omega d_h}{2c}$$

$$= \sqrt{\frac{\mu\omega}{2\rho c^2}}\sin^2\vartheta + (\gamma - 1)\sqrt{\frac{K\omega}{2\rho c^2 C_p}}$$

are the effective acoustic conductance and susceptance of the boundary layer, to first order in the small quantities d_h and d_v.

This is to be compared with Eq. (6.3.5). As mentioned previously, the correction for heat conduction acts exactly the same as an additional acoustic admittance of the surface, but the viscosity correction has a different dependence on angle of incidence; to first order there is no viscous absorption for a normally incident wave $\vartheta = 0$. Reference to Eq. (6.3.6) shows that the energy absorbed per unit area of surface, from a plane wave of intensity $I = |A|^2/\rho c$, incident at angle ϑ, during its reflection, is

$$I(1 - |C_r|^2) \cos \theta \simeq \frac{4I(\xi + \xi_b) \cos^2 \vartheta}{(\cos \vartheta + \xi + \xi_b)^2 + (\sigma + \sigma_b)^2} \qquad (6.4.42)$$

where $\rho c/z = \xi - i\sigma$. This also is valid only to first order in d_v and d_h.

If the surface is absolutely rigid, so that $\rho c/z = 0$, the first-order approximation to the energy absorbed per unit area is

$$4I\xi_b = \frac{4|A|^2}{\rho c} \xi_b \simeq \frac{|p(x,0,t)|^2}{\rho c} \xi_b = L_{bv} + L_{bh}$$

This formula checks with those of Eqs. (6.4.37) and (6.4.38).

Molecular energy equipartition

The foregoing discussion has covered, sufficiently for the needs of this book, the energy losses caused by shear friction and thermal diffusion. The magnitude, the dependence on frequency and temperature of these losses, and their relation with the kinetics of molecular motion have been set forth. A similar discussion cannot be given for liquids; molecular detail regarding the heat and momentum diffusion in liquids is not well understood. All we can do is to use experimentally determined constants in the formulas we have derived, or, in many cases, to use the formulas backward to determine empirical values for μ, γ, l_h, and l_v from measurements of energy loss. However, in the case of the energy loss related to the bulk modulus η of gases, present in Eqs. (6.4.8) and (6.4.20), additional discussion is desirable.

As we have seen in Eq. (6.4.24), the effect of heat conduction on sound propagation can be expressed in terms of a complex compressibility. Its imaginary part is proportional to the ratio between mean-free-path and wavelength, in other words, to the ratio between the period of the sound oscillation and the mean-free-time between molecular collisions. The pressure change resulting from a reduction of volume occupied is not quite in phase with the change in density, and energy is thereby changed to heat. Through interaction with the surrounding gas the compressed portion loses heat by diffusion, and a corresponding "relaxation" of the sound pressure results.

The effect of viscosity involves a similar diffusion of momentum from the moving part to the rest of the gas. These effects are irreversible transfers of potential and kinetic energy from the organized motion of the sound wave to the disorganized translational motion of the molecules known as heat. The attenuation caused by viscosity and heat conduction is thus often referred to as translational relaxation effects. Measurements of sound propagation in monatomic gases have shown that the attenuation is in agreement with that predicted by translational relaxation, as presented in the foregoing portion of this section.

In diatomic and polyatomic gases, however, viscosity and heat conduction are only part of the sound-attenuating mechanism. In addition, there is an irreversible energy transfer from the collective motion of the sound wave to other degrees of freedom of the molecules, other than the translational ones. It is easy to see qualitatively how this attenuation will come about. Imagine a volume element of gas to be compressed quickly. The translational motion of the molecules will adjust nearly instantaneously, and since the pressure of the gas is determined solely by this translational motion, the pressure adjusts immediately to the density change. After the compression is completed, some of the translational energy of the molecules goes more slowly into rotational and vibrational energy, and the pressure decreases accordingly. If the gas is allowed to expand to its initial volume, the work delivered by the gas during the expansion is less than that absorbed during the compression. Therefore, in each such cycle, there will be a net energy transfer to the internal degrees of freedom of the molecules, which will heat the gas and attenuate the sound.

The magnitude of this energy loss will depend on the ratio between the length of the cycle and the "relaxation time" of the transfer from translational to internal energy of the molecules. In the extreme case, when the gas is expanded immediately after compression, before such transfer can take place, the gas will behave as a monatomic gas, and only viscosity and heat conduction will cause attenuation. On the other hand, if both compression and expansion take place very slowly, so that thermal equilibrium between all internal modes of motion is continuously established, relaxational attenuation is again absent. Thus the relaxational attenuation will be important in some intermediate range of frequencies.

Frequency dependence of the specific heat

The effect of the delay in distribution of energy among the various degrees of freedom of the molecules is most clearly displayed in the frequency dependence of the heat capacity of the gas. Thence, via the ratio γ of Eqs. (6.4.19), it can be related to the compressibility of a gas subjected to simple-harmonic pressure variations. The specific heat at constant volume (strictly speaking, specific heat is the heat capacity of a mole of the gas; the quantity C_v used here is the heat capacity of a *unit mass* of the gas) is the rate of change of the internal energy of the gas with temperature, $(\partial E/\partial T)_v$, at constant volume (i.e., constant density).

In thermodynamics we assume that the temperature is changed slowly enough so that all the degrees of freedom of the molecules can take up their share of the increased energy. In this case the energy absorbed during a unit rise in T by the nth degree of freedom of a unit mass of gas molecules can be written $(R/2M)F_n(T)$, where R is the gas constant and M is the molecular weight. Function F_n is zero at $T = 0$, rises rapidly at temperatures near T_n, the characteristic temperature for the nth degree of freedom, and for $T \gg T_n$ reaches an asymptotic value of unity; in the limit each mode absorbs $\frac{1}{2}R$ per mole per degree rise.

In the range of temperatures of interest to us, the three F_n's corresponding to the translational degrees of freedom of the molecules (those contributing to the pressure) have long since reached their asymptotic values, whereas those for the electronic motions have not yet begun to rise from near zero. Therefore we can write the heat capacity at constant volume of the gas as

$$C_V = \left(\frac{\partial E}{\partial T}\right)_v = \frac{R}{M}\left[\tfrac{3}{2} + \tfrac{1}{2}\sum_n F_n(T)\right]$$

where the sum over n includes all the internal degrees of freedom of the molecule having to do with the nuclear motions, the rotational ones (zero, two, or three of them, depending on whether the molecules are mono-, di-, or polyatomic), and the vibrational ones, corresponding to the various normal modes of vibration of the nuclei, as discussed in Chap. 3 (the number of such degrees of freedom being zero for a monatomic, one for a diatomic, and $3N - 6$ for an N-atomic molecule). For all molecules except hydrogen, the F_n's for the rotational degrees of freedom (if any) have also reached their unit asymptotic values by 0°C. Usually, the F_n's for the vibrational modes are less than unity at room temperatures.

Any book on thermodynamics will demonstrate that the difference between the specific heats at constant pressure and at constant volume, for a perfect gas, is R, the gas constant, the difference being caused by the expansion of the gas when heated at constant pressure, the expansion requiring extra energy. Therefore, for a unit mass of gas,

$$C_P = C_V + \frac{R}{M} = \frac{R}{M}\left[\tfrac{5}{2} + \tfrac{1}{2}\sum_n F_n(T)\right]$$

and the ratio between these heat capacities, the quantity γ entering the equation for the velocity of sound, is

$$\gamma = \frac{C_P}{C_V} = \frac{\tfrac{5}{2} + \tfrac{1}{2}\sum F_n(T)}{\tfrac{3}{2} + \tfrac{1}{2}\sum F_n(T)} \tag{6.4.43}$$

Once one works out, from statistical mechanics, the expressions for the F_n's as functions of T, then the dependence of γ on T, for reversible processes, can be calculated.

As pointed out earlier, however, the gas during a cycle of a sound wave does not have time to come to thermodynamic equilibrium. The increase in density immediately increases the energy of translational motion, but the internal degrees of freedom, represented by the F_n's, take time to absorb their component shares of energy. The nth degree of freedom has its characteristic time of response, called its *relaxation time* t_n, to reach equilibrium. The rotational modes, being more affected by collisions, reach equilibrium fairly quickly, but the values of t_n for the vibrational modes are considerably longer, some of them being of the order of magnitude of a few milliseconds for standard pressures and temperatures.

Statistically, the rate of approach of the energy E_n of the nth degree of freedom to its equilibrium value E_n^e is proportional to the difference $E_n - E_n^e$, the proportionality factor being the reciprocal of the relaxation time.

$$\frac{dE_n}{dt} = -\frac{1}{t_n}(E_n - E_n^e) \qquad (6.4.44)$$

Thus, when left alone, the energy approaches E_n^e exponentially:

$$E_n = E_n^e + (E_n^i - E_n^e)e^{-t/t_n}$$

relaxing from its initial, nonequilibrium value E_n^i to E_n^e in a time roughly equal to t_n. In the presence of a simple-harmonic sound wave, however, the gas is not left alone; energy is alternately given and withdrawn at a rate determined by the factor $e^{-i\omega t}$. As long as the acoustic pressure p is small compared with the equilibrium pressure P, the situation is not far from equilibrium and it is not a bad approximation to assume that $E_n^e = E_n^0 + \epsilon_n e^{-i\omega t}$, where ϵ_n is the amplitude of the energy change in the nth degree of freedom which would be produced by the sound wave *if* equilibrium were attained at each instant of the cycle. Since the internal modes cannot follow so rapidly, the actual energy content E_n of the nth mode during the cycle is obtained from Eq. (6.4.44).

$$E_n = E_n^0 + \frac{\epsilon_n}{1 - i\omega t_n} e^{-i\omega t}$$

If the "driving force" is a periodic temperature change $\tau = \tau_0 e^{-i\omega t}$, the equilibrium amplitude ϵ_n of the internal energy change would be C_v or C_p times τ, and the amplitude of the actual change would be this product times $(1 - i\omega t_n)^{-1}$. Thus we can talk about a *dynamic* heat capacity, the ratio of internal energy to temperature change for a simple-harmonic temperature fluctuation in a gas,

$$C_V = \frac{R}{M}\left(\tfrac{3}{2} + \tfrac{1}{2}\sum_n \frac{F_n}{1 - i\omega t_n}\right)$$

and

$$C_P = \frac{R}{M}\left(\tfrac{5}{2} + \tfrac{1}{2}\sum_n \frac{F_n}{1 - i\omega t_n}\right)$$

and finally, a ratio of specific heats appropriate for such fluctuations,

$$\gamma(T,\omega) = \frac{5 + \sum [F_n(T)/(1 - i\omega t_n)]}{3 + \sum [F_n(T)/(1 - i\omega t_n)]} = \frac{\gamma + \frac{1}{2}(\gamma - 1) \sum \dfrac{i\omega t_n F_n}{1 - i\omega t_n}}{1 + \frac{1}{2}(\gamma - 1) \sum \dfrac{i\omega t_n F_n}{1 - i\omega t_n}} \qquad (6.4.45)$$

where γ is the equilibrium value defined in Eq. (6.4.43).

Relaxation attenuation

This dynamic value of γ should be used in computing the compressibility $\kappa_s = \kappa_T/\gamma$ for the processes in the sound wave and in computing the speed of sound $c = \sqrt{1/\rho\kappa_s}$. The attenuation of a plane wave of sound is then the imaginary part of the wavenumber $k = \omega/c$. In the present derivation of c and k, we have neglected the effects of viscosity and of thermal conduction; therefore, for the whole expression for the propagational mode, we should combine the results of Eqs. (6.4.23) and (6.4.45), obtaining for the square of the wavenumber

$$k^2 \simeq \gamma \left(\frac{\omega}{c_0}\right)^2 \frac{1 + \frac{1}{2}(\gamma - 1) \sum \dfrac{i\omega t_n F_n}{1 - i\omega t_n}}{\gamma + \frac{1}{2}(\gamma - 1) \sum \dfrac{i\omega t_n F_n}{1 - i\omega t_n}} \left[1 + i\frac{\omega}{c}l'_v + i(\gamma - 1)\frac{\omega}{c}l_h\right] \qquad (6.4.46)$$

where

$$c_0{}^2 = \frac{\gamma}{\kappa_T \rho} = \frac{\gamma P}{\rho} \qquad \text{(for a perfect gas)}$$

and where

$$l'_v = \frac{\eta + \frac{4}{3}\mu}{\rho c} \qquad \text{and} \qquad l_h = \frac{K}{\rho c C_p}$$

The formula for a plane wave in the x direction would be $e^{ikx - i\omega t}$; so the imaginary part of k is the attenuation constant a of Eq. (6.4.14), the reciprocal of the distance within which the wave amplitude is attenuated by a factor e^{-1}. The ratio between ω and the real part of k is the phase velocity of the wave, to be discussed in detail in the latter part of Sec. 9.1. For low frequencies, such that $\omega t_n \ll 1$ for all n's, the fraction in Eq. (6.4.46) can be approximated; to the first order in ωt_n, l'_v, and l_h, the expression for the wavenumber is

$$k \simeq \frac{\omega}{c_0}\left\{1 + \frac{i\omega}{2\rho c_0{}^2}\left[\frac{(\gamma - 1)^2}{2\kappa_T} \sum_n t_n F_n(T) + \eta + \frac{4}{3}\mu + (\gamma - 1)\frac{K}{C_p}\right]\right\} \qquad (6.4.47)$$

the attenuation constant being proportional to the quantity in the square brackets, and the phase velocity differing from c_0 only to second order in the small quantities. At low frequencies all effects enter linearly, each degree of freedom making its contribution independently of the others.

Thus the quantity

$$\eta_r = \frac{(\gamma - 1)^2}{2\kappa_T} \sum t_n F_n(T)$$

enters the formulas for attenuation on a par with the bulk modulus η. Since neither η_r nor η enters the shear and conductivity effects near a boundary except in second order, it is hard to distinguish between η_r and η. Factor η is an assumed internal friction of compression, and η_r comes from the hysteresislike loss to internal modes during a compression cycle. Of course, there are no internal degrees of freedom for monatomic gases; so η_r should be zero in this case. But the internal-frictional term η for a monatomic gas is so small that it is doubtful whether it is really present; and with di- and polyatomic gases, one cannot distinguish η_r from η. Thus it is doubtful whether there should be a separate factor η; we might as well consider η_r to *be* the η introduced earlier in this section.

If we insert the expressions for μ and for K/C_p in terms of molecular mean-free-path l, given in Eqs. (6.4.2) and (6.4.5), into the formula for the nondimensional quantity $a\lambda = (c_0/\omega)\mathrm{Im}\, k$, the attenuation per wavelength,

$$a\lambda \simeq \pi \frac{(\gamma - 1)^2}{2\gamma} \sum_n \omega t_n F_n + \tfrac{4}{3}\pi \frac{\omega}{\sqrt{\gamma}} \frac{l}{c_0} + \pi \frac{\gamma - 1}{\gamma^{\frac{3}{2}}} \omega \frac{l}{c_0}$$

we see that the viscosity and thermal effects are quite similar to the relaxation effects. The mean time between collisions l/c_0 (since c_0 is roughly equal to the mean speed of the molecules; see page 229) is equivalent to a relaxation time; indeed, it *is* the relaxation time for the translational degrees of freedom of the molecules. Since the relaxation time for the rotational degrees of freedom is not much longer than the time between collisions, and since F_n for the rotational terms is nearly unity at room temperatures, the rotational terms in the sum over n are of the same order of magnitude as the viscosity and thermal-diffusion terms (i.e., the translational terms). At standard conditions of temperature and pressure, the time between collisions in air is roughly 10^{-9} sec; so the rotational and translational contributions to acoustic attenuation are extremely small for frequencies less than about a megacycle.

Vibration-relaxation attenuation

As was stated on page 277, we thus find that the attenuation caused by viscosity, thermal diffusion, and rotational relaxation, away from boundaries, can usually be neglected (the boundary effects, as we have seen, are several orders of magnitude larger). This is not true for the vibrational modes, however, particularly for polyatomic gases. Although the factors F_n for these modes are small, between 0.1 and 0.01 for room temperatures, the relaxation times are much longer than the rotational times; it takes many collisions to excite some vibrations. For the oxygen molecule, for example,

t_n is about 10^{-3} sec. However, the presence of an impurity, such as water vapor, is known to decrease t_n so its inverse is in the acoustic frequency range.

Since the F_n's for the vibrational modes are small at room temperatures, we can expand the fraction of Eq. (6.4.46) for these terms also, but we cannot neglect the higher orders of ωt_n. The complete expression for kc_0/ω, using appropriate approximations, is

$$\frac{kc_0}{\omega} \simeq 1 - \frac{(\gamma - 1)^2}{4\gamma} \sum_{\text{vib}} \frac{(\omega t_n)^2 F_n}{1 + (\omega t_n)^2} + i\frac{(\gamma - 1)^2}{4\gamma} \sum_{\text{vib}} \frac{\omega t_n F_n}{1 + (\omega t_n)^2}$$

$$+ i\frac{(\gamma - 1)^2}{4\gamma} \sum_{\text{rot}} \omega t_n F_n + \tfrac{4}{3}i\frac{\omega\mu}{\rho c_0^2} + i(\gamma - 1)\frac{\omega K}{\rho c_0^2 C_p} \quad (6.4.48)$$

where the sum has been split into a rotational and a vibrational part, as labeled below the summation signs.

The last three terms can be neglected, at standard conditions, if there are any vibrational terms present. In fact, for air containing moderate amounts of molecules such as H_2O or CO_2, the vibrational terms for O_2 and N_2 vibrations cannot be neglected but, in reality, provide the largest contributions in Eq. (6.4.48) (multiplied, of course, by the factor corresponding to their concentration, which can be included in the factors F_n).

The first two terms in this expression are real and represent the effect of the vibration relaxation on the speed of sound. The phase velocity is

$$\frac{\omega}{\operatorname{Re} k} \simeq c_0 \left[1 + \frac{(\gamma - 1)^2}{4\gamma} \sum_{\text{vib}} \frac{(\omega t_n)^2 F_n}{1 + (\omega t_n)^2}\right] \quad (6.4.49)$$

At frequencies well below $1/t_n$ for all n's this is equal to $c_0 = \sqrt{\gamma P/\rho}$, the velocity in dry air. As the frequency is increased, this phase velocity increases, approaching a somewhat higher value asymptotically. Thus moist air is a dispersive medium for sound waves (see the latter part of Sec. 9.1).

The attenuation per wavelength

$$2\pi \operatorname{Im} \frac{kc_0}{\omega} \simeq \pi \frac{(\gamma - 1)^2}{2\gamma} \sum_{\text{vib}} \frac{\omega t_n F_n}{1 + (\omega t_n)^2} \quad (6.4.50)$$

also is small at low frequencies, rising linearly with ω. But it reaches a maximum, for each vibrational mode, at $\omega = 1/t_n$; when the period of oscillation of the sound is equal to 2π times the relaxation time, a maximum amount of energy is converted into heat because of the relaxation hysteresis. For still higher frequencies the attenuation decreases again, approaching zero as $\omega \to \infty$, as mentioned on page 295.

Problems

1   (a) A plane wave in air has an intensity of 100 ergs per sec per sq cm.   What is the magnitude of the temperature fluctuation in the air, caused by the wave? (b) At $t = 0$ a sound wave has the form of a step function, with $P = P_0 + p$ for $x < 0$ and $P = P_0$ for $x > 0$.   Explain qualitatively how heat conduction would change the shape of this step function as it travels in the positive x direction.

2   As shown in Eq. (6.4.14), the attenuation of a plane wave depends on the size of the shear modulus μ.   Explain how shear stresses can arise in a plane wave.

3   Suppose the air, which was initially at rest at the point x_0, y_0, z_0 has displacement $X - x_0 = AF(ct - x_0)$ in the x direction, where

$$F(q_0) = \begin{cases} 0 & q_0 < 0 \\ \alpha q_0 & 0 < q_0 < \dfrac{1}{\alpha} \\ 2 - \alpha q_0 & \dfrac{1}{\alpha} < q_0 < \dfrac{2}{\alpha} \\ 0 & \dfrac{2}{\alpha} < q_0 \end{cases} \qquad \text{where} \quad \begin{aligned} q &= ct - x \\ q_0 &= ct - x_0 \end{aligned}$$

Show that the displacement of the portion of the air, which happens to be at point x at time t, is

$$D(q) = A \begin{cases} 0 & 0 > q \\ \dfrac{\alpha q}{1 - \alpha A} & 0 < q < \dfrac{1}{\alpha} - A \\ \dfrac{2 - \alpha q}{1 + \alpha A} & \dfrac{1}{\alpha} - A < q < \dfrac{2}{\alpha} \\ 0 & \dfrac{2}{\alpha} < q \end{cases}$$

Show that the fluid velocity at point x at time t is $+c\alpha A$ for $0 < q < (1/\alpha) - A$ and is $-c\alpha A$ for $(1/\alpha) - A < q < (2/\alpha)$; zero otherwise.   Show that the pressure there is $\rho c^2(\partial D/\partial q)$.   What happens to the displacement and pressure wave when $\alpha A \to 1$?   Plot D, u, and p as functions of x for $t = 0$ and $\alpha A \to 1$. What is the physical meaning of this result?   What happens when $\alpha A > 1$?

4   By the usual methods of the calculus of variations, show that Eq. (6.2.14) is the Lagrange-Euler equation for the Hamilton principle

$$\delta \int dt \iiint L(\psi, \psi_x, \psi_y, \psi_z, \psi_t) \, dx \, dy \, dz = 0$$

where the subscript denotes differentiation, as in Eq. (6.2.14), and when the variation of ψ does not alter ψ at the limits of the integration in space or time. Show that adding to L the divergence of a vector $\mathbf{A}$, which goes to zero at the limits of integration, leaves Eq. (6.2.14) unchanged.

5   Show that if a Lagrange density L is a function of second derivatives ψ_{xy}, etc., as well as of ψ and its first derivatives, the Lagrange-Euler equation for the

Hamilton principle, $\int L \, dt \, dV = 0$, is

$$\frac{\partial L}{\partial \psi} - \sum_i \frac{\partial}{\partial x_i} \frac{\partial L}{\partial \psi_i} + \sum_{i,j} \frac{\partial^2}{\partial x_i \partial x_j} \frac{\partial L}{\partial \psi_{ij}} = 0$$

instead of Eq. (6.2.14), where $\psi_1 = \partial \psi / \partial x$, $\psi_{24} = \partial^2 \psi / \partial y \, \partial t$, etc. Show that, if L is a quadratic function of four functions p, its adjoint p^a, τ, and its adjoint τ^a [see discussion of Eq. (6.2.28)] and their first and second derivatives, there will be four Lagrange-Euler equations, one

$$\frac{\partial L}{\partial p^a} - \sum_i \frac{\partial}{\partial x_i} \frac{\partial L}{\partial p_i{}^a} + \sum_{i,j} \frac{\partial^2}{\partial x_i \partial x_j} \frac{\partial L}{\partial p_{ij}{}^a} = 0$$

for function p, and three more, one for p^a, one for τ, and one for its adjoint τ^a. Write out the formulas for the elements W_{ij} of the stress-energy tensor, and show how they are related by equations of continuity.

6 Use the results of Prob. 5 to show that Eqs. (6.4.22) are Lagrange-Euler equations for p and τ, corresponding to a Lagrange density function

$$L = -\frac{\gamma}{2c^2} \frac{\partial p}{\partial t} \frac{\partial p^a}{\partial t} + \frac{\gamma^2 \alpha^2 l_h}{2(\gamma - 1)c^2} \frac{\partial \tau}{\partial t} \frac{\partial \tau^a}{\partial t}$$

$$+ \tfrac{1}{2} \operatorname{grad} p \cdot \operatorname{grad} p^a - \frac{\gamma l_v'}{4c} \left(\frac{\partial p}{\partial t} \nabla^2 p^a - \frac{\partial p^a}{\partial t} \nabla^2 p \right)$$

$$- \frac{\alpha^2 \gamma^2 l_h}{4(\gamma - 1)c} \left(\frac{\partial \tau}{\partial t} \nabla^2 \tau^a - \frac{\partial \tau^a}{\partial t} \nabla^2 \tau \right) + \frac{\alpha \gamma}{2c^2} \left(\frac{\partial p}{\partial t} \frac{\partial \tau^a}{\partial t} + \frac{\partial p^a}{\partial t} \frac{\partial \tau}{\partial t} \right)$$

$$+ \frac{\alpha \gamma l_h}{4c} \left(\frac{\partial p}{\partial t} \nabla^2 \tau^a - \frac{\partial p^a}{\partial t} \nabla^2 \tau + \frac{\partial \tau}{\partial t} \nabla^2 p^a + \frac{\partial \tau^a}{\partial t} \nabla^2 p \right)$$

which takes into account viscosity and heat conduction (note the arrangement of signs in the last parentheses). Compute the stress-energy tensor W_{ij} [Eqs. (6.2.15)]. Discuss the physical significance of the formulas for W_{tt} and the vector $\mathbf{I}$ with x component W_{tx}.

7 One method of measuring the normal impedance of a boundary (as a function of the angle of incidence of the sound wave) is the following: Imagine the (complex) sound-pressure amplitude p_∞, that is, magnitude and phase (relative to some reference phase) to be measured at the surface of a perfectly reflecting surface (infinite impedance). Keeping the sound source the same, the surface is then replaced by the surface to be measured, and the complex surface pressure is again measured. Denote this by p_ζ. The complex number $w = p_\zeta / p_\infty$ is thus determined from the experiments.

(a) Show that the relationship between the normal impedance ζ (to be determined) and the measured quantity w is

$$\zeta = \frac{w}{1 - w \cos \vartheta} \frac{1}{}$$

where ϑ is the angle of incidence of the wave.

(b) If we set $\zeta \cos \vartheta = u + iv$ and $w = |w|e^{i\gamma}$, show that the relation between $\zeta \cos \vartheta$ and w can be represented as a set of orthogonal circles in the uv plane, circles for constant $|w|$ and constant γ. These circles are, with $[|w|^2/(1 - |w|^2)] = A$,

$$(u - A)^2 + v^2 = A(1 + A) \qquad \text{(constant } |w| \text{ circles)}$$
$$(v - \tfrac{1}{2} \cot \gamma)^2 + (u + \tfrac{1}{2})^2 = \tfrac{1}{4} \sin^{-2} \gamma \qquad \text{(constant } \gamma \text{ circles)}$$

8 The region $y < 0$ is filled with porous material of porosity Ω and flow resistance Φ; the region for $y > 0$, with air of characteristic impedance ρc and wave velocity c. Show that the effective acoustic impedance $z(\vartheta) = -(p/u_y)$, which the interface $y = 0$ presents to a plane wave of sound of frequency $\omega/2\pi$, incident at angle of incidence ϑ, is

$$\rho_p c_p \left[\frac{1 + i(\Phi/\rho_p\omega)}{1 - (c_p/c)^2 \sin^2 \vartheta}\right]^2$$

where ρ_p and c_p are defined in connection with Eq. (6.2.24). Show that, when $c_p \ll c$ and $\Phi \ll \rho_p\omega$, the interface behaves like a locally reacting surface of acoustic resistance $\rho_p c_p$ and stiffness reactance $-(c_p\Phi/2\omega)$.

9 The xz plane is a rigid wall; the region between $y = 0$ and $y = l$ is occupied by a material with acoustic parameters ρ_e and c_e, which may be complex quantities. Show that the ratio of reflected to incident intensity, for a normally incident wave, is

$$|C_r|^2 = \left|\frac{\rho_e c_e + i\rho c \tan (\omega l/c_e)}{\rho_e c_e - i\rho c \tan (\omega l/c_e)}\right|^2$$

10 A panel of porous material 3 cm thick is glued on one side to a plane rigid wall; the other side is open to the air. The acoustic parameters of the material, as defined in connection with Eq. (6.2.24), are $\rho_p = 2\rho$, $c_p = \tfrac{1}{2}c$, $\Phi/\rho c = 100 \text{ m}^{-1}$. A plane wave of frequency $\omega/2\pi$ is normally incident on this coated wall. Plot the fraction of incident acoustic energy which is reflected, as a function of frequency, from 0 to 10,000 cps.

11 Plot against angle of incidence ϑ the fraction of energy $|C_r|^2$ reflected from the plane interface between the air (ρ,c) and a medium of density $\tfrac{2}{5}\rho$ and sound velocity $\tfrac{5}{4}c$, when the incident wave comes from the air side. Plot $|C_r|^2$ when $\rho_w = \tfrac{8}{5}\rho$, $c_w = \tfrac{5}{4}c$.

12 Suppose a unit-pulse pressure wave $\delta(T_i - t_0)$, where

$$T_i = t - (1/c)(x \sin \vartheta - y \cos \vartheta)$$

impinges on the surface $(y = 0)$ of a porous material of the sort discussed in Prob. 8, where $z = \rho_p c_p + i(c_p\Phi/2\omega)$. Show that the reflected pressure wave is

$$\delta(T_r - t_0) + \frac{\rho c c_p \Phi \cos \vartheta}{(\rho c + \rho_p c_p \cos \vartheta)^2} \exp\left[\frac{-c_p \Phi \cos \vartheta}{2\rho c + 2\rho_p c_p \cos \vartheta} (T_r - t_0)\right]$$

when $t > 0$, where $T_r = t - (1/c)(x \sin \vartheta + y \cos \vartheta)$. Discuss the physical significance of the dependence on ϑ, t, and Φ/ρ_p.

13 A fluid having viscosity and thermal conductivity is confined between two rigid plates, one at $y = +\tfrac{1}{2}a$, the other at $y = -\tfrac{1}{2}a$. A sound wave is sent in the x direction in the fluid, with frequency $\omega/2\pi$. Show that the three types of waves, which together fit the boundary conditions $\tau = 0$ and $\mathbf{u} = 0$ at $y = \pm\frac{1}{2}a$, are, to first order in $kd_v = \sqrt{\frac{1}{2}kl_v}$ and $kd_h = \sqrt{\frac{1}{2}kl_h}$ [see Eq. (6.4.31)],

$$p_p = P \exp(i\mu x - i\omega t) \cosh(\beta y) \qquad \tau_p \simeq \frac{\gamma - 1}{\gamma\alpha} p_p$$

$$u_{px} = \frac{\mu}{\rho ck} p_p \qquad\qquad\qquad\qquad\qquad \text{(propagational wave)}$$

$$u_{py} = -\frac{i\beta P}{\rho ck} \exp(i\mu x - i\omega t) \sinh(\beta y)$$

$$p_h \simeq -i\gamma\alpha kl_h \tau_h$$

$$\tau_h \simeq \frac{\gamma - 1}{\gamma\alpha} P \frac{\cosh[(1 - i)y/d_h]}{\cosh[(1 - i)a/2d_h]} \exp(i\mu x - i\omega t) \cosh\frac{\beta a}{2}$$

$$u_{hx} \simeq 0 \qquad\qquad\qquad\qquad\qquad\qquad \text{(thermal wave)}$$

$$u_{hy} \simeq -(1 - i)kd_h \frac{\gamma - 1}{2\rho c} P \frac{\sinh[(1 - i)y/d_h]}{\cosh[(1 - i)a/2d_h]} \exp(i\mu x - i\omega t) \cosh\frac{\beta a}{2}$$

$$u_{tx} \simeq \frac{-\mu P}{\rho ck} \frac{\cosh[(1 - i)y/d_v]}{\cosh[(1 - i)a/2d_v]} \exp(i\mu x - i\omega t) \cosh\frac{\beta a}{2} \quad \text{(shear wave)}$$

$$u_{ty} \simeq -(1 - i)\frac{\mu^2 d_v}{\rho ck} P \frac{\sinh[(1 - i)y/d_v]}{\cosh[(1 - i)a/2d_v]} \exp(i\mu x - i\omega t) \cosh\frac{\beta a}{2}$$

where $k = \omega/c$, and the wavenumber $\mu = \sqrt{k^2 + \beta^2}$. Show that, in order that $u_x = 0$ at $y = \pm\frac{1}{2}a$,

$$\beta^2 a^2 \simeq k^2 a^2 \frac{(\gamma - 1)F(d_h/a) + F(d_v/a)}{1 - F(d_v/a)}$$

if βa is smaller than about $\frac{1}{2}$, where

$$F(x) = (1 + i)x \tanh\frac{1 - i}{2x} \rightarrow \begin{cases} (1 + i)x & x \ll 1 \\ 1 + \dfrac{i}{6x^2} & x \gg 1 \end{cases}$$

14 A plane sheet of cloth has flow resistance R such that the air velocity through it equals $1/R$ times the pressure difference on the two sides of the sheet. This sheet is suspended along the xz plane, the surface of a rigid wall is a distance behind it in the plane $y = -l$, with air in the interspace. Show that the reflection coefficient of Eqs. (6.3.5) and (6.3.13) is

$$C_r = \frac{R\cos\vartheta + i\rho c \cot(kl\cos\vartheta) - \rho c}{R\cos\vartheta + i\rho c \cot(kl\cos\vartheta) + \rho c}$$

Discuss the dependence of the fraction of incident energy which is absorbed on kl, R, and ϑ.

15 Discuss the similarities and differences between the C_r of Prob. 14 and the C_r for a covered wall of the sort described in Prob. 10, with arbitrary values of ρ_p, c_p, Φ, and panel thickness l. What are the advantages and disadvantages of the two sorts of acoustic covering?

16 A plane sound wave with angle of incidence $\vartheta = 60°$ is reflected from a plane wall of acoustic impedance z, and the sound pressure at the surface is measured in amplitude and phase. The experiment is then repeated, with the absorptive wall replaced by a perfectly rigid one. The mean-square pressure at the hard wall is found to be twice that for the absorbing wall, but the phase difference is zero. Use the results of Prob. 7 to compute the impedance of the absorbing wall. What fraction of the incident energy does it absorb?

17 Suppose the sheet of Prob. 14 is light enough so that its motion as a whole must be included. Show that the normal air velocity at its surface is $(1/R) + (i/\omega c)$ times the pressure drop across the sheet. Compute C_r and compare its dependence on kl, ϑ, and R with that of Prob. 14.

18 It is known that the bulk viscosity of nitrogen is about 0.8 times the ordinary shear viscosity. From this information and the discussion on page 299 determine the relaxation time of the rotational modes of motion in nitrogen. At 20°C and 1 atm the shear viscosity coefficient in nitrogen is $\mu = 1.78 \times 10^{-4}$ cgs units.

19 Carbon dioxide, CO_2, has its atoms along a line. The characteristic (Debye) temperature T_n (see page 296) for the transverse vibrational mode of motion of CO_2 is $T_n = 960°K$. (The longitudinal modes have considerably higher Debye temperatures.) Calculate and plot as functions of frequency the speed of sound and the attenuation per wavelength in CO_2 resulting from the transverse mode when the relaxation frequency $\frac{1}{2}\pi t_n$ of this mode at 20°C and 1 atm is known to be about 30,000 Hz. For a vibrational mode the specific heat function F_n of Eq. (6.4.43) is known from statistical mechanics to be

$$ F_n = \left(\frac{x}{e^x - 1} \right)^2 e^x \qquad \text{vibration} $$

where $x = T_n/T = hv/k_B$, $v = $ vibrational frequency
$\qquad h = $ Planck's constant
$\qquad k_B = $ Boltzmann's constant
$\qquad T = $ absolute temperature

20 With reference to the discussion leading to Eqs. (6.4.48) to (6.4.50), calculate and plot as functions of frequency the speed of sound and attenuation per wavelength in air at some different relative humidities. It is known that the average relaxation time of vibration in air depends on the relative humidity H as given by $v_n = (2\pi t_n)^{-1} \simeq 1 + 30H^2$, where H is measured in percent at a temperature of 20°C. (At 20°C the fraction of H_2O molecules in air is $H/4,320$.) In particular, consider $H = 18$ (corresponding to $v_n \simeq 10,000$ Hz), $H = 50$, and $H = 80$. On the basis of the corresponding curves discuss the role of water vapor on acoustic dispersion in the atmosphere as compared with the effects of viscosity and heat conduction.

CHAPTER 7

THE RADIATION OF SOUND

7.1 POINT SOURCES

Sound waves are generated by the vibration of any solid body in contact with the fluid medium, or by vibratory forces acting directly on the fluid, or by the violent motion of the fluid itself, as from a jet, or by oscillatory thermal effects, as would be produced by a modulated laser beam. In each case energy is transferred from the source to the fluid. The characteristics of the source determine the frequency and directional characteristics of the generated sound field; reciprocally, the directional properties of the field can be used to shed light on the nature of the source.

Of course, any motion of one portion of the fluid medium is transmitted to other parts of the medium; the motion in the frontal region of a sound wave at one instant may be regarded as the "source" of the subsequent wave motion, farther along in the medium. But in this case the energy of the wave is just being transmitted from one part of the medium to another; no new energy is being introduced. In the cases mentioned in the first paragraph, however, energy originally not acoustic is being changed to acoustic energy at the source, to be radiated outward and lost to the source. In the case of the jet, for example, the high-speed, organized motion of the jet is transformed by nonlinear processes into heat and acoustic energy, the sound being transmitted outward into the otherwise quiescent surrounding medium. From the point of view of acoustics, therefore, a *source* is a region of space, in contact with the fluid medium, where new acoustic energy is being generated, to be radiated outward as sound waves.

In this chapter we shall not go into the details of the nonlinear effects which sometimes generate the acoustic energy, but shall devote our attention to the generated sound waves as they move outward from the source. We shall assume that the fluid medium outside the source region is initially uniform and at rest and that the acoustic pressure generated outside the source is small enough so that the first-order equations of sound, derived in the preceding chapter, are valid in the region outside the source. We also shall concentrate on wave propagation in an infinite medium; sound propagation in ducts and rooms will be taken up in a later chapter.

Curvilinear coordinates

Since the wave motion near a source is seldom a plane wave, we must extend the form of our first-order equations (6.2.7) and (6.2.8) to more appropriate coordinate systems. The most useful, aside from the usual rectangular system x, y, z, are the spherical coordinates r, ϑ, φ, where

$$x = r \sin \vartheta \cos \varphi \qquad y = r \sin \vartheta \sin \varphi \qquad z = r \cos \vartheta$$

These are appropriate when the source region is concentrated in a small region of space. The circular cylindrical coordinates w, ϕ, z, where

$$x = w \cos \phi \qquad y = w \sin \phi \qquad z = z$$

are appropriate when the source region is extended in one dimension and concentrated in the other two. (When both spherical and cylindrical coordinates are used, we shall use w for the cylindrical radius; when spherical coordinates are not introduced, we shall use r for the cylindrical radius.)

As with the two-dimensional coordinates for the membrane, discussed in connection with Eq. (5.3.3), the differential operators, expressing the divergence, gradient, and Laplacian of the properties of the fluid, have different forms in the different coordinates. For example, the gradient $\operatorname{grad} f \equiv \boldsymbol{\nabla} f$ of a scalar quantity, such as the pressure or the velocity potential, which enters the basic equations

$$\rho \frac{\partial \mathbf{u}}{\partial t} = -\operatorname{grad} p \qquad \text{and} \qquad \mathbf{u} = -\operatorname{grad} \Psi$$

is a vector with components $\nabla_x f$, $\nabla_y f$, $\nabla_z f$ along the rectangular axes, components $\nabla_w f$, $\nabla_\phi f$, $\nabla_z f$ along the cylindrical coordinate axes, and components $\nabla_r f$, $\nabla_\vartheta f$, $\nabla_\varphi f$ along the radius from the origin, along a perpendicular to r in the rz plane and along a perpendicular to the rz plane, respectively. The expressions for these components are

$$\operatorname{grad}_x(\ ) = \frac{\partial}{\partial x} \qquad \operatorname{grad}_y(\ ) = \frac{\partial}{\partial y} \qquad \operatorname{grad}_z(\ ) = \frac{\partial}{\partial z}$$

(in rectangular coordinates)

$$\operatorname{grad}_w(\ ) = \frac{\partial}{\partial w} \qquad \operatorname{grad}_z(\ ) = \frac{\partial}{\partial z} \qquad \operatorname{grad}_\phi(\ ) = \frac{1}{w} \frac{\partial}{\partial \phi}$$

(in cylindrical coordinates)

$$\operatorname{grad}_r(\ ) = \frac{\partial}{\partial r} \qquad \operatorname{grad}_\vartheta(\ ) = \frac{1}{r} \frac{\partial}{\partial \vartheta} ; \qquad \operatorname{grad}_\varphi(\ ) = \frac{1}{r \sin \vartheta} \frac{\partial}{\partial \varphi}$$

(in spherical coordinates)

(7.1.1)

The second of the basic equations (6.2.7) involves the divergence of the vector **u**. In the three coordinate systems the divergence of vector **A** is

$$\frac{\partial A_x}{\partial x} + \frac{\partial A_y}{\partial y} + \frac{\partial A_z}{\partial z} \qquad \text{(in rectangular coordinates)}$$

$$\frac{1}{w}\frac{\partial}{\partial w}(wA_w) + \frac{1}{w}\frac{\partial}{\partial \phi}(A_\phi) + \frac{\partial}{\partial z}(A_z) \qquad \text{(in cylindrical coordinates)}$$

$$\frac{1}{r^2}\frac{\partial}{\partial r}(r^2 A_r) + \frac{1}{r\sin\vartheta}\frac{\partial}{\partial\vartheta}(A_\vartheta \sin\vartheta) + \frac{1}{r\sin\vartheta}\frac{\partial}{\partial\varphi}(A_\vartheta)$$

$$\text{(in spherical coordinates)}$$

(7.1.2)

where A_r, A_ϑ, A_φ are the components of **A** along the three axial directions in spherical coordinates, etc.

The Laplace operator $\nabla^2 \equiv \text{div grad}$ thus has three forms in the three different systems:

$$\nabla^2 = \begin{cases} \dfrac{\partial^2}{\partial x^2} + \dfrac{\partial^2}{\partial y^2} + \dfrac{\partial^2}{\partial z^2} & \text{(in rectangular coordinates)} \\[3mm] \dfrac{1}{w}\dfrac{\partial}{\partial w}\left(w\dfrac{\partial}{\partial w}\right) + \dfrac{1}{w^2}\dfrac{\partial^2}{\partial\phi^2} + \dfrac{\partial^2}{\partial z^2} & \text{(in cylindrical coordinates)} \\[3mm] \dfrac{1}{r^2}\dfrac{\partial}{\partial r}\left(r^2\dfrac{\partial}{\partial r}\right) + \dfrac{1}{r^2\sin\vartheta}\dfrac{\partial}{\partial\vartheta}\left(\sin\vartheta\dfrac{\partial}{\partial\vartheta}\right) \\[3mm] \qquad\qquad + \dfrac{1}{r^2\sin^2\vartheta}\dfrac{\partial^2}{\partial\varphi^2} & \text{(in spherical coordinates)} \end{cases}$$

(7.1.3)

The first-order equations of motion and the second-order equations for the acoustic energy density and intensity of Eqs. (6.2.5) are given in standard form in terms of these operators:

$$\rho\frac{\partial \mathbf{u}}{\partial t} = -\text{grad } p \qquad \mathbf{u} = -\text{grad } \Psi$$

$$\frac{\partial p}{\partial t} = -\text{div } \mathbf{u} \qquad p = \rho\frac{\partial \Psi}{\partial t}$$

$$\frac{\partial^2 p}{\partial t^2} = c^2\nabla^2 p \qquad \frac{\partial^2 \Psi}{\partial t^2} = c^2\nabla^2\Psi$$

$$w = \tfrac{1}{2}\rho\left(|\text{grad }\Psi|^2 + \frac{1}{c^2}\left|\frac{\partial \Psi}{\partial t}\right|^2\right) = \tfrac{1}{2}\rho\,|\mathbf{u}|^2 + \frac{1}{2}\frac{1}{\rho c^2}\,|p|^2$$

$$\mathbf{I} = -\rho\,\text{Re}\left(\frac{\partial \Psi^*}{\partial t}\,\text{grad }\Psi\right) = \text{Re}\,(\mathbf{u}^* p)$$

They can be expressed in different form for the different coordinate systems by using Eqs. (7.1.1) to (7.1.3).

In extension of the remarks on page 193 on the physical meaning of the Laplacian, we should point out that the Laplacian in three dimensions measures the negative of the concentration of a quantity. The value of $\nabla^2 p$ at a point is proportional to the difference between the average pressure near the point and the pressure right at the point. When this is negative, there is a concentration of pressure at the point; when it is positive, there is a lack of concentration there. The wave equation simply states, in a quantitative manner, that if there is a concentration of pressure at some point, the pressure there will tend to diminish.

The simple source

If the source region is compact and the generating motion has no preferred direction, it will produce a wave which spreads spherically outward. If the medium is infinite in extent, the waveform will depend on the distance r from the center of the source, but will be independent of the spherical angles ϑ and φ, as though the wave were generated by a sphere, centered at the origin, uniformly expanding and contracting. The wave equation in this case is

$$\frac{1}{r^2}\frac{\partial}{\partial r}\left(r^2\frac{\partial p}{\partial r}\right) = \frac{1}{c^2}\frac{\partial^2 p}{\partial t^2}$$

A general solution of this equation which is finite everywhere except at $r = 0$ is

$$p = \frac{1}{r}F(r - ct) + \frac{1}{r}f(r + ct)$$

consisting of a wave of arbitrary form going outward from the center and another wave focusing in on the center.

This solution, except for the factor $1/r$, is similar to Eq. (4.1.2) for waves on a string and to the equation for plane waves of sound given in Eq. (6.3.3). This means that spherical waves are more like plane waves than they are like cylindrical waves. Plane waves do not change shape or size as they travel; spherical waves do not change shape as they spread out, but they do diminish in amplitude because of the factor $1/r$; whereas cylindrical waves change both shape and size as they go outward, leaving a wake behind. Figure 5.8 shows that if a cylinder sends out a pulse of sound, the wave as it spreads out has a sharp beginning but no ending; the pressure at a point r from the axis is zero until a time $t = r/c$ after the start of the pulse, but the pressure does not settle back to its equilibrium value after the crest has gone by. With both plane and spherical waves the wave for a pulse has a sharp beginning and ending, the pressure settling back to equilibrium value after the pulse has gone past.

This behavior is another example of the general law (proved in books on the mathematics of wave motion) that waves in an odd number of dimensions (one, three, five, etc.) leave no wake behind them, whereas waves in an even number of dimensions (two, four, etc.) do leave wakes.

Spherical waves do resemble circular waves on a membrane, however, in that they have infinite values at $r = 0$. As we have seen on page 196, this simply means that the size of the source must be taken into account; every actual source of sound has a finite size, so that the wave motion never extends in to $r = 0$ where it would be infinite.

Suppose that a sphere of average radius a is expanding and contracting so that the radial velocity of its surface is everywhere the same function of time $U(t)$. The rate of flow of air away from the surface of the sphere, in every direction, is $4\pi a^2 U(t) = S(t)$. To obtain an expression for the pressure wave radiated from the sphere, we write Newton's equation as $\rho(\partial u_r/\partial t) = -(\partial p/\partial r)$. If p is chosen to be an arbitrary outgoing wave $p = P(r - ct)/r$, the requirement at the surface of the sphere,

$$\frac{1}{r}\frac{\partial P}{\partial r} - \frac{P}{r^2} = -\frac{\rho}{4\pi a^2}\frac{\partial S}{\partial t} \qquad \text{at } r = a$$

serves to determine the shape of the wave P.

If the vibrating sphere is very small (more specifically, if a is small compared with the wavelength of the sound radiated), the sphere is called a *simple source* of sound. In this case P/r is much larger than dP/dr at $r = a$, and $P \simeq \dfrac{\rho}{4\pi}\dfrac{dS}{dt}$ at $r = a$. The pressure wave at a distance r from the center of the simple source is therefore

$$p \simeq \frac{\rho}{4\pi r} S'\left(t - \frac{r}{c}\right) \tag{7.1.4}$$

where $S'(z) = (d/dz)S(z)$. The function S gives the instantaneous value of the total flow of air away from the center of the source. The pressure at the distance r is proportional to the rate of change of this flow a time r/c earlier. If, for example, the flow outward started suddenly, being zero for $t < 0$ and 1 for $t > 0$, the generated pressure wave would be a pulse, $(\rho/4\pi r)\,\delta[t - (r/c)]$; a steady flow outward produces no sound; only a change in flow can be heard.

The periodic simple source

In the special case when the source is simple-harmonic, so that the flow outward from the origin is $S_\omega e^{-i\omega t}$, the resulting wave motion, its mean-square energy and intensity [Eqs. (6.2.21)], and total power radiated Π are

$$p = p_s(\omega)e^{-i\omega t} = -\frac{ik\rho c}{4\pi r} S_\omega e^{ik(r-ct)} \qquad k = \frac{\omega}{c} = \frac{2\pi}{\lambda}$$

$$u_r = \frac{-1}{4\pi r^2}(ikr - 1)S_\omega e^{ik(r-ct)} = \frac{p}{\rho c}\left(1 + \frac{i\lambda}{2\pi r}\right)$$

$$w = \rho\left(\frac{1}{4\pi r^2}\right)^2 |S_\omega|^2 [(kr)^2 + \tfrac{1}{2}] = \rho\left(\frac{1}{2\lambda r}\right)^2 |S_\omega|^2 \left[1 + \frac{1}{2}\left(\frac{\lambda}{2\pi r}\right)^2\right] \quad (7.1.5)$$

$$I_r = \rho c\left(\frac{1}{2\lambda r}\right)^2 |S_\omega|^2 = \frac{|p|^2}{\rho c} \qquad I_\vartheta = I_\varphi = 0$$

$$\Pi = (4\pi r^2)I_r = \rho c\,\frac{\pi}{\lambda^2}|S_\omega|^2 = \frac{\rho\omega^2}{4\pi c}|S_\omega|^2$$

The pressure amplitude is inversely proportional to the distance r from the center of the sphere; the fluid velocity is entirely radial, being in phase with the pressure at distances large compared with λ and out of phase in the region $r < \lambda$. The time-average intensity $\mathbf{I}$ is also radial, proportional to the square of the pressure amplitude, and thus independent of time; its divergence is zero except at $r = 0$. The total outflow of energy per second from the source is $\pi\rho c(|S_\omega|/\lambda)^2$, independent of r; thus, for a given source strength S_ω, the power radiated is proportional to the square of the frequency.

At distances r from the source which are large compared with the wavelength, the relationships between p, $\mathbf{u}$, and $\mathbf{I}$ approach the simple ones given in Eqs. (6.3.3) for a plane wave, $u_r = p/\rho c$, $w_r = |p|^2/\rho c^2$, and $I_r = |p|^2/\rho c$. The sound field in this region is called the *far field*. Here the fluid velocity is in phase with the pressure, so that all the energy density is radiant, moving outward with the speed of sound, since $I_r = cw_r$.

In contrast, the *near field*, in the region $r \ll \lambda$, has a large velocity component out of phase with the pressure, which is responsible for the presence of additional mean-square energy density $w_x = \frac{1}{2}\rho(|S|/4\pi r^2)^2$, reactive energy which does not radiate outward. The radiant energy in a shell of unit thickness around the source is $\rho k^2 |S_\omega|^2/4\pi$, independent of the radius of the shell. The reactive energy stays close to the source; the total amount outside a radius r_m is $\dfrac{\rho |S_\omega|^2}{8\pi}\dfrac{1}{r_m}$, inversely proportional to r_m but independent of ω.

Two additional properties of the sound field from a simple source will be of utility in our later discussions. The first is the obvious fact that the source need not be at the origin of coordinates. If the *source point* is at $\mathbf{r}_0 = (x_0, y_0, z_0)$, then the pressure at the *measurement point* $\mathbf{r} = (x, y, z)$ is

$$p_\omega(\mathbf{r}\,|\,\mathbf{r}_0)e^{-i\omega t} = -ik\rho c S_\omega g_\omega(\mathbf{r}\,|\,\mathbf{r}_0)e^{-i\omega t} \qquad (7.1.6)$$

where

$$g_\omega(\mathbf{r}\,|\,\mathbf{r}_0) = \frac{1}{4\pi R}e^{ikR}$$

and

$$R^2 = |\mathbf{r} - \mathbf{r}_0|^2 = (x - x_0)^2 + (y - y_0)^2 + (z - z_0)^2$$

The other quantities in Eq. (7.1.5) can be obtained by substituting for r the distance R between source and measurement point. The quantity $g_\omega(\mathbf{r} \mid \mathbf{r}_0)$ is the Green's function for an infinite medium [Eq. (5.1.4)]; we shall discuss its properties and uses later.

The second point to note about the field from a simple source is that it is symmetric with respect to interchange of $\mathbf{r}$ and $\mathbf{r}_0$. In other words, the pressure at the measurement point $\mathbf{r}$, caused by a source at $\mathbf{r}_0$, is equal to the pressure which would be measured at $\mathbf{r}_0$ if the source were placed at $\mathbf{r}$. The Green's function is said to conform to the *principle of reciprocity*.

Dipole and quadrupole sources

To generate the sound field we have just described requires that we, somehow, periodically introduce and withdraw fluid from a small region of space, the source point, or else locate there a sphere which can alternately expand and contract. Although this condition is approximately fulfilled by many musical instruments and some loudspeakers (as we shall see later), in many other cases sound is produced by moving a portion of the fluid back and forth, no new fluid being introduced. The simplest arrangement of this sort can be simulated by two simple sources, close together, opposite in sign and equal in magnitude. In this way fluid is being sucked in by one as it is being ejected by the other, as though it were simply being moved from the one source point to the other. Such a source is appropriately called a *dipole source*.

The resulting radiation field can be derived from that for a simple source. For the simple-harmonic case, if the source of strength S_ω is at $\frac{1}{2}\mathbf{d}$ and the one of strength $-S_\omega$ is at $-\frac{1}{2}\mathbf{d}$, and if $\mathbf{d}$, the vector distance between the two sources, is very small compared with λ, then, from Eq. (7.1.6) by Taylor's theorem,

$$p = p_d(\omega)e^{-i\omega t} = -ik\rho c S_\omega[g_\omega(\mathbf{r} \mid \tfrac{1}{2}\mathbf{d}) - g_\omega(\mathbf{r} \mid -\tfrac{1}{2}\mathbf{d})]e^{-i\omega t}$$

$$= -ik\rho c D_\omega \{\mathrm{grad}_0 [g_\omega(\mathbf{r} \mid \mathbf{r}_0)]\}_{r_0=0} e^{-i\omega t}$$

$$= -k^2 D_\omega \frac{\rho c}{4\pi r} \cos \vartheta \left(1 + \frac{i}{kr}\right)e^{ikr-i\omega t}$$

$$u_r = -\frac{k^2 D_\omega}{4\pi r} \cos \vartheta \left(1 + \frac{2i}{kr} - \frac{2}{k^2 r^2}\right)e^{ikr-i\omega t}$$

$$u_\vartheta = i\frac{k D_\omega}{4\pi r^2} \sin \vartheta \left(1 + \frac{i}{kr}\right)e^{ikr-i\omega t} \qquad (7.1.7)$$

$$w = \rho\left(\frac{k^2 \mid D_\omega\mid}{4\pi r}\right)^2 \left[\cos^2 \vartheta + \frac{1}{2}\left(\frac{1}{kr}\right)^2 + \frac{1}{2}\left(\frac{1}{kr}\right)^4 (1 + 3\cos^2 \vartheta)\right]$$

$$I_r = \rho c\left(\frac{k^2 \mid D_\omega\mid}{4\pi r}\right)^2 \cos^2 \vartheta \qquad I_\vartheta = I_\varphi = 0$$

$$\Pi = 2\pi r^2 \int_0^\pi I_r \sin \vartheta \, d\vartheta = \rho c \frac{4\pi^2}{3\lambda^4} \mid D_\omega\mid^2 = \frac{\rho \omega^4}{12\pi c^3} \mid D_\omega\mid^2$$

where $\mathbf{D}_\omega$, the vector *dipole strength*, equals $S_\omega\mathbf{d}$, the strength of the component simple sources times the vector distance between them, and ϑ is the angle between $\mathbf{D}_\omega$ and the radius vector $\mathbf{r}$ to the measurement point.   As would be expected, this field has directionality, represented by the dependence on angle ϑ.

Even the far field from a dipole has directionality; when $kr \gg 1$, we have $p \rightarrow -\rho c(k^2 D_\omega/4\pi r) \cos \vartheta e^{ikr-i\omega t}$, the amplitude varying inversely with r, as with the far field from the simple source, but having its maximum along the dipole axis and dropping to zero at $\vartheta = \frac{1}{2}\pi$, in the plane through the dipole, normal to $\mathbf{D}_\omega$.   However, the far field still has the simple relationship valid for a plane wave; to the first order in $1/kr$, $u_r \rightarrow p/\rho c$ and $u_\vartheta \rightarrow u_\varphi \rightarrow 0$; to the second order, $w \rightarrow |p|^2/\rho c^2 = \rho |u_r|^2$ and $I_r \rightarrow wc$.   At these large distances the energy density is all radiant, traveling with velocity c.   The total power radiated is more strongly dependent on frequency than is that from the simple source, being proportional to ω^4.

The near field is more complicated in form than that for the simple source; the fluid motion is not purely radial, and the reactive energy is relatively larger and more concentrated near the origin.   The radiant energy in a shell of unit thickness is $\rho k^4 |D_\omega|^2/12\pi$, independent of r but proportional to ω^4. The total reactive energy outside radius r_m is $\rho |D_\omega|^2/12\pi r_m{}^3$ when $kr_m \ll 1$, independent of ω but inversely proportional to $r_m{}^3$.

If the dipole is centered at the point $\mathbf{r}_0 = (x_0, y_0, z_0)$, instead of at the origin, and if the axis of spherical coordinates is not pointed along the dipole axis, the dipole vector $\mathbf{D}_\omega$ has components D_x, D_y, D_z, and the pressure wave at point $\mathbf{r}$ can be written in terms of the radius vector $\mathbf{R} = \mathbf{r} - \mathbf{r}_0$ and the spherical angles ϑ_R, φ_R of vector $\mathbf{R}$.

$$p = -ik\rho c\mathbf{D}_\omega \cdot \mathrm{grad}_0 \, [g_\omega(\mathbf{r} \,|\, \mathbf{r}_0)]e^{-i\omega t} = -ik\rho c\mathbf{D}_\omega \cdot \mathbf{g}'_\omega(\mathbf{r} \,|\, \mathbf{r}_0)e^{-i\omega t} \quad (7.1.8)$$

where

$$g'_x = \sin \vartheta_R \cos \varphi_R |g'_\omega| \qquad g'_y = \sin \vartheta_R \sin \varphi_R |g'_\omega| \qquad g'_z = \cos \vartheta_R |g'_\omega|$$

$$|g'_\omega| = -\frac{ik}{4\pi R}\left(1 + \frac{i}{kR}\right)e^{ikR}$$

There can also be more complex point sources, made up of assemblages of dipoles.   For example, there can be a combination of four simple sources, two of strength $+Q$, placed at the points $x = \frac{1}{2}d$, $y = \frac{1}{2}d$, $z = 0$, and $(-\frac{1}{2}d, -\frac{1}{2}d, 0)$ and two of strength $-Q$ at the points $(+\frac{1}{2}d, -\frac{1}{2}d, 0)$ and $(-\frac{1}{2}d, +\frac{1}{2}d, 0)$, which correspond to two dipoles of strength Qd, one pointed along the positive x axis with its center at $(0, +\frac{1}{2}d, 0)$, the other pointed along the negative x axis with its center at $(0, -\frac{1}{2}d, 0)$.   Extension of the argument resulting in Eq. (7.1.7) shows that if $d \ll \lambda$,

$$p = -ik\rho c Q_{xy}\left[\frac{\partial^2}{\partial x_0 \, \partial y_0} g_\omega(\mathbf{r} \,|\, \mathbf{r}_0)\right]_{r_0=0} e^{-i\omega t}$$

$$= i\rho c k^3 Q_{xy}\frac{xy}{4\pi r^3}\left[1 + \frac{3i}{kr} - 3\left(\frac{1}{kr}\right)^2\right]e^{ikr-i\omega t} \quad (7.1.9)$$

where $Q_{xy} = Qd^2$. Such a combination of sources is called a *tesseral quadrupole* of strength Q_{xy} [it has dimensions of (length)^5 per time]. There can be three such arrangements, $Q_{xy} = Q_{yx}$, $Q_{xz} = Q_{zx}$, and $Q_{yz} = Q_{zy}$, lying in the three rectangular coordinate planes and having similar expressions for p, except for the factors xy, xz, or yz and for the magnitude of the strengths.

There can also be a pair of dipoles, pointed along the x axis and spaced along the x axis, which would correspond to a simple source of strength $-2S$ at the origin and two of strength S, placed at $(+\tfrac{1}{2}d, 0, 0)$ and $(-\tfrac{1}{2}d, 0, 0)$. This would produce a pressure wave of the form

$$p = -ik\rho c Q_{xx}\left[\frac{\partial^2}{\partial x_0^2}g_\omega(\mathbf{r}\mid\mathbf{r}_0)\right]_{r_0=0}e^{-i\omega t}$$

$$= i\rho c k^3 Q_{xx}\frac{1}{4\pi r}\left[\left(\frac{x}{r}\right)^2 + \frac{3x^2 - r^2}{r^2}\left(\frac{i}{kr} - \frac{1}{k^2 r^2}\right)\right]e^{ikr-i\omega t} \quad (7.1.10)$$

where $Q_{xx} = Sd^2$. This is called an axial *quadrupole source* of strength Q_{xx}. Similar arrangements along the other axes would produce Q_{yy} and Q_{zz} with similar pressure waves, using y^2 and z^2 instead of x^2.

We thus see that the possible quadrupole source strengths can be arranged in tensor form.

$$\mathfrak{Q}_\omega = \begin{pmatrix} Q_{xx} & Q_{xy} & Q_{xz} \\ Q_{yx} & Q_{yy} & Q_{yz} \\ Q_{zx} & Q_{zy} & Q_{zz} \end{pmatrix}$$

There are only six independent quadrupole components, however, for the tensor is symmetric, Q_{xy} being equal to Q_{yx}, etc. As a matter of fact, there are only five distinct quadrupole waves, for a sum of three axial quadrupoles Q_{xx}, Q_{yy}, Q_{zz}, of equal magnitude, produces (outside the source region) a wave which is indistinguishable from a simple-source wave. This can be seen by adding three expressions like that of Eq. (7.1.10) and remembering that $x^2 + y^2 + z^2 = r^2$, or else by recognizing that

$$\left(\frac{\partial^2}{\partial x_0^2} + \frac{\partial^2}{\partial y_0^2} + \frac{\partial^2}{\partial z_0^2}\right)g_k(\mathbf{r}\mid\mathbf{r}_0) = \nabla_0^2 g_k = -k^2 g_k(\mathbf{r}\mid\mathbf{r}_0)$$

from the wave equation. Of course, g_k itself is the spatial factor for the wave from a simple source, as given by Eq. (7.1.6). Because of the extra k^2 factor, the magnitude of the simple-source wave coupled to the quadrupole wave is small if the dimensions of the quadrupole are small.

There can be poles of still higher order, though they are of less practical interest. Considering the simple source (monopole) to be a pole of zero order, a dipole to have order $m = 1$, and a quadrupole to be a pole of order $m = 2$, then a pole of order three would be called an octupole, and so on, the number of simple sources required to generate a wave of order m being 2^m.

Source impedances

To produce any of these elementary waveforms, some mechanism, of dimensions very small compared with the wavelength, must be pushing the fluid about rhythmically. To produce a simple source wave, S_ω volume units of fluid is pushed out in all directions from the source, and then back again; for the dipole the fluid is pushed back and forth along some axis; for the quadrupole there must be two opposed motions. To produce such motion requires a driving force, acting somehow on the fluid, capable of contributing energy to the wave. The exact details will depend on the particular mechanism which is producing the wave, but for each elementary type we can calculate, from the near field, a typical driving pressure which could produce such a wave. The effective driving force and responding motion can be related in several ways; an rms pressure and the corresponding flux (volume per second) or an effective force (pressure times area) and a corresponding velocity. In one case the ratio is an analogous impedance [Eq. (9.1.10)]; in the other case it is a mechanical impedance [Eq. (2.3.2)]; in either case the real part of the product (pressure or force) times flux or velocity will equal the time average of the power radiated.

For example, in the simple source the flow is uniformly radial, so that, at the surface of the sphere of radius $a \ll \lambda$, just enclosing the generating mechanism, the radial flux is the source strength S_ω times the phase factor $e^{ik(a-ct)}$, and the radial velocity U_s is equal to this flux divided by the surface area $4\pi a^2$. The driving force F_s at $r = a$ required to produce this radial velocity is the pressure P_s at $r = a$ times the surface area $4\pi a^2$. Thus, from Eq. (7.1.5), the simple-source driving force, velocity, mechanical impedance, and total power radiated are

$$F_s = -i\rho c k a S_\omega e^{ik(a-ct)} \qquad U_s = \frac{S_\omega}{4\pi a^2}(1 - ika)e^{ik(a-ct)}$$

$$Z_s = \frac{F_s}{U_s} = -ik\rho c(4\pi a^3)(1 - ika)^{-1}$$

$$\simeq -i\omega(4\pi\rho a^3) + \rho c(4\pi a^2)(ka)^2 = -iX_s + R_s$$

(7.1.11)

$$\Pi_s = \mathrm{Re}\,(F_s U_s^*) = R_s\,|U_s|^2 = \frac{\rho\omega^2}{4\pi c}\,|S_\omega|^2$$

where the approximate values are, for $ka \ll 1$, appropriate for these elementary sources. This equivalent mechanical impedance has a reactive term equivalent to the mass load three times the mass of a sphere of radius a filled with fluid. The mechanical resistance is $4\pi a^2$ times ρc, the characteristic impedance of the medium, times the small quantity $(ka)^2$, but this small term cannot be neglected, for it is the real part of the impedance and thus is responsible for the power radiated.

Alternatively, we could have used the pressure P_s and the flux S_s and obtained an analogous impedance Z_{as}.

$$P_s = -\frac{i\rho cka}{4\pi a^2} S_\omega e^{ik(a-ct)} \qquad S_s = S_\omega(1 - ika)e^{ik(a-ct)}$$

$$Z_{as} = \frac{P_s}{S_s} = -\frac{ik\rho c}{4\pi a}(1 - ika)^{-1} \simeq -i\omega\frac{\rho}{4\pi a} + \rho c\frac{k^2}{4\pi}$$

The power radiated, $\mathrm{Re}\,(P_s S_s^*)$, is equal to the Π_s given in Eq. (7.1.11), as of course it must.

In the case of the dipole, the flow at the surface $r = a$ is back and forth in the direction of the dipole axis, so that both radial velocity and pressure are proportional to the cosine of the angle ϑ between $\mathbf{r}$ and $\mathbf{D}_\omega$. The force required to push the sphere of radius a to and fro is the integral of the component of the surface force along $\mathbf{D}_\omega$, over the sphere's surface,

$$\mathbf{F}_d = -ik\rho c\mathbf{D}_\omega(1 - ika)\,\tfrac{1}{2}\int_0^\pi \cos^2\vartheta\,\sin\vartheta\,d\vartheta\,e^{ik(a-ct)}$$

$$= -\tfrac{1}{3}ik\rho c\mathbf{D}_\omega(1 - ika)e^{ik(a-ct)}$$

and the corresponding velocity $\mathbf{U}_d$ is the maximum radial velocity at the sphere's surface (for $\vartheta = 0$). It thus is the velocity the sphere would have if it were producing the dipole by moving to and fro bodily. If the dipole is at an angle with respect to the coordinate axes, with components D_x, D_y, D_z, the pressure wave is

$$p_d = -ik\rho c\frac{D_\omega}{4\pi a^2}(1 - ika)\left(\frac{x}{r}D_x + \frac{y}{r}D_y + \frac{z}{r}D_z\right)$$

In this case we find the component along D_ω of the force acting on the sphere by multiplying p_d by the directionality factor

$$\frac{1}{D_\omega}\left(\frac{x}{r}D_x + \frac{y}{r}D_y + \frac{z}{r}D_z\right)$$

and integrating over the surface of the sphere. Since

$$D_\omega = (D_x^2 + D_y^2 + D_z^2)^{\frac{1}{2}}$$

is the magnitude of the dipole strength, we end up with the same result as given above.

Similarly with the quadrupole source. We define the magnitude of such a source as $Q_\omega = [(Q_{xy} + Q_{yx})^2 + (Q_{xz} + Q_{zx})^2 + (Q_{yz} + Q_{zy})^2 + \frac{1}{3}(2Q_{zz} - Q_{xx} - Q_{yy})^2 + (Q_{xx} - Q_{yy})^2]$ so as to combine the axial terms to eliminate the simple source part [see also Eq. (7.2.24)]. We then multiply the expression, obtained from Eqs. (7.1.9) and (7.1.10), for the total quadrupole pressure at $r = a$,

$$-3\frac{ik\rho c}{4\pi a^3}\left\{(1 - ika - \tfrac{1}{3}k^2a^2)\left[(Q_{xy} + Q_{yx})\frac{xy}{a^2} + \cdots + \tfrac{1}{2}(Q_{xx} - Q_{yy})\frac{x^2 - y^2}{a^2}\right.\right.$$

$$\left.\left. + (Q_{zz} - \tfrac{1}{2}Q_{xx} - \tfrac{1}{2}Q_{yy})\frac{z^2 - \tfrac{1}{3}a^2}{a^2}\right] + \tfrac{1}{3}k^2a^2(Q_{xx} + Q_{yy} + Q_{zz})\right\}e^{ik(a-ct)}$$

(where $x^2 + y^2 + z^2 = a^2$) by the "directionality factor"

$$\frac{1}{Q_\omega}\left[(Q_{xy} + Q_{yx})\frac{xy}{a^2} + \cdots + \tfrac{1}{2}(Q_{xx} - Q_{yy})\frac{x^2 - y^2}{a^2}\right.$$

$$\left. + (Q_{zz} - \tfrac{1}{2}Q_{xx} - \tfrac{1}{2}Q_{yy})\frac{z^2 - \tfrac{1}{3}a^2}{a^2}\right]$$

and integrate over the surface of the sphere, obtaining the "component of the applied force" in the direction of the quadrupole. The cross terms integrate to zero; only the squared terms remain, giving a factor $Q_\omega{}^2/Q_\omega = Q_\omega$. The final results for the effective force, velocity, impedance, and power radiated, for dipole and quadrupole, are

$$F_d = -\tfrac{1}{3}ik\rho c D_\omega(1 - ika)e^{ik(a-ct)}$$

$$U_d = \frac{D_\omega}{2\pi a^3}(1 - ika - \tfrac{1}{2}k^2a^2)e^{ik(a-ct)}$$

$$Z_d \simeq -i\omega\rho\frac{2\pi a^3}{3}[1 + \tfrac{1}{2}(ka)^2] + \rho c(4\pi a^2)\frac{(ka)^4}{12}$$

$$\Pi_d = R_d|U_d|^2 \simeq \frac{\rho\omega^4}{12\pi c^3}|D_\omega|^2$$

$$F_q = -i\frac{k\omega c}{5a}Q_\omega(1 - ika - \tfrac{1}{3}k^2a^2)e^{ik(a-ct)} \qquad (7.1.12)$$

$$U_q = \frac{9}{4\pi a^4}Q_\omega(1 - ika - \tfrac{4}{9}k^2a^2) + \tfrac{1}{9}ik^3a^3)e^{ik(a-ct)}$$

$$Z_q \simeq -i\omega\rho\frac{4\pi a^3}{45}[1 + \tfrac{1}{9}(ka)^2 + \tfrac{4}{81}(ka)^4] + \rho c(4\pi a^2)\frac{(ka)^6}{1{,}215}$$

$$\Pi_q \simeq \frac{\rho\omega^6}{60\pi c^5}|Q_\omega|^2$$

The reactive load for the monopole of Eq. (7.1.11) is the mass of a volume $4\pi a^3$ of the fluid, three times the volume of the sphere of radius a; that for the dipole is $\tfrac{1}{2}$ of the sphere's mass, and that of the quadrupole is $\tfrac{1}{15}$ of it. For a pole of order m, the reactive load is $3/[(m + 1) 1 \cdot 3 \cdot 5 \cdots (2m + 1)]$ of the sphere's mass, for $ka \ll 1$ (the order of the monopole is $m = 0$, that for the dipole is $m = 1$, etc.). The resistive load, the first real term in the expansion of Z in powers of ka, diminishes much more rapidly than this; it is

$[\rho c(4\pi a^2)][(ka)^{2m}/(m+1)1\cdot 3\cdot 5\cdots(2m+1)]$.   These expressions are valid when the multipoles radiate into unbounded space.   If the fluid is confined within boundaries, which are a finite distance away, both the waveform and the value of the acoustic impedance are changed, as will be shown later.

Radiation from nonperiodic multipoles

Now that we have worked out the radiation from simple-harmonic multipole sources, we can use the Fourier transform to calculate the radiation from multipoles with strengths which vary in an arbitrary way with time, of the sort already mentioned for the simple source.   We consider the periodic sources just discussed, with strengths S_ω, $\mathbf{D}_\omega$, etc., to be the components of the general source which has frequency $\omega/2\pi$, and we add all the components of different frequencies together to make up the general wave.   In other words, we assume that the strengths S_ω, $\mathbf{D}_\omega$, etc., are the Fourier transforms of the general time-varying sources of strengths $S(t)$, $\mathbf{D}(t)$, $\mathfrak{Q}(t)$, according to the procedure of Eqs. (1.3.16).

$$S_\omega = \frac{1}{2\pi}\int_{-\infty}^{\infty} S(t)e^{i\omega t}\,dt \qquad \mathbf{D}_\omega = \frac{1}{2\pi}\int_{-\infty}^{\infty}\mathbf{D}(t)e^{i\omega t}\,dt \qquad \cdots$$

Having calculated the radiation from the component simple-harmonic sources, we can then obtain the complete field by taking the inverse Fourier transform of the spatial part $p(\omega)$,

$$p_s(\mathbf{r},t) = \int_{-\infty}^{\infty} p_s(\omega)e^{-i\omega t}\,d\omega \qquad p_d(\mathbf{r},t) = \int_{-\infty}^{\infty} p_d(\omega)e^{-i\omega t}\,d\omega \qquad \cdots$$

To carry this out we note that, if $F(\omega)$ is the Fourier transform of $f(t)$, as given in Eqs. (1.3.16), then the inverse transform of $-i\omega F(\omega)$ is the time derivative $f'(t)$ of f, $-\omega^2 F(\omega)$ transforms to the second time derivative $f''(t)$, and so on. From Eqs. (7.1.5) we can thus obtain

$$p_s(\mathbf{r},t) = \frac{\rho}{4\pi r}\int_{-\infty}^{\infty}(-i\omega)S_\omega\exp\left(-i\omega t + i\frac{\omega}{c}r\right)d\omega = \frac{\rho}{4\pi r}S'\left(t-\frac{r}{c}\right)$$

which corresponds to Eq. (7.1.4) for the nonperiodic simple source.

Similarly, from Eqs. (7.1.7) to (7.1.10), we can compute the pressure waves from the nonperiodic dipole and quadrupole sources.

$$p_d(\mathbf{r},t) = \frac{\rho}{4\pi cr}\frac{\mathbf{r}}{r}\cdot\left[\mathbf{D}''\left(t-\frac{r}{c}\right)+\frac{c}{r}\mathbf{D}'\left(t-\frac{r}{c}\right)\right]$$

$$\tag{7.1.13}$$

$$p_q(\mathbf{r},t) = \frac{\rho}{4\pi c^2 r}\left\{\frac{\mathbf{r}}{r}\cdot\left[\mathfrak{Q}'''\left(t-\frac{r}{c}\right)+\frac{3c}{r}\mathfrak{Q}''\left(t-\frac{r}{c}\right)+3\left(\frac{c}{r}\right)^2\mathfrak{Q}'\left(t-\frac{r}{c}\right)\right]\cdot\frac{\mathbf{r}}{r}\right.$$
$$\left.-\left|\frac{c}{r}\mathfrak{Q}''\left(t-\frac{r}{c}\right)+\left(\frac{c}{r}\right)^2\mathfrak{Q}'\left(t-\frac{r}{c}\right)\right|\right\}$$

where vector $\mathbf{r}$ has the components (x,y,z), so that $(\mathbf{r}/r) \cdot \mathbf{D}''$, for example, is $(x/r)D''_x + (y/r)D''_y + (z/r)D''_z$, and

$$\frac{\mathbf{r}}{r} \cdot \mathbf{Q} \cdot \frac{\mathbf{r}}{r} = \left[\left(\frac{xx}{r^2}\right)Q_{xx} + \left(\frac{xy}{r^2}\right)Q_{xy} + \cdots + \left(\frac{yz}{r^2}\right)Q_{yz} + \left(\frac{zz}{r^2}\right)Q_{zz} \right]$$

Q_{xx}, Q_{xy}, etc., being the elements of the matrix $\mathbf{Q}$ of quadrupole strengths. The symbol $|\mathbf{Q}''|$ around matrix $\mathbf{Q}''$ denotes the sum of the diagonal elements, $Q''_{xx} + Q''_{yy} + Q''_{zz}$, etc.

Thus we see that all the waves are propagated with velocity c and that, far from the source, the wave is propagated with unchanged shape but with amplitude inversely proportional to r. The shape of the far field from the simple source is proportional to the first time derivative of the source strength $S(t)$, as we noted earlier. The shape of the far field from the dipole is proportional to the second time derivative of $\mathbf{D}(t)$, and that from the quadrupole is proportional to the third time derivative of the components of the matrix $\mathbf{Q}(t)$. The near field for the dipole and quadrupole has terms proportional to the first derivative of the source strengths. The dipole field is oriented along the direction of the dipole, and the far field from the quadrupole has the angular symmetry of $\mathbf{Q}$, but the near field has an additional isotropic term, coming from the diagonal terms. Another aspect of the property of the quadrupole field was mentioned on page 314.

The Green's function

Before applying these results to specific situations of interest, we should say more about the function $g_\omega(\mathbf{r} \mid \mathbf{r}_0)$ of Eq. (7.1.6). In the first place, it is a solution, for the unbounded medium, of the equation

$$\nabla^2 g_\omega(\mathbf{r} \mid \mathbf{r}_0) + k^2 g_\omega(\mathbf{r} \mid \mathbf{r}_0) = -\delta(\mathbf{r} - \mathbf{r}_0) \qquad (7.1.14)$$

where $k = \omega/c$, and $\delta(\mathbf{r} - \mathbf{r}_0) = \delta(x - x_0)\,\delta(y - y_0)\,\delta(z - z_0)$ is the delta function for three dimensions [Eq. (1.3.24)]. Thus g_ω is the spatial factor for a wave from a unit, simple-harmonic, point source at $\mathbf{r}_0$. Its form, as we have mentioned earlier, is $(1/4\pi R)e^{ikR}$, where $R = |\mathbf{r} - \mathbf{r}_0|$.

To verify this we first note that g_ω is a solution of the homogeneous equation $\nabla^2 g + k^2 g = 0$ everywhere except at the point $\mathbf{r}_0$. Next we integrate both sides of Eq. (7.1.14) (in the coordinates x, y, z of $\mathbf{r}$) over the volume of a sphere of very small radius a, centered at $\mathbf{r}_0$. The integral of the right-hand side, because of the properties of the delta function, is -1. The integral of the $k^2 g_\omega$ term goes to zero as $a^3 \to 0$. The integral of the $\nabla^2 g_\omega$ term can be obtained by remembering that $\nabla^2 g_\omega$ is the divergence of the gradient of g_ω. Now the volume integral of the divergence of a vector is equal to the surface integral of the outward, normal component of the vector, over the surface bounding the volume. Therefore the integral of $\nabla^2 g_\omega$ over the volume of a sphere, with $\mathbf{r}_0$ the center, is equal to the radial gradient of g_ω, integrated over the surface of the sphere.   This radial gradient, with center at r_0, is $\partial g_\omega / \partial R = [-(1/4\pi R^2) + (ik/4\pi R)]e^{ikR}$, which is constant in value over the spherical surface $R = a$.   Thus the volume integral of $\nabla^2 g_\omega$ over the interior of the sphere is equal to $(\partial g_\omega / \partial R)_{R=a}$ times the surface area $4\pi a^2$ of the sphere. When $a \to 0$ this approaches the value -1, which thus equals the right-hand side, verifying the form of g_ω.

We should note that the most general solution of Eq. (7.1.14) is $g_\omega(r \mid r_0)$ plus any solution $\chi(r)$ of the homogeneous equation $\nabla^2 \chi + k^2 \chi = 0$.   We have chosen the particular form g_ω as being appropriate for outgoing radiation in an unbounded medium.   If the medium is bounded by some surface on which boundary conditions are imposed, we may have to use the function

$$G_\omega(r \mid r_0) = g_\omega(r \mid r_0) + \chi(r) \qquad (7.1.15)$$

in order to satisfy these boundary conditions.   Thus G_ω is a solution of Eq. (7.1.14) which satisfies appropriate boundary conditions along boundaries within finite distance of the source; if no such boundaries exist, $\chi = 0$.   We shall later show that this more general Green's function G_ω also satisfies the principle of reciprocity, $G_\omega(r \mid r_0) = G_\omega(r_0 \mid r)$.

To demonstrate the utility of the Green's function G_ω, we consider the following problem: We wish to solve the equation

$$\nabla^2 p_\omega + k^2 p_\omega = -f_\omega(r) \qquad (7.1.16)$$

for the spatial factor of a pressure wave from a simple-harmonic, distributed source, represented by the function $f_\omega(r)e^{-i\omega t}$.   We multiply Eq. (7.1.14) for G_ω by p and subtract from it G_ω times the equation for p, obtaining

$$G_\omega(r \mid r_0) \, \nabla^2 p_\omega(r) - p_\omega(r) \, \nabla^2 G_\omega(r \mid r_0) = p_\omega(r) \, \delta(r - r_0) - G_\omega(r \mid r_0)f_\omega(r)$$

Now interchange r and r_0, utilize the symmetry of G_ω and $\delta(r - r_0)$ with respect to this interchange, and integrate with respect to x_0, y_0, z_0 over the whole volume occupied by the medium, obtaining

$$\iiint [G_\omega(r \mid r_0) \, \nabla_0^2 p_\omega(r_0) - p_\omega(r_0) \, \nabla_0^2 G_\omega(r \mid r_0)] \, dv_0$$

$$= \iiint p_\omega(r_0) \, \delta(r - r_0) \, dv_0 - \iiint f_\omega(r_0)G_\omega(r \mid r_0) \, dv_0$$

where the zero subscripts indicate differentiation and integration with respect to the x_0, y_0, z_0 coordinates.

The quantity in brackets is the divergence of the vector $G_\omega \, \text{grad}_0 \, p_\omega - p_\omega \, \text{grad}_0 \, G_\omega$, so that this volume integral can be changed to a surface integral, over the surface enclosing the medium, of the component of this vector, normal to the surface, pointing outward, *away from the medium*.   Writing this normal component of grad_0 as $\partial / \partial n_0$ and utilizing Eq. (1.3.24) for the delta function, we obtain

$$p_\omega(\mathbf{r}) = \iiint f_\omega(\mathbf{r}_0)G_\omega(\mathbf{r} \mid \mathbf{r}_0) \, dv_0$$

$$+ \iint \left[G_\omega(\mathbf{r} \mid \mathbf{r}_0) \frac{\partial}{\partial n_0} p_\omega(\mathbf{r}_0) - p_\omega(\mathbf{r}_0) \frac{\partial}{\partial n_0} G_\omega(\mathbf{r} \mid \mathbf{r}_0) \right] dS_0 \quad (7.1.17)$$

which is an equation for the spatial factor $p_\omega(\mathbf{r})$ of the pressure wave *within and on the surface* bounding the medium, in terms of a volume integral of the source function f_ω over the bounded volume and a surface integral of the boundary values of p_ω and its outward-pointed normal gradient over the boundary surface. Because we integrated over the bounded volume, the equation is not valid for $\mathbf{r}$ outside this volume. This formula is not very mysterious. Since $G_\omega(\mathbf{r} \mid \mathbf{r}_0)$ is the field at $\mathbf{r}$ due to a point source at $\mathbf{r}_0$, it states that the total field at r is the summation of the fields from all the elementary sources $f_\omega \, dv_0$ plus the waves reflected by the boundary surface.

This integral expression will be used frequently in later sections when we consider the effects of boundaries on the wave. In this section we are considering waves radiated out into an unbounded medium, in which case $G_\omega = g_\omega = (1/4\pi R)e^{ikR}$, and if p_ω consists solely of outward-going waves, the surface integral vanishes. Thus, for the unbounded medium, the equation becomes

$$p_\omega(\mathbf{r}) = \iiint f_\omega(\mathbf{r}_0)g_\omega(\mathbf{r} \mid \mathbf{r}_0) \, dx_0 \, dy_0 \, dz_0$$

All the simple-source waves, from the various elementary sources $f_\omega(\mathbf{r}_0) \, dv_0$, add together to produce the resultant field at $\mathbf{r}$, the measurement point.

At times we may wish to calculate the radiation from a source distribution without first analyzing the frequency components f_ω of the source. We can then utilize the result of Prob. 23 of Chap. 1. If, for example, $f_\omega(\mathbf{r})$ is the Fourier transform $(1/2\pi) \int f(\mathbf{r},t)e^{i\omega t} \, dt$ of the source function $f(\mathbf{r},t)$, the equation for the total pressure wave radiated from the distributed source is

$$p(\mathbf{r},t) = \iiint dv_0 \int_{-\infty}^{\infty} f(\mathbf{r}_0,t_0)g(\mathbf{r}, t \mid \mathbf{r}_0, t_0) \, dt_0$$

where $g(\mathbf{r}, t \mid \mathbf{r}_0, t_0) = (1/2\pi)g(\mathbf{r}; \mathbf{r}_0; t - t_0)$ is $1/2\pi$ times the inverse transform of the Green's function for frequency $\omega/2\pi$.

$$\frac{1}{2\pi} g(\mathbf{r}; \mathbf{r}_0; \tau) = \frac{1}{2\pi} \int_{-\infty}^{\infty} \frac{1}{4\pi R} e^{i(\omega/c)R - i\omega\tau} \, d\omega = \frac{1}{4\pi R} \delta\left(\tau - \frac{R}{c}\right)$$

using Eq. (1.3.25) for the inverse transform of the delta function.

Therefore we can, if more convenient, compute the total radiation from a nonperiodic source distribution by carrying out the following integration in space *and* time:

$$p(\mathbf{r},t) = \iiiint f(\mathbf{r}_0,t_0)g(\mathbf{r}, t \mid \mathbf{r}_0, t_0) \, dx_0 \, dy_0 \, dz_0 \, dt_0$$

$$g(\mathbf{r}, t \mid \mathbf{r}_0, t_0) = \frac{1}{4\pi R} \delta\left(t - t_0 - \frac{R}{c}\right) \qquad R = \mathbf{r} - \mathbf{r}_0 \quad (7.1.18)$$

This time-dependent Green's function is the solution of the equation

$$\nabla^2 g - \frac{1}{c^2}\frac{\partial^2 g}{\partial t^2} = -\delta(x - x_0)\, \delta(y - y_0)\, \delta(z - z_0)\, \delta(t - t_0)$$

for the pulse-wave from a point pulse source at $\mathbf{r}_0$, t_0. The pulse travels outward with velocity c. Note the relationship with Eq. (2.3.9) and the fact that Eq. (7.1.4) can be obtained directly from Eq. (7.1.18).

Radiation from a region of violent fluid motion

As a first illustration of the power of the integral equation (7.1.17), let us work out the radiation, out into unbounded space, from a small region of the fluid which is in violent motion. Inside this region the motions of the air are great enough so that the linear equations of wave motion no longer hold; energy may be fed into the medium by the introduction of new fluid, by forces acting directly on portions of the fluid, and by the nonlinear conversion of mass flow into turbulent motion. Within this small region, therefore, the equations of fluid motion must include all the nonlinear terms we neglected when we obtained the linear wave equation. It is still possible, however, to neglect the damping effects of viscosity and thermal diffusion, which were treated in Sec. 6.4, as long as rigid boundaries are not present.

The sources of sound, which feed energy into the acoustic field, are usually of three sorts, monopole, dipole, and quadrupole, corresponding to three ways of feeding energy into the field. The monopole sources correspond to the introduction of new fluid into the disturbed region, which may occur when an explosion or combustion changes a solid or liquid into a gas. Or they may correspond to the introduction of heat (as when a modulated laser or other focused radiation heats the fluid), which expands the fluid irregularly. Or else they may be caused by time or space variations of the compressibility, produced by turbulence or by irregular heating.

The equation of continuity (6.1.15) for mass flow is $(\partial\rho/\partial t) + \text{div}\,(\rho\mathbf{u}) = \rho q(\mathbf{r},t)$, where q is the rate of introduction of new fluid volume into the region, and thus ρq is the rate of introduction of new fluid mass. Using Eqs. (6.4.21), relating change of ρ to change of pressure and temperature, results in

$$\frac{\partial\rho}{\partial t} \equiv \frac{\partial\delta}{\partial t} \simeq \gamma\rho\kappa_s \frac{\partial}{\partial t}(p - \alpha\tau)$$

If heat energy $\epsilon(t)$ per unit mass is introduced into the fluid, Eq. (6.4.18) becomes

$$T \frac{\partial \sigma}{\partial t} \simeq K \nabla^2 \tau + \epsilon(t) \qquad \text{where } \sigma = \frac{C_p}{T}\left(\tau - \frac{\gamma - 1}{\alpha \gamma} p\right)$$

Since we can neglect thermal conduction in the main body of the fluid, we have

$$\frac{\partial}{\partial t}(p - \alpha \tau) \simeq \frac{1}{\gamma} \frac{\partial p}{\partial t} - \frac{\alpha \epsilon}{C_p}$$

and the equation of continuity for mass flow becomes

$$\rho \kappa \frac{\partial P}{\partial t} + \operatorname{div}(\rho \mathbf{u}) = \rho q + \frac{\alpha \gamma \rho \kappa}{C_p} \epsilon \qquad (7.1.19)$$

where the subscript s on κ is understood; we use only the adiabatic compressibility from now on. Also, but in this discussion only, we shall use ρ instead of $\rho + \delta$ as the actual density of the fluid, and P as its total pressure instead of $P + p$; of course, $\partial P/\partial t$ equals the $\partial p/\partial t$ used elsewhere.

By using the thermodynamic relations of Eqs. (6.4.19), the heat-energy term can be alternatively written

$$\frac{\alpha \gamma \rho \kappa}{C_p} \epsilon = \frac{\beta}{C_p} H = (\gamma - 1) \frac{H}{T \beta c^2} \rightarrow -(\gamma - 1) \frac{H}{c^2}$$

where H is the heat energy added per unit volume ($= \rho \epsilon$). The last expression is the limiting form for a perfect gas.

The equation of continuity (6.2.11) for momentum flow can be written as

$$\frac{\partial \rho \mathbf{u}}{\partial t} = \mathbf{F} - \operatorname{grad} P - \boldsymbol{\nabla} \cdot (\mathfrak{J} + \mathfrak{D}) \qquad (7.1.20)$$

where $\mathbf{F}(\mathbf{r}, t)$ is the force (electric or gravitational or otherwise) acting on a unit volume of fluid at $\mathbf{r}$, t, and $\mathfrak{J}$ is the momentum-flux tensor of Eq. (6.2.10), with typical component $J_{xy} = \rho u_x u_y$. We have separated the pressure P out of the stress tensor $\mathfrak{P}$ of Eq. (6.2.10), so that, as given in Eq. (6.4.9),

$$-\mathfrak{D} = P\mathscr{I} - \mathfrak{P} \qquad -D_{xx} = P - P_{xx} \qquad -D_{xy} = -P_{xy} \qquad \cdots$$

since we expect to obtain an equation for P. The elements of the tensor $\mathfrak{D}$ are the components of the stress of viscosity in the violently moving fluid, discussed in Sec. 6.4. The total tensor in the last term of Eq. (7.1.20),

$$\mathfrak{T} = \mathfrak{J} + \mathfrak{D} \qquad T_{xx} = P_{xx} - P + \rho u_x u_x \qquad T_{xy} = P_{xy} + \rho u_x u_y = T_{yx} \qquad \cdots$$

is the stress-momentum tensor, exclusive of fluid pressure defined in Eqs. (6.2.11) and (7.1.20). In fluids such as air and water, its components are small whenever $|\mathbf{u}|^2$ can be neglected compared with $|\mathbf{u}|$.

As before, to obtain a wave equation for P, we differentiate Eq. (7.1.19) with respect to t, take the divergence of Eq. (7.1.20), and then eliminate the $\rho\mathbf{u}$ term between them. In a region of the sort we are discussing, neither ρ nor κ will be uniform with respect to either space or time; in fact, the composition of the fluid may be nonuniform. Therefore the time derivative of the first term of Eq. (7.1.19) is

$$\rho\kappa \frac{\partial^2 P}{\partial t^2} + \frac{\partial P}{\partial t}\left[\kappa \frac{\partial \delta}{\partial t} + \rho\left(\frac{\partial \kappa}{\partial P}\right)_T \frac{\partial}{\partial t}(p - \alpha\tau)\right] + \rho\left(\frac{\partial \kappa}{\partial t}\right)_{PT} \frac{\partial P}{\partial t}$$

$$= \rho\kappa \frac{\partial^2 P}{\partial t^2} + \frac{\rho}{\gamma}\left[\kappa^2 + \left(\frac{\partial \kappa}{\partial P}\right)_T\right]\left(\frac{\partial P}{\partial t} - \frac{\alpha\gamma\epsilon}{C_p}\right)\frac{\partial P}{\partial t} + \rho\left(\frac{\partial \kappa}{\partial t}\right)_{PT} \frac{\partial P}{\partial t}$$

and the time derivative of the last term is

$$\frac{\alpha\gamma\rho\kappa}{C_p} \frac{\partial \epsilon}{\partial t} + \frac{\alpha\gamma\rho\epsilon}{C_p}\left(\frac{\partial \kappa}{\partial t}\right)_{TP} + \alpha\rho\left[\kappa^2 + \left(\frac{\partial \kappa}{\partial P}\right)_T\right]\left(\frac{\partial P}{\partial t} - \frac{\alpha\gamma\epsilon}{C_p}\right)\frac{\epsilon}{C_p}$$

We have assumed that the compressibility κ at $\mathbf{r}$ may depend explicitly on time because the fluid composition at $\mathbf{r}$ may change with time, in addition to the implicit dependence of κ on t through the variation of P and T when the composition does not change.

Thus the final wave equation for P, with the thermal and viscous terms neglected (except for the heat source term ϵ), is

$$\nabla^2 P - \rho\kappa \frac{\partial^2 P}{\partial t^2} \simeq -f(\mathbf{r},t) + \text{div } \mathbf{F}(\mathbf{r},t) - \boldsymbol{\nabla}\cdot\mathfrak{T}(\mathbf{r},t)\cdot\boldsymbol{\nabla} \qquad (7.1.21)$$

where the scalar quantity

$$\boldsymbol{\nabla}\cdot\mathfrak{T}\cdot\boldsymbol{\nabla} = \frac{\partial^2}{\partial x^2} T_{xx} + \frac{\partial^2}{\partial x\,\partial y}(T_{xy} + T_{yx}) + \cdots$$

$$\simeq \frac{\partial^2}{\partial x^2} \rho u_x^{\,2} + \frac{\partial^2}{\partial x\,\partial y}(\rho u_x u_y + \rho u_y u_x) + \cdots$$

and where the monopole source term is

$$f(\mathbf{r},t) \equiv \frac{\partial}{\partial t}\rho s(\mathbf{r},t) = \frac{\partial}{\partial t}\rho q + \frac{\alpha\gamma\rho\kappa}{C_p} \frac{\partial \epsilon}{\partial t} - \rho\left(\frac{\partial \kappa}{\partial t}\right)_{PT}\left(\frac{\partial P}{\partial t} - \frac{\alpha\gamma\epsilon}{C_p}\right)$$

$$- \frac{\rho}{\gamma}\left[\kappa^2 + \left(\frac{\partial \kappa}{\partial P}\right)_T\right]\left(\frac{\partial P}{\partial t} - \frac{\alpha\gamma\epsilon}{C_p}\right)^2 \qquad (7.1.22)$$

The function $s(\mathbf{r},t)$ is the effective monopole source-strength density, the generalization of the point source strength S of Eq. (7.1.4) for a distributed source. The last two terms are usually small compared with the first two. We thus see that, even if no new fluid is introduced (even if q is zero), the monopole source term is not necessarily zero in a region where heat energy is being introduced or where the composition of the fluid is nonuniform in space and time.

This equation, with its higher-order terms on the right-hand side, is usually impossible to solve exactly. But for our purposes here, we can consider the three terms on the right of Eq. (7.1.21), the monopole-source term f, the force term $\mathbf{F}$, and the stress-momentum term in $\mathfrak{T}$, as source terms, either determined by measurement or by calculation. They are non-negligible only in the region of violent fluid motion; in all the rest of the infinite region they are zero, and Eq. (7.1.21) is the usual linear wave equation for the acoustic pressure p. Here we shall assume that the values of the functions on the right-hand side are known; they are the source of new energy being injected into the sound field.

Radiation from a small region

To solve for the radiation field outside the source region, we first take the Fourier transform of Eq. (7.1.21), defining

$$p_\omega(\mathbf{r}) = \frac{1}{2\pi} \int_{-\infty}^{\infty} P(\mathbf{r},t)e^{i\omega t}\, dt \qquad P(\mathbf{r},t) = \int_{-\infty}^{\infty} p_\omega(\mathbf{r})e^{-i\omega t}\, d\omega$$

with similar relationships between $f_\omega(\mathbf{r})$ and $f(\mathbf{r},t)$, $\mathbf{F}_\omega(\mathbf{r})$ and $\mathbf{F}(\mathbf{r},t)$, and $\mathfrak{T}_\omega$ and $\mathfrak{T}$, according to Eq. (1.3.6). Then the equation for the Fourier transform of P, the quantity which becomes the acoustic pressure outside the source region, is

$$\nabla^2 p_\omega + k^2 p_\omega = -f_\omega + \text{div }\mathbf{F}_\omega - \boldsymbol{\nabla}\cdot\mathfrak{T}_\omega\cdot\boldsymbol{\nabla} \qquad k = \frac{\omega}{c} \qquad (7.1.23)$$

where $f_\omega = -i\omega\rho s_\omega$ is defined in Eq. (7.1.22). The source terms, on the right-hand side of Eq. (7.1.23), are assumed to be known; they are zero except within the small region of violent motion, which we have called the source region.

This equation has the same form as Eq. (7.1.16) and thus has the solution given in Eq. (7.1.17) for an unbounded medium,

$$p_\omega(\mathbf{r}) = \iiint [f_\omega(\mathbf{r}_0) - \text{div}_0\,\mathbf{F}_\omega(\mathbf{r}_0) + \boldsymbol{\nabla}_0\cdot\mathfrak{T}_\omega\cdot\boldsymbol{\nabla}_0]g_\omega(\mathbf{r}\mid\mathbf{r}_0)\, dv_0 \qquad (7.1.24)$$

where g_ω is the Green's function $(1/4\pi R)e^{ikR}$, and the source terms in the brackets represent the injection of energy by the injection of heat and the creation of fluid, or by forces acting on the fluid, or by the nonlinear effects of fluid motion (Bernoulli effect) and viscosity. Each of these terms sends a wave of different form out into the surrounding medium, as can be most easily seen if we work out the radiation from a source region small compared with the wavelength $2\pi/k$.

In the first place, we can integrate the second and third source terms in Eq. (7.1.24) by parts. For example,

$$g_\omega \text{ div}_0\,\mathbf{F}_\omega = -\mathbf{F}_\omega\cdot\text{grad}_0\,g_\omega + \text{div}_0(g_\omega\mathbf{F}_\omega)$$

and since $\mathbf{F}_\omega$ is zero outside the source region, the integral of $g_\omega \operatorname{div}_0 \mathbf{F}_\omega$ over the whole source region is equal to minus the integral of $\mathbf{F}_\omega \cdot \operatorname{grad}_0 g_\omega$ over the source region. Thus we can rewrite Eq. (7.1.24) in the more useful form

$$p_\omega(\mathbf{r}) = \iiint [f_\omega(\mathbf{r}_0) g_\omega(\mathbf{r} \mid \mathbf{r}_0) + \mathbf{F}_\omega(\mathbf{r}_0) \cdot \nabla_0 g_\omega(\mathbf{r} \mid \mathbf{r}_0)$$
$$+ \mathfrak{T}_\omega : (\nabla_0 g_\omega \nabla_0)] \, dv_0 \quad (7.1.25)$$

where the scalar products of the gradient and tensor gradient of g_ω are

$$\mathbf{F}_\omega \cdot \nabla g_\omega = F_x \frac{\partial}{\partial x} g_\omega + F_y \frac{\partial}{\partial y} g_\omega + F_z \frac{\partial}{\partial z} g_\omega$$

and

$$\mathfrak{T}_\omega : \nabla g_\omega \nabla = T_{xx}{}^\omega \frac{\partial^2}{\partial x^2} g_\omega + T_{xy}{}^\omega \frac{\partial^2}{\partial x \, \partial y} g_\omega + \cdots$$

where

$$T_{xy}{}^\omega = \frac{1}{2\pi} \int_{-\infty}^{\infty} \rho u_x u_y e^{i\omega t} \, dt \quad \cdots$$

We now take the integral of Eq. (7.1.25) over a sphere of radius a, which is just large enough to enclose the whole source region (we are assuming that $ka \ll 1$) and set the origin of coordinates at the center of the sphere. In this case we are making little error if we assume that g_ω and its derivatives are nearly constant throughout the sphere, being equal to their values at $\mathbf{r}_0 = 0$. Thus if, and only if, $ka \ll 1$, the g factor can be taken outside the integral, in which case the resulting field takes on a particularly simple form.

For example, the first source term, that due to the introduction of new fluid, produces a spherically symmetric wave outside the source region.

$$p_s(\mathbf{r}) \simeq g_\omega(\mathbf{r} \mid 0) \iiint f_\omega(\mathbf{r}_0) \, dv_0 = -ik \frac{\rho c}{4\pi r} \left(\iiint s_\omega \, dv_0 \right) e^{ikr} \quad ka \ll 1$$

$$(7.1.26)$$

Comparison with Eq. (7.1.5) shows that the integral in parentheses equals S_ω, the strength of the monopole source produced by the periodic injection or expansion of fluid in the region $r < a$.

The second term in the integral of Eq. (7.1.25) becomes, when $ka \ll 1$, the product of the integral of vector $\mathbf{F}_\omega$ and the gradient of g_ω.

$$p_d(\mathbf{r}) \simeq \left[\iiint \mathbf{F}_\omega(\mathbf{r}_0) \, dv_0 \right] \cdot [\operatorname{grad}_0 g_\omega(\mathbf{r} \mid \mathbf{r}_0)]_{r_0=0}$$

$$= -k^2 D_\omega \frac{\rho c}{4\pi r} \left(1 + \frac{i}{kr} \right) e^{ikr} \cos \vartheta \quad (7.1.27)$$

where $\mathbf{D}_\omega = (i/k\rho c)\int \mathbf{F}_\omega \, dv_0$ is the vector strength of the dipole source produced by the forces acting on the fluid in the source region, and ϑ is the angle between $\mathbf{D}_\omega$ and the radius vector $\mathbf{r}$ to the measurement point.

The third term in the integral of Eq. (7.1.25) results in a quadrupole wave when $ka \ll 1$.

$$p_q(\mathbf{r}) \simeq -ik\rho c\left\{Q_{xx}\left[\frac{\partial^2}{\partial x_0{}^2}g_\omega(\mathbf{r}\mid\mathbf{r}_0)\right]_{r_0=0}\right.$$

$$\left. + (Q_{xy} + Q_{yx})\left[\frac{\partial^2}{\partial x_0\,\partial y_0}g_\omega(\mathbf{r}\mid\mathbf{r}_0)\right]_{r_0=0} + \cdots\right\} \quad (7.1.28)$$

where

$$Q_{xx} = \frac{i}{\rho ck}\iiint T_{xx}(\mathbf{r}_0)\,dv_0 = \frac{i}{\rho ck}\iiint(P_{xx} - P + \rho u_x{}^2)\,dv_0$$

and the form of the wave is given in Eq. (7.1.10), where

$$Q_{xy} = \frac{i}{\rho ck}\iiint(P_{xy} + \rho u_x u_y)\,dv_0$$

and the waveform is that given in Eq. (7.1.9), and so on. Thus the integral, over the source region, of the tensor

$$\frac{i}{\rho ck}\mathfrak{T} = \frac{i}{\rho ck}(\mathfrak{P} - P\mathcal{I} + \mathfrak{J})$$

is the tensor strength of the quadrupole source produced by the nonlinear effects of the fluid motion in the source region, if this region is small enough.

The complete wave radiated from the small region of violent fluid motion,

$$p_\omega(\mathbf{r}) = p_s(\mathbf{r}) + p_d(\mathbf{r}) + p_q(\mathbf{r}) \quad (7.1.29)$$

is thus a combination of monopole, dipole, and quadrupole waves, with relative strengths determined by the relative magnitudes of the fluid injection, fluid force, and stress-momentum characteristics of the small source region. Corresponding expressions for nonperiodic sources can be obtained from Eq. (7.1.13).

Random source functions

It should be pointed out that when the disturbed region is not small in dimensions compared with a wavelength, the angular distribution of the radiation is not as simply organized as this. Sound generated by the various portions of the region can interfere destructively in some directions and constructively in others. It is possible to have a distribution of simple sources which radiates like a dipole or quadrupole, or even a combination with higher multipoles. For example, if the radiating region is a sphere of radius a and the simple-source function $s_\omega(\mathbf{r}_0)$ of Eq. (7.1.26) is distributed within the sphere so that s_ω is positive for $x_0 > 0$ and negative for $x_0 < 0$, then, when ka is not negligibly small, part of the radiation will be dipole in character. Likewise, a quadrupole source distribution can produce some simple-source and some dipole radiation if the radiating region is large enough. We shall have more to say about this under the next heading and in Chap. 11.

To simplify Eq. (7.1.29), we calculate the acoustic field at large distance r from a region, inside a sphere of radius a which is small compared with r but not small compared with λ, where the monopole and quadrupole source densities $s(\mathbf{r},t)$ and $\mathfrak{T}(\mathbf{r},t)$ fluctuate in a random manner with respect to both space and time. In this case the first and last terms in the integral of Eq. (7.1.25) are important, but there is no fluctuating gravitational or electric field; so the middle term is missing.

To compute the far field we note that, when $r \gg r_0$, the Green's function approaches the simple form

$$g_\omega(\mathbf{r} \mid \mathbf{r}_0) \rightarrow \frac{e^{ikr}}{4\pi r} e^{-i\mathbf{k}\cdot\mathbf{r}_0}$$

where vector $\mathbf{k}$, with components $k_x = k \sin \vartheta \cos \varphi$, $k_y = k \sin \vartheta \cos \varphi$, $k_z = k \cos \vartheta$, has magnitude k and is pointed in the direction of the radius vector $\mathbf{r}$ to the measurement point. Inserting this in Eq. (7.1.25) and omitting the term in $\mathbf{F}_\omega$, we obtain

$$p_\omega(\mathbf{r}) \rightarrow \frac{e^{ikr}}{r} \psi_\omega(\vartheta,\varphi)$$

where

$$\psi_\omega(\vartheta,\varphi) = \frac{1}{4\pi} \iiint [-ik\rho c s_\omega(\mathbf{r}_0) - \mathbf{k}\cdot\mathfrak{T}_\omega(\mathbf{r}_0)\cdot\mathbf{k}]e^{-i\mathbf{k}\cdot\mathbf{r}_0}\, dv_0 \quad (7.1.30)$$

is the function determining the angle distribution of the component of the wave having frequency $\omega/2\pi$.

If the region of disturbance is very small compared with the wavelength $\lambda = 2\pi/k$, then the exponential $e^{i\mathbf{k}\cdot\mathbf{r}_0}$ will be practically equal to 1 over the region of integration, and the wave radiated will be a combination of simple-source and quadrupole waves. As pointed out earlier, if we define the constants

$$S_\omega^0 = \iiint s_\omega(\mathbf{r}_0)\, dV_0 \qquad Q_{\omega xx}^0 = \frac{i}{\rho c k} \iiint T_{\omega xx}\, dv_0 \qquad \cdots$$

the long-wavelength radiation has the simple angle distribution

$$\psi_\omega \rightarrow \frac{-ik\rho c}{4\pi} S_\omega^0 + \frac{ik^3\rho c}{4\pi} (Q_{\omega xx}^0 \sin^2 \vartheta \cos^2 \varphi$$
$$+ Q_{\omega xy}^0 \sin^2 \vartheta \cos \varphi \sin \varphi + \cdots)$$

a combination of a pure simple source and a pure quadrupole.

But for somewhat higher frequencies, when the exponential must be written $e^{-i\mathbf{k}\cdot\mathbf{r}_0} \simeq 1 - i(k_x x_0 + k_y y_y + k_z z_0)$, integrals like

$$D_{\omega x}{}^s = \iiint s_\omega(\mathbf{r}_0)x_0\,dv_0 \qquad O^q_{\omega xyz} = \frac{i}{\rho ck} \iiint T_{\omega xy}(\mathbf{r}_0)z_0\,dv_0 \qquad \cdots$$

also enter. The fluid-injection distribution s_ω (if it is not symmetrically distributed) will contribute some dipole radiation to ψ_ω of the form $-(k^2\rho c/4\pi)(D_{\omega x}{}^s \sin\vartheta\cos\varphi + D_{\omega y}{}^s \sin\vartheta\sin\varphi + D_{\omega z}{}^s \cos\vartheta)$, and the momentum-tensor terms will contribute octupole radiation with terms in ψ_ω such as $(k^4\rho c/4\pi)O^q_{\omega xyz} \sin^2\vartheta\cos\vartheta\sin\varphi\cos\varphi$, etc. At still higher frequencies the radiation becomes quite complex, and there is no particular advantage in separating it into monopole, dipole, etc., components. A more useful separation is into various spherical harmonics, as will be discussed in Sec. 7.2.

The autocorrelation function

In such a disturbed region, if both monopole and quadrupole source densities fluctuate in a highly random manner, both in space and time, the methods of Sec. 1.3 are most appropriate for their description. The mean rate of introduction of heat, for example, averaged over a long interval of time T and over the whole volume V of the active region, produces no noise, but the difference between the actual rate of production at point $\mathbf{r}_0$ at time t and this constant mean value may be a highly fluctuating quantity, containing a wide spread of frequencies and wavelengths. In accord with Eq. (1.3.18), we call $s(\mathbf{r}_0,t)$ this difference between actual and average source density, taken only in the time interval $-\frac{1}{2}T < t < \frac{1}{2}T$ and in the volume V of the active region. By an obvious generalization of Eq. (1.3.20) it can be described in terms of its multiple Fourier transform

$$S(\mathbf{K},\omega) = \frac{1}{16\pi^4} \int \cdots \int s(\mathbf{r}_0,t)e^{i\omega t - i\mathbf{K}\cdot\mathbf{r}_0}\,dv_0\,dt \tag{7.1.31}$$

where the integration over t is from $-\frac{1}{2}T$ to $+\frac{1}{2}T$, and the integration over x_0, y_0, z_0 is over volume V. We note that, according to Eq. (7.1.22),

$$s_\omega = \frac{1}{2\pi} \int_{-\frac{1}{2}T}^{\frac{1}{2}T} s(r_0,t)e^{i\omega t}\,dt$$

But we can also describe the monopole source density in terms of the autocorrelation function

$$\Upsilon_s(\mathbf{d},\tau) = \frac{1}{VT} \int \cdots \int s(\mathbf{r},t)s(\mathbf{r}+\mathbf{d},\,t+\tau)\,dv\,dt \tag{7.1.32}$$

where the integration is again over T and V. In accord with Eq. (1.3.20), we see that the Fourier transform of Υ_s is proportional to the square of the Fourier transform of s.

$$|S(\mathbf{K},\omega)|^2 = \frac{VT}{(2\pi)^8} \int \cdots \int \Upsilon_s(\mathbf{d},\tau) e^{i\omega\tau - i\mathbf{K}\cdot\mathbf{d}} \, d\tau \, dv_d \qquad (7.1.33)$$

The intensity of sound of frequency $\omega/2\pi$, caused by gas expansion or production, measured at $\mathbf{r}$, if $r \gg a$, the "radius" of the disturbed region, is

$$I_{s\omega} = \frac{1}{\rho c r^2} |\psi_s(\vartheta,\varphi)|^2 = \frac{k^2 \rho c}{64\pi^4 r^2} \left| \int \cdots \int s(\mathbf{r}_0,t) e^{i\omega t - i\mathbf{k}\cdot\mathbf{r}_0} \, dv_0 \, dt \right|^2$$

$$= \frac{4\pi^4 k^2 \rho c}{r^2} |S(\mathbf{k},\omega)|^2$$

$$= \frac{VT\rho c k^2}{64\pi^4 r^2} \int \cdots \int \Upsilon_s(\mathbf{d},\tau) e^{i\omega\tau - i\mathbf{k}\cdot\mathbf{d}} \, d\tau \, dv_d \qquad (7.1.34)$$

which enables us to compute the total radiation produced by the monopole source in the active region, averaged over the time interval T.

The correlation function Υ_s will have its maximum value at $\mathbf{d}$ and $\tau = 0$, where its value will be the mean square of $s(\mathbf{r},t)$, as shown by Eq. (7.1.32). This difference $s(\mathbf{r},t)$ between the value of the $s(\mathbf{r},t)$ of Eq. (7.1.22) and its average value has dimensions volume per second per volume $= \sec^{-1}$; its mean square, the *variance* of this density, averaged over V and T, written $\langle\Delta_s{}^2\rangle$, has dimensions $\sec^{-2}$. According to the discussion of Eq. (2.3.12), if all frequencies are present in the fluctuation between 0 and some cutoff frequency $\omega_s/2\pi = 1/2\pi\tau_s$, Υ_s will be fairly constant for $0 < \tau < \tau_s$ and then will fall off to zero as τ is further increased. Thus there is little time correlation over a longer time interval than τ_s, called the *time correlation interval*. Likewise, there will be a *spatial correlation distance* w_s; for d^2 larger than $w_s{}^2$, function Υ_s will drop rapidly to zero. An obvious generalization of Eq. (2.3.13) would lead us to expect that

$$\Upsilon_s(\mathbf{d},\tau) = \langle\Delta_s{}^2\rangle \exp\left[-\frac{1}{2}\left(\frac{d}{w_s}\right)^2 - \tfrac{1}{2}(\omega_s\tau)^2 \right]$$

would be a good approximation to the correlation function if the monopole fluctuations are isotropically random.

Inserting this in Eq. (7.1.34), we find for the intensity at frequency $\omega/2\pi$

$$I_{s\omega} = \rho c \langle\Delta_s{}^2\rangle \frac{k^2 VT}{16\pi^2 r^2} \frac{w_s{}^3}{\omega_s} \exp\left\{ -\tfrac{1}{2}\omega^2 \left[\left(\frac{w_s}{c}\right)^2 + \left(\frac{1}{\omega_s}\right)^2 \right] \right\} \qquad (7.1.35)$$

This increases proportionally to the square of the frequency below the limiting value $(\omega_s/2\pi)(1 + k_s{}^2 w_s{}^2)^{-\frac{1}{2}}$, where $k_s = \omega_s/c = 2\pi/\lambda_s$, λ_s being the wavelength corresponding to the cutoff frequency $\omega_s/2\pi$. For frequencies greater than this, the intensity at $\mathbf{r}$ falls off to zero.

To obtain the total intensity, for all frequencies, we use Eq. (1.3.17) to convert an average over the time interval T into an integral over all frequencies. Since

$$\frac{1}{\rho c T}\int_{-\frac{1}{2}T}^{\frac{1}{2}T}|p(\mathbf{r},t)|^2\,dt = \frac{2\pi}{\rho c T}\int_{-\infty}^{\infty}|p_\omega(\mathbf{r})|^2\,d\omega$$

the total intensity at $\mathbf{r}$, averaged over interval T, is

$$I_s = \frac{\rho c V}{\sqrt{32\pi}\,r^2}\,\langle\Delta_s^2\rangle\,\frac{k_s^2 w_s^3}{(1+k_s^2 w_s^2)^{\frac{3}{2}}} \qquad (7.1.36)$$

When the space-correlation distance w_s for the monopole fluctuations is larger than c times the time-correlation interval τ_s of these fluctuations, the total radiated intensity is about $1/40r^2$ of the mean variance $\langle\Delta_s^2\rangle$ of these fluctuations, times the volume V of the active region, times the cutoff wavelength λ_s, times ρc. When w_s is smaller than $c\tau_s = \lambda_s/2\pi$, the resulting intensity is reduced by the factor $(2\pi w_s/\lambda_s)^3$.

Exactly the same sort of calculation can be carried out for the radiation from turbulent motion, corresponding to the second term in Eq. (7.1.30). When the gas has low viscosity, the largest part of each term in the matrix $\mathfrak{T}$ is the kinetic-energy term $\rho u_x u_y$, etc.; these exhibit large fluctuations in a region of turbulence, as will be shown in Sec. 11.4. The term in the integral of Eq. (7.1.30) is the component of this matrix in the direction of $\mathbf{k}$, that is, in the direction of the measurement point. Thus $\mathbf{k}\cdot\mathfrak{T}(\mathbf{r}_0,t)\cdot\mathbf{k}$ is equal to k^2 times twice the kinetic energy of turbulent motion in the $\mathbf{k}$ direction (which is $\frac{1}{3}$ of the mean kinetic energy if the turbulence is isotropic). The mean value of this, averaged over time and over V, produces no radiation, but the deviation from this mean, $\Delta_e = \frac{1}{2}T_{kk}(\mathbf{r}_0,t) - \frac{1}{2}\langle T_{kk}\rangle$, is responsible for the sound of turbulence at $\mathbf{r}$. It can be described in terms of its Fourier transform $\Delta_e(\mathbf{K},\omega)$ or in terms of its correlation function

$$\Upsilon_e(\mathbf{d},\tau) = \langle\Delta_e^2\rangle\exp\left[-\frac{1}{2}\left(\frac{d}{w_e}\right)^2 - \frac{1}{2}(\omega_e t)^2\right]$$

having a Fourier transform proportional to $|\Delta_e(\mathbf{k},\omega)|^2$ [see Eq. (11.4.15) for further discussion]. Constant w_e is the spatial correlation distance for the turbulent fluctuations, and $\tau_e = 1/\omega_e = 1/ck_e = \lambda_e/2\pi c$ is the time-correlation interval.

Thus, as with the monopole sources, we obtain an expression for the intensity of the turbulent sound at point $\mathbf{r}$, over all frequencies.

$$I_e = \frac{3V\langle\Delta_e^2\rangle}{\sqrt{2\pi}\,\rho c r^2}\,\frac{k_e^4 w_e^3}{(1+k_e^2 w_e^2)^{\frac{5}{2}}} \qquad (7.1.37)$$

If the turbulence has some orientation along an axis, $k_e w_e$ probably is independent of angle, but $k_e \langle \Delta_e{}^2 \rangle$ has the general form $A + B \cos^2 \vartheta$, where ϑ is the angle between $\mathbf{r}$ and the axis of symmetry. Further discussion of this radiation will be found in Sec. 11.4.

The total intensity from both effects is the sum of the two, if the fluctuations are statistically independent. Then

$$I_r = \frac{V}{\sqrt{2\pi}\, r^2} \left\{ \frac{\rho c \langle \Delta_s{}^2 \rangle}{4 k_s} \frac{(k_s w_s)^3}{[1 + (k_s w_s)^2]^{\frac{3}{2}}} + \frac{3 \langle \Delta_e{}^2 \rangle}{\rho c w_e} \frac{(k_e w_e)^4}{[1 + (k_e w_e)^2]^{\frac{5}{2}}} \right\} \quad (7.1.38)$$

When the products kw are small, i.e., when the spatial correlation is more restricted than c times the time-correlation interval τ, then the total radiation is weak and the simple-source term predominates. The intensity at $\mathbf{r}$ is equal to that of a simple source of strength roughly equal to $(2\pi)^{\frac{3}{4}}(w_s{}^3 V \langle \Delta_s{}^2 \rangle)^{\frac{1}{2}}$ unit volumes per second and frequency $\omega_s/2\pi$ cps. When the products kw are both large, i.e., when the spatial correlation is more widespread than the temporal correlation times c, then the turbulence term is more important, and the intensity is roughly equal to that of a source of radius w_e and frequency $\omega_e/2\pi$, with driving force F_s [Eq. (7.1.11)] of magnitude $(288\pi)^{\frac{1}{4}}(V \lambda_e \langle \Delta_e{}^2 \rangle)]^{\frac{1}{2}}$ newtons. Translational motion of the active region with respect to the surrounding air distorts this wave in a manner which will next be discussed in Chap. 11.

7.2 RADIATION FROM SPHERES

Sound waves can also be generated by the motion of solid bodies, of course. To correlate with the discussion of the preceding section, we shall first discuss the sound waves caused by the motion of a sphere, centered at the origin. Such waves can most easily be described in terms of spherical coordinates. Using Eqs. (7.1.4) and (7.1.3), we see that the first-order wave equation in these coordinates is (omitting dependence on φ)

$$\frac{1}{r^2} \frac{\partial}{\partial r}\left(r^2 \frac{\partial p}{\partial r}\right) + \frac{1}{r^2 \sin \vartheta} \frac{\partial}{\partial \vartheta}\left(\sin \vartheta \frac{\partial p}{\partial \vartheta}\right) = \frac{1}{c^2} \frac{\partial^2 p}{\partial t^2}$$

If p is set equal to the factored quantity $p = R(r)P(\vartheta)e^{-i\omega t}$, the equation can be separated, becoming

$$\frac{1}{R} \frac{d}{dr}\left(r^2 \frac{dR}{dr}\right) + k^2 r^2 = -\frac{1}{P} \frac{1}{\sin \vartheta} \frac{d}{d\vartheta}\left(\sin \vartheta \frac{dP}{d\vartheta}\right) \qquad k = \frac{\omega}{c}$$

Each side of this equation must equal the same constant, C. On the other hand, if we wish to include dependence also on the axial angle φ, we use the factored form $p = R(r)P(\vartheta)\Phi(\varphi)e^{-i\omega t}$, giving

$$\frac{d^2\Phi}{d\varphi^2} + n^2\Phi = 0 \qquad \Phi = \cos{(n\varphi)} \text{ or } \sin{(n\varphi)}$$

$$\frac{1}{\sin\vartheta}\frac{d}{d\vartheta}\left(\sin\vartheta\frac{dP}{d\vartheta}\right) + \left(C - \frac{n^2}{\sin^2\vartheta}\right)P = 0$$

$$\frac{1}{r^2}\frac{d}{dr}\left(r^2\frac{dR}{dr}\right) + \left(k^2 - \frac{C}{r^2}\right)R = 0$$

(7.2.1)

where k^2, n^2, and C are the separation constants.

Legendre functions

The dependence on the axial coordinate φ is sinusoidal, as it is for the circular membrane of Eq. (5.2.18). In order that Φ have no discontinuity at $\varphi = 0, 2\pi$, n must be an integer. In a large number of cases there is axial symmetry, so that the wave is independent of φ, in which case $n = 0$ and the equation for the ϑ factor is

$$(1 - \eta^2)\frac{d^2P}{d\eta^2} - 2\eta\frac{dP}{d\eta} + CP = 0 \qquad (7.2.2)$$

where we have made the substitution $\eta = \cos\vartheta$ for the independent variable. This solution for $n = 0$ is doubly important, for the solutions for $n \neq 0$ can be expressed in its terms. In fact, it is not difficult to show that the first two equations of (7.2.1) are satisfied by the solutions

$$\cos \text{ine or sine of } (n\varphi) \text{ times } (1 - \eta^2)^{\frac{1}{2}n}\frac{d^n}{d\eta^n}P(\eta) \qquad (7.2.3)$$

where $P(\eta) = P(\cos\vartheta)$ is a solution of Eq. (7.2.2). Therefore we concentrate first on solving this equation, for the axially symmetric solutions.

We solve it by setting P equal to an arbitrary power series in η, substituting this in Eq. (7.2.2), and solving for the coefficients. The result is

$$P = a_0\left[1 - \frac{C}{2!}\eta^2 - \frac{C(6 - C)}{4!}\eta^4 - \frac{C(6 - C)(20 - C)}{6!}\eta^6 - \cdots\right]$$

$$+ a_1\left[\eta + \frac{2 - C}{3!}\eta^3 + \frac{(2 - C)(12 - C)}{5!}\eta^5 + \cdots\right]$$

These two series diverge at $\eta = \pm 1$ if the series are infinite ones, with all the powers of η. The only way to avoid this is to have one or the other series break off at a finite number of terms, i.e., to have $C = 0$ or $C = 6$ or $C = 20$, etc., and have $a_1 = 0$, or else to have $C = 2$ or $C = 12$, etc., and have $a_0 = 0$. Thus the allowed values of C are

$$C = m(m + 1) \qquad m = 0, 1, 2, 3, \ldots$$

The solution of Eq. (7.2.2) which is finite over the range of x from -1 to $+1$, corresponding to $C = m(m + 1)$, is labeled $P_m(\eta)$. It can be obtained by substituting the proper value for C in the series given above, making one a equal to zero and giving the other a the value that makes $P_m(1) = 1$. The resulting solutions are

$$
\begin{array}{lll}
m = 0, & C = 0: & P_0(\eta) = 1 \\
 & & P_0(\cos \vartheta) = 1 \\
m = 1, & C = 2: & P_1(\eta) = \eta \\
 & & P_1(\cos \vartheta) = \cos \vartheta \\
m = 2, & C = 6: & P_2(\eta) = \tfrac{1}{2}(3\eta^2 - 1) \qquad\qquad (7.2.4) \\
 & & P_2(\cos \vartheta) = \tfrac{1}{4}(3 \cos 2\vartheta + 1) \\
m = 3, & C = 12: & P_3(\eta) = \tfrac{1}{2}(5\eta^3 - 3\eta) \\
 & & P_3(\cos \vartheta) = \tfrac{1}{8}(5 \cos 3\vartheta + 3 \cos \vartheta)
\end{array}
$$

. . . .

The function P_m is called a *Legendre function* of order m. It can be shown to have the following properties:

$$(\eta^2 - 1)\frac{d^2 P_m}{d\eta^2} + 2\eta \frac{dP_m}{d\eta} - m(m + 1)P_m = 0 \qquad \eta = \cos \vartheta$$

$$P_m(\eta) = \frac{1}{2^m m!} \frac{d^m}{d\eta^m} (\eta^2 - 1)^m$$

$$(2m + 1)\eta P_m(\eta) = (m + 1)P_{m+1}(\eta) + mP_{m-1}(\eta) \qquad (7.2.5)$$

$$(2m + 1)P_m(\eta) = \frac{d}{d\eta}[P_{m+1}(\eta) - P_{m-1}(\eta)]$$

$$\int_{-1}^{1} P_n(\eta)P_m(\eta)\, d\eta = \begin{cases} 0 & n \neq m \\ \dfrac{2}{2m + 1} & n = m \end{cases}$$

Values of some of these functions are given in Table VI.

The last equation of (7.2.5) shows that the functions $P_m(\eta)$ constitute a set of orthogonal characteristic functions. Any function of η in the range from

$\eta = 1$ to $\eta = -1$ can be expanded in terms of a series of these functions:

$$F(\eta) = \sum_{m=0}^{\infty} B_m P_m(\eta) \qquad B_m = (m + \tfrac{1}{2}) \int_{-1}^{1} F(\eta) P_m(\eta) \, d\eta \qquad (7.2.6)$$

The expression for the coefficients B_m can be obtained by the method discussed on page 123.

Having defined the axially symmetric angle factors for waves from spheres, we now turn to Eq. (7.2.3) to obtain the solutions which depend on φ. We define the *spherical harmonics*

$$Y_m(\vartheta) = Y_{m0}{}^1(\vartheta,\varphi) = P_m(\eta) \qquad Y_{m0}{}^{-1} = 0; \qquad \eta = \cos \vartheta$$

$$Y_{mn}{}^1(\vartheta,\varphi) = \cos (n\varphi)(1 - \eta^2)^{\frac{1}{2}n} \frac{d^n}{d\eta^n} P_m(\eta)$$

$$0 < n \leqslant m$$

$$Y_{mn}{}^{-1}(\vartheta,\varphi) = \sin (n\varphi)(1 - \eta^2)^{\frac{1}{2}n} \frac{d^n}{d\eta^n} P_m(\eta)$$

$$\int_0^{2\pi} d\varphi \int_0^{\pi} Y_m(\vartheta) Y_l(\vartheta) \sin \vartheta \, d\vartheta = \frac{4\pi}{2m + 1} \delta_{ml} \qquad (7.2.7)$$

$$\int_0^{2\pi} d\varphi \int_0^{\pi} Y_{mn}{}^{\sigma}(\vartheta,\varphi) Y_{lk}{}^{\tau}(\vartheta,\varphi) \sin \vartheta \, d\vartheta = \frac{2\pi}{2m + 1} \frac{(m + n)!}{(m - n)!} \delta_{ml} \delta_{nk} \delta_{\sigma\tau}$$

$$P_m[\cos \vartheta \cos \theta - \sin \vartheta \sin \theta \cos (\varphi - \phi)]$$

$$= \sum_{m,n,o} \epsilon_n \frac{(m - n)!}{(m + n)!} Y_{mn}{}^{\sigma}(\vartheta,\varphi) Y_{mn}{}^{\sigma}(\theta,\phi)$$

where σ, τ have the values $+1$ or -1; m, n are positive integers or zero; and $Y_{mn} = 0$ if $n > m$. Also, the coefficient δ_{ml} equals 1 if $m = l$ and equals zero if $m \neq l$, and coefficient ϵ_n equals 1 if $n = 0$ and equals 2 if $n > 0$. The last equation gives the relation between a Legendre function of one side of a spherical triangle in terms of spherical harmonics of the other two sides and the angle between them; the order m of the harmonic is not changed by a change of axis. The integrals show that any function of the angles ϑ, φ on the surface of a sphere can be expressed in terms of a generalization of series (7.2.6), using the spherical harmonics $Y_{mn}{}^{\sigma}(\vartheta,\varphi)$.

The radial factor

We must now solve the equation for the radial factor R.

$$\frac{d}{dr}\left(r^2 \frac{dR}{dr}\right) + k^2 r^2 R = CR = m(m + 1)R \qquad k = \frac{\omega}{c}$$

where m must be zero or a positive integer (the order of the harmonic) for the angle factor Y to be finite everywhere. Changing scale, to get rid of k, we set

$\zeta = kr$ and obtain

$$\frac{d^2R}{d\zeta^2} + \frac{2}{\zeta}\frac{dR}{d\zeta} + [\zeta^2 + m(m+1)]\frac{R}{\zeta^2} = 0 \tag{7.2.8}$$

This equation looks very much like the Bessel equation (1.2.4); in fact, solutions of the equation are $(1/\sqrt{\zeta})J_{m+\frac{1}{2}}(\zeta)$. The solutions (for m an integer) can be expressed more simply in terms of the trigonometric functions, however. We have already shown, in Eq. (7.1.5), that a solution for $m = 0$ is $R = e^{ikr}/r$. We therefore try the expression $R = (e^{i\zeta}/\zeta)F_m(\zeta)$ and obtain the equation

$$\frac{d^2F_m}{d\zeta^2} + \left(2 - \frac{2m}{\zeta}\right)\frac{dF_m}{d\zeta} - \frac{2m}{\zeta}F_m = 0$$

Assuming a power series for F_m soon demonstrates that it must be a polynomial in powers of $1/\zeta$ when m is an integer. Direct substitution demonstrates that

$$F_m(\zeta) = \sum_{s=0}^{m} \frac{(m+s)!}{s!\,(m-s)!}\left(\frac{i}{2\zeta}\right)^s \to \begin{cases} 1 & \zeta \to \infty \\ \dfrac{(2m)!}{m!}\left(\dfrac{i}{2\zeta}\right)^m & \zeta \to 0 \end{cases}$$

Therefore we define the solution of Eq. (7.2.8), the radial-wave factor to use for outgoing waves, as

$$h_m(\zeta) = \frac{i^{-m}}{i\zeta}\sum_{s=0}^{m}\frac{(m+s)!}{s!\,(m-s)!}\left(\frac{i}{2\zeta}\right)^s e^{i\zeta} \to \begin{cases} \dfrac{(2m)!}{im!\,\zeta}\left(\dfrac{1}{2\zeta}\right)^m & \zeta \to 0 \\ \dfrac{i^{-m-1}}{\zeta}e^{i\zeta} & \zeta \to \infty \end{cases} \tag{7.2.9}$$

This function is called the *spherical Hankel function* of order m. The first few such functions are

$$h_0(\zeta) = \frac{e^{i\zeta}}{i\zeta} \qquad h_1(\zeta) = -\frac{e^{i\zeta}}{\zeta}\left(1 + \frac{i}{\zeta}\right)$$

$$h_2(\zeta) = \frac{ie^{i\zeta}}{\zeta}\left(1 + \frac{3i}{\zeta} - \frac{3}{\zeta^2}\right) \qquad \cdots \qquad \text{where} \quad \zeta = kr$$

We note that the simple source of Eq. (7.1.5) has $h_0(kr)$ for its radial factor and that $h_1(kr)$ is the radial factor for the dipole radiation of Eq. (7.1.7). In fact, the three dipole components, along the x, y, and z coordinates, can be expressed in terms of the three solutions for $m = 1$: $Y_{11}{}^1h_1 = \cos\varphi\sin\vartheta$ $P_1'(\cos\vartheta)h_1(kr)$, $Y_{11}{}^{-1}h_1 = \sin\varphi\sin\vartheta\, P_1'h_1(kr)$, $Y_{10}{}^1h_1 = P_1(\cos\vartheta)h_1(kr)$, where $P_1'(\eta) = dP_1/d\eta = 1$, since $P_1(\eta) = \eta$. Thus the three possible solutions for $m = 1$ correspond to the three orientations of dipole radiation.

Similarly, the set for $m = 2$ corresponds to quadrupole radiation. For example, the xy component of Eq. (7.1.9) is proportional to

$$2 \cos \varphi \sin \varphi \sin^2 \vartheta \, P_2''(\cos \vartheta) h_2(kr) = 3 \sin (2\varphi) \sin^2 \vartheta \, h_2(kr)$$

which is $Y_{22}^{-1}(\vartheta, \varphi) h_2(kr)$, and the xz term is proportional to $Y_{21}^{1}(\vartheta, \varphi) h_2(kr)$. As noted earlier, one combination of the axial quadrupole waves yields the simple source wave $h_0(kr)$. Two others, however, are waves of order $m = 2$, for

$$\frac{1}{2r^2} (3z^2 - r^2) h_2(kr) = P_2(\cos \vartheta) h_2(kr) = Y_{20}^{1} h_2$$

$$\frac{3}{r^2} (x^2 - y^2) h_2(kr) = Y_{22}^{1}(\vartheta, \varphi) h_2(kr)$$

Thus the five independent quadrupole waves correspond to the five possible spherical-wave solutions for $m = 2$, one for $n = 0$, two for $n = 1$, and two for $n = 2$.

Spherical Bessel functions

The spherical Hankel functions $h_m(kr)$ correspond to outgoing waves, appropriate for situations where acoustic energy is radiating outward into an unbounded medium. Such a wave goes to infinity at $r = 0$ as r^{-m-1}, which does not preclude using it for this purpose since no physical source has zero radius, and the point $r = 0$ is not included in a function valid only outside the source. It does preclude our using the spherical Hankel function for a wave in the interior of a spherical enclosure, where r does go to zero; so we must find other solutions of Eq. (7.2.8) which are finite at $r = 0$. These are obtained by taking the real part of $h_m(kr)$. We define the *spherical Bessel function* of order m, $j_m(\zeta)$, as the real part of $h_m(\zeta)$, and while we are at it, we define the *spherical Neumann function* of order m, $n_m(\zeta)$, as the imaginary part of h_m.

$$h_m(\zeta) = j_m(\zeta) + i n_m(\zeta) \qquad j_m, n_m \text{ real} \qquad (7.2.10)$$

It turns out that j_m is finite at $r = 0$ but n_m is not. Since h_m is a solution of Eq. (7.2.8), so also are both j_m and n_m. Their specific expressions, for the first values of m, are

$$j_0(z) = \frac{\sin z}{z} \qquad\qquad n_0(z) = -\frac{\cos z}{z}$$

$$j_1(z) = \frac{\sin z}{z^2} - \frac{\cos z}{z} \qquad n_1(z) = -\frac{\sin z}{z} - \frac{\cos z}{z^2} \qquad (7.2.11)$$

$$j_2(z) = \left(\frac{3}{z^3} - \frac{1}{z}\right) \sin z - \frac{3}{z^2} \cos z \qquad n_2(z) = -\frac{3}{z^2} \sin z - \left(\frac{3}{z^3} - \frac{1}{z}\right) \cos z$$

. . . .

These functions have the following properties:

$$j_m(\zeta) = \sqrt{\frac{\pi}{2\zeta}} J_{m+\frac{1}{2}}(\zeta) \qquad n_m(\zeta) = \sqrt{\frac{\pi}{2\zeta}} N_{m+\frac{1}{2}}(\zeta)$$

$$j_m(\zeta) \xrightarrow[\zeta \to 0]{} \frac{\zeta^m}{1 \cdot 3 \cdot 5 \cdots (2m+1)}$$

$$n_m(\zeta) \xrightarrow[\zeta \to 0]{} -\frac{1 \cdot 1 \cdot 3 \cdot 5 \cdots (2m-1)}{\zeta^{m+1}}$$

$$j_m(\zeta) \xrightarrow[\zeta \to \infty]{} \frac{1}{\zeta} \cos\left(\zeta - \frac{m+1}{2}\pi\right) \qquad n_m(\zeta) \xrightarrow[\zeta \to \infty]{} \frac{1}{\zeta} \sin\left(\zeta - \frac{m+1}{2}\pi\right) \quad (7.2.12)$$

$$\int j_0^2(\zeta)\zeta^2 \, d\zeta = \frac{\zeta^3}{2} [j_0^2(\zeta) + n_0(\zeta) j_1(\zeta)]$$

$$\int n_0^2(\zeta)\zeta^2 \, d\zeta = \frac{\zeta^3}{2} [n_0^2(\zeta) - j_0(\zeta) n_1(\zeta)]$$

$$h_m'(\zeta) j_m(\zeta) - j_m'(\zeta) h_m(\zeta) = \frac{i}{\zeta^2} \qquad j' = \frac{dj}{d\zeta}$$

and have the following properties, for either j_m, n_m, or h_m:

$$j_{m-1}(\zeta) + j_{m+1}(\zeta) = \frac{2m+1}{\zeta} j_m(\zeta)$$

$$\frac{d}{d\zeta} j_m(\zeta) = \frac{1}{2m+1} [m j_{m-1}(\zeta) - (m+1) j_{m+1}(\zeta)]$$

$$\frac{d}{d\zeta} [\zeta^{m+1} j_m(\zeta)] = \zeta^{m+1} j_{m-1}(\zeta) \qquad \frac{d}{d\zeta} [\zeta^{-m} j_m(\zeta)] = -\zeta^{-m} j_{m+1}(\zeta) \quad (7.2.13)$$

$$\int j_1(z) \, dz = -j_0(z) \qquad \int j_0(z) z^2 \, dz = z^2 j_1(z)$$

$$\int j_m^2(z) z^2 \, dz = \frac{z^3}{2} [j_m^2(z) - j_{m-1}(z) j_{m+1}(z)] \qquad m > 0$$

In all Eqs. (7.2.13) the function n can be substituted for j. Values of some of the functions j and n are given in Table VII.

Radiation from a general spherical source

We take next the general case of a sphere, not necessarily small, whose surface vibrates with a velocity $U(\vartheta)e^{-2\pi i v t}$, where U is any sort of function of ϑ. We first express the velocity amplitude $U(\vartheta)$ in terms of a series of Legendre functions

$$U(\vartheta) = \sum_{m=0}^{\infty} U_m P_m(\cos \vartheta)$$

$$U_m = (m + \tfrac{1}{2}) \int_0^\pi U(\vartheta) P_m(\cos \vartheta) \sin \vartheta \, d\vartheta \qquad (7.2.14)$$

To correspond to this we also express the radiated pressure wave in a series

$$p = \sum_{m=0}^{\infty} A_m P_m(\cos \vartheta) h_m(kr) e^{-i\omega t}$$

where the values of the coefficients A_m must be determined in terms of the known coefficients U_m. The radial velocity of the air at the surface of the sphere is

$$u_a = \frac{1}{\rho c} \sum_{m=0}^{\infty} A_m B_m P_m(\cos \vartheta) e^{i\delta_m - i\omega t}$$

where

$$\frac{d}{d\zeta} h_m(\zeta) = i B_m(\zeta) e^{i\delta_m(\zeta)} \qquad \zeta = ka = \frac{2\pi a}{\lambda}$$

Therefore, using Eqs. (7.2.12) and (7.2.13),

$$m n_{m-1}(ka) - (m + 1) n_{m+1}(ka) = (2m + 1) B_m \cos \delta_m$$
$$(m + 1) j_{m+1}(ka) - m j_{m-1}(ka) = (2m + 1) B_m \sin \delta_m$$

When $ka \gg m + \frac{1}{2}$,

$$B_m \simeq \left(\frac{1}{ka}\right) \qquad \delta_m \simeq ka - \tfrac{1}{2}\pi(m + 1)$$

When $ka \ll m + \frac{1}{2}$, (7.2.15)

$$B_0 \simeq \left(\frac{1}{ka}\right)^2 \qquad \delta_0 \simeq \tfrac{1}{3}(ka)^3$$

$$B_m \simeq \frac{1 \cdot 3 \cdot 5 \cdots (2m - 1)(m + 1)}{(ka)^{m+2}} \qquad m > 0$$

$$\delta_m \simeq \frac{-m(ka)^{2m+1}}{1^2 \cdot 3^2 \cdot 5^2 \cdots (2m - 1)^2(2m + 1)(m + 1)}$$

Values of the amplitudes B_m and the phase angles δ_m are given in Table VIII.

The radial velocity of the air at $r = a$ must equal that of the surface of the sphere, and equating coefficients of the two series, term by term, we obtain equations for the coefficients A_m in terms of U_m:

$$A_m = \frac{\rho c U_m}{B_m} e^{-i\delta_m}$$

The pressure and radial velocity at large distances (i.e., many wavelengths) from the sphere can then be expressed

$$u_r \simeq U_0 \frac{a}{r} e^{ik(r - ct)} \psi(\vartheta) \qquad p \simeq \rho c u_r$$

(7.2.16)

$$\psi(\vartheta) = \frac{1}{ka} \sum_{m=0}^{\infty} \frac{U_m}{U_0 B_m} P_m(\cos \vartheta) e^{-i\delta_m - \frac{1}{2}i\pi(m+1)}$$

where we have multiplied and divided by U_0, the average velocity of the surface of the sphere. The air velocity near the sphere is, of course, not in phase with the pressure, nor is it entirely radial; but far from the sphere the velocity is radial and in phase with the pressure.

When kr is very large, the intensity at the point (r,ϑ) and the total power radiated are

$$I_r \simeq \rho c U_0{}^2 \frac{a^2}{r^2} F_r(\vartheta) \qquad F_r(\vartheta) = |\psi(\vartheta)|^2$$

$$F_r(\vartheta) = \left(\frac{1}{kaU_0}\right)^2 \sum_{m,n=0}^{\infty} \frac{U_m U_n}{B_m B_n} P_m(\cos\vartheta) P_n(\cos\vartheta)$$
$$\times \cos\left[\delta_m - \delta_n + \tfrac{1}{2}\pi(m-n)\right] \qquad (7.2.17)$$

$$\Pi = \int_0^{2\pi} d\phi \int_0^{\pi} r^2 I_r \sin\vartheta \, d\vartheta = \frac{\rho c^3}{4\pi\nu^2} \sum_{m=0}^{\infty} \frac{U_m{}^2}{(2m+1)B_m{}^2}$$

The function $F_r(\vartheta)$ is called the *angle-distribution function* for radiation from a sphere.

When $ka = 2\pi a/\lambda$ is quite small, all but the first terms in these expansions can be neglected, since B_m then increases quite rapidly with m. Consequently, the pressure and intensity a distance r from the center of the sphere, at any angle, are

$$p \simeq -\frac{ik\rho c}{4\pi r}(4\pi a^2 U_0)e^{ik(r-ct)} \qquad I_r \simeq \rho c\,|U_0|^2 \frac{k^2 a^4}{r^2}$$

equivalent to those of a simple source of strength $S_\omega = 4\pi a^2 U_0$.

On the other hand, when ka is quite large, B_m is equal to $1/ka$ for all values of m from zero up to a value roughly equal to ka, then drops sharply in value for $m > ka$. Suppose the series (7.2.14), for the surface velocity $U(\vartheta)$, includes terms for m up to some value M, but all coefficients U_m for $m > M$ are negligible. Then, when $ka > M$ $(2\pi a > M\lambda)$, all the non-negligible terms in the series for the angle-distribution factor ψ are simply proportional to U_m and $\psi \simeq (1/U_0)U(\vartheta)e^{-ika}$. In this case the pressure and the radiated intensity at great distances r from the sphere are

$$p(r,\vartheta,\varphi) \simeq \rho c U(\varphi)\frac{a}{r}e^{ik(r-a-ct)} \qquad I_r \simeq \rho c\,|U(\vartheta)|^2\left(\frac{a}{r}\right)^2 \qquad r \gg a$$

A spherical harmonic Y_{mn} of order m has m nodes and loops on the sphere's surface, and therefore a "wavelength" of about $2\pi a/m$. What we have just said is that when the wavelength of the radiation is smaller than the "fine structure" of the velocity distribution on the surface of the sphere, the pressure at r, ϑ is proportional to the velocity at a, ϑ, as though the sound traveled radially outward from the surface, the intensity above a region of large motion being greater than that above a quiet region. At $r = a$ the impedance opposing the surface motion is ρc and the intensity is $\rho c\,|U(\vartheta)|^2$. This is the region of "geometrical acoustics." At longer wavelengths the pattern of the velocity distribution $U(\vartheta)$ becomes blurred out in the radiated wave and the intensity decreases, until, for long wavelengths, the sphere radiates like a simple source.

Of course, if $U_0 = 0$, this limiting expression is invalid. For example, suppose the wave is caused by a rigid sphere, of radius a, moving to and fro along the z axis with velocity $U_1 e^{-i\omega t}$. The radial velocity of the fluid at the surface $r = a$ is then $U_1 \cos \vartheta e^{-i\omega t}$, and the resulting pressure wave is $(\rho c U_1/B_1)e^{-i\delta_1} \cos \vartheta\, h_1(kr)e^{-i\omega t}$. This is, of course, a dipole wave, of dipole strength $D_\omega = (4\pi/k^3 B_1)U_1$; in the limit of $ka \ll 1$, this strength becomes $2\pi a^3 U_1$, the cross section πa^2 of the sphere times $2aU_1$. In this case, since there is no $m = 0$ term present, the limiting form of the wave is that given in Eq. (7.1.7).

Radiation from a point source on a sphere

To show graphically how the radiation changes from symmetrical to directional as the frequency of the radiated sound increases, we shall work out the details of the radiation problem for two cases.

The first case is that of a point source at the point $\vartheta = 0$ on the surface; i.e., the surface velocity of the sphere is zero except for a small circular area of radius Δ around $\vartheta = 0$. The definition of $U(\vartheta)$ is

$$U(\vartheta) = \begin{cases} u_0 & 0 \leqslant \vartheta < \dfrac{\Delta}{a} \\[2mm] 0 & \dfrac{\Delta}{a} < \vartheta \leqslant \pi \end{cases}$$

and the coefficients U_m are

$$U_m = (m + \tfrac{1}{2})u_0 \int_0^{\Delta/a} P_m(\cos \vartheta) \sin \vartheta \, d\vartheta \xrightarrow[\Delta \to 0]{} \tfrac{1}{2}(m + \tfrac{1}{2})u_0 \left(\frac{\Delta}{a}\right)^2$$

since $P_m(1) = 1$. The intensity and power radiated can be obtained from Eqs. (7.2.17):

$$I_r \simeq \rho c u_0{}^2 \frac{\Delta^4}{16a^2 r^2} F_r(\vartheta) \tag{7.2.18}$$

$$F_r(\vartheta) = \frac{1}{\mu^2} \sum_{m,n=0}^{\infty} \frac{(2m+1)(2n+1)}{B_m B_n} P_m(\cos \vartheta) P_n(\cos \vartheta)$$
$$\times \cos[\delta_m - \delta_n + \tfrac{1}{2}\pi(m-n)]$$

$$\Pi = \rho c u_0{}^2 \frac{\Delta^4}{16a^2} \frac{4\pi}{\mu^2} \sum_{m=0}^{\infty} \frac{2m+1}{B_m{}^2} \qquad \mu = \frac{2\pi a}{\lambda}$$

At very low frequencies, only the first terms in these series are important, and the pressure wave again has the familiar form for the simple source

$$p \simeq -i\omega \frac{\rho}{4\pi r} (\pi \Delta^2 u_0) e^{ik(r-ct)}$$

with a strength equal to the velocity u_0 times $\pi \Delta^2$, the area of the radiating element.

Figure 7.1 gives curves for the distribution in angle of the intensity radiated from a point source on a sphere for different ratios of wavelength to sphere circumference. We see again the gradual change from radiation in all directions to a sharply directional pattern as the frequency is increased.

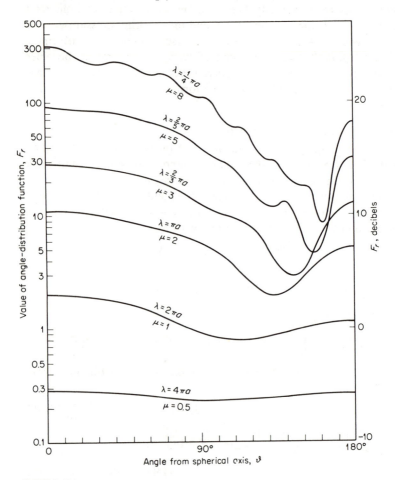

FIGURE 7.1
Distribution in angle of intensity radiated from a point source set in the surface of a sphere of radius a, for different values of $\mu = 2\pi a/\lambda$. Curves also give mean-square pressure at point (a,ϑ) on surface of sphere due to incident plane wave traveling in negative x direction.

These curves are of particular interest because of their dual role, as required by the *principle of reciprocity*, mentioned in Sec. 7.1. As computed, the curves give intensity or pressure amplitude squared at a point (r,ϑ), a considerable distance r from the sphere (when $kr \gg 1$, $|p| = \sqrt{\rho c I_r}$) at an angle

ϑ with respect to the line from the center of the sphere through the radiating element of area at point $(a,0)$. But the principle of reciprocity says that the pressure at a point Q due to a unit simple source at point P is equal to the pressure at point P due to a unit source at point Q. Consequently, the curves of Fig. 7.1 also represent the square of the pressure amplitude at a point $(a,0)$ on the surface of the sphere due to a point source of strength $\pi\Delta^2 u_0$ at the point (r,ϑ). Therefore the curves are useful as an indication of the directional properties of the ear plus head, or of a microphone in a roughly spherical housing.

Radiation from a piston set in a sphere

The other example to be worked out is that of a piston of radius $a \sin \vartheta_0$ set in the side of a rigid sphere. As long as ϑ_0 is not too large, this corresponds fairly closely to the following distribution of velocity on the surface of the sphere:

$$U(\vartheta) = \begin{cases} u_0 & 0 \leqslant \vartheta < \vartheta_0 \\ 0 & \vartheta_0 < \vartheta \leqslant \pi \end{cases}$$

The general formulas (7.2.17) can be used, with

$$U_m = (m + \tfrac{1}{2})u_0 \int_{\cos\vartheta_0}^1 P_m(x)\, dx = \tfrac{1}{2}u_0[P_{m-1}(\cos\vartheta_0) - P_{m+1}(\cos\vartheta_0)]$$

where, for the case $m = 0$, we consider $P_{-1}(x) = 1$.

These expressions for U_m can be substituted in Eq. (7.2.17) to give series for intensity and power. Curves for I as function of ϑ are given in Fig. 7.2. When the wavelength is long compared with $2\pi a$, the pressure and intensity are those for a simple source of strength $4\pi a^2 u_0 \sin^2(\vartheta_0/2)$.

The radiation impedance for the piston set in a sphere can be computed by integrating the expression for the pressure at $r = a$ over the surface of the piston. After quite a little algebraic juggling, we find

$$Z_p = R_p - iX_p = \rho c 4\pi a^2 \sin^2\left(\frac{\vartheta_0}{2}\right)(\theta_p - i\chi_p)$$

$$\theta_p = \frac{1}{4}\sum_{m=0}^{\infty} \frac{[P_{m-1}(\cos\vartheta_0) - P_{m+1}(\cos\vartheta_0)]^2}{\mu^2(2m+1)B_m^2 \sin^2(\vartheta_0/2)} \qquad (7.2.19)$$

$$\chi_p = \frac{1}{4}\sum_{m=0}^{\infty} \frac{[P_{m-1}(\cos\vartheta_0) - P_{m+1}(\cos\vartheta_0)]^2}{(2m+1)B_m \sin^2(\vartheta_0/2)}$$

$$\times [j_m(\mu)\sin(\delta_m) - n_m(\mu)\cos(\delta_m)]$$

where $\mu = ka = 2\pi a/\lambda$.

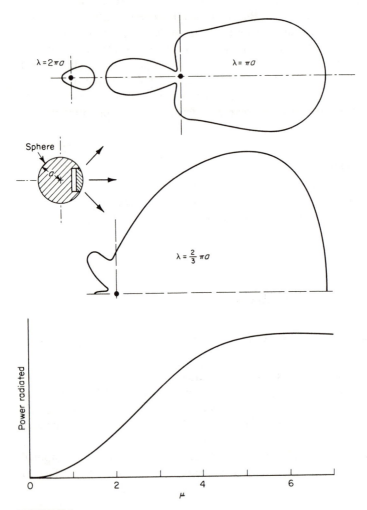

FIGURE 7.2
Distribution in angle of intensity and total power radiated
from a piston set in a sphere.

Since $4\pi a^2 \sin^2 (\vartheta_0/2) = \pi a_p{}^2$ is the area of the piston (a_p thus defined being the effective piston radius), the dimensionless quantity $\theta_p - i\chi_p$ is the impedance per unit area, divided by the characteristic impedance ρc of the medium. The specific resistance θ_p and reactance χ_p are plotted in Fig. 7.3 as functions of the ratio of equivalent diaphragm circumference to wavelength, $2\pi a_p/\lambda = (\omega a_p/c)$ [$a_p = 2a \sin (\vartheta_0/2)$], for several different values of $\vartheta_0 = 2 \sin^{-1} (a_p/2a)$. We notice that the resistive terms all are small at low frequencies, increasing as ω^2; rising to a value of approximately unity

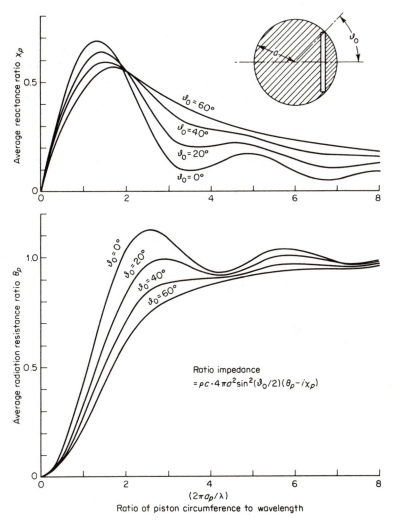

FIGURE 7.3
Values of radiation resistance and reactance ratios $(R_r, X_r)/\pi a_p^2 \rho c$, where $a_p = 2a \sin(\vartheta_0/2)$, as function of $2\pi a_p/\lambda = \omega a_p/c$ for different values of ϑ_0, for radiation of sound from a piston set in a sphere of radius a.

at wavelengths a little smaller than one-third times the equivalent diaphragm circumference; and then, for very high frequencies, approaching the usual limit of 1 ($z = \rho c$). The reactance is always positive, representing a mass load. It first increases linearly with frequency, as the reactance due to a constant mass would, but then it reaches a maximum at $\lambda \simeq 2\pi a$, and for higher frequencies it diminishes more or less rapidly.

These curves indicate the sort of radiation load one would expect on the diaphragm of a dynamic loudspeaker, set in a spherical case. The diaphragm of the dynamic speaker is not usually the surface of a sphere, but often has the shape of an inverted cone in order to increase its mechanical strength. The radiation from such a cone would naturally differ from that from a section of a sphere; but it turns out that the average radiation impedance on a piston is approximately the same, no matter what its shape, as long as its circumference is not changed and as long as the volume of the mounting case is not changed. Therefore the radiation reaction on the diaphragm of a dynamic speaker of outer circumference $2\pi a_p$, set in a cabinet of volume $V_p = 4\pi a^3/3$, is approximately given by $\pi a_p{}^2 \rho c$ times the curves of Fig. 7.3 for $\vartheta_0 = 2\sin^{-1}[a_p(\pi/6V_p)^{\frac{1}{3}}]$, with the frequency scale equal to $2\pi \nu a_p/c$.

Multipoles and Legendre functions

Since the Green's function $g_\omega(\mathbf{r}\,|\,\mathbf{r}_0)$ of Eqs. (7.1.6) and (7.1.14) may be expressed in spherical coordinates, it should be possible to express it in terms of Legendre functions and spherical Hankel functions. First we discuss the derivatives of g_ω with respect to x_0, y_0, z_0 at $\mathbf{r}_0 = 0$, the basis for our earlier discussion of multipole radiation.

The Green's function itself is

$$g_\omega(\mathbf{r}\,|\,0) = \frac{ik}{4\pi}\,h_0(kr) = \frac{1}{4\pi r}\,e^{ikr} \tag{7.2.20}$$

in terms of the spherical Hankel function, defined in Eqs. (7.2.9). The three dipole waves are proportional to the x, y, z components of the gradient of g_ω at $\mathbf{r}_0 = 0$.

$$g_x'(\mathbf{r}\,|\,0) = \left[\frac{\partial}{\partial x_0}\,g_\omega(\mathbf{r}\,|\,\mathbf{r}_0)\right]_{r_0=0} = \frac{ik^2}{4\pi}\,Y_{11}{}^1(\vartheta,\varphi)h_1(kr)$$

$$g_y'(\mathbf{r}\,|\,0) = \frac{ik^2}{4\pi}\,Y_{11}{}^{-1}(\vartheta,\varphi)h_1(kr) \qquad g_z'(\mathbf{r}\,|\,0) = \frac{ik^2}{4\pi}\,Y_1(\vartheta)h_1(kr) \tag{7.2.21}$$

involving all three of the spherical harmonics of order $m = 1$. The five spherical harmonics of order $m = 2$ all represent quadrupole radiation, as defined in Eqs. (7.1.9) and (7.1.10). Thus

$$g_{xy}''(\mathbf{r}\,|\,0) = \left[\frac{\partial^2}{\partial x_0\,\partial y_0}\,g_\omega(\mathbf{r}\,|\,\mathbf{r}_0)\right]_{r_0=0} = \frac{ik^3}{24\pi}\,Y_{22}{}^{-1}(\vartheta,\varphi)h_2(kr)$$

$$g_{xz}''(\mathbf{r}\,|\,0) = \frac{ik^3}{12\pi}\,Y_{21}{}^1(\vartheta,\varphi)h_2(kr) \qquad g_{yz}'' = \frac{ik^3}{12\pi}\,Y_{21}{}^{-1}(\vartheta,\varphi)h_2(kr) \tag{7.2.22}$$

as mentioned earlier. The axial solutions, as mentioned, contain a certain amount of the $m = 0$ wave as well:

$$g_{zz}''(\mathbf{r} \,|\, 0) = \frac{ik^3}{24\pi} [4 Y_{20}^1(\vartheta,\varphi)h_2(kr) - 2 Y_{00}^1(\vartheta,\varphi)h_0(kr)]$$

$$g_{xx}''(\mathbf{r} \,|\, 0) = \frac{ik^3}{24\pi} [Y_{22}^1 h_2 - 2 Y_{20}^1 h_2 - 2h_0] \qquad (7.2.23)$$

$$g_{yy}''(\mathbf{r} \,|\, 0) = \frac{ik^3}{24\pi} [- Y_{22}^1 h_2 - 2 Y_{20}^1 h_2 - 2h_0]$$

with the Y's as defined in Eqs. (7.2.7). Of course, $Y_{00}^1 = 1$ and $Y_{20}^1 = P_2(\cos \vartheta)$. Therefore the multipole fields of Eqs. (7.1.5) to (7.1.10) may be expressed in terms of spherical harmonics and spherical Hankel functions as follows:

$$p_s(\mathbf{r}) = \frac{\rho c k^2}{4\pi} S_\omega Y_{00}^1(\vartheta,\varphi)h_0(kr)$$

$$p_d(\mathbf{r}) = \frac{\rho c k^3}{4\pi} (D_x Y_{11}^1 + D_y Y_{11}^{-1} + D_z Y_{10}^1)h_1(kr)$$

$$\begin{aligned}
p_q(\mathbf{r}) = \frac{\rho c k^4}{24\pi} \{ &-2(Q_{xx} + Q_{yy} + Q_{zz})h_0(kr) \\
&+ [2(Q_{zz} - Q_{xx} - Q_{yy}) Y_{20}^1(\vartheta,\varphi) \\
&+ 2(Q_{xz} + Q_{zx}) Y_{21}^1 + 2(Q_{yz} + Q_{zy}) Y_{21}^{-1} \\
&+ (Q_{xx} - Q_{yy}) Y_{22}^1 + (Q_{xy} + Q_{yx}) Y_{22}^{-1}]h_2(kr)\}
\end{aligned} \qquad (7.2.24)$$

In accord with page 314, there are only five distinct quadrupole distributions corresponding to the five different spherical harmonics of order 2.

Sphere with randomly vibrating surface

It should be evident by now that we are using the sphere as a simplified model of an object which has roughly the same width, height, and depth. The details of radiation from a more complicated shape differ somewhat from those computed for the sphere, but the general acoustical behavior will be pretty much the same. As we mentioned previously, a loudspeaker in a cubical cabinet will produce a sound field which is not markedly different from that portrayed in Figs. 7.2 and 7.3.

Another type of sound field which also can be approximated by using the spherical model is the sort of random noise which surrounds a piece of machinery in violent vibration, or the sort of roar which is generated when highly turbulent air reacts with the surface of a large object placed in it (or moving through it). In many cases the machine action (or the turbulence) produces vibrations of the outer surface of the object which are random in nature, containing components of all frequencies and having little correlation between the motion of one area of the surface and another. Such a source of sound can be approximated by a spherical surface in random oscillation.

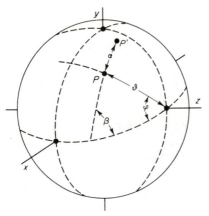

FIGURE 7.4
Angles defining the correlation function on the surface of a sphere.

To describe such motion we go back again to Eqs. (1.3.20) and (2.3.11) [see also Eqs. (7.1.33) to (7.1.38)], relating the Fourier transform of the motion and the autocorrelation function of the motion.   We first take the Fourier transform, with respect to time, of the surface velocity, obtaining the component $U_\omega(\vartheta,\varphi)$, for the frequency $\omega/2\pi$, of the radial velocity of the part of the spherical surface at the point specified by the angles ϑ, φ.

Even the single-frequency motion is random in regard to its spatial distribution over the sphere, so that we use an autocorrelation function in space to describe this degree of randomness.   We define this function by the integral

$$\Upsilon_\omega(\alpha) = \frac{1}{4\pi} \int_0^{2\pi} d\varphi \int_0^\pi U_\omega(P) U_\omega(P') \sin \vartheta \, d\vartheta \qquad (7.2.25)$$

where point P has coordinates ϑ, φ as shown in Fig. 7.4, and point P' is an angular distance α from P, in a direction at angle β with respect to the xz meridian.   If the motion is isotropically random, Υ_ω will be independent of β.   The value of $\Upsilon_\omega(0)$ is, of course, $|U_\omega^0|^2$, the mean-square radial velocity of the surface at frequency $\omega/2\pi$, the analogue of the spectrum density of Eq. (2.3.11).   For random motion, $\Upsilon_\omega(\alpha)$ will decrease as α is increased, dropping to zero when α is larger than the "mean correlation length," divided by the radius a of the sphere [see the discussion of Eq. (2.3.13)].

A rough approximation to such a correlation function would thus be

$$\Upsilon_\omega(\alpha) = \begin{cases} (\cos \alpha)^{N^2-1} |U_\omega^0|^2 & 0 \leqslant \alpha \leqslant \tfrac{1}{2}\pi \\ 0 & \tfrac{1}{2}\pi \leqslant \alpha \leqslant \pi \end{cases} \qquad (7.2.26)$$

which equals $|U_\omega^0|^2$ at $\alpha = 0$ and drops to zero (more rapidly the larger N is), with no correlation assumed between opposite halves of the sphere. The mean angular width of the peak in α is roughly $2\pi/N$, so that the *correlation length*, the mean distance along the surface within which the motion is more or less in phase, is $w \simeq 2\pi a/N$, one-Nth of the perimeter of the sphere.   We assume for the moment that N is independent of ω.

If this were a plane-surface problem [as that of Eq. (7.4.27) will be], so that the velocity and its correlation function could be expressed in terms of Fourier integrals, we could use Eq. (1.3.20) to find the transform of the velocity with respect to its wavenumber $k = 2\pi/\lambda$. Since the surface is a sphere, the analogue to the Fourier transform of $U(\vartheta,\varphi)$ is the coefficient $U_{mn}{}^\sigma$ in the expansion of $U(\vartheta,\varphi)$ in terms of the spherical harmonics $Y_{mn}{}^\sigma(\vartheta,\varphi)$, the discrete variable m being more or less the analogue of the continuous variable k. We should expect that the mean-square values of $U_{mn}{}^\sigma$ will be independent of n and σ, but will decrease slowly with m, corresponding to the decrease in component amplitude as the wavelength of the component decreases. The rate of this decrease should bear some relation to the correlation length w, analogous to the relationship between the dependence of Υ on τ and of $|F|^2$ on ω shown in Eq. (2.3.13).

The correlation function of Eq. (7.2.26) can be expanded in spherical harmonics; the coefficients are products of gamma functions, but when N is large, the approximate form,

$$\Upsilon_\omega(\alpha) \simeq \sum_{m=0}^\infty \frac{2\,|U_\omega^0|^2}{2m+1}\, N^{-2} e^{-2m(m+1)/N^2} P_m(\cos\alpha)$$

is a fairly good approximation. The coefficients for $n > 0$ do not appear because of the choice of axis centering around P for α; if the last of Eqs. (7.2.7) is used, the series can be transformed to one in $Y_{mn}{}^\sigma(\vartheta,\varphi)$. The coefficients of this series are the analogues of the Fourier transform of the correlation function, and thus should be proportional to the square of the radial velocity coefficient $U_{mn}{}^\sigma$. Consequently, the series for the radial velocity of the surface for frequency $\omega/2\pi$ is

$$U_\omega(\vartheta,\varphi) = \frac{U_\omega^0}{N} \sum_{m,n,\sigma} \left[\frac{2}{2m+1} \frac{(m-n)!}{(m+n)!} \epsilon_n \right]^{\frac{1}{2}} \exp\left[-\frac{m(m+1)}{N^2} + i\gamma_{mn} \right]$$
$$\times\, Y_{mn}(\vartheta,\varphi) \quad (7.2.27)$$

where we have taken the square root of the coefficient for Υ_ω, multiplied by a factor depending on n, to make the mean square of the $n > 0$ terms equal to that for $n = 0$. We have also included a phase γ_{mn} to allow for the fact that the phase of each term is a random angle between 0 and 2π, corresponding to the random nature of the velocity. The series for Υ_ω does not specify these angles.

This series is an obvious generalization of the series of Eq. (7.2.14), and is thus ready for us to calculate the radiated intensity and power. From Eq. (7.2.17) the intensity at the point r, ϑ, φ, a considerable distance from the sphere, is

$$I_r \to \tfrac{1}{2}\rho c\, |U_\omega^0|^2 \left(\frac{a}{r}\right)^2 |\psi(\vartheta,\varphi)|^2$$

where the angle-distribution function $|\psi|^2$ contains squares and cross products of the coefficients $U_{mn}{}^\sigma$. Because of the random nature of the phase relations between the $U_{mn}{}^\sigma$'s, and because these phase relations are changing with time (since the time variation is also random), the time average of $|\psi|^2$ will be the sum of the squared terms only.

$$|\psi|^2 \simeq (ka)^{-2} \sum_{m,n,\sigma} \frac{1}{2m+1} \frac{(m-n)!}{(m+n)!} \epsilon_n \frac{e^{-2m(m+1)N^2}}{N^2 B_m{}^2} (Y_{mn}{}^\sigma)^2$$

$$= \frac{1}{N^2 k^2 a^2} \sum_{m=0}^{\infty} \frac{e^{-2m(m+1)/N^2}}{(2m+1)B_m{}^2} \qquad N \gg 1 \qquad (7.2.28)$$

where the sum over n, σ has been accomplished by use of the last of Eqs. (7.2.7) for $\vartheta = 0$, $\varphi = \phi$. Thus the radiated intensity is independent of angle, as it should be if the random motion is isotropic.

This averaged series can be still further simplified by referring to the approximate formulas for B_m given in Eqs. (7.2.15). When m is smaller than $ka = 2\pi a/\lambda$, $(1/B_m)^2 \simeq (ka)^2$ and, as m is made larger than ka, $(1/kaB_m)^2$ drops rapidly to zero. Consequently, an approximate series for $|\psi|^2$, valid for $N^2 \gg 1$, is

$$|\psi|^2 \simeq \frac{1}{N^2} \sum_{m=0}^{ka} \frac{e^{-2m(m+1)/N^2}}{2m+1} \simeq 1 - e^{-(ka/N)^2} \qquad (7.2.29)$$

where

$$\frac{ka}{N} = \frac{2\pi a}{\lambda} \frac{w}{2\pi a} = \frac{w}{\lambda} = \frac{w\omega}{2\pi c}$$

For very small values of this parameter, $|\psi|^2$ is approximately $(ka/N)^2$; for very large values, $|\psi|^2$ is approximately 1. The last expression for $|\psi|^2$ is a very crude approximation to the earlier series, but it should be satisfactory in view of the approximations we have been making on account of the random nature of the motion. It shows that when the correlation length w is small compared with λ, the intensity radiated at this wavelength, $\frac{1}{2}\rho c |U_\omega|^2 (wa/\lambda r)^2$, is as if a circular patch of the sphere's surface, of radius w, is sending energy to the point r, ϑ, φ, the radiation from all the rest of the surface interfering to zero. On the other hand, when w is large compared with λ (i.e., for the high-frequency part of the radiation), the radiated intensity $\frac{1}{2}\rho c |U_\omega|^2 (a^2/r^2)$ is as if the whole spherical surface were radiating with surface velocity equal to the root-mean-square velocity U_ω. When w is small, I_r is proportional to $\pi(wa/\lambda)^2$; when w is large, I_r is proportional to the total area πa^2 visible by the observer.

To bring in the time dependence of the motion, we must invert the Fourier analysis of ψ in frequency. The surface velocity is also random in time, with a time-correlation function $Y_f(\tau)$, which decreases as τ increases, like the Y_f of Eq. (2.3.13). The spectrum level $|F(\omega)|^2$ for the surface velocity, the square of the Fourier transform of the root-mean-square velocity of the surface, should be some function which goes to zero when ω

becomes larger than some upper-limit value ω_n. Equation (2.3.13) shows that the correlation time interval τ_n is the reciprocal of ω_n.

Suppose the time-correlation function for the surface velocity has the familiar form

$$\Upsilon_f(\tau) = |U_0|^2\, e^{-\frac{1}{2}\omega_n{}^2\tau^2} \qquad \tau_n = \frac{1}{\omega_n}$$

where $|U_0|^2$ is the mean-square surface velocity, averaged over the surface of the sphere and also over time T. According to Eq. (1.3.20), the spectrum density $|F(\omega)|^2$ for velocity is proportional to the Fourier transform of Υ_f,

$$|F(\omega)|^2 = \frac{T}{4\pi^2} \int_{-\infty}^{\infty} \Upsilon_f(\tau) e^{i\omega\tau}\, d\tau = \frac{T\,|U_0|^2}{\omega_n(2\pi)^{\frac{3}{2}}}\, e^{-\frac{1}{2}(\omega/\omega_n)^2}$$

Therefore the spectrum density for intensity at r, ϑ, φ is $\frac{1}{2}\rho c(a^2/r^2)\,|F(\omega)|^2\,|\psi|^2$, where ψ is the angle-distribution factor of Eqs. (7.2.28) and (7.2.29). The mean value of the intensity, averaged over time and surface, is then [according to Eq. (1.3.19)]

$$\langle I_r \rangle = \frac{2\pi}{T} \int_{-\infty}^{\infty} I_r(\omega)\, d\omega$$

$$= -\rho c \left(\frac{a}{r}\right)^2 \frac{|U_0|^2}{\omega_n\sqrt{2\pi}} \int_{-\infty}^{\infty} e^{-\frac{1}{2}(\omega/\omega_n)^2}[1 - e^{-(w\omega/2\pi c)^2}]\, d\omega \qquad (7.2.30)$$

$$= \tfrac{1}{2}\rho c \left(\frac{a}{r}\right)^2 |U_0|^2 \frac{2w^2}{\lambda_n{}^2 + 2w^2} \qquad \lambda_n = \frac{2\pi c}{\omega_n}$$

When the wavelength λ_n, corresponding to upper limit ω_n of the spectrum density, is small compared with the correlation length w of velocity distribution over the spherical surface, the intensity is as though the whole sphere were vibrating in phase, with a velocity amplitude equal to the root-mean-square velocity U_0. When λ_n is large compared with w, that is, when the cutoff frequency is low or the correlation length w is very small, the intensity is much smaller, as though a patch of the sphere, of radius aw/λ_n, were doing the radiating. In many cases encountered in practice, neither of these limiting cases holds, λ_n being one to four times w.

Many modifications and improvements could be made in this calculation; a different dependence on ω of the spectrum level could be used, or we could have taken into account possible variation of the correlation length on ω, or we could have used a better approximation for the factor $|\psi|^2$ of Eq. (7.2.29). The resulting formulas would then have had a somewhat different appearance, but if the surface vibration is random and isotropic, the relationship between I_r and the spatial and temporal correlation intervals, w and $1/\omega_n$, will be more or less that set forth in Eq. (7.2.30).

Green's function in spherical coordinates

To compute the radiation from a volume distribution of sources which is not small compared with the wavelength, or to develop an alternative way of computing the radiation from a vibrating sphere of moderate size, we need the expansion of the Green's function $g_\omega(\mathbf{r} \mid \mathbf{r}_0)$ in terms of the coordinates r, ϑ, φ of the measurement point and r_0, ϑ_0, φ_0 of the source point. This expansion is

$$g_\omega(\mathbf{r} \mid \mathbf{r}_0) = \frac{ik}{4\pi} h_0(kR)$$

$$= \frac{ik}{4\pi} \sum_{m,n,\sigma} (2m+1)\epsilon_n \frac{(m-n)!}{(m+n)!} Y_{mn}{}^\sigma(\vartheta,\varphi)$$

$$\times Y_{mn}{}^\sigma(\vartheta_0,\varphi_0) \begin{cases} j_m(kr_0)h_m(kr) & r > r_0 \\ j_m(kr)h_m(kr_0) & r < r_0 \end{cases} \quad (7.2.31)$$

which exhibits the reciprocity between measurement and source points.

For example, if a region of violent motion is confined inside a sphere of radius a (not necessarily small compared with λ), the radiation outside $r = a$ can be obtained by inserting series (7.2.31) in Eq. (7.1.25). The required gradients of the series for g_ω can be worked out from the formulas of (7.2.5) and (7.2.13). After a lot of algebra we obtain

$$\frac{\partial}{\partial z} Y_{mn}{}^\sigma(\vartheta,\varphi)j_m(kr) = \frac{k}{2m+1} [(m+n) Y_{m-1,n}^\sigma(\vartheta,\varphi)j_{m-1}(kr)$$

$$- (m-n) Y_{m+1,n}^\sigma(\vartheta,\varphi)j_{m+1}(kr)]$$

$$\frac{\partial}{\partial x} Y_{mn}{}^\sigma j_m = \frac{k}{(2m+1)\epsilon_n} [(m+n)(m+n-1) Y_{m-1,n-1}^\sigma j_{m-1}$$

$$- Y_{m-1,n+1}^\sigma j_{m-1} \qquad\qquad (7.2.32)$$

$$+ (m-n+1)(m-n+2) Y_{m+1,n-1}^\sigma j_{m+1} - Y_{m+1,n+1}^\sigma j_{m+1}]$$

$$\frac{\partial}{\partial y} Y_{mn}{}^\sigma j_m = \frac{-\sigma k}{(2m+1)\epsilon_n} \{[(m+n)(m+n-1) Y_{m-1,n-1}^{-\sigma}$$

$$+ Y_{m-1,n+1}^{-\sigma}]j_{m-1}$$

$$+ [(m-n+1)(m-n+2) Y_{m+1,n-1}^{-\sigma} + Y_{m+1,n+1}^{-\sigma}]j_{m+1}\}$$

When r is considerably larger than a, large enough so that $h_m(kr) \to (i^{-m-1}/kr)e^{ikr}$, the pressure wave can be expressed in terms of the integrals of the various source functions over the region of violent motion ($r < a$).

$$S^\sigma(m,n) = \frac{i}{\rho c k} \int f_\omega(\mathbf{r}) Y_{mn}{}^\sigma(\vartheta,\varphi)j_m(kr) \, dv$$

$$D_x{}^\sigma(m,n) = \frac{i}{\rho c k} \int F_x(\mathbf{r}) \frac{\partial}{\partial x} [Y_{mn}{}^\sigma(\vartheta,\varphi)j_m(kr)] \, dv \quad \cdots$$

$$Q_{xy}{}^\sigma(m,n) = \frac{i}{\rho c k} \int T_{xy}(\mathbf{r}) \frac{\partial^2}{\partial x \, \partial y} (Y_{mn}{}^\sigma j_m) \, dv \quad \cdots$$

Referring to the discussion following Eqs. (7.2.12) and (7.2.15), we see that when $ka \to 0$, all these coefficients go as $(ka)^m$ to zero, except for $S^1(0,0) \to S_\omega$, the strength of the equivalent monopole source. Also, the largest D is $D_x^{\ 1}(0,0) \to D_x$, etc., for the three dipole components, and the largest Q's are $Q_{xy}^{\ 1}(0,0) \to Q_{xy}$, etc. When ka is not small, more terms in the series must be used. The full series is

$$p_\omega(\mathbf{r}) \simeq -\frac{ik\rho c}{4\pi r} e^{ikr}[\Phi_s(\vartheta,\varphi) - ik\Phi_d(\vartheta,\varphi) - k^2\Phi_q(\vartheta,\varphi)] \quad (7.2.33)$$

for $r > a$ and $kr \gg 1$, where

$$\Phi_s = \sum_{mn\sigma} i^{-m}\epsilon_n (2m + 1) \frac{(m-n)!}{(m+n)!} S^\sigma(m,n) Y_{mn}^{\ \sigma}(\vartheta,\varphi)$$

The first term in the series is a spherically symmetric wave of strength $S^1(0,0)$. This is the *only* term which appears if $f_\omega(\mathbf{r}_0)$ is spherically symmetric, independent of ϑ_0 and φ_0. But if f_ω does depend on ϑ_0 and/or φ_0, some of the coefficients $S^\sigma(m,n)$ will differ from zero, and the radiation from the distribution of simple sources, in a region not small compared with λ, will not be spherically symmetric; some higher multipole radiation will be present. The amplitude of the mth multipole is proportional to $(ka)^m$, becoming negligible when ka is very small.

Likewise, the radiation from the driving-force distribution $\mathbf{F}_\omega(\mathbf{r})$ will not be a pure dipole wave if ka is not very small. For example, the wave from the z component of the force, once we have inserted Eqs. (7.2.31) and (7.2.32) in the appropriate part of (7.1.25), is the combination

$$-\frac{ik\rho c}{4\pi r} e^{ikr} \sum_{mn\sigma} i^{-m}\epsilon_n \frac{(m-n)!}{(m+n)!} (2m + 1) Y_{mn}^{\ \sigma}(\vartheta,\varphi)$$

$$\times \iiint F_z \frac{\partial}{\partial z_0} Y_{mn}^{\ \sigma}(\vartheta_0,\varphi_0) j_m(kr_0)\, dv_0$$

$$= \frac{ik^2\rho c}{4\pi r} e^{ikr} \sum i^{-m}\epsilon_n \frac{(m-n)!}{(m+n)!} Y_{mn}^{\ \sigma}(\vartheta,\varphi)$$

$$\times \left[(m - n) \iiint F_z Y_{m+1,n} j_{m+1}\, dv_0 - (m + n) \iiint F_z Y_{m-1,n} j_{m-1}\, dv_0\right]$$

$$(7.2.34)$$

If $F_z(\mathbf{r}_0)$ is independent of ϑ_0 and φ_0, only the $m = 1$, $n = 0$, $\sigma = 1$ term differs from zero, since it is the only one having the nonzero integral $-F_z Y_{00}^{\ 1} j_0\, dv_0$. The angle distribution of this term is $Y_{10}^{\ 1}(\vartheta,\varphi)$, typical of a dipole pointed along the z axis. But if $F_z(\mathbf{r}_0)$ is a function of ϑ_0 and/or φ_0, other integrals can be nonzero, and the radiation will contain multipole terms. This is also true of the radiation from the matrix components of $\mathfrak{T}$. Only when all components are independent of ϑ_0 and φ_0 will the radiation be purely quadrupole if ka is not small. This result was anticipated in our discussion of Eq. (7.1.30).

Spherical-surface effects

The Green's function $g_\omega(\mathbf{r} \mid \mathbf{r}_0)$, defined in Eq. (7.1.14) and expanded in series (7.2.31), is appropriate for calculating acoustic radiation in an unbounded medium. When there are boundaries enclosing the medium at which boundary conditions must be satisfied, it is usually convenient to modify g_ω by the addition of a free standing wave $\chi(\mathbf{r})$ of such a form that $G_\omega = g_\omega + \chi$ satisfies some boundary condition on the surface. The general equation (7.1.17), for p_ω in terms of integrals of G_ω, indicates the alternatives. As it stands, we should have to specify both the value of p and of its normal gradient on the surface S_0 in order to compute the surface integral. But if we arranged that G_ω was zero on S_0, then we should need only to specify p_ω, or alternatively, if we chose χ so that the normal gradient of G_ω was zero on S_0, we should need only to specify the normal gradient of p on S_0 in order to determine p uniquely elsewhere, in the region occupied by the medium. Since we are often interested in the effects of the motion of a part of the boundary surface, and since, for simple-harmonic motion, $\mathbf{u} = (1/ik\rho c)\,\mathrm{grad}\,p$, it appears that we should prefer a G_ω which has a zero normal gradient on the surface. When this choice is made, Eq. (7.1.17) becomes

$$p_\omega(\mathbf{r}) = \iiint f_\omega(\mathbf{r}_0) G_\omega(\mathbf{r} \mid \mathbf{r}_0)\, dv_0 + ik\rho c \iint u_n(\mathbf{r}_0{}^s) G_\omega(\mathbf{r} \mid \mathbf{r}_0{}^s)\, dS_0 \quad (7.2.35)$$

where $u_n(\mathbf{r}_0{}^s) = \dfrac{1}{ik\rho c} \dfrac{\partial p_\omega}{\partial n_0}$ is the normal velocity of the boundary surface S_0 at the surface point $\mathbf{r}_0{}^s$. The positive direction of u_n is outward, away from the medium; the equation is valid for any point $\mathbf{r}$ within or on the boundary surface S_0.

To demonstrate the use of this equation, we first consider the fluid enclosed in a hollow rigid sphere of internal radius a. Here we must add to the series of Eq. (7.2.31) a solution of $(\nabla^2 + k^2)\chi = 0$ which has no poles within or on the surface of the sphere, such that the radial gradient of the sum $g_\omega + \chi$ is zero at $r = a$. Symbolically, this is not difficult to do. If $j_m'(\zeta) = dj_m/d\zeta$, the function

$$G_\omega(\mathbf{r} \mid \mathbf{r}_0) = \frac{ik}{4\pi} \sum_{m=0}^{\infty} (2m+1) \sum_{n,\sigma} \epsilon_n \frac{(m-n)!}{(m+n)!}\, Y_{mn}{}^\sigma(\vartheta,\varphi)$$

$$\times Y_{mn}{}^\sigma(\vartheta_0,\varphi_0) \left[H_m - \frac{h_m'(ka)}{j_m'(ka)}\, j_m(kr)\, j_m(kr_0) \right] \qquad \text{for } r, r_0 \leqslant a$$

$$(7.2.36)$$

where
$$H_m = \begin{cases} j_m(kr_0) h_m(kr) & r > r_0 \\ j_m(kr) h_m(kr_0) & r < r_0 \end{cases}$$

has a radial gradient zero at $r = a$ (when $r_0 \leqslant a$) and also at $r_0 = a$ (when $r \leqslant a$). The series with H_m alone is g_ω; the additional series is finite and continuous inside r, $r_0 = a$ and modifies g_ω to make $\partial G_\omega / \partial r = 0$ at $r = a$; it represents the wave reflected by the rigid spherical surface.

To use Eq. (7.2.35), we note that S_0 is the sphere of radius a and that the positive direction of the normal velocity is outward, in the positive r direction. If the sphere is rigid, the surface integral is zero, and p_ω equals the volume integral of G_ω times the source function $f_\omega(\mathbf{r}_0)$, integrated over the interior of the sphere. We note that when ka is such that $j_m'(ka)$ is zero, the reflected wave χ becomes infinite; the source frequency has coincided with one of the resonance frequencies of the hollow sphere. Green's functions for enclosures of finite size exhibit this resonance property, as with the Green's function of Eq. (4.4.14) for the string between two rigid supports a finite distance apart.

On the other hand, if the region of interest is between the outer surface of a sphere of radius a and the sphere at infinity, we must add to g_ω a function χ such that the radial gradient of the sum is zero at $r = a$ (when $r_0 \geqslant a$). The combination which does this is

$$
\begin{aligned}
G_\omega &= \frac{ik}{4\pi} \sum_{m=0}^{\infty} (2m+1) \sum_{n,\sigma} \epsilon_n \frac{(m-n)!}{(m+n)!} Y_{mn}{}^\sigma(\vartheta,\varphi) \\
&\quad \times Y_{mn}{}^\sigma(\vartheta_0,\varphi_0) \left[H_m - \frac{j_m'(ka)}{h_m'(ka)} h_m(kr)h_m(kr_0) \right] \\
&\to \frac{e^{ikr}}{4\pi r} \sum_{m=0}^{\infty} (2m+1)i^{-m} \sum_{n,\sigma} \epsilon_n \frac{(m-n)!}{(m+n)!} Y_{mn}{}^\sigma(\vartheta,\varphi) \\
&\quad \times Y_{mn}{}^\sigma(\vartheta_0,\varphi_0) \left[\frac{h_m'(ka)j_m(kr_0) - j_m'(ka)h_m(kr_0)}{h_m'(ka)} \right] \qquad kr \to \infty \quad (7.2.37)
\end{aligned}
$$

Since $h_m'(ka)$ has no zeros for real values of ka, this Green's function has no resonances, in contrast to that of Eq. (7.2.36).

If the only source of acoustic energy, in the region outside $r = a$, comes from the motion of the spherical surface itself, the only nonzero integral in Eq. (7.2.35) is the surface integral

$$
\begin{aligned}
p_\omega(\mathbf{r}) &= ik\rho c a^2 \int_0^{2\pi} d\varphi_0 \int_0^{\pi} u_n G_\omega \sin \vartheta_0 \, d\vartheta_0 \\
&= \rho c k^2 a^2 \sum_{mn\sigma} U_{mn}{}^\sigma \left[\frac{h_m'(ka)j_m(ka) - j_m'(ka)h_m(ka)}{h_m'(ka)} \right] Y_{mn}{}^\sigma(\vartheta,\varphi) h_m(kr)
\end{aligned}
$$

where

$$
U_{mn}{}^\sigma = \frac{\epsilon_n}{4\pi} (2m+1) \frac{(m-n)!}{(m+n)!} \int_0^{2\pi} d\varphi_0 \int_0^{\pi} [-u_n(\vartheta_0,\varphi_0)] Y_{mn}{}^\sigma(\vartheta_0,\varphi_0) \sin \vartheta_0 \, d\vartheta_0
$$

since the normal velocity u_n away from the fluid is here pointed in toward the origin; thus $-u_n$ is the radial velocity of the surface of the sphere.

From Eq. (7.2.12) we see that $h'_m j_m - j'_m h_m = i/k^2 a^2$, and from Eq. (7.2.15) we see that $h'_m(ka) = iB_m e^{i\delta_m}$, so that

$$p_\omega(\mathbf{r}) = \rho c \sum_{mn\sigma} \frac{U_{mn}{}^\sigma}{B_m} e^{-i\delta_m} Y_{mn}{}^\sigma(\vartheta,\varphi) h_m(kr)$$

$$\rightarrow -\frac{i\rho c}{kr} e^{ikr} \sum_{mn\sigma} i^{-m} \frac{U_{mn}{}^\sigma}{B_m} e^{-i\delta_m} Y_{mn}{}^\sigma(\vartheta,\varphi) \qquad kr \rightarrow \infty \qquad (7.2.38)$$

which is an obvious generalization of Eq. (7.2.16). Thus the integral formula (7.2.35) can be used to compute radiation from any source, whether it be a moving surface or the motion of the fluid itself.

7.3 RADIATION FROM CYLINDERS

The other simple geometrical surface which is often encountered in acoustics is the circular cylinder. Waves of cylindrical shape are obvious generalizations, to three dimensions, of the waves on a circular membrane, already discussed in Sec. 5.2. We showed there that circular waves have a more complicated shape than plane or spherical waves; they leave a "wake" behind them as they spread out. Cylindrical waves show the same behavior. As with the rest of this chapter, we shall concentrate on simple-harmonic waves; solutions with more general time dependence can be computed by use of the Fourier transform.

The general solution

The equation $\nabla^2 p + k^2 p = 0$ $(k = \omega/c)$ can be expressed in the cylindrical coordinates w, ϕ, z by using Eq. (7.1.3). The separated equations for the factored solution $p = R(w)\Phi(\phi)Z(z)e^{-i\omega t}$ are

$$\frac{d^2 Z}{dz^2} + k_z{}^2 Z = 0 \qquad \frac{d^2 \Phi}{d\phi^2} + m^2 \Phi = 0$$

$$\frac{1}{w}\frac{d}{dw}\left(w\frac{dR}{dw}\right) + \left(k_w{}^2 - \frac{m^2}{w^2}\right)R = 0 \qquad k^2 = k_w{}^2 + k_z{}^2 \tag{7.3.1}$$

The general solution for Z is a combination of $e^{ik_z z}$ and $e^{-ik_z z}$; the solution for Φ is $\cos(m\phi)$ or $\sin(m\phi)$, where m must be an integer for the solution to be continuous at $\phi = 0 = 2\pi$. The solution of the radial equation is a combination of the Bessel function $J_m(k_w w)$, defined in Eqs. (5.2.20) and (5.2.21), and the Neumann function $N_m(k_w w)$, mentioned in Eq. (5.3.15) as being the second solution of the Bessel equation for R. They have the following properties:

$$\frac{1}{z}\frac{d}{dz}\left(z\frac{dN_m}{dz}\right)+\left(1-\frac{m^2}{z^2}\right)N_m = 0$$

$$N_m(z)\xrightarrow[z\to\infty]{}\sqrt{\frac{2}{\pi z}}\sin\left(z-\frac{2m+1}{4}\pi\right)$$

$$N_0(z)\xrightarrow[z\to 0]{}\frac{2}{\pi}\ln(0.890536z)=\frac{2}{\pi}(\ln z-0.11593) \qquad (7.3.2)$$

$$N_m(z)\xrightarrow[z\to 0]{}-\frac{(m-1)!}{\pi}\left(\frac{2}{z}\right)^m \qquad m>0$$

$$N_{m-1}(z)J_m(z)-N_m(z)J_{m-1}(z)=\frac{2}{\pi z}$$

The properties given in Eqs. (5.2.21) for the functions $J_m(z)$ are also true for the corresponding functions $N_m(z)$.

We note that $N_m(k_w w)$ goes to infinity as $w\to 0$, whereas $J_m(k_w w)$ is finite at $w=0$. We also note that the combination $H_m^{(1)}(k_w w)=J_m(k_w w)+iN_m(k_w w)$ represents an outgoing wave, for

$$J_m(k_w w)+iN_m(k_w w)\to\sqrt{\frac{2}{\pi k_w w}}\exp[ik_w w-\tfrac{1}{2}\pi i(m+\tfrac{1}{2})] \qquad k_w w\gg 1$$

For $k_w w$ near zero, in the region of the near field, the combination is not just a simple exponential; in fact, the imaginary part of the combination goes to infinity as $w\to 0$. $H_m^{(1)}$ is called a *Hankel function* of mth order.

Uniform radiation

For waves spreading uniformly out from a cylinder, we use the function for $k_z=0$ and $m=0$, which represents outgoing waves,

$$p = A\left[J_0\left(\frac{2\pi vw}{c}\right)+iN_0\left(\frac{2\pi vw}{c}\right)\right]e^{-2\pi ivt}$$

$$\xrightarrow[w\to\infty]{}A\sqrt{\frac{2}{\pi kw}}\,e^{ik(w-ct)-i(\pi/4)} \qquad k=\frac{2\pi v}{c}=\frac{2\pi}{\lambda} \qquad (7.3.3)$$

$$\xrightarrow[w\to 0]{}i\frac{2A}{\pi}\ln(w)e^{-2\pi ivt}$$

Suppose that we have a long cylinder of radius a which is expanding and contracting uniformly in such a manner that the velocity of the surface of the cylinder is $u_0 = U_0 e^{-2\pi ivt}$. The constant A, to correspond to the radiated wave, must be chosen so that the velocity of the air perpendicular to the cylinder surface $u_w = \dfrac{1}{2\pi i v \rho}\dfrac{\partial p}{\partial w}$ is equal to u_0 at $w = a$. If a is small compared with the wavelength, this velocity is $(A/\pi^2 v \rho a)e^{-2\pi i v t}$, so that A must equal $\pi^2 v \rho a U_0$. The pressure and velocity at large distances from the cylinder are then

$$p \to \pi \rho a U_0 \sqrt{\frac{cv}{w}}\, e^{ik(w-ct)-i(\pi/4)}$$

$$u_w \to \pi a U_0 \sqrt{\frac{v}{cw}}\, e^{ik(w-ct)-i(\pi/4)}$$

The product of the real part of each of these expressions gives the flow of energy outward per second per square centimeter, and the average value of this,

$$I \simeq \pi^2 \rho a^2 U_0^2 \frac{v}{w} \qquad a \ll \frac{c}{\omega} \tag{7.3.4}$$

is the intensity of the sound at a distance w from the cylinder's axis. The total energy radiated in ergs per second per centimeter length of the cylinder is

$$\Pi \simeq 2\pi^3 \rho v a^2 U_0^2$$

Radiation from a vibrating wire

A somewhat more complicated wave is generated by a cylinder of radius a vibrating back and forth in a direction perpendicular to its axis, with a velocity $U_0 e^{-2\pi i v t}$. If the plane of vibration is taken as the reference plane for ϕ, the velocity of the part of the cylinder's surface at an angle ϕ from the plane of vibration has a component $U_0 \cos \phi\, e^{-2\pi i v t}$ perpendicular to the surface. In this case we take the radiated wave to be

$$p = A \cos \phi [J_1(kw) + iN_1(kw)]e^{-2\pi i v t} \qquad k = \frac{2\pi v}{c}$$

$$\xrightarrow[w \to 0]{} -i\frac{AC}{\pi^2 v w} \cos \phi\, e^{-2\pi i v t}$$

$$\xrightarrow[w \to \infty]{} A\sqrt{\frac{c}{\pi^2 v w}}\, e^{ik(w-ct)-i(3\pi/4)} \cos \phi$$

If a is small, the radial velocity at $w = a$ is

$$\frac{Ac}{2\pi^3 v^2 \rho a^2} \cos \phi\, e^{-2\pi i v t}$$

which must equal the radial velocity of the surface, so that $A = 2\pi^3 v^2 \rho a^2 U_0/c$. The radial component of the particle velocity at large distances from the cylinder is

$$u_w \xrightarrow[w \to \infty]{} \frac{A}{\rho c}\sqrt{\frac{c}{\pi^2 v w}}\, e^{ik(w-ct)-i(3\pi/4)} \cos \phi$$

There is a component of particle velocity perpendicular to the radius w, but it diminishes as $w^{-\frac{3}{2}}$ at large distances, and so is negligible there compared with u_w. The intensity at large distances and the total power radiated per centimeter length of vibrating wire are

$$I \simeq \frac{4\pi^4 \nu^3 \rho a^4 U_0^2}{c^2 r}\cos^2\phi \qquad \Pi \simeq \frac{4\pi^5 \nu^3 \rho a^4 U_0^2}{c^2} \qquad (7.3.5)$$

The amount of sound energy radiated by a vibrating string therefore diminishes rapidly as the frequency of vibration of the string decreases and diminishes very rapidly if the string's thickness is decreased. As we mentioned at the end of Chap. 4, a vibrating string is a very inefficient radiator of sound.

The reaction of the air back on the vibrating wire is obtained from the expression for the pressure at $w = a$.

$$p \simeq -i\omega\rho a U_0 \cos\phi\, e^{-i\omega t}$$

The net reaction force on the wire per unit length in the direction of its motion is

$$a \int_0^{2\pi} p \cos\phi\, d\phi = -F_0 e^{-i\omega t} \simeq -i\omega\rho(\pi a^2) U_0 e^{-i\omega t}$$

The ratio of this to the velocity of the wire is the mechanical impedance per unit length of wire, due to sound radiation,

$$\frac{F_0}{U_0} \simeq -i\omega(\pi a^2 \rho) \qquad a \ll \frac{c}{\omega}$$

which is equivalent to the reactance of a mass of air of volume equal to that of the wire. The resistive part R of this impedance is too small to be included in this approximation when a is small. We can find the resistive part from Eq. (7.3.5) for the power radiated, for of course $\Pi = R U_0^2$,

$$Z_{\text{rad}} \simeq -i\omega(\pi a^2 \rho) + \frac{\pi^2 \omega^3 \rho a^4}{2c^2} \qquad a \ll \frac{c}{\omega} \qquad (7.3.6)$$

In the discussion of Eqs. (10.1.9) and (10.1.19) it will be shown, however, that a string of finite length behaves quite differently.

Radiation from an element of a cylinder

To solve more complicated problems, where the velocity of the surface of the cylinder is a less simple function of ϕ than the preceding examples, it is convenient first to solve the problem where only a single line element on the surface of the cylinder does the vibrating. Suppose that the radial velocity of the surface $w = a$ is

$$u_a = \begin{cases} U e^{-2\pi i\nu t} & -\dfrac{d\alpha}{2} < \phi < +\dfrac{d\alpha}{2} \\[3mm] 0 & +\dfrac{d\alpha}{2} < \phi < 2\pi - \dfrac{d\alpha}{2} \end{cases}$$

The Fourier-series expansion for this function of ϕ is

$$u_a = \frac{U\, d\alpha}{\pi}\left[\tfrac{1}{2} + \sum_{m=1}^{\infty} \cos (m\phi)\right] e^{-2\pi i v t}$$

To fit this distribution of velocity at the surface, we choose a general sort of outgoing wave.

$$p = \sum_{m=0}^{\infty} A_m \cos (m\phi)[J_m(kw) + iN_m(kw)]e^{-2\pi i v t} \qquad k = \frac{2\pi v}{c}$$

The corresponding radial particle velocity at $w = a$ is

$$u_a = \frac{1}{2\pi i v \rho_0} \frac{\partial p}{\partial w} = \left[\frac{A_0 E_0}{2\rho_0 c} e^{i\gamma_0} + \sum_{m=1}^{\infty} \frac{A_m E_m}{\rho c} e^{i\gamma_m} \cos (m\phi)\right] e^{-2\pi i v t}$$

where

$$2\frac{d}{d\mu}[J_0(\mu) + iN_0(\mu)] = iE_0 e^{i\gamma_0}$$

$$\frac{d}{d\mu}[J_m(\mu) + iN_m(\mu)] = iE_m e^{i\gamma_m} \qquad m > 0$$

and where $\mu = ka$. Therefore, using Eqs. (5.2.20),

$$J_1(ka) = \tfrac{1}{2}E_0 \sin \gamma_0 \qquad N_1(ka) = -\tfrac{1}{2}E_0 \cos \gamma_0$$

$$J_{m+1}(ka) - J_{m-1}(ka) = 2E_m \sin \gamma_m$$

$$N_{m-1}(ka) - N_{m+1}(ka) = 2E_m \cos \gamma_m \qquad m > 0$$

Note the additional factor of 2 in E_0, to anticipate the factor $\tfrac{1}{2}$ in the sum for $m = 0$. Limiting values of the amplitudes E_m and phase angles γ_m are given by the following approximate formulas:

when

$$ka = \frac{2\pi a}{\lambda} \gg m + \tfrac{1}{2}$$

$$E_0 \simeq \sqrt{\frac{8}{\pi ka}} \qquad \gamma_0 \simeq ka - \frac{\pi}{4}$$

$$E_m \simeq \sqrt{\frac{2}{\pi ka}} \qquad \gamma_m \simeq ka - \tfrac{1}{2}\pi(m + \tfrac{1}{2}) \qquad m > 0$$

(7.3.7)

when

$$ka = \frac{2\pi a}{\lambda} \ll m + \tfrac{1}{2}$$

$$E_0 \simeq \frac{4}{\pi ka} \qquad \gamma_0 \simeq \pi\left(\frac{ka}{2}\right)^2$$

$$E_m \simeq \frac{m!}{2\pi}\left(\frac{2}{ks}\right)^{m+1} \qquad \gamma_m \simeq -\frac{\pi m}{(m!)^2}\left(\frac{ka}{2}\right)^{2m} \qquad m > 0$$

Values of some of the E's and γ's are given in Table V.

To fit the expression for u_w at $w = a$ to the expression for the velocity of the cylinder u_a, we must make

$$A_m = \frac{\rho c U\, d\alpha}{\pi E_m} e^{-i\gamma_m}$$

Since, at very large distances from the cylinder,

$$J_m(kw) + iN_m(kw) \simeq \sqrt{\frac{2}{\pi kw}} e^{i\left[kw - \frac{\pi}{4}(2m+1)\right]}$$

we have the following expressions for the pressure, particle velocity, and intensity of sound at the point w, ϕ (when w is a large number of wavelengths) and for the total power radiated by the element per unit length of cylinder:

$$u_w \simeq U\, d\alpha \sqrt{\frac{a}{w}} e^{ik(w-ct)}\psi(\phi) \qquad p \simeq \rho c u$$

$$\psi(\phi) = \sqrt{\frac{2}{\pi^3 ka}} \sum_{m=0}^{\infty} \frac{\cos(m\phi)}{E_m} e^{-i\left[\gamma_m + \frac{\pi}{4}(2m+1)\right]}$$

$$\tag{7.3.8}$$

$$I \simeq \frac{\rho c^2 (U\, d\alpha)^2}{\pi^4 v w} \sum_{m,n=0}^{\infty} \frac{\cos(m\phi)\cos(n\phi)}{E_m E_n} \cos\left[\gamma_m - \gamma_n + \tfrac{1}{2}\pi(m-n)\right]$$

$$\Pi = \frac{\rho c^2 (U\, d\phi\alpha)^2}{\pi^3 v} \left(\frac{2}{E_0{}^2} + \sum_{m=1}^{\infty} \frac{1}{E_m{}^2}\right)$$

To find the intensity, we have, of course, multiplied the real parts of u_w and p together and averaged over time. The total power Π per unit length of cylinder is the integral $w \int_0^{2\pi} I\, d\phi$.

Long- and short-wave limits

When the wavelength is quite long compared with $2\pi a$, we can use the second part of Eqs. (7.3.7) to compute the radiation. The largest values of $1/E_m$ are for $m = 0$, so that, to the first approximation,

$$\psi(\phi) \simeq \sqrt{\frac{ka}{8\pi}} e^{-i\pi/4} \tag{7.3.9}$$

$$I \simeq \frac{\rho v a^2}{4w} (U\, d\alpha)^2 \qquad \Pi \simeq \frac{\pi \rho v a^2}{2} (U\, d\alpha)^2$$

At these low frequencies the sound radiates out with equal intensity *in all directions*, and the amount radiated is small. The expression for intensity is the same as that given in Eq. (7.3.4) for a uniformly expanding cylinder, if we substitute for U_0 in the earlier expression the *average velocity* $U\, d\alpha/2\pi$ of the surface.

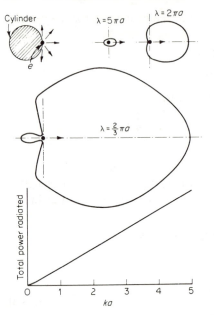

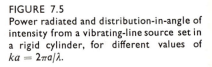

FIGURE 7.5
Power radiated and distribution-in-angle of
intensity from a vibrating-line source set in
a rigid cylinder, for different values of
$ka = 2\pi a/\lambda$.

Values of I and Π are plotted in Fig. 7.5. Polar curves of the intensity
are shown for different values of $\eta = ka = 2\pi a/\lambda$, and a curve is given for Π
as a function of η. We notice that at long wavelengths very little power is
radiated, and the intensity has very little directionality. As the wavelength is
decreased, more power is radiated and the intensity has more directionality;
the cylinder begins to cast a "shadow," and a smaller proportion of the energy
is sent out on the side of the cylinder opposite the radiating element. For
very short waves the intensity is large and uniform from $\phi = -(\pi/2)$ to
$\phi = +(\pi/2)$ and is zero from $\phi = \pi/2$ to $\phi = 3\pi/2$, in the shadow. In the
intermediate range of η, where the wavelength is about the same size as a,
interference effects are noticeable. The polar curve for $\eta = 3$ shows that a
fairly intense beam is sent out from the cylinder in a direction diametrically
opposite to the position of the line source ($\phi = 180°$).

The general properties illustrated by this set of curves are characteristic
of all wave motion when it strikes an obstacle. When the wavelength is large
compared with the size of the obstacle, the wave pays hardly any attention to
its presence. The first polar curve in Fig. 7.5 shows that for long waves the
intensity is distributed in approximately the same manner as it would be if the
line source were not in the side of a cylinder but were all by itself, radiating
into free space. On the other hand, when the wavelength is very small
compared with the size of the obstacle, the motion resembles the motion of
particles, the waves traveling in straight lines and the obstacles casting sharp-
edged shadows. Light waves have this raylike property in most cases;
geometrical optics is a valid approximation because the light waves are very
much shorter than the size of most of the obstacles they encounter.

When the wavelength is about the same size as the obstacle, complicated interference effects can sometimes occur, and the analysis of the behavior of the waves becomes quite involved.

Radiation from a cylindrical source of general type

If the line source is not at $\phi = 0$ on the surface of the cylinder but is at $\phi = \alpha$, the pressure and velocity at large distances are, by Eqs. (7.3.8),

$$p \to \rho c (U_0 \, d\alpha) \sqrt{\frac{a}{w}} \, \psi(\phi - \alpha) e^{ik(w-ct)}$$

$$u_w \to (U_0 \, d\alpha) \sqrt{\frac{a}{w}} \, \psi(\phi - \alpha) e^{ik(w-ct)}$$

the axis of the polar diagram being turned through an angle α. If several sources are distributed over the surface of the cylinder, each for a different value of α, the resulting radiated wave will be the sum of all the waves for the individual sources taken separately. This fact can be used to express the radiation from a cylinder whose surface vibrates with any arbitrary distribution of velocity amplitude. If the distribution is such that the surface at $\phi = \alpha$ has the radial velocity $U(\alpha)e^{-2\pi i \nu t}$, the wave may be considered to be the result of an infinite sequence of line sources, the one at $\phi = \alpha$ having the velocity amplitude $U(\alpha)$, etc. The pressure and radial velocity at large distances are then obtained by integrating the expressions given above for a single line source:

$$p \simeq \rho c \sqrt{\frac{a}{w}} \, e^{ik(w-ct)} \int_0^{2\pi} \psi(\phi - \alpha) U(\alpha) \, d\alpha$$

$$u \simeq \frac{p}{\rho c}$$

$$(7.3.10)$$

For instance, if a section of the cylinder between $\alpha = -\alpha_0$ and $\alpha = +\alpha_0$ is vibrating, so that $U(\alpha) = U_0$ for $-\alpha_0 < \alpha < \alpha_0$, and is zero for the rest of the values of α, then the pressure wave at large distances from the cylinder is

$$p = \rho c U_0 \sqrt{\frac{a}{w}} \, e^{ik(w-ct)} \int_{-\alpha_0}^{\alpha_0} \psi(\phi - \alpha) \, d\alpha$$

$$= \frac{2\rho c U_0}{\pi^2} \sqrt{\frac{c}{\nu w}} \, e^{ik(w-ct)} \sum_{m=0}^{\infty} \frac{\sin(m\alpha_0) \cos(m\phi)}{m E_m} e^{-i\left[\gamma_m + \frac{\pi}{4}(2m+1)\right]} \quad (7.3.11)$$

where we use the convention that $[\sin(m\alpha_0)]/m = \alpha_0$ when $m = 0$.

The intensity and total power radiated are

$$I = \frac{4\rho c_2 U_0^2}{\pi^4 \nu w} \sum_{m,n=0}^{\infty} \frac{\sin(m\alpha_0)\sin(n\alpha_0)\cos(m\phi)\cos(n\phi)}{mnE_m E_n}$$

$$\times \cos\left[\gamma_m - \gamma_n + (m-n)\frac{\pi}{2}\right]$$

$$\Pi = \frac{4\rho c^2 U_0^2}{\pi^3 \nu}\left[\frac{2\alpha_0^2}{E_0^2} + \sum_{m=1}^{\infty}\frac{\sin^2(m\alpha_0)}{m^2 E_m^2}\right] \qquad (7.3.12)$$

When the frequency is very small, so that λ is much larger than a, the intensity and power radiated are

$$I \rightarrow \frac{\rho \nu a^2 U_0^2 \alpha_0^2}{w} \qquad \Pi \rightarrow 2\pi\rho\nu a^2 U_0^2 \alpha_0^2$$

to the first approximation. The expression for I is quite similar to Eq. (7.3.4) for a uniformly expanding and contracting cylinder. Even though the velocity of the surface of the cylinder, in the present case, is not symmetrical about the axis, nevertheless, the radiation at very long wavelengths is symmetrical, behaving as though it came from a uniform cylindrical source whose velocity amplitude is $U_0\alpha_0/\pi$ (the average velocity amplitude of the actual surface). This is another example of the fact, mentioned above, that wave motion is insensitive to details smaller in size than the wavelength.

The Green's function

The Green's function of Eqs. (7.1.14) and (7.2.31) can also be expressed in cylindrical coordinates. To show this most easily we first obtain a Fourier transform of the Green's function in rectangular coordinates. The equation it must satisfy is

$$\left(\frac{\partial^2}{\partial x^2} + \frac{\partial^2}{\partial y^2} + \frac{\partial^2}{\partial z^2} + k^2\right)g_\omega = -\delta(x-x_0)\,\delta(y-y_0)\,\delta(z-z_0) \quad (7.3.13)$$

so the relation between $g_\omega(\mathbf{r}\mid\mathbf{r}_0)$ and its transform $G_\omega(\mathbf{K})$ is the three-dimensional extension of Eq. (1.3.16) (for x, y, z instead of t).

$$G_\omega = (2\pi)^{-3}\iiint g_\omega e^{i\mathbf{K}\cdot\mathbf{r}}\,dv_r \qquad g_\omega = \iiint G_\omega e^{-i\mathbf{K}\cdot\mathbf{r}}\,dv_K$$

where $\mathbf{K}$ is the vector with components K_x, K_y, K_z and volume element $dv_K = dK_x\,dK_y\,dK_z$; $\mathbf{r}$ is the vector (x,y,z) with volume element $dv_r = dx\,dy\,dz$. The transform of the term on the right of Eq. (7.3.13) is $-(2\pi)^{-3}e^{i\mathbf{K}\cdot\mathbf{r}_0}$, and the transform of the left-hand side is $(k^2 - K^2)G_\omega(\mathbf{K})$, so that the Green's function is given by the triple Fourier integral

$$g_\omega(\mathbf{r}\mid\mathbf{r}_0) = (2\pi)^{-3}\iiint \frac{\exp[i\mathbf{K}\cdot(\mathbf{r}_0 - \mathbf{r})]}{K^2 - k^2}\,dv_K \qquad (7.3.14)$$

where $k = \omega/c$, and $K^2 = K_x{}^2 + K_y{}^2 + K_z{}^2$. The contour for integration in the complex K plane goes above the pole at $K = -k$ and below the pole at $K = +k$ in order that the wave may be an outgoing one.

To convert this integral to one in cylindrical coordinates w, ϕ, z, we first change to cylindrical coordinates μ, β, K_z in K space, where $K_x = \mu \cos \beta$, $K_y = \mu \sin \beta$, and $K^2 = K_z{}^2 + \mu^2$. Then $dv_K = dK_z\,d\beta\mu\,d\mu$, $\mathbf{K \cdot r} = \mu w \cos (\beta - \phi) + z K_z$, and $\mathbf{K \cdot r_0} = \mu w_0 \cos (\beta - \phi_0) + z_0 K_z$. We first integrate over K_z, using Eqs. (1.2.15) to evaluate the integral in terms of the poles at $K_z = \pm\sigma = \pm\sqrt{k^2 - \mu^2}$ (going around the pole at $+\sigma$ when $z_0 - z$ is positive and around the pole at $-\sigma$ when $z - z_0$ is positive), so that

$$g_\omega(\mathbf{r} \mid \mathbf{r_0}) = -\frac{i}{8\pi^2} \int_0^{2\pi} d\beta \int_0^\infty \exp [i\mu w_0 \cos (\beta - \phi_0)$$
$$- i\mu w \cos (\beta - \phi)] \frac{\mu\,d\mu}{\sigma} e^{i\sigma|z - z_0|}$$

Next we use the series expansion of Eq. (1.2.9) to convert the exponential into a pair of Fourier series.

$$\exp [-] = \sum_{m,n=0}^\infty \epsilon_m \epsilon_n i^{m-n} \cos m(\beta - \phi_0) \cos n(\beta - \phi) J_m(\mu w_0) J_n(\mu w)$$

Integration over β removes all but the terms for which $m = n$; the final result is

$$g_\omega(\mathbf{r} \mid \mathbf{r_0}) = \sum_{m=0}^\infty \frac{i\epsilon_m}{4\pi} \cos m(\phi - \phi_0) \int_0^\infty J_m(\mu w_0) J_m(\mu w) \frac{\mu\,d\mu}{\sigma} e^{i\sigma|z - z_0|} \quad (7.3.15)$$

where $\sigma = \sqrt{k^2 - \mu^2}$ when $0 \leqslant \mu < k$ and equals $i\sqrt{\mu^2 - k^2}$ when $0 \leqslant k < \mu$.

This expansion for the free-field Green's function in cylindrical coordinates w, ϕ, z will be useful later in this chapter. When $w_0 = 0$, this function must become the spherically symmetric function $g_\omega(\mathbf{r} \mid 0)$ of Eq. (7.1.6). Since $J_0(0) = 1$ and $J_m(0) = 0$ for $m > 0$, the series reduces to

$$g_\omega(\mathbf{r} \mid 0) = \frac{e^{ikr}}{4\pi r} = \frac{i}{4\pi} \int_0^\infty J_0(\mu w) e^{i|z|\sqrt{k^2 - \mu^2}} \frac{\mu\,d\mu}{\sqrt{k^2 - \mu^2}} \quad (7.3.16)$$

where of course $r^2 = w^2 + z^2$. Thus series (7.3.15) is identical with the function defined in Eq. (7.1.6).

We shall return to other acoustical situations which require the use of cylindrical coordinates, later in this chapter, and also in Chap. 9, when we take up the generation and transmission of sound inside a hollow cylinder.

Occasionally, we shall have use for the Green's function in two dimensions, the solution of the equation

$$\nabla^2 g_\omega + k^2 g_\omega = -\delta(x - x_0)\,\delta(y - y_0)$$

which can also be considered to be the three-dimensional wave generated by a unit line source at $x = x_0$, $y = y_0$, $-\infty < z_0 < +\infty$. By changing to coordinates centered on the source and by methods paralleling the discussion of Eq. (7.1.14), we can see that the solution of

$$\frac{1}{W}\left[\frac{\partial}{\partial W}\left(W\frac{\partial g}{\partial W}\right)\right] + k^2 g = \frac{\delta(W)}{W}$$

which represents an outgoing wave, is

$$g_\omega(x,y \mid x_0 y_0) = \frac{i}{4}\left[J_0(kW) + iN_0(kW)\right] = \frac{i}{4} H_0^{(1)}(kW) \qquad (7.3.17)$$

where $W^2 = (x - x_0)^2 + (y - y_0)^2$. By using the Fourier transform it is then possible to show that this function can be expanded in the polar coordinates w, ϕ.

$$g_\omega(w,\phi \mid w_0,\phi_0) = \frac{i}{4}\sum_{m=0}^{\infty}\epsilon_m \cos m(\phi - \phi_0)\begin{cases} J_m(kw)H_m^{(1)}(kw_0) & w < w_0 \\ H_m^{(1)}(kw)J_m(kw_0) & w > w_0 \end{cases}$$

$$(7.3.18)$$

where the function $H_m^{(1)}$ is the outgoing-wave solution of the Bessel equation (1.2.4).

$$H_m^{(1)}(z) = J_m(z) + iN_m(z) \rightarrow i^{-m}\sqrt{\frac{2}{\pi i z}}\, e^{iz} \qquad z \rightarrow \infty$$

7.4   RADIATION AND REFLECTION FROM A PLANE SURFACE

So far we have been considering the radiation of sound into an infinite medium, unbounded by any barriers. This is a rather unrealistic arrangement, though the computed results do approximate the sound field generated by aircraft or by a source in an anechoic chamber. The arrangements discussed in this section, sources in the presence of an infinite plane barrier, so that the medium is confined to one side of the plane, are somewhat more realistic. They will serve to introduce the subject of the effects of boundaries which generate, reflect, or absorb sound energy.

Image sources

Suppose the plane is perfectly rigid. Then the boundary condition at its surface is that the normal fluid velocity is zero there, and thus [by Eq. (6.2.7)] that the normal gradient of the pressure is zero at the surface. This can most easily be accomplished by imagining that the boundary plane is replaced by a continuation of the medium into the region back of the boundary. Since the gradient of the field is still to be zero at the plane where the boundary had been, the field must be symmetric with respect to this plane; its value on one side of the plane should be the reflection of its value on the other side. Thus we may replace the effect of the boundary plane by a set of image sources, symmetrically placed with respect to the boundary plane, with both source and its image radiating into unbounded space. Of course, only the region outside the boundary plane contains the medium and carries acoustic energy. The part of the field from the original source, which is inside the boundary, has no physical reality (as is the case with the wave in the bounded string of Fig. 4.7). Conversely, the portion of the wave from the image source, outside

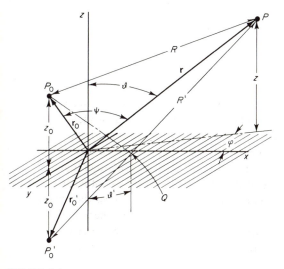

FIGURE 7.6
Angles and distances for the field at P from a source at P_0 and image, in the xy plane, at P_0'.

the boundary, does have physical reality. It constitutes the reflected wave and carries the energy reflected from the boundary surface.

For example, suppose the boundary surface is the xy plane, with the medium in the region $z > 0$. If the source is a simple source at point P_0 shown in Fig. 7.6, located by vector $\mathbf{r}_0$ from the origin, the image source will also be a simple source, of equal strength and phase, at point P_0', a vertical distance $2z_0$ below P_0, located by vector $\mathbf{r}_0'$. The acoustic pressure at P, located by vector $\mathbf{r}$, above the xy plane, for a simple-harmonic source, is then

$$-i\rho c k S_\omega \left(\frac{1}{4\pi R} e^{ikR} + \frac{1}{4\pi R'} e^{ikR'} \right) e^{-i\omega t} \qquad (7.4.1)$$

This represents, in the region $z > 0$, the original wave plus a reflected wave, which automatically satisfies the boundary conditions at $z = 0$. Along the boundary surface, $R' = R$ and the pressure is just twice what it would have been if the boundary had not been present.

At distances R from the source, which are much larger than the source is from the origin (i.e., for $R \gg r_0$), the two waves combine to form what appears to be a single nonsimple source at the origin.   To see this we note that when $r \gg r_0$,

$$R = |\mathbf{r} - \mathbf{r}_0| \simeq r - \left(\frac{\mathbf{r}}{r}\right) \cdot \mathbf{r}_0 = r - r_0 \cos \psi$$

and

$$R' \simeq r + 2z_0 \cos \vartheta - r_0 \cos \psi$$

Thus, neglecting higher-order terms in the small quantity r_0/r, the pressure field at $\mathbf{r}$ is

$$p \simeq - \frac{i \rho c k}{2\pi r} S_\omega \cos (kz_0 \cos \vartheta) \exp [ik(r - r_0 \cos \psi + z_0 \cos \vartheta) - i\omega t]$$

$$(7.4.2)$$

If kz_0 and $kr_0 \ll 1$, that is, if the source is considerably less than a wavelength above the boundary surface, the far field is very weakly dependent on ϑ or ψ, thus resembling the far field from a simple source, at the origin, of strength $2S_\omega$.   The effective strength is doubled because the reflected wave adds to the initial wave in this case; the source and image are close enough together to be considered as a single source of double strength.   On the other hand, when $kz_0 = \frac{1}{2}\pi$, when the source is a quarter wavelength above the surface, the source and its image are a half wavelength apart, and there is destructive interference between the two waves.   Directly above the source ($\vartheta = 0$), the interference is complete and the far-field pressure is zero; near the surface ($\vartheta = \frac{1}{2}\pi$), the source and image still add together.   Thus the effect of reflection, when $z_0 > 1/k$, is to provide the radiated pressure wave with interference bands.

Comparison with Eq. (7.1.6) indicates that the appropriate Green's function for a rigid boundary at the xy plane is

$$G_\omega(\mathbf{r} \mid \mathbf{r}_0) = g_\omega(\mathbf{r} \mid \mathbf{r}_0) + g_\omega(\mathbf{r} \mid \mathbf{r}_0') \qquad (7.4.3)$$

where $\mathbf{r}_0 = (x_0, y_0, z_0)$ and $\mathbf{r}_0' = (x_0, y_0, -z_0)$.

If the source at P_0 is a dipole of strength $\mathbf{D}_\omega$, inclined at the angles θ, ϕ with respect to the axes related to the boundary plane, the mirror image will have the same x and y components as the source, but will have a D_z opposite in sign from that of the source.   Therefore the field is

$$p = -ik\rho c \mathbf{D}_\omega \cdot [\mathrm{grad}_0\, g_\omega(\mathbf{r} \mid \mathbf{r}_0) + \mathrm{grad}_0\, g_\omega(\mathbf{r} \mid \mathbf{r}_0')]e^{-i\omega t}$$

$$\simeq -k^2 \frac{\rho c}{2\pi r} \left[\left(\frac{x}{r} D_x + \frac{y}{r} D_y\right) \cos (kz_0 \cos \vartheta) \right.$$

$$\left. + i\frac{z}{r} D_z \sin (kz_0 \cos \vartheta) \right] e^{ik(r - r_0 \cos \psi - z_0 \cos \vartheta) - i\omega t} \quad (7.4.4)$$

the last expression being the far field, for $r \gg r_0$. The source and its image thus combine, at large distances and for long wavelengths, into a dipole of components $2D_x$ and $2D_y$, parallel to the xy plane, plus an axial quadrupole, oriented along the z axis, of strength $Q_{zz} = z_0 D_z$. As kz_0 is increased, interference between the waves from the horizontal and vertical components appears. At no time does the far field appear to be that from a single isolated dipole (except when $z_0 = 0$).

The Green's function

It might be well to examine more carefully the reasons for our choice of the Green's function of Eq. (7.4.3). It is, of course, to be inserted in Eq. (7.1.17) to compute the radiation in the presence of a rigid boundary in the xy plane. The boundary condition in this case is that $\partial p/\partial n_0$ is zero at $z_0 = 0$. The reason we choose G_ω to satisfy the same boundary condition is that, if this is done, the surface integral of Eq. (7.1.17) vanishes and the pressure wave in the medium ($z > 0$) is

$$p_\omega(\mathbf{r}) = \iiint f_\omega(\mathbf{r}_0) G_\omega(\mathbf{r} \mid \mathbf{r}_0) \, dv_0 \qquad (7.4.5)$$

where f_ω is the source function in the region outside the boundary surface. The difference between this solution and that of Eq. (7.1.18), for the unbounded medium, is that function G_ω, in contrast to g_ω, contains a term representing the wave reflected from the boundary plane.

As we have already shown under the preceding heading, the term to add to $g_\omega(\mathbf{r} \mid \mathbf{r}_0)$ to make $\partial G_\omega/\partial n = 0$ at $z = 0$ is one containing a wave from an image source, $g_\omega(\mathbf{r} \mid \mathbf{r}_0')$. Since the image part comes from a source inside the surface, this term is a solution of the homogeneous wave equation $(\nabla^2 + k^2)\chi = 0$ everywhere outside the surface, in the medium. Thus, when the rigid boundary is the xy plane,

$$G_\omega(\mathbf{r} \mid \mathbf{r}_0) = \frac{1}{4\pi R} e^{ikR} + \frac{1}{4\pi R'} e^{ikR'} \qquad (7.4.6)$$

where

$$R = [(x - x_0)^2 + (y - y_0)^2 + (z - z_0)^2]^{\frac{1}{2}}$$

and

$$R' = [(x - x_0)^2 + (y - y_0)^2 + (z + z_0)^2]^{\frac{1}{2}}$$

The geometry of this function is shown in Fig. 7.6. The first term is the pressure at point P, $\mathbf{r} = (x,y,z)$, caused by a unit source at P_0, $\mathbf{r}_0 = (x_0,y_0,z_0)$, in the absence of a boundary. The second term is the pressure at P, caused by an image source at P_0', $\mathbf{r}_0' = (x_0, y_0, -z_0)$, outside the limits of the medium. The first term has a singularity at $\mathbf{r} = \mathbf{r}_0$, but the second term has no singularity in the region $z \geqslant 0$, occupied by the medium. This second term, of course, represents the wave reflected by the rigid boundary at $z = 0$. Function G_ω satisfies Eq. (7.1.14) when z and z_0 are both positive.

It will occasionally be convenient to write this Green's function in other forms, related to those of Eqs. (7.3.14) and (7.3.15) for g_ω. For example, the Fourier integral expression for G_ω of Eq. (7.4.6) is

$$G_\omega(x,y,z \mid x_0,y_0,z_0) = \frac{1}{4\pi^3} \int\!\!\!\int\!\!\!\int_{-\infty}^{\infty} \exp\left[iK_x(x_0 - x) + iK_y(y_0 - y) + iK_z z\right]$$

$$\times \frac{\cos(K_z z_0)\, dK_x\, dK_y\, dK_z}{K_x^2 + K_y^2 + K_z^2 - k^2} \quad (7.4.7)$$

It can also be expressed in terms of cylindrical coordinates w, ϕ, z, with the cylindrical axis perpendicular to the boundary plane.

$$G_\omega = \sum_{m=0}^{\infty} \frac{i\epsilon_m}{2\pi} \cos m(\phi - \phi_0) \int_0^{\infty} J_m(\mu w)J_m(\mu w_0) \frac{\mu\, d\mu}{\kappa} \begin{cases} e^{i\kappa z_0}\cos(\kappa z) \\ e^{i\kappa z}\cos(\kappa z_0) \end{cases} \quad (7.4.8)$$

where $\kappa^2 = k^2 - \mu^2$, and the upper term after the brace is used when $0 \leqslant z \leqslant z_0$ and the lower term is used when $0 \leqslant z_0 \leqslant z$. This is to be compared with the integral of Eq. (7.3.15).

The effect of boundary impedance

If the boundary surface at $z = 0$ is not perfectly rigid, but is still passive, yielding somewhat to the incident acoustic pressure, the reflected wave is modified, so that the image term of Eq. (7.4.6) must be modified. If the surface is one of local reaction (see page 260), the formulas of Eqs. (6.3.5) and (6.3.6), relating the reflected wave to the acoustic impedance $z(\omega) = p/u_n$ of the surface, may be used to find the reflected wave. The most convenient form to use is the Fourier integral of Eq. (7.3.14). The integrand of this expression represents a plane wave which strikes the surface $z = 0$ at an angle of incidence $\cos^{-1}(K_z/K)$. Using Eq. (6.3.7), we see that the presence of the plane surface of impedance $z(\omega)$ at $z = 0$ introduces an extra term into the integrand, corresponding to a reflected wave

$$\exp\left[iK_x(x_0 - x) + iK_y(y_0 - y) + iK_z(z_0 + z)\right]$$

with a factor

$$C_r = \frac{(K_z/K) - [\rho c/z(\omega)]}{(K_z/K) + [\rho c/z(\omega)]}$$

to take into account the change of phase and amplitude on reflection.

In the discussion of this and succeeding chapters it will be convenient to deal with the *acoustic admittance* of the surface, the reciprocal of the impedance. We define the dimensionless parameter, the *specific admittance* of the surface, by

$$\beta = \frac{\rho c}{z(\omega)} = \xi - i\sigma \qquad z(\omega) = \frac{p}{u_n} = R - iX$$

$$\xi = \rho c \frac{R}{R^2 + X^2} \qquad \sigma = -\rho c \frac{X}{R^2 + X^2} \quad (7.4.9)$$

The parameter ξ is the acoustic conductance of the surface, in units of the characteristic admittance $1/\rho c$ of the medium; σ is the corresponding susceptance. If the wall is perfectly rigid, $\beta = 0$, $C_r = 1$, and the integral reduces to that of Eq. (7.4.7). If the wall is quite soft, β is large, and the integral differs appreciably from that of Eq. (7.4.7).

Assuming that β is uniform over the whole boundary surface $z = 0$, we can transform the Fourier integral into one in the cylindrical coordinates w, ϕ, z [a natural extension of Eq. (7.4.8)],

$$G_\omega = \sum_{m=0}^{\infty} \frac{i\epsilon_m}{4\pi} \cos m(\phi - \phi_0) \int_0^{\infty} J_m(\mu w)J_m(\mu w_0)$$

$$\times \left[e^{i\kappa|z-z_0|} + \frac{\kappa - k\beta}{\kappa + k\beta} e^{i\kappa(z+z_0)} \right] \frac{\mu \, d\mu}{\kappa} \quad (7.4.10)$$

which will be of use later in this section.

An exact representation of this Green's function in the form of Eq. (7.4.6) cannot be obtained, but a very good approximation, valid as long as the measurement point P is not closer than a half wavelength to the boundary surface, is

$$G_\omega \simeq \frac{e^{ikR}}{4\pi R} + \frac{\cos \vartheta' - \beta}{\cos \vartheta' + \beta} \frac{e^{ikR'}}{4\pi R'} \quad (7.4.11)$$

where the quantities involved are defined in Fig. 7.6. To this approximation there is an image at the point P_0' (as with the rigid boundary), but the wave it radiates has an amplitude which depends on the angle ϑ' which the line R' makes with the z axis, normal to the boundary. The physical significance of the dependence is easily seen if we realize that the ray of acoustic radiation which reaches the measurement point P from the source point P_0, by reflection, strikes the reflecting surface at Q at an angle ϑ' of incidence and reflection. Thus, according to Eq. (6.3.5), its reflection factor C_r should be the one given in Eq. (7.4.11). The wave which goes directly from P_0 to P without reflection, given by the first term, is independent of the inclination of R; the reflected wave, given by the second term, depends on the angle of reflection.

From these expressions, by methods which are by now familiar, we can compute the radiation from a multipole source in the presence of a plane surface of specific admittance β. For example, the counterpart to Eq. (7.4.2), for the far field from a simple-harmonic, simple source of strength S_ω, is

$$p \simeq -\frac{i\rho ck}{2\pi r} S_\omega \frac{\cos \vartheta \cos (kz_0 \cos \vartheta) - i\beta \sin (kz_0 \cos \vartheta)}{\cos \vartheta + \beta}$$

$$\times \exp \left[ik(r - r_0 \cos \psi + z_0 \cos \vartheta) - i\omega t\right] \quad (7.4.12)$$

which reduces to that of Eq. (7.4.2) when $\beta = 0$. A parallel formula can be worked out for a dipole source, the generalization of Eq. (7.4.4).

Boundary effects on multipole impedances

To calculate the effects of the boundary on the radiation load of a multipole source, we must compute the distortion of the near field produced by the boundary. For ease of computation, let us place the origin at the center of the radiating multipole. The boundary surface, of specific acoustic admittance β, is then the plane $z = -z_0$, and the image source is at $z = -2z_0$. Using the expansion of the Green's function in spherical harmonics, given in Eq. (7.2.31), we have

$$G_\omega = \frac{ik}{4\pi} \left[h_0(kr) + C_r \sum_{m=0}^{\infty} (-1)^m (2m+1) P_m(\cos \vartheta) j_m(kr) h_m(2kz_0) \right] \quad (7.4.13)$$

where we have assumed that $r < 2z_0$. For the case under consideration, actually, r is considerably smaller than z_0, so that the $\cos \vartheta'$ in the formula for C_r differs from unity only in terms of the second order in r/z_0. Therefore, to compute pressures and velocities at the surface of the sphere $r = a$, we can consider C_r to be the constant

$$C_r \simeq \frac{1-\beta}{1+\beta} = \frac{z(\omega) - \rho c}{z(\omega) + \rho c}$$

A simple source at the origin would produce a pressure wave $-ik\rho c S_\omega G_\omega e^{-i\omega t}$. The integral of the pressure over the surface of the small sphere $r = a$, just outside the source, the driving force for the source, is

$$F_s = -i\rho cka S_\omega \left[e^{ika} + C_r \frac{\sin(ka)}{k} \frac{e^{2ikz_0}}{2z_0} \right] e^{-i\omega t}$$

$$\rightarrow -i\rho cka S_\omega \left[1 + C_r \frac{a}{2z_0} e^{2ikz_0} \right] e^{-i\omega t} \quad ka \rightarrow 0$$

where we have substituted from Eqs. (7.2.9) and (7.2.11) for the various spherical Bessel functions, and we have taken advantage of the fact that the integral over the sphere of $P_m(\cos \vartheta)$ is zero except for $m = 0$. Comparison with Eq. (7.1.11) shows that the presence of the boundary has distorted the near field. As long as ka is small, the distortion is not large, however. The radial velocity $\frac{1}{ik\rho c} \frac{\partial p}{\partial r}$ at $r = a$ is also somewhat nonisotropic, but its mean value is

$$U_s \rightarrow \frac{S_\omega}{4\pi a^2} \left(1 - ika + \tfrac{1}{3} C_r k^2 a^2 \frac{a}{2z_0} e^{2ikz_0} \right) e^{-i\omega t}$$

the distorting effect being smaller by a factor $k^2 a^2$ than the correction term for the pressure.

Therefore the mechanical impedance of a simple source a distance z_0 away from a plane wall of acoustic impedance $\rho c/\beta$ is [see Eq. (7.1.11)]

$$Z_s = -i\omega\rho(4\pi a^3)\left\{1 + \frac{a}{2z_0}[R_r\cos(2kz_0) - X_r\sin(2kz_0)]\right\}$$

$$+ \rho c(4\pi a^2)(ka)^2\left\{1 + \frac{1}{2kz_0}[R_r\sin(2kz_0) + X_r\cos(2kz_0)]\right\}$$

where

$$R_r = \text{Re } C_r \rightarrow 1 - 2\xi \qquad X_r = \text{Im } C_r \rightarrow 2\sigma \qquad |\beta| \ll 1$$

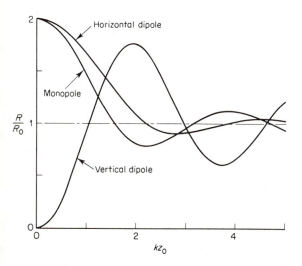

FIGURE 7.7
Ratio of effective radiation resistance R, for a multipole of frequency $\omega/2\pi = kc/2\pi$, a distance z_0 above a rigid plane, to the effective radiation resistance R_0 in the absence of the plane.

The wall correction to the mechanical reactance is small because a/z_0 is small; this correction can usually be neglected. But the correction to the small resistive term is proportionally larger and, if $\beta \rightarrow 0$, can be equal to the resistive term itself. In fact, for a rigid wall ($\beta = 0$), the impedance becomes

$$Z_s \simeq -i\omega\rho(4\pi a^3) + \rho c(4\pi a^2)(ka)^2\left[1 + \frac{\sin(2kz_0)}{2kz_0}\right] \qquad (7.4.14)$$

The real part of this impedance is a measure of the power radiated by the monopole. We note that when $kz_0 = 2\pi z_0/\lambda$ is small (either the source is close to the wall or the wavelength is long), this resistive term is twice that for a source in an unbounded medium. At these long wavelengths the image source reinforces the primary source. As kz_0 is increased, the factor in brackets approaches 1 in an oscillatory manner, as shown in Fig. 7.7. For $\lambda \rightarrow 0$ the source radiates as though the boundary were not present.

To find the corresponding change in the dipole impedance we need to compute $\text{grad}_0\, G_\omega$, but in Eq. (7.4.13) we have removed the r_0 coordinates by placing the origin at the source point. However, since $\mathbf{R} = \mathbf{r} - \mathbf{r}_0$ in the first term, we can set grad_0 equal to $-\text{grad}$; in the second term we remove the subscript zero by changing the sign of the x and y derivatives but keeping $\partial/\partial z_0 = \partial/\partial z$. This, together with Eqs. (7.2.32) (which hold for Hankel as well as Bessel functions), results in

$$p = -i\rho ck\Big(\frac{1 - ikr}{4\pi r^2}\, e^{ikr}(D_x Y_{11}{}^1 + D_y Y_{11}{}^{-1} + D_z Y_{10}{}^1)$$

$$+ C_r \frac{ik^2}{4\pi} \sum_{m=0}^{\infty} (-1)^m h_m(2kz_0)\{D_x(Y^1_{m-1,1}j_{m-1} + Y^1_{m+1,1}j_{m+1})$$

$$+ D_y(Y^{-1}_{m-1,1}j_{m-1} + Y^{-1}_{m+1,1}j_{m+1})$$

$$+ D_z[m Y_{m-1,0}j_{m-1} - (m + 1) Y_{m+1,0}j_{m+1}]\}\Big)e^{-i\omega t}$$

where the spherical harmonics Y are functions of the spherical angles ϑ, φ of the radius vector $\mathbf{r}$, and the Bessel functions j are functions of kr.

Both direction and magnitude of the net force on the dipole are modified by the presence of the boundary surface. If the components of the dipole would be D_x, D_y, D_z if the plane were not present, the components in the presence of the plane are $1 + \frac{1}{3}k^3a^3(h_0 + h_2)$ times D_x and D_y and $1 + \frac{1}{3}k^3a^3(-h_0 + 2h_2)$ times D_z, where the Hankel functions are functions of $2kz_0$. If the direction angles of the undistorted dipole are θ, ϕ, so that $D_x = D_\omega \sin\theta \cos\phi$, $D_y = D_\omega \sin\theta \sin\phi$, and $D_z = D_\omega \cos\theta$, the magnitude of the driving force is

$$F_d = -\tfrac{1}{3}ik\rho cD_\omega\{(1 - ika)e^{ika} - \tfrac{1}{3}ik^3a^3[(h_0 - \tfrac{1}{2}h_2)\cos 2\theta + \tfrac{1}{2}h_2]\}e^{-i\omega t}$$

for the case $\beta = 0$ $(C_r = 1)$, and the radial fluid velocity in the dipole direction is

$$U_d = \frac{D_\omega}{2\pi a^3}\{(1 - ika - \tfrac{1}{2}k^2a^2)e^{ika} + \tfrac{1}{6}ik^3a^3[(h_0 - \tfrac{1}{2}h_2)\cos 2\theta + \tfrac{1}{2}h_2]\}e^{-i\omega t}$$

The ratio between these is the effective mechanical impedance of the dipole in the presence of the rigid wall. As with the monopole source, the effect on the reactive term is negligible, but the effect of the presence of the wall on the resistive term is of the same order in ka as is the unmodified term. This part of the correction term involves the real part of the Hankel functions [Eq. (7.2.10)], so that, for $C_r = 1$,

$$Z_a \simeq -i\omega\rho\, \frac{2\pi a^3}{3}\,(1 + \tfrac{1}{2}k^2a^2)$$

$$+ \rho c(4\pi a^2)\frac{k^4a^4}{12}\{1 + [\tfrac{1}{2}j_2(2kz_0) - j_0(2kz_0)]\cos 2\theta + \tfrac{3}{2}j_2(2kz_0)\} \quad (7.4.15)$$

is the effective impedance of a dipole placed a distance z_0 from a rigid plane wall and inclined at angle θ to its normal. The modification required for a nonrigid wall is not difficult and will be left for a problem.

We see that the interference effects, in the resistance term, caused by the presence of the wall, depend on the orientation of the dipole. If the dipole is vertical ($\theta = 0$ or π), the image dipole is opposed and long-wavelength interference is destructive; the factor in braces goes to zero for $kz_0 \to 0$. On the other hand, if the dipole is parallel to the wall, the interference is constructive; the factor goes to 2 as $kz_0 \to 0$. The behavior of this factor, and that for the monopole, are shown in Fig. 7.7.

Radiation from the boundary

Sound waves in the bounded medium can also be produced by motion of the boundary surface itself, the energy entering the medium from some area of its surface. With a plane rigid boundary at $z = 0$, some portion of the plane may be moving in the z direction. If the motion is simple-harmonic $u_z = u_\omega e^{-i\omega t}$, the boundary condition in this area is that the normal gradient of the pressure is $ik\rho c u_\omega(x,y)e^{-i\omega t}$, whereas, for the rest of the plane, $\partial p/\partial z$ is zero.

An obvious means of finding the pressure field is to apply the Green's function of Eq. (7.1.17) again, this time using the surface integral; since there is no sound source within the medium, the volume integral is zero. The surface integral involves both p and its outward-pointing normal gradient (that is, $-\partial p/\partial z$) at $z = 0$, but our description of the physical situation has specified only the normal gradient. However, if we use the Green's function of Eqs. (7.4.6) to (7.4.8), which has been adjusted so that its normal gradient is zero at $z = 0$, the resulting surface integral then only includes the specified normal gradient of p, and we have

$$p_\omega(x,y,z) = -ik\rho c \iint u_\omega(x_0,y_0)G_\omega(x,y,z \mid x_0,y_0,0)\, dx_0\, dy_0$$

$$= -2ik\rho c \iint u_\omega(x_0,y_0)g_\omega(x,y,z \mid x_0,y_0,0)\, dx_0\, dy_0 \quad (7.4.16)$$

since, at $z = 0$, $R' = R$, and thus $G_\omega = 2g_\omega$.

Each moving element of area of the plane boundary acts like a simple source of *twice* the strength $u_\omega\, dx_0\, dy_0$ it would have if the rest of the boundary were not there. The reason is that this element radiates only into the half space $z > 0$; the source and its image have coalesced to form a source of double strength.

If the boundary plane, at $z = 0$, is not completely rigid, but reacts to an incident pressure wave with point impedance $z(\omega) = \rho c/\beta$, then every portion of the plane is moving, and we must devise a more sophisticated method of distinguishing the active regions of the boundary, where energy is fed into the medium, from the passive areas. For the passive areas the relationship between p and its normal derivative follows the definition of the acoustic impedance, $-u_z = -\dfrac{1}{ik\rho c}\dfrac{\partial p}{\partial z} = \dfrac{\beta}{\rho c}p$, at $z = 0$. For the active regions the normal fluid velocity, plus $(\beta/\rho c)p$, at $z = 0$, must differ from zero; we shall call this difference the *driving velocity* u_ω,

$$\left(u_z + \frac{\beta}{\rho c}p\right)_{z=0} = \left(\frac{1}{ik\rho c}\frac{\partial p}{\partial z} + \frac{\beta}{\rho c}p\right)_{z=0} = u_\omega e^{-i\omega t} \qquad (7.4.17)$$

To simplify the form of the surface integral of Eq. (7.1.17), we choose a Green's function such that $\partial G_\omega/\partial z_0 = -ik\beta G_\omega$. For then the surface integral over the passive area of the boundary is zero, and the integral over the active area is, with $\partial/\partial n_0 = -\partial/\partial z_0$,

$$p_\omega = -\iint\left(\frac{\partial p}{\partial z_0}G_\omega - p_\omega\frac{\partial G_\omega}{\partial z_0}\right)_{z_0=0} dx_0\,dy_0$$

$$= -\iint\left(\frac{\partial p_\omega}{\partial z_0} + ik\beta p_\omega\right)G_\omega\,dx_0\,dy_0$$

$$= -ik\rho c\iint u_\omega(x_0,y_0)G_\omega(x,y,z \mid x_0,y_0,0)\,dx_0\,dy_0 \qquad (7.4.18)$$

which has the same form as before. But now the Green's function, satisfying these requirements at $z = 0$, is the one we already wrote down in Eqs. (7.4.10) and (7.4.11) for reflection from a plane surface of impedance $\rho c/\beta$. The equation for the pressure wave, in cylindrical and spherical coordinates, is

$$p_\omega(\mathbf{r}) = \frac{k\rho c}{2\pi}\sum_{m=0}^{\infty}\epsilon_m\int_0^{2\pi}\cos m(\phi - \phi_0)\,d\phi_0\int_0^a u_\omega(w_0,\phi_0)w_0\,dw_0$$

$$\times\int_0^\infty J_m(\mu w)J_m(\mu w_0)\frac{\mu\,d\mu}{\kappa + k\beta}e^{i\kappa z}$$

$$\simeq -\frac{ik\rho c}{2\pi}\iint u_\omega(x_0,y_0)\frac{\cos\vartheta'}{\cos\vartheta' + \beta}\frac{e^{ikR}}{R}\,dx_0\,dy_0 \qquad (7.4.19)$$

where $\kappa^2 = k^2 - \mu^2$, $R^2 = (x - x_0)^2 + (y - y_0)^2 + z^2$, and $\cos\vartheta' = z/R$. The integral is, of course, over the active area A of the boundary, which is assumed to be confined within a circle of radius a about the origin. When $\beta = 0$ the boundary is rigid and the formulas reduce to those of Eq. (7.4.16). When $\beta > 0$ the second expression is not accurate for the near field, but is valid for the far field.

The far field

When the distance r, from the source region on the boundary, is much larger than a, the radius of the circle enclosing the source area, the last integral expression of Eq. (7.4.19) can be further simplified. Neglecting terms of order a/r and smaller, we have $\cos \vartheta' \simeq \cos \vartheta = z/r$.

$$R \simeq r - (x_0 \cos \varphi + y_0 \sin \varphi) \sin \vartheta \qquad \frac{1}{R} \simeq \frac{1}{r}$$

where $\tan \varphi = y/x$. Therefore

$$p_\omega(\mathbf{r}) \simeq -ik\rho c \, \frac{\cos \vartheta}{\beta + \cos \vartheta} \, \frac{e^{ikr}}{2\pi r} \iint u_\omega(x_0,y_0) e^{-i\mathbf{k}\cdot\mathbf{w}_0 \sin \vartheta} \, dx_0 \, dy_0$$

where vector $\mathbf{w}_0$ lies in the xy plane, having components x_0, y_0; and vector $\mathbf{k}$ points along $\mathbf{r}$, having components $k_x = k \cos \varphi \sin \vartheta$, $k_y = k \sin \varphi \sin \vartheta$, $k_z = k \cos \vartheta$.

This is a very interesting formula, for it shows that the angular distribution of the far field is proportional to the two-dimensional Fourier transform of the distribution of driving velocity in the active region A of the boundary plane. By analogy with Eqs. (1.3.16), we define

$$U_\omega(K_x,K_y) = \frac{1}{4\pi^2} \int\limits_{-\infty}^{\infty}\!\!\int u_\omega(x_0,y_0) e^{-iK_x x_0 - iK_y y_0} \, dx_0 \, dy_0$$

$$\hspace{7cm} (7.4.20)$$

$$u_\omega(x_0,y_0) = \int\limits_{-\infty}^{\infty}\!\!\int U_\omega(K_x,K_y) e^{iK_x x_0 + iK_y y_0} \, dK_x \, dK_y$$

In terms of this Fourier transform, the far field can be written in the particularly simple form

$$p_\omega(\mathbf{r}) \simeq -ik\rho c \, \frac{e^{ikr}}{4\pi r} f_\omega(\vartheta,\varphi)$$

$$f_\omega(\vartheta,\varphi) = \frac{8\pi^2 \cos \vartheta}{\beta + \cos \vartheta} \, U_\omega(k \cos \varphi \sin \vartheta, \, k \sin \varphi \sin \vartheta) \qquad (7.4.21)$$

where $k = \omega/c$. The angle-distribution factor $f_\omega(\vartheta,\varphi)$ measures the relative amplitude of the acoustic radiation in the direction ϑ, φ from the center of the active area of the boundary, at great distances from this area. It is the strength of the radiation in the direction ϑ, φ from the radiating area, measured in equivalent simple-source strengths.

The formula is useful in a number of ways. For example, for wavelengths long enough so that $ka \ll 1$, $U(K_x,K_y) \simeq (A_r/4\pi^2)\langle u_\omega \rangle$, where A_r is the area of the active portion of the boundary and $\langle u_\omega \rangle$ is the mean value of the driving velocity of that portion. We can write $A_r \langle u_\omega \rangle = S_\omega$ as the strength of the simple source equivalent to the active region and write

$$f_\omega(\vartheta,\varphi) \simeq \frac{2\cos\vartheta}{\beta + \cos\vartheta} S_\omega \qquad ka \ll 1$$

When $\beta \ll 1$ the angle-distribution function f_ω is practically independent of ϑ, until ϑ is close enough to $\frac{1}{2}\pi$ so that $\cos\vartheta = |\beta|$; from that value to $\vartheta = \frac{1}{2}\pi$ it drops rapidly to zero. Thus, for a hard, but not perfectly rigid, wall, a low-frequency wave spreads out isotropically from the radiating area, except for angles closer than about $|\beta|$ radians to the grazing angle $\frac{1}{2}\pi$. That part of the wave which stays close to the boundary is absorbed by the boundary, unless it is perfectly rigid.

Another way of looking at Eq. (7.4.21) is afforded by rotating the coordinate axes on the surface from x_0, y_0 to X_0, Y_0, where $x_0 = X_0 \cos\varphi - Y_0 \sin\varphi$ and $y_0 = X_0 \sin\varphi + Y_0 \cos\varphi$; so that the X_0 axis is the intersection of the rz plane with the xy plane, and the Y_0 axis is in the xy plane, perpendicular to the rz plane. We then see that

$$U_\omega(k\cos\varphi\sin\vartheta, k\sin\varphi\sin\vartheta) = \frac{1}{4\pi^2} \int e^{-ikX_0\sin\vartheta}\, dX_0 \int u_\omega(X_0, Y_0)\, dY_0$$

(7.4.22)

is the one-dimensional Fourier transform, in the X_0 direction, of the average $u_\varphi(\omega,X_0) = (1/2\pi)\int u_\omega(X_0, Y_0)\, dY_0$ of the velocity $u_\omega(X_0, Y_0)$ in the Y_0 direction. This quantity $u_\varphi(\omega,X_0)$ is a function of but one surface coordinate, X_0, the distance in the direction of the measurement point (i.e., along the rz plane); it is also a function of φ, the direction of the rz plane, and of the frequency $\omega/2\pi$.

From this expression we can go on to find the far field from a region of the boundary which is moving with a normal velocity $u_z(x,y,t)$, an arbitrary function of x, y and of time t. We first note that the $u_\omega(x,y)$ of Eqs. (7.4.20) and (7.4.22) is the time-dimension Fourier transform of $u_z(x,y,t)$; next we note [Eq. (2.3.5)] that $-i\omega u_\omega(x,y)$ is the Fourier transform of $\partial u_z/\partial t$. Thus the pressure field at the point $r = (r,\vartheta,\varphi)$ (when $r \to \infty$) is the inverse transform of p_ω of Eq. (7.4.21).

$$p(\mathbf{r},t) \simeq \frac{\rho}{4\pi r} \int_{-\infty}^{\infty} f_\omega(\vartheta,\varphi) \exp\left[-i\omega\left(t - \frac{r}{c}\right)\right] d\omega$$

This integral is tedious to compute if the boundary admittance β is strongly dependent on ω.

But if β can be considered to be constant over the range of ω for which f_ω is nonnegligible, the far field from an arbitrary velocity distribution $u_z(x,y,t)$ of the boundary can be expressed in simple form. First we rotate coordinate axes to the X_0, Y_0 just defined, expressing u_z in these coordinates as $u_z(X_0, Y_0, t)$. Then, from the definitions (1.3.16) of the Fourier transform, we have

$$p(\mathbf{r},t) \simeq \frac{\rho}{2\pi r} \frac{\cos\vartheta}{\beta + \cos\vartheta} \int_{-\infty}^{\infty} -i\omega e^{-i\omega\left(t-\frac{r}{c}\right)} d\omega \int_{-\infty}^{\infty} e^{i(\omega/c)X_0 \sin\vartheta} dX_0$$

$$\times \int_{-\infty}^{\infty} dY_0 \frac{1}{2\pi} \int_{-\infty}^{\infty} e^{i\omega\tau} u(X_0,Y_0,\tau)\, d\tau$$

$$= \frac{\rho}{2\pi r} \frac{\cos\vartheta}{\beta + \cos\vartheta} \frac{\partial}{\partial t}\left[\int dX_0 \int dY_0\, u\left(X_0, Y_0, t + \frac{1}{c} X_0 \sin\vartheta - \frac{r}{c}\right)\right]$$

$$(7.4.23)$$

This result demonstrates several properties of the far field. First, the field depends on an appropriate summation of the normal *accelerations* of the various portions of the active area of the boundary surface. Second, the contribution to the pressure at $\mathbf{r}$ at time t, from the line X_0, $-\infty < Y_0 < \infty$, depends on the normal accelerations along this line at time $t + (1/c)X_0 \sin\vartheta - (r/c)$, a time $(r/c) - (X_0/c)\sin\vartheta$ earlier, this being the time required for the wave to travel from the line to the point $\mathbf{r}$. The dependence of $p(\mathbf{r},t)$ on φ is inherent in the choice of the coordinates X_0, Y_0; its dependence on ϑ comes from the $(X_0/c)\sin\vartheta$ term, and also from the boundary-admittance term outside the integral (for example, see Fig. 7.10). A typical modification of this result, when β depends strongly on ω, is given as a problem.

Since, for the far field, $p \simeq \rho c u_r$, the intensity of the acoustic radiation at great distances from the active area is all radial, with time-averaged magnitude $(|p|^2/\rho c)$.

Far field from a randomly vibrating area

The formulas derived in the last subsection can be used to compute the mean intensity at r, ϑ, φ produced by an area A of the boundary which is vibrating with a random distribution of velocity. Motion of this sort could be caused by the impinging of turbulent air on the surface, or it could be that a portion of the boundary is attached to some vibrating machinery. If we consider the velocity distribution $u_z(x,y,t)$ of the active area as a function of position as well as of time, u_z can be described either in terms of its auto-correlation function Υ or its spectrum density $|U_\omega|^2$, where

$$\Upsilon(\xi,\eta,\tau) = \frac{1}{AT} \iiint u_z^*(x,y,t)u_z(x + \xi, y + \eta, t + \tau)\, dx\, dy\, dt$$

$$(7.4.24)$$

$$U_\omega(K_x,K_y) = \frac{1}{8\pi^3} \iiint u_z(x,y,t)e^{i\omega t - iK_x x - iK_y y}\, dx\, dy\, dt$$

where the integration in x, y is over the active area A; in t, is over the long interval of time T. As we have seen, a simple extension of Eq. (1.3.20) shows that the space-time spectrum level $|U_\omega|^2$ is the Fourier transform of the autocorrelation function:

$$|U_\omega(K_x,K_y)|^2 = \frac{AT}{64\pi^6} \iiint\limits_{-\infty}^{\infty} \Upsilon(\xi,\eta,\tau) e^{i\omega\tau - iK_x\xi - iK_y\eta}\, d\xi\, d\eta\, d\tau \quad (7.4.25)$$

However, we have already encountered the velocity transform U_ω in the expression (7.4.21) for the far field. In fact, the time-averaged intensity for the radiation of frequency $\omega/2\pi$, at large distances from the radiating area,

$$I_r(\omega) \to \frac{|p|^2}{\rho c} \simeq \frac{\rho c}{r^2} \frac{4\pi^2 k^2 \cos^2\vartheta}{|\beta(\omega) + \cos\vartheta|^2} |U_\omega(k\sin\vartheta\cos\varphi, k\sin\vartheta\sin\varphi)|^2$$

$$(7.4.26)$$

is proportional to the spectrum density, and so can be expressed in terms of the threefold autocorrelation function of the surface velocity. The total intensity at r, ϑ, φ, averaged over the time interval T, is $2\pi/T$ times the integral of $I_r(\omega)$ over ω, in accord with Eq. (1.3.19).

If the motion of the active area is completely random, we should expect that the autocorrelation function Υ falls off smoothly as τ, ξ, or η is made larger than zero; so that a reasonable assumption for its form would be

$$\Upsilon(\xi,\eta,\tau) = |U_0|^2 \exp\left[-\frac{1}{2w_c^2}(\xi^2 + \eta^2) - \tfrac{1}{2}(\omega_n\tau)^2\right] \quad (7.4.27)$$

where $|U_0|^2$ is the mean-square velocity of the active area, averaged over time interval T and area A; w_c is the correlation distance along the surface, as in Eq. (7.2.29); and $1/\omega_n$ is the correlation time interval, as in Eq. (2.3.13). Therefore $\omega_n/2\pi$ is the upper frequency limit of the power spectrum, below which all frequencies are represented and above which $|U_\omega|^2$ drops rapidly to zero.

If this is a good approximation for Υ, the corresponding power spectrum is

$$I_r(\omega) \simeq \rho c \frac{AT}{r^2} (2\pi)^{-\frac{5}{2}} \frac{\omega^2 w_c^2 |U_0|^2 \cos^2\vartheta}{c^2\omega_n |\beta + \cos\vartheta|^2} \exp\left[-\frac{1}{2}\left(\frac{\omega w_c}{c}\right)^2 \sin^2\vartheta - \frac{1}{2}\left(\frac{\omega}{\omega_n}\right)^2\right]$$

To find the total intensity at $\mathbf{r}$ we must integrate this over ω and multiply by $2\pi/T$. The integrand has a maximum at $\omega = \sqrt{2}\,\omega_n(1 + k_n^2 w_c^2 \sin^2\vartheta)^{-\frac{1}{2}}$, where $k_n = \omega_n/c = 2\pi/\lambda_n$, λ_n being the lower limit of wavelength in the power spectrum. If $\beta(\omega)$, the specific acoustic admittance of the boundary surface, does not change much with ω near this maximal frequency, then β can be taken outside the integral, and

$$I_r(\mathbf{r}) = \frac{2\pi}{T} \int_{-\infty}^{\infty} I_r(\omega)\, d\omega \simeq \rho c \frac{A}{2\pi r^2} \frac{|U_0|^2 \cos^2\vartheta}{|\beta + \cos\vartheta|^2} \frac{(k_n w_c)^2}{[1 + (k_n w_c)^2 \sin^2\vartheta]^{\frac{3}{2}}}$$

$$(7.4.28)$$

where β is evaluated at $\omega = \sqrt{2}\,\omega_n(1 + k_n^2 w_c^2 \sin^2\vartheta)^{-\frac{1}{2}}$.

This is an interesting, and fairly simple, formula. The mean intensity at **r** is proportional to ρc, the characteristic impedance of the medium, times $|U_0|^2$, the mean-square velocity of the active area, times the ratio between the area A which is active and the area of the half sphere of radius r which is in the medium, times an angle-distribution factor, which depends in part on the ratio between the spatial-correlation length w_c and the wavelength λ_n of the cutoff frequency of the power spectrum, and in part on the admittance of the boundary surface. If the motion is caused by turbulence of the medium close to the surface, then $k_n w_c$ is of the order of unity; if the motion is caused by vibrations transmitted through a stiff boundary wall, then $k_n w_c$ may be larger than unity. In the latter case the intensity has a fairly strong preference for the normal direction, $\vartheta = 0$. If the wall is stiff, so that β is small, the factor involving β is nearly unity, except for near-grazing directions, $\vartheta \to \frac{1}{2}\pi$.

Radiation from a circular piston

As an interesting example of the radiation from a moving portion of a plane boundary, we shall compute the far field from a circular piston of radius a, set in the boundary plane $z = 0$. This is, of course, a special case of the formulas we have been examining in this section; it is worth carrying through in detail because of its practical interest. A cone loudspeaker, set in a plane baffle, radiates sound roughly like a plane piston set in a plane baffle of infinite extent. We shall take as our model the case where its normal velocity is $u_\omega e^{-i\omega t}$ inside a circle of radius a, centered at the origin, and is zero over the rest of the plane $z = 0$ [if the surface has admittance β, the difference $u_z + (\beta/\rho c)p$ at the surface is $u_\omega e^{-i\omega t}$ for $r < a$].

Returning to Eq. (7.4.21), changing from coordinates x_0, y_0 to polar coordinates w_0, ϕ_0 in the integral for U_ω, we find, using Eqs. (1.3.11) and (5.2.21),

$$p_\omega(r) \simeq -ik\rho c \, \frac{e^{ikr}}{2\pi r} \frac{\cos\vartheta}{\beta + \cos\vartheta} u_\omega \int_0^{2\pi} d\phi_0 \int_0^a w_0 \, dw_0$$

$$\times \exp\left[-ikw_0 \cos(\phi_0 - \phi)\sin\vartheta\right]$$

$$= -ik\rho c \, \frac{e^{ikr}}{4\pi r} f_\omega(\vartheta) \qquad f_\omega = 2u_\omega \pi a^2 \left[\frac{2J_1(ka\sin\vartheta)}{ka\sin\vartheta} \frac{\cos\vartheta}{\beta + \cos\vartheta}\right]$$

$$\text{(7.4.29)}$$

The equivalent strength in direction ϑ is thus equal to twice the total outflow amplitude $\pi a^2 u_\omega$ times the factor in brackets, which is nearly unity at $\vartheta = 0$, when $\beta \ll 1$.

The corresponding intensity at **r** is

$$I_r \simeq \frac{|p|^2}{\rho c} = \tfrac{1}{4}\rho c \, |u_\omega|^2 \, k^2 a^2 \left(\frac{a}{r}\right)^2 \frac{\cos^2\vartheta}{|\beta + \cos\vartheta|^2} \left[\frac{2J_1(ka\sin\vartheta)}{ka\sin\vartheta}\right]^2 \quad \text{(7.4.30)}$$

Values of $[2J_1(x)/x]$ can be obtained from Table IX. It is unity when x is zero, remains nearly unity until x is about $\pi/2$, goes to zero at x about 1.2π, falls to about -0.13 at x about 1.7π, goes to zero again at x about 2.2π, and so on, having a sequence of maxima and minima which diminishes in size.

When the wavelength λ of the sound radiated is longer than the circumference $2\pi a$ of the piston, the value of $\mu \sin \vartheta$ is less than $\pi/2$ even for

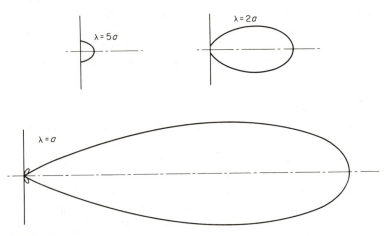

FIGURE 7.8
Polar diagrams of distribution-in-angle of radiated intensity from a piston set in a plane wall, for different ratios between the wavelength λ and the radius of the piston a. In the curve for $\lambda = a$, the small loops at the side are diffraction rings.

$\vartheta = 90°$ (i.e., even along the wall), and the term in brackets is practically unity for all values of ϑ. Therefore long-wavelength sound spreads out uniformly in all directions from the piston, with an intensity four times that due to a simple source of strength $\pi a^2 u_0$. If the wall were not present and the "piston" were the open end of a pipe, this end would act like a simple source of strength $\pi a^2 u_0$ for long wavelengths, so that the wall, or baffle plate, produces a fourfold increase in intensity. The sound reflected from the baffle reinforces the sound radiated outward, thereby doubling the amplitude of the wave, and thus quadrupling the intensity, which depends on the square of the amplitude. Of course, to have the baffle give this considerable increase in amplitude, it must be considerably larger than the wavelength of the sound radiated, so that it will act as though it were infinite in extent.

If λ is smaller than $2\pi a$, then the reflected sound still reinforces the sound radiated straight ahead, and the intensity has its maximum value

$\rho v^2 (\pi a^2 u_0)^2 / 2cr^2$ at $\vartheta = 0$. At points off the axis, however, the reflected sound interferes with that radiated directly, and I diminishes in value as ϑ increases, falling to zero when $\sin \vartheta$ is about equal to $0.6(\lambda/a)$; then rises to a secondary maximum (where there is a little reinforcement) of about 0.02 times the intensity for $\vartheta = 0$; then goes again to zero; and so on. Therefore high-frequency sound is chiefly sent out perpendicular to the wall, with little spread of the beam. Most of the intensity is inside a cone whose axis is along the axis of the piston and whose angle is about $\sin^{-1}(\lambda/2a)$. This main beam is surrounded by *diffraction rings*, secondary intensity maxima, whose magnitude diminishes rapidly as we go from one maximum to the next away from the main beam.

Values of the intensity as a function of ϑ are given in polar diagrams in Fig. 7.8, for different values of $\mu = 2\pi v a/c$. The increased directionality with increase in frequency is apparent. It is this directionality that makes it difficult to use a single loudspeaker in outdoor public-address systems: although the low-frequency sound is spread out in all directions, only the people standing directly in front of the loudspeaker will hear the high-frequency sound. In rooms of small size, the sound is scattered so much by the walls that the directionality does not matter particularly (unless it is very marked), and one loudspeaker is usually adequate.

Radiation impedance of the piston

To find the reaction of the fluid back on the driving piston we must calculate the near field. For this purpose the first of Eqs. (7.4.19) is useful. Setting u_ω constant for $r < a$ and then setting $z = 0$ gives

$$P_\omega(\omega) = \rho c k u_\omega \int_0^a w_0 \, dw_0 \int_0^\infty J_0(\mu w_0) J_0(\mu w) \frac{\mu \, d\mu}{k\beta + \sqrt{k^2 - \mu^2}}$$

$$= \rho c k a u_\omega \int_0^\infty J_0(\mu w) J_1(\mu a) \frac{d\mu}{k\beta + \sqrt{k^2 - \mu^2}}$$

since u_ω is independent of w_0 and ϕ_0. This pressure is not constant across the face of the piston. But since the piston is presumed rigid, the quantity of practical interest is the total reaction force on the whole piston. In the case of $\beta = 0$, this can be expressed in closed form.

$$F_\omega = 2\pi \int_0^a p_\omega(w) w \, dw = \pi \rho c k a^2 u_\omega \int_0^\infty [J_1(\mu a)]^2 \frac{d\mu}{\mu \sqrt{k^2 - \mu^2}}$$

$$= \pi \rho c k a^2 u_\omega \left\{ \int_0^k [J_1(\mu a)]^2 \frac{d\mu}{\mu \sqrt{k^2 - \mu^2}} - i \int_k^\infty [J_1(\mu a)]^2 \frac{d\mu}{\mu \sqrt{\mu^2 - k^2}} \right\}$$

and thus the mechanical impedance of the radiation load on the piston is

$$Z_r = \frac{F_\omega}{u_\omega} = \pi a^2 \rho c(\theta_0 - i\chi_0) = \pi a^2 (R - iX)$$

$$\theta_0(\gamma) = 1 - \frac{2}{\gamma} J_1(\gamma) \qquad \gamma = 2ka = \frac{4\pi a}{\lambda} = \frac{2\omega a}{c}$$

$$\chi_0(\gamma) = M_1(\gamma) = \frac{4}{\pi} \int_0^{\pi/2} \sin(\gamma \cos \alpha) \sin^2 \alpha \, d\alpha$$

(7.4.31)

$$\theta_0 \to \begin{cases} \frac{1}{2}(ka)^2 & ka \to 0 \\ 1 & ka \to \infty \end{cases} \qquad \chi_0 \to \begin{cases} \dfrac{8ka}{3\pi} & ka \to 0 \\ \dfrac{2}{\pi ka} & ka \to \infty \end{cases}$$

See Table IX for values of the functions θ_0 and χ_0. Curves are shown in Fig. 7.9. Comparison with Fig. 7.3 shows that change from plane to spherical baffle makes little change in the average impedance load on the piston, though comparison between Figs. 7.8 and 7.2 indicates that the directionality pattern does differ considerably. Sound is sent out in different directions from the two sources, but its reaction back on the piston depends more on $2\pi a/\lambda$ than on baffle shape. At high frequencies the impedance is resistive, equal to piston area times the characteristic impedance ρc.

We might here say a few things about the radiation reaction to the general motion $u_\omega(x_0, y_0)$, which produces the far field of Eq. (7.4.21). In this case the pressure at the point w, ϕ on the surface of the moving boundary is

$$p_\omega(w, \phi) = \frac{k\rho c}{2\pi} \sum_{m=0}^{\infty} \epsilon_m \int_0^{2\pi} \cos m(\phi_0 - \phi) \, d\phi_0$$

$$\times \int_0^\infty J_m(\mu w) \frac{\mu \, d\mu}{k\beta + \sqrt{k^2 - \mu^2}} \int_0^a u_\omega(w_0, \phi_0) J_m(\mu w_0) w_0 \, dw_0$$

where we have expressed $u_\omega(x_0, y_0)$ as a function of the polar coordinates w_0, ϕ_0. This set of integrals must be worked out numerically in most cases, but in the two limiting cases $ka \ll 1$ and $ka \gg 1$, approximate formulas can be obtained.

In the first place, reference to Eq. (1.3.27) indicates that the last integral is the Hankel transform of u_ω, so that

$$u_\omega(w, \phi) = \sum_{m=0}^{\infty} \frac{\epsilon_m}{2\pi} \int_0^{2\pi} \cos m(\phi_0 - \phi) \, d\phi_0 \int_0^\infty J_m(\mu w) U_{m\omega}(\mu, \phi_0) \mu \, d\mu$$

$$U_{m\omega}(\mu, \phi_0) = \int_0^a u_\omega(w_0, \phi_0) J_m(\mu w_0) w_0 \, dw_0$$

where the upper limit to the integral for the transform $U_{m\omega}$ could just as well have been ∞, since u_ω is zero for $w_0 > a$. The quantity $U_{m\omega}$, as a function of μ, goes to zero as $\mu \to 0$, for $m > 0$, because $J_m(z) \to z^m/2^m m!$ as $z \to 0$.

For $m = 0$ the limiting value

$$U_{0\omega}(\mu,\phi_0) \rightarrow \int_0^a u_\omega(w_0,\phi_0)w_0 dw_0 \qquad \mu a \ll 1$$

is proportional to the mean value of u_ω over the interior of the circle $w = a$, in the ϕ_0 direction. It also can be demonstrated that $U_{m\omega} \rightarrow 0$ for all m when μ becomes considerably larger than $1/a$.

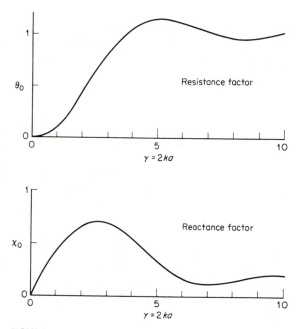

FIGURE 7.9
Radiation impedance of a circular piston in a plane baffle equals area of piston times $(\theta_0 - i\chi_0)\rho c$. Resistance and reactance factors plotted as a function of frequency $\omega/2\pi = c\gamma/4\pi a$.

Now if $ka \gg 1$, so that the radiated wavelength is small compared with a, and also small enough so that u_ω does not change appreciably in a distance of a wavelength, then, over the range of integration of μ within which $U_{m\omega}$ differs appreciably from zero, the denominator of the integral over μ is approximately $k(1 + \beta)$. In this case the pressure at point w, ϕ is

$$p_\omega(w,\phi) \simeq \frac{\rho c}{1 + \beta} \sum_{m=0}^\infty \frac{\epsilon_m}{2\pi} \int_0^{2\pi} \cos m(\phi - \phi_0) \, d\phi_0 \int_0^\infty J_m(\mu w) U_{m\omega}(\mu,\phi_0)\mu \, d\mu$$

$$\rightarrow \rho c \frac{u_\omega(w,\phi)}{1 + \beta} \qquad ka \gg 1 \tag{7.4.32}$$

Therefore, for very short wavelength radiation from the surface, the pressure at any point on the active part of the surface equals the normal velocity there times $\rho c/(1 + \beta) = [(1/\rho c) + (1/z)]^{-1}$, where $z(\omega) = \rho c/\beta$ is the acoustic impedance of the surface, and ρc is the characteristic impedance of the medium. The radiation load on each unit area of the active surface, for $ka \gg 1$, is thus equivalent to the characteristic impedance of the medium, in parallel with the acoustic impedance of the surface. At these short wavelengths each portion of the surface radiates separately and is separately loaded.

On the other hand, if $ka \ll 1$, the denominator of the integral over μ for p_ω is approximately $i\mu$ in the range $1/a > \mu > k \ (\ll 1/a)$, where $U_{m\omega}$ is large. In this range $J_0 \to 1$ and $J_m(\mu w) \to 0 \ \ (m > 0)$ for $w < a$. Therefore

$$p_\omega \simeq -\frac{ik\rho c}{2\pi} \int_0^{2\pi} d\phi_0 \int_0^{1/a} d\mu \int_0^a u_\omega(w_0,\phi_0) w_0 \, dw_0 = -\tfrac{1}{2}i\rho cka\langle u_\omega \rangle \qquad ka \ll 1$$

where

$$\langle u_\omega \rangle = \frac{S_\omega}{\pi a^2} = \frac{1}{\pi a^2} \int_0^{2\pi} d\phi_0 \int_0^a u_\omega(w_0,\phi_0) w_0 \, dw_0$$

is the mean velocity of the active area of the boundary, and S_ω is the amplitude of the total outflow from the surface (the equivalent strength of the surface as a simple source).

Thus, at long wavelengths, the pressure is uniform across the active area, being proportional to the mean velocity amplitude $\langle u_\omega \rangle$ of the area. To this approximation the ratio between pressure and mean velocity is imaginary, indicating that the power radiated is proportional to a higher power of $ka \ll 1$. The real part of this ratio can be obtained by using Eq. (7.4.21) to compute the total power radiated. Since, for $ka \ll 1$, the factor U_ω in f_ω is just S_ω, a constant, the far field and the total power are

$$p_\omega \simeq -ik\rho c \, \frac{e^{ikr}}{2\pi r} \, \frac{\cos\vartheta}{\beta + \cos\vartheta} S_\omega \qquad \beta = \xi - i\sigma$$

$$\Pi \to \frac{1}{\rho c} \int_0^{2\pi} d\varphi \int_0^\pi |p_\omega|^2 \sin\vartheta \, d\vartheta = \frac{\rho ck^2}{2\pi} |S_\omega|^2 \, \Omega(\beta)$$

where

$$\Omega = 1 + (\xi^2 + \sigma^2) \tan^{-1}\left(\frac{\sigma}{1 + \xi^2 + \sigma^2}\right) - \xi \ln\frac{(1 + \xi)^2 + \sigma^2}{\xi^2 + \sigma^2}$$

But the power radiated must equal the power leaving the active area, which is the pressure (constant over the area at these wavelengths) p_ω times the part of the mean outflow S_ω which is in phase with p_ω; we already know the part out of phase with p_ω. Therefore the pressure at $z = 0$, $w < a$, for $ka \ll 1$, is, approximately,

$$p_\omega \simeq \tfrac{1}{2}\rho cka(-i + ka\Omega)\langle u_\omega \rangle = \frac{1}{2}\left(-i\omega \frac{\rho}{\pi a} + \rho c \frac{k^2 a^2}{\pi}\Omega\right) S_\omega \qquad (7.4.33)$$

For long wavelengths the active area becomes the equivalent of twice a simple source of strength S_ω (when $\beta = 0$) radiating uniformly into the half space $z > 0$, and the pressure over the active area is constant, proportional to the mean velocity of the active area. The area radiates as a whole, in contrast to the situation for $ka \gg 1$.

Transient radiation from a piston

The general equation (7.4.23) for radiation from an elementary source can be used to compute the transient pressure wave from a piston in a plane wall. Suppose that the velocity of the piston is $U(t)$ and its acceleration is

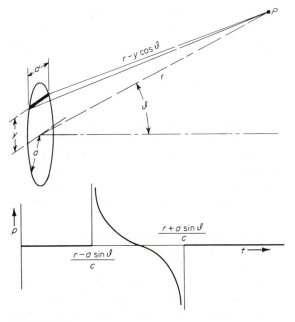

FIGURE 7.10
Radiation of a transient from a piston. Sound reaching point P came from strip d a time $(1/c)(r - y \sin \vartheta)$ earlier. Lower curve shows pressure fluctuation at P due to a velocity pulse $U = \delta(t)$ of the piston.

$A(t) = dU/dt$. Study of Fig. 7.10 shows that a part of the pressure wave arriving at the point (r, ϑ) at the time t is that due to the acceleration of the strip of length $d = 2\sqrt{a^2 - y^2}$ and width dy a time $(1/c)(r - y \sin \vartheta)$ earlier (if r is much larger than a). The amplitude of this component of the pressure pulse is, from Eq. (7.4.23),

$$dp \simeq \frac{\rho}{2\pi r} \sqrt{a^2 - y^2}\, A\left[t - \frac{1}{c}(r - y \sin \vartheta)\right] dy$$

The total pressure is obtained by integrating this over the surface of the piston, from $-a$ to $+a$. Change of integration variable and integration by parts yields the general formulas

$$p = \frac{\rho c}{2\pi r \sin^2 \vartheta} \int_e^f \sqrt{a^2 \sin^2 \vartheta - (c\tau + r - ct)^2}\, A(\tau)\, d\tau$$

$$= \frac{\rho c^2}{2\pi r \sin^2 \vartheta} \int_e^f \frac{(r + c\tau - ct)U(\tau)\, d\tau}{\sqrt{a^2 \sin^2 \vartheta - (r + c\tau - ct)^2}}$$

where the limits of integration are $e = t - (1/c)(r + a \sin \vartheta)$ and $f = t - (1/c)(r - a \sin \vartheta)$.

This is a very interesting formula, for it shows, perhaps more clearly than Eq. (7.4.30), the dependence of the pressure wave on the angle ϑ. Directly ahead ($\vartheta = 0$) the pressure wave reproduces the piston *acceleration* exactly:

$$p = \frac{\rho a^2}{4r} A\left(t - \frac{r}{c}\right) \qquad \vartheta = 0$$

The force on a microphone diaphragm directly ahead of the piston is therefore proportional to the piston acceleration; and if the diaphragm is mass-controlled, its acceleration is proportional to the force, so that the accelerations, velocities, and displacements of piston and diaphragm are proportional. As ϑ is increased, however, the integral covers a larger and larger interval of time τ, so that more and more of the piston motion gets blurred together in the pressure wave arriving at P.

A simple example of this is for the case when the piston suddenly moves outward a distance Δ. In this case the displacement of the piston is $\Delta u(t)$, where $u(t)$ is the step function defined in Eq. (1.3.22); the velocity is $\Delta \delta(t)$, proportional to the impulse function; and the acceleration $\Delta \delta'(t)$ is formally proportional to the derivative of the impulse function, a "pathological" function going first to plus infinity and then to minus infinity in an infinitesimal period of time. Its integral properties are $\int_{-\infty}^t \delta'(\tau - a)f(\tau)\, d\tau = -\{[df(a)/da]u(t - a)\}$. The resulting pressure at (r, ϑ) is

$$p = \begin{cases} 0 & ct < r - a \sin \vartheta \\[2mm] \dfrac{\rho c^2 \Delta}{2\pi r \sin^2 \vartheta} \dfrac{r - ct}{\sqrt{a^2 \sin^2 \vartheta - (r - ct)^2}} & r - a \sin \vartheta < ct < r + a \sin \vartheta \\[2mm] 0 & ct > r + a \sin \vartheta \end{cases}$$

This pulse, shown in Fig. 7.10, is a stretched-out version of the δ' function; the greater the angle ϑ, the greater the stretch. Only directly ahead of the piston, at $\vartheta = 0$, is the pressure pulse as instantaneous as the piston pulse.

The freely suspended disk

The next two examples considered in this section could have been treated in Sec. 7.3, for in neither case is a boundary plane present. They have been postponed to this section, however, because their solution involves techniques which have been developed in this section. The first is that of a sheet of metallic foil, in the form of a disk of radius a, somehow suspended in an unbounded medium, driven by electromagnetic forces perpendicular to its surface. Large loudspeakers, of a form roughly corresponding to this idealized model, have been constructed.

The setup can be approximated by imagining that the fluid itself is acted on by a force F per unit volume, pointed in the positive z direction, of a magnitude

$$F = \frac{F_\omega}{\pi a^2} e^{-i\omega t} \begin{cases} \delta(z) & w < a \\ 0 & w > a \end{cases}$$

where w is the distance from the center of the disk, in the xy plane, and where F_ω is the total force amplitude, driving the whole disk. Using Eq. (7.1.25), we obtain an expression for the pressure.

$$p_\omega(\mathbf{r}) = \frac{F_\omega}{\pi a^2} \int_0^{2\pi} d\phi_0 \int_0^a \left[\frac{\partial}{\partial z_0} g(w,\phi,z \mid w_0,\phi_0,z_0) \right]_{z_0=0} w_0\, dw_0 \quad (7.4.34)$$

To compute the far field from this source we use the approximation

$$g(r,\vartheta,\varphi \mid w_0,\phi_0,z_0) \simeq \frac{1}{4\pi r} \exp\{ik[r - w_0 \sin\vartheta \cos(\phi_0 - \phi) - z_0 \cos\vartheta]\}$$

$$(7.4.35)$$

valid for $r \gg r_0$. The pressure at large distances from the free-floating disk is then

$$p_\omega(\mathbf{r}) \simeq -ik \frac{e^{ikr}}{4\pi r} \frac{F_\omega}{\pi a^2} \cos\vartheta \int_0^a w_0\, dw_0 \int_0^{2\pi} e^{-ikw_0 \sin\vartheta \cos(\phi_0 - \phi)}\, d\phi_0$$

$$= -ikF_\omega \frac{e^{ikr}}{4\pi r} \left[\frac{2J_1(ka \sin\vartheta)}{ka \sin\vartheta} \right] \cos\vartheta \quad (7.4.36)$$

where the factor in brackets is the same as that entering the expression for the far field from a piston in a plane. In this case, however, there is an extra directionality factor $\cos\vartheta$, which ensures antisymmetry about the xy plane.

For long wavelengths the factor in brackets, as we saw earlier, is approximately unity. Comparison with Eq. (7.1.7) shows that at low frequencies the disk radiates like a dipole source of strength $D_z = i(F_\omega/\rho c k)$. The angle distribution for the radiated intensity has an additional factor $\cos^2\vartheta$, compared with that for the piston in a plane; thus there is no radiation in the plane of the disk.

Using Eq. (7.3.15) for the near field, we can calculate the pressure on the front surface of the disk and the normal velocity of the portion of the disk a distance $w < a$ from its center.

$$p_+ = \frac{F_\omega}{2\pi a^2} \int_0^\infty J_0(\mu w)\mu \, d\mu \int_0^a J_0(\mu w_0)w_0 \, dw_0 = \frac{F_\omega}{2\pi a^2} \qquad w < a$$

$$u_z = \frac{F_\omega}{2\pi a^2 \rho c k} \int_0^\infty J_0(\mu w)\mu\sqrt{k^2 - \mu^2} \, d\mu \int_0^a J_0(\mu w_0)w_0 \, dw_0$$

$$(7.4.37)$$

where we have used Eq. (1.3.27) to compute the pressure. The pressure at the back of the disk is $p_- = -(F_\omega/2\pi a^2)$, so that the total force required to move the fluid ahead and behind the disk is $\pi a^2(p_+ - p_-) = F_\omega$, as it was assumed to be.

When ka is large, the radical in the integral for u_z is equal to k over the range of μ for which the integrand is large, so that Eq. (1.3.27) produces $u_z \simeq F_\omega/\rho c 2\pi a^2$ for $w < a$. For long wavelengths, on the other hand, the integral for u converges poorly because of the large tangential velocity near the edge of the disk. However, we can use Eqs. (7.1.12) to compute an effective normal velocity, an appropriate average over the surface of the disk. The expressions for the pressure at the disk surface, the limiting forms of the disk velocity, and the total power radiated are

$$p_+ = \frac{F_\omega}{2\pi a^2} = -p_- \qquad \langle u_z \rangle = F_\omega Y(\omega)$$

$$Y(\omega) \to \frac{1}{2\pi a^2 \rho c} \qquad ka \gg 1$$

$$\to \frac{1}{6\pi a^3 \rho \omega} [i(1 - \tfrac{1}{2}ik^2 a^2) + \tfrac{1}{2}i(ka)^3] \qquad ka \ll 1 \qquad (7.4.38)$$

$$\Pi = |F_\omega|^2 \operatorname{Re} Y \to \begin{cases} |F_\omega|^2 \dfrac{1}{2\pi a^2 \rho c} = 2\pi a^2 \dfrac{|p_+|^2}{\rho c} & ka \gg 1 \\[2ex] |F_\omega|^2 \dfrac{k^2 a^2}{12\pi a^2 \rho c} & ka \ll 1 \end{cases}$$

where Y is the radiative admittance of the disk. Its reciprocal,

$$Z_r = \frac{F_\omega}{u_z} \to \begin{cases} 2\pi a^2 \rho c & ka \gg 1 \\ 3\pi a^2 \rho c(-2ika + k^4 a^4) & ka \ll 1 \end{cases}$$

has much the same form, for intermediate values of ka, as that of the impedance Z_r of Eqs. (7.4.31) for the piston in a plane baffle (there is an extra factor 2 in the present case, since the disk radiates to $z \to -\infty$ as well as to $z \to \infty$).

Radiation from an aircraft propeller

These same techniques can be used to compute the far field from an airplane propeller, whirling in the xy plane with an angular velocity ω. If there are n propeller blades, each of length a, with center at the origin, the air in the neighborhood of the disk of radius a in the xy plane is subjected to n successive pushes during each rotation of the propeller axle. Part of this push is in the $-z$ direction, but also, because of turbulence drag, there will be an accompanying force in the ϕ direction, perpendicular to the polar radius w. As with the free disk, we can assume that this driving force is concentrated in the xy plane, for $w < a$, and is greater near the propeller tips at $w = a$. A very rough approximation to the two components of the force on the medium is

$$F_z = -\left[\frac{(n+1)F}{2\pi a^{n+2}}\right] w^n e^{in(\phi - \omega t)}\, \delta(z) u(a - w)$$

$$F_\phi = \frac{(n+1)\alpha F}{2\pi a^{n+3}} w^{n+1} e^{in(\phi - \omega t)}\, \delta(z) u(a - w)$$

where F is approximately equal to the total backward thrust of the propeller, α is the drag-to-lift ratio (usually less than 1), and u is the step function of Eq. (1.3.21), being unity when $w < a$, zero when $w > a$.

To find the far field we insert this and the g of Eq. (7.4.35) into Eq. (7.1.25) (f_ω and $\mathfrak{T}_\omega = 0$), obtaining

$$p \simeq \frac{ik}{4\pi r} \left\{\frac{(n+1)F}{2\pi a^{n+2}}\right\} e^{ik(r - ct)} \left\{\cos \vartheta \int_0^a w_0^{n+1}\, dw_0 \int_0^{2\pi} d\phi_0 e^{in\phi_0 - ikw_0 \sin \vartheta \cos (\phi_0 - \varphi)}\right.$$

$$\left. + \frac{\alpha}{a} \sin \vartheta \int_0^a w_0^{n+2}\, dw_0 \int_0^{2\pi} d\phi_0 \sin (\phi_0 - \varphi) e^{in\phi_0 - ikw_0 \sin \vartheta \cos (\phi_0 - \varphi)}\right\}$$

$$= i^{-n-1} \frac{kF}{4\pi r} \left(\frac{\alpha n}{ka} - \cos \vartheta\right) \frac{(ka \sin \vartheta)^n}{2^{n+1} n!}$$

$$\times \left[\frac{2^{n+1}(n+1)!\, J_{n+1}(ka \sin \vartheta)}{(ka \sin \vartheta)^{n+1}}\right] e^{ik(r - ct) + in\varphi} \qquad (7.4.39)$$

where we must remember that, in this case, $k = n\omega/c$, since we are here using ω as the angular velocity of the propeller; the frequency of the radiation is $n\omega/2\pi$ since there are n blades. The quantity in brackets is roughly unity for $0 < ka \sin \vartheta < n$. Since, usually, the propeller tips are traveling somewhat slower than the speed of sound, $\omega a/c = ka/n$ is somewhat less than 1; thus the factor in brackets is nearly 1 for all values of ϑ, and the radiated intensity is, approximately,

$$I_r \simeq \frac{a^{2n} k^{2n+2} |F|^2}{2^{2n+6} \pi^2 r^2 (n!)^2 \rho c} \left(\frac{\alpha n}{ka} - \cos \vartheta\right)^2 \sin^{2n} \vartheta \qquad (7.4.40)$$

Aside from the distorting factor in parentheses, this is the radiation from a tesseral multipole of order $n + 1$, with a strong maximum in the xy plane. Because of the distorting factor the maximum is just aft of the plane of the propeller, the peak being the more pronounced, the greater the number of blades n. The factor in parentheses comes about because of the relationship between lift and drag; the term $\alpha n/ka$ is usually somewhat less than unity, so that the intensity is noticeably greater in the backward direction (the direction of the push) than in the forward direction. Also, at the angle $\cos^{-1}(\alpha n/ka)$ to the axis in the forward direction, no sound is radiated. The whole angular pattern is further distorted when the propeller is moving with respect to the air [Eq. (11.2.34)]. In addition, the actual force is not sinusoidal, but more pulselike, with harmonic terms $e^{imn(\phi - \omega t)}$ ($m = 2, 3, \ldots$). For these higher-frequency terms the factor in brackets in Eq. (7.4.39) imposes a preference for the axial rather than the equatorial directions. Thus the higher harmonics are louder along the axis than in the plane of the propeller.

$$\Pi \simeq \frac{a^{2n}k^{2n+2}|F|^2}{2^{n+4}\pi n! \, \rho c} \frac{1}{1 \cdot 3 \cdot 5 \cdot \, \cdots \, \cdot (2n + 3)} \left[(2n + 3)\left(\frac{\alpha n}{ka}\right)^2 + 1 \right] \quad (7.4.41)$$

Comparison of the expressions for Π given in Eqs. (7.1.11) and (7.1.12) indicates that the power radiated from a multipole of order m, of strength M_m, is

$$\Pi_m = \frac{\rho c k^{2m}|M_m|^2}{4\pi \cdot 1 \cdot 3 \cdot \, \cdots \, \cdot (2m + 1)}$$

Therefore the rotating n-blade propeller radiates as though it were a combination of a tesseral multipole of order n and strength M_n, minus another of order $n + 1$ and strength M_{n+1}, where

$$M_n \simeq \frac{Fa^n}{2\rho c\sqrt{2^n n!}} \frac{\alpha n}{ka} \qquad M_{n+1} \simeq \frac{F(a^n/k)}{2\rho c\sqrt{2^n n!}}$$

the n-pole, symmetric about the xy plane, corresponding to the effects of the propeller drag and the $(n + 1)$-pole, antisymmetric about the xy plane, caused by the propeller thrust in the negative z direction. This problem will be discussed again in Chap. 11.

The rectangular piston

Since they will be of use later, we shall work out the formulas for a flat piston of rectangular shape, of sides a and b, set in a plane rigid wall. The integrals for the far field and for the piston impedance are similar to those for the circular piston; only the boundaries of integration are now between $-\frac{1}{2}a$ and $\frac{1}{2}a$ for x_0 and between $-\frac{1}{2}b$ and $\frac{1}{2}b$ for y_0. The integral for the far field, analogous to Eq. (7.4.29) for $\beta = 0$, is

$$p_\omega(\mathbf{r}) \simeq -ik\rho c u_\omega \frac{e^{ikr}}{2\pi r} \int_{-\frac{1}{2}a}^{\frac{1}{2}a} dx_0 \int_{-\frac{1}{2}b}^{\frac{1}{2}b} dy_0 \exp\left[-ik\sin\vartheta\,(x_0\cos\varphi + y_0\sin\varphi)\right]$$

$$= 2ik\rho c u_\omega ab S(\tfrac{1}{2}ka\sin\vartheta\cos\varphi)S(\tfrac{1}{2}kb\sin\vartheta\sin\varphi)\frac{e^{ikr}}{4\pi r} \qquad (7.4.42)$$

where

$$S(z) = \frac{\sin z}{z}$$

The radiated sound intensity is concentrated in a cone ahead of the piston, as is the sound from the circular piston; only now the cone is elliptic in cross section, with its wider axis in the direction of the *smaller* dimension (a or b). Function $S(z)$ has much the same behavior as $2J_1(z)/z$, the corresponding function for the circular piston, though it falls off rather less rapidly as z increases and its zeros come for different values of z.

The radiation impedance of the rectangular piston is obtained by using Eq. (7.4.7) for the Green's function, instead of Eq. (7.4.10), in Eq. (7.4.18) (with $\beta = 0$). Setting both z and z_0 zero, the pressure at the surface of the piston is

$$p_\omega(x,y) = \frac{-i}{4\pi^3}\rho c k u_\omega ab \iiint_{-\infty}^{\infty} \frac{dK_x\,dK_y\,dK_z}{K_x{}^2 + K_y{}^2 + K_z{}^2 - k^2} S(\tfrac{1}{2}K_x a)S(\tfrac{1}{2}K_y b)e^{-iK_z x - iK_y y}$$

$$= \frac{1}{4\pi^2}\rho c k u_\omega ab \iint_{-\infty}^{\infty} \frac{dK_x\,dK_y}{\sqrt{k^2 - K_x{}^2 - K_y{}^2}} S(\tfrac{1}{2}K_x a)S(\tfrac{1}{2}K_y b)e^{iK_z x + iK_y y}$$

and the total reaction force on the piston is

$$F_\omega = \frac{1}{4\pi^2}\rho c k u_\omega a^2 b^2 \int_0^\infty \frac{\mu\,d\mu}{k^2 - \mu^2} \int_0^{2\pi} S^2(\tfrac{1}{2}\mu a\cos\theta)S^2(\tfrac{1}{2}\mu b\sin\theta)\,d\theta \qquad (7.4.43)$$

where $\mu^2 = K_x{}^2 + K_y{}^2$. By expanding the functions S^2 in power series and integrating term by term, a power series for the integral over θ can be computed. The integral over μ can then be calculated numerically, for the appropriate values of ka and kb. Comparison of series, however, shows that a rough approximation to F_ω, valid for all values of k as long as a/b is neither very large nor very small, is

$$F_\omega \simeq \rho c u_\omega ab\,\frac{a^2\theta_\square(ka) - b^2\theta_\square(kb) - ia^2\chi_\square(ka) + ib^2\chi_\square(kb)}{a^2 - b^2}$$

$$\to \rho c u_\omega ab[\theta_0(ka) - i\chi_0(ka)] \qquad b \to a \qquad (7.4.44)$$

where

$$\theta_\square(z) = \frac{2}{z^2}\int_0^z \theta_0(y)y\,dy = \left[1 - 4\frac{1 - J_0(z)}{z^2}\right] \to \begin{cases} \frac{1}{16}z^2 & z \ll 1 \\ 1 & z \gg 1 \end{cases}$$

$$\chi_\square(z) = \frac{2}{z^2}\int_0^z \chi_0(y)y\,dy = \frac{8}{\pi z}\left[1 - \frac{\pi}{2z}M_0(z)\right] \to \begin{cases} \dfrac{8}{9\pi}z & z \ll 1 \\ \dfrac{8}{\pi z} & z \gg 1 \end{cases}$$

and $M_0(x) = (2/\pi)\displaystyle\int_0^{\pi/2}\sin(x\cos u)\,du$ is the Struve function of zero order.

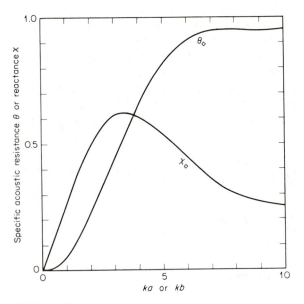

FIGURE 7.11
Impedance functions for the rectangular piston. See Eq. (7.4.44).

The functions θ_0, χ_0 are defined in Eqs. (7.4.31), plotted in Fig. 7.9, and given in Table IX. Functions $\theta_\square$ and $\chi_\square$ are plotted in Fig. 7.11. Approximate limiting values for the radiation impedance of the piston are thus

$$Z_r = \frac{F_\omega}{u_\omega} \to \begin{cases} \frac{1}{16}k^2(a^2 + b^2) - \dfrac{8i}{9\pi}k\,\dfrac{a^2 + ab + b^2}{a + b} & ka \ll 1 \\ 1 - i\dfrac{8}{\pi k(a + b)} & ka \gg 1 \end{cases} \qquad (7.4.45)$$

for values of b/a which are neither very large nor very small.

Problems

1 An airtight box, with four loudspeakers set in the front panel as shown, is suspended in free space. Each loudspeaker has an area A and oscillates with velocity $U_0 e^{-i\omega t}$. Determine the power radiated from this system at low

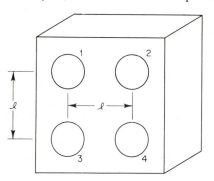

frequencies when the wavelength is much larger than the dimensions of the box:

   (a) When all speakers oscillate in phase
   (b) When speakers 1 and 2 are in phase but 180° out of phase with 3 and 4
   (c) When 1 and 4 are in phase and 180° out of phase with 2 and 3

2 The portion of the surface of a long cylinder of radius a which is between $-\phi_0$ and $+\phi_0$ vibrates with velocity normal to the surface, and the rest of the cylinder is rigid. Obtain a series analogous to Eqs. (7.2.19) for the average acoustic impedance over the vibrating surface.

3 A sphere and a circular disk are both driven in one-dimensional motion with a force $F_x = F_0 \exp(-i\omega t)$. The disk is driven normal to its plane. If the sphere and the disk have the same mass m and radius a, show that the velocity amplitudes at low frequencies are

$$u_s = \frac{F_0}{\omega(m + m_s)} \quad \text{and} \quad u_d = \frac{F_0}{\omega(m + m_d)}$$

where

$$m_s = \frac{2\pi}{3}\rho a^3 \quad \text{and} \quad m_d = \frac{8\pi}{3}\rho a^3$$

where ρ is the density of the surrounding fluid. For an arbitrary body the mass reaction of the fluid must in general be expressed as a tensor $\mathfrak{M}$ such that $iF_i = \omega \sum_j M_{ij}u_j$. What, then, is the ratio between the velocity amplitude and the amplitude of the driving force? What are the elements of the tensor $\mathfrak{M}$ for the sphere and for the disk?

4 Compute the impedance for an elementary dipole a distance l from a plane locally reacting wall of acoustic impedance $\rho c/\beta$ when the dipole axis is parallel to the wall. What is the impedance when the dipole axis is perpendicular to the wall?

5 A monopole is a distance h above an infinite horizontal locally reacting boundary having acoustic impedance $\rho c \zeta$. The monopole strength at frequency $\omega/2\pi$ is $S(\omega)$. Calculate the sound pressure as a function of the distance r

from the source in the horizontal plane of the source. Show that if the source height is larger than λ, the sound-pressure amplitude exhibits maxima and minima in a region given approximately by $0 < r < h^2/\lambda$ and decreases monotonically with r for sufficiently large r $(>h^2/\lambda)$. If ζ is finite, show that this monotonic decrease approaches proportionality to $1/r^2$ for large r. Estimate the value of r, in terms of $|\zeta|$, above which this is approximately true.

6 Show that a point force $F(t)$, applied at $r = 0$ in a gas, produces a sound pressure field

$$p(\mathbf{r},t) = \frac{1}{4\pi rc}\left(\frac{\partial F}{\partial t} + \frac{cF}{r}\right)\cos\vartheta$$

where ϑ is the angle between $\mathbf{F}$ and $\mathbf{r}$ (assuming that the direction of $\mathbf{F}$ is constant). What is the result if the direction as well as the magnitude of $\mathbf{F}$ changes with time? (Parentheses indicate retarded time.)

7 A pulsating sphere of radius a has a radial surface velocity $U_0 e^{-i\omega t}$, where U_0 is independent of ϑ and φ as well as of t. Show that the acoustic radiation impedance at the surface is $\rho c \zeta_r$, where

$$\zeta_r = \frac{k^2a^2}{1 + k^2a^2} - i\frac{ka}{1 + k^2a^2}$$

where $kc = \omega$. Is this valid for any value of ka?

8 The monopole source density $s_\omega(\mathbf{r})$ of Eq. (7.1.30) has the uniform value $S_\omega/\pi a^2 l$ throughout the inside of a cylinder of radius a, with axis along the x axis and between the planes $x = \pm\frac{1}{2}l$, and is zero outside the cylinder. Change the factor $e^{-i\mathbf{k}\cdot\mathbf{r}_0}\,dv_0$ in the integral to $\exp[-ikw_0\sin\vartheta\cos(\phi - \phi_0) - ikx_0\cos\vartheta]\,w_0\,d\phi_0\,dw_0\,dx_0$, where $w_0^2 = y_0^2 + z_0^2$, and use the formulas

$$\int_0^{2\pi} e^{iq\cos\theta}\cos(m\theta)\,d\theta = 2\pi i^m J_m(q) \quad\text{and}\quad \int u^{m+1}J_m(u)\,du = u^{m+1}J_{m+1}(u)$$

to show that the pressure wave a distance r from the origin is

$$-ik\,\rho c S_\omega\,\frac{\sin(\frac{1}{2}kl\cos\vartheta)}{\frac{1}{2}kl\cos\vartheta}\,\frac{J_1(ka\sin\vartheta)}{\frac{1}{2}ka\sin\vartheta}\,\frac{e^{ikr-i\omega t}}{4\pi r} \qquad r \gg a, l$$

where ϑ is the angle between r and the x axis. Plot the dependence of the intensity at r on ϑ when $ka = 4$ and $kl = 3$. When kl and ka are small but not negligible, show that the far-field pattern is similar to that from a combination of a point monopole S_ω minus three quadrupoles of strengths

$$Q_{xx} = \frac{1}{24}l^2 S_\omega \qquad Q_{yy} = Q_{zz} = \frac{1}{8}a^2 S_\omega$$

all at the origin. *Hint*: use the expansion

$$J_m(x) = \frac{(x/2)^m}{m!}\left[1 - \frac{(x/2)^2}{m+1} + \cdots\right]$$

9 A spherical cloud of ionized gas is acted on by an oscillating electric field so that the force $\mathbf{F}$ of Eq. (7.1.24) is in the z direction and

$$F_z = \begin{cases} A(a^2 - r^2)e^{-i\omega t} & r < a \\ 0 & r > a \end{cases}$$

Show that the pressure wave produced outside the sphere is

$$p(r,\vartheta,\varphi) = \tfrac{1}{3}ia^3 A\, j_2(ka)\cos\vartheta\, h_1(kr)$$

Use the integral relationship $\int z^{n+2} j_n(z)\, dz = z^{n+2} j_{n+1}(z)$.

10 The fluid inside a sphere of radius a is in violent motion so that the velocity components with frequency $\omega/2\pi$ are

$$u_x = \begin{cases} -\tfrac{1}{3}Q_\omega y^2 e^{-i\omega t} & r < a \\ 0 & r > a \end{cases} \qquad u_y = \begin{cases} Q_\omega xy e^{-i\omega t} & r < a \\ 0 & r > a \end{cases}$$

and $u_z = 0$. Use Eqs. (7.1.25) and (7.2.31) to show that the pressure wave for $r > a$, generated by this motion, has frequency $\omega/2\pi$ and amplitude

$$i\,|Q|^2\, a^4 j_3(2ka)\sin^2\vartheta\,\cos 2\varphi\, h_2(2kr) \qquad k = \omega/c$$

What is the pressure amplitude when $r \to \infty$? When $ka \to 0$?

11 A region of volume V contains ionized gas, which is acted on by a randomly fluctuating electromagnetic field, so that the vector $\mathbf{F}$ in Eq. (7.1.25) is not zero. The mean-square amplitudes $\langle|F_x|^2\rangle$ and $\langle|F_y|^2\rangle$ are equal, but are not equal to $\langle|F_z|^2\rangle$; the correlation length w_d and correlation time $\tau_d = 1/\omega_d$ are the same for all three components. Show that the rms intensity at r, for all frequencies, averaged over a long time interval T, is

$$I_d \to \frac{V\lambda_d}{\sqrt{32\pi}\,\rho c r^2}\frac{w_d^{\,3}}{w_d^{\,2} + \lambda_d^{\,2}}\left(\langle|F_y|^2\rangle\cos^2\vartheta + \langle|F_x|^2\rangle\sin^2\vartheta\right)$$

for $r \to \infty$, where $\lambda_d = c/\omega_d$, ϑ is the angle between vector r and the z axis.

12 A long rigid cylinder of diameter d, placed in a transverse airstream, experiences a transverse oscillatory force because of the flow instability produced in the stream (see Chap. 11). As a result, sound is radiated (Aeolian tone) at a characteristic frequency $\omega/2\pi$. If the correlation length of the force along the cylinder is L, calculate the rms power radiated per unit length of the cylinder.

13 Suppose the specific acoustic admittance β of the surface of Eq. (7.4.21) does depend on frequency, so that the Fourier transform of the admittance term is

$$b(t) = \int_{-\infty}^{\infty}\frac{\omega\cos\vartheta}{\beta + \cos\vartheta}\,e^{-i\omega t}\,d\omega$$

By use of Prob. 23 of Chap. 1, show that the pressure wave at r, t (r very large), caused by a velocity distribution $u(x,y,t)$, is

$$\frac{i\rho}{4\pi^2 r}\int dX_0\int dY_0\int b(\tau)u\,[X_0,\,Y_0,\,T(X_0) - \tau]\,d\tau$$

where $T(X_0) = t - (r/c) + (X_0/c)\sin\vartheta$, and u, X_0, Y_0 are as defined in connection with Eq. (7.4.23).

14 The surface of Prob. 13 has acoustic impedance $R + i(K/\omega)$, so that $\beta = \rho c\omega/(\omega R + iK)$. Show that

$$b(t) = -2\pi i\left(\frac{K}{R}\right)^2\frac{(\rho c/R)\cos^2\vartheta}{[(\rho c/R) + \cos\vartheta]^3}\exp\left(-Wt\right)u(t)$$

where

$$W = \frac{(K/R)\cos\vartheta}{(\rho c/R) + \cos\vartheta}$$

and $u(t)$ is the unit step function of Eq. (1.3.22). If the radiating area extends over X_0 from 0 to l, over Y_0 from 0 to a (i.e., if the rz plane is parallel to two sides of the rectangular radiating area), and if the velocity of the area is a pulse $u(X, Y, t) = \Delta_0 \delta(t)$, show that the pressure wave at $\mathbf{r}$, t is

$$\frac{pc\,\Delta_0 a}{2\pi r}\frac{K}{R}\frac{(pc/R)\cos\vartheta}{[(pc/R) + \cos\vartheta]^2}$$

$$\times \begin{cases} \exp\left[-W\left(t - \dfrac{r}{c}\right)\right] - \exp\left[-WT(l)\right] & r < ct \\ 1 - \exp\left[-WT(l)\right] & r > ct > r - l\sin\vartheta \\ 0 & ct < r - l\sin\vartheta \end{cases}$$

where $T(x) = (x/c)\sin\vartheta + t - (r/c)$. Discuss the physical significance of this result; plot it as function of t for appropriate values of the constants.

15 A circular membrane of radius a, under tension, is set in an otherwise rigid plane wall. Suppose the membrane is vibrating with its (m,n)th mode [Eq. (5.2.25)] so that the u of Eq. (7.4.19) is $U_m \cos(m\phi)J_m(\pi\beta_{mn}r/a)\exp(-ik_{mn}ct)$. Show that the asymptotic form of the pressure wave a distance r from the center of the membrane is, when $r \gg a$,

$$-i\pi ka^2 pc\beta_{mn}U_{mn}J_m(\pi\beta_{mn})\frac{J_m(ka\sin\vartheta)}{k^2a^2\sin^2\vartheta - \pi^2\beta_{mn}^2}\frac{1}{r}\exp\left[ik_{mn}(r - ct)\right]$$

where k_{mn} equals $\pi\beta_{mn}/a$ times the ratio between the wave velocity in the membrane and that in air.

16 A spherical source of radius a consists of a uniform monopole distribution of source strength $Q_0 e^{-i\omega t}$ per unit volume. Show that the radiated pressure is $(-i\omega Q_0 V/4\pi r)F(ka)e^{ik(r-a)-i\omega t}$, where $k = \omega/c$, $V = 4\pi a^3/3$, and

$$F(x) = \frac{6}{x^3}(x - \sin x) + \frac{3i}{x^3}(x^2 - 2 + 2\cos x)$$

17 By using Eqs. (7.4.38), determine by what factor the low-frequency output of a loudspeaker can be increased by providing it with a baffle box, in effect to convert it from a dipole to a monopole source. In making the comparison, let the amplitude of oscillation of the speaker disk be the same for the two cases. Also assume that the wavelength is larger than the dimensions of the box.

18 A small gas bubble of radius a in a liquid medium acts as a monopole source of sound when set into radial oscillation. Starting with the equilibrium conditions for the bubble, $P_i = P_0 + (2\sigma/a)$, where P_i is the pressure inside the bubble, P_0 the static pressure in the liquid, and σ the surface tension, take into account the radiative mass load on the bubble and determine the resonance frequency of the radial vibrations. Assuming that $P_i \gg P_0$, and that the compression of the gas in the bubble is adiabatic, show that, in the long-wavelength limit, when a is small compared with λ, the resonance frequency is

$$\frac{1}{2\pi}\sqrt{(3\gamma - 1)\frac{2\sigma}{a^3\rho}}$$

where $\gamma = C_p/C_v$ for the gas in the bubble, and ρ is the density of the liquid.

19 As a result of the Alaskan earthquake of March, 1964, surface waves (Rayleigh waves) of considerable amplitude passed over the United States.   These waves radiated sound into the atmosphere.   At infrasonic experimental stations in Washington and in Boulder, Colo., the vertical velocity amplitude of the earth surface was found to be about $\frac{1}{2}$ cm per sec, with a period of oscillation of 25 sec.   If the speed of the surface waves was five times the speed of sound in air, determine the direction of propagation and the intensity of the radiated sound wave (assumed to be a plane wave).   From conservation of energy of the sound wave, determine the resulting amplitude of oscillation of the ionosphere.   The gas pressure in the ionosphere is about 10^{-8} atm.

CHAPTER

8

THE SCATTERING OF SOUND

8.1 SIMPLE EXAMPLES OF SCATTERING

When a sound wave encounters an obstacle, some of the wave is deflected from its original course. It is usual to define the difference between the actual wave and the undisturbed wave, which would be present if the obstacle were not there, as the *scattered* wave. When a plane wave, for instance, strikes a body in its path, in addition to the undisturbed plane wave there is a scattered wave, spreading out from the obstacle in all directions, distorting and interfering with the plane wave. If the obstacle is very large compared with the wavelength (as it usually is for light waves and very seldom is for sound), half of this scattered wave spreads out more or less uniformly in all directions from the scatterer, and the other half is concentrated behind the obstacle in such a manner as to interfere destructively with the unchanged plane wave behind the obstacle, creating a sharp-edged shadow there. This is the case of geometrical optics; in this case the half of the scattered wave spreading out uniformly is called the *reflected* wave, and the half responsible for the shadow is called the *interfering* wave. If the obstacle is very small compared with the wavelength (as it often is for sound waves), all the scattered wave is propagated out in all directions, and there exists no sharp-edged shadow. In the intermediate cases, where the obstacle is about the same size as the wavelength, a variety of curious interference phenomena can occur.

In this chapter, since we are studying sound waves, we shall be interested in the second and third cases, where the wavelength is longer than, or at least the same size as, the obstacle. We shall not encounter or discuss sharply defined shadows. So much of the scattered wave will travel in a different direction from the plane wave that destructive interference will be unimportant, and we shall be able to separate all the scattered wave from the undisturbed plane wave. We shall be interested in the total amount of the wave that is scattered, in the distribution in angle of this wave, and in the effect of this scattered wave on the pressure at various points on the surface of the obstacle.

Scattering from a cylinder

Let us first compute the scattering, by a cylinder of radius a, of a plane wave traveling in a direction perpendicular to the cylinder's axis. If the plane wave has intensity I, the pressure wave, if the cylinder were not present, would be

$$p_p = P_0 e^{ik(x-ct)} = P_0 e^{ik(r\cos\phi-ct)} \qquad P_0 = \sqrt{\rho c I} \qquad k = \frac{2\pi}{\lambda}$$

where the direction of the plane wave has been taken along the positive x axis.

In Eq. (1.2.9) we expressed this plane wave in terms of cylindrical waves:

$$p_p = P_0 e^{ik(r\cos\phi-ct)} = P_0 \left[J_0(kr) + 2\sum_{m=1}^{\infty} i^m \cos(m\phi) J_m(kr) \right] e^{-2\pi i v t} \qquad (8.1.1)$$

The radial velocity corresponding to this wave is

$$u_{pr} = \frac{P_0}{\rho c} \left\{ iJ_1(kr) + \sum_{m=1}^{\infty} i^{m+1} [J_{m+1}(kr) - J_{m-1}(kr)] \cos(m\phi) \right\} e^{-2\pi i v t}$$

When the cylinder is present with its axis at $r = 0$, the wave cannot have the form given by the above series, for the cylinder distorts the wave. There is present, in addition to the plane wave, a scattered outgoing wave of such a size and shape as to make the radial velocity of the combination zero at $r = a$, the surface of the cylinder. We shall choose the form of this outgoing wave to be the general series

$$p_s = \sum_{m=1}^{\infty} A_m \cos(m\phi) [J_m(kr) + iN_m(kr)] e^{-2\pi i v t}$$

$$u_{sr} = \frac{1}{\rho c} \left\{ iA_0[J_1(kr) + iN_1(kr)] + \frac{i}{2} \sum_{m=1}^{\infty} A_m \cos(m\phi)[J_{m+1}(kr) \right.$$
$$\left. - J_{m-1}(kr) + iN_{m+1}(kr) - iN_{m-1}(kr)] \right\} e^{-2\pi i v}$$

The combination $J + iN$ has been chosen because it ensures that all the scattered wave is outgoing.

Our first task is to find the values of the coefficients A which make the combination $u_{pr} + u_{sr}$ equal zero at $r = a$. Equating u_{sr} to $-u_{pr}$ at $r = a$ term by term, we obtain

$$A_m = -\epsilon_m P_0 i^{m+1} e^{-i\gamma m} \sin\gamma_m \qquad P_0 = \sqrt{\rho c I}$$

$$\tan\gamma_0 = -\frac{J_1(ka)}{N_1(ka)} \qquad \tan\gamma_m = \frac{J_{m-1}(ka) - J_{m+1}(ka)}{N_{m+1}(ka) - N_{m-1}(ka)} \qquad (8.1.2)$$

where $\epsilon_0 = 1$ and $\epsilon_m = 2$ for all values of m larger than unity. These phase angles γ_m have already been defined in Eqs. (7.3.7), in connection with the radiation of sound from a cylinder. Values of some of them are given in Table V. The behavior of these phase angles completely determines the behavior of the scattered wave. It is interesting to notice the close connection between the waves scattered by a cylinder and the waves radiated by the same cylinder when it is vibrating. The quantities needed to compute one are also needed to compute the other.

The pressure and radial velocity of the scattered wave at large distances from the cylinder are

$$p_s \simeq -\sqrt{\frac{2\rho c Ia}{\pi r}}\,\psi_s(\phi)e^{ik(r-ct)} \qquad u_s \simeq \frac{p_s}{\rho c}$$

$$\psi_s(\phi) = \frac{1}{\sqrt{ka}} \sum_{m=0}^{\infty} \epsilon_m \sin(\gamma_m)e^{-i\gamma m} \cos(m\phi)$$

The intensity of the scattered part, at the point $(r,\phi)(kr \gg 1)$, is therefore

$$\Upsilon_s \simeq \frac{Ia}{\pi r}|\psi_s(\phi)|^2 \tag{8.1.3}$$

$$|\psi_s|^2 = \frac{1}{ka}\sum_{m,n=0}^{\infty}\epsilon_m\epsilon_n \sin\gamma_m \sin\gamma_n \cos(\gamma_m - \gamma_n)\cos(m\phi)\cos(n\phi)$$

where $\epsilon_0 = 1$, $\epsilon_m = 2$ $(m > 0)$. This intensity is plotted as a function of ϕ on a polar plot in Fig. 8.1, for different values of $\eta = 2\pi va/c = 2\pi a/\lambda$.

It is interesting to notice the change in directionality of the scattered

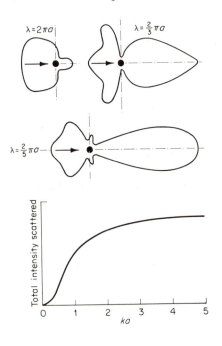

FIGURE 8.1
The scattering of sound waves from a rigid cylinder of radius a. Polar diagrams show the distribution in angle of the intensity of the scattered wave; lower graph shows the dependence of the scattered intensity on $\eta = 2\pi a/\lambda$.

wave as the wavelength is changed. For very long wavelengths (η small) but little is scattered, and this is scattered almost uniformly in all the backward directions. As the frequency is increased, the distribution in angle becomes more and more complicated, diffraction peaks appearing and moving forward, until for very short wavelengths (much shorter than those shown in Fig. 8.1), one-half of the scattered wave is concentrated straight forward (the interfering beam), and the other half is spread more or less uniformly over all the other directions, giving a polar plot which is a cardioid, interrupted by a sharp, very high peak in the forward direction, as shown in Eq. (8.1.4) below.

For very long wavelengths only the two cylindrical waves corresponding to $m = 0$ and $m = 1$ are important in the scattered wave. As shown in Eq. (7.3.7),

$$\gamma_0 \simeq -\gamma_1 \simeq \frac{\pi k^2 a^2}{4} \gg -\gamma_2, \; -\gamma_3, \; - \ldots \qquad \text{when } ka \to 0$$

The first-order approximation for the scattered intensity at long wavelengths is therefore

$$I_s \simeq \frac{\pi \omega^3 a^4}{8 c^3 r} I(1 - 2 \cos \phi)^2 \qquad \omega a \ll c$$

Short-wavelength limit

For wavelengths very small compared with the circumference of the cylinder, the relatively simple approximation of "geometrical acoustics" is valid, with the "scattered" wave dividing into two parts, the "reflected" and the "shadow-forming" waves. However, the process of demonstrating that the series of Eqs. (8.1.3) really does behave in a simple manner involves mathematical manipulation of considerable intricacy.

In the first place, the series of Eqs. (8.1.3), which is in a form useful for calculating scattering at longer wavelengths, is not particularly suitable for showing in detail the interference between part of the scattered wave and the primary wave to form the shadow. In optics, one differentiates between *Fraunhofer diffraction*, where intensities are measured at distances so large that the angle subtended by the diffracting object is small compared with the ratio $(\lambda/2\pi a) = 1/ka$, and *Fresnel diffraction*, for distances large compared with the wavelength but not extremely large compared with $2\pi a$. The Fresnel-diffraction formulas show the shadow with its related diffraction bands, but at the great distances involved in the Fraunhofer-diffraction formulas, the shadow has become blurred again. Series (8.1.3) is for distances corresponding to Fraunhofer diffraction, so that what can be demonstrated is the separation of the scattered wave into a reflected wave and a shadow-forming wave, but not the details of the interference with the incident wave, which are characteristic of the Fresnel formulas.

In the second place, the simplicity of the formulas for very short wavelengths only appears as an average; the scattered intensity varies rapidly with angle in a complicated sort of way, and only the average intensity per degree (or per minute) varies smoothly. This rapid fluctuation is seldom measured, however, for any small change of frequency or of position of the cylinder will blur it out, leaving only the average intensity. Consequently, our calculations should separate the rapid fluctuations of intensity from the average behavior of the reflected wave.

When all the necessary manipulations are made, the expression for the scattered intensity at short wavelengths is

$$\frac{I_s}{I} \simeq \frac{a}{r} \sin\left(\frac{\phi}{2}\right) + \frac{1}{\pi k r} \cot^2 \frac{\phi}{2} \sin^2 (ka \sin \phi)$$

$$+ \text{(rapidly fluctuating terms)} \qquad kr \gg ka \gg 1 \quad (8.1.4)$$

The first term of this expression constitutes the reflected intensity, which, for a cylinder, reflects more in the backward direction ($\phi \simeq \pi$) than in a forward direction ($\phi \simeq 0$). The second term is the shadow-forming beam, concentrated in the forward direction within an angle $\pi/ka = \lambda/2a$ which is smaller the smaller λ is compared with a. The third term contains rapidly fluctuating quantities that average to zero, and so may be neglected.

Total scattered power

The total power scattered by the cylinder per unit length is obtained by multiplying I_s by r and integrating over ϕ from 0 to 2π. The cross terms in the sum of Eqs. (8.1.3) disappear, because of the integral properties of the characteristic functions cos $(m\phi)$, leaving the result

$$\Pi_s = 4aI \frac{1}{ka} \sum_{m=0}^{\infty} \epsilon_m \sin^2 (\gamma_m) \simeq \begin{cases} \dfrac{6\pi^5 a^4}{\lambda^3} I & \lambda \gg 2\pi a \\ 4aI & \lambda \ll 2\pi a \end{cases} \quad (8.1.5)$$

The limiting value for total *scattered* power for very short wavelengths is the power contained in a beam *twice* as wide as the cylinder (4a). This is due to the fact, discussed above, that the scattered wave includes *both* the reflected *and* the shadow-forming waves, the first and second terms of Eq. (8.1.4). The integral of r times the first term is the total *reflected* power, which is just $2aI$; and the integral of r times the second term is also $2aI$, showing that the shadow-forming wave has enough power just to cancel the primary wave behind the cylinder.

The quantity $\Pi_s/4aI$ is plotted in Fig. 8.1 as a function of $ka = 2\pi a/\lambda$. Notice that in spite of the various peculiarities of the distribution in angle of the intensity, the total scattered intensity turns out to be a fairly smooth function of ka. We also should note that Π_s is not usually measured experimentally, because of the difficulty of separating primary from scattered waves at small angles of scattering. What is usually measured is more nearly the quantity

$$\Pi_{\text{exp}} \simeq 2\int_{\Delta}^{\pi-\Delta} I_s r \, d\phi$$

where Δ is a small angle, if the experimental conditions are good, but never zero. It turns out that Π_{exp} is very nearly equal to Π_s for the longer wavelengths; but as the wavelength is made smaller, the shadow-forming beam [the second term in Eq. (8.1.4)] is less and less included in the integral, until for very short wavelengths Π_{exp} is equal to $\frac{1}{2}\Pi_s$. The transition from Π_s to $\frac{1}{2}\Pi_s$ comes at values of ka near $\pi/2\Delta$, at wavelengths λ near $4a\Delta$.

The force on the cylinder

Returning now to the expression for the total pressure due to both the undisturbed plane wave and the scattered wave, we find, after some involved juggling of terms [the use, for instance, of the last of Eqs. (7.3.2)], that the total pressure at the surface of the cylinder at an angle ϕ from the x axis is

$$p_a = (p_p + p_s)_{r=a} = \frac{4P_0}{\pi ka} e^{-2\pi i \nu t} \sum_{m=0}^{\infty} \frac{\cos(m\phi)}{E_m} e^{i[-\gamma_m+(\pi m/2)]} \quad (8.1.6)$$

where the quantities E_m are defined in Eq. (7.3.7). This expression is proportional to the expression in Eq. (7.3.8), giving the pressure at some distance from the cylinder due to a vibrating line element on the cylinder, when we make the necessary change from ϕ to $\pi - \phi$ (since now $\phi = 0$ is in the direction *opposite* to the source). This is an example of the *principle of reciprocity*. The pressure at a point A due to a source at point B is equal to the pressure at B due to a source at A, everything else being equal. Therefore the polar curves of Fig. 7.5 show the distribution of intensity about a cylinder having a line source, and they also show the distribution of the square of the pressure on the surface of the cylinder due to a line source at very large distances away from the cylinder (distances so large that the wave has become a plane wave by the time the wave strikes the cylinder).

When η is very small ($\lambda \gg a$), the expression for the pressure at $r = a$ reduces to

$$p \xrightarrow[ka\to 0]{} P_0(1 + 2ika\cos\phi)e^{-2\pi i \nu t}$$

which approaches in value the pressure $P_0 e^{-2\pi i \nu t}$ of the plane wave alone as η goes to zero.

The net force on the cylinder per unit length is in the direction of the plane wave and is

$$F = a\int_0^{2\pi} p\cos\phi \, d\phi = 4aP_0 \frac{1}{kaE_1} e^{-i\omega t - i\gamma_1 + i\pi/2}$$

$$\simeq \begin{cases} i\omega \dfrac{4\pi^2 a^2}{c} P_0 e^{-i\omega t} & \omega \ll \dfrac{c}{a} \\ \sqrt{4a\lambda}P_0 e^{i(\omega/c)(a-ct)+i\pi/4} & \lambda \ll 2\pi a \end{cases} \quad (8.1.7)$$

This force lags behind the pressure which the plane wave would have at $r = 0$ if the cylinder were not there by an angle $-\gamma_1 + (\pi/2)$.

The quantity $|F/P_0|$ is the net force on the cylinder per unit length per unit pressure of the plane wave. This quantity, divided by $2\pi a$, is plotted in Fig. 8.2 as a function of $ka = \eta$. We note that for small frequencies the force is proportional to the frequency (that is, to η), but that when η becomes larger than unity (i.e., when the wavelength becomes smaller than the circumference of the cylinder), the linear relation breaks down and the force diminishes with increasing frequency thenceforth. This result is of interest

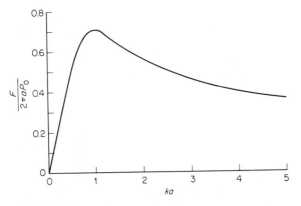

FIGURE 8.2
Amplitude of sideward force per unit length F on a cylinder of radius a due to the passage of a plane wave of pressure amplitude P_0, plotted as a function of $2\pi a/\lambda$.

in connection with the so-called velocity-ribbon microphone, which consists of a light metal strip more or less open to the atmosphere, pushed to and fro by the sound wave. The ribbon is in a transverse magnetic field, so that the motion induces an emf along the ribbon, which actuates an amplifier. The net force on the strip is, of course, not exactly the same function of η as that given in Eq. (8.1.7) for the cylinder, but the behavior will be the same, in general. The force on the strip will be proportional to the frequency for small frequencies, but this linear dependence will break down when the wavelength becomes smaller than twice the width of the strip.

There is an approximate method of finding the net force on the cylinder which gives the correct result for wavelengths longer than the circumference of the cylinder. If the pressure in the plane wave is $P_0 e^{ik(x-ct)}$, the pressure at the surface of the cylinder due to the plane wave is $P_0 e^{i\eta \cos \phi - 2\pi i vt}$. If η is small, this can be expanded into $p_p \simeq P_0(1 + i\eta \cos \phi)e^{-2\pi i vt}$. This is the pressure due to the plane wave; there is also a scattered wave, enough of a wave to make the radial velocity at the surface come out to zero. This scattered wave contributes a term $i\eta P_0 \cos \phi e^{-2\pi i \nu t}$ to this approximation to the pressure, so that the net pressure is that given in Eq. (8.1.6), and the net force is the limiting value given in (8.1.7).   We thus see that even for very long waves the distortion of the plane wave due to the presence of the cylinder contributes a factor 2 to the net force on the cylinder.

Scattering from an inhomogeneity

The two-dimensional example just concluded has illustrated a number of general characteristics of scattered sound; they can, however, be more generally displayed in terms of the Green's function.   A sound wave is scattered, not only by a solid object, but also by a region in which the acoustic properties of the medium, such as its density or compressibility, differ from their values in the rest of the medium.   Turbulent air scatters, as well as generates, sound.   A rough patch on a plane wall scatters, as well as reflects, sound.   Fog particles in air, and bubbles in water, scatter sound.

When an object or region scatters sound, some of the energy carried by the incident wave is dispersed.   The energy lost to the incident wave may be absorbed by the scatterer, or it may simply be deflected from its original course.   The amount of energy lost per second by an incident plane wave, divided by the incident wave's intensity (energy per second per unit area), is called the *cross section* Σ of the object or region.   It is the cross-sectional area of that portion of the incident wave which is removed by the scatterer. If the energy is absorbed by the object, the ratio is called the absorption cross section; if the sound is simply deflected, so it goes to infinity in a different direction, it is called the scattering cross section; the sum is the total cross section.   In any case, the incident plane wave is reduced in intensity because of the loss.   If there are, on the average, N equivalent scatterers per unit volume of the medium, then, as the plane wave progresses, each unit volume removes a fraction $N\Sigma$ of the wave's intensity.   In other words, the intensity is attenuated by the fraction $N\Sigma$ per unit length; thus the intensity of a plane wave traversing a scatterer-filled medium is $I_0 e^{-N\Sigma x}$, where x is the distance of penetration.

If the scattering object is a region in which the medium differs in acoustic properties from the rest of the medium, the wave solution can be written in terms of the volume integral of Eq. (7.1.17).   For example, suppose, in the region R, which can be inscribed in a sphere of radius a at the origin, the density of the fluid is ρ_e and its compressibility is $\kappa_e = 1/\rho_e c_e^2$ (so that the wave velocity in R is c_e), whereas these quantities have values ρ and $\kappa = 1/\rho c^2$ throughout the rest of the medium.

To obtain the appropriate equations of motion from the general equations (6.2.6) and (6.2.11) or (6.4.1) to (6.4.22), in the present case, we must distinguish between the variations of pressure and density caused by the sound wave and the variations of density and compressibility inherent in the inhomogeneous nature of the medium.   If these inhomogeneities are caused by "foreign bodies" in a fluid of density ρ and compressibility κ, then, in the absence of a sound wave, the density, as a function of $\mathbf{r}$ and t, will be $\rho + \delta_e(\mathbf{r},t)$, where δ_e is zero except inside one of the "foreign bodies." These bodies move as wholes. They may be fog droplets in air or bubbles in water, and in the meaning of Eq. (6.1.9), the total time derivative of $\rho + \delta_e$ is zero.

$$0 = \frac{d}{dt}(\rho + \delta_e) = \frac{\partial \delta_e}{\partial t} + \mathbf{u} \cdot \operatorname{grad} \delta_e \qquad \rho \text{ is constant}$$

Now we add a sound wave, with its velocity $\mathbf{u}$, its pressure p, and its additional density change δ, related by the usual first-order equation $\delta = (\rho + \delta_e)\kappa_e p$, where κ_e is equal to the constant value κ in the homogeneous part of the fluid but is different from κ in each inhomogeneous region. The equation of continuity for density (6.2.6) thus becomes

$$\frac{\partial}{\partial t}(\rho + \delta_e + \delta) \equiv \frac{\partial}{\partial t}(\delta_e + \delta) = -\operatorname{div}[(\rho + \delta_e + \delta)\mathbf{u}]$$
$$\simeq -(\rho + \delta_e)\operatorname{div}\mathbf{u} - \mathbf{u} \cdot \operatorname{grad}\delta_e$$

where we have neglected the two terms involving both δ and $\mathbf{u}$ as being much smaller than the others. Applying the previous two equations, we eventually obtain

$$\rho_e \kappa_e \frac{\partial p}{\partial t} = -\rho_e \operatorname{div}\mathbf{u} \qquad \text{or} \qquad \operatorname{div}\mathbf{u} = -\kappa_e \frac{\partial p}{\partial t} \qquad (8.1.8)$$

where $\rho_e = \rho + \delta_e$ is the fluid density, equal to the constant value ρ outside the foreign bodies; κ_e likewise is κ outside, in the homogeneous region.

The argument regarding the force equation (6.2.11) in the absence of turbulence is rather similar, and if we neglect viscosity and heat conduction, the result is

$$\frac{\partial \mathbf{u}}{\partial t} = -\frac{1}{\rho_e}\operatorname{grad}p \qquad (8.1.9)$$

The combination of these two equations, eliminating $\mathbf{u}$, is

$$\operatorname{div}\left(\frac{1}{\rho_e}\operatorname{grad}p\right) - \kappa_e \frac{\partial^2 p}{\partial t^2} = 0 \qquad (8.1.10)$$

If turbulence is present, so that the fluid velocity is the sum of $\mathbf{U}$, the turbulent velocity, and $\mathbf{u}$, the small acoustic velocity, then the right-hand side of the equation, instead of being zero, is equal to $(1/\rho)\nabla \cdot \mathfrak{J} \cdot \nabla$, where $\mathfrak{J}$ is the kinetic tensor of Eq. (6.2.10). The velocity components entering the elements of this tensor are of two kinds: the components of the acoustic velocity $\mathbf{u}$ of Eq. (8.1.8), which are usually quite small compared with c, the velocity of sound, and the components of the turbulent velocity $\mathbf{U}$, which may not be small compared with c. Thus in the region where $\mathbf{U}$ differs from zero, the kinetic term to insert on the right in Eq. (8.1.10) is

$$\frac{1}{\rho}\nabla\cdot\mathfrak{J}\cdot\nabla \simeq \frac{1}{\rho}\left[\frac{\partial^2}{\partial x^2}U_x{}^2 + 2\frac{\partial^2}{\partial x\,\partial y}U_x U_y + \cdots\right]$$

$$+\frac{2}{\rho}\left[\frac{\partial^2}{\partial x^2}R_{xx} + \frac{\partial^2}{\partial x\,\partial y}(R_{xy}+R_{yx}) + \cdots\right]$$

$$R_{xx} = \rho u_x U_x \qquad R_{xy} = \tfrac{1}{2}\rho(u_x U_y + u_y U_x) \quad \cdots$$

where we have omitted the very small terms $u_x u_y$, etc.

The $U_i U_j$ terms in $\mathfrak{J}$ are independent of the acoustic wave; if they vary with time, they represent sources of sound, of the sort discussed in Chap. 7. We are interested, at present, in the scattering of sound, coming from some other source, by the inhomogeneities in R, not by the generation of sound within R. Therefore we drop the terms $U_i U_j$ from our discussion, assuming their effects have been computed by the methods of Chap. 7. The portion of the $\nabla\cdot\mathfrak{J}\cdot\nabla$ term which is of importance here is that arising from the cross products, which may be written $(2/\rho)\nabla\cdot\mathfrak{R}\cdot\nabla$, with $\mathfrak{R}$ having elements R_{xy}, etc., as already defined. See Sec. 11.4 for further discussion.

Thus the equation for the acoustic pressure, either inside the region R or outside, where ρ_e and κ_e have the constant values ρ and κ, can be written in the form

$$\nabla^2 p - \frac{1}{c^2}\frac{\partial^2 p}{\partial t^2} = \frac{1}{c^2}\frac{\partial^2 p}{\partial t^2}\gamma_\kappa(\mathbf{r},t) + \mathrm{div}\,[\gamma_\rho(\mathbf{r},t)\,\mathrm{grad}\,p] - 2\nabla\cdot\mathfrak{R}\cdot\nabla \quad (8.1.11)$$

where

$$\gamma_\kappa = \begin{cases}\dfrac{\kappa_e - \kappa}{\kappa} & \text{inside } R \\[2mm] 0 & \text{outside } R\end{cases} \qquad \gamma_\rho = \begin{cases}\dfrac{\rho_e - \rho}{\rho_e} & \text{inside } R \\[2mm] 0 & \text{outside } R\end{cases}$$

and where $c^2 = 1/\rho\kappa$. We have added the terms $(1/\rho)\nabla^2 p - \rho\kappa(\partial^2 p/\partial t^2)$ on both sides of Eq. (8.1.10), and then multiplied by ρ to obtain it in a form appropriate for changing into an integral equation. We have omitted the source terms involving $U_i U_j$ for the reasons given two paragraphs above.

Equation (8.1.11) is homogeneous in the unknown quantity p, though we have put the terms representing the effect of R on the right, as though they were some sort of source term. In a sense they are source terms, for they represent sources of scattered sound, produced by the interaction between the sound wave p and the nonuniformities in R; but they do not represent any new energy being introduced into the sound field, as would terms on the right-hand side which were independent of p, discussed in Chap. 7.

If the irregularities in R are stationary in time, as they would be if R were a "foreign particle" in the fluid, such as a fog droplet in the air or a bubble in water (in these cases $\mathfrak{R}$ is usually zero), we can profitably consider the case where the acoustic motion has a single frequency $\omega/2\pi$ and use the time-independent equation for the pressure amplitude,

$$\nabla^2 p_\omega + k^2 p_\omega = -k^2 \gamma_\kappa(\mathbf{r})p_\omega - \text{div} \, [\gamma_\rho(\mathbf{r}) \, \text{grad} \, p_\omega] \qquad \text{where } k = \frac{\omega}{c} \quad (8.1.12)$$

From this equation we can obtain steady-state single-frequency solutions and transient solutions, if needed, by the usual Fourier transform procedures.

The integral equation

As with radiation problems, we shall find it much more useful to deal with an integral equation for the pressure wave, rather than with a differential equation plus boundary conditions. To obtain the integral equation we return to Eq. (7.1.17), giving p_ω in terms of a volume integral of $f_\omega G_\omega$ plus a surface integral of p_ω and G_ω. For the present case, with an unbounded medium, the Green's function is $g_\omega = e^{ikr}/4\pi r$. We cannot neglect the surface integral over the sphere at infinity in the present case, for we are now feeding energy into the medium, from infinity, via the incident wave. In fact, it turns out that the surface integral of Eq. (7.1.17), over the sphere at infinity, just produces $p_i(\mathbf{r})$, the incident wave, which comes from infinity and is scattered by region R.

Thus the integral equation for the scattering of an incident wave $p_i(\mathbf{r})$ by a single region R of different ρ and κ is

$$p_\omega(\mathbf{r}) = p_i(\mathbf{r}) + \iiint\limits_R \{k^2\gamma_\kappa(\mathbf{r}_0)p_\omega(\mathbf{r}_0) - \text{div} \, [\gamma_0(\mathbf{r}_0)\nabla_0 p_\omega(\mathbf{r}_0)]\}g_\omega(\mathbf{r} \mid \mathbf{r}_0) \, dv_0$$

$$= p_i(\mathbf{r}) + \iiint\limits_R (k^2\gamma_\kappa p_\omega g_\omega + \gamma_\rho\nabla_0 p_\omega \cdot \nabla_0 g_\omega) \, dv_0 \qquad (8.1.13)$$

where

$$g_\omega(\mathbf{r} \mid \mathbf{r}_0) = \frac{1}{4\pi \mid \mathbf{r} - \mathbf{r}_0 \mid} \exp{(ik \mid \mathbf{r} - \mathbf{r}_0 \mid)}$$

To obtain the second form from the first we have integrated by parts, as we did with Eq. (7.1.25). In fact, our discussion of this integral equation can go very much as it did for Eq. (7.1.25). The integral here represents the scattered wave produced by the interaction between the incident wave and the irregularities of ρ and κ inside R. The first term comes from the fact that R has not the same compressibility as the surrounding medium, which results in a monopole source distribution of source density $(ik/\rho c)\gamma_\kappa(\mathbf{r})p_\omega(\mathbf{r})$. The second term gives rise to dipole scattered wavelets from each dv_0; because of the difference in density, region R does not move in response to the force grad p_ω with the same velocity as does the surrounding medium; this produces a dipole source density $(1/ik\rho c)\gamma_\rho(\mathbf{r})$ grad $p_\omega(\mathbf{r})$. Note that these sources depend on p_ω, the actual pressure field inside R, not on the incident pressure p_i.

This is an inhomogeneous integral equation; instead of the volume term being the inhomogeneous term, as it was in the case of radiation, the incident wave $p_i(\mathbf{r})$ (a specified function) is the inhomogeneous term. We have transformed a differential equation, which is homogeneous because it represents the fact that no energy is being added mechanically in the finite part of space, into an integral equation, which is inhomogeneous because energy is being added from infinity, via the incident wave p_i.

The scattered wave

At large distances from the scattering region R ($r \gg a$, the radius of the sphere enclosing R), the Green's function takes on the simple form

$$g_\omega(\mathbf{r} \mid \mathbf{r}_0) \to \frac{1}{4\pi r} \exp(ikr - i\mathbf{k}_s \cdot \mathbf{r}_0)$$

where vector $\mathbf{k}_s$ has magnitude k, and is pointed in the direction of the radius vector $\mathbf{r}$ from the origin, inside R, to the point of observation. If, for example, the incident wave were a plane wave of amplitude A and wavenumber $k = \omega/c$, the exact form for the asymptotic behavior of the pressure wave would be

$$p_\omega(\mathbf{r}) \to A e^{i\mathbf{k}_i \cdot \mathbf{r}} + A \left(\frac{e^{ikr}}{r}\right) \Phi_e(\mathbf{k}_s) \tag{8.1.14}$$

$$\Phi_s(\mathbf{k}_s) = \frac{k^2}{4\pi A} \iiint\limits_R \left[\gamma_\kappa(\mathbf{r}_0) p_\omega(\mathbf{r}_0) - i\gamma_\rho(\mathbf{r}_0) \frac{\mathbf{a}_r}{k} \cdot \mathbf{\nabla}_0 p_\omega\right] \exp(-i\mathbf{k}_s \cdot \mathbf{r}_0) \, dv_0$$

where $\mathbf{a}_r = \mathbf{k}_s/k$ is the unit vector in the direction of the observer. The second term for p_ω is, of course, a scattered wave, radiating outward from the region R, with an amplitude depending on vector $\mathbf{k}_s$, which points in the direction of the observer. Thus, if we knew the exact form of the sound field p_ω in R, we could compute the exact form of the scattered wave at large distances.

Function $\Phi_s(\mathbf{k}_s)$ is called the *angle-distribution factor* for the scattered wave, or sometimes the *scattered amplitude*. Its square $|\Phi_s|^2$, divided by $\rho c r^2$, is the intensity of the scattered wave at the observation point $\mathbf{r}$ ($r \gg a$). Ordinarily, the scattered wave must be measured a considerable distance from the scatterer in order to separate the scattered wave from the incident wave; so the function Φ_s is usually the goal of our calculations. An important point is that Φ_s is the three-dimensional Fourier transform of the function

$$\varphi_s(\mathbf{r}) = \frac{2\pi^2 k^2}{A} \left[\gamma_\kappa(\mathbf{r}) p_\omega(\mathbf{r}) - i\gamma_\rho(\mathbf{r})(\mathbf{a}_r/k) \cdot \mathbf{\nabla} p_\omega(\mathbf{r})\right]$$

$$= \iiint \Phi_s(\mathbf{K}) e^{i\mathbf{K} \cdot \mathbf{r}} \, dv_K \tag{8.1.15}$$

$$\Phi_s(\mathbf{K}) = \frac{1}{8\pi^3} \iiint \varphi_s(\mathbf{r}) e^{-i\mathbf{K} \cdot \mathbf{r}} \, dv_r$$

$$dv_K = dK_x \, dK_y \, dK_z \qquad dv_r = dx \, dy \, dz$$

for the special value $\mathbf{K} = \mathbf{k}_s$. This property of the scattered amplitude will be of considerable utility in our analyses.

A few other general properties of Φ_s, resulting from the Fourier transform relationship, merit exposition here. For example, we can generalize the convolution integral of Prob. 23 of Chap. 1 to obtain an alternative form for Φ_s. If we define the spatial Fourier transforms of p_ω and γ as

$$P(\mathbf{K}) = \frac{1}{8\pi^3}\iiint p_\omega(\mathbf{r})e^{-i\mathbf{K}\cdot\mathbf{r}}\,dv_r, \qquad \Gamma(\mathbf{K}) = \frac{1}{8\pi^3}\iiint \gamma(\mathbf{r})e^{-i\mathbf{K}\cdot\mathbf{r}}\,dv_r$$

and if we remember that the Fourier transform of grad p_ω is $i\mathbf{K}P(\mathbf{K})$, then the angle-distribution factor can be written

$$\Phi_s(\mathbf{k}_s) = \frac{2\pi^2 k^2}{A}\iiint P(\mathbf{K})\left[\Gamma_\kappa(\mathbf{k}_s - \mathbf{K}) + \frac{\mathbf{K}\cdot\mathbf{a}_r}{k}\Gamma_\rho(\mathbf{k}_s - \mathbf{K})\right]dv_K \quad (8.1.16)$$

the convolution in this case being in wavenumber space rather than in distance space. If, for example, in the region R, p_ω resembles a traveling wave $A\exp(i\mathbf{k}_i\cdot\mathbf{r})$, its Fourier transform will have a sharp peak at $\mathbf{K} = \mathbf{k}_i$, resembling the transform of $A\exp(i\mathbf{k}_i\cdot\mathbf{r})$, which is $A\,\delta(\mathbf{K} - \mathbf{k}_i)$. In this case $\Phi_s(\mathbf{k}_s)$ would be approximately equal to

$$(2\pi^2 k^2/A)[\Gamma_\kappa(\mathbf{k}_s - \mathbf{k}_i) - \Gamma_\rho(\mathbf{k}_s - \mathbf{k}_i)\cos\vartheta]$$

where ϑ is the angle between $\mathbf{k}_s$ and $\mathbf{k}_i$. The dependence on ϑ comes explicitly on the cosine factor, but also implicitly, because the vector $\mathbf{k}_s - \mathbf{k}_i$ depends, both in magnitude and direction, on ϑ, the angle of scattering.

The intensity of the scattered wave, at large distances from R, is given by a double integral which can be transformed into the Fourier transform of an autocorrelation function.

$$I_s = \frac{k^4}{16\pi^2\rho c r^2}\iiint_R f_\omega(\mathbf{r}_0,\mathbf{a}_r)e^{-i\mathbf{k}_s\cdot\mathbf{r}_0}\,dv_0\iiint_R f_\omega^*(\mathbf{r}_1,\mathbf{a}_r)e^{i\mathbf{k}_s\cdot\mathbf{r}_1}\,dv_1$$

$$= \frac{k^4 V}{16\pi^2\rho c r^2}\iiint_R e^{-i\mathbf{k}_s\cdot\mathbf{l}}\Upsilon_\omega(\mathbf{l})\,dv_l \qquad (8.1.17)$$

where

$$f_\omega(\mathbf{r},\mathbf{a}_r) = \gamma_\kappa(\mathbf{r})p_\omega(\mathbf{r}) - i\gamma_\rho(\mathbf{r})\frac{\mathbf{a}_r}{k}\cdot\nabla p_\omega(\mathbf{r})$$

$$\Upsilon_\omega(\mathbf{l}) = \frac{1}{V}\iiint_R f_\omega(\mathbf{r},\mathbf{a}_r)f_\omega^*(\mathbf{r} + \mathbf{l},\,\mathbf{a}_r)\,dv_r \qquad \mathbf{k}_s = k\mathbf{a}_r$$

V being the volume of the scattering region R, and $\mathbf{l} = \mathbf{r}_0 - \mathbf{r}_1$. We notice that Υ_ω is zero when $l > 2a$, twice the radius of the sphere enclosing R. Thus, when $ka \ll 1$, the scattered intensity depends only on the volume integral of Υ_ω, and its directionality depends only on the directionality of f_ω.

The Born approximation

We have seen that, once the exact form of the pressure distribution p_ω is determined inside the scattering region R, then all the properties of the sound field outside R can be computed, including the asymptotic behavior of the scattered wave. The integral equation (8.1.13) can occasionally be solved exactly; a few examples will be given in the next section. It may be solved approximately by variational methods; an example will be given in the problems. Or it may be solved by successive approximations. Suppose we take, for our first approximation, the field produced when the unknown p_ω in the integral of Eq. (8.1.13) is replaced by the known function p_i, the incident wave.

$$p_\omega(\mathbf{r}) \simeq p_i(\mathbf{r}) + p_1(\mathbf{r})$$
$$p_1(\mathbf{r}) = \iiint\limits_R [k^2 \gamma_\kappa(\mathbf{r}_0) p_i(\mathbf{r}_0) g_\omega(\mathbf{r} \mid \mathbf{r}_0) + \gamma_\rho(\mathbf{r}_0) \boldsymbol{\nabla}_0 p_i \cdot \boldsymbol{\nabla}_0 g_\omega] \, dv_0 \qquad (8.1.18)$$

If $p_1(\mathbf{r})$ turns out to be considerably smaller than $p_i(\mathbf{r})$ throughout R, then this approximation, called the *Born approximation* [Eq. (4.5.16)], will give us a satisfactory representation of the behavior of the scattered sound. If this is not the case, we might add higher terms in the approximation series.

$$p_\omega(\mathbf{r}) = p_i(\mathbf{r}) + p_1(\mathbf{r}) + p_2(\mathbf{r}) + p_3(\mathbf{r}) + \cdots$$
$$p_{n+1}(\mathbf{r}) = \iiint\limits_R [k^2 \gamma_\kappa(\mathbf{r}_0) p_n(\mathbf{r}_0) g_\omega(\mathbf{r} \mid \mathbf{r}_0) + \gamma_\rho(\mathbf{r}_0) \boldsymbol{\nabla}_0 p_n \cdot \boldsymbol{\nabla}_0 g_\omega(\mathbf{r} \mid \mathbf{r}_0)] \, dv_0 \qquad (8.1.19)$$

For the sorts of functions γ we encounter in acoustics, this series converges, but it usually converges slowly unless the γ's are quite small (in which case p_1 is sufficient). Thus, when the Born approximation is inadequate, we turn to variational methods, or else to exact solutions, when these are possible.

When the γ's are small and the Born approximation is valid, the calculation of the scattered wave is fairly simple. For example, when the incident wave is a plane wave, $p_i(\mathbf{r}) = A \exp{(i\mathbf{k}_i \cdot \mathbf{r})}$, the angle-distribution function is nearly equal to

$$\Phi_b(\boldsymbol{\mu}) = \frac{k^2}{4\pi} \iiint\limits_R [\gamma_\kappa(\mathbf{r}_0) + \gamma_\rho(\mathbf{r}_0) \cos\vartheta] \exp{(i\boldsymbol{\mu} \cdot \mathbf{r}_0)} \, dv_0$$
$$= 2\pi^2 k^2 [\Gamma_\kappa(\boldsymbol{\mu}) + \Gamma_\rho(\boldsymbol{\mu}) \cos\vartheta] \qquad \boldsymbol{\mu} = \mathbf{k}_s - \mathbf{k}_i \qquad (8.1.20)$$

where $\cos\vartheta = \mathbf{k}_i \cdot \mathbf{k}_s / k^2$ is the cosine of the angle between $\mathbf{k}_s$ and $\mathbf{k}_i$ (i.e., the angle of scattering between the direction of the incident plane wave and the radius vector $\mathbf{r}$ to the observer), and the Γ's are the spatial Fourier transforms of the γ's, as defined in Eq. (8.1.16).

If R is a sphere of radius a and both the γ's depend only on the radius r_0, then the angle coordinates of dv_0 can be integrated over and

$$\Phi_b(\mu) = \frac{k^2}{\mu} \int_0^a [\gamma_\kappa(r_0) + \gamma_\rho(r_0) \cos \vartheta] \sin (\mu r_0) r_0 \, dr_0$$

$$\rightarrow \tfrac{1}{3} k^2 a^3 \left(\left\langle \frac{\kappa_e - \kappa}{\kappa} \right\rangle + \left\langle \frac{\rho_e - \rho}{\rho_e} \right\rangle \cos \vartheta \right) \qquad \mu a \ll 1 \quad (8.1.21)$$

where $\mu = |\mathbf{k}_s - \mathbf{k}_i| = 2k \sin (\vartheta/2)$. Thus, at long wavelengths, the dipole wave, proportional to the mean value $\langle \gamma_\rho \rangle$ over R, has the same order of dependence on ka as the monopole wave, which is proportional to the mean value $\langle \gamma_\kappa \rangle$ over R. Other properties of the Born approximation to the scattered wave will be taken up in the problems.

Scattering from turbulence

Our second example of the use of the Born approximation is the case where the γ's and $\mathfrak{R}$ are small, but vary in a random manner with time and position within R. If the rms values are small, the approximate solution should be satisfactory. This time we shall not analyze the wave first into simple-harmonic components, but shall use the time-dependent Green's function of Eq. (7.1.18), with a source function

$$f(\mathbf{r},t) \simeq -\frac{1}{c^2} \gamma_\kappa(\mathbf{r},t) \frac{\partial^2 p_i}{\partial t^2} - \mathrm{div} \, [\gamma_\rho(\mathbf{r},t) \, \mathrm{grad} \, p_i(\mathbf{r},t)] + 2\nabla \cdot \mathfrak{R} \cdot \nabla \quad (8.1.22)$$

where the components of $\mathbf{u}$ in the elements of $\mathfrak{R}$ are set equal to the components of $\mathbf{u}_i$, related to p_i via the equation $-(\partial \mathbf{u}_i/\partial t) = (1/\rho) \, \mathrm{grad} \, p_i$.

Again we specialize by assuming that the incident wave is a plane wave, $p_i = A \exp (i\mathbf{k}_i \cdot \mathbf{r} - ik_i ct)$, and since we are interested in the asymptotic form of the scattered wave, we use the asymptotic form of the Green's function,

$$g(\mathbf{r},t \mid \mathbf{r}_0,t_0) \rightarrow \frac{1}{4\pi r} \delta \left[t_0 - t + \frac{r}{c} - \frac{1}{c} (\mathbf{r}_0 \cdot \mathbf{a}_r) \right]$$

where $\mathbf{a}_r$ is a unit vector pointed along $\mathbf{r}$. Thus $\mathbf{r}_0 \cdot \mathbf{a}_r$ equals r_0 times the cosine of the angle between $\mathbf{r}_0$ and $\mathbf{r}$, and $k_i r_0 \cdot \mathbf{a}_r$ equals $\mathbf{k}_s \cdot \mathbf{r}_0$, where $\mathbf{k}_s = k_i \mathbf{a}_r$, as previously. Dealing first with the γ_κ term, we can write the integral for the asymptotic form of this part of the scattered wave, eliminating the delta function by integrating over t_0.

$$p_{s\kappa}(\mathbf{r},t) \rightarrow \frac{Ak_i^2}{4\pi r} \int dt_0 \iiint dv_0 \, \gamma_\kappa(\mathbf{r}_0,t_0) \, \delta \left(t - t_0 - \frac{r}{c} + \frac{1}{c} \mathbf{r}_0 \cdot \mathbf{a}_r \right)$$

$$\times \exp [i\mathbf{k}_i \cdot \mathbf{r}_0 - ik_i ct_0]$$

$$(8.1.23)$$

$$= \frac{Ak_i^2}{4\pi r} e^{ik_i r} \iiint dv_0 \, \gamma_\kappa \left(\mathbf{r}_0, t - \frac{r}{c} + \frac{1}{c} \mathbf{r}_0 \cdot \mathbf{a}_r \right)$$

$$\times \exp [i(\mathbf{k}_i - \mathbf{k}_s) \cdot \mathbf{r}_0 - ik_i ct]$$

To demonstrate that this approximate formula for the scattered wave does indeed display a frequency spread, we now carry out a frequency analysis by taking the Fourier transform of $p_{s\kappa}$ for the frequency $\omega/2\pi$ (not necessarily equal to $k_i c/2\pi$).

$$P_{s\kappa}(\mathbf{r},\omega) \to \frac{Ak_i^2}{4\pi r} e^{ik_i r} \iiint dv_0 \int dt \gamma_\kappa \left(\mathbf{r}_0, t - \frac{r}{c} + \frac{1}{c}\mathbf{r}_0 \cdot \mathbf{a}_r \right)$$

$$\times \exp\left[i(\mathbf{k}_i - \mathbf{k}_s) \cdot \mathbf{r}_0 + i(\omega - k_i c)t\right]$$

This is the four-dimensional Fourier transform of the fluctuating function $\gamma_\kappa(\mathbf{r},t)$.

$$\Gamma_\kappa(\mathbf{K},\omega_\gamma) = (2\pi)^{-4} \int dt_1 \iiint dv_1\, \gamma_\kappa(\mathbf{r}_1,t_1) \exp\left(-i\mathbf{K} \cdot \mathbf{r}_1 + i\omega_\gamma t_1\right)$$

for a specific value of $\mathbf{K}$ and ω_κ, multiplied by a phase factor, which arises because the proper integration variable is not t but $t_1 = t - (r/c) + (1/c)\mathbf{r}_0 \cdot \mathbf{a}_r$. Carrying out the integrals in turn, we obtain

$$P_{s\kappa}(\mathbf{r},\omega) \to \frac{Ak_i^2}{8\pi^3 r} e^{ik_i r} \int dt_1 \iiint dv_0\, \gamma_\kappa(\mathbf{r}_0,t_1)$$

$$\times \exp\left[i(\mathbf{k}_i - \mathbf{k}_s) \cdot \mathbf{r}_0 + i(\omega - k_i c)\left(t_1 + \frac{r}{c} - \frac{1}{c}\mathbf{r}_0 \cdot \mathbf{a}_r\right)\right]$$

$$= \frac{2\pi^2 A}{r} k_i^2 e^{i\omega r/c} \Gamma_\kappa\left(\frac{\omega}{c}\mathbf{a}_r - \mathbf{k}_i, \omega - k_i c\right) \qquad (8.1.24)$$

Because of the random motion within R, the transform of γ is not a delta function $\delta(\omega - k_i c)$ exactly at the frequency $k_i c/2\pi$ of the incident wave, but has a spread in frequency roughly equal to the reciprocal of the correlation time interval $2\pi\tau_c = 2\pi/\omega_c$. Thus the scattered wave can have frequencies $\omega/2\pi$ and wavenumber ω/c which differ from the incident values $k_i c/2\pi$ and k_i. The transform Γ_κ also has a broader peak at $(\omega/c)\mathbf{a}_r \simeq \mathbf{k}_i$, which produces a dependence on the angle between $\mathbf{a}_r$ and $\mathbf{k}_i$ in the scattered wave.

The other parts of the scattered wave may be computed by similar procedures, though the calculation is further complicated because of the divergence operators, which give difficulty when integrated with a delta function. The safe way is to integrate by parts, obtaining derivatives of the delta function, such as $\delta'(x)$, which has the property that $\int f(x)\delta'(x - a)\,dx = f'(a)$, where the prime indicates differentiation with respect to the argument. One then can integrate over t_0 and eventually obtain for the ω component of the complete scattered wave

$$p_s(\mathbf{r},\omega) \to \frac{2\pi^2 A}{r} e^{i\omega r/c} \left\{ k_i^2 \Gamma_\kappa(\boldsymbol{\mu}_s, \Delta_s) \right.$$

$$\left. - \cos\vartheta \left[k_i \frac{\omega}{c} \Gamma_\rho(\boldsymbol{\mu}_s, \Delta_s) - 2\left(\frac{\omega}{c}\right)^2 \Omega_r(\boldsymbol{\mu}_s, \Delta_s) \right] \right\} \qquad (8.1.25)$$

where

$$\mu_s = \frac{\omega}{c}\, \mathbf{a}_r - \mathbf{k}_i \qquad \Delta_s = \omega - k_i c$$

and ϑ is the angle between $\mathbf{a}_r$ and $\mathbf{k}_i$ (the scattering angle). Functions Γ_ρ and Ω_r are the Fourier transforms, respectively, of γ_ρ and of $1/c$ times the component U_r, along $\mathbf{r}$, of the turbulent velocity $\mathbf{U}$. The details of these calculations will be taken up in the problems.

Because of the random motion, the phases of Γ_κ, Γ_ρ, and Ω_r are random functions of their arguments. Since, usually, γ_κ, γ_ρ, and U_r/c vary independently, when we square p_s to obtain the intensity of the scattered wave, the cross terms average out, and the intensity involves just the sum of the squares of Γ_κ, Γ_ρ, and Ω_r separately. Again we go into the details in regard to Γ_κ here, leaving the rest of the calculation for a problem. Thus the spectrum level of the sound scattered by γ_κ, analyzed over the time interval at T, is

$$I_\kappa(\mathbf{r}) \to \frac{4\pi^4\,|A|^2\,k_i^{\,4}}{\rho c r^2}\left|\Gamma_\kappa\!\left(\frac{\omega}{c}\,\mathbf{a}_r - \mathbf{k}_i,\,\omega - k_i c\right)\right|^2 \tag{8.1.26}$$

where $|\Gamma_\kappa(\mu_s,\Delta_s)|^2$ is the spatial-temporal spectrum density, for wavenumber vector μ_s and frequency difference $\Delta_s/2\pi$, of the inhomogeneity factor $\gamma_\kappa(\mathbf{r},t)$ in R. If the fluctuations in R are isotropic, $|\Gamma_\kappa|^2$ will depend on the magnitude of μ_s, not its direction, and if its fluctuations have equal contributions for frequencies up to $\omega_c/2\pi$ and from wavenumbers of magnitudes up to $1/l_c$ (i.e., if the correlation time is $1/\omega_c$ and the correlation length is l_c), then $|\Gamma_\kappa(\mathbf{K},\omega_\gamma)|^2$ will be maximum at $\mu_s = 0$ and $\Delta_s = 0$ and will drop to zero when $|\mu_s|$ is larger than $1/l_c$, when $|\Delta_s|$ is larger than ω_c.

Therefore Eq. (8.1.26) shows that the greatest amount of this scattered intensity, measured at $\mathbf{r}$, is for a frequency equal to $k_i c/2\pi = \omega_i/2\pi$ and for a direction $\mathbf{a}_r$ for which $(\omega/c)\mathbf{a}_r = \mathbf{k}_i$; it drops to zero for frequencies such that $|\omega - \omega_i|/2\pi$ is greater than $\omega_c/2\pi$ and when $|(\omega/c)\mathbf{a}_r - \mathbf{k}_i|$ is greater than $1/l_c$. To make these points a little clearer, we work out the case for which the autocorrelation function for γ_κ has the simple form

$$\Upsilon_\kappa(\mathbf{l},\tau) \equiv \frac{1}{VT}\int_{-T}^{0} dt \iiint_R \gamma_\kappa(\mathbf{r},t)\gamma(\mathbf{r}+\mathbf{l},\,t+\tau)\,dv_r$$

$$= \langle \gamma_\kappa^{\,2}\rangle \exp\!\left[-\frac{1}{2}\left(\frac{l}{l_c}\right)^2 - \tfrac{1}{2}\omega_c^{\,2}\tau^2\right]$$

where $\langle \gamma_\kappa^{\,2}\rangle$ is the mean square of γ_κ, averaged over region R and over time interval T. Generalization of Eq. (2.3.13) indicates that an autocorrelation function of this form will have a spectrum density

$$|\Gamma_\kappa(\mathbf{u},\Delta)|^2 = \frac{VT_c^{\,3}l}{(2\pi)^6\omega_c}\,\langle \gamma_\kappa^{\,2}\rangle \exp\!\left[-\tfrac{1}{2}\mu^2 l_c^{\,2} - \frac{1}{2}\left(\frac{\Delta}{\omega_c}\right)^2\right]$$

Therefore the intensity of the sound scattered at frequency $\omega/2\pi$ from a plane wave, of amplitude A and frequency $\omega_i/2\pi$, going in the direction $\mathbf{k}_i$, by the region R with irregularity γ_κ characterized above, is

$$I_\kappa(\mathbf{r}) \simeq \frac{|A|^2 k_i^4 V T l_c^3}{16\pi^2 \rho c \omega_c r^2} \langle \gamma_\kappa^2 \rangle \exp\left[-\tfrac{1}{2}l_c^2 \left| \frac{\omega}{c} \mathbf{a}_r - \mathbf{k}_i \right|^2 - \frac{(\omega - k_i c)^2}{2\omega_c^2} \right] \quad (8.1.27)$$

which exhibits the dependence on frequency and on the angle between the direction $\mathbf{k}_i$ of the incident wave and the direction $\mathbf{a}_r$ to the observer.

Finally, Eq. (1.3.17) shows that the total intensity at $\mathbf{r}$, averaged over a time interval T, is the same as the integral of $I_\kappa(\mathbf{r})$ over ω.

$$I_{\kappa t}(\vartheta) \equiv \frac{1}{T} \int I_t(t)\, dt = \frac{2\pi}{T} \int I_\kappa\, d\omega \simeq \frac{|A|^2 V k_i^4}{\sqrt{32\pi}\,\rho c r^2} \frac{l_c^3}{\sqrt{1 + k_c^2 l_c^2}} \langle \gamma_\kappa^2 \rangle$$

$$\times \exp\left[\frac{-k_i^2 l_c^2}{2(k_c^2 l_c^2 + 1)} (4 \sin^2 \tfrac{1}{2}\vartheta + k_c^2 l_c^2 \sin^2 \vartheta) \right] \quad (8.1.28)$$

where $k_c = \omega_c/c$, and ϑ is the angle of scattering, between $\mathbf{a}_r$ and $\mathbf{k}_i$. Integration of this, times $2\pi r^2 \sin \vartheta\, d\vartheta$ over ϑ from 0 to π, would equal the total power scattered by the inhomogeneous term γ_λ. Again we see that the scattered intensity falls off as the scattering angle increases. For high frequencies (k_i large) it falls off rapidly, particularly if $k_c l_c = l_c/c\tau_c$ is large. When the correlation length l_c is small compared with c times the correlation time τ_c, and the scattered peak in the k_i direction is broader and at lower frequencies ($k_i l_c \ll 1$), the scattered intensity is nearly isotropic.

More calculation of the same sort indicates that the total spectrum density of sound scattered by random fluctuations in compressibility, density, and turbulent velocity is

$$I_\omega(r) \rightarrow \frac{|A|^2 V T l_c^3}{16\pi^2 \rho c \omega_c r^2} \left[k_i^4 \langle \gamma_\kappa^2 \rangle + k_i^2 \left(\frac{\omega}{c}\right)^2 \langle \gamma_\rho^2 \rangle \cos^2 \vartheta \right.$$

$$\left. + 4\left(\frac{\omega}{c}\right)^4 \langle M_t^2 \rangle \cos^2 \vartheta \right] \exp\left[-\tfrac{1}{2}\mu_s^2 l_c^2 - \frac{1}{2}\left(\frac{\Delta_s}{\omega_c}\right)^2 \right] \quad (8.1.29)$$

where γ_κ and γ_ρ are defined in Eq. (8.1.11), and $M_t = U_r/c$, so that $\langle M_t^2 \rangle$ is the mean-square Mach number of the turbulent motion, independent of ϑ, if the turbulence is isotropic [see Eq. (11.4.15) for further discussion]. The quantities μ_s and Δ_s are given in Eq. (8.1.25). The total scattered intensity, for all wavelengths, is then

$$I_s(\mathbf{r}) \rightarrow \frac{|A|^2 V}{\sqrt{32\pi}\,\rho c l_c r^2} \frac{\alpha^4}{\sqrt{1 + \alpha^2}} \left\{ \eta^4 \langle \gamma_\kappa^2 \rangle + \eta^2 \langle \gamma_\rho^2 \rangle \frac{1 + \eta^2(1 + \alpha^2 \cos \vartheta)^2}{1 + \alpha^2} \right.$$

$$\left. + 12\langle M_t^2 \rangle \frac{1 + 2\eta^2(1 + \alpha^2 \cos \vartheta)^2 + \tfrac{1}{3}\eta^4(1 + \alpha^2 \cos \vartheta)^4}{(1 + \alpha^2)^2} \right\}$$

$$\times \exp\left[\frac{-\eta^2 \alpha^2}{2(1 + \alpha^2)} (4 \sin^2 \tfrac{1}{2}\vartheta + \alpha^2 \sin^2 \vartheta) \right] \quad (8.1.30)$$

where $\eta = k_i/k_c = \omega_i/\omega_c$, and $\alpha = k_c l_c$. We have assumed that although γ_κ, γ_ρ, and U_r/c fluctuate independently, their correlation lengths and times are roughly equal.

Ordinarily, in a turbulent region, the fluctuations of κ and of ρ are proportional to M_t^2, so that if the turbulent Mach number is small, the first

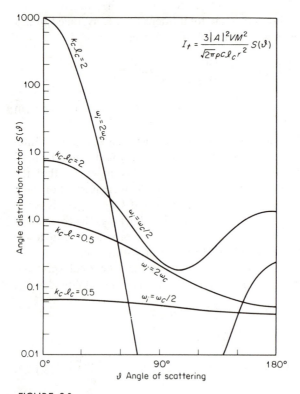

$$I_f = \frac{3|A|^2 V M^2}{\sqrt{2\pi} \rho c \mathscr{l}_c r^2} S(\vartheta)$$

FIGURE 8.3
Angle distribution of total intensity scattered by isotropic turbulence of Mach number M.

two terms are small compared with the last one in the braces. The angle distribution for this last term, assuming γ_κ and γ_ρ are negligible, is plotted in Fig. 8.3, to display the very large variety in the directionality of the scattering from a turbulent region.

8.2 SCATTERING BY SPHERES

To obtain more accurate formulas for sound scattering we must specialize our calculations for objects of relatively simple shape. In the preceding section we worked out the scattering from circular cylinders; in this section we take up the other simple shape, the sphere.   This case is of considerable practical importance; many scattering objects are more or less spherical. Scattering from a few other shapes can be computed exactly, from disks and ellipsoids, for example.   But these computations are much more complex and introduce no new principles; so we omit their discussion.

If the sphere is composed of material of quite different acoustical properties from those of the medium, it will turn out to be advantageous to fit boundary conditions at the surface of the sphere, rather than to use the volume integral of Eq. (8.1.13).   First let us go through the simplest case, that of a rigid, motionless sphere of radius a, centered at the origin.

The rigid sphere

The analysis follows the same lines as that for the cylinder.   The expression for a plane wave traveling to the right along the polar axis is

$$p_p = Ae^{ik(r\cos\vartheta - ct)} = A \sum_{m=0}^{\infty} (2m + 1)i^m P_m(\cos\vartheta)j_m(kr)e^{-2\pi ivt} \quad (8.2.1)$$

where $A = \sqrt{\rho_0 cI}$, and the factors P_m and j_m are defined in Eqs. (7.2.4) and (7.2.11), respectively.   The expression for the wave scattered from a sphere of radius a whose center is the polar origin is

$$p_s = -A \sum_{m=0}^{\infty} (2m + 1)i^{m+1}e^{-i\delta_m} \sin\delta_m P_m(\cos\vartheta)$$
$$\times [j_m(kr) + in_m(kr)]e^{-2\pi ivt} \quad (8.2.2)$$

where the angles δ_m have been defined in Eqs. (7.2.15) in connection with the radiation from a sphere.   The values of some of them are given in Table VIII.

The intensity of the scattered wave and the total power scattered are

$$I_s \simeq \frac{a^2 I}{r^2} \frac{1}{k^2 a^2} \sum_{m,n=0}^{\infty} (2m + 1)(2n + 1)$$
$$\times \sin\delta_m \sin\delta_n \cos(\delta_m - \delta_n)P_m(\cos\vartheta)P_n(\cos\vartheta)$$
$$\simeq \begin{cases} \dfrac{16\pi^4 v^4 a^6 I}{9c^4 r^2}(1 - 3\cos\vartheta)^2 & ka \ll 1 \\[4mm] I\left[\dfrac{a^2}{4r^2} + \dfrac{a^2}{4r^2}\cot^2\left(\dfrac{\vartheta}{2}\right)J_1^2(ka\sin\vartheta)\right] & ka \gg 1 \end{cases} \quad (8.2.3)$$

$$\Pi_s = 2\pi a^2 I \frac{2}{k^2 a^2} \sum_{m=0}^{\infty} (2m + 1)\sin^2\delta_m \simeq \begin{cases} \dfrac{256\pi^5 a^6}{9\lambda^4}I & \lambda \gg 2\pi a \\[4mm] 2\pi a^2 I & \lambda \ll 2\pi a \end{cases}$$

The discussion concerning the short-wavelength limit for the scattering from a sphere is similar to that preceding and following Eq. (8.1.4) for the cylindrical case.   The total power scattered is that contained in an area of primary beam equal to twice the cross section πa^2 of the sphere.   Half of this is reflected equally in all directions from the sphere (the first term in the last expression for I); and the other half is concentrated into a narrow beam which tends to interfere with the primary beam and cause the shadow

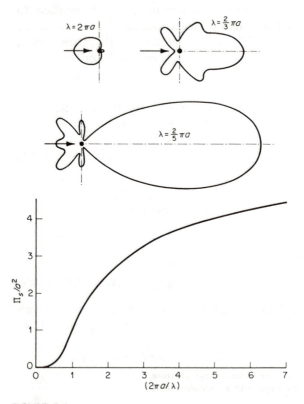

FIGURE 8.4
Distribution-in-angle of intensity scattered from a sphere of radius a and total power scattered Π, per unit incident intensity.

(the second term in the last expression).   If the experimentally measured "total scattered power" includes everything from $\vartheta = \Delta$ to $\vartheta = \pi - \Delta$, Π_{exp} will equal Π_s for wavelengths longer than $4a/\Delta$ and will approach $\frac{1}{2}\Pi_s$ for wavelengths much shorter than $4a/\Delta$.

Figure 8.4 shows polar curves of the scattered intensity as a function of the angle of scattering ϑ, for different values of $\eta = ka$, and a curve Π_s as function of η.   As with the cylinder, the directionality of the scattered wave increases as the frequency increases.

The force on the sphere

The total pressure at a point on the sphere an angle ϑ from the polar axis (note that the point $\vartheta = 0$ is the point farthest away from the source of the sound) turns out to be

$$p_a = Ae^{-i\omega t}\left(\frac{1}{ka}\right)^2 \sum_{m=0}^{\infty} \frac{2m+1}{B_m} P_m(\cos\vartheta)e^{-i(\delta_m - \frac{1}{2}\pi m)}$$ (8.2.4)

$$\simeq (1 + \tfrac{3}{2}ika\cos\vartheta)Ae^{-i\omega t} \qquad ka \ll 1$$

As with the cylinder, this expression is proportional to that for the pressure, at large distances, due to a simple source set in the sphere. The curves of Fig. 7.1 therefore show the dependence of $p_a{}^2$ on ϑ.

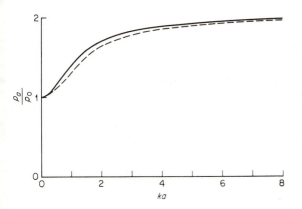

FIGURE 8.5
Ratio of pressure amplitude at a point on a sphere to pressure amplitude of the plane wave striking the sphere, plotted as function of $\eta = 2\pi a/\lambda$. Solid line is the pressure at a point facing the incident wave; dotted line is the average pressure for a circular area around this point with an angular radius of 30°.

The amplitude of the pressure at the point nearest the source of sound ($\vartheta = \pi$), for a plane wave of unit pressure amplitude, is plotted in Fig. 8.5 as a function of η. We notice that for wavelengths long compared with the circumference of the sphere, the pressure at $\vartheta = \pi$ equals the pressure of the plane wave, but that for shorter wavelengths the distortion of the wave due to the presence of the sphere makes p_a differ from A. This general fact will be true for obstacles of other than spherical shape, even though the dependence of p_a on η will be somewhat different from that given in Eq. (8.2.4). Therefore a microphone measures the pressure of the wave striking it only as long as its circumference is smaller than the wavelength of the wave. For smaller wavelengths a correction must be made for the distortion due to the presence of the microphone.

By using Eqs. (7.2.5) we can find the average value of the pressure on that portion of the surface of the sphere contained between the angles $\vartheta = \pi$ and $\vartheta = \pi - \vartheta_0$:

$$
\begin{aligned}
P_{\text{avg}} &= \frac{1}{1 - \cos \vartheta_0} \int_{\pi - \vartheta_0}^{\pi} p_a \sin \vartheta \, d\vartheta \\
&= A e^{-i\omega t} \sum_{m=0}^{\infty} \frac{i^m e^{-i\delta m}}{k^2 a^2 B_m} \frac{P_{m-1}(\cos \vartheta_0) - P_{m+1}(\cos \vartheta_0)}{P_{-1}(\cos \vartheta_0) - P_1(\cos \vartheta_0)} \qquad (8.2.5) \\
&\simeq A e^{-i\omega t}[1 + \tfrac{3}{4} i k a(1 + \cos \vartheta_0)] \qquad ka \ll 1
\end{aligned}
$$

where $P_{-1}(\cos \vartheta_0) = 1$. The dotted line in Fig. 8.5 gives the values of $P_{\text{avg}}/A_0 e^{-i\omega t}$ as function of $\eta = ka$ for the case $\vartheta_0 = 30°$.

Nonrigid sphere

To compute the scattering from a sphere which is not perfectly rigid, we shall find it convenient to use the full machinery of the Green's function integral. The fluid medium fills all space outside the surface of the sphere of radius a centered at the origin. In this region the volume integral of Eq. (7.1.17) is zero, and the surface integral is over the sphere at infinity and that at $r = a$. The integral over the sphere at infinity produces the incident wave p_i, as mentioned in connection with Eq. (8.1.13). Thus Eq. (7.1.17) becomes

$$
\begin{aligned}
p_\omega(r,\vartheta,\varphi) = p_i + a^2 \int_0^{2\pi} d\varphi_0 \int_0^{\pi} \sin \vartheta_0 \, d\vartheta_0 \bigg[p_\omega(a,\vartheta_0,\varphi_0) \frac{\partial}{\partial a} g_\omega \\
- g_\omega(r,\vartheta,\varphi \mid a,\vartheta_0,\varphi_0) \frac{\partial}{\partial a} p_\omega(a,\vartheta_0,\varphi_0) \bigg] \qquad (8.2.6)
\end{aligned}
$$

where the incident wave p_i is usually the plane wave $A e^{ikz}$ and where we have replaced $\partial/\partial n_0$ in the integral by $-(\partial/\partial a)$. The appropriate series for the Green's function is given in Eq. (7.2.31). To make the integral equation specific, we need to determine the ratio between p_ω and $\partial p_\omega/\partial r$ at the surface of the sphere $r = a$. This ratio is, of course, fixed by the acoustic properties of the material inside $r = a$ [which is not a part of the region in which Eq. (8.2.6) holds]. If the material is such that its surface at $r = a$ is one of local reaction, then $u_r = -(\beta/\rho c)p$ or $\partial p/\partial r = -ik\beta p$ just outside $r = a$, where $\rho c/\beta$ is the point impedance, and β is the specific admittance of the surface.

On the other hand, if the material of the scattering sphere allows sound waves to propagate through it, the surface at $r = a$ will be one of extended reaction, and we shall have to determine the ratio between u_r and p at $r = a$ by investigating the wave motion inside $r = a$. Suppose the material of the sphere has density ρ_e and compressibility κ_e, so that wave motion propagates with velocity $c_e = 1/\sqrt{\rho_e \kappa_e}$ in it. In this case the acoustic pressure inside $r = a$ can be represented by a series of terms of the form $Y_{mn}{}^\sigma(\vartheta,\varphi)j_m(k_e r)$, where $k_e = \omega/c_e$; $Y_{mn}{}^\sigma$ and j_m are the spherical harmonics and spherical Bessel functions discussed in Sec. 7.2. We must use j_m, not n_n, because this wave cannot go to infinity at $r = 0$.

For each spherical harmonic of order m, the relation between u_r and p just inside $r = a$ is then

$$u_r = \frac{1}{i\omega\rho_e}\left(\frac{\partial p}{\partial r}\right)_{r=a} = \frac{1}{i\rho_e c_e}\,j'_m(k_e a) = -\frac{\beta_m}{\rho c}\,j_m(k_e a) \qquad (8.2.7)$$

where

$$\beta_m = i\,\frac{\rho c}{\rho_e c_e}\left[\frac{j'_m(k_e a)}{j_m(k_e a)}\right]$$

and where j'_m represents the derivative of j_m with respect to its argument. We have expressed the ratio between the portions of u_r and p, which belong to the spherical harmonic $Y_{mn}{}^\sigma$, in terms of an effective admittance β_m, by analogy with the locally reacting case. Here, however, β_m is a function of the order m of the spherical harmonic, whereas for the locally reacting case β is independent of m.

In either case we shall have to expand the solution $p_\omega(r,\vartheta,\varphi)$ outside $r = a$ into a spherical harmonic series

$$p_\omega(r,\vartheta,\varphi) = \sum_{mn\sigma} Y_{mn}{}^\sigma(\vartheta,\varphi)q_{mn}{}^\sigma(kr) \qquad k = \frac{\omega}{c} \qquad (8.2.8)$$

where $q_{mn}{}^\sigma$ is a combination of $j_m(kr)$ and $h_m(kr)$, which fits the boundary conditions at $r = a$. In fact, the boundary condition is that, just outside $r = a$,

$$u_r = \frac{1}{i\omega\rho}\left(\frac{\partial q}{\partial r}\right)_{r=a} = \frac{1}{i\sigma\rho c}\,q_{mn}{}^{\sigma\prime}(ka) = -\frac{\beta_m}{\rho c}\,q_{mn}{}^\sigma(ka)$$

by definition of β_m. Quantity $q_{mn}{}^{\sigma\prime}(\zeta) = dq_{mn}{}^\sigma(\zeta)/d\zeta$. However, the ratio between u_r and p at $r = a$, required by the acoustic properties of the material inside $r = a$, is given by combining the equations of the foregoing paragraphs. We see that

$$q_{mn}{}^{\sigma\prime}(ka) = -i\beta_m q_{mn}{}^\sigma(ka) \qquad \text{where } \beta_m = i\,\frac{\rho c}{\rho_e c_e}\left[\frac{j'_m(k_e a)}{j_m(k_e a)}\right] \qquad (8.2.9)$$

for a surface of extended reaction at $r = a$. For a surface of local reaction, of point impedance $\rho c/\beta$, the relation between q'_{mn} and q_{mn} is the same, but $\beta_m = \beta$, independent of m, instead of being given by the second equation. This fixes the relationship between the first and second terms in the brackets in Eq. (8.2.6), for either kind of surface reaction.

Solving the integral equation

We can now carry out the surface integral of Eq. (8.2.6), using the expansion (7.2.31) for the Green's function, the expansion (8.2.8) for p_ω, and Eq. (8.2.9). Because of the integral relationships between the spherical harmonics, given in Eqs. (7.2.7), the double summation reduces to a single sum, and the integral simplifies considerably. For example, the first term in the surface integral turns out to be

$$\frac{ik^2a^2}{4\pi} \int_0^{2\pi} d\varphi_0 \int_0^\pi \sin \vartheta_0 \, d\vartheta_0 \sum_{str} q_{st}{}^{\tau\prime}(ka) \, Y_{st}{}^\tau(\vartheta_0,\varphi_0)$$

$$\times \sum_{mn\sigma} (2m+1) \frac{(m-n)!}{(m+n)!} \, Y_{mn}{}^\sigma(\vartheta_0,\varphi_0) \, Y_{mn}{}^\sigma(\vartheta,\varphi) j_m'(ka) h_m(kr)$$

$$= ik^2a^2 \sum_{mn\sigma} Y_{mn}{}^\sigma(\vartheta,\varphi) h_m(kr) q_{mn}{}^\sigma(ka) j_m'(ka)$$

where we have chosen the form of the Green's function appropriate for $r \geqslant a$.

Thus Eq. (8.2.6) becomes

$$p_\omega(r,\vartheta,\varphi) = p_i(r,\vartheta,\varphi) + ik^2a^2 \sum_{mn\sigma} [j_m'(ka)$$
$$+ i\beta_m j_m(ka)] q_{mn}{}^\sigma(ka) \, Y_{mn}{}^\sigma(\vartheta,\varphi) h_m(kr) \quad \text{(8.2.10)}$$

which will enable us to evaluate the pressure anywhere outside the sphere, as soon as we have computed the values of the constants $q_{mn}(ka)$. The first term on the right-hand side is the incident wave; the second term, made up of outgoing Hankel functions, is the scattered wave. We compute the values of $q_{mn}(ka)$ by requiring that Eq. (8.2.10) hold on the boundary $r = a$. The equation

$$p_i(a,\vartheta,\varphi) = \sum_{mn\sigma} q_{mn}{}^\sigma(ka) \, Y_{mn}{}^\sigma(\vartheta,\varphi)\{1 - ik^2a^2 h_m(ka)[j_m'(ka) + i\beta_m j_m(ka)]\}$$

simply states that the total wave, given by Eq. (8.2.8), minus the scattered wave, at $r = a$, equals the incident wave.

One further simplification can be made in this equation. The last of Eqs. (7.2.12) states that $j_m'(ka) h_m(ka) = j_m(ka) h_m'(ka) - (i/k^2a^2)$. Therefore the equation determining the coefficients $q_{mn}(ka)$, the equivalent of the integral equation (8.2.6), becomes

$$p_i(a,\vartheta,\varphi) = -ik^2a^2 \sum_{m,n,\sigma} q_{mn}{}^\sigma(ka) \, Y_{mn}{}^\sigma(\vartheta,\varphi) j_m(ka)[h_m'(ka) + i\beta_m h_m(ka)]$$

$$\text{(8.2.11)}$$

The incident wave also may be expanded in terms of spherical harmonics.

$$p_i(r,\vartheta,\varphi) = \sum_{m,n,\sigma} A_{mn}{}^\sigma \, Y_{mn}{}^\sigma(\vartheta,\varphi) j_m(kr)$$

Since the incident wave is the complete solution *in the absence* of the scattering sphere, it must be finite at $r = 0$, and thus only the Bessel functions $j_m(kr)$ can be used. If the incident wave is a plane wave of amplitude $A = \sqrt{\rho c I}$, intensity I, and wavelength $\lambda = 2\pi/k$, traveling in the direction of the spherical axis (z axis), the form is

$$p_i = A \sum_{m=0}^{\infty} (2m + 1)i^m P_m(\cos \vartheta)j_m(kr) \quad \text{or} \quad A_{mn} = A(2m + 1)i^m \delta_{\sigma 1} \delta_{n0}$$

(8.2.12)

since $Y_{mo}{}^1(\vartheta, \varphi) = P_m(\cos \vartheta)$.

Therefore the coefficients of expansion of p_ω on the surface $r = a$ have values

$$q_{mn}{}^\sigma(ka) = i \frac{A_{mn}{}^\sigma}{k^2 a^2} [h'_m(ka) + i\beta_m h_m(ka)]^{-1}$$

$$= \frac{(2m + 1)i^{m+1}A}{h'_m(ka) + i\beta_m h_m(ka)} \frac{\delta_{\sigma 1}\delta_{n0}}{k^2 a^2} \quad \text{(for a plane wave)} \quad \text{(8.2.13)}$$

For the rest of this chapter we shall concentrate exclusively on the case of an incident plane wave, for which the second expansion holds for $q_{mn}{}^\sigma(ka)$. The more general case will be taken up later.

Absorption and scattering

Inserting the second form of Eq. (8.2.13) into Eq. (8.2.10), we obtain, for the total sound field outside the sphere, for an incident plane wave,

$$p_\omega(r,\vartheta) = Ae^{ikr \cos \vartheta} - A \sum_{m=0}^{\infty} (2m + 1)i^m P_m(\cos \vartheta)$$

$$\times \frac{j'_m(ka) + i\beta_m j_m(ka)}{h'_m(ka) + i\beta_m h_m(ka)} h_m(kr)$$

$$= A \sum_m (2m + 1)i^m P_m(\cos \vartheta)[j_m(kr) - \tfrac{1}{2}(1 + R_m)h_m(kr)]$$

$$\to \frac{iA}{2kr} \sum_m (2m + 1)P_m(\cos \vartheta)[R_m e^{ikr} + (-1)^m e^{-ikr}] \quad r \to \infty$$

(8.2.14)

In the last expression we have used the asymptotic formulas for j_m and h_m given in Eqs. (7.2.9) and (7.2.12). The last two formulas have been simplified by defining a *reflection coefficient* R_m.

$$1 + R_m = 2\frac{j'_m(ka) + i\beta_m j_m(ka)}{h'_m(ka) + i\beta_m h_m(ka)} \qquad R_m = \frac{h'^*_m(ka) + i\beta_m h^*_m(ka)}{h'_m(ka) + i\beta_m h_m(ka)}$$

with $h'^*_m(ka) = j'_m(ka) - in'_m(ka)$, $h^*_m = j_m - in_m$ being the complex conjugates of the Hankel functions $h'_m = j'_m + in'_m$ and $h_m = j_m + in_m$, respectively. These results deserve detailed discussion.

The asymptotic expression indicates that the pressure wave outside the sphere can be expressed as a series of waves traveling radially outward from the origin (characterized by the factor e^{ikr}) plus another series traveling radially inward (with the factor e^{-ikr}). The waves traveling inward are unaffected by the nature of the spherical surface at $r = a$; they have been determined solely by the shape of the incident plane wave. The outward-bound waves, on the other hand, have amplitudes proportional to the factors R_m, which depend on the surface admittances β_m. We may call this factor, the ratio between the mth outgoing and incoming waves, the *reflection coefficient* for the mth partial wave. If β_m is purely imaginary (reactive), R_m is the ratio between the quantity $h'_m + i\beta_m h_m$ and its complex conjugate, so that $|R_m| = 1$. In this case each wave is reflected from the surface $r = a$ with no change in amplitude, but a change in phase; no energy is lost at the sphere's surface.

On the other hand, if β_m has a real part, $|R_m|$ is less than unity. Less energy is reflected than is incident on the sphere; some of the energy is absorbed. Finally, if the material inside the sphere has the same compressibility and density as the medium outside, so that $\beta_m = i[j'_m(ka)/j_m(ka)]$, then $1 + R_m = 0$, $R_m = -1$ for all m, there is no scattering or absorption, and the incident plane wave is undistorted.

The asymptotic form of the scattered wave is

$$p_s(r,\vartheta,\varphi) = -\tfrac{1}{2}A \sum_m (2m + 1)i^m(1 + R_m)P_m(\cos\vartheta)h_m(kr)$$

$$\to A\frac{e^{ikr}}{r}\Phi(\vartheta) \qquad \Phi(\vartheta) = \frac{i}{2k}\sum_m (2m + 1)(1 + R_m)P_m(\cos\vartheta)$$

$$(8.2.15)$$

Factor Φ, the angle-distribution factor for the scattered wave, is equal to $1/A$ times the factor Φ_s of Eq. (8.1.14). The total power scattered, divided by the incident intensity $I = |A|^2/\rho c$, is equal to the *scattering cross section* [see discussion preceding Eq. (8.1.8)].

$$\Sigma_s = 2\pi\int_0^\pi |\Phi(\vartheta)|^2 \sin\vartheta\, d\vartheta = \frac{4\pi}{k^2}\sum_m (2m + 1)\left|\frac{j'_m(ka) + i\beta_m j_m(ka)}{h'_m(ka) + i\beta_m h_m(ka)}\right|^2$$

$$= \frac{\pi}{k^2}\sum_m (2m + 1)|1 + R_m|^2 \to \frac{4\pi}{k^2}\sum_m (2m + 1)\sin^2\delta_m \qquad \text{when } \beta \to 0$$

$$(8.2.16)$$

[see Eqs. (8.2.3)]. If β has a real part, energy is absorbed by the sphere as well as scattered to infinity. To calculate the power absorbed, we compute the energy flux, $-\mathrm{Re}\,(pu_r^*)$ at $r = a$, into the sphere. Equations (8.2.8) and (8.2.13) enable us to calculate the pressure at the surface, whereas Eq. (8.2.9) shows that

$$(u_r)_{r=a} = -\frac{A}{\rho ck^2a^2}\sum_m \frac{(2m + 1)i^m\beta_m}{h'_m(ka) + i\beta_m h_m(ka)}P_m(\cos\vartheta)$$

If the admittance β_m has a real part ξ_m and imaginary part σ_m, so that $\beta_m = \xi_m - i\sigma_m$, then the *absorption cross section* Σ_a is the power absorbed by the sphere, divided by I_i.

$$\Sigma_a = \frac{k^4 a^4}{4\pi a^2} \sum_m \frac{(2m+1)\xi_m}{[h'_m(ka) + i\beta_m h_m(ka)]^2} = \frac{\pi}{k^2} \sum_m (2m+1)(1 - |R_m|^2) \quad (8.2.17)$$

Quite a bit of algebra is required to show that the two series are equal.

Having written out the two expressions for the power withdrawn from the incident beam, we can write down the sum, called the *total cross section*.

$$\Sigma = \Sigma_s + \Sigma_a = \frac{2\pi}{k^2} \sum_m (2m+1)(1 + \operatorname{Re} R_m) = \frac{4\pi}{k} \operatorname{Im}[\Phi(0)] \quad (8.2.18)$$

since the imaginary part of the angle-distribution factor Φ, for $\vartheta = 0$, is proportional to the sum of the real parts of $(2m+1)(1 + R_m)$. We note that the forward-directed scattering function Φ is proportional to the *total cross section*. Since $\operatorname{Im}[\Phi(0)]$ forms the shadow, it is not surprising that it measures the total power abstracted from the incident wave, whether absorbed or scattered.

Long wavelengths

To show how these formulas can be used and to demonstrate the range of validity of the Born approximation [Eq. (8.1.20)] for the case of a spherical scatterer, we shall work out the limiting forms for the angle-distribution function Φ and for the cross section Σ when ka is small (i.e., when λ is larger than $2\pi a$). Insertion of Eqs. (7.2.9) and (7.2.12) into Eqs. (8.2.9), (8.2.13), and (8.2.15) produces approximate formulas for angle-distribution function, cross sections, and pressure and velocity distributions at the surface of the sphere, to the lowest order in the small quantity ka.

$$\Phi(\vartheta) \simeq \tfrac{1}{3}k^2 a^3 \left(\frac{\kappa_e - \kappa}{\kappa} + \frac{3\rho_e - 3\rho}{2\rho_e + \rho} \cos \vartheta \right)$$

$$p_\omega(a,\vartheta) \simeq A\left(1 + ika \frac{3\rho_e}{2\rho_e + \rho} \cos \vartheta \right) \qquad p_s(r,\vartheta) \to A \frac{e^{ikr}}{r} \Phi(\vartheta)$$

$$u_r(a,\vartheta) \simeq \frac{A}{\rho c}\left(\tfrac{1}{3}ika \frac{\kappa_e}{\kappa} + \frac{3\rho}{2\rho_e + \rho} \cos \vartheta \right) \qquad (8.2.19)$$

$$\Sigma_s \simeq \tfrac{4}{9}\pi a^2 (ka)^4 \left(\left|\frac{\kappa_e - \kappa}{\kappa}\right|^2 + \frac{1}{3}\left|\frac{3\rho_e - 3\rho}{2\rho_e + \rho}\right|^2 \right)$$

$$\Sigma_a \simeq \tfrac{4}{3}\pi a^2 (ka) \operatorname{Im}\left(\frac{\kappa_e - \kappa}{\kappa} + \frac{3\rho_e - 3\rho}{2\rho_e + \rho} \right)$$

These formulas also deserve extended discussion. For a heavy, incompressible sphere ($\kappa_e \ll \kappa$, $\rho_e \gg \rho$) the angle distribution of the scattered intensity has the factor $(-1 + \frac{3}{2} \cos \vartheta)^2$, with 25 times as much scattered backward $(\vartheta = \pi)$ as forward $(\vartheta = 0)$ and with a node of zero intensity at $\vartheta = \cos^{-1} \frac{2}{3} \simeq 48°$. The radial velocity of the surface of the sphere in this case is zero, of course, and the pressure at the surface is $A(1 + \frac{3}{2}ika \cos \vartheta)$, whereas the incident pressure at $r = a$ is $A(1 + ika \cos \vartheta)$. This indicates the error made by the Born approximation, which uses the incident pressure. The modified approximation of Eq. (8.1.21) checks with the present exact solutions when $\rho_e - \rho$ is small compared with ρ_e. But for a very heavy, incompressible sphere the modified Born approximation would predict an angle distribution of $(-1 + \cos \vartheta)^2$, with nothing scattered forward, instead of the $(-1 + \frac{3}{2} \cos \vartheta)^2$ of the exact solution. The unmodified Born approximation, obtained by substituting p_i for p_ω in the integral of Eq. (8.1.14), would be much less satisfactory, for the effect of ρ_e would be the same as the effect of κ_e, and there would be *no* cosine term, of second order in ka, in the resulting $\Phi(\vartheta)$.

For a light, compressible sphere $(\kappa_e > \kappa, \rho_e \ll \rho)$ the angle factor for the scattered intensity is $[(\kappa_e/\kappa) - 3 \cos \vartheta]^2$, which again has a predominance in the backward direction. The pressure at the surface of the sphere is A, with negligible dependence on ϑ, but the radial velocity at $r = a$ is large, being $(A/\rho c)[ika(\kappa_e/\kappa) + 3 \cos \vartheta]$; the sphere moves back and forth along the z axis with three times the velocity of the surrounding fluid; so it is not surprising that the scattering cross section is large. This is the case with air bubbles in water. The absorption cross section differs from zero only if the compressibility or density of the material of the sphere, or both, are complex, corresponding to frictional effects.

Porous sphere

In this connection it is interesting to work out the scattering from a sphere of porous material, of porosity Ω and flow resistance Φ_e, as discussed in connection with Eq. (6.2.23). We first assume that the material of the pore walls is stationary, the air flowing through the pores as described in Sec. 6.2. From Eq. (6.2.25) we see that the effective compressibility and density of the fluid in the pores, for sound of frequency $\omega/2\pi$, are

$$\kappa_e = \kappa_p \Omega \qquad \rho_e = \rho_p + i \frac{\Phi_e}{\omega}$$

To this approximation the compressibility has no imaginary part, though inclusion of thermal transfer would introduce a small imaginary term.

For many porous acoustic materials in air, $\Phi_e/\rho_p\omega$ ranges from 1 to 10 for frequencies of 1,000 cps and is proportionally larger for lower frequencies; so the imaginary part of ρ_e is often larger than its real part. When this is true we can write

$$\frac{3\rho_e - 3\rho}{2\rho_e + \rho} \simeq \frac{3}{2} + \frac{3}{4}i \frac{\rho ck}{\Phi_e}$$

If, in addition, $\kappa_p\Omega$ is not very different from κ, the expressions of Eq. (8.2.19) become

$$\Phi(\vartheta) \simeq \tfrac{1}{2}k^2a^3\left[1 + \tfrac{1}{2}i\,\frac{\rho ck}{\Phi_e}\right]\cos\vartheta$$

$$p_\omega(a,\vartheta) \simeq A\left[1 + ika\left(\tfrac{3}{2} - \tfrac{3}{4}i\,\frac{\rho ck}{\Phi_e}\right)\cos\vartheta\right]$$

$$u_r(a,\vartheta) \simeq \frac{A}{\rho c}\left[\tfrac{1}{3}ika - \tfrac{3}{2}i\,\frac{\rho ck}{\Phi_e}\cos\vartheta\right] \qquad (8.2.20)$$

$$\Sigma_s \simeq \tfrac{1}{3}\pi a^2(ka)^4\left[1 + \frac{1}{4}\left(\frac{\rho ck}{\Phi_e}\right)^2\right]$$

$$\Sigma_a \simeq \pi a^2(ka)\,\frac{\rho ck}{\Phi_e}$$

Note that the angle-distribution function $\Phi(\vartheta)$ is not to be confused with flow resistance Φ_e.

When the effective compressibility is nearly equal to that of air, the scattered wave is dipolelike; the air flows back and forth through the pores, the flow resistance ensuring that the dipole is out of phase with the driving pressure. The absorption cross section is larger than the scattering cross section if ka is small.

When the framework of the pore walls is not held at rest, yet another correction must be added because of the motion of the solid material. If the pore-wall material is effectively incompressible, the radially symmetric term ($m = 0$) in the series for the pressure inside the sphere is unaffected by this motion, and the term $(\kappa_e - \kappa)/\kappa$ in Eqs. (8.2.19) is unchanged. However, the $m = 1$ term, representing linear motion along the z axis, is changed because the solid part of the sphere as a whole will move back and forth (as well as the air in the pores) unless the pore-wall material has infinite mass. Suppose this material has density ρ_m, large but not infinite. Then the mass of a unit volume of the sphere, exclusive of the fluid contained in the pores, is $\rho_m(1 - \Omega)$, where $1 - \Omega$ is the fraction of the volume occupied by the solid material. This is usually larger than ρ_e.

When this is the case, for the $m = 1$ term (proportional to $\cos\vartheta$), the impedance to the fluid flow is equivalent to a parallel circuit, the resistance Φ_e of the porous material being shunted by the reactance $-i\omega\rho_m(1 - \Omega)$ of the solid material, which is forced into motion by the drag of the air moving through the pores. In other words, the Φ_e of Eqs. (8.2.20), for the $m = 1$ term only, must be replaced by

$$\Phi_p = \left[\frac{1}{\Phi_e} + \frac{i}{\omega\rho_m(1 - \Omega)}\right]^{-1}$$

and the factor $\rho c k / \Phi_e$ in the first three of Eqs. (8.2.20) must be replaced by

$$\frac{\rho c k}{\Phi_e} + i \frac{\rho}{\rho_m(1 - \Omega)}$$

where the susceptance term is usually 1/2 to 1/10 of the magnitude of the conductance term. The spherically symmetric $(m = 0)$ components are unaffected, to this approximation, the velocity being out of phase with the pressure at $r = a$. The axial $(m = 1)$ components, on the other hand, are changed by the motion of the solid material, both in regard to phase relations and to magnitude ratios.

The fourth and fifth of Eqs. (8.2.20) then become

$$\Sigma_s = \tfrac{1}{3}(\pi a^2)(ka)^4 \left\{ 1 + \frac{1}{4}\left(\frac{\rho c k}{\Phi_e}\right)^2 + \frac{1}{4}\left[\frac{\rho}{\rho_m(1 - \Omega)}\right]^2 \right\}$$

$$\Sigma_a = \pi a^2 (ka) \frac{\rho c k}{\Phi_e}$$

(8.2.21)

indicating that the absorption cross section is not affected by the motion of the solid material (to this approximation) and the scattering cross section is not changed much when $\rho/\rho_m(1 - \Omega)$ is smaller than $\rho c k / \Phi_e$.

For larger spheres or shorter wavelengths these approximate formulas are not valid, and the exact formulas of Eqs. (8.2.15) to (8.2.17) must be used.

Viscosity and thermal conduction

Finally, we can combine these equations with Eqs. (6.4.17) and (6.4.25) to find the effect of viscosity and thermal conduction on the scattered wave. The incident-plus-scattered wave is, of course, a propagational one; in air the attenuation effects can be neglected, except at the surface of the scatterer. Thus the incident plane wave p_i is still given by Eq. (8.2.1), with $k = \omega/c$, and the wave scattered by the sphere can still be written

$$p_s = A\sum_m B_m P_m(\cos\vartheta) h_m(kr) \to A \frac{e^{ikr}}{ikr} \sum_m i^m B_m P_m(\cos\vartheta)$$

where $B_m = -\tfrac{1}{2}(2m + 1)i^m(1 + R_m)$, R_m being the reflection coefficient of Eq. (8.2.14). In addition, in the boundary layer just outside the sphere, there are thermal and shear waves, of an amplitude sufficient to satisfy the boundary conditions on the temperature and tangential velocity at the surface. These also modify the pressure and radial-velocity conditions somewhat, so that coefficients B_m, and thus the reflection coefficients R_m, have not the values they would have if these small effects were neglected.

Usually, these effects can be neglected, particularly if the sphere is fairly large or is "soft." They are proportionally more important for small scatterers and for a cloud of small spheres (such as fog or rain); the effect can be quite noticeable. For a sphere of radius a, large compared with the boundary layer thicknesses d_h and d_v of Eq. (6.4.31), the effects can be approximated by adding to the surface admittance β_m of Eq. (8.2.7) the additional term

$$\beta_{sm} = \tfrac{1}{2}(1 - i)\left[(\gamma - 1)kd_h + \frac{m(m + 1)}{a^2}kd_v\right]$$

for each value of m, with the sums $\beta_m + \beta_{sm}$ to be used instead of β_m in Eqs. (8.2.14) and elsewhere. According to the discussion following Eq. (6.4.36), as well as Eq. (6.4.41), this additional admittance will account for the effects of thermal and viscous losses whenever a is large compared with d_h and d_v. Since the β_{sm}'s are small, they can be neglected whenever the β_m's are larger.

When a is small, however, the effects cannot be expressed in terms of effective admittances; the shear and thermal waves must be calculated, and the boundary conditions must be worked out in detail. Fortunately, when a is small, we usually need only to consider the $m = 0$ and $m = 1$ terms in the series, since a will then usually be small compared with the wavelength. As an example of this situation we consider the material inside the sphere to have much greater viscosity and thermal conduction than does air; so we can neglect the thermal and shear waves inside the sphere (this would be the case for a water droplet suspended in air, for instance). The fluid inside the sphere has effective density ρ_e, compressibility κ_e, and wave velocity $c_e = 1/\sqrt{\rho_e \kappa_e}$. Thus the wave inside the sphere will be [Eq. (8.2.7)]

$$p_e = A \sum_m B_m{}^e P_m(\cos \vartheta) j_m(k_e r) \qquad r \leqslant a; \qquad k_e = \frac{\omega}{c_e}$$

When $k_e a$ is small we can neglect all but the first two terms of this series and can use the first terms in the series for the Bessel functions.

$$j_0(z) \simeq 1 - \tfrac{1}{6}z^2 \qquad j_1(z) \simeq \tfrac{1}{3}z \qquad h_0(z) \simeq 1 - \frac{i}{z} \qquad h_1(z) \simeq -\frac{i}{2} - \frac{i}{z^2}$$

The temperature fluctuation accompanying these incident and scattered waves is $\tau_p = [(\gamma - 1)/\gamma\alpha](p_i + p_s)$ for $r \geqslant a$ and is zero for $r \leqslant a$.

The thermal scattered wave, just outside the sphere, is obtained by using the k of Eq. (6.4.25) in the spherical Hankel functions

$$\tau_h = A \sum_m E_m P_m(\cos \vartheta) h_m \frac{(1 + i)r}{d_h}$$

For frequencies around a kilocycle, in air at standard conditions, d_h is roughly equal to 10^{-4} cm. Thus, even for spheres as small as 0.01 cm in radius, the argument of this Hankel function is large enough so that its asymptotic form, $h_m(z) \to (i^{-m-1}/z)e^{iz}$, can be used, at least for $m = 0$ and 1. Therefore we write the first two terms of the series, with the corresponding pressure and radial velocity, as

$$\tau_h \simeq -(1-i)\frac{Ad_h}{2r}(iE_0 + E_1 \cos \vartheta)e^{(i-1)r/d_h}$$

$$p_h \simeq i\gamma\alpha k^2(d_h{}^2 - d_v'^2)\frac{\tau_h}{2}$$

$$u_{hr} \simeq \frac{\gamma\alpha A}{2i\rho c}\frac{kd_h{}^2}{r}(iE_0 + E_1 \cos \vartheta)e^{(i-1)r/d_h}$$

Because $d_h{}^2$ can be neglected compared with d_h, we can neglect p_h and $u_{h\vartheta}$ at $r = a$; but τ_h and u_{hr} cannot be neglected.

There can be no shear wave which is spherically symmetric; the $m = 0$ terms have no tangential velocity components, and no shear wave is needed to fit boundary conditions. Thus the $m = 0$ terms satisfy the following conditions at $r = a$:

p continuous:

$$B_0^e j_0(k_e a) = j_0(ka) + B_0 h_0(ka)$$

or

$$B_0^e \simeq 1 - \frac{iB_0}{ka}$$

u_r continuous:

$$\frac{1}{i\rho_e c_e}B_0^e j_0'(k_e a) = \frac{1}{i\rho c}\left[j_0'(ka) + B_0 h_0'(ka) + \frac{i\alpha\gamma kd_h{}^2 C_0}{2a} \right]$$

or

$$-\frac{\rho c}{3\rho_e c_e}B_0^e k_e a \simeq -\tfrac{1}{3}ka + \frac{iB_0}{k^2 a^2} + \frac{i\alpha\gamma kd_h{}^2 C_0}{2a}$$

τ zero:

$$\frac{\gamma - 1}{\gamma\alpha}\left[j_0(ka) + B_0 h_0(ka) \right] - (1+i)\frac{d_h C_0}{2a} = 0$$

or

$$\frac{\gamma - 1}{\gamma\alpha}\left(1 - \frac{iB_0}{ka}\right) - (1+i)\frac{d_h C_0}{2a} \simeq 0$$

where $C_0 = E_0 e^{(i-1)r/d_h}$.

The approximate solution of this trio of equations, valid for ka and d_h/a small, is

$$B_0 \simeq \tfrac{1}{3}(ka)^3\left[i\gamma_\kappa - 3(1+i)(\gamma - 1)\frac{d_h}{a} \right]$$

$$B_0^e \simeq 1 + \tfrac{1}{3}(ka)^3 \qquad C_0 \simeq (1-i)\left[(\gamma - 1)\frac{a}{\alpha\gamma d_h} \right]$$

where $\gamma_\kappa = (\kappa_e - \kappa)/\kappa$ from Eq. (8.1.11), and $d_h = \sqrt{2K/\rho\omega C_p}$ from Eq.

(6.4.31).   We note that the reflection coefficient

$$R_0 \simeq -\left[1 - (1+i)(ka)^3(\gamma - 1)\frac{d_h}{a} - \tfrac{2}{3}i(ka)^3\gamma_\kappa\right]$$

has a magnitude squared (to the first order in kd_h and γ_κ),

$$|R_0|^2 \simeq 1 - 2(\gamma - 1)(ka)^2kd_h$$

which is smaller than unity, representing energy absorption.

To satisfy boundary conditions for the $m = 1$ terms, we must include a shear wave in the boundary layer just outside the sphere, a solution of Eqs. (6.4.17) and (6.4.29) with $k_v^2 = 2i/d_v^2$, d_v as defined in Eqs. (6.4.31).   The r and ϑ components of the shear wave required to fit the boundary conditions are

$$u_{tr} = \frac{2AV}{k_v r}\cos\vartheta\, h_1(k_v r) \to 2iAV\cos\vartheta\,\frac{kd_h^2}{2k_v r^2}e^{(i-1)r/d_v}$$

$$u_{t\vartheta} = \frac{-AV}{k_v r}\frac{d}{dr}[rh_1(k_v r)]\sin\vartheta \to (1+i)AV\sin\vartheta\,\frac{d_v}{2r}e^{(i-1)r/d_v}$$

where u_ϑ is the velocity component tangential to the sphere in the rz plane. Again we can use the asymptotic form for $h_1(k_v r)$ because d_v is so small. That this wave is purely transverse can be demonstrated by showing that

$$\text{div }\mathbf{u} \equiv \frac{1}{r^2}\frac{\partial}{\partial r}(r^2 u_r) + \frac{1}{r\sin\vartheta}\frac{\partial}{\partial\vartheta}(\sin\vartheta u_\vartheta) = 0$$

It also can be shown that curl (curl $\mathbf{u}$) $= k_v^2\mathbf{u}$.

The boundary conditions at $r = a$ for these waves are p continuous:

$$B_1^e j_1(k_e a) = 3ij_1(ka) + B_1 h_1(ka)$$

or

$$\tfrac{1}{3}k_e aB_1^e \simeq ika - \frac{i}{k^2 a^2}B_1$$

τ zero:

$$\frac{\gamma - 1}{\gamma\alpha}\left[3ij_1(ka) + B_1 h_1(ka)\right] - (1-i)\frac{C_1 d_h}{2a} = 0$$

or

$$ika\frac{\gamma - 1}{\gamma\alpha}\left(1 - \frac{B_1}{k^3 a^3}\right) \simeq (1-i)\frac{C_1 d_h}{2a}$$

u_r continuous:

$$\frac{1}{i\rho_e c_e}B_1^e j_1'(k_e a) = \frac{1}{i\rho c}\left[3ij_1'(ka) + B_1 h_1'(ka) + \alpha\gamma\frac{kd_h^2}{2a}C_1 - \frac{d_v^2}{a^2}D_1\right]$$

or

$$\frac{1}{3}\frac{\rho c}{\rho_e c_e}B_1^e \simeq i + \frac{2iB_1}{k^3 a^3} + \alpha\gamma\frac{kd_h^2}{2a}C_1 - \frac{d_v^2}{a^2}D_1$$

where
$$C_1 = E_1 e^{(i-1)a/d_h} \quad \text{and} \quad D_1 = \rho c V e^{(i-1)a/d_v}$$

u_9 continuous:

$$\frac{-B_1^e}{i\rho_e c_e k_e a} j_1(k_e a) = \frac{-1}{i\rho c k a} [3ij_1(ka) + B_1 h_1(ka) - \tfrac{1}{2}(1-i)kd_v D_1]$$

or

$$\frac{1}{3}\frac{\rho c}{\rho_e c_e} B_1^e \simeq i - i\frac{B_1}{k^2 a^2} - \tfrac{1}{2}(1-i)\frac{d_v}{a} D_1$$

These suffice to calculate the various coefficients; the approximate formulas for ka small are

$$B_1 \simeq -\tfrac{1}{3}(ka)^3 \gamma'_\rho \left[1 - (1+i)\frac{3\rho_e}{2\rho_e + \rho}\frac{d_v}{a}\right]$$

$$B_1^e \simeq \frac{9i\rho_e}{2\rho_e + \rho}\frac{c_e}{c}\left[1 - \tfrac{1}{3}(1+i)\gamma'_\rho \frac{d_v}{a}\right]$$

$$D_1 \simeq -(1-i)\frac{\delta'_\rho a}{d_v}\left[1 - \tfrac{1}{3}(1+i)\frac{d_v}{a}\gamma'_\rho\right]$$

$$C_1 \simeq (1+i)\frac{\gamma-1}{\gamma\alpha}\frac{3i\rho_e}{2\rho_e + \rho}\frac{ka^2}{d_h}\left[1 - \tfrac{1}{3}(1+i)\gamma'_\rho \frac{d_v}{a}\right]$$

where
$$\gamma'_\rho = \frac{3\rho_e - 3\rho}{2\rho_e + \rho}$$

We note that the thermal wave for $m = 1$ can be neglected; its magnitude C_1 and its contribution to the scattered wave are another order of magnitude smaller than the viscous effects for $m = 1$. This is in contrast to the $m = 0$ wave, for which the viscous wave is nonexistent and the only energy loss comes from thermal diffusion. The $m = 0$ wave is primarily a compressive mode, to which thermal diffusion responds; the $m = 1$ wave is primarily one of motion of the sphere back and forth along the z axis, which produces viscous energy loss. Next we note that the viscous loss cannot be computed by using Eq. (8.2.17) to determine the absorption cross section, as can be done for the thermal loss; the viscous loss is caused by shear motion, and it must be computed by using Eq. (6.4.38). Using the expressions for u_t and D_1, we see that the viscosity contribution to the absorption cross section $\Sigma_a = \rho c/|A|^2$ (power lost) is

$$\frac{1}{2}\frac{\rho c}{|A|^2}\rho\omega d_v \iint |u_{t3}|^2 \, dA \simeq \tfrac{9}{2}kd_v \gamma'^2_\rho 2\pi a^2 \int \sin^3 \vartheta \, d\vartheta = 12\pi ka^2 d_v \gamma'^2_\rho$$

Therefore the expressions for pressure and velocity at the surface of the sphere, the scattered wave, and the cross sections, appropriate for long wavelengths, are

$$p_s(r,\vartheta) \to A \frac{e^{ikr}}{r} \Phi(\vartheta) \qquad \gamma_\kappa = \frac{\kappa_e - \kappa}{\kappa} \qquad \gamma_\rho' = \frac{3\rho_e - 3\rho}{2\rho_e + \rho}$$

$$\Phi(\vartheta) \simeq \tfrac{1}{3}k^2a^3 \left\{ \gamma_\kappa - 3(1-i)(\gamma-1)\frac{d_h}{a} \right.$$
$$\left. + \gamma_\rho' \cos\vartheta \left[1 - (1+i)\frac{3\rho_e}{2\rho_e + \rho}\frac{d_v}{a} \right] \right\}$$

$$\Sigma_s \simeq \tfrac{4}{9}\pi a^2(ka)^4 \left\{ (\gamma_\kappa)^2 - 6(\gamma-1)\gamma_\kappa \frac{d_h}{a} + \tfrac{1}{3}(\gamma_\rho')^2 \left[1 - \frac{6\rho_e}{2\rho_c + \rho}\frac{d_v}{a} \right] \right\}$$

$$\Sigma_a \simeq 2\pi a^2(\gamma-1)kd_h + \tfrac{4}{3}\pi a^2(\gamma_\rho')^2 kd_v \qquad (8.2.22)$$

$$p_\omega(a,\vartheta) \simeq A + ikaA \frac{3\rho_e}{2\rho_e + \rho} \left[1 - \tfrac{1}{3}(1+i)\gamma_\rho'\frac{d_v}{a} \right] \cos\vartheta$$

$$u_r \simeq \frac{A}{\rho c} \left[\tfrac{1}{3}ika\frac{\kappa_e}{\kappa} - (\gamma-1)kd_h \left\{ \begin{matrix} 1+i \\ -(1+3i) \end{matrix} \right\} \right]$$
$$+ \frac{A\cos\vartheta}{\rho c} \left[\frac{3\rho}{2\rho_e + \rho} + (1+i)\frac{\gamma_\rho'}{2\rho_e + \rho}\frac{d_v}{a} \left\{ \begin{matrix} (2\rho_e) \\ (-\rho) \end{matrix} \right\} \right]$$

$$u_\vartheta \simeq \frac{A\sin\vartheta}{\rho c} \left[\frac{-3}{2\rho_e + \rho} + (1+i)\frac{\gamma_\rho'}{2\rho_e + \rho}\frac{d_v}{a} \right] \left\{ \begin{matrix} \rho_e \\ \rho \end{matrix} \right\}$$

where $d_v = \sqrt{2\mu/\rho\omega}$ and $d_h = \sqrt{2K/\rho\omega C_p}$, μ being the viscosity of the air, K its thermal conductivity, and C_p its specific heat at constant pressure.   For the two velocity components, the upper factors in the braces are the values just outside the boundary layer, and the lower factors are to be used in computing the velocity at the surface $r = a$.

We see that the $m = 0$ quantities are modified by the thermal conduction, and the $m = 1$ terms, by the viscosity of the medium outside the sphere. The angle distribution of the scattered wave is only slightly modified from that given in Eqs. (8.2.19), which neglects these losses, but the absorption cross section is of course entirely dependent on their presence [unless ρ_e or κ_e is complex, in which case these terms must be added, as in Eqs. (8.2.19)]. The sphere expands and contracts with an amplitude proportional to κ_e/κ. It also moves back and forth along the z axis with a velocity amplitude

$$U \simeq \frac{3A}{(2\rho_e + \rho)c} \left[1 - \tfrac{1}{3}(1+i)\gamma_\rho'\frac{d_v}{a} \right]$$

this amplitude being slightly diminished because of the viscosity.   The air clings to it as it moves.   The part of the pressure there which is proportional to $\cos\vartheta$ is in phase with the part of the radial velocity which is proportional to $\cos\vartheta$, but the spherically symmetric parts of the surface pressure and radial velocity at the surface are not in phase, so that energy is lost, both in the thermal boundary layer and inside the sphere itself (by thermal conduction into the sphere). The tangential motion of the air increases through the viscous boundary layer, corresponding to the shear action there. Just outside the layer the amplitude of the tangential velocity, relative to the moving sphere, is proportional to $\rho_e - \rho$: the air outside moves faster than the sphere if $\rho_e > \rho$; slower than the sphere if the medium's density is greater than that of the sphere. The viscous-loss term in the absorption cross section is therefore proportional to $(\gamma'_\rho)^2$, that is, to $(\rho_e - \rho)^2$.

Cloud of scatterers

When sound traverses a cloud of small objects, such as air bubbles in water or fog droplets in air, each object produces a scattered wave, and these wavelets reinforce in some directions and interfere in some others. The wave incident on each scatterer is affected because of the presence of the others, which in turn affects the shape of the scattered wave. This gives rise to coherent, incoherent, and multiple scattering and, in cases where the scatterers are regularly spaced, to diffraction.

Suppose the dimensions of the scattering objects are all quite small compared with the wavelength λ; furthermore, that the mean distance between scatterers is not larger than λ. Next, suppose that the dimensions of the region R, inhabited by the scatterers, are considerably larger than λ and, finally, that the distance from R to the point of measurement is much larger than the dimensions of R (and thus is very much larger than λ). All this is usually the case with mists or with bubble clouds. Also suppose that the scatterers are all spheres; that there are, on the average, N of them per unit volume inside R; that the nth sphere has radius a_n, compressibility κ_n, density ρ_n; and that its center is at the point represented by vector $\mathbf{r}_n$. The fluid outside R and in the space between scatterers inside R has compressibility κ_0 and density ρ_0. The incident wave will be the plane wave $Ae^{ikz} = Ae^{i\mathbf{k}_i \cdot \mathbf{r}}$, where k_i points along the z axis.

In the integral of Eqs. (8.1.19) the quantities $(\kappa_n/\kappa_0) - 1$ and $1 - (\rho_0/\rho_n)$ (we are not considering turbulence here; so $\Re = 0$) differ from zero inside each of the scattering spheres inside R. The scattered wave at r will thus be a sum of waves arising from the various small spheres, each having a relative phase and amplitude determined by the particular sphere's size and position. If the spheres are regularly placed, these phases will all add up, in certain directions, to produce diffracted beams, as with the Bragg diffraction of x-rays from crystals. However, the objects scattering sound waves are usually distributed at random, in which case no reinforcement occurs except in the direction of the incident wave.

In this forward direction the reinforced scattering modifies the incident wave as though it were traveling through a region R of different index of refraction. To see this, let us separate the compressibility and density factors γ_κ and γ_ρ in the integrand of Eqs. (8.1.19) into their mean values, throughout the region R, plus the variation of these factors from their mean values. The mean values are equal to the sums $\sum [(\kappa_n/\kappa_0) - 1] \cdot \frac{4}{3}\pi a_n{}^3$ and $\sum [1 - (\rho_0/\rho_n)] \cdot \frac{4}{3}\pi a_n{}^3$ over each of the N spheres per unit volume in R.

These sums are responsible for the coherent scattering. Considering only the average terms for γ_κ and γ_ρ for the moment, Eq. (8.1.13), inside region R, has the form

$$p_\omega(\mathbf{r}) = p_i(\mathbf{r}) + \left[\sum_n \tfrac{4}{3}\pi a_n{}^3 \left(\frac{\kappa_n}{\kappa_0} - 1 \right) \right] \iiint k^2 p_\omega(\mathbf{r}_0) g_\omega(\mathbf{r} \mid \mathbf{r}_0) \, dv_0$$

$$- \left[\sum_n \tfrac{4}{3}\pi a_n{}^3 \left(1 - \frac{\rho_0}{\rho_n} \right) \right] \iiint g_\omega(\mathbf{r} \mid \mathbf{r}_0) \, \nabla_0{}^2 p_\omega(\mathbf{r}_0) \, dv_0$$

where $k^2 = \rho_0\kappa_0\omega^2$. This is the equation for a wave, induced by the incident wave p_i, which has an equation of motion $\nabla^2 p_\omega + k^2 p_\omega = 0$ in the region outside R and an equation of motion

$$\nabla^2 p_\omega + k^2 p_\omega = \left[\sum_n \left(\frac{\rho_0}{\rho_n} - 1 \right) \tfrac{4}{3}\pi a_n{}^3 \right] \nabla^2 p_\omega + \left[\sum_n \left(\frac{\kappa_n}{\kappa_0} - 1 \right) \tfrac{4}{3}\pi a_n{}^3 \right] k^2 p_\omega$$

inside R. This last equation is that for sound waves in a medium of density and compressibility given by

$$\rho_R = \left[\frac{1}{\rho_0} + \sum_{n=1}^N \tfrac{4}{3}\pi a_n{}^3 \left(\frac{1}{\rho_n} - \frac{1}{\rho_0} \right) \right]^{-1} \qquad \kappa_R = \kappa_0 + \sum_{n=1}^N \tfrac{4}{3}\pi a_n{}^3 (\kappa_n - \kappa_0)$$

$$(8.2.23)$$

The effective compressibility is the average of the compressibility over each unit volume of R; the effective density is the reciprocal of the average value of $1/\rho$ over each unit volume.

Thus, as far as the average terms in the integrand are concerned, region R behaves as though it had a uniform density and compressibility which differ from that of the medium outside R. This difference will produce scattering (or reflection or refraction, whichever term seems more appropriate) of sound incident on R. For example, if R is a spherical region of radius a_R, Eq. (8.2.15) can be used, with Eq. (8.2.9) used to calculate the effective admittances β_m in terms of a_R, ρ_R, and κ_R. In the very long wavelength limit of Eqs. (8.2.19), the scattered intensity is proportional to the squares of the summations for ρ_R and κ_R in Eqs. (8.2.23), in other words, to the square of the number N of scattering particles per unit volume in R. This cooperative portion of the scattered wave, in which each scatterer adds its contribution to the wave amplitude (and thus the intensity is proportional to N^2), is called the *coherent* scattered wave. The rest of the scattering, the intensity of which we shall see is proportional to N rather than to N^2, is called the *incoherent* scattered wave.

Incoherent scattering

The incoherent scattering is, of course, generated by the variable parts of the functions κ and ρ, after the constant average values have been taken into account by the change of index of refraction just worked out and given in Eqs. (8.2.23). The nth little sphere will produce its scattered wave as though it were a sphere of radius a_n, with density and compressibility ρ_n and κ_n, in a medium of density ρ_R and κ_R throughout R. The incident wave will have been refracted by the change of index of refraction as it enters R, but if R is large and the interface is a plane, it will have the form $A_R e^{ik_R z}$ inside R, where $k_R^2 = \rho_R \kappa_R \omega^2$. The wave scattered from the nth sphere will have the form $A_R(e^{ik_R r}/r)\Phi_n(\vartheta)$ inside R, but by the time it has emerged from R and reached the measurement point, its wavenumber will have changed back from k_R to $k = \omega\sqrt{\rho_0\kappa_0}$, with corresponding (but very small, if k_R does not differ much from k) change in amplitude A.

The easiest way to compute the incoherent scattering, if the N scatterers are distributed at random throughout R, is to use the correlation function of the variable parts of the compressibility and density factors γ_κ and γ_ρ. If all scatterers are spheres, we can use Eqs. (8.2.19) to ensure accuracy even when κ_n and ρ_n differ greatly from κ_0 and ρ_0 (as long as $a_n \ll \lambda$) and obtain an expression for the pressure at very large distances from R.

$$p(\mathbf{r}) \to A_i e^{ikz} + \sum_n \tfrac{4}{3}\pi a_n^3(\gamma_{\kappa n} + \gamma_{\rho n}\cos\vartheta)k_R^2 p(\mathbf{r}_n)(e^{ik_R R_n}/4\pi R_n)$$

where

$$\gamma_{\kappa n} = \frac{\kappa_n - \kappa_R}{\kappa_R} \qquad \gamma_{\rho n} = \frac{3\rho_n - 3\rho_R}{2\rho_n + \rho_R}$$

$\mathbf{r}_n$ is the radius vector to the center of the nth sphere from the center of R, and $R_n = |\mathbf{r} - \mathbf{r}_n|$. This expresses the incoherent scattering as the sum of wavelets scattered from each sphere. But it would be more convenient, for our present purposes, to express the scattered wave as an integral of the form of Eq. (8.1.14).

We have already included the average values of the factors γ_κ and γ_ρ in the index of refraction of region R, which gives rise to the coherent scattering. The variable parts of the γ's may be expressed in terms of the function

$$\delta(\mathbf{r}) = \begin{cases} \dfrac{\kappa_n - \kappa_R}{\kappa_R} + \dfrac{3\rho_n - 3\rho_R}{2\rho_n + \rho_R}\cos\vartheta & (\mathbf{r}\ \text{inside the } n\text{th sphere}) \\[3mm] \dfrac{\kappa_0 - \kappa_R}{\kappa_R} + \dfrac{3\rho_0 - 3\rho_R}{2\rho_0 + \rho_R}\cos\vartheta & (\text{inside } R, \text{ outside a sphere}) \end{cases} \qquad (8.2.24)$$

which has average value zero over R, by definition, since we have subtracted the average values of the γ's. This function can be defined in terms of its correlation function

$$\Upsilon(\mathbf{d}) = \frac{1}{V}\iiint_R \delta(\mathbf{r})\delta(\mathbf{r} + \mathbf{d})\,dv$$

where V is the volume of region R. As usual, the value of $\Upsilon(0)$ is the mean-square value of function $\delta(\mathbf{r})$,

$$\langle \delta^2 \rangle \simeq N(\tfrac{4}{3}\pi a^3)\,|\gamma_\kappa + \gamma_\rho \cos \vartheta|^2 \tag{8.2.25}$$

where N is the mean number of spheres per unit volume in R, a is their mean radius, and γ_κ and γ_ρ are the mean values of their factors $(\kappa_n - \kappa_0)/\kappa_0$ and $(3\rho_n - 3\rho_0)/(2\rho_n + \rho_0)$, respectively. However, since the mean value of $\delta(\mathbf{r})$ is zero, it must be that the integral of $\Upsilon(\mathbf{d})$ over R is zero. Thus $\Upsilon(\mathbf{d})$ cannot simply fall off exponentially to zero as d increases; it must go negative to a sufficient extent so that its average value is zero. The simplest form satisfying these requirements is

$$\Upsilon(\mathbf{d}) = \langle \delta^2 \rangle \left(1 - \frac{d^2}{3a^2}\right) \exp\left[-\frac{1}{2}\left(\frac{d}{a}\right)^2\right] \tag{8.2.26}$$

where a is a scale factor proportional to, perhaps nearly equal to, the mean radius of the scattering spheres.

If the density of scattering particles is sufficiently small so that the chance of multiple scattering is negligible, we can use the Born approximation, as with Eq. (8.1.18). The incoherently scattered wave at the observation point is

$$p_{is}(\mathbf{r}) \to \frac{A_i e^{ikr}}{4\pi r}\,k_R^2 \iiint\limits_R \delta(\mathbf{r}_0) e^{-i\boldsymbol{\mu} \cdot \mathbf{r}_0}\,dv_0 \qquad \mu = 2k_R \sin \frac{\vartheta}{2}$$

and thus the incoherently scattered intensity at $\mathbf{r}$, divided by the incident intensity $|A_i|^2/\rho c$, is $(4\pi^4 k_R^4/r^2)\,|\Delta(\boldsymbol{\mu})|^2$, where $\Delta(\mathbf{K})$ is the Fourier transform

$$\Delta(\mathbf{K}) = \frac{1}{8\pi^3} \iiint\limits_R \delta(\mathbf{r}) e^{-i\mathbf{K}\cdot\mathbf{r}}\,dv \tag{8.2.27}$$

Note that $\Delta(0) = 0$. We know, from Eq. (8.1.26), that $|\Delta(\mathbf{K})|^2$ is proportional to the Fourier transform of the correlation function of $\delta(\mathbf{r})$.

$$|\Delta(\mathbf{K})|^2 = \frac{V}{64\pi^6} \iiint \delta(\mathbf{d}) e^{-i\mathbf{K}\cdot\mathbf{d}}\,dv_d$$

$$= \langle \delta^2 \rangle \frac{VK^2 a^5}{3(2\pi)^{\frac{9}{2}}}\, e^{-\frac{1}{2}K^2 a^2} \tag{8.2.28}$$

Therefore the incoherently scattered intensity, per unit incident intensity, at angle ϑ to the incident wave, for frequency $\omega/2\pi$, is (for $K = \mu$)

$$\frac{I_{is}}{I_i} \simeq NV \frac{\sqrt{2\pi}}{9} \frac{k_R^4 a^6}{r^2}\,|\gamma_\kappa + \gamma_\rho \cos \vartheta|^2\, (\tfrac{1}{8}k_R^2 a^2 \sin^2 \tfrac{1}{2}\vartheta) \exp\left(-2k_R^2 a^2 \sin^2 \tfrac{1}{2}\vartheta\right)$$

$$\tag{8.2.29}$$

to the approximation which neglects multiple scattering. If thermal and viscous losses are included, we use the quantity in braces in the Φ of Eqs. (8.2.22) instead of $(\gamma_\kappa + \gamma_\rho \cos \vartheta)$.

If we had just used Eqs. (8.2.19), assuming that each of the NV scatterers in R scattered sound independently of the others and that each wavelet did not interfere with any other, so that we could add intensities, the ratio of scattered

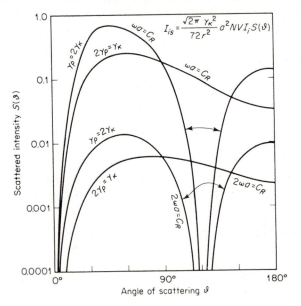

$$I_{is} = \frac{\sqrt{2\pi}\,\gamma_\kappa^2}{72\,r^2}\,a^2 NVI_i S(\vartheta)$$

FIGURE 8.6
Incoherent scattering from a cloud of particles of radius a, with compressibility factors γ_κ and density factors γ_ρ.

to incident power at r would have the value

$$\left(\frac{I_{is}}{I_i}\right)_0 \simeq \tfrac{1}{9} NV \frac{k_0^4 a^6}{r^2}\,|\gamma_\kappa + \gamma_\rho \cos\vartheta|^2$$

Equations (8.2.29) show that the interference between the wavelets modifies the distribution-in-angle of the scattered wave, reducing it in the forward direction and augmenting it to the side and backward, so that the power fraction scattered is

$$\left(\frac{I_{is}}{I_i}\right)_0 \cdot \tfrac{1}{8}\sqrt{2\pi}\left(\frac{k_R}{k_0}\right)^4 (k_R\,a\,\sin\tfrac{1}{2}\vartheta)^2 \exp\left(-2k_R^2 a^2 \sin^2\tfrac{1}{2}\vartheta\right)$$

Plots of the incoherent scattering are shown in Fig. 8.6 ($k_R = \omega/c_R$).

This formula holds only if region R is small enough, or is so sparsely populated with scatterers that the total power scattered out of the incident beam is a small fraction of the power incident on R. According to the discussion on page 407, the power scattered away from the incident beam in traversing a distance dz of R is I_i times $N\Sigma_s\,dz$, and the power absorbed by the scatterers is $I_i N\Sigma_a\,dz$. Thus the incident intensity a distance z inside R is not I_{i0}, the intensity at the start of penetration, but $I_{i0}\exp\left[-Nz(\Sigma_s + \Sigma_a)\right]$. But Eq. (8.2.29) was derived assuming that I_i is constant throughout R, which means that we have assumed that $Nl(\Sigma_s + \Sigma_a)$, where l is the distance along the beam from one side of R to the other, is small compared with unity. If this is not the case, much of the scattered wave will be rescattered before it emerges from R; multiple scattering will be important. Considerations of space persuade us to omit this complex subject.

Whether multiple scattering is important or not, the attenuation of the *incident* wave in traversing a distance z of region R is given by the factor $\exp\left[-Nz(\Sigma_s + \Sigma_a)\right]$, with Σ_s and Σ_a given by Eqs. (8.2.19), (8.2.20), or (8.2.22) for small scatterers, by Eq. (8.2.18) for larger ones, and with N the mean number of scatterers per unit volume of R. In a distance dz the fraction of *incident* energy absorbed by the scatterers is $N\Sigma_a\,dz$, and the fraction which is scattered in directions other than the incident direction is $N\Sigma_s\,dz$. When multiple scattering is important, some energy is absorbed by the scattered wave before it emerges from R, and some of the scattered wave is rescattered, but these formulas still tell what happens to the incident wave, the part of the wave which has not yet been affected by a scatterer.

8.3 SCATTERING FROM SURFACE IRREGULARITIES

When a plane wave of sound impinges on a surface, large in extent with respect to a wavelength but irregular either in shape or in regard to its reaction to sound, the wave returning from the surface will be distorted because of the irregularities. The simplest example of these effects occurs for a plane reflecting surface, as shown in Fig. 8.7. We saw in Sec. 6.3 that a plane wave of amplitude P_i, incident at angle ϑ_i on a smooth boundary plane of uniform point impedance $\rho c/\beta$, is reflected as a plane wave at angle

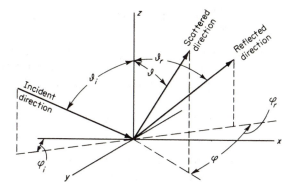

FIGURE 8.7
Directions and angles for reflection and scattering of a plane wave incident on the xy plane.

$\vartheta_r = \vartheta_i$ to the z axis, with amplitude $P_i C_r$, where

$$C_r = \frac{\cos \vartheta - \beta}{\cos \vartheta_i + \beta}$$

If either the specific admittance β changes from point to point along the plane or if the boundary itself is not a perfect plane, the reflected wave will not be plane.   If the surface deviation is not large, the wave reflected may be represented as a reflected plane wave plus a scattered wave, which measures the degree of distortion.

Irregularities in surface admittance

Suppose the xy plane has point impedance $\rho c / \beta_0$ everywhere except over an area A, which can be circumscribed by a circle of radius a, with center at the origin.   Within A the surface admittance is $\beta(x,y)$, different from β_0.   As before, we carry out the analysis by means of the Green's function of Fig. 7.6 and Eqs. (7.4.10) and (7.4.11), corresponding to a source point at (x_0,y_0,z_0) (with $z_0 \geqslant 0$) plus an image point at $(x_0, y_0, -z_0)$, producing a reflected wave of amplitude modified by the factor $(\cos \vartheta' - \beta_0)/(\cos \vartheta' + \beta_0)$, corresponding to the uniform admittance β_0 over the whole xy plane.   This Green's function satisfies the boundary condition

$$(\partial G/\partial z)_{z=0} = -ik\beta_0 G(x,y,0 \mid x_0,y_0,z_0)$$

The acoustic pressure wave, on the other hand, must satisfy the boundary conditions $(\partial p/\partial z)_{z=0} = -ik\beta p_{z=0}$ within area A; $(\partial p/\partial z)_{z=0} = -ik\beta_0 p_{z=0}$ over the rest of the boundary plane.

If there were no irregularity in surface admittance, β being β_0 everywhere, the complete solution, for the case of a plane incident wave at angle ϑ_i, in a plane at angle φ_i to the xz plane, would be the combination of incident and reflected waves,

$$p_i(r) = P_i \exp \left[ik \sin \vartheta_i(x \cos \varphi_i + y \sin \varphi_i) - ikz \cos \vartheta_i\right]$$

$$+ P_i \frac{\cos \vartheta_i - \beta_0}{\cos \vartheta_i + \beta_0} \exp \left[ik \sin \vartheta_i(x \cos \varphi_i + y \sin \varphi_i) + ikz \cos \vartheta_i\right] \quad (8.3.1)$$

times the factor $e^{-i\omega t}$.   This is an obvious generalization of Eq. (6.3.7). When $\beta \neq \beta_0$ inside A, the complete solution may be obtained from Eq. (7.1.17) as p_i plus a surface integral over area A.

$$p(\mathbf{r}) = p_i(\mathbf{r}) + ik \iint_A (\beta - \beta_0)p(x_0,y_0,0)G(x,y,z \mid x_0,y_0,0) \, dx_0 \, dy_0 \quad (8.3.2)$$

This is an inhomogeneous integral equation for p, which must usually be solved by approximation techniques.

The far field

In order to measure the scattered wave, represented by the integral, separately from the incident-reflected wave, the measurement point r, ϑ, φ must be far from the area A, usually far enough so that the approximation

$$G \simeq \frac{e^{ikr}}{4\pi r} \left[e^{-ikz_0 \cos \vartheta} + \frac{\cos \vartheta - \beta_0}{\cos \vartheta + \beta_0} e^{ikz_0 \cos \vartheta} \right]$$

$$\times \exp\left[-ik \sin \vartheta \left(x_0 \cos \varphi + y_0 \sin \varphi\right)\right] \quad (8.3.3)$$

is valid. To compute the asymptotic form for the scattered wave, we need only insert this form for G, times a good approximation for the pressure $p(x_0,y_0,0)$ on the area A, into the integral.

Well outside A this surface pressure will be approximately equal to $p_i(x_0,y_0,0)$, having amplitude $2P_i \cos \vartheta_i/(\cos \vartheta_i + \beta_0)$ and phase factor $\exp[ik \sin \vartheta_i(x \cos \varphi_i + y \sin \varphi_i)]$, whereas, if A is very large, the surface pressure well inside A will be approximately $2P_i \cos \vartheta_i/(\cos \vartheta_i + \beta_a)$ times the same phase factor, if β inside A does not differ much from its mean value β_a inside A. Probably a good approximation is to use this last expression for $p(x_0,y_0,0)$ everywhere inside A.

Thus an approximate formula for the scattered wave, at large distances from the scattering area, is

$$p_s(\mathbf{r}) \to P_i \Phi(\vartheta_i,\varphi_i \mid \vartheta,\varphi) \frac{e^{ikr}}{r} \qquad r \gg a$$

$$\Phi \simeq \frac{ik \cos \vartheta \cos \vartheta_i}{2\pi(\cos \vartheta + \beta_0)(\cos \vartheta_i + \beta_a)} \iint_A (\beta - \beta_0) e^{i\mu_x x_0 + \mu_y y_0} \, dx_0 \, dy_0 \tag{8.3.4}$$

where

$$\mu_x = k\left(\sin \vartheta_i \cos \varphi_i - \sin \vartheta \cos \varphi\right)$$

$$\mu_y = k(\sin \vartheta_i \sin \varphi_i - \sin \vartheta \sin \varphi)$$

$$k^2\gamma^2 = \mu_x{}^2 + \mu_y{}^2 = k^2[\sin^2 \vartheta_i + \sin^2 \vartheta - 2\sin \vartheta_i \sin \vartheta \cos (\varphi - \varphi_i)]$$

Thus the scattered amplitude is approximately proportional to the two-dimensional Fourier transform of the admittance difference $\beta - \beta_0$ for area A. If we define vector $\mathbf{k}_r = (k \sin \vartheta_i \cos \varphi_i, k \sin \vartheta_i \sin \varphi_i, k \cos \vartheta_i)$ as the wavenumber vector, of magnitude k, in the direction of propagation of the reflected wave, and $\mathbf{k}_s = (k \sin \vartheta \cos \varphi, k \sin \vartheta \sin \varphi, k \cos \vartheta)$ as the wavenumber vector in the direction of the point of observation, then the quantity in the exponent in the integral is $i\boldsymbol{\mu} \cdot \mathbf{r}_0$, where $\mathbf{r}_0 = (x_0,y_0,0)$ is the radius vector to a point on the scattering area A, and $\boldsymbol{\mu} = \mathbf{k}_r - \mathbf{k}_s$ is the recoil vector of Eq. (8.1.20). Thus $k\gamma$ is the length of the projection of vector $\boldsymbol{\mu}$ on the xy plane.

In general (but not always), the integral, and thus the amplitude of the scattered wave, is largest when $\boldsymbol{\mu} = 0$, that is, in the direction of the reflected wave.   When the dimensions of A are small compared with the wavelength, the scattered wave is nearly isotropic; when A is large compared with λ, the scattered wave is (usually) concentrated in the direction of the reflected wave.

To illustrate these generalities we compute the angle-distribution factor $\Phi(\vartheta_i, \varphi_i \mid \vartheta, \varphi)$ for a few simple distributions of admittance.   The first is a rectangular patch of length l in the x direction and width w in the y direction, with β constant (and different from β_0) over the whole area lw.   The angle-distribution factor is

$$\Phi_\beta \simeq \frac{iklw(\beta - \beta_0)\cos\vartheta_i\cos\vartheta}{2\pi(\cos\vartheta_i + \beta)(\cos\vartheta + \beta_0)}\left[\frac{\sin(\mu_x l/2)}{(\mu_x l/2)}\frac{\sin(\mu_y w/2)}{\mu_y w/2}\right] \quad (8.3.5)$$

Referring to the definition of $\boldsymbol{\mu}$ and of γ, we see that if both kl and kw are less than $\frac{1}{2}\pi$, the quantity in brackets changes very little with ϑ or φ, so the scattered wave is nearly independent of ϑ and φ.   On the other hand, if $kl \gg \frac{1}{2}\pi$ and $kw < \frac{1}{2}\pi$, then the scattered amplitude is only large when $\gamma_x = \sin\vartheta_i\cos\varphi_i - \sin\vartheta\cos\varphi$ is much smaller than unity, i.e., when the vector $\boldsymbol{\mu} = \mathbf{k}_r - \mathbf{k}_s$ is nearly perpendicular to the x axis (the long axis of the rectangle); in other words, the scattered wave is confined mainly to the plane defined by the reflection vector k_r and the y axis.   If both l and w are large compared with a wavelength, the scattered wave is concentrated mostly in the direction of reflection.

If the patch of admittance β is a circular area of radius a, the angle factor is

$$\Phi_\beta \simeq \frac{ik\pi a^2(\beta - \beta_0)\cos\vartheta_i\cos\vartheta}{2\pi(\cos\vartheta_i + \beta)(\cos\vartheta + \beta_0)}\left[\frac{2J_1(k\gamma a)}{k\gamma a}\right] \quad (8.3.6)$$

the only difference from Eq. (8.3.5) being in the factor in brackets, which is more symmetric, depending on $\varphi - \varphi_i$ instead of on φ and φ_i separately. Figure 8.8 shows plots of $|\Phi_\beta|$ for the incident wave at 45° on a circular "soft spot" of radius a.   Contours of constant amplitude of Φ_β are plotted on the ϑ, φ surface, for two different values of $ka = 2\pi a/\lambda$, to show the variation of scattered pressure amplitude with angle of scattering.   For the longer wavelength, the scattered wave is not much dependent on ϑ or φ; for $ka = 4$, the scattered wave is concentrated in a peak around the direction of the reflected wave, at $\varphi = 0°$, $\vartheta = 45°$, indicated by the small circle on both plots.   In the latter case the scattered wave has a node ($|\Phi_\beta| = 0$) and one diffraction band (where Φ_β changes sign).

An example where the scattered wave is not most intense in the direction of the reflected wave is that where β is periodic, forming a "diffraction grating,"

$$\beta - \beta_0 = B\cos\frac{2\pi x}{d} \quad \text{for } -\tfrac{1}{2}l < x < \tfrac{1}{2}l, \ -\tfrac{1}{2}w < y < \tfrac{1}{2}w$$

In this case, integration of Eq. (8.3.4) produces

$$\Phi_\beta \simeq \frac{iklwB \cos \vartheta_i \cos \vartheta}{2\pi(\cos \vartheta_i + \beta_0)(\cos \nu + \beta_0)} \left\{ \frac{\sin \left[\frac{1}{2}\mu_x l - (\pi l/d)\right]}{\frac{1}{2}\mu_x l - (\pi l/d)} \right.$$
$$\left. + \frac{\sin \left[\frac{1}{2}\mu_x l + (\pi l/d)\right]}{\frac{1}{2}\mu_x l + (\pi l/d)} \right\} \frac{\sin (\mu_y w/2)}{\mu_y w/2} \quad (8.3.7)$$

Here the strongest scattered wave is not in the direction of reflection, but rather in the directions of the diffracted beams, for which $\mu_x/2\pi \equiv (1/\lambda) \times (\sin \vartheta_i \cos \varphi_i - \sin \vartheta \cos \varphi) = \pm(1/d)$.

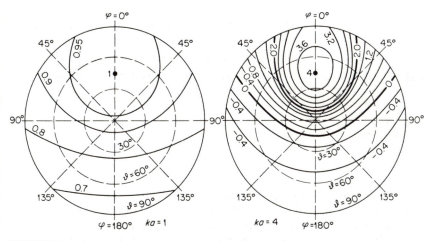

FIGURE 8.8
Contours of equal scattered pressure amplitude, plotted against angles of scattering ϑ, φ for plane wave incident at angles $\vartheta_i = 45°$, $\varphi_i = 0°$, on an absorbing disk of radius a, set in an otherwise rigid plane. See Eq. (8.3.6).

Surface roughness

If the irregularity in area A is in actual shape rather than in admittance, the scattering effect will come from the fact that the boundary condition is now applied for a surface which differs from the xy plane. Suppose the reflecting surface is not $z = 0$ for all values of x and y, but is $z = \xi(x,y)$ within A; suppose also that this surface, at x, y, has admittance $\beta(x,y)$, which may differ from the constant admittance β_0 of the flat surface outside A. The integral equation for the pressure wave is then

$$p(x,y,z) = p_i(x,y,z) + \iint \left[G(x,y,z \mid x_0,y_0,\xi) \frac{\partial}{\partial n_0} p(x_0,y_0,\xi) \right.$$
$$\left. - p(x_0,y_0,\xi) \frac{\partial}{\partial n_0} G(x,y,z \mid x_0,y_0,\xi) \right] dx_0 \, dy_0 \quad (8.3.8)$$

where p_i is the incident-reflected wave of Eq. (8.3.1), G is as given in Eq. (8.3.3), and the normal gradient $\partial/\partial n_0$ is normal to the surface $\xi(x_0, y_0)$.

$$\frac{\partial}{\partial n_0} = -\frac{\partial}{\partial z_0} + \frac{\partial \xi}{\partial x_0}\frac{\partial}{\partial x_0} + \frac{\partial \xi}{\partial y_0}\frac{\partial}{\partial y_0} \qquad \text{if } \operatorname{grad}_0 \xi \ll 1$$

If the displacement ξ of the surface from the xy plane is everywhere less than a wavelength, we can expect the Born approximation to be valid. In this case we can use the approximations

$$p \simeq P_i \frac{2\cos \vartheta_i}{\cos \vartheta_i + \beta}(1 - ik\beta z_0) \exp \left[ik \sin \vartheta_i (x_0 \cos \varphi_i + y_0 \sin \varphi_i)\right]$$

$$G \simeq \frac{e^{ikr}}{4\pi r}\frac{2\cos \vartheta}{\cos \vartheta + \beta_0}(1 - ik\beta_0 z_0) \exp \left[-ik \sin \vartheta (x_0 \cos \varphi + y_0 \sin \varphi)\right]$$

inside the integral; the approximate expression for the incident plus reflected plus scattered wave becomes $p = p_i(\mathbf{r}) + p_s(\mathbf{r})$, where p_i is given in Eq. (8.3.1), and $p_s(\mathbf{r}) = P_i(e^{ikr}/r)\Phi(\vartheta, \varphi)$, with

$$\Phi(\vartheta, \varphi) \simeq \frac{i}{\pi}\frac{\cos \vartheta_i \cos \vartheta}{(\cos \vartheta_i + \beta)(\cos \vartheta + \beta_0)}\iint \left[k(\beta - \beta_0) + \mu_x \frac{\partial \xi}{\partial x_0}\right.$$
$$\left. + \mu_y \frac{\partial \xi}{\partial y_0}\right]e^{i(\mu_x x_0 + \mu_y y_0)}\, dx_0 \, dy_0 \qquad (8.3.9)$$

The components of vector $\boldsymbol{\mu}$ are given in Eq. (8.3.4). This equation for the angle distribution of the scattered wave is a generalization of Eq. (8.3.4) when the area A is rough, as well as uneven in admittance.

Since this integral can be manipulated just as that of Eq. (8.3.4) was, we shall content ourselves with a single example, that of a rectangular, raised area of length l, width w, and height h. In this case the expressions for the components of the gradient of ξ are given in terms of delta functions at the edges.

$$\frac{\partial \xi}{\partial x_0} \simeq h[\delta(x_0 + \tfrac{1}{2}l) - \delta(x_0 - \tfrac{1}{2}l)] \qquad \text{for } -\tfrac{1}{2}w < y_0 < \tfrac{1}{2}w$$

$$\frac{\partial \xi}{\partial y_0} \simeq h[\delta(y_0 + \tfrac{1}{2}w) - \delta(y_0 - \tfrac{1}{2}w)] \qquad \text{for } -\tfrac{1}{2}l < x_0 < \tfrac{1}{2}l$$

The scattering arising from the difference in admittance has already been computed. The additional scattering coming from the change in shape of the boundary is

$$\Phi_\xi(\vartheta_i, \varphi_i \mid \vartheta, \varphi) \simeq \frac{ikwkh \cos \vartheta_i \cos \vartheta}{\pi(\cos \vartheta_i + \beta)(\cos \vartheta + \beta_0)}\left[\gamma^2 \frac{\sin (\mu_x l/2)}{\mu_x l/2}\frac{\sin (\mu_y w/2)}{\mu_y w/2}\right]$$
$$(8.3.10)$$

which does not differ much from the expression of Eq. (8.3.5). Aside from the dimensionless factor kh, measuring the change in shape, which replaces the dimensionless factor $\beta - \beta_0$, there is an additional directional factor,

$$\gamma^2 = \sin^2 \vartheta_i + \sin^2 \vartheta - 2 \sin \vartheta_i \sin \vartheta \cos (\varphi_i - \varphi) \qquad (8.3.11)$$

proportional to the square of the length of the projection of vector $\boldsymbol{\mu}$ on the xy plane. This is small when $\boldsymbol{\mu}$ is small, i.e., for scattering in a direction close to the direction of the reflected wave, and large only when $\boldsymbol{\mu}$ is large and nearly horizontal, i.e., in a direction near that of the negative of the incident wave. Thus the factor γ^2 is small when the rest of the factors in brackets are large, and vice versa, so that this part of the scattering is small for all directions. This is not surprising, for the "roughness" in the example we have chosen is only at the edges of the raised panel; the rest of area A is flat and contributes nothing to this part of the scattering.

The effect of random roughness

A situation of some practical interest involves the scattering of sound incident on a patch of rough surface which also has random variations in surface admittance. In area A the quantities $\beta - \beta_0$ and ξ exhibit random fluctuations, with correlation lengths w_β and w_ξ, respectively. Area A may be the only area of the surface in which $\beta - \beta_0$ and ξ differ from zero, or it may be that area A is the only part of the surface which is "illuminated" by the beam of sound sent out by a distant source. If the beam covers a solid angle Ω, if the axis of the beam is at an angle ϑ_i to the z axis, and if the distance r_s to the source is great, then the area $A = \Omega r_s^2/\cos \vartheta_i$ and the scattered sound will arise from this area. Figure 8.9 shows the quantities involved.

An obvious modification of Eq. (8.3.9) gives, for the scattered pressure reaching point $\mathbf{r}$ at angle ϑ, φ, if the source is producing a steady wave of

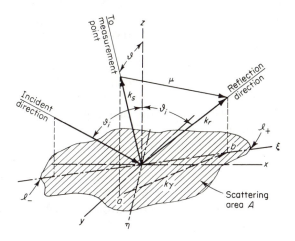

FIGURE 8.9
Directions and axes involved in calculating the scattering from a rough surface..

frequency $\omega/2\pi$, having a pressure amplitude P_i by the time it reaches A,

$$p_s(\mathbf{r},\omega) = P_i \frac{e^{ikr}}{\pi r} \frac{\cos \vartheta_i \cos \vartheta}{(\cos \vartheta_i + \beta_0)(\cos \vartheta + \beta_0)} \iint [ikb(\mathbf{r}_0)$$
$$+ k^2\gamma^2\xi(\mathbf{r}_0)]e^{i\mu_x x_0 + i\mu_y y_0}\, dx_0\, dy_0$$

By now the rest of the procedure is familiar. The function $b = \beta - \beta_0$ is the deviation of the admittance at point $\mathbf{r}_0$ on the surface from the mean value β_0, and ξ is the normal displacement of the boundary surface from its mean shape, the xy plane. Their correlation functions will be, approximately,

$$\Upsilon_\beta(\mathbf{d}) = \langle b^2 \rangle e^{-\frac{1}{2}(d^2/w_\beta{}^2)} \qquad \Upsilon_\xi(\mathbf{d}) = \langle \xi^2 \rangle e^{-\frac{1}{2}(d^2/w_\xi{}^2)}$$

if their fluctuations are random and isotropic. Quantities $\langle b^2 \rangle$ and $\langle \xi^2 \rangle$ are the mean-square values of b and of ξ over the illuminated area A. If, in addition, the two variations are statistically independent, the intensity of the scattered sound turns out to be

$$I_s \simeq 4\pi^2 I_i \frac{A}{r^2} \left| \frac{\cos \vartheta_i \cos \vartheta}{(\cos \vartheta_i + \beta_0)(\cos \vartheta + \beta_0)} \right|^2 \{k^2 |B(k\gamma)|^2 + k^4\gamma^4 |Z(k\gamma)|^2\}$$

$$= I_i \frac{A}{2\pi r^2} \left| \frac{\cos \vartheta_i \cos \vartheta}{(\cos \vartheta_i + \beta_0)(\cos \vartheta + \beta_0)} \right|^2 \{k^2 w_\beta{}^2 \langle b^2 \rangle e^{-\frac{1}{2}k^2\gamma^2 w_\beta{}^2}$$
$$+ k^4\gamma^4 w_\xi{}^2 \langle \xi^2 \rangle e^{-\frac{1}{2}k^2\gamma^2 w_\xi{}^2}\} \quad (8.3.12)$$

where $B(\mathbf{K})$ is the two-dimensional spatial Fourier transform of $b(\mathbf{r})$, and $Z(\mathbf{K})$ is the corresponding transform of $\xi(\mathbf{r})$. The angles and quantities involved are shown in Fig. 8.9. $I_i = |P_i|^2/\rho c$ is the intensity incident on A; the dependence of γ^2 on the angles of incidence and of scattering is given in Eq. (8.3.11). For scattering in the direction of reflection, $\gamma = 0$; for scattering back to the source, $\gamma = 2 \sin \vartheta_i$.

The first term in braces, the scattering caused by the admittance variation, has its maximum in the direction of the reflected wave, as mentioned earlier. When the incident wavelength is large compared with the admittance correlation length w_β, this scattering is nearly isotropic. The second term in braces, the scattering caused by roughness, avoids the direction of reflection and, for very long wavelengths, produces little scattering. For shorter wavelengths the scattered intensity concentrates in a cone around the reflection direction; this term is zero when $\gamma = 0$ and has its maximum value for $k\gamma w_\xi = 2$, that is, at an angle roughly $\lambda/\pi w_\xi$ radians from the reflection direction. Figure 8.10 shows a typical example, with the admittance scattering given in one plot, the roughness scattering in the other. Note that the roughness scattering avoids the reflection direction. Compare with Fig. 8.8.

If we work out the scattering of a pulse-wave from this area, we should obtain a result quite similar to that of Prob. 11. The scattered wave would not be a pulse (except in the reflection direction), but would be a rumble, reproducing in time the spatial irregularities of admittance and shape. The spectrum density of this scattered sound would equal the I_s of Eq. (8.3.12).

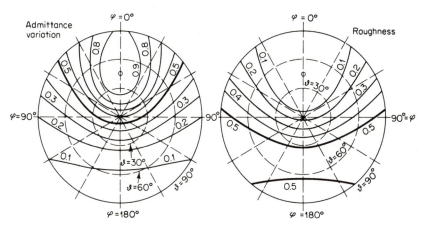

FIGURE 8.10
Contours of equal amplitude of wave scattered from a patch of random variation in admittance or from a rough patch, as function of angles ϑ, φ of scattering for $kw = 2\pi w/\lambda = 0.707$, w being the correlation length for admittance or for roughness. Reflection direction is $\vartheta = 45°$, $\varphi = 0°$, position of maximum for admittance scattering, of zero scattering for roughness. Compare with Fig. 8.8.

8.4 DIFFRACTION

When the scattering object is large compared with the wavelength of the scattered sound, we usually say the sound is reflected and diffracted, rather than scattered. The effects are really the same, but the relative magnitudes differ enough so that there seems to be a qualitative difference. Behind the object there is a shadow, where the pressure amplitude is vanishingly small; in front or to the side, in the "illuminated" region, there is a combination of the incident wave and the wave reflected from the surface of the scattering object. At the edge of the shadow the wave amplitude does not drop discontinuously from its value in the illuminated region to zero; the amplitude oscillates about its illuminated value, with increasing amplitude, reaching its maximum just before the edge of the shadow, and then dropping monotonically, approaching zero well inside the shadow.

These fluctuations of amplitude, near the shadow edge, are called *diffraction bands*. Their angular spacing depends on the ratio between the wavelength of the incident sound and the distance from the observation point to the line on the scattering object separating "light" from "shadow." This behavior was touched on in connection with the scattering from a cylinder [see discussion of Eq. (8.1.4)] and from a sphere. To study it in detail we shall compute the diffraction of an incident plane wave from a rigid half-plane. Diffraction patterns also arise in the reflection of a plane wave from a plane surface, one large area of which differs in impedance from the rest of the plane.

Diffraction from a knife-edge

The simplest diffraction problem is that of a plane wave traveling in the negative x direction, striking a rigid half-plane extending from the line $x = y = 0$, $-\infty < z < \infty$, to infinity in the direction $\frac{3}{2}\pi - \Psi$, as shown in Fig. 8.11. The "shadow line" is the negative x axis; the region III, $\pi < \phi < \frac{3}{2}\pi - \Psi$, is in shadow. In region I, $-2\Psi < \phi < \pi$, the incident wave $Ae^{-ikx} = Ae^{-ikr\cos\phi}$ predominates, and in region II, $-\frac{1}{2}\pi - \Psi < \phi < -2\Psi$, there is, in addition to the incident wave, a reflected wave $Ae^{ikr\cos(\phi+\Psi)}$. The details of how these solutions join together at the edges of the regions must be obtained from the solution of the boundary-value problem.

This solution is most easily obtained from the function

$$U(r,\phi) = \tfrac{1}{2} \sum_{m=0}^{\infty} \epsilon_m(-i)^{\frac{1}{2}m} \cos(\tfrac{1}{2}m\phi) J_{\frac{1}{2}m}(kr) \tag{8.4.1}$$

Comparison with Eq. (8.1.1) shows that this is not the expansion of a plane wave in polar coordinates because it includes half-integer orders as well as integral orders of J. In fact, the plane wave is

$$e^{-ikr\cos\phi} = \sum_{n=0}^{\infty} \epsilon_n(-i)^n \cos(n\phi) J_n(kr) = U(r,\phi) + U(r,\phi + 2\pi)$$

Reference to Eq. (7.3.1) indicates that $U(r,\phi)$ is a solution of the wave equation; but it is not periodic in ϕ with period 2π; it is periodic with period 4π.

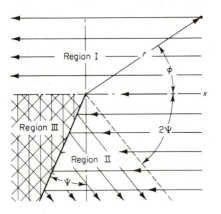

FIGURE 8.11
Diffraction of a plane wave by the edge of a rigid half-plane.

Thus, to use U for a solution in free space would be forbidden, for U is two-valued in ϕ if ϕ is unlimited.

In the problem we are now solving, however, ϕ is limited in range; its greatest positive value is $\frac{3}{2}\pi - \Psi$, and its largest negative value is $-\frac{1}{2}\pi - \Psi$ (Ψ can be negative, of course; its allowed range of values is $-\frac{1}{2}\pi < \Psi < \frac{1}{2}\pi$). Within this range of values of ϕ, we can use a combination of U's to satisfy boundary conditions at the half-plane. If the plane is rigid, the condition is that $\partial p/\partial \phi = 0$ at $\phi = \frac{3}{2}\pi - \Psi$ or $-\frac{1}{2}\pi - \Psi$. This is satisfied if we use the following combination for the pressure:

$$p(r,\phi) = AU(r,\phi) + AU(r, 3\pi - 2\Psi - \phi) \tag{8.4.2}$$

Since U is periodic with period 4π, the two U's do not cancel everywhere; their ϕ gradients do cancel on both sides of the plane barrier, however.

In order to find out the details of this solution, we should find a closed form for the series U. This can be done in several ways. By using the first of Eqs. (5.2.21) and combining the series, we can show that

$$\frac{\partial U}{\partial \phi} - ikr \sin \phi \, U = -\sqrt{\frac{kr}{2\pi i}} \, e^{ikr} \sin \tfrac{1}{2}\phi$$

Solution of this, plus the fact that $U(0,\phi) = \frac{1}{2}$, shows that

$$U(r,\phi) = e^{ikr.\cos \phi} \, E(\sqrt{2kr} \cos \tfrac{1}{2}\phi) \tag{8.4.3}$$

where

$$E(z) = \frac{1}{\sqrt{i\pi}} \int_{-\infty}^{z} e^{it^2} dt \to \begin{cases} 1 - \sqrt{\dfrac{i}{4\pi z^2}} \, e^{iz^2} & z \to +\infty \\[3mm] \dfrac{1}{2} & z \to 0 \\[3mm] -\sqrt{\dfrac{i}{4\pi z^2}} \, e^{iz^2} & z \to -\infty \end{cases}$$

Alternatively, the integral expression for U can be expanded in a Fourier series and shown to be equal to the series of Eq. (8.4.1).

For intermediate values of $z = \sqrt{u}$, we can express $E(z)$ in terms of Fresnel integrals

$$C(u) = \sqrt{\frac{2}{\pi}} \int_{0}^{\sqrt{u}} \cos t^2 \, dt \qquad S(u) = \sqrt{\frac{2}{\pi}} \int_{0}^{\sqrt{u}} \sin t^2 \, dt \tag{8.4.4}$$

as follows:

$$E(\sqrt{u}) = \begin{cases} 1 - \tfrac{1}{2}[1 - C(u) - S(u)] - \tfrac{1}{2}i[C(u) - S(u)] & \sqrt{u} > 0 \\[2mm] \tfrac{1}{2}[1 - C(u) - (u)] + \tfrac{1}{2}i[C(u) - S(u)] & \sqrt{u} < 0 \end{cases}$$

$$\tag{8.4.5}$$

The functions $1 - C(u) - S(u)$ and $C(u) - S(u)$ are plotted in Fig. 8.14. Tables of the functions are not hard to find.

Returning to Eq. (8.4.2) for the solution, we see that

$$p(r,\phi) = Ae^{-ikr\cos\phi}\, E[\sqrt{2kr}\cos\tfrac{1}{2}\phi]$$
$$+ Ae^{ikr\cos(\phi+2\Psi)}\, E[\sqrt{2kr}\cos(\tfrac{3}{2}\pi - \Psi - \tfrac{1}{2}\phi)] \quad (8.4.6)$$

$$p \xrightarrow[kr\to\infty]{}
\begin{cases}
AD(r,\phi) & \text{(in region III)} \\
Ae^{-ikr\cos\phi} + AD(r,\phi) & \text{(in region I)} \\
Ae^{-ikr\cos\phi} + Ae^{ikr\cos(\phi+2\Psi)} + AD(r,\phi) & \text{(in region II)}
\end{cases}$$
$$(8.4.7)$$

where

$$D(r,\phi) = \frac{e^{ikr}}{\sqrt{8\pi ikr}}\left[\frac{1}{\sin(\Psi+\tfrac{1}{2}\phi)} - \frac{1}{\cos\tfrac{1}{2}\phi}\right]$$

The first expression in the asymptotic formula is for the shadow region; in this region the only wave is a cylindrical wave represented by function D, emanating as if from the knife-edge, with amplitude which diminishes as one goes deeper into the shadow as $\phi - \pi$ increases; the asymptotic form is not valid for ϕ very close to π or very close to -2Ψ. In region I the predominant term is the incident wave, but the cylindrical wave is also present. Finally, in region II, the wave reflected from the half-plane is present, as well as the incident wave and the cylindrical "diffraction wave" $D(r,\phi)$.

Looking more closely at the behavior of the wave, let us calculate the mean-square pressure in the region near the shadow line dividing region III from region I. Suppose the pressure is measured along a line $x = -l$, for different values of y: $y = 0$ being the edge of the geometric shadow, y negative being in shadow, and y positive being in region I. We define

$$\alpha = \pi - \phi \qquad y = r\sin\alpha \qquad l = r\cos\alpha$$
$$u = kr(1 + \cos\phi) = k(\sqrt{l^2 + y^2} - l) \qquad\qquad (8.4.8)$$
$$v = kr[1 + \cos(2\Psi - \alpha)] = u + kl(1 + \cos 2\Psi) + ky\sin 2\Psi$$

Then the asymptotic expression for the mean-square pressure as function of y is

$$|p|^2 \to |A|^2
\begin{cases}
\dfrac{1}{4\pi}[(v)^{-\frac{1}{2}} - (u)^{-\frac{1}{2}}]^2 & y < -\sqrt{\dfrac{l}{k}} \\[2ex]
1 + \dfrac{1}{\sqrt{\pi}}[(v)^{-\frac{1}{2}} - (u)^{-\frac{1}{2}}]\cos(u + \tfrac{1}{4}\pi) & y > \sqrt{\dfrac{l}{k}}
\end{cases}$$
$$(8.4.9)$$

This formula predicts the monotonic decrease in $|p|^2$ inside the shadow as $-y$ increases. It also indicates that $|p|^2$ oscillates about its mean value $|A|^2$ as y is increased in the illuminated region I. But it is not valid close to the edge of the shadow, where $|y| < \sqrt{l/k}$, because we have used the asymptotic form of Eq. (8.4.3) in Eq. (8.4.6) for both the functions E. A more accurate formula, using (8.4.5) for $E(\sqrt{2kr}\cos\frac{1}{2}\phi)$ and the asymptotic form for the other E, since it is valid throughout the region near $\phi = \pi$ as long as $kr > 20$, is

$$
\left|\frac{p}{A}\right|^2 \simeq
\begin{cases}
[\frac{1}{2} + \frac{1}{2}C(u) + \frac{1}{2}S(u) + (4\pi v)^{-\frac{1}{2}}\cos(u + \frac{1}{4}\pi)]^2 \\
\quad + [\frac{1}{2}C(u) - \frac{1}{2}S(u) - (4\pi v)^{-\frac{1}{2}}\sin(u + \frac{1}{4}\pi)]^2 & y \geqslant 0 \\
[\frac{1}{2} - \frac{1}{2}C(u) - \frac{1}{2}S(u) + (4\pi v)^{-\frac{1}{2}}\cos(u + \frac{1}{4}\pi)]^2 \\
\quad + [\frac{1}{2}C(u) - \frac{1}{2}S(u) + (4\pi v)^{-\frac{1}{2}}\sin(u + \frac{1}{4}\pi)]^2 & y \leqslant 0
\end{cases}
\tag{8.4.10}
$$

plotted in Fig. 8.12 for $kl = 50$, as function of y/l. It shows that the transition from illumination to shadow indicates that $|p/A|^2$ is about $\frac{1}{4}$ at $y = 0$, the geometric shadow line, and that its maximum value is about $\frac{4}{3}$ at a point about two wavelengths inside the illuminated region. It also shows that the asymptotic formula (8.4.9) is adequate for $|y|$ greater than about three wavelengths. These results would be somewhat altered for different values of kl or of Ψ. The effect of Ψ on the results is small, however; the curve for $|p/A|^2$ versus y/l for $kl = 50$ and $\Psi = 45°$ would be too close to the curve for $\Psi = 0°$, which is plotted, to be able to distinguish clearly between the two.

The transition from region I to region II is provided with similar diffraction bands, further complicated by the interference between the incident and reflected waves. Equations (8.4.5) and (8.4.6) enable one to compute $|p/A|^2$ here also.

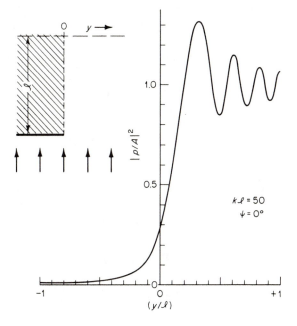

FIGURE 8.12
Mean-square pressure of diffracted plane wave at point l beyond obstructing half-plane and distance y from "shadow line," for $l \simeq 8\lambda$.

$kl = 50$
$\psi = 0°$

Diffraction of a reflected wave

Similar diffraction effects occur when a sound wave reflects from an infinite plane wall, one half of which is rigid, the other half soft.   Suppose the wall is the xy plane, the half of the wall for x positive having a uniform, specific acoustic admittance β, and the half for x negative having $\beta = 0$. If the incident-wave direction is given by the angles ϑ_i, φ_i of Fig. 8.7, we can separate off the factor $\exp\left[i(\omega/c)\sin\vartheta_i\sin\varphi_i\right]$ from all the equations, leaving a two-dimensional formula for the motion in the xz plane.   For example, the wave equation for p can be written

$$e^{iy(\omega/c)\sin\vartheta_i\sin\varphi_i}\left(\frac{\partial^2}{\partial x^2} + \frac{\partial^2}{\partial z^2} + k^2\right)p = 0$$

where

$$k^2 = \left(\frac{\omega}{c}\right)^2(1 - \sin^2\vartheta_i\sin^2\varphi_i)$$

and the equation for the Green's function is

$$e^{iy(\omega/c)\sin\vartheta_i\sin\varphi_i}\left(\frac{\partial^2}{\partial x^2} + \frac{\partial^2}{\partial z^2} + k^2\right)G = \delta(x - x_0)\delta(z - z_0)e^{iy(\omega/c)\sin\vartheta_i\sin\phi_i}$$

$$(8.4.11)$$

In terms of the component wavenumber k in the xz plane, the x and z components are $k_x = k\sin\theta_i$, $k_z = k\cos\theta_i$, where

$$\sin\theta_i = \frac{\sin\vartheta_i\cos\varphi_i}{\sqrt{1 - \sin^2\vartheta_i\sin^2\varphi_i}} \qquad \cos\theta_i = \frac{\cos\vartheta_i}{\sqrt{1 - \sin^2\vartheta_i\sin^2\varphi_i}} \qquad (8.4.12)$$

We can thus neglect the y component of the wave and work out the problem as though the incident wave were pointed in the xz plane, with a wavenumber $k < \omega/c$ and at an angle of incidence θ_i.

Referring to Eqs. (8.3.1) and (8.3.2), we see that the integral equation for the pressure wave in the region $z > 0$ is

$$p(x,z) = 2P_i e^{ikx\sin\theta_i}\cos(kz\cos\theta_i) + ik\beta\int_0^\infty p(x_0,0)G_k(x,z\mid x_0,z_0)\,dx_0$$

$$(8.4.13)$$

The Green's function is the two-dimensional one satisfying the boundary condition $\partial G/\partial z = 0$ at $z = 0$.   It can be expressed in a number of different forms [see Eqs. (7.3.17) and (7.4.7)],

$$G_k(x,z\mid x_0,z_0) = \frac{i}{4}H_0^{(1)}(kR) + \frac{i}{4}H_0^{(1)}(kR')$$

$$= \frac{1}{2\pi^2}\int_{-\infty}^\infty dK_x\int_{-\infty}^\infty dK_z\,\frac{\cos(K_z z_0)e^{iK_x(x-x_0)+iK_z z}}{K_x^2 + K_z^2 - k^2} \qquad (8.4.14)$$

where $R^2 = (x - x_0)^2 + (z - z_0)^2$, $R'^2 = (x - x_0)^2 + (z + z_0)^2$, and the function $H_0^{(1)}$ is the Hankel function [see discussion following Eq. (7.3.2)]

$$H_0^{(1)}(z) = J_0(z) + iN_0(z) \to \sqrt{\frac{\pi}{2ikr}} \exp{(ikr)} \qquad kr \to \infty \qquad (8.4.15)$$

The integral equation (8.4.13) uses $\beta_0 = 0$ as the "average" admittance (in the sense of page 445), so the incident-plus-reflected wave is for a completely rigid wall. We could, of course, have chosen the reflected wave and the Green's function for the wall admittance β, in which case the integral in Eq. (8.4.13) changes sign and the integration is from $-\infty$ to 0. We know,

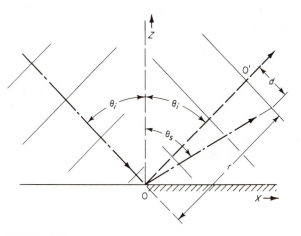

FIGURE 8.13
Incident and reflected plane waves from a half-rigid–half-soft plane at $z = 0$. Diffraction pattern is exhibited in reflected wavefront near point O'.

of course, that well to the left of point O' in Fig. 8.13 (i.e., for d large and negative) the amplitude of the reflected wave is P_i, since this part of the wave has been reflected from the rigid part of the wall; whereas well to the right of O' (d large and positive) the reflected-wave amplitude is $P_i[(\cos\theta_i - \beta)/(\cos\theta_i + \beta)]$, since this part is reflected from the region of admittance β [see Eq. (6.3.5)]. In between, for $|d|$ small, a diffraction-pattern curve connects the two different values.

Integral equation (8.3.13) can be solved exactly using the Wiener-Hopf technique. The problem is clearly related to that of the diffraction of waves by a straightedge, which has just been discussed. The result, of course, involves the Fresnel functions defined in Eqs. (8.4.4). The exact solution produces the Fourier transform of the pressure distribution on the soft part of the reflecting plane, and this, according to Eq. (8.3.3), is proportional to the shape of the reflected wave a large distance from the origin. In fact, the exact solution predicts that the reflected amplitude, as a function of d, along the wavefront, when kr is very large, is

$$P_{\text{ref}} \rightarrow \begin{cases} P_i - \dfrac{P_i\beta/2}{\cos\theta_i + \beta}\{[1 - C(u) - S(u)] - i[C(u) - S(u)]\} & d < 0 \\ \\ P_i\dfrac{\cos\theta_i - \beta}{\cos\theta_i + \beta} + \dfrac{P_i\beta/2}{\cos\theta_i + \beta}\{[1 - C(u) - S(u)] \\ \qquad\qquad\qquad\qquad\qquad - i[C(u) - S(u)]\} & d > 0 \end{cases} \quad (8.4.16)$$

where $u = (kd/\sqrt{2kr})^2$. The functions $(1 - C - S)$ and $(C - S)$ are plotted at the top of Fig. 8.14; we see that they go asymptotically to zero as $u \rightarrow \infty$. Therefore, for d large and negative, the reflected amplitude is P_i, the amplitude of a wave reflected from a rigid plane; for d large and positive, the amplitude is

$$P_i\frac{\cos\theta_i - \beta}{\cos\theta_i + \beta}$$

the amplitude of a wave reflected from a plane of admittance β everywhere. For $|d|$ small, the functions $(1 - C - S)$ and $(C - S)$ produce the changes in amplitude and phase typical of a diffraction pattern.

Although the exact form of the Fourier transform of the surface pressure distribution is obtained, the inverse transform, the pressure itself, cannot be expressed in closed form. Therefore we shall be satisfied with an approximate solution of the integral equation, obtained from Eq. (8.4.13) by setting $z = 0$. We set into the right-hand integral the first approximation to p,

$$p(x_0,0) \simeq \begin{cases} 2P_i e^{ikx_0 \sin\theta_i} & x_0 < 0 \\ \\ 2P_i\dfrac{\cos\theta_i}{\cos\theta_i + \beta} e^{ikx_0 \sin\theta_i} & x_0 > 0 \end{cases}$$

to obtain, as a second-order approximation,

$$p(x,0) \simeq 2P_i e^{ikx \sin\theta_i} - \frac{P_i\beta\cos\theta_i}{\cos\theta_i + \beta}\int_0^\infty e^{ikx_0 \sin\theta_i} H_0(k\,|x - x_0|)\,d(kx_0)$$

$$\simeq 2P_i e^{ikx \sin\theta_i} - \frac{P_i\beta\, e^{ikx \sin\theta_i}}{\cos\theta_i + \beta}\begin{cases} [X(\theta_i,-kx) + iY(\theta_i,-kx)] & x < 0 \\ \\ [2 - X(-\theta_i, kx) - iY(-\theta_i,kx)] & x > 0 \end{cases}$$

$$(8.4.17)$$

where

$$X(\theta,u) + iY(\theta,u) = \cos\theta\int_u^\infty e^{iw \sin\theta}[J_0(w) + iN_0(w)]\,dw$$

$$\rightarrow \begin{cases} 1 + \dfrac{2\theta}{\pi} & u \rightarrow 0 \\ \\ 0 & u \rightarrow \infty \end{cases}$$

These functions, for $\theta = 0$, are plotted in the lower half of Fig. 8.14. They are sufficiently similar to the functions $(1-C-S)$ and $(C-S)$ to lead us to believe that Eq. (8.4.17) is a good approximation, at least for $|\beta| < 1$. And the integrals X and Y are much easier to compute than is the exact solution for $\theta \neq 0$.

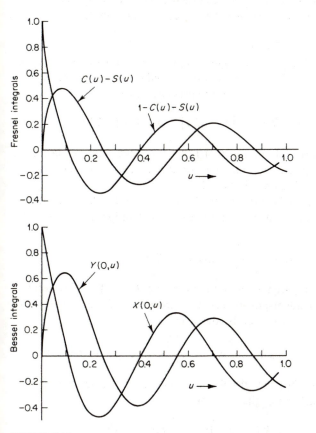

FIGURE 8.14
Comparison of Fresnel integrals, defined in Eqs. (8.4.4), and Bessel integrals, defined in Eq. (8.4.17).

The power absorbed by the wall, per unit area, a distance x from the edge of the soft region, is $\xi |p|^2/\rho c$ for $x > 0$ and is zero for $x < 0$ (note that $\beta = \xi - i\sigma$). Therefore the strip of width about $\lambda/2\pi$, next to the edge, absorbs more energy than the average; the next strip, of width about $\lambda/2$, absorbs less than the average; and so on. Since the integral of $X + iY$ from $x = 0$ to $x = \infty$ is zero, the mean rate of energy loss per unit area for a strip much wider than $\lambda = 2\pi/k$ nearly equals $|2P_i \cos \theta_i/(\cos \theta_i + \beta)|^2(\xi/\rho c)$ if the admittance $|\beta| < 1$. Since this would be the loss rate if the whole wall had admittance β [see Eq. (6.3.6)], we see that no edge correction is necessary in calculating the average energy lost near the edge of an absorbing area of dimensions large compared with λ; to first order in $|\beta|$ the diffraction fluctuations in p near the edge average out, and even the second-order correction is a small fraction times $|\beta|^2$. Under the next heading we shall see that this is not true when two edges are close together, i.e., when the smallest dimension of the absorbing area is not large compared with λ.

Scattering from an absorbing strip

Suppose the wall at $z = 0$ is rigid except for a strip from $x = -\tfrac{1}{2}a$ to $x = +\tfrac{1}{2}a$, within which the specific acoustic impedance is β, independent of x and y. In this case the equation for the pressure wave is

$$p(x,z) = 2P_i e^{ikx \sin \theta_i} \cos (kz \cos \theta_i) - \tfrac{1}{2}k\beta \int_{-\frac{1}{2}a}^{\frac{1}{2}a} p(x_0,0)H_0^{(1)}(kR) \, dx_0 \quad (8.4.18)$$

using the same notation as in Eq. (8.4.13). An exact solution, using the Wiener-Hopf technique, is not possible for this case; so we shall use the variational procedure discussed in Sec. 4.5.

First we shall use the asymptotic form for the Hankel function of Eq. (7.3.3) to obtain a formula for the wave at a point designated by the polar coordinates r, θ_s,

$$H_0^{(1)}(kR) \rightarrow \sqrt{\frac{2}{i\pi kr}} \exp (ikr - ikx_0 \sin \theta_s)$$

when kr is large compared with both unity and kx_0. The first term in Eq. (8.4.18) is the incident-plus-reflected combination of plane waves; the second term is the wave scattered by the strip,

$$p_{\theta_i}(x,z) \rightarrow 2P_i e^{ikx \sin \theta_i} \cos (kz \cos \theta_i) - \sqrt{\frac{1}{2\pi i kr}} e^{ikr} F_{\theta_i}(k \sin \theta_s) \quad (8.4.19)$$

The function F describes the distribution-in-angle of the scattered wave; it is the Fourier transform of the function

$$f_\theta(x) = 2\pi k\beta(x)p_\theta(x,0) = \int_{-\infty}^{\infty} F_\theta(\kappa)e^{i\kappa x} \, d\kappa$$

$$F_\theta(\kappa) = k\beta \int_{-\frac{1}{2}a}^{\frac{1}{2}a} p_\theta(x_0,0)e^{-i\kappa x_0} \, dx_0 \quad (8.4.20)$$

where $\beta(x)$ equals the constant β in the range $-\tfrac{1}{2}a < x < +\tfrac{1}{2}a$ and equals zero for other values of x.

The integral equation which $p(x,0)$ must satisfy is

$$p_{\theta_i}(x,0) = 2P_i e^{ikx \sin \theta_i} - \tfrac{1}{2}\beta \int_{-\frac{1}{2}a}^{\frac{1}{2}a} p_{\theta_i}(x_0,0)H_0^{(1)}(k |x - x_0|) \, d(kx_0) \quad (8.4.21)$$

If this can be solved for p in the range $-\tfrac{1}{2}a < x < +\tfrac{1}{2}a$, then p, for any value of θ_i, of x and of z, can be computed from Eq. (8.4.18) or, asymptotically, from Eq. (8.4.19).   We might try to solve this equation on the average by multiplying it by $\beta P(x,0)$, integrating over x from $-\frac{1}{2}a$ to $+\frac{1}{2}a$, and adjusting the average value of p in this range to satisfy the resulting equation.   But we can do better than this if we notice that the integral of the first term on the right in Eq. (8.4.21), times βp, is related to the angle-distribution factor F; in fact, it is the F for the angle of incidence changed from θ_s to $-\theta_i$, as though the incident wave originated from the observation point r, θ_s and the scattered wave were being observed at the point r, $-\theta_i$.

Suppose we multiply Eq. (8.4.21) for $\theta_i = \theta$ by $\beta p_\theta{}^a(x,0)$ and integrate from $x = -\frac{1}{2}a$ to $x = +\frac{1}{2}a$, where we shall have to determine the relation between p^a and p later.   Simplifying our notation in an obvious manner, we have the equation

$$0 = 2P_i k \int_{-\frac{1}{2}a}^{\frac{1}{2}a} e^{ikx \sin \theta_i} \beta p_\theta{}^a \, dx$$

$$- \tfrac{1}{2}k^2 \int \beta p_\theta{}^a \, dx \int H_0^{(1)}(k \, |x - x_0|) \beta p_\theta \, dx_0 - k \int p_\theta{}^a \beta p_\theta \, dx$$

We also write down the equation for F for $\theta_s = \theta_i = \theta$ in the form

$$2P_i F_\theta(k \sin \theta) = 2P_i k \int e^{-ikx \sin \theta} \beta p_\theta \, dx$$

and add the equations to obtain a symmetric expression.

$$2P_i F_\theta(k \sin \theta) = 2P_i k \left(\int e^{ikx \sin \theta} \beta p_\theta{}^a \, dx + \int e^{-ikx \sin \theta} \beta p_\theta \, dx \right)$$

$$- k \int p_\theta{}^a \beta p_\theta \, dx - \tfrac{1}{2}k^2 \int \beta p_\theta{}^a \, dx \int H_0^{(1)}(k \, |x - x_0|) \beta p_\theta \, dx_0 \quad (8.4.22)$$

This equation for F can serve as a basis for our variational procedure.

For example, if we vary the shape of p^a slightly, so that it becomes $p^a + \delta p^a$, the variation of $2P_i F$ is

$$k \int \beta \delta p_\theta{}^a \, dx \left[2P_i e^{ikx \sin \theta} - \tfrac{1}{2}k \int H_0^{(1)}(k \, |x - x_0|) \beta p_\theta \, dx_0 - p_\theta \right]$$

If p_θ is a solution of Eq. (8.4.21), this variation is zero.   Likewise, if p is varied, the change in $2P_i F$ proportional to δp is zero if p^a is a solution of the equation resulting when θ is changed to $-\theta$ in Eq. (8.4.21) [i.e., $p_\theta{}^a$ is the surface pressure when source point and observation point are interchanged; this equation and its solution are said to be *adjoint* to Eq. (8.4.21) and its solution].

Generalizing the arguments of page 156, we can say that if the exact solution p_θ and its adjoint $p_\theta{}^a$ are inserted in the right-hand side of Eq. (8.4.22), the result will equal $2P_i$ times the correct angle-distribution function $F_\theta(k \sin \theta)$ for the angle of incidence θ, but if trial functions $\chi(\alpha)$ and $\chi^a(\alpha)$

are used in the integrals, and if these functions differ from p and p^a by small quantities of the order of δ, the resulting expression will differ from $2P_iF$ by a constant of the order of δ^2. Furthermore, if the shape of $\chi(\alpha)$ can be varied by changing the value of the parameter α, then the shape of $\chi(\alpha)$ for which the derivative $\partial/\partial\alpha$ of the variational expression is zero is the shape most closely corresponding to p, and the value of the variational expression for this minimal value of α is most nearly equal to $2P_iF(k \sin \theta)$.

For example, suppose we assume that the pressure over the soft strip is proportional to $e^{ikx \sin \theta}$ (as it would if $\beta = 0$) but that its magnitude is not $2P_i$ but A. We should obtain a Born approximation for F if we inserted $2P_ie^{ikx \sin \theta}$ for p in Eq. (8.4.20) for F; this would be correct to first order in β. We expect a second-order approximation if we insert the trial function $\chi = Ae^{ikx \sin \theta}$ instead of p_θ and its adjoint $Ae^{-ikx \sin \theta}$ for $p_\theta{}^a$ into Eq. (8.4.22) and vary the value of A to obtain a zero slope. The resulting variational expression is

$$4P_ika\beta A - ka\beta A^2 - \tfrac{1}{2}k^2\beta^2A^2 \int e^{-ikx \sin \theta}\, dx \int H_0^{(1)}(k\,|x - x_0|)e^{ikx \sin \theta}\, dx_0$$

$$(8.4.23)$$

The derivative of this with respect to A vanishes when

$$A = \frac{2P_i}{1 + \dfrac{k\beta}{2a}\displaystyle\int e^{-ikx \sin \theta}\, dx \int H_0^{(1)}(k\,|x - x_0|)e^{ikx_0 \sin \theta}\, dx_0}$$

This can be expressed in terms of the functions X and Y of Eq. (8.4.17) for

$$Z(\theta,u) \equiv \cos \theta \int_0^u e^{iw \sin \theta} H_0^{(1)}(w)\, dw$$

$$= \begin{cases} -1 - \dfrac{2\theta}{\pi} + X(\theta,u) + iY(\theta,u) & u < 0 \\[3mm] 1 + \dfrac{2\theta}{\pi} - X(\theta,u) - iY(\theta,u) & u > 0 \end{cases} \qquad (8.4.24)$$

so that

$$k \cos \theta \int_{-\frac{1}{2}a}^{\frac{1}{2}a} H_0^{(1)}(k\,|x - x_0|)e^{ikx \sin \theta}\, dx_0$$

$$= e^{ikx \sin \theta}[Z(\theta, \tfrac{1}{2}ka - kx) + Z(-\theta, \tfrac{1}{2}ka + kx)]$$

and

$$\frac{1}{a}\int_{-\frac{1}{2}a}^{\frac{1}{2}a} e^{-ikx \sin \theta}\, dx \int_{-\frac{1}{2}a}^{\frac{1}{2}a} H_0^{(1)}(k\,|x - x_0|)e^{ikx_0 \sin \theta}\, dx_0$$

$$= \frac{1}{ka}\int_0^{ka}[Z(\theta,-u) + Z(-\theta,u)]\, du \equiv W(\theta,ka) + W(-\theta,ka)$$

The real and imaginary parts of Z and W, for $\theta = 0$, are plotted in Fig. 8.15.

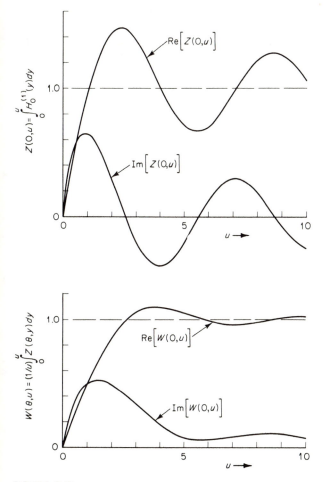

FIGURE 8.15
Functions entering the calculation of pressure distribution across a strip of absorbing material.

In terms of the Z's and the W's, the variational solution is

$$A = \frac{2P_i \cos \theta}{\cos \theta + \frac{1}{2}\beta[W(\theta,ka) + W(-\theta,ka)]}$$

$$p_{\theta_i}(x,0) \simeq 2P_i\left\{1 - \frac{1}{2}\beta\frac{Z(\theta_i, \frac{1}{2}ka - kx) + Z(-\theta_i, \frac{1}{2}ka + kx)}{\cos \theta_i + \frac{1}{2}\beta[W(\theta_i,ka) + W(-\theta_i,ka)]}\right\}e^{ikx \sin \theta_i}$$

$$\rightarrow \begin{cases} 2P_i e^{ikx \sin \theta_i} & |x| \gg a \\ 2P_i e^{ikx \sin \theta_i}\dfrac{\cos \theta_i}{\cos \theta_i + \beta} & |kx| \ll ka \gg 1 \end{cases} \qquad (8.4.25)$$

$$F_{\theta_i}(k \sin \theta_s) \simeq \frac{2P_i ka\beta \cos \theta_i}{\cos \theta_i + (\beta/2)[W(\theta_i,ka) + W(-\theta_i,ka)]}\frac{\sin\left[\frac{1}{2}ka(\sin \theta_s - \sin \theta_i)\right]}{\frac{1}{2}ka(\sin \theta_s - \sin \theta_i)}$$

All these expressions are good to second order in the mean-square departure of the surface pressure on the soft strip from its mean value there (i.e., the mean-square difference between the true p and $Ae^{ikx\sin\theta}$). The formula for $p_{\theta_i}(x,0)$ is obtained by inserting the "best" trial function into the integral on the right of Eq. (8.4.21); that for the angle-distribution function

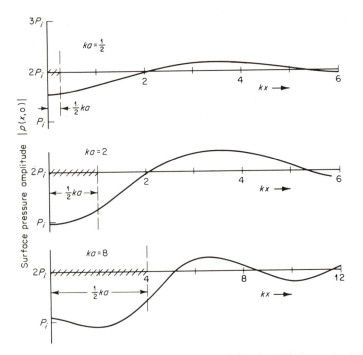

FIGURE 8.16
Pressure amplitude on wall having strip of admittance $\beta = 1$ of width a, otherwise rigid, produced by normally incident plane wave of wavelength $2\pi/k$ and amplitude P_i, as function of distance x from center of strip, for different values of k. Vertical dashed line indicates edge of strip.

for the scattered wave by setting the same "best" function into Eq. (8.4.20). The best value of the variational expression is, of course, $F_{\theta_i}(k\sin\theta_i)$, the scattered amplitude at the angle of reflection $\theta_s = \theta_i$.

As an indication of the adequacy of this solution, we plot in Fig. 8.16 the amplitude $p(x,0)$ of the surface pressure from Eq. (8.4.25) as function of x, for $\theta_i = 0$, for $\beta = 1$, and for several different values of k. We see that, for $|x| < \frac{1}{2}a$, the pressure amplitude is not far from being constant, and thus that the curves are not far from those of the exact solution. For $|\beta| < 1$ the fit would be even better.

We note that when ka is very small, the pressure on the soft strip is not much different from the $2P_i$ which would be present if the wall were all rigid; the longer wavelength cannot accommodate itself to the narrow strip; so the dip in pressure is shallow and extends well beyond the edge of the strip. As k is increased, the pressure at the center of the strip approaches the value P_i which it would have if the strip were infinitely wide (for $\beta = 1$ there is no reflected wave from a normally incident wave; the only pressure there is the incident pressure P_i). For $ka \gg 1$ the pressure amplitude approaches a step function, going from P_i for $|x| < \frac{1}{2}a$ to $2P_i$ for $|x| > \frac{1}{2}a$, with a diffraction pattern noticeable at the strip edge.

Problems

1 Write out the long-wavelength expressions for the angle distribution of the pressure wave scattered from a rigid cylinder and a rigid sphere from Eqs. (8.1.1) and (8.2.2). What are the ratios between the forward-, sideward-, and backward- scattered intensity? In the case of the sphere, what are the separate contributions from monopole and dipole scattering?

2 When the scattering sphere of radius a is immovable and incompressible, the surface integral of Eq. (7.1.17) can be used, and the integral equation for the incident plus scattered wave takes the form

$$p_\omega(\mathbf{r}) = p_i(\mathbf{r}) + a^2 \iint \left[p_\omega(\mathbf{r}_0) \frac{\partial}{\partial r_0} g_\omega(\mathbf{r} \mid \mathbf{r}_0) \right]_{r_0 = a} d\Omega_0$$

where $d\Omega_0$ is the element of solid angle in the $\mathbf{r}_0$ coordinates. Show that the asymptotic form of the scattered wave is

$$p_s(\mathbf{r}) \to -\frac{ika^2}{4\pi r} e^{ikr} \iint [p_\omega(\mathbf{r}_0)]_{r_0 = a} \, e^{-ika \cos \Psi} \cos \Psi \, d\Omega_0$$

where Ψ is the angle between $\mathbf{r}$ and $\mathbf{r}_0$. Show that the Born approximation to this asymptotic form, if the incident wave is the plane wave $A \exp(i\mathbf{k}_i \cdot \mathbf{r})$, is

$$p_s(\mathbf{r}) \simeq -\frac{ia^2 A}{4\pi r} e^{ikr} \iint \exp\left(-i\boldsymbol{\mu} \cdot \mathbf{a}_0\right) \left[(\boldsymbol{\mu} + \mathbf{k}_i) \cdot \mathbf{a}_0 \right] d\Omega_0$$

$$= \frac{aA}{2r} e^{ikr} (\mu a) j_1(\mu a) \qquad \boldsymbol{\mu} = \mathbf{k}_s - \mathbf{k}_i \qquad \mu = 2k \sin\frac{\vartheta}{2}$$

where $\mathbf{a}_0$ is the unit vector in the direction of the element of solid angle $d\Omega_0$ (*Hint*: use the direction of $\boldsymbol{\mu}$ as the axis for the angles of $d\Omega_0$). Show that this does not agree, even at long wavelengths, with the exact solution of Eqs. (8.2.19), for $\kappa_e = 0$ and $\rho_e \to \infty$. Why is this?

3 In a heavy rain the mean radius of the raindrops is 2 mm and there is, on the average, one drop per 1,000 cu cm. Use Eqs. (8.2.22) to calculate the fraction of energy removed from a plane wave, per meter, at 100 and at 1,600 cps, assuming that $\gamma_\kappa = -1$ and $\gamma_\rho = \frac{3}{2}$. Include both absorption and scattering cross sections. Carry out the same calculations for a mist in which the mean radius of the droplets is 0.1 mm but the same amount of water per cubic centimeter is present as in the rain.

4 The material inside an infinite cylinder of radius a, with axis along the z axis, has density ρ_e and compressibility κ_e. It is surrounded by a medium of density ρ and compressibility κ. An incident plane wave of sound

$$p_i = \exp\,(ik_z z + ik_i x - i\omega t) = \exp\,(ik_z z + ik_i r \cos\phi - i\omega t)$$

where $k_z^2 + k_i^2 = \rho\kappa\omega^2$, is incident on the cylinder, producing a scattered wave. Show that the pressure amplitude inside and outside the cylinder can be expressed in terms of the cylindrical coordinates z, r, ϕ by the series

$$p_\omega = \begin{cases} e^{ik_z z} \sum \epsilon_m i^m B_m \cos\,(m\phi)J_m(k_e r) & r < a \\ e^{ik_z z} \sum \epsilon_m i^m \cos\,(m\phi)\,[J_m(k_i r) + D_m H_m^{(1)}(k_i r)] & a < r \end{cases}$$

where $k_e^2 = (\rho_e \kappa_e / \rho\kappa)k_i^2$. Show that requirements of continuity of p and u_r across the cylindrical surface require that

$$D_m = \frac{J_m(k_e a)J_m'(k_i a) - \alpha J_m'(k_e a)J_m(k_i a)}{\alpha J_m'(k_e a)H_m^{(1)}(k_i a) - J_m(k_e a)H_m^{(1)\prime}(k_i a)} \qquad \alpha^2 = \frac{\kappa_e \rho}{\kappa \rho_e}$$

where the primes indicate differentiation with respect to the argument. Show that the asymptotic form of the scattered wave is

$$p_s \to \sqrt{\frac{2}{\pi i k r}}\, e^{ik_z z + ik_i r} \sum \epsilon_m D_m \cos\,(m\phi)$$

5 Using the results of Prob. 4, plus the series expansions of J_m and N_m, show that the long-wavelength form for the wave scattered from the cylinder, to second order in $k_i a$, is

$$p_s \to \sqrt{\frac{2}{\pi i k_i r}}\, e^{ik_z z + ik_i r}\, \frac{i\pi k_i^2 a^2}{4}\left(\frac{\kappa_e - \kappa}{\kappa} + \frac{2\rho_e - 2\rho}{\rho_e + \rho}\cos\phi\right)$$

6 Use the Born approximation of Eq. (8.1.18) and the two-dimensional Green's function $g_\omega(\mathbf{r}\,|\,\mathbf{r}_0) = (i/4)H_0^{(1)}(k\,|\mathbf{r} - \mathbf{r}_0|)$ to calculate the scattered wave of Prob. 4. If γ_κ and γ_ρ are functions of the cylindrical coordinate r only, show that

$$p_s \simeq \tfrac{1}{2}\pi i \sqrt{\frac{2}{\pi i k_i r}}\, e^{ik_z z + ik_i r} k^2 \int_0^a [\gamma_\kappa(r) + \gamma_\rho(r)\cos\phi]J_0\,(\mu r)r\,dr$$

where $\mu = 2k_i \sin\,(\vartheta/2)$. When γ_κ and γ_ρ are constant inside $r = a$ and zero outside, show that the integral is approximately equal to $(a^2/2)(\gamma_\kappa + \gamma_\rho \cos\phi)$. Compare this result with that of Prob. 5. What does it indicate regarding the range of validity of the Born approximation?

7 The cylinder of Prob. 4 contains material of large viscosity and thermal conduction, but the surrounding medium has values μ, K, and C_P, which cannot be neglected. Show that, at long wavelengths, if the incident wave has amplitude A and moves normal to the cylindrical axis ($k_i = k = \omega/c$), the scattered wave is, approximately,

$$p_s \to A\sqrt{\frac{i\pi}{8kr}}\, e^{ikr} k^2 a^2 \left\{\frac{\kappa_e - \kappa}{\kappa} - 2(1 - i)(\gamma - 1)\frac{d_h}{a}\right.$$
$$\left. + \frac{2\rho_e - 2\rho}{\rho_e + \rho}\left[1 - (1 + i)\frac{2\rho_e}{\rho_e + \rho}\frac{d_v}{a}\right]\cos\phi\right\}$$

Compute the energy lost per unit length of cylinder.

8 Show that, neglecting the difference between γ_p and $(3\rho_e - 3\rho)/(2\rho_e + \rho)$, the results of Eqs. (8.2.22) can be obtained from the Born approximation if we assume that the quantity in braces in the first integral of Eq. (8.1.13) is

$$k^2 \gamma_\kappa(\mathbf{r}_0) p_i(\mathbf{r}_0) - \mathrm{div}_0[\gamma_\rho(\mathbf{r}_0)\,\boldsymbol{\nabla}_0 p_i] + (i-1)(\gamma-1)\,d_h\,\mathrm{div}\,[p_i(0)\mathbf{a}_{r_0}\,\delta(r_0-a)]$$

$$+ \tfrac{1}{2}(i+1)\gamma_\rho d_v\,\mathrm{div}_0[\mathbf{a}_{\theta_0}(\mathbf{a}_{\theta_0}\cdot\boldsymbol{\nabla}_0 p_i)\,\delta(r_0-a)]$$

where $r_0 = 0$ is the center of the sphere, γ_κ and γ_ρ are as defined in Eq. (8.1.11), and γ is the ratio of specific heats, where δ is the Dirac delta function.   Vector $\mathbf{a}_{r_0}$ is the unit vector in the direction of $\mathbf{r}_0$, and $\mathbf{a}_{\theta_0}$ is the unit vector perpendicular to $\mathbf{a}_{r_0}$ in the $r_0 z$ plane, and thus $\mathbf{a}_{\theta_0}(\mathbf{a}_{\theta_0}\cdot\boldsymbol{\nabla}_0 p_i)$ is the component of the gradient of p_i perpendicular to $\mathbf{r}_0$.   Show that these terms correspond to forces acting on the surface of the sphere.   Discuss the relationship between these forces and the physics of the heat conduction and viscous flow.

9 Suppose the sphere of radius a, centered at the origin, contains material of the same density ρ as the medium in the rest of space, but has compressibility $\kappa_e = \kappa(1 + \gamma_\kappa)$ differing from the κ for the medium outside.   Show that the integral equation for p_ω, for an incident plane wave,

$$p_\omega(\mathbf{r}) = e^{i\mathbf{k}_i\cdot\mathbf{r}} + \iiint k^2\,\gamma_\kappa(\mathbf{r}_0)p_\omega(\mathbf{r}_0)g_\omega(\mathbf{r}\mid\mathbf{r}_0)\,dv_0$$

can be converted into the variational expression

$$4\pi\Phi(\mathbf{k}_s) = k^2\iiint\gamma_\kappa(\mathbf{r})p(\mathbf{r})e^{-i\mathbf{k}_s\cdot\mathbf{r}}\,dv$$

$$+ k^2\iiint\gamma_\kappa(\mathbf{r})p^a(\mathbf{r})e^{i\mathbf{k}_i\cdot\mathbf{r}}\,dv - k^2\iiint\gamma_\kappa(\mathbf{r})p^a(\mathbf{r})p(\mathbf{r})\,dv$$

$$+ k^4\iiint\gamma_\kappa(\mathbf{r})p^a(\mathbf{r})\,dv\iiint g_\omega(\mathbf{r}\mid\mathbf{r}_0)\gamma_\kappa(\mathbf{r}_0)p(\mathbf{r}_0)\,dv_0$$

where p^a is the adjoint function, the solution of the integral equation, with $\mathbf{k}_i$ changed to $-\mathbf{k}_s$.   Show that, as the shape of p inside $r = a$ is varied, the stationary value of Φ comes when p is the solution of the integral equation, and that when the chosen shape of p differs somewhat from the correct solution p_ω, the resulting value of Φ differs from the correct Φ_s to second order.   Use $p = Ae^{i\mathbf{k}_i\cdot\mathbf{r}}$ and $p^a = Ae^{-i\mathbf{k}_s\cdot\mathbf{r}}$ as trial functions and show that the best values of A and Φ are

$$A \simeq \left\{1 - k^2\,\frac{\displaystyle\iiint\gamma_\kappa e^{-i\mathbf{k}_s\cdot\mathbf{r}}\,dv\iiint g_\omega\gamma_\kappa e^{i\mathbf{k}_i\cdot\mathbf{r}_0}\,dv_0}{\displaystyle\iiint\gamma_\kappa(\mathbf{r})\exp[i(\mathbf{k}_i-\mathbf{k}_s)\cdot\mathbf{r}]\,dv}\right\}$$

$$\Phi_s \simeq \frac{k^2 A}{4\pi}\iiint\gamma_\kappa(\mathbf{r})\exp[i(\mathbf{k}_i-\mathbf{k}_s)\cdot\mathbf{r}]\,dv$$

10 Suppose γ_κ in Prob. 9 is constant throughout the sphere.   Show that the integrals in the expression for A and Φ are

$$\iiint\gamma_\kappa e^{-i\mathbf{k}_s\cdot\mathbf{r}}\,dv\iiint g_\omega\gamma_\kappa e^{i\mathbf{k}_i\cdot\mathbf{r}_0}\,dv_0$$

$$= \frac{8\pi}{15}\gamma_\kappa^2 a^5\left[1 + \tfrac{5}{6}ika\,\tfrac{1}{21} - k^2 a^2\,(13 - \cos\vartheta)\cdots\right]$$

$$\iiint\gamma_\kappa\exp[i(\mathbf{k}_i-\mathbf{k}_s)\cdot\mathbf{r}]\,dv = \frac{4\pi}{3}\gamma_\kappa a^3[1 - \tfrac{1}{10}k^2 a^2 + \tfrac{1}{280}k^4 a^4\cdots]$$

Calculate the angle-distribution factor Φ and compare with the Born approximation and with the exact solution for the same case.

II Show that the backscattered sound of frequency $\omega/2\pi$ from a patch of variable acoustic admittance β, in an otherwise rigid wall, is, approximately,

$$p_{\omega b} \to \frac{ik}{2\pi r} e^{ikr} \int B(x_0) \exp(2ikx_0 \sin \theta) \, dx_0$$

where the angle of incidence is θ and the angle of scattering is $\vartheta = -\theta$ for backscattering, the direction of the incident wave being in the xz plane, and where $B(x) = \int \beta(x,y) \, dy$. Show that the component of the pulsed incident-plus-reflected wave

$$p_{pi} = A\delta\left(t + \frac{z}{c}\cos\theta - \frac{x}{c}\sin\theta\right) + A\delta\left(t - \frac{z}{c}\cos\theta - \frac{x}{c}\sin\theta\right)$$

which has frequency $\omega/2\pi$, is

$$p_{\omega i} = \frac{A}{2\pi}[\exp(-ikz\cos\theta + ikx\sin\theta) + \exp(ikz\cos\theta + ikx\sin\theta)]$$

which is the incident wave producing the $p_{\omega b}$ above. Show thus that the backscattered wave produced by the incident pulse is

$$p_{pb} \simeq \frac{A}{4\pi r \sin\theta} \frac{d}{dt} B\left(\frac{ct - r}{2\sin\theta}\right)$$

Explain in simple terms why the variations in space of B have turned into variations in time of p. If the material on the wall is laid out in periodic bands, parallel to the y axis, so that $B \simeq B_0 \sin(2\pi x/l)$ for $-\frac{1}{2}d < x < +\frac{1}{2}d$ $(d \gg l)$, what frequency sound will be backscattered from the pulse-wave?

CHAPTER

9

SOUND WAVES IN DUCTS AND ROOMS

9.1 ACOUSTIC TRANSMISSION LINES

A large number of sound-generating devices are tubular in shape, sound waves of large amplitude being set up inside the tube, and some of this stored-up energy being radiated out into the open. Other tubular devices, such as mufflers, are designed to attenuate the sound passing through them.

When a sound wave is transmitted along the interior of a tube, and when the wavelength is large compared with the transverse dimensions of the tube, the fluid motion is predominantly parallel to the tube axis, and the wave motion is very nearly one-dimensional. The mean fluid velocity through a cross-sectional area a distance x from one end of the tube, at time t, can be written $u(x,t)$, and the mean acoustic pressure there is $p(x,t)$. If the first-order wave equation (6.2.4) is to be valid for this motion, the fluid should be at rest and in thermal equilibrium when no sound is present, and u and p must be small compared with c, the wave velocity, and with P, the equilibrium pressure, respectively. For these waves the tube is analogous to an electric transmission line, and ac circuit theory can be used. Constrictions or openings in the tube can be considered as lumped-circuit elements, and uniform yielding of the tube walls can be considered to be contributing to the line impedance.

Standing waves

A long rigid tube of uniform internal cross section is a useful apparatus with which to measure the acoustic properties of a surface placed at the end of the tube. A vibrating piston, generating sound of frequency $\omega/2\pi$, is placed at the end $x = 0$, and a sample of the surface to be tested closes the other end of the tube, at $x = l$. By measurement of the pressure amplitude of the standing wave within the tube, the acoustic impedance of the surface can be determined. To see what measurements are to be made, we must study the nature of the standing wave set up in the tube.

Suppose two simple-harmonic waves are traveling along the tube, one, $p_+ = A_+ e^{ikx-i\omega t}$, going in the positive x direction with frequency $\omega/2\pi$ and

467

wavelength $\lambda = 2\pi/k = 2\pi c/\omega$, and the other, $p_- = A_- e^{-ikx-i\omega t}$, with the same frequency and wavelength, but going in the negative x direction. The actual acoustic pressure p at any x and t is of course given by the sum $p_+ + p_-$, as long as the amplitudes are small compared with P. Using Eqs. (6.2.3), we find that the fluid velocity at x, t is

$$u(x,t) = \frac{1}{i\omega\rho}\frac{\partial p}{\partial x} = \frac{1}{\rho c}[p_+(x,t) - p_-(x,t)] \qquad (9.1.1)$$

where

$$p(x,t) = p_+(x,t) + p_-(x,t)$$

As long as the tube walls are completely rigid and no energy is absorbed as the waves progress, the wave velocity $c = \omega/k = \sqrt{1/\rho\kappa}$ is real and equal to the wave velocity in the open (i.e., in the same fluid, of infinite extent, with no tube walls present). Also, the amplitudes $|p_+|$ and $|p_-|$ are independent of x and t. On the other hand (as we shall see), if the tube walls are not perfectly rigid, or if the viscous and thermal-conduction effects are non-negligible, the wave number k will have a positive imaginary part, corresponding to wave attenuation. Usually, this part is small compared with the real part of k, small enough so that the relationship $\operatorname{Re} k = 2\pi/\lambda$, valid when k is real, is still a good approximation. Thus we can write

$$k = \frac{2\pi}{\lambda} + i\pi\gamma \qquad (9.1.2)$$

The acoustic quantities may be conveniently related by means of the ratio (p/u) at point x for a simple-harmonic wave of frequency $kc/2\pi$, which can be called the impedance z at x for that frequency. The value z_0 at $x = 0$ determines the relation between the piston velocity and the pressure at the piston surface; the value z_l must equal the acoustic impedance of the material closing the end of the tube. These impedances are most compactly expressed in terms of the hyperbolic tangent function of a complex quantity $\pi(\alpha - i\beta)$, where α and β turn out to be linear functions of x (note that this β is *not* the specific acoustic admittance of Chap. 6). The relationship between α and β and the real and imaginary parts of the hyperbolic tangent are shown graphically in Plates I and II.

With these goals in mind, we write out the expressions for the pressure and velocity, in a plane wave of frequency $kc/2\pi$, as

$$p = A_+ e^{ikx} + A_- e^{-ikx} = A \cosh(ikx + \Phi_0) = A \cosh(ikd - \Phi_l)$$

$$u = \frac{A}{\rho c}\sinh(ikx + \Phi_0) = -\frac{A}{\rho c}\sinh(ikd - \Phi_l)$$

$$A_+ = \tfrac{1}{2}Ae^{\Phi_0} \qquad A_- = \tfrac{1}{2}Ae^{-\Phi_0} \qquad d = l - x \qquad \Phi_l = \Phi_0 + ikl \quad (9.1.3)$$

$$\Phi_0 = \pi\alpha_0 - i\pi(\beta_0 + \tfrac{1}{2}) \qquad \Phi_l = \pi\alpha_l - i\pi(\beta_l + \tfrac{1}{2})$$

$$ik = -\pi\gamma + i\pi\frac{2}{\lambda} \qquad \alpha_0 = \alpha_l + \gamma l \qquad \beta_0 = \beta_l + \frac{2l}{\lambda}$$

where the value of A is determined by the velocity amplitude of the piston at $x = 0$, and the values of α_0 and β_0 are determined by the acoustic reaction of the material closing the duct at $x = l$. This reaction is measured by the phase angle Φ_l; the ratio between reflected- and incident-wave amplitudes at $x = l$ is $\exp(-2\,\mathrm{Re}\,\Phi_s) = e^{-2\pi\alpha_l}$, and the phase change on reflection is $-2\,\mathrm{Im}\,\Phi_s = \pi(2\beta_s + 1)$.

If the reaction of the end material can be expressed in terms of an impedance $z_l = (p/u)_l = \rho c \zeta_l = \rho c(\theta - i\chi)$ [Eq. (6.3.4)], the equation determining α_l and β_l is

$$\zeta_l = \coth \Phi_l = \coth \pi(\alpha_l - i\beta_l - \tfrac{1}{2}i) = \tanh \pi(\alpha_l - i\beta_l) \quad (9.1.4)$$

since $\coth x = \tanh(x - \tfrac{1}{2}\pi i)$. If the specific acoustic impedance ζ_l is known, α_l and β_l can be computed graphically from Plates I and II, and then the impedance of the fluid in the tube on the piston

$$\zeta_0 = \coth \Phi_0 = \tanh \pi(\alpha_0 - i\beta_0) = \tanh \pi\left[(\alpha_l + \gamma l) - i\left(\beta_l + \frac{2l}{\lambda}\right)\right]$$

$$(9.1.5)$$

can also be computed from the same charts. A typical example is shown in Fig. 9.1.

Measurement of impedance

On the other hand, if the impedance $\rho c \zeta_l$ is not known, it can be measured by measuring the pressure amplitude in the tube as a function of x. The mean-square pressure a distance d from the material at $x = l$ is

$$|A|^2\,|\cosh(ikd - \Phi_l)|^2 = |A|^2 \left|\cosh \pi\left[(\alpha_l + \gamma d) - i\left(\beta_l + \frac{2d}{\lambda} + \tfrac{1}{2}\right)\right]\right|^2$$

$$= |A|^2\left[\cosh^2 \pi(\alpha_l + \gamma d) - \cos^2 \pi\left(\beta_l + \frac{2d}{\lambda}\right)\right] \quad (9.1.6)$$

As with any standing wave, this has a sequence of minima, a half-wavelength apart, separated by maxima, halfway between. The minima come at a distance $d_{\min}$ from the $x = l$ end, where

$$\beta_l + \frac{2d_{\min}}{\lambda} = n \quad \text{(an integer)} \quad \text{or} \quad d_{\min}(n) = -\lambda(n - \beta_l)$$

with the rms pressure there

$$p_{\min}(n) = |A| \sinh \pi[\alpha_l + \tfrac{1}{2}\gamma\lambda(n - \beta_l)]$$

The maxima come at $d_{\max}(n + \tfrac{1}{2}) = \tfrac{1}{2}\lambda(n + \tfrac{1}{2} - \beta_l)$, where n again is an integer; the rms pressure there is

$$p_{\max}(n + \tfrac{1}{2}) = |A| \cosh \pi[\alpha_l + \tfrac{1}{2}\gamma\lambda(n + \tfrac{1}{2} - \beta_l)]$$

The larger the argument of the sinh, the less sharp is the pressure minimum and the less is the difference between maxima and minima. If there is no friction (i.e., if α_l and γ are both zero), the pressure minimum is zero, and the standing wave is "perfect." Usually, γ is small; if the material

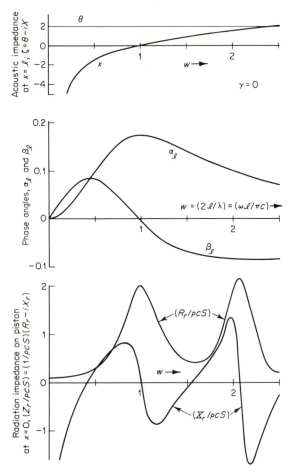

FIGURE 9.1
Specific acoustic impedance ζ of material at $x = l$ end of tube, phase angle Φ_l of standing wave there, and corresponding radiation impedance Z_r on piston at $x = 0$.

at $x = l$ is fairly rigid, α_l also is small, and the pressure minima are quite pronounced. On the other hand, if the impedance at $x = l$ is equal to the characteristic impedance ρc of the medium, then $\zeta = 1$ and $\alpha_l \rightarrow \infty$, in which case $|p_{\min}| \rightarrow |p_{\max}|$; no wave is reflected, and there is no interference; so there can be no minima.

To measure ζ_l, we first determine the spacing between pressure minima and the distance between the nearest minimum, $n = 1$, and the termination, $x = l$. The spacing is equal to $\frac{1}{2}\lambda$, and $\beta_l = 1 - [2d_{\min}(1)/\lambda]$. The value of γ is determined by measuring the rms pressure at two successive minima, say, for n and $n + 1$, and at the intermediate maximum. The difference between the pressure amplitudes at the two minima is

$$2\,|A|\,\cosh \pi[\alpha_l - \tfrac{1}{2}\gamma\lambda(n + \tfrac{1}{2} - \beta_l)]\,\sinh \frac{\gamma\lambda}{4}$$

and the ratio between this and the amplitude at the maximum,

$$\frac{|p_{\min}(n)| - |p_{\min}(n+1)|}{|p_{\max}(n + \tfrac{1}{2})|} = 2\sinh \frac{\gamma\lambda}{4} \to \frac{\gamma\lambda}{2} \qquad \gamma\lambda \ll 1 \qquad (9.1.7)$$

serves to determine the attenuation factor γ.

To determine α_l, we compute the ratio of the mean value of $|p_{\min}|$ at the successive minima to the intermediate $|p_{\max}|$.

$$\frac{|p_{\min}(n)| + |p_{\min}(n+1)|}{2\,|p_{\max}(n + \tfrac{1}{2})|} = \cosh \frac{\gamma\lambda}{4}\,\tanh \pi[\alpha_l - \gamma d_{\max}(n + \tfrac{1}{2})] \qquad (9.1.8)$$

Knowing γ, λ, and $d_{\max}(n + \tfrac{1}{2})$, this equation allows one to obtain the value of α_l. Then, from Plate I or II, we can compute $\rho c \zeta_l$, the impedance of the material at $x = l$ and $\rho c S \zeta_0$, the acoustic load of the air in the tube on the driving piston.

Open tube end

When the end of the tube at $x = l$ is open to free space, the conditions of flow and the distribution of pressure across the opening are not uniform, even when the wavelength is much longer than the tube diameter. In this long-wavelength limit, however, the only wave which can propagate inside the tube is the plane wave, the fundamental mode, which is the situation under discussion in this section. When this is the situation, it is possible to ignore the nonuniformity of velocity across a tube cross section and to assume that the air in the open end, at $x = l$, acts like a piston, radiating sound out into the open, as well as reflecting some back toward the driving piston at $x = 0$. At very long wavelengths little is radiated out of the open end, a strong standing wave is built up inside the tube, and the simple approximation we are discussing is close to the actual behavior. But as the driving frequency is increased, more and more energy leaks out of the open end, and the simple model loses validity.

For a circular tube of length l and inside radius a, with end $x = 0$ provided with a driving piston and the end $x = l$ flush with a rigid plane wall (i.e., the open end is provided with a large, rigid flange), the air in the open end may be considered, for long wavelengths, to be a piston of zero mass, radiating some energy into the open and returning some back down the tube.

To this approximation we can talk about an acoustic impedance of the open end, the ratio of the mean pressure at $x = l$ to the mean axial velocity there. The reactive part of this impedance would produce reflection back to the driving piston at $x = 0$; the resistive part would represent energy lost from the tube by being radiated out of the open end.   Reference to Eq. (7.4.31) indicates that this open-end impedance, for a flanged open end, would be the function $\theta_0 - i\chi_0$ (plotted in Fig. 7.9 and shown in Table IX) multiplied by ρc.

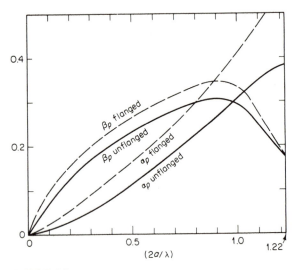

FIGURE 9.2
Impedance parameters $\Phi_p/\pi = \beta_p + i\alpha_p$ for open end of circular tube.   Curves marked *unflanged* are for open end in infinite space; curves marked *flanged* are for opening in plane wall, radiation confined to one side of wall.

Therefore the standing-wave parameters α_l and β_l for the wave inside the tube can be determined.   We define the quantities $\alpha_p = \alpha_l$ and $\beta_p = \beta_l$ such that

$$\theta_0 - i\chi_0 = \tanh \pi(\alpha_p - i\beta_p) \quad \text{(see Table IX)}$$

The radiation load on the piston at $x = 0$ is then easily computed from Eq. (9.1.8).

$$Z_r = R_r - iX_r = \rho c S \tanh \pi \left[\alpha_p + \gamma l - i\left(\beta_p + \frac{2l}{\lambda} \right) \right] \quad (9.1.9)$$

where $S = \pi a^2$ is the cross-sectional area of the tube. R_r and X_r can thus be calculated by using Plates I or II.   The parameters α_p and β_p for a tube with flanged open end are plotted as the dashed lines marked *flanged*, as functions of $2a/\lambda = a\omega/\pi c$, in Fig. 9.2.   We see that, for $\lambda > 10a$, the majority of the energy stays inside the tube (the ratio of reflected to incident intensity is $e^{-4\pi\alpha}$, and for $\alpha < 0.05$, this ratio is greater than 0.5, indicating that a minority of the energy coming to the open end, from the piston, escapes from the tube). We also see that the reactive effect of the open end is considerable (this reactive effect is measured by the phase change $e^{2\pi i\beta_p}$ of the pressure wave on reflection, and β_p is larger than 0.1 unless $\lambda > 20a$).

In many cases the tube is not flanged, but radiates into free space (as with an open organ pipe, for example), in which case the impedance functions θ_0 and χ_0 given in the dashed curves of Fig. 9.2 are not appropriate. The equivalent impedance of an unflanged circular tube can be computed exactly for $\lambda > 2a$ [see Levine and Schwinger, *Phys. Rev.*, **73**: 383 (1948)]. It requires the solution of a Wiener-Hopf integral equation, a subject beyond the scope of this book. The results may be translated into values of α_p and β_p, and are included in Fig. 9.2 as the solid lines marked *unflanged*. We note that, because the escaping sound must spread out over the whole solid angle 4π, instead of the 2π in front of the flange in the previous case, the *unflanged* curves indicate a smaller reactive effect and less energy radiated than for the flanged tube at the same frequency (a larger solid angle corresponds to a poorer impedance match between inside and outside). The general trend of the curves is the same, however. The calculations for the *unflanged* case are invalid for $2a/\lambda$ larger than 1.22, above which a higher-mode symmetric wave can be transmitted inside the tube, as will be shown in Sec. 9.2.

Having values of α_p and β_p, at least for the longer wavelengths, we now can use Eq. (9.1.9) to compute the radiation impedance on the driving piston at $x = 0$. A single example of the results is displayed in Fig. 9.3, for an unflanged tube twice as long as its diameter. The lower curve gives the real part, and the upper curve the imaginary part, of the hyperbolic tangent. The resonances are clearly indicated, particularly at the lower frequencies. If the boundary condition at $x = l$ had been that p is zero there (as is sometimes assumed in calculating resonance frequencies for open tubes), α_p, β_p, and R_r would have been zero, and X_r would have been $\rho cS \tan(2\pi l/\lambda)$. The resonances would have come at $2l/\lambda = \frac{1}{2}, \frac{3}{2}, \ldots$. Figure 9.3 shows that the actual resonances are for somewhat smaller values, for $2l/\lambda = 0.43, 1.33, 2.25, 3.21$. As the frequency is increased, more and more energy is lost through the open end (α_p increases); so the resonance peaks get broader and less high until, for $2l/\lambda > 4.88$, the plane-wave approximation loses all validity. Obviously, a much narrower tube, in relation to its length, would have many more sharp resonances, coming at values of $2l/\lambda$ much closer to half-integer values, at least for the lower resonances.

If the tube is used as a musical instrument, the "driving piston" is either a vibrating reed or vibrating jet of air, which generates sound of all frequencies. In this case the energy radiated will be more or less proportional to the radiation resistance R_r at the piston. This radiated energy is concentrated in narrow frequency bands, corresponding to the peaks in the curve of R_r. For a long narrow tube, these peaks come for frequencies such that $2l/\lambda \simeq n + \frac{1}{2}$, the usual formula for resonance in an open-ended tube. For the higher harmonics, however, when n begins to be as large as l/a, the peaks drop off in height and narrowness, and the resonances differ more and more from the simple-harmonic ratio $2n + 1$ to the fundamental.

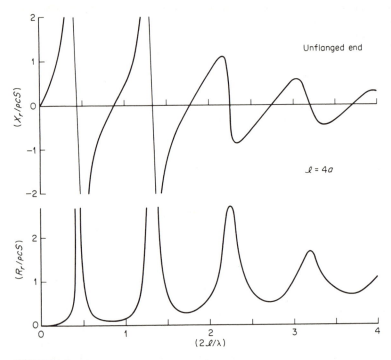

FIGURE 9.3
Radiation resistance R and reactance X of a piston driving air in a circular tube of radius a and length l, the other end open to free space, as functions of $kl/\pi = 2l/\lambda$.

Analogous impedance

When the wavelength of the sound wave in the tube is much greater than the cross-sectional dimensions of the tube, the wave motion is quite analogous to the flow of electric current in a transmission line. If the tube has cross-sectional area S, the total flux of fluid, $U = Su$, is the analogue of the electric current I, and the sound pressure p is the analogue of the voltage E. The inductance L_l per unit length of transmission line is defined, in circuit theory, by saying that the magnetic energy per unit length of line is $\frac{1}{2}L_l I^2$. The analogous inductance L_a, per unit length of the pipe, is likewise defined by saying that the kinetic energy of the fluid, per unit length of pipe, is $\frac{1}{2}L_a U^2$, so that

$$L_a = \frac{2}{U^2} \text{KE} = \frac{2}{U^2} \tfrac{1}{2} S \rho u^2 = \frac{\rho}{S} \tag{9.1.10}$$

from Eq. (6.1.7), since the volume of a unit length of pipe is S.

Similarly, we can define the analogous capacitance C_a per unit length of pipe by saying that the potential energy per unit length is $\tfrac{1}{2} C_a p^2$. Again from Eq. (6.1.7),

$$C_a = \frac{2}{p^2} \text{PE} = \frac{2}{p^2} \tfrac{1}{2} S \kappa p^2 = S \kappa = \frac{S}{\rho c^2} \tag{9.1.11}$$

Thus the analogous characteristic impedance of the acoustic transmission line and the velocity of wave motion along it are

$$\sqrt{\frac{L_a}{C_a}} = \frac{\rho c}{S} \qquad \sqrt{\frac{1}{L_a C_a}} = c$$

and the power transmitted along the wave is $pU = (\rho c/S)U^2 = S(\rho c u^2)$, where $\rho c u^2$ is the intensity of the sound wave.

This is all that needs to be said if the tube walls are perfectly rigid, if we are willing to neglect energy losses at the tube walls, and if the tube cross section is independent of x. But if we desire to include the effects of heat conduction and viscosity at the tube walls, we must also include an analogous resistance R_a per unit length of tube. By analogy the resistance would be defined by stating that the power loss per unit length is $R_a |U|^2$. According to Eqs. (6.4.37) and (6.4.38), for a wave moving tangentially to the surface ($k_t = \omega/c = k$), the power loss per unit area of surface is $[kd_v + (\gamma - 1)kd_h](|p|^2/2\rho c)$, where d_v and d_h are defined in Eqs. (6.4.31). If D is the perimeter of the duct's internal cross section, and since $|p| \simeq \rho c |u| = (\rho c/S) |U|$ on the average, we can write the equation for the analogous resistance per unit length of duct of uniform cross-sectional area S and perimeter D as

$$R_a = \frac{\rho c D}{2S^2} [kd_v + (\gamma - 1)kd_h] \tag{9.1.12}$$

This resistance is proportional to the square root of the driving frequency; for ducts of cross-sectional area larger than 1 sq cm and for frequencies of a few kilocycles or less, it can ordinarily be neglected.

Ducts with yielding walls

The equation of motion given in Eqs. (6.2.3) for free space may be written for a plane wave in a duct in terms of acoustic pressure and flux $U = Su$, in accord with the electric-circuit analogy,

$$L_a \frac{\partial U}{\partial t} + R_a U = -\frac{\partial p}{\partial x} \tag{9.1.13}$$

where, as we have just said, we can often neglect the R_a term.   If the duct walls yield somewhat to the internal pressure, the equation of continuity, relating the time rate of change of pressure to the gradient of U, must be modified.

The term $-(\partial U/\partial x)$ represents a rate of influx of gas into a unit length of duct, and this corresponds to a change of density, and thus of pressure there.   If the duct walls are rigid, this influx equals the compressibility $\kappa = 1/\rho c^2$ of the gas times the time rate of change of pressure.   But if the duct walls yield, so that an increase of pressure produces a fractional change of cross-sectional area, this dilutes somewhat the increased compression of the gas, as though the two effects acted in parallel.   Thus the equation of continuity would be written

$$-\frac{\partial U}{\partial x} = S\left(\kappa + \frac{1}{K_w}\right)\frac{\partial p}{\partial t}$$

where K_w, the *wall stiffness constant*, is the ratio between the pressure p and the fractional change in cross-sectional area of the duct, produced by p.

The pressure p is not static, however, and we also should take into account the inertia of the duct walls.   In a stiffness-controlled system [Eqs. (2.3.4)] the impedance is $K/-i\omega$, where $\omega/2\pi$ is the frequency of the wave.   If mass is to be considered, the impedance is $-i\omega M + (K/-i\omega) = Z$. Therefore the equation of continuity for a duct with yielding walls is

$$-\frac{\partial U}{\partial x} = S\left(\kappa + \frac{1}{-i\omega Z_w}\right)\frac{\partial p}{\partial t} = S\left[\frac{1}{\rho c^2} + \frac{1}{M_w(\omega_0{}^2 - \omega^2)}\right]\frac{\partial p}{\partial t}$$

where $\omega_0{}^2 = K_w/M_w$, $\omega_0/2\pi$ being the resonance frequency of transverse expansion oscillation of the duct walls, unencumbered by the fluid inside. Thus the second of Eqs. (6.2.3) becomes

$$-\frac{\partial U}{\partial x} = S\kappa\frac{\omega_1{}^2 - \omega^2}{\omega_0{}^2 - \omega^2}\frac{\partial p}{\partial t} \to S\kappa\frac{\partial p}{\partial t} \qquad \text{as} \qquad K_w \to \infty \qquad (9.1.14)$$

where $\omega_1{}^2 = \omega_0{}^2 + (\rho c^2/M_w)$, $\omega_1/2\pi$ being the resonance frequency of the walls when stiffened by the air inside it.   It should be noted that this analysis assumes that the duct walls are *locally reacting*, in the sense of page 260; that no expansive wave motion is produced along the duct by the wave motion inside (see Prob. 1 of Chap. 10 for waves in the duct walls).

In an electric transmission line the corresponding equation is $-(\partial I/\partial x) = C(\partial V/\partial t)$, where C is the capacitance per unit length of the line.   Therefore, for a tube with locally yielding walls, Eq. (9.1.11) for the analogous capacitance per unit length of duct becomes

$$C_a = \frac{S}{\rho c^2}\frac{\omega_1{}^2 - \omega^2}{\omega_0{}^2 - \omega^2} \qquad\qquad (9.1.15)$$

The wave velocity in the duct is then

$$c_\varphi = \frac{1}{\sqrt{L_a C_a}} = c\left(\frac{\omega_0^2 - \omega^2}{\omega_1^2 - \omega^2}\right)^{\frac{1}{2}} \qquad c^2 = \frac{1}{\rho\kappa} \qquad (9.1.16)$$

as long as the wavelength is long compared with the tube perimeter D. The effect of the wall motion is to make this velocity a function of frequency. If the walls are "soft," so that ω_0 and ω_1 are in the range of frequencies for which the transmission-line analogy holds (i.e., for which $\omega_0 D/c$ and $\omega_1 D/c$ are much smaller than unity), the wave velocity inside the tube differs greatly from the speed c of sound in the open.

As ω is increased from zero, the wave velocity first diminishes, becoming zero when $\omega = \omega_0$, when the wave energy is entirely diverted to exciting tube oscillations. For $\omega_0 < \omega < \omega_1$, the wave velocity is imaginary, indicating that acoustic motion is so strongly attenuated that no true wave motion can exist (this is identical with the anomalous dispersion and absorption in optical media). Finally, when ω is larger than ω_1, the wave velocity is again real, but larger than c. It decreases as ω increases further, approaching c as $\omega \to \infty$. Of course, when ω becomes larger than c/d, the wavelength of the sound is no longer much greater than the tube diameter, and our simple one-dimensional theory is no longer valid.

Thus the wave velocity in the tube is a function of the frequency of the wave. Optical media, in which the light velocity is a function of frequency, exhibit *dispersion*, so that a medium in which the wave velocity is frequency-dependent is called a *dispersive medium*. A pipe with yielding walls is a dispersive medium for longitudinal acoustic waves. Waves of arbitrary shape will not maintain their shape as they travel along the interior of the tube.

Phase and group velocity

With dispersive waves, when c is a function of ω, we must distinguish between a number of different wave velocities. First there is the *phase velocity* c_φ, the velocity with which the phase angle $\omega[t - (x/c_\varphi)]$ of a simple-harmonic wave of frequency $\omega/2\pi$ progresses along the tube. This is the wave velocity of the wave of Eq. (9.1.16).

Then there is the *group velocity*, the velocity of progress of the "center of gravity" of a group of waves that differ somewhat in frequency. For example, we might combine a sequence of simple-harmonic waves, each of the form $\exp[i(\omega/c_\varphi)(x - c_\varphi t)]$, with pressure amplitudes which are maximum for $\omega = \omega_a$ and fall off on either side, as given by the "error-function" formula, $(A/\delta\sqrt{2\pi}) \exp[-(\omega - \omega_a)^2/2\delta^2]$, so that the acoustic pressure is

$$p = \frac{A}{\delta\sqrt{2\pi}} \int_{-\infty}^{\infty} \exp\left[-i\omega t + ix\frac{\omega}{c_\varphi(\omega)} - \left(\frac{1}{2\delta^2}\right)(\omega - \omega_a)^2\right] d\omega \quad (9.1.17)$$

Here $\omega_a/2\pi$ is the mean frequency of the group, and $\delta/2\pi$ is the mean deviation in frequency of any one component of the wave from this average. If the phase velocity c_φ were independent of ω, this group of waves would have the form

$$p = A \exp\left[-i\omega_a\left(t - \frac{x}{c_\varphi}\right) - \tfrac{1}{2}\delta^2\left(t - \frac{x}{c_\varphi}\right)^2\right]$$

which is a pulse extending, roughly, from $x = c_\varphi t - (c_\varphi/\delta)$ to $c_\varphi t + (c_\varphi/\delta)$ in space, traveling with velocity c_φ and oscillating with frequency $\omega_a/2\pi$ as it travels along.

But if c is a function of ω, we must take this into account in performing the integration. We can, for example, expand the quantity

$$\frac{\omega}{c_\varphi(\omega)} = \frac{\omega_a}{c_a} + (\omega - \omega_a)\frac{1}{c_g} + \cdots$$

where

$$c_a = c_\varphi(\omega_a) \qquad \frac{1}{c_g} = \left[\frac{d}{d\omega}\frac{\omega}{c_\varphi(\omega)}\right]_{\omega=\omega_a} \tag{9.1.18}$$

in a Taylor's series about the average value ω_a. If δ, the "spread" in frequency of the group, is small compared with ω_a, we can usually neglect all the terms higher than the first order in this expansion, and the exponent of the integrand of Eq. (9.1.17) becomes

$$-i\omega_a\left(t - \frac{x}{c_a}\right) - i(\omega - \omega_a)\left(t - \frac{x}{c_g}\right) - \frac{1}{2\delta^2}(\omega - \omega_a)^2$$

The integration can then be carried out, giving

$$p = A \exp\left[-i\omega_a\left(t - \frac{x}{c_a}\right) - \tfrac{1}{2}\delta^2\left(t - \frac{x}{c_g}\right)^2\right] \tag{9.1.19}$$

This represents a pulse of spatial length c_g/δ and of frequency $\omega_a/2\pi$. The velocity of the surfaces of constant phase angle in this group wave is $c_a = c_\varphi(\omega_a)$, the phase velocity of the component of average frequency. But the velocity of the wave peak, the point where the various components are all in phase, is not c_a, but $c_g = 1/[d(\omega/c_\varphi)/d\omega]_{\omega_a}$, which is the group velocity of any group of waves of frequency centered around ω_a. If c_φ is independent of ω, $c_g = c_a$, but if c_φ does vary with ω,

$$c_g = c_a\left[1 - \frac{\omega_a}{c_a}\left(\frac{\partial c_\varphi}{\partial \omega}\right)_{\omega_a}\right]^{-1}$$

which is smaller than c_a if $dc/d\omega$ is negative (as it is for acoustic waves in a yielding tube).

Finally, there is the *signal velocity*, which is the velocity of the *front* of a group of waves, the speed at which one end of the tube first learns that a signal has been started from the other end.   Since a wave, to have a beginning, must be a combination of waves of different frequencies, it is obvious that the signal velocity (if there be one) is some sort of average value of c_φ. But if there are waves with phase velocity of any value, no matter how large (as there are with waves in a nonrigid tube), it is not obvious that a wavefront can be maintained.   In fact, it would seem probable that the components of high phase velocity would inevitably draw ahead of the rest and that eventually there would be no sharp wavefront.

This is the case with the diffusion of heat, for example.   If the temperature at the end of a rod is suddenly increased at time $t = 0$, the temperature at a distance x along the rod begins *immediately* to rise, slowly at first but continuously, so that at no time after $t = 0$ can we say that the "heat front" has suddenly arrived.   There is no wavefront, and thus no signal velocity for heat flow (in fact, there is no true wave motion for it).   One can ask whether there can be such a thing as a sharp wavefront (and hence a signal velocity) in a dispersive medium, such as a pipe with yielding walls.

It turns out that the crucial test is not whether the phase velocity goes to infinity for some intermediate value of ω, but whether c has a finite limiting value as $\omega \to \infty$.   By extension of Eqs. (1.3.16) and (2.3.6), we see that the pressure at point x at time t in a one-dimensional dispersive wave, for which the pressure is $p_0(t)$ at $x = 0$, is

$$p(x,t) = \int_{-\infty}^{\infty} P(\omega) \exp\left(i\frac{\omega}{c_\varphi}x - i\omega t\right) d\omega$$

where

$$P(\omega) = \frac{1}{2\pi}\int_{-\infty}^{\infty} p_0(t)\, e^{i\omega t}\, dt \qquad (9.1.20)$$

and c_φ is some function of ω.   Now, if $p_0(t)$ is zero for $t < 0$, then all the poles and other singularities of $P(\omega)$ will be *on or below the real axis* of the variable ω, and the contour of integration of ω will be just above its real axis. Also, the singularities and zeros of $c_\varphi(\omega)$ are on or below the real axis of ω [Eq. (9.1.16)], so that the contour of integration for ω lies above all the singularities of the integrand.   Thus, according to Eqs. (1.2.15), the acoustic pressure $p(x,t)$ is zero for $t - [x/c(\infty)] < 0$, and so, if $c_\varphi(\omega)$ approaches a finite value, $c_\varphi(\infty)$, as $\omega \to \infty$, nothing will occur at point x until time $t = x/c_\varphi(\infty)$; the wave will have a true wavefront which travels with the signal velocity $c_\varphi(\infty)$.

The velocity of simple-harmonic "waves" of heat is proportional to $\sqrt{\omega}$; thus $c_\varphi(\omega) \to \infty$ as $\omega \to \infty$, and heat conduction has no wavefront, as we pointed out earlier.   On the other hand, for waves in a pipe, if we assume that Eq. (6.4.21) holds over the whole range of ω, then $c_\varphi(\infty)$ is just equal to c, the velocity of sound waves in free space; thus the signal velocity of waves in a tube exists and equals c.

Lumped-circuit elements

Sudden changes of duct cross section produce sudden changes of pressure, and in the long-wavelength approximation, their effects are equivalent to the lumped-circuit impedances, capacitances, and resistances familiar in electric circuits. For example, a constriction in a duct gives rise to a concentration of flow through the constriction, with consequent increase in kinetic-energy density. If the net increase is concentrated in a region small compared with the wavelength, the total increase can be represented by a lumped inductance. If the constriction also produces an increase in viscous-energy loss, this addition can be represented as a lumped resistance.

The various types of acoustical circuit elements are shown in Fig. 9.4, together with their electrical analogues. The closed volume V_D at D behaves like a capacitance, since additional volume of fluid, introduced into D, increases the pressure there. The potential energy of this compression is approximately $\frac{1}{2}V_D\kappa p^2$; so the analogous capacitance corresponding to D is [see also Eq. (9.1.11)]

$$C_D \simeq \frac{V_D}{\rho c^2} \tag{9.1.21}$$

To calculate the analogous impedance and resistance of a constriction like the openings at C or B of Fig. 9.4, we compute the extra kinetic energy and viscous-energy loss, produced by steady flow through the opening. The simplest case is that of a circular opening of radius b in a thin but rigid plate. If the area πb^2 is considerably smaller than the area S of the duct cross section, we can consider the flow pattern through the hole to be unaffected by the duct walls and calculate the flow through a circular hole in an infinite plate.

As shown in Fig. 9.5, the lines of flow are hyperbolas, $\eta = \text{constant}$, with foci the traces of the sharp edge of the hole on the rx plane. The surfaces of constant velocity potential, $\xi = \text{constant}$, are oblate spheroids, with elliptical sections in the rz plane, having the same foci. Thus ξ, η, and

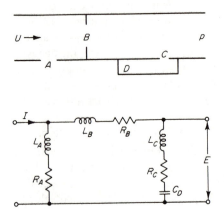

FIGURE 9.4
Illustration of electric-circuit elements analogous to an acoustical transmission line with openings, constrictions, and shunt volumes.

ϕ, the angle around the x axis, the appropriate set of coordinates to describe the motion, are related to x, y, and z by the equations

$$x = b\xi\eta$$
$$y = b\sqrt{(\xi^2 + 1)(1 - \eta^2)} \cos \phi$$
$$z = b\sqrt{(\xi^2 + 1)(1 - \eta^2)} \sin \phi$$

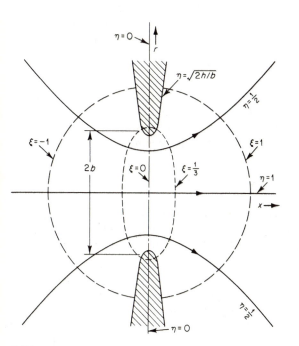

FIGURE 9.5
Lines of flow, $\eta = $ const, and velocity-potential surfaces, $\xi = $ const, for steady flow through a circular orifice.

It can be shown that the magnitude u of the fluid velocity at the point x, y, z or ξ, η, ϕ is $(U/2\pi b^2)[(1 + \xi^2)(\xi^2 + \eta^2)]^{-\frac{1}{2}}$, going to zero at infinite distances from the hole ($\xi \to \infty$) and going to infinity at the edge of the hole ($\xi \to \eta \to 0$) if the plate is infinitely thin. Factor U is the total rate of flow through the opening, and thus is the value of the analogue to the current, as already used. The volume element in the ξ, η, ϕ coordinates is

$$b^3(\xi^2 + \eta^2) \, d\xi \, d\eta \, d\phi$$

So the total kinetic energy of the fluid is

$$\tfrac{1}{2}\rho\left(\frac{U}{2\pi b^2}\right)^2 \int_0^{2\pi} d\phi \int_0^1 d\eta \int_{-\infty}^{\infty} \frac{b^3(\xi^2 + \eta^2) \, d\xi}{(\xi^2 + 1)(\xi^2 + \eta^2)} = \frac{\rho U^2}{2\pi b} \int_0^{\infty} \frac{d\xi}{1 + \xi^2} = \frac{\rho}{4b} U^2$$

Thus, in accord with Eq. (9.1.10), when πb^2 is much smaller than S, the lumped-circuit inductance of the opening is

$$L_a \simeq \frac{\rho}{2b}$$

If we had chosen a mechanical, instead of electrical, analogue, we should have talked about the air in and near the opening as equivalent to a solid plug of material, moving back and forth through the opening with velocity $u = U/\pi b^2$ and kinetic energy $\frac{1}{2}M_a u^2$. In this case the effective mass would be

$$M_a \simeq \tfrac{1}{2}\pi^2 b^3 \rho = (\pi b^2)^2 L_a$$

which is equivalent to a slug of fluid of density ρ, of cross section equal to $S_0 = \pi b^2$, the area of the opening, and of length $l_0 = \frac{1}{2}\pi b$ equal to one-quarter of the perimeter of the opening. This leads us to a way of writing these formulas, in terms of the area S_0 of the opening and its perimeter D_0.

$$M_a \simeq \tfrac{1}{4}D_0 S_0 \rho \qquad L_a \simeq \frac{\rho D_0}{4 S_0} \qquad (9.1.22)$$

which we have just shown is valid for circular holes, as long as $S_0 \ll S \ll \lambda^2$. We should expect such formulas to apply reasonably well to square holes or holes of other shape, as long as $(D_0/4)^2$ is of the same order of magnitude as S_0 (i.e., as long as the opening is about as long as it is wide).

To compute the analogous resistance of the circular hole, we use Eq. (6.4.39) to compute the additional power lost because of the increase in velocity near the edge of the opening. Equation (9.1.12) gives the analogous resistance per unit length of duct for the uniform duct. The tangential velocity near the sharp edges of the hole is much larger than that tangential to the duct walls. Thus, though the thermal-conduction loss rate is not increased much, the viscous loss may be quite large near the edge of the hole. In fact, if we assume that the surface of the plate corresponded to the surface $\eta = 0$ (i.e., the plate were of zero thickness), the tangential velocity next to the plate would be $U/2\pi b^2 \sqrt{\xi(1 + \xi^2)}$, which goes to infinity at $\xi = 0$, at the infinitely sharp edge of the opening.

This formula for the tangential velocity next to the plate is incorrect for two reasons. First, the plate has a finite thickness, and the edge has a finite radius of curvature, not zero. It would be more accurate to approximate the surface of the plate by the surface $\eta_s = \sqrt{2h/b}$, where $h \ll b$ is the radius of curvature of the edge of the hole, and thus $2h$ is roughly equal to the thickness of the plate near its edge; h is small but not zero. In the second place, the formula of Eq. (6.4.39) has to do with tangential velocity just *outside* the viscous boundary layer; the tangential velocity at the surface of the plate is zero. The thickness $d_h = \sqrt{2\mu/\rho\omega} \simeq 0.21/\sqrt{\nu}$ cm for air at standard conditions applies only when d_h is smaller than the radius of curvature of the surface.   Only when this is true can we assume that the u_{tan} of Eq. (6.4.39) is the u_{tan} calculated for the surface when viscosity is neglected, the way Eq. (6.4.39) has been used.   Thus the h in the formula $\eta_s = \sqrt{2h/b}$ is either the half-thickness of the plate near the edge of the hole, or else is the thickness d_v of the viscous boundary layer, whichever is *larger*.

Using Eq. (6.4.39), the power lost by flow through a circular hole of radius b is the integral of $-\rho\omega d_v(u_{tan})^2$ over the surface $\eta_s = \sqrt{2h/b}$.

$$\tfrac{1}{2}\rho\omega d_v\left(\frac{U}{2\pi b^2}\right)^2\int_0^{2\pi} d\phi \int_{-\infty}^{\infty} \frac{\sqrt{\xi^2 + (2h/b)}}{(\xi^2 + 1)\left(\xi^2 + \dfrac{2h}{b}\right)}\, b^2\, d\xi \simeq \frac{\rho\omega d_v}{4\pi b^2}\, U^2 \ln \frac{2b}{h}$$

when $h \ll b$.   Therefore the analogous resistance of a circular hole of radius $b \ll \sqrt{S/\pi}$ is

$$R_a \simeq \frac{\rho\omega d_v}{4\pi b^2}\ln \frac{2b}{h} = \frac{1}{4S_0}\sqrt{2\rho\omega\mu}\,\ln \frac{2b}{h} \qquad (9.1.23)$$

where $h \ll b$ is either the half-thickness of the plate near the hole edge (i.e., is the mean radius of curvature of the edge) or else is the viscous-boundary-layer thickness d_v, whichever is larger.   (When b becomes the same size as d_v, the boundary layers merge in the hole, and the exact solutions of Sec. 6.4 must be used; fortunately, such a minute hole is of no great practical interest.) The resistance R_a is in series with the lumped-circuit inductance L_a of Eq. (9.1.22) for the hole, as shown in Fig. 9.4.   At moderate and large intensities, turbulence around the edges modifies R; see Eq. (11.3.36).

Two-dimensional flow

To determine the lumped-circuit impedance and resistance of a constricted opening which is considerably longer than it is wide, we resort to a technique which can be used whenever the problem is a two-dimensional one. If the duct is rectangular in cross section, with two walls parallel to the xy plane, everywhere a distance d apart, the distance between the other two walls varying with x, then the fluid motion in the duct will be parallel to the xy plane, and p and u will be functions of x and y only (except for the time dependence).   We then can use the complex notation (Fig. 1.1) to assist us. Both velocity and pressure can be expressed in terms of the velocity potential Ψ of Eq. (6.2.13) (for convenience here we shall use minus the potential, $\psi = -\Psi$).

When the wavelength is long compared with the duct width, the function $\psi(x,y)$ will approximate a solution of the Laplace equation, $(\partial^2\psi/\partial x^2) + (\partial^2\psi/\partial y^2) \simeq 0$.   In this case ψ will turn out to be the real part of a function of the complex variable $z = x + iy$:

$$F(z) = \psi(x,y) + i\theta(x,y) \qquad z = x + iy \qquad z^* = x - iy$$

$$F^* = \psi - i\theta \qquad \frac{\partial}{\partial x} = \frac{\partial}{\partial z} + \frac{\partial}{\partial z^*} \qquad \frac{\partial}{\partial y} = i\frac{\partial}{\partial z} - i\frac{\partial}{\partial z^*}$$

and the velocity potential ψ and the *flow function* θ, so defined, will be related by the Cauchy relations

$$\frac{\partial \psi}{\partial x} = \frac{\partial \theta}{\partial y} \qquad \frac{\partial \psi}{\partial y} = -\frac{\partial \theta}{\partial x}$$

We can then use the geometric representation of a complex quantity, pictured in Fig. 1.1, to define the vector velocity of the fluid, the real part, $\partial \psi / \partial x$, being the x component, and the imaginary part, $\partial \psi / \partial y$, being the y component, so that

$$\frac{\partial \psi}{\partial x} + i\frac{\partial \psi}{\partial y} = \frac{1}{2}\frac{\partial}{\partial x}(\psi - i\theta) + \frac{1}{2}i\frac{\partial}{\partial y}(\psi - i\theta) = \frac{1}{2}\frac{\partial}{\partial x}F^* + \frac{1}{2}i\frac{\partial}{\partial y}F^*$$

$$= \frac{1}{2}\frac{\partial}{\partial z}F^* + \frac{1}{2}\frac{\partial}{\partial z^*}F^* - \frac{1}{2}\frac{\partial}{\partial z}F^* + \frac{1}{2}\frac{\partial}{\partial z^*}F^* = \frac{\partial F^*}{\partial z^*}$$

Since F is a function of z alone (not z^*), F^* will be a function of z^* alone, and $\partial F^*/\partial z^*$ will be the complex conjugate of the complex quantity dF/dz. Thus the complex quantity representing the magnitude and direction of the fluid flow at the point $x + iy$ within the duct is given by the complex conjugate of the derivative dF/dz, where the real part of $F(z) = \psi + i\theta$ is the negative velocity potential.

$$u_x = \text{Re}\,\frac{dF}{dz} \qquad u_y = -\text{Im}\,\frac{dF}{dz}$$

$$p \simeq i\rho\omega\,\text{Re}\,F \qquad |u|^2 = \left|\frac{dF}{dz}\right|^2$$

(9.1.24)

Inductance and resistance of a slit

If the traces of the duct walls perpendicular to the xy plane consist of portions of straight lines, we can use the Schwartz-Christoffel transformation to find F as a function of z (actually, we find z as a function of F, but this is only moderately embarrassing). As an example of the method, we consider the duct of Fig. 9.6, having depth d perpendicular to the xy plane and having constant width a in the y direction, except for a diaphragm at $x = 0$ with an open slit of width b symmetrically located.

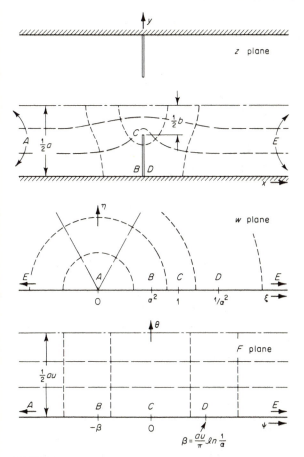

FIGURE 9.6
Conformal transformations from the z plane of the duct to the velocity-potential F plane, via the w plane, for half a rectangular duct with slit of width b. Dashed lines are streamlines, $\theta = \text{const}$; dotted lines, velocity-potential contours, $\psi = \text{const}$.

To find the steady-state flow, we first transform the interior of the half-duct, between the lines $y = 0$ and $y = -\frac{1}{2}a$ to the upper half of the $w = \xi + i\eta$ plane, by the Schwartz-Christoffel transformation

$$\frac{dz}{dw} = \frac{a}{2\pi w}\frac{w-1}{\sqrt{(w-\alpha^2)(w-1/\alpha^2)}}$$

$$z = \frac{a}{\pi}\ln\frac{\sqrt{w-\alpha^2}+\sqrt{w-(1/\alpha)^2}}{\sqrt{(1/\alpha)^2-(1/w)}+\sqrt{\alpha^2-(1/w)}} \rightarrow \begin{cases} \dfrac{a}{2\pi}\ln\dfrac{4w}{Q} & w\rightarrow\infty \\[2mm] \dfrac{a}{2\pi}\ln\dfrac{wQ}{4} & w\rightarrow 0 \end{cases}$$

$$(9.1.25)$$

where $Q = (1/\alpha)^2 + \alpha^2 + 2 = [(1/\alpha) + \alpha]^2$, and where, in order that point C ($w = 1$) correspond to $z = (i/2)(a - b)$, the quantity α must equal $\tan(\pi b/4a)$.

The transformation from the w to the F plane is simple.

$$w = e^{2\pi F/au} \qquad F = \frac{au}{2\pi} \ln w$$

if the fluid velocity at $x = \pm\infty$ is u, so that the total flux is $uad = U$. Thus at point E ($x \to \infty$ or $w \to \infty$ or $\psi \to \infty$), the fluid velocity, in magnitude and direction, is the limiting value of the complex conjugate of

$$\frac{dF}{dz} = \frac{dF}{dw}\frac{dw}{dz} = \frac{au}{2\pi w}\frac{2\pi w}{a(w - 1)}\sqrt{(w - \alpha^2)\left[w - \left(\frac{1}{\alpha}\right)^2\right]} \to u \qquad w \to \infty$$

The limiting value is real, indicating that the vector velocity is parallel to the x axis, and has magnitude u, as it should.

To calculate the lumped-circuit inductance of the slit at $x = 0$, we need to calculate the total kinetic energy of the fluid in the duct between $x = -l$ and $x = +l$, where $l \gg a$. This, minus the quantity $\frac{1}{2}\rho u^2(2lad)$ (which would be the kinetic energy if there were no constriction), must equal $\frac{1}{2}L_a U^2$, where L_a is the analogous inductance and $U = uad$. To compute the corresponding resistance we must integrate $|u|^2$ along the surface of the plate, from points B to C to D (or from B to C times 2). From Eq. (6.4.39), the analogous resistance R_a of the slit is $\rho\omega d_v/U^2$ times the integral of $|dF/dz|^2$ from B to C.

To transform the integral from the z to the F plane, we use the relation $dx\, dy = |dz/dF|^2\, d\psi\, d\theta$. For the inductance we integrate over half the duct and multiply by 2.

$$L_a = \frac{4}{U^2}\left(\frac{1}{2}\rho d \int_{-l}^{l} dx \int_{0}^{a/2} dy \left|\frac{dF}{dz}\right|^2 - \frac{1}{2}\rho ladu^2\right)$$

$$= \frac{2\rho d}{(uad)^2}\left(\int_{\psi_-}^{\psi_+} d\psi \int_{0}^{\frac{1}{2}au} d\theta - lau^2\right) = \frac{\rho}{aud}\left[(\psi_+ - \psi_-) - 2lu\right]$$

where ψ_+ is the value of ψ at $x = +l$, and ψ_- is its value at $-l$. Since $w_+ = e^{2\pi\psi_+/au}$ and $w_- = e^{2\pi\psi_-/au}$, we can use the second line of Eqs. (9.1.25) to relate ψ_+ and ψ_- to $\pm l$.

$$l \simeq \frac{\psi_+}{u} - \frac{a}{\pi}\ln\left(\frac{\alpha}{2} + \frac{1}{2\alpha}\right) \qquad -l \simeq \frac{\psi_-}{u} + \frac{a}{\pi}\ln\left(\frac{\alpha}{2} + \frac{1}{2\alpha}\right)$$

Thus, finally, the lumped-circuit inductance of the slit, of width b, in a duct of cross section a by d, is

$$L_a = \frac{2\rho}{\pi d} \ln \left(\tfrac{1}{2}\alpha + \tfrac{1}{2}\alpha^{-1} \right) = \frac{2\rho}{\pi d} \ln \left(\tfrac{1}{2} \tan \frac{\pi b}{4a} + \tfrac{1}{2} \cot \frac{\pi b}{4a} \right)$$

$$\to \begin{cases} \dfrac{2\rho}{\pi d} \ln \dfrac{2a}{\pi b} & b \ll a \\[2mm] \dfrac{\pi\rho}{4d} \left(\dfrac{a-b}{a} \right)^2 & a - b \ll a \end{cases} \tag{9.1.26}$$

Since the Schwartz-Christoffel transformation includes the duct sides, this formula is valid for all values of b in the range $0 \leqslant b \leqslant a$, as long as λ is much larger than a. If the slit is wide, the obstruction being but a pair of fins sticking out from the walls, the inductive effect is small, being proportional to the square of the relative extension $(a - b)/a$. For a relatively narrow slit, the analogous inductance goes to infinity logarithmically as slit width goes to zero, in contrast to that of the circular hole, which is inversely proportional to the radius of the hole when this radius is small compared with a.

The effective mass of the high-speed air in the slit is

$$M_a = (bd)^2 L_a = \frac{2}{\pi} \rho b^2 d \ln \left(\tfrac{1}{2} \tan \frac{\pi b}{4a} + \tfrac{1}{2} \cot \frac{\pi b}{4a} \right)$$

When $b \ll a$, this is equivalent to the mass of a slug of air of cross section equal to that of the slit bd and of length $(2b/\pi) \ln (2a/\pi b)$, which is longer and longer, compared with its width b, as b goes to zero.

According to Eqs. (6.4.39) and (9.1.23), the additional power lost because of the constriction, $R_a U^2$, should be $2d$ times the integral of $\tfrac{1}{2}\rho\omega d_v (u_{\tan})^2 \, dz$ over the surface, in the z plane, from B to C to D. The magnitude of u^2 equals $|dF/dz|^2 = |dF/dw|^2 |dw/dz|^2$, and if we transform to the w plane, $dz = |dz/dw| \, dw$. Therefore

$$R_a = \frac{2\rho\omega d_v}{a^2 \, du^2} \int_{\alpha^2}^{1} \left| \frac{dF}{dw} \right|^2 \left| \frac{dw}{dz} \right| dw = \frac{\rho\omega d_v}{\pi a d} \int_{\alpha^2}^{1} \left| \frac{(w - \alpha^2)[(1/\alpha)^2 - w]}{w^2(1 - w)^2} \right|^{\frac{1}{2}} dw$$

If we assume that the fins forming the constriction have zero thickness, the path of integration will be along the real axis in the w plane, and the integral will diverge. But if we assume that the fins have small but finite thickness, with maximum radius of curvature h at their edges, the integration will be along the line $\eta_s = \{(4\pi h/a)[(1/\alpha) - \alpha]\}^{\frac{1}{2}}$ above the axis in the $w = \xi + i\eta$ plane. The result, if h is small compared with a, is the same as if we integrate along the ξ axis from α to $1 - \eta_s$:

$$R_a \simeq \frac{\rho \omega d_v}{2\pi a d}\left(\frac{1}{\alpha} - \alpha\right)\ln\left\{\frac{a}{\pi h}\frac{(1/\alpha) - \alpha}{[(1/\alpha) + \alpha]^2}\right\}$$

$$\rightarrow \begin{cases} \dfrac{2\rho \omega d_v}{\pi^2 b d}\ln\dfrac{b}{4h} & 4h \ll b \ll a \\[3mm] \dfrac{\rho \omega d_v(a - b)}{2a^2 d}\ln\dfrac{a - b}{4h} & a - b \ll a \end{cases}$$

(9.1.27)

where $\alpha = \tan(\pi b/4a)$, and $d_v = \sqrt{2\mu/\rho\omega}$, as before. Also as before, with Eq. (9.1.23), h is either the half-thickness of the plate near the slit edge, or else it is d_v, the viscous-boundary-layer thickness, whichever is larger. When the slit width b becomes as small as d_v, the formula no longer holds. Turbulence also modifies these results, as shown in Eq. (11.3.36).

Change of duct width

The same analysis can be applied to the case of a rectangular duct of depth d and width b for $x < 0$ and a for $x > 0$, as shown in Fig. 9.7. The details will be left for a problem; the results are given here. The lumped-circuit inductance and resistance corresponding to the sudden increase in cross section are

$$L_a = \frac{\rho}{\pi d}\left[\frac{(a - b)^2}{2ab}\ln\frac{a + b}{a - b} + \ln\frac{(a + b)^2}{4ab}\right]$$

$$\rightarrow \begin{cases} \dfrac{\rho}{\pi d}\left(1 + \ln\dfrac{a}{4b}\right) & d_v \ll b \ll a \\[3mm] \dfrac{\rho}{\pi d}\left(\dfrac{a - b}{2a}\right)^2\left(1 + \ln\dfrac{2a}{a - b}\right) & b \rightarrow a \end{cases}$$

$$R_a = \frac{\rho \omega d_v}{2ad}\frac{a - b}{b}\left(1 + \frac{a^2 - b^2}{\pi ab}\ln\frac{a + b}{a - b}\right)$$

(9.1.28)

$$\rightarrow \begin{cases} \dfrac{\rho \omega d_v}{2bd}\dfrac{2 + \pi}{\pi} & d_v \ll b \ll a \\[3mm] \dfrac{\rho \omega d_v}{2ad}\dfrac{a - b}{b} & b \rightarrow a \end{cases}$$

Both L_a and R_a vanish when $a = b$ (i.e., when there is no change in width). When b is much smaller than a, both L_a and R_a are fairly large since the excess velocity at shoulder C becomes large. The resistance increases as the inverse of b, whereas the inductance increases only as the logarithm of $a/4b$. Since the angle at c is only $90°$ (instead of the $180°$ at the edge of the plates of Figs. 9.5 and 9.6), the integral of $(u_{\tan})^2$ does not diverge, and we need not introduce the minimal radius of curvature h.

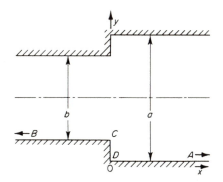

FIGURE 9.7
Section of a rectangular duct with sudden
change of width in the y direction.

Acoustical circuits

Returning to Fig. 9.4, the effective impedance of opening A, which
allows energy to escape from the duct, would be a combination of the local
mass and viscous loads, represented by Eqs. (9.1.22) and (9.1.23), and the
radiation impedance of the opening, represented by the functions $\theta_0 - i\chi_0$
used in Eq. (9.1.9) and plotted in Fig. 7.9 (or else the impedance corresponding to the α_p, β_p marked "unflanged" in Fig. 9.1). Since reactance χ_0
corresponds to the mass load outside the hole, the reactance of Eq. (9.1.22)
should be divided by 2, to represent only the concentration of flow inside the
duct, near the hole. Therefore the analogous impedance of opening A of
Fig. 9.4 is

$$Z_A = R_A - i\omega L_A \qquad L_A \simeq \frac{\rho}{S_A}\left[\tfrac{1}{8}D_A + \frac{\lambda}{2\pi}\chi_0\left(2k\sqrt{\frac{S_A}{\pi}}\right)\right]$$

$$R_A \simeq \frac{\rho c}{S_A}\left[\tfrac{1}{8}kd_v \ln\left(\frac{4S_A}{\pi h_A{}^2}\right) + \theta_0\left(2k\sqrt{\frac{S_A}{\pi}}\right)\right]$$

(9.1.29)

where $k = \omega/c = 2\pi/\lambda$; S_A and D_A are the area and perimeter, respectively,
of the opening A; h_A is the half-thickness of the duct wall; and θ_0, χ_0 are the
functions shown in Table IX, with argument $w = 2k\sqrt{S_A/\pi}$.

The impedance of opening B is, of course, $R_B - i\omega L_B$, where

$$L_B \simeq \frac{\rho D_B}{4S_B} \qquad R_B \simeq \frac{\rho c k d_v}{8S_B}\ln\frac{4S_B}{\pi h_B{}^2}$$

unless B is a slit, extending from one side of the duct to the other, in which
case Eqs. (9.1.26) and (9.1.27) can be used, with d and b equaling the length
and width, respectively, of the slit, and a equaling the cross-sectional area
of the duct S divided by d (that is, $S = ad$). All these formulas are valid
only as long as the wavelength of the sound is considerably longer than the
square root of the cross-sectional area of the duct, S.

The opening C and volume D of Fig. 9.4 constitute a resonating acoustical system, called a *Helmholtz resonator*. The analogous elements C_D,

R_C, and L_C are given in Eqs. (9.1.21), (9.1.22), and (9.1.23).   The resonance frequency is $\omega_0/2\pi$, where

$$\omega_0 = \frac{1}{\sqrt{L_C C_D}} \simeq 2c\sqrt{\frac{S_C}{D_C V_D}} \tag{9.1.30}$$

The Q of the resonator [Eq. (2.2.4)] is

$$Q = \sqrt{\frac{\omega L_C}{R_C}} \simeq \left[\frac{2D_C}{d_v \ln(4S_C/\pi h_C{}^2)}\right]^{\frac{1}{2}}$$

This defines the rate of damping of free oscillations of the resonator and the width of the resonance peak.   If opening C looks out into open space instead of the duct interior, the inductance and resistance of the analogous circuit will be given by Eqs. (9.1.29), since radiation resistance must be added. The resonance frequency is then the solution of the implicit equation

$$\omega_0 \simeq 2c\sqrt{\frac{S_C}{D_C V_D}}\left[\frac{1}{2} + \frac{8}{D_C}\sqrt{\frac{S_C}{\pi}}\frac{1}{w_0}\chi_0(w_0)\right]^{-\frac{1}{2}} \qquad w_0 = 2\frac{\omega_0}{c}\sqrt{\frac{S_C}{\pi}}$$

and the Q of the resonator will be

$$Q \simeq \left[\frac{\chi_0(w_0) + w_0(D_C/16)\sqrt{\pi/S_C}}{\theta_0(w_0) + w_0(d_v/16)\sqrt{\pi/S_C}\ln(4S_C/\pi h_C{}^2)}\right]^{\frac{1}{2}}$$

which is smaller than the Q of Eq. (9.1.30), since it is reduced by the added radiation resistance.

Periodic structures

The acoustical elements can, finally, be arranged sequentially to form a filter network.   For example, a duct with equally spaced constrictions, as shown in Fig. 9.8, is equivalent to a network of series inductances and shunt capacitances, making a low-pass filter.   The equations of motion for flow through the nth constriction are

$$p_n - p_{n+1} = (R_a - i\omega L_a)U_n = Z_a U_n$$

for low frequencies, where R_a and L_a are given by Eqs. (9.1.22) and (9.1.23),

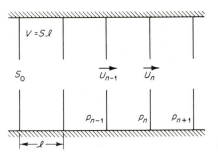

FIGURE 9.8
Acoustical low-pass filter circuit.

or else (9.1.26) and (9.1.27), if the opening is a slit. The equation of continuity, relating the compressibility times the pressure change in the $(n + 1)$st enclosure to the net influx to this volume, is

$$U_n - U_{n+1} = \kappa V \frac{\partial p_{n+1}}{\partial t} = -i\omega C_a p_{n+1}$$

where C_a is given in Eq. (9.1.21).

Expressing U_{n+1} and p_{n+1} in terms of U_n and p_n, we have

$$p_{n+1} = p_n - Z_a U_n$$
$$U_{n+1} = i\omega C_a p_n + (1 - i\omega C_a Z_a)U_n \qquad (9.1.31)$$

For wave motion propagated in the positive n direction, p_{n+1} and U_{n+1} must both be a constant factor times p_n and U_n, respectively. If we write this factor $e^{i\Phi}$, the real part of Φ will be the phase change from element to element in a simple-harmonic wave, and the imaginary part will measure the attenuation of the wave. Solving Eq. (9.1.31), with $p_{n+1} = e^{i\Phi}p_n$ and $U_{n+1} = e^{i\Phi}U_n$, we find that

$$e^{i\Phi} = 1 - \tfrac{1}{2}i\omega C_a Z_a \pm \sqrt{-i\omega C_a Z_a - \tfrac{1}{4}(\omega C_a Z_a)^2}$$

or

$$\cos\Phi = 1 - \tfrac{1}{2}i\omega C_a Z_a = 1 - \frac{1}{2}\left(\frac{\omega}{\omega_0}\right)^2 - \frac{1}{2}i\left(\frac{\omega}{\omega_0 Q^2}\right) \qquad (9.1.32)$$

where

$$\omega_0{}^2 = \frac{1}{L_a C_a}, \qquad \text{and} \qquad Q^2 = \frac{\omega_0 L_a}{R_a} = \left(\frac{1}{R_a}\right)\sqrt{\frac{L_a}{C_a}}.$$

The real and imaginary parts of Φ can be determined graphically from Plate III.

For frequencies considerably below the resonance frequency $\omega_0/2\pi$, the relation

$$\Phi \simeq \frac{\omega}{\omega_0} + \frac{1}{2}i\frac{1}{Q^2} \simeq \frac{l\omega}{2c}\sqrt{\frac{SD_0}{S_0 l}} + i\frac{d_v\omega}{16c}\sqrt{\frac{S_0 D_0}{Sl}} \qquad (9.1.33)$$

is valid. It provides a basis for calculating the wave velocity and attenuation in the filter sequence. Since the pressure in the nth enclosure, roughly, a distance $x = nl$ from the beginning, is $P_0 e^{in\Phi}$, the pressure a distance x from the beginning is

$$P_0 e^{ix\Phi/l} \simeq P_0 \exp\left[i\omega\left(\frac{x}{2c}\sqrt{\frac{SD_0}{S_0 l}} - t\right) - \frac{x}{l}\frac{kd_v}{16}\sqrt{\frac{S_0 D_0}{Sl}}\right]$$

$$\text{for } \omega \ll \omega_0 \qquad Q \gg 1 \quad (9.1.34)$$

where $\quad S = $ cross-sectional area of duct

$\quad l = $ distance between successive constrictions

$\quad S_0 = $ area of constricted opening

$\quad D_0 = $ perimeter of opening

Thus we see that the effective phase velocity c_φ of the wave along the network, and the damping constant γ, the inverse of the distance within which the amplitude decreases by a factor e^{-1}, are

$$c_\varphi \simeq c\sqrt{\frac{4S_0 l}{SD_0}} \quad \text{and} \quad \gamma \simeq \frac{kd_v}{16}\sqrt{\frac{S_0 D_0}{Sl}}$$

As the driving frequency is increased to $\omega_0/2\pi$ and greater, these formulas are no longer valid, and one must solve Eq. (9.1.32) graphically or numerically. A sudden change comes at the frequency ω_0/π, particularly when $Q \gg 1$. For at driving frequencies greater than ω_0/π, the real part of $\cos \Phi$ is less than -1, and the imaginary part of Φ, which had been small for $\omega < 2\omega_0$, now increases rapidly with frequency.

$$\Phi \simeq \pi + i\cosh^{-1}\left[\frac{1}{2}\left(\frac{\omega}{\omega_0}\right)^2 - 1\right] \quad \omega > 2\omega_0 \quad Q^2 \gg 1 \quad (9.1.35)$$

Each enclosure oscillates out of phase with its neighbors, and the wave is rapidly damped out with increasing n. Above this cutoff frequency ω_0/π, true wave motion is impossible.

Other sequential acoustical circuits can be devised, to act as high-pass or bandpass filters. The details will be left to the problems.

9.2   HIGHER MODES IN DUCTS

It is possible to consider a sound wave in a pipe as a one-dimensional wave, somewhat like a wave in a string, only when the wavelength of the sound transmitted is much longer than the transverse dimensions of the pipe. This limiting case was dealt with in detail in the preceding section; it is now time to take up the higher-frequency waves, where this simplification is not valid. In addition to the fundamental wave, traveling in the axial direction and having wavefronts which are uniform across the duct cross section, there are waves which reflect back and forth from the duct walls as they travel along, for which the pressure distribution is not uniform across the duct.

These higher modes of wave transmission have phase velocities which are greater than the velocity $c = \sqrt{1/\kappa\rho}$ of sound in free space. As the frequency is decreased, the phase velocity of each higher mode increases, to become infinite at some cutoff frequency, characteristic of the mode considered. Below the cutoff, true transmission cannot take place for this mode; the wave attenuates rapidly. Thus, at low frequencies, only the plane wave, the fundamental mode, is transmitted. At higher frequencies some of the higher modes also are propagated, and these additional waves must be included in our calculations.

Internal reflections and higher modes

Thus we can consider a wave inside the duct as a combination of plane-wave components, incident on and reflecting from the duct walls. When the wavelength $\lambda = 2\pi c/\omega$ in free space is small enough, in addition to the plane wave traveling in the direction of the tube axis, there will be other waves, which reflect from side to side as they travel along. They cannot strike the sides at any angle; the incidence must be such as to avoid destructive interference between successively reflected waves.

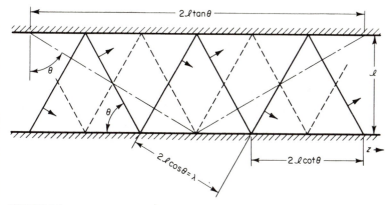

FIGURE 9.9
Reflected-wave components making up the first higher mode propagating between two parallel plates. Solid lines represent pressure maxima of the component wave; dotted lines, pressure minima. Arrows, representing direction of propagation of the components, are normal to the wavefronts.

To illustrate these points we first consider a two-dimensional model, that of wave motion between plane parallel plates, as shown in Fig. 9.9. Reference to Eq. (6.3.5) indicates that if the duct surfaces are rigid, the phase of the incident and reflected waves must be equal at the plate surface, maximum coinciding with maximum and minimum with minimum. If the alternate reinforcements are a single wavelength apart, so that $2l \cos \theta = \lambda$, as shown, the wave is the first transverse mode; if they are m wavelengths apart, so that $2l \cos \theta = m\lambda$, the wave is the mth transverse mode. The wave for $\theta = \frac{1}{2}\pi$ $(m = 0)$ is, of course, the fundamental mode. Thus the distance between successive maxima on the sides of the duct is

$$\frac{2l}{m} \cot \theta = \frac{\lambda}{\sqrt{1 - (m\lambda/2l)^2}}$$

and since $\omega/2\pi = c/\lambda$ of them must pass a given point in a second, the speed of the mth mode down the duct is

$$\text{Phase velocity } c_\varphi = \frac{c}{\sin \theta_m} = \frac{c}{\sqrt{1 - (m\lambda/2l)^2}}$$

$$(9.2.1)$$

$$\text{Angle of incidence} = \theta_m = \cos^{-1} \frac{m\lambda}{2l} = \cos^{-1} \frac{\pi m c}{\omega l}$$

The maximum λ for the mth mode, corresponding to $\theta_m = 0°$, is $\lambda_m = 2l/m$. At the frequency $\omega_m/2\pi = mc/2l$, corresponding to this maximum λ, the phase velocity of the mth mode is infinite; below this mth *cutoff frequency* the mth mode cannot propagate along the tube. Finally, by noting where pressure maxima cross pressure minima of the reflected waves, we can convince ourselves that the mth mode has m nodal surfaces (where p is zero) parallel to the duct axis.

This same geometrical analysis can also provide an approximate answer when the acoustic admittance of the walls is not zero. For example, in the case of the parallel plates of Fig. 9.9, if the lower plate has specific acoustic admittance β_1 and the upper one has admittance β_2 (with $\beta = \xi - i\sigma$ in both cases), then Eq. (6.3.5) indicates that at each reflection the wave is changed in amplitude and phase by the factor

$$C_{sm} = \frac{1 - (\beta_s/\cos \theta_m)}{1 + (\beta_s/\cos \theta_m)} \simeq \exp \frac{-2\xi_s + 2i\sigma_s}{\sqrt{m\lambda/2}} \quad \text{if} \quad m\lambda \gg 4l \,|\beta_s| \quad (9.2.2)$$

the subscript s being 1 for reflection from the lower plate, and 2 for reflection from the upper plate. For $m\lambda \gg |4l\beta_s|$, the formula for θ_m in Eqs. (9.2.1) is still valid. Thus, at each reflection, the amplitude of the wave is reduced by a factor roughly equal to $\exp(-2\xi_s/\cos \theta_m)$, and the phase of the wave is advanced by $2\sigma_s/\cos \theta_m$ radians, so that incident maxima do not exactly coincide with reflected maxima at the sides of the duct.

Since in two reflections, one from each side, the acoustic ray has progressed a distance $2l \tan \theta_m$ down the duct, the attenuation of the wave per unit distance is given by the factor

$$\exp\left(-\frac{2\xi_1 + 2\xi_2}{\cos \theta_m} \frac{1}{2l \tan \theta_m}\right) = \exp\left(-\frac{\xi_1 + \xi_2}{l \sin \theta_m}\right) \quad (9.2.3)$$

and the phase of the wave is advanced by $(\sigma_1 + \sigma_2)/l \sin \theta_m$ radians per unit length, because of the reflections. However, even if the wall susceptances σ_1 and σ_2 were zero, the phase would be advanced in a unit length by an amount equal to the angular velocity $\omega = 2\pi c/\lambda$ divided by the phase velocity c_φ of Eqs. (9.2.1). Thus the total change of phase per unit length along the duct is, roughly, $(2\pi/\lambda) \sin \theta_m + [(\sigma_1 + \sigma_2)/l \sin \theta_m]$, and the dependence of the pressure on the distance z along the duct is via the factor

$$\exp\left\{\left[\frac{2\pi i}{\lambda}\sqrt{1 - \left(\frac{m\lambda}{2l}\right)^2} + i\frac{\sigma_1 + \sigma_2}{\sqrt{l^2 - \frac{1}{4}m^2\lambda^2}} - \frac{\xi_1 + \xi_2}{\sqrt{l^2 - \frac{1}{4}m^2\lambda^2}}\right]z\right\} \quad (9.2.4)$$

when $|\beta| \ll m\lambda/4l$. Compare these results with the exact solution, Eqs. (9.2.14) and (9.2.16).

An approximate answer is all we can expect from this mixture of wave and geometrical acoustics. Furthermore, the effect of reflections, when the duct walls are not planes, is not easily computed this way. On both counts it is desirable to work out the exact solution for a wave traveling along the inside of a uniform duct.

General formulas for transmission

We set the z axis parallel to the axis of the duct. The interior cross section, parallel to the xy plane, is some area S, bounded by a perimeter D, the intersection of the cross-sectional plane and the inner duct surface. The wave equation for simple-harmonic waves

$$\left(\frac{\partial^2}{\partial x^2} + \frac{\partial^2}{\partial y^2} + \frac{\partial^2}{\partial z^2}\right)p + \left(\frac{\omega}{c}\right)^2 p = 0 \quad (9.2.5)$$

thus allows the z factor to be separated off, $p = \Psi(x,y)e^{ik_z z}$, where we have, as usual, omitted the time-dependent factor $e^{-i\omega t}$. The equation for the transverse factor Ψ is then

$$\left(\frac{\partial^2}{\partial x^2} + \frac{\partial^2}{\partial y^2}\right)\Psi + \varkappa^2\Psi = 0 \qquad \left(\frac{\omega}{c}\right)^2 = k^2 = k_z^2 + \varkappa^2 \quad (9.2.6)$$

Note that the $\varkappa$ of this section is *not* the compressibility κ.

This equation for Ψ is the same as Eq. (5.2.3) for the vibrations of a membrane, except that now it is for a factor of the pressure, instead of transverse displacement. As with the earlier problem, the solutions which fit the appropriate boundary conditions occur only for a discrete set of values of the constant $\varkappa$, these values being called characteristic values (or eigenvalues), and the corresponding solutions Ψ being called characteristic functions (or eigenfunctions). Since the boundary conditions in the present case differ from that for the membrane, the functions and values differ in the two cases, but many of the general properties are the same.

For the membrane within rigid boundaries, the displacement has to be zero along the perimeter. For a duct with rigid walls, the fluid velocity normal to the surface wall must be zero. Since the vector velocity is related to the pressure according to the equation

$$u = \frac{1}{i\rho c k}\,\text{grad}\, p \qquad \omega = ck$$

the boundary condition for rigid walls is that the component of grad Ψ, normal to the perimeter (which we can write $\partial\Psi/\partial n$), be zero at every point on the perimeter.

On the other hand, if the duct walls are locally reactive to the pressure, so that we can ascribe to the surface an acoustic impedance $z = p/u_n$ or a specific acoustic admittance $\beta = \xi - i\sigma = \rho c/z$, then the boundary condition, to be satisfied everywhere on the perimeter, is

$$\frac{\partial\Psi}{\partial n} = ik\beta\Psi \qquad k = \frac{\omega}{c} \qquad (9.2.7)$$

where the normal is directed into the wall, away from the interior of the duct. The discrete set of eigenvalues of $\varkappa$ for which this boundary condition is satisfied may, for simplicity, be distinguished from each other by the index number n (though we shall see later that it is often more convenient to use a pair of index numbers). The characteristic function satisfying boundary condition (9.2.7), corresponding to $\varkappa_n$, will be denoted $\Psi_n(x,y)$.

If the impedance of the duct surface is purely reactive, so that β is pure-imaginary, the eigenvalues $\varkappa_n$ will all be real numbers. If, however, β has a real part, then $\varkappa_n$ will have a negative-imaginary part (as will be demonstrated later), and thus the wavenumber $k_z \equiv k_n = [(\omega/c)^2 - \varkappa_n^2]^{\frac{1}{2}}$ for this mode will have a positive-imaginary part. The corresponding characteristic functions Ψ_n have the same general orthogonality properties as indicated in Eqs. (5.2.16) and (5.2.26). These general properties can be summarized as follows:

$$P_\omega = \sum_n A_n \Psi_n(x,y) e^{ik_n z - i\omega t}$$

$$\left(\frac{\partial^2}{\partial x^2} + \frac{\partial^2}{\partial y^2}\right)\Psi_n \equiv \nabla_2^2\Psi_n = -\varkappa_n^2\Psi_n$$

Ψ_n satisfies Eq. (9.2.7) along the perimeter D,

$$(9.2.8)$$

$$k_n \equiv \frac{2\pi}{\lambda_n} + i\pi\gamma_n = \sqrt{\left(\frac{\omega}{c}\right)^2 - \varkappa_n^2}$$

$$\iint_S \Psi_n\Psi_m \, dx \, dy = \begin{cases} 0 & m \neq n \\ S\Lambda_n & m = n \end{cases}$$

where the subscript 2 is used to specify that the Laplace operator ∇_2^2 is for the two-dimensional area of cross section. The complex dimensionless factor Λ_n is the mean value of Ψ_n^2 averaged over the duct cross section S. Note that if the duct surface is *not* locally reactive, the β of Eq. (9.2.7) will depend on m and the set of functions Ψ_m will *not* be mutually orthogonal.

The separation of k_n into its real and imaginary parts shows that the wavelength along the duct axis of the nth mode is λ_n, and thus that its phase velocity is $c_n = \omega\lambda_n/2\pi$, and that the pressure amplitude attenuates by a factor of e^{-1} in a distance $1/\pi\gamma_n$. Since $|\varkappa_n|$ increases as n increases, we note that λ_n tends to decrease as n increases. Therefore, for any specified frequency, there is a value of n for which $\mathrm{Re}\,\varkappa_n{}^2 < (\omega/c)^2 < \mathrm{Re}\,\varkappa_{n+1}^2$. At this frequency the imaginary part of k_n, for all values of n greater than this, is much larger than the imaginary parts for smaller n's, indicating that all these higher modes are strongly attenuated. To put it another way, each mode has a *cutoff frequency* c/λ_n; for driving frequencies above this the mode propagates along the duct with (usually) quite small attenuation, but for driving frequencies less than this the attenuation is very large, so large that the mode cannot be said to propagate as a wave.

When the duct walls are rigid ($\beta = 0$), the lowest eigenvalue $\varkappa_0$ is zero, for which $\Psi_0 = 1$, so that $\Lambda_0 = 1$. This is the fundamental mode; when $\beta = 0$, its phase velocity is c, the velocity of sound in free space; it propagates as a plane wave. For frequencies below the cutoff frequency of the lowest higher mode (i.e., when $\omega < 2\pi c/\lambda_1$), only this fundamental mode is propagated without attenuation (if $\beta = 0$); all higher modes are strongly attenuated. (This is the justification of the assumptions made in Sec. 9.1.) As ω is increased, when it becomes larger than $2\pi c/\lambda_1$, the first higher mode also propagates with zero attenuation (if $\beta = 0$), though at a phase velocity higher than c. And so on; as the frequency is increased, successively higher modes can be transmitted along the duct, and the phase velocity of each decreases, coming closer to the constant value c of the fundamental mode. Compare this with the discussion of Eqs. (9.2.1); the quantity $\varkappa_n$ replaces the quantity $\pi m/l$ of the simple model, and for $\beta = 0$, the effective angle of incidence θ_n of the simple model is equivalent to $\cos^{-1}(\varkappa_n c/\omega)$ of the exact solution.

This same behavior is also true when the specific admittance β is not zero but is pure-imaginary, except that here the phase velocity of even the fundamental mode differs from c; all modes exhibit dispersion, as discussed on page 477. If the specific conductance ξ of the duct walls is not zero, all modes are attenuated at all frequencies, and thus the distinction between no attenuation above the cutoff frequency and no transmission below it becomes less clear-cut. As long as ξ is small compared with unity, however, it still is true that above its cutoff frequency the attenuation of the nth mode is slight; below it the mode is sharply attenuated.

We should also mention that, unless the tube is very narrow, the surface effects of thermal conduction and viscosity may be expressed in terms of the equivalent surface admittance

$$\beta_s \simeq (1 - i)\left[(\gamma - 1)kd_h + \left(\frac{k_t}{k}\right)^2 kd_v\right]$$

of Eq. (6.4.36) (see also page 292), which may be added to the admittance of the surface itself to obtain the β of Eq. (9.2.7). We shall show later, for various duct shapes, how to compute the value of the ratio (k_t/k) for each mode. Calculation of the effects in a very narrow circular tube will be given at the end of this section.

Dispersion of the higher modes

Thus we see that wave propagation, for all modes except the fundamental mode in a duct with rigid walls (and neglecting thermal and viscous effects), is dispersive in the sense of the discussion of page 477. For a duct with rigid walls, the phase velocity of the nth mode is

$$c_\varphi = \frac{\omega}{k_n} = \frac{c}{\sqrt{1 - (c\varkappa_n/\omega)^2}}$$

where c is the velocity of sound in free space. Below the cutoff frequency $c\varkappa_n/2\pi$, the phase velocity is imaginary, indicating no propagation. Just above the cutoff frequency the phase velocity is very large and real, and as $\omega \to \infty$, c_φ approaches c from above. The group velocity, for a group of waves centering at frequency $\omega_a/2\pi$, is, according to Eq. (9.1.18),

$$c_g = \left(\frac{\partial k_n}{\partial \omega}\right)^{-1} = c\sqrt{1 - \left(\frac{c\varkappa_n}{\omega_a}\right)^2} = \frac{c^2}{c_\varphi}$$

which is less than c for the frequencies above the cutoff.

The signal velocity $c_\varphi(\infty)$ is c, in spite of the fact that the phase velocity is greater than c for all frequencies for which the wave transmits. For example, by use of Eq. (9.1.20) and an extended table of Fourier transforms, we can show that a delta-function pulse, for the nth mode, started at $z = 0$ at $t = 0$, becomes the wave

$$p = \frac{1}{2\pi}\Psi_n(x,y)\int_{-\infty}^{\infty} \exp\left[i\frac{z}{c}\sqrt{\omega^2 - (\varkappa_n c)^2} - i\omega t\right] d\omega$$

$$= \Psi_n(x,y)\left\{\delta\left(t - \frac{z}{c}\right) - u\left(t - \frac{z}{c}\right)\varkappa_n z \frac{J_1[\varkappa_n c\sqrt{t^2 - (z/c)^2}]}{\sqrt{t^2 - (z/c)^2}}\right\}$$

a time t later (where u is the unit step function). The pulse, represented by the $\delta(t - z/c)$ term, travels with the free-space velocity c; trailing behind it is a "wake," represented by the second term. Although the dispersive propagation spreads out the wave behind the pulse, nothing reaches the observer at z until time z/c, when the pulse arrives.

For ducts with nonrigid walls even the fundamental mode is dispersive, and the expressions for phase and group velocities are more complicated than the ones just discussed; but as long as the duct walls have stiffness and mass, the signal velocity $c_\varphi(\infty)$ is equal to the free-space value c.

Generation of sound in a duct

Sound can be generated by placing a vibrating piston at one end of the duct. Suppose the duct extends from $z = 0$ to $z = \infty$, and suppose the driving surface at $z = 0$ has velocity in the z direction, at point x, y on its surface, equal to $U(x,y)e^{-i\omega t}$. As with Eq. (5.2.17), we use the properties of the eigenfunctions given in Eqs. (9.2.8) to work out the series expansions for the wave in the duct.

$$u_z = \frac{1}{i\rho\omega}\frac{\partial p}{\partial z} = \sum_n U_n\Psi_n e^{ik_n z - i\omega t} \qquad U_n = \frac{1}{S\Lambda_n}\iint\limits_S U(x,y)\Psi_n\,dx\,dy$$

$$\tag{9.2.9}$$

$$p = \rho c \sum_n \frac{\omega}{ck_n} U_n\Psi_n e^{ik_n z - i\omega t} = \rho c \sum_n \left[1 - \left(\frac{c\varkappa_n}{\omega}\right)^2\right]^{-\frac{1}{2}} U_n\Psi_n e^{ik_n z - i\omega t}$$

All modes are present (unless U_n is identically zero) close to the driving surface, but those modes having cutoff frequencies greater than $\omega/2\pi$ die out rapidly, leaving only the lower modes to transmit energy any appreciable distance along the tube. The wave cannot transmit all the details of the piston motion, as represented by the function $U(x,y)$; fine detail of dimensions smaller than $\lambda = 2\pi c/\omega$ is lost in transmission.

If the duct wall is rigid and we neglect thermal and viscous losses at the walls, the lowest characteristic function is $\Psi_0 = 1$, and $\varkappa_0$ is zero, as mentioned earlier. Thus, by the last of Eqs. (9.2.8), the integrals $\iint \Psi_n\,dx\,dy$, for all $n > 0$, are zero. In this case, if the driving surface at $z = 0$ is a perfect piston, with velocity U independent of x and y, U_0 will equal U, all U_n for $n > 0$ will be zero, and the propagated wave

$$u_z = Ue^{i(\omega/c)z - i\omega t} \qquad p = \rho c Ue^{i(\omega/c)z - i\omega t}$$

will be the plane wave discussed in Sec. 9.1, with constant ratio between pressure and velocity. If the driving surface is not a perfect piston, however, some of the higher modes will be excited, and the ratio between p and u_z at the driving surface will not be exactly ρc. In fact, for those modes for which $\varkappa_n > k$, k_n will be positive-imaginary and the radical in the sum for p will be imaginary.

$$\rho c \left[1 - \left(\frac{c\varkappa_n}{\omega}\right)^2\right]^{-\frac{1}{2}} = \frac{\rho\omega}{k_n} = -i\omega\rho\left[\varkappa_n^2 - \left(\frac{\omega}{c}\right)^2\right]^{-\frac{1}{2}}$$

and the corresponding radiation impedance at the driving surface will contain a mass-reactance term.

If the duct walls are not perfectly rigid, or if thermal and viscous effects at the surface are not neglected, some of the higher modes will be excited, even if the driving surface is a perfect piston. For in this case Ψ_0 will be a function of x and y, and therefore the integrals $\iint \Psi_n\,dx\,dy$ $(n > 0)$ will not be zero, and the radiation impedance will have reactive terms even for a perfect piston. This effect will be accentuated if the driving frequency happens to be just below some cutoff frequency, so that the factor $[\varkappa_n^2 - (\omega/c)^2]^{-\frac{1}{2}}$ is large for that n.

The Green's function

If the sound generator is a simple source, at point x_0, y_0, z_0 within a duct which extends from $z = -\infty$ to $z = +\infty$, the resulting wave can be computed from the Green's function for an infinite duct, the solution of the equation

$$\left[\frac{\partial^2}{\partial x^2} + \frac{\partial^2}{\partial y^2} + \frac{\partial^2}{\partial z^2} + \left(\frac{\omega}{c}\right)^2\right] g_\omega(x,y,z \mid x_0,y_0,z_0)$$
$$= -\delta(x - x_0)\, \delta(y - y_0)\, \delta(z - z_0)$$

which satisfies the boundary conditions of Eq. (9.2.7). It can be expressed in terms of a series of the eigenfunctions Ψ_n.

$$g_\omega = \sum_n F_n(z)\Psi_n(x,y)$$

Insertion in the differential equation, multiplication by Ψ_n, and integration of both sides over S results in a set of equations

$$\left(\frac{d^2}{dz^2} + k_n^2\right)F_n = -\frac{\Psi_n(x_0,y_0)}{S\Lambda_n}\delta(z - z_0) \qquad k_n^2 = \left(\frac{\omega}{c}\right)^2 - \varkappa_n^2$$

Reference to the derivation of Eq. (4.4.11) from Eq. (4.4.9) shows that

$$F_n = \frac{i\Psi_n(x_0,y_0)}{2k_n S\Lambda_n}\exp(ik_n |z - z_0|)$$

where, when Re $\varkappa_n^2 > (\omega/c)^2$, k_n must equal i (not $-i$) times the radical $[\varkappa_n^2 - (\omega/c)^2]^{\frac{1}{2}}$ in order that the higher modes attenuate, rather than augment, with increasing values of $|z - z_0|$.

Thus the Green's function for the infinite duct is

$$g_\omega(x,y,z \mid x_0,y_0,z_0) = \frac{i}{2S}\sum_n \frac{\Psi_n(x,y)\Psi_n(x_0,y_0)}{\Lambda_n k_n}\exp(ik_n |z - z_0|) \quad (9.2.10)$$

As mentioned before, if we can assume that the duct walls are rigid, $\varkappa_0 = 0$ and $\Psi_0 = 1$, eigenvalues $\varkappa_n$ are real, nonnegative quantities, and the Ψ_n's are real, satisfying the condition $\partial\Psi/\partial n = 0$ along the perimeter of S. In this case we can separate the attenuated upper modes (for which $n > N$, where $\varkappa_N < \omega/c < \varkappa_{N+1}$) from the propagated upper modes for $n \leqslant N$.

$$g_\omega = \frac{ic}{2\omega S}e^{i(\omega/c)|z-z_0|} + \frac{i}{2S}\sum_{n=1}^{N}\frac{\Psi_n(x,y)\Psi_n(x_0,y_0)}{\Lambda_n k_n}e^{ik_n|z-z_0|}$$

$$+ \frac{1}{2S}\sum_{n=N+1}^{\infty}\frac{\Psi_n(x,y)\Psi_n(x_0,y_0)}{\Lambda_n K_n}e^{-K_n|z-z_0|} \quad (9.2.11)$$

$$k_n = \sqrt{\left(\frac{\omega}{c}\right)^2 - \varkappa_n^2} \qquad K_n = \sqrt{\varkappa_n^2 - \left(\frac{\omega}{c}\right)^2}$$

This series looks very different from the simple expression $(1/4\pi R)e^{ikR}$ of Eq. (7.1.6) for the Green's function in free space. Indeed it is; the reflections from the duct walls have so modified the wave that the radiated energy is channeled outward only along the $\pm z$ axis, instead of going out in all directions.   Of course, if the source point is a large number of wavelengths away from the nearest portion of the duct wall, the interior of the duct will look almost like free space to the source.   In this limiting case, when N is very large, k_n will be real for a large number of terms in the series. Physical intuition thus tells us that, for x, y, z close to x_0, y_0, z_0 ($R < \lambda$), the series of Eq. (9.2.10) approaches $(1/4\pi R)e^{ikR}$ when the driving frequency is larger than the cutoff frequencies of a large number of modes of the duct. This can be proved mathematically, though the manipulations required are too lengthy to go into here.

Impedance of a simple source

When the wavelength is roughly the same size as the transverse dimensions of the duct, the radiated pressure wave is appreciably different from the free-space Green's function, even in the region close to the source point. In this case the presence of the duct walls appreciably modifies the reaction of the medium back on the simple source, even more than the presence of a nearby plane surface [Eq. (7.4.14)].   To show this more specifically, we suppose the sound generator to be a simple source, of strength S_ω (not to be confused with S, the cross-sectional area of the duct) and of driving frequency $\omega/2\pi$, of the sort discussed in Sec. 7.1.   The pressure wave inside a duct with rigid walls and outside the sphere of radius $R = a[R^2 = (x - x_0)^2 + (y - y_0)^2 + (z - z_0)^2]$ is $-i\omega\rho S_\omega g_\omega(\mathbf{r}_0 + \mathbf{R} \mid \mathbf{r}_0)$.   For wavelengths of the same size as the duct dimensions, or longer, ω/c is smaller than $\varkappa_n$ for all but the first N modes, with N small, and all but a few of the terms in series (9.2.11) are nearly equal to

$$\frac{1}{2S\varkappa_n\Lambda_n}\Psi_n(x,y)\Psi_n(x_0,y_0)e^{-\varkappa_n|z-z_0|} \qquad n > N$$

It is possible to show that, close to the source point,

$$\frac{1}{4\pi R} \to -\frac{|z - z_0|}{2S} + \frac{1}{2S}\sum_{n=1}^{\infty}\frac{\Psi_n(x,y)\Psi_n(x_0,y_0)}{\Lambda_n\varkappa_n}e^{-\varkappa_n|z-z_0|} \qquad \text{as } R \to 0$$

For the longer wavelengths (N small), therefore, the terms in series (9.1.11) for $n > N$ are approximately equal to the corresponding terms in the series for $1/4\pi R$, for R small.

Adding and subtracting this comparison series, we finally obtain an approximate formula for g_ω, valid for the longer wavelengths.

$$g_\omega \simeq \left[\frac{1}{4\pi R} + \frac{i}{2Sk}e^{ik|z-z_0|} + \frac{|z - z_0|}{2S}\right.$$
$$\left. + \sum_{n=1}^{N}\frac{\Psi_n(x,y)\Psi_n(x_0,y_0)}{2S\Lambda_n}\left(\frac{ie^{ik_n|z-z_0|}}{k_n} - \frac{e^{-\varkappa_n|z-z_0|}}{\varkappa_n}\right)\right]e^{-i\omega t}$$

The mean pressure $-i\omega\rho g_\omega$ on the small sphere of radius a of the source is, approximately,

$$P_s \simeq \frac{\rho\omega}{2S}\left\{\frac{-iS}{4\pi a} + \frac{1}{k} + \sum_{n=1}^{N}\frac{[\Psi'_n(x_0,y_0)]^2}{\Lambda_n}\left(\frac{1}{k_n} + \frac{i}{\varkappa_n}\right)\right\}S_\omega e^{-i\omega t}$$

The fluid flux at the surface of the sphere is, by definition,

$$S_\omega(1 - ika)e^{ika-i\omega t} \simeq S_\omega e^{-i\omega t}$$

Therefore the limiting forms for the mechanical force, velocity, and impedance of a simple source in a duct are

$$F_s \simeq \left\{-i\rho\omega a\left[1 - \frac{2\pi a}{S}\sum_{n=1}^{N}\frac{\Psi'_n{}^2(x_0,y_0)}{\Lambda_n\varkappa_n}\right]\right.$$

$$\left. + \rho c\,\frac{2\pi a^2}{S}\left[1 + \sum_{n=1}^{N}\frac{k\Psi'_n{}^2(x_0,y_0)}{\Lambda_n k_n}\right]\right\}S_\omega e^{-i\omega}$$

$$U_s \simeq \frac{S_\omega}{4\pi a^2}e^{-i\omega t} \tag{9.2.12}$$

$$Z_s \simeq -i\omega(4\pi\rho a^3)\left[1 - \frac{2\pi a}{S}\sum_{n=1}^{N}\frac{\Psi'_n{}^2(x_0,y_0)}{\Lambda_n\varkappa_n}\right]$$

$$+ \rho c\,\frac{8\pi^2 a^4}{S}\left[1 + \sum_{n=1}^{N}\frac{k}{k_n}\frac{\Psi'_n{}^2(x_0,y_0)}{\Lambda_n}\right]$$

for $ka \ll 1$ and for N small.

This is an interesting formula, particularly in comparison with Eqs. (7.1.11), the analogue for a simple source in free space. The first term in the reactive part is the same as the reactive part in Eqs. (7.1.11); the presence of the duct walls reduces the mass load on the source by a small amount, proportional to the ratio between the radius a of the source and the mean dimension $\sqrt{S}$ of the duct cross section, the exact value depending on the placement of the source.

In contrast, the resistive term in Eqs. (9.2.12) is larger than the resistance for a source in free space by the large factor $(1/ka)^2$. The radiation resistance does not go to zero as $(ka)^2$ for ka small as it does for waves spreading out radially in three dimensions; it approaches a constant value, independent of ka, as it does for plane waves moving in one direction. The first term, corresponding to the fundamental (plane-wave) mode, is equal to the radiation resistance on a plane piston of area $8\pi^2 a^4/S$ (much smaller than S) radiating plane waves normal to its surface. The additional terms, one for each higher mode which propagates at the driving frequency, have a more complicated frequency dependence. When the driving frequency is just larger than the cutoff frequency of a particular mode (say, the nth one), its radiation resistance is larger than that of the fundamental mode [unless the source is at a node of the nth mode, i.e., unless $\Psi'_n{}^2(x_0,y_0)$ is zero].

As the frequency is increased, the radiation resistance increases stepwise, each new step corresponding to a new mode becoming propagational, being introduced by a narrow peak of cross-duct resonance for that mode. The steps get closer and closer together, and the peaks get narrower and narrower and, if we average out the narrow peaks, the curve approaches the value $\rho c (4\pi a^2) k^2 a^2$, which is the radiation resistance of the source in free space.

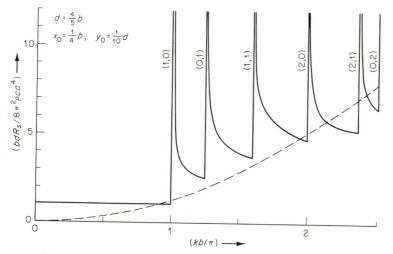

FIGURE 9.10

Radiation resistance R_s of a simple source of radius a at point $x_0, y_0, 0$ in a rectangular duct of sides b, d. Numbers by each peak indicate the m, n of the mode resonating. Dashed curve is radiation resistance for same source in free space.

Figure 9.10 is a plot of the real part R_s of the Z_s of Eqs. (9.2.12) for a rectangular duct of width b and depth d, with the source at x_0, y_0 as noted. The stepwise character of the curve is apparent. The dashed curve is the corresponding free-space resistance.

The impedance of a simple dipole in a duct differs even more from its counterpart in free space, as described in Eqs. (7.1.12). For example, if the dipole is oriented transverse to the duct axis, there is no radiation below the first cutoff frequency; at low frequencies the radiation impedance is purely reactive. However, the general formulas do not merit writing down here.

Rectangular ducts

To make further progress we must deal with specific forms of duct cross section. The simplest case is the rectangular duct, of breadth b in the x direction and depth d in the y direction. If the duct walls are rigid, we choose the z axis to run along one corner of the duct, so that the boundary conditions are

$$\frac{\partial p}{\partial x} = 0 \quad \text{at } x = 0, x = b \qquad \frac{\partial p}{\partial y} = 0 \quad \text{at } y = 0, y = d$$

The characteristic functions and values for the transverse factors of the various modes are then

$$\Psi_{mn} = \cos\frac{\pi m x}{b} \cos\frac{\pi n y}{d} \qquad \Lambda_{mn} = \frac{1}{\epsilon_m \epsilon_n} \qquad \epsilon_m = \begin{cases} 1 & m = 0 \\ 2 & m > 0 \end{cases}$$

$$\varkappa_{mn} = \sqrt{\left(\frac{\pi m}{b}\right)^2 + \left(\frac{\pi n}{d}\right)^2} \qquad k_{mn}{}^2 = \left(\frac{\omega}{c}\right)^2 - \varkappa_{mn}{}^2$$

(9.2.13)

Contrast these functions with those of Eqs. (5.2.15) for the displacement of a rectangular membrane, which must go to zero at the boundary. The fundamental mode is, of course, the $m = n = 0$ case. In the change from the two-index notation used here to the one-index notation used earlier, the particular sequence of (m,n) pairs, which will form a monotonically increasing sequence of $\varkappa_{mn}$'s, is one which depends on the value of b/d. The cutoff frequency for the (m,n)th mode is $c\varkappa_{mn}/2\pi$.

Incidentally, these results should be compared with the simple reflection model of Fig. 9.9. Since the earlier model considers reflections from one pair of planes, we should compare with the modes for $n = 0$. The phase velocity $c_\varphi = \omega/k_{mo}$ is equal to $c/\sqrt{1 - (\pi mc/\omega b)^2}$, from Eq. (9.2.13), which checks with Eq. (9.2.1) exactly. Therefore, in general, the angle of incidence on the sidewalls of the (m,n)th mode is $\cos^{-1}\sqrt{(\pi mc/\omega b)^2 + (\pi nc/\omega d)^2}$. The number of nodal surfaces (where p is zero) parallel to the x walls is m; the number parallel to the y walls is n.

If the duct walls are not rigid, but have uniform specific acoustic admittance, β_x for the two walls perpendicular to the x axis and β_y for the two y walls, then it is somewhat more convenient to run the z axis along the central line of the pipe, so that the boundary condition is

$$\frac{\partial p}{\partial x} = \pm ik\beta_x p \qquad \text{at } x = \pm\tfrac{1}{2}b \qquad \frac{\partial p}{\partial y} = \pm ik\beta_y p \qquad \text{at } y = \pm\tfrac{1}{2}d$$

In this case we must distinguish between even and odd symmetry in each direction. We set

$$\Psi_{mn} = Cs_m\left(q_m, \frac{x}{b}\right) Cs_n\left(q_n, \frac{y}{d}\right)$$

where

$$Cs_m(q_m, u) = \begin{cases} \cos(\pi q_m u) & \text{when } n = 0, 2, 4, \dots \\ \sin(\pi q_m u) & \text{when } n = 1, 3, 5, \dots \end{cases}$$

The boundary conditions for the x factor are met, for the even functions, by adjusting q_m so that

$$\mp \frac{\pi q_m}{b} \sin \frac{\pi q_m}{2} = \pm ik\beta_x \cos \frac{\pi q_m}{2}$$

or

$$\tfrac{1}{2}\pi q_{xm} \tan (\tfrac{1}{2}\pi q_{xm}) = -\tfrac{1}{2}i\beta_x kb \qquad \text{for } m \text{ an even integer}$$

Those for the odd functions require that q_m satisfy

$$\frac{\pi q_m}{b} \cos \frac{\pi q_m}{2} = ik\beta_x \sin \frac{\pi q_m}{2}$$

or

$$\tfrac{1}{2}\pi q_{xm} \cot (\tfrac{1}{2}\pi q_{xm}) = \tfrac{1}{2}i\beta_x kb \qquad \text{for } m \text{ an odd integer}$$

Two similar equations relate q_{yn} to β_y.

In many cases of practical interest, the wall admittance is small enough so that $\beta kb = (2\pi b/\lambda)\beta$ is considerably smaller than unity for both the x and the y walls. When this is true we can use first approximations to the solutions. For the x walls,

$m = 0$:

$$i\pi q_{x0} \equiv i\pi(\mu_{x0} - i\nu_{x0}) \simeq \sqrt{2i\beta_x kb}$$

or

$$\mu_{x0} \simeq \frac{1}{\pi} \text{Im} [\sqrt{(\sigma_x + i\xi_x)2kb}] \qquad \nu_{x0} \simeq \frac{1}{\pi} \text{Re} [\sqrt{(\sigma_x + i\xi_x)2kb}]$$

$m > 0$:

$$iq_{xm} \equiv i\mu_{xm} + \nu_{xm} \simeq im + 2\beta_x \frac{kb}{\pi^2 m} \qquad\qquad (9.2.14)$$

or

$$\mu_{xm} \simeq m - \frac{2\sigma_x kb}{\pi^2 m} \qquad \nu_{xm} \simeq \frac{2\xi kb}{\pi^2 m}$$

and the x factor in the characteristic function, with its normalizing factor, is

m even:

$$N_{xm} = \frac{1}{b} \int_{-b/2}^{b/2} \cos^2 \left(\frac{\pi q_{xm}x}{b}\right) dx \simeq \begin{cases} 1 + \tfrac{1}{6}(\sigma_x + i\xi_x)kb & m = 0 \\[2mm] \tfrac{1}{2} - (\sigma_x + i\xi_x)\dfrac{kb}{\pi^2 m^2} & m > 0 \end{cases}$$

$$Cs_m\left(q_{xm}, \frac{x}{b}\right) \simeq \cos \frac{\pi \mu_{xm}x}{b} + i \frac{\pi \nu_{xm}x}{b} \sin \frac{\pi \mu_{xm}x}{b}$$

m odd:

$$N_{xm} = \frac{1}{b} \int_{-b/2}^{b/2} \sin^2 \left(\frac{\pi}{b} q_{xm}x\right) dx \simeq \tfrac{1}{2} + (\sigma_x + i\xi_x)\frac{kb}{\pi^2 m^2} \qquad (9.2.15)$$

$$Cs_m\left(q_{xm}, \frac{x}{b}\right) \simeq \sin \left(\frac{\pi \mu_{xm}x}{b}\right) - i \frac{\pi \nu_{xm}x}{b} \cos \frac{\pi \mu_{xm}x}{b}$$

as long as $\beta_x kb \ll 1$. We note that, since the wall conductance must be positive, ν_{xm} is positive for all values of m.

The y factors are constructed similarly, with d instead of b in the equations, and with different values of β if it should happen that the y walls differ in admittance from the x walls. Thus the characteristic functions, their normalizing factors, and their propagation constants may be written

$$\Psi_{mn} = Cs_m\left(q_{xm}, \frac{x}{b}\right)Cs_n\left(q_{yn}, \frac{y}{d}\right) \qquad \Lambda_{mn} = N_{xm}N_{yn}$$

$$\varkappa_{mn}{}^2 = \left(\frac{\pi q_{xm}}{b}\right)^2 + \left(\frac{\pi q_{yn}}{d}\right)^2 \simeq \left(\frac{\pi m}{b}\right)^2 + \left(\frac{\pi n}{d}\right)^2 - \frac{2i\epsilon_m\beta_x k}{b} - \frac{2i\epsilon_n\beta_y k}{d} \quad (9.2.16)$$

$$k_{mn} \simeq k^0_{mn} + \frac{k}{k^0_{mn}}\left(\frac{\epsilon_m\sigma_x}{b} + \frac{\epsilon_n\sigma_y}{d}\right) + i\frac{k}{k^0_{mn}}\left(\frac{\epsilon_m\xi_x}{b} + \frac{\epsilon_n\xi_y}{d}\right)$$

where $(k^0_{mn})^2 = k^2 - (\pi m/b)^2 - (\pi n/d)^2$; the approximate formulas are for $kb\beta$ small; and as always, $\epsilon_m = 1$ when $m = 0$, $\epsilon_m = 2$ when $m > 0$.

It is interesting to note that for a two-walled duct the transmission factor would be

$$k_m \simeq k\sqrt{1 - \left(\frac{\pi m}{kb}\right)^2} + \frac{\epsilon_m(\sigma + i\xi)}{\sqrt{b^2 - (\pi m/k)^2}}$$

and therefore, to this approximation, the factor $e^{ik_m z}$ corresponds to the factor of Eq. (9.2.4), obtained from the geometrical model for the higher modes ($m > 0$, $\epsilon_m = 2$) when $\beta_1 = \beta_2 = \beta$. The reflection model, however, gave an indeterminate answer for the fundamental mode, whereas the present, more detailed analysis shows that the fundamental mode is affected just half as much ($\epsilon_0 = 1$) as are the higher modes, to this approximation.

Thus, when $|\beta|$ is small compared with the ratio between the wavelength $\lambda = 2\pi/k$ and the perimeter $2(b + d)$ of the duct, the attenuation factor Im k_{mn} equals the weighted mean $(\epsilon_m\xi_x/b) + (\epsilon_n\xi_y/d)$ of the specific conductances of the walls, divided by the sine of the angle of incidence, $\sin\theta_{mn} = [1 - (\pi m/kb)^2 - (\pi n/kd)^2]^{\frac{1}{2}}$. The numerator is the same (to this approximation) for all modes except those which travel parallel to a pair of walls (i.e., for which m and/or n are zero). For the fundamental ($m = n = 0$), the numerator would be $(\xi_x/b) + (\xi_y/d)$; for waves tangential to the x walls ($m = 0$, $n > 0$), it would be $(\xi_x/b) + (2\xi_y/d)$; and so on. This is, of course, true only for those waves which propagate at the frequency $\omega/2\pi$; for those modes having cutoff frequencies higher than this, k^0_{mn} is imaginary, and no true propagation is possible.

When the duct is fitted with a piston at $z = 0$ which drives the air in the region $z > 0$, with a uniform velocity amplitude $U(x,y) = U_0$, the formulas of Eqs. (9.2.9) hold. In this case the approximate formulas for the amplitude factors U_{mn} of the transmitted wave turn out to be (if $\beta_x = \beta_y = \beta$)

$$U_{00} \simeq U_0[1 - \tfrac{1}{12}(\sigma + i\xi)k(b + d)] \qquad U_{mn} = 0 \text{ if } m \text{ and/or } n \text{ is odd}$$

$$U_{0,2n} \simeq U_0(-1)^{n-1}(\sigma + i\xi)\frac{kd}{\pi^2 n^2}$$

$$U_{2m,0} \simeq U_0(-1)^{m-1}(\sigma + i\xi)\frac{kb}{\pi^2 m^2} \qquad\qquad\qquad (9.2.17)$$

$$U_{2m,2n} \simeq U_0(-1)^{n+m}(\sigma + i\xi)^2\frac{k^2 bd}{(\pi^2 mn)^2} \qquad m, n > 0$$

Thus, even though the piston is perfectly rigid, so that only the fundamental mode would be radiated if the duct walls were rigid, when the walls are nonrigid, the higher modes are also excited to some extent. And when the driving frequency happens to be just greater than some cutoff frequency, so that some k_{mn} is very small, Eqs. (9.2.9) indicate that this mode will have a large amplitude and a phase velocity markedly different from the other components of the pressure wave. Measurements made in such a tube, if interpreted on the assumption that only the fundamental mode is present, will appear to produce erroneous results, since both the phase velocity and attenuation constants of these higher modes may differ, by fairly large factors, from those of the fundamental.

Finally, we can see, by reference to page 519 and to Eq. (6.4.36), that the thermal and viscous effects can be approximated by ascribing to the duct walls an additional specific acoustic admittance β_h and, for the (m,n)th mode, admittances β_{vxm} and β_{vyn} for the x and y walls, respectively, where

$$\beta_h \simeq (1 - i)(\gamma - 1)kd_h \qquad \beta_{vxm} \simeq (1 - i)\left[1 - \left(\frac{\pi m}{kb}\right)^2\right]kd_v$$

$$\beta_{vyn} \simeq (1 - i)\left[1 - \left(\frac{\pi n}{kd}\right)^2\right]kd_v \qquad\qquad\qquad (9.2.18)$$

Large wall admittances

When $|\beta|kb$ is not a small quantity, the tangent-cotangent equations for q_m must be solved numerically. A graphical method is illustrated by Plate IV. We set $q_m = -2ig$, so that $g = \tfrac{1}{2}\nu_m + \tfrac{1}{2}i\mu_m$, and we set

$$\frac{\beta kb}{2\pi} = \frac{\rho c}{|z|}\frac{b}{\lambda}e^{i\varphi} = he^{i\varphi}$$

so that

$$h = \frac{\rho c}{|z|}\frac{b}{\lambda} = \frac{|\beta|kb}{2\pi}$$

where $z = R - iX = |z|e^{-i\varphi}$ is the acoustic impedance of the duct walls,

$|z|$ its magnitude, and φ its phase.   Then the equations giving g in terms of h and φ are

$$g \tanh (\pi g) = ihe^{i\varphi} \quad \text{for } m \text{ even} \qquad g \coth (\pi g) = ihe^{i\varphi} \quad \text{for } m \text{ odd}$$

The conformal transformation between $v + i\mu$ and $he^{i\varphi}$ is plotted for these two cases in Plate IV.   Knowing $|z|$ and φ, b and λ, we can determine h and φ, and the corresponding values of μ and v can be picked off the chart, for both the x and the y walls.   Then the transverse function for that mode is

$$\Psi_{mn}(x,y) = Cs_m\left(\mu_{xm} - iv_{xm}, \frac{x}{b}\right)Cs_n\left(\mu_{yn} - iv_{yn}, \frac{y}{d}\right)$$

and its propagation factor k_{mn} is given by

$$k_{mn}^2 = \left(\frac{\omega}{c}\right)^2 - \left(\frac{\pi\mu_{xm}}{b}\right)^2 + \left(\frac{\pi v_{xm}}{b}\right)^2 - \left(\frac{\pi\mu_{yn}}{d}\right)^2 + \left(\frac{\pi v_{yn}}{d}\right)^2$$

$$+ 2i\frac{\pi\mu_{xm}}{b}\frac{\pi v_{xm}}{b} + 2i\frac{\pi\mu_{yn}}{d}\frac{\pi v_{yn}}{d} \quad (9.2.19)$$

The values of μ and v for m (or n) $= 0$ lie in the region marked "first even mode" in the chart to the right; those for $m = 1$ are in the region marked "first odd mode" in the chart to the left; and so on.   These values can then be used in Eqs. (9.2.15) and (9.2.16) in place of the approximate formula given there.

Conversely, if the real and imaginary parts of k_{mn}^2 can be measured (for a single mode), the values of μ_x, v_x, μ_y, and v_y may be computed (if $b = d$, $q_x = q_y$ for $m = n$, for example), and the charts can be used to determine h, and therefrom the wall admittance β.   For the reasons given in the discussion of Eqs. (9.2.17), it is usually difficult to isolate a single mode, even the fundamental, so that this procedure for measuring wall impedance has not been very successful.

When three sides of the duct are rigid, the fourth side (say, the side at $x = b$) having admittance β, the same charts can be used to compute μ_x and v_x.   We set the z axis along a corner of the duct, away from the soft wall.   Then

$$\Psi_{mn} = \cos\frac{\pi q_{xm}x}{b}\cos\frac{\pi ny}{b}$$

so that Eq. (9.2.7) is automatically satisfied for the two y walls and for the rigid x wall and becomes $(\pi q)\tan(\pi q) = -i\beta kb$ for the soft wall.   Setting

$$q_{xm} = -ig \quad \text{and} \quad h = |\beta|\frac{kb}{\pi} = \frac{\rho c}{|z|}\frac{2b}{\lambda}$$

produces the equation

$$g \tanh (\pi g) = ihe^{i\varphi}$$

which is plotted on the right-hand side of Plate IV.   Thus, introducing the extra factor 2 in the calculation of h and multiplying the values read from the chart by $\frac{1}{2}$ to obtain μ_{xm} and ν_{xm} enables one to use the same chart to work out this case.   The corresponding propagation constant can be obtained from

$$k_{mn}^{2} = \left(\frac{\omega}{c}\right)^{2} - \left(\frac{\pi n}{d}\right)^{2} - \left(\frac{\pi \mu_{xm}}{b}\right)^{2} + \left(\frac{\pi \nu_{xm}}{b}\right)^{2} + 2i\,\frac{\pi \mu_{xm}}{b}\,\frac{\pi \nu_{xm}}{b} \quad (9.2.20)$$

It is obvious that the behavior of the sound waves, even of the principal wave, is quite complicated when the admittance ratio of the walls is not small compared with unity.   If one adds to this the fact that many porous acoustic materials have impedances that vary considerably with change in frequency, it becomes apparent that very few sweeping generalities can be made concerning the behavior of sound in ducts with highly absorbent walls.   An examination of Plate IV shows that, for negative phase angles (stiffness reactance) and for large values of h (small values of impedance and/or high frequencies), the value of ν can become quite large, and therefore even the principal wave can be highly damped.   Further examination shows that in these cases the principal wave is far from a uniform plane wave, the negative reactance of the walls having in some way pulled most of the energy of the wave away from the center of the duct to the periphery, where it is more quickly absorbed as it travels along.   A positive phase angle (mass reactance) has the opposite effect.   The change of acoustic behavior with frequency is greatest when h and ϑ are close to one of the "branch points" of the equation for g, shown as circles on the dashed lines separating the modes.

Figure 9.11 gives a typical example for a fairly "soft" duct wall (with acoustic impedance similar to a number of sound-absorbent materials). The real and imaginary parts of $k_{mn} = (2\pi/\lambda_{mn}) + i\pi\gamma_{mn}$ [Eq. (9.2.8)] have physical significance for the transmission properties of the (m,n)th mode.   As discussed on page 497, the reciprocal of the real part is proportional to the phase velocity of this mode, $c_{mn} = c(k/k_{mn})$, and the imaginary part is the attenuation factor $\pi\gamma_{mn}$ per unit length of duct. Since a reduction of $|p|$ by a factor of e^{-1} corresponds to a power reduction of 8.686 db, the power attenuation of the mode in decibels per unit length is $8.686\pi\gamma_{mn} = 27.3\gamma_{mn}$.   The lower sets of curves of Fig. 9.11 show how the phase velocity of the (1,1) mode rises for frequencies just above its cutoff. They also show the very large attenuation of this mode, even for frequencies well above the cutoff, for duct walls as soft as this.

Ducts of circular cross section

When the duct cross section is a circle of radius b, the transverse, eigenfunctions and values are

$$\Psi_{mn}^{\sigma} = \frac{\cos}{\sin}\,(m\phi)J_{m}\!\left(\frac{\pi q_{mn}r}{b}\right) \qquad \varkappa_{mn} = \frac{\pi q_{mn}}{b} \quad (9.2.21)$$

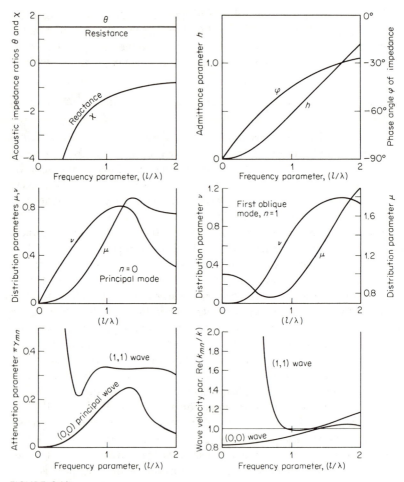

FIGURE 9.11
Transmission of sound along a square duct of width l covered with acoustic material of acoustic impedance $\rho c(\theta - i\chi)$, for principal (0,0)-mode and a higher (1,1)-mode wave. Quantities proportional to attenuation per unit length and phase velocity are plotted against frequency. Sudden rise of γ_{mn} and k_{mn} for (1,1)-mode illustrates the fact that higher modes cannot be transmitted at low frequencies.

where $\sigma = +1$ for the cosine ϕ factor and -1 for the sine factor, and r, ϕ are polar coordinates about the central axis of the tube. The properties of the Bessel functions are given in Eqs. (5.2.20) and (5.2.21). The boundary condition (9.2.7) is that

$$\pi q_{mn} J'_m(\pi q_{mn}) \equiv \pi q_{mn} J_{m-1}(\pi q_{mn}) - m J_m(\pi q_{mn})$$
$$= i\beta k b J_m(\pi q_{mn})$$

J'_m being the derivative of J_m with respect to its argument. For each integral value of m there is a discrete set of values of q which satisfies this equation; the smallest will be labeled q_{m0}, the next q_{m1}, and so on. When all Ψ'_{mn}'s satisfy this condition, their normalizing constants are

$$\int_0^{2\pi} d\phi \int_0^b \Psi'_{mn}{}^\sigma \Psi'_{lk}{}^\tau r\, dr = \begin{cases} 0 & \sigma \neq \tau \text{ or } l \neq m \text{ or } k \neq n \\ \pi b^2 \Lambda_{mn} & \sigma = \tau \text{ and } l = m \text{ and } k = n \end{cases} \quad (9.2.22)$$

where

$$\Lambda_{mn} = \frac{1}{\epsilon_m}\left[1 - \frac{m^2 + (\beta k b)^2}{(\pi q_{mn})^2}\right] J_m{}^2(\pi q_{mn})$$

and $\epsilon_m = 1$ when $m = 0$, $= 2$ when $m > 0$. The (m,n)th wave has m plane nodal surfaces, extending radially out from the axis, and n cylindrical nodes concentric with the axis.

For a rigid duct the boundary condition is $J'_m(\pi q) = 0$ or $\pi q J_{m-1}(\pi q) = m J_m(\pi q)$. Solutions of this equation will be called $q = \alpha_{mn}$, to distinguish them from the roots β_{mn} of Eqs. (5.2.23), solutions of $J_m(\pi\beta) = 0$ for the circular membrane. A few values and corresponding values of $J_m(\pi\alpha)$ are

$$
\begin{array}{llll}
\alpha_{00} = 0.0000 & \alpha_{01} = 1.2197 & \alpha_{02} = 2.2331 & \cdots \\
\alpha_{10} = 0.5861 & \alpha_{11} = 1.6970 & \alpha_{12} = 2.7140 & \cdots \\
\alpha_{20} = 0.9722 & \alpha_{21} = 2.1346 & \alpha_{22} = 3.1734 & \cdots
\end{array}
$$

$$\alpha_{mn} \rightarrow n + \tfrac{1}{2}m - \tfrac{3}{4} \qquad m < n \gg 1 \qquad (9.2.23)$$

$$
\begin{array}{lll}
J_0(\pi\alpha_{00}) = 1.0000 & J_0(\pi\alpha_{01}) = -0.4028 & J_0(\pi\alpha_{02}) = 0.3001 \\
J_1(\pi\alpha_{10}) = 0.5819 & J_1(\pi\alpha_{11}) = -0.3461 & J_1(\pi\alpha_{12}) = 0.2733 \\
J_2(\pi\alpha_{20}) = 0.4865 & J_2(\pi\alpha_{21}) = -0.3135 & J_2(\pi\alpha_{22}) = 0.2547
\end{array}
$$

$$J_m(\alpha_{mn}) \simeq \sqrt{\frac{2}{\pi^2 \alpha_{mn}}} \qquad m < n \gg 1$$

As expected, the fundamental $(0,0)$ mode for the rigid tube has a zero characteristic value and thus is a plane wave, propagating at the free-space velocity c. The cutoff frequency for the (m,n)th mode is $\alpha_{mn}c/2b$. Thus the first modes to become propagational are $(1,0)$, then $(2,0)$ (each being a pair of modes, one for $\sigma = +1$, the other for $\sigma = -1$), and only after these is the second symmetric mode $(0,1)$ propagated, as the frequency is increased.

Various properties of cylindrical tubes will be worked out in later sections, but we can now use the formulas just developed to illustrate some of the statements made regarding the radiation reaction on a simple source in a duct. If the source, a sphere of radius $a \ll b$, has its center at the point $\phi = 0$, $r = r_0$, $z = 0$, then the frequency-dependent factor in Eqs. (9.2.12) for the radiation resistance, the bracketed factor in the real part of Z_s, becomes

$$\sum_{n,m}^{\pi\alpha < kb} \frac{\epsilon_m [J_m(\pi\alpha_{mn}r_0/b)/J_m(\pi\alpha_{mn})]^2}{[1 - (m/\pi\alpha_{mn})^2]\sqrt{1 - (\pi\alpha_{mn}/kb)^2}}$$

where we have absorbed the unity term (for $m = n = 0$) into the sum, which is over all pairs of values of m and n for which $\alpha_{mn} < kb/\pi$. The graph for this sum is quite similar to that shown in Fig. 9.10, with the successive peaks coming at $kb = \pi\alpha_{mn}$. As before, the radiation resistance jumps discontinuously as the driving frequency reaches the cutoff frequency of another mode. Of course, if the tube were not perfectly rigid (or if we took thermal and viscous surface losses into account), the peaks would be rounded off and finite in height, becoming smaller and closer together as ω is increased, until the curve roughly equals the smooth curve for $\frac{1}{2}(kb)^2$, the corresponding radiation-resistance term for an isolated source. The differences, at lower frequencies, illustrate how much the reaction on the simple source is affected by the walls of the tube which surround it.

When the wall admittance is not zero but is small compared with $\lambda/2\pi b$, one can develop approximate formulas, similar to Eqs. (9.2.14) to (9.2.16), by using the equations

$$J'_m(\pi\alpha_{mn}) = 0 \qquad \text{and} \qquad J''_m(\pi\alpha_{mn}) = -[1 - (m/\pi\alpha_{mn})^2]J_m(\pi\alpha_{mn})$$

To the first approximation, in the small quantity $|\beta|\, kb$, we have

$$i\pi q_{00} \simeq \sqrt{2(i\beta kb)} \qquad k_{00} \simeq k + \frac{1}{2b}(\sigma + i\xi)$$

$$\pi q_{mn} \simeq \pi\alpha_{mn}\left[1 - \frac{(\sigma + i\xi)kb}{(\pi\alpha_{mn})^2 - m^2}\right] \qquad k_{mn} \simeq k^0_{mn} + \frac{(\sigma + i\xi)k}{k^0_{mn}b[1 - (m/\pi\alpha_{mn})^2]}$$

$$\Psi_{00} \simeq 1 + (\sigma + i\xi)\frac{kr^2}{4b} \qquad \Lambda_{00} \simeq 1 - \tfrac{3}{4}(\sigma + i\xi)kb \qquad (9.2.24)$$

$$\Psi_{mn} \simeq \genfrac{}{}{0pt}{}{\cos}{\sin}(m\phi)\left\{J_m\left(\frac{\pi}{b}\alpha_{mn}r\right) - \frac{(\sigma + i\xi)\pi\alpha_{mn}}{(\pi\alpha_{mn})^2 - m^2}\left[\frac{m}{\pi\alpha_{mn}}J_m\left(\frac{\pi}{b}\alpha_{mn}r\right)\right.\right.$$

$$\left.\left. - J_{m-1}\left(\frac{\pi}{b}\alpha_{mn}r\right)\right]\right\}$$

$$\Lambda_{mn} \simeq \tfrac{1}{2}\left[1 - \left(\frac{m}{\pi\alpha_{mn}}\right)^2\right]\left\{1 - \frac{2(\sigma + i\xi)kb}{[(\pi\alpha_{mn})^2 - m^2]^2}\right\}J_m^{\,2}(\pi\alpha_{mn})$$

where $k^0_{mn} = \sqrt{k^2 - (\pi\alpha_{mn}/b)^2}$ or $i\sqrt{(\pi\alpha_{mn}/b)^2 - k^2}$, depending on whether the mode's cutoff frequency is below or above the driving frequency.

The lowest mode of any axial symmetry ($n = 0$ for a given m) exhibits the greatest attenuation, for frequencies above its cutoff frequency, because the factor $1 - (m/\pi\alpha_{mn})^2$ in the denominator for k_{mn} is smallest, for a given m, for $n = 0$ (note that $m > \pi\alpha_{m0}$ for all m). This is in contrast to the case of the rectangular duct, where the lowest modes (m or $n = 0$) have a *lower* attenuation than the rest. In the rectangular duct, of course, the higher-mode waves reflect back and forth from plane-parallel walls, so that there can be a whole set of modes which do not reflect at all from one pair of walls (the ones $m = 0$, $n > 0$ or $m > 0$, $n = 0$). But for the cylindrical duct, the only wave which does not reflect from the curved walls is the fundamental mode, which has an attenuation constant $\xi/2b$, less than half the value $\xi/b\sqrt{1 - (\pi\alpha_{mn}/kb)^2}[1 - (m/\pi\alpha_{mn})^2]$ for any other mode (to this approximation).

In the case of the circular pipe the waves have in general two possibilities, when the wavelength is small compared with b: they can reflect more or less normally from the surface and thus focus in to the center, with circular nodal surfaces for its standing waves (this corresponds to n large and m small), or they can move in a spiral path, roughly parallel to the wall, with a preponderance of radial nodes (which would correspond to m large and n small). In the latter case the waves would not avoid reflecting from the wall, as they would in the rectangular duct; their angle of incidence is large, and they tend to "cling" to the curved wall. As a matter of fact, the modes for m large and n small have very small amplitude near the tube axis and differ appreciably from zero only for r nearly equal to b. Thus the majority of these modes' energy is close to the duct wall, where it is quickly absorbed; these "spiral-tangential" waves are *more* strongly attenuated than the others, in contrast to the truly tangential waves in the rectangular duct, which are *less* strongly attenuated than the others.

If $|\beta|\,kb$ is not small, the boundary condition

$$-i\beta kb = m - \pi q\,\frac{J_{m-1}(\pi q)}{J_m(\pi q)} \qquad J_{-1} = -J_1 \text{ for } m = 0$$

must be solved exactly. If we set $q = -iw$, so that $w = v + i\mu$, and set $h = |\beta|\,kb/\pi$, with $\beta = |\beta|\,e^{i\varphi} = \rho c/|z|\,e^{-i\varphi}$ as before, the equation for w in terms of h and φ for the J_m modes is

$$ihe^{i\varphi} = -\frac{m}{\pi} - iw\,\frac{J_{m-1}(-i\pi w)}{J_m(-i\pi w)} \tag{9.2.25}$$

Plate V plots this conformal transform for $m = 0$ and $m = 1$. There is a general similarity to the corresponding plots of Plate IV for the rectangular duct, though the exact locations of branch points, zeros, etc., differ somewhat.

As an example, to compare with Fig. 9.11, we have calculated the values of v and μ for the fundamental mode $(0,0)$ in a circular pipe, with walls having a unit specific acoustic resistance and a stiffness reactance inversely proportional to the frequency, as in the case of Fig. 9.11. The real and imaginary parts of the propagation parameter

$$k_{00} = \frac{2\pi}{\lambda_{00}} + i\pi\gamma_{00} = \left[k^2 - \left(\frac{\pi q_{00}}{b}\right)^2\right]^{\frac{1}{2}}$$

can then be computed and are plotted in Fig. 9.12. As mentioned earlier, the real part of k_{00}/k is the ratio of the velocity of sound in free space to the phase velocity of the mode in the tube, and its imaginary part is proportional to the attenuation, and thus to the decibel drop in intensity as the wave travels along the tube. These quantities are plotted in Fig. 9.12, for two

different wall stiffness constants.   The impedances in the two examples
were chosen to display large attenuations.   On the w plane of Plate V the
dashed curve goes to the right of the branch point at $\eta = 0.41$, $\mu = 0.94$;
the solid curve goes to the left.   The solid curves resemble those of Fig. 9.11
for the fundamental mode (note that the left-hand curve of Fig. 9.11 is for
$\pi\gamma_{00}$, whereas the upper curves of 9.12 are proportional to $\lambda\gamma_{00}$).

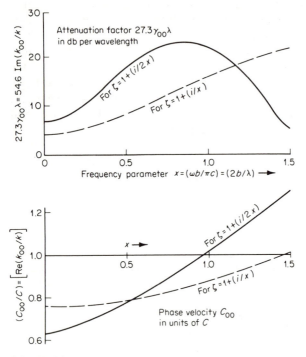

FIGURE 9.12
Attenuation and phase velocity of lowest mode in a circular
tube, with walls having specific acoustic impedance $\zeta = Z/\rho c$,
as function of frequency, for two different values of stiffness
reactance.

Representation of randomly distributed sources

Occasionally, it is of interest to calculate the generation of sound from a
region inside the duct, where fluid is being produced (or expanded by radiant
heat) in a random manner or where the fluid is in violent turbulence.   This
problem was discussed in Sec. 7.1 for a turbulent region in free space.   The
confining effect of the duct walls introduces additional complications, in
part because it is less easy to define a random function in a restricted region
than in free space.   To demonstrate these difficulties and to show how they
may be overcome, we consider a region of fluctuating fluid generation inside
a rectangular duct with rigid walls.

Our first task is to devise a measure of the variability of some fluctuating function $f(x,y)$ within the finite area $0 < x < b$, $0 < y < d$. In unconfined space this would be given by the autocorrelation function for f, generalizing the definition of Eq. (1.3.13). But if we define

$$\Upsilon_f(\xi,\eta) = \frac{1}{bd} \int_0^b dx \int_0^d f(x,y) f(x + \xi, y + \eta) \, dy \qquad (9.2.26)$$

as the correlation function for f, we must somehow decide what values we should use for f when $x + \xi$ exceeds b or $y + \eta$ exceeds d.

We could, of course, say that f is zero outside the limited area, but this leads to computational difficulties. It might seem more "physical" to consider the walls as mirrors and to consider f outside the duct cross section to be the mirror image of the wave inside, somewhat as was done in the case of the string (Fig. 4.7). Another way of saying the same thing is to expand f in terms of the eigenfunctions of the cross section

$$f(x,y) = \sum_{m,n} C_{mn} \Psi_{mn}(x,y) \qquad C_{mn} = \frac{1}{bd\Lambda_{mn}} \int_0^b \int_0^d f \Psi_{mn} \, dx \, dy \quad (9.2.27)$$

and say that f, outside $0 < x < b$, $0 < y < d$, is simply the analytic continuation of the series, over the wider range of the variables. If the walls are rigid, function f will be symmetric about the boundary lines.

If we do this, it is then easy to calculate the expansion of the autocorrelation function in terms of the normal modes of the duct cross section. Inserting series (9.2.27) twice into Eq. (9.2.26), using the formulas

$$\cos\left[(\pi m/b)(x + \xi)\right] = \cos(\pi m x/b) \cos(\pi m \xi/b)$$

$$- \sin(\pi m x/b) \sin(\pi m \xi/b), \text{ etc.}$$

and remembering that the integral of $\cos(\pi m x/b) \sin(\pi m x/b)$ over x from 0 to b is zero, we obtain

$$\Upsilon_f(\xi,\eta) = \sum_{m,n} \Lambda_{mn} |C_{mn}|^2 \Psi_{mn}(\xi,\eta) \qquad (9.2.28)$$

Thus the coefficients in the expansion of the correlation function of f are the squares of the coefficients of the expansion of f itself, a simple extension of the relationship of Eq. (1.3.21). The coefficients C_{mn} can be considered as transforms of function f, with respect to the set of eigenfunctions Ψ_{mn}. Alternatively,

$$|C_{mn}|^2 = \frac{1}{bd\Lambda_{mn}^2} \int_0^b d\xi \int_0^d d\eta \, \Upsilon_f(\xi,\eta) \Psi_{mn}(\xi,\eta) \qquad (9.2.29)$$

Now suppose the region R_s of the duct from $z = -\frac{1}{2}L$ to $z = \frac{1}{2}L$ is filled with fluctuating production of fluid, at a rate $s(\mathbf{r},t)$ units of volume per second per unit volume, s being a function which is randomly variable in region R_s, having a mean-square magnitude $\langle s^2 \rangle$ there, a correlation distance in space of w, and a correlation interval in time of $\tau_c = 1/\omega_0$, with $\omega_0/2\pi$ the upper limit of the frequency distribution. Then, from Eqs. (1.3.20), (7.1.33), and (9.1.28), we can write

$$C_{mn}{}^s(\omega,K) = \frac{1}{4\pi^2 bd\Lambda_{mn}} \int \cdots \int e^{i\omega t - iKz}\Psi_{mn}(x,y)s(\mathbf{r},t)\,dx\,dy\,dz\,dt$$

$$|C_{mn}{}^s(\omega,K)|^2 = \frac{LT}{16\pi^4\Lambda_{mn}{}^2 bd} \int \cdots \int \Upsilon_s(d,\tau)e^{i\omega\tau - iKd_z}\Psi_{mn}(d_x,d_y)\,d\tau\,dv_d$$

$$\Upsilon_s(d,\tau) = \frac{1}{VT} \int \cdots \int s(\mathbf{r},t)s(\mathbf{r}+\mathbf{d},\,t+\tau)\,dt\,dv \tag{9.2.30}$$

$$= \frac{4\pi^2}{LT} \sum_{mn} \Lambda_{mn}\Psi_{mn}(d_x,d_y) \iint |C_{mn}{}^s(\omega,K)|^2\,e^{-i\omega\tau + iKd_z}\,d\omega\,dK$$

where the integrations are over the cross section of the duct, over z or d_z from $-\tfrac{1}{2}L$ to $\tfrac{1}{2}L$, and over t or τ from $-\tfrac{1}{2}T$ to $\tfrac{1}{2}T$, and the final integral over ω and K is from $-\infty$ to $+\infty$.

The component of the pressure wave of frequency $\omega/2\pi$, at point $\mathbf{r} = (x,y,z)$, to the right of region R_s, can be obtained from the Green's function of Eq. (9.2.10) by integrating $-ik\rho c g_\omega$ times the component of $s(\mathbf{r},t)$ having the frequency $\omega/2\pi$.

$$p_\omega(\mathbf{r}) = \frac{k\rho c}{4\pi bd} \sum_{mn}{}' \frac{\Psi_{mn}(x,y)e^{ik_{mn}z}}{\Lambda_{mn}k_{mn}} \int \cdots \int e^{i\omega t - ik_{mn}z_0}\,\Psi_{mn}(x_0,y_0)s(\mathbf{r}_0,t)\,dv_0\,dt$$

$$= \pi k\rho c \sum_{mn}{}' \frac{C_{mn}{}^s(\omega,k_{mn})}{k_{mn}} \Psi_{mn}(x,y)e^{ik_{mn}z} \tag{9.2.31}$$

Also

$$u_z(\mathbf{r}) = \pi \sum_{mn}{}' C_{mn}{}^s(\omega,k_{mn})\Psi_{mn}(x,y)e^{ik_{mn}z} \qquad \text{for } z \gg \tfrac{1}{2}L$$

is the axial component of fluid velocity for the frequency $\omega/2\pi$. The sums include all those modes for which the eigenvalue $\varkappa_{mn}{}^2 = (\pi m/b)^2 + (\pi n/d)^2$ is less than $k^2 = (\omega/c)^2$, so that $k_{mn}{}^2 = k^2 - \varkappa_{mn}{}^2$ is positive, and the mode propagates (which is the meaning of the prime over the summation sign). The intensity at $\mathbf{r}$ in the unit ω band at ω is then the real part of $(p_\omega u_z^*)$, and the total power radiated to the right from R_s is

$$W_\omega = \int (p_\omega u_z^*)\,dx\,dy = \pi^2\rho cbd \sum_{mn}{}' \frac{k}{k_{mn}} |C_{mn}{}^s(\omega,k_{mn})|^2 \tag{9.2.32}$$

Now suppose the source distribution has an autocorrelation function which can be approximated by the usual formula

$$\Upsilon_s(\mathbf{d},\tau) = \langle s^2 \rangle \exp\left[-\frac{1}{2}\left(\frac{d}{w}\right)^2 - \tfrac{1}{2}(\omega_0\tau)^2 \right] \tag{9.2.33}$$

where $\langle s^2 \rangle$ is the mean-square value of the source function s in region R_s, and w is the correlation length and $1/\omega_0$ the correlation time interval for s. By using Eqs. (9.2.30), if $w \ll L$, b, and d and $(1/\omega_0) \ll T$, we can write out an approximate formula for $|C|^2$ and for the power-spectrum level W_ω.

$$|C_{mn}{}^s(\omega,k_{mn})|^2 \simeq \frac{LTw^3\langle s^2\rangle}{16\pi^2\Lambda_{mn}{}^2 bd\omega_0} \exp\left[-\tfrac{1}{2}(kw)^2 - \frac{1}{2}\left(\frac{\omega}{\omega_0}\right)^2 \right]$$

$$W_\omega \simeq \frac{LTw^3\rho c}{16\omega_0} \langle s^2\rangle \sum_{mn}{}' \frac{\exp\left[-\tfrac{1}{2}(kw)^2 - \tfrac{1}{2}(\omega/\omega_0)^2\right]}{\sqrt{1-(\varkappa_{mn}/k)^2}} \tag{9.2.34}$$

The total power radiated to the right, for all frequencies, is obtained by using Eq. (2.3.13).

$$W = \frac{2\pi}{T}\int_{-\infty}^{\infty} W_\omega\, d\omega = \frac{\pi^{\frac{3}{2}}Lw^3\rho c}{8[(k_0w)^2+1]} \langle s^2\rangle \sum_{mn}{}' \exp\left[-\tfrac{1}{2}\varkappa_{mn}{}^2\left(w^2+\frac{1}{k_0{}^2}\right) \right]$$

$$\simeq \frac{\sqrt{\pi}}{16}\rho c\, \frac{V_s w(k_0w)^2}{[(k_0w)^2+1]^2} \langle s^2\rangle \qquad \text{if } w \text{ and } \frac{1}{k_0} \ll \frac{1}{\varkappa_{10}} = \frac{b}{\pi} \tag{9.2.35}$$

where $k_0 = \omega_0/c$, and $V_s = Lbd$ is the volume of the radiating region R_s. The second formula is valid when the exponential in the sum m, n varies so slowly that we can substitute for the sum an integral over n and m. The quantity $k_0w = 2\pi w/\lambda_0$ is 2π times the ratio between the spatial-correlation length w and the wavelength λ_0 corresponding to the upper limit $\omega_0/2\pi$ of the frequencies present in the random source. The total power radiated is maximum when this quantity is unity.

A similar calculation can be made for the radiation from a turbulent region. Instead of the source function $s(\mathbf{r},t)$, the kinetic-energy function

$$T(\mathbf{r},t) = \frac{\partial^2}{\partial x^2}(\rho U_x{}^2) + \frac{\partial^2}{\partial y^2}(\rho U_y{}^2) + \frac{\partial^2}{\partial z^2}(\rho U_z{}^2)$$

$$+ 2\frac{\partial^2}{\partial x\, \partial y}(\rho U_x U_y) + 2\frac{\partial^2}{\partial x\, \partial z}(\rho U_x U_z) + 2\frac{\partial^2}{\partial y\, \partial z}(\rho U_y U_z)$$

is the generating term, where U_x, U_y, U_z are the components of the turbulent velocity at $\mathbf{r}$, t. An autocorrelation function Υ_t can be formed from this function, and the expansion coefficients $C_{mn}{}^t$ can be calculated, having relationships to T and Υ_t similar to Eqs. (9.2.30) for $C_{mn}{}^s$. Integration by parts produces the alternative formula

$$C_{mn}{}^t(\omega,K) = \frac{\epsilon_m \epsilon_n}{4\pi^2 bd} \int \cdots \int \rho \left\{ -\left[U_x{}^2 \left(\frac{m\pi}{b}\right)^2 + U_y{}^2 \left(\frac{n\pi}{d}\right)^2 + U_z{}^2 K^2 \right] \right.$$

$$\times \cos \frac{m\pi x}{b} \cos \left(\frac{n\pi y}{d}\right)$$

$$+ 2U_x U_y \frac{m\pi}{b} \frac{n\pi}{d} \sin \frac{\pi m x}{b} \sin \left(\frac{\pi n y}{d}\right)$$

$$+ 2i U_x U_z K \frac{m\pi}{b} \sin \frac{\pi m x}{b} \cos \left(\frac{\pi n y}{d}\right)$$

$$\left. + 2i U_y U_z K \frac{n\pi}{d} \cos \frac{\pi m x}{b} \sin \frac{\pi n y}{d} \right\} e^{i\omega t - iKz} \, dx \, dy \, dz \, dt$$

The rest of the calculation proceeds as for the function s.

Extension of these formulas to ducts of other shape is made difficult by the complications of relating the expansion coefficients C to an autocorrelation function. For example, for a cylindrical duct of radius a, for which a random function $f(x,y)$ is represented by a series of Bessel functions,

$$f(x,y) = \sum_{mn} [C_{mn}{}^e \cos(m\phi) + C_{mn}^0 \sin(m\phi)] J_m \left(\frac{\pi \alpha_{mn} r}{a}\right) \qquad C_{0n}^0 = 0$$

$$C_{mn}{}^e = \frac{1}{\Lambda_{mn} \pi a^2} \int_0^{2\pi} d\phi \int_0^a f(r \cos\phi, r \sin\phi) \cos(m\phi) J_m \left(\frac{\pi \alpha_{mn} r}{a}\right) r \, dr$$

and similarly for C_{mn}^0. If the random behavior of f is isotropic, the autocorrelation function should be independent of ϕ. It can be shown, in that case, that

$$\Upsilon(\delta) \equiv \frac{1}{\pi a^2} \int_0^{2\pi} d\phi \int_0^a f(r \cos\phi, r \sin\phi)$$

$$\times f(r \cos\phi + \delta \cos\theta, r \sin\phi + \delta \sin\theta) r \, dr$$

$$= \sum_{mn} (\Lambda_{mn}{}^e |C_{mn}{}^e|^2 + \Lambda_{mn}^0 |C_{mn}^0|^2) J_m \left(\frac{\pi \alpha_{mn} \delta}{a}\right)$$

which is *not* an expansion in terms of the axially symmetric eigenfunctions $J_0(\pi \alpha_{0n} \delta/a)$. Alternatively, one could define a rotational correlation function

$$\Upsilon_r(\theta) \equiv \frac{1}{\pi a^2} \iint f(r \cos\phi, r \sin\phi) f[r \cos(\phi + \theta), r \sin(\phi + \theta)] r \, dr \, d\phi$$

$$= \sum_{mn} (\Lambda_{mn}{}^e |C_{mn}{}^e|^2 + \Lambda_{mn}^0 |C_{mn}^0|^2) \cos(m\theta)$$

expressible in terms of the angle functions, but it is not certain what use this function would have. About all we can do is to assume that $|C_{mn}{}^e|^2 = |C_{mn}^0|^2$ has, roughly, the form given in Eqs. (9.2.34), and continue as before.

Losses at the duct walls

Viscous and thermal losses at the duct walls tend to attenuate the waves. If the cross-sectional dimensions are large compared with the boundary-layer thickness, these losses can be approximately allowed for in the boundary conditions on the propagational waves, as shown in Eq. (6.4.36). That equation indicates that these effects act as though the surface had, in addition to its usual specific acoustic admittance $\beta = \rho c/z$, an effective viscothermal admittance

$$\beta^v \simeq \tfrac{1}{2}(1 - i)\left[(\gamma - 1)kd_h + \left(\frac{k_t}{k}\right)^2 kd_v\right] \qquad (9.2.36)$$

where $k = \omega/c$, and d_h and d_v, defined in Eqs. (6.4.31), are the thermal- and viscous-boundary-layer thicknesses, respectively. The square of the tangential component k_t of the wavenumber is k^2 minus the portion of $\varkappa_n{}^2$ coming from the factor normal to the surface. If β is not very small, the effects of β^v can usually be neglected, but if the duct walls are quite rigid, β^v may be the largest part of the effective wall admittance.

We have already seen, in Eq. (9.2.18), that the effective viscothermal admittance for the (m,n)th mode in a rectangular duct is

$$\beta_{xm}{}^v \simeq \tfrac{1}{2}(1 - i)\left\{(\gamma - 1)kd_h + \left[1 - \left(\frac{\pi m}{kb}\right)^2\right]kd_v\right\} \qquad \text{(for the } x \text{ walls)}$$

$$\qquad\qquad (9.2.37)$$

$$\beta_{yn}{}^v \simeq \tfrac{1}{2}(1 - i)\left\{(\gamma - 1)kd_h + \left[1 - \left(\frac{\pi n}{kd}\right)^2\right]kd_v\right\} \qquad \text{(for the } y \text{ walls)}$$

These quantities, plus the usual wall admittances, are then used in Eqs. (9.2.15) and (9.2.16) to compute the waveforms and transmission factors for the (m,n)th mode.

Similarly, Eqs. (9.2.24) can be used for circular ducts with rigid walls, by using $k_t{}^2 = k^2 - (\pi \alpha_{mn}/b)^2$ and thus assuming an effective wall conductance and susceptance

$$\xi_{mn}{}^v = \sigma_{mn}{}^v = \tfrac{1}{2}(\gamma - 1)kd_h + \tfrac{1}{2}\left[1 - \left(\frac{\pi \alpha_{mn}}{kb}\right)^2\right]kd_v \qquad (9.2.38)$$

for the viscothermal part of the wall admittance for the (m,n)th mode.

When the tube is so narrow that the boundary layers merge, the approximation yielding an equivalent admittance breaks down, and we must return to the equations for thermal and viscous waves to find a solution satisfying the boundary conditions of normal and tangential velocity and temperature variation, all going to zero at the wall. As an example of the calculations, we shall work out the formulas for the fundamental mode in a circular tube of radius b which is not very large compared with d_v or d_h.

The propagational wave [see Eq. (6.4.32)] for $m = n = 0$ has acoustic pressure, temperature, and radial and tangential velocity

$$p_p = J_0\!\left(\frac{\pi q r}{b}\right)e^{ik_t z} \qquad \tau_p = \frac{\gamma - 1}{\alpha \gamma} J_0\!\left(\frac{\pi q r}{b}\right)e^{ik_t z}$$

$$u_{pr} = \frac{-\pi q}{i\omega\rho b} J_1\!\left(\frac{\pi q r}{b}\right)e^{ik_t z} \qquad u_{pz} = \frac{k_t}{\rho c k} J_0\!\left(\frac{\pi q r}{b}\right)e^{ik_t z}$$

(9.2.39)

where $k_t{}^2 = k^2 - (\pi q/b)^2$; all of this is in accord with Eq. (9.2.21). Constant q is to be adjusted to fit the boundary conditions. One of the conditions is that the acoustic temperature τ is to be zero at $r = b$. To fit this we must introduce a thermal wave [Eq. (6.4.25)]

$$\tau_h = -\frac{\gamma - 1}{\alpha \gamma} B_h J_0\!\left[(1 + i)\frac{r}{d_h}\right]e^{ik_t z} \qquad B_h = \frac{J_0(\pi q)}{J_0\!\left[(1 + i)\dfrac{b}{d_h}\right]}$$

(9.2.40)

$$u_{hr} \simeq (1 + i)\frac{\gamma - 1}{2\rho c} k d_h B_h J_1\!\left[(1 + i)\frac{r}{d_h}\right]$$

The pressure and tangential velocity for this wave are of the order of magnitude of $(k d_h)^2$, and may be neglected for air at normal pressures and temperatures for frequencies less than 100 kc.

This wave cancels out the temperature fluctuations of the propagational wave at $r = b$, but it does not cancel the tangential velocity there. To satisfy this condition we must add a shear wave, solution of the last of Eqs. (6.4.17), with tangential velocity equal to $-u_{pz}$ at $r = b$. The appropriate solution is not the plane wave of Eq. (6.4.30), but is an appropriate constant times

$$\mathbf{U} = \text{curl curl}\left\{\mathbf{a}_z J_0\!\left[(1 + i)\frac{r}{d_v}\right]e^{ik_t z}\right\}$$

where $\mathbf{a}_z$ is the unit vector in the z direction. This is a solution of the equation

$$\text{curl curl } \mathbf{U} = \left(\frac{2i}{d_v{}^2} - k_t{}^2\right)\mathbf{U} \simeq \left(\frac{2i}{d_v{}^2}\right)\mathbf{U}$$

which automatically has zero divergence and the required pattern of tangential velocity at $r = b$. The appropriate constant to multiply $\mathbf{U}$ is

$$\tfrac{1}{2}i\,\frac{k_t}{2\rho c k} B_v d_v{}^2 \qquad \text{where} \qquad B_v = \frac{J_0(\pi q)}{J_0\!\left[(1 + i)\dfrac{b}{d_v}\right]}$$

Thus the shear wave has velocity components

$$u_{vr} \simeq \frac{1 + i}{2\rho c}\frac{k_t{}^2 d_v}{k} B_v J_1\!\left[(1 + i)\frac{r}{d_v}\right]e^{ik_t z}$$

$$u_{vz} \simeq -\frac{k_t}{\rho c k} B_v J_0\!\left[(1 + i)\frac{r}{d_v}\right]e^{ik_t z}$$

with zero pressure and temperature fluctuations.

The sum of the three waves has zero acoustic temperature and tangential velocity u_z at the tube surface $r = b$. The acoustic pressure and normal velocity are

$$p \simeq p_p = J_0(\pi q)e^{ik_l z}$$

$$u_r = u_{pr} + u_{hr} + u_{vr} \simeq \frac{-\pi q}{i\omega \rho b} J_1(\pi q) + \frac{1+i}{2\rho c} J_0(\pi q)$$

$$\times \left\{ (\gamma - 1)k d_h \frac{J_1[(1+i)b/d_h]}{J_0[(1+i)b/d_h]} + \frac{k_l^2 d_v}{2\rho c} \frac{J_1[(1+i)b/d_v]}{J_0[(1+i)b/d_v]} \right\}$$

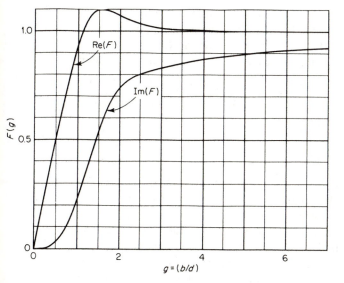

FIGURE 9.13
Real and imaginary parts of function $F(g)$ defined in Eq. (9.2.41) for thermal and viscous boundary conditions at inner surface of a circular pipe.

and q must be adjusted so that $u_r = 0$ there.   The resulting equation for q is

$$\pi q \frac{J_1(\pi q)}{J_0(\pi q)} \simeq -\frac{1}{2}\left[k^2 - \left(\frac{\pi q}{b}\right)^2 \right] b d_v F\left(\frac{b}{d_v}\right) - \frac{1}{2}(\gamma - 1)k^2 b d_h F\left(\frac{b}{d_h}\right) \quad (9.2.41)$$

where

$$F(g) = (1 - i)\frac{J_1[(1+i)g]}{J_0[(1+i)g]} \rightarrow \begin{cases} g + \frac{1}{4}ig^3 & g \ll 1 \\ 1 + i & g \gg 1 \end{cases}$$

The real and imaginary parts of function $F(g)$ are plotted in Fig. 9.13. When $k^2 b d_v$ and $k^2 b d_h$ are both small, then πq is small, and the Bessel functions on the left-hand side of this equation may be approximated by the first terms in their series expansions. The resulting equation for q is

$$(\pi q)^2 \simeq -\left[k^2 - \left(\frac{\pi q}{b}\right)^2\right]bd_v F\left(\frac{b}{d_v}\right) - (\gamma - 1)k^2 bd_h F\left(\frac{b}{d_h}\right)$$

or

$$(\pi q)^2 \simeq \frac{-k^2 bd_v F(b/d_v) - (\gamma - 1)k^2 bd_h F(b/d_h)}{1 - (d_v/b)F(b/d_v)} \qquad k^2 bd \ll 1 \quad (9.2.42)$$

When $b/d_v \gg 1$, this equation becomes $(\pi q)^2 \simeq -(1 + i)k^2 b[d_v + (\gamma - 1)d_h]$, which is the same as the expression for πq_{00} in Eqs. (9.2.24), with ξ_{00} and σ_{00} as given in Eq. (9.2.38); the effective impedance approximation is valid in this range.

When b/d_v and b/d_h are not very large, but $k^2 bd_v$ and $k^2 bd_h$ are still small, Eq. (9.2.42) can still be used, but the F's are no longer equal to $1 + i$. Formulas (9.2.24) can still be used (for the case $m = n = 0$ which we are considering here), but the effective surface admittance is now

$$\beta_{00} = \xi_{00} - i\sigma_{00} \simeq -\tfrac{1}{2}i\frac{kd_v F(b/d_v) + (\gamma - 1)kd_h F(b/d_h)}{1 - (d_v/b)F(b/d_v)}$$

$$\rightarrow \begin{cases} 2\gamma k\dfrac{d_v^2}{b} & b < d_v, d_h \\[2mm] \tfrac{1}{2}(1 - i)k[d_v + (\gamma - 1)d_h] & b > d_v, d_h \end{cases} \qquad (9.2.43)$$

Figure 9.13 shows that the range of transition between the limiting forms is small: for b/d less than 1, the $b < d$ approximation can be used; for b/d greater than 3, the $b > d$ form is valid. For capillary tubes $b < d$, the thermal loss exhibits itself only in the factor γ; the equivalent surface admittance is purely real and larger than the conductance for $b > d$ by the factor $4(d_v/b)$. The attenuation factor $\text{Im } k_{00} = \pi\gamma_{00}$ is equal to $\gamma k(d_v/b)^2$, which increases rapidly as the radius of the tube is decreased. The dependence on frequency comes only through the square-root factor inherent in d_v.

When $k^2 bd_v$ is not small, none of these approximations can be used, but Eq. (9.2.41) can be solved graphically, using Plate V. Since, in this case, k^2 is usually much larger than $|\pi q/b|^2$, the equation can be solved by successive approximations, first solving for $q = \mu - iv$ by neglecting $(\pi q/b)^2$ compared with k^2 in the brackets and then setting this solution into the right-hand side of the equation to solve again for q, and so on until the sequence produces no further change in the computed value of q.

9.3   REFLECTION FROM DUCT DISCONTINUITIES

Calculation of acoustic transmission along infinite ducts is a necessary preliminary, but infinite ducts seldom occur in practice. In many cases the continuity of the duct is broken by a change in cross-sectional shape or size, or by a septum partially obstructing the wave, or else by an "elbow" occasioned by a change in direction of the duct axis.   In still other cases the duct is terminated, the wave emerging into whole- or half-space.   Each of these discontinuities has its effect on a wave encountering it; part of the wave energy is reflected, and part is transmitted past the obstruction.

The situation is somewhat similar to that described in Sec. 5.2 for waves on a string, with reflected and transmitted waves arising from the discontinuity.   The analogy was the basis of the transmission-line approximation developed in Sec. 9.1, where we neglected all modes but the fundamental. But the actual behavior of the wave in the duct is more complicated than this, because of the possibility of several modes of wave transmission; both reflected and transmitted waves may contain modes other than the incident one, or ones.   We shall work out the implications of this coupling of modes by using the general notation of Eqs. (9.2.8), and then will consider several specific problems in detail.   The details are rather forbidding, but the general problem of the acoustic coupling of one region with another is never a simple task to solve, though it often is of considerable practical interest. The coupling of two parts of a duct, which we shall concentrate on in this section, is "simple" enough to serve as an introduction; the more difficult coupling problems will be taken up in the next chapter.

Reflection from an impedance surface

If the duct is semi-infinite, extending to $z \to -\infty$ and being terminated at $z = 0$ with a plane wall of uniform acoustic impedance, the modes will not "mix up" on reflection; if an nth mode wave is sent from $-\infty$, an nth-mode wave will be reflected back.   In line with the procedure of Eq. (9.1.3), we can relate pressure, axial velocity, and terminal impedance z_a by the equations

$$p = A\Psi_m(x,y) \cosh(\Phi_m + ik_m z) \qquad k_m{}^2 = k^2 - \varkappa_m{}^2$$

$$u_z = \frac{1}{i\rho ck}\frac{\partial p}{\partial z} = \frac{Ak_m}{\rho ck}\Psi_m(x,y)\sinh(\Phi_m + ik_m z) \qquad (9.3.1)$$

$$z_a = \rho c\xi = \left(\frac{p}{u_z}\right)_{z=0} = \rho c\frac{k}{k_m}\coth\Phi_m$$

or

$$\beta = \frac{\rho c}{z_a} = \frac{k_m}{k}\tanh\Phi_m \qquad \nabla_2{}^2\Psi_m + \varkappa_m{}^2\Psi_m = 0$$

where the complex phase angle Φ_m can be determined from the value of z_a or β, by using Plate I.

If the wall at $z = 0$ has impedance $\rho c\xi(x,y)$ which is a function of position on the wall, or if there is some other nonuniform obstruction at $z = 0$, the obstruction will tend to couple several modes.   If a single mode is sent on from $-\infty$, several different modes will be reflected.   To study this coupling we need to devise a set of standard incident-plus-reflected combinations, which we can combine to fit various requirements of interest. The most convenient combination turns out to be a single-mode incident wave plus its related multiple-mode reflected wave. Such a combination has the form, for $z < 0$,

$$u_m^- = \Psi_m(x,y) \sinh(\Phi_m + ik_m z) + \sum_{n \neq m} U_n^m \Psi_n(x,y) e^{-ik_n z}$$

$$p_m^- = \rho c \left[\frac{k}{k_m} \Psi_m \cosh(\Phi_m + ik_m z) - \sum_{n \neq m} \frac{k}{k_n} U_n^m \Psi_n e^{-ik_n z} \right] \quad (9.3.2)$$

This is a unit wave; so u_m^- and the U's and Ψ's are dimensionless, and p_m^- has the dimensions of acoustic impedance. Any actual wave would then have an amplitude factor, with dimensions of velocity, multiplying the combination. The incident mode m will be called the *primary* mode of the combination; the other modes will be called *subsidiary* modes. The constants Ψ_m, U_n^m are to be adjusted so that $u_m^- = (p_m^-/\rho c)\beta(x,y)$ at every point on the terminal surface $z = 0$. Note that the coupling coefficients U_n^m and the phase angle Φ_m are related to the axial velocity $u_m^-(x,y,0) = u_m^0(x,y)$ at the terminal surface by the equations

$$\sinh \Phi_m \equiv U_m^m = \frac{1}{S\Lambda_m} \iint_S u_m^0 \Psi_m \, dx \, dy$$

$$U_n^m = \frac{1}{S\Lambda_n} \iint_S u_m^0 \Psi_n \, dx \, dy \quad (9.3.3)$$

The combination of Eqs. (9.3.2) has a number of advantages, which will become apparent as we work with it. It has one disadvantage, namely, that the incident amplitude is not unity, but is $\frac{1}{2}e^{\Phi_m}$. Thus the reflected amplitude of the nth mode, for *unit* incident amplitude of the primary wave, is not U_n^m but $2e^{-\Phi_m}U_n^m$ for $n \neq m$ and $e^{-2\Phi_m}$ for $n = m$. All the modes participate in the combination, including those having cutoff frequencies greater than $\omega/2\pi$ (i.e., those for which $n > N$). These modes decay exponentially as $z \to -\infty$, but contribute to the shape of the wave near the terminal surface $z = 0$. As $-z$ increases, however, the only surviving modes are those for $n \leqslant N$ (this, of course, presupposes that $m \leqslant N$, for the primary wave must propagate to be incident).

Reflections and standing waves

This example need not be pursued further; we took it up because it provided a link with Sec. 9.1, and thus was a natural means of introducing the set of modal combinations given in Eqs. (9.3.2). It turns out that the effects of any discontinuity in a duct, at $z = 0$, can be worked out in terms of similar modal combinations, where the phase angles Φ_m for the primary reflected wave and the coupling coefficients $U_n{}^m$ for the nth subsidiary waves are determined by appropriate boundary conditions at $z = 0$. Before we investigate these conditions, for several situations of practical interest, we should discuss the general properties of such combinations.

As Eqs. (9.3.2) indicate, each combination consists of a single primary mode traveling toward the obstruction, plus a combination of primary and subsidiary reflected waves. The intensity of the sound in the duct is, of course, Re (pu^*), and the total flow of energy in the axial direction for the combination having the mth mode as primary is (if we assume that the duct walls are rigid, so that the eigenfunctions Ψ_m are real)

$$\int\!\!\int \text{Re}\,(p_m{}^- u_m{}^{-*})\,dx\,dy = \frac{\rho c k}{2k_m} S\Lambda_m \sinh{(2\,\text{Re}\,\Phi_m)} - \sum_{n \neq m}^{N} \frac{\rho c k}{k_n} S\Lambda_n\,|U_n{}^m|^2$$

where N is the upper limit of the propagating waves, $c\varkappa_N < \omega < c\varkappa_{N+1}$.

Since the amplitude of the incident pressure wave is $\rho c(k/k_m)\Psi_m e^{\Phi_m}$, this whole expression can be separated into a single term representing the energy flowing to the right, incident on the obstruction, minus terms representing flow to the left, energy reflected from the obstruction.

$$\text{Incident power} = \frac{\rho c k}{4k_m} S\Lambda_m \exp{(2\,\text{Re}\,\Phi_m)}$$

$$(9.3.4)$$

$$\text{Reflected power} = \frac{\rho c k}{4k_m} S\Lambda_m \exp{(-2\,\text{Re}\,\Phi_m)} + \sum_{u \neq m}^{N} \frac{\rho c k}{k_n} S\Lambda_n\,|U_n{}^m|^2$$

Thus, by analogy with Eqs. (9.3.1), the effective admittance of the obstruction at $z = 0$ for the primary wave is $\beta_m = (k_m/k)\tanh{\Phi_m}$. This measures the phase and amplitude of the reflected primary wave, though it does not, in this case, tell the whole story in regard to the reflected wave. Inspection of Eqs. (9.3.2), compared with Eq. (6.3.5), suggests that we can define a set of reflection factors for this situation. For the primary wave the situation is simple; the ratio, in amplitude and phase, between the reflected and incident waves is the exponential $\exp{(-2\Phi_m)}$. This could be called the reflection coefficient $R_m{}^m$ for the primary wave; the square of its magnitude is the fraction of the incident power reflected by means of the primary mode, as shown in Eqs. (9.3.4). By analogy the ratio between the reflected amplitude of the nth subsidiary mode and the incident amplitude could be called the reflection coefficient $R_n{}^m$ for the nth subsidiary mode, when the mth mode is primary. Thus

$$R_m{}^m = e^{-2\Phi_m} \qquad R_n{}^m = 2U_n{}^m e^{-\Phi_m} \qquad (9.3.5)$$

The fraction of power reflected in the nth mode is then $(k_m\Lambda_n/k_n\Lambda_m)\,|R_n{}^m|^2$.

In most cases of practical interest, more than one of the combinations of Eqs. (9.3.2) will be present, their relative magnitudes being determined by the conditions at the other end of the duct. For example, suppose a piston closes the end, at $z = -l$, moving in such a manner that its normal velocity is $V_M \Psi_M(x,y)e^{-i\omega t}$. If there were no obstruction in the duct, this motion would generate a single-mode wave, traveling to the right. But with an obstruction at $z = 0$, waves of several modes will be reflected in addition to the primary, all of which will be re-reflected from the piston, and so on. The resulting steady-state wave in the region near $z = -l$ should be a combination of combinations:

$$u \to \sum_{m=0}^{N} A_m{}^M u_m = \sum_{m=0}^{N} A_m{}^M [\Psi_m \sinh (\Phi_m + ik_m z) + \sum_{n \neq m}^{N} U_n{}^m \Psi_n e^{-ik_n z}]$$

$$z \to -l$$

where we have assumed that l is large enough so that the effects of the non-propagational modes $(n > N)$ can be neglected at $z = -l$. We are to choose values of the coefficients $A_n{}^M$ so that, at $z = -l$, the velocity (though not the pressure) is just the driving mode, $u = V_M \Psi_M(x,y)$. This gives rise to the following equations:

$$A_m{}^M \sinh (\Phi_m - ik_m l) + \sum_{n \neq m}^{N} A_n{}^M U_m{}^n e^{-ik_m l} = V_M \delta_{Mm} \qquad m = 0, 1, \ldots, N$$

for the $N + 1$ coefficients $A_m{}^M$. When the frequency is low enough so that only the lowest mode propagates, there is just one equation; M and m must be zero, and the summation term does not exist. This case has been treated in Sec. 9.1.

The solution of these equations is not difficult, once we know the values of the Φ's and the $U_n{}^m$'s, which have been determined in terms of the boundary conditions at the obstruction at $z = 0$. If the coupling coefficients $U_n{}^m$ are all small, an approximate solution is

$$A_M{}^M \simeq \left[\frac{V_M}{Q_M} \sinh (\Phi_M - ik_M l) \right] \qquad 0 \leqslant M \leqslant N$$

$$A_n{}^M \simeq - \frac{A_M{}^M U_n{}^M e^{-ik_n l}}{\sinh (\Phi_n - ik_n l)} \qquad 0 \leqslant n \leqslant N; n \neq M$$

$$Q_M = 1 - \sum_{n \neq M}^{N} \frac{U_n{}^N U_N{}^n \exp [i(k_n + k_M)l]}{\sinh (\Phi_M - ik_M l) \sinh (\Phi_n - ik_n l)}$$

Referring to Eqs. (9.3.2), we see that the pressure wave just in front of the piston is

$$p = \rho c \sum_{m=0}^{N} \frac{k}{k_m} \Psi_m \left[A_m{}^M \cosh (\Phi_m - ik_m l) - e^{ik_m l} \sum_{n \neq m}^{N} A_n{}^M U_m{}^n \right]$$

$$= \sum_{m=0}^{N} P_m \Psi_m$$

The component of this pressure, corresponding to the driving mode, is the coefficient P_M in this series. The ratio of P_M to V_M may be considered as an acoustic impedance for the driving piston.

$$Z_M = \frac{P_M}{V_M} \simeq \rho c \frac{k}{k_M} \coth (\Phi_M - ik_M l) \frac{H_M}{Q_M}$$

$$H_M = 1 + \sum_{n \neq m}^{N} \frac{U_n{}^M U_M{}^n \exp [i(k_N + k_n)l]}{\cosh (\Phi_M - ik_m l) \sinh (\Phi_n - ik_n l)}$$

(9.3.6)

The relation to the simpler solution (9.1.4) is obvious; the phase angle Φ_M takes the place of the earlier $\pi(\alpha_l - i\beta_l - \frac{1}{2}i)$.

When the coupling constants U are small, the factors H_M and Q_M are nearly unity, and the impedance reduces to the single-mode expression which is a generalization of Eq. (9.1.9). In fact, for frequencies low enough so that only the lowest mode propagates, this is the exact expression (as long as $l \gg 1/\varkappa_{N+1}$) and is the justification of the approximations made in Sec. 9.1. With a single mode present, the impedance varies periodically with $k_0 l$. As soon as the next mode can propagate, however, this periodicity disappears; the upper mode or modes have wavelengths different from that of the fundamental.

Boundary conditions at a constriction

Further discussion cannot be so general; details of procedure must depend on the particular problem. One class of problems is concerned with uniform infinite ducts, partially obstructed by a rigid wall, in the xy plane, which is pierced by a window (as exemplified in Fig. 9.14), coupling the region for $z < 0$ to the region for $z > 0$. The interior cross-sectional area of the duct is S; the area of the window in the barrier plate is S_0 (smaller than S); and the area of the wall, which obstructs the free passage of the wave, is $S_b = S - S_0$.

If an mth-mode wave is sent to the right from $z = -\infty$, this partial barrier will reflect waves of several different modes to the left and will also allow waves of several modes to propagate on to $z = +\infty$. In the region to the left of the barrier the general form of the wave motion will be the combination given in Eqs. (9.3.2). Since the barrier is rigid, the z component of the velocity, $u_m^0(x,y)$, at $z = 0$, will be zero over the area S_b, so

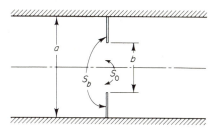

FIGURE 9.14
Obstruction of rectangular duct by partial barriers, leaving an open slit of width b.

that the integrals of Eqs. (9.3.3) for the coefficients U will be over the open area S_0 only. The wave motion to the right of the barrier will have the form

$$u_m{}^+ = \sum_{n=0}^{\infty} U_n{}^m \Psi_n(x,y) e^{ik_n z} \qquad p_m{}^+ = \rho c \sum_{n=0}^{\infty} \frac{k}{k_n} U_n{}^m \Psi_n e^{ik_n z} \quad (9.3.7)$$

for $z > 0$ where $U_m{}^m = \sinh \Phi_m$, as before. Since the axial velocity must be continuous across the opening S_0, the coefficients $U_n{}^m$ must be equal to those given in Eqs. (9.3.3) and used in Eqs. (9.3.2) for the reflected wave. In the region $z > 0$ the transmitted modes go to the right; in the region $z < 0$ the incident wave goes to the right, the reflected modes go to the left. The coupling between the various waves is usually determined by the requirement that the fluid pressure, as well as the velocity, be continuous across the area of the opening S_0.

The use of coefficients $U_n{}^m$ in Eqs. (9.3.7) ensures continuity of velocity. Continuity of pressure (if this is assumed) is satisfied by requiring that $p_m{}^-$ be equal to $p_m{}^+$ in the opening. This corresponds to the equation

$$\frac{k}{k_m} \Psi_m \cosh \Phi_m - \sum_{n \neq m} \frac{k}{k_n} U_n{}^m \Psi_n = \frac{k}{k_m} \Psi_m \sinh \Phi_m + \sum_{n \neq m} \frac{k}{k_n} U_n{}^m \Psi_n$$

The equality holds just over the area S_0 of the opening; it does not hold over the remaining area S_b, where the rigid plate stops the motion and produces a pressure difference between its two sides. This equation is really an integral equation for u_m^0, as can be seen by using Eqs. (9.3.3) to eliminate the U's. We can write

$$0 = \Psi_m(x,y) \cosh \Phi_m - \iint_{S_0} G_m(xy \mid x_0 y_0) u_m^0(x_0 y_0) \, dx_0 \, dy_0 \quad (9.3.8)$$

where

$$G_m = \frac{1}{S\Lambda_m} \Psi_m(xy) \Psi_m(x_0 y_0) + \sum_{n \neq m}^{\infty} \frac{2k_m}{Sk_n \Lambda_n} \Psi_n(xy) \Psi_n(x_0 y_0)$$

is symmetric to interchange of xy and $x_0 y_0$.

Power flow

We now can conclude our discussion of power flow for this situation. For unit incident amplitude of the mth mode, the reflected amplitudes are the quantities $R_n{}^m$ of Eq. (9.3.5). The related amplitudes of the modes transmitted through the opening and on to $z = +\infty$ are

$$T_m{}^m = 2U_m{}^m e^{-\Phi_m} = 1 - e^{-2\Phi_m} \qquad T_n{}^m = 2U_n{}^m e^{-\Phi_m} \quad (9.3.9)$$

The total power transmitted to the right, for the combination of Eqs. (9.3.7),

$$\frac{\rho c k}{4k_m} S\Lambda_m [e^{2 \operatorname{Re} \Phi_m} + e^{-2 \operatorname{Re} \Phi_m} - 2 \cos (2 \operatorname{Im} \Phi_m)] + \sum_{n \neq m}^{N} \frac{\rho c k}{k_n} S\Lambda_n |U_n{}^m|^2$$

is to be compared with Eq. (9.3.4). Inasmuch as the assumption that the U_n^m's of Eqs. (9.3.7) are equal to the U_n^m's of Eqs. (9.3.2) is equivalent to the assumption that no energy is lost at the opening (velocity and pressure are continuous across the opening), we must conclude that the sum of the power reflected and the power transmitted equals the incident power, which corresponds, after dividing through by $\rho c k S \Lambda_m / 4 k_m$, to the equation

$$e^{2 \operatorname{Re} \Phi_m} = e^{2 \operatorname{Re} \Phi_m} + 2 e^{-2 \operatorname{Re} \Phi_m} - 2 \cos (2 \operatorname{Im} \Phi_m) + 8 \sum_{n \neq m}^{N} \frac{k_m \Lambda_n}{k_n \Lambda_m} |U_n^m|^2$$

This power-balance equation can be verified directly from the integral equation (9.3.8) or its equivalent in terms of the U's.

$$\Psi_m e^{-\Phi_m} = 2 \sum_{n \neq m}^{\infty} \frac{k_m}{k_n} U_n^m \Psi_n \qquad \text{(across } S_0)$$

where we have used the identity $\cosh \Phi - \sinh \Phi = e^{-\Phi}$. If the duct walls are rigid, the Ψ_n's are real, and we can multiply by the complex conjugate of $2 u_m^0(xy)$ and integrate over S_0 to obtain

$$4 \sum_{n \neq m}^{\infty} \frac{k_m \Lambda_n}{k_n \Lambda_m} |U_n^m|^2 = 2 e^{-\Phi_m^*} \sinh \Phi_m = e^{-2i \operatorname{Im} \Phi_m} - e^{-2 \operatorname{Re} \Phi_m}$$

In the summation every term for the propagated modes ($0 \leqslant n \leqslant N$) is real, and every term for the higher modes is imaginary; so we can separate real and imaginary parts and equate each separately.

$$\cos (2 \operatorname{Im} \Phi_m) - e^{-2 \operatorname{Re} \Phi_m} = 4 \sum_{n \neq m}^{N} \frac{k_m \Lambda_n}{k_n \Lambda_m} |U_n^m|^2$$

$$-i \sin (2 \operatorname{Im} \Phi_m) = 4 \sum_{n=N+1}^{\infty} \frac{k_m \Lambda_n}{k_n \Lambda_m} |U_n^m|^2$$

The first of these equations, multiplied by 2, is just the power-balance equation written earlier. The second may be used to calculate the phase shift of the primary wave on reflection.

Viscous losses at the constriction

To be meticulous, we should take into account the additional energy loss, at the constriction, caused by the viscous friction of the fluid in its higher-speed flow over the surface S_b of the constriction. When this is done the sum of the reflected and transmitted energies should not exactly equal the incident energy, but should equal the incident energy less the integral of $\frac{1}{2} \rho c k d_v |u_{\tan}|^2$ over the surface of the constriction (i.e., over $2S_b$), as indicated by Eq. (6.4.39). This result, because of some interesting properties of the U's, can be represented in a very simple manner.

The square of the magnitude of the tangential velocity at $z = 0$ is $(1/\rho \omega)^2 |\operatorname{grad}_t p|^2$, where grad_t is the component of the gradient perpendicular to the z axis. This tangential velocity is 0 at the walls of the duct (i.e., the outer perimeter of S_b), rises to a large magnitude at the sharp edge of the hole (the perimeter of S_0), and then is very small over the area of the opening S_0, where the velocity is chiefly in the z direction. However, since

$$|\text{grad}_t \, p|^2 \equiv \left(\frac{\partial p}{\partial x}\right)^2 + \left(\frac{\partial p}{\partial y}\right)^2 = \text{div}\,(p^* \, \text{grad}_t \, p) - p^* \nabla_2^2 p$$

where

$$\nabla_2^2 p \equiv \frac{\partial^2 p}{\partial x^2} + \frac{\partial^2 p}{\partial y^2} = -\rho c \sum_{n=0}^{\infty} \frac{k \varkappa_n^{\,2}}{k_n} U_n^{\,m} \Psi_n \qquad \text{at } z = 0$$

and since the integral of div $(p^* \, \text{grad}_t \, p)$ over S is zero (because $\text{grad}_t \, p$ goes to zero at the perimeter of the duct), the integral of $\rho c k d_v \, |u_{\text{tan}}|^2$ over S can be changed as follows:

$$\frac{d_v}{\rho c k} \iint_S |\text{grad}_t \, p|^2 \, dx \, dy = -\frac{d_v}{\rho c k} \iint_S p^* \nabla_2^2 p \, dx \, dy = \rho c k d_v \sum_{n=0}^{\infty} S\Lambda_n \left|\frac{\varkappa_n U_n^{\,m}}{k_n}\right|^2$$

But we have just seen that the integrand of the first integral is large only over the area S_b of the obstructing plate, and is very small in the area S_0; therefore the first integral is practically equal to the integral of $(d_v/\rho c k)$ $|\text{grad}_t \, p|^2$ over S_b, which equals the viscous power loss along both sides of the obstruction. On the other hand, the last sum approximates the integral of $|u_m^0|^2$ over the area of the opening S_0. For when $\varkappa_n$ is larger than k, $|k_n/\varkappa_n|^2 = 1 - (k/\varkappa_n)^2 \to 1$, and also, from Eqs. (9.3.3),

$$\iint_S |u_m^0|^2 \, dx \, dy = \sum_{n=0}^{\infty} S\Lambda_n \, |U_n^{\,m}|^2$$

so that, approximately,

Power lost at constriction $\equiv \Pi_a = \dfrac{d_v}{\rho c k} \iint |\text{grad}_t \, p|^2 \, dx \, dy$

$$\simeq \rho c k d_v \left(\iint |u_m^0|^2 \, dx \, dy + \sum_{n=0}^{N} S\Lambda_n \frac{2\varkappa_n^{\,2} - k^2}{k^2 - \varkappa_n^{\,2}} \, |U_n^{\,m}|^2\right) \qquad (9.3.10)$$

When k is smaller than $\varkappa_1$ (i.e., when $N = 0$), this last term in the parentheses becomes $-S \, |U_0^{\,m}|^2$. For the rest of this discussion we shall carry through the calculations for this long-wavelength case. The additional corrections, when $N > 0$, are easily added.

The first term in the parentheses corresponds to the power loss which would occur at the constriction if the area S_0 were not completely open, but had across it a permeable membrane of flow resistance $\rho c k d_v$. In that case the pressure drop across it, and thus the difference between the p^- of Eqs. (9.3.2) and the p^+ of Eqs. (9.3.7), would be $\rho c k d_v u_m^0(x,y)$ at point xy, and the total power loss caused by the pressure drop would be the first term of Eq. (9.3.10). The correction term, for long wavelengths, modifies this simple picture slightly; the loss of Eq. (9.3.10) corresponds to a pressure drop across the plane of the opening S_0,

$$\rho c k d_v [u_m^0(x,y) - U_0^m] = \rho c k d_v \sum_{n=1}^{\infty} U_n^m \Psi_n(x,y)$$

proportional to the difference between the axial velocity u_m^0 at xy and its average value U_0^m across S_0. The total loss is then the integral of this drop times u_m^0 itself. When the barrier vanishes, $S_0 \to S$ and $u_m^0 \to U_0^m$, and there is no viscous loss.

Thus, when viscosity is taken into account, the incident power will not exactly equal the sum of the reflected and transmitted power, as was the case with the formulas following Eqs. (9.3.9). Both reflected and transmitted power will have one-half of the small term Π_a of Eq. (9.3.10) subtracted from the quantities given in Eqs. (9.3.4) and (9.3.9). The integral equation (9.3.8) also must be slightly modified. For example, for long wavelengths and for a fundamental incident wave, the integral equation will be

$$0 = \cosh \Phi_0 - \iint_{S_0} G_0^v(xy \mid x_0 y_0) u_0^0(x_0 y_0) \, dx_0 \, dy_0 \qquad (9.3.11)$$

where

$$G_0^v = G_0 - \frac{kd_v}{S} + kd_v \delta(x_0 - x)\delta(y_0 - y) = G_0 + \frac{kd_v}{S} \sum_{n=1}^{\infty} \frac{\Psi_n(xy)\Psi_n(x_0 y_0)}{\Lambda_n}$$

$$G_0(xy \mid x_0 y_0) = \frac{1}{S} + \sum_{n=1}^{\infty} \frac{2k}{Sk_n \Lambda_n} \Psi_n(xy)\Psi_n(x_0 y_0)$$

The inclusion of viscous losses does not change the shape of the solution for u_m^0 very much, as we shall demonstrate. It does, of course, diminish the magnitudes of the reflected and the transmitted power by the small amount $\frac{1}{2}\Pi_a$, as mentioned.

A variational procedure

Equation (9.3.8), or (9.3.11) if we wish to include the effects of viscosity, is a Fredholm integral equation of the first kind, with a symmetric kernel G_m. It must be solved either by successive approximations or else by a variational procedure. In the present case the variational method is easier.

We multiply Eq. (9.3.8) by $u_m^0(xy)$ and integrate over S_0, obtaining

$$0 = \cosh \Phi_m \iint_{S_0} u_m^0(xy) \Psi_m \, dx \, dy$$

$$- \iint_{S_0} u_m^0(xy) \, dx \, dy \iint_{S_0} G_m(xy \mid x_0 y_0) u_m^0(x_0 y_0) \, dx_0 \, dy_0$$

the second term of which is symmetric to exchange of xy and $x_0 y_0$. Next we utilize the first of Eqs. (9.3.3) to write

$$S\Lambda_m \cosh \Phi_m \sinh \Phi_m = \cosh \Phi_m \iint_{S_0} u_m^0(xy) \Psi_m(xy) \, dx \, dy$$

Adding the two equations and at the same time simplifying the notation, to save space, we obtain

$$S\Lambda_m \cosh \Phi_m \sinh \Phi_m = 2 \cosh \Phi_m \int u_m^0 \Psi_m \, dS - \int u_m^0 \, dS \int G_m u_m^0 \, dS_0$$

(9.3.12)

which is the basis for our variational procedure.

We discussed variational expressions like this in Secs. 4.5 and 8.3. According to the discussion following Eq. (4.5.22), if the quantity u_m^0 in Eq. (9.3.12) is equal to the correct form for the axial velocity distribution in the area S_0, then inserting it in the integrals will provide the correct value of Φ_m. But if, instead of u_m^0, we use in the integrals a function $u(xy) = u_m^0(xy) + \lambda v(xy)$ which differs little from the correct form u_m^0 (i.e., if λ is small), then the value of Φ, computed therefrom, will differ from the correct value Φ_m by an amount proportional to λ^2, second order in the small difference. Furthermore, if we use a function $u(\lambda;xy)$, with a parameter λ which changes the form of u, then if we adjust the value of λ so that the derivative of the left-hand side of Eq. (9.3.12) with respect to λ is zero, the corresponding form of u will be closest to the correct u_m^0, and the value of Φ will be as close to the correct value Φ_m as the chosen form permits.

The variational expression (9.3.12) can be used in a number of ways. If we are clever enough to choose a form which is close to the correct form, we can use it simply to determine the "best" amplitude. For example, we can set $u_m^0 = A\chi(xy)$, where χ has a form we hope is close to the correct one, and magnitude A is the variable parameter. In this case the variational expression is

$$2A \cosh \Phi_m \int \chi \Psi_m \, dS - A^2 \int \chi \, dS \int G_m \chi \, dS_0$$

since G is symmetric. Setting the derivative of this with respect to A equal to zero, we obtain for the "best" value of A and of the other quantities of interest

$$A = \frac{\int \chi \Psi_m \, dS}{\int \chi \, dS \int G_m \chi \, dS_0} \cosh \Phi_m \qquad \tanh \Phi_m \simeq \frac{\left| \int \chi \Psi_m \, dS \right|^2}{S\Lambda_m \int \chi \, dS \int G_m \chi \, dS_0}$$

$$u_m^0 \simeq \frac{S\Lambda_m}{\int \chi \Psi_m \, dS} \sinh \Phi_m \, \chi(xy)$$

(9.3.13)

$$U_n{}^m \simeq \frac{S\Lambda_m \int \chi \Psi_n \, dS}{S\Lambda_n \int \chi \Psi_m \, dS} \sinh \Phi_m$$

If the form of u_m^0 needs to be determined with still greater accuracy, we can try expressing it as a sum of appropriately chosen functions χ_l.

$$u_m^0 = \cosh \Phi_m \sum_{l=1}^{m} A_l{}^m \chi_l(xy)$$

Inserting this in Eq. (9.3.12) and dividing by $(S\Lambda_m)^2 \cosh^2 \Phi_m$, the variational expression becomes

$$2 \sum_l C_l{}^m A_l{}^m = \sum_{ls} A_l{}^m B_{ls}{}^m A_s{}^m$$

where

$$C_l{}^m = \int \chi_l \Psi_m \, dS \quad \text{and} \quad B_{ls}{}^m = \int \chi_l \, dS \int G_m \chi_s \, dS_0 = B_{sl}{}^m$$

since G_m is symmetric. Setting the partial of this with respect to each $A_l{}^m$ equal to zero produces a set of simultaneous equations,

$$\sum_{s=1}^{M} B_{ls}{}^m A_s{}^m = C_l{}^m \qquad l = 1, 2, 3, \ldots, M$$

which can be solved for the A's. In terms of these solutions the "best" value of $\tanh \Phi_m$ and the "best" form for u_m^0 are then

$$\tanh \Phi_m \simeq \sum_{l=1}^{M} A_l{}^m C_l{}^m \qquad u_m^0 \simeq S\Lambda_m \cosh \Phi_m \sum_{l=1}^{M} A_l{}^m \chi_l(xy) \quad (9.3.14)$$

Since we shall use them later, we write out the solutions for $M = 2$.

$$\tanh \Phi_m \simeq \frac{P_m}{Q_m} \qquad \sinh \Phi_m \simeq \frac{P_m}{R_m}$$

$$P_m = (C_1{}^m)^2 B_{22}{}^m + (C_2{}^m)^2 B_{11}{}^m - 2 C_1{}^m C_2{}^m B_{12}{}^m$$

$$Q_m = S\Lambda_m [B_{11}{}^m B_{22}{}^m - (B_{12}{}^m)^2] \qquad R_m{}^2 = -P_m{}^2 + Q_m{}^2 \qquad (9.3.15)$$

$$u_m^0 \simeq \frac{S\Lambda_m}{R_m} [(C_1{}^m B_{22}{}^m - C_2{}^m B_{12}{}^m)\chi_1(xy) + (C_2{}^m B_{11}{}^m - C_1{}^m B_{12}{}^m)\chi_2(xy)]$$

$$U_m{}^m = \sinh \Phi_m \qquad U_n{}^m \simeq \frac{\Lambda_m}{R_m \Lambda_m} [(C_1{}^m B_{22}{}^m - C_2{}^m B_{12}{}^m)C_1{}^n$$

$$+ (C_2{}^m B_{11}{}^m - C_1{}^m B_{12}{}^m)C_2{}^n]$$

For ducts with rigid walls, the constants $C_n{}^m$ are real quantities, since Ψ_m and (presumably) the trial functions χ_n are real. The series for G_m [Eq. (9.3.8)], however, includes terms for the nonpropagating modes, for which k_n is positive-imaginary; thus the constants $B_{ls}{}^m$ have negative-imaginary parts. Both Φ_m and u_m^0 are therefore complex quantities, indicating that the fluid velocity is not in phase with the pressure at $z = 0$ and that the effective impedance $\rho c \coth \Phi_m$ of the obstruction is complex. The real part of this impedance corresponds to the energy transmitted past the obstruction. The imaginary part is proportional to the added mass and stiffness produced by the distorted flow around the obstruction; at low frequencies this is masslike.

Reflection from a slit

We shall work out two examples of these formulas. The simplest case is that of a uniform rectangular duct with rigid walls, of width a in the x direction, d in the y direction, partly obstructed by a rigid barrier having a slit in it, parallel to the y axis, of width b, as shown in Fig. 9.14. We shall consider the case of the fundamental mode being the primary wave and will take the case of the slit being symmetrically placed. In this case, because of the symmetry, all modes reflected and transmitted from the slit are independent of y and are symmetric with respect to the central yz plane. Thus we can consider the problem to be a two-dimensional one, can take the z axis as the central axis of the duct, and can use the combinations

$$u_z = \begin{cases} \sinh(\Phi_0 + ikz) + \sum_{n=1}^{\infty} U_n^0 \cos\left(2\pi n \frac{x}{a}\right) e^{-ik_n z} & z < 0 \\[2em] \sinh(\Phi_0) e^{ikz} + \sum_{n=1}^{\infty} U_n^0 \cos\left(2\pi n \frac{x}{a}\right) e^{+ik_n z} & z > 0 \end{cases} \qquad (9.3.16)$$

$$\Psi_n = \cos\left(2\pi n \frac{x}{a}\right) \qquad S\Lambda_n = \frac{a}{\epsilon_n} \qquad k_n^2 = k^2 - \left(\frac{2\pi n}{a}\right)^2$$

The kernel G_0 is the sum

$$G_0(x \mid x_0) = \frac{1}{a}\left[1 + \sum_{n=1}^{\infty} \frac{4k}{k_n} \cos\left(2\pi n \frac{x}{a}\right) \cos\left(2\pi n \frac{x_0}{a}\right)\right]$$

The area S_0 is the region at $z = 0$ between $x = -\frac{1}{2}b$ and $x = +\frac{1}{2}b$.

The crucial decision in applying the variational procedure is the choice of trial function for u_0^0, the axial velocity in the slit. For steady flow the discussion of Fig. 9.5 shows that this velocity is approximately proportional to $[1 - (2x/b)^2]^{-\frac{1}{2}}$, with the velocity fairly uniform across most of the opening but becoming very large close to the sharp edges of the slit. For slits considerably narrower than a wavelength, this is probably a good form to use for χ, but when b is not much smaller than a and when the frequency is large enough so that more than one mode propagates, it is unlikely to be good enough. Of course, when $b = a$, there is no obstruction, and the velocity at $z = 0$ is uniform if the $m = 0$ mode is the primary one.

Thus a reasonable assumption for a trial function would be

$$u_0^0 \simeq a \cosh \Phi_0\left(A_0 + \frac{A_2}{\sqrt{1 - (2x/b)^2}}\right) \qquad -\tfrac{1}{2}b < x < \tfrac{1}{2}b$$

which we can use in Eqs. (9.3.15) (we use the subscript 0 instead of 1 for reasons which will eventually become apparent). The constants C are easy to calculate (we omit the superscripts, since we are only computing the

$m = 0$ case):

$$C_0 = \int_{-\frac{1}{2}b}^{\frac{1}{2}b} dx = b \qquad C_2 = \int_{-\frac{1}{2}b}^{\frac{1}{2}b} \frac{dx}{\sqrt{1 - (2x/b)^2}} = \frac{\pi b}{2}$$

The constants B_{ns}, involving the kernel G_0, are less simple. Two types of integrals are involved; the first, for χ_{0}, is

$$\int_{-\frac{1}{2}b}^{\frac{1}{2}b} \cos\left(2\pi n \frac{x}{a}\right) dx = \frac{a}{\pi n} \sin\left(\pi n \frac{b}{a}\right) = (-1)^{n-1} \frac{a}{\pi n} \sin\left(\pi n \frac{a-b}{a}\right)$$

where the second form of the result is useful when $a - b$ is small. The second type of integral, involving χ_2, is

$$\int_{-\frac{1}{2}b}^{\frac{1}{2}b} \cos\left(2\pi n \frac{x}{a}\right) \frac{dx}{\sqrt{1 - (2x/b)^2}} = \frac{b}{2} \int_{-\frac{1}{2}b}^{\frac{1}{2}b} \cos\left(\pi n \frac{b}{a} \sin\theta\right) d\theta = \frac{\pi b}{2} J_0\left(\frac{\pi n b}{a}\right)$$

where we have utilized the integral representation for the Bessel function

$$\int_{-\frac{1}{2}\pi}^{\frac{1}{2}\pi} \cos(z \sin\phi) \cos(2m\phi) \, d\phi = \pi J_{2m}(z)$$

obtained by taking the real part of Eq. (1.3.11). Also, we have

$$B_{00} = \frac{b^2}{a} + \frac{4a}{\pi^2} \sum_{n=1}^{N} \frac{\sin^2(\pi n b/a)}{n^2 \sqrt{1 - (2\pi n/ka)^2}} - \frac{2ika^2}{\pi^3} \sum_{n=N+1}^{\infty} \frac{\sin^2(\pi n b/a)}{n^3 \sqrt{1 - (ka/2\pi n)^2}}$$

where $N < ka/2\pi < N + 1$ (the Nth mode is the highest one which propagates). If b is nearly equal to a, we can use $(a - b)/a$ instead of b/a in the sine functions.

For moderate frequencies the first series (which is real) consists of only a few terms, which can easily be summed. The infinite series (the imaginary part of B_{00}) becomes tedious if either b or $a - b$ is small compared with a. But in this case we can add and subtract $2ika^2/\pi^3$ times the comparison series

$$\sum_{n=1}^{\infty} \frac{\sin^2(\pi n b/a)}{n^3} = 2 \int_0^{\pi b/a} dx \int_0^x \sum_{n=1}^{\infty} \frac{\cos(2ny)}{n} \, dy$$

$$= -2 \int_0^{\pi b/a} dx \int_0^x \ln(2 \sin y) \, dy \rightarrow \begin{cases} \left(\frac{\pi b}{a}\right)^2 \left(\frac{1}{2} - \ln \frac{\pi b}{a}\right) & \pi b \ll a \\ 1.0518 & b = \frac{1}{2}a \end{cases}$$

[Again, when $b - a \ll a$, we exchange $(a - b)/a$ for (b/a) in these formulas.] This trick is particularly useful at low frequencies, when just the fundamental mode propagates; so the real series is not present, and the imaginary series is practically equal to $2ika^2/\pi^3$ times the comparison series. Then, as $b \to 0$,

$$aB_{00} \to b^2\left[1 - \frac{2ika}{\pi}\left(\tfrac{1}{2} - \ln\frac{\pi b}{a}\right)\right] \to 0$$

or if $b \to a$,

$$aB_{00} \to b^2 - \frac{2ika}{\pi}(a - b)^2\left[\tfrac{1}{2} - \ln\left(\pi\frac{a - b}{a}\right)\right] \to b^2$$

The constant B_{22},

$$B_{22} = \frac{1}{a}\left(\frac{\pi b}{2}\right)^2\left[1 + 4\sum_{n=1}^{N}\frac{J_0^2(\pi nb/a)}{\sqrt{1 - (2\pi n/ka)^2}} - \frac{2ika}{\pi}\sum_{n=N+1}^{\infty}\frac{J_0^2(\pi nb/a)}{n\sqrt{1 - (ka/2\pi n)^2}}\right]$$

also can be provided with a comparison series, for

$$\sum_{n=1}^{\infty}\frac{J_0^2(\pi nb/a)}{n} = \frac{2}{\pi}\int_0^{\pi/2}\sum_{n=1}^{\infty}\frac{1}{n}J_0\left(2\pi n\frac{b}{a}\sin\theta\right)d\theta$$

$$= \frac{4}{\pi^2}\int_0^{\pi/2}d\theta\int_0^{\pi/2}\sum\frac{1}{n}\cos\left(2\pi n\frac{b}{a}\sin\theta\sin\phi\right)d\phi$$

$$= -\frac{4}{\pi^2}\int_0^{\pi/2}d\theta\int_0^{\pi/2}\ln\left(\pi\frac{b}{a}\sin\theta\sin\phi\right)d\phi$$

$$\to -\ln\frac{\pi b}{2a} \qquad \pi b \ll a$$

where we have used the formulas

$$J_m^2(z) = \frac{2}{\pi}\int_0^{\pi/2}J_{2m}(2z\cos\theta)\,d\theta \qquad \text{and} \qquad \int_0^{\pi/2}\ln(\sin\phi)\,d\phi = -\frac{\pi}{2}\ln 2$$

in addition to the previous tricks. Finally, the constant

$$B_{02} = \frac{\pi b^2}{2a} + \sum_{n=1}^{N}\frac{2b\sin(\pi nb/a)}{n\sqrt{1 - (2\pi n/ka)^2}}J_0\left(\frac{\pi nb}{a}\right)$$

$$- \frac{ikab}{\pi}\sum_{n=N+1}^{\infty}\frac{\sin(\pi nb/a)J_0(\pi nb/a)}{n^2\sqrt{1 - (ka/2\pi n)^2}}$$

may also be approximated, for $b \ll a$, by the series

$$\sum_{n=1}^{\infty} \frac{\sin(\pi nb/a)J_0(\pi nb/a)}{n^2}$$

$$= \sum_{n=1}^{\infty} \frac{1}{\pi} \int_0^{\pi/2} \frac{d\theta}{n^2} \left[\sin\left(2\pi n \frac{b}{a} \cos^2\frac{\theta}{2}\right) + \sin\left(2\pi n \frac{b}{a}\sin^2\frac{\theta}{2}\right)\right]$$

$$\rightarrow \begin{cases} -\dfrac{\pi b}{a} \ln \dfrac{\pi b}{2a} & \pi b \ll a \\[2mm] \dfrac{2(a-b)}{\pi a} \ln 2 & a - b \ll a \end{cases}$$

These expressions can then be inserted in Eqs. (9.3.15) to obtain the effective admittance of the slit and the coupling constants $U_n{}^m$. A number of special cases should be considered. When $b \ll a$, it turns out that $C_0 B_{22}$ is almost exactly equal to $C_2 B_{02}$, so that the coefficient of $\chi_0 = 1$ in Eqs. (9.3.15) is very small. Hence the "best" form for u_0^0 in this case is the pure form χ_2, as earlier predicted. The variational solution is

$$\coth \Phi_0 \simeq 1 + 4 \sum_{n=1}^{N} \frac{J_0^2(\pi nb/a)}{\sqrt{1 - (2\pi n/ka)^2}} - \frac{2ika}{\pi} \sum_{n=N+1}^{\infty} \frac{J_0^2(\pi nb/a)}{n\sqrt{1 - (ka/2\pi n)^2}}$$

$$\rightarrow 1 - \frac{2ika}{\pi} \ln \frac{2a}{\pi b} \qquad ka < \tfrac{1}{2}\pi \qquad\qquad (9.3.17)$$

$$U_n^0 \simeq \epsilon_n J_0\left(\frac{\pi nb}{a}\right) \sinh \Phi_0 \rightarrow \frac{\pi \epsilon_n}{2ka} \frac{J_0(\pi nb/a)}{\ln(2a/\pi b)} \frac{\sqrt{\tfrac{1}{2}\beta - \tfrac{1}{2}} - i\sqrt{\tfrac{1}{2}\beta + \tfrac{1}{2}}}{\beta}$$

where

$$\beta^2 = 1 + \left(\frac{\pi}{ka} \ln \frac{2a}{\pi b}\right)^2 \qquad |R_0^0| = |e^{-2\Phi_0}| \rightarrow \frac{1}{\beta} \qquad ka < \tfrac{1}{2}\pi$$

Viscous losses

If viscosity is included, we must use the kernel $G_0{}^v$ of Eq. (9.3.11). The additional term in the integral B_{22} is

$$kd_v \int \chi_2(x)\, dx \int \left[\delta(x_0 - x) - \frac{1}{a}\right] \chi_2(x_0)\, dx_0$$

$$= kd_v \left[2\int_0^{b/2} \frac{dx}{1 - (2x/b)^2} - \frac{1}{a}\left(\frac{\pi b}{2}\right)^2\right]$$

The first term in the last expression is infinite if we carry the integration to $x = \tfrac{1}{2}b$. But of course we must stop short of the boundary layer, or else allow for the fact that the edge of the slit is not infinitely sharp. As with Eqs. (9.1.23) and (9.1.27), we carry the integral to $\tfrac{1}{2}b - h$, where h is either the half-thickness of the plate forming the slit edge or else is d_v, whichever is larger. The additional term becomes

$$\frac{kd_v}{a}\left(\frac{\pi b}{2}\right)^2\left(\frac{2a}{\pi^2 b}\ln\frac{b}{h}-1\right)$$

and the variational solution, for $b < \frac{1}{2}a$ and $ka < \frac{1}{2}\pi$, including the viscous effects, is

$$\coth\Phi_0 \simeq 1 + kd_v\left[\frac{2a}{\pi^2 b}\ln\left(\frac{b}{h}\right)-1\right]-\frac{2ika}{\pi}\ln\frac{2a}{\pi b}=1-\tfrac{1}{2}\gamma^2$$

$$u_0^0 \simeq \frac{2a}{\pi b\gamma}\left[1-\left(\frac{2x}{b}\right)^2\right]^{-\frac{1}{2}}\qquad \gamma^2 = 2kd_v\left(\frac{2a}{\pi^2 b}\ln\frac{b}{h}-1\right)-\frac{4ika}{\pi}\ln\frac{2a}{\pi b}$$

$$U_n^0 \simeq \frac{\epsilon_n}{\gamma}J_0\left(\frac{\pi n b}{a}\right)$$

(9.3.18)

$$R_0^0 \simeq \frac{\gamma^2}{4+\gamma^2}=\frac{\ln(2a/\pi b)-i(\pi d_v/2a)+i(d_v/\pi b)\ln(b/h)}{\ln(2a/\pi b)+i(\pi/ka)-i(\pi d_v/2a)+i(d_v/\pi b)\ln(b/h)}$$

The effective impedance $\rho c \coth\Phi_0$ at these long wavelengths is thus equivalent to ρc plus a lumped inductance of magnitude $(2\rho/\pi d)\ln(2a/\pi b)$ and a lumped resistance of magnitude $(2\rho\omega d_v/\pi^2 bd)[\ln(b/h)-(\pi^2 b/a)]$, where d is the width of the duct in the y direction. This is to be compared with the approximate results of Eqs. (9.1.26) and (9.1.27). The present formulas, primarily developed for the higher frequencies, thus join smoothly on to those which were developed in Sec. 9.1, which hold only for very long wavelengths.

The fraction of incident power reflected in the fundamental mode is $|R_0^0|^2$, which approaches unity, for a given value of k (i.e., of frequency) as b is made small, such that $2a/\pi b$ is quite large but b/h is not yet as small as unity. This is not surprising; as the slit is made smaller more and more of the incident wave should be reflected. But the slowness of the approach to unity, proportional to the reciprocal of the logarithm of $1/b$, may be surprising. Evidently, a narrow slit still transmits a lot of energy. This is even more apparent when we hold b constant (and small) and reduce frequency (i.e., reduce ka). No matter how small b is (as long as it is not zero), the reflection coefficient can be made as small as one likes by reducing frequency sufficiently. Incidentally, $|R_0^0|^2$ is the fraction of power reflected, and $|T_0^0|^2$ of Eqs. (9.3.9) is the fraction transmitted, in the primary wave, without the additional subtraction of the correction term $\frac{1}{2}\Pi_a$ of Eq. (9.3.10). The present formulas have included the viscous effects in the variational calculations, not as an afterthought. However, the effect is usually small enough so that, in the rest of this section, we neglect viscosity in the variational calculations and assume it can be allowed for by using Eq. (9.3.10) after u_m^0 is computed.

Wide slit and shorter wavelengths

For the case where b is not much smaller than a, particularly when $a - b$ is small, the single trial function χ_2 is not good enough and the formulas (9.3.15) must be resorted to, using both χ_0 and χ_2. For intermediate values of b, the series expressions for B_{00}, B_{02}, and B_{22} converge reasonably well, and the computation is straightforward, though tedious. When $b \to a$ and the obstruction reduces to a couple of fins sticking out from the duct walls, the series again converge slowly and the comparison series, worked out earlier, must be used.

For frequencies not larger than $3c/2a$ ($ka < 3\pi$), so that only the lowest mode, or else only the lowest two modes, propagate, the formulas, to first order in the small quantity $(a - b)/a$, are

$$\coth \Phi_0 \simeq 1 + \frac{a - b}{b}\frac{S}{R} \to 1 + \frac{a - b}{b}\frac{2 \ln 2}{\pi^2 \ln (2a/\pi b)} \qquad ka < 1$$

$$\frac{A_2}{A_0} \simeq -2\frac{a - b}{b}\frac{S}{R} \qquad R_0^0 \simeq \frac{a - b}{2b}\frac{S}{R}$$

$$R = \frac{J_0^2(\pi b/a)}{\sqrt{1 - (2\pi/ka)^2}} - i\frac{ka}{2\pi}\left[\ln\left(\frac{2a}{\pi b}\right) - J_0^2\left(\frac{\pi b}{a}\right)\right] \to -i\frac{ka}{2\pi}\ln\frac{2a}{\pi b} \qquad ka < 1$$

$$S = \frac{J_0(\pi b/a)}{\sqrt{1 - (2\pi/ka)^2}} - i\frac{ka}{2\pi}\left[\frac{2}{\pi^2}\ln 2 - J_0\left(\frac{\pi b}{a}\right)\right] \to -i\frac{ka}{\pi^3}\ln 2 \qquad ka < 1$$

$$2e^{-\Phi_0}U_0^0 \simeq 1 - 2\frac{a - b}{b}\frac{S}{R}$$

$$2e^{-\Phi_0}U_n^0 \simeq (-1)^{n-1}\frac{a}{\pi n b}\sin\left(\pi n\frac{a - b}{a}\right) - \frac{a - b}{b}\frac{S}{R}J_0\left(\pi n\frac{b}{a}\right) \qquad n > 0$$

$$(9.3.19)$$

The effective impedance $\rho c \coth \Phi_0$ is real, to the first order in $(a - b)/a$, when ka is small. This checks with the "exact" solution for very long wavelengths of Eq. (9.1.26), which indicates that $L_a \simeq (\pi\rho/4d)[(a - b)/a]^2$ as $b \to a$; thus the reactance is a second-order term, which we have neglected. As $b \to a$, the acoustic impedance approaches ρc, the characteristic impedance of the unobstructed duct; also, all the reflected modes, except the fundamental, go to zero ($U_n^0 \to 0$ for $n > 0$). As we surmised earlier, the relative amplitude A_2/A_0 of the χ_2 trial function goes to zero as $b \to a$, so that the velocity in the slit opening becomes uniform across the duct as $b \to a$. Of course, when $b = a$, no waves are reflected at all, and Φ_0 becomes infinite; our choice of modal combination (9.3.2) reveals its major deficiency by making the incident amplitude go to infinity in an attempt to keep a finite reflected amplitude. However, the reflected and transmitted amplitudes per unit amplitude incident wave, $2e^{-\Phi_0}U_n^0$, are finite and behave as expected when $b \to a$.

One further complication is apparent from Eqs. (9.3.17), for all values of $0 < b < a$. When the frequency happens to be equal to one of the modal cutoff frequencies (when ka equals $2\pi M$, say), coth Φ_0 becomes infinite, and the term corresponding to the Mth mode in u_0^0 (the value of U_M^0) is infinite. At just this frequency it appears that the mode, which is resonating across the duct, is excited with infinite amplitude by the obstruction. These infinities are not really present, of course; what has happened is that neither χ_0 nor χ_2 is an appropriate trial function for the narrow range of frequency within which the resonance occurs. Reference to the integral equation (9.3.8) and the related kernel G_m shows what the trouble is. When $ka \rightarrow 2\pi M$, the factor k_M in the series for G_m goes to zero, and the Mth term in G_m becomes very large, trying to make the integral $\int G_m u_m^0 \, dS_0$ equal to a large quantity times Ψ'_M. However, the integral equation requires this integral to be a finite factor times Ψ'_m, the transverse factor for the primary wave. Since the primary wave must be a propagating one, M must be larger than m. The integral equation thus requires that u_m^0, at this resonance frequency, must have a form such that the integral of u_m^0 times Ψ'_M, over the opening, be zero. Since our previous trial functions χ_0 and χ_2 do not have this property, they are not good enough for frequencies very close to the Mth cutoff.

To see what can be done, let us again take the plane-wave fundamental mode as the incident wave and consider the narrow range of frequencies close to the cutoff for the first symmetric mode, i.e., for ka near 2π. We must include, in our set of trial functions, a function χ_1 such that $\int \chi_1(x) \cos(2\pi x/a) \, dx = 0$, the integral being over S_0, for x from $-\frac{1}{2}b$ to $+\frac{1}{2}b$. A possible form is

$$\chi_1(x) = \cos \frac{2\alpha x}{a}$$

where

$$\alpha \tan \frac{\alpha b}{a} = \pi \tan \frac{\pi b}{a} \quad \text{or} \quad \alpha \simeq \frac{\pi a}{b} + \frac{\pi b}{a} \quad \text{when } b \ll a \quad (9.3.20)$$

Constant α is large enough so that function χ_1 changes sign once in the range $0 < x < \frac{1}{2}b$. Using this function as a trial function, we can calculate the necessary integrals from the basic one.

$$C_1{}^n = \int \chi_1 \Psi_n \, dx = \pi a \frac{\tan(\pi b/a) - n \tan(\pi n b/a)}{\alpha^2 - (\pi n)^2} \cos \frac{\alpha b}{a} \cos \frac{\pi n b}{a}$$

$$\rightarrow \pi \frac{b^2}{a} \frac{n \sin(\pi n b/a) - \tan(\pi b/a) \cos(\pi n b/a)}{1 - (b/a)^2(n^2 - 1)} \quad \text{when } b \ll a$$

$$C_1{}^n = 0 \quad \text{for } n = 1$$

This term, squared, will be multiplied by k/k_n to form one term in the sum for B_{11}; we have chosen χ_1, so that the unruly $n = 1$ term is missing from the sum.

From here on we carry out the calculations for $b \ll a$; the case is easier to write out, and it illustrates the procedure. We also shall assume that the frequency is very close to the cutoff for the $n = 1$ mode, so that all higher modes are nonpropagating, and $1 - (ka/2\pi n)^2 \simeq 1$ for $n > 1$. The constants to use in Eqs. (9.3.15), for trial functions χ_1 and χ_2, are then

$$C_1 \simeq -\frac{b^3}{a^2} \qquad C_2 = \frac{\pi b}{2}$$

$$B_{11} \simeq \frac{b^6}{a^5}\left[1 - i\left(\frac{a}{b}\right)^4 Z\right] \qquad B_{12} \simeq -\frac{\pi b^4}{2a^3}\left[1 + i\left(\frac{a}{b}\right)^2 Y\right]$$

$$B_{22} \simeq \frac{\pi^2 b^2}{2a}\left[1 + \frac{4J_0^2(\pi b/a)}{\sqrt{1 - (2\pi/ka)^2}} - iX\right]$$

$$X = \frac{2ka}{\pi}\left[\ln\left(\frac{2a}{\pi b}\right) - J_0^2\left(\frac{\pi b}{a}\right)\right] \tag{9.3.21}$$

$$Y = \frac{2ka}{\pi}\sum_{n=2}^{\infty}\frac{J_0(\pi nb/a)}{n}\frac{n^2(a/\pi nb)\sin(\pi nb/a) - \cos(\pi nb/a)}{(a/b)^2 - n^2 + 1}$$

$$Z = \frac{2ka}{\pi}\sum_{n=2}^{\infty}\frac{[n^2(a/\pi nb)\sin(\pi nb/a) - \cos(\pi nb/a)]^2}{n[(a/b)^2 - n^2 + 1]^2}$$

where we have separated off the term in B_{22} which will go to infinity when $ka \to 2\pi$ (χ_1 has been chosen so that no such term appears in either B_{11} or B_{12}). The series in Y and Z are slow to converge when b/a is small, but they have been defined so that their magnitude is of the order of unity. For example, the terms in the series for Z, for $n < (a/b)$, are each roughly equal to $n(b/a)^2$; whereas for $n > (a/b)$, the terms drop off rapidly in size (and are negative). Therefore a very rough value of Z is $(b/a)^2$ times the sum of n from $n = 1$ to $n =$ the integer nearest a/b, which sum is roughly $\frac{1}{2}(a/b)^2$; so $Z \simeq \frac{1}{2}$ when $b \ll a$. Thus X, Y, and Z are roughly equal in magnitude.

If ka is not nearly equal to 2π, the factor multiplying χ_2 in u_0^0 of Eqs. (9.3.15),

$$C_2 B_{11} - C_1 B_{12} \simeq -i\frac{\pi b^3}{2a}\left[Z + \left(\frac{b}{a}\right)^2 Y\right]$$

is roughly equal to the factor multiplying χ_1.

$$C_1 B_{22} - C_2 B_{12} = i\frac{\pi^2 b^3}{4a}\left[Y + \left(\frac{b}{a}\right)^2 X + 4i\left(\frac{b}{a}\right)^2\frac{J_0^2(\pi b/a)}{\sqrt{1 - (2\pi/ka)^2}}\right]$$

because Y is roughly equal to Z. Thus function χ_1 helps in improving the accuracy of the solution outside the resonance range (almost any two trial functions are better than either alone), but it does not make an outstanding improvement. When $ka \to 2\pi$, however, the last term in brackets in the factor for χ_1 becomes quite large, indicating that in the resonance range, χ_1 is a much better trial function than χ_2. The effective impedance in this range is ρc times

$$\coth \Phi_0 \simeq 1 + \frac{Z[4J_0^2(\pi b/a)/\sqrt{1 - (2\pi/ka)^2}]}{Z + i(b/a)^4[4J_0^2(\pi b/a)/\sqrt{1 - (2\pi/ka)^2}]} \quad (9.3.22)$$

Very close to the cross-duct resonance for the first subsidiary mode the second term in the denominator is larger than the first term and $\coth \Phi_0 \simeq 1 - i(a/b)^4 Z$, which is a quite different value from that computed without including χ_1 in the trial function [as given in Eqs. (9.3.17), for example]. The range of frequency over which χ_1 is important is very narrow if $b \ll a$ (the width of the range of ka is proportional to b^8/a^8). Within this narrow range the form u_0^0 of the velocity distribution at the slit is close to χ_1, which is orthogonal to the resonating-mode function Ψ'_1. Thus, instead of going to infinity as $ka \to 2\pi$ [as falsely predicted by Eqs. (9.3.17)], $\coth \Phi_0$ becomes large but finite (a/b is large) and nearly pure-imaginary, corresponding to a large mass load. Outside this narrow resonance range, however,

$$\coth \Phi_0 \simeq 1 + \frac{4J_0^2(\pi b/a)}{\sqrt{1 - (2\pi/ka)^2}}$$

which corresponds to Eqs. (9.3.17) when we reinclude the additional terms which could be neglected near the resonance but which cannot be neglected outside this range.

Thus, for a narrow slit, χ_2 is a good approximation for the form of the velocity shape in the slit opening, and Eqs. (9.3.17) give a good approximation to its reflection and transmission properties, with the exception of a set of very narrow frequency bands around each cutoff frequency, where the corresponding mode has cross-duct resonance. Within each of these bands the velocity shape u_0^0 tends to be orthogonal to the resonating mode, and the effective impedance $\rho c \coth \Phi_0$ just at the resonance frequency is large and imaginary, instead of going to infinity. When $b \ll a$ these bands are very narrow; outside them Eqs. (9.3.17) are adequate.

This finishes our first example of the use of the variational technique in connection with a coupling problem. We have gone into fairly complete detail with this one, so that its weaknesses, as well as its power, can be demonstrated. The computational complexities are considerable, but all coupling problems of any practical importance are inherently complicated, and attempts to find solutions by straightforward series expansions (such as trying to solve for the U's by direct solution of equations) lead to further proliferation of poorly convergent series, and series of series. When used with firmness and understanding, the variational method can achieve excellent results (note the circles of Fig. 4.20, for example). Of course, if poor choices are used for the trial function, absurd or misleading results may be obtained, as we have just seen.

We could pursue this example further. The case of the nonsymmetric slit, or an asymmetric incident wave, can be worked out by these general methods, using appropriately chosen trial functions. These results would be useful in some engineering problems; they would not demonstrate any new principles. We shall, instead, go on to consider a different sort of obstacle.

Transmission through a membrane

Suppose a flexible membrane, of density σ per unit area, is stretched with tension T across the duct at $z = 0$. The incident wave, from $z = -\infty$, sets the membrane into motion, which produces waves reflected back to $-\infty$ and transmitted on to $+\infty$. Thus two integral equations, one relating the pressure $p^- - p^+$ across the membrane to its velocity distribution u^0, and the other relating this velocity to the amplitudes of the various reflected and transmitted waves in the duct, must be combined and solved. There are two sorts of resonances, those of the standing waves in the membrane and the cross-duct resonances of the modes in the duct, at the cutoff frequencies. If the speed c of sound in air is larger than the speed $c_\sigma = \sqrt{T/\sigma}$ of wave motion in the membrane (in absence of air), the membrane resonances will begin at a lower frequency than the lowest cutoff frequency of the duct, and it may be possible to excite one or more membrane resonances at frequencies so low that only the fundamental plane-wave duct mode is propagated.

The equation for the motion of the membrane can be set up in accord with the discussion of Sec. 5.2. Suppose its modes of transverse vibration are the set $\phi_n(xy)$, which satisfies the boundary condition that ϕ be zero along the perimeter of the membrane (assumed to be coincident with the duct walls at $z = 0$) and which satisfies the equations

$$\nabla^2 \phi_n + K_n{}^2 \phi_n = 0 \qquad \iint \phi_m \phi_n \, dS = S \Sigma_n \delta_{mn}$$

where the integral is over the whole area of the membrane (of the duct cross section). The resonance frequency of the nth membrane mode is thus $K_n c_\sigma/2\pi = K_n c\gamma/2\pi$, where $\gamma = c_\sigma/c$, the ratio between the velocity of waves on the membrane and the speed of sound in air, is usually smaller than unity.

The equation of motion for the displacement η of the membrane from the equilibrium xy plane, produced by the pressure difference on its two sides, is

$$\sigma \frac{\partial^2 \eta}{\partial t^2} - T\nabla^2 \eta = p^- - p^+ \qquad \text{or} \qquad \nabla^2 \eta + \left(\frac{\omega}{c_\sigma}\right)^2 \eta = -\frac{1}{T}(p^- - p^+)$$

$$(9.3.23)$$

if the pressure difference is simple-harmonic, with frequency $\omega/2\pi$. If an mth-mode wave is incident on the left-hand side of the membrane, the sound waves in the duct can be represented by the combinations given in Eqs. (9.3.2) and (9.3.7). The driving force on the membrane can be written as $\rho c P_m^0(xy) = (p_m{}^- - p_m{}^+)$ at $z = 0$, and the membrane transverse velocity is equal to $u_m^0(xy)$, the fluid velocity at $z = 0$.

We next express the membrane velocity and displacement as series of normal modes.

$$u_m^0 = \sum_{n=1}^{\infty} V_n{}^m \phi_n(xy) \qquad V_n{}^m = \frac{1}{S\Sigma_n} \iint u_m^0 \phi_n \, dS$$

$$\eta_m = \frac{i}{\omega} \sum_n V_n{}^m \phi_n(xy)$$

(9.3.24)

where the series coefficients are to be determined. The series must satisfy the equation of motion (9.3.23). Therefore, using the equation for the ϕ_n, we obtain

$$\frac{i}{\omega} \sum_n \left[\left(\frac{\omega}{c_\sigma} \right)^2 - K_n{}^2 \right] V_n{}^m \phi_n = -\frac{\rho c}{T} P_m^0(xy)$$

or

$$V_n{}^m = \frac{ik\rho/S\Sigma_n \sigma}{k^2 - (K_n\gamma)^2} \iint_S P_m^0(xy)\phi_n(xy) \, dx \, dy \qquad (9.3.25)$$

where $\rho c P_m^0 = p_m{}^- - p_m{}^+$; $k = \omega/c$; $\gamma = c_\sigma/c$. Thus the integral equation relating the velocity of the membrane surface (and of the air next to it) to the pressure difference across its surface is

$$u_m^0(xy) = \iint_S G_\sigma(xy \mid x_0 y_0) P_m^0(x_0 y_0) \, dx_0 \, dy_0 \qquad (9.3.26)$$

where

$$G_\sigma = \frac{ik\rho}{S\sigma} \sum_n \frac{\phi_n(xy)\phi_n(x_0 y_0)}{\Sigma_n[k^2 - (K_n\gamma)^2]} \qquad \iint \phi_n{}^2 \, dx \, dy = S\Sigma_n$$

The effect of the membrane resonances is contained in the denominators of the terms in G_σ; when $k \equiv \omega/c = K_n\gamma \equiv K_n(c_\sigma/c)$, the nth membrane mode resonates.

The equation relating P_m^0 to u_m^0 is obtained from Eqs. (9.3.2) and (9.3.7).

$$P_m^0(xy) = \frac{k}{k_m} \Psi_m(xy) \cosh \Phi_m - \frac{k}{k_m} U_m{}^m \Psi_m(xy) - 2 \sum_{n \neq m} \frac{k}{k_n} U_n{}^m \Psi_n(xy)$$

$$= \frac{k}{k_m} \Psi_m(xy) \cosh \Phi_m - \iint_S G_d(xy \mid x_0 y_0) u_m^0(x_0 y_0) \, dx_0 \, dy_0 \qquad (9.3.27)$$

where

$$G_d = \frac{k}{k_m S\Lambda_m} \Psi_m(xy)\Psi_m(x_0 y_0) + 2 \sum_{l \neq m} \frac{k}{k_l S\Lambda_l} \Psi_l(xy)\Psi_l(x_0 y_0)$$

As before, the functions Ψ_l are the normal transverse modes for sound waves in the duct, with the boundary condition that the normal gradient of Ψ at the perimeter be zero (if the duct walls are rigid). Factor $k_l{}^2 = k^2 - \varkappa_l{}^2$, the mean-square value of Ψ_m is Λ_m, and the cutoff frequency of the lth mode is $\varkappa_l c/2\pi$. Thus the k/k_l factors contain the effects of the duct-mode resonances.

These two integral equations must be solved simultaneously to find u_m^0 and P_m^0, and thence to find the displacement of the membrane as well as the amplitudes of the reflected and transmitted waves. All this could presumably be found by expansions, either in terms of the Ψ's or the ϕ's, but such a procedure would lead to an infinite number of equations in an infinite number of unknowns, with more than usual convergence difficulties. Again it is better to use the variational technique.

To obtain a variational expression, we multiply Eq. (9.3.27) by u_m^0 and integrate over S, obtaining

$$0 = \frac{k}{k_m} \cosh \Phi_m \int \Psi_m u_m^0 \, dS - \int u_m^0 \, dS \int G_d u_m^0 \, dS_0 - \int u_m^0 P_m^0 \, dS$$

We also recall that $U_m{}^m = \sinh \Phi_m$, or

$$\frac{k}{k_m} S\Lambda_m \sinh \Phi_m \cosh \Phi_m = \frac{k}{k_m} \cosh \Phi_m \int \Psi_m u_m^0 \, dS$$

Adding the two equations and substituting from Eq. (9.3.26) for some of the u_m^0's, we obtain our variational expression,

$$\frac{k}{k_m} S\Lambda_m \sinh \Phi_m \cosh \Phi_m = \frac{k}{k_m} \cosh \Phi_m \left(\int \Psi_m u_m^0 \, dS + \int \Psi_m \, dS_0 \int G_\sigma P_m \, dS \right)$$

$$- \int u_m^0 \, dS \int G_d \, dS_0 \int G_\sigma P_m^0 \, dS_1 - \int u_m^0 P_m^0 \, dS \quad (9.3.28)$$

This is an asymmetric expression, relating the two adjoint functions u_m^0 and P_m^0. It has been chosen so that, when the expression on the right is varied, by varying the form of u_m^0, the first-order variation will be zero if P_m^0 satisfies Eq. (9.3.27). Specifically, in calculating the effect of a small change in u_m^0, as was done on page 459, the coefficient of δu, inside the integral over dS, is

$$\frac{k}{k_m} \cosh (\Phi_m) \Psi_m - \int G_d \, dS_0 \int G_\sigma P_w^0 \, dS_1 - P_m^0$$

If this is zero, P_m^0 satisfies Eq. (9.3.27), as can be seen by using Eq. (9.3.26) to substitute for $\int G_\sigma P_m^0 \, dS_1$.

Also, if we vary the value of the expression on the right of Eq. (9.3.28) by changing the form of P_m^0, the coefficient of δP_m^0 in the integral is

$$\frac{k}{k_m} \cosh \Phi_m \int G_\sigma \Psi_m \, dS_0 - \int G_\sigma \, dS_0 \int G_d u_m^0 \, dS_1 - u_m^0 = \int G_\sigma P_m^0 \, dS_0 - u_m^0$$

if we use Eq. (9.3.27) to eliminate the first term. This coefficient is zero if u_m^0 satisfies Eq. (9.3.26). Thus expression (9.3.28) embodies both integral equations and, if used properly, is the basis of a variational procedure for either P_m^0 or u_m^0, with the stationary value of the phase angle Φ_m differing from its correct value by an error which is to second order in the error in the trial function, whether it be for u_m^0 or P_m^0. It should be noted that it would be improper to use the expression obtained by multiplying Eq. (9.3.27) by P_m^0 (instead of by u_m^0) and integrating, for the resulting expression,

$$2 \frac{k}{k_m} \cosh \Phi_m \int \Psi_m P_m^0 \, dS - \int P_m^0 \, dS \int G_d \, dS_0 \int G_0 P_m^0 \, dS_1 - \int (P_m^0)^2 \, dS$$

has a kernel

$$K(xy \mid x_1 y_1) = \int G_d(xy \mid x_0 y_0) G_\sigma(x_0 y_0 \mid x_1 y_1) \, dS_0$$

which is not symmetric to interchange of xy and $x_1 y_1$. Therefore the coefficients of the two δP's in the variation will not be equal, and we shall not obtain Eq. (9.3.27) from the variation. Only by combining the two unknown functions, as we did, in the single expression, can we combine the two equations.

By now we are aware of the advantages and dangers of the variational procedure and are prepared to discuss the physical situation before we decide what trial functions should be used.

Choice of trial function

At low frequencies, below all resonances of the membrane and cutoff frequencies of the duct, the incident wave must be a plane wave, since we are assuming that the duct walls are rigid. It thus would be reasonable to assume that the pressure P_m^0 across the membrane should be reasonably uniform across the duct. The velocity u_m^0, of course, would not be uniform; it must go to zero at the perimeter of the membrane. By the time the frequency had reached the first membrane resonance ($k = \gamma K_1$), the velocity shape u_0^0 would be, primarily, the shape ϕ_1 of the first mode of vibration of the membrane. According to Eq. (9.3.26), this means that, to avoid infinities, the integral of $P_m^0 \phi_1$ must go to zero at $k = K_1 \gamma$; that is, at this frequency, P_m^0 must be orthogonal to ϕ_1. On the other hand, at the first duct resonance ($k = \varkappa_1$), the velocity u_m^0 must be orthogonal to the resonating duct mode Ψ_1.

If, in a given range of frequency, we have decided that the pressure drop across the membrane is approximated by the trial function $\vartheta(xy)$, we can determine its "best" magnitude by setting $P_m^0 = A\vartheta(xy)$ into Eq. (9.3.28), using Eq. (9.3.26) to express u_m^0 in terms of A and ϑ, and then finding the A for which the resulting expression is stationary to change of A. The variational expression becomes

$$2A \frac{k}{k_m} \cosh (\Phi_m) \int \Psi_m \int G_\sigma \vartheta - A^2 \int \vartheta \int G_\sigma \int G_d \int G_\sigma \vartheta - A^2 \int \vartheta \int G_\sigma \vartheta$$

where we have drastically simplified the notation, for the time being, in order to condense the formulas. It should be noted that the kernels in the quadratic terms of this *are* symmetric. Setting the derivative of this, with respect to A, equal to zero, we obtain

$$\coth \Phi_m \simeq \frac{k_m S \Lambda_m}{k} \frac{\int \vartheta \int G_\sigma \vartheta + \int \vartheta \int G_\sigma \int G_d \int G_\sigma \vartheta}{[\int \Psi_m \int G_\sigma \vartheta]^2} \qquad P_m^0 \simeq A\vartheta(xy)$$

$$A = \frac{k}{k_m} \cosh \Phi_m \frac{\int \Psi_m \int G_\sigma \vartheta}{\int \vartheta \int G_\sigma \vartheta + \int \vartheta \int G_\sigma \int G_d \int G_\sigma \vartheta} \qquad u_m^0 \simeq A \int G_\sigma \vartheta \, dS_0$$

$$(9.3.29)$$

On the other hand, if we decide to try $u_m^0 = B\chi$, the variational expression is

$$\left(\frac{k}{k_m} \cosh \Phi_m\right)^2 \int \Psi_m \int G_\sigma \Psi_m - 2B\frac{k}{k_m} \cosh \Phi_m \int \Psi_m \int G_\sigma \int G_d \chi$$

$$+ B^2 \int \chi \int G_d \chi + B^2 \int \chi \int G_d \int G_\sigma \int G_d \chi$$

The kernels of the quadratic terms of this expression also are symmetric. Differentiating with respect to B, we find

$$\tanh \Phi_m \simeq \frac{1}{S \Lambda_m} \int \Psi_m \int G_\sigma \Psi_m - \frac{k}{k_m S \Lambda_m} \frac{[\int \Psi_m \int G_\sigma \int G_d \chi]^2}{\int \chi \int G_d \chi + \int \chi \int G_d \int G_\sigma \int G_d \chi}$$

$$B = \frac{k}{k_m} \cosh \Phi_m \frac{\int \Psi_m \int G_\sigma \int G_d \chi}{\int \chi \int G_d \chi + \int \chi \int G_d \int G_\sigma \int G_d \chi} \qquad u_m^0 \simeq B\chi(xy) \quad (9.3.30)$$

$$P_m^0 \simeq \frac{k}{k_m} \cosh (\Phi_m \Psi_m)xy - B\int G_d \chi \, dS_0$$

Incidentally, if a stiff diaphragm were used, instead of a membrane, across the duct at $z = 0$, Eqs. (9.3.29) and (9.3.30) would be unchanged in form, but G_σ would be modified by the insertion of the normal-mode functions and eigenvalues of Sec. 5.3 for a thin plate, instead of ϕ_n and K_n in the series, and the coefficient would be modified to replace c_σ by the transverse wave velocity in the plate.

Variational calculations for a circular membrane

The circular duct of cross-sectional radius a, obstructed by a circular membrane at $z = 0$, is a useful example to work out. It is particularly convenient because we can drastically simplify the expression $\int G_\sigma \vartheta$ if we choose an appropriate form for ϑ. The duct and membrane functions, for motion symmetric about the duct axis, are

$$\Psi_l = J_0\left(\frac{\pi \alpha_{0l} r}{a}\right) \qquad \varkappa_l = \frac{\pi \alpha_{0l}}{a} \qquad \Lambda_l = J_0^2(\pi \alpha_{0l})$$

$$(9.3.31)$$

$$\phi_n = J_0\left(\frac{\pi \beta_{0n} r}{a}\right) \qquad K_n = \left(\frac{\pi \beta_{0n}}{a}\right) \qquad \Sigma_n = J_1^2(\pi \beta_{0n})$$

where

$$J_0(\pi\beta_{0n}) = 0 \quad \text{and} \quad -J_0'(\pi\alpha_{0l}) \equiv J_1(\pi\alpha_{0l}) = 0 \quad k_l^2 = k^2 - \left(\frac{\pi\alpha_{0l}}{a}\right)^2$$

In this case we can neglect the angle φ of the polar coordinates, considering it to be a one-dimensional problem, and thus $S = \frac{1}{2}a^2$. The values of some of the α's are given in Eq. (9.2.23), and of some of the β's in Eq. (5.2.23) (see also Table X). If the incident wave is not independent of φ, but has the factor $\cos(m\varphi)$, then all terms would have this factor. After dividing out the $\cos(m\varphi)$, the procedure to be discussed would also carry through, with J_m's instead of J_0's, α_{ml}'s instead of α_{0l}'s, and so on. The integrals occurring in the variational solution can be worked out from the general formulas of Eq. (5.2.21).

If we choose the trial function $A\vartheta = AJ_0(qr/a)$ for P_m^0, a closed form can be found for the integral $A\int G_\sigma \vartheta = u_m^0$. This is a fairly flexible trial function; by varying A and/or q we can have a constant pressure ($q = 0$) or else a function of r which may be equal to or orthogonal to any of the Ψ's or ϕ's. Of course, not all values of q are appropriate, since ϑ should satisfy the boundary condition of zero slope at $r = a$, which the duct pressure should satisfy.

Since

$$\left[\nabla^2 + \left(\frac{\omega}{c_\sigma}\right)^2\right]J_0\left(\frac{qr}{a}\right) = \left[\left(\frac{k}{\gamma}\right)^2 - \left(\frac{q}{a}\right)^2\right]J_0\left(\frac{qr}{a}\right)$$

we see we can solve Eq. (9.3.23) directly, rather than use Eq. (9.3.26). Since $u = -i\omega\eta$, the solution of

$$\nabla^2 u + \left(\frac{k}{\gamma}\right)^2 u = \frac{i\omega\rho c}{T}P_m^0 = \frac{ik\rho}{\sigma\gamma^2}AJ_0\left(\frac{qr}{a}\right)$$

which goes to zero at $r = a$, is the combination

$$\frac{iAka(\rho a/\sigma)}{k^2a^2 - q^2\gamma^2}\left[J_0\left(\frac{qr}{a}\right) - \frac{J_0(q)}{J_0(ka/\gamma)}J_0\left(\frac{kr}{\gamma}\right)\right] = u_m^0 \qquad (9.3.32)$$

and therefore must be the velocity distribution u_m^0. Thus we have obtained a closed form for the integral

$$\int_0^a G_\sigma(r \mid r_0)J_0\left(\frac{qr_0}{a}\right)r_0\,dr_0 = \frac{i(\rho a/\sigma)ka}{k^2a^2 - q^2\gamma^2}\left[J_0\left(\frac{qr}{a}\right) - \frac{J_0(q)}{J_0(ka/\gamma)}J_0\left(\frac{kr}{\gamma}\right)\right]$$

$$(9.3.33)$$

which greatly reduces the number of infinite series entering the formulas, as long as we use the trial function $\vartheta = J_0(qr/a)$.

Membrane resonances appear in this closed form for u_m^0 when ka/γ equals one of the roots $\pi\beta_{0n}$ at which $J_0(ka/\gamma)$ goes to zero. When $ka \to q\gamma$, both the denominator and the square bracket in the numerator go to zero, and

$$\int G_\sigma J_0\left(\frac{qr_0}{a}\right) = \frac{i\rho a}{2\gamma\sigma}\left[\frac{r}{a}J_1\left(\frac{qr}{a}\right) - \frac{J_1(q)}{J_0(q)}J_0\left(\frac{qr}{a}\right)\right] \qquad ka \to q\gamma$$

The closed expression (9.3.33) for u_m^0 is a function of position r/a on the membrane; in addition, it is a function of four dimensionless parameters. Parameter q is one of the variational parameters used to choose a "best" form for ϑ. Parameter ka is proportional to the frequency of the incident wave. Parameter $\gamma = c_\sigma/c$ is the ratio between wave velocities, in the membrane and in the air; in most cases it is smaller than unity. Finally, parameter $\rho a/\sigma$ is equal to the ratio of the mass of air in a length of duct equal to the radius a to the mass of the membrane; it also is usually smaller than unity.

Therefore, when $J_0(qr/a)$ is used for the trial function, the general solution (9.3.29) becomes, for an incident plane wave ($m = 0$),

$$\coth \Phi_0 \simeq 1 - \frac{i\gamma^2 J_0(ka/\gamma)}{(\rho a/\sigma)ka}\frac{Y(q)}{X_0^2(q)} - 2ika\sum_{l=1}^{\infty}\frac{1}{\mu_l}\left[\frac{X_l(q)}{X_0(q)}\right]^2$$

$$P_0^0(r,q) \simeq \frac{2[(ka)^2 - (q\gamma)^2]J_0(ka/\gamma)X_0(q)\cosh\Phi_0 J_0(qr/a)}{J_0(ka/\gamma)Y(q) + 4(\rho a/\sigma)ka\left[iX_0^2 + 2ka\sum_{l=1}^{\infty}(X_l^2/\mu_l)\right]} \qquad (9.3.34)$$

$$u_0^0(r,q) \simeq \frac{2i(\rho a/\sigma)kaX_0\cosh\Phi_0[J_0(ka/\gamma)J_0(qr/a) - J_0(q)J_0(kr/\gamma)]}{J_0(ka/\gamma)Y(q) + 4(\rho a/\sigma)ka\left[iX_0^2 + 2ka\sum_{l=1}^{\infty}(X_l^2/\mu_l)\right]}$$

where

$$\mu_l^2(ka) = (\pi\alpha_{0l})^2 - (ka)^2$$

$$Y(q) = q^2[J_0(q)J_2(q) - J_1^2(q)]J_0\left(\frac{ka}{\gamma}\right) - \left(\frac{ka}{\gamma}\right)^2\left[J_0^2(q)J_2\left(\frac{ka}{\gamma}\right) - J_1^2(q)J_0\left(\frac{ka}{\gamma}\right)\right]$$

$$X_0(q) = J_2(q)J_0\left(\frac{ka}{\gamma}\right) - J_0(q)J_2\left(\frac{ka}{\gamma}\right) = \frac{2}{q}J_1(q)J_0\left(\frac{ka}{\gamma}\right) - 2\frac{\gamma}{ka}J_0(q)J_1\left(\frac{ka}{\gamma}\right)$$

$$X_l(q) = \frac{2qJ_1(q)J_0(ka/\gamma)}{q^2 - (\pi\alpha_{0l})^2} - \frac{2(ka/\gamma)J_0(q)J_1(ka/\gamma)}{(ka/\gamma)^2 - (\pi\alpha_{0l})^2}$$

We have written it for $ka < \pi\alpha_{01}$, but if any of the higher modes are propagating ($ka > \pi\alpha_{01}$, say), the corresponding μ_1 in the denominators can be changed to $-i[(ka)^2 - (\pi\alpha_{01})^2]^{\frac{1}{2}}$.

As pointed out earlier, for frequencies well below the first membrane-resonance, it should be satisfactory to let $q = 0$, that is, $\vartheta = 1$. In this case the formulas become much simpler.

$$\coth \Phi_0 \simeq 1 + \frac{ika J_0(ka/\gamma)}{(\rho a/\sigma) J_2(ka/\gamma)} - \frac{2ika}{J_2^2(ka/\gamma)} \sum_{l=1}^{\infty} \frac{1}{\mu_l} \left[\frac{(2ka/\gamma) J_1(ka/\gamma)}{(ka/\gamma)^2 - (\pi\alpha_{0l})^2} \right]^2 \equiv Q$$

$$P_0^0(0) \simeq i \frac{ka J_0(ka/\gamma) \cosh \Phi_0}{(\rho a/\sigma) Q J_2(ka/\gamma)} \to 1 \qquad ka \ll \gamma$$

$$u_0^0(0) \simeq \frac{J_0(kr/\gamma) - J_0(ka/\gamma)}{Q J_2(ka/\gamma)} \cosh \Phi_0 \to -i \frac{\rho c \omega a^2}{4T} \left[1 - \left(\frac{r}{a}\right)^2 \right]$$

$$(9.3.35)$$

$$U_0^0 = \sinh \Phi_0 \to -i \frac{\rho c \omega a^2}{8T} \qquad \mu_l^2(ka) = (\pi\alpha_{0l})^2 - (ka)^2$$

$$U_l^0 \simeq \frac{(2ka/\gamma) J_1(ka/\gamma) \cosh \Phi_0}{J_2(ka/\gamma) J_0(\pi\alpha_{0l})[(\pi\alpha_{0l})^2 - (ka/\gamma)^2]} \to -i \frac{\rho a}{\sigma} \frac{(ka)^3}{8\gamma^4(\pi\alpha_{0l})^2}$$

$$Q \to \frac{8iT}{\rho c \omega a^2} \qquad ka \ll \gamma \qquad \mu_l = -ik_l a \qquad ka > \pi\alpha_{0l}$$

First, we notice that, for frequencies below the lowest membrane resonance, the effective acoustic impedance of the membrane has the simple form

$$\rho c \coth \Phi_0 \to \rho c + \frac{i}{\omega} \frac{8T}{a^2} - i\omega \rho a \sum_{l=1}^{\infty} \frac{128}{(\pi\alpha_{0l})^5} \qquad ka \ll \gamma$$

where the sum equals 0.150. This corresponds to a circuit with three components in series: first, the characteristic impedance ρc of the uniform duct to the right of the membrane; second, the stiffness reactance of the membrane below its first resonance; and third, a mass reactance produced by the nonuniform velocity distribution at the membrane and represented by the higher (nonpropagating) duct modes. At these low frequencies the system is stiffness-controlled, the pressure is uniform across the surface of the membrane, and the membrane velocity has a parabolic distribution, maximum at the center and zero at $r = a$.

Membrane and duct resonances

To be consistent, we should vary parameter q until $\partial \coth (\Phi_m)/\partial q$ is zero. Then we should have a "best" form for ϑ and a "best" value for $\coth \Phi_m$ and the other quantities, within the limits of our choice of $J_0(qr/a)$ for ϑ. But this policy implies numerical evaluation and differentiation of the complicated expressions of Eqs. (9.3.34), which would hardly clarify the physics of the situation. It also would involve the use of values of q for which

$d\vartheta/dr \neq 0$ at $r = a$, not appropriate for a pressure wave in a duct. We can see, by using the series expansions [Eqs. (5.2.20)] of $J_m(q)$, that for q small, coth $[\Phi_0(q)]$ equals coth $[\Phi_0(0)]$ [the value given in Eqs. (9.3.35)] plus q^2 times an expression independent of q, plus terms in still higher powers of q. Thus $q = 0$ is one root of the equation ∂ coth $(\Phi_0)/\partial q = 0$. But there are other roots to this equation, since $Y(q)$ and the $X_l(q)$'s are oscillating functions of q. These correspond to the values $q = \pi\alpha_{0n}$, for which $d\vartheta/dr = 0$ and $\partial\Phi_0/\partial q = 0$. Which root is the "best" one will be most easily determined by looking at the physics of the situation. Certainly, for low frequencies ($ka \ll \gamma$), it makes sense to assume that the pressure distribution across the membrane is the same as the distribution in the incident primary wave (i.e., that $\vartheta = 1$ and $q = 0$), and thus to use Eqs. (9.3.35) as the "best" expressions for Φ_0, P_0^0, and u_0^0.

To see the sequence of events as the frequency is increased, we note that the first root of $J_0(x) = 0$ is $\pi\beta_{01} = 2.40$; the first common root (aside from $x = 0$) of $J_0'(x) = 0$ and $J_1(x) = 0$ is $\pi\alpha_{01} = \pi\beta_{11} = 3.83$; the first root of $J_2(x) = 0$ is $\pi\beta_{21} = 5.14$; and the second root of $J_0(x) = 0$ is $\pi\beta_{02} = 5.52$. Thus the first resonance, if $\gamma < 1$, is at $ka = \gamma\pi\beta_{01}$, where $J_0(ka/\gamma) = 0$ and the membrane has its first resonance. Here the second term in coth Φ_0 in Eqs. (9.3.35) is zero; but this is to be expected at a membrane resonance. This second term is proportional to the membrane impedance, which should go to zero at membrane resonance. Study of Eqs. (9.3.34), moreover, shows that the second term goes to zero no matter what value of q we choose; therefore the choice of $q = 0$ and the use of Eqs. (9.3.35) is still a reasonable choice at a membrane resonance. Also, P_0^0 goes to zero at a membrane resonance ($ka = \gamma\pi\beta_{0l}$) no matter what value q has, indicating that the primary pressure wave has a node at $z = 0$. Even though it is at resonance the amplitude u_0^0 of the membrane velocity is finite; this is because the membrane is not oscillating in a vacuum, but is impeded in its motion by the air on both sides.

Before the frequency reaches the second membrane resonance, $ka = \gamma\pi\beta_{02}$, the frequency parameter passes through the value $\gamma\pi\beta_{21}$, when $J_2(ka/\gamma) = 0$. Here there is a real difference between Eqs. (9.3.35) and (9.3.34); the impedance factor coth Φ_0 goes to infinity for $\vartheta = 1$ ($q = 0$), but for $q > 0$ it does not. The difficulty here is that, at $ka/\gamma = \pi\beta_{21}$, the shape of the membrane's motion, when excited by a uniform pressure, is such that it cannot excite the primary reflected wave. Thus, *if* the pressure at $z = 0$ is independent of r, no reflected or transmitted fundamental mode can be produced, and our combination tries to compensate for this by increasing P_0^0 to infinity. Obviously, the choice $q = 0$ is not a good one for this frequency.

The next choice, which satisfies the boundary condition that $d\vartheta/dr = 0$ at $r = a$, is $\vartheta = J_0(\pi\alpha_{01}r/a)$, corresponding to the first higher duct mode, orthogonal to $\vartheta = 1$, which does permit excitation of the fundamental.

Inserting $q = \pi\alpha_{01}$ in Eqs. (9.3.34), we obtain

$$\coth \Phi_0 \simeq 1 + \frac{ika J_0(ka/\gamma)}{4(\rho a/\sigma) J_1^2(ka/\gamma)}\left[(\pi\alpha_{01})^2 J_0\left(\frac{ka}{\gamma}\right) + \left(\frac{ka}{\gamma}\right)^2 J_2\left(\frac{ka}{\gamma}\right)\right]$$

$$- 2ika \sum_{l=1}^{\infty} \frac{1}{\mu_l}\left[\frac{(ka/\gamma)^2}{(ka/\gamma)^2 - (\pi\alpha_{0l})^2}\right]^2$$

after judicious application of Eqs. (5.2.21) and of the fact that $J_2(\pi\beta_{21}) = 0$. This expression is finite at $ka = \gamma\pi\beta_{21}$, but it goes to infinity at $ka = \gamma\pi\alpha_{01}$ or $\gamma\pi\alpha_{02}$.

A still better choice for trial function would of course be $\vartheta = A + B J_0(\pi\alpha_{01}r/a)$. Rough calculations indicate that if we do this, B will be large compared with A only in a narrow frequency range near $ka = \gamma\pi\beta_{21}$ (and B would be zero for $ka = \gamma\pi\alpha_{01}$ or $\gamma\pi\alpha_{02}$). Thus we can use Eqs. (9.3.34) over most of the frequency range, except very close to the values $\gamma\pi\beta_{2n}$, where $J_2(ka/\gamma) = 0$. In these narrow bands the impedance has a peak, but does not go to infinity, as Eqs. (9.3.34) would predict. The peak value at ka exactly equal to $\gamma\pi\beta_{21}$, and the corresponding pressure distribution, are

$$\coth \Phi_0 \simeq 1 + \frac{i\gamma(\pi\alpha_{01})^2}{(\rho a/\sigma)\pi\beta_{21}} - 8i\gamma(\pi\beta_{21})^3 \sum_{l=1}^{\infty} \frac{1/\mu_l(\gamma\pi\beta_{21})}{[(\pi\beta_{21})^2 - (\pi\alpha_{0l})^2]^2}$$

$$P_0^0 \simeq 2\gamma^2[(\pi\beta_{21})^2 - (\pi\alpha_{01})^2] \cosh \Phi_0 \frac{J_0(\pi\alpha_{01}r/a)}{H J_0(\pi\alpha_{01})}$$

$$H = (\pi\alpha_{01})^2 - 4i\left(\frac{\rho a}{\sigma}\right)(\gamma\pi\beta_{21})$$

$$\quad\quad (9.3.36)$$

$$- 8\frac{\rho a}{\sigma}(\gamma\pi\beta_{21})^2 \sum_{l=1}^{\infty} \frac{1}{\mu_l(\gamma\pi\beta_{21})}\left[\frac{(\pi\beta_{21})^2}{(\pi\beta_{21})^2 - (\pi\alpha_{0l})^2}\right]^2$$

$$\mu_l^2(x) = (\pi\alpha_{0l})^2 - x^2 \quad\quad \mu_l = -ik_l a$$

after extended manipulation of Bessel-function properties. The velocity distribution u_0^0 is given by the form of Eq. (9.3.32), with $q = \pi\alpha_{01}$. At higher frequencies B again goes to zero, and Eqs. (9.3.35) can again be used until close to the next membrane "antiresonance" frequency, $ka = \gamma\pi\beta_{22}$.

It might appear that the third term for $\coth \Phi_0$ in Eqs. (9.3.35) becomes infinite when $ka/\gamma = \pi\alpha_{0n}$. This is not the case, however, for $J_1(ka/\gamma)$ becomes zero at this point, all but the $l = n$ term in the series vanish, and the limiting value of this third term, when $ka/\gamma = \pi\alpha_{0n}$, is just $-2i[\gamma/\sqrt{1 - \gamma^2}]$.

We have not yet discussed the cross-duct resonances, which occur when $ka = \pi\alpha_{0l}$, because we have been assuming that $\gamma = c_a/c$ is small and that several membrane resonances take place before the frequency reaches the first cutoff frequency for the duct. When $ka = \pi\alpha_{0m}$, the mth term in the third term for $\coth \Phi_0$ in Eqs. (9.3.34) is infinite unless $X_m(q)$ is zero, which suggests that, in the narrow frequency band near $ka = \pi\alpha_{0m}$, we set $q = \pi\alpha_{0m}/\gamma$, and thus set $\vartheta = J_0(\pi\alpha_{0m}r/a\gamma)$. Physically, this corresponds to saying that, in this frequency band, the pressure drop across the membrane adjusts itself so that the resulting membrane motion cannot excite the mth resonant duct mode.   The rest of the discussion of this frequency range proceeds analogously to the corresponding discussion of the transmission through a slit, near a duct cutoff frequency.

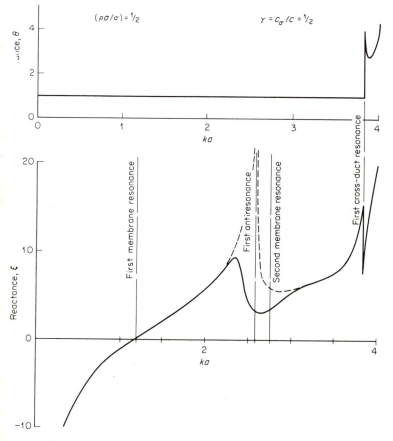

FIGURE 9.15
Specific impedance of membrane in circular tube of radius a to plane incident wave.   From Eqs. (9.3.35), except in the range $2.3 < ka < 2.9$; in that range, dotted-line plots Eqs. (9.3.35); solid line is corrected according to Eqs. (9.3.36).

A typical plot of real and imaginary parts of the effective impedance of the membrane-duct system for an incident plane wave, $\coth \Phi_0$, is plotted in Fig. 9.15 against the frequency parameter ka, for a typical case.   We have chosen γ less than unity, so that two membrane resonances are reached before a cross-duct resonance is reached.   Also, $\rho a/\sigma$ is not very small, so that the membrane reactance (the second term in the expressions for coth Φ_0) does not completely mask the wave-distortion reactance (the third term). Above the first cross-duct resonance a second mode can transmit energy, so that the resistance suddenly increases.   The dotted line is the continuation of Eqs. (9.3.35) in the region of antiresonance, where it is not a good approximation; the solid line in this range of ka follows the better approximation of Eqs. (9.3.36).   This theory can be applied to the operation of the human ear canal and drum.

9.4   STANDING WAVES IN CAVITIES

When sound waves are produced in a region completely enclosed by walls, rigid or otherwise, all wave motion is standing-wave motion, and the acoustic-energy content of the cavity is determined by the nature of its walls. For cavities smaller than, or of the same size as, a wavelength, it will be more convenient to analyze this motion in terms of the normal modes of the enclosure; for cavities much larger than a wavelength, it is often more convenient to use geometrical acoustics, following the "rays" of sound as they reflect back and forth from various portions of the wall surface.   Rooms are cavities for acoustic waves, usually of complex shape and with nonuniform wall response.   For most rooms the geometric analysis turns out to be more fruitful, except for the long wavelengths.   The geometric analysis is dealt with in numerous books on room acoustics; so it will not be discussed here in any detail.   This section will be primarily concerned with acoustic cavities of smaller size than most rooms, or with the very low frequency acoustic behavior of ordinary rooms.   Of course, at frequencies so low that the wavelength is much larger than the dimensions of the cavity, the air in the enclosure can only exhibit stiffness reactance, expanding and contracting in phase, in response to a driving source, as was assumed in connection with the Helmholtz resonator of Eq. (9.1.30).   When the ratio of wavelength to cavity dimension is between about $\frac{1}{3}$ and 3, the normal-wave analysis is most useful.   For shorter wavelengths the geometrical analysis is preferable.

Forced motion; normal modes

Suppose the walls of the cavity are locally reacting so that we can, with good approximation, assign a specific acoustic admittance $\beta(\omega,\mathbf{r}_s) = \chi - i\sigma = \rho c(u_n/p)$ to each point $\mathbf{r}_s$ on the wall surface for each frequency $\omega/2\pi$ of the sound.   Suppose the wall surface S completely encloses region R, which has volume V.   Then the acoustic response to a source of sound in R

$$\nabla^2\Psi_n \equiv \frac{\partial^2\Psi_n}{\partial x^2} + \frac{\partial^2\Psi_n}{\partial y^2} + \frac{\partial^2\Psi_n}{\partial z^2} = -K_n{}^2\Psi_n$$

$$\iiint\limits_R \Psi_n(\omega,\mathbf{r})\Psi_m(\omega,\mathbf{r})\,dV = V\Lambda_n(\omega)\delta_{mn}$$

(9.4.1)

can be expressed in terms of the normal modes of the cavity, $\Psi_m(\omega, \mathbf{r})$, where where $dV = dx\, dy\, dz$, and Ψ_n must satisfy the boundary condition that, for each point $\mathbf{r}_s$ on S, the component $\partial \Psi_n / \partial n_s$ of the gradient of Ψ_n in the direction of the outward-pointing (away from R) normal to S at that point be equal to $i(\omega/c)\beta(\omega, \mathbf{r}_s)$ times the value of Ψ_n at that point.   If any portion of S has nonzero conductance, the eigenvalues K_n will be complex, with negative-imaginary parts.   The index n stands for a trio of numbers, as required for a three-dimensional standing wave.   An important fact to keep in mind is that the boundary conditions depend on frequency (except in the limiting case of rigid walls, $\beta = 0$) so that the set of eigenfunctions Ψ_n and eigenvalues K_n also are functions of ω.   Thus our analysis must first be made for steady-state situations, for a specific frequency $\omega/2\pi$; transient behavior will be computed by use of the Fourier transform.

As usual, the first task is to calculate the Green's function $G_\omega(\mathbf{r} \mid \mathbf{r}_0)$, representing the spatial distribution of the radiation from a point source of frequency $\omega/2\pi$ at point $\mathbf{r}_0$ in R.   As indicated in Eq. (7.1.14), the Green's function satisfies

$$\nabla^2 G_\omega + \left(\frac{\omega}{c}\right)^2 G_\omega = -\delta(x - x_0)\, \delta(y - y_0)\, \delta(z - z_0)$$

Expanding G_ω in a series of normal modes, we can solve for the coefficients and obtain

$$G_\omega(\mathbf{r} \mid \mathbf{r}_0) = \sum_n \frac{\Psi_n(\omega, \mathbf{r})\Psi_n(\omega, \mathbf{r}_0)}{V\Lambda_n[K_n^2(\omega) - (\omega/c)^2]} \tag{9.4.2}$$

Very close to the source this series approaches in value the free-space Green's function $g_\omega = e^{i\omega R/c}/4\pi R$ of Eq. (7.1.14); farther away it differs from g_ω because of the waves reflected from the cavity walls.   In the present case the wave has a resonance whenever ω is equal to c times the real part of one of the eigenvalues $K_n(\omega)$.   At the nth resonance the nth standing wave Ψ_n predominates, having an amplitude inversely proportional to the imaginary part of K_n for that frequency.

Suppose a root of the equation $cK_n(\omega) = \omega$ is $\omega_n - i\gamma_n$, with ω_n and γ_n positive quantities.   Because of the symmetry of the admittance function $\beta(\omega)$ about the imaginary ω axis, there will be another root at $-\omega_n - i\gamma_n$, so that the two roots of the equation $cK_n(\omega) = \omega$ are

$$\omega = \pm\omega_n - i\gamma_n = cK_n(\pm\omega_n - i\gamma_n) \tag{9.4.3}$$

with γ_n usually much smaller than ω_n.   The denominator in the nth term of the series for G_ω, for $\omega = \omega_n + \epsilon$ (driving frequency $\omega/2\pi$ close to ω_n), is

$$-2\omega_n(\epsilon + i\gamma_n)(1 - cK_n') \frac{V\Lambda_n}{c^2}$$

to first order in the small quantities ϵ and γ_n, where K_n' is the derivative of K_n with respect to ω at $\omega = \omega_n - i\gamma_n$.   The square of the magnitude of this quantity has its minimum at $\epsilon = 0$ ($\omega = \omega_n$) and rises to twice its minimal value for $\epsilon \simeq \pm\gamma_n$ ($\omega \simeq \omega_n \pm \gamma_n$); thus the "width" of the nth resonance peak of G_ω is γ_n. When the separation between successive resonances comes closer together than γ_n, the response of the air in the room to a single-frequency source merges into a more or less uniform response, as the frequency is changed, instead of a response punctuated by separate resonance peaks.

The mean number ΔN of resonance peaks in an enclosure of volume V, within the range of values of ω between ω and $\omega + \Delta\omega$, is

$$\Delta N \simeq \frac{V\omega^2}{2\pi^3 c^3} \Delta\omega \tag{9.4.4}$$

This will be demonstrated [see Eq. (9.5.13)] later for a rectangular enclosure. Therefore the mean distance between resonances, on the ω scale, is $\Delta\omega/\Delta N \simeq 2\pi^2 c^3/V\omega^2$. When this distance becomes smaller than the width of the average peak, i.e., when

$$\frac{2\pi^2 c^3}{V\omega^2} \ll \gamma_{\text{avg}} \qquad \text{or when} \qquad \omega^2 \gg \frac{2\pi^2 c^3}{V\gamma_{\text{avg}}} \tag{9.4.5}$$

the room response becomes a smoothly varying function of ω, and the geometric-acoustics analysis of Sec. 9.5 will usually be adequate. At present, however, we are discussing the response at lower frequencies, where the resonance peaks are separated from each other and the standing-wave analysis is necessary.

Impedance of simple source

Following Eq. (7.4.1), the pressure wave $\mathbf{r}$, produced by a simple source of frequency $\omega/2\pi$ and strength S_ω at point $\mathbf{r}_0$ in R, is $p_s = -i\omega\rho S_\omega G_\omega(\mathbf{r} \mid \mathbf{r})e^{-i\omega t}$ inside R. If the source is a sphere of radius $a \ll \lambda$, the mean pressure at its surface is $1/4\pi a^2$ times the integral of p_s, in the $\mathbf{r}$ coordinates, over the surface of the sphere of radius a, centered at $\mathbf{r}_0$. The series for G_ω diverges as $\mathbf{r} \to \mathbf{r}_0$, so that, as $a \to 0$, the mean value of G_ω over the sphere will increase without limit. If a is small, the series for the mean will converge very slowly, so that the majority of its value will come from the terms for n large; thus we can calculate the limiting value of the sum by dealing with the terms of large n. These terms simplify in form considerably, and by a rather involved argument, it can be demonstrated that the mean value of G_ω at very low frequencies, close to the source, is

$$\frac{1}{4\pi} \int_0^{2\pi} d\varphi \int_0^\pi G_\omega^1(x_0 + a \sin\vartheta \cos\varphi, y_0 + a \sin\vartheta \sin\varphi, z_0 + a \cos\vartheta) \sin\vartheta \, d\vartheta$$

$$\to \frac{1}{4\pi a} \qquad \text{as } \omega \to 0 \text{ and } a \to 0$$

where G_ω^1 is the series of Eq. (9.4.2), with the term $n = 0$ omitted.

Thus we can write the mean value of the pressure on the small sphere in the form

$$\langle p_s \rangle \to -i\omega\rho S_\omega \frac{e^{-i\omega t}}{4\pi a} - ik^3\rho c S_\omega \sum_{n=1}^{\infty} \frac{\Psi_n^2(\mathbf{r}_0)e^{-i\omega t}}{V\Lambda_n K_n^2(K_n^2 - k^2)} \qquad a \to 0$$

somewhat like the form in Eq. (9.2.12) for the simple source in a tube. The effective driving force is then $F_s = 4\pi a^2 \langle p_s \rangle$, and the net fluid velocity at $|\mathbf{r} - \mathbf{r}_0| = a$ is $U_s \simeq (S_\omega/4\pi a^2)e^{-i\omega t}$, as in Eq. (7.1.11), the equation for the simple source in free space. The ratio between the two quantities

$$Z_s \simeq -i\omega\rho(4\pi a^3) - i\omega\rho(4\pi a^3)\left\{4\pi k^2 a^2 \sum_n \frac{(a^3/V)\Psi_n^2(\mathbf{r}_0)}{\Lambda_n(K_n a)^2[(K_n a)^2 - (ka)^2]}\right\}$$

$$(9.4.6)$$

is the effective mechanical impedance of the simple source.

Comparing this with Eq. (7.1.11), we see that the first term is the mass reactance, which is present when the simple source is in free space; it is the mass load of the air just outside the sphere. The second term is purely reactive if the dimensionless function in braces is real; this is real if the walls are rigid or the wall impedance is purely reactive. If the wall conductance is nonzero, however, both K_n and Ψ_n are complex quantities, and there is a real part to Z_s; some energy disappears. The series in the braces converges satisfactorily, because of the additional $(K_n a)^2$ term in the denominators, arising because the series for $1/4\pi a$ has been subtracted from series (9.4.2). In fact, if the driving frequency is not close to a resonance frequency, the term with the series is a small correction to the first, reactive term, since by definition a^3/V is small.

If, however, ω is near one of the values ω_n of Eq. (9.4.3) (the one for $n = N$, for instance), the Nth term in the series becomes

$$\rho\omega(4\pi a^3)\left(\frac{\omega}{\omega_N - i\gamma_N}\right)^3 \left\{\frac{a^3\Psi_N^2(\mathbf{r}_0)}{V\Lambda_N k^2 a^2}\left[\frac{2\pi\omega}{\gamma_N - i(\omega - \omega_N)}\right]\right\}$$

$$= \rho c(8\pi^2 a^2)\frac{k^3 a^3}{K_N^4 a^4}\frac{a^3\Psi_N^2(\mathbf{r}_0)\omega_N}{V\Lambda_N\gamma_N} \qquad \text{when } \omega = \omega_N$$

Because, usually, ω_N is considerably larger than γ_N, this term is usually considerably larger than any other term in the series of Eq. (9.4.6) [unless it happens that the source is at a node of the Nth standing wave and $\Psi_N(\mathbf{r}_0)$ is zero]. Exactly at resonance ($\omega = \omega_N$), the term is almost purely resistive, though there is some reactive part, because K_N and Ψ_N have imaginary parts. At high frequencies the resonance peaks overlap, and the second term of (9.4.6) approaches the smooth function $\rho c(4\pi a^2)(ka)^2$, typical of the simple source in free space, in a manner similar to that discussed in connection with Fig. 9.10.

Transient effects; reverberation

To work out the acoustic response of the cavity to point sources of nonperiodic sound we can utilize again the discussion resulting in Eq. (7.1.13). If the outflow of air from the source at $\mathbf{r}_0$ is $S(t)$, then the component S_ω for the frequency $\omega/2\pi$ is the Fourier transform of $-i\rho\omega S_\omega G_\omega$. For example, suppose the source function $S(t)$ is the step function $u(t)$, the source being quiescent until $t = 0$, after which it expels a unit volume of air per second. According to Eqs. (1.3.22), the Fourier transform of $u(t)$ is $i/2\pi\omega$, so the wave generated by the source is

$$p(t) = -\frac{\rho c^2}{2\pi} \int_{-\infty}^{\infty} \sum_n \frac{\Psi_n(\omega,\mathbf{r})\Psi_n(\omega,\mathbf{r}_0)e^{-i\omega t}}{V\Lambda_n(\omega)[\omega^2 - c^2 K_n^2(\omega)]} \, d\omega \qquad (9.4.7)$$

which can be evaluated by calculating residues, as indicated in Eqs. (1.2.15). According to Eq. (9.4.3), the poles of the integrand are at $\omega = \pm\omega_n - i\gamma_n$. Because of the symmetry of the wall impedance functions with respect to reversal of sign of ω, $\Psi_n(-\omega_n - i\gamma_n, \mathbf{r})\Psi_n(-\omega_n - i\gamma_n, \mathbf{r}_0)/\Lambda_n(-\omega_n - i\gamma_n)$ is the complex conjugate of $\Psi_n(\omega_n - i\gamma_n, \mathbf{r})\Psi_n(\omega_n - i\gamma_n, \mathbf{r}_0)/\Lambda_n(\omega_n - i\gamma_n)$. Therefore the pressure wave from an explosive source at $\mathbf{r}_0$ in the cavity is given by

$$p_s(\mathbf{r},t) = \begin{cases} 0 & t < 0 \\ \rho c^2 \sum_n A_n e^{-\gamma_n t} \sin(\omega_n t + \Phi_n) & t > 0 \end{cases}$$

where

$$\frac{\Psi_n(\omega_n - i\gamma_n, \mathbf{r})\Psi_n(\omega_n - i\gamma_n, \mathbf{r}_0)}{(\omega_n - i\gamma_n)V\Lambda_n(\omega_n - i\gamma_n)} = A_n e^{i\Phi_n} \qquad A_n, \Phi_n \text{ real}$$

The resulting wave in the cavity, after $t = 0$, consists of a series of standing waves, $\Psi_n(\omega_n - i\gamma_n, \mathbf{r})$, each oscillating with its own natural frequency ω_n and each damping out in time with its own damping factor γ_n. This kind of wave, started at $t = 0$ and subsequently dying out exponentially, is called *reverberation*. The length of time it takes the mean energy of the wave to reduce to a millionth part of its initial mean value is called the *reverberation time*, a much used measure of the room's acoustical properties. It will be discussed again in the next section.

If the point source generated a simple-harmonic wave from $t = -\infty$ to $t = 0$ and then stopped generating sound, reverberation would be present after $t = 0$. Suppose $S(t) = \sin(\omega_g t)$ from $t = -\infty$ to $t = 0$ and is zero from $t = 0$ to $t = +\infty$; the Fourier transform of $S(t)$ would be $\omega_g/2\pi(\omega^2 - \omega_g^2)$, where the contour of integration over ω goes slightly *below* the points $\omega = \pm\omega_g$, so that the poles are included in the integration only when $t < 0$. Then the resulting pressure wave is

$$p_s(\mathbf{r},t) = \frac{\rho c^2}{2\pi i} \omega_g \int_{-\infty}^{\infty} \sum_n \frac{\Psi_n(\omega,\mathbf{r})\Psi_n(\omega,\mathbf{r}_0)e^{-i\omega t}\omega \, d\omega}{(\omega^2 - \omega_g^2)V\Lambda_n(\omega)[\omega^2 - c^2 K_n^2(\omega)]}$$

where the contour goes *above* the points $\omega = \pm\omega_n - i\gamma_n$. The result is

$$p_s(\mathbf{r},t) = \begin{cases} \rho c^2 \omega_g \sum_n B_n \cos(\omega_g t - \Gamma_n) & t < 0 \\ \rho c^2 \omega_g \sum_n C_n e^{-\gamma_n t} \cos(\omega_n t + \Omega_n) & t > 0 \end{cases}$$

where

$$B_n e^{i\Gamma_n} = \frac{\Psi_n(\omega_g,\mathbf{r})\Psi_n(\omega_g,\mathbf{r}_0)}{[\omega_g{}^2 - c^2 K_n{}^2(\omega_g)]V\Lambda_n(\omega_g)}$$

and

$$C_n e^{-i\Omega_n} = \frac{\Psi_n(\omega_n - i\gamma_n, \mathbf{r})\Psi_n(\omega_n - i\gamma_n, \mathbf{r}_0)}{[\omega_g{}^2 - (\omega_n - i\gamma_n)^2]V\Lambda_n(\omega_n - i\gamma_n)}$$

Here the wave is periodic before $t = 0$, with frequency $\omega_g/2\pi$; after $t = 0$, when the source has ceased generating sound, the wave has the typical reverberant form, each individual standing wave oscillating with its own natural frequency and damping out in time with its own factor γ_n. It might appear from the two expressions that the pressure and related fluid velocity have discontinuities in value at $t = 0$. This is not so, as a detailed study of the relationships between the amplitudes B_n, C_n and the phase angles Γ_n, Ω_n will demonstrate. Specific examples of this behavior, for a room of particularly simple shape, will be worked out later in this section.

Nonuniform surface admittance

The normal modes in a room of simple shape, such as a perfect parallelepiped or a simple cylinder, have quite simple forms, with regularly spaced nodal surfaces and, for the cylindrical rooms, regions of concentration or avoidance of the wave. Such rooms have poor acoustic qualities, because pronounced echoes and focal regions can occur and because the nodal surfaces of a number of standing waves can coincide in certain regions, with consequent lack of certain components of speech or music. Such rooms can be improved by introducing nonuniformities of shape and/or of admittance of the boundary surface, thus breaking up the echoes and the regularity of spacing of the nodal surfaces. These nonuniformities can be considered as coupling the simple standing waves, so that each normal mode of the irregular room is a combination of several of the modes for the regular room, with a corresponding shift in resonance frequency and damping factor. The phenomenon is one of scattering, which was considered in Chap. 8. Here, however, the effects of interest are the shift in eigenvalues and change in shape of the normal modes, rather than the behavior of the scattered sound at large distances from the nonuniformity.

To compute these effects we start with a room of regular shape, for which we can compute the normal modes. For simplicity we assume the walls are all rigid; so the eigenfunctions $\phi_n(\mathbf{r})$ and eigenvalues $\eta_n{}^2$ are real quantities, independent of ω:

$$\nabla^2\phi_n + \eta_n^2\phi_n = 0 \qquad \text{within } R$$

$$\iiint \phi_n^2 \, dv = V\Lambda_n^0 \qquad \frac{\partial\phi_n}{\partial n} = 0 \text{ on } S \qquad (9.4.8)$$

We now ask what changes in the eigenfunctions and eigenvalues are produced if the boundary surface S is no longer rigid but has a specific acoustic admittance $\beta(\omega,\mathbf{r}_s)$ which may vary from point to point on S. The boundary condition now is that, at point $\mathbf{r}_s$ on S, the outward-pointing normal gradient of p must equal $ik\beta(\omega,\mathbf{r}_s)$ times p itself.

The most direct way of answering this question is by use of a Green's function of the general form of Eq. (9.4.2) for a uniform room with rigid walls. But we wish to use this Green's function to obtain a set of solutions of Eqs. (9.4.1) for a nonuniform room with nonrigid walls, for a frequency $\omega/2\pi \equiv kc/2\pi = K_Nc/2\pi$, where K_N is an eigenvalue for the nonuniform room. The equation for the Green's function is therefore

$$\nabla^2 G_K + K_N^2 G_K = -\delta(x - x_0)\,\delta(y - y_0)\,\delta(z - z_0) \qquad \frac{\partial G_K}{\partial n} = 0 \text{ on } S$$

Expressing this function in terms of the eigenfunctions ϕ_n of Eqs. (9.4.8), which satisfy the same boundary conditions, we find that

$$G_K(\mathbf{r} \mid \mathbf{r}_0) = \sum_n \frac{\phi_n(\mathbf{r})\phi_n(\mathbf{r}_0)}{V\Lambda_n(\eta_n^2 - K_N^2)} \qquad (9.4.9)$$

The equation relating the solution Ψ_N of Eq. (9.4.1) and this Green's function is obtained from Eqs. (7.1.17), the general relation between a Green's function and a solution of the corresponding homogeneous equation. In this case there is no source, so that the volume integral does not occur; and since $\partial G_K/\partial n = 0$ on S, only one of the surface-integral terms remains, resulting in an integral equation for Ψ_N, the eigenfunction for the room with nonrigid walls,

$$p(\mathbf{r}) = \iint_S G_K(\mathbf{r} \mid \mathbf{r}_0) \frac{\partial p}{\partial n_0} \, dS_0 = ik \iint_S G_K(\mathbf{r} \mid \mathbf{r}_0)\beta(\mathbf{r}_0)p(\mathbf{r}_0) \, dS_0 \quad (9.4.10)$$

where we have introduced the boundary conditions which p must satisfy on S.

This is a homogeneous integral equation, which includes both the differential equation and also the boundary conditions which p must satisfy. Its exact solution would yield the correct form for the eigenfunction Ψ_N, but also the correct value of the corresponding eigenvalue K_N. Unless K_N has this value, the only possible solution of Eq. (9.4.10) is $p = 0$ everywhere inside R.

The general properties of this equation and its solution can be more clearly demonstrated by a modification of its form. Suppose we imagine going from the simple room with rigid walls, for which the solution is ϕ_N and η_N, to the nonuniform room under consideration by multiplying $\beta(\mathbf{r}_s)$ by some parameter λ and letting λ go from 0 to 1. The eigenfunctions and eigenvalues will change smoothly from ϕ_N, η_N to the final ones Ψ_N, K_N as λ is increased. We separate out the term $n = N$ in the series for G_K and write Eq. (9.4.10) as

$$p(\mathbf{r}) = ik \iint G_N(\mathbf{r} \mid \mathbf{r}_s)\beta(\mathbf{r}_s)p(\mathbf{r}_s)\,dS + ik\left[\frac{\iint \phi_N(\mathbf{r}_s)\beta(\mathbf{r}_s)p(\mathbf{r}_s)\,dS}{V\Lambda_N^0(\eta_N{}^2 - K_N{}^2)}\right]\phi_N(\mathbf{r})$$

where

$$G_N(\mathbf{r} \mid \mathbf{r}_0) = \sum_{n \neq N} \frac{\phi_n(\mathbf{r})\phi_n(\mathbf{r}_0)}{V\Lambda_n^0(\eta_n{}^2 - K_N{}^2)}$$

G_N being the Green's function, for frequency $cK_N/2\pi$, for the rigid-walled room, with the term $n = N$ omitted.

Of course, when $\beta = 0$, $\eta_N{}^2 - K_N{}^2$ must go to zero and $p \to \phi_N$; in other words, p is zero unless $K_N = \eta_N$ when $\beta = 0$. When β is not zero, the first term on the right is not zero, and p differs from ϕ_N. The equation is homogeneous; so p can be multiplied by any constant factor and still be a solution. For convenience we choose that factor which allows the expression in square brackets to be equal to $1/ik$; we shall call this particular solution Ψ_N. Thus

$$\iint \phi_N(\mathbf{r}_s)\beta(\mathbf{r}_s)\Psi_N(\mathbf{r}_s)\,dS = \frac{1}{ik}V\Lambda_N^0(\eta_N{}^2 - K_N{}^2) \tag{9.4.11}$$

and the integral equation for $p \equiv \Psi_N$ becomes

$$\Psi_N(\mathbf{r}) = \phi_N(\mathbf{r}) + ik\iint G_N(\mathbf{r} \mid \mathbf{r}_s)\beta(\mathbf{r}_s)\Psi_N(\mathbf{r}_s)\,dS \tag{9.4.12}$$

It is obvious now that when $\beta \to 0$, then $\Psi_N \to \phi_N$, and $K_N{}^2 = \eta_N{}^2$.

Variational calculation

A first-approximation solution can be obtained for Ψ_N by setting $\Psi_N \simeq \phi_N$ in the integral, but a better approximation can be obtained by using a variational procedure. We multiply both sides of Eq. (9.4.12) by $\Psi_N(r)\beta(r)$ and integrate over S, obtaining

$$0 = \int \phi_N\beta\Psi_N + ik\int \Psi_N\beta\int G_N\beta\Psi_N - \int \Psi_N\beta\Psi_N$$

in our usual shorthand. Adding Eq. (9.4.11) to this, we obtain

$$V\Lambda_N^0(\eta_N{}^2 - K_N{}^2) = 2ik\int \phi_N\beta\Psi_N - k^2\int \Psi_N\beta\int G_N\beta\Psi_N - ik\int \Psi_N\beta\Psi_N \tag{9.4.13}$$

If the correct form for Ψ'_N is inserted in this equation the value of K_N computed from the result will be the correct value of the corresponding eigenvalue for the nonuniform room. Note that this is an implicit relationship, for $ik\beta$ is a function of ω and, for the correct solution, the ω used in $ik\beta$ must be $K_N c = \omega_N - i\gamma_N$ in accord with Eq. (9.4.3).

If Ψ'_N is varied on the right-hand side, the coefficient of Ψ'_N will be zero if Ψ'_N satisfies Eq. (9.4.12); the argument then proceeds as on page 532. A reasonable trial function for Ψ'_N when β is small would be $A\phi_N$, where A is adjusted so that the derivative of the right-hand side of (9.4.13) is zero. This results in a value for K_N^2 and a form for Ψ'_N, good to second order in the small quantity β. Application of the variational procedure gives

$$A \simeq \left(1 - ik\frac{\int \phi_N\beta \int G_N\beta\phi_N}{\int \phi_N\beta\phi_N}\right)^{-1} \qquad \Psi_N \simeq \phi_N + ikA\int G_N\beta\phi_N$$

$$K_N^2 \simeq \eta_N^2 - \frac{ikA}{V\Lambda_N^0}\int \phi_N\beta\phi_N \tag{9.4.14}$$

or

$$K_N \simeq \eta_N\left[1 - \frac{ik\int \phi_N\beta\phi_N}{2V\Lambda_N^0\eta_N^2} + \frac{k^2\int \phi_N\beta \int G_N\beta\phi_N}{2V\Lambda_N^0\eta_N^2} + \frac{1}{8}\left(\frac{k\int \phi_N\beta\phi_N}{V\Lambda_N^0\eta_N^2}\right)^2\right]$$

If a first-order approximation is sufficient, we can use

$$K_N(\omega) \simeq \eta_N - \frac{k}{2V\Lambda_N^0\eta_N}\iint \phi_N^2(\mathbf{r}_s)[\sigma(\mathbf{r}_s) + i\xi(\mathbf{r}_s)]\,dS \tag{9.4.15}$$

$$\Psi_N(\omega,\mathbf{r}) \simeq \phi_N(\mathbf{r}) + k\iint G_N(\mathbf{r}\,|\,\mathbf{r}_s)(\sigma + i\xi)\phi_N(\mathbf{r}_s)\,dS$$

where $\beta = \xi - i\sigma$ is a function of ω and of $\mathbf{r}_s$. Note that these are second or first approximations to the eigenfunctions and eigenvalues defined in Eqs. (9.4.1) and used in Eqs. (9.4.2), (9.4.6), and (9.4.7). Since $k\beta$ is a function of ω, the resonance frequency and damping factor γ_N for free vibration of the Nth standing wave in the room are obtained by solving Eq. (9.4.3), using the $K_N(\omega)$ of Eqs. (9.4.14) or (9.4.15). When used to represent simple-harmonic forced motion, as in Eq. (9.4.6) or (9.4.7), ω equals 2π times the driving frequency.

We see that, to the first approximation, the change in the eigenvalue K_N, caused by the fact that $\beta \neq 0$, is proportional to the average value of $i\beta$ over the walls, weighted by ϕ_N^2 so that the wall areas, where the Nth standing wave is large, are emphasized. The imaginary part of cK_N, which equals the damping factor γ_N when the ω in $k = \omega/c$ and in β is set equal to $(\omega_N - i\gamma_N)$, is proportional, to the first order, to the mean value of the wall conductance ξ, similarly weighted. The effectiveness of the distribution of absorbing material in coupling other normal modes to the Nth one is measured by the magnitudes of the integrals $\iint \phi_n \beta \phi_N \, dS$ entering $\iint G_N \beta \phi_N \, dS$ in the expression for Ψ'_N. If β is nearly uniform, most of the integrals $\iint \phi_n \beta \phi_N \, dS \, (n \neq N)$ will be quite small, and Ψ'_N will not differ much from the uniform eigenfunction ϕ_N. We shall work out examples of these equations shortly.

An added complication occurs when state N is *degenerate*, i.e., when s different eigenfunctions $\phi_N, \phi_{N+1}, \ldots, \phi_{N+s-1}$, all have the *same* eigenvalue, $\eta_N = \eta_{N+1} = \cdots = \eta_{N+s-1}$. This can occur when the cavity is particularly symmetrical in shape. In such a case, the separation of the Nth term in G_K, to form the G_N function of Eq. (9.4.12), does not separate off all the terms with the small quantity $\eta_N^2 - K_N^2$ in the denominator; G_N still has terms with $n = N + 1, N + 2, \ldots, N + s - 1$, which are as large as the term separated off.

When this happens the appropriate procedure is to form s linear combinations,

$$\chi_\nu(\mathbf{r}) = A_{\nu 1} \phi_N(\mathbf{r}) + A_{\nu 2} \phi_{N+1}(\mathbf{r}) + \cdots + A_{\nu s} \phi_{N+s-1}(\mathbf{r}) \qquad \nu = 1, 2, \ldots, s$$

such that

$$\iint \chi_\nu(\mathbf{r}_s) \beta(\mathbf{r}_s) \chi_\mu(\mathbf{r}_s) \, dS = 0 \qquad \mu \neq \nu$$

and then to use the χ_ν's instead of the $\phi_N, \ldots, \phi_{N+s-1}$ in G_N and in Eqs. (9.4.14). The χ's are just as good eigenfunctions as the ϕ's, for they all have the same eigenvalue, η_N. But the χ's have the property that all the terms in the integral $\int \chi_\nu \beta \int G_N \beta \chi_\nu$, involving the products $\chi_\mu(\mathbf{r}) \chi_\mu(\mathbf{r}_0)$ $(\mu \neq \nu)$ in G_N, are zero, thus eliminating the embarrassingly large terms in this integral.

The coefficients for the different linear combinations are found by solving the system of equations

$$\sum_{i=1}^{s} A_{\nu i} \iint \phi_{N+i-1} \beta \phi_{N+j-1} \, dS = B_\nu A_{\nu j} \qquad j = 1, 2, \ldots, s$$

for the s different roots $B_1, \ldots, B_s$ and corresponding sets of coefficients $A_{1j}, \ldots, A_{sj}$ $(j = 1, \ldots, s)$. Methods of solution and properties of the solutions will be found in any text on matrix analysis. The χ's are then used instead of the ϕ's in Eqs. (9.4.14).

An interesting physical interpretation of the equation for the damping factor γ_N arises from the fact that $V\Lambda_n^0$ is the integral of the square of the pressure amplitude throughout the rigid room, and is thus approximately proportional to the total energy of the Nth standing wave. The quantity $\xi(\mathbf{r}_s) \phi_N^2(\mathbf{r}_s)$ is similarly proportional to the power lost per unit area at $\mathbf{r}_s$ on the surface. Therefore, according to Eqs. (9.4.15), the damping factor γ_N is proportional to the fraction of the total energy of the standing wave which is lost to the walls per second, a not unreasonable result.

Change of room shape

The room can be made nonuniform by modifying its shape, instead of (or in addition to) making the wall admittance nonuniform. For the appropriate equation we return to the integral equation (7.1.17) for p, in terms of the Green's function of Eq. (9.4.9) for the uniform room. In this case the surface over which the integration is made, S_1, is at least in part different from S. In fact, part of S_1 may be a completely separate closed surface, corresponding to the insertion of an obstacle to the sound waves in the interior of the room.

The integral equation, taking into account the admittance as well as the change in shape, is then

$$p(\mathbf{r}) = \iint\limits_{S_1} p(\mathbf{r}_s)\left[ik\beta(\mathbf{r}_s)G_K(\mathbf{r}\,|\,\mathbf{r}_s) - \frac{\partial}{\partial n_s} G_K(\mathbf{r}\,|\,\mathbf{r}_s)\right] dS$$

Operating on this as we did on Eq. (9.4.10), we obtain, for the normal mode corresponding to ϕ_N in the uniform room,

$$\Psi_N(\omega,\mathbf{r}) = \phi_N(\mathbf{r}) + \iint\limits_{S_1} \Psi_N(\omega,\mathbf{r}_s)\left[ik\beta(\omega,\mathbf{r}_s)G_N(\mathbf{r}\,|\,\mathbf{r}_s)\right.$$

$$\left. - \frac{\partial}{\partial n_s} G_N(\mathbf{r}\,|\,\mathbf{r}_s)\right] dS$$

$$\eta_N{}^2 - K_N{}^2 = \frac{1}{V\Lambda_N^0} \iint\limits_{S_1} \Psi_N(\omega,\mathbf{r}_s)\left[ik\beta(\omega,\mathbf{r}_s)\phi_N(\mathbf{r}_s)\right.$$

$$\left. - \frac{\partial}{\partial n_s} \phi_N(\mathbf{r}_s)\right] dS$$

(9.4.16)

where the value of K_N which satisfies the second equation is the eigenvalue $K_N(\omega)$, corresponding to $\Psi_N(\omega,\mathbf{r})$, for the driving frequency $\omega/2\pi$.

One should note that the integral equations (9.4.12) and (9.4.16) have the same form as integral equations (8.3.2) and (8.3.8) for the scattering of sound from admittance and/or shape irregularities in an infinite plane wall. In the present case the "incident" wave p_i is the unperturbed wave ϕ_N, which is produced by the rigid walls of the uniform room; the Green's function G_N differs from the free-space function g_ω because of the reflections from the enclosing walls. Thus the scattered wave, the second term on the right of the integral equation, corresponds to the scattering of the combination of incident and reflected waves which make up the normal mode ϕ_N, instead of the single incident-plus-reflected plane wave dealt with in Sec. 8.3. In the present case we are not as much interested in the form of the scattered wave as we are in the difference between the eigenvalue K_N for the nonuniform cavity and the value η_N for the uniform cavity, which is given by an integral involving the value of the correct eigenfunction Ψ_N at the scattering surface.

The effect of a scattering object

The effect of a small object in the interior of the room on the eigenfunctions and eigenvalues is even more clearly related to the scattering formulas of Chap. 8. Suppose, for example, the acoustic properties of this foreign object (a drape hung away from a wall or an object hung from the ceiling, say) can be expressed in terms of its effective density ρ_e and compressibility κ_e, as was done in Sec. 8.1 and 8.2. In this case the equation for the pressure distribution has a volume, as well as a surface integral.

In this example the walls of the room are rigid and regular, so that the standing waves, in the absence of the scattering object, are the ϕ_N and η_N of Eqs. (9.4.8). The scattering object, of density ρ_e and compressibility κ_e, occupies region R_1, of volume V_1, considerably smaller than the volume V of the room. The integral equation for this situation is obtained as Eq. (8.1.13) was; the surface integrals over the rigid walls vanish, and the remaining integral is the volume integral over R_1.

$$p(\mathbf{r}) = \iiint\limits_{R_1} [k^2 \delta_\kappa p(\mathbf{r}_1) G_K(\mathbf{r} \mid \mathbf{r}_1) + \delta_\kappa (\mathrm{grad}_1 p) \cdot (\mathrm{grad}_1 G_K)] \, dv_1$$

with G_K given by Eq. (9.4.9), and $\delta_\kappa = [(\kappa_e - \kappa)/\kappa]$, $\delta_\rho = [(\rho_e - \rho)/\rho_e]$.

To simplify the intermediate equations we define the symbol $Q(f,F)$, for an arbitrary function $F(\mathbf{r})$ and for a function $f(\mathbf{r})$ which is a solution of the equation $\nabla^2 f + (\omega^2 \kappa_e \rho_e) f = 0$ inside R_1, as the integral

$$Q(f,F) = \iiint\limits_{R_1} [k^2 \delta_\kappa f(\mathbf{r}_1) F(\mathbf{r}_1) + \delta_\rho (\mathrm{grad}_1 f) \cdot (\mathrm{grad}_1 F)] \, dv_1$$

$$= \iiint\limits_{R_1} [k^2 \delta_c f(\mathbf{r}_1) + (\mathrm{grad}_1 f) \cdot (\mathrm{grad}_1 \delta_\rho)] F(\mathbf{r}_1) \, dv_1 \qquad (9.4.17)$$

where $\delta_c = (c^2 - c_e^2)/c_e^2 = (\rho_e \kappa_e / \rho \kappa) - 1$, and $k^2 = (\omega/c)^2 = \omega^2 \rho \kappa$.

The second form is obtained from the first by integration by parts, using

$$\mathrm{div}\,(\delta_\rho F \,\mathrm{grad}\, f) = \delta_\rho (\mathrm{grad}\, F) \cdot (\mathrm{grad}\, f) + F(\mathrm{grad}\, \delta_\rho) \cdot (\mathrm{grad}\, f)$$

$$- k^2 \frac{\rho_e \kappa_e}{\rho \kappa} \delta_\rho f F$$

and extending the integration far enough beyond R_1 so that δ_ρ is zero and the surface integral arising from the divergence term is zero. Thus the second form includes an integral over the surface region of R_1, where the density and compressibility change suddenly from their exterior values ρ, κ to their interior values ρ_e, κ_e. Integrals Q have the dimension of length if f and F are dimensionless.

In terms of these symbols we can write the integral equation for the pressure wave in the room,

$$p(\mathbf{r}) = \sum_n \frac{\phi_n(\mathbf{r})Q(\phi_n,p)}{V\Lambda_n^0(\eta_n{}^2 - K^2)} = \iiint\limits_{R_1} G_K(\mathbf{r}\,|\,\mathbf{r}_1)(k^2\delta_c p + \operatorname{grad} p \cdot \operatorname{grad} \delta_\rho)\,dv_1$$

and carry out the manipulations we developed earlier, to obtain equations for Ψ'_N, K_N, the eigenfunction and eigenvalue for the room containing the scattering object in region R_1. As before, we define the magnitude of Ψ'_N by an equation, which then becomes the means of computing K_N, after we solve the second equation.

$$V\Lambda_N^0(K_N{}^2 - \eta_N{}^2) = Q(\phi_N,\Psi'_N)$$

and

$$\Psi'_N(\mathbf{r}) = \phi_N(\mathbf{r}) + \sum_{n \neq N} \frac{\phi_n(\mathbf{r})Q(\phi_n,\Psi'_N)}{V\Lambda_n^0(\eta_n{}^2 - K_N{}^2)} \tag{9.4.18}$$

$$= \phi_n(\mathbf{r}) + \iiint\limits_{R_1} G_N(\mathbf{r}\,|\,\mathbf{r}_1)(k^2\delta_c\Psi'_N + \operatorname{grad}\Psi'_N \cdot \operatorname{grad}\delta_\rho)\,dv_1$$

where G_N is as defined earlier.

This pair of equations is amenable to approximate solution by variational means. Since $G_N(\mathbf{r}\,|\,\mathbf{r}_1)$ is symmetric in $\mathbf{r}$, $\mathbf{r}_1$, we multiply both sides of the second equation by

$$H(\Psi'_N) = [k^2\delta_c\Psi'_N(\mathbf{r}) + \operatorname{grad}\Psi'_N \cdot \operatorname{grad}\delta_\rho]$$

and integrate over R_1 in the $\mathbf{r}$ coordinates, obtaining

$$0 = Q(\Psi'_N,\phi_N) + \sum_{n \neq N} \frac{[Q(\Psi'_N,\phi_n)]^2}{V\Lambda_n^0(\eta_n{}^2 - K_N{}^2)} - Q(\Psi'_N,\Psi'_N)$$

$$= \int \phi_N H(\Psi'_N) + \int H(\Psi'_N)\int G_N H(\Psi'_N) - \int \Psi'_N H(\Psi'_N)$$

Adding the first of Eqs. (9.4.18) to this, we obtain the variational expression

$$V\Lambda_N^0(\eta_N{}^2 - K_N{}^2) = 2Q(\Psi'_N,\phi_N) + \sum_{n \neq N} \frac{[Q(\Psi'_N,\phi_n)]^2}{V\Lambda_n^0(\eta_n{}^2 - K_N{}^2)} - Q(\Psi'_N,\Psi'_N)$$

$$= 2\int \phi_N H(\Psi'_N) + \int H(\Psi'_N)\int G_N H(\Psi'_N) - \int \Psi'_N H(\Psi'_N) \tag{9.4.19}$$

Since Q, in its first form in Eq. (9.4.17), is seen to be symmetric in f and F, both integrals which are quadratic in Ψ'_N are symmetric, and the variational techniques we have used before can be used here.

For example, if the scattering object is a compact shape, like a sphere or a cube, we could assume that the pressure wave inside it is not too different from the unperturbed normal mode ϕ_N, except for a constant factor, and use $A\phi_N$ for the trial function instead of Ψ'_N in the integrals. In this case the variational solution is

$$A \simeq \left[1 - \sum_{n \neq N} \frac{[Q(\phi_N,\phi_N)]^2}{V\Lambda_n^0(\eta_n^2 - \eta_N^2)Q(\phi_N,\phi_N)} \right]^{-1}$$

$$K_N^2 - \eta_N^2 \simeq \frac{A}{V\Lambda_N^0} Q(\phi_N,\phi_N) = \frac{A}{V\Lambda_N^0} \iiint_{R_1} (k^2\delta_\kappa \phi_N^2 + \delta_\rho \,|\mathrm{grad}\,\phi_N|^2)\, dv$$

$$\tag{9.4.20}$$

$$\Psi'_N(\mathbf{r}) \simeq \phi_N(\mathbf{r}) + A \sum_{n \neq N} \frac{Q(\phi_N,\phi_n)}{V\Lambda_n^0(\eta_n^2 - \eta_N^2)} \phi_n(\mathbf{r})$$

where we have inserted for K_N^2 its "zeroth approximation" η_N^2 in the denominators of the (presumably) small terms in the summations. The solution is quite accurate when the dimensions of the scattering region R_1 are not larger than a half wavelength. For larger objects the error may be considerable unless δ_κ and δ_ρ are small. We have shown in Sec. 8.2 that if the object is a sphere, some improvement in accuracy can be attained if we substitute $(3\rho_e - 3\rho)/(2\rho_e + \rho)$ instead of δ_ρ in the equations.

The equation for the resonance frequency and damping factor of Eq. (9.4.3), in the presence of the scattering object, is more clearly stated in terms of the rms pressure-amplitude distribution $\phi_N(\mathbf{r})$ and the velocity-amplitude distribution $\mathbf{U}_N(\mathbf{r}) = (1/k\rho c)\,\mathrm{grad}\,\phi_n$ of the Nth unperturbed wave,

$$K_N^2 \simeq \eta_N^2 \left[1 - \frac{k^2 A}{\eta_N^2 V\Lambda_N^0} \iiint_{R_1} (\delta_\rho \,|\rho c\mathbf{U}_N|^2 + \delta_\kappa \phi_N^2)\, dv \right]$$

$$\rightarrow \eta_N^2 - \frac{k^2 V_1}{V\Lambda_N^0} \langle \rho^2 c^2 U_N^2 \rangle \delta_\rho - \frac{k^2 V_1}{V\Lambda_N^0} \langle \phi_N^2 \rangle \delta_\kappa \qquad V_1 \ll \lambda^3 \quad (9.4.21)$$

where

$$\langle f \rangle = (1/V_1) \iiint_{R_1} f\, dv$$

The second expression, valid for wavelengths longer than the dimensions of R_1, indicates that the effect of the scattering object on the wavenumber K_N of the Nth mode is proportional to the ratio between the volume V_1 of the scatterer and the volume V of the room. One part of the correction is proportional to the density difference $\delta_\rho = (\rho_e - \rho)/\rho_e$ times the mean-square velocity of the unperturbed wave at the location of the scatterer; the other part is proportional to the fractional compressibility difference $\delta_\kappa = (\kappa_e - \kappa)/\kappa$ times the mean square of the pressure there. If the object is heavy and incompressible ($\rho_e > \rho$, $\kappa_e < \kappa$), the first part tends to *decrease* K_N

because the object's immobility causes the air near it to move faster to get around it, which increases the effective mass of the wave; the second part tends to *increase* K_N because the object is stiffer than the displaced air, which increases the effective stiffness of the wave. If the object is near a pressure node (where $\rho c U_N > \phi_N$), the former predominates and $K_N < \eta_N$; if the object is near a pressure maximum for the unperturbed standing wave, the latter effect predominates and $K_N > \eta_N$. Unless ρ_e or κ_e is complex, K_N is real.

If the object is porous, with porosity Ω and flow resistance Φ, we can use the results of Eqs. (8.2.20) and (8.2.21) and substitute $(\kappa_p \Omega - \kappa)/\kappa$ and $\frac{3}{4} + \frac{3}{4}i(\rho c k/\Phi)$ for δ_κ and δ_ρ, respectively, in Eq. (9.4.21) (if the object is roughly spherical). We then see that, for a small object, the imaginary part of K_N is

$$\text{Im } K_N \simeq \frac{3}{8} \frac{\rho c k^3}{\eta_N \Phi} \frac{V_1}{V \Lambda_N^0} \langle \rho^2 c^2 U_N{}^2 \rangle \qquad (9.4.22)$$

According to Eq. (9.4.3), the damping constant for free vibration of the Nth standing wave, in the presence of the porous object, would be approximately c times the value of this expression when k is set equal to η_N. Since a porous object absorbs energy because of the air flow through its pores, it should be placed at a velocity maximum if this damping constant is to be as large as possible.

If there are a number of similar scattering regions placed at random about the room, their effect on K_N will be additive as long as their total volume is small compared with V. For a large number of objects, multiple scattering effects arise, which produce equivalents to the coherent and incoherent scattering discussed at the end of Sec. 8.2. The coherent effect is to modify the effective density and compressibility of the air in the room as a whole; the incoherent scattering is a differential effect, producing distortion of the standing waves.

Finally, we see that the presence of the scattering object tends to couple the normal modes of the unperturbed room, so that each eigenfunction $\Psi_N(\mathbf{r})$ is a mixture of many of the simple waves $\phi_N(\mathbf{r})$. To the first approximation the coefficient of coupling between the Nth mode and the nth simple wave is

$$\frac{Q(\phi_N, \phi_n)}{V \Lambda_n^0 (\eta_n{}^2 - \eta_N{}^2)} \rightarrow \frac{k^2 V_1}{V \Lambda_N^0 (\eta_n{}^2 - \eta_N{}^2)} [\delta_\rho \langle \rho^2 c^2 \mathbf{U}_n \cdot \mathbf{U}_N \rangle \\ + \delta_\kappa \langle \phi_n \phi_N \rangle] \qquad V_1 \ll \lambda^3 \quad (9.4.23)$$

the compressibility-coupling term being proportional to the product of the pressure amplitudes of the two coupled waves, and the density-coupling term being proportional to the scalar product of the two standing-wave velocities, both at the position of the scatterer.

Scattering by a porous panel

If the scattering object is a panel of thickness h and area S_1, oriented perpendicular to a unit normal vector $\mathbf{a}_1$ (this could be a piece of drapery, for example, hung away from the walls), the variational expression (9.4.19) still is applicable, but a different choice of trial function is required. In this case there is a tendency for a pressure drop from one side of the panel to the other, particularly if the panel is denser than the air, and thus does not move as easily. The mean pressure, on a plane halfway between the surfaces, may be roughly proportional to the pressure which would be there if no panel were present, but the pressure at either surface, a distance $\pm \frac{1}{2}h$ from the central plane, may be larger or smaller than the unperturbed pressure there.

A simple form which has this property, inside the panel, is the trial function $(A + Bx_1)\phi_N$, where x_1 is the distance, in the direction of the unit normal vector $\mathbf{a}_1$, from the central plane. This is not a good representation of the pressure distribution throughout the room, but the integrals of Eq. (9.4.19) are concerned only with the pressures inside the scattering region, where this function is a reasonable, as well as a simple, representation. Although the range of x_1 is only from $-\frac{1}{2}h$ to $+\frac{1}{2}h$ (with h small), the pressure difference between the two sides may be large; so B may be large. Thus we can expect that the gradient of the trial function,

$$\text{grad}\,[(A + Bx_1)\phi_N] = (A + Bx_1)\,\text{grad}\,\phi_N + B\phi_N\mathbf{a}_1$$

would have the second term, $B\phi_N\mathbf{a}_1$, much larger than the first; i.e., we should expect that the motion of the panel material (or of the air in its pores) would be predominantly in the direction of the vector $\mathbf{a}_1$, in response to the pressure drop between the surfaces.

Thus the integrals entering into the variational expression will have the approximate values, to the second order in h,

$$Q(\Psi_N,\phi_n) \simeq Ak^2h\delta_\kappa \iint \phi_N\phi_n\,dS_1 + Bh\delta_\rho\mathbf{a}_1 \cdot \iint \phi_N\,\text{grad}\,\phi_n\,dS_1$$

$$Q(\Psi_N,\Psi_N) \simeq A^2k^2h\delta_\kappa \iint \phi_N{}^2\,dS_1 + B^2h\delta_\rho \iint \phi_N{}^2\,dS_1 \qquad (9.4.24)$$

where the integrals over dS_1 are over the area of the central plane of the panel. The terms in h^3 and higher powers are neglected. The variational expression thus becomes

$$V\Lambda_N^0(\eta_N{}^2 - K_N{}^2) \simeq (2A - A^2)k^2h\delta_\kappa S_N + h\delta_\rho(2B\mathbf{a}_1 \cdot \mathbf{W}_N - B^2S_N)$$
$$+ A^2(k^2h\delta_\kappa)^2\Sigma_{11} + 2ABk^2h^2\delta_\kappa\delta_\rho\Sigma_{12} + B^2h^2\delta_\rho{}^2\Sigma_{22}$$

where

$$S_N = \iint \phi_N{}^2 \, dS_1 \qquad \mathbf{W}_N = \iint \phi_N \operatorname{grad} \phi_N \, dS_1$$

$$\Sigma_{11} = \sum_{n \neq N} \frac{(\iint \phi_N \phi_n \, dS_1)^2}{V\Lambda_n^0(\eta_n{}^2 - \eta_N{}^2)}$$

$$\Sigma_{12} = \sum_{n \neq N} \frac{\mathbf{a}_1 \cdot \iint \phi_N \operatorname{grad} \phi_n \, dS_1}{V\Lambda_n^0(\eta_n{}^2 - \eta_N{}^2)} \iint \phi_N \phi_n \, dS_1$$

$$\Sigma_{22} = \sum_{n \neq N} \frac{(\mathbf{a}_1 \cdot \iint \phi_N \operatorname{grad} \phi_n \, dS_1)^2}{V\Lambda_n^0(\eta_n{}^2 - \eta_N{}^2)}$$

all the integrals being over the area S_1 of the central plane of the panel.

Differentiating this with respect to both A and B and equating both derivatives to zero, then solving for A and B, we finally obtain the second-order solution.

$$A \simeq \frac{1}{\Delta} [S_N{}^2 - h\delta_\rho(S_N \Sigma_{22} - \mathbf{a}_1 \cdot \mathbf{W}_N \Sigma_{12})]$$

$$B \simeq \frac{1}{\Delta} [S_N \mathbf{a}_1 \cdot \mathbf{W}_N - k^2 h \delta_\kappa(\mathbf{a}_1 \cdot \mathbf{W}_N \Sigma_{11} - S_N \Sigma_{12})]$$

$$\Delta = (S_N - k^2 h \delta_\kappa \Sigma_{11})(S_N - h\delta_\rho \Sigma_{22}) - k^2 h^2 \delta_\kappa \delta_\rho \Sigma_{12} \qquad (9.4.25)$$

$$K_N{}^2 \simeq \eta_N^2 - \frac{A}{V\Lambda_n^0} k^2 h \delta_\kappa S_N + \frac{B}{V\Lambda_n^0} h\delta_\rho \mathbf{a}_1 \cdot \mathbf{W}_N$$

$$\Psi_N \simeq \phi_N(\mathbf{r}) + Ak^2 h \delta_\kappa \sum_{n \neq N} \frac{\iint \phi_N \phi_n \, dS_1}{V\Lambda_n^0(\eta_n{}^2 - \eta_N{}^2)} \phi_n(\mathbf{r})$$
$$+ Bh\delta_\rho \mathbf{a}_1 \cdot \sum_{n \neq N} \frac{\iint \phi_N \operatorname{grad} \phi_n \, dS_1}{V\Lambda_n^0(\eta_n{}^2 - \eta_N{}^2)} \phi_n(\mathbf{r})$$

When the terms in h^2 can be neglected, the expressions for A, B, and K_N become more manageable.

$$A \simeq 1 + \frac{k^2 h \delta_\kappa}{S_N} \Sigma_{11} + \frac{h\delta_\rho \mathbf{a}_1 \cdot \mathbf{W}_N}{S_N{}^2} \Sigma_{12}$$

$$B \simeq \frac{\mathbf{a}_1 \cdot \mathbf{W}_N}{S_N} + \frac{k^2 h \delta_\kappa}{S_N} \Sigma_{12} + \frac{h\delta_\rho \mathbf{a}_1 \cdot \mathbf{W}_N}{S_N{}^2} \Sigma_{22} \qquad (9.4.26)$$

$$K_N \simeq \eta_N \left[1 - \frac{hS_1 k^2}{2V\Lambda_N^0 \eta_N^2} \frac{\delta_\kappa \langle \phi_N{}^2 \rangle^2 + (\delta_\rho/k^2)(\mathbf{a}_1 \cdot \langle \phi_N \operatorname{grad} \phi_N \rangle)^2}{\langle \phi_N{}^2 \rangle} \right]$$

where $\langle F \rangle$ is the average value of F over the central plane of the panel. Finally, if the panel is small enough with respect to the wavelength, or is located so that $\langle \phi_N \operatorname{grad} \phi_N \rangle \simeq \langle \phi_N \rangle \langle \operatorname{grad} \phi_N \rangle$ and $\langle \phi_N{}^2 \rangle \simeq \langle \phi_N \rangle^2$, we have

$$K_N{}^2 \simeq \eta_N^2 \left[1 - \frac{hS_1 k^2}{V\Lambda_N^0 \eta_N^2} (\delta_\kappa \langle \phi_N \rangle^2 + \delta_\rho \cos^2 \vartheta_N \, |\langle \rho c U_N \rangle|^2) \right] \qquad (9.4.27)$$

where $\langle U_N \rangle$ is the mean value of the velocity amplitude of the Nth unperturbed wave, averaged over the central plane of the panel; $|\langle U \rangle|$ is the magnitude of this vector, and ϑ_N is the angle it makes with the unit vector $\mathbf{a}_1$, normal to the surface.

Thus the major difference between this equation, for the panel, and Eq. (9.4.21) for a near-spherical object lies in the factor $\cos^2 \vartheta_N$. If the panel is parallel to the air flow in the unperturbed wave, there will be no pressure drop across it and no motion normal to its surface; its only effect will be that caused by the compressibility difference δ_κ. The maximum effect of the δ_ρ term comes when the panel is perpendicular to the motion of the air in the Nth standing wave ϕ_N. When the panel is porous, with porosity Ω and flow resistance Φ, $\delta_\kappa \simeq (\kappa_p \Omega - \kappa)/\kappa$ and $\delta_\rho \simeq 1 + \frac{1}{2}i(\rho c k/\Phi)$. The contribution to the imaginary part of K_N is then

$$\operatorname{Im} K_N \simeq -\frac{1}{4}\frac{\rho c k^3}{\eta_N \Phi}\frac{h S_1}{V \Lambda_n^0}\langle|\rho c U_N|^2 \cos^2 \vartheta_N\rangle \qquad (9.4.28)$$

The rectangular room

The general formulas we have just developed will be illustrated by one example, the case when the unperturbed room is a rectangular parallelepiped of sides l_x, l_y, l_z, with rigid walls. The characteristic values and functions, which we use to build solutions for modified rooms, are

$$\phi_n(\mathbf{r}) = \cos\frac{\pi n_x x}{l_x}\cos\frac{\pi n_y y}{l_y}\cos\frac{\pi n_z z}{l_z}$$

$$\eta_n^2 = \left(\frac{\pi n_x}{l_x}\right)^2 + \left(\frac{\pi n_y}{l_y}\right)^2 + \left(\frac{\pi n_z}{l_z}\right)^2 \qquad (9.4.29)$$

$$V = l_x l_y l_z \qquad \Lambda_n^0 = \frac{1}{\epsilon_{n_x}\epsilon_{n_y}\epsilon_{n_z}} \qquad n_x, n_y, n_z = 0, 1, 2, \ldots$$

where the origin of coordinates is at one corner of the room.

If the walls are not rigid but have uniform specific admittance $\beta = \xi - i\sigma$, we can use the formulas of Eq. (9.2.14) and Plate IV to obtain eigenfunctions and eigenvalues for the driving frequency $\omega/2\pi = kc/2\pi$.

$$\Psi_n(\mathbf{r}) = \frac{\cos}{\sin}\left(\frac{\pi q_{xn_x}x}{l_x}\right)\frac{\cos}{\sin}\left(\frac{\pi q_{yn_y}y}{l_y}\right)\frac{\cos}{\sin}\left(\frac{\pi q_{zn_z}z}{l_z}\right)$$

$$K_n^2 = \left(\frac{\pi q_{xn_x}}{l_x}\right)^2 + \left(\frac{\pi q_{yn_y}}{l_y}\right)^2 + \left(\frac{\pi q_{zn_z}}{l_z}\right)^2 \qquad (9.4.30)$$

$$\Lambda_n = L_x L_y L_z \qquad L_x = \tfrac{1}{2} \pm \frac{1}{2\pi q_{xn_x}}\sin\left(\pi q_{xn_x}\right)$$

where, in this case, the origin is at the center of the room. The cosine function is used in the first line and the plus sign in the third line when the corresponding n_x, n_y, or n_z is an even integer; the sine function and the minus sign when the corresponding n is an odd integer.   The values of $q = \mu - i\nu$ for each of the three pairs of walls are found graphically in Plate IV by determining the point corresponding to the values of $h = |\beta|\, kl/2\pi$, φ in the region corresponding to the choice of n (for $n = 0$ use the lower part of the right-hand chart; for $n = 1$ use the lower part of the left-hand chart; for $n = 2$ use the upper part of the right-hand chart, and so on).   Here $\beta = \xi - i\sigma = \beta e^{i\varphi}$ is the wall admittance at the frequency $\omega/2\pi$, and φ is the phase angle for the impedance ($\varphi = 90°$ if the impedance is masslike and $\varphi = -90°$ if the impedance is stiffness-controlled).

If β is constant over each wall and βkl is a small quantity, the results of Eq. (9.2.15) can be extended to provide a first-order solution.

$$q_{x0} \simeq \frac{1}{\pi i}\sqrt{ikl_x(\beta_{x0} + \beta_{x1})} \qquad q_{xn} \simeq n - i\frac{kl_x}{\pi^2 n}(\beta_{x0} + \beta_{x1})$$

$$K_n^2 \simeq \left(\frac{\pi q_{xn_x}}{l_x}\right)^2 + \left(\frac{\pi q_{yn_y}}{l_y}\right)^2 + \left(\frac{\pi q_{zn_z}}{l_z}\right)^2$$

$$K_n \simeq \eta_n - \frac{ik}{2\eta_n}\left(\epsilon_{n_x}\frac{\beta_{x0} + \beta_{x1}}{l_x} + \epsilon_{n_y}\frac{\beta_{y0} + \beta_{y1}}{l_y} + \epsilon_{n_z}\frac{\beta_{z0} + \beta_{z1}}{l_z}\right) \qquad (9.4.31)$$

$$\Psi_n \simeq \cos\left(q_{xn_x}\frac{\pi x}{l_x} + i\beta_{x0}\frac{kl_x}{\pi q_{xn_x}}\right)$$

$$\times \cos\left(q_{yn_y}\frac{\pi y}{l_y} + i\beta_{y0}\frac{kl_y}{\pi q_{yn_y}}\right)\cos\left(q_{zn_z}\frac{\pi z}{l_z} + i\beta_{z0}\frac{kl_z}{\pi q_{zn_z}}\right)$$

where we have relocated the origin at a corner of the room, and β_{x0} is the specific admittance of the wall $x = 0$, β_{x1} the admittance of wall $x = l_x$, etc. The effect of the factors ϵ_{n_x}, etc., is to make the first-order effect of the admittance of a given wall just half as large for those waves moving parallel to that wall ($n = 0$) as it is for those waves which reflect from the wall ($n > 0$).

Note the correspondence with the approximate solutions for waves in a rectangular duct, given in Eq. (9.2.16).   The reason for the factors ϵ_n is that, to first order, the mean energy contained in the mode $(0,n_y,n_z)$, for example, is twice that contained in a mode ($n_x > 0$, n_y, n_z), because the mean square of 1 is 1, whereas the mean square of $\cos(\pi n x/l_x)$ is $\frac{1}{2}$.   Since the rate of withdrawal of energy from the x walls is the same in both cases, the fractional rate of withdrawal when $n_x = 0$ is half that when $n_x > 0$.

We note also that if the β's are small, the normal modes are roughly equal to the simple forms $\phi_n(\mathbf{r})$ of Eqs. (9.4.29) and that, if the wave in the room is the single mode ϕ_n, the rate of energy loss to the walls is the integral over the wall area of ϕ_n^2 times the acoustic conductance $\xi/\rho c$ of the wall, which equals

$$\frac{V\Lambda_n^0}{\rho c}\left(\epsilon_{n_x}\frac{\xi_{x0} + \xi_{x1}}{l_x} + \epsilon_{n_y}\frac{\xi_{y0} + \xi_{y1}}{l_y} + \epsilon_{n_z}\frac{\xi_{z0} + \xi_{z1}}{l_z}\right) \qquad (9.4.32)$$

where Λ_n^0 is the mean-square value of the pressure throughout the room, and the factors ϵ give the ratio between the mean-square pressure at the walls to that throughout the room.   Factor V/l_x is the area of one of the walls perpendicular to the x axis, etc.   This quantity is, to the first order in the ξ's, proportional to the imaginary part of K_n, as one sees from Eqs. (9.4.31).

Room with randomly distributed impedance

In most cases, however, the impedance of the walls is irregularly distributed, and various approximate methods must be used to calculate the normal modes and their dynamic behavior.   One example of such a calculation will be given before we take up the high-frequency acoustics of rooms.

Suppose the specific acoustic admittance of the $z = 0$ wall is randomly distributed, with a mean value $\langle \beta \rangle$ and a variance $\Delta_\beta^2 = \langle \beta^2 \rangle - \langle \beta \rangle^2$.   Its variability can be measured by its autocorrelation function

$$\Upsilon_\beta(\xi,\eta) = \frac{1}{l_x l_y} \iint \beta(x,y)\beta(x + \xi, y + \eta)\, dx\, dy$$
$$\simeq \langle \beta \rangle^2 + \Delta_\beta^2 \exp\left[-\frac{1}{2}\left(\frac{\xi}{w}\right)^2 - \frac{1}{2}\left(\frac{\eta}{w}\right)^2 \right] \quad (9.4.33)$$

where w is the correlation length of the variation.   Alternatively, its variability can be measured by its double Fourier series, in terms of the eigenfunctions of the wall.

$$\beta(x,y) = \sum_{mn} B_{mn} \cos \frac{\pi m x}{l_x} \cos \frac{\pi n y}{l_y}$$
$$(9.4.34)$$
$$B_{mn} = \frac{\epsilon_m \epsilon_n}{l_x l_y} \iint \beta \cos \frac{\pi m x}{l_x} \cos \frac{\pi n y}{l_y}\, dx\, dy$$

where $B_{00} = \langle \beta \rangle$.   But from Eq. (9.2.29) we see that

$$B_{mn}{}^2 = \frac{\epsilon_m \epsilon_n}{l_x l_y} \iint \Upsilon_\beta(\xi,\eta) \cos \frac{\pi m \xi}{l_x} \cos \frac{\pi n \eta}{l_y}\, d\xi\, d\eta$$
$$\simeq \frac{\pi \epsilon_m \epsilon_n w^2}{2 l_x l_y} \Delta_\beta^2 \exp\left[-\frac{1}{2}\left(\frac{\pi m w}{l_x}\right)^2 - \frac{1}{2}\left(\frac{\pi n w}{l_y}\right)^2 \right] + \langle \beta \rangle^2 \delta_{m0}\delta_{n0} \quad (9.4.35)$$

and that

$$\Delta_\beta^2 = \sum_{mn} B_{mn}{}^2 - B_{00}{}^2$$

Although the squares of the quantities B_{mn} are smoothly varying functions of w and m and n, the quantities themselves, the square root of the expressions in Eqs. (9.4.35), have plus and minus signs distributed at random as function of m and n, because of the random distribution of β.   Therefore, in the calculation of the integrals of Eqs. (9.4.14), to be used in a variational calculation of the effect of the absorbing material,

$$\int \phi_N \beta \phi_N = \tfrac{1}{4} l_x l_y \left(B_{00} + \frac{1}{\epsilon_{N_x}} B_{2N_x,0} + \frac{1}{\epsilon_{N_y}} B_{0,2N_y} + \frac{1}{\epsilon_{N_x}\epsilon_{N_y}} B_{2N_x,2N_y} \right)$$

the second, third, and fourth terms, being smaller and randomly distributed in sign, can be neglected, except for the cases N_x and/or $N_y = 0$. Therefore

$$\int \phi_N \beta \phi_N \simeq \frac{l_x l_y}{\epsilon_{N_x}\epsilon_{N_y}} \langle \beta \rangle$$

If we choose Ψ'_N to be $A\phi_N$, we can use the formulas of (9.4.14). The Green's function integral will involve integrals of the sort

$$\left(\int \phi_N \beta \phi_n \right)^2 = (\tfrac{1}{16} l_x l_y)^2 (B_{n_x-N_x,n_y-N_y} + B_{n_x-N_x,n_x+N_y}$$
$$+ B_{n_x+N_x,n_y-N_y} + B_{n_x+N_x,n_y+N_y})^2 \qquad \text{for } n_x, n_y \neq N_x, N_y$$

Because of the random distribution of signs among the B's, the cross terms cancel out, on the average, leaving the sum of squares, which then reduces to

$$\left(\int \phi_n \beta \phi_N \right)^2 \simeq \frac{\pi}{128} w^2 l_x l_y \Delta_\beta{}^2 \left\{ \exp\left[-\frac{1}{2}\left(\frac{\pi w}{l_x}\right)^2 (n_x - N_x)^2 \right] \right.$$
$$\left. + \exp\left[-\frac{1}{2}\left(\frac{\pi w}{l_x}\right)^2 (n_x + N_x)^2 \right] \right\} \left\{ \exp\left[-\frac{1}{2}\left(\frac{\pi w}{l_y}\right)^2 (n_y - N_y)^2 \right] \right.$$
$$\left. + \exp\left[-\frac{1}{2}\left(\frac{\pi w}{l_y}\right)^2 (n_y + N_y)^2 \right] \right\} \simeq \left(\frac{l_x l_y}{4}\right)^2 \langle \beta \rangle^2$$
$$n_x = N_x > 0; \; n_y = N_y > 0 \quad (9.4.36)$$

Thus the Green's function term in (9.4.14) will be

$$A^2 \int \phi_N \beta \int G_N \beta \phi_N \simeq A^2 \frac{\pi w^2}{16 l_z} \Delta_\beta{}^2$$

$$\times \sum_{n \neq N} \frac{\exp\{-\frac{1}{2}[\pi(n_x - N_x)w/l_x]^2 - \frac{1}{2}[\pi(n_y - N_y)w/l_y]^2\}}{(\pi/l_x)^2(n_x{}^2 - N_x{}^2) + (\pi/l_y)^2(n_y{}^2 - N_y{}^2) + (\pi/l_z)^2(n_z{}^2 - N_z{}^2)}$$

$$+ A^2 \frac{l_x l_y l_z}{2\pi^2} \frac{\langle \beta \rangle^2}{n_z{}^2 - N_z{}^2}$$

where the sum is over n_x and n_y from $-\infty$ to $+\infty$ and n_z from 0 to $+\infty$, except for the term $n_x, n_y, n_z = N_x, N_y, N_z$. If $N_x w, N_y w$ are the same order of magnitude or larger than l_x, l_y, the largest terms in this series will be for n_x near N_x and n_y near N_y.

Changing from summation over n_x, n_y to integration, an approximate value for the sum is

$$A^2 \int \phi_N \beta \int G_N \beta \phi_N \simeq A^2 \frac{l_x l_y l_z}{8\pi^2} (\Delta_\beta^2 + 4\langle\beta\rangle^2) \sum_{n_z \neq N_z} \frac{1}{n_z^2 - N_z^2}$$

$$\simeq -A^2 \frac{V}{16\pi^2 N_z^2} (\Delta_\beta^2 + 4\langle\beta\rangle^2) \qquad N_x, N_y, N_z \text{ large} \quad (9.4.37)$$

Therefore the variational expression is

$$\tfrac{1}{8} V(\eta_N^2 - K_N^2) \simeq \tfrac{1}{2} i k l_x l_y \langle\beta\rangle A - \tfrac{1}{4} i k l_x l_y \langle\beta\rangle A^2 + \frac{k^2 V}{16\pi^2 N_z^2} (\Delta_\beta^2 + 4\langle\beta\rangle^2) A^2$$

with a solution

$$A \simeq \left(1 + \frac{ikl_z}{4\pi^2 N_z^2} \frac{\Delta_\beta^2 + 4\langle\beta\rangle^2}{\langle\beta\rangle}\right)^{-1}$$

$$\qquad\qquad\qquad\qquad\qquad\qquad\qquad\qquad (9.4.38)$$

$$K_N \simeq \eta_N - \frac{ik\langle\beta\rangle}{\eta_N l_z} + \frac{k^2 \langle\beta\rangle^2}{2\eta_N^3 l_z^2} - \frac{k^2(\Delta_\beta^2 + 4\langle\beta\rangle^2)}{4\pi^2 \eta_N N_z^2}$$

for large values of N_x, N_y, N_z.

Though the coupling coefficients have random sign, we can write out an approximate expression for the square of their magnitudes.

$$\Psi_N(\mathbf{r}) \simeq \phi_N(\mathbf{r}) + \sum_{n \neq N} V_n^N \phi_n(\mathbf{r})$$

$$|V_n^N|^2 \simeq \frac{\pi k^2 w^2 |\Delta_\beta^2| |A^2|}{16 V l_z (\eta_n^2 - \eta_n^2)} \exp\left\{-\tfrac{1}{2}(\pi w)^2 \left[\left(\frac{n_x - N_x}{l_x}\right)^2 + \left(\frac{n_y - N_y}{l_y}\right)^2\right]\right\}$$

$$\qquad\qquad\qquad\qquad\qquad\qquad\qquad\qquad (9.4.39)$$

which drop off in value rapidly as $|n_x - N_x|$ and $|n_y - N_y|$ become greater than l_x/w or l_y/w, respectively.

These formulas, of course, are for just one of the walls, perpendicular to the z axis, to have nonzero β. The extension of the formulas to the case with all walls nonrigid is not difficult. To the first order in the quantities $\langle\beta\rangle/kl$, the effects of the different walls add, and we can write

$$K_N \simeq \eta_N - \frac{ik}{\eta_N} \left(\frac{\langle\beta_{x0} + \beta_{x1}\rangle}{l_x} + \frac{\langle\beta_{y0} + \beta_{y1}\rangle}{l_y} + \frac{\langle\beta_{z0} + \beta_{z1}\rangle}{l_z}\right) \quad (9.4.40)$$

for N_x, N_y, N_z large; β_{x0} is the admittance of the wall at $x = 0$, β_{x1} the admittance of the wall at $x = l_x$, and so on.

It is interesting to note that this first-order formula for K_N is the same as that given in Eqs. (9.4.31) for a room with uniformly distributed admittance on its walls, the average values of the β's occurring instead of the uniform values. The difference comes, not in the first-order changes in the eigenvalues, but in the fact that the fluctuating admittance couples modes together to give a much more randomly distributed standing wave, with little regularity in its nodal surfaces. In Eqs. (9.4.31) the coupling coefficients V_n^N are zero; the generalization of the $|V_n{}^N|^2$ of Eqs. (9.4.39) is a sum of similar terms for each wall, each term being proportional to the variance $|\Delta_\beta{}^2|$ of the admittance on each wall; the greater this variance, the more is ϕ_N modified by a random assortment of ϕ_n's for n near N, and thus the more random is the distribution of fluid velocity in the room. In contrast, the uniform-admittance room has normal modes with regularly placed nodal surfaces and with velocity distributions which are far from random. Of course, this same coupling of modes can be attained by altering the shape of the walls in a random manner, or by placing scattering objects at random positions inside the room.

9.5 ROOM ACOUSTICS

The analysis of the preceding section is useful and conceptually fruitful in cases when the wavelength of the sound is larger than about one-third of the shortest dimension of the room. At these lower frequencies only a few standing waves are excited, so that the series expansions converge fairly rapidly, the resonances are distinct, and the waveforms are simple. As the last example of the preceding section indicates, however, at higher frequencies the resonances merge, the series expansions contain dozens or hundreds of nonnegligible terms, and the whole procedure of using normal modes of vibration to portray the acoustic behavior of the room no longer has either conceptual simplicity or provides ease of computation. To recapture these simplicities for the higher frequencies we must turn to a description in terms of acoustic rays and average intensities, analogous to geometric optics, with its rays of light and intensity of light rays. By analogy we shall call this conceptual model *geometrical acoustics.*

Assumptions of geometrical acoustics

The basic assumption of geometrical acoustics is that the room walls are irregular enough so that the acoustic energy density w is distributed uniformly through the room. For this to be true, a large number of standing waves must be involved, so that the nodal surfaces of each will be "filled in" by the others. Since each standing wave can be considered to be made up of a number of plane traveling waves, reflecting from the walls at appropriate angles, the sound in the room at the point characterized by the vector $\mathbf{r}$ can be represented by an assemblage of plane waves, each going in a direction specified by the angles φ, ϑ, each with pressure amplitude $A(\mathbf{r} \mid \varphi, \vartheta)$ and intensity $|A(\mathbf{r} \mid \varphi, \vartheta)|^2/\rho c$. In other words, we represent the pressure at point $\mathbf{r}$ in the room by the combination of plane waves

$$p(\mathbf{r}) = \int_0^{2\pi} d\varphi \int_0^{\pi} A(\mathbf{r} \mid \varphi, \vartheta) e^{i\mathbf{k}\cdot\mathbf{r} - i\omega t} \sin \vartheta \, d\vartheta$$

where $\mathbf{k}$ is the vector of magnitude ω/c pointed in the direction denoted by the angles φ, ϑ.

Since the energy density of each plane wave is $|A|^2/\rho c^2$, the mean energy density must then be

$$w(\mathbf{r}) = \frac{1}{\rho c^2} \int_0^{2\pi} d\varphi \int_0^{\pi} |A(\mathbf{r} \mid \varphi, \vartheta)|^2 \sin \vartheta \, d\vartheta$$

and the power incident on a unit area normal to the φ, ϑ axis (which can be in any direction) would be

$$I(\mathbf{r}) = \frac{1}{\rho c} \int_0^{2\pi} d\varphi \int_0^{\pi/2} |A(\mathbf{r} \mid \varphi, \vartheta)|^2 \cos \vartheta \sin \vartheta \, d\vartheta$$

The basic assumptions of geometric acoustics are that the plane waves constituting p are distributed so much at random that the energy density $w(\mathbf{r})$ in the room is *independent* of $\mathbf{r}$ and the energy flow I is *isotropic*; in other words, the average value of $|A(\mathbf{r} \mid \varphi, \vartheta)|^2$ over a small region of space is independent of $\mathbf{r}$, φ, or ϑ. When this is true, the simple relation

$$I = \tfrac{1}{4}cw \tag{9.5.1}$$

holds between mean energy density w and power flux I incident on any unit area in the room, independent of its position and orientation.

Geometric acoustics next makes the assumption that the acoustic absorption properties of the wall surface can be adequately represented by an *absorption coefficient* $\alpha(\mathbf{r}_s)$, the fraction of incident acoustic power which the region of wall near the surface point $\mathbf{r}_s$ absorbs [the value of $1 - |C_r|^2$ for the C_r of Eq. (6.3.5), for example], averaged over all directions of incidence, for isotropic distribution of incidental rays. With these assumptions the rate of loss of acoustic energy at the walls of the room would be

$$aI = \tfrac{1}{4}acw \qquad \text{where} \qquad a = \iint \alpha(\mathbf{r}_s) \, dS \tag{9.5.2}$$

the integration being over the wall area S of the room. Quantity a, called the *absorption* of the room, has the dimensions of area; if the area is given in square feet, a is said to be given in units of *sabins*.

If and when all these simplifying assumptions obtain (and we shall have to see when they are likely to do so), there results an elementary relationship between the rate of change of the total acoustic energy $Vw = 4VI/c$ in the room and the difference $\Pi(t) - aI$ between the power $\Pi(t)$ introduced into the room by some source of sound and the power aI absorbed by the walls,

$$\frac{d}{dt}\frac{4VI(t)}{c} = \Pi(t) - aI(t) \tag{9.5.3}$$

from which one can compute the mean sound intensity $I(t)$ anywhere in the room at time t.

Reverberation

The solution of this equation is

$$I = \frac{c}{4V} e^{-act/4V} \int_{-\infty}^{t} e^{ac\tau/4V} \, \Pi(\tau) \, d\tau \qquad (9.5.4)$$

indicating that the intensity at a given instant depends on the power output Π for the previous $4V/ac$ sec, but depends very little on the power output before that time (due to the exponential inside the integral). If the power Π fluctuates slowly, changing markedly in a time long compared with $4V/ac$, then the intensity I will be roughly proportional to Π, and Eq. (9.5.4) reduces to

$$I \simeq \frac{\Pi(t)}{a} \qquad \frac{d}{dt} \Pi(t) \ll \frac{ac}{4V} \Pi$$

$$(9.5.5)$$

$$\text{Intensity level} \simeq 10 \log\left(\frac{\Pi}{a}\right) + 90 \text{ db} \qquad \text{above } 10^{-16} \text{ watt/cm}^2$$

if Π is in ergs per second and a in square centimeters. If Π is in watts and a in square feet, the equation is

$$\text{Intensity level} \simeq 10 \log\left(\frac{\Pi}{a}\right) + 130 \text{ db}$$

This result is easily seen from Eq. (9.5.3), for if $d\Pi/dt$ is small, then $d(4VI/c)/dt$ can be neglected and $aI \simeq \Pi$. The intensity is thus inversely proportional to the room absorption a, so that, for steady-state intensity to be large, a should be small.

On the other hand, if Π varies widely in a time short compared with $4V/ac$, then the intensity will not follow the fluctuations of Π and the resulting sound will be "blurred." If the sound is shut off suddenly at $t = 0$, for instance, the subsequent intensity will be

$$I = I_0 e^{-act/4V}$$

$$\text{Intensity level} = 10 \log I_0 + 90 - 4.34 \frac{act}{4V} \qquad \text{db} \qquad (9.5.6)$$

The "blurring" of rapid fluctuations of speaker power is known as *reverberation*. It is related to the fact that the intensity level in the room does not immediately drop to zero when the power is shut off, but drops off linearly, with a slope $-4.34ac/4V$ db per sec. This linear dependence of intensity level on time is typical of rooms with uniform sound distribution. We shall discuss, later in this section, cases that have more complex behavior.

Reverberation time

The slope of the decay curve (the intensity level plotted against time after the power is shut off) indicates the degree of fidelity with which the room follows transient fluctuations in speaker output. The length of time for the level to drop 60 db is used as a measure of this slope, and is called the *reverberation time T.* If lengths are measured in centimeters, this time is

$$T = 60 \frac{4V}{4.34 \, ac} \quad \text{sec}$$

When lengths are measured in feet, and for air at normal conditions of pressure and temperature, the reverberation time is

$$T = 0.049 \frac{V}{a} = \frac{0.049 V}{\Sigma \alpha_s A_s} \quad \text{sec} \qquad (9.5.7)$$

When the speaker output changes slowly compared with T, then the intensity follows the output; but when the speaker output changes markedly in a time less than one-tenth of the reverberation time, the fluctuations will not be followed.

Therefore, in order that the room transmit transient sound faithfully, the reverberation time should not be large. For this requirement a should be large, in contradiction to the requirement that a be kept small to keep the steady-state intensity large. A compromise must be worked out between these opposing requirements, a compromise that varies with the size of the room. For a small room ($V \simeq 10,000$ cu ft) T can be as small as 1 sec, and the average intensity will still be satisfactorily high; but for a large room ($V \simeq 1,000,000$ cu ft) T may need to be as large as 2 sec for the intensity to be high enough throughout the room. If the room is used primarily for speech, which fluctuates rapidly, the reverberation time should be about two-thirds of this, for if the hall is large, the intensity can be increased by a public-address system. If the room is used chiefly for music, we can allow more reverberation without detriment (in fact, the music does not sound "natural" unless there is a certain amount of reverberation).

Thus an analysis of an extremely simplified example of sound in a room indicates the sort of compromise between reinforcement and absorption that must be reached for any sort of room, even if the sound is not uniformly distributed throughout its extent. The analysis has also indicated that a useful criterion to indicate the degree of uniformity of the sound distribution is the shape of the decay curve for the sound after the source is shut off. If this is a straight line (on a decibel scale), the chances are that the sound is fairly evenly spread throughout the room; but if it is a curve, it is certain that the sound is not uniformly distributed, either in space or in direction of propagation or both.

Absorption coefficient and acoustic impedance

Before we finish our discussion of the idealized case of uniform distribution of sound, we must compute the relationship between the specific acoustic impedance of the wall material and the absorption coefficient α.

As stated above, this quantity is the average fraction of power absorbed by the wall when sound is falling on it equally from all directions.

To obtain this average, we go back to our discussion of the amplitudes $A(\varphi, \vartheta)$. Suppose that we choose φ and ϑ so that the polar axis is perpendicular to the wall (assumed plane) and ϑ is the angle of incidence of the wave of amplitude $A(\varphi, \vartheta)$. In the case we are at present considering, A is independent of φ and ϑ, so that the power falling on a unit area of wall is

$$ I = \frac{1}{2\rho c} \int_0^{2\pi} d\varphi \int_0^{\pi/2} |A|^2 \cos \vartheta \sin \vartheta \, d\vartheta = \frac{\frac{1}{2}\pi}{\rho c} |A|^2 $$

But from Eq. (6.3.6) we see that the fraction of power lost by a wave of angle of incidence ϑ, on reflection from a plane surface of specific acoustic admittance $\beta = \xi - i\sigma$, is

$$ \alpha(\vartheta) = 1 - \left| \frac{\cos \vartheta - \beta}{\cos \vartheta + \beta} \right|^2 = \frac{4\xi \cos \vartheta}{(\xi + \cos \vartheta)^2 + \sigma^2} $$

if the wall is a locally reacting surface (Sec. 6.3).

Therefore the average value of α, which is to be used in the case of uniform and isotropic sound distribution, is given in terms of the acoustic conductance and susceptance of the wall by the formula

$$ \alpha = \frac{1}{2\rho c I} \int_0^{2\pi} d\varphi \int_0^{\pi} \alpha(\vartheta) \, |A|^2 \cos \vartheta \sin \vartheta \, d\vartheta $$

$$ = 8\xi \left[1 + \frac{\xi^2 - \sigma^2}{\sigma} \tan^{-1} \frac{\sigma}{\sigma^2 + \xi^2 + \xi} - \xi \ln \frac{(\xi + 1)^2 + \sigma^2}{\xi^2 + \sigma^2} \right] $$

$$ \rightarrow 8\xi \qquad |\beta| \ll 1 \tag{9.5.8} $$

Values of this quantity, in terms of magnitude and phase of β or $\rho = 1/\beta$, can be obtained from Plate VI. This plot shows that the maximum value of the absorption coefficient ($\alpha = 0.96$) comes when the wall impedance is a pure resistance, a little bit larger than $\rho c(\xi \simeq 1.55)$. As the impedance is decreased or increased from this value, the absorption coefficient diminishes, and at very large values of ζ, α is approximately equal to eight times the specific conductance ξ of the wall. The formula for α is more complicated if the wall is a surface of extended reaction (see page 266).

Therefore, when the room walls are irregular enough and the frequency is high enough so that the acoustical energy is fairly uniformly distributed throughout the room volume, the acoustical characteristics are given by Eqs. (9.5.3) to (9.5.7), and the slowness of response to transient sounds is measured by the reverberation time. The absorption coefficient entering these formulas is given in terms of the wall impedance (*if* the wall is locally reactive) by Eq. (9.5.8). If the sound is not uniformly distributed, however, Eqs. (9.5.3) to (9.5.7) will not be valid, and Eq. (9.5.8) for α will not be applicable; in fact, the term absorption coefficient will have no meaning.

Simple-source geometrical acoustics

When the power input, represented by the term $\Pi(t)$ of Eq. (9.5.3), is provided by an acoustic source of small size, placed at some point $\mathbf{r}_0$ inside the room, the assumption of uniformity of energy density and isotropy of power flux cannot be the case everywhere in the room. Though it may be uniform in most of the room, close to the source it will not be uniform, for most of the sound there is radiating outward from the source, and the intensity varies with the inverse square of the distance, as from a simple source. At larger distances the outward radiation is lost beneath the randomly scattered waves, which have more or less uniform intensity everywhere in the room. If Π is the power output of the source in watts, r the distance from the source in feet, a the room absorption in square feet [Eq. (9.5.2)], the intensity close to the source in watts per square centimeter is $\Pi/4,000\pi r^2$, where the factor 1,000 is approximately the number of square centimeters in a square foot. For large values of r, the intensity is $\Pi/1,000a$.

At this point we must halt to point out the difference between our definition of intensity and the way sound "intensity" is usually measured. We have defined intensity as sound power falling on one side of a square centimeter of area. This can conceivably be measured, but in many cases the result may depend on the orientation of the area. Close to the source the intensity is all flowing outward, so that we must arrange that the square centimeter be placed perpendicular to the radius r if our measurement is to equal $\Pi/4,000\pi r^2$; if it were placed parallel to r, the intensity measured would not be at all as large. On the other hand, throughout the rest of the room, according to our assumption, the intensity flows equally in all directions, and the intensity-measuring device need not have any special orientation to measure the predicted amount.

In actual practice, sound intensity is rarely measured directly; what is measured is *mean-square pressure*, as was mentioned in connection with Eq. (6.3.3). This quantity is simply related to the average energy density w by the relation $p_{rms}^2 = \rho c^2 w$, but it is not simply related to the sound intensity. If the intensity is flowing in only one direction, the relation is $p_{rms}^2 = \rho c I$, but if it is flowing equally in all directions, the relation is $p_{rms}^2 = 4\rho c I$, as was shown in Eq. (9.5.1). Consequently, the quantity to compute, which can be checked directly with measurement, is the mean-square pressure, rather than the intensity, or else the pressure level [which is $20 \log (p_{rms}) + 74 = 10 \log (w) + 136$] rather than the intensity level. See page 229.

In terms of these quantities the statements made above become

$$\text{Pressure level} \simeq \begin{cases} 10 \log (\Pi) - 20 \log (r) + 49 \text{ db} & r^2 < \dfrac{a}{50} \\[2mm] 10 \log (\Pi) - 10 \log (a) + 66 \text{ db} & r^2 > \dfrac{a}{50} \end{cases}$$

for the statistical case, where Π is source power in watts, r the distance from the source in feet, and a the room absorption in square feet. The criterion for range of validity (r^2 versus $a/50$) is obtained by equating the two formulas. If the power Π is measured in ergs per second, r in centimeters, and a in square centimeters, the formulas for mean-square pressure are

$$p_{\text{rms}}^2 \simeq \begin{cases} \dfrac{\rho c \Pi}{4\pi r^2} & r^2 < \dfrac{a}{50} \\[3mm] \dfrac{4\rho c \Pi}{a} & r^2 > \dfrac{a}{50} \end{cases}$$

$$a = \sum_s \alpha_s A_s$$

Close to the source, the first expression is larger and is used; far from the source, the second term, representing the random sound, is larger and is valid.

If the sound generator is a simple source of strength S_ω, then Eq. (7.1.5) shows that the power generated will be $\rho \omega^2 |S_\omega|^2/4\pi c$. Consequently, over most of the volume of the room the mean-square pressure from the simple source is

$$p_{\text{rms}}^2 = \frac{\rho^2 \omega^2}{\pi a} |S_\omega|^2 \tag{9.5.9}$$

This equation is written here, for the case of the room for which geometric acoustics is valid, so as to compare it with the expression we shall obtain for a regular rectangular room, where the assumption of uniform distribution is not valid.

Relation between wave and geometric acoustics

To determine the frequency range and type of room for which geometric acoustics is valid, we must extend the normal-mode analysis of the previous section to higher frequencies, to see where it begins to correspond to the simple geometric formulas we have just derived. And to do this we must first obtain approximate formulas for the number of standing-wave resonance frequencies in some frequency band. We shall do this in some detail for a rectangular room, and only mention the result for other room shapes.

For a rectangular room of sides l_x, l_y, l_z, with rigid walls, the resonance frequencies are obtained from Eqs. (9.4.29).

$$\nu_n{}^2 = \frac{\omega_n{}^2}{4\pi^2} = \left(\frac{\eta_n c}{2\pi}\right)^2$$

$$= \left(\frac{n_x c}{2l_x}\right)^2 + \left(\frac{n_y c}{2l_y}\right)^2 + \left(\frac{n_z c}{2l_z}\right)^2 \tag{9.5.10}$$

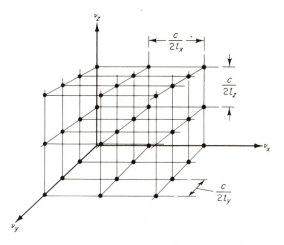

FIGURE 9.16
Distribution of allowed frequencies in "frequency
space" for a rectangular
room of sides l_x, l_y, and l_z.
The length of the vector
from the origin to one of
the lattice points is an
allowed frequency, and the
direction of the vector gives
the direction cosines of the
corresponding standing wave
in the room.

This suggests that we can consider ν_n to be the magnitude of a vector having components $n_x c/2l_x$, $n_y c/2l_y$, $n_z c/2l_z$. The direction of the vector gives the direction of the plane wave producing the standing wave; the length of the vector is the frequency of the standing wave. A normal mode of oscillation can therefore be considered as a point in "frequency space," whose x component is an integral number of unit lengths $c/2l_x$, whose y component is an integer times $c/2l_y$, etc. The length of the line joining this point and the origin is the frequency of the normal mode, and the direction of this line is the direction of the wave that can be used to generate the standing wave. Some of these "characteristic points" are shown in Fig. 9.16, and it can be seen that they correspond to the intersections of a rectangular lattice with x, y, and z spacings equal to $c/2l_x$, $c/2l_y$, $c/2l_z$, respectively. It can also be seen that all the normal modes are included among the points in the octant of space between the positive ν_x, ν_y, and ν_z axes, for any of the waves of frequency ν with directions corresponding to

$$(cn_x/2l_x,\ cn_y/2l_y,\ cn_z/2l_z)$$
$$(-cn_x/2l_x,\ cn_y/2l_y,\ cn_z/2l_z)$$
$$(cn_x/2l_x,\ -cn_y/2l_y,\ cn_z/2l_z)$$
$$\cdots\cdots\cdots\cdots\cdots\cdots$$

will generate, by reflection, the same standing wave.

This picture of a lattice of characteristic points in frequency space is extremely useful in discussing the number and type of normal modes having frequencies within a given frequency range. For instance, since there are $8V/c^3$ lattice cells per unit volume of frequency space ($V = l_x l_y l_z$), there will be, on the average, $\dfrac{8V}{c^3}\dfrac{\pi\nu^3}{6}$ normal modes having frequency equal to or less than ν (the factor $\pi\nu^3/6$ being the volume of an eighth of a sphere of radius ν). The actual number of modes having frequency less than ν varies in an irregular manner as ν increases, being zero until ν equals the smallest of the three quantities $c/2l_x$; $c/2l_y$, $c/2l_z$, when it suddenly jumps to unity, and so on.

Axial, tangential, and oblique waves

Referring to Sec. 9.4, and anticipating the results of this section a little, we note that waves traveling "parallel" to a wall are affected by the wall (are absorbed by it, for instance) to a lesser extent than waves having oblique incidence. Therefore we separate our standing waves into three categories and seven classes:

Axial waves (for which two n's are zero):

x-axial waves, parallel to the x axis $(n_y, n_z = 0)$
y-axial waves, parallel to the y axis $(n_x, n_z = 0)$
z-axial waves, parallel to the z axis $(n_x, n_y = 0)$

Tangential waves (for which one n is zero):

y, z-tangential waves, parallel to the yz plane $(n_x = 0)$
x, z-tangential waves, parallel to the xz plane $(n_y = 0)$
x, y-tangential waves, parallel to the xy plane $(n_z = 0)$

Oblique waves (for which no n is zero)

It will turn out that, even in the first approximation, waves of different classes have different reverberation times and, to the first approximation, waves of the same class (with ν's approximately equal) have the same reverberation time.

Consequently, it will be quite important to count the number of standing waves of a given class having frequency less than ν. The representation in a lattice system is again useful here, for the axial waves have their lattice points on the corresponding axis in "frequency space," and the tangential waves have their points in the corresponding coordinate planes. Again, the number of lattice points can be counted, or a "smoothed-out" average number can be computed.

To take an example of practical interest: Suppose that the source sends out a pulse of sound of frequency ν_0 and duration Δt. According to Eq. (1.3.16), if this pulse is to be transmitted in the room without serious distortion of shape, there must be a sufficient number (it turns out that a "sufficient number" is more than 10) of standing waves with frequencies within a frequency band between $\nu_0 - (\Delta\nu/2)$ and $\nu_0 + (\Delta\nu/2)$, where $\Delta\nu = 1/\Delta t$, in order to "carry" the sound. If, for instance, we should wish to have the room transmit adequately a pulse of length $\frac{1}{10}$ sec, we should be interested in counting the number of resonance frequencies of the room between $\nu_0 - 5$ and $\nu_0 + 5$.   If this number is less than 10 for a certain value of ν_0, then a pulse of frequency ν_0 and of duration $\frac{1}{10}$ sec would not be transmitted with fidelity in the room.   If the number is larger than about 10, and if in addition the reverberation time of each of the standing waves involved

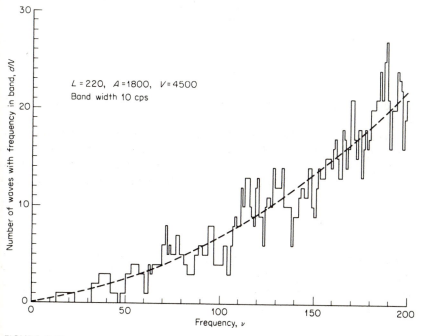

FIGURE 9.17
Number of standing waves with frequencies between $\nu - 5$ and $\nu + 5$ in a room 10 by 15 by 30 ft.   Irregular solid line gives exact values; dashed smooth curve is plot of Eq. (9.5.12), giving approximate values of dN.

is less than about a second, the pulse will be transmitted with reasonable fidelity.   Figure 9.17 shows a curve (solid irregular line) representing a count of this sort for a room 10 by 15 by 30 ft.   It indicates that a pulse of $\frac{1}{10}$ sec duration would not be reproduced adequately unless its frequency were larger than about 150 cps.

Average formulas for numbers of allowed frequencies

It is quite tedious to count the individual allowed frequencies less than a given frequency or in a given frequency band, so that it is useful to obtain "smoothed-out" formulas for average values of the counts.   This can be done by considering that each lattice point "occupies" a rectangular block of dimensions $c/2l_x$, $c/2l_y$, $c/2l_z$, in frequency space, with the actual lattice point at the center of the block. Then the average number of points can be obtained by dividing the volume of frequency space considered by the volume $c^3/8V$ ($V = l_x l_y l_z$) of each block.

As an example, we can count up the numbers of different classes of waves having frequencies less than ν. The average number of x-axial waves is just ν divided by the lattice spacing in the ν_x direction, $2\nu l_x/c$ (i.e., it is the number of blocks in a rod of cross section $c^2/4l_y l_z$ and length ν), and the average number of all axial waves with frequencies less than ν is

$$N_{ax} \simeq \frac{\nu L}{2c}$$

where $L = 4(l_x + l_y + l_z)$ is the sum of the lengths of all the edges of the room.

The average number of y, z-tangential waves is the number of blocks in a quarter of a disk, of thickness $c/2l_x$ and of radius ν, minus a correction to allow for the axial waves, which have been counted separately. This correction in volume is *one-half* the space "occupied" by the y and the z axial lattice points, viz.,

$$\frac{\nu c^2}{8V}(l_y + l_z)$$

The factor $\frac{1}{2}$ comes in because only one-half the volume "occupied" by the axial lattice points is inside the angular sector formed between the y and z axes, which bounds the quarter disk. Therefore the average number of y, z-tangential waves having frequencies less than ν is

$$N_{ta,yz} = \frac{\pi \nu^2}{c^2} l_y l_z - \frac{\nu}{c}(l_y + l_z)$$

and the average number of all tangential waves with frequencies less than ν is

$$N_{ta} \simeq \frac{\pi \nu^2 A}{2c^2} - \frac{\nu L}{2c}$$

where $A = 2(l_x l_y + l_x l_z + l_y l_z)$ is the total wall area. We are neglecting the corrections for the overlapping regions at the origin, $\nu = 0$, for they are independent of ν and are small in magnitude.

The volume "occupied" by the lattice points for the oblique waves of frequency less than ν is the volume of one-eighth sphere minus the volume already counted for the other classes of wave:

$$N_{ob} \simeq \frac{4\pi \nu^3 V}{3c^3} - \frac{\pi \nu^2 A}{4c^2} + \frac{\nu L}{8c}$$

where $V = l_x l_y l_z$ is the volume of air in the room. Therefore the total number of standing waves of all classes which have frequencies less than ν is

$$N \simeq \frac{4\pi\nu^3 V}{3c^3} + \frac{\pi\nu^2 A}{4c^2} + \frac{\nu L}{8c} \qquad (9.5.11)$$

The correct value for N fluctuates above and below this average value, but is seldom more than one or two units away, unless the room is too symmetrical; this will be discussed later.

Average number of frequencies in band

The number of standing waves with frequencies in a band of width $d\nu$ is obtained by differentiating the formulas given above:

$$dN_{ax,x} \simeq \frac{2l_x}{c} d\nu, \cdots$$

$$dN_{ax} \simeq \frac{L}{2c} d\nu$$

$$dN_{ta,yz} \simeq \left[\frac{2\pi\nu}{c^2} l_y l_z - \frac{1}{c}(l_y + l_z)\right] d\nu, \cdots$$

$$dN_{ta} \simeq \left(\frac{\pi\nu A}{c^2} - \frac{L}{2c}\right) d\nu \qquad (9.5.12)$$

$$dN_{ob} \simeq \left(\frac{4\pi\nu^2 V}{c^3} - \frac{\pi\nu A}{2c^2} + \frac{L}{8c}\right) d\nu$$

$$dN \simeq \left(\frac{4\pi\nu^2 V}{c^3} + \frac{\pi\nu A}{2c^2} + \frac{L}{8c}\right) d\nu$$

$$L = 4(l_x + l_y + l_z) \qquad A = 2(l_x l_y + l_x l_z + l_y l_z) \qquad V = l_x l_y l_z$$

The value of dN, obtained from this formula, for $d\nu = 10$ and for appropriate values of the other constants, is shown as the dashed line in Fig. 9.17. It is seen that this curve is a good "smoothed-out" approximation to the correct step curve. At very high frequencies just the term proportional to ν^2 is important.

We notice that the average number of allowed frequencies in a band increases with the square of the frequency at the higher frequencies. If we assume that the average intensity of sound in a room (for a constant output source) is proportional to the number of standing waves that carry the sound (i.e., the number with frequency inside the band characteristic of the driver), the intensity in the room increases as the square of the frequency, for high frequencies, according to Eqs. (9.5.12). This is very interesting, because the power output into free space from a simple source is proportional to v^2, according to Eq. (7.1.5). Therefore the power transmitted from source to receiver in a room varies, on the average, with frequency, as it does in the open; but superimposed on the smooth rise are fluctuations (as shown in Fig. 9.17) because of the fluctuations of the number of standing waves in the frequency band of the driver. These irregularities of response are more pronounced, the more symmetric the shape of the room, or the narrower the frequency band of the sound source.

When dN, as given by Eqs. (9.5.12), becomes 2 or less, the fluctuations become so large that they appreciably reduce the fidelity of transmission. For the room referred to in Fig. 9.17, this lower limit, for a bandwidth of 10 cps, is about 50 cps; for a bandwidth of 5 cps, it is about 100 cps; etc.

The effect of room symmetry

We have mentioned several times in the preceding pages that the response curve of a room, as evidenced by the exact curves for dN, is more irregular when the room is more symmetrical. This is due to the increase in the number of *degenerate* modes, standing waves with different n's that have the same frequency. As an example of the effect of degeneracies, we can consider the sequence of rooms of dimensions l_x, $l_y = ql_x$, $l_z = l_x/q$, so chosen that the volume $V = (l_x)^3$ remains the same but the relative dimensions change as we change q. The natural frequencies are

$$v = \frac{c}{2l_x}\sqrt{n_x^2 + \frac{n_y^2}{q^2} + q^2 n_z^2}$$

which change in spacing as we change q. We shall write down the lowest allowed frequencies for two rooms of this sequence: one for $q^2 = 1$, a cubical room, the most symmetric, and one for $q^2 = 2$, which is not symmetric.

Table 9.1 gives the allowed frequencies and the combinations of integers (n_x, n_y, n_z) which label the corresponding characteristic functions for these two cases. We notice immediately the tendency of all the characteristic frequencies to "clump together" in the cubical-room case. Threefold, and even sixfold, degeneracies are common even at these low frequencies (for instance, 2.236 and 3.000). These result in large ranges of frequency within which there is no characteristic value, so that the response is very irregular. In contrast, the case of $q^2 = 2$ never gives more than twofold degeneracies in the frequency range considered, and the allowed frequencies are therefore more evenly spaced along the scale. We note that, because the room volumes are equal, there are approximately the same *number* of frequencies equal to

$3c/2l$ or less (28 in one case, 27 in the other), but the particular values of (n_x,n_y,n_z) included, and their order on the frequency scale differs.   If we had picked an incommensurate value for q^2 (the cube root of 5, for instance), we should have had no degeneracies at all, and the allowed frequencies would

TABLE 9.1

Characteristic frequency parameters $(2lv/c)$ and corresponding quantum numbers for standing waves in a cubical room of side l and in a room of dimensions $l, l\sqrt{2}, l/\sqrt{2}$

CUBICAL ROOM				NONCUBICAL ROOM			
$2lv/c$	n_x	n_y	n_z	$2lv/c$	n_x	n_y	n_z
	1	0	0	0.707	0	1	0
1.000	0	1	0	1.000	1	0	0
	0	0	1	1.225	1	1	0
	1	1	0	1.414	0	0	1
1.414	1	0	1		0	2	0
	0	1	1	1.581	0	1	1
1.732	1	1	1	1.732	1	0	1
	2	0	0		1	2	0
2.000	0	2	0	1.871	1	1	1
	0	0	2	2.000	2	0	0
	2	1	0		0	2	1
	1	2	0	2.121	2	1	0
	2	0	1		0	3	0
2.236	1	0	2	2.236	1	2	1
	0	2	1	2.345	1	3	0
	0	1	2	2.449	2	0	1
	2	1	1		2	2	0
2.449	1	2	1	2.550	2	1	1
	1	1	2		0	3	1
	2	2	0	2.739	1	3	1
2.828	2	0	2	2.828	0	0	2
	0	2	2		0	4	0
	3	0	0	2.915	0	1	2
	0	3	0		2	3	0
	0	0	3		1	0	2
3.000	2	2	1	3.000	3	0	0
	2	1	2		1	4	0
	1	2	2				

have been still more more evenly spaced along the scale.   We can never get absolutely uniform spacing with a rectangular room, of course, because the lattice in frequency space is always rectangular.   A room with irregular walls would correspond to a more random arrangement of lattice points in frequency space and, perhaps, to a more uniform response.

Nonrectangular rooms

Our analysis of standing waves has depended to some extent on the fact that we have chosen to study rectangular rooms. This is not a serious limitation, for most rooms approximate a rectangular form. Nevertheless it would be more satisfactory if it could be shown that Eqs. (9.5.11) and (9.5.12) hold for all room shapes. This cannot be done, for several reasons. In the first place, although it is not difficult to generalize the quantities V (room volume) and A (area of walls) to rooms of other shape, the quantity L (total length of edge) becomes a problem. (For example, if L for a cylindrical room is just $4\pi R$, what is L for a room of octagonal floor plan, or with a floor plan that is a polygon of a large number of sides—approaching a circular form—and what is L for a spherical enclosure?)

In the second place, it becomes progressively more difficult to define axial and tangential waves as the room shape is made more complex (this, of course, is another aspect of the reciprocal relationship between uniformity of room shape and uniformity of wave behavior: all waves are oblique waves if the room is irregular enough).

We have discussed some of this, subsequent to Eqs. (9.2.24), in connection with a cylindrical duct. A plot of dN for a cylindrical room would be roughly similar to Fig. 9.17, but the allowed frequencies would be less evenly spaced than the rectangular case shown there. Because of the greater symmetry of the cylindrical room, there are many modes having nearly the same resonance frequency, with large gaps in between, resulting in greater irregularity of response.

The distribution in frequency of the standing waves in a spherical enclosure is even less regular than in a cylinder, and still higher frequencies must be used before the actual curve for dN is smoothed out and approaches the average curve. Such an enclosure would not be satisfactory for use as a room, because of the fluctuation in its resonating characteristics as the frequency is changed. In fact, we can state as a general rule that the more symmetrical an enclosure is, the larger will be the range of frequency over which the resonance properties fluctuate, and the less desirable will it be for use as an auditorium. The curve for dN for a room of the same volume as that used for Fig. 9.17, but having very irregular walls, will approach the smooth average curve still more rapidly than the curve shown does.

Irregular walls also serve to spread out the sound energy more or less uniformly over the room. Most of the high-frequency standing waves in a rectangular enclosure have an average amplitude that is nearly the same everywhere in the room, but many of the standing waves in a spherical room have larger amplitudes near the center than near the wall. In rooms having smooth concave surfaces, focal points of considerable excess intensity may occur to render the room undesirable as an auditorium.

No matter what shape the room has, however, it can be shown that the first term in the equation for dN in (9.5.12), the term which preponderates at high frequencies, is still correct. The number of resonance frequencies in the frequency band dv approaches the number

$$dN \simeq \frac{4\pi v^2 V}{c^3}\, dv = \frac{\omega^2 V}{2\pi^2 c^3}\, d\omega \qquad \text{as } v \to \infty \qquad (9.5.13)$$

for a room of volume V, no matter what its shape.

Steady-state response

We now have assembled the formulas which enable us to follow the transition from the irregular acoustical-response characteristic of low frequencies to the statistically uniform response of the geometrical-acoustics formulation. First we return to the representation of the pressure wave at $\mathbf{r}$, within a rectangular room, when excited by a simple source of strength S_ω and frequency $\omega/2\pi$ at $\mathbf{r}_0$. According to Eqs. (7.4.1) and (9.4.2), this is

$$p_s(\mathbf{r} \mid \mathbf{r}_0) = -\frac{i\rho\omega}{V}\, S_\omega \sum_n \frac{\Psi_n(\omega,\mathbf{r})\Psi_n(\omega,\mathbf{r}_0)}{\Lambda_n(K_n^2 - k^2)} \qquad n = n_x, n_y, n_z \qquad (9.5.14)$$

where $\Psi_n(\omega,\mathbf{r})$ and K_n^2 are the eigenfunction and eigenvalue for the nth standing wave in the rectangular room. If the room walls are uniformly covered with material of acoustic admittance $\beta_{x0}, \beta_{x1}, \beta_{y0}$, etc., these are given by Eqs. (9.4.30). If acoustical material is applied in a random manner to the walls, the formulas (9.4.36) to (9.4.40) can be used. If, in addition, sound-absorbing material is placed in the room interior, formulas such as (9.4.21) and (9.4.27) can be used.

We note that K_n^2 has the dimensions of inverse area and that, if the room walls are fairly stiff, the real part of K_n^2 is nearly equal to η_n^2, the corresponding eigenvalue for the room with rigid walls, and the imaginary part of K_n^2 is approximately equal to $-(k/l_x l_y l_z)$ times a quantity proportional to the sum of wall areas times specific acoustic conductance $\mathrm{Re}\,\beta = \xi$ of the wall. For example, both Eqs. (9.4.31) and (9.4.40) show that, for either uniform or random distribution of wall admittance, the first-order expression for K_n^2 can be written

$$K_n^2 \simeq \eta_n^2 - \frac{k}{V}\, [\epsilon_{n_x}\langle\sigma_{x0} + \sigma_{x1} + i\xi_{x0} + i\xi_{x1}\rangle l_y l_z + \cdots]$$

To relate this to the absorption coefficients of geometrical acoustics, we recall that the room absorption a is equal to the sum of the wall areas times their respective absorption coefficients and, from Eq. (9.5.8), that for fairly stiff walls (and for oblique waves), the absorption coefficient α of Eq. (9.5.2) should be roughly equal to eight times the conductance ξ of the wall. Therefore, if we define g_n and a_n by the equation

$$K_n^2 = \eta_n^2 - \frac{k}{4V}\, [g_n(\omega) + ia_n(\omega)] \qquad (9.5.15)$$

we can see that, for $|\beta| \ll 1$, quantity a_n is given to the first order by the formula

$$a_n \simeq 4[\epsilon_{n_x}\langle \xi_{x0} + \xi_{x1}\rangle l_y l_z + \epsilon_{n_y}\langle \xi_{y0} + \xi_{y1}\rangle l_x l_z$$
$$+ \epsilon_{n_z}\langle \xi_{z0} + \xi_{z1}\rangle l_x l_y] \qquad |\beta| \ll 1 \quad (9.5.16)$$

which does equal the expression for a given in Eq. (9.5.2) in the limit of $|\beta| \ll 1$, when $\alpha \to 8\xi$ (and also when $n > 0$). In this limit the value of a_n is the same for all oblique waves, in the sense of the definition of page 584. For tangential and axial waves, one or more of the factors ϵ_n are unity rather than 2, so that these a_n's have smaller values, even in the limit of $|\beta| \ll 1$.

For larger values of β, the quantity a_n, as defined in Eq. (9.5.15), still has meaning, although its value is not given by the first-order formula (9.5.16); it still is equal to $-(4V/k)$ times the imaginary part of $K_n{}^2$. In fact, as the irregularities of room shape and impedance distribution are increased, the various normal modes get more and more "mixed up," their properties become harder and harder to distinguish in a simple manner (such as the position and orientation of nodal surfaces), and the quantities a_n, as defined in Eq. (9.5.15), approach each other in value. For a room with uniform wall impedances, corresponding to Eqs. (9.4.30), the absorption constants for the nth standing wave,

$$a_n = \frac{8\pi^2 l_x l_y l_z}{k}\left(\frac{\mu_{n_x}\nu_{n_x}}{l_x{}^2} + \frac{\mu_{n_y}\nu_{n_y}}{l_y{}^2} + \frac{\mu_{n_z}\nu_{n_z}}{l_z{}^2}\right) \qquad \beta \text{ uniform} \quad (9.5.17)$$

do not merge in value but remain separate, some being small, some large, as ξ is increased. In contrast, for a room with randomly situated irregularities in wall shape and impedance, so that the expression for Ψ_N contains a large number of ϕ_n's [as in Eq. (9.4.39)], the higher-order formulas for the a's have similar mixtures of terms which differ less and less from each other as the coupling is increased.

Returning to Eq. (9.5.14), we know from the discussion preceding Eq. (9.4.6) that the pressure wave close to the simple source is a spherical outgoing wave, but that far from the source the pressure amplitude is more or less uniform. Since the mean-square value of Ψ_n is Λ_n, by definition, we see that the mean-square pressure, far from the source, is equal to

$$|p_s{}^2| = \left(\frac{\omega\rho}{V}\right)^2|S_\omega{}^2|\sum_n \frac{[|\Psi_n{}^2(\omega,\mathbf{r}_0)|/\Lambda_n]}{[\eta_n{}^2 - k^2 - (kg_n/4V)]^2 + (ka_n/4V)^2} \quad (9.5.18)$$

The quantity in square brackets in the denominator is the resonance term; when it is zero, the nth mode is resonating. Since $a_n/4V\eta_n$ is usually quite small, these resonance peaks are usually high and narrow; the width of the frequency band within which the response is greater than half its maximum value is $\Delta\nu_n = (c/2\pi)\Delta k = ca_n/8\pi V$, which is called the *width* of the resonance peak.

At low frequencies these resonance peaks are spaced much farther apart than their width; only one standing wave is excited.   In such a case we cannot realistically average over $\mathbf{r}$ to obtain a mean-square pressure; at the nth resonance (if it is distinct) the square of the pressure amplitude at r would be

$$|p_s^2| \simeq \left(\frac{4\rho c}{a_n \Lambda_n}\right)^2 |S_\omega^2| \, |\Psi_n^2(c\eta_n,\mathbf{r})| \, |\Psi_n^2(c\eta_n,\mathbf{r_0})|$$

which exhibits all the nodes and peaks of the individual standing wave [the formula, of course, is not valid for $\mathbf{r}$ close to $\mathbf{r_0}$ where other terms in series (9.5.18) are in phase, cross products do not cancel out, and the square of the series adds up to a spherical outgoing wave].

The behavior of the mean-square pressure is illustrated by the curves of Fig. 9.18, which are response curves for a rectangular room with the source at one corner of the room, so that $|\Psi_n^2(\omega,\mathbf{r_0})| \simeq 1$ for all values of n.   The upper curve is the average response, according to Eq. (9.5.18), and the lower curves are the mean-square pressures, according to Eq. (9.5.14), for two different positions in the room.   At the low frequencies the response at specific points can differ considerably from the average response; at higher frequencies the fractional differences diminish, and all curves approach the dotted curve, which is the geometrical-acoustics formula.   These curves should be compared with those of Fig. 9.17, which are for a room with the same dimensions.

As the driving frequency is increased, the resonance peaks begin to overlap, and eventually, when the peak width is larger than the mean distance between resonances, the response curve becomes fairly smooth, as shown in the upper curve of Fig. 9.18.   The mean frequency spacing between resonance peaks is given by Eqs. (9.5.12).   When $\lambda = c/\nu$ is less than about one-quarter of the smallest dimension of the room, the number of oblique waves in any frequency band is considerably larger than the number of tangential or axial waves, and the simple formula

$$dN \simeq \frac{4\pi\nu^2 V}{c^3}\, d\nu = \frac{\omega^2 V}{2\pi^2 c^3}\, d\omega = \frac{k^2 V}{2\pi^2}\, dk \qquad (9.5.19)$$

[Eq. (9.5.13)] is a fair estimate of the total number of resonances within the frequency band $d\nu$, the ω band $d\omega$, or the wavenumber band dk.   Thus the mean frequency spacing between resonance peaks is approximately

$$\frac{d\nu}{dN} \simeq \frac{c^3}{4\pi\nu^2 V}$$

We can therefore say that when this mean spacing is less than about one-eighth of the mean width of the resonance peak, i.e., when

$$\nu^2 > 16\frac{c^2}{\langle a \rangle} \qquad \text{or} \qquad \lambda < \tfrac{1}{4}(\langle a \rangle)^{\frac{1}{2}} \qquad (9.5.20)$$

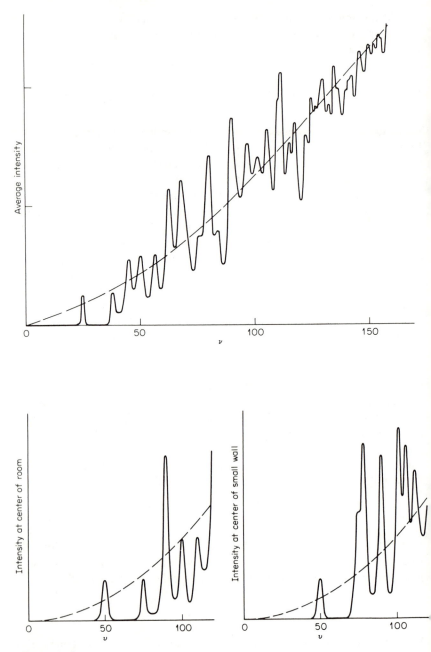

FIGURE 9.18
Response curves for a room 15 by 30 by 10 ft with reverberation time 1 sec. Dotted lines correspond to approximate formula (9.5.18), which the actual curves approach
asymptotically.

more than eight normal modes will be strongly excited by a single-frequency simple source. The symbol $\langle a \rangle$ stands for the mean value of a_n for those modes resonating at frequency ν.

In this range of frequency we can use Eq. (9.5.18) as a good approximation for the mean-square pressure measured at any point not closer than a wavelength from the source and not closer than a half wavelength from a wall. Furthermore, we can approximate the factor $|\Psi_n{}^2(\omega,\mathbf{r}_0)|/\Lambda_n$ by its mean value $E_n = \epsilon_{n_x}\epsilon_{n_y}\epsilon_{n_z}\langle\Psi_n{}^2(\omega,\mathbf{r}_0)\rangle$, which is unity if the source is placed at random in the room, more than a wavelength away from the walls, but is $\epsilon_{n_x}\epsilon_{n_y}\epsilon_{n_z}$ if the source is placed at a corner of the room. Thus the equation for the mean-square pressure, not too close to source or walls, is

$$\langle p_\omega{}^2 \rangle = \left(\frac{\rho\omega}{V}\right)^2 |S_\omega|^2 \sum_n \frac{E_n(\mathbf{r}_0)}{[\eta_n{}^2 - k^2 - (kg_n/4V)]^2 + (ka_n/4V)^2}$$

For frequencies in the range defined by Eq. (9.5.20) there are enough excited modes so that the sum can be changed to an integral over the wave-number variable η, using $dN = (\eta^2 V/2\pi^2)\, d\eta$. Thus

$$\langle p_\omega{}^2 \rangle \simeq \frac{\omega^2\rho^2}{2\pi^2} |S_\omega{}^2| E(\mathbf{r}_0) \int_0^\infty \frac{\eta^2\, d\eta}{[\eta^2 - k^2 - (kg_n/4V)]^2 + (ka_n/4V)^2}$$

$$\simeq \frac{\rho^2\omega^2}{\pi\langle a \rangle} |S_\omega{}^2| E(\mathbf{r}_0) \qquad \begin{matrix} 16\lambda^2 < \langle a \rangle \\ \langle a \rangle \text{ and } \langle g \rangle < 4kv \end{matrix} \qquad (9.5.21)$$

where $E(\mathbf{r}_0)$ is 1 if the source is well away from all walls, is 2 if the source is on a wall, is 4 if the source is along an edge, and is 8 if the source is at a corner of the room. This formula is identical with the geometrical formula (9.5.9), except that now a_n has been defined as $-(4V/k)$ times the imaginary part of $K_n{}^2$, sometimes computable from Eq. (9.5.16) or (9.5.17), and $\langle a \rangle$ is the mean value of a_n, averaged over a frequency band of width $c\langle a \rangle/8\pi V$. Thus we see that, at frequencies higher than $4c(\langle a \rangle)^{-\frac{1}{2}}$, the relationship between the mean-square pressure in the room and the source strength S_ω of the driving source does have the form predicted by geometrical acoustics. However, the value of the effective absorption coefficient $\langle a \rangle$, which enters the formula, is only approximately given by Eq. (9.5.2); only when the coupling between the normal modes is large because of room irregularities can we ascribe an effective absorption coefficient to each area of the wall, independent of the mode being absorbed, and only when the wall admittance is small is this coefficient equal to eight times the wall conductance ξ, as prescribed in Eq. (9.5.16).

Transient behavior; reverberation

When the power source driving a particular standing wave is shut off, the wave will damp out exponentially. As indicated in the discussion following Eq. (9.4.15), the damping factor γ_n of Eq. (9.4.3) is proportional to the ratio between the power lost at the walls by the nth mode and the total energy of the standing wave. Since the power lost per unit area of wall at $\mathbf{r}_s$ is $|\Psi'_n{}^2(\mathbf{r}_s)|/\rho c$ times the specific conductance ξ at $\mathbf{r}_s$, and since the energy density at point $\mathbf{r}$ is $|\Psi'_n{}^2(\mathbf{r})|/\rho c^2$, we see that

$$2\gamma_n = \frac{Ac}{V} \frac{\text{mean value of } \xi\, |\Psi'_n{}^2| \text{ over all walls}}{\text{mean value of } |\Psi'_n{}^2| \text{ over room volume}} \tag{9.5.22}$$

where A is the wall area, and V the room volume as before.

Equation (9.4.7) enables us, however, to calculate the damping factor γ_n in terms of the imaginary part of the eigenvalue $K_n{}^2$. Suppose the acoustical source has excited all the standing waves having resonance frequencies between $\nu - \tfrac{1}{2}d\nu$ and $\nu + \tfrac{1}{2}d\nu$ (i.e., all those with K_n between $\eta - \tfrac{1}{2}d\eta$ and $\eta + \tfrac{1}{2}d\eta$, where $\eta = 2\pi\nu/c$). Then, when the source is shut off, the pressure wave in the room will be, according to Eq. (9.4.7),

$$p_s(\mathbf{r},t) = \sum_n{}' A_n e^{-\gamma_n t} \cos(\omega_n t + \Phi_n) \tag{9.5.23}$$

where the sum is over the normal modes, with K_n between $\eta - \tfrac{1}{2}d\eta$ and $\eta + \tfrac{1}{2}d\eta$; the magnitudes of the coefficients A_n are roughly equal in this range of n; and γ_n is the imaginary part of the solution of Eq. (9.4.3), $cK_n(\omega) = \omega \equiv \omega_n - i\gamma_n$.

For a rectangular room with uniform wall admittance, with K_n given by Eq. (9.4.30), we can calculate γ_n by finding a self-consistent solution of the equation $cK_n(\omega) = \omega$, computing K_n from Plate IV. In this case, if $\gamma_n \ll \omega_n$ (as it usually is), we have

$$\omega_n \simeq \pi c \left[\left(\frac{\mu_{n_x}}{l_x}\right)^2 + \left(\frac{\mu_{n_y}}{l_y}\right)^2 + \left(\frac{\mu_{n_z}}{l_z}\right)^2 \right]$$

$$\gamma_n \simeq \frac{\pi^2 c}{K_n} \left[\frac{\mu_{n_x}\nu_{n_x}}{l_x^2} + \frac{\mu_{n_y}\nu_{n_y}}{l_y^2} + \frac{\mu_{n_z}\nu_{n_z}}{l_z^2} \right] \tag{9.5.24}$$

where $a_n = 8VK_n\gamma_n/\omega$ is given by Eq. (9.5.17). Thus the decay factor $e^{-2\gamma_n t}$ for the mean-square pressure of the nth mode has the form of the factor in Eq. (9.5.6), obtained from geometrical acoustics; only now the quantity a_n is not given by Eq. (9.5.2) but by Eq. (9.5.17). For the case we are now considering, a rectangular room with uniform wall admittance, each different standing wave has a different value of γ_n, and thus of a_n, so that

$$\langle p^2 \rangle = \sum A_n{}^2 e^{-2\gamma_n t} \tag{9.5.25}$$

is a combination of exponentials, not a single constant times a single exponential, as the simple geometric analysis would lead us to expect.

As a matter of fact, even when $|\beta|$ is small, the intensity decay will not be a straight line on a logarithmic plot against time, since Eq. (9.4.31) [see also Eq. (9.5.16)] indicates that the effective absorption for oblique waves (see discussion on page 592) differs from that for axial or tangential waves, because of the factors ϵ in the formulas. Thus, for $|\beta| \ll 1$, uniform wall admittance,

$$a_{ob}(\omega) \simeq 8[l_y l_z(\xi_{x0} + \xi_{x1}) + l_x l_z(\xi_{y0} + \xi_{y1}) + l_x l_y(\xi_{z0} + \xi_{z1})]$$

$$a_{\tan,yz} \simeq 8[\tfrac{1}{2}l_y l_z(\xi_{x0} + \xi_{x1}) + l_x l_z(\xi_{y0} + \xi_{y1}) + l_x l_y(\xi_{z0} + \xi_{z1})] \cdots \quad (9.5.26)$$

$$a_{ax,x} \simeq 8[l_y l_z(\xi_{x0} + \xi_{x1}) + \tfrac{1}{2}l_x l_z(\xi_{y0} + \xi_{y1}) + \tfrac{1}{2}l_x l_y(\xi_{z0} + \xi_{z1})]$$

Thus, using Eqs. (9.5.12) to approximate the relative numbers of oblique, tangential, and axial waves in the frequency band between $k - \tfrac{1}{2}dk$ and $k + \tfrac{1}{2}dk$, the mean-square pressure in the decaying sound, for a rectangular room with uniform, not very absorbent walls is

$$\langle |p^2| \rangle \simeq \left[\frac{k^2 V}{2\pi^2} \exp\left(\frac{-ca_{ob}t}{4V} \right) + \frac{kl_y l_z}{2\pi} \exp\left(\frac{-ca_{\tan,yz}t}{4V} \right) \right.$$

$$\left. + \frac{l_x}{\pi} \exp\left(\frac{-ca_{ax,x}t}{4V} \right) + \cdots \right] \langle |A^2| \rangle \, dk \quad (9.5.27)$$

$$\simeq I_0 D_x(t) D_y(t) D_z(t)$$

$$D_x = e^{-(ca_x/4V)t} + \frac{\pi}{kl_x} e^{-(ca_x/8V)t} + \cdots$$

$$a_x = 8l_y l_z(\xi_{x0} + \xi_{x1}) \cdots$$

For this case the sound-decay curve (intensity level vs. time) is a sum of three terms of the form $10 \log D_x(t)$. Each of these terms starts at $t = 0$ as a straight line with negative slope proportional to a_x, etc. A time $(8V/ca_x) \times \ln(c/2\nu l_x)$ later the curve has a break, ending up, beyond this, as a straight line with slope proportional to $\tfrac{1}{2}a_x$, *half* the initial slope. Since each of the additive terms has its "break" at a different time (unless the room is completely symmetrical), the resulting decay curve is quite far from being the straight line that was indicated in Eq. (9.5.6) for a uniform distribution of sound in an irregular room.

At high enough frequencies, the "breaks" come late enough so that the first 20 or 30 db of the curve is nearly straight, with a slope and indicated "reverberation time" corresponding to the oblique waves. If, by chance, this result were assumed to correspond to that given in Eq. (9.5.6) for an irregular room with diffuse sound, the quantities 8ξ (where ξ is the wall-conductivity ratio) would be presumed to equal the absorption coefficient α. We have seen, however, that this is an inaccurate correlation, which may work fairly well for very stiff walls [ξ very small; see the comments on Eq. (9.4.30)] but which fails for more absorptive walls, when the break in slope of the decay curve is more pronounced.

If most of the absorbing material is concentrated on the two opposite walls of a room, those standing waves that do not reflect from the absorbing walls will take about twice as long to die out as do all other waves. When a sound with a "spread" of frequency is used to excite a number of standing waves at the same time, the dying out of these waves after the source is shut off is a rather complicated phenomenon. When only two or three standing waves are excited, these waves as they die out may alternately reinforce and interfere with each other, because they are of slightly different frequency. The intensity will then fluctuate instead of decreasing uniformly, the sort of fluctuations obtained depending on the position of source and microphone in the room and on the manner of starting the sound.

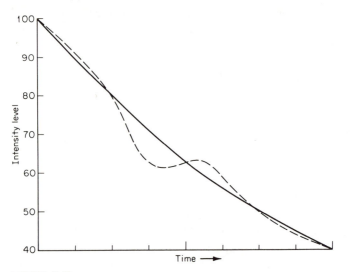

FIGURE 9.19
The decay of sound in a room with two opposite walls more absorbent than the rest. Solid curve shows the average decay of a large number of normal modes; dotted line shows the interference effects possible when only two normal modes have been excited.

If more than three standing waves have been excited, these fluctuations will be more or less averaged out, and the resulting intensity will first diminish uniformly at a rate dependent on the reverberation time of the standing waves which are reflected from all six walls. After these waves have died out, the rest of the sound, due to waves not striking the most absorbent walls, remains and dies out more slowly. The intensity level as a function of time approximates a broken line, the steeper initial part corresponding to most of the standing waves, and the less steep later part due to the waves that do not strike some walls.

In such cases the term "reverberation time" has a specific meaning only in connection with the damping out of single normal modes, i.e., in connection with the slopes of the two portions of the broken line of intensity level against time. The actual length of time that it takes for the intensity level to drop 60 db will depend on the relative amounts of energy possessed by the rapidly and the slowly damped standing waves.

Figure 9.19 illustrates these points; it shows curves of intensity level in a room 10 by 15 by 30 ft as a function of the length of time after the sound is shut off.   The two smallest walls are supposed to be much more absorbent than the other four.   The solid line is the curve for intensity when the room has been excited by a tone with wide enough frequency "spread" to excite 10 standing waves about equally, so that a smooth decay results, and the "break" in the curve is apparent.   The dotted line is the curve when only two standing waves are excited, the resulting fluctuation, due to interference, masking the exponential decay.

These interference oscillations and breaks in the curve for decay of intensity level are present even when all the walls are about equally absorbent, but they are less pronounced.

On the other hand, if the irregularities of wall shape and admittance distribution are great enough so that each standing wave is a more or less random combination of a number of simple modes of the kind given in Eqs. (9.4.39), so that there is no sharp separation between oblique, tangential, and axial waves and no corresponding difference between the mean value of $|\Psi'_n{}^2|$ over the walls or throughout the volume, from one n to another, the values of γ_n for all the waves in the frequency band $d\nu$ will be nearly equal, and the mean-square pressure, during a transient decay, will all attenuate according to the same exponential factor $e^{-i\gamma t}$.   In this case and, in addition, if the frequency band is wide enough [or the frequency satisfies Eq. (9.5.20)], *only then* can we assume that geometric acoustics holds, and we can talk about a room absorption a, for the frequency ν, and set it equal to $8V\gamma/c$.

Problems

1   A uniform tube of length L and inner cross section S is closed at both ends. The tube is vibrated with harmonic motion along its axis, with amplitude X_0 and frequency $\omega/2\pi$, thus producing a sound wave inside the tube.   Show that the pressure amplitude in the tube is $\omega X_0 \rho c \sin [k(x - \tfrac{1}{2}L)]/\cos (kL/2)$ and that the force amplitude required to maintain this field (excluding the mass of the tube) is $2S\omega X_0 \rho c \tan (kL/2)$, where $k = \omega/c$.

2   A uniform rigid duct is terminated at $x = L$ by a plane with specific acoustic admittance $\beta = \xi - i\sigma$ and is driven at the $x = 0$ end by a piston, producing plane waves.   Show that the real and imaginary parts of the pressure ratio (p_0/p_L) can be obtained by the following means: (a) Mark off the angles $\beta = -\tan^{-1} \sigma$ and kL on the unit circle as shown in the figure (points A and B,

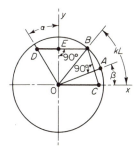

respectively) ($k = \omega/c$). (b) Draw a straight line from B normal to the line OA. The intersection C on the x axis measures the real part OC of the ratio p_0/p_L. (c) Mark off the angle $\alpha = \tan^{-1}\xi$ as shown, and draw the line BD normal to the y axis; DE is then the imaginary part of p_0/p_L.   Show also that a similar construction can be used to find the real and imaginary parts of the velocity ratio (u_0/u_L), where the corresponding angles are given in terms of the impedance, rather than the admittance.

3 A vibrating piston is placed at one end ($x = 0$) of a tube whose cross section is 10 sq cm, and a second piston whose mechanical impedance is to be measured is placed at the other end ($x = 30$).   When the pressure due to the sound wave is measured at different points along the tube, it is found that the pressure amplitude is a maximum at the points $x = 3$, $x = 15$, $x = 27$, having an amplitude of 10 at these points.   The pressure amplitude is a minimum at $x = 9$, $x = 21$, with a value of 6.57.   From these data find the mechanical impedance of the driven piston, the frequency of the sound used, and the amplitude of vibration of the driving piston.

4 A vibrating piston is set in one end of a tube whose length is 86 cm and whose cross section is 10 sq cm.   Closing the other end of the tube is a diaphragm whose overall mechanical impedance is to be measured.   The measured radiation impedance of the driving piston (not including the mechanical impedance of the piston itself) has the following values:

v	100	200	300	400	500	600	700	800
R_p	0	42	147	420	189	840	67	1,260
X_p	−71	1,260	−365	0	126	−840	76	−1,680

What are the real and imaginary parts of the mechanical impedance of the diaphragm at these frequencies?   Plot a curve of the magnitude of the impedance.

5 Two similar pistons, each of mass 10 g, slide freely in opposite ends of a uniform tube of length 34.4 cm and cross section 10 sq cm.   One piston is driven by a force $100e^{-2\pi ivt}$ dynes.   Plot the amplitude of motion of the other piston as a function of v from $v = 0$ to $v = 500$.

6 A Helmholtz resonator has a cylindrical open neck 1 cm long and 1 cm in diameter.   If the resonating vessel is spherical, what must its radius be to have the resonance frequency be equal to 400 cps?

7 The back of a loudspeaker diaphragm looks into a "tank" of volume V, which connects with free space by a constriction of negligible length, and area S of opening.   Set up the equivalent circuit for the analogous impedance at the back of the diaphragm; and give the formula for the additional mechanical impedance load on the diaphragm.   Over what range of frequencies will the motion of air in the constriction be in phase with the motion of the diaphragm *with respect to the outside* (i.e., move out as the diaphragm moves out; be sure to express the phase relations between input current in the equivalent circuit and diaphragm motion correctly) and be larger than the diaphragm motion?   If the area of the constriction is 100 cm², what volume must the "tank" have in order to have this reinforcement of the diaphragm motion come at and below about 100 cps?   Above what frequency will the equivalent circuit be invalid?

8 An air-conditioning system has a circulation fan that produces noise of frequency chiefly above 200 cps.   Design a low-pass acoustic filter, consisting of two vessels and three narrow tubes in series, which will filter out the noise. The narrow tubes cannot be less than 5 cm in diameter, and the vessels cannot have a volume larger than 30,000 cu cm.

9 A duct has a square cross section 34.4 cm on a side.   Use Plate IV to calculate the optimum wall impedance to give maximum attenuation per length of duct at 400 cps for the least attenuated mode.   Repeat the calculations for 1,000 cps.

10 A square duct, $2a$ on a side, has its inner walls coated with material of specific acoustic impedance $R + (iK/\omega)$, with the values of R and K such that the quantities h and φ of Eq. (9.2.19) and Plate IV are given by $he^{i\varphi} = (x^3 - ix^2)/(x^2 + 1)$, where $x = ka/\pi$.   Plot μ and ν, the real and imaginary parts of k_{mn}/k for the lowest two modes (as in Fig. 9.11), against x, from $x = 0$ to $x = 2$.

11 A circular duct of radius a has walls with impedance the same as those of Prob. 10.   Use Plate V to plot μ, ν, real and imaginary parts of k_{mn}/k for the first two modes of wave motion for $0 < x < 2$.   Compare with Fig. 9.12.

12 Use the Schwartz-Christoffel transformation

$$\frac{dz}{dw} = \frac{a}{2\pi w}\sqrt{(w-1)/(w-\gamma)} \qquad \gamma = \frac{a^2}{b^2}$$

$$z = \frac{a}{2\pi}\cosh^{-1}\frac{2w-\gamma-1}{\gamma-1} - \frac{b}{2\pi}\cosh^{-1}\frac{(\gamma-1)w-2\gamma}{(\gamma-1)w}$$

to transform the lower half of Fig. 9.7 onto the real w axis, with points $A \to w = \infty$, $B \to w = 0$, $C \to w = 1$, $D \to w = \gamma$.   Then use the further transformation $w = \exp[(\pi/ua)(\psi + i\theta)]$ to obtain the expressions for analogous impedance and resistance given in Eqs. (9.1.28).

13 Use the Schwartz-Christoffel transformation

$$\frac{dz}{dw} = \frac{b}{\pi(1-w)}\sqrt{\frac{\beta-w}{w}} \qquad \beta = 1 + \left(\frac{a}{b}\right)^2$$

$$z = \frac{b}{\pi}\cos^{-1}\frac{\beta-2w}{\beta} + \frac{a}{\pi}\cosh^{-1}\frac{\beta(w+1)-2w}{\beta(1-w)}$$

to transform the elbow duct shown in the figure onto the real axis of the w plane.   Show that a further transformation, $w = 1 + \exp[(\pi/ua)(\psi + i\theta)]$,

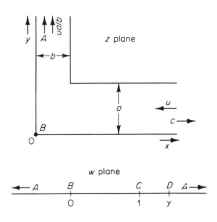

will specify a velocity potential ψ corresponding to fluid coming in from C with velocity u, going irrotationally around the corner and eventually going out to A with velocity ua/b. Show that the analogous impedance of the corner region for long-wavelength sound in the duct is

$$L_a = \frac{\rho}{aud}\left[\psi_+(y) - \psi_-(x) - u(x - b) - \frac{ua}{b}(y - a)\right] \qquad x, y \to \infty$$

$$= \frac{\rho}{\pi d}\left[\left(\frac{a}{b} + \frac{b}{a}\right)\cos^{-1}\left(\frac{b^2 - a^2}{b^2 + a^2}\right) + \frac{\pi a}{b} + 2\ln\frac{b^2 + a^2}{4ab}\right] \qquad b > a$$

$$= \frac{\rho}{\pi d}\left[\pi\left(\frac{2a}{b} + \frac{b}{a}\right) - \left(\frac{a}{b} + \frac{b}{a}\right)\cos^{-1}\left(\frac{a^2 - b^2}{a^2 + b^2}\right) + 2\ln\frac{b^2 + a^2}{4ab}\right] \qquad a > b$$

where d is the depth of the duct in the z direction.

14 A square duct 2 ft on a side carries cooled air; it also carries unwanted 200-cps fan noise. If the duct walls were covered with acoustic material of specific impedance, $z/\rho c = 2 + 5i$ at 200 cps, what length of duct must be covered in order to reduce the noise by 60 db? Suppose the noise is to be attenuated by attaching a sequence of Helmholtz resonators, of the sort shown in C and D of Fig. 9.4, the holes C being a foot apart along the duct, each of the volumes V_D being $\frac{1}{4}$ cu ft, and the openings C being adjusted in size so that the resonance frequency is 200 cps. What length of duct must have this treatment in order to attenuate the sound 60 db?

15 If the cross-sectional area S of the duct changes slowly with x, show that the equations for long-wavelength sound are

$$\frac{\partial p}{\partial t} = -\frac{\rho c^2}{S}\frac{\partial Su}{\partial x} \qquad \rho\frac{\partial u}{\partial t} = -\frac{\partial p}{\partial x} \qquad \frac{1}{S}\frac{\partial}{\partial x}\left(S\frac{\partial p}{\partial x}\right) = \frac{1}{c^2}\frac{\partial^2 p}{\partial t^2}$$

Show that when $S = S_0[y(x)]^2$, if we set $p = (A/y)\exp[i(\omega/c)\int\tau\,dx - i\omega t]$, the wave equation will be approximately satisfied if we set

$$\tau^2 = 1 - \left(\frac{c}{\omega}\right)^2\frac{1}{y}\frac{d^2y}{dx^2}$$

and if we can neglect $d\tau/dx$ compared with ω/c (this is called the WKB approximation). Show that, when $y(x) = e^{x/h}$, τ is a constant and the phase velocity is $c/\sqrt{1 - (c/\omega h)^2}$. What happens when ω is smaller than c/h?

16 A dipole of strength D_ω, radius a, and frequency $\omega/2\pi$ is placed at the point x_0, y_0 in the plane $z_0 = 0$ in a duct of cross section S. Show that if the dipole is pointed in the x direction, transverse to the duct axis, its impedance [Eqs. (7.1.12) and (9.2.12)] is

$$Z_d \simeq -i\omega\left(\frac{2\pi\rho a^3}{3}\right)\left[1 - \frac{2\pi a^3}{S}\sum_{n=1}^{N}\frac{1}{\Lambda_n\varkappa_n}\left(\frac{\partial\Psi_n}{\partial x}\right)^2_{x_0y_0}\right]$$

$$+ \rho c\left(\frac{2\pi^2 a^6}{3S}\right)\sum_{n=1}^{N}\frac{k}{\Lambda_n k_n}\left(\frac{\partial\Psi_n}{\partial x}\right)^2_{x_0y_0}$$

and that if it is pointed in the z direction, along the axis, it is

$$Z_d \simeq -i\omega\left(\frac{2\pi\rho a^3}{3}\right)\left[1 - \frac{3\pi a^2}{S}\sum_{n=1}^{N}\frac{\varkappa_n a}{\Lambda_n}\Psi'_n{}^2(x_0 y_0)\right]$$

$$+ \rho c\left(\frac{2\pi a^2}{S}\right)k^2 a^2\left[1 + \sum_{n=1}^{N}\frac{k_n a}{\Lambda_n}\Psi'_n{}^2(x_0 y_0)\right]$$

17   Plot the real parts of the impedances, times $b^4/2\rho c a^4$, of Prob. 16, for the dipole on the axis of a circular duct of inner radius b, as functions of kb/π from 0 to 3. Compare the results with the real part of Z_d of Eqs. (7.1.12) for free space.

18   A rigid wall, coincident with the xy plane, has a circular hole of radius a in it, centered at the origin.   This hole is the open end of a cylindrical pipe of inside radius a, extending to $z = -\infty$.   A plane wave of sound, of frequency $kc/2\pi$, comes along the inside of the pipe from $-\infty$.   Part of the wave radiates out of the open end, into the unbounded region $z > 0$; part is reflected back to $z = -\infty$.   Show that the pressure wave is

$$p(r,z) = \begin{cases} \rho c\cosh(\Phi + ikz) - \rho c\displaystyle\sum_{n=1}^{\infty}\frac{k}{k_n}U_n J_0\left(\frac{\pi\alpha_{0n}r}{a}\right)e^{-ik_n z} & z < 0 \\ \displaystyle\int_0^\infty J_0(\mu r)e^{iqz}\,\mu\,d\mu\int_0^a u_0(r_0)J_0(\mu r_0)r_0\,dr_0 & z > 0 \end{cases}$$

where $k_n{}^2 = k^2 - (\pi\alpha_{0n}/a)^2$, $q^2 = k^2 - \mu^2$, $u_0(r)$ is the axial velocity of the fluid in the open end of the tube at $z = 0$, and

$$U_n = \frac{2}{a^2 J_0{}^2(\pi\alpha_{0n})}\int_0^a u_0(r)J_0\left(\frac{\pi\alpha_{0n}r}{a}\right)r\,dr$$

Set up a variational expression for u_0 and show that, if we use $u_0 = A\chi(r)$ as a trial function, the "best" value for the effective impedance of the open end of the tube is

$$\rho c\coth\Phi = \frac{\rho c a^2}{Q}\int_0^a \chi(r)r\,dr\int_0^a\left[\frac{2}{a^2}\sum_{n=1}^{\infty}\frac{k}{k_n}\frac{J_0(\pi\alpha_{0n}r_0/a)J_0(\pi\alpha_{0n}r/a)}{J_0{}^2(\pi\alpha_{0n})}\right.$$

$$\left. + k\int_0^\infty J_0(\mu r)J_0(\mu r_0)\frac{\mu}{q}\,d\mu\right]\chi(r_0)r_0\,dr_0$$

where $Q = \left(2\displaystyle\int_0^a \chi r\,dr\right)^2$.   Show that, if we choose $\chi = 1$, the impedance, to this order of approximation, is $\rho c(\theta_0 - i\chi_0)$, the impedance function of Eqs. (7.4.31).

19   A duct consists of the space between two concentric cylinders, the outer one of radius a, the inner one of radius b.   Show that the normal modes in the annular space between cylinders have the forms

$$\Psi_{mn} = {\cos\atop\sin}(m\varphi)\phi_{mn}(r)\exp(iK_{mn}z)$$

where

$$\phi_{mn}(r) = J_m\left(\frac{\pi\gamma_{mn}r}{a}\right) - \frac{J_m'(\pi\gamma_{mn}b/a)}{N_m'(\pi\gamma_{mn}b/a)} N_m\left(\frac{\pi\gamma_{mn}r}{a}\right)$$

$$K_{mn}^2 = k^2 - \varkappa_{mn}^2 \qquad \varkappa_{mn} = \frac{\pi\gamma_{mn}}{a}$$

and where, if the cylinders are rigid, γ_{mn} is determined by

$$\frac{J_m'(\pi\gamma_{mn})}{N_m'(\pi\gamma_{mn})} = \frac{J_m'(\pi\gamma_{mn}b/a)}{N_m'(\pi\gamma_{mn}b/a)}$$

A table of values of $[1 - (b/a)]\gamma_{mn}$ is as follows:

$\dfrac{a}{b}$	$m = 0$		$m = 1$			$m = 2$		
	$n = 1$	$n = 2$	$n = 0$	$n = 1$	$n = 2$	$n = 0$	$n = 1$	$n = 2$
1	1.000	2.000	0.000	1.000	2.000	0.000	1.000	2.000
2	1.018	2.009	0.215	1.045	2.022	0.427	1.114	2.059
3	1.041	2.023	0.327	1.119	2.060	0.622	1.283	2.164
4	1.062	2.038	0.392	1.195	2.103	0.718	1.478	2.324

For $b \to a$:

$$\gamma_{mn} \to \frac{na}{a-b} + \frac{4m^2 + 3}{8\pi^2 nb}(a - b) + \cdots$$

For the fundamental mode $\gamma_{00} = 0$ and $\phi_{00} = 1$, the constant $\Sigma_{00} = 1 - (b/a)^2$.

Show that

$$\int_b^a (\Psi_{mn})^2 r \, dr = \tfrac{1}{2}a^2 \Sigma_{mn}$$

where

$$\Sigma_{mn} = \phi_{mn}^2(a) - \left(\frac{b}{a}\right)^2 \phi_{mn}^2(b) - \left(\frac{m}{\pi\gamma_{mn}}\right)^2 [\phi_{mn}^2(a) - \phi_{mn}^2(b)]$$

For the fundamental mode, where $\gamma_{00} = 0$ and $\phi_{00} = 1$, show that $\Sigma_{00} = 1 - (b/a)^2$. List the m, n pairs in order of increasing $\varkappa_{mn}$, for the first four modes, and for $a = 4b$, compute $\varkappa_{mn}a$, Σ_{mn}, and the coefficient of N_m in ϕ_{mn} for these four modes.

20 A concentric circular duct of the sort described in Prob. 19 extends from $z = 0$ to $z = -\infty$; it is joined to a simple circular duct of radius a, which extends from $z = 0$ to $z = \infty$, so that the outer wall is continuous and the inner cylinder of radius b stops at $z = 0$. Call the transverse modes for the annular duct $(z < 0)$ $\cos(m\varphi)\phi_{mn}(r)$ as defined in Prob. 19; call the modes for $z > 0$ $\cos(m\varphi)\Psi_{ms}(r)$, where

$$\Psi_{ms}(r) = J_m\left(\frac{\pi\alpha_{ms}r}{a}\right) \qquad J_m'(\pi\alpha_{ms}) = 0 \qquad k_{ms}^2 = k^2 - \left(\frac{\pi\alpha_{ms}}{a}\right)^2$$

$$\int_0^a \Psi_{ms}^2 r \, dr = -a^2 \Lambda_{ms} \qquad \Lambda_{ms} = \left[1 - \left(\frac{m}{\pi\alpha_{ms}}\right)^2\right] J_m^2(\pi\alpha_{ms})$$

A wave, $\cos(m\varphi)\phi_{mn}(r) \exp(iK_{mn}z)$, comes from $z = -\infty$ and strikes the discontinuity at $z = 0$; part is reflected, and part transmitted to $z = \infty$. Show that the wave motion is

$$
p = \begin{cases}
\left[\dfrac{pck}{K_{mn}} \phi_{mn} \cosh (\Phi_{mn} + iK_{mn}z) - \sum_{l \neq n} \dfrac{pck}{K_{ml}} U_l{}^n \phi_{ml} e^{-ik_{ml}z} \right] \\
\hspace{6cm} \times \cos (m\varphi) \qquad z < 0 \\[2mm]
\cos (m\varphi) \sum_s \dfrac{pck}{k_{ms}} V_s{}^n \Psi_{ms}(r) e^{-ik_{ms}z} \hspace{2.2cm} z > 0
\end{cases}
$$

where $U_l{}^n$ and $V_s{}^n$ are integrals involving the axial velocity in the plane $z = 0$. Show that a variational expression, suitable for varying the magnitude and shape of this velocity distribution, becomes, when the amplitude variation has been carried out,

$$
\coth \Phi_{mn} \simeq \tfrac{1}{2} a^2 \Sigma_{mn} \frac{\int \chi \int G_{mn}{}^+ \chi + \int \chi \int G_{mn}{}^- \chi}{(\int \chi \phi_{mn})^2}
$$

$$
A \simeq \frac{(\int \chi \phi_{mn}) \cosh \Phi_{mn}}{\int \chi \int G_{mn}{}^+ \chi + \int \chi \int G_{mn}{}^- \chi} \qquad
U_l{}^n \simeq \frac{\Sigma_{mn} \int \chi \phi_{mn}}{\Sigma_{mn} \int \chi \phi_{ml}} \sinh \Phi_{mn}
$$

$$
V_s{}^n \simeq \frac{\Sigma_{mn} \int \chi \Psi_{ms}}{\Lambda_{ms} \int \chi \Psi_{mn}} \sinh \Phi_{mn} \qquad \text{Trial function } u_n^0 = A\chi(r)
$$

where

$$
G_{mn}{}^- = \sum_{l \neq n}^{\infty} \frac{K_{mn}}{K_{ml}} \frac{\phi_{ml}(r)\phi_{ml}(r_0)}{\tfrac{1}{2} a^2 \Sigma_{ml}} \qquad
G_{mn}{}^+ = \sum_{s=0}^{\infty} \frac{K_{mn}}{k_{ms}} \frac{\Psi_{ms}(r)\Psi_{ms}(r_0)}{\tfrac{1}{2} a^2 \Lambda_{ms}}
$$

What reasons can you give for choosing $\chi = (r/a)^m + (a/r)^m$ for trial function, if the incident wave is the lowest mode for a given m ($m0$ mode)? Show that, if it is used, the "best" value of the impedance is (assume that $\pi \gamma_{m0} < ka < \pi \gamma_{m1}$ and $\pi \alpha_{mN} < ka < \pi \alpha_{m, N+1}$)

$$
\coth \Phi_{m0} \simeq \sum_{s=0}^{N} \frac{M_{m0}}{N_{ms}} \frac{\Sigma_{m0}}{\Lambda_{ms}} \left(\frac{\pi \gamma_{m0} Y_{ms}}{\pi \alpha_{ms} X_{m0}} \right)^2 - ika \sum_{S=N+1}^{\infty} \frac{M_{m0} \Sigma_{m0} (\pi \gamma_{m0})^2}{Q_{ms} \Lambda_{ms} (\pi \alpha_{ms})^3} \left(\frac{Y_{ms}}{X_{m0}} \right)^2
$$

$$
- ika \sum_{l=1}^{\infty} \frac{M_{m0} \Sigma_{m0} (\pi \gamma_{m0})^2}{L_{ml} \Sigma_{ml} (\pi \gamma_{ml})^3} \left[\frac{X_{ml}(b/a)}{X_{m0}(b/a)} \right]^2
$$

where

$$
M_{m0}{}^2 = 1 - \left(\frac{\pi \gamma_{m0}}{ka} \right)^2 \qquad
L_{ml}{}^2 = 1 - \left(\frac{ka}{\pi \gamma_{ml}} \right)^2
$$

$$
N_{ms}{}^2 = 1 - \left(\frac{\pi \alpha_{ms}}{ka} \right)^2 \qquad
Q_{ms}{}^2 = 1 - \left(\frac{ka}{\pi \alpha_{ms}} \right)^2
$$

and

$$
\int_b^a \chi \Psi_{ms} r \, dr = \frac{a^2}{\pi \alpha_{ms}} \left[\left(\frac{a}{b} \right)^{m-1} J_{n-1}\left(\pi \alpha_{ms} \frac{b}{a} \right) - \left(\frac{b}{a} \right)^{m+1} J_{m+1}\left(\pi \alpha_{ms} \frac{b}{a} \right) \right]
$$

$$
\equiv \frac{a^2}{\pi \alpha_{ms}} Y_{ms} \frac{b}{a}
$$

$$
\int_b^a \chi \phi_{ml} r \, dr = \frac{a^2}{\pi \gamma_{ml}} \left(\frac{a}{b} \right)^{m-1} \left[J_{m-1}\left(\pi \gamma_{ml} \frac{b}{a} \right) + R_{ml} N_{m-1}\left(\pi \gamma_{ml} \frac{b}{a} \right) \right]
$$

$$
\times \left[1 - \left(\frac{b^2}{a^2} \right)^{m+1} \right]
$$

$$
\equiv \frac{a^2}{\pi \gamma_{ml}} X_{ml} \frac{b}{a}
$$

where

$$R_{ml} = \frac{J_{m-1}(\pi\gamma_{ml}) - J_{m+1}(\pi\gamma_{ml})}{N_{m+1}(\pi\gamma_{ml}) - N_{m-1}(\pi\gamma_{ml})}$$

Does the expression for $\coth \Phi_{m0}$ go to infinity? For what values of ka? Should it? What can be done to improve the result?

21 Show, by use of Plates IV and V, that the approximate formulas for q of Eqs. (9.2.14) and (9.2.24) hold reasonably well (within about 10 percent) when βkb is less than unity.

22 Show that, if the walls of a cavity are mass-controlled (susceptance σ negative, conductances negligible), a low-frequency nodeless ($n = 0$) standing wave is possible, with frequency approximately proportional to $-\sigma$. For a cubical enclosure, of length b on a side, show that this frequency is $\omega_{000}/2\pi$, where $\omega_{000} \simeq (6c/b)(-\sigma)$, and that the corresponding pressure amplitude is $\Psi_{000} \simeq \cos(-2\sigma x/b)\cos(-2\sigma y/b)\cos(-2\sigma z/b)$ ($-\sigma$ small) if the origin is placed at the center of the cube.

23 A rectangular room, with sides b, $b\sqrt{\tfrac{1}{2}}$, $b\sqrt{\tfrac{2}{3}}$ has all its walls locally reacting, with specific impedance $\zeta = 5 + i(20/kb)$. Use Eq. (9.4.31) to compute the real and imaginary parts of the eigenvalue $K_n b$, for all the standing waves having $\eta_n b$ less than 7, and the expressions for the corresponding pressure amplitudes Ψ_n. A simple source of strength S_ω is placed near one corner. Plot the response of the room, $(V/\rho c S_\omega b)^2 |p|^2$, at the center of the room and at the corner opposite the source, against kb from 0 to 10.

24 The radius of the simple source of Prob. 23 is a. Plot $Vb/4\pi a^6 \rho c$ times the real part of the impedance of Eq. (9.4.6), against kb, from 0 to 7, when the source is close to one corner of the room; when the source is at the center of the room.

25 Placing the origin at the center of the room of Prob. 23, with the x axis in the direction of the longest dimension and the z axis in the direction of the shortest dimension, the wall at $x = \tfrac{1}{2}b$ has a band of acoustic material of admittance $\beta = \tfrac{1}{2}$ between the lines $y = \pm\tfrac{1}{4}b\sqrt{\tfrac{2}{3}}$ and another band of the same material on the wall at $z = b\sqrt{\tfrac{1}{8}}$ between the lines $x = \pm\tfrac{1}{4}b$; the remaining walls have zero admittance. Use Eqs. (9.4.15) to compute the values of $K_n b$ for the first 10 standing waves. As a measure of the degree of "mixing" of the modes, calculate the magnitude of the largest coefficient of ϕ_n in the expression for Ψ_N ($n \neq N$), for each of these waves.

26 A cylindrical room has radius b and height b. Duplicate the calculations of Probs. 23 and 24 for this room, for the range of $\eta_n b$ from 0 to 5.

27 A room with cylindrical walls, of radius 5 m, has a flat floor and ceiling 4 m apart. Plot the number of allowed frequencies in the room between ν and $\nu + 5$ as a function of ν from $\nu = 0$ to $\nu = 50$. Above what frequency will this curve become fairly uniform?

28 A rectangular corridor is 2 m wide, 3 m high, and 10 m long. Plot the number of allowed frequencies in the enclosure between ν and $\nu + 5$ as a function of ν from $\nu = 0$ to $\nu = 100$. Above what frequency will this curve be fairly uniform?

29 A cubical room 5 m on a side has an average absorption coefficient for floor and ceiling of 0.2; for the walls, a value of 0.04. What is the reverberation time for those waves which strike floor and ceiling? For those waves which do not strike floor and ceiling? List all the allowed frequencies between 0 and

100 cps and give the position of the nodal planes and the reverberation times of each corresponding standing wave.

30 List frequencies, nodes, and reverberation times of the normal modes between $\nu = 0$ and $\nu = 100$ for the room of Prob. 29 when walls, floor, and ceiling all have an average absorption coefficient of 0.1.

31 The air in the room of Prob. 29 is started into vibration so that all the normal modes between $\nu = 98$ and $\nu = 102$ are set into motion with equal initial amplitudes. What normal modes are excited? Plot the decay curve of intensity level against time after the source is shut off at the midpoint of the room; at the midpoint of one wall; at a point 167 cm out from two walls and 250 cm up from the floor.

CHAPTER 10

COUPLING OF ACOUSTICAL SYSTEMS

10.1 SYSTEMS OF INFINITE EXTENT

In the earlier chapters of this book we studied the vibrations and wave motions of homogeneous systems of infinite, and also of finite, extent. In Chaps. 4 and 5, for example, we discussed the traveling and standing waves in uniform strings, membranes, and plates in vacuum; in Chap. 6 we took up the study of the wave motion in a homogeneous fluid of infinite extent, uncoupled to any vibrating solid system. Even in Chap. 7 we assumed that the radiated waves were produced by vibrating sources having known velocities, so that, though the waves in the medium were produced by the vibrations of the solid source, we effectively neglected to include the reaction of the medium back on the source.

In many cases this neglect is justifiable, particularly if the source is small in size compared with the wavelength and is stiff in comparison with the medium. But if the vibrating solid system is extended in size and/or if it also supports wave motion, which can be induced by the reaction of the medium, the coupling between solid and fluid systems should be included from the start in our theory. Many problems of practical interest involve the interaction between two or more wave-carrying systems, coupled together so strongly that the motion of each is noticeably affected by the other. Membranes or plates do not usually vibrate *in vacuo*; they affect and are affected by the air (or water) in contact with their surfaces; the velocity of their transverse waves is appreciably modified by the contact.

In this chapter we shall take up a few examples of this coupled motion, indicate the characteristic results of the coupling, and illustrate some of the methods by which these characteristics can be calculated. Some systems are of sufficient size so that we can observe traveling, rather than standing, waves; these will be discussed in the first section. We shall see that the coupled motion involves two (or more) different sorts of waves, each kind involving motion of both systems, but with the apportionment of energy between the systems different for the different kinds. We shall also see that the

608

characteristics of these different kinds of wave will depend markedly on the relationship between the wave velocities in the two (or more) separate systems.

There are two possible traveling, transverse waves on a membrane in contact with air, for example, one wave having velocity near that for the same membrane *in vacuo*, and the other with velocity near that for sound in air. Both waves involve the motion of both membrane and air. If the wave velocity of the membrane *in vacuo* is less than that of air, the former wave, but not the latter, can propagate along the membrane without attenuation. On the other hand, if the sound velocity in air is not as great as that of transverse waves on the membrane *in vacuo*, the situation is reversed; the latter kind of wave will "cling" to the membrane and propagate along it without attenuation, whereas the former kind will continually radiate energy away from the membrane into the air; the membrane wave of this kind will lose energy as it travels along, and thus will be attenuated.

Elastic, porous solid

The simplest example of coupled wave motion is that of two copenetrating media which couple together at every point in space. A porous solid, such as a piece of sound-absorbing material, for instance, couples to the air in its pores. The solid portion of the system can be considered as an interlocking mesh of elastic fibers, with a characteristic compressibility and effective mass density. *In vacuo* we could consider this mesh, for waves considerably longer than pore size, as an elastic medium of mean density ρ_s and volume compressibility κ_s (s meaning "solid" here, not isentropic). It also would support shear waves, but we are not interested in following this example further than discussing the coupling effects in the compressional waves.

The motion of the air in the pores has been discussed in Sec. 6.2; it also has an effective density ρ_a, which may not be equal to the density of air in free space, and an effective bulk modulus κ_a, which was written as $\Omega\kappa_p$ in Eq. (6.2.22). These two systems may have different displacements from equilibrium and different Eulerian velocities, defined so that $\mathbf{u}_a$ is the volume of air flowing per unit time per unit area through the pores, and $\rho_a\mathbf{u}_a$ is the fluid momentum per unit volume; $\mathbf{u}_s$ is the velocity, and $\rho_s\mathbf{u}_s$, the momentum per unit volume for the solid portion of the system.

Next we note that compressional force can be applied to the two parts of the system independently. Pressure can be applied to the air, to change its density without appreciably changing the density of the solid fibers; also, forces can be applied to the fiber network which will compress the net without appreciably affecting the interspersed air. This latter force, in the large, can be considered to have a mean compressional component p_s which will change the fiber spacing and thus the mean solid density, the gradient of which will produce acceleration of the network (there also will be shear stresses, but these are not pertinent to our discussion). True, compression of the solid network will reduce the porosity Ω somewhat and thus will change the density of the air in the pores, but if Ω is not very small and if the compressibility of the fibers is considerably smaller than that of the air, this coupling effect will be of second order and can be neglected for small-amplitude waves.

The major part of the coupling between the two systems will come in the frictional force arising when the velocities of the solid net and the entrapped air are not equal. In Eq. (6.2.23), when $\mathbf{u}_s$ was considered to be zero, we defined this force in terms of a measurable flow resistance Φ. The generalization is obvious: the coupling force is $\Phi(\mathbf{u}_a - \mathbf{u}_s)$, acting with opposite sign on the two systems. Thus, to first order, the equations of continuity of the two portions are separate, but the equations of motion are coupled by the flow resistance.

$$\kappa_a \frac{\partial p_a}{\partial t} = -\text{div } \mathbf{u}_a \qquad \kappa_s \frac{\partial p_s}{\partial t} = -\text{div } \mathbf{u}_s$$

$$\rho_a \frac{\partial \mathbf{u}_a}{\partial t} = -\text{grad } p_a + \Phi(\mathbf{u}_s - \mathbf{u}_a) \tag{10.1.1}$$

$$\rho_s \frac{\partial \mathbf{u}_s}{\partial t} = -\text{grad } p_s + \Phi(\mathbf{u}_a - \mathbf{u}_s)$$

If no air is present ($\rho_a \to 0$, $\kappa_a \to \infty$), p_a is zero, $\mathbf{u}_a$ equals $\mathbf{u}_s$, and the solid network will transmit compressional waves with a wave velocity $c_s = \sqrt{1/\rho_s \kappa_s}$ (κ_s often is complex, because of internal friction in the fiber net). Likewise, if the density and compressibility of the fiber structure were equal to that of the air, $\mathbf{u}_s$ again would equal $\mathbf{u}_a$, and the wave velocity would equal that of air, $c_a = \sqrt{1/\rho_a \kappa_a}$. For intermediate cases the coupling will modify the wave velocity of both parts of the system.

To demonstrate this we ask for the plane-wave solutions, setting

$$p_a = P_a e^{ikx - i\omega t} \qquad u_a = \frac{\omega \kappa_a}{k} p_a = \frac{\omega}{\rho_a c_a^2 k} p_a$$

$$p_s = P_s e^{ikx - i\omega t} \qquad u_s = \frac{\omega \kappa_s}{k} p_s = \frac{\omega}{\rho_s c_s^2 k} p_s$$

and solving for k as a function of ω, the κ's, and ρ's.

$$i\omega \Phi \kappa_s P_s = (\omega^2 \alpha_a^2 - k^2) P_a \quad \text{and} \quad i\omega \Phi \kappa_a P_a = (\omega^2 \alpha_s^2 - k^2) P_s$$

where

$$\alpha_s^2 = \kappa_s \rho_s + i \frac{\kappa_s \Phi}{\omega} \quad \text{and} \quad \alpha_a^2 = \kappa_a \rho_a + i \frac{\kappa_a \Phi}{\omega}$$

or

$$k^2 = \tfrac{1}{2}\omega^2(\alpha_s^2 + \alpha_a^2) \pm \tfrac{1}{2}[\omega^4(\alpha_s^2 - \alpha_a^2)^2 - 4\omega^2 \kappa_a \kappa_s \Phi^2]^{\frac{1}{2}} \tag{10.1.2}$$

When $|c_a{}^2 - c_s{}^2| \gg 4\Phi^2/\omega^2\rho_a\rho_s$, the approximation

$$k^2 \simeq \left(\frac{\omega}{c_s}\right)^2 + i\Phi \frac{\omega\rho_s}{c_s{}^2} + \frac{\Phi^2}{\rho_s\rho_a} \frac{1}{c_a{}^2 - c_s{}^2}$$

or

$$\simeq \left(\frac{\omega}{c_a}\right)^2 + i\Phi \frac{\omega\rho_a}{c_a{}^2} + \frac{\Phi^2}{\rho_s\rho_a} \frac{1}{c_s{}^2 - c_a{}^2}$$

is valid, indicating two possible waves, with two velocities,

$$\frac{\omega}{k} \simeq c_a\left(1 + \frac{\Phi^2}{2\omega^2\rho_a\rho_s} \frac{c_a{}^2}{c_a{}^2 - c_s{}^2}\right) \quad \text{or} \quad c_s\left(1 + \frac{\Phi^2}{2\omega^2\rho_a\rho_s} \frac{c_s{}^2}{c_s{}^2 - c_a{}^2}\right)$$

The result is reminiscent of that of Eqs. (3.1.3) for a pair of simple coupled oscillators. The velocity of the faster wave is increased by the coupling; that of the slower wave is diminished. Inserting the result into the equation relating the magnitudes P_a and P_s shows that the wave with velocity near c_a has most of its energy in air motion; that with velocity near c_s carries most of its energy in motion of the solid network.

This particular example will not be carried further; all that was intended was to point out the parallel with the coupled oscillators, to demonstrate that there will be as many different kinds of waves as there are different systems interacting, and that the particular wave which has its velocity near that of one of the component systems will entrust its energy chiefly to that component. In the more interesting and more complicated examples taken up in the rest of this section, these general characteristics will appear, though somewhat modified by the complications of the energy flow, which can be concentrated close to a component system, traveling only in one direction, or else can radiate outward to infinity.

Flexible string

As an example of somewhat greater practical interest, we take up next the effect of the surrounding medium on a flexible string of diameter $2a$, of infinite length, under tension T. According to the discussion of Chap. 4, the velocity of transverse waves in such a string, if coupling to the surrounding medium is ignored, is $c_s = \sqrt{T/\pi a^2\rho_s}$, where ρ_s is the density of the string material, usually much larger than ρ, the density of the surrounding medium. Thus we have two dimensionless parameters, $\alpha = c/c_s$, the ratio of acoustic wave velocity in the medium to the string-wave velocity, and $\mu = \rho/\rho_s$, the ratio of density of the medium to that of the string material. There also are two wavenumbers, $k = \omega/c$ for the medium, and K for the string, which equals ω/c_s *in vacuo* but which differs from this in the presence of the medium.

We first look for possible free vibrations of the coupled system, to see whether there can be a simple-harmonic wave propagating along the string, for example. If such a wave were possible, the dependence of all parts of the system on z, the distance along the string, and on time t, would be via the factor $\exp(iKz - i\omega t)$. Presumably, this is possible only when there is a particular relationship between frequency $\omega/2\pi = kc/2\pi$ and the string wavenumber K, which must be determined. The ratio (ω/K) is then the phase velocity of the wave along the string.

Suppose the displacement $y(z,t)$ of the string, from its equilibrium position along the z axis, is in the yz plane. If the wave is to have the characteristics we have assumed, the displacement should be

$$\eta = A \exp(iKz - i\omega t)$$

and the velocity, normal to the string surface $(r = a)$ at the surface, is

$$u_r = -i\omega A \cos\varphi \exp(iKz - i\omega t)$$

where r, φ are the polar coordinates in the xy plane, φ being the angle between r and the yz plane of displacement of the string.

The pressure wave in the medium outside the string, produced by the string's motion, would then be

$$p(r,\varphi,z,t) = P_0 \cos\varphi \, H_1^{(1)}(qr) \exp(iKz - i\omega t)$$

where $H_1^{(1)}(u) = J_1(u) + iN_1(u) \rightarrow -i\sqrt{\pi/2iu}\, e^{iu}$ $(u \rightarrow \infty)$ is the Hankel function for outgoing waves for $m = 1$, and $q = \sqrt{k^2 - K^2}$ when $k = \omega/c > K$ and $= i\sqrt{K^2 - k^2}$ when $K > k$. If $K > k$, the argument of the Hankel function is imaginary, and the wave attenuates exponentially; no acoustic energy radiates to $r \rightarrow \infty$.

The radial velocity of the medium at $r = a$ is

$$u_r = \frac{1}{ik\rho c}\left(\frac{\partial p}{\partial r}\right)_{r=a} = -i\frac{P_0}{\rho c}\frac{q}{k}\cos\varphi\, H_1'(qa)\exp(iKz - i\omega t)$$

where

$$H_1'(u) = \frac{\partial H_1^{(1)}(u)}{\partial u} = \tfrac{1}{2}H_0(u) - \tfrac{1}{2}H_2(u) = H_0(u) - \frac{1}{u}H_1(u)$$

omitting the superscripts for convenience. This must equal the radial velocity of the string surface, which provides a relationship between P_0 and A, the amplitude of motion of the string,

$$P_0 = \frac{k^2\rho c^2}{qH_1'(qa)}A \tag{10.1.3}$$

The wave motion in the medium reacts back on the string. The net force in the y direction, per unit length of string, is

$$F(z,t) = \int_0^{2\pi} d\varphi\, a \cos\varphi\, p(a,\varphi,z,t) = \pi a P_0 H_1(qa)\exp(iKz - i\omega t)$$

$$= \frac{\pi a k^2\rho c^2}{q}\frac{H_1(qa)}{H_1'(qa)}A\exp(iKz - i\omega t) \tag{10.1.4}$$

The equation of motion of the string is therefore

$$\pi a^2 \rho_s \frac{\partial^2 \eta}{\partial t^2} = T \frac{\partial^2 \eta}{\partial z^2} + F \qquad T = \frac{\pi a^2 \rho_s}{c_s^2}$$

or

$$(k^2\alpha^2 - K^2)A = \frac{k^2\alpha^2\mu}{qa} \frac{H_1(qa)}{H_1'(qa)} A \qquad (10.1.5)$$

where $\alpha = c/c_s$ and $\mu = \rho/\rho_s$. This is the equation which relates K and k. Usually, the wavelength of sound in the medium is much longer than the string diameter $2a$. When this is the case we can use the first few terms in the series expansion for the Hankel function.

$$H_1(u) \simeq -\frac{2i}{\pi u} + \tfrac{1}{2}u\left[1 - \frac{i}{\pi} + \frac{2i}{\pi}\ln{(Cu)}\right] \qquad C = 0.8905$$

$$\frac{H_1(u)}{H_1'(u)} \simeq -u[1 - u^2\ln{(Cu)} + \tfrac{1}{2}\pi i u^2] \qquad \text{if } u \text{ is real}$$

If u is positive-imaginary, the imaginary term in brackets in the last equation vanishes, and the logarithm is of the magnitude of u. Since $u = qa$ and $u^2 = k^2a^2 - K^2a^2$, we obtain an approximate relation between K, the wavenumber in the z direction, and k, the wavenumber in the medium, ω/c, determined by the frequency.

$$K \simeq \begin{cases} k\alpha\{1 + \mu - (1 - \alpha^2 - \mu\alpha^2)k^2a^2\mu[\ln{(Cka\sqrt{1 - \alpha^2 - \mu\alpha^2})} - \tfrac{1}{2}\pi i]\}^{\frac{1}{2}} \\ \qquad\qquad\qquad\qquad\qquad\qquad\qquad \text{when } (1 + \mu)\alpha^2 < 1 \\ k\alpha\{1 + \mu + (\alpha^2 + \mu\alpha^2 - 1)k^2a^2\mu\ln{(Cka\sqrt{\alpha^2 + \mu\alpha^2 - 1})}\}^{\frac{1}{2}} \\ \qquad\qquad\qquad\qquad\qquad\qquad\qquad \text{when } (1 + \mu)\alpha^2 > 1 \end{cases}$$

$$(10.1.6)$$

When the string is heavy and the tension is not great enough to make $c_s = \sqrt{T/\pi a^2\rho_s}$ larger than c, quantity $\alpha^2(1 + \mu)$ is larger than unity. In this case the wavelength $2\pi/K$ of transverse waves on the string is smaller than the wavelength $2\pi/k$ of waves in the medium, and an outgoing wave, carrying energy from the string to $r \to \infty$, cannot be fitted onto the wave in the string. The radial wavenumber component $q \simeq ik\sqrt{\alpha^2(\mu + 1) - 1}$ is pure-imaginary, and the function $H_1(qr)$ is a real exponential, attenuating to zero as $r \to \infty$. None of the energy possessed by the string or by the motion of the fluid close to the string can be radiated away, except along the string to $z \to \infty$. The phase velocity of this wave in the z direction is [see discussion of Eq. (9.1.17)]

$$c_\varphi = \frac{kc}{K} \simeq c_s\{1 - \tfrac{1}{2}\mu[1 + (\alpha^2 + \mu\alpha^2 - 1)k^2a^2\ln{(0.8905ka\sqrt{\alpha^2 + \mu\alpha^2 - 1})}]\}$$

$$\text{when } \alpha^2(1 + \mu) > 1 \quad (10.1.7)$$

a speed somewhat less than c_s, and thus definitely less than c. The effect is as though the string were loaded with the additional mass of some of the neighboring medium. Being a mass load, however, no energy is lost and K is real, implying no attenuation of the wave as it traverses the string.

The situation is quite different if the string is light enough and the tension great enough so that $1/\alpha^2 \equiv (c_s/c)^2 > 1 + \mu$. In this case q is real, energy can radiate away to the medium, and the wavenumber K has an imaginary part, representing attenuation in the z direction. To the first approximation, in the small quantity μ, the attenuation constant and the phase velocity for the string wave are

$$\operatorname{Im} K \simeq ka(1 - \alpha^2)\frac{\pi k^2 a^2 \mu}{4} \qquad \mu \ll 1;\ \alpha^2(1 + \mu) < 1$$

$$(10.1.8)$$

$$c_\varphi = \frac{kc}{\operatorname{Re} K} \simeq c_s\{1 - \tfrac{1}{2}\mu[1 - k^2 a^2(1 - \alpha^2)\ln{(Cka\sqrt{1 - \alpha^2})}]\}$$

The transverse wave in the string is then

$$\eta = A \exp{(iKz - i\omega t)} = A e^{-z\,\operatorname{Im}{(K)}} \exp{[i\operatorname{Re} K\,(z - c_\varphi t)]}$$

The wave in the medium is a conical one, propagating in a direction at an angle Ψ, such that

$$\cos\Psi = \frac{\operatorname{Re} K}{k} \simeq \alpha\sqrt{1 + \mu}\,[1 - \tfrac{1}{2}(1 - \alpha^2)k^2 a^2 \mu \ln{(Cka\sqrt{1 - \alpha^2})}]$$

$$\sin\Psi \simeq \sqrt{1 - \alpha^2(\mu + 1)}[1 + \tfrac{1}{2}k^2 a^2 \alpha^2 \mu \ln{(Cka\sqrt{1 - \alpha^2})}] \qquad \mu \ll 1$$

The wave, and its asymptotic form, is

$$p = P_0 \cos\varphi\, H_1(qr) \exp{(iKz - i\omega t)}$$

$$\xrightarrow[r \to \infty]{} -iP_0 \cos\varphi\sqrt{\frac{2}{\pi i q r}} \exp{[ik(r\sin\Psi + z\cos\Psi - ct)}$$
$$- \tfrac{1}{4}k\alpha(1 - \alpha^2)\pi k^2 a^2 \mu z_0] \qquad (10.1.9)$$

where $q \simeq k\sin\Psi$, and P_0 is related to A by Eq. (10.1.3). Quantity $z_0 = z - r\cot\Psi$ is the point on the string where the conical wave reaching point r, φ, z started from, as shown in Fig. 10.1. When $\alpha^2(\mu + 1) < 1$, the radial

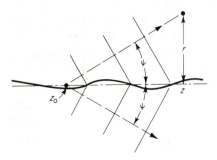

FIGURE 10.1
Conical radiation from traveling wave on a string.

wavenumber q has a negative-imaginary part, and K has a positive-imaginary part. Therefore the wave attenuates as we increase z, holding r constant, but increases exponentially as we increase r, holding z constant, since the wave comes from farther to the left on the string (z_0 greater negative), and thus from a portion of the string which has not attenuated so much.

Incidentally, we can verify Eq. (7.3.6) for the radiation impedance per unit length of the string, at least for very long string wavelengths. We insert the approximate formula for the Hankel function in Eq. (10.1.4) and substitute $U/-i\omega$ for A, where U is the velocity amplitude of the string. The ratio of F to U is the desired impedance

$$Z_{\text{rad}} \simeq -i\omega(\pi a^2 \rho)[1 - a^2(k^2 - K^2) \ln (Ca\sqrt{k^2 - K^2})] + \tfrac{1}{2}\pi^2 \rho c a^4 k(k^2 - K^2)$$

$$(10.1.10)$$

(if $K > k$, omit the resistive term and use $K^2 - k^2$ in the second reactive term). This is equivalent to Eq. (7.3.6) if we neglect the logarithmic term and set $K \to 0$. Formula (10.1.10) shows, however, that the radiation resistance is strongly dependent on the wavelength $2\pi/K$ of the wave on the string, vanishing completely if this wavelength is shorter than the wavelength $2\pi/k$ of the sound in the medium. Under the next heading we shall show that the radiation resistance of a string of finite length differs even more from that given by Eq. (7.3.6), unless c_s is very much greater than c.

Arbitrary waveshape

The wave motion portrayed by Eqs. (10.1.8) is valid for the simple-harmonic wave in a string of infinite extent. Because of the attenuation of the wave for $\alpha^2(1 + \mu) < 1$, it implies an infinite amplitude at $z \to -\infty$. A more realistic situation would be that of a string of infinite extent, with only a finite portion of the string allowed to move transversally. We continue to neglect the effect of the outer boundaries of the acoustic medium, but we couple this infinite medium to a string of finite length, fastened rigidly at both ends, for example. We could consider the string as being a very narrow cylinder of finite length, but this would introduce difficulties at each end; it would be easier to continue to consider the string as having infinite extent but held motionless over all but a finite portion. The added motionless part makes it easier to fit boundary conditions, but does not appreciably affect the wave motion if the string radius is small compared with the wavelength.

Thus our next task is to survey the relationship between the transverse motion of a string, only a portion of which moves, to the coupled wave motion in the medium. We assume that the string displacement $\eta(z)e^{-i\omega t}$ is entirely in the yz plane and is simple-harmonic with frequency $\omega/2\pi$; $\eta(z)$ is nonzero over only a finite range of z. Adapting Eq. (7.1.17), using a

Green's function with zero radial gradient at $r = a$, we have

$$p_\omega(r) = -a \iint \frac{\partial p}{\partial r_0} G_\omega(r,\varphi,z \mid a,\varphi_0,z_0) \, d\varphi_0 \, dz_0$$

where we have omitted the common factor $e^{-i\omega t}$. For motion in the yz plane this becomes

$$P_\omega(r) = -\rho a \omega^2 \int_0^{2\pi} \cos \varphi_0 \, d\varphi_0 \int_{-\infty}^\infty \eta(z_0) G_\omega(r,\varphi,z \mid a,\varphi_0,z_0) \, dz_0$$

The appropriate Green's function is

$$G_\omega = \frac{i}{8\pi} \sum_{m=0}^\infty \epsilon_m \cos m(\varphi - \varphi_0) \int_{-\infty}^\infty e^{iK(z-z_0)} \, dK$$

$$\times H_m(qr) \left[J_m(qr_0) - \frac{J'_m(qa)}{H'_m(qa)} H_m(qr_0) \right] \quad (10.1.11)$$

when $r_0 < r$, where $q^2 = k^2 - K^2$ and $k = \omega/c$. Inserting this in the integral for p and using the formula

$$H'_m(u)J_m(u) - J'_m(u)H_m(u) = \frac{2i}{\pi u}$$

derived from the last of Eqs. (7.3.2), we have for the radiated wave and the reaction force on the string,

$$p_\omega = \frac{\rho\omega^2}{2\pi} \cos \varphi \int_{-\infty}^\infty e^{iKz} \frac{H_1(qr)}{H'_1(qa)} \frac{dK}{q} \int_{-\infty}^\infty e^{-iKz_0}\eta(z_0) \, dz_0$$

$$\to -i\rho\omega^2 \cos \varphi \sqrt{\frac{2}{\pi i r}} \int_{-\infty}^\infty e^{iKz+iqr} \left[\frac{Y(K)}{q^{\frac{3}{2}}H'_1(qa)} \right] dK \quad kr \to \infty \quad (10.1.12)$$

$$F_\omega(z) = -\pi\rho c^2 k^2 a^2 \int_{-\infty}^\infty e^{iKz} \left[\frac{H_1(qa)}{qaH'_1(qa)} \right] Y(K) \, dK$$

where $Y(K)$ is the Fourier transform of the string displacement

$$Y(K) = \frac{1}{2\pi} \int_{-\infty}^\infty \eta(z_0)e^{-iKz_0} \, dz_0 \qquad \eta(z) = \int_{-\infty}^\infty Y(K)e^{iKz} \, dK$$

If η were a traveling wave $Ae^{i\varkappa z}$, then $Y(K)$ would be the delta function $A\delta(K - \varkappa)$, and F would have the form of Eq. (10.1.4), with $\varkappa$ instead of K. If, however, η were some function which goes to zero rapidly enough so that the integral of $|\eta|^2$ over z from $-\infty$ to $+\infty$ is finite, then, according to Eq. (1.3.17), $Y(K)$ goes to zero, as $|K| \to \infty$, rapidly enough so that the integral of $|Y|^2$, over the whole range of K, is finite; in fact, so that

$$\int_{-\infty}^\infty |\eta(z)|^2 \, dz = 2\pi \int_{-\infty}^\infty |Y(K)|^2 \, dK$$

The spread of $Y(K)$ in K depends on the degree of "jaggedness" of η. If η is smoothly varying so that components of wavelength shorter than λ_{min} are unimportant, then $Y(K)$ will be negligible for values of $|K|$ larger than $2\pi/\lambda_{min}$. In particular, if there is no "kink" or rapid change in slope of η with a radius of curvature smaller than about $20a$ (where a is the radius of the string), then Y is effectively zero for $|K|$ larger than about $1/3a$.

When such is the case, the quantity in brackets in the integral for $F_\omega(z)$ in Eqs. (10.1.12) can be approximated by the first term in its series expansion.

$$\frac{H_1(qa)}{qaH_1'(qa)} \simeq -1 + a^2(k^2 - K^2)[\ln(Ca\sqrt{k^2 - K^2}) - \tfrac{1}{2}\pi i] \qquad |K| < k$$

$$\simeq -1 - a^2(K^2 - k^2)\ln(Ca\sqrt{K^2 - k^2}) \qquad |K| > k$$

$$\text{(10.1.13)}$$

(where $C = 0.8905$) over the range of K, for which $Y(K)$ is nonnegligible. If the string radius is very small, this approximation is valid even for string shapes with a fair amount of "jaggedness." For example, if η is zero for $|z| > \tfrac{1}{2}l$, as it would be if we were representing the motion of a string of finite length, held rigid at $\pm\tfrac{1}{2}l$, as long as $l \gg a$, the "kinks" at the two supports will not be perfectly sharp in actual practice, but will be curves instead of cusps, with a curvature radius of about $20a$ or greater. If this is the case, the integral for the reactive force can be greatly simplified.

$$F_\omega(z) \simeq \pi\rho c^2 k^2 a^2 \int_{-\infty}^{\infty} e^{iKz} Y(K)[1 + a^2(K^2 - k^2)\ln(Ca\sqrt{K^2 - k^2})]\, dK$$

$$\text{(10.1.14)}$$

where the quantity in brackets is real and positive at $K \to \infty$, and the contour of integration (C in Fig. 10.2) goes just above the real axis of K from $-\infty$ to 0 and just below the real axis from 0 to $+\infty$ in order that p_ω be an outward-traveling wave.

The first term in the brackets can be integrated immediately, for it is the inverse transform of Y. Thus, as was indicated in Eq. (10.1.10), the largest part of the reaction force is $\pi a^2 \rho \omega^2 \eta(z)$, an inertial reaction, proportional to $-\omega^2$ times the displacement η, the added mass load being equal to an amount of fluid medium equal to the volume of the string [this corresponds to the μ term in braces in Eqs. (10.1.6)]. The logarithmic term is more difficult; in fact, it can be worked out only when we know the formula for $Y(K)$ (or for η). It is important, however, for it is responsible for the resistive part of the reaction, the part producing energy loss. We of course do not know the exact form of η (in fact, we are in the process of determining its form).

But suppose we assume that this part of the reaction does not change the shape of the vibrating string very much, that its shape is roughly the same as if it were vibrating in a vacuum, between supports at $\pm\tfrac{1}{2}l$, that is, that

$$\eta(z) \simeq \begin{cases} 0 & z < -\tfrac{1}{2}l \\ A\cos\dfrac{\pi z}{l} & -\tfrac{1}{2}l < z < \tfrac{1}{2}l \\ 0 & z > \tfrac{1}{2}l \end{cases}$$

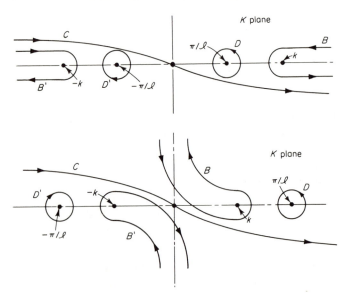

FIGURE 10.2
Contours of integration for the reaction integral.

The Fourier transform for this function is

$$Y(K) = -\frac{A}{2l}\frac{e^{iKl/2} + e^{-iKl/2}}{K^2 - (\pi/l)^2} \tag{10.1.15}$$

In this case the transverse force the medium exerts on the string, per unit length of string, in the range $-\tfrac{1}{2}l < z < \tfrac{1}{2}l$ (it does not matter what force is exerted on the rest of the string; it is being held motionless), becomes

$$F_\omega(z) \simeq \pi a^2 \rho \omega^2 A \cos\frac{\pi z}{l} - \frac{\pi a^2 \rho \omega^2 A a^2}{2l}\int_C \frac{K^2 - k^2}{K^2 - (\pi/l)^2}$$

$$\times\; (e^{iKl/2} + e^{-iKl/2})\ln\,(aC\sqrt{K^2 - k^2})e^{iKz}\,dK$$

where the contour of integration is the one marked C in Fig. 10.2.

For the exponential $e^{iK(\frac{1}{2}l+z)}$, if $z > -\frac{1}{2}l$, the contour must be closed by a semicircle around the upper half of the K plane, which can be shrunk down to the contour D around the pole of the integrand at π/l, plus the infinite loop B around the branch point at $K = k$. For the exponential $e^{-iK(\frac{1}{2}l-z)}$, if $z < \frac{1}{2}l$, the closing semicircle must be around the lower half-plane; it can be reduced to the circle D' around the pole at $-(\pi/l)$ plus the loop B' around the branch point $-k$. The shape of the loops B, B' depends on whether k is larger or smaller than π/l, in other words, whether the wavelength $\lambda_s = 2\pi/\alpha k$ of the wave in the string (if vibrating *in vacuo*) is smaller or larger than $\lambda = 2\pi/k$, the wavelength of sound in the medium, i.e., whether the string wave velocity $c_s = c/\alpha$ is smaller or larger than the medium velocity c. If $c_s > c$ $(\alpha < 1)$, then $\lambda_s = 2l$ will be greater than $\lambda = 2\pi/k$ and $\pi/k < l$; in this case the contours will be those shown in the upper half of Fig. 10.2. Conversely, if c_s is smaller than c, then k is smaller than π/l, and the contours to use are shown in the lower half of Fig. 10.2.

In either case, Eq. (1.2.15) shows that the integrals D and D', around the poles, when $|z| < \frac{1}{2}l$, are

$$D + D' = -\frac{\pi^2 i \rho \omega^2 A a^2}{l} \left(\frac{ie^{i\pi z/l}}{2\pi/l} + \frac{ie^{-i\pi z/l}}{2\pi/l} \right) \left[\left(\frac{\pi}{l} \right)^2 - k^2 \right] \ln \left[aC \sqrt{\left(\frac{\pi}{l} \right)^2 - k^2} \right]$$

If $z > \frac{1}{2}l$, the contour D' does not exist, and the $e^{-i\pi z/l}$ term is missing; if $z < -\frac{1}{2}l$, the $e^{i\pi z/l}$ term vanishes. If $k > \pi/l$, the logarithmic factor has an imaginary part, $-(i\pi/2)$.

This part of the reaction force is proportional to $\cos(\pi z/l)$, that is, proportional to η (in the range $|z| < l/2$). But the integrals B and B' result in a function of z which is not proportional to $\cos(\pi z/l)$. The difference in value of the logarithmic factor between the incoming and outgoing parts of B is $i\pi$; all other factors are unchanged as the path goes around $K = k$. Thus, if $k > \pi/l$,

$$B = -\frac{i\pi^2 a^2 \rho \omega^2 A a^2}{2l} \int_k^\infty \frac{K^2 - k^2}{K^2 - (\pi/l)^2} e^{iK(\frac{1}{2}l+z)} \, dK$$

and

$$B' = -\frac{i\pi^2 a^2 \rho \omega^2 A a^2}{2l} \int_k^\infty \frac{K^2 - k^2}{K^2 - (\pi/l)^2} e^{iK(\frac{1}{2}l-z)} \, dK$$

These integrals can be made to converge more rapidly by changing the path of integration to one from k to $+i\infty$. It can be seen that $B + B'$ is not proportional to $\cos(\pi z/l)$, and thus that the total expression for F is not proportional to η. Therefore $A \cos(\pi z/l)$ is not an exact solution for the shape of a string vibrating between fixed supports, when in contact with an acoustic medium.

Variational calculation of frequency

Since an exact solution in closed form is not possible, we set up a variational procedure to find a close approximation, at least for the natural frequency of vibration of the string. We are considering a string of infinite extent, held fixed along the z axis except for the portion between $-\tfrac{1}{2}l$ and $+\tfrac{1}{2}l$, which can move transversally in the yz plane, with a displacement $\eta(z)e^{-i\omega t}$. The motion is subject to the string tension $T = \pi a^2 \rho_s c_s^2 = \pi a^2 \rho c^2(1/\mu\alpha^2)$ $(\mu = \rho/\rho_s,\ \alpha = c/c_s)$ and to the reaction force $F_\omega(z)e^{-i\omega t}$ of the surrounding acoustic medium. We wish to determine the shape of the string when it vibrates under the influence of these forces, and more important, we wish to compute the natural frequency $\omega_n = k_n/c$ of this motion. The real part of $\omega_n/2\pi$ is the frequency of free vibration; the imaginary part of ω_n (which must be negative) is the damping constant, related to the loss of energy by radiation. What we have seen in the foregoing discussion is that the damping constant depends strongly on α, the ratio of wave velocities in medium and string.

We first set down the equation of motion for the free portion of the string. Rewriting the equation for F somewhat, to assist in the subsequent manipulations, we have

$$-\pi a^2 \rho_s \omega^2 \eta = T \frac{\partial^2 \eta}{\partial z^2} + F_\omega(z) \qquad -\tfrac{1}{2}l < z < \tfrac{1}{2}l$$

or

$$\omega^2 \eta(z) = -\left(\frac{c}{a}\right)^2 \left[\frac{\partial^2 \eta}{\partial z^2} + \frac{\mu\omega^2}{2\pi}\int_{-\frac{1}{2}l}^{\frac{1}{2}l} G_\omega(z \mid z_0)\eta(z_0)\, dz_0\right] \qquad (10.1.16)$$

where

$$G_\omega = \int_{-\infty}^{\infty} e^{iK(z-z_0)}\left[\frac{H_1(qa)}{qaH_1'(qa)}\right] dK$$

We multiply this equation by $\eta(z)$ and integrate over z from $-\tfrac{1}{2}l$ to $+\tfrac{1}{2}l$, obtaining

$$\omega^2 = \frac{1}{\int \eta^2\, dz}\left[\left(\frac{c}{a}\right)^2 \int \left(\frac{\partial \eta}{\partial z}\right)^2 dz + \frac{\mu\omega^2}{2\pi}\int \eta(z)\, dz \int G_\omega(z \mid z_0)\eta(z_0)\, dz_0\right] \qquad (10.1.17)$$

with all integrations from $-\tfrac{1}{2}l$ to $+\tfrac{1}{2}l$. We have integrated the first term by parts, utilizing the fact that $\eta = 0$ at the end points of the integration.

The expression on the right of this equation is a variational expression; if we vary the shape of η, the coefficient of $\delta\eta$ inside the integral is zero if η is a solution of Eq. (10.1.16); so for shapes near this, the variation of the expression is second order. Therefore, if we insert a trial function $\chi(\gamma)$ instead of η in the expression of Eq. (10.1.17) and adjust γ so the expression's derivative with respect to γ is zero, the difference between χ and the correct η will be minimal, and the expression's value will differ from the correct ω^2 to the second order of this difference. We know, of course, that there are an infinite number of exact solutions and a corresponding infinity of allowed values of ω, corresponding to the fundamental and the higher harmonics of the vibrating string. The lowest allowed value of ω^2 is that for the fundamental, so, as we vary χ, the absolute minimum of the variational expression is the square of the ω for the fundamental mode. We shall see later how to determine the higher harmonics. Note that the variational expression is independent of the magnitude of χ, since η^2 (or χ^2) enters both numerator and denominator. Since Eq. (10.1.16) is linear and homogeneous in η, a constant times a solution is a solution.

If the coupled medium is light enough ($\mu = \rho/\rho_s \ll 1$) so that the radiative reaction is small compared with the effect of string tension, we should expect that η does not differ much from the shape of the same string *in vacuo*. For the fundamental mode, therefore, we can insert $\cos (\pi z/l)$ for χ in the variational expression and expect that the resulting value would differ very little from the correct value of ω^2 for the fundamental. Thus a second-order approximation to the fundamental eigenvalue is

$$\omega^2 \simeq \frac{2}{l} \left\{ \left(\frac{\pi c}{\alpha l}\right)^2 \frac{l}{2} + 2\pi\mu\omega^2 \int_{-\infty}^{\infty} |Y(K)|^2 \left[\frac{H_1(qa)}{qaH_1'(qa)}\right] dK \right\}$$

where $\alpha = c/c_s$; $q^2 = k^2 - K^2$; and $Y(K)$ is the Fourier transform of $\cos (\pi z/l)$, given in Eq. (10.1.15), without the factor A.

If $l \gg a$, we can assume, as before, that Y^2 is negligible in the range of integration, where approximation (10.1.13) is no longer valid, and the simpler form can be used for the quantity in brackets in the integral. Then we can use Eq. (1.3.17) for the first term of the approximate form, and the variational expression becomes

$$\omega^2 \simeq \left(\frac{\pi c_s}{l}\right)^2 - \mu\omega^2 - \frac{\pi\mu\omega^2 a^2}{l^3} \int_{-\infty}^{\infty} (K^2 - k^2)$$

$$\times \left[\frac{e^{iKl/2} + e^{-iKl/2}}{K^2 - (\pi/l)^2}\right]^2 \ln (aC\sqrt{K^2 - k^2}) \, dK \quad (10.1.18)$$

the contour being the one marked C in Fig. 10.2, as before. Also as before, this can be broken into the set $B + D$ for the terms $1 + e^{iKl}$ in the denominator, and the set $B' + D'$ for the terms $1 + e^{-iKl}$. The integrals D and D' around the second-order poles are evaluated by means of Eq. (1.2.13), and we have

$$\omega^2 \simeq \left(\frac{\pi c_s}{l}\right)^2 - \mu\omega^2 - \mu\omega^2 a^2 \left[\left(\frac{\pi}{l}\right)^2 - k^2\right] \ln \left[aC\sqrt{\left(\frac{\pi}{l}\right)^2 - k^2}\right] + B + B'$$

for $k < \pi/l$. If $k > \pi/l$, the logarithmic factor becomes

$$\ln \left[aC\sqrt{k^2 - \left(\frac{\pi}{l}\right)^2}\right] - \tfrac{1}{2}i\pi$$

These first three terms for the eigenvalue are the same as those in Eq. (10.1.6) for a traveling wave of infinite extent. If k is smaller than π/l (i.e., if $c + \mu c > c_s$), these first three terms are real; a wave of infinite extent, if it moves slower than the speed of sound in the medium (times $1 + \mu$), loses no energy by radiation to $r \to \infty$, as was noted earlier. For the finite string, however, some energy is lost even when $c(1 + \mu) > c_s$, because of the sudden change of slope of η at $z = \pm\frac{1}{2}l$. This contribution to the imaginary part of ω^2 comes from the contours B and B' of Fig. 10.2, which arise because of the form of $Y(K)$, that is, because $\eta(z)$ is zero for $|z| > \frac{1}{2}l$. The form of these integrals differs, depending on whether k is greater or less than π/l.

When $k < \pi/l$ ($l < \frac{1}{2}\lambda$ or $c + \mu c > c_s$), the contours in the lower part of Fig. 10.2 are to be used. The difference in the logarithmic factor, between the incoming and outgoing part of B, is $i\pi$; that for B' is $-i\pi$. The resulting integrals, for those parts going from $K = 0$ to $K = \pm i\infty$, cancel each other; the parts from 0 to $\pm k$ add, so that

$$
\begin{aligned}
B + B' &\simeq -\frac{2i\pi^2\mu\omega^2 a^2}{l^3} \int_0^k \frac{k^2 - K^2}{[(\pi/l)^2 - K^2]^2}(1 + e^{iKl})\, dK \\
&= -\frac{2i\mu\omega^2 k^3 a^2}{\pi^2} \int_0^{\pi/2} \frac{\sin^3 \theta(1 + e^{ikl\cos\theta})}{[1 - (kl/\pi)^2 \cos^2 \theta]^2}\, d\theta \\
&\to -\frac{2i\mu\omega^2 k^3 a^2}{\pi^2}\left(\frac{4}{3} + \frac{i}{4}kl\right) \qquad 2l \gg \lambda
\end{aligned}
$$

Thus, when the wave velocity in the string is very much smaller than that in the medium, the formula for the resonance frequency of the fundamental mode, correct to first order in the small quantity $\mu\pi^2 a^2/l^2$, is

$$
\omega_1 \simeq \frac{\pi c \gamma_s}{l}\left\{1 - \frac{\mu/2}{1 + \mu}\left(\frac{\pi a}{l}\right)^2\left[(1 - \gamma_s^2)\ln\left(\frac{a\pi C}{l}\sqrt{1 - \gamma_s^2}\right) + \frac{8i}{3\pi}\gamma_s^3\right]\right\}
$$

$$(10.1.19)$$

when $c_s \ll c$, where $\gamma_s^2 = [c_s^2/c^2(1 + \mu)] = [1/\alpha^2(1 + \mu)]$, and $C = 0.8905$.

The last term in brackets is the only imaginary part of ω_1; when $c_s < c$ the sole reason for loss of energy is the finite extent of the string displacement. This result differs radically from that of the simple Eq. (7.3.6); energy loss from a string of finite extent is quite different from energy loss from a string of infinite extent, particularly when c_s is less than c. Of course, with the strings used in musical instruments, the coupling with the air is not primarily direct, but is transmitted through the string supports, which are not rigid, to a sounding board, which is a much more efficient acoustic radiator than is the string itself. Consequently, Eq. (10.1.19) is not applicable for the calculation of the damping constant of a violin or piano string, for example; it applies only to the case of the string with truly rigid supports.

In the other limit, when $c_s \gg c$ $(kl \gg \pi)$, the contours B and B' are those shown in the upper half of Fig. 10.2, with the integrations carried from k to $\pm i\infty$. The approximate result is

$$
\omega_1 \simeq \frac{\pi c \gamma_s}{l} \left\{ 1 + \frac{\mu/2}{1+\mu} \left(\frac{\pi a}{l} \right)^2 (\gamma_s^2 - 1) \ln \left(\frac{\pi a C}{l} \sqrt{\gamma_s^2 - 1} \right) \right.
$$
$$
\left. - \frac{i\mu}{1+\mu} \left(\frac{\pi a}{l} \right)^2 \left[\frac{1+\gamma_s^2}{\pi} \coth^{-1}(\gamma_s) - \frac{\gamma_s}{\pi} + \frac{\pi}{2} (\gamma_s^2 - 1) \right] \right\} \qquad c_s \gg c
$$

$$(10.1.20)$$

which has a larger damping constant than does the case for $c_s \ll c$. For the intermediate range, when c_s is roughly equal to c, the integrals B and B' must be computed more accurately.

Calculations for the higher modes can be carried out by using $\chi = \cos(\pi n z / l)$ for n odd or $\chi = \sin(\pi n z / l)$ for n even in Eq. (10.1.17). But these are subject to increasing uncertainties, as the following discussion will indicate. We can write Eq. (10.1.16) as $\omega^2 \eta = \mathfrak{G}_\omega \eta$, where $\mathfrak{G}_\omega$ is the differential and integral operator written on the right of Eq. (10.1.16). When the exact solution η_n for the nth harmonic is used, the quantity $\mathfrak{G}_\omega \eta_n$ will equal $\omega_n^2 \eta_n$. The eigenvalues ω_n^2 are complex because of the energy loss to the medium, but their real parts are all positive and can be arranged in an ordered sequence so that $\mathrm{Re}\,\omega_{n+1}^2 > \mathrm{Re}\,\omega_n^2$, with ω_1^2 the eigenvalue for the fundamental. Furthermore, it is not difficult to show that the η_n's are mutually orthogonal; i.e.,

$$
\int_{-l/2}^{l/2} \eta_n(z) \eta_m(z)\, dz = \begin{cases} 0 & n \neq m \\ N_n & n = m \end{cases}
$$

where the complex quantity N_m is the normalizing constant for η_m.

Choosing an arbitrary form for η is equivalent to choosing an arbitrary expansion $\eta = \sum A_n \eta_n$ in terms of the eigenfunctions. Inserting this into Eq. (10.1.17) shows that the variational equation is equivalent to

$$
\frac{\sum A_n^2 N_n \omega_n^2}{\sum A_n^2 N_n} = W - iV
$$

where $V \ll W$ in most cases. Of course, we do not know the η_n's or ω_n's (otherwise we should not be using the variational procedure), but this representation of the variational expression is useful because it tells us more about the limitations of the method. For example, since $\mathrm{Re}\,\omega_1^2$, though positive, is smaller than the real part of any other ω_n^2, we can see that W can never be smaller than $\mathrm{Re}\,\omega_1^2$. Thus, as we vary the A_n's by varying the shape of χ, the absolute minimal value of W is $\mathrm{Re}\,\omega_1^2$, which only occurs when $\chi = \eta_1$ ($A_n = 0$, for all n except $n = 1$).

To find ω_2^2 by means of the variational procedure, we must choose a series for χ which has $A_1 = 0$; in other words, we must arrange that the trial function, though variable in form, always is orthogonal to η_1 (i.e., that $\int \chi \eta_1 \, dz = 0$). When $A_1 = 0$, the argument used for $\omega_1{}^2$ works for $\omega_2{}^2$; the minimal value of W for all shapes orthogonal to η_1 is Re $\omega_2{}^2$, which occurs only when $\eta = \eta_2$. And so on. But we do not know η_1 exactly; so we cannot find the family of trial functions exactly orthogonal to η_1. If we make the trial functions for η_2 orthogonal to the best approximation we can find for η_1, these trial functions may have a small component of η_1 in them (A_1 is not exactly zero), and thus we cannot guarantee that the minimal value of W for these functions is Re $\omega_2{}^2$; it may be somewhat smaller than this. And if we then go on to $\omega_3{}^2$ by using a set of trial functions orthogonal to the approximations for η_1 and η_2, we are compounding our errors.

Thus the variational procedure used here (sometimes called the Rayleigh-Ritz variational procedure) is most accurate in determining the fundamental mode, and rapidly loses utility when it is used to compute higher modes. In contrast, the procedure used in Eq. (9.4.19), though it involves more computation, loses efficacy more slowly as n is increased. This will be illustrated in a problem.

In the case of the string, we have illustrated the point, mentioned at the beginning of this section, that the behavior of the wave representing coupled motion of string and medium depends markedly on the ratio α of wave velocities in the two systems, when uncoupled. The other point mentioned earlier, that there are always two kinds of waves, with two velocities, has not been emphasized in this example. The wave of Eq. (10.1.9) is obviously related to that of the string *in vacuo*. There are, of course, waves in the air, going with velocity c, which move the string slightly; these would be the other set of waves. In the case of the string, however, the coupling is so small that the characteristics of this second set are not appreciably affected by the presence of the string (at least in regard to wave velocity); we can calculate their space distribution by use of Sec. 8.1.

If the interacting systems involve greater coupling, such as a membrane plus surrounding air, the second class of waves may have its velocity also modified by the coupling. In fact, as we shall see, for some wave-velocity ratios, waves in the medium may be "trapped" by the plate; so they travel parallel to the plate surface, the majority of the energy being carried by the medium near the plate. This can be demonstrated by solving the problem of the coupled motion of a thin plate with the medium surrounding it.

Free motion of an infinite plate

We consider a thin plate of thickness $2h$ and density ρ_p, with modulus of elasticity Q and Poisson ratio s (usually equal to $\frac{1}{3}$). The analogous, but simpler, example of a membrane in contact with a medium will be given as a problem. According to Eq. (5.3.9), the equation of motion for the plate's transverse displacement, when driven by a simple-harmonic force $F_\omega(x,y)e^{-i\omega t}$, is

$$\left(\frac{\partial^2}{\partial x^2} + \frac{\partial^2}{\partial y^2}\right)^2 \eta - \gamma^4 \eta = \frac{\gamma^4}{2kh\rho_p c\omega} F_\omega \qquad (10.1.21)$$

where $\gamma^4 = 3\omega^2 \rho_p (1 - s^2)/Qh^2$ and $k = \omega/c$ as before, ρ and c being the density and sound velocity, respectively, of the medium. As an example of magnitudes, for a sheet of steel 2 mm thick, with $\rho_p = 8$ and $Q = 2 \times 10^{12}$ dynes per sq cm, parameter γ is approximately equal to $\sqrt{\nu/4{,}400}$ (ν being the frequency), and the ratio of the wave velocity of the plate ω/γ to the velocity of sound c in the air is $(c_p/c) \simeq \sqrt{\nu/10^4}$. The plate, of course, is a dispersive medium for transverse waves.

First let us ask what sort of free sinusoidal waves can travel along the plate in contact with the medium. We assume that the transverse displacement of the plate and its transverse velocity are

$$\eta = A \exp(iKx - i\omega t) \qquad u_z = -i\omega\eta$$

where we have placed the xy plane at the mid-plane of the plate when in equilibrium. The pressure waves above and below the plate must have the same dependence on x; because of symmetry the two are equal and opposite in sign at the plate surface.

$$p_+ = P \exp(iKx + izq - i\omega t) \qquad p_- = -P \exp(iKx - izq - i\omega t)$$

$$u_z = \pm \frac{q}{\rho ck} p \qquad q = \sqrt{k^2 - K^2} \quad \text{or} \quad i\sqrt{K^2 - k^2}$$

where we have assumed that h is so small compared with λ that we can assume that the boundary conditions are at $z = 0$. The reaction force of these two waves on the plate is $F_\omega = (p_- - p_+)_{z=0}$. Therefore there are two equations relating A and P, one equating the normal velocity of plate and neighboring air, the other with $-2P \exp(iKx - i\omega t)$ for F_ω in Eq. (10.1.21). The combination relates the amplitude A of plate motion to amplitude P of acoustic wave and also fixes the relationship between plate wavenumber and frequency $kc/2\pi$ at which there is free vibration.

$$-i\omega A = \left(\frac{q}{\rho ck}\right) P \qquad K^4 - \gamma^4 = i\frac{\rho\gamma^4}{\rho_p qh} = i\frac{\mu\gamma^4}{q} \qquad (10.1.22)$$

where $\mu = \rho/\rho_p h$.

First let us find the various roots of the equation for K, regardless of whether the plate wave goes to infinity at $x = \pm\infty$, but insisting that the radiated wave is outgoing and noninfinite at $z = \pm\infty$ (i.e., that only the sheet of $\sqrt{K^2 - k^2}$ with positive-real or positive-imaginary part be used). There are four such roots, two real and two imaginary, when $k < \gamma$ (when the plate velocity is smaller than that in air). When both μ/γ and k/γ are quite small, approximate expressions for K are

$$K = \pm K_+ \qquad K_+ \simeq \gamma\left[1 + \frac{\mu}{4(\gamma^2 - k^2)^{\frac{1}{2}}}\right]$$

$$q_+ \simeq i\sqrt{\gamma^2 - k^2}\left[1 + \frac{\mu\gamma^2}{4(\gamma^2 - k^2)^{\frac{3}{2}}}\right]$$

$$K = \pm iK_- \qquad K_- \simeq \gamma\left[1 + \frac{i\mu}{4(\gamma^2 + k^2)^{\frac{1}{2}}}\right] \qquad (10.1.23)$$

$$q_- \simeq \sqrt{\gamma^2 + k^2}\left[1 + \frac{i\mu\gamma^2}{4(\gamma^2 + k^2)^{\frac{3}{2}}}\right] \qquad \mu = \frac{\rho}{\rho_p h}$$

For the steel plate 2 mm thick, μ/γ is roughly equal to $\sqrt{1/100\nu}$ when the plate is in air, and is $\sqrt{10,000/\nu}$ when it is in water. The other roots, near $K = \pm k$, are not allowed, since they correspond to $\sqrt{K^2 - k^2}$ being negative-imaginary, when $k < \gamma$, which is not allowed. When μ is not small compared with γ, these formulas are not valid, but the roots still are two real and two nearly pure-imaginary.

The first pair of roots corresponds to wave propagation along the plate with no attenuation; the corresponding wave in air attenuates with increasing $|z|$, so that no energy is radiated away to $|z| \to \infty$. The second pair, being imaginary, cannot correspond to free-wave motion in an infinite plate; they are used only to fit boundary conditions in a finite or semi-infinite plate. We should note, however, that the corresponding q is nearly real, corresponding to energy loss to $z \to \pm\infty$; the presence of a boundary induces loss of energy at the edge even for the case $k < \gamma$. The wave velocity for the roots $\pm K_+$ is close to ω/γ, the wave velocity for the plate in vacuum, when $\mu < \gamma$. Most of the energy is in the plate.

At higher frequencies, when $k > \gamma$, all six roots are allowed. Formulas for small values of μ/γ are

$$K = \pm K_+ \qquad K_+ \simeq \gamma\left[1 + \frac{i\mu}{4(k^2 - \gamma^2)^{\frac{1}{2}}}\right]$$

$$q_+ \simeq \sqrt{k^2 - \gamma^2}\left[1 - \frac{i\mu\gamma^2}{4(k^2 - \gamma^2)^{\frac{3}{2}}}\right]$$

$$K = \pm iK_- \qquad K_- \simeq \gamma\left[1 + \frac{i\mu}{4(k^2 + \gamma^2)^{\frac{1}{2}}}\right] \qquad (10.1.24)$$

$$q_- \simeq \sqrt{k^2 + \gamma^2}\left[1 + \frac{i\mu\gamma^2}{4(k^2 + \gamma^2)^{\frac{3}{2}}}\right]$$

$$K = \pm K_s \qquad K_s \simeq k\left[1 + \frac{1}{2}\left(\frac{\mu}{k}\right)^2\left(\frac{\gamma^4}{k^4 - \gamma^4}\right)^2\right] \qquad q_s \simeq i\mu\frac{\gamma^4}{k^4 - \gamma^4}$$

The first root again corresponds to the free wave of the plate in vacuum. This time $(c_p > c)$, some of its energy can radiate away to $|z| \to \infty$. The pressure wave above the plate is, approximately,

$$p_+ \simeq -i\rho c^2 \frac{kA}{\sqrt{1 - (\gamma/k)^2}} \exp\left[ik(x \cos \Psi + z \sin \Psi) - \frac{\rho \cot \Psi}{4\rho_p h} x_0 - i\omega t \right]$$

$$(10.1.25)$$

where $\cos \Psi = \gamma/k$, and $x_0 = x - z \cot \Psi$. This is a plane wave, moving in a direction at angle Ψ to the x axis, with velocity $\omega/k = c$, with amplitude proportional to the plate amplitude at the point x_0, the point at which the wave which arrives at (x,z) was generated. The geometry is similar to that of Fig. 10.1, with (z,r) changed to (x,z) and with the present wave a plane wave, instead of the conical wave produced by the string. Since high-frequency wave motion can radiate to $|z| \to \infty$, the plate wave attenuates with x, and the acoustic wave attenuates with x_0.

In contrast to this wave is the *surface wave*, corresponding to the third pair of roots (only allowed when $k > \gamma$). This wave has most of its energy carried in the medium; since q is imaginary and small, the ratio of P to A is large, and the disturbance in the medium attenuates slowly as $|z|$ increases. This wave has velocity slightly less than the acoustic wave velocity in the medium. The plate moves just enough to cause the sound wave to "cling" to the plate, with its wavefronts normal to the plate and with wave amplitude largest near the plate, diminishing exponentially as $|z|$ increases. This sort of wave, produced by plate-medium coupling when $c_p > c$, has no attenuation in the x direction; thus no energy is lost to $|z| \to \infty$. Such a wave does not exist for a plate in vacuum, nor does it exist in a medium without a plate; it requires the interaction of both.

Linear driving force

If the plate is subject to a transverse driving force $F(K)e^{iKx - i\omega t}$ per unit area of plate, in addition to the radiation reaction of the surrounding medium, the steady-state waves in the plate and medium can be obtained from Eqs. (10.1.21) and (10.1.22).

$$\eta = A(K)e^{iKx - i\omega t} \quad \left(K^4 - \gamma^4 - \frac{\mu\gamma^4}{\sqrt{K^2 - k^2}}\right) A(K) = \frac{\gamma^4}{2h\rho_p\omega^2} F(K)$$

$$p = \pm P(K) \exp\left(iKx \mp z\sqrt{K^2 - k^2} - i\omega t\right) \quad P(K) = -\frac{\rho\omega^2}{\sqrt{K^2 - k^2}} A(K)$$

$$(10.1.26)$$

where the upper signs in the expression for p are used for $z > 0$, the lower signs for $z < 0$. For the p wave to have the right asymptotic behavior, the square root must be in the first quadrant of the complex plane; we take it to be real and positive as K approaches $+\infty$.

If the driving force is simple-harmonic but not a sinusoidal function of x, the force shape can be built up as a Fourier integral.

$$f(x) = \int_{-\infty}^{\infty} F(K)e^{iKx}\, dK \qquad F(K) = \frac{1}{2\pi} \int_{-\infty}^{\infty} f(x)e^{-iKx}\, dx$$

The steady-state driven motion is then $\eta = a(x)e^{-i\omega t}$, and $p = \pm p(x, \pm z)e^{-i\omega t}$, where

$$a(x) = \int_{-\infty}^{\infty} A(K)e^{iKx}\, dK = \frac{\gamma^4}{2h\rho_p\omega^2} \int_C \frac{F(K)\exp(i\kappa x)\sqrt{\kappa^2 - k^2}}{(K^4 - \gamma^4)\sqrt{K^2 - k^2} - \mu\gamma^4}\, dK$$

$$p(x,z) = \tfrac{1}{2}\mu\gamma^4 \int_C \frac{F(K)\exp(iKx - z\sqrt{K^2 - k^2})}{(K^4 - \gamma^4)\sqrt{K^2 - k^2} - \mu\gamma^4}\, dK \qquad z > 0$$

$$(10.1.27)$$

where the contour of integration is the one marked C in Fig. 10.3, and $\mu = \rho/\rho_p h$. When x is positive this contour must be completed by a semicircle in the upper half of the K plane. It then surrounds the roots K_+ and iK_- of Eq. (10.1.23) if $k < \gamma$, or the roots K_+, iK_-, and K_s of Eq. (10.1.24) if $k > \gamma$. It also must avoid the branch cut, which starts from $K = k$ and goes to infinity in the first quadrant.

If the driving force $f(X)$ is a line force $F_0\delta(x)$ along the line $x = 0$, the transform is simply $F(K) = F_0/2\pi$, having no additional singular points in the K plane. In this case the integrals reduce to circuits around the two or three poles of the integrand, plus a U-shaped integral around the branch cut, as shown in Fig. 10.3. If $x < 0$, the circuit is completed around the lower half-plane, and the circuits are similar ones around the other poles, with a U around $-k$ going to $-i\infty$. To compute the integrals around the poles we must calculate the residues (see page 15) at each, by setting $K = K_+ + \epsilon$, etc., and finding the coefficient of $1/\epsilon$ as $\epsilon \to 0$. If $\mu \ll 1$ and $k > \gamma$, we can use the approximate formulas of Eqs. (10.1.24) and find the residues of $e^{iKx}/[(K^4 - \gamma^4)\sqrt{K^2 - k^2} - \mu\gamma^4]$ to be $(1/4\gamma^2 K_+)e^{iK_+x}$ at K_+, $(i/4\gamma^2 K_-)e^{-K_-x}$ at iK_-, and $(\mu^2/k\gamma^4)[\gamma^4/(k^4 - \gamma^4)]^3$ at K_s.

Therefore, for $\mu \ll 1$ and for $x > 0$, the integrals around the poles for $a(x)$ are

$$D + D' + D'' \simeq \frac{i\gamma^2 F_0}{2h\rho_p\omega^2}\left[\frac{e^{iK_+x}}{4K_+} + \frac{ie^{-K_-x}}{4K_-} + \frac{\mu^2}{k\gamma^2}\left(\frac{\gamma^4}{k^4 - \gamma^4}\right)^3 e^{iK_sx}\right]$$

$$\text{for } \gamma < k$$

The first two terms together have zero slope at $x = 0$ and thus fit smoothly onto the symmetrical pair for $x < 0$. The third term, corresponding to a surface wave of small amplitude, does not join symmetrically to its opposite number for $x < 0$; the integral around the branch point cancels out this discontinuity in slope. Remembering that $K^2 - k^2$ is real and positive

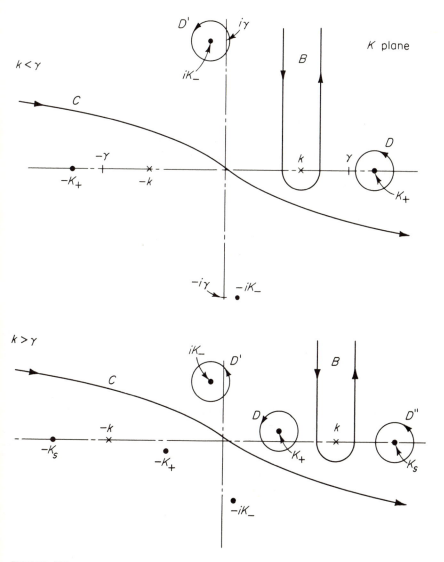

FIGURE 10.3
Contours for the integrals for η and p of Eq. (10.1.27).

when $K \to +\infty$, and thus that $\sqrt{K^2 - k^2} = e^{i\pi/4}\sqrt{u(2k + iu)}$ along the upgoing half of B (with $K = k + iu$) and $\sqrt{K^2 - k^2} = -e^{i\pi/4}\sqrt{u(2k + iu)}$ along the downgoing part, this integral, for $x > 0$, is

$$B_\eta \simeq \frac{-F_0}{2\pi\rho c^2 k^2} e^{ikx} \int_0^\infty e^{-ux} \frac{e^{3\pi i/4}\sqrt{u(2k + iu)}\, du}{1 - i(1/\mu\gamma^4)(k^4 + \cdots + u^4 - \gamma^4)^2 u(2k + iu)}$$

$$\to \frac{2\pi i F_0}{\rho c^2 (2\pi ikx)^{\frac{3}{2}}} e^{ikx} \quad x \to \infty \qquad \to -\frac{iF_0}{2\rho c^2}\left(\frac{\mu\gamma^4/k}{k^4 - \gamma^4}\right)^3 \quad x \to 0$$

$$(10.1.28)$$

This just cancels the integral D'' at $x = 0$, thus leaving no discontinuity in η there. The asymptotic formula indicates that this part of the wave does not attenuate exponentially, but falls off as $1/x^{\frac{3}{2}}$. This corresponds to a cylindrical wave in the medium, centered at the line source at $x = 0$, as will be demonstrated by computing p.

The integral around the branch point

The residues around the three poles in the integral for p are calculated the same way as were those for η. The integral around the branch cut from k to ∞ requires further discussion. We can write it

$$B_p = \frac{-i\mu\gamma^4 F_0}{4\pi} \int_k^{i\infty} \left[\frac{\exp(iKx + iz\sqrt{k^2 - K^2})}{(K^4 - \gamma^4)\sqrt{k^2 - K^2} - i\mu\gamma^4} \right.$$

$$\left. + \frac{\exp(iKx - iz\sqrt{k^2 - K^2})}{(K^4 - \gamma^4)\sqrt{k^2 - K^2} + i\mu\gamma^4} \right] dK \quad (10.1.29)$$

When x and/or z is large, both terms in the integrand oscillate rapidly over most of the path of integration. In such cases it is useful to find a point in the allowed region of the K plane, where the argument of the exponential has zero slope; close to this point the integrand ceases to oscillate rapidly, and the majority of the value of the integral arises from this region.

The derivative with respect to K of the exponent of the first term is zero when

$$K = \frac{kx}{\sqrt{x^2 + z^2}} = k\cos\phi$$

where $\phi = \tan^{-1}(z/x)$ is the angle between the radius vector $\mathbf{r}$ from the origin and the x axis. Near this point on the K plane, when $K = k\cos\phi + \epsilon$, the exponent of the first term has the form

$$iKx + iz\sqrt{k^2 - K^2} \simeq ikr - \frac{1}{2}i\frac{r^3}{kz^2}\epsilon^2 + \cdots \qquad r^2 = x^2 + z^2$$

If the path of integration is deformed so that it crosses the real K axis at $k \cos \phi$ at an angle of either 45 or 135° (i.e., if ϵ is either $u\sqrt{i}$ or $u\sqrt{-i}$, with u real), the integrand does not change phase, and thus does not oscillate near $u = 0$. But if $\epsilon = u\sqrt{i}$, the exponential increases as $|u|$ is increased; so the point $K = k \cos \phi$ is a minimum for the integrand in this region, whereas if $\epsilon = u\sqrt{-i}$, the point is a maximum. For this reason a point of zero derivative is called a *saddle point* of the function.

If we run the path of integration through the saddle point $K = k \cos \phi$ at 135°, as shown in Fig. 10.4, it is possible to route the rest of the path so that by far the largest contribution to the integral comes from the region close to the saddle point. And if kr is very large, the integrand is non-negligible only for small values of $|u|$. The first approximation to the integral for kr large is then obtained by using the first two terms in the expansion of the exponent and by extending the range of integration to $\pm\infty$. Thus the asymptotic approximation for the first term in B can be written

$$
B_p \simeq -\frac{i\sqrt{-i}\gamma^4\mu F_0 e^{ikr}}{4\pi[(k^4\cos^4\phi - \gamma^4)k\sin\phi - i\mu\gamma^4]}\int_{-\infty}^{\infty} e^{-(r/2k\sin^2\phi)u^2}\,du
$$

$$
\simeq -\frac{i\mu\gamma^4 k F_0 \sin\phi\, e^{ikr}}{2\sqrt{2\pi i kr}[(k^4\cos^4\phi - \gamma^4)k\sin\phi - i\mu\gamma^4]} \qquad kr,\, kz \gg 1
$$

$$
\to \frac{ikF_0 \tan\phi}{2\sqrt{\pi i kx}} e^{ikx} \qquad kr \to \infty;\ \phi \to 0
$$

$$
(10.1.30)
$$

There is no saddle point in the first quadrant for the second term in the brackets of Eq. (10.1.29); when $kr \gg 1$, the second integrand oscillates continuously, no matter what allowable path is taken from $K = k$ to $K = i\infty$. Thus the expression of Eq. (10.1.30) is the complete expression for the asymptotic form of integral B_p, as written. We note that it is a cylindrical wave, with origin the origin of the x, z coordinates (i.e., the location of the line force). Its amplitude dies out inversely as the square root of r; it is dipole radiation, since its amplitude goes to zero along the plate ($\phi = 0°$ and 180°).

Having reached this stage, we must point out that the arrangement of the branch cut to go through the saddle point $K = k \cos \phi$ has an important effect on the contour integral around the pole at $K = K_+$. If $K = k \cos \phi$ is greater than γ, the cut goes above the pole at K_+, as shown in the upper part of Fig. 10.4. In this case the denominator of the first integral goes to zero at $K = K_+$, as defined in Eq. (10.1.24); therefore K_+ is a pole for the integrand. But if $k \cos \phi$ is smaller than γ, the cut goes below K_+, factor $\sqrt{k^2 - K^2}$ in the numerator changes sign, and K_+ is no longer a pole for the integrand. Therefore the contour integral around K_+ is to be included in the formula for p *only as long as $k \cos \phi$ is larger than γ.*

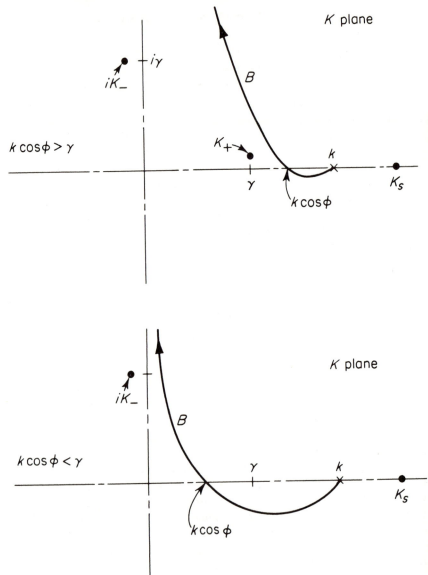

FIGURE 10.4
Path of integration along the branch cut B, going through
the saddle point at $K = k \cos \phi$.

Collecting all the expressions and carrying out the contour integrations, we can finally write out the asymptotic formulas for plate displacement and pressure wave, for x and z positive, for μ/γ small, and for $k > \gamma$.

$$
\eta_\omega \simeq \frac{i\mu\gamma F_0}{8\rho\omega^2} \left\{ \frac{\exp\left[i\gamma x - (\mu\gamma/4k_-)x\right]}{1 + i(\mu\gamma/4k_-)} + \frac{\exp\left[-\gamma x - i(\mu\gamma/4k_+)x\right]}{1 + i(\mu\gamma/4k_+)} \right.
$$
$$
\left. + \frac{4\mu^2}{k\gamma}\left(\frac{\gamma^4}{k^4 - \gamma^4}\right)^3 \exp\left(iK_s x\right) \right\} e^{-i\omega t} + B_\eta(x)e^{-i\omega t}
$$

$$
p_\omega \simeq \frac{\mu\gamma F_0}{8} \left\{ \frac{\exp\left[i\gamma x + ik_- z - (\mu\gamma x/4k_-) + (\mu\gamma^2 z/4k_-^2)\right]}{k_-[1 + (i\mu/4k_-) + (i\mu\gamma^2/4k_-^3)]} u(k\cos\phi - \gamma) \right.
$$
$$
+ i\frac{\exp\left[-\gamma x + ik_+ z - (i\mu\gamma x/4k_+) - (\mu\gamma^2 z/4k_+^2)\right]}{k_+[1 + (i\mu/4k_+) + (i\mu\gamma^2/4k_+^3)]}
$$
$$
\left. + \frac{4i\mu}{k\gamma}\left(\frac{\gamma^4}{k^4 - \gamma^4}\right)^2 \exp\left[iK_s x - \left(\frac{\mu\gamma^4}{k^4 - \gamma^4}\right)z\right] \right\} e^{-i\omega t} + B_p e^{-i\omega t}
$$
$$
\tag{10.1.31}
$$

where $k_-^2 = k^2 - \gamma^2$, $k_+^2 = k^2 + \gamma^2$, and $\mu = \rho/h\rho_p$; B_η and B_p are given in Eqs. (10.1.28) and (10.1.30), respectively, and γ is given in Eq. (10.1.21). Function $u(w)$ is the step function, being zero when $w < 0$ and unity when $w > 0$, and K_s is given in Eqs. (10.1.24).

The radiated wave

The radiated wave has an interesting, though not unexpected, shape. Starting from $\phi = 90°$, vertically above the driving force, at first, as we decrease ϕ, the predominant part of p for large r is the cylindrical wave represented by B_p of Eq. (10.1.30). The first term in braces in Eq. (10.1.31) is zero as long as $\cos\phi$ is less than γ/k; the second and third terms are negligibly small for large z. As long as ϕ is larger than $\Psi = \cos^{-1}(\gamma/k)$ [see Eq. (10.1.25)], the wave amplitude is small, because of the factor μ/γ. As ϕ approaches Ψ, this amplitude increases, as will be clear if we rewrite the asymptotic form of $|B_p|^2$ in terms of the angle Ψ.

$$
|B_p|^2 \simeq \frac{(\mu k F_0)^2}{16\pi kr} \frac{\sin^2\phi}{[(\cos\phi/\cos\Psi)^4 - 1]^2 \sin^2\phi + \mu^2}
$$

This has a sharp maximum when $\phi = \Psi$, the direction of the pressure wave of Eq. (10.1.25), arising from the free wave in the plate, having wavenumber K_+.

At angle $\phi = \Psi$, the first term in braces for p_ω suddenly enters. Rewriting the mean-square amplitude of this term,

$$
|D_p|^2 \simeq \left(\frac{\mu}{8}F_0\right)^2 \cot^2\Psi \frac{\exp\left[-(\mu r/2)\cot\Psi \csc\Psi \sin(\Psi - \phi)\right]}{1 + \mu^2\csc^4\Psi} \qquad \phi < \Psi
$$

we see that it represents a plane wave going in the Ψ' direction, with no attenuation at the angle $\phi = \Psi'$, but dropping off rapidly as $\Psi' - \phi$ increases. When $\mu r \gg 1$, this drop-off is sharp enough so that $|D_p|^2$ has a delta-function peak in the Ψ' direction.

$$|D_p|^2 \to \frac{\mu \cos \Psi' \, |F_0|^2}{32 r (1 + \mu^2 \csc^4 \Psi')} \, \delta(\Psi' - \phi)$$

For ϕ less than Ψ', the B_p term is again the largest, though even its amplitude drops as ϕ diminishes, first rapidly, then more slowly approaching zero as $\phi \to 0$. When ϕ is so close to zero that $\mu z < 1$ (kr still very large), the effect of the surface wave, the third term in braces in the formula for p in Eqs. (10.1.31), begins to be felt. This wave, as mentioned earlier, travels without attenuation, confined roughly to the region $|z| < (h\rho_p/\rho)[(k^4 - \gamma^4)/\gamma^4]$; so its asymptotic, mean-square amplitude is again a delta function

$$\frac{\mu^3 \, |F_0|^2}{8 k^2 r} \left(\frac{\gamma^4}{k^4 - \gamma^4} \right)^3 \delta(\phi)$$

Thus, although a small amount of the energy supplied by the linear driving force radiates away in all directions (the cylindrical wave B_p), most of the energy is concentrated into six sharp beams, the more intense being at the angles $\pm\Psi'$, $\pm(\pi - \Psi')$ to the plate, the two beams of lesser intensity being along the plate ($\phi = 0$, π), corresponding to the surface waves D_p. By the time $|x| \to \infty$, the majority of the energy, which was mostly possessed by the plate when $|x|$ was small, has been transferred to the medium and has radiated away. Involved in this coupled motion are all the waves mentioned at the beginning of this section: the waves with velocity near that of the plate *in vacuo* [the first term in braces in Eq. (10.1.31)]; the bound surface waves with velocity near c (the third term in braces); and the free-wave motion in the medium, caused by the discontinuity at $x = 0$, the line of application of the force (the integrals B). The second term in braces does not constitute a wave; it is the correction term required by the fourth-order equation for the plate, needed only to satisfy boundary conditions at the origin, large only near the origin.

The case of the plate driven by a line force at low frequency, when $k < \gamma$ ($c_p < c$), is much simpler. There is no surface wave, but the wave with wavenumber K_+ propagates along the plate without attenuation. Also, the integral along the branch cut has no high peak, and there is no cutoff of the integral D, since $\Psi' = \cos^{-1}(\gamma/k)$ is imaginary. Thus the power radiates fairly uniformly in all directions, but with a peak near the plate surface, and here the majority of the energy is carried by the plate. The details are left as a problem.

Incident plane wave

One way of forcing the plate into motion is to send a wave of sound in its direction. Suppose the wave is plane, propagated at an angle of incidence θ to the z axis, normal to the plate, in the xz plane. This wave will be reflected from the plate, will also set the plate into motion, and thus will generate a transmitted wave on the other side of the plate. The acoustic waves and the induced transverse displacement of the plate will have the forms

$$p = \begin{cases} P_i \exp\left[ik(x\sin\theta - z\cos\theta) - i\omega t\right] \\ \qquad\qquad + P_r \exp\left[ik(x\sin\theta + z\cos\theta) - i\omega t\right] & z > 0 \\ P_t \exp\left[ik(x\sin\theta - z\cos\theta) - i\omega t\right] & z < 0 \end{cases}$$

$$\eta = A\exp\left(ikx\sin\theta - i\omega t\right)$$

In this case neither the surface wave, of wavenumber K_s, nor the cylindrical wave, represented by the integrals B, is excited.

The equations of motion, with $(p_- - p_+)_{z=0}$ as F_ω and the two equations matching fluid and plate velocities on the two sides of the plate, are

$$-i\omega A = \frac{\cos\theta}{\rho c}(-P_i + P_r) = -\frac{\cos\theta}{\rho c}P_t$$

and

$$(k^4\sin^4\theta - \gamma^4)A = \frac{\gamma^4}{2h\rho_p\omega^2}(P_t - P_i - P_r)$$

with solutions

$$P_t = \frac{P_i}{1 - iQ(\theta)} \qquad P_r = \frac{-iQ(\theta)P_i}{1 - iQ(\theta)} \qquad A = \frac{-i(1/\rho c\omega)\cos\theta\, P_i}{1 - iQ(\theta)} \qquad (10.1.32)$$

where

$$Q = \left[1 - \left(\frac{k}{\gamma}\right)^4\sin^4\theta\right]\frac{k}{\mu}\cos\theta \qquad \mu = \frac{\rho}{\rho_p h}$$

If k/γ (which is roughly equal to $\sqrt{\nu/60,000}$ for the 2-mm steel plate in air) is much smaller than unity, then the quantity Q is large for all values of θ, the angle of incidence, except for grazing incidence, $\theta = \frac{1}{2}\pi$. In this case P_t, the amplitude of the transmitted wave, is small; most of the incident wave is reflected. If the frequency is great enough so that $k/\gamma > 1$, then Q will be zero when $\sin\theta = \gamma/k$; at this angle of incidence all the incident energy is transmitted through the plate. Put another way, when the frequency and angle of incidence are such that the horizontal component $k\sin\theta$ of the incident wavenumber is equal to the wavenumber γ of the transverse wave in the plate, the plate is transparent to the incident wave.

Now suppose the incident wave, instead of being sinusoidal, is a pulse, with Fourier transform $P_i(\omega)$.

$$p_i = P_i \delta(T) \qquad T = t - \frac{1}{c}(x \sin \theta - z \cos \theta)$$

$$P_i(\omega) = \frac{1}{2\pi} \int_{-\infty}^{\infty} p_i(T) e^{i\omega t}\, dt = \frac{P_i}{2\pi} e^{i(\omega/c)(x \sin \theta - z \cos \theta)}$$

Then the reflected and transmitted waves, and the displacement of the plate, will be corresponding Fourier transforms of the functions given in Eqs. (10.1.32). For example, the transmitted wave, in the region $z < 0$, is

$$p_t = \frac{P_i}{2\pi} \int_{-\infty}^{\infty} \frac{\exp\left[ik(x \sin \theta - z \cos \theta) - i\omega t\right] d\omega}{1 - i(k/\mu)[1 - (\omega/\omega_p)^2 \sin^4 \theta]\cos \theta} \qquad k = \frac{\omega}{c}$$

$$= \frac{-iP_i}{2\pi} \frac{\mu c \omega_p^2}{\cos \theta \sin^4 \theta} \int_{-\infty}^{\infty} \frac{e^{-i\omega T}\, d\omega}{\omega[\omega^2 - (\omega_p/\sin^2 \theta)^2] - i(\mu c \omega_p^2/\cos \theta \sin^2 \theta)}$$

$$\text{(10.1.33)}$$

where $\mu = \rho/h\rho_p$, and $\omega_p^2 = c^4\gamma^4/\omega^2 = 3c^4\rho_p(1 - s^2)/Qh^2$; for the 2-mm steel plate in air, $\mu \simeq 0.0016$ and $\omega_p \simeq 5 \times 10^4$.

There are three poles of this integrand, roots of the denominator; one of them is on the negative-imaginary axis of the ω plane; the other two are above the real axis, symmetrically placed with respect to the origin. We can label the roots of

$$\omega^3 - \omega \left(\frac{\omega_p}{\sin^2 \theta}\right)^2 - i\frac{\mu c \omega_p^2}{\cos \theta \sin^2 \theta} = 0$$

as

$$\omega_+ = w + iv \qquad \omega_- = -w + iv \qquad \omega_s = -2iv$$

where

$$w^2 - 3v^2 = \frac{\omega_p^2}{\sin^4 \theta} \qquad v(w^2 + v^2) = \frac{\mu c}{2 \cos \theta} \frac{\omega_p^2}{\sin^4 \theta} \qquad \text{(10.1.34)}$$

$$w \to \frac{\omega_p}{\sin^2 \theta} \qquad v \to \frac{\mu c}{2 \cos \theta} \qquad \text{when} \qquad \mu c \sin^2 \theta \ll \omega_p \cos \theta$$

the last line giving approximate values when $\mu c/\omega_p \equiv (\rho/c\rho_p)\sqrt{Q/3\rho_p(1 - s^2)}$ is smaller than $\cos \theta/\sin^2 \theta$ (it equals about 10^{-3} for the steel plate in air, about 0.4 for the same plate in water). When $\mu c/\omega_p$ is not small, w and v are given by Table 10.1.

The second line of Eq. (10.1.33) has been written so that the denominator of the integrand is

$$(\omega - \omega_+)(\omega - \omega_-)(\omega - \omega_s) = (\omega - iv - w)(\omega - iv + w)(\omega + 2iv)$$

According to Eqs. (1.2.15), the transmitted pressure wave is

$$p_t = \frac{P_i \mu c \omega_p{}^2}{\cos\theta \sin^4\theta} \begin{cases} \left(\cos wT - 3\dfrac{v}{w}\sin wT\right)\dfrac{e^{vT}}{w^2 + 9v^2} & T < 0 \\[3mm] \dfrac{e^{-2vT}}{w^2 + 9v^2} & T > 0 \end{cases} \tag{10.1.35}$$

where $T = t - (1/c)(x\sin\theta - z\cos\theta)$, $(1/c)(x\sin\theta - z\cos\theta)$ being the time the incident pulse-wave would arrive at the point x, z ($z < 0$) if the plate were not present.

In this case a "precursor" wave is present; the observer at point x, z ($z < 0$) hears sound before he would hear the pulse if the plate were not present (i.e., for $T < 0$). Because of the factor e^{vT}, the sound is not loud

TABLE 10.1

$\dfrac{\mu c \sin^2\theta}{\omega_p \cos\theta}$	$\dfrac{w \sin^2\theta}{\omega_p}$	$\dfrac{v \sin^2\theta}{\omega_p}$	$\dfrac{\mu c \sin^2\theta}{\omega_p \cos\theta}$	$\dfrac{w \sin^2\theta}{\omega_p}$	$\dfrac{v \sin^2\theta}{\omega_p}$
0.1	1.0037	0.0495	1.2	1.1975	0.3802
0.2	1.0138	0.0965	1.4	1.2312	0.4148
0.4	1.0462	0.1776	1.8	1.2938	0.4740
0.6	1.0848	0.2428	2.2	1.3504	0.5242
0.8	1.1239	0.2961	2.6	1.4026	0.5678
1.0	1.1616	0.3412	3.0	1.4505	0.6067

until $-T$ is small. The frequency heard is inversely proportional to the square of the sine of the angle of incidence (if $\theta \to 0$, $w \to \infty$, and there is no precursor wave). In addition, there is an exponential decay of pressure when T is positive, which attenuates most slowly when $\theta \to 0$ (since $v \to \mu c/2 \cos\theta$ as $\theta \to 0$). This result appears to conflict with the discussion of page 479, which indicated that no sound can arrive at the observer faster than the signal velocity can bring it. There is no contradiction, however; the velocity of transverse motion on a plate is proportional to the square root of the frequency ($c_p = \omega/\gamma = c\sqrt{\omega/\omega_p}$), so that there is no limiting velocity as $\omega \to \infty$, and therefore there is no signal velocity (or if you like, the signal velocity is infinite). The pulse, incident on the plate at angle θ, moves along the plate with a velocity $c/\sin\theta$; and $\omega_p/2\pi \sin^2\theta \simeq w/2\pi$ is the *frequency of the plate wave which has this velocity*.

What is happening is also illustrated by the expression for the plate displacement η as function of T. From Eqs. (10.1.32) and (10.1.33) we have

$$\eta = -\frac{P_i\mu c\omega_p{}^2}{2\pi\rho c \sin^4\theta}\int_{-\infty}^{\infty}\frac{e^{-i\omega T}\,d\omega}{\omega(\omega - w - iv)(\omega + w - iv)(\omega + 2iv)}$$

$$= \frac{-P_i\mu\omega_p{}^2}{\rho\sin^4\theta(w^2+v^2)(w^2+9v^2)}\begin{cases}\dfrac{e^{vT}}{w}\,[(w^2-3v^2)\sin wT \\ \qquad\qquad + 4wv\cos wT] & T<0 \\[2mm] \dfrac{1}{2v}\,[(w^2+9v^2)-(w^2+v^2)e^{-2vT}] & T>0\end{cases}$$

where $T = t - (x/c)\sin\theta$, since $z = 0$ for the plate. Quantity η is continuous in value and slope at $T = 0$; its greatest positive value is just before $T = 0$; after the pulse has passed $(T > 0)$, the plate approaches exponentially its final displacement, $-[P_i\mu\omega_p{}^2/2\rho v(w^2 + v^2)\sin^4\theta] = -(P_i\cos\theta/\rho c)$, which is thus its greatest negative displacement. At grazing incidence $(\theta \to \tfrac{1}{2}\pi)$, $v \to \tfrac{1}{2}(\mu c\omega_p{}^2/\cos\theta\sin^2\theta)^{\frac{1}{3}}$ and $w \to \sqrt{3}\,v$, both going to infinity as $\cos\theta \to 0$. In this limit all waves become pulse-waves, traveling parallel to the plate surface.

Incident spherical wave

Suppose a simple source of frequency $\omega/2\pi$ and strength S_ω is placed at $x = y = 0$, $z = l$, above a plane plate in the xy plane, the plate having density ρ_p, thickness $2h$, elastic modulus Q, and Poisson ratio s, as before. If the plate were not present, the pressure wave from the source would be

$$p_i = \frac{ik\rho c}{4\pi R}S_\omega e^{ikR-i\omega t} = -ik\rho c S_\omega g_k(r\,|\,r_0)e^{-i\omega t}$$

$$= \frac{k\rho c}{4\pi}S_\omega\int_0^{\infty} J_0(\beta r)e^{iw|z-l|}\frac{\beta\,d\beta}{w}e^{-i\omega t}$$

in terms of the cylindrical coordinates of Fig. 10.5, where $w = \sqrt{k^2 - \beta^2}$ or $= i\sqrt{\beta^2 - k^2}$. The first formula comes from Eqs. (7.1.5); the last is still another form of the Green's function, appropriate for our present use.

With the source on the z axis, all waves will be symmetric about this axis, independent of angle ϕ. Thus, from Eqs. (1.3.27), the plate displacement and the pressure waves below and above the plate can be written

$$\eta = e^{-i\omega t}\int_0^{\infty} A(\beta)J_0(\beta r)\beta\,d\beta \tag{10.1.36}$$

$$p = \begin{cases} p_i + p_r = e^{-i\omega t}\displaystyle\int_0^{\infty}\left[\frac{k\rho c S_\omega}{4\pi w}e^{iw|z-l|} + B(\beta)e^{iwz}\right]J_0(\beta r)\beta\,d\beta & z>0 \\[3mm] p_t = e^{-i\omega t}\displaystyle\int_0^{\infty} C(\beta)e^{-iwz}J_0(\beta r)\beta\,d\beta & z<0 \end{cases}$$

where A, B, and C must be determined from the boundary conditions at

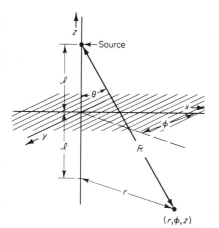

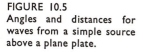

FIGURE 10.5
Angles and distances for
waves from a simple source
above a plane plate.

$z = 0$. To join velocities at $z = 0$, we set $-i\omega A = -(S_\omega/4\pi)e^{iwl} + (w/\rho ck)B = -(w/\rho ck)C$ or $(\rho ck S_\omega/4\pi w)e^{iwl} = B + C$ and $A = -(iw/\rho c^2 k^2)C$. The equation of motion (10.1.21), when applied to the integrands, with $F = p_t - p_i - p_r$, is

$$(\beta^4 - \gamma^4)A = \frac{\mu\gamma^4}{2\rho\omega^2}\left(C - B - \frac{k\rho c S_\omega}{4\pi w}e^{iwl}\right)$$

where $\mu = \rho/h\rho_p$, and $\gamma^4 = 3\omega^2\rho_p(1 - s^2)/Qh^2$, as before.

Solution of the simultaneous equations results in

$$A = -i\frac{w}{\rho w^2}C \qquad B = \frac{\rho\omega}{4\pi w}S_\omega e^{iwl} - C$$

$$C = \frac{-(i/4\pi)\mu\gamma^4\rho\omega S_\omega e^{iwl}}{w^2(\beta^4 - \gamma^4) - i\mu w\gamma^4} \qquad w^2 = k^2 - \beta^2$$

which then can be substituted in the integrals for η, p_r, and p_t. The integrals are similar to those of Eqs. (10.1.27), but the differences merit discussion. In the first place, the integrals over β are from 0 to ∞, not from $-\infty$ to ∞. Since the asymptotic form of J_0 is not simple, it is best to separate it into the two Hankel functions,

$$J_0(\beta r) = \tfrac{1}{2}H_0^{(1)}(\beta r) + \tfrac{1}{2}H_0^{(2)}(\beta r)$$

$$H_0^{(1)}(u) = J_0(u) + iN_0(u) \to \sqrt{\frac{2}{\pi iu}}e^{iu} \qquad u \to \infty \qquad (10.1.37)$$

$$H_0^{(2)}(u) = J_0(u) - iN_0(u) \to \sqrt{\frac{2i}{\pi u}}e^{-iu}$$

in accord with the discussion following Eqs. (7.3.2). The two Hankel functions have a logarithmic singularity at zero, but have simple asymptotic forms, $H_0^{(1)}(u)$ going exponentially to zero when the imaginary part of u is large and positive, $H_0^{(2)}(u)$ going to zero when the imaginary part of u is large and negative.

Thus the integrals of Eq. (10.1.36) can be broken into two parts, the term containing $H_0^{(1)}$ being integrated along contour C of Fig. 10.6, and that

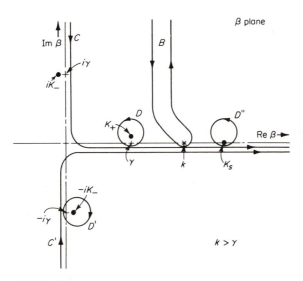

FIGURE 10.6
Contours of integration for Eqs. (10.1.37).

containing $H_0^{(2)}$ going along C'. This is possible because $H_0^{(1)}(iz) = -H_0^{(2)}(-iz)$ (z real), as an examination of Eqs. (5.3.15) will demonstrate. Therefore the integral of the $H_0^{(1)}$ part along the positive-imaginary portion of C is exactly canceled by the integral of the $H_0^{(2)}$ part along the negative-imaginary portion of C'; so these parts can be added to the parts along the 0 to ∞ portions of C and C' without changing the value of the integral sum. Contour C can then be completed by a quarter circle around the first quadrant at ∞, since the integrand is zero there; similarly, contour C' can be completed by surrounding the fourth quadrant.   Finally, these closed contours can be shrunk, C reducing to the circles about the poles $\beta = K_+$ and $\beta = K_s$ (if $k > \gamma$) plus the loop from $i\infty$ around the branch point $\beta = k$ and back to $i\infty$, and C' to the circle around the pole $-iK_-$. Approximate values of K_+, K_-, and K_s are given in Eqs. (10.1.24).

The calculation of the integrals proceeds as with the integrals of Eqs. (10.1.27).   For example, the wave p_t, transmitted through the plate to the region $z < 0$, for $k > \gamma$ and $R^2 = r^2 + (l - z)^2$ very large, is

$$p_t \to -\tfrac{1}{8}\rho ck S_\omega \mu \gamma^4 \left\{ u(k \sin \theta - \gamma) \frac{\exp\left[iK_+ r + iq_+(l - z)\right]}{\gamma^2 q_+ \sqrt{k^2 - \gamma^2}\sqrt{2\pi i K_+ r}} \right.$$

$$+ \frac{\exp\left[-K_- r + iq_-(l - z)\right]}{\gamma^2 q_- \sqrt{k^2 + \gamma^2}\sqrt{2\pi K_- r}} - 4iK_s \frac{\exp\left[iK_s r + iq_s(l - z)\right]}{k(k^4 - \gamma^4)\sqrt{2\pi i K_s r}}$$

$$\left. + \frac{4\sqrt{2}}{(k^4 \sin^4 \theta - \gamma^4)k \cos \theta - i\mu\gamma^4} \frac{e^{iKr}}{4\pi R} \right\} \quad (10.1.38)$$

Definitions and approximate value of K_+, K_-, K_s, and q_+, etc., are given in Eqs. (10.1.24) for $k > \gamma$.

When μ/γ is small, the exponential of the first term in braces can be rewritten

$$\exp\left\{i\gamma r + i(l - z)\sqrt{k^2 - \gamma^2} - \frac{\mu\gamma}{4(k^2 - \gamma^2)}\left[r\sqrt{k^2 - \gamma^2} - \gamma(l - z)\right]\right\}$$

$$= \exp\left\{ikR \cos(\tfrac{1}{2}\pi - \theta - \Psi) - \tfrac{1}{4}\mu R \sin(\tfrac{1}{2}\pi - \theta - \Psi)\cot\Psi \csc\Psi\right\}$$

where $r = R \sin \theta$; $l - z = R \cos \theta$; and $k \cos \Psi = \gamma$, Ψ being the angle of Eq. (10.1.25) between the plate and the direction of propagation of the radiation into the medium for the frequency $\omega/2\pi$. Because of the step function $u(k \sin \theta - \gamma)$, this wave has zero amplitude (for R large) for θ smaller than $\tfrac{1}{2}\pi - \Psi$. At angle $\theta = \tfrac{1}{2}\pi - \Psi$ to the z axis, there is no exponential attenuation of this part of the wave (though the wave is inversely proportional to $\sqrt{r}$), but the amplitude drops exponentially with $\sin(\tfrac{1}{2}\pi - \Psi - \theta)$ for larger values of θ. This conical wave arises from the spherical incident wave. Since, as we saw earlier, the plate is transparent, at this frequency, only to waves with angles of incidence close to $\tfrac{1}{2}\pi - \Psi$, the transmitted wave (this part of it, at any rate) is concentrated on the surface of a cone of angle $\tfrac{1}{2}\pi - \Psi$ with respect to the z axis and vertex at the source point.

The second term in Eq. (10.1.38) is negligible for large R; it comes from the exponentiallike solution $J_0(i\gamma r)$ required to maintain continuity of slope at $r = 0$ [see discussion of Eqs. (5.3.15)]. The third term corresponds to the surface wave induced by the incident wave; it represents a radial flow of energy outward in a region close to the xy plane of the plate. An incident spherical wave produces such a surface wave (though of small amplitude); an incident plane wave does not, as Eqs. (10.1.32) show.

The last term in Eq. (10.1.38), coming from the integral around the branch cut, is a spherical wave emanating from the source point, with a pronounced dependence on the angle θ. When $\theta = 0$ (along the $-z$ axis) and when $\theta = \tfrac{1}{2}\pi$ (out along the xy plane), the amplitude is small because of the μ in the factor outside the braces. But the amplitude of this part of the wave has a sharp maximum, larger by the factor k/μ, at $\theta = \frac{1}{2}\pi - \Psi$, the angle at which a plane wave can penetrate the plate without loss. Thus all the wave types discussed earlier are produced by the incident spherical wave.

At low frequencies, when $c_p < c$ $(k < \gamma)$, there is no surface wave and no direction of strong penetration, since $\Psi = \cos^{-1}(\gamma/k)$ is complex. In the expression for p_t, corresponding to Eq. (10.1.38), the only term which does not die out exponentially as $z \to \infty$ is the integral around the branch point, representing a spherical transmitted wave of small amplitude, increasing monotonically as θ increases from 0 to $\frac{1}{2}\pi$ and having no maximum. Details are left for a problem.

10.2 A FINITE PANEL

When all the coupled systems are infinite in extent, as they were in the problems discussed in Sec. 10.1, the wave motion can extend to infinity, and reflections can be neglected. But when one or more of the systems are finite in size, then standing waves, and therefore resonances, add their complications to an already intricate problem. The calculations required for a solution are often involved, and the solutions, usually approximations, are necessarily complex in order to take into account all the resonances, antiresonances, transparencies, and other effects typical of coupled resonating systems. Space is available, in this chapter and its problems, for only a few examples. In this section a single example is worked out of a triple system, one element of which is finite. A problem deals with a double system, a simplified Helmholtz resonator.

Circular membrane in rigid wall

We consider a rigid wall coincident with the xy plane, with a circular opening in it of radius a, centered at the origin. This opening has a membrane, of areal density σ, stretched across it under tension T, so that the velocity of transverse waves is $c_p = \sqrt{T/\sigma} = c/\alpha$, c being the acoustic velocity in the medium on either side of the wall. The transverse vibration of the membrane is subject to the equation

$$\frac{1}{r}\frac{\partial}{\partial r}\left(r\frac{\partial \eta}{\partial r}\right) + \frac{1}{r^2}\frac{\partial^2 \eta}{\partial \phi^2} + k^2\alpha^2\eta = -\frac{\alpha^2}{\sigma c^2}P \qquad r < a \qquad (10.2.1)$$

where $\eta = 0$ when $r = a$
 $\quad k = \omega/c$
r and ϕ = polar coordinates in xy plane
 $\quad \eta$ = displacement of membrane in z direction
 $\quad P$ = pressure difference $p_- - p_+$ tending to move membrane in positive z direction

According to Eq. (5.2.22) [see also Eq. (9.3.31)], the eigenfunctions for free vibration of the membrane, in the absence of the medium, are

$$\begin{matrix} \cos \\ \sin \end{matrix} (m\phi)J_m\left(\frac{j_{mn}r}{a}\right) \qquad \text{where } J_m(j_{mn}) = 0 \qquad (10.2.2)$$

Thus $j_{mn} = \pi\beta_{mn}$ is the nth zero of J_m (see Table X). By methods similar to those used in deriving Eq. (9.4.2), we can show that the Green's function for the membrane, solution of $(\nabla_2{}^2 + k^2\alpha^2)G = -\delta(x - x_0)\delta(y - y_0)$, is

$$G(r,\phi \,|\, r_0,\phi_0) = \sum_{m=0}^{\infty} \frac{\epsilon_m}{\pi} \cos m(\phi - \phi_0) \sum_{n=1}^{\infty} \frac{J_m(j_{mn}r/a)J_m(j_{mn}r_0/a)}{J_{m-1}^2(j_{mn})[j_{mn}{}^2 - (k\alpha a)^2]}$$

$$\text{for } 0 \leqslant r, r_0 \leqslant a \quad (10.2.3)$$

Thus the displacement of the membrane under the influence of a net pressure difference $P(r,\phi)e^{-i\omega t}$ is

$$\eta(r,\phi) = \frac{\alpha^2}{\sigma c^2} \int_0^{2\pi} d\phi_0 \int_0^a G(r,\phi \,|\, r_0,\phi_0)P(r_0,\phi_0)r_0 \, dr_0 \qquad (10.2.4)$$

times the factor $e^{-i\omega t}$.

The integral is a Fourier series in ϕ, because of the form of G; thus we have a sequence of equations for the Fourier components of η. If we set

$$\eta = \sum_m \cos(m\phi)Y_m(r) \qquad \text{and} \qquad P = \sum_m \cos(m\phi)\Phi_m(r)$$

then

$$Y_m(r) = \frac{\alpha^2}{\sigma c^2} \int_0^a G_m(r \,|\, r_0)\Phi_m(r_0)r_0 \, dr_0 \qquad (10.2.5)$$

where

$$G_m(r \,|\, r_0) = \sum_{n=1}^{\infty} \frac{2}{J_{m-1}^2(j_{mn})} \frac{J_m(j_{mn}r/a)J_m(j_{mn}r_0/a)}{j_{mn}{}^2 - (k\alpha a)^2}$$

If Φ_m happens to be proportional to the Bessel function $J_m(qr)$, a closed form for Y_m can be found [as was done with Eq. (9.3.33)]. Since the function

$$f(r,\phi) = A \cos(m\phi)\left[J_m(qr) - \frac{J_m(qa)}{J_m(k\alpha a)}J_m(k\alpha r)\right]$$

satisfies the boundary condition $f(a,\phi) = 0$ and also satisfies the equation $(\nabla_2{}^2 + k^2\alpha^2)f = A \cos(m\phi)(k^2\alpha^2 - q^2)J_m(qr)$, we can show that

$$\int_0^r G_m(r \,|\, r_0)J_m(qr_0)r_0 \, dr_0 = \frac{1}{q^2 - k^2\alpha^2}\left[J_m(qr) - \frac{J_m(qa)}{J_m(k\alpha a)}J_m(k\alpha r)\right]$$

$$= \frac{1}{2k\alpha}\left[rJ_{m-1}(k\alpha r) - \frac{J_{m-1}(k\alpha a)}{J_m(k\alpha a)}aJ_m(k\alpha r)\right] \qquad \text{when } q = k\alpha \quad (10.2.6)$$

These formulas will be useful later.

Reaction of the medium

The air, or other medium, on either side of the wall, has density ρ and transmits acoustic waves with velocity c. The Green's function for the region $z > 0$ is most appropriately given in Eq. (7.4.8) for cylindrical coordinates r, ϕ, z and for a rigid wall.

$$g_+ = \sum_{m=0}^{\infty} \frac{i\epsilon_m}{2\pi} \cos m(\phi - \phi_0) \int_0^{\infty} J_m(\mu r) J_m(\mu r_0) \frac{\mu \, d\mu}{\varkappa} \begin{cases} \cos \varkappa z \, e^{i\varkappa z_0} & z_0 \geqslant z \geqslant 0 \\ \cos \varkappa z_0 \, e^{i\varkappa z} & z \geqslant z_0 \geqslant 0 \end{cases}$$
$$(10.2.7)$$

where

$$\varkappa = \sqrt{k^2 - \mu^2} \quad \text{or} \quad i\sqrt{\mu^2 - k^2}$$

The Green's function g_- for the region $z \leqslant 0$ is obtained from g_+ by reversing the signs of z and z_0.

Motion of the membrane produces waves in the medium, and this produces a reaction back on the membrane. Suppose a plane wave $p = A \exp (ikz \cos \theta + ikr \cos \phi \sin \theta - i\omega t)$, with direction of propagation in the xz plane at angle of incidence θ with respect to the z axis, comes from $-\infty$ to strike the negative side of the wall at $z = 0$. If the whole wall were rigid, this wave would be reflected with no change of amplitude or phase, and the pressure in the region $z < 0$ would be (we omit the time factor)

$$p_i = 2P_i \cos (kz \cos \theta) \exp (ikr \cos \phi \sin \theta)$$
$$= 2P_i \cos (kz \cos \theta) \sum_m \epsilon_m i^m \cos (m\phi) J_m(kr \sin \theta) \quad (10.2.8)$$

by using Eq. (1.2.9). If the membrane displacement is expressed as a Fourier series $\eta = \sum \cos (m\phi) Y_m(r)$, the pressure wave in region $z < 0$ is modified to become

$$p_- = p_i + \rho\omega^2 \int_0^{2\pi} d\phi_0 \int_0^a g_-(r,\phi,z \mid r_0,\phi_0,0)\eta(r_0,\phi_0)r_0 \, dr_0$$
$$\xrightarrow[z \to 0]{} \sum_m \left[2P_1 \epsilon_m i^m J_m(kr \cos \theta) + \rho\omega^2 \int_0^a g_m(r \mid r_0) Y_m(r_0)r_0 \, dr_0 \right] \cos (m\phi)$$
$$(10.2.9)$$

where

$$g_m(r \mid r_0) = i \int_0^{\infty} J_m(\mu r) J_m(\mu r_0) \frac{\mu \, d\mu}{\varkappa} \qquad \varkappa^2 = k^2 - \mu^2$$

as obtained from Eqs. (10.2.7) and (7.4.16), taking into account the fact that the membrane velocity u_z is equal to $-i\omega\eta$, and for the negative side of the wall, the expression of Eq. (7.4.16) for p_- is $+i\rho\omega$, rather than $-i\rho\omega$, times the first integral.

If there is no incident wave in the region $z > 0$, the only sound there will be that produced by the motion of the membrane.

$$P_+ = -\rho\omega^2 \int_0^{2\pi} d\phi_0 \int_0^a g_+(r,\phi,z \mid r_0,\phi_0,0)\eta(r_0,\phi_0)r_0\,dr_0$$

$$\xrightarrow[z \to 0]{} -\rho\omega^2 \sum_m \cos(m\phi) \int_0^a g_m(r \mid r_0) Y_m(r_0)r_0\,dr_0 \quad (10.2.10)$$

Taking the second lines of Eqs. (10.2.9) and (10.2.10), we have a second integral equation relating P and η, and thus a second set of equations relating the Φ's with the Y's.

$$P = (p_- - p_+)_0 = p_i + 2\rho\omega^2 \int_0^{2\pi} d\phi_0 \int_0^a g(r,\phi,0 \mid r_0,\phi_0,0)\eta(r_0,\phi_0)r_0\,dr_0$$

or

$$(10.2.11)$$

$$\Phi_m(r) = 2P_i\epsilon_m i^m J_m(kr\sin\theta) + 2\rho\omega^2 \int_0^a g_m(r \mid r_0) Y_m(r_0)r_0\,dr_0$$

for the range $0 \leqslant r \leqslant a$ only. The simultaneous solutions to these and the counterpart equations (10.2.5) will be called $\Phi_m(\theta \mid r)$ and $Y_m(\theta \mid r)$. From them one can compute the energy transmitted through the membrane, as well as the angular distribution of the transmitted wave in region $z > 0$ and the distortion of the reflected wave produced by the membrane motion.

The transmitted wave

Knowing $Y_m(\theta \mid r)$, Eq. (10.2.10) can be used to compute the wave in the region $z > 0$. Its asymptotic form, for z very large, can be computed by using the alternative form for g, given in Eq. (7.4.6). Using the spherical coordinates R, ϕ, ϑ instead of cylindrical coordinates r, ϕ, z, the asymptotic form for g_+ is

$$g_+(R,\phi,\vartheta \mid r_0,\phi_0,0) \to \frac{1}{2\pi}\frac{e^{ikR}}{R} \exp[-ikr_0\cos(\phi - \phi_0)\sin\vartheta]$$

Inserting this in Eq. (10.2.10) and using Eq. (1.3.11), we obtain

$$p_+ \to \frac{e^{ikR}}{R} \sum_m \Theta_m(\theta \mid \vartheta) \cos(m\phi)$$

$$(10.2.12)$$

$$\Theta_m(\theta \mid \vartheta) = \rho\omega^2 i^{-m} \int_0^a J_m(kr_0\sin\vartheta) Y_m(\theta \mid r_0)r_0\,dr_0 \qquad 0 \leqslant \vartheta \leqslant \tfrac{1}{2}\pi$$

Thus the sum $\sum \Theta_m(\theta \mid \vartheta) \cos(m\phi)$ gives the asymptotic dependence on the spherical angles ϑ, ϕ of the wave transmitted through the membrane, generated by a plane wave incident at angle θ on the other side of the wall.

The total power transmitted through the membrane could be obtained by integrating $|p_+|^2/\rho c$ over the hemisphere of large radius, from (10.2.12).

But we can also calculate it by integrating $\mathrm{Re}\,(p_+ u_z^*)$ over the surface of the membrane, with $u_z = -i\omega\eta$. Since, for $z = 0$, $p_- - p_+ = P$ and $p_- + p_+ = p_i$, we can write

$$p_+ = \tfrac{1}{2}p_i - \tfrac{1}{2}P = -\rho\omega^2 \sum_m \cos(m\phi) \int_0^a g_m(r \mid r_0) Y_m(\theta \mid r_0) r_0\, dr_0$$

from Eq. (10.2.10). Since $u_z = -i\omega \sum \cos(m\phi) Y_m(r)$, we have for the total power transmitted

$$\Upsilon = 2\pi\rho\omega^3\,\mathrm{Re}\left[\sum_m \frac{-i}{\epsilon_m}\int_0^a Y_m(\theta \mid r) r\, dr \int_0^a g_m(r \mid r_0) Y_m^*(\theta \mid r_0) r_0\, dr_0\right]$$

$$= 2\pi\rho\omega^3\,\mathrm{Re}\left[\sum_m \frac{1}{\epsilon_m}\int_0^a Y_m(\theta \mid r) r\, dr \right.$$

$$\left. \times \int_0^a Y_m^*(\theta \mid r_0) r_0\, dr_0 \int_0^\infty J_m(\mu r)J_m(\mu r_0)\frac{\mu\, d\mu}{\sqrt{k^2 - \mu^2}}\right] \quad (10.2.13)$$

A possible method of solution is to expand Y_m into a series of the eigenfunctions $J_m(j_{mn}r/a)$ $(n = 1, 2, \ldots)$.

$$Y_m(\theta \mid r) = \sum_{n=1}^\infty A_n{}^m(\theta)J_m\left(\frac{j_{mn}r}{a}\right)$$

This can be inserted in the left-hand side of Eq. (10.2.5), to equate with the coefficients of the expansion of G_m, obtaining

$$A_n{}^m(\theta) = \frac{2\alpha^2}{\sigma c^2 J_{m-1}^2(j_{mn})}\,\frac{1}{j_{mn}{}^2 - (k\alpha a)^2}\int_0^a \Phi_m(\theta \mid r_0)J_m\left(\frac{j_{mn}r_0}{a}\right) r_0\, dr_0$$

Substituting Eq. (10.2.11) for Φ_m and again using the expansion of Y_m produces a set of simultaneous equations from which the coefficients $A_n{}^m$ can be computed. Using the equation for g_m, we have

$$A_n{}^m(\theta) = \frac{\alpha^2 a^2/\sigma c^2}{j_{mn}{}^2 - (k\alpha a)^2}\left\{2P_i\epsilon_m i^m F_n{}^m(ka\sin\theta)\right.$$

$$\left. + \tfrac{1}{2}i\rho\omega^2 a \sum_v A_v{}^m(\theta)J_{m-1}^2(j_{nv})[\theta_{nv}{}^m(ka) - i\chi_{nv}{}^m(ka)]\right\} \quad (10.2.14)$$

where

$$F_n{}^m(x) = \frac{2j_{mn}J_m(x)}{J_{m-1}(j_{mn})(x^2 - j_{mn}{}^2)} \to 1 \qquad x \to j_{mn}$$

and

$$\theta_{nv}{}^m(x) - i\chi_{nv}{}^m(x) = \int_0^\infty F_n{}^m(\mu)F_v{}^m(\mu)\frac{\mu\, d\mu}{\sqrt{x^2 - \mu^2}}$$

We have used the last of Eqs. (5.2.21), which in the present case becomes

$$\int_0^a J_m\left(\frac{j_{mn}r}{a}\right) J_m(\mu r) r\, dr = \tfrac{1}{2}a^2 J_{m-1}^2(j_{mn})F_n{}^m(\mu a)$$

to carry out the integrations over the Bessel-function products.

The functions $F_n{}^m(ka \sin \theta)$ measure the direct effect of the incident wave on the mth component of the transmitted wave. Some of these functions are plotted in Fig. 10.7; in general, they are large only in the range $j_{m,n-1} < x < j_{m,n+1}$, and only those for $m = 0$ are large near $x = 0$. The

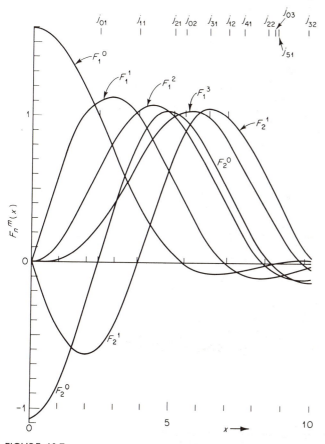

FIGURE 10.7
Angle-dependence functions $F_n{}^m(x)$ for wave transmitted through circular membrane. See Eqs. (10.2.14) and (10.2.15) and Table XI.

quantities $\theta_{nv}{}^m(x) - i\chi_{nv}{}^m(x)$ measure the interaction between the various eigenfunctions of free-membrane vibration, via the reaction of the medium on both sides of the membrane. The θ and χ are resistance and reactance factors analogous to the factors of Fig. 7.9 for the circular piston. Some of these factors are plotted in Fig. 10.8, more are given in Table XII, the F's in Table XI. In general, they have large magnitudes only near j_{mn} and j_{mv}, for n different from v, and the diagonal terms $\theta_{nn}{}^m$ and $\chi_{nn}{}^m$ are larger than the nondiagonal terms, for $n \neq v$.

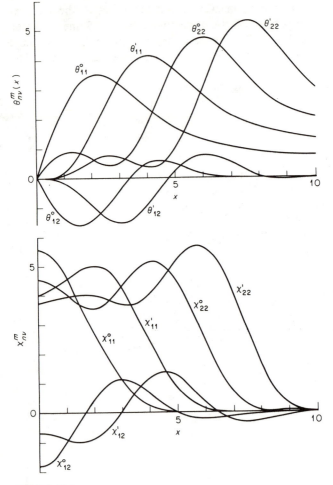

FIGURE 10.8
Resistance and reactance factors for sound transmission
through a circular membrane. See Eq. (10.2.14) and Table XII.

Once the A's are determined, the items of physical interest can be computed. For example, the asymptotic form for the transmitted wave, on the
side opposite the incident wave, is given in terms of the functions Θ_m of
Eqs. (10.2.12).

$$\Theta_m(\theta \mid \vartheta) = \tfrac{1}{2} i^{-m} \rho \omega^2 a^2 \sum_{n=1}^{\infty} A_n{}^m(\theta) J_{m-1}^2(j_{mn}) F_n{}^m(ka \sin \vartheta) \quad (10.2.15)$$

The power transmitted through the membrane becomes

$$\Upsilon = \tfrac{1}{2} \pi \rho \omega^3 a^3 \sum_{m=0}^{\infty} \frac{1}{\epsilon_m} \sum_{n,\nu} A_n{}^m A_\nu{}^m J_{m-1}^2(j_{mn}) J_{m-1}^2(j_{m\nu}) \theta_{n\nu}{}^m(ka) \quad (10.2.16)$$

But to solve for the A's by successive approximations involves the summation of many series. The variational procedure will usually achieve satisfactory accuracy and will result in more compact formulas. First, however, we should see what Eq. (10.2.14) predicts in various limiting situations.

Approximate solutions

When $k\alpha a$ is close in value to j_{mn}, the (m,n)th normal mode of the membrane is strongly excited, and $A_n{}^m(\theta)$ becomes the dominant coefficient in the expansion for η (as long as $-\rho\omega^2 a$ is small compared with $\sigma c^2/\alpha^2 a^2$). For frequencies well below the lowest resonance ($k\alpha a \ll j_{01}$), only A_1^0 is nonnegligible; the pressure near the membrane is nearly independent of position and of angle of incidence. Approximate formulas for the quantities of interest, in this case, are

$$\eta \simeq 4\frac{\alpha^2 a^2}{\sigma c^2}\frac{P_i}{j_{01}{}^3 J_1(j_{01})}J_0\left(\frac{j_{01}r}{a}\right)$$

$$p_+ \to \frac{4\rho\alpha^2 k^2 a^4}{\sigma j_{01}{}^4}P_i\frac{e^{ikR}}{R} \qquad R \to \infty \qquad (10.2.17)$$

$$\Upsilon \simeq 8\pi a^2\left(\frac{\rho a}{\sigma}\right)^2 k^3 a^3 \alpha^4\frac{|P_i|^2}{\rho c}\frac{J_1{}^2(j_{01})}{j_{01}{}^6}\theta_{11}^0(ka)$$

The power transmitted equals $2\pi R^2$ times $1/\rho c$ times the square of the asymptotic amplitude of p_+ (as it should), since $\theta_{11}^0(ka) \to 4ka/j_{01}{}^2 J_1{}^2(j_{01})$ when $ka \to 0$. We note that the transmitted power is proportional to the square of the ratio $(\rho a/\sigma)$ between the density of the medium and that of the membrane; at these low frequencies the transmission is mass-controlled.

As the frequency of the incident wave is increased to the resonance frequency of the (0,1)st normal mode of the membrane, A_1^0 remains the dominant coefficient,

$$A_1^0 \simeq \frac{(2\alpha^2 a^2 P_i/\sigma c^2)F_1^0(ka\sin\theta)}{j_{01}{}^2 - (k\alpha a)^2 - (\rho k^2 a^3/2\sigma)J_1{}^2(j_{01})[\chi_{11}^0(ka) + i\theta_{11}^0(ka)]} \qquad (10.2.18)$$

becoming quite large, but not infinite, at resonance ($k\alpha a = j_{01}$). The asymptotic form of the transmitted wave is, approximately,

$$p_+ \to \frac{aP_i F_1^0(ka\sin\theta)F_1^0(ka\sin\vartheta)(e^{ikR}/R)}{[2\sigma/\rho k^2 a^3 \alpha^2 J_1{}^2(j_{01})][j_{01}{}^2 - (k\alpha a)^2] - \chi_{11}^0(ka) - i\theta_{11}^0(ka)} \qquad (10.2.19)$$

when $R \to \infty$. The symmetry between the angle of incidence θ and the angle of transmission ϑ is a consequence of the principle of reciprocity.

Close to the (1,1) resonance, the coefficient A_1^1 predominates, and when $k\alpha a$ is exactly equal to j_{11}, the membrane displacement is

$$\eta \simeq \frac{4P_i F_1^1(ka\sin\theta)}{\rho\omega^2 a J_1{}^2(j_{01})[\chi_{11}^1(ka) + i\theta_{11}^1(ka)]}\cos\phi\, J_1\left(\frac{j_{11}r}{a}\right)$$

and so on.

Variational formulas

For purposes of computation, the variational procedure is more useful than the series expansion. As indicated in connection with Eq. (9.3.28), we can work either with the shape of Φ_m, the pressure component, or with that of Y_m, the membrane-displacement component. We have just seen that Y_m changes shape radically as the frequency approaches one of the resonance frequencies of the membrane. The driving pressure Φ_m also alters near membrane resonance, but if ρa is small compared with σ, the first term on the right-hand side of Eq. (10.2.11), representing the direct effect of the incident wave, remains the dominant term (or at least does not become negligible compared with the second term) at resonance. Thus it will be easier to devise a form for Φ_m which is satisfactory over a wide range of frequency than it will be to discover and work with one for Y_m.

We start by multiplying Eq. (10.2.11) by $Y_m(\theta \mid r)$ and integrating.

$$0 = 2P_i \epsilon_m i^m \int J_m(kr \sin \theta) Y_m(\theta \mid r) r \, dr - \int \Phi_m Y_m(\theta \mid r) r \, dr$$
$$+ 2\rho\omega^2 \int Y_m(\theta \mid r) \int g_m(r \mid r_0) Y_m(\theta \mid r_0) r_0 \, dr_0 \, r \, dr$$

This can be changed to an expression in terms of the Φ_m's by using Eq. (10.2.5) several times. From Eq. (10.2.12) we see that the integral in the first term is proportional to the function Θ_m, which determines the asymptotic form of the transmitted wave,

$$\int J_m(k \sin \theta) Y_m(\theta \mid r) r \, dr = \frac{1}{\rho\omega^2} i^m \Theta_m(\theta \mid \theta)$$

Adding this to the preceding equation and using Eq. (10.2.5) to eliminate the functions Y_m, we obtain a symmetric expression which can be used to obtain a variational solution,

$$(-1)^m \frac{2P_i \epsilon_m}{\rho\omega^2} \Theta_m(\theta \mid \theta) = 4P_i \epsilon_m i^m \frac{\alpha^2}{\sigma c^2} \int \Phi_m \int G_m J_m(kr \sin \theta)$$
$$- \frac{\alpha^2}{\sigma c^2} \int \Phi_m \int G_m \Phi_m + \frac{2\rho\omega^2 \alpha^4}{\sigma^2 c^4} \int \Phi_m \int G_m \int g_m \int G_m \Phi_m \quad (10.2.20)$$

in our shorthand notation. As usual, we check this by noting that if the shape of Φ_m is varied (inserting $\Phi_m + \delta\Phi_m$ for Φ_m), the coefficient of $\int G_m \, \delta\Phi_m$, inside the integral, is zero if Φ_m satisfies Eqs. (10.2.5) and (10.2.11).

Thus, if we choose to approximate $\Phi_m(\theta \mid r)$ by the trial function $B_m U_m(r)$, the optimal value of B_m will be

$$B_m = \frac{2P_i \epsilon_m i^m \int J_m(kr \sin \theta) \int G_m U_m}{\int U_m \int G_m U_m - (2\rho k^2 \alpha^2/\sigma) \int U_m \int G_m \int g_m \int G_m U_m} \quad (10.2.21)$$

from which we can calculate $\Theta_m(\theta \mid \vartheta)$, the transmitted power and any other desired part of the solution. If U_m is well chosen, the resulting quantities will be close to the exact solution. Moreover, if U_m is a Bessel function $J_m(qr)$, we can take advantage of Eq. (10.2.6) to obtain a closed form, rather than a series, for the integral of $G_m U_m$. For example, we define the function $H_m(uvw)$, finite for all real values of u, v, w and symmetric to interchange of u, v, w, as

$$
\begin{aligned}
H_m(uvw) &= \frac{J_m(w)}{a^4} \int_0^a J_m\left(u\frac{r}{a}\right) r\, dr \int_0^a G_m(r \mid s) J_m\left(v\frac{s}{a}\right) s\, ds \\
&= -\frac{u J_{m-1}(u) J_m(v) J_m(w)}{(u^2 - v^2)(u^2 - w^2)} - \frac{v J_{m-1}(v) J_m(u) J_m(w)}{(v^2 - u^2)(v^2 - w^2)} \\
&\quad - \frac{w J_{m-1}(w) J_m(u) J_m(w)}{(w^2 - u^2)(w^2 - v^2)} \\
&\xrightarrow[uvw \to 0]{} \frac{(uvw)^m}{2^{3m+3} m!\,(m+1)!\,(m+2)!} \\
&\xrightarrow[v \to u]{} \frac{J_m(w)}{2(u^2 - w^2)}\left[J_m^2(u) - J_{m-1}(u) J_{m+1}(u)\right] \\
&\qquad + \frac{J_m(u)}{(u^2 - w^2)^2}\left[u J_{m-1}(u) J_m(w) - w J_{m-1}(w) J_m(u)\right] \\
&\xrightarrow[u \to 0]{} \frac{u^{2m} J_{m+2}(w)}{w^2 2^{2m+1} m!\,(m+1)!} \\
&\xrightarrow[uv \to w]{} \frac{1}{4w^2}\{J_m(w)[m J_m^2(w) - (m-1) J_{m-1}(w) J_{m+1}(w)] \\
&\qquad - w J_{m-1}(w)[J_m^2(w) - J_{m-1}(w) J_{m+1}(w)]\}
\end{aligned}
$$
(10.2.22)

where $u = ka \sin\theta$ and $w = \alpha ka$, as indicated in Eq. (10.2.6). This provides a closed form, rather than an infinite series, for the integral in the numerator of B_m (if U_m is a Bessel function) and also for the first term in the denominator. Functions H_m are plotted in Fig. 10.9, given in Table XIII.

The same procedure may be used for the second integral in the denominator of B_m, but this would result in a set of integrals over the variable μ in g_m, different from those of Eq. (10.2.14), and thus would require a lot of additional numerical integration. Therefore the series expansion for the G's is preferable here, especially since the result,

$$
\begin{aligned}
S_m(uw \mid x) &= \frac{J_m^2(w)}{a^7} \int J_m\left(u\frac{r}{a}\right) \int G_m \int g_m \int G_m J_m\left(u\frac{s}{a}\right) \\
&= \sum_{n,\nu} \frac{[J_{m-1}(j_{mn}) J_{m-1}(j_{m\nu})]^3}{16 j_{mn} j_{m\nu}} F_n^m(u) F_\nu^m(u) F_n^m(w) F_\nu^m(w) \\
&\qquad \times [\chi_{n\nu}^m(x) + i\theta_{n\nu}^m(x)] \\
&\equiv I_m(uw \mid x) + i Q_m(uw \mid x)
\end{aligned}
$$
(10.2.23)

involves functions already defined and tabulated, converges rapidly, and is symmetric to interchange of u and w. These functions are plotted in Fig. 10.10.

Setting $U_m = J_m(qr/a)$ for trial function yields

$$B_m = \frac{2P_i \epsilon_m i^m J_m(w) H_m(uqw)}{J_m(w) H_m(qqw) - 2\gamma w^2 S_m(qw \mid x)}$$

where $w = \alpha ka$, $x = ka$; and $\gamma = \rho a/\sigma$ measures the coupling between membrane and medium. Since both γ and S_m are usually small, the second

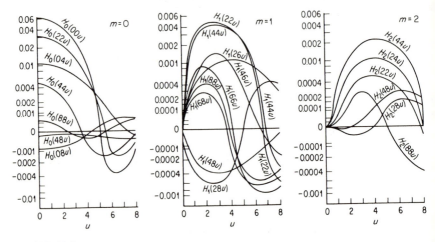

FIGURE 10.9
Typical plots of the angle-distribution function H_m. Note that $H_m(uvw) = H_m(vuw) = H_m(wvu)$. See Eq. (10.2.22) and Table XIII.

term in the denominator is usually much smaller than the first, except when $J_m(w) = 0$ (membrane resonance) or $H_m(qqw) = 0$ (antiresonance). Determining the "best" value of q cannot be done analytically, but graphical determination indicates that $q = u$ is ordinarily a good choice. In nearly all cases the variational expression of Eq. (10.2.20) changes slowly near $q = u$; the exact value at which its derivative with respect to q is zero depends a little on the value of γ, but as long as the second term in the denominator is not large, the choice $q = u = ka \sin \theta$ is quite satisfactory. To obtain a decidedly better trial function, we should have to use a two-term $BJ_m(ur/a) + CJ_m(qr/a)$, with $q \neq u$. We shall show later that this complication is seldom necessary.

When we set $q = u$, the variational solution for the mth component of the pressure difference across the membrane becomes

$$\Phi_m(\theta \mid r) \simeq 2P_i \epsilon_m i^m J_m(w) \frac{i J_m(kr \sin \theta)}{Z_m(uw \mid x)}$$

where

$$Z_m(uw \mid x) \equiv R_m(uw \mid x) - iX_m(uw \mid x) = |Z_m| \, e^{-i\Omega_m} \quad (10.2.24)$$

$$= iJ_m(w) - 2i\gamma w^2 \, \frac{S_m(uw \mid x)}{H_m(uuw)}$$

may be considered the impedance of the mth component of the membrane-plus-radiation load to the incident plane wave. As before, $u = ka \sin \theta$,

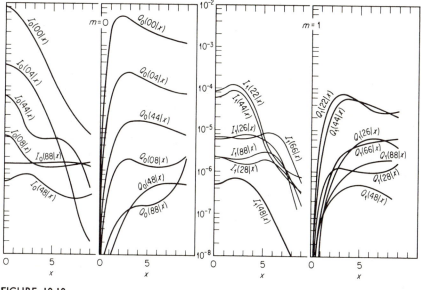

FIGURE 10.10
Typical curves of the radiation resistance and reactance functions I_m and Q_m. Note that $I_m(uw \mid x) = I_m(wu \mid x)$ and $Q_m(uw \mid x) = Q_m(wu \mid x)$.

$w = \alpha ka$, and $x = ka$; $\gamma = \rho a/\sigma$ is the coupling constant. When γ goes to zero there is no radiation load, Z_m becomes simply $iJ_m(w)$, and Φ_m reduces to the mth component of the incident wave. The real part of Z_m, proportional to the θ's of Eq. (10.2.14) and Fig. 10.8, is small if γ is small. The reactive part, X_m, includes the reactance $-iJ_m(w)$ of the unloaded membrane, in addition to the radiation reactance.

The mth component of the membrane's displacement is, from Eqs. (10.2.5) and (10.2.6),

$$Y_m(\theta \mid r) \simeq \frac{2a^2 \epsilon_m i^m P_i}{i\sigma c_p^2 (u^2 - w^2) Z_m(uw \mid z)} \left[J_m(w) J_m\!\left(u \frac{r}{a}\right) - J_m(u) J_m\!\left(w \frac{r}{a}\right) \right]$$

$$(10.2.25)$$

This function goes to zero at $r = a$, as it should; it does not become infinite at $u = w$, as the limiting form of Eq. (10.2.6) shows. Its amplitude is inversely proportional to the impedance Z_m. The angle-dependence factor Θ_m for the transmitted wave and the resulting transmitted intensity are then

$$\Theta_m(\theta \mid \vartheta) \simeq 2a\gamma\epsilon_m P_i w^2 \frac{H_m(uvw)}{iZ_m(uw \mid x)}$$

$$\frac{1}{\rho c}|p_+|^2 \xrightarrow[R \to \infty]{} \frac{4a^2}{R^2} \gamma^2 |P_i|^2 \frac{w^4}{\rho c} \sum_{m,l} \epsilon_m \epsilon_l \frac{H_m(uvw)H_l(uvw)}{|Z_m(uw \mid x)Z_l(uw \mid x)|}$$

$$\times \cos(m\phi)\cos(l\phi)\cos(\Omega_m - \Omega_l) \quad (10.2.26)$$

The transmitted intensity varies with the azimuth ϕ and with the zenith angle ϑ through the variable $v = ka\sin\vartheta$; thus the functions $H_m(uvw)$ determine the dependence of the transmitted wave on angle of incidence θ, on the zenith angle ϑ of the radius to the observer, on the wave-velocity ratio $\alpha = c/c_p$, and on the coupling constant γ. Because of the symmetry of the functions $H_m(uvw)$, all three factors, $\sin\theta$, $\sin\vartheta$, and α, have the same effect. The symmetry between θ and ϑ is another way of stating the reciprocity principle; the symmetry between $\sin\theta$ and α is related to the transparency of the infinite membrane or plate which occurs when $\sin\theta$ equals α [see discussion of Eq. (10.2.30)].

Finally, the total acoustic power transmitted through the membrane is

$$\Upsilon \simeq 8\pi\alpha a^2\gamma^2 \frac{|P_i|^2}{\rho c} \sum_m \epsilon_m \frac{w^3 Q_m(ka\sin\theta, ka\alpha \mid ka)}{|Z_m(ka\sin\theta, ka\alpha \mid ka)|^2} \quad (10.2.27)$$

where the factors Q_m involve the radiation-resistance factors $\theta_n{}^m(ka)$, plotted in Fig. 10.10. We note that $(\pi a^2 |P_i|^2/\rho c)\cos\theta$ is the power incident on the membrane, and thus that the dimensionless quantity

$$T \simeq 2\alpha\gamma^2 \sum_m \epsilon_m w^2 \operatorname{Re} \frac{1}{Z_m(ka\sin\theta, ka\alpha \mid ka)} \quad (10.2.28)$$

may be called the *transmission factor*, the ratio between the power actually transmitted through the membrane to the power carried in an area πa^2 of the incident wavefront. If the membrane transmitted all the incident power, then T would equal $\cos\theta$. It is usually considerably smaller than this, for many reasons, which our variational solution is accurate enough to demonstrate.

Transmission; resonance and antiresonance

The presence of the membrane impedes the flow of incident energy through the circular aperture in the wall, because of its mass and tension, as indicated by the presence of the factors $\alpha = c/c_p = c\sqrt{\sigma/T}$ and $\gamma = \rho a/\sigma$ in the formula for Υ. The transmission factor is inversely proportional to the square root of the membrane tension and the three-halves power of the membrane mass density σ.   These general effects are modified, as we change frequency and angle of incidence, by the various membrane resonances and by the results of the coupling between membrane and medium.

Well below the lowest membrane resonance ($\alpha ka \ll 1$), all terms for $m > 0$ are negligible, and Eqs. (10.2.25) to (10.2.27) become

$$\eta \simeq \frac{a^2 P_i}{2\sigma c_p{}^2}\left[1 - \left(\frac{r}{a}\right)^2\right]$$

$$P_+ \to \frac{\rho a}{8\sigma}(ka\alpha)^2 P_i \frac{ae^{ikR}}{R} \qquad R \to \infty$$

$$\Upsilon \simeq 8\pi a^2\left(\frac{\rho a}{\sigma}\right)^2 \frac{|P_i|^2}{\rho c}\left(\frac{ka\alpha}{j_{01}}\right)^4$$

which are similar to, but better approximations than, Eqs. (10.2.17).

To separate the various effects which enter as the incident frequency is increased, let us first look at the case of normal incidence.   When $\theta = 0$, parameter $u = ka \sin \theta$ is zero, there is no dependence on the axial angle ϕ, and consequently only the $m = 0$ term in the sums over m differs from zero. The simplified formulas become

$$H_0(0vw) = \frac{vJ_0(v)J_1(w) - wJ_0(w)J_1(v)}{vw(w^2 - v^2)} \xrightarrow{v \to w} \frac{1}{2w^2}[J_1{}^2(w) - J_0(w)J_2(w)]$$

$$H_0(00w) = \frac{1}{2w^2}J_2(w) \qquad F_n^0(0) = \frac{2}{j_{0n}J_1(j_{0n})}$$

$$S_0(0w \mid x) = \sum_{n,v}\left[\frac{J_1(j_{0n})J_1(j_{0v})}{2j_{0n}j_{0v}}\right]^2 F_n^0(w)F_v^0(w)[\chi_{nv}^0(x) + i\theta_{nv}^0(x)]$$

$$\eta = Y_0(0 \mid r) \simeq \frac{2a^2 P_i}{\sigma c_p{}^2} \frac{J_0(wr/a) - J_0(w)}{w^2 Z_0(0w \mid x)}$$

$$P_+ \to \Theta_0(0 \mid \vartheta)\frac{e^{ikR}}{R} \simeq 2\gamma w^2 P_i \frac{H_0(0vw)}{Z_0(0w \mid x)}\frac{ae^{ikR}}{R} \qquad R \to \infty \qquad (10.2.29)$$

$$\Upsilon \simeq 2\pi a^2\alpha\gamma^2 \frac{|P_i|^2}{\rho c}\operatorname{Re}\frac{1}{Z_0(0w \mid x)}$$

where

$$Z_0(0w \mid x) = iJ_0(w) + 4\gamma\frac{w^4}{J_2(w)}\sum_{n,v}\left[\frac{J_1(j_{0n})J_1(j_{0v})}{2j_{0n}j_{0v}}\right]^2$$

$$\times F_n^0(w)F_v^0(w)[\theta_{nv}^0(x) - i\chi_{nv}^0(x)]$$

or $v = ka \sin \vartheta$, $w = \alpha ka$, and $x = ka$.

In Fig. 10.11 four curves for $T = \rho c \Upsilon / \pi a^2 |P_i|^2$ are plotted as functions of ka for two different values of $\alpha = c/c_p$ and of $\gamma = \rho a/\sigma$. Increase of γ increases the coupling between the medium and the membrane, thus increasing "throughput" and raising T closer to unity, the value for a completely

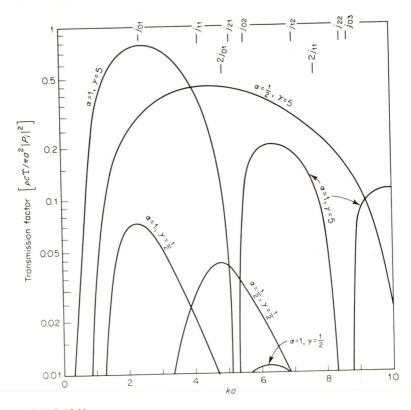

FIGURE 10.11
Power Υ transmitted through a circular membrane of radius a in a rigid wall, from a normally incident plane wave of frequency $kc/2\pi$, for different membrane-medium coupling constants $\gamma = \rho a/\sigma$ and wave-velocity ratios $(c/c_p) = \alpha$.

"transparent" membrane. It also broadens the membrane-resonance peaks. Increase of $1/\alpha$ proportionally increases all the membrane resonances in terms of $x = ka$, since a membrane resonance occurs when $w = \alpha ka = j_{0n}$.

Thus, taking the $\alpha = 1$, $\gamma = 5$ curve, T increases from zero at $ka = 0$ to reach a maximum at $ka \simeq j_{01}$, where $X_0(0w \mid x) = 0$. This is the lowest symmetric resonance of the membrane, at which nearly 80 percent of the incident power is transmitted. As ka is further increased, T decreases, until when $ka = j_{21}$, T is zero again. At this frequency, called the *antiresonance* frequency, $H_0(00w)$ is zero; the membrane shape is such that no energy i radiated to infinity, and the membrane behaves as though it were rigid. To the degree of approximation of these formulas, the effective impedance $Z_0(0w \mid x)$ becomes infinite. A better trial function, chosen so that not all parts of the membrane would vibrate in phase [as it must with the form $BJ_0(ur/a)$ for Φ_0], would remove the zero, but there still would be a pronounced minimum at $w = j_{21}$. When $\theta > 0$, the other components, which do not have minima at the same place, would cover the minimum; so it is not usually worthwhile to go to the trouble to find an improved value at the minimum for $m = 0$.

Just above this antiresonance frequency the membrane has its next $m = 0$ resonance. Because the antiresonance is close below, the resonance peak is distorted, the maximum coming somewhat above $w = j_{02}$. As ka is still further increased, there comes another antiresonance at $w = j_{22}$; just beyond it X_0 goes to zero at the next $m = 0$ resonance; and so on. At higher frequencies the peaks and valleys come closer together on a logarithmic scale (and with a better trial function, the valleys diminish in depth), until at high frequencies the curve is an irregularly fluctuating one, with a general trend of decrease with increasing frequency, typical of mass-controlled response. At other angles of incidence, the $m > 0$ terms add their differently placed peaks and valleys; so the curve for T is even less irregular, only the first few peaks and valleys being distinguishable.

Thus the membrane resonance maxima and antiresonance minima, both typical of the finite size of the membrane boundaries, serve to differentiate these phenomena from the reflection and transmission of sound through an unbounded membrane. Appropriate modification of the discussion of Eq. (10.1.32) for the case of a membrane, rather than a plate (taken up in Prob. 7), provides the transmission formulas for an infinite membrane. If the membrane has areal density σ and tension $T = \sigma c_p{}^2$, an incident plane wave of amplitude P_i and angle of incidence θ, coming from $-\infty$, would produce a plane transmitted wave in region $z > 0$, traveling in the same direction as the incident wave, with an amplitude

$$P_t = P_i\left[1 + \tfrac{1}{2}i\,\frac{\sigma}{\rho a}\,ka\cos\theta\left(\frac{1}{\alpha^2}\sin^2\theta - 1\right)\right]^{-1}$$

Thus the infinite membrane sends all its transmitted energy in one direction ($\phi = 0$, $\vartheta = \theta$), instead of in all directions, as Eqs. (10.2.26) indicate for the finite membrane. The equivalent, for the infinite membrane, of the transmission factor T of Eqs. (10.2.29) and Fig. 10.11 would be the power transmitted through an area πa^2 of the membrane, divided by πa^2 times the incident intensity,

$$T = \frac{\cos\theta}{1 + (ka/2\gamma)^2\cos^2\theta[(1/\alpha^2)\sin^2\theta - 1]^2} \qquad \text{(unbounded membrane)}$$

$$(10.2.30)$$

the cosine in the numerator coming from the inclination of the area πa^2 of the membrane to the direction of the incident-transmitted wave. This formula has a strong maximum at $\sin \theta = \alpha$, when the velocity of the pressure wavefronts at the membrane equals the velocity of transverse waves on the membrane.

To compare this with the transmission factor of Eq. (10.2.28), for the bounded membrane, we have plotted Υ as a function of the angle of incidence θ, for several different frequencies. We choose $\alpha = \frac{1}{2}$ so that the peak of transmission for the infinite membrane will come at $\theta = 30°$; we choose the coupling constant $\gamma = 1$ large enough to expect an appreciable fraction of the incident energy to be transmitted at the angle of transparency. The choice of frequency is not easy if we are trying to exhibit a simple relationship between the transmissions for infinite and finite membranes; the finite membrane has too many resonances and antiresonances to be able to avoid them all (except at very low frequencies, where very little energy gets through anyway). We have chosen $ka = 2$, 4, and 8 to illustrate some of the complexities; other frequencies would have exhibited other peculiarities.

Thus Fig. 10.12 has three plots of T, as computed from Eq. (10.2.28), as function of angle of incidence θ, for $ka = 2$, 4, and 8. The dashed lines are T from Eq. (10.2.30) for the unbounded membrane for the same values of α, γ, and ka, for comparison. At $ka = 2$ the driving frequency is well below any resonance or antiresonance, but the incident wavelength $2\pi/k$ is longer than the diameter of the membrane, and one would not expect much directionality, and thus much similarity, with the infinite-membrane behavior. The transmitted wave is more or less independent of ϕ, and less than one percent of the incident power is transmitted—somewhat less at $\theta = 90°$ than at $\theta = 0°$.

The frequency corresponding to $ka = 4$ is near the first resonance for $m = 0$ $(j_{01}/\alpha = 4.8)$. Thus the curve of T for $ka = 4$ reflects the fact that the $m = 0$ component of Υ is exaggerated; the transmitted power is large for $\theta = 0$ and falls off as θ increases. The components for $m > 0$, which would contribute to a "transparency" peak at $\theta = 30°$, are almost drowned out by the strong $m = 0$ contribution; one can see a small "bump" near $\theta = 30°$ where the $m = 1$ and $m = 2$ components have their maxima.

On the other hand, the frequency corresponding to $ka = 8$ is close to a membrane resonance for $m = 1$ $(j_{11}/\alpha = 7.7)$, so that the $m = 1$ component is exaggerated. Also, both $H_0(uuw)$ and $H_1(uuw)$ have zeros, corresponding to antiresonances, the one for H_0 coming at θ about $30°$ and the one for H_1 at θ about $50°$. Thus the $m = 0$ contribution is nearly nonexistent at $\theta = 30°$, the angle at which the infinite membrane is transparent. The peak of Υ which is exhibited, corresponding to an increase in transmission by a factor of nearly 10 from $\theta = 0$ to $\theta = 20°$, comes predominantly from the $m = 1$ term. This term goes to an antiresonance zero at $50°$, leaving the small contributions from the $m = 0$, 2, and 3 components.

Beyond 50° the transmission has another small peak; the resulting curve has little resemblance to the dashed curve.

The angle distribution of the transmitted wave also exhibits some of the complexities arising from the wave reflections from the finite-membrane

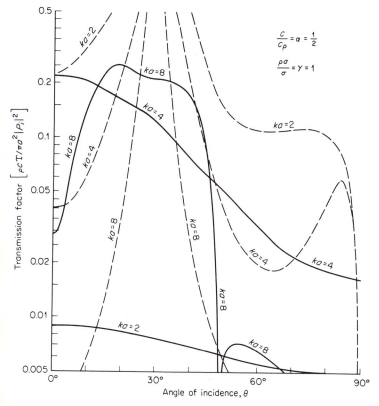

FIGURE 10.12

Power Υ transmitted through a circular membrane of radius a in a rigid wall as a function of angle of incidence θ of a plane wave of frequency $kc/2\pi$. Dashed lines would be the transmission factor if the circular membrane were to behave as though it were of infinite extent, with no wave reflection at its edges.

boundaries, which produce the large differences from the simple behavior of the infinite membrane. In Fig. 10.13 are plotted contours of intensity of the transmitted wave, a large number of wavelengths from the origin on the $z > 0$ side, for a plane incident wave in the direction $\phi = 0$, $\theta = 30°$ (the direction of maximum transmission for an infinite membrane) for the same values of α, γ, and ka as for Fig. 10.12. The transmitted wave for an unbounded membrane would be a plane wave in the same direction as the incident wave, which would be represented by a delta-function peak at $\phi = 0$, $\vartheta = 30°$, shown by the cross on each contour plot.

In the left-hand plot, for $ka = 4$, we see again the effect of the $m = 0$ resonance. The transmitted energy radiates away in a pattern nearly symmetric with respect to the normal ($\vartheta = 0$); the intensity peak is deflected a

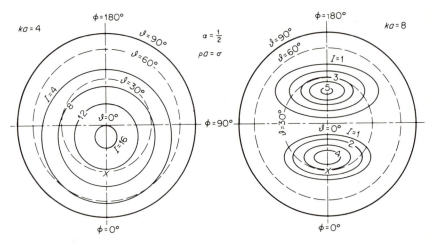

FIGURE 10.13
Angle distribution of transmitted sound for conditions of Fig. 10.12, for angle of incidence of 30°. Contours of I, proportional to $|p_+|^2$, at large distance R from origin, plotted against spherical angles ϑ and ϕ.

small amount away from the zenith, but is certainly not centered on the cross. The peak is quite broad, typical of long wavelengths. In contrast, the radiation pattern for $ka = 8$ is typical of the $m = 1$ resonance. Since the driving frequency is higher than the $m = 1$ resonance, the $m = 1$ component is out of phase with the others, which results in more being radiated in the $\phi = 180°$ direction than in the $\phi = 0$ direction of the incident wave. The $m = 0$ component is very small, as mentioned earlier; the interaction between the $m = 1$ and $m = 2$ components broadens the peaks parallel to the $\phi = 90$–$270°$ axis (makes them "ridges," rather than peaks); it, plus the $m = 0$, $m = 1$ interaction, accounts for the asymmetry with respect to this axis, and the fact that B_1 is negative accounts for the fact that the $\phi = 180°$ ridge is higher than the $\phi = 0$ one. The peak maxima are not at $\vartheta = 30°$, though they are closer than the peak for $ka = 4$.

Other calculations, for other values of ka, α, γ, and θ, show a wide variety of forms, corresponding to the manifold combination of resonances, antiresonances, and transparencies, which can result in all possible combinations of amplitudes and phases for each component.   In such an exploration of a truly complex system, the results of a variational calculation are of great value, for they are usually accurate enough to indicate the correct order of magnitude of the various effects, and though the answers may differ by 10 or 20 percent from the correct values, they do not go to infinity or to zero, and thus fail completely, at crucial points.

Somewhat better results can be obtained by using the two-term trial function $AJ_m(ur/a) + BJ_m(qr/a)$, with q adjusted so that $J_m(w)H_m(qqw) = 2\gamma w^2 I_m(qw \mid x)$.   In this case the "best" values of A and B turn out to be

$$A_m = \frac{2P_i\epsilon_m i^m J_m(w)H_m(uqw)}{J_m(w)H_m(uqw) - 2\gamma w^2 S_m(uqw \mid x)}$$

$$B_m = 2P_i\epsilon_m i^m J_m(w)H_m(uqw)\frac{-2\gamma w^2 S_m(uw \mid x)}{[J_m(w)H_m(uqw) - 2\gamma w^2 S_m(uqw \mid x)]^2}$$

where

$$S_m(uqw \mid x) = \frac{J_m^2(w)}{a^7}\int J_m\left(\frac{ur}{a}\right)\int G_m\int g_m\int G_m J_m\left(\frac{qr}{a}\right)$$

is an obvious generalization of the function of Eq. (10.2.23).   When the coupling constant γ is small, B_m is small compared with A_m; so this improvement is slight, except near an antiresonance.   Because $S_m(uw \mid x)$ is complex, B_m is out of phase with A_m, so that the resulting shape of η, for each m, has some traveling waves as well as standing waves, as would be required for a damped membrane.   As mentioned before, the values of the component terms at their antiresonances are not usually of much interest, so that it is not usually worthwhile to carry out this more arduous calculation. Thus, in practice, as long as γ is less than about 10, the one-term trial function $BJ_m(ur/a)$, with the resulting formulas (10.2.24) to (10.2.27), is sufficient.

A similar calculation could be carried out for a plate set in an otherwise rigid wall.   The computations would be somewhat more laborious than the ones sketched here for the membrane, but the general characteristics would turn out to be much the same.   The functions F, H, θ, χ, and S would involve combinations of the Bessel functions of real and imaginary arguments required for a disk clamped at its edge, but their general shapes would be similar to those plotted for the membrane, with a different frequency scale to correspond to the different resonance and antiresonance frequencies.   A similar series of peaks and valleys would result for the curves of the transmission factor T.   Some of this will be demonstrated in the problems.

10.3 PERIODICALLY SUPPORTED SYSTEMS

A system of considerable practical interest consists of a plane partition, strengthened by linear studs or beams, periodically placed. The complete details of finite structures of this type cannot be represented by any soluble model, but many of their properties can be understood by studying a model of infinite extent, with the struts replaced by linear impedance loads, attached at equal spacing to the wall, each load having stiffness, mass, and resistance per unit length. A completely satisfactory model would have a thin, uniform plate for the partition. Such a model can be worked out (some aspects are given as problems), but the algebraic complexities hinder an understanding of the results.

Membrane with regularly spaced supports

We shall, instead, replace the partition by a flexible membrane (see Sec. 5.2) of mass σ per unit area and with transverse wave velocity $c_p = c/\alpha$, c being the wave velocity in the medium (air or water) on both sides of the partition. This is not a bad approximation for simple-harmonic motions, as long as the wavelength of the transverse waves in the wall is smaller than the spacing between struts. In this case the results computed for the membrane model, particularly in regard to the coupled waves in the medium, are much the same as for a model using a plate with equal σ and c_p. Transient effects will differ, of course, because the velocity of transverse waves in a plate is proportional to $\sqrt{\omega}$, instead of being independent of ω, as with a membrane. With this limitation in mind, we proceed with the simpler model, leaving the modifications produced by the stiffness of the plate to be taken up in the problems.

Thus our model is an infinite membrane, the equilibrium plane being the xy plane, of mass σ per unit area, under tension $T = \sigma c_p{}^2$, with transverse wave velocity $c_p = c/\alpha$ (thus defining the dimensionless constant α). Attached to this membrane, along lines parallel to the y axis, are line impedances, of magnitude Z per unit length (dimensions of force per unit length per unit velocity) along the lines $x = nl$ $(n = \ldots, -1, 0, 1, 2, \ldots)$.

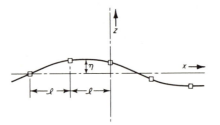

FIGURE 10.14
Transverse displacement of membrane, periodically loaded with linear supports, each having impedance Z per unit length in the y direction.

These are roughly equal to struts or beams, spaced a distance apart. In many cases Z is larger than σc_p, the characteristic impedance of the membrane. Thus our problem, in a sense, involves coupling three systems, the membrane, the set of parallel struts, and the medium on both sides of the wall. Figure 10.14 shows a section of the membrane, with its supporting struts.

Motion in a vacuum

First we deal with the coupling of a membrane-and-strut system, without the added complications of the radiation reaction of the surrounding fluid. We omit also the complication of wave motion in the y direction, since this depends on the longitudinal stiffness of the struts; we consider only wave motion in the x direction. In this case the struts move as a set of coupled oscillators, and the strips of membrane between each pair move in combinations of moving and standing waves coupling the struts. The motion is related to, but more complex than, the motion of coupled oscillators, treated in Sec. 3.3.

As an easy introduction, we can study the behavior of the system under the influence of a transverse driving force $Fe^{i\varkappa x - i\omega t}$ per unit area. The steady-state membrane displacement η must also have the factor $e^{-i\omega t}$ and must repeat itself in each successive strip, with the exception of an additional phase factor $e^{i\varkappa l}$ per strip. In other words, $\eta(x + l)$ should equal $e^{i\varkappa l}\eta(x)$.

In addition to the driving force, there will be, at $x = nl$, the linear reaction of the strut impedance, $-Z(\partial\eta/\partial t) = ikc\eta Z$ times the delta function $\delta(x - nl)$. Thus the equation of motion for the membrane will be

$$\frac{\sigma c^2}{\alpha^2}\frac{\partial^2\eta}{\partial x^2} + \sigma k^2 c^2 \eta = -F_x e^{i\varkappa x} - ikcZ\eta\sum_{n=-\infty}^{\infty}\delta(x - nl) \qquad (10.3.1)$$

Except at the points $x = nl$, where there is a discontinuity in slope, η must be a combination of a solution of the homogeneous equation plus a solution of the inhomogeneous equation, $-F_x e^{i\varkappa x}$ being the inhomogeneous term. Thus the solution, except at $x = nl$, will be the combination

$$\frac{\alpha^2 F_x/\sigma c^2}{\varkappa^2 - k^2\alpha^2}(e^{i\varkappa x} + Be^{ik\alpha x} + Ce^{-ik\alpha x}) \qquad 0 < x < l$$

with constants C and B so adjusted that $Be^{ik\alpha x} + Ce^{-ik\alpha x}$ at $x = l$ is $e^{i\varkappa l}$ times its value $B + C$ at $x = 0$. To achieve this, B must equal $A(e^{i\varkappa l} - e^{-ik\alpha l})$, and C must equal $A(e^{ik\alpha l} - e^{i\varkappa l})$, where A must then be adjusted to satisfy the strut-reaction requirements at $x = nl$.

Thus the solution is

$$\eta = \frac{\alpha^2 F_x/\sigma c^2}{\varkappa^2 - k^2\alpha^2}(e^{i\varkappa x} + \Omega)$$

where

$$\Omega = A \begin{cases} [\sin \alpha k(x + l) - e^{-i\varkappa l} \sin \alpha kx] & -l < x < 0 \\ [e^{i\varkappa l} \sin \alpha kx + \sin \alpha k(l - x)] & 0 < x < l \\ [e^{2i\varkappa l} \sin \alpha k(x - l) + e^{i\varkappa l} \sin \alpha k(2l - x)] & l < x < 2l \end{cases}$$

with A adjusted so that the discontinuity in slope at $x = 0$ is $-ik\alpha\zeta\eta(0)$, where $\zeta = Z/\sigma c_p = \alpha Z/\sigma c$ is the dimensionless strut-impedance parameter. In other words,

$$A[\alpha k e^{i\varkappa l} - \alpha k \cos (\alpha kl) - \alpha k \cos (\alpha kl) + \alpha k e^{-i\varkappa l}]$$
$$= -i\alpha k \zeta[1 + A \sin (\alpha kl)]$$

The solution of this is $A = -i\zeta/Q$, where

$$Q = 2 \cos (\varkappa l) - 2 \cos (\alpha kl) + i\zeta \sin (\alpha kl)$$

The steady-state solution may be written in the form just shown, or as $e^{i\varkappa x}$ times a Fourier series, the coefficients of which may be obtained by substitution in Eq. (10.3.1). Contrary to appearances, η does not become infinite when $\varkappa = k\alpha$. Various forms for η are given here:

$$\eta = \frac{\alpha^2 F_\varkappa/\sigma c^2}{\varkappa^2 - k^2\alpha^2} \begin{cases} \left\{ e^{i\varkappa x} - i\dfrac{\zeta}{Q} [\sin \alpha k(x + l) + e^{-i\varkappa l} \sin \alpha kx] \right\} & -l < x < 0 \\ \\ \left\{ e^{i\varkappa x} - i\dfrac{\zeta}{Q} [e^{i\varkappa l} \sin \alpha kx + \sin \alpha k(l - x)] \right\} & 0 < x < l \end{cases}$$

$$\xrightarrow[\varkappa \to \alpha k]{} \frac{i\alpha F_\varkappa/2k\sigma c^2}{\sin (\alpha k)} \left\{ \left[\left(x + \frac{2l}{\zeta}\right) e^{i\alpha k x} \sin \alpha kl - l e^{i\alpha kl} \sin \alpha kx \right] \right\} \quad 0 < x < l$$

$$= \frac{\alpha^2 F_\varkappa/\sigma c^2}{\varkappa^2 - k^2\alpha^2} e^{i\varkappa x} \left\{ 1 + \frac{2i\zeta\alpha kl}{Q} [\cos (\varkappa l) - \cos (\alpha kl)] \right.$$
$$\left. \times \sum_{m=-\infty}^{\infty} \frac{e^{(2\pi im/l)x}}{(\varkappa l + 2\pi m)^2 - k^2\alpha^2 l^2} \right\}$$

$$\xrightarrow[\varkappa \to \alpha k]{} \frac{\alpha^2 F_\varkappa}{\sigma c^2} e^{i\alpha k x} \left[\frac{1 - \alpha kl \cot (\alpha kl) + 4i(\alpha kl/\zeta)}{4k^2\alpha^2} \right.$$
$$\left. - \sum_{m=-\infty}^{\infty} (1 - \delta_{0m}) \frac{l^2}{2\pi m} \frac{e^{(2\pi im/l)x}}{2\alpha kl + 2\pi m} \right] \quad (10.3.2)$$

the factor $1 - \delta_{0m}$ in the last form ensuring that the term $m = 0$ is omitted from the sum.

The Fourier series expansions are obtained by substituting $\eta = \sum B_n \exp(i\varkappa x + 2\pi i n x/l)$ in Eq. (10.3.1). To find the coefficients B_m, we multiply by $\exp(-i\varkappa x - 2\pi i m x/l)$ and integrate over x from $-\epsilon$ to $l - \epsilon$ ($\epsilon \to 0$). The integral over the last term on the right is $-ikcZ \sum B_n$, for any value of m; thus we obtain an equation for B_m in terms of $F_\varkappa$ and of $\sum B_n$. Summation over m results in an equation for $\sum B_n$, which involves a sum which can be expressed in closed form.

$$\sum_n \frac{ik\alpha l\zeta}{(\varkappa l + 2\pi n)^2 - (k\alpha l)^2} = -\tfrac{1}{2}i\zeta \frac{\sin(\alpha k l)}{\cos(\varkappa l) - \cos(\alpha k l)}$$

Further algebra results in the series form given in Eq. (10.3.2). Alternatively, it may be obtained by expanding the first form of Eq. (10.3.2) in a Fourier series in the usual manner.

The first form of Eq. (10.3.2) indicates the division between the traveling wave $e^{i\varkappa x}$ and the membrane wave motion, at wavenumber $k\alpha$, induced in each strip by the motion of the struts, each successive strip differing in phase by the factor $e^{i\varkappa l}$. The limiting form shows that η does not become infinite when $\varkappa \to k\alpha$; it becomes infinite only if *both* $\varkappa$ and $k\alpha$ equal $\pi n/l$ or when the quantity Q is zero. The Fourier-series form, valid for all values of x, shows that η is equal to $e^{i\varkappa x}$ times a function which is periodic in x with period l, as required.

Free-wave motion

The free-wave motion of the strut-membrane system is obtained by finding the values of $\varkappa$ for which $\eta/F_\varkappa$ becomes infinite, for then η can be finite even though $F_\varkappa \to 0$. This occurs when Q, defined earlier, becomes zero. Suppose the root of $Q = 0$ with the smallest, positive, real part is $\varkappa_q + i\xi$. Then, because of the periodicity of Q, all the possible roots of $Q = 0$ are

$$\varkappa_n = \varkappa_q + \frac{2\pi n}{l} + i\xi \quad \text{and} \quad -\varkappa_q + \frac{2\pi n}{l} - i\xi \qquad (10.3.3)$$

where n is any integer, positive or negative. Since the addition or subtraction of $2\pi n/l$ to $\varkappa$ makes no difference to the value of either $e^{i\varkappa l}$ or $\cos \varkappa l$, there are really only two different solutions for the free-wave motion of the system.

$$\eta = \begin{cases} A\{\exp[\pm(i\varkappa_q - \xi)l]\sin(\alpha k x) + \sin[\alpha k(l - x)]\} & 0 < x < l \\ A\{\exp[\pm 2(i\varkappa_q - \xi)l]\sin[\alpha k(x - l)] \\ \qquad + \exp[\pm(i\varkappa_q - \xi)l]\sin[\alpha k(2l - x)]\} & l < x < 2l \end{cases}$$

$$= -iA\sin(\alpha k l)\sum_{n=-\infty}^{\infty} \frac{\exp\{\pm i[\varkappa_0 + (2\pi n/l)]x\}}{(\varkappa_0 l + 2\pi n)^2 - (\alpha k l)^2} \qquad \text{(all values of } x\text{)} \quad (10.3.4)$$

where $\varkappa_0 = \varkappa_q + i\xi$. The solution with the plus signs represents a wave moving to the right; the one with minus signs, a wave going to the left; each with speed $c_q = \omega/\varkappa_q$ and with attenuation constant ξ. The Fourier-series form shows the nature of the periodicity.

A plot of the root $\varkappa_q$ of the equation $Q = 0$ is given in Fig. 10.15 for purely reactive strut impedances, $Z = -iX$. Contours for the wavenumber parameter $\varkappa_q l/\pi = u_q$ are plotted against different values of $v = \alpha k l/\pi$ and

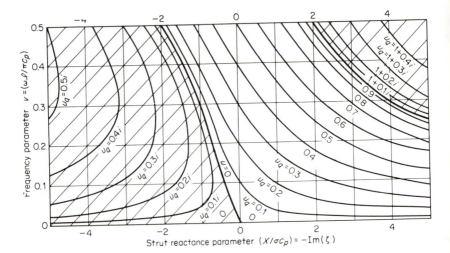

FIGURE 10.15
Relation between system wavenumber parameter $u_q = \varkappa_q l/\pi$, frequency parameter $v = (\omega l/\pi c_p)$, and strut reactance parameter $-\mathrm{Im}\,(\zeta)$ for traveling waves in a periodically supported membrane with purely reactive support struts ($Z = R - iX$, $R = 0$).

of $-\mathrm{Im}\,\zeta$. In this case, when $\mathrm{Re}\,\zeta = 0$, there is a band of values of v and of $-\mathrm{Im}\,\zeta$ for which the parameter u_q is real ($\xi = 0$), and the wave does not attenuate, but travels at a speed $c_q = \omega l/\pi u_q$. Outside this band, in the shaded regions of the plot, $\varkappa$ has an imaginary part ($\xi \neq 0$) and the wave is strongly damped. This is a general property of wave propagation in periodic structures [see, for example, the discussion of Eq. (9.1.35)]; the systems behave like bandpass filters. Because of the periodicity of the equation $Q = 0$, the plot can be extended upward and downward. The range $0 > v > -\frac{1}{2}$ can be covered by changing the signs of u, v, and X, in effect rotating the chart 180° about the origin and changing the sign of u. The range $\frac{1}{2} < v < 1$ is obtained by changing u and v to $1 - u^*$ and $1 - v$ and reversing the sign of X, in effect rotating the chart 180° about the point $X = 0$, $v = \frac{1}{2}$. We note that, for ω small, massy struts ($X > 0$) allow wave propagation, stiff struts ($X < 0$) attenuate the waves.

If Z is not purely reactive, the wavenumber parameter u will be complex over the whole area of Fig. 10.15, but if Re ζ is small, there will be a band, roughly the same as the unshaded area of Fig. 10.15, in which ξ is small and outside of which ξ increases rapidly. When the strut reactance is large compared with the characteristic impedance of the membrane (i.e., when $|\zeta| \gg 1$), the roots of Eq. (10.3.3) are, approximately,

$$\varkappa_q l \simeq \cos^{-1} [1 - \tfrac{1}{2}g \sin (\alpha k l)] \qquad \xi l \simeq \frac{f \sin (k \alpha l)}{2 \sin (\varkappa_q l)}$$

where $\zeta = f - ig$. These formulas are for the small range of values of $k\alpha l$ for which $\sin (k\alpha l)$ lies between zero and $2/f$ (f is large); for the rest of the range,

$$\varkappa_q l \simeq \frac{g}{f} |\cos (k \alpha l)|^{\frac{1}{2}} \qquad \xi l \simeq \ln [g \sin (k \alpha l)]$$

Forced motion *in vacuo*

If, for example, a single transverse line force $F_0 e^{-i\omega t} \delta(x)$ acts on the membrane and strut at $x = 0$, the steady-state forced motion is

$$\eta(x) = \begin{cases} \cdots\cdots\cdots\cdots\cdots\cdots\cdots\cdots\cdots\cdots\cdots\cdots\cdots \\ A\{-e^{(i\varkappa_q - \xi)l} \sin (\alpha k x) + \sin [\alpha k(l + x)]\}e^{-i\omega t} & -l < x < 0 \\ A\{e^{(i\varkappa_q - \xi)l} \sin (\alpha k x) + \sin [\alpha k(l - x)]\}e^{-i\omega t} & 0 < x < l \\ \cdots\cdots\cdots\cdots\cdots\cdots\cdots\cdots\cdots\cdots\cdots\cdots\cdots \end{cases}$$

(10.3.5)

where

$$A = \frac{\alpha^2 F_0/\sigma c^2}{2e^{(i\varkappa_q - \xi)l} - 2\cos (\alpha k l) + i\zeta \sin (\alpha k l)}$$

the free wave for positive x going to the right, that for negative x going to the left. If the strut impedance is that of a simple oscillator, $Z = i[(K/\omega) - \omega M] = -iX$, the strut resonance will be at $\omega_0 = \sqrt{K/M}$. Since Z is purely reactive, we can use the plot of Fig. 10.15 to work out the behavior of the system. At very low driving frequencies, X is large and negative and the wave is strongly damped; no power is transmitted to $x = \infty$ by the wall. As ω is increased, the point on the $(X/\sigma c_p, v)$ plane, corresponding to the system, moves along the line $X/\sigma c_p = (M\pi/\sigma l)[v - (v_0^2/v)]$ (the dashed line of Fig. 10.16), where $v_0 = \omega_0 l/c_p \pi$. If $v_0 < \tfrac{1}{2}$, this line will cross the heavy line marked $u = 0$ at a value of v less than $\tfrac{1}{2}$, at point a in Fig. 10.16, for example. At this point (for this driving frequency), the wavenumber $\varkappa_0$ is zero, the attenuation constant ξ is zero, and the amplitude A is equal to infinity. For a somewhat higher frequency, $\omega_0/2\pi$, the strut impedance is zero, but $\varkappa_0$ by this time is positive and real (it equals $k\alpha$ at this point), indicating that true wave motion is carrying power away from the driving force to $\pm\infty$. There is no resonance here because the membrane is resisting the force, even if the struts are not.

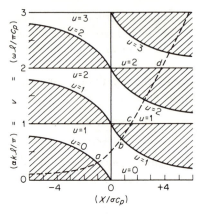

FIGURE 10.16
Illustration of computation of behavior of membrane periodically loaded with simple-oscillator struts. Dashed line is strut reactance curve; when it is in shaded regions the wave motion is attenuated.

As the frequency is further increased, $\varkappa_0$ increases further, until at point b, where the dashed line crosses the curve marked $u = 1$, resonance occurs again, with alternate struts moving in opposite directions, instead of being all in phase, as they were when $\varkappa_0 = 0$. Further increase of frequency brings the curve into a shaded region, when the wave motion is attenuated. At still higher frequency, at point c, ξ again becomes zero, and the system begins again to carry unattenuated waves. For this case, and for the similar points such as d, when *both* u and v equal n, the amplitude of motion does not go to infinity, as it does at points like a and b, where $\varkappa_0$ equals $n\pi$ but does not equal $k\alpha l$. Of course, the amplitude would never be infinite if we were to take into account the reaction of the air or other medium surrounding the membrane-strut system.

Thus, for some ranges of driving frequency, wave motion can carry power along the membrane; for other ranges, where the curve for strut reactance passes through a shaded region of Fig. 10.16, wave motion is severely attenuated.

Green's functions

To calculate the effect of the presence of an acoustic medium on the system we have just been describing, it will be useful to set up appropriate Green's functions, for both the strut-membrane system and the acoustic media on either side of the membrane. If we are interested in a traveling wave of frequency $\omega/2\pi$, both Green's functions should have a phase factor $e^{i\varkappa x - i\omega t}$. Because of the periodicity of the system in x, we should expect an additional factor, periodic in x with period l, just as the solution of Eq. (10.3.2) has. We might as well make our Green's functions periodic also, by having them correspond to a regularly spaced sequence of line sources, rather than just one.

Thus the equation for the Green's function for the medium, for two-dimensional motion (still assuming no motion in the y direction), would be for a line source, not only at x_0, z_0, but also ones at $x_0 \pm nl$, z_0, each source being out of phase by an amount $\varkappa l$ radians from the source next to the left. The equation for such a periodic Green's function is

$$\left(\frac{\partial^2}{\partial x^2} + \frac{\partial^2}{\partial z^2}\right)g + k^2 g = -\delta(z - z_0) \sum_{n=-\infty}^{\infty} e^{i n \varkappa l}\, \delta(x - x_0 - nl)$$

so that $g(x + l, z \mid x_0, z_0) = e^{i \varkappa l} g(x, z \mid x_0, z_0)$ and also $g(x, z \mid x_0 + l, z_0) = e^{-i \varkappa l} g(x, z \mid x_0, z_0)$. Then, instead of integrating $g(\partial p/\partial n_0)$, etc., over x_0 from $-\infty$ to $+\infty$, we need only integrate from 0 to l, since this moves each one of the series of sources from one end of its period to the other end, thus covering the whole range of x_0.

Put another way, if we multiply the equation for g by p, and the equation for p, $(\partial^2 p/\partial x_0^2) + (\partial^2 p/\partial z_0^2) + k^2 p = 0$, by g, subtract, and integrate over x_0 from 0 to l, over z_0 from 0 to ∞, we integrate over but one of the delta functions, the n for which $nl < x < (n + 1)l$. Thus the integral reduces to an integral over the delta function and a surface integral, so that

$$e^{i n \varkappa l} p(x - nl, z) = \iint \left[g \frac{\partial p}{\partial n_0} - p \frac{\partial g}{\partial n_0} \right] dS_0$$

with the surface integral over the x_0 axis from $x_0 = 0$ to l, over the line $x_0 = l$ from $z_0 = 0$ to $+\infty$, over the line at $z_0 = \infty$ from $x_0 = l$ to 0, and back to the origin from $z_0 = \infty$ to 0 along the z_0 axis. Since p also satisfies the condition $p(x_0 + l, z_0) = e^{i \varkappa l} p(x_0, z_0)$, the $e^{i \varkappa l}$ for the p cancels the $e^{-i \varkappa l}$ for the x_0 dependence of g, and the quantity in brackets is periodic in x_0, with period l. Therefore the integral up the line $x_0 = l$ is exactly canceled by the integral down the line $x_0 = 0$, and if the wave is entirely outgoing the integral over the segment at infinity is zero. What is left is the integral along the x_0 axis from 0 to l.

If now we arrange that $\partial g/\partial z_0 = 0$ for $z_0 = 0$, and if we remember that $e^{i n \varkappa l} p(x - nl, z) = p(x, z)$, we obtain

$$p_+(x, z) = -\int_0^l g_+(x, z \mid x_0, 0) \left(\frac{\partial p}{\partial z_0}\right)_{z_0 = 0} dx_0$$

Since $(\partial p/\partial z)_{z=0} = i\rho\omega u = \rho\omega^2 \eta(x)$, where u is the transverse velocity of the membrane and η is its displacement, we finally arrive at the equation relating the pressure wave in the region $z > 0$ to the membrane displacement,

$$p_+(x, z) = -\rho\omega^2 \int_0^l g_+(x, z \mid x_0, 0) \eta(x_0)\, dx_0 \qquad (10.3.6)$$

if η also satisfies the requirement $\eta(x + l) = e^{i \varkappa l} \eta(x)$. This is, of course, quite similar in form to Eq. (10.2.10); the difference lies in the differences between the Green's functions and the limits of integration.

To find the expression for the function g_+, we note that the series

$$g_+ = \sum_{m=-\infty}^{\infty} f_m(z, z_0, x_0) e^{i k_m x} \qquad k_m = \varkappa + \frac{2\pi m}{l}$$

is as easy to handle as a simple Fourier series, for

$$\int_0^l e^{ik_m x} e^{-ik_n x}\, dx = l\delta_{nm}$$

Therefore we can manipulate the equation for g as easily as was done to obtain Eqs. (4.4.14) and (7.4.8), eventually obtaining

$$g_+ = \sum_{m=-\infty}^{\infty} \frac{i \exp\left[ik_m(x - x_0)\right]}{lq_m} \begin{cases} \cos\left(q_m z\right)e^{iq_m z_0} & 0 \leqslant z \leqslant z_0 \\ \cos\left(q_m z_0\right)e^{iq_m z} & 0 \leqslant z_0 \leqslant z \end{cases} \quad (10.3.7)$$

where $q_m = (k^2 - k_m{}^2)^{\frac{1}{2}} = i(k_m{}^2 - k^2)^{\frac{1}{2}}$. The corresponding q_- for the region $z < 0$ is obtained from g_+ by reversing the signs of z and z_0. The integral relation between p_- and η is similar to Eq. (10.3.6), with $-g_-$ substituted for $+g_+$. Therefore the pressure difference at $z = 0$, which produces transverse acceleration of the membrane, is

$$P(x) \equiv p_-(x,0) - p_+(x,0) = 2\rho\omega^2 \int_0^l g(x \mid x_0)\eta(x_0)\, dx_0 \quad (10.3.8)$$

where

$$g(x \mid x_0) = \sum_{m=-\infty}^{\infty} \frac{i}{lq_m} \exp\left[ik_m(x - x_0)\right]$$

This also satisfies the periodicity condition $P(x + l) = P(x)e^{ixk}$ if η does.

The Green's function for the membrane-strut system is more complicated. Not only must it satisfy the periodicity condition; it must account for the strut impedances as well as a periodic array of unit sources. The equation is

$$\left(\frac{d^2}{dx^2} + k^2\alpha^2\right)G(x \mid x_0) = -\sum e^{inxl}\, \delta(x - x_0 - nl)$$
$$- \frac{ik\alpha^2 Z}{c} \sum \delta(x - nl)G(x \mid x_0)$$

where $\alpha = c/c_p$. After a great deal of algebra one can write out the appropriate Green's function in any of the following forms:

$$G(x \mid x_0) = \begin{cases} \cdots\cdots\cdots\cdots\cdots\cdots\cdots\cdots\cdots\cdots\cdots\cdots \\[4pt] \dfrac{1}{2k\alpha Q}\{2e^{-ixl}\sin \alpha k(x - x_0) - 2\sin \alpha k(x + l - x_0) \\ \qquad + i\zeta[\cos \alpha k(x + x_0 - l) - \cos \alpha k(x + l - x_0)]\} \\ \hfill 0 < x < x_0 < l \\[8pt] \dfrac{1}{2k\alpha Q}\{2\sin \alpha k(x - l - x_0) - 2e^{ixl}\sin \alpha k(x - x_0) \\ \qquad + i\zeta[\cos \alpha k(x + x_0 - l) - \cos \alpha k(x_0 + l - x)]\} \\ \hfill 0 < x_0 < x < l \\[4pt] \cdots\cdots\cdots\cdots\cdots\cdots\cdots\cdots\cdots\cdots\cdots\cdots \end{cases}$$

$$= \frac{l}{\pi^2} \sum_m \frac{e^{ik_m x}}{\Delta_m} \Psi_m(-x_0) = \frac{l}{\pi^2} \sum_m \frac{e^{-ik_m x_0}}{\Delta_m} \Psi_m(x) \quad (10.3.9)$$

where

$$\alpha = \frac{c}{c_p} \qquad k = \frac{\omega}{c} = \frac{2\pi}{\lambda} \qquad \zeta = \frac{\alpha Z}{\sigma c} = \frac{Z}{\sigma c_p}$$

$$u = \frac{\varkappa l}{\pi} \qquad v = \frac{k\alpha l}{\pi} = \frac{2\alpha l}{\lambda} \qquad k_m = \varkappa + \frac{2\pi m}{l}$$

$$u_m = \frac{k_m l}{\pi} = u + 2m \qquad \Delta_m = u_m{}^2 - v^2 \qquad \mu = \frac{2\rho}{\pi\sigma} v^2$$

$$q_m = (k^2 - k_m{}^2)^{\frac{1}{2}} \qquad w_m = \frac{lq_m}{\pi} = \left[\left(\frac{v}{\alpha}\right)^2 - u_m{}^2\right]^{\frac{1}{2}} = iW_m \qquad (10.3.10)$$

$$W_m = \left[u_m{}^2 - \left(\frac{v}{\alpha}\right)^2\right]^{\frac{1}{2}} \qquad C(u) = \cos(\pi u) - \cos(\pi v)$$

$$Q(u) = 2\cos(\pi u) - 2\cos(\pi v) + i\zeta\sin(\pi v)$$

$$B(u) = \frac{2\zeta v}{\pi} \frac{C(u)}{Q(u)}$$

to list all the abbreviations to be used in the following discussion. The functions

$$\Psi_m(x) = e^{ik_m x} - \frac{i\xi}{Q}\begin{cases} \cdots\cdots\cdots\cdots\cdots\cdots\cdots\cdots\cdots \\ \sin[\alpha k(l + x)] - e^{-i\varkappa l}\sin(\alpha k x) \qquad -l < x < 0 \\ e^{i\varkappa l}\sin(\alpha k x) + \sin[\alpha k(l - x)] \qquad 0 < x < l \\ e^{2i\varkappa l}\sin[\alpha k(x - l)] + e^{i\varkappa l}\sin[\alpha k(2l - x)] \\ \qquad\qquad\qquad\qquad\qquad\qquad\qquad l < x < 2l \\ \cdots\cdots\cdots\cdots\cdots\cdots\cdots\cdots\cdots \end{cases}$$

$$= e^{ik_m x} + iB(u)\sum_{n=-\infty}^{\infty}\frac{e^{ik_n x}}{\Delta_n(u)} \qquad (10.3.11)$$

are generalizations of the forced waves in vacuum, of Eq. (10.3.2), written in more compact form.

Using this Green's function, the transverse displacement of the membrane-strut system, under the influence of a force $F_\varkappa(x)e^{-i\omega t}$ which satisfies the periodicity condition $F_\varkappa(x + l) = e^{i\varkappa l}F_\varkappa(x)$, is

$$\eta(x) = \frac{\alpha^2}{\sigma c^2}\int_0^l G(x \mid x_1)F_\varkappa(x_1)\,dx_1$$

$$= \frac{\alpha^2}{\pi^2\sigma c^2}\sum\frac{\Psi_m(x)}{\Delta_m(u)}\int_0^l e^{-ik_m x_1}F_\varkappa(x_1)\,dx_1 \qquad (10.3.12)$$

where again the introduction of an array of sources in the equation for G has meant that we need only to integrate over a single period of x_1.

Free-wave motion, fluid present

If we wish to investigate the free-wave motion of the system, we set the force F_x equal to the pressure P of the medium caused by the membrane motion, given in Eq. (10.3.8). Inserting this for F_x in Eq. (10.3.12), we obtain the integral equation for free motion of a periodically supported membrane, in contact on both sides with a medium of density ρ and acoustic velocity c,

$$\eta(x) = \frac{2\rho\alpha^2\omega^2}{\sigma c^2} \int_0^l G(x \mid x_1)\, dx_1 \int_0^l g(x_1 \mid x_0)\eta(x_0)\, dx_0$$

$$= i\mu \sum \frac{\Psi'_m(x)}{w_m\Delta_m} \int_0^l e^{-ik_mx_0}\eta(x_0)\, dx_0 \qquad (10.3.13)$$

when we substitute the series for G and g and carry out the integration over x_1 [see Eqs. (10.3.10) for definitions of symbols].

In the present case an exact solution can be worked out because the component functions $\Psi'_m(x)$ consist of a term e^{ik_mx} plus a term which is independent of m. We assume that $\eta(x)$ is a series $\Sigma A_n e^{ik_nx}$, and proceed to calculate the relative magnitudes of the A's. The procedure is complicated by the fact that the Ψ'_n's are *not* orthogonal; so we cannot just multiply by $\Psi'_m(-x)$ and integrate, but must expand everything in series of terms e^{ik_mx}, which are orthogonal. The steps go as follows:

$$\sum_i A_i e^{ik_ix} = i\mu l \sum_j \frac{1}{w_j\Delta_j}\left[e^{ik_jx} + iB(u)\sum_n \frac{e^{ik_nx}}{\Delta_n}\right]A_j$$

with all summations from $-\infty$ to $+\infty$. Multiplying both sides by e^{-ik_mx} and integrating over x from 0 to l, we obtain

$$A_m = -\frac{w_m}{w_m\Delta_m - i\mu l}\left[\mu lB(u)\sum_n \frac{A_n}{w_n\Delta_n}\right] \qquad (10.3.14)$$

where the quantity in brackets is independent of m but is a function of $u = \kappa l/\pi$.

This is an infinite homogeneous set of simultaneous equations for the A_m's, which has nonzero solutions only for certain values of u, corresponding to the various sorts of free waves possible for the system. An equation of self-consistency can be obtained by multiplying the equation for A_m by $1/w_m\Delta_m$ and summing, then equating the coefficients of $\Sigma(A_m/w_m\Delta_m)$ on both sides. The result is

$$0 = 1 + \mu lB(u)\sum_m \frac{1}{\Delta_m(w_m\Delta_m - i\mu l)}$$

or

$$1 = \frac{4i\xi\rho lv^3}{\pi^2\sigma}\Gamma(u) \qquad (10.3.15)$$

where

$$\Gamma(u) \equiv \frac{C(u)}{Q(u)} \sum_m \frac{1}{\Delta_m(W_m \Delta_m - \mu l)} \equiv \frac{\cos(\pi u) - \cos(\pi v)}{2\cos(\pi u) - 2\cos(\pi v) + i\zeta \sin(\pi v)}$$

$$\times \sum_{m=-\infty}^{\infty} [(u+2m)^2 - v^2]^{-1} \left\{ \left[(u+2m)^2 - \left(\frac{v}{\alpha} \right)^2 \right]^{\frac{1}{2}} \right.$$

$$\left. \times [(u+2m)^2 - v^2] - \left(\frac{2\rho l}{\pi\sigma} \right) v^2 \right\}^{-1}$$

with the proviso that, whenever $v/\alpha > u + 2m$, the radical in the summation becomes $-i[(v/\alpha)^2 - (u+2m)^2]^{\frac{1}{2}}$, *not* $+i$ times the radical.

In our manipulations to obtain Eq. (10.3.15), we have, in effect, multiplied by Δ_m. Thus, when $u_m = v$, *both* $C(u)$ and Δ_m go to zero and, in the limit, Eq. (10.3.15) is satisfied. These roots do not correspond to solutions of Eq. (10.3.13). The other roots of Eq. (10.3.15), however, do correspond to values of $u = \varkappa l/\pi$ (and thus to values of $\varkappa$) which *are* solutions of Eq. (10.3.13), and therefore represent possible free motions of the membrane-strut-medium system. There are two such sets of roots, one for the wave motion which has most of its energy in the membrane-strut part of the system, the other where most of the energy is in the medium.

We note that function $\Gamma(u)$ is periodic in u with period 2, that is, $\Gamma(u + 2n) = \Gamma(u)$. Thus we need determine only those roots of the equation in the first period of u; all other roots are obtained by subtracting or adding an even integer to these. Moreover, an examination of Eq. (10.3.14) shows that the other solutions are really the same, that the series for η for $u = u_0$, one root of Eq. (10.3.15) in the first period, is exactly the same as the series for η for $u = u_0 + 2n$; only the coefficients and terms are relabeled, the $(m + n)$th term of the first series being equal to the mth term of the second series. Thus examination of the four roots of Eq. (10.3.15) in the first period (solutions in pairs, corresponding to the two possible directions of propagation) covers the complete capabilities for wave motion of the triply coupled system.

The membrane-strut waves

When $2\rho l/\pi\sigma$, the mass of a unit area of blanket of the medium, of thickness $2l/\pi$, divided by the mass σ of a unit area of the membrane, is small compared with unity (as it is if the medium is air), then $\Gamma(u)$ must be quite large in order that Eq. (10.3.15) may be satisfied. Thus, to find solutions valid to first order in $2\rho l/\pi\sigma$, we look for singularities of $\Gamma(u)$ in the first period of u. In the range $0 < \mathrm{Re}\, u < 1$, $Q(u)$ has one root, the one plotted in Fig. 10.15 for wave motion of the membrane-strut system *in vacuo* (we have called this root u_q). Therefore, when $2\rho l/\pi\sigma$ is small, there is a solution of Eq. (10.3.15) corresponding to wave motion with velocity near $\omega l/\pi u_q$. These waves involve the greatest amount of motion of the struts; so we call them the membrane-strut waves.

Call the first-period solution of $Q(u) = 0$ by the subscript u_q (if the strut impedance Z is reactive, u_q can be real for certain values of v, as shown in Fig. 10.15). Setting $u = u_q + \epsilon$ and expanding $\Gamma(u)$ to first order in ϵ, we obtain, for one root of Eq. (10.3.15),

$$u_l \simeq u_q - \frac{\zeta^2 \rho l v^3}{\pi^3 \sigma} \frac{\sin(\pi v)}{\sin(\pi u_q)} \left\{ D(u_q) \right.$$

$$\left. + \frac{1}{(u_q^2 - v^2)[(u_q^2 - v^2)\sqrt{u_q^2 - (v/\alpha)^2} - (2\rho l/\pi \sigma)v^2]} \right\} \quad (10.3.16)$$

where the second term in the braces is the term for $m = 0$ in Γ, and the term $D(u_q)$ is all the other terms, for $u = u_q$. We do this because, if $v = \alpha k l/\pi = 2l/\lambda_m$ (λ_m being the wavelength of waves on the membrane at frequency $\omega/2\pi$) is small compared with unity, the series D can be approximated by

$$D(u) \simeq \frac{1}{32} \sum_{m=1}^{\infty} \left[\frac{1}{(m + \frac{1}{2}u)^5} + \frac{1}{(m - \frac{1}{2}u)^5} \right]$$

which is a small correction to the second term. At higher frequencies several terms in the series must be separated off, and the computation becomes more tedious, but the series always converges satisfactorily.

As with any wave-supporting surface exposed to an acoustic medium, the properties of the surface wave motion depend on whether the wave velocity of the surface *in vacuo* is larger or smaller than the speed c of sound in the medium. In the present case the criterion is whether $v/\alpha = kl/\pi = 2l/\lambda = \omega l/\pi c$ is larger or smaller than $\text{Re } u_q = x_q l/\pi = \omega l/\pi c_q$, where $c_q = \omega/x_q$ is the effective speed of transverse wave motion in the strut-membrane system *in vacuo*. When $c_q < c$ ($u_q > v/\alpha$) (this occurs to the right of the v axis in Fig. 10.15, for example), the whole quantity in braces in Eq. (10.3.16) is real, and if the strut impedance is either purely reactive or purely resistive, u_l is real to this order of approximation (if ζ is purely reactive, u_l is real to all orders in this case). Thus, as usual, when the velocity c_q is less than c, the acoustic wave motion cannot couple to the wall motion, and the wave propagates along the wall without attenuation [see also Eqs. (10.1.23)]. On the other hand, if $c_q > c$ ($u_q < v/\alpha$), the radical in the second term in braces in Eq. (10.3.16) is $-i[(v/\alpha)^2 - u_q^2]^{\frac{1}{2}}$, u_l has an imaginary part (which turns out to be positive-imaginary), and there is some energy transmitted through the medium to $z = \pm\infty$, and thus some attenuation of the wall wave.

Returning to the equations for η and for p, we see that this sort of wave can be written

$$\eta(x) = A_l \sum_{m=-\infty}^{\infty} \frac{W_m(u_l)e^{i(u_l+2m)(\pi x/l)}}{W_m(u_l)\Delta_m(u_l) - i\mu l}$$

and

$$p_+(x,z) = -i\omega A_l \frac{pck}{\pi}$$

$$\times \sum \frac{\exp\{i(u_l + 2m)(\pi x/l) + (i\pi z/l)[(\omega l/\pi c)^2 - (u_l + 2m)^2]^{\frac{1}{2}}\}}{w_m(u_l)\Delta_m(u_l) - i\mu l}$$

(10.3.17)

The expression for p_- is obtained by changing the sign of z and also of the whole expression. The series for η is not the same as for $\Psi_0(x)$, the wave *in vacuo*, as given in Eq. (10.3.5), which in our present notation is

$$\sum \frac{\exp[i(u_q + 2m)(\pi x/l)]}{\Delta_m(u_q)}$$

If $\rho l/\pi\sigma$ is small, however, the difference is not great. Root u_l is not much different from u_q, and the effect on the denominator of each term in the series for η lies in the term $i\mu l/\omega_m$, the radiation impedance load on the membrane. Thus, when we include the load of the medium, there will be no infinite amplitudes for η at points such as a and b of Fig. 10.16. The terms in the series for p for which $|u_l + 2m| > \omega l/\pi c$ attenuate as z increases and only serve to modify p near the struts. If $u_l > \omega l/\pi c$, all terms so attenuate, and no energy escapes to $|z| \to \infty$.

There is also a root of $Q = 0$ at $2 - u_l$, within the first period of u; this corresponds to the root $-u_l$, for wave motion in the negative direction. All other roots are for $u = \pm u_l \pm 2m$. However, adding or subtracting $2m$ to u_l in Eqs. (10.3.17) does not change η or p; so really there are only two different membrane-strut waves.

The surface wave

In addition to the wave motion with velocity near that of the wall *in vacuo*, just discussed, there may occur a surface wave which travels at a velocity slightly smaller than the wave velocity c in the unbounded medium, but in which the energy flow is parallel to the surface, being mostly in the part of the medium near the surface. Such a surface wave can exist only if the velocity c_p of transverse wave in the membrane is greater than c [see also the discussion of Eq. (10.1.24)]. The possibility of such a solution arises from the fact that $\Gamma(u)$ has a sequence of poles at points where $[u^2 - (v/\alpha)^2]^{\frac{1}{2}}$ is equal to $(2\rho l/\pi\sigma)[v^2/(u^2 - v^2)]$, the square root being small, not $u^2 - v^2$. Set $u = (v/\alpha) + \epsilon$; if $(v/\alpha)^2 > v^2$, that is, if $c_p > c$, a real value of ϵ is possible. For $\rho l/\pi\sigma$ small, it is approximately given by

$$\sqrt{\frac{2\epsilon v}{\alpha}} \simeq \frac{2\rho l\alpha^2}{\pi\sigma(1 - \alpha^2)} + \frac{4i\zeta\rho l\alpha^4 C(v/\alpha)}{\pi^2\sigma v(1 - \alpha^2)^2 Q(v/\alpha)}$$

leading to

$$u_s \simeq \frac{kl}{\pi} + \frac{\alpha}{2v}\left(\frac{2\rho l}{\pi\sigma}\right)^2\left(\frac{\alpha^2}{1-\alpha^2}\right)^2\left[1 + \frac{2i\zeta}{\pi v}\frac{C(v/\alpha)}{Q(v/\alpha)}\frac{\alpha^2}{1-\alpha^2}\right]^2 \quad (10.3.18)$$

corresponding to a velocity $\omega l/\pi u_s$ a little smaller than c, the acoustic velocity of the medium. If the strut impedance is imaginary, u_s is real.

Since w_0 is small for this wave, we see from Eqs. (10.3.17) (substituting u_s for u_l in the formulas) that η is small and p large, in comparison with the waves for $u = u_l$, when the majority of the energy is carried by the wall. All components of the pressure wave attenuate with increase of $|z|$, because all coefficients of z in the exponent are negative and real. The smallest coefficient, that for $m = 0$, is

$$i\frac{\pi}{l}\left[\left(\frac{v}{\alpha}\right)^2 - u_s^2\right]^{\frac{1}{2}} \simeq -\frac{2\rho}{\sigma}\frac{\alpha^2}{1-\alpha^2}\left[1 + \frac{2i\zeta}{\pi v}\frac{C(v/\alpha)}{Q(v/\alpha)}\frac{\alpha^2}{1-\alpha^2}\right]$$

The disturbance extends out from the wall to a distance of the order of magnitude of the reciprocal of this, a considerable distance if σ is much larger than ρ. All other components, for $|m| > 0$, attenuate more sharply.

When $\alpha > 1$ $(c_p < c)$, ϵ becomes imaginary, and there can be no root of Eq. (10.3.15) which satisfies the physical requirement that p remain finite at $|z| \to \infty$. Thus no surface wave exists when the velocity of transverse waves on the membrane *in vacuo* is less than the acoustic velocity c in the medium.

We can also carry out the calculations for a driving force $Fe^{i\varkappa x - i\omega t}$ applied transversally to the wall, including the reaction of the medium. This can be manipulated, by means of the Fourier integral in $\varkappa$, to obtain the membrane shape and pressure wave arising when the strut at $x = 0$ is driven by a simple-harmonic line force. But this exercise will be left as a problem. The one example of forced motion deserving attention here is when the driving force is produced by an incident plane wave of sound, producing reflected and transmitted waves.

Wave penetration through the wall

If the pressure wave incident on the strut-membrane system is A exp $(ikx \sin\theta - ikz \cos\theta - i\omega t)$, θ being the angle of incidence (in the xz plane), then the integral equation for the pressure above the xy plane is a combination of Eqs. (8.3.2) and (10.3.6).

$$p_+(x,z) = 2Ae^{ikx \sin\theta}\cos(kz\cos\theta) - \rho\omega^2\int_0^l g_+(x,z \mid x_0,0)\eta(x_0)\,dx_0$$

where the $k_0 \equiv \varkappa$ in the g_+ of Eq. (10.3.7) is now $k \sin\theta$. Since this Green's function is chosen to have its normal gradient equal to zero at $z = 0$, the inhomogeneous term in the integral equation must satisfy the same condition, including what would be the reflected wave if the wall were rigid.

We next insert Eq. (10.3.12) for η, with $F_\kappa = (p_- - p_+)_{z-0} \equiv P(x)$. Therefore p_+, p_-, and P are

$$p_+(x,z) = 2Ae^{ikx \sin \theta} \cos(kz \cos \theta)$$

$$- \frac{\rho k^2 \alpha^2}{\sigma} \int_0^l g_+(x,z \mid x_0,0) \, dx_0 \int_0^l G(x_0 \mid x_1)P(x_1) \, dx_1$$

$$p_-(x,z) = \frac{\rho k^2 \alpha^2}{\sigma} \int_0^l g_-(x,z \mid x_0,0) \, dx_0 \int_0^l G(x_0 \mid x_1)P(x_1) \, dx_1 \qquad (10.3.19)$$

$$P(x) = -2Ae^{ikx \sin \theta} + \frac{2\rho k^2 \alpha^2}{\sigma} \int_0^l g(x \mid x_0) \, dx_0 \int_0^l G(x_0 \mid x_1)P(x_1) \, dx_1$$

where g_+ is given in Eq. (10.3.7), g_- is obtained by reversing the signs of z and z_0, and g is given in Eq. (10.3.8) and G in Eq. (10.3.9). In all these we set $k_m = k \sin \theta + (2\pi m/l)$.

It is obvious from the form of the series for g and G that P will be expressed as a series of the form $\Sigma P_m e^{ik_m x}$. Once this series is known, the series for p_+ and p_- can be determined, for from Eqs. (10.3.19),

$$p_- - p_+ = P \qquad \text{and} \qquad p_- + p_+ = 2Ae^{ik_0 x} \qquad \text{at } z = 0$$

Therefore the wave transmitted through the strut-membrane system into the region $z < 0$ is, at the surface of the membrane,

$$p_-(x,0) = Ae^{ikx \sin \theta} + \tfrac{1}{2}P(x)$$

We can express k_m and $q_m = (k^2 - k_m{}^2)^{\frac{1}{2}}$ in terms of angles θ_m of transmission of the various waves with respect to the negative z axis by defining

$$\sin \theta_m = \sin \theta + \frac{2\pi m}{kl} \qquad \cos \theta_m = (1 - \sin^2 \theta_m)^{\frac{1}{2}} \qquad (10.3.20)$$

If $|\sin \theta_m| > 1$, the requirement of finiteness at $z \to -\infty$ is that $\cos \theta_m$ be positive-imaginary; if $|\sin \theta_m| < 1$, $\cos \theta_m$ must be positive-real. In these terms the transmitted wave in region $z < 0$ is expressible in terms of the coefficients of the series for P.

$$p_- = Ae^{ikx \sin \theta - ikz \cos \theta} + \tfrac{1}{2} \sum P_m e^{ikx \sin \theta_m - ikz \cos \theta_m}$$

$$= \sum_{m=-\infty}^{\infty} T_m \exp (ikx \sin \theta_m - ikz \cos \theta_m) \qquad (10.3.21)$$

Depending on the size of $2\pi/kl = \lambda/l$, there are one or more values of m for which $|\sin \theta_m| < 1$; corresponding to these values there will be one or more plane waves, carrying energy to $z \to -\infty$ at the corresponding angles θ_m (which can be positive or negative). These are the waves diffracted by the periodically spaced struts, according to the usual expression for the transmission angles θ_m from a diffraction grating. The waves for which $|\sin \theta_m| > 1$ will die out exponentially with z; their effect is to adapt the shape of the wave near $z = 0$ to the motion of the membrane near each strut, as the whole moves in response to the pressure of the incident wave. Thus, if we can solve Eqs. (10.3.19) for P to obtain the coefficient P_m, we can compute the amplitudes of the various diffracted waves.

The equation for P is quickly solved. We substitute Eqs. (10.3.7) and (10.3.9) and the series for P into the last of Eqs. (10.3.19) and integrate over x_0 and x_1 to obtain

$$P_m e^{ik_m x} = -2Ae^{ikx_0} + \frac{2i\rho\alpha^2}{k\sigma} \sum_m \frac{P_m e^{ik_m x}}{(\sin^2 \theta_m - \alpha^2)\cos\theta_m}$$

$$- \frac{2\rho\zeta\alpha^3 C}{k^2\sigma l Q} \sum_m \frac{e^{ik_m x}}{(\sin^2 \theta_m - \alpha^2)\cos\theta_m} \sum_n \frac{P_n}{\sin^2\theta - \alpha^2}$$

which can be solved exactly as Eq. (10.3.14) was. The functions C and Q are defined in Eqs. (10.3.10), with $\pi u = kl \sin\theta$ and $\pi v = k\alpha l$. Referring to Eq. (10.3.21), we see that the amplitude T_0 of the directly transmitted wave, propagating at the angle $\theta = \theta_0$, and the amplitude T_m of the mth diffracted wave, going at angle θ_m as defined in Eqs. (10.3.20), are

$$T_0 = \frac{-igA}{(\sin^2\theta - \alpha^2)\cos\theta - ig} + B_0 \qquad T_m = B_m$$

where

$$B_m = \frac{A\cos\theta}{[(\sin^2\theta - \alpha^2)\cos\theta - ig][(\sin^2\theta_m - \alpha^2)\cos\theta_m - ig]}$$

$$\times \left\{ \Upsilon \frac{2\cos(kl\sin\theta) - 2\cos(k\alpha l) + i\zeta\sin(k\alpha l)}{\cos(kl\sin\theta) - \cos(k\alpha l)} \right. \tag{10.3.22}$$

$$\left. + \sum_n \frac{1}{(\sin^2\theta_n - \alpha^2)[(\sin^2\theta_n - \alpha^2)\cos\theta_n - ig]} \right\}^{-1}$$

and where $\alpha = c/c_p$; $\Upsilon = k^2 l\sigma/2\rho\zeta\alpha^3$, and $g = 2\rho\alpha^2/k\sigma$. The only waves which propagate to $z \to -\infty$ are those for which $|\sin\theta_m| < 1$.

Usually, constant g is small, because $\lambda\rho$ is usually smaller than σ. Where this is the case and where $\alpha < 1$ ($c_p > c$), the wall becomes transparent to the incident wave when the angle of incidence θ is equal to $\sin^{-1}\alpha$, that is, when the horizontal phase velocity $(\omega/k)\sin\theta$ is equal to the membrane wave velocity $c_p = \omega/k\alpha$. In this case the amplitude of the directly transmitted wave ($m = 0$) is A, the incident amplitude; and all the diffracted beams vanish ($B_m = 0$). When the angle of incidence θ is such that $\sin\theta_m = \alpha$ ($m \neq 0$), the diffracted beams also disappear, though the directly transmitted wave ($m = 0$) is in this case not equal to the incident wave; part of the incident wave is reflected to $z \to +\infty$.

If the strut impedance $Z = \sigma c_p \xi$ is zero, all the B_m's for $m \neq 0$ are zero, and the result reduces to that of a wave incident on a uniform membrane. If the struts are very stiff or heavy ($|\xi| \gg 1$), the B_m's are largest when $\sin(k\alpha l) = 0$, which is when the membrane strip between each strut resonates; at these frequencies the diffracted beams are most intense. Except at these resonance angles or frequencies, the transmission through the strut-membrane system is mass-controlled (proportional to $g = 2\rho c^2 / k\sigma c_p^2$) for the first part of T_0, and is strut-impedance-controlled (proportional to $\gamma = \Upsilon l / \xi \alpha g$) for the B_m's governing the diffracted waves.

10.4 COUPLED CAVITIES

Finally, we shall work out the phenomena which arise when two resonating cavities are coupled together. Suppose cavity a has volume V_a and boundary surface S_a; cavity b has volume V_a and wall surface S_b. The two are joined by an opening S_0, a surface common to both S_a and S_b, as shown in Fig. 10.17, through which energy can pass from V_a to V_b, and vice versa.

The integral equations

To solve this problem we first compute the normal modes for each room separately, with the coupling area S_0 rigidly closed; in fact, we assume that the whole boundary area S_a is rigid, and similarly for S_b. In this case the eigenfunctions $\phi_n{}^a$ and eigenvalues $\eta_n{}^a$ and mean-square values $\Lambda_n{}^a$, as defined in Eqs. (9.4.8), for room a, are all real, and the corresponding quantities $\phi_m{}^b(r)$, $\eta_m{}^b$, and $\Lambda_m{}^b$ for isolated room b are also real and independent of frequency. The corresponding Green's functions for driving frequency $\omega / 2\pi$,

$$G_a = \frac{1}{V_a} \sum_n \frac{\phi_n{}^a(\mathbf{r})\phi_n{}^a(\mathbf{r}_0)}{\Lambda_n{}^a[(\eta_n{}^a)^2 - (\omega/c)^2]}$$

$$G_b = \frac{1}{V_b} \sum_m \frac{\phi_m{}^b(\mathbf{r})\phi_m{}^b(\mathbf{r}_0)}{\Lambda_m{}^b[(\eta_m{}^b)^2 - (\omega/c)^2]}$$

(10.4.1)

also have no imaginary parts. The series for function G_a is valid only

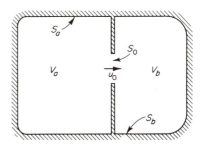

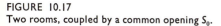

FIGURE 10.17
Two rooms, coupled by a common opening S_0.

within V_a and on S_a; the series for G_b holds only within V_b and on S_b; thus only on S_0 are both series valid.

The problem we wish to solve is that for which S_0 is not rigid but is an open space coupling the two rooms (another example, where S_0 is a flexible membrane, is considered in the problems). Across this surface the normal velocity is $u_0(r^s) = \dfrac{1}{i\omega\rho}\dfrac{\partial p}{\partial n}$ (for convenience, we assume that u_0 is positive for flow from a to b) and the acoustic pressure is $p_0(r^s)$. Utilizing Eq. (7.1.17) for each room separately, the equations expressing the pressure in each room are

$$p(\mathbf{r}) = \begin{cases} i\rho\omega \displaystyle\iint G_a(\mathbf{r}\mid \mathbf{r}_0{}^s)u_0(\mathbf{r}_0{}^s)\,dS_0 & \mathbf{r} \text{ in } V_a \text{ or on } S_a \\[2mm] -i\rho\omega \displaystyle\iint G_b(\mathbf{r}\mid \mathbf{r}_0{}^s)u_0(\mathbf{r}_0{}^s)\,dS_0 & \mathbf{r} \text{ in } V_b \text{ or on } S_b \end{cases} \qquad (10.4.2)$$

where both integrations are over the common surface S_0. Both equations hold simultaneously *only* when $\mathbf{r}$ is on S_0, in which case p is the common pressure p_0. For the solution which approaches the Nth mode in room a as S_0 is reduced to zero, the procedures which produced Eqs. (9.4.11) and (9.4.12) give, for the pressure in the opening and for the eigenvalue when S_0 is open,

$$0 = \Psi_N{}^a(\mathbf{r}^s) + \iint G_{aN}(\mathbf{r}^s\mid\mathbf{r}_0{}^s)u_N^0(\mathbf{r}_0{}^s)\,dS_0 + \iint G_b(\mathbf{r}^s\mid\mathbf{r}_0{}^s)u_N^0(\mathbf{r}_0{}^s)\,dS_0 \quad (10.4.3a)$$

where

$$G_{aN} = \frac{1}{V_a}\sum_{n\neq N}\frac{\phi_n{}^a(\mathbf{r})\phi_n{}^a(\mathbf{r}_0)}{\Lambda_n{}^a[(\eta_n{}^a)^2-(\omega/c)^2]} \qquad (10.4.3b)$$

and where

$$\Lambda_N{}^a V_a\left[(\eta_N{}^a)^2 - \left(\frac{\omega}{c}\right)^2\right] = \iint \phi_N{}^a(\mathbf{r}_0{}^s)u_N^0(\mathbf{r}_0{}^s)\,dS_0 \qquad (10.4.3c)$$

Equation (10.4.3a) is one for the coupling velocity u_N^0 over S_0. Once it is solved, the allowed frequency of free vibration $\omega_N/2\pi$ can be obtained from Eq. (10.4.3c), and the pressure distribution in the two rooms can be obtained from Eq. (10.4.2). The integral equation is amenable to variational calculation; the function to be varied is

$$\Lambda_N{}^a V_a\left[(\eta_N{}^a)^2 - \left(\frac{\omega}{c}\right)^2\right] = 2\int\phi_N{}^a u_N^0 + \int u_N^0\int G_{aN}u_N^0 + \int u_N^0\int G_b u_N^0 \quad (10.4.4)$$

If we use the very simple trial function $u_N^0 = A$, the "best" values for $k = \omega/c$ and for A are given by the equations

$$A = \frac{-\int\phi_N{}^a\,dS}{\int dS\int(G_{aN}+G_b)\,dS_0} \qquad k^2 \simeq (\eta_N{}^a)^2 + \frac{(\int\phi_N{}^a\,dS)^2}{\Lambda_N{}^a V_a\int dS\int(G_{aN}+G_b)\,dS_0}$$

The resonance frequency $c\eta_N{}^a/2\pi$ of room a, uncoupled, may be close to a resonance frequency $c\eta_M{}^b/2\pi$ of room b, uncoupled; if so, it may be useful also to separate off the term in G_b having $(\eta_M{}^b)^2 - (\omega/c)^2$ in the denominator. The equations for k^2 and for A then can be rearranged into the forms

$$\frac{(\int \phi_N{}^a \, dS)^2/\Lambda_N{}^a V_a}{k^2 - (\eta_N{}^a)^2} + \frac{(\int \phi_M{}^b \, dS)^2/\Lambda_M{}^b V_b}{k^2 - (\eta_M{}^b)^2} \simeq \int dS \int G(\mathbf{r} \mid \mathbf{r}_0) \, dS_0$$

$$G(\mathbf{r} \mid \mathbf{r}_0) = G_{aN} + G_{bM} \qquad (10.4.5)$$

$$A \simeq \frac{\Lambda_N{}^a V_a}{\int \phi_N{}^a \, dS} [(\eta_N{}^a)^2 - k^2] = \left(\int \phi_N{}^a \, dS \right)$$

$$\times \frac{\Lambda_M{}^b V_b [k^2 - (\eta_M{}^b)^2]}{(\int \phi_M{}^b \, dS)^2 - \Lambda_M{}^b V_b [k^2 - (\eta_M{}^b)^2] \int dS \int G \, dS_0}$$

where all integrations are over the coupling area S_0, and G_{aN} and G_{bM} are the series of Eqs. (10.4.1), with the Nth and the Mth terms, respectively, omitted. The first equation, for k^2, is now symmetric with respect to the two nearest resonance frequencies, one for each room. The equation for A is not symmetric because the normalizing equation for u_N^0, Eq. (10.4.3c), is in terms of the normal mode $\phi_N{}^a$ of room a, which is not symmetric. The ratios between the pressures in the two rooms, calculated from Eq. (10.4.2), will be symmetric, however.

Analogy with coupled oscillators

The equation for k^2 has the same form as that of two coupled linear oscillators. For example, Eqs. (3.1.1) can be written in the form

$$(\omega_{01}{}^2 - \omega^2)A = \omega_{c1}{}^2(B - A) \qquad (\omega_{02}{}^2 - \omega^2)B = \omega_{c2}{}^2(A - B)$$

where

$$x = Ae^{-i\omega t} \qquad y = Be^{-i\omega t} \qquad \omega_{01}{}^2 = \frac{K_1}{m_1} \qquad \omega_{c1}{}^2 = \frac{K_3}{m_1}$$

$$\omega_{02}{}^2 = \frac{K_2}{m_2} \qquad \text{and} \qquad \omega_{c2}{}^2 = \frac{K_3}{m_2}$$

The natural frequencies of the two oscillators, without the presence of the coupling spring of stiffness K_3, are $\omega_{01}/2\pi$ and $\omega_{02}/2\pi$, respectively, and the quantities ω_1, ω_2, and μ of Eq. (3.1.2) are given by

$$\omega_1{}^2 = \omega_{01}{}^2 + \omega_{c1}{}^2 \qquad \omega_2{}^2 = \omega_{02}{}^2 + \omega_{c2}{}^2 \qquad \mu^2 = \omega_{c1}\omega_{c2}$$

The equation for ω^2, giving the natural frequencies of the coupled oscillators, is obtained by eliminating $A - B$ from Eqs. (3.1.1). Written in the form

$$\frac{\omega_{c1}^2}{\omega^2 - \omega_{01}^2} + \frac{\omega_{c2}^2}{\omega^2 - \omega_{02}^2} = 1$$

it resembles the first of Eqs. (10.4.5). Thus we can write

$$(K_{cN}{}^a)^2 = \frac{(\int \phi_N{}^a \, dS)^2 / \Lambda_N{}^a V_a}{\int dS \int G(\mathbf{r} \mid \mathbf{r}_0) \, dS_0} \quad \text{and} \quad (K_{cM}{}^b)^2 = \frac{(\int \phi_M{}^b \, dS)^2 / \Lambda_M{}^b V_b}{\int dS \int G(\mathbf{r} \mid \mathbf{r}_0) \, dS_0}$$

and obtain the equation

$$\frac{(K_{cN}{}^a)^2}{k^2 - (\eta_N{}^a)^2} + \frac{(K_{cM}{}^b)^2}{k^2 - (\eta_M{}^b)^2} \simeq 1 \tag{10.4.6}$$

for the natural frequencies of the coupled rooms which are nearest to the Nth and Mth frequencies of the uncoupled rooms.

The solution of this equation for the natural frequencies of the coupled rooms is $ck_{NM}{}^+/2\pi$ and $ck_{NM}{}^-/2\pi$, where

$$(k_{NM}{}^+)^2 = \tfrac{1}{2}[(\eta_N{}^a)^2 + (\eta_M{}^b)^2] + \tfrac{1}{2}\{[(\eta_N{}^a)^2 - (\eta_M{}^b)^2]^2 + 4\mu_{NM}{}^2\}^{\frac{1}{2}}$$

$$\to (\eta_N{}^a)^2 + \frac{\mu_{NM}{}^4}{(\eta_N{}^a)^2 - (\eta_M{}^b)^2} \tag{10.4.7}$$

$$(k_{NM}{}^-)^2 = \tfrac{1}{2}[(\eta_N{}^a)^2 + (\eta_M{}^b)^2] - \tfrac{1}{2}\{[(\eta_N{}^a)^2 - (\eta_M{}^b)^2]^2 + 4\mu_{NM}{}^2\}^{\frac{1}{2}}$$

$$\to (\eta_N{}^a)^2 - \frac{\mu_{NM}{}^4}{(\eta_N{}^a)^2 - (\eta_M{}^b)^2}$$

where

$$\mu_{NM}{}^2 = K_{cN}{}^a K_{cM}{}^b = \Lambda_N{}^a \Lambda_M{}^b V_a V_b \frac{\int \phi_N{}^a \, dS \int \phi_M{}^b \, dS}{\int dS \int G(\mathbf{r} \mid \mathbf{r}_0) \, dS_0}$$

The limiting forms, indicated by the arrows, hold when $\mu^2 \ll (\eta_N{}^a)^2 - (\eta_M{}^b)^2$ (we assume, with no loss of generality, that $\eta_N{}^a > \eta_M{}^b$). In the rare cases when $\mu^2 \gg (\eta_N{}^a)^2 - (\eta_M{}^b)^2$, the approximate solution is

$$(k_{NM}{}^\pm)^2 \to \tfrac{1}{2}[(\eta_N{}^a)^2 + (\eta_M{}^b)^2] \pm \mu_{NM}{}^2$$

Thus, for frequencies nearest the Nth resonance $(c\eta_N{}^a/2\pi)$, of room a, uncoupled, and the Mth resonance $(c\eta_M{}^b/2\pi)$ of uncoupled room b, the coupled rooms behave like two coupled oscillators with coupling constants $K_{cN}{}^a$ and $K_{cM}{}^b$ and $\mu_{NM}{}^2 = K_{cN}{}^a K_{cM}{}^b$. As with the simpler system, the resonance frequencies are pushed apart by the coupling.

When constant $\mu_{NM}{}^2$ is much smaller than $(\eta_N{}^a)^2 - (\eta_M{}^b)^2$, the particular standing wave for the coupled rooms which oscillates with the frequency

$ck_{NM}{}^+/2\pi$ has the following waveforms in the two rooms [obtained by setting the u_0 of Eq. (10.4.2) equal to the A of Eqs. (10.4.5)]:

$$p_a(\mathbf{r}) \simeq (i\rho c k_{NM}{}^+)\left[\phi_N{}^a(\mathbf{r}) + \sum_{n\neq N} \frac{\Lambda_N{}^a \int \phi_n{}^a \, dS}{\Lambda_n{}^a \int \phi_N{}^a \, dS} \frac{(\eta_N{}^a)^2 - (k_{NM}{}^+)^2}{(\eta_n{}^a)^2 - (k_{NM}{}^+)^2} \phi_n{}^a(\mathbf{r})\right]$$

$$p_b(\mathbf{r}) \simeq (i\rho c k_{NM}{}^+)\left(\frac{\Lambda_N{}^a V_a}{\Lambda_M{}^b V_b}\right)^{\frac{1}{2}} \frac{\mu_{NM}{}^2}{(k_{NM}{}^+)^2 - (\eta_M{}^b)^2} \tag{10.4.8}$$

$$\times \left[\phi_M{}^b(\mathbf{r}) + \sum_{m\neq M} \frac{\Lambda_M{}^b \int \phi_m{}^b \, dS}{\Lambda_m{}^b \int \phi_M{}^b \, dS} \frac{(\eta_M{}^b)^2 - (k_{NM}{}^+)^2}{(\eta_m{}^b)^2 - (k_{NM}{}^+)^2} \phi_m{}^b(\mathbf{r})\right]$$

each expression being multiplied by the common time factor $\exp(-ik_{NM}{}^+ct)$. Thus, for the resonance frequency nearest $c\eta_N{}^a/2\pi$, the standing wave is largest in room a, the ratio between p_b and p_a being approximately the small quantity

$$\sqrt{\frac{\Lambda_N{}^a V_a}{\Lambda_M{}^b V_b}} \frac{\mu_{NM}{}^+}{(k_{NM}{}^+)^2 - (\eta_M{}^b)^2}$$

For the resonance frequency $ck_{NM}{}^-/2\pi$, nearest the room b natural frequency, the formulas are reversed, and the sound is mostly concentrated in room b.

When either or both rooms have sound-absorbent walls, the eigenvalues and eigenfunctions $\eta_n{}^a$, $\phi_n{}^a$, etc., become complex and frequency-dependent, so that Eq. (10.4.6) is an implicit equation for k^2. The roots $k_{NM}{}^+$ and $k_{NM}{}^-$ are then complex, corresponding to the fact that the free vibrations of the air in the coupled rooms lose energy to the walls.

The coupling constant

The quantity most difficult to compute in these formulas is the function

$$\int dS \int G(\mathbf{r} \mid \mathbf{r}_0) \, dS_0 = \sum_{n\neq N} \frac{[\int \phi_n{}^a(\mathbf{r}^s) \, dS]^2}{\Lambda_n{}^a V_a[(\eta_n{}^a)^2 - k^2]} + \sum_{m\neq M} \frac{[\int \phi_m{}^b(\mathbf{r}^s) \, dS]^2}{\Lambda_m{}^b V_b[(\eta_m{}^b)^2 - k^2]}$$

These series converge, though slowly. If the walls are absorbing, the result has an imaginary, as well as a real, part. If the area S_0 of the opening is small compared with S_a and S_b, we can use the following argument to obtain an approximate formula for $\int dS \int G(\mathbf{r} \mid \mathbf{r}_0) \, dS_0$, which is at least as accurate as the trial function we have used for u_0.

We first point out that if we add the missing $n = N$ and $m = M$ terms, the expression is proportional to the radiation impedance on a rigid piston placed in the opening S_0, radiating sound into both rooms. Referring to Eqs. (7.4.31) or Eq. (7.4.44), we see that this impedance would be

$$Z_r = \frac{F_a + F_b}{u_0} = -i\rho c k \iint_{S_0} dS \iint_{S_0} [G_a(\mathbf{r}^s \mid \mathbf{r}_0{}^s) + G_b(\mathbf{r}^s \mid \mathbf{r}_0{}^s)] \, dS_0$$

If the rooms are nonabsorbent, Z_r will be pure-imaginary, indicating that all the energy radiated is reflected back to the piston, chiefly in the form of the most strongly resonant waves (the $n = N$ and $m = M$ terms, which were omitted in the sums). But if the room walls are absorbent, so that much of the sound emitted by the piston never returns to the piston, and if the resonant terms are omitted, the resulting expression should be approximately equal to the piston impedance for completely absorbent rooms; and this should be approximately equal to the impedance load on a piston set in a plane wall, radiating into infinite space on both sides of the wall (as long as S_0 is not near a corner of either room).

What has been just said is that if both rooms are at least moderately absorptive (8 Re β greater than the ratio of S_0 to $S_a + S_b$, as will be shown), the quantity $\int dS \int G(\mathbf{r} \mid \mathbf{r}_0) \, dS_0$ is approximately proportional to the impedances given in Eqs. (7.4.31) or Eq. (7.4.44); in fact, that

$$\int dS \int G(\mathbf{r} \mid \mathbf{r}_0) \, dS_0 \simeq 2i \, \frac{Z}{\rho c k} = \frac{2S_0}{k} (\chi + i\theta) \qquad (10.4.9)$$

where, if the opening S_0 is circular, of radius a, $\chi = \chi_0(2ka)$, $\theta = \theta_0(2ka)$; if the opening is rectangular with sides a and b, $\chi = [ab/(a^2 - b^2)][\chi_\square(ka) - \chi_\square(kb)]$ and $\theta = [ab/(a^2 - b^2)][\theta_\square(ka) - \theta_\square(kb)]$; in any case, χ will be the specific reactance, and θ the specific resistance $(R/\rho c S_0)$, per unit area of a piston of the shape and size of the opening, radiating into free space on both sides.

To this approximation the coupling constant becomes

$$\mu_{NM}{}^2 \simeq \frac{\int \phi_N{}^a \, dS \int \phi_M{}^b \, dS}{S_0{}^2 \sqrt{\Lambda_N{}^a \Lambda_M{}^b}} \frac{kS_0/2}{(\chi + i\theta)\sqrt{V_a V_b}} \qquad (10.4.10)$$

Thus the coupling constant is roughly equal to the product of the ratio between the mean values of $\phi_N{}^a$ and of $\phi_M{}^b$ across S_0 to the rms values of $\phi_N{}^a$ and $\phi_M{}^b$ over V_a or V_b, and the ratio between the factor $kS_0/2(\chi + i\theta)$ and the geometric mean of the room volumes, a small quantity having the dimensions of k^2. We note that if a nodal surface of either $\phi_N{}^a$ or $\phi_M{}^b$ cuts across S_0, so that either of the area integrals is zero, then, to this degree of approximation, there is no coupling between the rooms for this frequency range. In these cases a better approximation can be worked out (if it is worth the large amount of algebra!) by setting $u_0 = A\phi_N{}^a + B\phi_M{}^b$ in expression (10.4.4) and carrying out the variational calculation for best values of A and B.

Forced motion of the coupled rooms

If an acoustic source of frequency $\omega/2\pi$, representable by a distributed source strength $S_\omega(\mathbf{r})$, is placed in room a, the integral equations for the pressure in the two rooms are

$$
p(\mathbf{r}) = \begin{cases} -ik\rho c \iiint G_a(\mathbf{r} \mid \mathbf{r}_0) S_\omega(\mathbf{r}_0) \, dv_0 \\[2mm] \quad + ik\rho c \iint G_a(\mathbf{r} \mid \mathbf{r}_0{}^s) u_0(\mathbf{r}_0{}^s) \, dS_0 \qquad \mathbf{r} \text{ in } V_a \text{ or on } S_a \qquad (10.4.11) \\[2mm] -ik\rho c \iint G_b(\mathbf{r} \mid \mathbf{r}_0{}^s) u_0(\mathbf{r}_0{}^s) \, dS_0 \qquad \mathbf{r} \text{ in } V_b \text{ or on } S_b \end{cases}
$$

The volume integral is the pressure which would be in room a if S_0 were rigidly closed; it could be called the *driving pressure* and could be symbolized as $P(\mathbf{r})$ ($\mathbf{r}$ in V_a or on S_a). We can now set up a variational expression for the integral,

$$
T_\omega \equiv \iint u_0(\mathbf{r}_0{}^s) P(\mathbf{r}_0{}^s) \, dS_0 \tag{10.4.12}
$$

the real part of which is the power that the driving pressure gives to the coupling area S_0, which is transmitted through to room b. Equating $p(\mathbf{r})$ on S_0 from the two parts of Eq. (10.4.11), multiplying by u_0 and integrating over S_0, and then adding Eq. (10.4.12), we obtain the variational expression

$$
T_\omega = 2 \int u_0 P \, dS + ik\rho c \int u_0 \, dS \int (G_a + G_b) u_0 \, dS_0
$$

Inserting the trial value $u_0 = A$, as before, we obtain

$$
A \simeq \frac{-\int P \, dS}{ik\rho c \int dS \int (G_a + G_b) \, dS_0} \qquad T_\omega \simeq \frac{-(\int P \, dS)^2}{ik\rho c \int dS \int (G_a + G_b) \, dS_0} \tag{10.4.13}
$$

with corresponding solutions for the steady-state pressure in the two rooms, obtained from Eq. (10.4.11).

Thus the sound pressure in room a is the pressure P, the pressure present if S_0 were closed, with a correction term because S_0 is open; the pressure in room b is the amount which has come through S_0 from a. If all walls are rigid, the transmission factor T_ω is pure-imaginary; in steady state no further power is being transmitted from room a to room b. If the walls of both rooms have nonzero wall conductance, T_ω will have a real part corresponding to the power which flows through S_0, to be absorbed on the walls of room b.

As an example, suppose both rooms are rectangular, having walls of specific acoustic admittance $\beta_a = \xi_a - i\sigma_a$ and $\beta_b = \xi_b - i\sigma_b$, respectively, with magnitude small but not zero. Then Eq. (9.3.41) indicates that, instead of $\eta_n{}^a$ in Eqs. (10.4.1) et seq., we should use $K_n{}^a$, where

$$
(K_n{}^a)^2 \simeq (\eta_n{}^a)^2 - 2(\sigma_a + i\xi_a) \frac{S_n{}^a k}{V_a} \tag{10.4.14}
$$

where

$$
2S_n{}^a = \epsilon_{n_x} S_x{}^a + \epsilon_{n_y} S_y{}^a + \epsilon_{n_z} S_z{}^a - \epsilon_{n_x} S_0
$$

and $\eta_n{}^a$ is as given in Eq. (9.4.29); $S_x{}^a = 2l_y{}^a l_z{}^a$ is the area of the two walls perpendicular to the x axis, etc.; and we assume that opening S_0 is in an x wall. There is a similar expression for $K_m{}^b$.

Now suppose the driving frequency is small, in the range where the resonance peaks are separated, and suppose we excite the Nth mode of room a, ω being set equal to the real part of $K_N{}^a$. Then one term in G_a will be larger than all the rest of $G_a + G_b$ (unless room b happens to have an exactly coinciding resonance). The rest of G_a and all G_b will produce the coupling term of Eq. (10.4.9); so the acoustic wave in room a and the power transmitted through S_0 to room b will be approximately

$$p_a(\mathbf{r}) \simeq B_N \phi_N{}^a(\mathbf{r}) \frac{L_w{}^a}{L_w{}^a + \frac{1}{2}L_c{}^a} \qquad \mathrm{Re}\, T_\omega \simeq |B_N|^2 L_c{}^a \frac{L_w{}^a}{L_w{}^a + \frac{1}{2}L_c{}^a} \qquad (10.4.15)$$

where

$$L_w{}^a = \frac{2\xi_a S_N{}^a \Lambda_N{}^a}{\rho c} \qquad L_c{}^a = \frac{(\iint \phi_N{}^a \, dS)^2 \theta}{2\rho c S_0(\theta^2 + \chi^2)} \qquad B_N = \frac{\rho c \iiint S_\omega \phi_N{}^a \, dv}{2\xi_a S_N{}^a \Lambda_N{}^a}$$

with θ and χ as given in Eq. (10.4.9).

These formulas have a fairly simple physical interpretation. According to Eq. (9.4.32), $L_w{}^a$ is the power which standing wave $\phi_N{}^a$ loses to the walls of room a. Quantity $L_c{}^a$ is the power lost by wave $\phi_N{}^a$ to coupling area S_0, transferred to room b. Thus the energy density in room a (proportional to $|p|^2$) is reduced, below the level $(B_n \phi_N{}^a)^2$ it would have if S_0 were closed, by the ratio between the energy $L_w{}^a$ lost to the walls alone and the total energy lost. The amount transmitted to room b is proportional to $L_w{}^a$, if there is no coincidence of resonance.

Alternatively, when the driving frequency of the source in room a is equal to one of the resonance frequencies of room b but room a does not resonate, the pressure in room b is correspondingly increased. The pressure distribution in room a, when S_0 is closed, can still be called $P(\mathbf{r})$, though it will have a more complex form than $B\phi_N{}^a$, which it had at resonance. In this case the resonant term in G_b is larger than G_a and the rest of G_b, and the wave in room b is

$$p_b(\mathbf{r}) \simeq \phi_M{}^b(\mathbf{r}) \frac{\iint P \, dS}{\iint \phi_M{}^b \, dS} \left[\frac{(\iint \phi_M{}^b \, dS)^2}{2\rho c S_0 |\theta - i\chi|} \right] \frac{1}{L_w{}^b + \frac{1}{2}L_c{}^b} \qquad (10.4.16)$$

Since the coupling factor in brackets is of the order of magnitude of $L_c{}^b = (\iint \phi_M{}^b \, dS)^2 \theta / 2\rho c S_0 |\theta - i\chi|^2$, and thus is much smaller than $L_w{}^b = \xi_b S_M{}^b \Lambda_M{}^b / \rho c$, we see that p_b is smaller than P, but is much larger than the p_b, for the situation of Eqs. (10.4.15).

The analysis starting with Eqs. (10.4.1) can also be carried through, when S_0 is covered by a membrane or thin plate which can vibrate and thus transmit acoustic power, by adapting the procedures discussed in the beginning of Sec. 10.2. The quantity to be varied is not the velocity in S_0, but the pressure difference across S_0, which can be considered constant across

S_0 if S_0 is small.   We are led to a double integral, involving two Green's functions, one for the air in the rooms and one for the membrane motion. The formulas are thus more involved, but satisfactory approximations can be developed.   This will be left to the problems.

Higher frequencies

When the frequency is high enough for geometrical acoustics to be valid, the formulas of (10.4.15) and (10.4.16) no longer apply.   In this frequency range the simple formulation of Sec. 9.5, using energy density w and mean intensity I, applies.   The intensity in room a is $I_a = |p_a|^2/4\rho c = cw_a/4$, and that in room b is I_b.   Quantity a_a, the wall absorption for room a, is given by Eq. (9.5.2) or (9.5.16); a_b is the corresponding wall absorption for room b. If coupling area S_0 is small, the wavelength of the sound may be small enough, compared with room dimensions, so that geometrical acoustics can be used, yet large enough so that transmission through S_0 is still that of a plane wave striking the opening, and the formulas (7.4.31) or (7.4.44) can still be used in computing the impedance $2\rho c S_0(\theta - i\chi)$ of the opening.

If the pressure is fairly uniform across S_0, the effective acoustic impedance of the air on the other side of the opening is $\rho c S_0(\theta - i\chi)$, and in accord with Eq. (10.4.12), the power removed from room a (and therefore transmitted into room b) is $\epsilon_{n_x} S_0 |p_a|^2 \theta/\rho c(\theta^2 - \chi^2)$.   The factor ϵ_{n_x}, which is 2 for most high-frequency modes, is the ratio of the mean-square pressure at the wall containing S_0 to the mean-square pressure throughout the room a. Thus the power transmitted from room a to room b is $2[S_0 |p_a|^2 \theta/\rho c(\theta^2 + \chi^2)]$, and that going in the opposite direction is $2[S_0 |p_b|^2 \theta/\rho c(\theta^2 + \chi^2)]$ [we assume that the phase relations between p_a and p_b are random so that the mean value of Re $(p_a p_b^*)$ is zero].   The total power lost to room a is then $I_a a_a + 8S_0 I_a[\theta/(\theta^2 + \chi^2)]$, and the power gain, by transmission through S_0 from room b, is $8S_0 I_b[\theta/(\theta^2 + \chi^2)]$.   The total acoustic energy in room a is $w_a V_a = 4V_a I_a/c$, and if a source of acoustic power Π is located in room a, the equations relating the rate of change of intensity to the power lost and gained in the two rooms are

$$\frac{4V_a}{c}\frac{dI_a}{dt} + a_a I_a + 8S_0 C_0 I_a = 8S_0 C_0 I_b + \Pi$$

$$(10.4.17)$$

$$\frac{4V_b}{c}\frac{dI_b}{dt} + a_b I_b + 8S_0 C_0 I_b = 8S_0 C_0 I_a$$

where $C_0 = \theta/(\theta^2 + \chi^2)$ is the specific conductance of the opening.   For steady state the solution is

$$I_a = \frac{\Pi(a_b + 8S_0 C_0)}{(a_a + 8S_0 C_0)(a_b + 8S_0 C_0) - (8S_0 C_0)^2} \simeq \frac{\Pi}{a_a + 8S_0 C_0} \qquad S_0 C_0 \ll a_a$$

$$(10.4.18)$$

$$I_b \simeq \frac{8\Pi S_0 C_0}{(a_a + 8S_0 C_0)(a_b + 8S_0 C_0)} \qquad S_0 C_0 \ll a_b$$

When opening S_0 is closed, the intensity in room a is Π/a_a, and the mean-square pressure, which can be written $|P|^2$, is equal to $4\rho c\Pi/a_a$. Thus we can write, to the first order in the small quantity S_0C_0/a_a,

$$|p_a|^2 \simeq |P|^2 \frac{a_a}{a_a + 8S_0C_0}$$

$$|p_b|^2 \simeq |P|^2 \frac{8S_0C_0}{a_b + 8S_0C_0} \frac{1}{1 + (8S_0C_0/a_a)}$$

(10.4.19)

for the mean-square pressure in the two rooms. This can now be compared with the formulas of Eqs. (10.4.15) and (10.4.16), obtained for resonance conditions at lower frequencies. Equations (9.5.16) and (10.4.15) indicate that $a_a = 8\xi_a S_N{}^a$ when ξ_a, the specific conductance of the walls of room a, is small. Multiplying top and bottom of Eqs. (10.4.19) by $\Lambda_N{}^a/4\rho c$ and remembering that $2\Lambda_N{}^a \simeq (\iint \phi_N{}^a\, dS)^2$ on the average, we see that, very approximately, the geometrical equation for $|p_a|^2$ is equivalent to

$$|p_a|^2 \simeq |P|^2 \frac{L_w{}^a}{L_w{}^a + L_c{}^a}$$

which is the same as the square of Eqs. (10.4.15) when $L_c{}^a \ll L_w{}^a$. A similar comparison can be made between the second of Eqs. (10.4.19) and the square of Eq. (10.4.16).

Although the subject of coupled systems, and the various tricks useful in working out their behavior, could be continued well-nigh indefinitely, we must turn to other matters. Between the text and the problems of this chapter, the more useful results and techniques have been at least touched on. Next we take up acoustic propagation in moving media.

Problems

I   A flexible tube under tension is filled with fluid of density ρ and adiabatic compressibility $\kappa = 1/\rho c_a{}^2$. It has inside radius a when at equilibrium with the static pressure P of the fluid, the longitudinal tension, and the tube's transverse elasticity. When an acoustic wave travels in the fluid, the acoustic pressure $p\ (\ll P)$ stretches the tube so that its radius is $a + \eta(\eta \ll a)$. Alternatively, a dilation wave can travel along the tube walls; the resulting change of radius η will expand or compress the fluid inside. Show that the first-order equations of motion relating the acoustic pressure p, the net flow of fluid $U = \pi(a + \eta)^2 u$ along the tube, and the change η in radius of the tube, for wave motion of wavelength long compared with a, are

$$2\pi a \epsilon \frac{\partial^2 \eta}{\partial t^2} = 2\pi a T \frac{\partial^2 \eta}{\partial z^2} + \pi a^2 p - 2\pi a K \eta$$

or

$$\frac{\partial^2 \eta}{\partial z^2} = \frac{1}{c_t{}^2} \frac{\partial^2 \eta}{\partial t^2} + \left(\frac{\omega_0}{c_t}\right)^2 \eta - \frac{a}{2T} p$$

$$-\rho \frac{\partial U}{\partial t} = \pi a^2 \frac{\partial p}{\partial z} \qquad -\frac{\partial U}{\partial z} = \pi a^2 \kappa \frac{\partial p}{\partial t} + 2\pi a \frac{\partial \eta}{\partial t}$$

or

$$\frac{\partial^2 p}{\partial z^2} = \frac{1}{c_a^2}\frac{\partial^2 p}{\partial t^2} + 2\frac{\rho}{a}\frac{\partial^2 \eta}{\partial t^2}$$

where ϵ, T, and K are the effective mass per unit length, tension, and transverse stiffness, respectively, of the tube walls; $c_t = \sqrt{T/\epsilon}$, the velocity of the dilation wave in an empty tube; $\omega_0 = \sqrt{K/\epsilon}$ is 2π times the frequency of dilation oscillation of the empty tube; and $c_a = \sqrt{1/\rho\kappa}$ is the acoustic velocity of the free fluid. Show that two simple-harmonic waves, of form $e^{i(\omega/c)z - i\omega t}$, are possible in the filled tube; that if $\omega c_t^2/\epsilon\omega^2 \ll 1$, the phase velocities of the two waves are, approximately,

$$c_a \left\{ \frac{\omega^2[1 - (c_t/c_a)^2] - \omega_0^2}{\omega^2[1 - (c_t/c_a)^2] - \omega_0^2 \mp (\rho c_a^2/\epsilon)} \right\}^{\frac{1}{2}}$$

Compare this with the formulas of Eq. (9.1.16), when $c_t \ll c_a$. Show that in the case of the minus sign in the denominator, most of the energy is carried by the fluid; for the plus sign, the tube walls carry most of the energy.

2 Suppose the flexible wall of the tube of Prob. 1, in addition to having fluid inside, is in contact with the same fluid outside the tube, extending to infinity. Show that the equation for $\partial^2\eta/\partial z^2$ has an additional term $(a/2T)p_0$, where p_0 is the pressure just outside the tube. Show also that the radial fluid velocity just outside the tube, $\partial\eta/\partial t$, must equal $\dfrac{1}{i\omega p}\dfrac{\partial p_0}{\partial r}$. For a simple-harmonic wave, such that $p = D \exp[i(\omega/c)z - i\omega t]$, $\eta = E \exp[i(\omega/c)z - i\omega t]$, and $p_0 = GH_0^{(1)}(\varkappa r) \exp[i(\omega/c) - i\omega t]$, where $\varkappa^2 = (\omega/c_a)^2 - (\omega/c)^2$, show that the formulas for phase velocity given in Prob. 1 are modified by changing the terms $1 - (c_t/c_a)^2$ to

$$1 - \left(\frac{c_t}{c_a}\right)^2 + \frac{\rho a^2}{4\epsilon}\left\{\ln\frac{1 - (c_t/c_a)^2 - (\omega_0/\omega)^2}{\rho a^2/\epsilon C^2} + i\pi\right\}$$

or

$$1 - \left(\frac{c_t}{c_a}\right)^2 - \frac{\rho a^2}{4\epsilon}\ln\left\{\frac{C^2\omega^2 a^2}{c_t^2}\frac{\omega^2[1 - (c_t^2/c_a^2)] - \omega_0^2 - (\rho c_a^2/\epsilon)}{\omega^2[1 - (c_t^2/c_a^2)] - \omega_0^2} - \frac{C^2\omega^2 a^2}{c_a^2}\right\}$$

for the fluid wave or the tube wave, respectively. [For the value of C, see Eq. (10.1.6).] Explain the presence of the imaginary term in the first expression; why should the fluid wave attenuate, and why should the tube wave not attenuate? Show that there is a third wave, also nonattenuating, thus with an imaginary value for $\varkappa$, with a velocity $c = [(1/c_a)^2 + (\varkappa/\omega^2 a^2)]^{-\frac{1}{2}}$, and with x the solution of $x = (1/c) \exp\{(4/\rho a^2)[1 - (c_t/a\omega)^2 x - (c_t/c_a)^2 - (\omega_0/\omega)^2 + (a^2\rho/\epsilon x)]\}$.

3 Show that an alternative solution of Eq. (10.1.16) is

$$\eta(z) = -\frac{\mu a^2 k^2}{2\pi}\int_0^l g(z \mid z_0)\, dz_0 \int_0^l G_\omega(z_0 \mid z_1)\eta(z_1)\, dz_1$$

where

$$G_\omega(z_0 \mid z_1) = -\int_{-\infty}^{\infty} e^{iK(z_0 - z_1)}[1 + a(K^2 - k^2)\ln(Ca\sqrt{K^2 - k^2})]\, dK$$

and $g(z \mid z_0)$ is the Green's function for the string of Eqs. (4.4.14) and (4.5.39) (we have shifted the origin from the center of the string to one end). Show that this homogeneous integral equation may be turned into a variational expression for the Nth eigenfunction by setting $(\omega l/c_s)^2 - (\pi N)^2 = (\mu\alpha^2 k^2 l/\pi)\Phi_N$, where

$$\Phi_N = \int_0^l dz_0 \sin\left(\frac{\pi N x_0}{l}\right) \int_0^l G_\omega(z_0 \mid z_1)\eta(z_1)\,dz_1$$

Show that the variational expression for Φ_N [see Eq. (9.4.13)] is

$$\Phi_N = 2\int \sin\left(\frac{\pi N z}{l}\right) \int G_\omega(z \mid z_0)\eta(z_0) - \int \eta(z) \int G_\omega(z \mid z_0)\eta(z_0)$$

$$- \frac{1}{\pi}\mu\alpha^2 k^2 l \int \eta(z) \int G_\omega(z \mid z_1) \int g_N(z_1 \mid z_2) \int G_\omega(z_2 \mid z_0)\eta(z_0)$$

where

$$g_N(z \mid z_0) = \sum_{n \neq N} \frac{\sin(\pi n z/l)\sin(\pi n z_0/l)}{(\pi n)^2 - (\omega l/c_s)^2}$$

Show that, if we choose η to be approximated by $A \sin(\pi N z/l)$ and vary A, the resulting approximate solution is

$$\Phi_N \simeq \int \sin\left(\frac{\pi N z}{l}\right) \int G_\omega(z \mid z_0) \sin\left(\frac{\pi N z_0}{l}\right)\left\{1 + \frac{\mu\alpha^2 k^2 l/\pi}{\int \sin(\pi N z/l) \int G_\omega \sin(\pi N z_0/l)}\right.$$

$$\left. \times \int \sin\left(\frac{\pi N z}{l}\right) \int G_\omega \int g_N \int G_\omega \sin\left(\frac{\pi N z_0}{l}\right)\right\}$$

Compare this result with Eq. (10.1.19), coming from the Rayleigh-Ritz variational procedure. Would the present method be more dependable for the higher modes?

4 Show that the asymptotic form for the pressure wave radiated by the plate of Eqs. (10.1.27), in the region $x > 0$, $z > 0$, for $k < \gamma$, is

$$p_\omega \to \frac{i\mu\gamma F_0}{8\sqrt{\gamma^2 - k^2}} \left\{ \exp(i\gamma x - z\sqrt{\gamma^2 - k^2}) \right.$$

$$+ \sqrt{\frac{\gamma^2 - k^2}{\gamma^2 + k^2}} \exp\left[-\gamma x + iz\sqrt{\gamma^2 + k^2} - \frac{\mu\gamma^2 z}{4(\gamma^2 + k^2)}\right]$$

$$\left. + \frac{4\gamma^3\sqrt{\gamma^2 - k^2}\,e^{ikr}}{\sqrt{2\pi i k r}(\gamma^4 - k^4\cos^4\phi)k\sin\phi} \right\}$$

Compare this with the expressions of Eqs. (10.1.30) and (10.1.31) for $\gamma < k$. What is the plate displacement for $\gamma > k$?

5 An infinite membrane of areal density σ, under tension T per unit length, is in contact on both sides with a medium of infinite extent, of density ρ, and acoustic velocity c. Show that linear simple-harmonic transverse waves of displacement $\eta = A \exp(iKx - i\omega t)$ are possible on the membrane if the wavenumber K has the value

$$K \simeq k_m\left(1 + \frac{\rho}{\sigma\sqrt{k_m^2 - k^2}}\right) \quad \text{when} \quad \frac{T}{\sigma} \equiv c_m^2 < c^2, \quad \frac{2\rho}{\sigma\sqrt{k_m^2 - k^2}} \ll 1$$

and where $k = \omega/c$, and $k_m = \omega/c_m$. What is the phase velocity of the wave?

What is the pressure distribution in the medium above the membrane? Show that, when $c_m > c$, two waves are possible, with two different wavenumbers,

$$K \simeq k_m \left(1 + \frac{i\rho}{\sigma \sqrt{k^2 - k_m^2}}\right) \quad \text{and} \quad K_s \simeq k\left[1 + \frac{(2\rho k_m^2/\sigma k)^2}{(k^2 - k_m^2)^2}\right]$$

What is the physical significance of the imaginary term in the first expression? What is the pressure distribution in the medium in both cases? Why is the K_s possible when $c_m > c$, and not possible when $c_m < c$?

6 The membrane of Prob. 5 is driven by a line force $F_0 \delta(x) e^{-i\omega t}$. Show that the asymptotic form for the pressure wave radiated in the region $z > 0$, $x > 0$, when $2\rho/\sigma\sqrt{(k_m^2 - k^2)} \ll 1$, is

$$p_+ \simeq \frac{ik_m \rho F_0}{\sigma\sqrt{k_m^2 - k^2}} \left\{ \exp(ik_m x - z\sqrt{k_m^2 - k^2}) \right.$$

$$\left. + \frac{2k_m \sqrt{k_m^2 - k^2}\, e^{ikr}}{\sqrt{2\pi ikr}(k_m^2 - k^2 \cos^2\phi)} \right\} e^{-i\omega t} \qquad c_m < c$$

$$\simeq \frac{k_m \rho F_0}{\sigma\sqrt{k^2 - k_m^2}} \left\{ u(k\cos\phi - k_m)\exp\left[ik_m x - \frac{\rho k_m x}{\sigma\sqrt{k^2 - k_m^2}} + iz\sqrt{k^2 - k_m^2}\right.\right.$$

$$\left. + \frac{\rho k_m^2 z}{\sigma(k^2 - k_m^2)}\right] - \frac{4i\rho k_m^3}{k\sigma(k^2 - k_m^2)}\exp\left[iK_s x - \frac{2\rho k_m^2 z}{\sigma(k^2 - k_m^2)}\right]$$

$$\left. - \frac{4ikk_m\sqrt{k^2 - k_m^2}\sin(\phi)\,e^{ikr}}{\sqrt{2\pi ikr}[(k^2\cos^2\phi - k_m^2)k\sin\phi - i(2\rho k_m^2/\sigma)]} \right\} e^{-i\omega t} \qquad c_m > c$$

where $c_m^2 = T/\sigma$, $c^2 = 1/\rho\kappa$, $k = \omega/c$, $k_m = \omega/c_m$, $r^2 = x^2 + z^2$, and $\tan\phi = z/x$; $u(w) = 0$ when $w < 0$, $= 1$ when $w > 0$; and K_s is as given in Prob. 5. What changes in the formulas are needed for $x < 0$? For $z < 0$? Compute the transverse displacement of the membrane. Discuss the behavior of the radiated wave for r large, $c_m > c$, and ϕ near $\psi = \cos^{-1}(k_m/k)$.

7 The membrane of Prob. 5 has incident on it a plane sound wave

$$p_i = P_i \exp[ik(x\sin\theta - z\cos\theta) - i\omega t]$$

at angle of incidence θ. Show that the reflected and transmitted wave amplitudes P_r and P_t and the amplitude A of transverse motion of the membrane are given by Eq. (10.1.32). except that now quantity Q is $[1 - (c_m/c)^2 \sin^2\theta] \times (\sigma k/2\rho)\cos\theta$. At what angles of incidence is the membrane transparent to the wave?

8 Show that the expression for the p_t of Eq. (10.1.38), for $k < \gamma$, is

$$p_t \to \frac{\rho ck S_\omega \mu \gamma^2}{8(\gamma^2 - k^2)} \left\{ \frac{\exp[i\gamma r - (l - z)\sqrt{\gamma^2 - k^2}]}{\sqrt{2\pi i\gamma r}} \right.$$

$$+ \frac{\gamma^2 - k^2}{\gamma^2 + k^2}\frac{\exp[-\gamma r + i(l - z)\sqrt{\gamma^2 - k^2}]}{\sqrt{2\pi\gamma r}}$$

$$\left. + \frac{4\sqrt{2}\gamma^2(\gamma^2 - k^2)(e^{ikR}/4\pi R)}{(\gamma^4 - k^4\sin^4\theta)k\cos\theta} \right\}$$

9 The spherical wave from a simple source S_ω and frequency $\omega/2\pi$ at $z = l$, $x = y = 0$, falls on the membrane of Prob. 5.   Show that the asymptotic form of the transmitted wave ($z < 0$) is

$$p_t \to \frac{k_m{}^2 \rho S_\omega}{\sigma(k_m{}^2 - k^2)} \left\{ \frac{1}{2\pi i k_m r} \exp\left[ik_m r - (l - z)\sqrt{k_m{}^2 - k^2}\right] \right.$$
$$\left. + \frac{(k_m{}^2 - k^2)(4e^{ikR}/2\pi R)}{(k_m{}^2 - k^2\cos^2\theta)k\sin\theta + i(2\rho k_m{}^2/\sigma)} \right\} e^{-i\omega t} \qquad c_m < c$$

$$\to \frac{-k_m{}^2 \rho S_\omega}{\sigma(k^2 - k_m{}^2)} \left\{ \frac{u(k\cos\theta - k_m)}{\sqrt{2\pi i k_m r}} \exp\left[ik_m r - \frac{\rho k_m r}{\sigma\sqrt{k^2 - k_m{}^2}} \right.\right.$$
$$\left. + i(l - z)\sqrt{k^2 - k_m{}^2} + \frac{\rho k_m{}^2(l - z)}{\sigma(k^2 - k_m{}^2)} \right]$$
$$+ \frac{2}{\sqrt{2\pi i k r}} \exp\left[ikr - \frac{2\rho k_m{}^2(l - z)}{\sigma(k^2 - k_m{}^2)} \right]$$
$$\left. + \frac{4(k^2 - k_m{}^2)(e^{ikR}/2\pi R)}{(k^2\cos^2\theta - k_m{}^2)k\sin\theta - i(2\rho k_m{}^2/\sigma)} \right\} e^{-i\omega t} \qquad c < c_m$$

where $r^2 = x^2 + y^2$, $R^2 = x^2 + y^2 + z^2$, and $\cos\theta = r/R$.   Compare this result with the solution of Eq. (10.1.38) for transmission through a plate.

10 Suppose the circular plate of Eq. (5.3.6) is substituted for the membrane of Eq. (10.2.1).   Write out the resulting modification of the integral equation and Green's function of Eq. (10.2.5).   Use the results of Eq. (5.3.10) to obtain the appropriate modification of Eq. (10.2.6).   What are the new expressions for the auxiliary functions H_m and S_m of Eqs. (10.2.22) and (10.2.23)?   In terms of them, write out the modified formulas (10.2.24), (10.2.25), and (10.2.27). What are the differences from the membrane in spacing of the first few resonances and antiresonances?

11 For $c/c_p = \frac{1}{2}$, $\gamma = \rho a/\sigma = 1$, plot the transmission factor of Fig. 10.12 against θ for $ka = 3$ and $ka = 6$.   Plot the angle distribution of Fig. 10.13 for $ka = 6$. Discuss the effects of the various resonances and antiresonances.

12 A Helmholtz resonator, consisting of a hollow sphere of radius a, centered at the origin, with a circular hole of angular aperture θ_0, is shown in the figure.

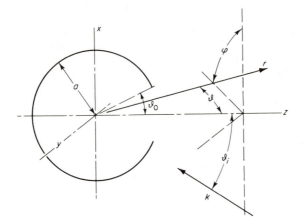

A plane wave of frequency $\omega/2\pi$ is incident on the sphere, its direction being in the xz plane, in the negative direction along a line at angle θ_i to the z axis. If the radial component of the air velocity in the hole is $u_a(\vartheta,\varphi)$, show that the pressure inside the sphere is, for $r < a$,

$$p_- = \sum_{m=0}^{\infty} \epsilon_m \cos (m\varphi) \sum_{n=m}^{\infty} (2n + 1) \frac{(n - m)!}{(n + m)!} S_n{}^m(\theta_i) P_n{}^m(\cos \vartheta) \frac{j_n(kr)}{j_n'(ka)}$$

where the spherical harmonics $Y_{mn}{}^1(\vartheta,\varphi) = \cos (m\varphi) P_n{}^m(\cos \vartheta)$ are defined in Eqs. (7.2.7). Show that the incident plane wave is

$$p_i = A \sum_m \epsilon_m \cos (m\varphi) \sum_n (-i)^n (2n + 1) \frac{(n - m)!}{(n + m)!} P_n{}^m (\cos \theta_i) P_n{}^m (\cos \vartheta) j_n(kr)$$

and thus that the total pressure field outside the sphere is

$$p_+ = \sum_m \epsilon_m \cos (m\varphi) \sum_n (2n + 1)(-i)^n \frac{(n - m)!}{(n + m)!} P_n{}^m (\cos \vartheta)$$

$$\times \left\{ A P_n{}^m (\cos \theta_i) \left[j_n(kr) - \frac{j_n'(ka)}{h_n'(ka)} h_n(kr) \right] + i^n S_n{}^m \frac{h_n(kr)}{h_n'(ka)} \right\}$$

$$\rightarrow p_i + i \frac{e^{ikr}}{kr} \sum_{m,n} \epsilon_m (2n + 1) \frac{(n - m)!}{(n + m)!} \left\{ A(-1)^n \frac{j_n'(ka)}{h_n'(ka)} P_n{}^m (\cos \theta_i) \right.$$

$$\left. - (-i)^n \frac{S_n{}^m(\theta_0)}{h_n'(ka)} \right\} \cos (m\varphi) P_n{}^m (\cos \vartheta)$$

The effect of viscosity is approximately to produce a pressure difference $p_- - p_+ = \rho c k d_v u_a$ across the surface of the opening.

If now we assume that $u_a = (U/\sqrt{2}) \sin (\tfrac{1}{2}\theta_0) (\cos \vartheta - \cos \theta_0)^{-\frac{1}{2}}$ (why make this assumption?), the coefficients $S_n{}^m$ become

$$S_n{}^m \equiv \frac{i\rho c}{4\pi} \int_0^{2\pi} \cos (m\varphi) \, d\varphi \int_0^{\theta_0} u_a(\varphi,\vartheta) P_n{}^m(\cos \vartheta) \sin \vartheta \, d\vartheta$$

$$= \frac{i\rho c U}{2n + 1} \sin (\tfrac{1}{2}\theta_0) \sin [(n + \tfrac{1}{2})\theta_0)] \, \delta_{m0}$$

The constant U could be determined by a variational procedure, but if we take the cruder approximation, requiring that the equation $p_- - p_+ = \rho c k d_v u_a$ at $\vartheta = 0$, show that the equation determining U is

$$- \frac{\rho c}{k^2 a^2} U \sin (\tfrac{1}{2}\theta_0) \sum_n \sin [(n + \tfrac{1}{2})\theta_0)] \left[\frac{1}{j_n'(ka) h_n'(ka)} + k^2 a^2 k d_v \right]$$

$$= A\Psi(\theta_i) \equiv \frac{iA}{k^2 a^2} \sum_n (2n + 1) \frac{(-i)^n P_n(\cos \theta_i)}{h_n'(ka)}$$

where

$$\Psi(\theta_i) = \sum_n \frac{2n + 1}{k^2 a^2} \frac{e^{-i\delta_n(ka)}}{B_n(ka)} (-i)^n P_n(\cos \theta_i) \xrightarrow[ka \to 0]{} 1 - \tfrac{3}{2} ika \cos \theta_i$$

Quantities B_n and δ_n are defined in Eqs. (7.2.15) and given in Table VIII. Series $A\Psi(\theta_i)$ is the pressure which would be present at the point $r = a$, $\vartheta = 0$, if there were no hole in the sphere. It can be considered to be an approximate expression for the driving pressure, forcing the flow U through the hole.

Therefore the coefficient of $-U$, on the first line, is a rough approximation to the impedance $\rho c \zeta$, per unit area, of the hole. Using the comparison series

$$\sum_n \sin\left[(n + \tfrac{1}{2})\theta_0\right] = \frac{1}{2 \sin\left(\tfrac{1}{2}\theta_0\right)}$$

$$\sum_n \frac{(2n + 1)\sin\left(n + \tfrac{1}{2}\right)\theta_0}{(n + \delta_{0n})(n + 1)} = (\pi - \theta_0)\cos\left(\tfrac{1}{2}\theta_0\right)$$

to approximately cancel out the terms in $1/j_n(ka)h_n'(ka)$, show that the impedance of the opening is $3(\rho c/ka) \sin^2(\tfrac{1}{2}\theta_0)(R - iX)$, where

$$X = \tfrac{1}{3}k^2a^2(\pi - \theta_0)\cot\left(\tfrac{1}{2}\theta_0\right) - \operatorname{Re} K$$

and

$$R = \tfrac{1}{6}k^2\, ad_v \csc^2\left(\tfrac{1}{2}\theta_0\right) - \operatorname{Im} K$$

where

$$K = \frac{1}{3ka} \sum_{n=0}^{\infty} \frac{\sin\left(n + \tfrac{1}{2}\right)\theta_0}{\sin\left(\tfrac{1}{2}\theta_0\right)}\left[\frac{\cot \delta_n}{B_n{}^2} + (ka)^3 \frac{2n + 1}{(n + 1)(n + \delta_{0n})} - \frac{i}{B_n{}^2}\right]$$

is a small quantity when $2\pi ka \ll 1$. Show that, at low frequencies, $K \simeq 1 - \tfrac{1}{15}k^2a^2 - \dfrac{i}{3}k^3a^3$, so that the impedance becomes

$$4\pi a^2 \sin^2\left(\tfrac{1}{2}\theta_0\right)[R_s - i\omega L_s + i(1/\omega C_s)]$$

where

$$R_s = \frac{\rho c}{4\pi a^2}\left[\tfrac{1}{2}kd_v \csc^2\left(\tfrac{1}{2}\theta_0\right) + k^2a^2\right] \qquad C_s = \frac{4\pi a^3}{3\rho c^2}\left(1 + \tfrac{1}{15}k^2a^2\right)$$

$$L_s = \frac{\rho}{4\pi a}(\pi - \theta_0)\cot\left(\tfrac{1}{2}\theta_0\right)$$

Point out the physical significance of each term in these analogous impedances. At what value of ω is the resonance? Is this the only resonance of the resonator (i.e., is this the only value of ω for which $\operatorname{Re} K = 0$)? Since the formulas for the velocity in the opening result in

$$S_n{}^m \simeq -\frac{Aka\Psi(\theta_i)}{3(2n + 1)}\frac{\sin\left(n + \tfrac{1}{2}\right)\theta_0}{\sin\left(\tfrac{1}{2}\theta_0\right)}\frac{\delta_{m0}}{X - iR}$$

write out the expression for the pressure amplitude at the center of the sphere. At what frequency is this maximum? What is its value then? Show that the scattering and absorption cross sections of the resonator at longer wavelengths are

$$\Sigma_s \simeq \tfrac{1}{3}\pi a^2(ka)^4(1 - \tfrac{3}{5}k^2a^2) + \frac{\pi a^2}{R^2 + X^2}\left(\tfrac{2}{9}(ka)^5X\frac{\sin\left(\tfrac{3}{2}\theta_0\right)}{\sin\left(\tfrac{1}{2}\theta_0\right)}\right.$$

$$\left. + (ka)^6\left\{1 + \tfrac{1}{5}X^2 + \frac{1}{27}\left[\frac{\sin\left(\tfrac{3}{2}\theta_0\right)}{\sin\left(\tfrac{1}{2}\theta_0\right)}\right]^2\right\} - \tfrac{1}{3}R^2\right)$$

$$\Sigma_a \simeq 4\pi a^2 \frac{kd_v\,|\Psi(\theta_i)|^2\,k^2a^2}{9(R^2 + X^2)} \csc^2(\tfrac{1}{2}\theta_0) \qquad ka \ll 1$$

Again explain the origin of each term. What are the peak values of Σ_s and Σ_a at resonance, when $X = 0$? Compare them with the geometrical cross section πa^2.

13 The thin plate of Eq. (10.1.21) is periodically loaded with line impedances, like the membrane of Sec. 10.3, so that its equation of motion, when driven by a force $F_x \exp(i\kappa x - i\omega t)$, is

$$\frac{\partial^4 \eta}{\partial x^4} - \gamma^4 \eta = \frac{\gamma^4 F_x}{2\omega^2 h \rho_p} e^{i\kappa x} + \frac{i\gamma^4 Z}{2\omega h \rho_p} \eta \sum_n \delta(x - nl)$$

Show that the steady-state displacement of the plate is

$$\eta = \frac{\gamma^4 F_x / 2\omega^2 h \rho_p}{\kappa^4 - \gamma^4} e^{i\kappa x - i\omega t} \left[1 + \frac{iZ(\gamma l)^4}{2\omega\, h \rho_p} \frac{C}{Q} \sum_{m=-\infty}^{\infty} \frac{e^{(2\pi i/l)x}}{(\kappa l + 2\pi m)^4 - (\gamma l)^4} \right]$$

where

$$C(\kappa l) = [\cos(\kappa l) - \cos(\gamma l)][\cos(\kappa l) - \cosh(\gamma l)]$$

$$Q(\kappa l) = \left[\cos(\kappa l) - \cos(\gamma l) + \left(\frac{i\gamma Z}{8\omega h \rho_p} \right) \sin(\gamma l) \right]$$

$$\times \left[\cos(\kappa l) - \cosh(\gamma l) - \left(\frac{i\gamma Z}{8\omega h \rho_p} \right) \sinh(\gamma l) \right] - \left(\frac{\gamma Z}{8\omega h \rho_p} \right)^2 \sin(\gamma l) \sinh(\gamma l)$$

Show that the free vibration of the plate occurs when $Q(\kappa l) = 0$. Sketch the plot, corresponding to Figs. 10.15 and 10.16, for $u_q = \kappa_q l / \pi$ versus $\pi X / 8\omega h \rho_v$ and $v = \gamma l / \pi$, for $Z = R - iX$ purely reactive ($R = 0$). Is the plot periodic in v?

14 Show that the Green's function for the periodically loaded plate of Prob. 13 is

$$G(x \mid x_0) = \frac{l^3}{\pi^4} \sum_m \frac{e^{ik_m x}}{\Delta_m} \psi_m(-x_0) = \frac{l^3}{\pi^4} \sum_m \frac{e^{-ik_m x_0}}{\Delta_m} \psi_m(x)$$

where

$$\psi_m(x) = e^{ik_n x} + iB \sum_n \frac{e^{ik_n x}}{\Delta_n} \qquad \Delta_n = u_n^2 - v^2 \qquad u_n = u + 2n = \frac{k_n l}{\pi}$$

$$k_n = \kappa + \frac{2\pi n}{l} \qquad u = \frac{\kappa l}{\pi} \qquad v = \frac{\gamma l}{\pi}$$

$$B = \frac{v^4 Z}{2\omega l h \rho_p} \frac{C(\kappa l)}{Q(\kappa l)}$$

functions C and Q being defined in Prob. 13.

15 Carry through the calculations of Eqs. (10.3.13) to (10.3.18) for the plate of Probs. 13 and 14, to obtain the wave velocities for the periodically loaded plate in contact with the medium.

16 Show that the displacement of the periodically supported membrane of Sec. 10.3, when in contact with the medium and when also acted on by a driving force $F(\kappa) \exp(i\kappa x - i\omega t)$, is

$$\eta_\kappa = \frac{\kappa^2 l F(\kappa)}{\pi^2 \sigma c^2} \left[\frac{w_0 e^{i\kappa x}}{\Delta_0 w_0 - i\mu l} + \sum_n \frac{iBw_n e^{ik_n x}}{\Delta_0(\Delta_n w_n - i\mu l)} \frac{1 + \dfrac{2i\rho l v^2 / \pi\sigma}{\Delta_0 w_0 - i\mu l}}{1 - (4i\zeta\rho l v^3 / \pi^2 \sigma)\Gamma(u)} \right] e^{-i\omega t}$$

where the quantities w_n, Δ_n, B, etc., are as defined in Eqs. (10.3.10), and $\Gamma(u)$ is as defined in Eq. (10.3.15). What is the pressure wave associated with this motion?

17 Plot the mean-square pressure amplitudes $|T_0|^2$ and $|T_m|^2$ of the transmitted and diffracted waves of Eq. (10.3.22), for those waves which go to $z \to -\infty$, for $\zeta = 10$, $2\pi/kl = \frac{1}{3}$, $g = \frac{1}{10}$, and $\alpha = \frac{1}{2}$, as function of the angle of incidence θ.

18 Carry out the calculations resulting in Eqs. (10.3.15) and (10.3.16) for the periodically loaded plate of Probs. 13 and 14.

19 Obtain the equivalent of Eq. (10.3.22) for the diffraction of sound through the periodically supported plate of Probs. 13 and 14.

20 A cylindrical enclosure is bounded by rigid walls at $r = a$, $z = b$, and $z = -d$. It is divided into two parts by a thin plate of the sort described in Sec. 5.3, along the plane $z = 0$.   If the coupled standing waves in the two volumes have common angular factors $\cos(m\phi)$ about the cylinder axis, show that the displacement of the plate, $Y_m(r)\cos(m\varphi)$, is related to the pressure difference $(p_- - p_+)_{z=0} = P_m(r)\cos(m\varphi)$ by the equation

$$Y_m(r) = \frac{\gamma^4}{2\omega^2 h \rho_p} \int_0^a G_m(r \mid r_0) P_m(r_0) r_0 \, dr_0$$

Write out the series for G_m. Show that a closed form for the integral is possible if P_m were to equal $J_m(qr)$.

$$\int_0^a G_m(r \mid r_0) J_m(qr) r_0 \, dr_0 = \frac{1}{(q^4 - \gamma^4)\Delta_m(\gamma a)} \left\{ \Delta_m J_m(qr) \right.$$
$$+ \left[\frac{q}{\gamma} J'_m(qa) I_m(\gamma a) - J_m(qa) I'_m(\gamma a) \right] J_m(\gamma r)$$
$$\left. + \left[J_m(qa) J'_m(\gamma a) - \frac{q}{\gamma} J'_m(qa) J_m(\gamma a) \right] I_m(\gamma r) \right\}$$

where

$$\Delta_m(\gamma a) = J_m(\gamma a) I'_m(\gamma a) - J'_m(\gamma a) I_m(\gamma a)$$

(resonance of the plate occurs when $\Delta_m = 0$). Show that P_m is given in terms of Y_m by

$$P_m(r) = -\frac{2\omega^2 \rho}{a^2} \int_0^a [g_b(r \mid r_0) + g_d(r \mid r_0)] Y_m(r_0) r_0 \, dr_0$$

$$g_b = \sum_{ij} \frac{\epsilon_j}{b\Lambda_{ij}} \frac{J_m(\pi\alpha_{mi} r/a) J_m(\pi\alpha_{mi} r_0/a)}{(\pi\alpha_{mi}/a)^2 + (\pi j/b)^2 - k^2} \qquad k = \frac{\omega}{c}$$

and g_d has d instead of b.   The quantities Λ and α are as defined in Eqs. (9.2.22) and (9.2.23).   Suppose that b does not differ much from d, so that the resonance frequencies for $i = n$, $j = l$ are nearly the same for the two cavities. Use the procedures of Sec. 10.4 to show that an approximate expression for the resonance frequencies $kc/2\pi$ closest to $c[(\alpha_{mn}/2a)^2 + (l/2b)^2]^{\frac{1}{2}}$ is

$$\frac{\gamma^4 a^4 \rho \epsilon_l}{2h\rho_p(\pi^4 \alpha_{mn}^4 - \gamma^4)} \left[\frac{4\gamma^3 a^3 J'_m(\gamma a) I'_m(\gamma a)}{(\pi^4 \alpha_{mn}^4 - \gamma^4)(1 - m^2/\pi^2 \alpha^2_{mn})\Delta_m(\gamma a)} - 1 \right]$$

$$\times \left[\frac{1/b}{k^2 - (\pi\alpha_{mn}/a)^2 - (\pi l/b)^2} + \frac{1/d}{k^2 - (\pi\alpha_{mn}/a)^2 - (\pi l/d)^2} \right] = 1$$

when $\gamma^4 a^5 \rho/2h\rho_p$ is small.   Work out expressions for the coupling coefficient, and discuss the effect of plate resonance on the resonances of the coupled rooms. What is the formula for the forced acoustic motion of the two enclosures?

21 Consider a "double wall" consisting of two parallel stretched membranes separated a distance d. A plane wave is incident at an angle of incidence θ on the wall. Determine the transmitted wave and the sound field between the membranes. Show that the transmitted wave amplitude P_t can be written in terms of the incident pressure amplitude P_i as

$$P_t = P_i \frac{1}{(1 + f_1)(1 + f_2) - f_1 f_2 \exp(2ikd \cos \theta)}$$

where

$$f = i \frac{\sigma \omega}{2 \rho c} \left[1 - \left(\frac{v}{c}\right)^2 \sin^2 \theta \right] \cos \theta$$

where σ = membrane mass per unit area
$\quad v = \sqrt{T/\sigma}$ = velocity of free transverse waves on membrane
$\quad \rho$ = density of air
$\quad c$ = speed of sound in air

The indices 1 and 2 refer to the two membranes which may have different values of ϵ and v.

CHAPTER

11

ACOUSTICS IN MOVING MEDIA

11.1 PLANE WAVES IN UNIFORM FLOW

Most of the material discussed so far has been devoted to the acoustics of a homogeneous fluid at rest, and this indeed covers a substantial portion of the problems of interest in acoustics. The restriction that the fluid be at rest may seem insignificant at first since the question of (uniform) motion or rest is merely a matter of a coordinate transformation. However, when we consider also the sound sources, receivers, and boundaries involved in a specific problem, it is clear that a relative motion between the fluid and the various material bodies involved cannot be eliminated by a coordinate transformation; a transformation that brings the medium to rest sets the sources and the boundaries in motion.

For example, let the fixed boundary be an ordinary straight-tube section open at both ends. If the tube length is L and if there is no flow, the time of travel for a plane-wave pulse back and forth in the tube is $T_1 = 2L/c$, and therefore the fundamental plane-wave resonance frequency of the tube is $f_1 = c/2L$. In the presence of flow, however, the round-trip time for a pulse is $T_1' = [L/(c + V)] + [L/(c - V)] = T_1/(1 - M^2)$, where $M = V/c$, and the corresponding resonance frequency is $f_1' = (1 - M^2)f_1$. In other words, the presence of flow lowers the resonance frequency of the tube. This simple example demonstrates the need for a reexamination of the problems of generation, propagation, and detection of sound when motion relative to the medium is involved. In some sense the motion of the medium makes it nonisotropic, since the speed of sound in the laboratory system depends on the direction of propagation with respect to the direction of motion of the medium. If, in addition, the medium is inhomogeneous, so that the speed of sound and other medium characteristics vary from one point to another as in the atmosphere and in the sea, various refraction and scattering effects must be considered. Some of these problems have already been discussed in previous chapters for a medium at rest. Therefore the main emphasis in this chapter will be on the influence of fluid motion on sound propagation.

Doppler shift

The changing pitch of the sound from, say, a fire-engine siren as it moves by at high speed is familiar to all; this effect, the Doppler effect, is one of the most obvious influences of relative motion between source and medium. Similarly, if a wave train of sound in the medium is received by an observer in motion with respect to the medium, a Doppler change of frequency also will result. If, for example, the observer moves with speed of the wave train, he will find a stationary pattern with zero frequency. The Doppler effect is a purely kinematic phenomenon, and can be studied without resort to the dynamical equations of motion of the wave.

Let us consider, first, the case when an observer A is at rest in the laboratory system S, recording the pressure fluctuations in the sound wave. The medium moves in the x direction with a velocity V and carries a plane sound wave that travels in the medium, also in the x direction, with the speed c of sound. In other words, an observer in a frame of reference S' moving with the medium observes a sound wave traveling in the x direction with a speed c; the frequency to him is, of course, $f' = c/\lambda$, where λ is the wavelength. With respect to an observer at rest, however, the speed of the wave is $c + V$, and since λ is the same to both observers, the observed frequency is $f = (c + V)/\lambda = f'(1 + M)$, where $M = V/c$ is the *Mach number* of the flow. (In what follows we always use *primes* to denote quantities measured by an *observer moving with* the medium; *unprimed quantities* are measured by an *observer at rest* in the "laboratory coordinates.")

Next let us turn to the source that generated the plane wave of wavelength λ. If the source moves with the medium, the source frequency to produce the wave is $f' = c/\lambda$. If the source is at rest in S, however, in the time of one period T_S of the source the wave will travel in the x direction a distance $(c + V)T_S$, which must equal the wavelength λ. The relation between the wavelength and the frequency of the source is then $f_S = 1/T_S = (c + V)/\lambda = f'(1 + M)$, where, as before, $f' = c/\lambda$ is the frequency observed in a coordinate system S' moving with the medium. We recall that the frequency recorded by a stationary observer is also $f'(1 + M)$, the same as the frequency of the stationary source.

If the medium and the source are both at rest in S, but the receiver is moving in the negative x direction with a velocity V, we have a situation that kinematically is equivalent to the previously considered case of a stationary receiver and a medium moving (with the source) in the positive x direction with a velocity V.

It is not difficult to extend this to three dimensions. What needs to be kept in mind is that a steady-state wave, such as a plane wave, is not like a projectile; it is a moving pattern of wavefronts. The distance between these fronts, for example the wavelength λ, and the direction normal to the fronts will be the same for *all* observers, whether they are moving with respect to the medium or not. Other quantities will change, such as measured frequency, particle velocity (of course), and sometimes pressure, but the geometry of the pattern is invariant.

Thus the wavenumber vector **k**, of magnitude $2\pi/\lambda$ and direction normal to the wavefronts of a plane wave, is the same in the laboratory system S as it is in the comoving system S', moving with the medium with a velocity **V** with respect to S. In the comoving system the pressure wave is proportional to

$$\exp{(i\mathbf{k} \cdot \mathbf{r'} - i\omega't)}$$

where $\mathbf{r'}$ and ω' are measured with respect to S', and $k = \omega'/c$. The transformation to the laboratory system is given by

$$\mathbf{r} = \mathbf{r'} + \mathbf{V}t \qquad t = t'$$

so that, to an observer at rest in the laboratory system, the pressure will be proportional to

$$\exp{[i\mathbf{k} \cdot \mathbf{r} - i(\omega' + \mathbf{k} \cdot \mathbf{V})t]} = \exp{(i\mathbf{k} \cdot \mathbf{r} - i\omega t)}$$

producing the three-dimensional Doppler formula, relating the frequency $\omega'/2\pi$ in the comoving system to the frequency $\omega/2\pi$ in the laboratory system,

$$\omega = \omega' + \mathbf{k} \cdot \mathbf{V} = \omega'(1 + M\cos\theta) \tag{11.1.1}$$

where θ is the angle between the direction of motion of the medium and the normal to the plane wavefronts. In the laboratory system k is *not* equal to ω/c.

The equations of motion

With the kinematical preliminaries out of the way, we next should examine the equations of wave motion. In the coordinate system S' moving with the fluid, a sound field is described by the usual equations for mass and momentum balance [see Eq. (6.2.7)], which we write in the form

$$\frac{\partial\rho}{\partial t'} + \rho_0 \sum_i \frac{\partial u_i}{\partial x_i'} = 0 \qquad \rho_0 \frac{\partial u_i}{\partial t'} = -\frac{\partial p}{\partial x_i'} \tag{11.1.2}$$

where the subscript i refers to x, y, and z, respectively. If the velocity of the fluid has the components V_i, the transformation of coordinates from the comoving system S' to the laboratory system S is given by

$$x_i = x_i' + V_i t' \qquad t = t' \qquad \mathbf{V} = \text{const}$$

Thus, for any function f of space and time, we have

$$\frac{\partial f}{\partial x_i'} = \frac{\partial f}{\partial x_i}\frac{\partial x_i}{\partial x_i'} + \frac{\partial f}{\partial t}\frac{\partial t}{\partial x_i'} = \frac{\partial f}{\partial x_i}$$

In other words, the space derivatives are invariant under the transformation; $\partial/\partial x_i' = \partial/\partial x_i$. However, for the time derivative, we get

$$\frac{\partial f}{\partial t'} = \frac{\partial f}{\partial t}\frac{\partial t}{\partial t'} + \sum_i \frac{\partial f}{\partial x_i}\frac{\partial x_i}{\partial t'} = \frac{\partial f}{\partial t} + \sum_i V_i \frac{\partial f}{\partial x_i}$$

or

$$\frac{\partial}{\partial t'} = \frac{\partial}{\partial t} + \sum_i V_i \frac{\partial}{\partial x_i} \equiv \frac{\partial}{\partial t} + \mathbf{V} \cdot \boldsymbol{\nabla}$$

[compare this with Eq. (6.1.14)]. Using these transformations in (11.1.2), we obtain the equations with respect to the laboratory system S.

$$\left(\frac{\partial}{\partial t} + \mathbf{V}\cdot\boldsymbol{\nabla}\right)\rho + \rho_0 \sum_i \frac{\partial u_i}{\partial x_i} = 0 \qquad \rho_0\left(\frac{\partial}{\partial t} + \mathbf{V}\cdot\boldsymbol{\nabla}\right)u_i = -\frac{\partial p}{\partial x_i} \quad (11.1.3)$$

From these two equations plus the relationship $dp = c^2\,d\rho$, the wave equation follows.

$$\frac{1}{c^2}\left(\frac{\partial}{\partial t} + \mathbf{V}\cdot\boldsymbol{\nabla}\right)^2 p = \nabla^2 p$$

or

$$\frac{\partial^2 p}{\partial t^2} + 2\sum_i V_i \frac{\partial}{\partial t}\frac{\partial p}{\partial x_i} + \sum V_i V_j \frac{\partial^2 p}{\partial x_i\,\partial x_j} = c^2\nabla^2 p \qquad (11.1.4)$$

For harmonic time dependence Eq. (11.1.4) reduces to

$$\nabla^2 p + \left(\frac{\omega}{c}\right)^2\left(1 + i\frac{1}{\omega}\mathbf{V}\cdot\boldsymbol{\nabla}\right)^2 p = 0$$

For a plane wave traveling in a direction specified by the vector wavenumber $\mathbf{k}$, where $k = \omega'/c$ and ω' is the angular frequency as measured in the co-moving coordinate system S', we have $\nabla^2 = -k^2$ and $\mathbf{V}\cdot\boldsymbol{\nabla} = i\mathbf{V}\cdot\mathbf{k}$, again arriving at the relation between ω and ω' given in Eq. (11.1.1).

So far we have considered only the Doppler shift in the case of a plane wave, in which case only the two values of the received frequency $f_1 = f_0/(1 + M)$ and $f_2 = f_0/(1 - M)$ are found. When we consider a spherical wave, emitted by a source that moves by a distant observer, the Doppler-shifted frequency varies continuously between the two values f_1 and f_2. This case will be discussed in some detail in Sec. 11.2.

Fluid motion and transducer response

An important question concerns the measurement of a sound field in a moving fluid. If the transducers used for pressure measurements move with the fluid, the true static P and acoustic pressure p are measured. If the same transducers are kept at rest, however, so that there is relative motion between them and the fluid, the quantities measured by the transducer are not necessarily P and p; the output from the transducers generally will depend on the

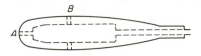

FIGURE 11.1
Streamlined transducer or transducer housing
in which the pressure-sensing area is located
at B for minimum influence of fluid motion
and turbulence.

relative motion. Although a detailed analysis of what actually is measured
by any one transducer in such a case is quite complex, it nevertheless is
possible to say something about this problem, by using some well-known
facts about the pressure distribution over a body in a moving fluid.

Suppose the pressure measured by a transducer moving with the medium
is P; we wish to measure the pressure when the medium is moving past it with
velocity V. If the transducer has a blunt front as illustrated in Fig. 11.1,
there will be one point A at the front of the body where the flow velocity is
zero. At this point, the stagnation point in the flow, Bernoulli's law [Eq.
(6.2.12)] for irrotational fluid motion predicts that the pressure would be

$$P_A = P + \frac{\gamma - 1}{\gamma} \frac{\rho V^2}{2} \qquad (11.1.5)$$

where γ is the ratio between the specific heats at constant pressure and
volume [see Eq. (6.1.3)].

If, on the other hand, we consider not the stagnation point, but a point
B on the side of the body, the flow velocity at this point is approximately equal
to the free-stream velocity V (if the body is sufficiently slender), and the pressure is then approximately equal to the free-stream pressure P.

$$P_B \simeq P \qquad (11.1.6)$$

In the presence of a plane incident wave with sound pressure p and
corresponding fluid velocity $\mathbf{u}$, in the bulk of the fluid away from the body
we have $P = P_0 + p$ and $\mathbf{V} = \mathbf{V_0} + \mathbf{u}$, where P_0 and $\mathbf{V_0}$ are the unperturbed
values of the pressure and velocity. If the wavelength of the sound is much
larger than the dimensions of the body, the sound pressure about the body
is approximately uniform and equal to the incident sound pressure. The
stagnation pressure P_A in (11.1.5) can then be expressed approximately as

$$P_A \simeq P_0 + p + \frac{\gamma - 1}{2\gamma} (\rho_0 + \delta) |\mathbf{V_0} + \mathbf{u}|^2 \qquad (11.1.7)$$

If the transducer has its pressure-sensing portion in the stagnation point,
we see that, if $V_0 \gg u$, the time-dependent portion of the pressure is

$$P_A \simeq p\left(1 + \frac{\gamma - 1}{\gamma} M_0 \cos \varphi\right)$$

where $M_0 = V_0/c$. We have introduced $\mathbf{V} \cdot \mathbf{u} = Vu \cos \varphi$ and $p = \rho cu$,
valid for a plane wave of sound; φ is the angle between $\mathbf{u}$ and $\mathbf{V}$. Thus the

relative motion between the transducer and the medium leads to an increase of the recorded pressure at A by an amount $p[(\gamma - 1)/\gamma]M_0 \cos \varphi$, which depends on the angle between $\mathbf{V}$ and the direction of the sound wave.

If, on the other hand, the pressure-sensing part of the microphone is at point B (assuming that the microphone has a slender shape), the time-dependent part of P_B is approximately equal to the sound pressure p, the same as the pressure measured by a transducer moving with the fluid.

If the medium is turbulent, the relative motion between the microphone and the medium will give rise to an additional time-dependent pressure on the microphone, a pressure which often "masks" the signal produced by the incident sound wave to be detected. The magnitude of these turbulent-pressure fluctuations depends on the location of the pressure-sensing area of the microphone. It is larger at the stagnation point A than at the point B in Fig. 11.1. In the following estimate of the influence of the turbulence we shall assume that the fluctuating turbulent velocity $v(x,y,z,t)$ is much smaller than the mean velocity V of the flow. The turbulent pressure field as recorded by a transducer moving with the average speed of the fluid is of the order of $\rho v^2(x',y',z',t)/2$. The turbulent pressure $p_{B,t}$ recorded at the position B of the stationary transducer will be of this same order of magnitude, $p_{B,t} \simeq \frac{1}{2}\rho v^2(x - Vt, y, z, t)$.

At A, on the other hand, the turbulent pressure is $p_{A,t} \simeq \rho \mathbf{V}_0 \cdot \mathbf{v}(x - Vt, y, z, t)$, assuming that $V \gg v$. This pressure is larger than $p_{B,t}$ by a factor of the order of $2V_0/v$.

If, instead of a pressure-sensitive transducer, we have one that responds to a velocity component, say, the x component, a microphone moving with a relative velocity V in the x direction in the turbulent fluid will record a velocity field $v_x(x - Vt, y, z, t)$.

The pressure fluctuations recorded by the moving transducer as a result of the turbulence in the fluid represent undesirable noise which interferes with the measurement of the sound pressure field. If the sound pressure field is p, the corresponding velocity field is $u = p/\rho c$, assuming plane-wave conditions. If we use a microphone with the pressure-sensing element at the point B, the *signal-to-noise ratio* between p and the turbulent noise pressure $p_{B,t}$ is of the order of $p/(\rho v^2/2) \simeq 2 cu/v^2$. The corresponding signal-to-noise ratio when referred to the stagnation point A is cu/vV. If, instead of a pressure transducer, we use a velocity-sensing transducer, the signal-to-noise ratio is u/v. Since in most cases of interest we have $c > V > v$, it follows that, when operating in turbulent flow, the pressure-type microphone measuring the pressure at point B is least influenced by turbulence, and the velocity transducer is the noisiest.

Most of the standard transducers, such as crystal, condenser, or moving-coil transducers, can be regarded as pressure transducers; when the pressure sensitive membrane points in the direction of the relative motion, the influence of turbulence is the largest. If the transducer is turned 90°, so that the membrane is parallel with the relative motion, the noise is reduced. The response to the sound pressure, on the other hand, is approximately independent of the orientation.

To get an idea of the order of magnitude of the turbulent-pressure fluctuations, consider a microphone in a turbulent airstream with an average velocity 30 m per sec. The rms value of the turbulent-velocity fluctuation is assumed to be 10 percent of the mean flow speed. At the stagnation point the order of magnitude of the turbulent-pressure fluctuation $\rho V v = 0.1 \rho V^2$ then will be about 900 dynes per sq cm, which corresponds to a sound-pressure level of 126 db. In other words, only very intense sound waves could be detected under these conditions. At point B the turbulence level would be 13 db below this, still quite loud.

In the discussion so far we have disregarded the pressure fluctuations induced by the transducer because of the turbulent flow it produces. As it moves through the fluid, a turbulent wake is produced behind the transducer, and the corresponding pressure fluctuations produced by the flow will be of the order of $\rho V^2/2$, where, as before, V is the mean relative speed between the microphone and the fluid. Actually, in the range of velocities corresponding to Reynolds numbers between 10^2 and 10^5, these fluctuations are practically periodic. This oscillation, which produces the familiar Kármán vortex street behind the body, has a frequency of the order of $0.2V/d$, where d is the transverse dimension of the wake. In the case of a cylinder at right angles to the flow, the corresponding oscillatory force per unit length of the cylinder, tending to shake the cylinder sidewise, is found to be of the order of $0.5d$ $(\rho V^2/2)$ per unit length of the cylinder, where d in this case is, roughly, the diameter of the cylinder. The corresponding pressure fluctuation recorded by the transducer is then a maximum when the membrane normal points in the direction transverse to the motion (i.e., is in position B). This is in contrast to the case when the pressure fluctuations are due to the turbulence in the incident flow, in which case, as we recall, the maximum noise is obtained when the membrane normal points in the direction of the flow (i.e., at point A). However, the self-induced turbulent-pressure fluctuations by the transducer can be reduced considerably by enclosing the transducer in a screen which alters the character of the wake behind the transducer.

Having made these general comments about the relation between the pressure actually recorded by various transducers, variously oriented and moving at various speeds through the medium, and the pressure P which would be measured at the same point by a transducer moving with the medium, in the rest of this chapter we shall assume that the necessary corrections have been made and will discuss P, the "true" pressure at a given point and time. Of course, we still have the basic difference between the Lagrange and Euler representations portrayed in Fig. 6.2, if the point in question is moving with respect to the medium. The pressure at point $\mathbf{r}$ at time t, in the laboratory system, in a medium moving with respect to this system, is the pressure measured by a comoving transducer which happens to be at point **r** at time t. If there is no turbulence, this is equal to the pressure measured at point B in a transducer at rest in the laboratory system. When there is turbulence, corrections of the sort just discussed must be made before the measurements will correspond to the pressure given by the formulas in this chapter.

Flow parallel to plane boundary

An important aspect of the effects of moving media is the interaction with the boundary. As a simple example we consider the boundary to be the yz plane, the fluid moving parallel to it with velocity V, as shown in Fig. 11.2. Further, we assume a sinusoidal transverse wave to be traveling along the boundary in the y direction, with velocity c_t. In other words, the boundary can be a membrane or plate, with transverse wave motion in it having wave velocity c_t with respect to its material, or it could be that the boundary is a rigid corrugated surface, in which case c_t would be zero. In any case, the relative velocity $c_t - V$ will produce wave motion in the medium.

We have already seen that velocities and frequencies are altered by relative motion, but that lengths and directions of normals to wavefronts are unaltered. This suggests that we deal with displacements and pressure (or density changes) in fitting boundary conditions, rather than joining velocities at the boundary, as has been done heretofore. As a first step, we describe the wave in the boundary surface in terms of its displacement from the yz plane and its wavelength L (or wavenumber $k_t = 2\pi/L$). Then the speed of this wave and its frequency in the laboratory system S will be c_t and $\omega/2\pi = c_t/L$, respectively. The transverse displacement of the boundary can thus be written, in the laboratory coordinates, as

$$\xi = \xi_0 \exp [ik_t(y - c_t t)] = \xi_0 \exp [i(\omega/c_t)y - i\omega t]$$

The relative motion between the wave in the boundary and that in the medium produces a plane wave in the medium. In the laboratory

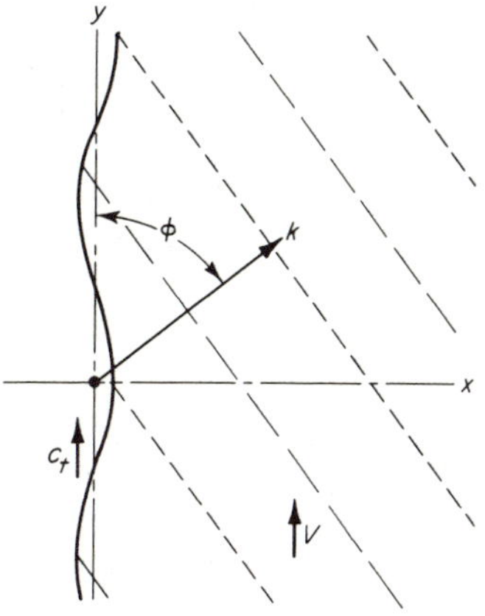

FIGURE II.2
A transverse wave in the boundary at $x = 0$ moves with velocity c_t; the medium to the right of the boundary moves with velocity V with respect to the laboratory coordinates x, y. Vector **k** is normal to the plane wavefronts in the medium.

coordinates x and y, the sound pressure (see the last paragraph under the preceding heading) will have the same frequency, $\omega/2\pi$, as that of the transverse wave in the boundary (in the same coordinates). Thus the form of the pressure wave, in terms of x, y, and t, will be

$$p = P \exp (ik_x x + ik_y y - i\omega t)$$
$$= P \exp [ik(x \sin \phi + y \cos \phi) - i\omega t] \tag{11.1.8}$$

where $\omega = 2\pi c_t/L$, and k_y must equal $k_t = \omega/c_t$ in order that the sound wave fit the boundary displacement wave along the xz plane.

For comparison, let us write down the results if the medium were at rest, if V were zero. In this case, k would equal ω/c, where c is the speed of sound in the medium. Equating $k \cos \phi$ to k_t yields

$$\cos \phi = \frac{c}{c_t} \quad \text{and therefore} \quad k_x = k \sin \phi = \frac{\omega}{c}\sqrt{1 - \left(\frac{c}{c_t}\right)^2} \quad V = 0$$

which determines the angle ϕ of the radiated wave. As we noted in Sec. 10.1, if the speed c of sound in the medium is greater than the speed c_t of the boundary waves, no true wave is propagated into the medium; k_x is imaginary.

The pressure amplitude P is related to the boundary displacement by using the equation $\rho(\partial^2 \xi/\partial t^2) = -(\partial p/\partial x)$, obtained from Eqs. (6.2.3). The result is (to first approximation—see page 238)

$$P = \frac{\rho\omega^2}{ik_x} \xi_0 = \frac{\rho c}{\sqrt{1 - (c/c_t)^2}} (-i\omega\xi_0) \quad V = 0$$

the impedance load of the medium per unit area of vibrating boundary surface thus being $\rho c/\sqrt{1 - (c/c_t)^2}$.

If, however, the medium is moving in the y direction with velocity V, these results must be modified. The form of the pressure wave in terms of the laboratory coordinates x and y is the same, that given in Eq. (11.1.8), but the values of P, ϕ, and k are changed by the motion. Instead of the simple wave equation $\partial^2 p/\partial t^2 = c^2 \nabla^2 p$, the pressure must satisfy Eq. (11.1.4), which in this case is

$$\left(\frac{\partial}{\partial t} + V\frac{\partial}{\partial y}\right)^2 p = c^2 \nabla^2 p$$

Substitution of Eq. (11.1.8) in this yields $(\omega - Vk \cos \phi)^2 = (ck)^2$, which, together with $k_y = k \cos \phi = \omega/c_t = 2\pi/L$, serves to determine both k and ϕ.

$$k = \frac{\omega}{c}\left(1 - \frac{V}{c_t}\right) = \frac{\omega}{c + V \cos \phi}$$

$$k_x = \frac{\omega}{cc_t}\sqrt{(c_t - V)^2 - c^2} \quad k_y = \frac{\omega}{c_t} \tag{11.1.9}$$

$$\cos \phi = \frac{c}{c_t - V} \quad \omega = \frac{2\pi c_t}{L}$$

Thus k is reduced by the factor $1 - (V/c_t)$, involving the difference $c_t - V$, and $\cos \phi$ is increased by the reciprocal of the same factor (since k_y cannot be changed by the motion of the medium). Thus, also, no energy is radiated away from the surface if c is greater than the velocity difference $|c_t - V|$. For example, if the "wave" in the boundary is motionless (the boundary being corrugated and $c_t = 0$), only if $V > c$ (i.e., only if the flow past the boundary is supersonic) will this flow produce a "wave."

Having determined the angle of propagation ϕ and the wavelength $\lambda = 2\pi/k$ of the wave, we next must determine its amplitude. As mentioned above, we must modify Eqs. (6.2.3) to take the fluid motion into account. The time-derivative terms in these equations are really total derivatives, the time rate of change of velocity or displacement measured in comoving coordinates. As long as the medium is at rest, the correction term $\mathbf{u} \cdot \boldsymbol{\nabla}$ is negligible because the $\mathbf{u}$ is the velocity caused by the sound wave, which is quite small. If the fluid moves with velocity $\mathbf{V}$, on the other hand, we cannot neglect the term $\mathbf{V} \cdot \boldsymbol{\nabla}$ of Eq. (6.1.14). Thus the equation relating P to ξ at the boundary, in laboratory coordinates, for the present case, is

$$\rho\left(\frac{\partial}{\partial t} + V\frac{\partial}{\partial y}\right)^2 \xi_x = -\frac{\partial p}{\partial x} \tag{11.1.10}$$

Inserting the expressions for ξ and for p, we obtain

$$pk_t^2(c_t - V)^2\xi_0 = ikP \sin \phi = i\frac{k_t}{c}(c_t - V)P \sin \phi$$

or

$$P = \rho c(-i\omega\xi_0)\frac{(c_t - V)^2}{c_t\sqrt{(c_t - V)^2 - c^2}} \qquad \omega = \frac{2\pi c_t}{L}$$

$$= \frac{-i\omega\rho c\xi_0}{(1 + M\cos\phi)\sin\phi} \qquad M = \frac{V}{c} \tag{11.1.11}$$

which goes to zero when the medium moves as fast as the wave on the boundary ($V = c_t$), and to infinity when $|c_t - V| = c$.

The ratio $P/{-i\omega\xi_0}$ is the impedance load z of the medium on the boundary. [We use the Eulerian $\partial\xi/\partial t$, rather than the Lagrangian $d\xi/dt$, for the denominator because the velocity in question is the rate of change of the point $(0,y)$ on the boundary, disregarding the sideward motion of the fluid; slippage of the fluid past the boundary is supposed to produce no force.] Thus

$$z = \frac{\rho c(c_t - V)^2}{c_t\sqrt{(c_t - V)^2 - c^2}} = \frac{\rho c}{(1 + M\cos\phi)\sin\phi} \tag{11.1.12}$$

where M is the Mach number V/c. This impedance is purely resistive when $|c_t - V|$ is greater than the acoustic velocity c; it is negative-imaginary (masslike) when the differential speed $|c_t - V|$ is subsonic. The acoustic power radiated per unit area of boundary is

$$\Upsilon = \mathrm{Re}\left(p^* \frac{\partial \xi}{\partial t}\right)_{x=0} = z\,|\omega\xi_0|^2$$

$$= 4\pi^2 \rho c \,\frac{c_t(c_t - V)^2}{L^2\sqrt{(c_t - V)^2 - c^2}} \qquad \text{when} \qquad |c_t - V| > c \quad (11.1.13)$$

where L is the wavelength of the sinusoidal wave in the boundary, c_t its velocity of propagation, and V the velocity of the fluid. We now see that when the "wave" in the boundary is a stationary corrugation ($c_t = 0$), no power is radiated, although a pressure "wave" is present in the medium, if $|V| > c$. In this case the "wave" is a pattern of alternately greater and lesser pressure which is stationary with respect to the laboratory system S, and the boundary is stationary, so that no power can be radiated. Also, when $|c_t - V| < c$, no power is radiated; in this case z is reactive.

Refraction at an interface

Next let us consider the transmission of a plane acoustic wave through an interface between two media in relative motion, as shown in Fig. 11.3. We start by investigating the kinematical conditions to be satisfied, from which we shall find a law of refraction analogous to Snell's law.

Consider two semi-infinite fluid regions separated by the yz plane and moving in the y direction with velocities V_1 and V_2, respectively. For simplicity let the propagation vector of the sound wave lie in the xy plane. The plane wave originates in region 1 and travels in a direction which makes an angle ϕ_1 with the y axis, as shown in Fig. 11.3. The sound is partially

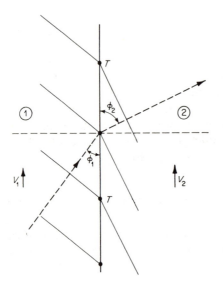

FIGURE 11.3
The trace velocity of the wave along the boundary between 1 and 2 (i.e., the velocity of T) must be the same in both media 1 and 2. $(c_1/\cos\phi_1) + V_1 = (c_2/\cos\phi_2) + V_2$.

transmitted into region 2 in a direction of propagation that makes an angle ϕ_2 with the y axis. To obtain the angle ϕ_2 in terms of the angle ϕ_1, we make use of the fact that the spatial variation of the fields along the interface must be the same in both regions. It follows that the motion of the intersection between a phase surface and the boundary must be the same regardless of whether the phase surface belongs to region 1 or 2. The velocity of such an intersection, indicated by T in Fig. 11.3, can be expressed in two ways, either as $(c_1/\cos \phi_1) + V_1$ or as $(c_2/\cos \phi_2) + V_2$; these, then, must be equal.

$$\frac{c_1}{\cos \phi_1} + V_1 = \frac{c_2}{\cos \phi_2} + V_2$$

or

$$\cos \phi_2 = \frac{c_2 \cos \phi_1}{c_1 - \Delta V \cos \phi_1} \qquad (11.1.14)$$

Here $\Delta V = V_2 - V_1$ is the relative velocity between the two regions, i.e., the velocity *increase* encountered when crossing the boundary in the direction of the sound.

If $V_2 > V_1$, so that ΔV is positive, which includes the case when region 1 is at rest and region 2 in motion, we note that $\phi_2 < \phi_1$. The critical angle of incidence ϕ_{1c}, corresponding to a refracted wave traveling along the boundary in region 2 (so that $\phi_2 = 0$), is given by

$$\cos \phi_{1c} = \frac{c_1}{c_2 + V} \qquad \phi_2 = 0$$

For angles ϕ_1 smaller than this critical angle, no plane traveling waves will penetrate into region 2, and total reflection occurs. (Under these conditions the sound pressure in region 2 decays exponentially with the distance from the boundary, as will be discussed later.) For the largest value of $\phi_1 = 180°$, the angle ϕ_2 is given by

$$-\cos \phi_{2,\text{max}} = \frac{c_2}{c_1 + \Delta V} \qquad \phi_1 = 180°$$

If $c_2 = c_1$, we note that $180 - \phi_{2,\text{max}} = \phi_{1c}$. The region $\phi_2 > \phi_{2,\text{max}}$ is

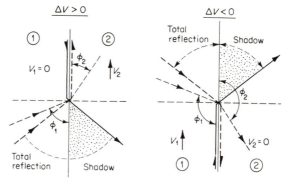

FIGURE 11.4
Wave transmission into a moving fluid, and from a moving fluid into one at rest. $\Delta V = V_2 - V_1$.

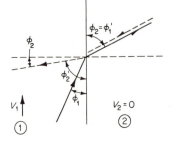

FIGURE 11.5
Reciprocity does not apply for refraction in moving media.

inaccessible for plane sound waves from region 1 and represents an acoustic shadow region. The behavior of the sound wave at the interface when traveling in a direction of increasing flow velocity is illustrated schematically in the first part of Fig. 11.4 for the case $c_2 = c_1$.

If, on the other hand, $V_1 > V_2$, so that ΔV *is negative*, which includes the case of transmission from a moving fluid into one at rest, we find from (11.1.14) that $\phi_2 > \phi_1$. The smallest possible value of ϕ_2 is obtained for $\phi_1 = 0$, and we find $\cos \phi_{2,\min} = c_2/(c_1 + |\Delta V|)$, and the region $\phi_2 < \phi_{2,\min}$ represents a shadow. The largest possible angle, $\phi_2 = 180°$, is obtained for the critical angle of incidence given by $-\cos \phi_{1c} = c_1/(c_2 + |\Delta V|)$. For angles of incidence larger than ϕ_{1c}, total reflection will occur, as indicated schematically in the right-hand part of Fig. 11.4.

It is important to note that if a wave that has been transmitted from region 1 to 2 is reversed in direction, it will not follow along the path of the initial ray; this is illustrated schematically in Fig. 11.5.

Reflection and transmission at the interface

We now turn to the dynamics of the problem in order to calculate the *reflection* and *transmission coefficients* at the interface between the two fluids in relative motion. As before, the interface is the yz plane, and the fluids move with uniform velocities V_1 and V_2 parallel with the yz plane. Let us first note that only the velocity components that lie in the plane of the incident wave (the plane defined by $\mathbf{k}$ and the normal to the interface) will affect the propagation and refraction of sound. This follows directly from the wave equation

$$\left(\frac{\partial}{\partial t} + \mathbf{V} \cdot \mathbf{\nabla}\right)^2 p = c^2 \nabla^2 p \qquad (11.1.15)$$

For a plane wave of the form $\exp(i\mathbf{k} \cdot \mathbf{r} - i\omega t)$ we have $\mathbf{V} \cdot \mathbf{\nabla} = i\mathbf{V} \cdot \mathbf{k} = iV_{11} \cdot k$, where V_{11} is the component in the plane of the wave. Since the main object now is to investigate the effect of flow, we therefore can, without lack of generality, let $\mathbf{V}_1$ and $\mathbf{V}_2$ be placed along the y axis in the plane of incidence of the wave. Then $\mathbf{k} = (k_x, k_y, 0)$, $\mathbf{V} = (0, V, 0)$, and $\mathbf{V} \cdot \mathbf{k} = Vk \cos \phi = Vk_y$, where ϕ is the angle between $\mathbf{V}$ and $\mathbf{k}$. With $k_y = k \cos \phi$ and

$k_x = k \sin \phi$, it follows from Eq. (11.1.15) that, for either medium,

$$k = \frac{\omega}{c + V \cos \phi}$$

(11.1.16)

$$k_x = k \sin \phi \qquad k_y = k \cos \phi$$

Using this expression for k_y in the two regions 1 and 2 (Fig. 11.3), together with the boundary condition that $k_{1y} = k_{2y}$,

$$\frac{c_1}{\cos \phi_1} + V_1 = \frac{c_2}{\cos \phi_2} + V_2$$

(11.1.17)

as found previously in Eq. (11.1.14).

If the incident and reflected pressure fields are p_i and p_r, the total field in region 1 is

$$p = p_i + p_r = P_1[e^{i(k_{1x}x + k_{1y}y)} + Re^{i(-k_{1x}x + k_{1y}y)}]e^{-i\omega t}$$

(11.1.18)

Region 2 contains only a transmitted field, which is of the form

$$p_2 = P_1 T e^{i(k_{2x}x + k_{2y}y) - i\omega t}$$

(11.1.19)

With reference to the previous discussion of boundary conditions in the problem of sound emission from a plane, the boundary conditions to be applied here are continuity of pressure and displacement ξ_x at the interface $x = 0$. The particle-displacement fields corresponding to (11.1.18) and (11.1.19) are obtained by integrating Eq. (11.1.10) and use of Eqs. (11.1.16) and (11.1.17). For example, the x displacement of medium 1 is

$$\xi_{1x} = \frac{iP_1 \sin \phi_1}{\rho_1 k_1 c_1^2} (e^{i(k_{1x}x + k_{1y}y)} - Re^{i(-k_{1x}x + k_{1y}y)})$$

Equating pressure at $x = 0$ gives us $1 + R = T$; equating displacements gives us $(\rho_1 c_1^2 k_1 \sin \phi_2)T = (\rho_2 c_2^2 k_2 \sin \phi_1)(1 - R)$. From these, plus the requirement that $k_1 \cos \phi_1 = k_2 \cos \phi_2$, an expression for the reflection coefficient is derived.

$$R = \frac{\rho_2 c_2^2 \sin (2\phi_1) - \rho_1 c_1^2 \sin (2\phi_2)}{\rho_2 c_2^2 \sin (2\phi_1) + \rho_1 c_1^2 \sin (2\phi_2)}$$

(11.1.20)

The corresponding transmission coefficient is $T = 1 + R$.

Generally, in discussions of refraction, one defines the angle of incidence and refraction as $\phi_1' = 90 - \phi_1$ and $\phi_2' = 90 - \phi_2$. The form of R is not altered by introducing these angles, but in problems involving flow it is more convenient to use the angles ϕ_1 and ϕ_2. For normal incidence, $\phi_1 = \phi_2 = 90°$, this relation reduces to the familiar result $(\rho_2 c_2 - \rho_1 c_1)/(\rho_2 c_2 + \rho_1 c_1)$ because it follows from (11.1.14) that, for normal incidence, we have $\sin (2\phi_1)/\sin (2\phi_2) = c_1/c_2$. In Fig. 11.6 we have plotted the reflection coefficient for the special case $\rho_1 = \rho_2$, $c_1 = c_2$, $V_2 = 0$, with $V_1 = c_1/2$ and c_1.

The corresponding relation between the angles ϕ_2 and ϕ_1 is also given. There are three angles at which there is no reflection from the interface. These angles are determined by $\sin(2\phi_1) = \sin(2\phi_2)$. With $\phi_1 = \phi_2$, from Eq. (11.1.17), we get $\phi_1 = 90°$. The two remaining angles, corresponding to $\phi_2 = 90 - \phi_1$ and $\phi_2 = 270 - \phi_1$, are determined from

$$\pm\cot\phi_1 = 1 + M_1 \cos\phi_1$$

One of these angles is less than $90°$; the other, larger than $90°$.

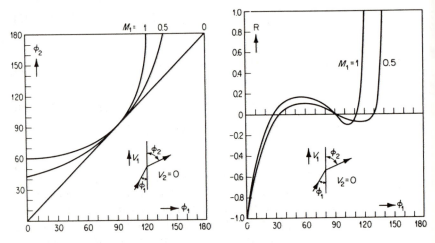

FIGURE 11.6
Angle of refraction and reflection coefficient for a plane wave going from a moving fluid to one at rest.

Reflection from a surface

The reflection coefficient of Eq. (11.1.20) may be derived in another way. According to Eq. (11.1.12), the impedance of medium 2 to being driven by medium 1 is

$$z_2 = \frac{\rho_2 c_2}{(1 + M_2 \cos\phi_2) \sin\phi_2}$$

Suppose we simply replace medium 2 by an equivalent impedance and calculate the reflection coefficient by requiring that the ratio between p_1 and $\partial\xi_1/\partial t = -i\omega\xi_1$ be z_2 at every point of the surface $x = 0$. Inserting the expressions for p_1 and ξ_1 produces

$$1 + R = \frac{z_2}{\rho_1 c_1}(1 + M_1 \cos\phi_1)(\sin\phi_1)(1 - R)$$

$$R = \frac{(z_2/\rho_1 c_1)(1 + M_1 \cos\phi_1)\sin\phi_1 - 1}{(z_2/\rho_1 c_1)(1 + M_1 \cos\phi_1)\sin\phi_1 + 1} \qquad (11.1.21)$$

where z_2 is the impedance of the boundary surface as it appears to medium 1.

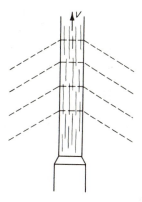

FIGURE 11.7
Sound emitted by a jet is refracted away from the jet at the jet boundary.

If, now, we insert the expression for the z_2 of medium 2, we obtain Eq. (11.1.20), after applying Eqs. (11.1.16) and (11.1.17). We may, however, use Eqs. (11.1.21) for the reflection of a plane wave from any plane surface of effective impedance z_2, when medium 1 is moving past the surface with relative velocity V (where V is in the same direction as the y component of the incident wave). Thus this is the generalization of Eq. (6.3.5) for the reflection coefficient (there called C_r, here called R) when relative motion of the medium is taken into account (we remember that $\cos \vartheta = \sin \phi_1$).

The boundaries between fluid regions in relative motion encountered in practice are of course not so well defined as in the idealized situations treated above. Nevertheless, the results obtained should be useful as a first approximation in more complex problems, such as are encountered, for example, in jets and in atmospheric sound "ducts." Thus, on the basis of the previous analysis, the sound field emitted from sources within a jet is expected to have an acoustic shadow inside a cone in the forward direction with an apex angle $2 \cos^{-1}[c_2/(c_1 + V)]$, where V is the velocity of the jet, and c_1 the speed of sound in the jet. In practice, one indeed finds a region in the forward direction of considerably reduced intensity. To obtain the actual field in this region, it is essential to account for diffraction effects, which of course do not enter when infinite plane boundaries are involved.

Another important problem concerns the propagation of sound in "ducts" formed by velocity and temperature gradients in the atmosphere. Such ducts can trap sound and serve as waveguides for sound over large

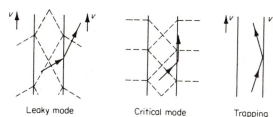

FIGURE 11.8
Acoustic waveguide formed by flow.

Leaky mode   Critical mode   Trapping

distances. Using the results indicated schematically in Fig. 11.4, the qualitative behavior of sound in these two problems of sound emission from a jet and wave trapping in an atmospheric duct is shown schematically in Figs. 11.7 and 11.8.

Ducts

To indicate at least one aspect of the influence of flow on sound transmission through a duct, we shall consider the simple case of propagation between two plane parallel boundaries. Let the boundary at $y = 0$ be rigid, that is, $z = \infty$, whereas the boundary at $y = d$ has a finite impedance $z = \rho c/\beta..$ The fluid between the boundaries moves in the x direction with a velocity V. The solution to the wave equation,

$$\left(\frac{\partial}{\partial t} + V_x \frac{\partial}{\partial x}\right)^2 p = c^2 \nabla^2 p$$

is then of the form

$$p = P_1 \cos (k_y y) e^{ik_x x} e^{-i\omega t} \tag{11.1.22}$$

where

$$k_x^2 + k_y^2 = k^2 \left(\frac{1 - Mk_x}{k}\right)^2$$

Employing the boundary condition for continuity of the particle displacement ξ_y at $y = d$, we have

$$k_y \tan k_y d = -i\beta \left(\frac{1 - Mk_x}{k}\right)^2 k$$

From these equations we obtain the propagation constants k_x and k_y in the x and y directions. For sufficiently small values of the admittance, we can set $\tan (k_y d) \simeq k_y d$ for the fundamental mode, and the corresponding value of k_x for the fundamental mode can be written in the form

$$k_x = k \frac{1}{1 - M^2 A} (\pm A^{\frac{1}{2}} - MA) \tag{11.1.23}$$

where $A = 1 + (i\beta/kd)$. The resulting pressure attenuation of the fundamental mode for propagation in the direction of the flow is then

$$\alpha \simeq \frac{\xi}{2d} \frac{1}{(1 + M)^2} \qquad \beta \ll 1$$

and for propagation in the opposite direction of the flow,

$$\alpha \simeq \frac{\xi}{2d} \frac{1}{(1 - M)^2} \qquad \beta \ll 1$$

where $\xi = \text{Re } \beta$. The same result clearly applies to the case where both planes have the impedance $z = \rho c \beta^{-1}$ if they are separated a distance $D = 2d$. When $|\beta| \ll 1$, the phase velocity downstream and upstream is $c + V$ and $c - V$, and is independent of β to the first order in β.

In this connection it is interesting to consider the case of a rigid duct wall in which $\beta = 0$. Then we obtain, for the two values of k_x corresponding to waves in the direction of the flow and against the flow,

$$k_x^+ = \frac{k}{1 + M} \qquad k_x^- = \frac{k}{1 - M}$$

In other words, the wavelength of the wave in the direction of the flow is stretched to the value $\lambda(1 + M)$, and in the opposite direction the wavelength is compressed to $\lambda(1 - M)$, where $\lambda = 2\pi/k = 2\pi c/\omega$ is the wavelength in a fluid at rest. Since the wave speeds in the direction with and against the flow are $c(1 + M)$ and $c(1 - M)$, respectively, it is clear that the frequency recorded by an observer at rest is the same regardless of the direction of travel of the wave. The evaluation of k_x and k_y under more general conditions requires numerical analysis. The solution presented here is a highly idealized one since it implies a laminar uniform flow in the duct and a point-reacting flexible boundary without turbulence and boundary layers.

In this example it is implied that there is no relative motion between the sound source and the observer; both are stationary in the laboratory system. This is implicit in the assumed form of the pressure wave given in Eq. (11.1.22), in which the frequency is given for the laboratory system S, independent of the flow velocity.

A moving plane-wave source

It was implied in the preceding analysis that the sound source was at rest in the duct. We did not concern ourselves further with the source conditions, but merely studied the behavior of the wave once it had been generated. Although we shall devote Sec. 11.2 to the problem of sound generation by moving sources, it is instructive at this point to consider a simple case involving sound generation in a duct by a plane-wave source when motion between the source and the medium is involved. The relative motion is obtained either by moving the fluid in the tube with the source stationary or by moving the source. Both situations are encountered in practice, particularly when the sound is produced by the heating of the neutral gas in a portion of the duct.

An interesting heating mechanism is encountered in an ordinary electric gas-discharge tube. In such a tube the electrons are kept at a considerably higher temperature than the neutral atoms by the action of the electric field in the tube, and the neutral gas is heated through the encounters with the electrons. If the electric field is time-dependent, the heating of the neutral gas will be time-dependent also, and as a result sound will be produced. Thus, if the gas in the tube is made to flow through the discharge region, we obtain a situation in which the source is stationary and the gas is moving. In a discharge tube, however, the reverse situation is more common. More often than not, there exist, in a gas discharge, traveling-wave disturbances, known as traveling striations, which represent perturbations of electron density and electron temperature, moving with a certain speed, generally from the anode to the cathode, in the discharge tube.   With reference to the neutral gas, these traveling waves represent a periodic source distribution with a frequency $\omega/2\pi = V/\lambda$, where V is the speed of the traveling wave and λ its wavelength.   The phase of the source varies along the tube in accordance with the time of travel of the source, and is expressed by the factor $\exp(i\omega x/V)$.

There are many similar situations in which the sound is produced by traveling disturbances, but we limit the discussion here to the one-dimensional problem in which the source function is a uniform monopole distribution traveling with the speed V in the x direction.

$$Q(x,t) = Q \exp\left[i\left(\frac{\omega}{V}\right)x - i\omega t \right] \qquad (11.1.24)$$

The amplitude Q of the source will be assumed constant in the regions $-L < x < L$ and zero outside, and the tube will be chosen long enough so that reflected sound can be ignored.   The wave equation $c^{-2}\,\partial^2 p/\partial t^2 - \partial^2 p/\partial x^2 = \partial Q/\partial t$, in the case of harmonic time dependence, reduces to $\partial^2 p/\partial x^2 + (\omega/c)^2 p = i\omega Q(x,\omega)$, where Q equals ρ times the s of Eq. (7.1.23), being the mass, rather than the volume generation.   We have seen in Eq. (4.4.11) that the Green's function $G(x,x_0,\omega)$ satisfying the equation $\partial^2 G/\partial x^2 + (\omega/c)^2 G = -\delta(x - x_0)$ is $G(x,x_0,\omega) = (i/2k)\exp(ik\,|x - x_0|)$, where $k = \omega/c$.   With the source function (11.1.24), the acoustic pressure field in the present case is

$$p(x,\omega) = \frac{-i}{2k}\,Q\int_{-L}^{L}(i\omega)e^{i\omega x_0/V}e^{ik|x-x_0|}\,dx_0$$

In the region $-L > x > L$, outside the source, where we have only outgoing waves, we get

$$p(x,\omega) = QcL\,\frac{\sin\{kL[1 \pm (1/M)]\}}{kL[1 \pm (1/M)]}\,e^{ikx-i\omega t} \qquad (11.1.25)$$

where $M = V/c$.   The minus sign refers to the region $x > L$ and represents the forward radiated sound from the traveling source.   We note that the pressure amplitude contributed per unit length of the source is a maximum when $V = c$, in which case the sound-pressure amplitude in the forward direction is QcL.   Under these conditions the sound from all regions of the source arrives in phase at the point of observation.   Such constructive interference cannot occur in the backward direction, for $x < -L$.

The sound pressure field inside the source region is made up of waves traveling in both positive and negative x directions.   We leave the calculation of this field as one of the problems.

11.2 SOUND EMISSION FROM MOVING SOURCES

The main object of the preceding section was to demonstrate the principal effects of fluid motion on sound transmission, and for this purpose only plane waves of sound were considered. The sound field from an actual sound source, however, is ordinarily not a plane wave; we must therefore inquire about the influence of flow on the sound field from such a source. The problem is not simple. For one thing, the flow field about a source of finite dimensions is generally turbulent, and the interaction of sound with this flow field must be accounted for in a detailed analysis. Furthermore, the corresponding static-pressure distribution and the turbulent-pressure fluctuations on the source in motion may influence the operation of the source considerably. To illustrate this point, we mention a design of a train whistle, which was found to operate perfectly well when tested in the laboratory, but which ceased to operate at all on the train at speeds exceeding 50 mph.

We shall not attempt an explanation of this and similar peculiar effects in moving media; we wish to study more general properties. Even then we have to idealize our problem by considering the radiation from point sources in relative motion with the medium. This avoids the question of the flow distortion about the source and the question of turbulence. In applying the results obtained to an actual source, however, we clearly have to be cautious, in particular at high frequencies, when the wavelength is of the order of the source dimensions.

Practical problems, where the influence of relative motion between the source and the medium is of interest, are found in numerous applications, particularly in aviation acoustics, and in treating some of these problems later in this chapter it is useful to be familiar with the idealized situations to be discussed in this section.

We start by considering the sound radiation from an acoustic point monopole source that moves along a trajectory in a fluid at rest. Our discussion is therefore an extension of the analysis in Chap. 7, to include relative motion between the source and the surrounding medium.

Kinematics

Before turning to a calculation of the sound pressure field, it is instructive to establish some purely kinematic properties of the problem, without using the dynamical equations of motion.

We consider a sound source moving along a path specified by the position vector $\mathbf{r}_s(t)$ with the components $x_s(t)$, $y_s(t)$, and $z_s(t)$. The point of observation O is at the position $\mathbf{r}$ with the coordinates x, y, z at rest with respect to the medium. The sound pressure observed at $\mathbf{r}$ at the time t was emitted by the source at a time $t_e = t - (R/c)$, at which time the source was

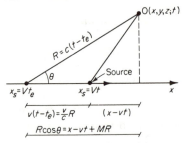

FIGURE 11.9
Source in uniform subsonic motion.

at $\mathbf{r}_s(t_e)$, the emission point E. The distance $\mathbf{R} = \mathbf{r} - \mathbf{r}_s(t_e)$ between E and the point of observation O is determined by the equation

$$R^2 = \left[x - x_s\left(t - \frac{R}{c}\right)\right]^2 + \left[y - y_s\left(t - \frac{R}{c}\right)\right]^2 + \left[z - z_s\left(t - \frac{R}{c}\right)\right]^2$$

(11.2.1)

To solve this equation for R, of course, requires specification of the path $\mathbf{r}_s(t)$. Depending on circumstances, it is possible that more than one value of R is found to satisfy this equation. For example, in the case of supersonic motion of the source along a straight line with constant speed, we shall find that there are two emission points and corresponding emission times which will produce simultaneous contributions to the sound field at the time t at the point of observation.

We shall start with the simplest possible situation, illustrated in Fig. 11.9, in which the source moves along a straight line, the x axis, with a constant speed V smaller than the speed of sound, so that the Mach number $M = V/c < 1$. Then, if we let the source pass the origin of the coordinate system at the time $t = 0$, we have $x_s(t) = Vt$ and $y_s = z_s = 0$. The equation for R in (11.2.1) then reduces to

$$R^2 = \left[x - V\left(t - \frac{R}{c}\right)\right]^2 + r^2 = [(x - Vt) + MR]^2 + r^2 \quad (11.2.2)$$

where $r^2 = y^2 + z^2$. This equation is satisfied by

$$R = \frac{M(x - Vt) \pm \sqrt{(x - Vt)^2 + (1 - M^2)r^2}}{1 - M^2}$$

(11.2.3)

If $M < 1$, it is clear from (11.2.3) that only the positive sign will give a positive value for R. If we introduce

$$R_1 = \sqrt{(x - Vt)^2 + (1 - M^2)r^2}$$

(11.2.4)

we have

$$R = \frac{M(x - Vt) + R_1}{1 - M^2}$$

(11.2.5)

If we denote by θ the angle between the direction of motion and the direction of $\mathbf{R}$ as indicated in Fig. 11.9, we have $x - x_e = x - Vt_e = R\cos\theta$, and with $R = c(t - t_e)$, we get $M(x - Vt) = M[x - Vt_e - V(t - t_e)] = MR(\cos\theta - M)$. Therefore it follows from (11.2.5) that

$$R_1 = R(1 - M\cos\theta) \qquad (11.2.6)$$

a result which will be used frequently later.

If $M = V/c > 1$, that is, when the speed of the source is supersonic, the plus and the minus signs in (11.2.3) are both acceptable, yielding positive values of R provided that $x - Vt < 0$, that is, when the point of observation is behind the source. We can then write

$$R_{\pm} = \frac{M(Vt - x) \pm (Vt - x)\sqrt{1 - (M^2 - 1)r^2/(Vt - x)^2}}{M^2 - 1}$$

If we introduce $r/(Vt - x) = \tan\theta_2$ and $\sin\theta_1 = 1/M$ as shown in Fig. 11.10, we get

$$R_{\pm} = \frac{M(Vt - x) \pm (Vt - x)\sqrt{1 - (\tan\theta_2/\tan\theta_1)^2}}{M^2 - 1} \qquad (11.2.7)$$

In other words, R_+ is real only when $\theta_2 < \theta_1$, that is, inside a cone with the vertex at the source and the vertex angle $2\theta_1$. If we consider a point of observation inside the cone as in Fig. 11.10, we can construct the points of emission E_+ and E_- corresponding to R_+ and R_- which contribute to the sound pressure at the point of observation O at a time t. This point O must

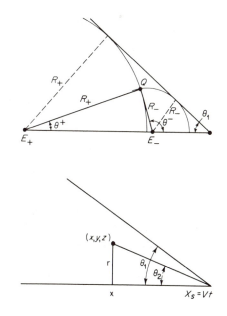

FIGURE 11.10
Kinematics of the wavefield from a point source in uniform supersonic motion.

lie on the intersection between two spherical phase surfaces which are both tangent to the cone and have their centers on the x axis. These centers are the emission points E_+ and E_- that correspond to O. The construction is illustrated in Fig. 11.10.

In the special case when the point of observation O is on the x axis, the distinction between the wave kinematics for subsonic and supersonic sources can be visualized in a simple manner in a space-time diagram as shown in Fig. 11.11. In the subsonic case a point O will be reached at a certain time

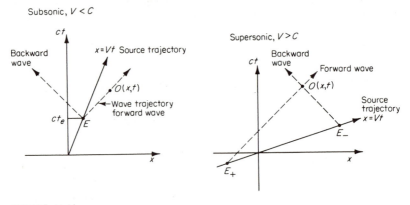

FIGURE 11.11
Space-time diagrams for the wave kinematics of sources in uniform subsonic and supersonic motion.

t only by the forward-going wave from the source corresponding to the emission point E. In the supersonic case, however, there is one forward- and one backward-going wave, reaching O simultaneously, although emitted at different times, corresponding to E_+ and E_-.

Figure 11.12 also gives a hint as to the intensity distribution of the sound. A point O_1 in front of the source receives the energy emitted during the time Δt in the time interval Δt_1 which is smaller than the time interval corresponding to a point of observation O_2 behind the source. In other words, we expect a source which has a symmetric radiation pattern at rest to focus its radiation in the direction of motion. This will be analyzed quantitatively a little later.

By examining the phase surfaces that correspond to a succession of emission points from the moving source, as illustrated in Fig. 11.9, we can easily obtain the Doppler shift noted by an observer at rest. In the forward direction the distance between phase surfaces emitted by the source with a time difference T is $cT - V_sT = T(c - V_s)$. Since the phase surfaces travel with a speed c relative to the observer at rest, he observes a time interval between these phase surfaces $T(c - V_s)/c = T[1 - (V_s/c)]$. The

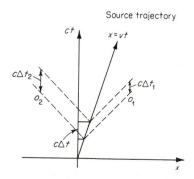

FIGURE 11.12
Relation between emission and reception time
intervals for points of observation in front of
and behind a moving source.

corresponding observed frequency is then larger than the source frequency
by a factor $(1 - M)^{-1}$, where $M = V_s/c$. For an observer at a point
corresponding to an angle θ between V and r measured from the emission
point, we find that the observed frequency is larger than the source frequency
by a factor $(1 - M \cos \theta)^{-1}$. We shall return for a derivation and detailed
discussion of this Doppler-shift formula.

The sound pressure field; subsonic motion

Turning now to the *dynamics* of the problem, we consider a monopole
point source of sound moving with uniform velocity V with respect to the
surrounding fluid. We recall that a monopole point source can be thought
of in terms of a small pulsating sphere, the strength of which is specified by q,
which represents the total rate of mass flux out of the source (thus q equals
ρ times the volume flux S of Sec. 7.1). The source-distribution density can
thus be expressed as $Q(\mathbf{r},t) = q(t)\delta(x - Vt)\delta(y)\delta(z)$. We can imagine this
point source monopole being produced, for example, by the heating and
expansion caused by some modulated radiation focused at a point that is
moved through the fluid, or else by the interaction between neutral atoms
and electrons, mentioned earlier.

In any event, the wave equation for the sound pressure field, produced
by the moving source distribution, is of the form

$$\nabla^2 p - \frac{1}{c^2}\frac{\partial^2 p}{\partial t^2} = -\frac{\partial}{\partial t} q(t)\,\delta(x - Vt)\,\delta(y)\,\delta(z)$$

Using a velocity potential ψ equaling ρ times the Ψ of Eq. (6.2.9), we have

$$p = \frac{\partial \psi}{\partial t}$$

and we get

$$\nabla^2 \psi - \frac{1}{c^2}\frac{\partial^2 \psi}{\partial t^2} = -q(t)\,\delta(x - Vt)\,\delta(y)\,\delta(z) \qquad (11.2.8)$$

This equation can be solved in many different ways. We shall here use a method based simply on a coordinate transformation which enables us to reduce the problem to that of radiation from a stationary source. Later in this chapter we obtain the solution using Fourier transforms.

If we perform a Galilean transformation of coordinates, we reduce the right-hand side of the equation to that corresponding to a stationary source, but then the left-hand side becomes a rather complicated differential form, as we have seen already, in Sec. 11.1. On the other hand, the left-hand side remains unchanged in form if, instead, we use a linear transformation of coordinates analogous to a Lorentz transformation. The transformed right-hand side of the equation hopefully can be written in the form of a stationary source. Thus we attempt the following transformation of coordinates:

$$x' = \gamma(x - Vt) \qquad y' = y \qquad z' = z$$

$$t' = \gamma\left(t - \frac{V}{c^2}x\right) \tag{11.2.9}$$

where
$$\gamma = \frac{1}{\sqrt{1 - M^2}} \qquad M = \frac{V}{c} \qquad c = \text{speed of sound}$$

If we denote by a prime differential operations with respect to the x_i' system, we obtain

$$\nabla'^2\psi - \frac{1}{c^2}\frac{\partial^2\psi}{\partial t'^2} = -q\left[\gamma\left(t' + \frac{x'V}{c^2}\right)\right]\delta\left(\frac{x'}{\gamma}\right)\delta(y')\,\delta(z')$$

Since $\delta(x'/\gamma) = \gamma\,\delta(x')$, this becomes

$$\nabla'^2\psi - \frac{1}{c^2}\frac{\partial^2\psi}{\partial t'^2} = -\gamma q(\gamma t')\,\delta(x')\,\delta(y')\,\delta(z')$$

To make the equivalence with a stationary source problem complete, we proceed to introduce $t'' = \gamma t'$, $x'' = \gamma x'$, $y'' = \gamma y'$, and $z'' = \gamma z'$, in which case $\nabla'' = \gamma^2\nabla'$, and the new equation becomes

$$\nabla''^2\psi - \frac{1}{c^2}\frac{\partial^2\psi}{\partial t''^2} = -\gamma^2 q(t'')\,\delta(y'')\,\delta(z'')\,\delta(x'') \tag{11.2.10}$$

The equation is now completely analogous to one representing the radiation from a stationary source with the strength $\gamma^2 q(t'')$. Using the results already obtained in Sec. 7.1, the solution can be written directly in terms of the coordinates (r'',t'').

$$\psi(r'',t'') = \gamma^2\frac{q[t'' \mp (r''/c)]}{4\pi r''} \tag{11.2.11}$$

Transforming back to the initial variables x, t, we obtain

$$t'' \mp \frac{r''}{c} = \gamma t' \mp \frac{\gamma r'}{c} = t - \frac{V(x - Vt)\gamma^2}{c^2} \mp \frac{\gamma}{c}\sqrt{(x - Vt^2)\gamma^2 + y^2 + z^2}$$

$$= t - \frac{M(x - Vt) \pm \sqrt{(x - Vt)^2 + (y^2 + z^2)(1 - M^2)}}{c(1 - M^2)} = t - \frac{R}{c}$$

where

$$R = \frac{M(x - Vt) \pm \sqrt{(x - Vt)^2 + (1 - \beta^2)(y^2 + z^2)}}{1 - M^2}$$

$$= \frac{M(x - Vt) \pm R_1}{1 - M^2} \tag{11.2.12}$$

From the discussion of the kinematics of the problem we recognize R as the distance between the emission point and the point of observation. The minus sign in (11.2.12), which corresponds to the plus sign in (11.2.11), is acceptable only when we have supersonic motion of the source, as discussed earlier.

In the subsonic case, only the plus sign can be used in (11.2.12). The corresponding value of R is labeled R^+, and the subsonic solution for ψ is then

$$\psi(\mathbf{r},t) = \frac{q[t - (R^+/c)]}{4\pi R_1} \tag{11.2.13}$$

where we have used $r'' = \gamma^2 R_1$.

$$R_1 = \sqrt{(x - Vt)^2 + (1 - M^2)(y^2 + z^2)}$$

Using the expression $R_1 = R(1 - M \cos \theta)$ given in (11.2.6), we can express this solution as

$$\psi(\mathbf{r},t) = \frac{q[t - (R/c)]}{4\pi R(1 - M \cos \theta)}$$

where, from Eq. (11.2.6) and Fig. 11.9, θ is the angle between the x axis and **R**. The corresponding sound pressure field, as we recall from (11.2.8), is $p = \partial\psi/\partial t$. Thus

$$p = \frac{\left(1 - \dfrac{1}{c}\dfrac{dR}{dt}\right)q'\left(t - \dfrac{R}{c}\right)}{4\pi R_1} - \frac{q\left(t - \dfrac{R}{c}\right)}{4\pi R_1^2}\frac{dR_1}{dt}$$

With R given in (11.2.12), we obtain

$$\frac{1}{c}\frac{dR}{dt} = -\frac{1}{1 - M^2}\left[M^2 + \frac{M(x - Vt)}{R_1}\right]$$

$$= -\frac{M}{1 - M^2}\left[M + \frac{R(\cos \theta - M)}{R(1 - M \cos \theta)}\right] = -\frac{M \cos \theta}{1 - M \cos \theta} \tag{11.2.14}$$

where we have used $x - Vt = x - Vt_e - V(t - t_e) = R \cos \theta - MR$ and $R = c(t - t_e)$. Hence

$$p = \frac{1}{4\pi} \frac{q'[t - (R/c)]}{R(1 - M \cos \theta)^2} + \frac{q}{4\pi} \frac{(\cos \theta - M)V}{R^2(1 - M \cos \theta)^2} \qquad (11.2.15)$$

The first of the two terms in (11.2.15) represents the radiation field with the corresponding pressure decreasing as $1/R$. We note that when $\theta = 90°$, the pressure in the far field is the same as when the source is at rest. As expected from the qualitative considerations in connection with Fig. 11.12, we now see that the pressure in the forward direction ($\theta = 0$) is larger than the pressure in the backward direction ($\theta = \pi$) by a factor $(1 + M)^2/(1 - M)^2$.

Simple-harmonic source

We now consider harmonic time dependence so that $q(t) = q_0 \sin(\omega_0 t)$. The pressure field, which contains the factor $q_0 \omega_0 \cos[\omega_0(t - R/c)]$, then has the phase

$$\phi = \omega_0\left(t - \frac{R}{c}\right)$$

which no longer is simply proportional to time since R now is time-dependent.

If the point of observation is on the line of motion of the source in front of the source, so that $x - Vt > 0$, we have $R = x - Vt$, and

$$\phi = \frac{\omega_0}{1 - M}t - \frac{\omega_0}{c(1 - M)}x$$

The observed frequency is then

$$\omega_1 = \frac{\omega_0}{1 - M}$$

and the corresponding wavelength obtained from the spatial dependence of the phase is $(1 - M)\lambda_0$.

Similarly, if the point of observation is behind the source, so that $x - Vt < 0$, we obtain in a similar manner the received frequency and wavelength

$$\omega_2 = \frac{\omega_0}{1 + M} \qquad \text{and} \qquad \lambda_2 = (1 + M)\lambda_0$$

When the point of observation is not on the line of motion, the frequency clearly will vary continuously as a function of time between the limits $\omega_0/(1 - M)$ and $\omega_0/(1 + M)$, which then correspond to $t = -\infty$ and $t = +\infty$. Actually, the concept of frequency is not so well defined, when, as in our case, the phase no longer is proportional to time. However, if we generalize the concept of frequency and define it as the time derivative of the phase, we obtain

$$\omega = \frac{d\phi}{dt} = \omega_0\left(1 - \frac{1}{c}\frac{dR}{dt}\right) = \omega_0\left(1 + \frac{M\cos\theta}{1 - M\cos\theta}\right) = \frac{\omega_0}{1 - M\cos\theta}$$

$$(11.2.16)$$

where we have used the expression for $(1/c)(dR/dt)$ in Eq. (11.2.14).   This is

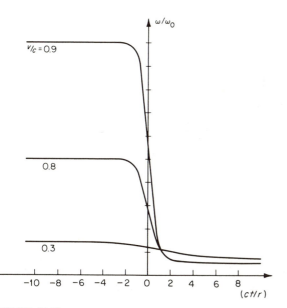

FIGURE 11.13
Time dependence of frequency $\omega/2\pi$ at point of observation located a distance r from the line of motion of the source. Frequency of source $= \omega_0/2\pi$, speed $= V$.

the general Doppler formula.   As the source moves past an observer, the angle θ varies between $\theta = \pi$ and $\theta = 0$, and the frequency varies between the values $\omega_0/(1 - M)$ and $\omega_0/(1 + M)$.

Thus we have seen that both the pressure amplitude and the frequency can be expressed in a very simple manner in terms of the angle θ as measured with respect to the source position at the retarded time $t - (R/c)$.   If, instead, we wish to express these quantities explicitly in terms of the time t, we have to introduce

$$(1 - M\cos\theta)^{-1} = \frac{R}{R_1} = (1 - M^2)^{-1}\left[1 - \frac{M\tau}{\sqrt{(1 - M^2) + \tau^2}}\right]$$

where $\tau = t/(r/V)$. R_1 and R are given in (11.2.12) and (11.2.13). In Figs. 11.13 and 11.14 we have shown the time dependence of the frequency and the amplitude at the position $x = 0$ a distance r from the line of motion of the source. We note that the main changes in the frequency and the amplitude both take place in a time of the order of r/V.

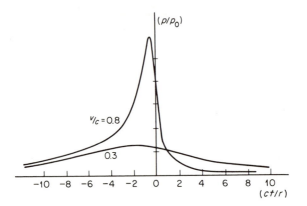

FIGURE 11.14
Time dependence of the pressure amplitude at the point of observation described in Fig. 11.13. The reference pressure p_0 is the pressure that would be observed when $V = 0$.

Supersonic motion

The sound pressure field in the case of supersonic motion is obtained by adding the contributions from the two source points E_+ and E_-.

$$p = \frac{\partial}{\partial t} \left\{ \frac{q[t - (R^+/c)]}{4\pi R_1} + \frac{q[t - (R^-/c)]}{4\pi R_1} \right\} \qquad (11.2.17)$$

where
$$R^{\pm} = \frac{M(Vt - x) \pm R_1}{M^2 - 1}$$

$$R_1 = \sqrt{(Vt - x)^2 - (M^2 - 1)r^2} \qquad (11.2.18)$$

Introducing $Vt - x = V(t - t_e) - (x - Vt_e) = MR - R\cos\theta$, we obtain

$$R_1 = R^+(M\cos\theta^+ - 1) = -R^-(M\cos\theta^- - 1)$$

where the θ^+ and θ^- indicate the directions of R^+ and R^- from the emission points E^+ and E^- to the point of observation, as previously indicated in

Fig. 11.10.   The time dependence of the two angles is given by

$$(M \cos \theta^+ - 1)^{-1} = \frac{R^+}{R_1} = \frac{1}{M^2 - 1} \frac{M\tau}{\sqrt{\tau^2 - (M^2 - 1)}} + 1$$

$$(M \cos \theta^- - 1)^{-1} = -\frac{R^-}{R_1} = -\frac{1}{M^2 - 1} \frac{M\tau}{\sqrt{\tau^2 - (M^2 - 1)}} - 1$$

where $\tau = (Vt - x)/r$.

   With

$$\frac{1}{c} \frac{dR^\pm}{dt} = \frac{M \cos \theta^\pm}{M \cos \theta^\pm - 1}$$

and considering (11.2.18), the pressure field can be expressed as

$$p = \frac{-q'[t - (R^+/c)]}{4\pi R^+ (M \cos \theta^+ - 1)^2} + \frac{q'[t - (R^-/c)]}{4\pi R^- (M \cos \theta^- - 1)^2}$$

$$- \frac{q[t - (R^+/c)]}{4\pi (R^+)^2} \frac{(M - \cos \theta^+)V}{(M \cos \theta^+ - 1)^3} + \frac{q[t - (R^-/c)](M - \cos \theta^-)}{4\pi (R^-)^2 (M \cos \theta^- - 1)^3} \quad (11.2.19)$$

If the point of observation is on the x axis, we have, simply, $R^+ = (Vt - x)/(M - 1)$ and $R^- = (Vt - x)/(M + 1)$.   Then, if the source has harmonic time dependence, so that $q'(t - R^+/c) = q_0\omega \cos \omega_0[t - (R^+/c)]$, we see that the phase becomes

$$\phi(t) = \omega_0\left[t - \frac{Vt - x}{c(M - 1)}\right] = -\frac{\omega_0}{M - 1} t + \frac{\omega_0 x}{c(M - 1)}$$

This phase corresponds to the signal received from the emission point E^+, in which case the source moves toward the point of observation.   We note that the frequency formally is negative, which may be interpreted as a result of the fact that the phase surfaces which were emitted last arrive first.   For the sound that arrives from the emission point E^-, the frequency has the same form as in the subsonic case, namely, $\omega_0/(1 + M)$.   For any arbitrary position, this frequency is $-\omega_0/(M \cos \theta - 1)$.

   An interesting feature of the supersonic field is the interference between the two contributions from the two emission points E^+ and E^-.   These waves, which have different frequencies and amplitudes, give rise to a sound-pressure amplitude which oscillates with time.   We leave it for one of the problems to study this interference field.

Power spectrum

The calculation of the frequency spectrum of the radiated field from a moving source can be carried out directly by Fourier transform of the pressure field in Eq. (11.2.15). However, it is instructive to start directly from the wave equation,

$$\nabla^2 p - \frac{1}{c^2}\frac{\partial^2 p}{\partial t^2} = -\frac{\partial}{\partial t}[q(t)\,\delta(x - Vt)\,\delta(y)\,\delta(z)]$$

and take its Fourier transform. Thus we multiply by $e^{i\omega t}$ and integrate over time. Rewriting the delta function as

$$\delta(x - Vt) = \delta\left[V\left(t - \frac{x}{V}\right)\right] = \frac{1}{V}\delta\left(t - \frac{x}{V}\right)$$

the Fourier transform of the right-hand side gives

$$\frac{1}{2\pi}\int_{-\infty}^{+\infty} e^{i\omega t}\frac{\partial}{\partial t}[q(t)\,\delta(x - Vt)]\,dt = -\frac{i\omega}{2\pi V}q\left(\frac{x}{V}\right)e^{i\omega(x/V)}$$

Similarly, the Fourier transform of the left-hand side of the equation is $\nabla^2 p(\mathbf{r},\omega) + (\omega/c)^2 p(\mathbf{r},\omega)$, where $p(\mathbf{r},\omega)$ is the Fourier transform of the pressure. We shall specialize further and consider harmonic time dependence of the source so that

$$q(t) = q_0\cos(\omega_0 t) = \tfrac{1}{2}q_0[\exp(-i\omega_0 t) + \exp(i\omega_0 t)] \qquad (11.2.20)$$

Consider, first, the contribution to the pressure field from $\exp(-i\omega_0 t)$ and denote this contribution by p^+. Then

$$\nabla^2 p^+ + \left(\frac{\omega}{c}\right)^2 p^+ = \frac{i\omega}{4\pi V}q_0 e^{i(\omega - \omega_0)(x/V)}\,\delta(y)\,\delta(z) \qquad (11.2.21)$$

Since the radiation field is symmetrical about the line of motion, the sound field will be of the form

$$p^+ = f(r)\exp\frac{i(\omega - \omega_0)x}{V}$$

where $r = \sqrt{y^2 + z^2}$. Inserting this expression into the wave equation, we obtain

$$\nabla^2_{y,z} f(r) + \left[\left(\frac{\omega}{c}\right)^2 - \frac{(\omega - \omega_0)^2}{V^2}\right]f(r) = \frac{i\omega}{4\pi V}q_0\,\delta(y)\,\delta(z) \qquad (11.2.22)$$

The free-space Green's function $g(\mathbf{r},\omega)$ to this Helmholtz equation in two dimensions is a solution to

$$\nabla^2_{x,y}g(\mathbf{r},\omega) + k^2 g(\mathbf{r},\omega) = -\delta(y)\,\delta(z)$$

and as we have found previously [Eq. (7.3.17)],

$$g(\mathbf{r},\omega) = \frac{i}{4} H_0^{(1)}(kr)$$

In this case

$$k^2 = \left(\frac{\omega}{c}\right)^2 - \left(\frac{\omega - \omega_0}{V}\right)^2$$

Since the field contributions from $e^{-i\omega_0 t}$ and $e^{i\omega_0 t}$ in the source term merely involved changes in the sign of ω_0, it is convenient to introduce the notation k^+ and k^-, where

$$(k^\pm)^2 = \left(\frac{\omega}{c}\right)^2 - \left(\frac{\omega \mp \omega_0}{V}\right)^2 \tag{11.2.23}$$

The solution to Eq. (11.2.22), then, is

$$f(r) = \frac{\omega q_0}{16\pi V} H_0^{(1)}(k^+ r)$$

and the pressure field is

$$p^+ = \frac{\omega q_0}{16\pi V} H_0^{(1)}(k^+ r) e^{i(\omega - \omega_0)x/V}$$

Inverting the Fourier transform, we find the space-time dependence of the sound pressure field p^+.

$$p^+ = \frac{q_0}{16\pi V} \int_{-\infty}^{+\infty} \omega H_0^{(1)}(k^+ r) e^{i(\omega - \omega_0)x/V} e^{-i\omega t}\, d\omega \tag{11.2.24}$$

The corresponding radial component of the particle velocity field obtained from $\rho(\partial u_r/\partial t) = -\partial p/\partial r$ is then

$$u_r^+ = -\frac{q_0}{16\pi V(i\rho)} \int_{-\infty}^{+\infty} k^+ H_1^{(1)}(k^+ r) e^{i(\omega - \omega_0)x/V} e^{-i\omega t}\, d\omega \tag{11.2.25}$$

where we have used $dH_0^{(1)}(z)/dz = -H_1^{(1)}(z)$.

The contributions p^- and u_r^- produced by the second term, $\exp(i\omega_0 t)$, in the source term (11.2.20) are obtained simply by changing the sign of ω_0 in Eqs. (11.2.24) and (11.2.25), and the total sound pressure field and the total radial particle velocity are then

$$p = p^+ + p^- \qquad u_r = u_r^+ + u_r^-$$

Before investigating these fields further, we shall study the frequency dependence of the propagation constants $k^\pm$ as given by (11.2.23). We have

$$(k^+)^2 = \left(\frac{\omega}{c}\right)^2 - \left(\frac{\omega - \omega_0}{V}\right)^2 = \frac{M^2 - 1}{V^2}\left(\omega - \frac{\omega_0}{1 - M}\right)\left(\omega - \frac{\omega_0}{1 + M}\right)$$

where $M = V/c$. By considering the signs of the two factors in this equation, we see that, for $M < 1$, the propagation constant k^+ will be real only in the range

$$\frac{\omega_0}{1 + M} < \omega < \frac{\omega_0}{1 - M} \qquad (11.2.26)$$

Similarly, for $M > 1$, there will be *two* intervals,

$$-\infty < \omega < -\frac{\omega_0}{M - 1} \qquad \frac{\omega_0}{M + 1} < \omega < \infty \qquad (11.2.27)$$

where k^+ is real.   The corresponding ranges for k^- are obtained by changing the sign of ω_0.

This behavior is consistent with our previous observations about the Doppler shift.   In the subsonic case, $M < 1$, the observed frequency was found to be $\omega_0/(1 - M \cos \theta)$, which has a maximum value $\omega_0/(1 - M)$ when $\theta = 0$ (emission point at $x = -\infty$) and a minimum value $\omega_0/(1 + M)$ when $\theta = \pi$ (emission point at $x = +\infty$).   Similarly, in the supersonic case, the frequency is $-\omega_0/(M \cos \theta - 1)$, which gives $-\omega_0/(M - 1)$ when $\theta = 0$ and goes to $-\infty$ when $\cos \theta = 1/M$.   Then ω switches to $+\infty$ as θ passes through the value $\theta = \cos^{-1}(1/M)$, and approaches $\omega_0/(1 + M)$ when $\theta \to 0 \, (x \to \infty)$.   As discussed earlier, the negative sign of the frequency in one of these regions expresses the fact that the signals that were emitted last from the source arrive first at the point of observation.   In the frequency intervals in which $(k^+)^2$ is negative, k^+ is imaginary, and the wave functions $H_0^{(1)}(k^+r)$ and $H_1^{(1)}(k^+r)$ will decay exponentially away from the line of motion of the source.   Thus these frequencies will not contribute to the sound field in the far-field region, and the corresponding frequency intervals can be excluded in the far field.

The acoustic power emitted by the moving source will be evaluated by integrating the radial component of the acoustic energy flux over the surface of an infinitely long cylinder enclosing the line of motion of the source. Actually, we shall determine the average power over one period, $2\pi/\omega_0$, of the source.   Thus, the radius of the cylinder being r, the acoustic power Π is expressed as

$$\Pi = \int_{-\infty}^{+\infty} 2\pi r \overline{pu_r} \, dx = \int_{-\infty}^{+\infty} 2\pi r \overline{(p^+ + p^-)(u_r{}^+ + u_r{}^-)} \, dx \quad (11.2.28)$$

where the bar indicates the time average over one period, $2\pi/\omega_0$.   In evaluating this integral the asymptotic form of the Hankel functions will be used,

$$H_0^{(1)}(kr) \simeq \sqrt{\frac{2}{\pi kr}} \, e^{ikr - \frac{1}{4}\pi i}$$

and $H_1^{(1)} \simeq -iH_0^{(1)}(kr)$.

In the far field only the frequency intervals in which k^- and k^+ are real will contribute to the pressure and velocity fields. To indicate this restriction in the range of integration, we introduce the function $S_+(\omega)$, which is unity when k^+ is real and zero elsewhere. Similarly, $S_-(\omega)$ is unity when k^- is real and zero elsewhere. Clearly, $S_-(-\omega) = S_+(\omega)$.

The integrals involved in (11.2.28) are then of the form

$$\int p^+ u^- \, dx = C \int_{-\infty}^{+\infty} dx \iint_{-\infty}^{+\infty} d\omega \, d\omega' \, S_+(\omega) S_-(\omega') e^{i(k^+ + k'^-)r}$$
$$\times \exp\left[(\omega + \omega')\left(\frac{x}{V} - t\right)\right]$$

where $k'^- = k^-(\omega')$, and $C = 4q_0{}^2/(16\pi V)^2 \rho$. Because $S_-(-\omega) = S_+(\omega)$, this integral turns out to be independent of t; thus an average over a period is not zero. A similar integral is obtained for $p^- u^+$. The contributions from $p^+ u^+$ and $p^- u^-$ will be zero when averaged over one period of the source because they both oscillate with time. The integration over x gives $2\pi V \delta(\omega + \omega')$, and therefore the integration over ω' merely involves replacing ω' by $-\omega$ in the integral. Using $S_-(-\omega) = S_+(\omega)$, the remaining integration over ω, then, is

$$\int p^+ u^- \, dx = (2\pi V) C \int_{-\infty}^{+\infty} S_+{}^2(\omega) \sqrt{\frac{k'^-}{k^+}} \, \omega e^{i(k^+ + k'^-)r} e^{-i\pi/2} \, d\omega$$

On account of $S_+{}^2(\omega)$, the integration reduces to the interval in which k^+ is real. It follows from (11.2.26) that in the range where k^+ is real (and positive), we have $k'^- = -k^+$ and $\sqrt{k'^-} = i\sqrt{k^+}$. Thus

$$\int p^+ u^- \, dx = (2\pi V) C \int_{-\infty}^{+\infty} S_+{}^2(\omega) \, \omega \, d\omega \qquad (11.2.29)$$

It is not difficult to see that $\int p^- u^+ \, dx$ is equal to $\int p^+ u^- \, dx$.

Thus, in the subsonic case in which the range of integration is $a = \omega_0/(1 + M) < \omega < \omega_0/(1 - M) = b$, we obtain for the radiated power

$$\Pi = \frac{h_0{}^2}{16\pi(\rho c) M} \int_a^b \omega \, d\omega = \frac{q_0{}^2 \omega_0{}^2}{8\pi\rho c} \frac{1}{(1 - M^2)^2} = \frac{\Pi_s}{(1 - M^2)^2} \qquad (11.2.30)$$

where

$$\Pi_s = \frac{(q_0 \omega_0)^2}{8\pi\rho c} = \frac{(\rho S_0 \omega_0)^2}{8\pi\rho c}$$

is the power radiated by a source at rest, which is identical with the result obtained in Eq. (7.1.5) (when we take into account that q_0 and S_0 here are maximum, not rms, values). Consequently, for a point monopole source the relative motion between the source and the fluid causes an increase of the acoustic power output by a factor $(1 - M^2)^{-2}$. The energy spectrum, defined by $\Pi = \int_0^\infty E_s(\omega)\, d\omega$, is seen to be

$$E_s(\omega) = \frac{q_0^2}{16\pi\rho c}\frac{\omega}{M} \qquad \frac{\omega_0}{1 + M} < \omega < \frac{\omega_0}{1 - M} \qquad M < 1$$

Although the source is simple-harmonic, the sound reaching the observer has a spread of frequency because the source is moving, another way of stating the Doppler effect.

In the supersonic case, $S_+(\omega)$ is different from zero in the two regions given in (11.2.27), and the integral in (11.2.29) becomes

$$\int_{-\infty}^\beta \omega\, d\omega + \int_\alpha^\infty \omega\, d\omega = \frac{\omega_0^2}{2}\left[\frac{1}{(M-1)^2} - \frac{1}{(M+1)^2}\right] = \frac{2\omega_0^2 M}{(M^2 - 1)^2}$$

where $\beta = -\omega_0/(M - 1)$ and $\alpha = \omega_0/(M + 1)$, and the power is

$$\Pi = \Pi_s\frac{1}{(M^2 - 1)^2} \qquad M > 1$$

Thus we see that, formally, the expressions for the total radiated power in the subsonic and supersonic cases are the same. The energy spectrum differs markedly, however; for $M < 1$, the frequencies are those below $\omega_0/2\pi(1 - M)$, whereas for $M > 1$, they are all those greater than $\omega_0/2\pi(1 + M)$, clear to infinity.

Dipole sources

The ideal point dipole is represented by an oscillating point force acting on a fluid. At sufficiently low frequencies many actual sound sources such as an oscillating piston are approximately equivalent to dipoles. An airplane in "bumpy" flight through turbulent air will act as an acoustic dipole source moving through the atmosphere, and in a similar manner a ship moving over the ocean waves will be a dipole source for low-frequency underwater sound. The sound radiation from propellers and turbines also can be described in terms of moving-dipole-source distributions, as will be discussed later in this section.

An ideal dipole source moving in the x direction with a velocity V is described by a force distribution per unit volume with components of the form

$$F_j = f_j(t)\delta(x - Vt)\delta(y)\delta(z)$$

The corresponding source term in the wave equation is $\partial F_j/\partial x_j$, as we have seen already in Eq. (7.1.23). Thus, if we introduce a vector $\mathbf{A}$ with the components A_j such that

$$p = \operatorname{div} \mathbf{A} \qquad\qquad (11.2.31)$$

it follows that the equation for A_j is

$$\nabla^2 A_j - \frac{1}{c^2} \frac{\partial^2 A_j}{\partial t^2} = f_j(t)\delta(x - Vt)\delta(y)\delta(z) \tag{11.2.32}$$

In complete analogy with (11.2.13), we obtain the solution, for $V < c$,

$$A_j = -\frac{f_j[t - (R^+/c)]}{4\pi R_1}$$

where R^+ and R_1 are defined in (11.2.18). The pressure then follows from (11.2.31). For future reference we express the fields from the transverse dipole with the force components $f_j = (0, f_2, 0)$.

$$p_2 = \frac{1}{4\pi R_1} \frac{y}{R_1} \left(\frac{1}{c} f_2' + \frac{1}{R_1} f_2 \right) \tag{11.2.33}$$

Here the prime signifies differentiation with respect to the argument. Similarly, the sound pressure field for a longitudinal dipole, in which the force components are $f_j = (f_1, 0, 0)$, is

$$p_1 = \frac{1}{4\pi R_1 c} f_1' \left[\frac{M}{1 - M^2} + \frac{x - Vt}{R_1(1 - M^2)} \right] + \frac{1}{4\pi R_1^2} \frac{x - Vt}{R_1} f_1 \tag{11.2.34}$$

The detailed study of the properties of these fields is left for some of the problems.

We turn here, instead, to the Fourier transform of A_j obtained from Eq. (11.2.32) when the force has harmonic time dependence $f_j(t) = f_{0j} \cos(\omega_0 t)$. Again, by analogy with (11.2.24), we get

$$A_j^+ = \frac{i f_{0j}}{16\pi V} \int_{-\infty}^{+\infty} H_0^{(1)}(k^+ r) e^{i(\omega - \omega_0)x/V} e^{-i\omega t} \, d\omega \tag{11.2.35}$$

The sound pressure field (11.2.31) and the corresponding radial velocity component then follow from A_j, and in much the same manner as for the monopole, we find that the power emitted from a longitudinal dipole (force component f_x) is

$$\Pi = \frac{f_{0x}^2 \omega_0^2}{16\pi V^3 \rho} \int_{\gamma_1}^{\gamma_2} \frac{(\gamma - 1)^2}{\gamma} \, d\gamma = \Pi_d G_1(M) \tag{11.2.36}$$

where

$$\gamma = \frac{\omega}{\omega_0} \qquad \gamma_1 = \frac{1}{1 + M} \qquad \gamma_2 = \frac{1}{1 - M}$$

$$\Pi_d = \frac{\omega_0^2 f_{0x}^2}{24\pi \rho c^3}$$

$$G_1(M) = \frac{3}{2} \frac{1}{M^3} \left[\log \left(\frac{1 + M}{1 - M} \right) + \frac{2M(2M^2 - 1)}{(1 - M^2)^2} \right]$$

We leave it as one of the problems to derive this and some similar expressions for other moving sources.

The quantity Π_d is the power emitted from a dipole at rest, in agreement with the result obtained in Eq. (7.1.7), for $f_0 = \omega D$, and allowing for the change from rms to maximum amplitude. The function $G_1(M)$ expresses the velocity dependence of the power. For the moving transverse dipole

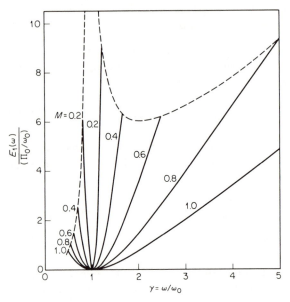

FIGURE 11.15
Power spectra of a longitudinal dipole in uniform subsonic motion for some different values of the velocity. $V/c = M$.

corresponding to the force component f_y (or f_z), the total power emitted is found to be

$$\Pi = \Pi_d G_2(M) \qquad (11.2.37)$$

where

$$G_2(M) = \frac{3}{4} \left[\frac{2}{M^2(1 - M^2)} - \frac{1}{M^3} \log \frac{1 + M}{1 - M} \right]$$

and Π_d is given in (11.2.36).

The power spectra, defined by $\Pi = \int_0^\infty E(\omega)\, d\omega$, for the longitudinal and transverse moving dipoles, are obtained directly from the integrals for Π; they are found to be

$$E_1(\omega) = \frac{\Pi_d}{\omega_0} \frac{3}{2M^3} \frac{(\gamma - 1)^2}{\gamma}$$

$$E_2(\omega) = \frac{\Pi_d}{\omega_0} \frac{3}{4M^3\gamma} [(M - 1)\gamma + 1][\gamma(M + 1) - 1]$$

$(11.2.38)$

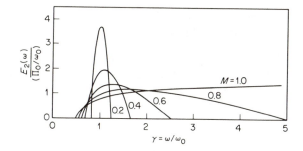

FIGURE 11.16
Power spectra of a transverse dipole in uniform subsonic motion for some different values of the velocity. $V/c = M$.

It may appear that the spectrum shape of the longitudinal moving dipole is independent of the speed. We should keep in mind, however, that the width of the spectrum determined by the extreme values of the Doppler-shifted frequencies does depend on M. These spectra are shown in Figs. 11.15 and 11.16 for some different values of M. For the longitudinal dipole the intensity is zero for $\omega = \omega_0$. This is understandable since, for a dipole lined up in the direction of motion, the direction of zero Doppler shift coincides with the direction of zero intensity in the dipole radiation pattern.

For the "transverse" dipole, on the other hand, the directions of zero intensity are along the line of motion, and on these lines the Doppler-shifted frequencies are $\omega_0/(1 - M)$ and $\omega_0/(1 + M)$, in front of and behind the source, respectively. Consequently, the energy spectrum is expected to be zero at these frequencies, and this is indeed the case.

Sound emission from a piston source in a moving medium

In many applications we encounter problems in which a sound source is at rest in the laboratory system and the fluid is moving. As already pointed out, however, the problem of sound emission from a point source moving in the positive x direction is equivalent to the emission from a stationary source in a fluid moving in the negative x direction. Therefore we should not need to say anything further about this case, except perhaps to point out that it is convenient to express the field in terms of coordinates attached to the source rather than the fluid. In addition, some comments should be made about the calculations of the sound field from an extended source, and we shall choose as an illustration the radiation from an oscillating piston in an infinite plane in contact with a fluid moving past it.

Thus, returning to the expression for the sound field from a moving monopole source in (11.2.13) or (11.2.15), we introduce the coordinates

$$\xi = x - Vt \qquad \eta = y \qquad \zeta = z$$

attached to the moving source. In terms of these coordinates, the solution represents the sound field from a stationary source in a flow field moving in the negative ξ direction. With reference to (11.2.15), in which we set $R_1 = R(1 - M \cos \theta)$ [see (11.2.6)], the sound pressure field from a monopole at the point (ξ_0, η_0, ζ_0), when the surrounding fluid moves with a velocity V in the negative ξ direction, is given by

$$p = \frac{1}{4\pi R_1}\left[\frac{R}{R_1}q'\left(t - \frac{R}{c}\right) + \frac{\xi V}{R_1{}^2}q\left(t - \frac{R}{c}\right)\right] \qquad (11.2.39)$$

where

$$R_1{}^2 = (\xi - \xi_0)^2 + (1 - M^2)[(\eta - \eta_0)^2 + (\zeta - \zeta_0)^2]$$

and

$$(1 - M^2)R = M(\xi - \xi_0) + R_1$$

For a source with harmonic time dependence $q = q_0 \exp(-i\omega t)$, this field can be expressed as

$$p(\mathbf{r};\mathbf{r}_0) = \frac{-i\omega}{4\pi R_1}e^{ikR}\left(\frac{R}{R_1} + i\frac{\xi - \xi_0}{R_1}\frac{M}{kR_1}\right)q_0 e^{-i\omega t} \qquad (11.2.40)$$

where

$$M = \frac{V}{c} \qquad k = \frac{\omega}{c}$$

With $q_0 = 1$ this corresponds to the free-space Green's function for a moving fluid (velocity V in the negative ξ direction).

We shall use this result for the calculation of the sound emission from a rectangular piston set in an otherwise rigid plane, coincident with the $\xi\eta$ plane, with the fluid moving in the negative ξ direction with a speed V. The piston, with the dimensions $2a$ and $2b$, has a velocity $U\exp(-i\omega t)$ in the ζ direction, and therefore a displacement $d_\zeta = (iU/\omega)\exp(-i\omega t)$. The corresponding ζ component of the *fluid* velocity at the plane is given by $[\partial/\partial t - V(\partial/\partial \xi)]\,d_\zeta = \{1 + i(V/\omega)[\delta(\xi - a) - \delta(\xi + a)]\}U\exp(-i\omega t)$. The δ-function contribution from the velocity arises when the fluid flows over the edge of the piston as it moves up and down.   Since we have assumed that the flow is in the $-\xi$ direction, these jumps are only at the $\xi = \pm a$ edges, not at the others.   With q_0 in (11.2.40) replaced by 2ρ times this fluid velocity, the sound field from the piston is obtained by integrating $p(r/r_0)$ in (11.2.40) over the region $-a < \xi_0 < +a$, $-b < \eta_0 < +b$.   The integral includes the δ-function contribution to the velocity from $\eta = -b$ to $\eta = +b$. Considering here only the far field, we use the first term in (11.2.40) and obtain for the pressure field

$$p \to -\frac{2i\rho cUkab}{\pi}\frac{e^{ikR'}}{R_1'}\frac{R'}{R_1'}\frac{\sin X}{X}\frac{\sin Y}{Y}\left(1 + M\frac{X}{ka}\right)$$

$$R_1' = \xi^2 + (1 - M^2)(\eta^2 + \zeta^2)$$

$$(1 - M^2)R' = M\xi + R_1'$$

$$X = \frac{ka[M - (\xi/R_1')]}{1 - M^2} \,;\; Y = kb\frac{\eta}{R_1'} \qquad (11.2.41)$$

The first term in the factor $1 + M(X/ka)$ represents the contribution from the surface of the piston, whereas the second term comes from the $\xi = \pm a$ edges of the piston. The surface of the piston represents an extended monopole source distribution, and it is interesting to see from (11.2.41) that, in contrast to the point source, this part of the emitted sound pressure field remains finite when $V = c$. The reason for the divergence for the point source when $V = c$ can be traced to the zero size of the source. At a given point of observation upstream from the source, the pressure-wave contributions from the point source all arrive in phase as $V \to c$, regardless of the emission time. But the wavelength of the sound in the fluid then approaches zero, and if the source has a nonzero size, however small, the difference in travel time for rays emitted from different parts of the source will be sufficient to cause destructive interference. This is what happens to the surface contribution from the piston, and as a result the field remains finite for $M = 1$.

The line sources at the edges of the piston, however, having zero size in the direction of motion of the flow, behave much like a point source, and the corresponding sound-field contribution does indeed diverge at $V = c$, as can be seen from Eqs. (11.2.41).

In practice, the behavior of the flow at the edges of the piston will not be steplike, as implied by the use of δ functions in the acoustic source function. Instead, the flow profile will be blurred because of turbulence, so that this contribution, also, will remain finite at $M = 1$.

11.3 SOUND EMISSION FROM RIGID-BODY FLOW INTERACTION

As already mentioned in Sec. 11.1 in the study of the effect of fluid motion on transducer response, the interaction between a body in motion relative to a fluid leads to pressure fluctuations on the body. These fluctuations result either from turbulence or inhomogeneities in the incident flow or from the turbulence produced by the body itself. To obtain the sound field produced in this manner, we must start by investigating the interaction between the body and the flow. A general analysis of this problem, starting from first principles, is hardly possible, and in this phase of the study we must rely upon experimental data relating the interacting forces with the parameters describing the flow field and the body. Such experimental data are available for a number of simple bodies such as spheres and cylinders and, in particular, for airfoils.

In this section we shall start with some comments about the sound from a rigid body in motion relative to a turbulent fluid. Next we consider the case when the incident flow is uniform and the turbulence is produced by the body itself. As a typical example, the Kármán vortex sound from a cylinder will be discussed. Actually, turbulence is not a necessary requirement for sound emission from a rigid body in a moving fluid. For example, if the motion of the body is supersonic, it is well known that a shock wave is produced, and if there is an array of bodies moving through a fluid, as in the case of a propeller, a periodic pressure field is generated which, under certain conditions, results in a radiated sound field. The important problem of propeller sound will be considered for the case of both uniform and nonuniform flow. Finally, some observations will be made relating to the problem of the effect of flow on a mechanical oscillator and on the acoustical properties of an orifice in a partition.

Sound from a propeller

As an example of sound radiation in Sec. 7.4, we considered briefly the sound field produced by a source distribution which, in its essentials, could

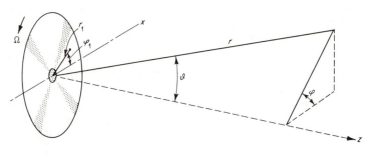

FIGURE 11.17
Sound radiation from a propeller.

serve as a model for a propeller. We shall now justify this model by considering the elementary aerodynamics of a body such as the propeller blade of Fig. 11.17 moving through a fluid, and include the case often encountered in practice when the propeller operates in an inhomogeneous fluid. For example, a ship propeller operates in the inhomogeneous flow in the wake of the ship, a compressor rotor interacts with the wakes of stationary guide vanes in front of the rotor, and an airplane propeller often moves through a turbulent atmosphere. It turns out that flow inhomogeneities give rise to sound radiation which generally far exceeds the radiation in a homogeneous flow. To study this aspect of the sound generation by a propeller, we shall start by investigating the interaction between a propeller blade and the surrounding fluid.

When a propeller blade moves through a homogeneous fluid, there will be a pressure distribution about the blade resulting in a net force distribution over the blade. It is customary to express the force per unit area of the blade in terms of a thrust and a drag component f_t and f_d, illustrated in Fig. 11.18, which are expressed as

$$f_t = c_t \frac{\rho V^2}{2} \qquad f_d = c_d \frac{\rho V^2}{2} \tag{11.3.1}$$

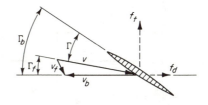

FIGURE 11.18
Airfoil (propeller blade) moving with a velocity
V_b through a fluid that moves with a velocity V_f.

The thrust and drag coefficients c_t and c_d depend strongly on the "angle of attack" Γ of the flow velocity (Fig. 11.17) and only weakly on the magnitude of the relative velocity V,

$$V = V_f - V_b$$

where V_f is the fluid velocity, and V_b the blade velocity.

The coefficients c_t and c_d must be determined empirically as functions of the position on the propeller blade. If the flow is uniform, the force on a blade is independent of the angular position of a blade as it moves through the fluid. Thus, in a coordinate system rotating with the angular velocity of the propeller, the force distribution over the plane of the propeller is time-independent. However, the effect of the propeller on the surrounding fluid as observed at a fixed point in the laboratory system is represented by a force which is periodic in time with an angular frequency equal to $\omega_1 = B\Omega$, where Ω is the angular velocity of the propeller shaft, and B the number of blades. The plane of the propeller, then, can be regarded as a sound source with a dipole source distribution corresponding to the force (11.3.1) on the portion occupied by the blades and zero between the blades. The coordinates in this plane are shown in Fig. 11.17.

Considering, first, the case of *uniform* flow, the amplitude of the periodic force on the fluid is independent of φ_1, but the phase is proportional to φ_1. Thus, if the z component of the force on the fluid at $\varphi_1 = 0$ is $f_z(t;r_1,0) = \sum_{-\infty}^{+\infty} A_s(r_1)e^{-is\omega_1 t}$, it is, at φ_1,

$$f_z(t;r_1,\varphi_1) = \sum A_s \exp\left[-is\omega_1\left(t - \frac{\varphi_1}{\Omega}\right)\right] = f_z^0 \Sigma \alpha_s e^{isB\varphi_1}e^{-is\omega_1 t} \qquad (11.3.2)$$

$$\omega_1 = B\Omega \qquad \alpha_s = \frac{A_s}{f_z^0}$$

Here the coefficients f_z^0 and α_s generally are functions of the radial coordinate r_1. For the φ_1 component of the force, we obtain

$$f_\varphi(t;r_1,\varphi_1) = f_\varphi^0 \sum_{-\infty}^{+\infty} \alpha_s e^{isB\varphi_1}e^{-is\omega_1 t} \qquad (11.3.3)$$

We have here assumed that the angular dependence of f_φ and f_z is the same but the magnitudes are different. Actually the time average value f_z^0 is considerably larger than f_φ^0 for a well-designed airfoil. Constants f_z^0 and f_φ^0 can

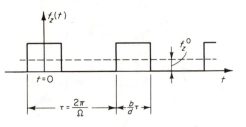

FIGURE 11.19
Example of time dependence of the dipole source function in the plane of the propeller.

be expressed in terms of the total thrust force F_t and the total torque T of the propeller from

$$F_t = \int f_z^0 r_1 \, dr_1 \, d\varphi_1 \quad \text{and} \quad T = \int (f_\varphi^0 r_1) r_1 \, dr_1 \, d\varphi_1 \qquad (11.3.4)$$

The coefficients α_s are obtained from

$$\alpha_s = \frac{1}{f_z^0} \frac{e^{-isB\varphi_1}}{\tau} \int_{-\tau/2}^{\tau/2} f_z(t,\varphi_1) e^{is\omega_1 t} \, dt \qquad (11.3.5)$$

If, as an illustration, we assume that the pressure distribution over a blade is uniform, the time dependence of f is represented by a rectangular function as shown in Fig. 11.19. Then we obtain from Eq. (11.3.5):

$$\alpha_0 = 1 \qquad 2\alpha_s = \frac{\sin (sb\pi/d)}{sb\pi/d} \qquad s > 0 \qquad (11.3.6)$$

where b is the width of the blade, and d the separation between two blades.

Nonuniform flow

If the flow is *nonuniform*, so that the velocity V in Eqs. (11.3.1) depends on r_1 and φ_1, the response of a propeller blade is considerably more complicated than in the homogeneous case, but as long as the flow is time-independent, the time dependence of the force distribution will still be periodic, with the fundamental frequency $\omega_1 = B\Omega$. We shall consider only this case of a time-independent (although inhomogeneous) flow field. In contrast to the homogeneous case, the amplitude factors f_z^0 and f_φ^0 are now functions of φ_1. If we express this φ_1 dependence in terms of the Fourier series,

$$f_z^0(\varphi_1) = f_z^0 \sum_{-\infty}^{+\infty} \beta_q e^{iq\varphi_1} \qquad (11.3.7)$$

we obtain, replacing f_z^0 by $f_z^0(\varphi_1)$ in (11.3.3),

$$f_z(t;\varphi_1,r_1) = f_z^0 \sum_{sq} \alpha_s \beta_q e^{i(sB+q)\varphi - is\omega_1 t} \qquad (11.3.8)$$

This expression includes, through the term $q = 0$, the case of uniform flow considered above. Similarly, the φ_1 component of the force becomes

$$f_\varphi(t;\varphi_1,r_1) = f_\varphi^0 \sum_{sq} \alpha_s \delta_q e^{i(sB+q)\varphi - is\omega_1 t} \qquad (11.3.9)$$

The relationship between the coefficients β_q, δ_q, and the flow velocity field will be quite complex in general, but in the case where the width of the propeller blade is sufficiently small, the expression (11.3.1) for the force should be a good approximation, even when V is a function of φ_1. Thus, if the nonuniform part of the flow velocity $\mathbf{V}_f$ is $\mathbf{v}(\varphi_1)$, the nonuniformity in the relative velocity $\mathbf{V}$ is also $\mathbf{v}(\varphi_1)$, since the blade velocity is constant. The corresponding nonuniformity in the force distribution is then obtained from Eq. (11.3.1) as

$$df_t = df_z = \frac{\rho}{2}\,[c_t d(V^2) + V^2\,dc_t] = \frac{\rho}{2}\left[c_t d(V^2) + V^2\frac{dc_t}{d\Gamma}\,d\Gamma\right]$$

The angle of attack Γ of the flow is $\Gamma = \Gamma_b - \Gamma_f$, where the angles Γ_b and Γ_f are defined in Fig. 11.18.

If the axial and tangential components of the flow velocity are V_{fz} and $V_{f\varphi}$ and the radial component is zero, the relative velocity can be expressed as

$$V = \sqrt{(V_b - V_{f\varphi})^2 + V_{fz}^{\,2}}$$

and the angle Γ_f is given by

$$\Gamma_f = \tan^{-1}\frac{V_{fz}}{V_b - V_{f\varphi}} \simeq \frac{V_{fz}}{V_b} \qquad \text{if } (V_{f\varphi}, V_{fz}) \ll V_b$$

If the components of the flow perturbation are v_z and v_φ and if $(v_z, v_\varphi) \ll V$, and furthermore, if $V \ll V_b$, we obtain $dV^2 \simeq -2V_b v_\varphi$ and $d\Gamma = -d\Gamma_f \simeq -v_z/V_b$. The corresponding expression above for the perturbation df_z in the z component of the force is then

$$-\frac{df_z}{f_z^0} = \frac{2v_\varphi}{V_b} + \frac{1}{c_t}\frac{dc_t}{d\Gamma}\frac{v_z}{V_b} \qquad (11.3.10)$$

In other words, instead of the constant force amplitude in (11.3.2), the expression for the total force is now

$$f_z = f_z^0\left[1 - \frac{2v_\varphi}{V_b} - \frac{1}{c_t}\frac{dc_t}{d\Gamma}\frac{v_z(\varphi)}{V_b}\right] \equiv f_z^0\sum_q \beta_q e^{iq\varphi_1} \qquad (11.3.11a)$$

with a similar expression for the drag force,

$$f_\varphi = f_\varphi^0\left[1 - \frac{2v_\varphi}{V_b} - \frac{1}{c_d}\frac{dc_d}{d\Gamma}\frac{v_z(\varphi)}{V_b}\right] \equiv f_\varphi^0\sum_q \delta_q e^{iq\varphi_1} \qquad (11.3.11b)$$

From this we can express the coefficients β_q and δ_q by a Fourier expansion of the flow field.

It is interesting in this connection to obtain the expressions (11.3.11) in a different and somewhat more direct manner. Since the force $\mathbf{f}$

(components f_i) depends on the relative velocity $\mathbf{V}$ (components V_j) between the propeller blade and the incident flow, it follows that, $\mathbf{V}$ being the only vector from which $\mathbf{f}$ can be constructed, the force components must have the form

$$f_i = \sum_j \alpha_{ij} \left(\frac{\rho V^2}{2}\right) \frac{V_j}{V} \tag{11.3.12}$$

where $\rho V^2/2$ is obtained from dimensional considerations. The dimensionless coupling coefficients α_{ij} may be functions of such dimensionless parameters as the Reynolds and the Mach numbers. These coefficients are assumed to be known from experiments. Splitting up the velocity $\mathbf{V}$ into a uniform part $\mathbf{V}^0$ and a perturbation $\mathbf{v}(\varphi) = (0,v_\varphi,v_z)$, it follows that the force f_z to first order in v becomes

$$f_z = \left[-\alpha_{zz}\frac{v_z}{V^0} + \alpha_{z\varphi}\left(1 - \frac{2v_\varphi}{V^0}\right)\right]\frac{\rho(V^0)^2}{2}$$

Also

$$f_\varphi = \left[-\alpha_{\varphi z}\frac{v_z}{V^0} + \alpha_{\varphi\varphi}\left(1 - \frac{2v_\varphi}{V^0}\right)\right]\frac{\rho(V^0)^2}{2}$$

Comparing these expressions with (11.3.11), we see that the coupling coefficients have the following meaning in terms of the thrust and drag coefficients:

$$\alpha_{z\varphi} = c_t \qquad \alpha_{\varphi\varphi} = c_d$$

$$\alpha_{zz} = \frac{dc_t}{d\Gamma} \qquad \alpha_{\varphi z} = \frac{dc_d}{d\Gamma} \tag{11.3.13}$$

The radiated sound

After having obtained the acoustic source function, we can now express the corresponding sound field, using the appropriate Green's function as demonstrated several times throughout this book. The acoustic source distribution is defined over the plane of the propeller over a circle of radius a, the radius of the propeller. A point source of unit strength placed at the position r_1, φ_1 in the plane produces the sound field $g_0 = \exp\left[ik\,|\mathbf{r} - \mathbf{r}_1|\right]/4\pi\,|\mathbf{r} - \mathbf{r}_1|$, where $k = \omega/c$ and $\mathbf{r}$ is the position vector of the point of observation. Expressing the position vectors in the spherical coordinates $\mathbf{r} = (r,\vartheta,\varphi)$ and $\mathbf{r}_1 = (r_1,\pi/2,\varphi_1)$, where $\vartheta = 0$ defines the axis of the propeller, we obtain in the far field $r \gg r_1$, $|\mathbf{r} - \mathbf{r}_1| \simeq r - r_1 \sin\vartheta\cos(\varphi - \varphi_1)$.

The sound field from the unit-strength harmonic point source at r_1 with a frequency ω is then

$$g_0 = \frac{1}{4\pi r}\exp\left[ikr - ikr_1\sin\vartheta\cos(\varphi - \varphi_1)\right]e^{-i\omega t} \qquad k = \frac{\omega}{c}$$

If we use the Fourier expansion of Eq. (1.2.9),

$$e^{iz \cos (\varphi - \varphi_1)} = \sum_{-\infty}^{+\infty} i^m J_m(z) e^{im(\varphi - \varphi_1)}$$

where J_m is the Bessel function of the mth order, the expression for g_0 can be written

$$g_0 = \frac{1}{4\pi r} e^{ikr} \sum_{-\infty}^{+\infty} i^m J_m(kr_1 \sin \vartheta) e^{im(\varphi - \varphi_1)} e^{-i\omega t}$$

As we have shown earlier, in Chap. 7, we can generate multipole fields from the monopole field simply by repeated differentiation with respect to the source coordinates. Thus the dipole-field sound field from a unit-strength point force in the z direction ($\vartheta = 0$) is simply $g_{1z} = -\partial g_0/\partial z = -(\partial g_0/\partial r) \cos \vartheta$.

$$g_{1z} = -ik \cos \vartheta \frac{e^{ikr}}{4\pi r} \sum_{m=-\infty}^{\infty} i^m J_m(kr_1 \sin \vartheta) e^{ik(\varphi - \varphi_1)} e^{-i\omega t} \quad (11.3.14a)$$

Similarly, the dipole field from a unit force in the φ_1 direction is

$$g_{1\varphi} = -\frac{1}{r_1} \frac{\partial g_0}{\partial \varphi} = \frac{i}{r_1} \frac{e^{ikr}}{4\pi r} \sum_m mi^m J_m(kr_1 \sin \vartheta) e^{im(\varphi - \varphi_1)} e^{-i\omega t} \quad (11.3.14b)$$

When applying these expressions for the calculation of the sound produced by the sth harmonic of the propeller-blade passage frequency $\omega_1/2\pi$, we have to set $\omega = s\omega_1$ and $k = s\omega_1/c = sk_1$ in the expressions for g_{1z} and $g_{1\varphi}$. The total sound field from the z component of the force distribution is then obtained as

$$p_z(r,\varphi,\vartheta) = \iint f_z(r_1,\varphi_1) g_{1z}(r_1,\varphi_1;r,\vartheta,\varphi) r_1 \, dr_1 \, d\varphi_1$$

with a similar expression for the field from the φ_1 component of the force. Inserting the expressions for f_z and g_{1z} as given by (11.3.8) and (11.3.14), we find that the integral over φ_1 is zero except when $m = sB + q$, in which case the integral is 2π, and the sound pressure becomes

$$p_\varphi(r,\varphi,\vartheta) = -\frac{ik_1 \cos \vartheta}{4\pi r} \sum_{sq} i^{sB+q} e^{isk_1 r} e^{i(sB+q)\varphi - is\omega_1 t}$$

$$\times \int_0^a sf_z^0 \alpha_s \beta_q J_{sB+q}(u) 2\pi r_1 \, dr_1 \, d\varphi_1 \qquad u = sk_1 r_1 \sin \vartheta \quad (11.3.15)$$

In general, f_z^0, α_s, and β_q are all functions of r_1. The quantities must be determined from the experimentally known aerodynamic characteristics of the propeller blades, and the integral in (11.3.15) must be evaluated numerically. However, formally, we can express it as

$$(\alpha_s \beta_q J_{sB+q})_{r_1 = a_z} \int f_z^0 2\pi r_1 \, dr_1$$

where a_z is an average value of the radial coordinate determined from

$$(\alpha_s B_q J_{sB+q})_{r_1=a_z} = (F_t)^{-1} \int_0^a f_z^0 \alpha \beta J_{sB+q} 2\pi r_1 \, dr_1 \qquad (11.3.16)$$

where

$$F_t = \int_0^a f_z^0 2\pi r_1 \, dr_1$$

is the total thrust force of the propeller. We can then rewrite (11.3.15) to express the sound pressure field from the z component of the force distribution in terms of the mean radius a_z and the total thrust force of the propeller.

$$p_z = (-ik_1 \cos \vartheta) F_t \frac{1}{4\pi r}$$

$$\times \sum s(\alpha_s \beta_q) J_{sB+q}(sk_1 a_z \sin \vartheta) \exp [isk_1 r + i(sB + q)(\varphi + \tfrac{1}{2}\pi) - is\omega_1 t]$$

In a completely similar manner we obtain the sound pressure p_φ produced by the φ component of the force

$$p_\varphi = \int f_\varphi g_{1\varphi} 2\pi r_1 \, dr_1$$

where f_φ and $g_{1\varphi}$ are given in Eqs. (11.3.11) and (11.3.14). There is one slight difference, inasmuch as the radial integration now contains an extra factor $(r_1)^{-1}$, the radial integral being

$$\int_0^a \alpha_s \delta_q f_\varphi^0 (r_1)^{-1} J_{sB+q} 2\pi r_1 \, dr_1$$

Again this integral has to be evaluated numerically, but we can always express it in terms of the total torque $T = \int (f_\varphi^0 r_1) 2\pi r_1 \, dr_1$ by introducing a mean radius a_φ such that

$$T \left(\frac{1}{a_\varphi}\right)^2 (\alpha_s \delta_q J_{zB+q})_{r_1=a\varphi} = \int_0^a \alpha_s \delta_q f_\varphi^0 (r_1)^{-1} J_{sB+q} 2\pi r_1 \, dr_1 \qquad (11.3.17)$$

It should be noted that the mean radii a_z and a_φ, determined from Eqs. (11.3.16) and (11.3.17), in general, are not the same. The sound field produced by the φ component of the force then can be expressed as

$$p_\varphi = \frac{T}{a_\varphi^2} \frac{1}{4\pi r} \sum_{s,q} (sB + q)(\alpha_s \delta_q)$$

$$\times J_{sB+q}(sk_1 a_d \sin \vartheta) \exp [isk_1 r + i(sB + q)(\varphi + \tfrac{1}{2}\pi) - is\omega_1 t]$$

If as an approximation we assume that the mean radii a_z and a_φ are the same, $(a_z, a_\varphi) = a_m$, the expression for the total sound pressure field $p = p_z + p_\varphi$ can be written in the form

$$p = \frac{1}{4\pi r} \sum_{s,q} \left(sk_1 F_t \cos \vartheta + T \frac{\delta_q}{\beta_q} \frac{sB + q}{a_m^2} \right) \alpha_s \beta_q J_{sB+q}(sk_1 a_m \sin \vartheta)$$

$$\times \exp [isk_1 r + i(sB + q)(\varphi + \tfrac{1}{2}\pi) - is\omega_1 t] \qquad (11.3.18)$$

Fundamental and harmonics

For numerical calculations and for comparison with experimental results, it is convenient to express the sound pressure field in real form. If we consider, first, the case of *uniform flow*, we have to set $q = 0$ and sum over all values of s. Thus, if n is a positive integer, we have to add terms corresponding to $s = n$ and $s = -n$ and then sum over n. Carrying out this summation and making use of $J_n(z) = J_{-n}(-z)$, we obtain for the pressure field

$$P_{q=0} = \frac{1}{4\pi r}\left(k_1 F_t \cos \vartheta + \frac{BT}{a_m{}^2}\right)$$

$$\times \sum_{n=1}^{\infty} n(2\alpha_n)J_{nB}(nk_1 a_m \sin \vartheta) \sin\left[nk_1 r + nB(\varphi + \tfrac{1}{2}\pi) - n\omega_1 t\right]$$

This sound field represents a pressure wave which spins with the angular velocity $\omega_1/B = \Omega$ of the propeller. The amplitude of the nth-harmonic component can be written

$$(p_n)_{q=0} = \frac{(2\alpha_n)k_1 F_t}{4\pi r}\left(\cos \vartheta + \frac{F_d}{F_t M}\right)nJ_{nB}(nBM \sin \vartheta)$$

where we have introduced the drag force $F_d = T/a_m$ and $M = a_m\Omega/c$, the Mach number of the propeller blade at the mean radius a_m. The value of the Fourier coefficients $2\alpha_n$ is obtained from Eq. (11.3.5), which in the special case when the pressure over the blade is assumed uniform results in Eqs. (11.3.6). For the first few harmonics, the value of $2\alpha_n$ is of the order of unity.

Ordinarily, the propeller blade at radius a_m is not traveling faster than the speed of sound, so that $M \sin \vartheta < 1$. In this case we can use the first term in the power series for the Bessel function [Eq. (5.2.20)] and can use the approximation $(nB)! \simeq \sqrt{2\pi nB}(nB)^{nB}e^{-nB-1}$ to obtain an approximate expression for the amplitude of the pressure wave radiated at frequency $n\Omega B/2\pi$.

$$(p_n)_{q=0} \simeq \frac{a_n\Omega}{2\pi cr}\sqrt{nB}\left(F_t \cos \vartheta + \frac{F_d}{M}\right)(\tfrac{1}{2}eM \sin \vartheta)^{nB} \qquad (11.3.19)$$

The factor $(\sin \vartheta)^{nB}$ peaks sharply at $\vartheta = 90°$, in the plane of the propeller. As long as $\tfrac{1}{2}eM \simeq 1.4M$ is less than unity, the amplitudes of the higher harmonics fall off fairly rapidly with increasing n, and if this were the sound produced by the usual airplane propeller, aircraft noise would not be the problem it is. The difficulty is that the flow past the propeller blade is *not* uniform, and we must include the terms for $q \neq 0$ in Eq. (11.3.18).

For nonuniform flow, we introduce the positive numbers n and l and sum the terms corresponding to $s = n$, $s = -n$, $q = l$, $q = -l$. In carrying out this addition, it is convenient to pair up the contributions from ($s = n$,

$q = l$) and ($s = -n,\ q = -l$) and from the terms ($s = n,\ q = -l$) and ($s = -n, q = l$).   Making use of $J_n(z) = J_{-n}(-z)$, we then find the following expression for the total sound pressure field:

$$p = \frac{1}{4\pi r} \sum_{n=1}^{\infty} (2nk_1\alpha_n) \left\{ \sum_{l=0}^{\infty} \left(\beta_l F_t \cos\vartheta + \delta_l \frac{nB - l}{nBM} F_d \right) J_{nB-l}(nBM \sin\vartheta) \right.$$

$$\times \sin[nk_1(r - ct) + (nB - l)(\varphi + \tfrac{1}{2}\pi)] + \sum_{l=0}^{\infty} \left(\beta_l F_t \cos\vartheta + \delta_l \frac{nB + l}{nBM} F_d \right)$$

$$\left. \times J_{nB+l}(nBM \sin\vartheta) \sin[nk_1(r - ct) + (nB + l)(\varphi + \tfrac{1}{2}\pi)] \right\} \quad (11.3.20)$$

We note that the sound field is composed of two distinct groups of waves, the first representing a pressure field which spins with an angular velocity, greater than Ω,

$$\omega_- = \frac{n\omega_1}{nB - l} = \frac{nB}{nB - l}\Omega \qquad (11.3.21)$$

in the positive φ direction.   The other wave spins with an angular velocity

$$\omega_+ = \frac{nB}{nB + l}\Omega \qquad (11.3.22)$$

which is always smaller than the angular velocity of the propeller.   This second kind of wave is negligible, compared with some of the first kind, as long as the propeller Mach number M is less than about 0.7, for the reasons we pointed out in connection with Eq. (11.3.19); Bessel functions of high order ($nB + l \gg 1$) and argument ($nBM \sin\vartheta$) smaller than order are quite small.

The first series over l, however, contains low-order Bessel functions; in fact, the term for $l = nB$ contains the factor $J_0(nBM \sin\vartheta)$, which has its maximum at $\vartheta = 0$ instead of $\vartheta = 90°$.   The phase of this wave is independent of φ; its amplitude is

$$\frac{1}{2\pi r} n\alpha_n k_1 \beta_{nB} F_t \cos\vartheta J_0(nBM \sin\vartheta)$$

which is maximum at $\vartheta = 0$ (along the axis of the propeller) and is zero in the plane of the propeller.   The amplitude of the component of the nth-harmonic wave, having a spin velocity $nB\Omega/m$ (where $m = nB - l$), on the other hand, is

$$\frac{\alpha_n nk_1}{2\pi r} \left(\beta_{nB-m} F_t \cos\vartheta + \delta_{nB-m} F_d \frac{m}{nBM} \right) J_m(nBM \sin\vartheta) \quad (11.3.23)$$

which is zero for $\vartheta = 0$, but does not peak sharply at $\vartheta = 90°$, at least when m is small.

Unless M is very close to 1, these terms are larger than the uniform-flow terms of Eq. (11.3.19), even if the nonuniformity factors β and δ are small, particularly in directions away from the equatorial plane of the propeller. Therefore, even a modest flow irregularity gives rise to a radiated sound pressure, which, at least at low Mach numbers, is considerably larger than the sound field from the uniform-flow component.

If an aircraft carrying such a propeller is moving through the air with velocity V_a, these formulas must be modified according to the discussion in Sec. 11.2. The motion modifies the directionality of the emitted sound, as measured by an observer at rest with respect to the fluid. For subsonic flight ($M_a = V_a/c < 1$), the angle ϑ in the foregoing formulas corresponds to the angle θ of Fig. 11.9; in other words, the ϑ for the sound which reaches an observer at O at time t is the angle between the direction of flight and the line of sight R measured at the time $t = R/c$, when this sound originated at the aircraft. All the directional factors will be modified by being multiplied by the factor $1/(1 - M \cos \theta)$, which accentuates the amplitudes ahead of the plane and diminishes those behind. In addition, of course, there are the Doppler changes of frequency and the further modifications of amplitude indicated in Eqs. (11.2.33) and (11.2.34).

Compressor noise

An axial-flow compressor such as in an aircraft jet engine is acoustically equivalent to a propeller which is operating in a cylindrical duct with rigid walls. The flow that enters through the rotating blade assembly of the compressor (the rotor) is generally not uniform, but is distorted by stationary guide vanes placed in front of each rotor. As a result, the flow velocity has a pronounced angular dependence, with a period determined by the number of guide vanes. Since the analysis of the sound produced in this case is quite analogous to the previous free-field radiation from a propeller in inhomogeneous flow, we leave the details of the analysis for the reader, and merely quote the result and make some comments about it.

Let the rotating blade assembly be located in a circular duct of radius r_0 in the plane $z = 0$, where z is the coordinate along the axis of the propeller. The angular frequency of the assembly shaft is Ω, and the fundamental angular frequency of the sound produced is

$$\omega_1 = B\Omega$$

where B is the number of blades. As before, we use Eqs. (11.3.7) and (11.3.9) to obtain the expression for the equivalent force distribution in the plane of the propeller. Thus for the axial force per unit area we have

$$f_z = f_{z0} \sum_{s,q=-\infty}^{\infty} \alpha_s \beta_q e^{i(q+sB)\varphi_1} e^{-is\omega t}$$

The tangential component has the same form, with f_z and f_{z0} replaced by f_φ and $f_{\varphi 0}$. The quantities f_{z0} and $f_{\varphi 0}$ are the time averages of the forces. The Fourier coefficients α_s relate to the force distribution over the propeller in uniform flow [see Eq. (11.3.5)] and the β_q's relate to the expansion of the flow inhomogeneity [see Eq. (11.3.11)].

Using the Green's function for the inside of a cylindrical tube (see Prob. 11.15), we find that the sound pressure produced by the force components f is

$$p = \sum_{s,q=-\infty}^{\infty} \sum_{n=0}^{\infty} \frac{1}{\pi r_0^2} \left(\pm F_t A_{mn}{}^{sq} + F_d B_{mn}{}^{sq} \frac{m}{\kappa_{mn} r_0} \right)$$

$$\tag{11.3.24}$$

$$\times \frac{J_m(\kappa_{mn} r) e^{ik_{mn} z}}{\left[1 - \left(\frac{m}{\kappa_{mn} r_0} \right)^2 \right] J_m{}^2(\kappa_{mn} r_0)} e^{im\varphi} e^{is\omega_1 t}$$

where, as before, $m = q + sB$. We have here introduced the total forces

$$F_t = 2\pi \int_0^{r_0} f_{z0}(r_1) r_1 \, dr_1 \quad \text{and} \quad F_d = 2\pi \int_0^{r_0} f_{\varphi 0} r_1 \, dr_1$$

and the coefficients $A_{mn}{}^{sq}$ and $B_{mn}{}^{sq}$ that express the coupling of these forces to the (m,n)th duct mode, as

$$A_{mn}{}^{sq} = \frac{1}{2F_t} \int_0^{r_0} \alpha_s \beta_q f_{z0} J_m(\kappa_{mn} r_1) r_1 \, dr_1$$

and a similar expression for $B_{mn}{}^{sq}$. The quantities κ_{mn} are the ratios $\pi \alpha_{mn}/r_0$ of Eq. (9.2.23).

Since the axial force produces an asymmetric field around the propeller plane $z = 0$ and the tangential force produces a symmetric one, we must use the plus sign in Eq. (11.3.24) for $z > 0$ and the negative sign for $z < 0$. (See Prob. 11.15 for a discussion of this equation.)

The propagation constant k_{mn}, of Eq. (9.2.6), is now

$$k_{mn} = \sqrt{\left(\frac{s\omega_1}{c} \right)^2 - \kappa_{mn}{}^2} \tag{11.3.25}$$

It follows that the (m,n)th mode will propagate if $|s\omega_1/c| > \kappa_{mn}$. For homogeneous flow, corresponding to $q = 0$ and $m = sB$, we have $J'_{sB}(\kappa_{mn} r_0) = 0$. For sufficiently large values of sB the smallest root κ_{m1} is approximately

$$\kappa_{m1} \simeq \frac{sB}{r_0}$$

Therefore the condition for the generation of a propagating mode in homogeneous flow is, approximately,

$$\frac{s\omega_1 r_0}{c} > sB \quad \text{or} \quad \frac{\Omega r_0}{c} > 1 \quad \text{or} \quad \frac{V}{c} > 1 \qquad (11.3.26)$$

where we have used $\omega_1 = \Omega B$. In other words, the condition is, roughly, that the tip speed $V = \Omega r_0$ of the rotor be larger than c. It comes as no surprise that this condition is much the same as that found in the study of sound generation from the moving corrugated board in Sec. 11.1. In most instances the tip speed of the rotor (with a radius equal to the duct radius) is indeed less than the speed of sound. Then, if the flow were strictly uniform across the duct, so that the load on all the blades was the same, the rotor would produce no propagating sound; all the wave modes would decay exponentially away from the rotor.

It is a well-known fact, however, that a jet-engine compressor *does* produce a very intense high-pitched screech which frequently dominates the noise of airports, so that the conclusion just reached about the absence of propagating modes is not applicable. In practice, however, the flow is never completely homogeneous, and modes other than those produced in homogeneous flow will be produced. Some of these modes represent spinning pressure patterns with angular velocities greater than the rotor velocity, as already indicated in Eq. (11.3.21). (The situation is much the same as when a wheel is photographed by a motion-picture camera. Depending upon the relationship between the angular velocity of the wheels, the number of spokes on the wheel, and the number of film frames per second, the projected motion of the wheel can have an angular velocity considerably different from the actual velocity of the wheel.)

Actually, according to this equation, the lth Fourier component in the angular expansion of the flow velocity gives rise to a spinning pressure pattern with an angular velocity $s\Omega B/(sB - l) = \omega_l$, where Ω is the angular velocity of the shaft, and s is the order of the harmonic of the emitted sound. The approximate condition for propagation of the corresponding acoustic mode produced by this pressure pattern, using $\omega_l r_0 > c$, then becomes

$$\frac{V}{c} > \left| \frac{sB - l}{sB} \right| = \left| 1 - \frac{l}{sB} \right|$$

In other words, in order for the mode to propagate, the tip speed V of the blade assembly need not exceed the speed of sound. In fact, the fundamental frequency $\omega = \omega_1$ corresponding to $s = 1$ will be produced at all speeds if $l = B$. Then a plane wave will be produced in the duct.

In the discussions of sound generation by a propeller, we have considered so far only the effect of the momentum transfer to the fluid corresponding to a dipole source distribution. Thus, if the pitch angle of the propeller blades is adjusted to give zero thrust of the propeller, this dipole source distribution will be eliminated. However, there remain other features of the interaction between the propeller and the surrounding fluid that give rise to sound. The fact that a blade occupies a certain volume means that when a blade moves into a region of the fluid, the volume of the fluid is displaced to make room for the blade. This "thickness effect" of the blade gives rise to an acoustic *monopole* source distribution.

To express this distribution in mathematical terms, we assign a propeller thickness function $S(r_1', \varphi_1')$ over the propeller plane, where the prime indicates coordinates attached to the propeller. This function is zero between the blades and equals the blade thickness at the position of the blades. When expressed in terms of the coordinates of the laboratory system, this function becomes $S(r_1, \varphi_1 - \Omega t)$. The rate of displaced air mass per unit area of the propeller plane at the position (r_1, φ_1) is then $Q(r_1 \varphi_1) = \rho(\partial S/\partial t) = -\Omega \rho S' = -\Omega \rho(\partial S/\partial \varphi_1')$. The corresponding acoustic source term in the wave equation for the sound pressure is then

$$\frac{\partial Q}{\partial t} = -\Omega \rho S'' = \Omega^2 \rho \frac{\partial^2 S}{\partial \varphi_1'^2} \qquad (11.3.27)$$

Once the source function is known, we can proceed to compute the radiated field in a manner completely analogous to the dipole field. We leave this calculation for one of the problems.

If we were to carry the analysis of the sound emission from a propeller even further, we should consider the influence of vibrations of the propeller blades. This effect, however, is ordinarily very small, and we shall not analyze it here. Instead, the problem of interaction of flow with an oscillating body will be discussed a little later, in terms of a simpler example.

In the preceding pages we have shown how the complicated flow field about a propeller can be described acoustically in terms of a monopole and a dipole source distribution that can be calculated from well-known characteristics of the propeller. Although these contributions are the dominant ones, and account quite well for experimental observations, there still remain other more complex contributions to the source function. One of them refers to the vortex shedding and turbulence about the blades. However, for an ordinary well-designed blade profile, the corresponding contribution to the sound field has been found to be small, and we shall not discuss it further in this context. Instead, we shall consider a situation in which the vortex sound actually is the predominant one.

Kármán vortex sound (Aeolian tone)

The interaction force between a rigid body in (subsonic) motion relative to the surrounding fluid is largely determined by the Reynolds number $R = VD/\nu$, where D is a characteristic dimension of the body transverse to the flow, V is the relative speed, and ν is the kinematic viscosity μ/ρ.

For sufficiently small Reynolds numbers, say, $R < 100$, the interaction force is mainly the viscous drag on the body which is proportional to the relative velocity. For example, the force on a sphere is given by the Stokes formula $f = 6\pi a \eta V$, which is familiar to physicists from the Millikan oil-drop experiment. Here η is the shear viscosity coefficient, V the velocity, and a the radius of the sphere. As the speed between the body and the fluid is increased, a value is reached at which the flow starts to separate from the body so that a wake is formed behind the body. The boundary between the wake and the surrounding fluid is a shear layer with a large transverse velocity gradient. This layer is unstable and breaks into turbulent motion farther down in the wake.

For large Reynolds numbers, say, $R > 10^5$, the fluid motion in almost the entire wake is random or turbulent. For lower Reynolds numbers, say, for $300 < R < 10^4$, strong periodic components of the fluid motion in the wake are frequently found. For example, the wake behind a cylinder exhibits a distinct transverse oscillatory motion in which vortices are shed off the cylinder alternately on one side and the other, forming the well-known Kármán vortex street in the wake. The trail of eddies is arranged in a zigzag pattern, and during the formation there is a periodic oscillatory transverse momentum component of the fluid motion in the immediate vicinity of the cylinder. (This phenomenon is not uncommon. We are reminded of the oscillatory or zigzag motion of a piece of paper falling through the air or of a flat stone or a lure moving in water. These motions no doubt are related to the Kármán vortex street.) As a result of the transverse oscillatory momentum transfer to the fluid, a sound field is produced equivalent to that of an oscillatory volume force acting on the fluid. The frequency and magnitude of this equivalent volume force can be determined by measuring the reaction force on the cylinder.

Experimentally, the frequency spectrum of the emitted sound in a free field has a predominant peak about a frequency given by

$$\frac{\omega}{2\pi} = \frac{\alpha V}{D} \qquad (11.3.28)$$

where $V =$ incident flow speed
$D =$ diameter of cylinder
$\alpha =$ so-called Strouhal number, approximately constant, $\alpha \simeq 0.2$, over a wide range of Reynolds numbers from 10^2 to 10^5

(Occasionally a slight indication of a second-harmonic component is found.) The width of the peak increases with the Reynolds number until $R \simeq 10^5$, by which value the spectrum becomes fairly uniform without any pronounced peaks. The fluid motion in the wake is then fully turbulent.

The magnitude of the transverse force per unit length of the cylinder is of the form

$$F_t \simeq \beta \frac{\rho V^2}{2} D$$

The value of β, found experimentally by various investigators, lies in the range between $\beta = 0.5$ and $\beta \simeq 2$. There can be several reasons for this scatter. In the first place, the degree of turbulence of the incident flow affects the formation of vortices about the cylinder. In fact, only if the incident flow is comparatively free from turbulence will there be a pronounced oscillatory motion. Second, the vortices shed off a cylinder have been found to be in phase over a comparatively short length of the cylinder, of the order of a few cylinder diameters. The correlation length Δ of the vortices has been found to depend on the total length of the cylinder and on the geometry of the flow system employed. Under such conditions some discrepancies between experimental data are to be expected. As a typical average value of β, we shall use $\beta \simeq 1$ in the following discussion. Correspondingly, a typical value of the coherence or correlation length Δ is about three to four times the diameter. The total acoustic power generated by the cylinder is then obtained approximately as the sum of the power contributions from each length Δ of the cylinder.

The wavelength of the emitted sound from the cylinder, using the frequency in (11.3.28), is

$$\lambda \simeq 5 \frac{c}{V} D$$

Since, ordinarily, $c/V \gg 1$, we have $\lambda \gg D$. Thus the wavelength ordinarily is much longer than the correlation length Δ, and the acoustic power emitted by a length Δ of the cylinder is approximately the same as that produced by an oscillatory point force

$$f_t = \beta \frac{\rho V^2}{2} D \Delta e^{-i\omega t} \tag{11.3.29}$$

To be somewhat more precise, f_t is the force applied to a length Δ of the cylinder by the turbulent layer of fluid surrounding it. If the cylinder does not move, this must equal the transverse force acting on the turbulence-free portion of fluid just outside this layer. If both D and Δ are small compared with λ, this can be equated to a point force f_t acting at the origin on moving fluid, with the cylinder absent. The corresponding sound pressure field can be obtained directly from the sound field of a moving dipole source already derived in (11.2.33). If in this expression we introduce the coordinates $\xi = x - Vt$, η, ζ, attached to the cylinder (cylinder along the ζ axis), we obtain the sound field from a stationary cylinder in a fluid moving in the negative ξ direction. The total emitted power per length Δ is then found by using the force (11.3.29) in (11.2.36):

$$\Pi_a = \frac{\pi}{12}\,\alpha^2\beta^2\left(\frac{\rho V^3}{2}\,\Delta^2\right)M^3 G_1{}^2(M) \qquad (11.3.30)$$

Ordinarily, the Mach numbers involved are so small that $G_1(M) \simeq 1$.

It is of interest to compare this acoustic power with the total flow-energy loss resulting from the steady drag force on the cylinder. Experimentally, this drag force has been found to be

$$f_d \simeq 1.3\,\frac{\rho V^2}{2}\,D\Delta$$

on a length Δ of the cylinder, and the corresponding rate of flow-energy loss (conversion into energy of turbulent flow) is

$$W_d \simeq 1.3\,\frac{\rho V^3}{2}\,D\Delta$$

The fraction of this energy that is converted into sound is, using Eq. (11.3.30),

$$\frac{W_a}{W_d} \simeq 0.01\,M^3\,\frac{\Delta}{D}$$

Effect of cylinder length

For a cylinder of length L, longer than the correlation length Δ of the oscillatory force along the cylinder, a somewhat more involved calculation is needed to obtain the radiated power. We assume a cylinder of diameter D and length L, its axis coinciding with the z axis and center at the origin, the fluid flow being in the x direction. The observer O (at rest with respect to the cylinder) is a distance r from the origin ($r \gg D, L$), the direction of r being given by the spherical angles ϑ (angle between r and the z axis) and φ (angle between the rz plane and the xz plane), as shown in Fig. 11.20. The driving force per unit length F_t is then pointed along the y axis. If M is small, Eqs. (11.2.6) and (11.2.33), combined with Eq. (11.3.30), indicate that, to the first order in M, the pressure wave at O, from an element of length dz of the cylinder at z, is

$$dp \simeq B(z)\,\frac{\sin\vartheta\,\sin\varphi}{(1 - M\cos\psi)^2}\,e^{ikr - ikz\cos\vartheta}$$

where $B(z) = (k\beta D/16\pi)\rho V^2 \exp\,[ig(z) - i\omega t]$, $k = 2\pi\alpha M/D$, $M = V/c$, and $\cos\psi = \sin\vartheta\cos\varphi$. Function $g(y)$ represents the fact that the phase of the force, and thence of p, is randomly distributed along the cylinder, with correlation length Δ. The total pressure p at O is the integral of this over z from $-\tfrac{1}{2}L$ to $+\tfrac{1}{2}L$, and the intensity at O is $|p|^2/\rho c$.

$$I = \frac{\sin^2\vartheta\,\sin^2\varphi}{\rho c(1 - M\sin\vartheta\cos\varphi)^4}\,\frac{1}{r^2}\int e^{\xi ik\cos\vartheta}\,d\xi\int B(z_1)B^*(z_1 + \xi)\,dz_1$$

where $\xi = z_2 - z_1$, and both integrals are over the length of the cylinder.

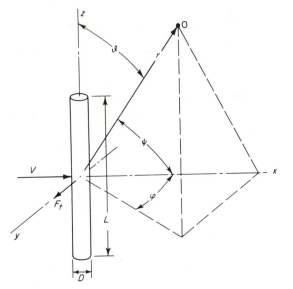

FIGURE 11.20
Cylinder generating an
Aeolian tone.

The second integral is, of course, L times the autocorrelation function Υ for B. If the correlation distance along z is Δ, a good approximation for Υ is $|B|^2 \exp(-\xi^2/2\Delta^2)$. Utilization of the familiar formula

$$\int_{-\infty}^{\infty} \exp\left(\frac{-\xi^2}{2\Delta^2} + iu\xi\right) d\xi = \sqrt{2\pi}\, \Delta \exp\left(-\tfrac{1}{2}u^2\Delta^2\right)$$

brings us to the desired equation for the intensity at O (assuming $L \gg \Delta$).

$$I \simeq \frac{\sqrt{2\pi}\alpha^2\beta^2}{32r^2} \frac{\rho V^3}{2} M^3 L\Delta \frac{\sin^2\vartheta \sin^2\varphi}{(1 - M\sin\vartheta\cos\varphi)^4} \exp\left[-\frac{1}{2}\left(\frac{2\pi\alpha M\Delta}{D}\right)^2 \cos^2\vartheta\right]$$

$$(11.3.31)$$

The integral of this over the sphere of radius r is the total power radiated.

$$\Pi \simeq \sqrt{\frac{\pi}{2}} \frac{\pi}{12} \alpha^2\beta^2\left(\frac{\rho V^3}{2} L\Delta\right) M^3 \equiv WLD(k\Delta)$$

When M and $k\Delta$ are small,

$$k\Delta = \frac{2\pi\alpha M\Delta}{D} \qquad W = \sqrt{\frac{\pi}{2}} \alpha\beta^2 \frac{\rho V^3}{48} M^2$$

which (aside from the numerical factor $\sqrt{\pi/2} \simeq 1.25$) differs from the result of Eq. (11.3.30), for L not large compared with Δ and λ, only by the factor L/Δ. For a long cylinder the power radiated is proportional to its length (not the square of its length), as well as to the correlation length.

When $2\pi\alpha M\Delta/D = k\Delta$ is less than unity (and M is small), the angle dependence of the intensity is, roughly, the same as for a point dipole pointed in the y direction.   This is the usual result, when the fluid flow past the cylinder has enough irregularities so that the correlation length Δ is roughly equal to the cylinder diameter D.   On the other hand, if the flow is smooth enough so that Δ is considerably longer than D, $k\Delta$ may be larger than unity, and the exponential function will concentrate the radiation near the equatorial plane ($\vartheta = 90°$).   When this is so,

$$\Pi \simeq \frac{\pi}{32}\,\alpha\beta^2\left(\frac{\rho V^3}{2}\,LD\right)M^2 = \frac{4}{3\sqrt{2\pi}}\,WLD \qquad M \text{ small} \qquad k\Delta \text{ large}$$

which is larger than the Π for $k\Delta$ small, but not as large as the ratio between the $k\Delta$'s.   In fact, when $k\Delta > 1$, Π becomes independent of $k\Delta$.

The occurrence of an oscillatory wake behind a body is not limited to cylinders only.   The basic condition for this phenomenon is the presence of a shear layer regardless of the shape of the body that produces the wake. The characteristic dimension D entering into the formulas (11.3.28) and (11.3.29) more appropriately should be the size of the wake (separation between two opposite shear layers).   Thus, for a plate inclined an angle θ with the flow direction, we have $D \simeq L \sin \theta$, where L is the width of the plate.

Stimulated vortex sound; whistles

The fluid oscillations about a cylinder and the corresponding sound field that we have just discussed concerned the conditions in free field, and may be referred to as *spontaneous* acoustic emission from the cylinder.   This is in contrast to what might be called *stimulated* generation of vortices and emission of sound when a sound field is present to reinforce the fluid motion about the cylinder.   The reinforcing sound may be externally produced, or it may be the sound from the cylinder itself, which has been reflected from surrounding surfaces.   Under proper conditions this reflected sound field will stimulate the formation of vortices and thus increase the intensity of the emitted sound.

Such an effect can be realized in a tube with fluid flow if a cylindrical rod is placed in the tube perpendicular to the axis.   If the flow speed is adjusted to a value such that the frequency of vortex shedding coincides with one of the transverse resonances in the tube, an intense sound field is produced.   The amplitude of the sound field is found to be considerably larger than would be expected on the basis of forced motion of the tube, using the free-field value of the oscillating force as a source function.   Therefore the sound emitted from the cylinder, after reflection from the tube walls, is not merely a matter of resonant excitation of a duct by a given dipole source, but rather the result of a reaction of the sound field back on the flow about the cylinder, drawing on the energy of the fluid flow.

Through this feedback mechanism the oscillatory force on the cylinder will depend on the amplitude of the sound, the system can become unstable, and large amplitudes can result.   The amplitude presumably will be limited mainly by acoustic losses due to viscous and conduction effects considered in Chap. 6.   Such a possibility for transforming the energy of steady flow into acoustic energy is of considerable practical interest, since it has been found to be responsible for the violent oscillations sometimes encountered in industrial test and processing facilities involving flow through tubes and ducts.   Not only will enormous sound emission result (sometimes several kilowatts of acoustic power), but mechanical failure of the duct systems frequently has been the result.

A similar instability will occur even in free field if, instead of being rigid, the cylinder is flexible in the transverse direction, with a resonance frequency coincident with the frequency of vortex shedding.   In principle, this situation is analogous to what occurred in the famous Tacoma Bridge disaster, in which enormous transverse oscillations occurred as the result of such an instability.

The conditions for stimulated sound emission from the vortex flow around a rod of diameter d set in a duct are easily obtained by setting the Kármán vortex frequency equal to the transverse resonance frequencies of the tube.   In the case of a rectangular duct with a transverse dimension D, the transverse resonance frequencies are given by $f_n = nc/2D$.   Therefore, using $f \simeq 0.2V/d$ for the vortex frequency as produced by the rod, given in Eq. (11.3.28), the condition for instability in the tube can be expressed as

$$\frac{V}{c} \simeq 2.5n \frac{d}{D} \quad \text{or} \quad \frac{V}{d} \simeq 2.5n \frac{c}{D} \qquad (11.3.32)$$

An additional condition is that the Reynolds number must lie in a region given approximately by $10^2 < R < 10^5$.   It is convenient here to express the Reynolds number as $R = Vd/\nu = \dfrac{V}{c}\dfrac{d}{l}$, where $l = \nu/c$ is of the order of the mean-free-path in the case of a gas, and $\nu = \mu/\rho$ is the kinematic viscosity. Then, introducing the value $V/c = 2.5nd/D$ from Eq. (11.3.32), we can express the additional condition for instability as

$$10^2 < 2.5n \frac{d^2}{Dl} < 10^5 \qquad (11.3.33)$$

For a gas at atmospheric pressure, $l \simeq 10^{-5}$ cm, and we get

$$10^{-3} < 2.5n \frac{d^2}{D} < 1 \qquad \text{cm}$$

As an example, a cylindrical rod with a diameter $d = 1$ in. is used in an air duct with $D \simeq 1$ ft; the first resonant instability ($n = 1$) will occur at a flow speed of about 24 ft per sec; and the frequency of oscillation is about 550 cps.

The $n = 2$ mode will be excited at a velocity of about 48 ft per sec with a frequency of about 1,100 cps, etc.   The spectrum of the sound in each case contains not only the fundamental frequency, but a number of intense harmonics as well (as many as 10 harmonics have been observed).

As an illustration of the second type of vortex-whistle mechanism, we shall consider the sound emitted from two or more identical cylindrical rods, with their axes perpendicular to the flow, placed one after the other along a line in the direction of the flow.   The vortex street produced by one rod interacts with the next rod.   This interaction gives rise to a pressure disturbance which affects the vortex shedding by the first rod.   If the pressure pulse arrives at the time of formation of a new vortex in a phase such as to stimulate the formation, the vortex is amplified, and through this feedback mechanism, strong acoustic emission results.

To obtain the condition for such a "resonance," it is important to realize that a vortex in the wake of a rod does not travel with the velocity V of the fluid, but somewhat slower.   Actually, for a cylindrical rod the vortex drift speed is $V_d \simeq 0.8V$.   If the distance between two rods is L, the vortex drift time between the rods is L/V_d.   The pressure pulse, produced when a vortex encounters the second rod, will reach the first rod after an additional time L/c, where c is the sound speed.   The total time elapsed between the formation of a vortex at the first rod and the arrival of the feedback pressure pulse from the second rod is then $(L/V_d) + (L/c) \simeq L/V_d$, where we have assumed $V_d < c$.   The feedback pulse has been found to stimulate vortex formation when this total time delay equals a multiple of the vortex shedding period, $T = 5d/V$, where d is the rod diameter.   The condition for resonance is then

$$\frac{L}{V_d} = 5n\,\frac{d}{V}$$

or with $V_d \simeq 0.8V$,

$$L \simeq 4nd \qquad\qquad (11.3.34)$$

Another way of interpreting this result is to say that resonance will occur when the distance L between two rods is a multiple of the distance between two consecutive vortices in the wake behind a rod.

In contrast to the first vortex-whistle mechanism, in which reflection of sound from the walls of the duct provided the feedback mechanism, the present mechanism relies upon the interaction of a vortex with a second object.   The characteristic length of significance in the feedback processes is the distance between the vortex source and the vortex receiver, and the characteristic velocity is the vortex drift speed.   Most vortex whistles can be described in terms of either one of these two types, or sometimes a combination of both.   For example, the well-known jet-edge whistle is analogous to the two-rod whistle.   Here a thin jet is emitted from a slit with sharp edges. Because of the shear layers produced, the jet is unstable for transverse perturbations, and the speed of these perturbations traveling down the jet is equivalent to the vortex drift speed in the previous example. The characteristic length involved is the distance between the slit and the edge which is placed in front of the slit. Resonance occurs when this distance is a multiple of the wavelength of the perturbations of the jet. For a given L, the whistle can operate in different modes, in which the distance between slit and edge is 1, 2, 3, etc., resonant wavelengths.

Similar oscillations occur also if, instead of a slit, we have an annular opening and a circular edge in front of the annulus.

Even the flow through a hole in a rigid partition, with a thickness comparable with the hole diameter, gives rise to feedback oscillations. The edge at the entrance of the hole serves as the vortex generator, and the edge at the other end of the hole as the vortex receiver, so that the characteristic length involved is the thickness of the partition. If a cavity resonator is added to this vortex generator, an intense sound can be produced.

The well-known excitation of cavity resonance by flow across the cavity opening involves a combination of the two basic mechanisms described above. The flow that passes the opening of the cavity defines the characteristic time V/d as in the jet-edge whistle. When this time equals the time of travel of an acoustic pulse being reflected from the walls of the cavity, strong resonance occurs; the approximate condition for oscillation in this case is $V/d = L/c$.

Effect of flow on acoustic impedances

To illustrate one effect of flow on an oscillating body, consider a one-dimensional mass-spring oscillator vibrating in a fluid which is moving in the direction of motion of the oscillator. If the velocity of the mass in the oscillatory motion is u, and the incident flow velocity is V, the force on the body caused by the flow is of the form

$$F = AC_d \frac{\rho(V - u)^2}{2} = AC_d \frac{\rho(V^2 + u^2)}{2} - AC_d \rho Vu \qquad (11.3.35)$$

where A is the area of the oscillating body, and C_d is the drag coefficient, discussed earlier in connection with the force on a propeller blade. Considering C_d to be independent of the velocity, it follows from Eq. (11.3.35) that the presence of the flow produces a time-dependent force on the oscillator which is equivalent to a friction force proportional to the velocity u of the oscillator. In addition, there is a force proportional to u^2 which contains twice the frequency of the oscillator, but if $u \ll V$, this force is of little importance. Nonlinear contributions to the force of this type will be discussed further in Chap. 13.

The acoustical characteristics of several other mechanical oscillators are influenced in a similar manner by flow. For example, the free oscillations of strings, membranes, and plates have been found to decay more rapidly when in relative motion with respect to the surrounding fluid. (There are conditions under which the flow can produce self-sustained oscillations of such systems, but this phenomenon will not be considered here.)

We return to the analysis of sound transmission through an aperture, discussed earlier in Chap. 9, to inquire about the effect of flow through an aperture on its acoustical impedance.   To study this question, we must first discuss the behavior of steady flow through the aperture.   As in the case of flow interaction with any other rigid body, discussed earlier, we have to resort to experimental facts.   At very low speeds the fluid motion no doubt is laminar, but for the velocities generally encountered in practice, the flow forms a turbulent jet on the downstream side of the aperture.   The critical Reynolds number $R = VD\rho/\mu$ at which flow separation and jet formation occur depends on the sharpness of the edges of the aperture, but generally the value is of the order of $R \simeq 100$.

The pressure in the jet at the hole is somewhat lower than the ambient pressure P_2 on the downstream side of the plate, but the jet pressure quickly adjusts to the ambient pressure as the jet becomes turbulent.

To get an idea of the effect of a periodic perturbation on the orifice flow, we shall consider sufficiently low frequencies so that we may neglect compressibility of the fluid in the aperture region.   The flow in the region between the upstream side of the plate and point 2 in Fig. 11.21, just before the flow becomes turbulent, is laminar, and we can apply the momentum equation in the form of Bernoulli's law for an incompressible fluid [Eq. (6.2.12)].

$$-\rho\,\frac{\partial\psi}{\partial t} + \frac{\rho}{2}\,V^2 + P = \text{const}$$

where ψ is the velocity potential, and $V = -\text{grad }\psi$.   Denote the incident flow speed and pressure by V_1 and P_1, and the velocity of the jet at point 2 by V_2.   The pressure at this point approximately equals the ambient pressure P_2 on the downstream side of the plate.   Furthermore, if s is the coordinate along a stream line, and V_s the corresponding velocity component (say, along the axis of the orifice), we have $\psi = -\int V_s\,ds$.   From Bernoulli's law, and

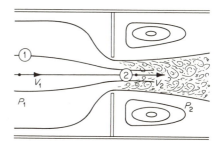

FIGURE 11.21
Flow through an aperture with a super-imposed sound field.

using the continuity relation $A_1 V_1 = A_2 V_2$,

$$\rho \int_1^2 \frac{\partial V_s}{\partial t} \, ds + \frac{\rho}{2} V_2^2 \left[1 - \left(\frac{A_2}{A_1} \right)^2 \right] = P_1 - P_2 \qquad (11.3.36)$$

where A_2 and A_1 are the cross-sectional areas at the stations 1 and 2, as indicated in Fig. 11.21.

Next we consider a small perturbation of the flow velocity, such that $V_s = V_s^0 + u_s$, with corresponding perturbations u_2, u_1, Δp in V_2, V_1, and $\Delta P = P_1 - P_2$. This perturbation represents a sound wave (incident) on the plate on the upstream side of the orifice. It follows from (11.3.36) that

$$\rho \int_1^2 \frac{\partial u_s}{\partial t} \, ds + \rho V_2^0 u_2 \left[1 - \left(\frac{A_2}{A_1} \right)^2 \right] = \Delta p$$

or

$$\frac{1}{\rho c} \frac{\Delta p}{u_1} = \frac{V_1^0}{c} \left(\frac{A_1}{A_2} - \frac{A_2}{A_1} \right) + \frac{\rho}{\rho c u_1} \int_1^2 \frac{\partial u_s}{\partial t} \, ds \qquad (11.3.37)$$

This expression represents the specific impedance of the orifice. The first term, which represents a resistance, accounts for the fact that the modulation of the jet by an oscillatory flow increases the rate of energy converted into turbulence. The second term is the mass reactance of the aperture, proportional to the M_a of Eq. (9.1.22). If we set $u_s = c(s)u_1$ and consider harmonic time dependence, we can express the integral as $-i\omega \rho l / \rho c$, where $l = \int_1^2 c(s) \, ds$ is a characteristic length that corresponds to the mass end correction of the aperture expressed in a different way in Sec. 9.1 [see, for example, Eqs. (9.1.26) and (9.1.28)]. The resistive term is proportional to the unperturbed flow velocity V_1^0. The reactive term corresponds to an end correction resulting from the convergence of the flow on the upstream side of the aperture. It is expected to be approximately one-half of the mass end correction in the absence of steady flow, because in the latter case the end corrections account for the flow convergence (and divergence) on *both* sides of the aperture. Measurements of orifice impedance in the presence of flow are in good (qualitative) agreement with the result in Eq. (11.3.37).

11.4 GENERATION OF SOUND BY FLUID FLOW

We have already considered some acoustical aspects of fluid motion that concerned the interaction of flow with a rigid body; a nonstationary flow impinging upon a body produces a time-dependent pressure distribution about the body, and as a result, sound is generated. Even in the absence of rigid bodies or boundaries, however, pressure fluctuation will be produced in the main body of a time-dependent flow field. This follows, for example, from Bernoulli's law, discussed in Chap. 6 [Eq. (6.2.12)]. The pressure fluctuations in the nonstationary fluid-flow region give rise to.(small) density fluctuations which propagate as sound to other regions of the fluid that may or may not be in motion. In any event, a nonstationary region of a fluid acts as a sound source, and the question that we wish to study in this section is the determination of the acoustic source strength and the corresponding sound field in terms of the flow field. If the fluid is turbulent, the velocity field and the pressure and density fluctuations are random functions of space and time, and only statistical properties (such as autocorrelation functions) of these quantities can be measured or predicted. Correspondingly, we can describe the sound field only in statistical terms. The results, of course, are similar to those carried out in Chap. 7 in connection with sound radiation from random source distributions.

Qualitative considerations

Before we get involved in mathematical analysis, however, we shall discuss those order-of-magnitude relationships between fluid motion and sound that can be understood from simple qualitative considerations. The mechanism of sound generation in the bulk of a turbulent fluid is different from most of the sources considered so far, inasmuch as the sound is not a direct result of the interaction of the fluid with vibrating boundaries or with external energy sources. In the present case the time dependence of the flow, required for sound emission, is caused by the instability of the fluid flow, so that even if the flow field is produced by a time-independent energy source, flow fluctuations and sound will be produced as a result of the flow instability.

A local fluctuation v in the fluid velocity will produce a local pressure fluctuation, which (from Bernoulli's law) will be of the order of ρv^2. This local-pressure-fluctuation field acts as a (monopole) source of sound and gives rise to a sound pressure, a distance R away, of the order of $p \simeq (d/R)\rho v^2$, where d is the size of the region over which the pressure fluctuation occurs. The corresponding radiated power $4\pi R^2(|p|^2/\rho c)$ then will be of the order of

$$\Pi_0 \approx \rho v^2 d^3 \frac{v}{c} \frac{v}{d} \tag{11.4.1}$$

Since $d/v = T$ is the characteristic time of fluctuation, this can be interpreted as the fraction v/c of kinetic energy $\rho v^2 d^3$ transformed into sound per unit time. Hence the "acoustic efficiency" of this flow monopole source is of the order of v/c, and the total acoustic output is proportional to the *fourth power* of the velocity fluctuation.

When we consider a larger volume V of the turbulent fluid rather than a (small) volume element d^3, this equation for Π_0, with V substituted for d^3, is valid only if there is a net pressure and corresponding density fluctuation in phase over the entire volume. If the turbulent region is isolated or surrounded by a quiescent region, however, there is no net mass transfer into the region, so that the net monopole strength will be zero. The pressure fluctuations in the fluid then will be distributed throughout the region, with positive and negative contributions which average to zero.

Next let us consider a pair of such flow monopoles, regions of pressure fluctuations with opposite signs. In Chap. 7 we have seen that two monopole sources of opposite sign represent a dipole source which produces a sound pressure of the order of d/λ times the monopole pressure, where d is the separation of the monopoles and λ the acoustic wavelength. For simplicity we have assumed this separation to be the same as the characteristic "size" of the monopole region used above. It follows, then, that the acoustic power emitted from a "flow dipole" will be of the order $(d/\lambda)^2$ times the power from the monopole. With a characteristic time of flow oscillations $T \simeq d/v$, we have $\lambda = c/T \simeq (c/v)d$, and we obtain for the acoustic power from the flow dipole, using (11.4.1),

$$\Pi_2 \approx \rho v^2 d^3 \left(\frac{v}{c}\right)^3 \frac{v}{d} \qquad (11.4.2)$$

We see that, in terms of the interpretation used for the monopole, the acoustic efficiency of the flow dipole is of the order of $(v/c)^3$, and the total emitted acoustic power is proportional to the *sixth power* of the velocity fluctuation.

This conclusion applies also to a finite region of the fluid if the region has a net flow-dipole strength. Since, as we have seen in Chap. 7, a dipole is equivalent to a force on the fluid, a finite dipole strength of the fluid region implies that there is momentum transfer from the region to the surrounding fluid. However, in the absence of external forces or boundaries acting on the fluid region, or if it is surrounded by a quiescent region of fluid, there will be no net momentum transfer and no net dipole strength. Therefore, in a region of velocity correlation in the fluid, two or more flow dipoles will occur, with opposite signs, so that the net dipole strength also is zero.

Thus, in the absence of both a net density fluctuation and a net momentum transfer to the (turbulent) flow region under consideration, the sound emission must be a result of quadrupole and higher-order sources (as was stated in Sec. 7.1). The acoustic pressure Π_4 from a flow quadrupole is of the order of $d/\lambda = d/(c/T) = v/c$ times the dipole pressure, and therefore the radiated power is of the order of $(v/c)^2$ times Π_2 in Eq. (11.4.2). Thus

$$\Pi_4 \approx \rho v^2 d^3 \left(\frac{v}{c}\right)^5 \frac{v}{d} \qquad (11.4.3)$$

This expression corresponds to the radiation from a single region of flow correlation (an "eddy") in the fluid of volume d^3. The acoustic efficiency is now proportional to $(v/c)^5$. In contrast to the monopole and dipole formulas, the quadrupole formula *can* apply to a finite region of the fluid, and was applied in obtaining Eqs. (7.1.28) and (7.1.37). The acoustic power from it is merely the sum of the power contributions from the various uncorrelated eddies in the fluid. The actual expression for this integrated power, although dimensionally of the form (11.4.3), involves additional constants which cannot be found from these simple considerations. For a region of isotropic turbulence we shall see that the constant factor to be used in (11.4.3) is of the order of 10^{-2}.

Thus, for subsonic velocity fluctuations with $M < 1$, the power removed from an eddy through sound radiation is generally very small. For example, with $v/c = 0.1$, the sound emission will remove about 10^{-5} percent of the kinetic energy of an eddy in one period of oscillation T. If sound were the only damping mechanism, the "lifetime" of an eddy would be large indeed. There are more important mechanisms, however, which damp out the eddy motion. For one thing, an eddy will feed energy to smaller eddies, which ultimately, through viscosity, go into heat.

Applying Eq. (11.4.3), we can get an idea about the sound emission from a turbulent jet. The fluctuating velocity component in the jet is known to be proportional to the average flow velocity V of the jet; typically, we have $v \approx 0.1V$. Furthermore, the size of the energy-carrying eddies in the flow is of the order of the diameter D of the jet. The quantity $\rho v^2 d^3 (v/d)$ in Eq. (11.4.3), then, is proportional to $\dfrac{\rho V^2}{2} \dfrac{\pi D^2}{4} V$, which is the kinetic energy per second being discharged from the jet nozzle. A certain fraction of this power will be radiated as sound, and it follows from (11.4.3) that the sound power is of the form

$$\Pi = (\text{const}) \, M^5 \frac{\rho V^3}{2} A \qquad M = \frac{V}{c} \qquad (11.4.4)$$

where A is the nozzle area. A large number of experiments have shown that the constant of proportionality in this formula is of the order of 10^{-4} for subsonic circular jets. The efficiency of conversion of jet energy into sound is then about $10^{-4}M^5$, which, for subsonic motion, is quite small. The jet energies involved, however, are often very large, and the acoustic powers emitted can be considerable, often of the order of kilowatts. This should be compared with the acoustic power from a human voice, which is about 10^{-6} watt.

Further analysis

We have already treated the radiation from a randomly fluctuating and turbulent region in Sec. 7.1, and we refer to that section as one approach to the mathematical analysis of our problem of sound generation by turbulence. We shall supplement that analysis by presenting here a somewhat different approach, plus some additional observations regarding this complex problem. There is no significant loss of generality if viscosity and heat conduction are left out of the discussion, and we shall do so. In the absence of an external force, viscosity, and heat conduction, the general equations of fluid motion

(6.1.13) and (6.2.11), expressing mass and momentum balance in the fluid, can be formulated as

$$\frac{\partial \rho}{\partial t} + \sum_i \frac{\partial(\rho V_i)}{\partial x_i} = 0$$

$$\frac{\partial(\rho V_i)}{\partial t} + \sum_j \frac{\partial(\rho V_i V_j)}{\partial x_j} = -\frac{\partial P}{\partial x_i}$$

(11.4.5)

where V_i ($i = 1, 2, 3$) stands for the components of the total fluid velocity, and ρ and P are the total fluid density and pressure. Under isentropic conditions the equation of state of the fluid can be expressed as $P = P(\rho)$, the pressure being a function of density only, and the compressibility of the fluid is given by

$$\kappa = \frac{1}{\rho} \frac{d\rho}{dP}$$

(11.4.6)

By differentiating the first of Eqs. (11.4.4) with respect to t and the second with respect to x_i, we obtain, using (11.4.6),

$$\rho \kappa \frac{\partial^2 P}{\partial t^2} - \sum_{ij} \frac{\partial^2(\rho V_i V_j)}{\partial x_i \partial x_j} = \sum \frac{\partial^2 P}{\partial x_i{}^2} = \nabla^2 P$$

(11.4.7)

where, as we recall from Eq. (6.2.4), $c^2 = 1/\rho\kappa$. In the absence of an external force field, viscosity, and heat conduction, this equation is a version of the general "wave equation" (7.1.21).

Both the compressibility κ and the density ρ, even in the absence of sound in the fluid, may be functions of position [Eq. (7.1.22)]. However, we have neglected a possible time dependence of $\rho\kappa = 1/c^2$. Before we proceed further, let us consider some general properties of Eq. (11.4.7). First we observe that if the compressibility is zero, the pressure field is fixed by the velocity field via the equation $\nabla^2 P = -\sum(\partial^2\rho V_i V_j/\partial x_i \partial x_j)$; the consequent pressure then is of the order of ρV^2. Under these conditions the ratio between the first and second term is of the order of $(L/T)^2/c^2$, where we have introduced a characteristic time T of pressure fluctuation in the fluid. In a turbulent fluid this characteristic time is of the order of $T \simeq L/V$, where L is the characteristic length L (correlation length), so that $(L/T)^2/c^2 \simeq (V/c)^2$. Under subsonic conditions, so that $V/c \ll 1$, compressibility has only a small effect, and the pressure field therefore is essentially the same as in an incompressible fluid. This pressure field we shall refer to as the *hydrodynamic*, or *aerodynamic*, pressure field. It is characterized by the property that the pressure fluctuation is proportional to the *square* of the velocity fluctuation. This is in contrast to a sound field, in which the sound pressure is proportional to the *first power* of the velocity fluctuation.

In a moving compressible fluid we generally have both hydrodynamic and acoustic pressure fluctuations; their relative strength ordinarily depends on the location of the point of observation. For example, inside a violent region of motion such as in a turbulent jet, the hydrodynamic pressure fluctuations predominate, whereas sufficiently far outside this region, the pressure in the sound field, emitted from the violent region of motion, is larger than the local hydrodynamic pressure fluctuations. In this far-field region the hydrodynamic motion of the fluid generally is very small, and the characteristic period of oscillation is long, compared with the periods in the sound field which has been generated by the rapid fluctuations in the violent region. If the typical angular frequency of the sound field is ω, the first term in Eq. (11.4.7) is of the order of $\omega^2 p/c^2 \simeq \omega^2 \delta$, where δ is the density fluctuation $\delta = p/c^2$. Similarly, the second term can be expressed as $\rho\Omega^2$, where we have introduced the characteristic frequency of the fluid motion $\Omega = V/L$. Consequently, the condition for predominance of the sound pressure field can be expressed as $\dfrac{\delta}{\rho}\left(\dfrac{\omega}{\Omega}\right)^2 \gg 1$.

But even if this condition is fulfilled, we cannot completely ignore the hydrodynamic motion of the fluid, because there will be some coupling between the fluid motion and the acoustic field, resulting in refraction and scattering of the sound field throughout the entire fluid (as discussed in Sec. 8.1). However, in many cases, as for a jet, there is a comparatively sharp boundary between the region of fluid motion and the outside "quiescent" region, and the refraction effects will be localized at the boundary. The refraction is then approximately the same as for a discontinuous change in fluid velocity, already considered in Sec. 11.1.

To account for all the inhomogeneities often encountered in, for example, turbulent jets, we let κ and ρ be functions of position. If the space average values are κ_0 and ρ_0, Eq. (11.4.7) can be written in the form

$$\rho_0\kappa_0 \frac{\partial^2 P}{\partial t^2} - \nabla^2 P = \rho_0(\kappa_0 - \kappa)\frac{\partial^2 P}{\partial t^2} + \left(\frac{\rho_0}{\rho} - 1\right)\nabla^2 P + \frac{\rho_0}{\rho}\frac{\partial^2 \rho V_i V_j}{\partial x_i \partial x_j} \equiv S(\mathbf{r}, t)$$

$$(11.4.8)$$

We leave the derivation for one of the problems [see also Eq. (8.1.11)].

Interaction between turbulence and sound

The foregoing discussion serves as a convenient starting point for most problems of sound generation and transmission in a loss-free inhomogeneous fluid, when external volume forces and heat sources are absent. Although it may be difficult to establish a clear separation between the "hydrodynamic" and "acoustic" contributions to the various field variables, the handling of this question in most cases of interest is fairly obvious. Suppose, for example, we are interested in the transmission of an acoustic signal from an external source through an inhomogeneous moving fluid, the state of which is time-independent. The sound wave then is simply represented by the time-dependent part of the field, and if the signal is sufficiently small, it can be treated as a perturbation to the fluid motion, as was done in Sec. 8.1.

The terms on the right-hand side in Eq. (11.4.8) then account for the scattering and refraction of sound from inhomogeneities in compressibility and density and from the motion of the fluid.   The latter effect is seen if we set $V_i = U_i + u_i$, where $\mathbf{U}$ is the unperturbed velocity, and $\mathbf{u}$ the perturbation.   The cross terms of the form $U_i u_j$ then provide the coupling between the sound wave and the fluid motion.

Another situation, in which the method of separating hydrodynamic and acoustic fields is fairly clear, refers to the generation of sound by a well-defined region of fluid motion, such as a turbulent jet outside which the fluid is practically at rest.   To obtain the sound field in the outside region, we integrate the source term multiplied by the appropriate Green's function over the source region.   In this region the hydrodynamic portions of the source term may be assumed predominant and known *a priori*.   The acoustic contribution to the source term in the source region accounts for the scattering and refraction of the emitted sound that it undergoes before reaching the outside region.   However, we shall ignore these effects in this discussion, and consider the source terms in Eq. (11.4.8) as known hydrodynamic variables in the source region, at least in a statistical sense.

Then, using an appropriate Green's function, we can express the radiated sound field directly in terms of integrals, as we have done several times before in this book.   For example, in Eq. (7.1.23), we worked with the Fourier-transformed wave equation, and in the absence of boundaries, the appropriate Green's function was g_ω, given in Eq. (7.1.6).   We shall now handle the problem a little differently and use the wave equation in its original form, without Fourier transformation [see also the derivation of Eq. (8.1.24)].   Then, in the absence of boundaries, the appropriate Green's function is as given in Eq. (7.1.18),

$$g(\mathbf{r},t \mid \mathbf{r}_0,t_0) = \frac{\delta\{t_0 - [t - (R/c)]\}}{4\pi R}$$

where $\mathbf{R} = \mathbf{r} - \mathbf{r}_0$.   If, for simplicity, we neglect inhomogeneity of ρ and κ in (11.4.8), the solution to this equation then can be written

$$p(\mathbf{r},t) = \int dt_0 \int dv_0 \left(\sum \frac{\partial^2 \rho V_i V_j}{\partial x_i \partial x_j} \right) \frac{\delta\{t_0 - [t - (R/c)]\}}{4\pi R} \qquad (11.4.9)$$

where the volume element dv_0 refers to the source region.   Through integration by parts, the far-field contribution in (11.4.9) can be expressed as

$$p(\mathbf{r},t) = \frac{1}{4\pi Rc^2} \frac{\partial^2}{\partial t^2} \int (\rho V_r^2) \, dv_0 \qquad (11.4.10)$$

which is analogous to Eq. (7.1.28).   Here $V_r = \Sigma x_i V_i / r$ is the component of the fluid velocity $\mathbf{V}$ in the direction $\mathbf{r}$ from the source region to the point of observation, and the parentheses following the integral indicate that the quantity inside is to be evaluated at the time $t - |\mathbf{r} - \mathbf{r}_0|/c$. It is interesting that only the velocity component V_r contributes to the radiated sound field at $\mathbf{r}$. This is related to the angular distribution of the sound field from the elementary quadrupoles in the region. Recall, for example, that the radiated sound field from a dipole is zero in a direction perpendicular to the direction of the dipole, and similar results apply to the quadrupoles.

If rigid boundaries are present so that $\partial P/\partial n = 0$, where $\mathbf{n}$ is the unit vector perpendicular to the boundary, it is advantageous to choose a Green's function G such that $\partial G/\partial n = 0$, as we have seen already in connection with (7.1.17). If we use this Green's function, there will be no contribution from integrals over the boundary. If, then, for example, we deal with a known turbulent flow field outside a boundary, we need not be concerned about the interaction between the flow and the boundary and the corresponding pressure distribution along the boundary. On the other hand, in some cases it may be of interest to relate the sound field to the pressure fluctuations over the boundary, and another choice of Green's function may then be appropriate. In this context we recall the discussion of sound generation from the Kármán vortex street in the wake behind a cylinder, in Sec. 11.3. In that case, the sound, in fact, is generated by the volume source distribution in the oscillatory flow in the vicinity of the rigid cylinder, but we replaced this source distribution by its net dipole strength, determined indirectly in terms of the reaction force on the cylinder. The validity of this equivalence can be formally proved, but we shall not pursue it here.

If, in Eq. (11.4.9), instead of expressing P in terms of $\mathbf{V}$ as we did earlier, we now express $\mathbf{V}$ in terms of P, using Eq. (11.4.7), the source term becomes a monopole distribution $\dfrac{1}{c^2}\dfrac{\partial^2 P}{\partial t^2} - \nabla^2 P$. Then, if we use the free-space Green's function of Eq. (11.4.10) and integrate by parts, the volume integral contributions to the far field from $\dfrac{1}{c^2}\dfrac{\partial^2 P}{\partial t^2}$ and $\nabla^2 P$ cancel each other, except for a term $\int \text{div}\,(g\,\text{grad}\,P)\,dv$, which can be transformed to a surface integral over the surface that encloses the source region. Then Eqs. (11.4.4) and (11.4.5) show that the sound pressure in the far field can be expressed as

$$p(\mathbf{r},t) = \frac{1}{4\pi r}\frac{d}{dt}\int (\rho V_n)\,dv_0 \qquad (11.4.11)$$

where V_n is the flow velocity component normal to the surface of integration. As before, the parentheses following the integral indicate that the quantity inside is to be evaluated at $t - |\mathbf{r} - \mathbf{r}_0|/c$. If the surface of integration is taken outside the turbulent region, the velocity component that enters is the acoustic velocity $\mathbf{u}$, and the formula then is merely an expression of Huygens' principle.

Description of turbulence

In a turbulent fluid the motion is random, and the velocity field, and hence the radiated sound pressure field, are stochastic functions; the detailed time dependence cannot be expressed in terms of simple functions. The procedure in calculating the corresponding emitted acoustic power was discussed in Chap. 7 for a particular form (Gaussian) of the correlation function of the velocity fluctuations. We shall extend the calculation here to consider the so-called (Kolmogorov) inertial subrange of isotropic turbulence, and obtain an expression, not only for the total intensity, but also the frequency spectrum. These quantities are conveniently obtained from the autocorrelation function for the sound pressure at the point of observation. This function is defined as

$$\Upsilon(\tau) = \langle p(\mathbf{r},t)p(\mathbf{r},t-\tau)\rangle \tag{11.4.12}$$

where the brackets indicate time average [see Eq. (1.3.13)]. From it we obtain the total acoustic intensity spectrum $E(\omega)$ from the Fourier transform

$$E(\omega) = 2\pi\rho c \int_{-\infty}^{+\infty} \Upsilon(\tau)e^{i\omega\tau}\,d\tau \tag{11.4.13}$$

and the total acoustic intensity is given by

$$I = \frac{\Upsilon(0)}{\rho c} = \frac{\langle p^2\rangle}{\rho c} = \int_{-\infty}^{+\infty} E(\omega)\,d\omega \tag{11.4.14}$$

If we express the sound pressure in terms of the fluid velocity field and introduce the notation

$$T_r = \rho V_r^2$$

the correlation function (11.4.12) can be expressed as [see Eq. (11.4.10)]

$$\Upsilon(\tau) = \left(\frac{1}{4\pi rc^2}\right)^2 \iint K_t(\mathbf{r}_0,\mathbf{r}_1,\tau)\,dv_0\,dv_1 \tag{11.4.15}$$

where

$$K_t(\mathbf{r}_0,\mathbf{r}_1,\tau) = \left\langle \frac{\partial^2 T_r(\mathbf{r}_0,t_0)}{\partial t^2}\frac{\partial^2 T_r(\mathbf{r}_1,t_1)}{\partial t^2}\right\rangle \tag{11.4.16}$$

and

$$t_0 = t - \frac{|\mathbf{r}-\mathbf{r}_0|}{c} \qquad t_1 = t - \tau - \frac{|\mathbf{r}-\mathbf{r}_1|}{c}$$

In other words, the acoustic emission from a turbulent fluid is uniquely specified by the function K, which in essence represents the correlation in the fluctuations of the kinetic energy at different locations. Equations (11.4.13) to (11.4.16) can be regarded as the formal solution to the problem of sound emission from fluid flow.

First let us consider a stationary region of turbulence with zero drift velocity. Under such conditions the characteristic period T of fluctuation v of the velocity is $T = L/v$, where L is the characteristic eddy size. In the presence of drift velocity $V_0 \gg v$, the corresponding relation would be $T \simeq L/V_0$, but if v is proportional to V, we may still set $T \simeq L/v$. To make a general estimate of the integrand K_t in Eq. (11.4.16), we set $\partial/\partial t \sim v/L$ and find, for the order of magnitude of K_t,

$$K_t \simeq \frac{\rho^2 v^8}{L^4}$$

We also observe, when carrying out the integration over r_1 in (11.4.15), that we shall get a contribution only for r_1 values within a distance of the correlation length L from r_0. Then, assuming that the total volume of the turbulent region is much larger than L^3, the integration over r_1 gives a value of the order of L^3, and the expression for the total emitted acoustic power from the turbulent region then can be expressed as

$$\Pi \simeq 4\pi r^2 I = \frac{4\pi r^2}{\rho c} \Upsilon(0) \sim \frac{\rho^2 v^8 L^2}{4\pi \rho c^5} \int \left(\frac{K_t}{\rho^2 v^8/L^4}\right)_{\tau=0} \frac{dv_0}{L^3} = C\left(\frac{v}{c}\right)^5 \frac{\rho v^3 L^2}{2}$$

$$(11.4.17)$$

which is consistent with (11.4.3) [compare with Eq. (7.1.38)]. The integral is dimensionless, and its value depends on the details of the flow in the turbulent region, but can always be expressed as the total volume of the region multiplied by some average value of the integrand. Therefore the constant factor in the last expression for Π can be regarded as proportional to the volume of the turbulent region, neglecting surface effects. It is interesting to compare this expression for the total radiated acoustic power with the mechanical power required to keep the turbulent region in motion. In the case of statistically stationary fluid motion, there must be a source which supplies mechanical power to the fluid, and this power, through the turbulent motion of the fluid, eventually is dissipated by viscosity and other loss mechanisms and by acoustic radiation. The mechanical power from the source first goes into kinetic energy of the large energy-carrying eddies, which are of the characteristic size of the driving mechanism. Here viscosity plays a minor role, and if the power transferred per unit mass of the fluid is ϵ, it follows from dimensional considerations that the velocity v is of the order of $(\epsilon L)^{\frac{1}{3}}$. Inserting this expression for the velocity in (11.4.17), and realizing that in this expression the constant C is proportional to the volume of the turbulent region divided by L^3, we see that the acoustic power Π emitted per unit mass of the fluid can be expressed as

$$\Pi \simeq (\text{const})M^5 \epsilon \qquad (11.4.18)$$

This shows that the efficiency of conversion of fluid-flow energy into sound via turbulent fluctuations is proportional to the fifth power of the Mach-number fluctuations in the flow. In the particular case of homogeneous isotropic turbulence, the dimensionless constant in Eq. (11.4.18) can be shown to be approximately equal to 38, but we shall not derive this result here.

Frequency distribution

These considerations referred only to the total emitted acoustic power. It is interesting to inquire also about the spectral distribution of this power. If we limit the discussion to statistically stationary turbulence, the time dependence of the function K involves only the time difference $t_0 - t_1$.

$$t_0 - t_1 = \tau - \left(\frac{|\mathbf{r} - \mathbf{r}_0|}{c} - \frac{|\mathbf{r} - \mathbf{r}_1|}{c}\right) \simeq \tau - \frac{\mathbf{Y} \cdot \mathbf{r}}{rc}$$

We have introduced here $\mathbf{Y} = \mathbf{r}_0 - \mathbf{r}_1$. If the fluid is homogeneous, the correlation function depends only on this difference, and not on the actual location of the points $\mathbf{r}_0$ and $\mathbf{r}_1$. In an inhomogeneous fluid this is not so, and it is then convenient to introduce, in addition to $\mathbf{Y}$, the variable $\mathbf{X} = \mathbf{r}_0 + \mathbf{r}_1$. Furthermore, if we introduce the time $\tau_0 = \mathbf{Y} \cdot \mathbf{r}/rc$ for sound to travel a distance Y, the expression for the correlation function takes the form

$$\Upsilon(\tau) = \left(\frac{1}{4\pi rc^2}\right)^2 \iiint d^3X \iiint K_t(\mathbf{Y}, \tau - \tau_0, \mathbf{X}) \, d^3Y \qquad (11.4.19)$$

The Fourier transform of the correlation function yields the intensity spectrum

$$E(\omega) = \frac{1}{16\pi^2 r^2 \rho c^5} \iiint d^3X \iiint \phi(\mathbf{Y}, \omega, \tau_0, \mathbf{X}) \, d^3Y \qquad (11.4.20)$$

where

$$\phi(\mathbf{Y}, \omega, \tau_0, \mathbf{X}) = \frac{1}{2\pi} \int_{-\infty}^{+\infty} K_t(\mathbf{Y}, \tau - \tau_0, \mathbf{X}) e^{i\omega\tau} \, d\tau$$

For homogeneous turbulence the function K is independent of $\mathbf{X} = \mathbf{r}_0 + \mathbf{r}_1$. The integral over d^3Y is then simply the volume of the turbulent region. Consequently, the intensity spectrum radiated per unit volume of the source region has the form

$$e(\omega) = \frac{1}{32\pi^3 r^2 \rho c^5} \iiint d^3Y \int_{-\infty}^{+\infty} K_t(\mathbf{Y}, \tau - \tau_0) e^{i\omega\tau} \, d\tau \qquad (11.4.21)$$

We recall that τ_0 represents the time of travel of sound a distance Y, or since K_t is substantially zero for Y larger than the typical eddy size L, we can say that τ_0 is approximately the time of travel across an eddy. Then, if the frequency of interest is such that the period of the sound wave is smaller than τ_0 (or the wavelength smaller than L), there will be interference between the sound emitted from different parts of an eddy. There is also the possibility that during the time τ_0 the state of an eddy has changed; i.e., the lifetime of the eddy is small compared with τ_0. Under these circumstances it is important to retain τ_0 in the equation. We can easily see, however, that if the fluctuating Mach number is considerably smaller than unity, the lifetime of an eddy is large compared with τ_0 (the characteristic time of the eddy is $T = L/v$). Then, if we restrict ourselves to the case of low Mach numbers, and also to wavelengths large compared with the typical eddy size, we are justified in setting $\tau_0 = 0$ in the equation.

With this simplification we turn to the question dealing with the form of the function K_t. If we have a statistically stationary turbulent region, it is clear that the particular form of the driving mechanism that maintains the turbulent flow must influence the structure of the turbulence, at least for eddies of the order of the size of the driving mechanism. These eddies receive their energy directly from the driver, and this energy is transmitted to smaller eddies through nonlinear coupling, and eventually the energy is degraded to the "microscale" eddies which give up their energy to heat. However, there is a range of eddy sizes, the so-called inertial subrange, for which the statistical properties of the flow are supposed to be independent both of the driving mechanism and of viscosity. The size of the eddies in this inertial, or similarity, range is smaller than the size L of the largest eddies, but is larger than the size l of the microscale eddies, which are influenced more by viscosity than by inertial effects. From dimensional considerations it follows that the size of the microeddies is of the order of $l \approx (\nu^2/\epsilon)^{\frac{1}{4}}$, where ν is the kinematic viscosity, $\nu = \mu/\rho$.

Thus, in the inertial subrange, the magnitude of Y will lie in the range $L > Y > l$. If in this range the statistical properties of the turbulence are independent of the geometry of the driving mechanism and of viscosity, the only quantities that should enter into the description of the motion are the eddy size R, the power ϵ transferred to the fluid per unit mass, and the density ρ of the fluid. (The power ϵ is transferred from larger eddies to smaller ones until it is transformed into heat.) Using these quantities, we can construct from dimensional arguments expressions for the quantities entering our acoustical problem. Thus, for the characteristic velocity, we obtain $v \sim (\epsilon Y)^{\frac{1}{3}}$, and for the characteristic time of oscillation of an eddy of size Y, we get $T = Y^{\frac{2}{3}}\epsilon^{-\frac{1}{3}}$. From dimensional considerations it follows, then, that the function K_t will be of the form $K_t \sim \rho^2(\epsilon Y)^{\frac{8}{3}}\phi$, where ϕ is a dimensionless function of the argument $\tau/T = \tau/Y^{\frac{2}{3}}\epsilon^{-\frac{1}{3}}$. Thus, if we use

$$K_t \sim \rho^2(\epsilon Y)^{\frac{8}{3}}\phi\left(\frac{\tau}{Y^{\frac{2}{3}}\epsilon^{-\frac{1}{3}}}\right)$$

in (11.4.21) to obtain the acoustic power spectrum $4\pi r^2 e_i(\omega)$ of the sound radiated from the inertial subrange, we get, with $dv_Y = 4\pi Y^2 dY$,

$$4\pi r^2 e_i(\omega) = \frac{4\pi}{8\pi^2\rho c^5}\int_l^{L_i} Y^2 dY \int_{-\infty}^{+\infty} K_i(Y,\tau)\phi(x) dx \qquad (11.4.22)$$

where L_i is the largest eddy size in the inertial range. The lower limit on this integral goes to zero as the kinematic viscosity goes to zero. With this choice of lower limit, and using the dimensionless variables $x = \tau/Y^{\frac{2}{3}}\epsilon^{-\frac{1}{3}}$ and $y = \omega Y^{\frac{2}{3}}\epsilon^{-\frac{1}{3}}$, we obtain

$$4\pi r^2 e_i(\omega) = \tfrac{4}{3}\pi\rho c^2 M^{\frac{21}{2}}\left(\frac{c}{\omega L}\right)^{\frac{1}{2}}\int_0^{y_m} y^{\frac{5}{2}} dy \int_{-\infty}^{+\infty} e^{iyx}\phi(x) dx$$

where $y_m = \omega L_i/cM$. To obtain the acoustic contribution from the rest of the eddies with $L_i < Y < L$, we must add to this integral a portion in which K_t does not have the similarity form but depends on the geometry of the driving mechanism. However, as $\omega \to \infty$, the upper limit in the integral goes to infinity, and the additional contribution from the larger eddies becomes negligible. Therefore, for $\omega \gg cM/L$, the acoustic spectrum should be "universal," in the sense that it does not depend on the geometry of the driving mechanism. The acoustic power spectrum in this range is

$$4\pi r^2 e(\omega) \simeq (\text{const})\rho c^2 M^{\frac{21}{2}}\left(\frac{c}{\omega L}\right)^{\frac{7}{2}} \qquad \omega \gg \frac{cM}{L} \qquad (11.4.23)$$

the constant involving the value of the double integral, with the integral over y extended to ∞.

This result indicates that the power-spectrum density increases as $M^{\frac{21}{2}}$ and decreases with frequency as $\omega^{-\frac{7}{2}}$ in the high-frequency range, whereas the total power, including the entire spectrum, is proportional to M^8.

A simplified line source

After these general considerations it is instructive to work through a simple specific example. Let us consider the sound emission from an idealized model flow field that may be used as a very simple model of a jet. When viewed from far away, a jet may be regarded approximately as an acoustic line source of finite length L, with a source strength that varies with position, and we shall use this model in the present example. The acoustic source strength per unit length of this source, as we have seen, is determined by the velocity fluctuation in the direction along the line of observation. For simplicity, we assume that the time dependence of v_r is a result of convection of a "frozen" pattern of flow irregularity. This means that in a coordinate system moving with the mean speed V_0 of the flow, the time dependence of the flow irregularity (eddies) is only that corresponding to the (sudden) creation at $x = 0$ and absorption or destruction of the eddies at $x = L$. The speed of convection is the mean velocity of the jet flow. To make the calculations simple, the spatial variation of v_r is assumed to be harmonic in the region $0 < x < L$, and zero elsewhere. The time dependence of v_r at a point x is then given by

$$v_r = \begin{cases} v_r = 0 & x < 0; \; x > L \\ v_0 \sin \dfrac{2\pi}{\Lambda}(x - V_0 t) \equiv v_0 \sin(Kx - \Omega t) & 0 < x < L \end{cases} \qquad (11.4.24)$$

where $K = 2\pi/\Lambda$ is the wavenumber corresponding to the "wavelength" Λ of the flow irregularity, and $\Omega/2\pi = V_0/\Lambda$ is the frequency of flow fluctuation as seen by an observer in the laboratory system. With $T_r = \rho v_r^2$, we have

$$\frac{\partial^2 T_r}{\partial t^2} = \frac{\rho v_0^2}{2} (2\Omega)^2 \cos\left[2(Kx - \Omega t)\right]$$

To evaluate the function K_t that occurs in (11.4.16), we have to evaluate $\partial^2 T_r / \partial t^2$ at (x_0, t_0) and (x_1, t_1), where

$$t_0 = t - \frac{|\mathbf{r} - \mathbf{r}_0|}{c} \simeq t - \frac{r}{c} + \frac{x_0 \cos\theta}{c}$$

$$t_1 = t - \tau - \frac{|\mathbf{r} - \mathbf{r}_1|}{c} \simeq t - \tau - \frac{r}{c} + \frac{x_1 \cos\theta}{c}$$

Here θ is the angle between $\mathbf{r}$ and the x axis. Using these expressions in Eq. (11.4.16), we obtain

$$K_t(x_0, x_1, \tau) = \left\langle \left(\frac{\partial^2 T_r}{\partial t^2}\right)_{r_0, t_0} \left(\frac{\partial^2 T_r}{\partial t^2}\right)_{r_1, t_1} \right\rangle$$

$$= \tfrac{1}{2} v_0^4 \Omega^4 \cos\left\{2[K(x_0 - x_1)(1 - M_0 \cos\theta) - \Omega\tau]\right\}$$

where $M_0 = V_0/c$. Next we have to integrate K_t over the range $0 < x_0 < L$, $0 < x_1 < L$. If the cross-sectional area of the jet is A, we have $d^3 r_0 = A\, dx_0$ and $d^3 r_1 = A\, dx_1$, and after integration over x_0 and x_1, we find

$$\Upsilon(\tau) = A^2 \int\!\!\int_0^L K_t \, dx_0 \, dx_1 = \left(\frac{AL}{4\pi rc^2}\right)^2 \frac{v^4 \Omega^4}{2} \cos(2\Omega\tau) \frac{\sin^2(\alpha KL)}{(\alpha KL)^2} \qquad (11.4.25)$$

$$\alpha = 1 - \frac{V_0}{c} \cos\theta$$

According to Eq. (11.4.14), the total emitted intensity radiated in the direction θ is

$$I(\theta) = \frac{\Upsilon(0)}{\rho c} = \frac{\pi^2}{r^2} \frac{A}{\Lambda^2} \left(\frac{L}{\Lambda}\right)^2 \frac{A\rho v_0^4 V_0^4}{c^5} \frac{\sin^2(\alpha KL)}{(\alpha KL)^2} \qquad (11.4.26)$$

The frequency spectrum of the emitted sound in this simple model is proportional to $\delta(\omega - 2\Omega)$, so that only the one frequency 2Ω is present. With $\Omega = 2\pi V_0/\Lambda$, we note that the intensity of the radiated sound is proportional to $v_0^4(\Omega)\Omega^4$, where we have indicated that the velocity amplitude v_0 generally will depend on Ω or Λ. The corresponding total radiated power is

$$\Pi = \int 2\pi r^2 I(\theta) \sin\theta \, d\theta$$

$$= 2\pi^3 \frac{A}{\Lambda^2} \left(\frac{L}{\Lambda}\right)^2 \frac{\rho v_0^4 V_0^4 A}{c^5} \frac{1}{KLM_0} \int_{(1-M)KL}^{(1+M)KL} \frac{\sin^2 y}{y^2} \, dy \qquad (11.4.27)$$

where we have assumed v_0 independent of θ. In this simple model the velocity fluctuations are correlated over the entire length L of the jet, but under more general conditions, the correlation length may be only some fraction of L. In that case we should obtain an expression of the form (11.4.26) for each length of the jet equal to the correlation length. The factor $\sin^2(\alpha KL)/(\alpha KL)^2$, which expresses the interference that takes place between sound emitted from different points along the correlation distance, has a maximum value of unity at an angle given by $\cos \theta = 1/M_0$. In other words, a true maximum in the angular distribution resulting from this factor occurs only if $M_0 > 1$. We also note that the maximum gets sharper as $KL = 2\pi L/\Lambda$ gets larger, i.e., as the frequency V_0/Λ of the emitted sound increases.

Turning to the total emitted power, we see that if $M_0 < 1$, the interference factor $\int (\sin^2 y/y^2)\,dy$ rapidly goes to zero as $L \to \infty$, and the power radiated per unit length of the jet then goes to zero. This behavior depends on the simplified model used here, in which there is no time dependence of the flow in the jet in a coordinate system moving with the mean velocity of the jet. On the other hand, if $M_0 > 1$, we see that, as $L/\Lambda \to \infty$, the interference factor $\int_0^\infty (\sin^2 y/y^2)\,dy$ equals $\pi/2$ for $M_0 > 1$, and the power then becomes *proportional to L*. If, on the other hand, the jet length is short, so that $L/\Lambda \ll 1$, the interference integral is proportional to L, and according to Eq. (11.4.20), the power then is *proportional to L^2*.

To simulate the conditions in a turbulent jet, we have to set $v_0 \sim V_0$ (typically, $v_0 < 0.1V_0$), and we see that, apart from the interference factor $\int (\sin^2 y/y^2)\,dy$, the acoustic power is proportional to V_0^8, consistent with the qualitative result in Eq. (11.4.4). As we have seen in this analysis, there is no radiation (per unit length of the jet) for a subsonic jet of infinite length; this is related to the fact that, in a coordinate system moving with the speed of the jet, the flow is time-independent. For a *finite* jet, however, there will be a radiated sound even at subsonic speed; this depends on the fact that, in a coordinate system moving with the speed of the jet, the flow is no longer time-independent, since there is a creation of a flow irregularity at $x = 0$ and a destruction thereof at $x = L$. The corresponding (single) "lifetime" of an eddy then can be considered to be $\tau \simeq L/V_0$. In practice, there is a distribution of lifetimes among the various eddies in the flow, and for each such lifetime there will be a corresponding length of travel $L_i = V_0\tau_i$ before the destruction of the eddy. Under these circumstances the flow always will be time-dependent, even in the coordinate system moving with the jet, since there is a continuous creation, decay, and destruction of eddies over the length of the jet.

The simple model of sound emission from a jet considered here can easily be extended to incorporate some of these features so that it may be applied to an arbitrary convected flow irregularity containing a continuous distribution of wavelengths and decay times. To put these observations in quantitative terms, we should introduce an appropriate correlation function for the jet flow and proceed to evaluate the correlation function for the sound pressure field as given in Eq. (11.4.15). Or we could follow the analysis in Sec. 7.1 dealing with the sound radiation from a randomly fluctuating fluid region. In attempting such an analysis, however, considerable difficulties are encountered because the jet flow is neither homogeneous nor isotropic. Furthermore, the mean flow velocity is not the same throughout the jet, and the large velocity gradients transverse to the jet axis couple with the turbulent fluctuations to give rise to large fluctuations in the momentum flux tensor. In addition, the mean velocity gradients cause refraction of the sound as it is transmitted from the jet to the surrounding quiescent atmosphere. The free-space Green's function used in Sec. 7.1 then is no longer appropriate. Qualitatively, the effect of refraction out from the jet is much the same as described in connection with Fig. 11.7.

Effect of translational motion of turbulence

A detailed analysis of the noise from a jet, accounting for the various effects mentioned above, will not be carried out here. Actually, for such an analysis to be meaningful, a better knowledge of the relevant flow correlations is required than is presently available. However, in view of the fact that, in Sec. 7.1, we have already treated the sound emission from a region of stationary turbulence, it is instructive in this context to determine the influence of translational motion of the turbulent region on the sound emission. We exclude the effect of refraction. The basic effect of the flow is to alter the autocorrelation function and the spectrum density of the excitation functions; they must be expressed in terms of coordinates at rest with respect to the observer. The transformation of coordinates, if the mean velocity of the turbulent region is U, in the direction of the z axis, is $x' = x$, $y' = y$, $z' = z - Ut$, and $t' = t$. The space and time correlation becomes mixed, and if, for example, $\Upsilon_s(\mathbf{d}, \tau)$ and $|S(\mathbf{K}, \omega)|^2$ of Eqs. (7.1.32) and (7.1.33) are the correlation function and spectrum density of the source function in coordinates moving with the jet at velocity U in the z direction, then the corresponding functions in coordinates at rest with respect to the observer and the surrounding air are

$$\Upsilon_s(d - U\tau, \tau) = (\Delta_s^2) \exp\left[-\frac{1}{2}\frac{(d - U\tau)^2}{w_s^2} - \frac{1}{2}(\omega_s\tau)^2\right]$$

$$S_s(\mathbf{K}, \omega - \mathbf{U}\cdot\mathbf{K}) = \frac{VT}{(2\pi)^6}\frac{w_s^2}{\omega_s}(\Delta_s^2) \exp\left[-\tfrac{1}{2}K^2 w_s^2 - \frac{(\omega - \mathbf{U}\cdot\mathbf{K})^2}{2\omega_s^2}\right]$$

(11.4.28)

In this system of coordinates the correlation between the same portion of fluid ($d' = 0$) at different times must be between the portion at $\mathbf{r}$ at $t = 0$ and the portion at $\mathbf{r} + \mathbf{U}\tau$ at time $t = \tau$. Correlation between times an interval τ apart, for $d = 0$, corresponds to comparison between two parts of the fluid a distance $U\tau$ apart in the moving fluid. The Fourier transform of this, the spectrum density $|S_s|^2$, has its frequency transformed to $\omega - \mathbf{U} \cdot \mathbf{K}$ because of the motion. Because of the motion, the frequency distribution of the radiation is shifted by an amount $\mathbf{U} \cdot \mathbf{K}$, in accordance with the Doppler shift in frequency of the moving source.

Equations (7.1.35) and (7.1.36) for the intensity at frequency $\omega/2\pi$ and the total intensity, at the distant observation point in quiet air outside the turbulent region, become

$$I_{s\omega} = \rho c \langle \Delta_s{}^2 \rangle \frac{\omega^2 VT}{16\pi^2 c^2 r^2} \frac{w_s{}^2}{\omega_s} \exp\left\{\frac{-\omega^2}{2\omega_s{}^2}\left[k_s{}^2 w_s{}^2 + \left(1 - \frac{U}{c}\cos\vartheta\right)^2\right]\right\}$$

$$\tag{11.4.29}$$

$$I_s = \frac{\rho c V}{\sqrt{32\pi} r^2} \langle \Delta_s{}^2 \rangle \frac{k_s{}^2 w_s{}^3}{\{k_s{}^2 w_s{}^2 + [1 - (U/c)\cos\vartheta]^2\}^{\frac{3}{2}}}$$

where ϑ is the angle between the direction of the velocity $\mathbf{U}$ of the jet flow and the vector $\mathbf{r}$ from the jet to the observation point. If the space variation of the random source distribution is more "fine-grained" than its time variation (i.e., if $w_s < c\tau_s = 4\pi c/\omega_s$), then $k_s w_s < 1$, and the radiated intensity is strongly dependent on the angle ϑ. The frequency distribution $I_{s\omega}$ has its maximum at

$$\omega_{max} = \sqrt{2}\omega_s\left[\left(1 - \frac{U}{c}\cos\vartheta\right)^2 + k_s{}^2 w_s{}^2\right]^{-\frac{1}{2}}$$

$$\simeq \sqrt{2}\,\frac{\omega_s}{1 - (U/c)\cos\vartheta} \qquad \text{if } k_s w_s \ll 1$$

When $U < c$, frequency $\omega_{max}/2\pi$ is greatest at $\vartheta = 0$, downstream of the turbulent region, and is least at $\vartheta = \pi$, upstream of the jet. The total intensity also decreases as ϑ is changed from 0 to π. Thus, as the observer moves in a circle of large radius r about the turbulent region, from a position away from the motion $\vartheta = \pi$ to a point in the direction of the motion $\vartheta = 0$, the mean frequency of the observed sound increases by an amount which is greater the nearer the jet velocity U is to the speed of sound c, and the total intensity also increases considerably.

If the time variation of the source function $s(\mathbf{r}',t)$ is rapid and the size of the turbulent cells is large, then $k_s w_s > 1$, and the radiation is not much dependent on the angle ϑ. In the limit the frequency distribution has its maximum at $\omega_{max} \simeq \sqrt{2}(c/w_s)$, and the total intensity at the observation point is

$$I_s \simeq \frac{\rho c^2 V}{\sqrt{32\pi}\omega_s r^2} \qquad k_s w_s \gg 1$$

independent of ϑ.

The same remarks can be made regarding the sound produced by the fluctuations of the stress-momentum tensor of Eq. (7.1.38). The factors $1 + k_s^2 w_s^2$ and $1 + k_e^2 w_e^2$ in Eqs. (7.1.36) and (7.1.38) are changed, respectively, to

$$\left[\left(1 - \frac{U}{c}\cos\vartheta\right)^2 + k_s^2 w_s^2\right] \qquad \text{and} \qquad \left[\left(1 - \frac{U}{c}\cos\vartheta\right)^2 + k_e^2 w_e^2\right]$$

The comments given above regarding the dependence of mean frequency and total intensity of the radiation on the angle ϑ between the mean velocity U and the radius vector to the observation point also apply to the combined radiation. The intensity I_e from the turbulent motion is somewhat more sharply directional than is the intensity I_s from the fluctuating monopoles.

When $U > c$, the maximum of the intensity, as well as the frequency ω_{max}, comes, not at $\vartheta = 0$, but at $\vartheta = \cos^{-1}(c/U)$.

Problems

1 Derive the expressions (11.2.36) and (11.2.37) for the power emitted from a moving longitudinal and transverse dipole. Also determine the corresponding power spectra given in (11.2.38) in the text. Discuss the relative magnitudes of the power emitted in the two cases.

2 An airplane travels through a turbulent atmosphere with a speed equal to half the sound speed and, as a result, experiences a fluctuation lift force with a characteristic period of the order of $T \simeq 1$ sec and with an amplitude of the order of the weight w of the airplane. As far as the influence on the surrounding atmosphere is concerned, regard the airplane as a traveling transverse dipole source, so that the pressure field produced can be determined from Eq. (11.2.33).

(*a*) Show that the integral of the near-field term in the pressure field over the ground surface (assumed plane) equals the weight of the airplane.

(*b*) What is the maximum amplitude of the sound pressure field at the ground surface if the airplane flies at a height of 5,000 ft and if $w = 10$ tons?

(*c*) What is the radiated acoustic power?

3 Derive the expressions (11.2.33) and (11.2.34) for the sound pressure field radiated from a moving transverse and longitudinal dipole source. Introduce in these expressions the distance R between the point of observation and the point of emission [Eqs. (11.2.6) and (11.2.12)] and the corresponding direction cosines, $\cos\theta_1$, $\cos\theta_2$, and $\cos\theta_3$, and show that the expressions for the pressure fields (11.2.33) and (11.2.34) can be written in the form

$$p_2 = \frac{1}{4\pi R}\frac{\cos\theta_2}{(1 - \beta\cos\theta_1)^2}\left[\frac{1}{c}f_2' + \frac{1}{R(1 - \beta\cos\theta_1)}f_2\right]$$

$$p_1 = \frac{f_1'}{4\pi cR}\frac{\cos\theta_1}{1 - \beta\cos\theta_1} + \frac{f_1}{4\pi R^2}\frac{\cos\theta_1 - \beta}{(1 - \beta\cos\theta_1)^3} \qquad \beta = \frac{V}{c}$$

4 A stationary acoustic source distribution in a duct has a uniform source strength $Q = Q_0 \exp(-i\omega t)$ per unit volume over a length L of the duct and is zero outside this region. The air in the duct is moving with a uniform velocity V along the duct. With reference to the discussion related to Eq. (11.1.25), calculate the sound field in the duct both outside and inside the source region.

5 A plane boundary is specified acoustically by a specific acoustic impedance ζ. The outside fluid moves parallel with the boundary with a velocity. V as shown in Fig. 11.3. A plane wave is incident on the boundary at an angle ϕ. Show that the pressure reflection coefficient is given by Eq. (11.2.21) and that the power absorption coefficient is

$$\alpha = \frac{4\mu(1 + M\cos\phi)\sin\phi}{[\xi + (1 + M\cos\phi)\sin\phi]^2 + \sigma^2}$$

where

$$\zeta = \frac{1}{\xi - i\sigma}$$

and $M = V/c$, c being the speed of sound in the fluid.

6 A thin plane partition of some kind, a membrane or a thin plate, separates two regions of a fluid in relative motion as indicated in Fig. 11.3. Generalize Eq. (11.1.20) for the reflection coefficient of a plane sound wave at the interface to account for the presence of a partition. The partition is specified acoustically by means of an impedance z, defined as the ratio between the pressure difference between the two sides of the partition and the velocity of the partition. Show that the reflection coefficient in the case of a stretched membrane may be obtained from (11.1.20) by setting

$$\zeta = -\frac{i\omega\varepsilon}{\rho_1 c_1}\left[1 - \frac{v_t^2}{c_1^2}\frac{\cos^2\phi_1}{(1 + M_1\cos\phi_1)^2}\right] + \frac{\rho_2 c_2}{\rho_1 c_1}\frac{1}{(1 + M_2\cos\phi_2)\sin\phi_2}$$

where ε is the mass per unit area of the membrane, and $v_t = \sqrt{T/\varepsilon}$, T = tension of membrane. For a thin plate with phase velocity of flexural waves v_p, replace the expression inside the brackets by $1 - (v_p/c_1)^4\cos^4\phi_1/(1 + M_1\cos\phi_1)^4$.

7 As mentioned in the discussion of the sound field from a supersonically moving monopole source [Eq. (11.2.19)], the sound field inside the sound cone can be regarded as a superposition of the field contributions from two emission points, E_- and E_+. In the case of harmonic time dependence of the sound source, this leads to the interesting effect of "standing-wave" formation inside the cone from a moving source. Thus, if the source strength is specified by $q = q_0\exp(-i\omega t)$, show that the sound-pressure amplitude on the axis of the sound cone a distance x from the vertex of the cone (i.e., from the sound source) is given by

$$p = \frac{\omega(2q_0)}{4\pi x(1 - \beta^2)}\left\{\sin^2\frac{\omega x}{c(1 - \beta^2)} + \beta^2\cos\frac{\omega x}{c(1 - \beta^2)}\right\}^{\frac{1}{2}}$$

8 Show that the d'Alembertian operator $\nabla^2 - (1/c^2)(\partial^2/\partial t^2)$ in the acoustic wave equation is invariant under the "Lorentz" transformation

$$x' = \gamma(x - Vt) \qquad y' = y \qquad \gamma = \frac{1}{\sqrt{1 - (V/c)^2}}$$

$$t' = \gamma\left(t - \frac{Vx}{c^2}\right) \qquad z' = z$$

9 Show that a Galilean transformation to a frame of reference moving with velocity V brings the ordinary acoustic wave equation into the form

$$\frac{1}{c^2}\left(\frac{\partial}{\partial t} + \mathbf{V}\cdot\nabla\right)^2 p - \nabla^2 p = 0$$

Show that this equation is consistent with Eq. (7.1.21). Is it valid when V depends on position?

10 In Chap. 6 the Lagrangian density in a sound field was shown to be $L = \frac{1}{2}\rho[(\nabla\Psi)^2 - c^{-2}(\partial\Psi/\partial t)^2]$, where Ψ is the velocity potential. This expression is valid in a coordinate system in which the unperturbed fluid is at rest. If the fluid is in motion with a uniform velocity V_0 in the x direction with respect to the laboratory system S, what is the Lagrangian density with respect to S? Use the results obtained in Chap. 6 to derive an expression for the acoustic intensity with respect to S.

11 A monopole source with a source strength $Q = Q_0 \exp(-i\omega t)$ moves along a circular path of radius a in the yz plane with an angular velocity Ω. Calculate the radiated sound field. Show that the spectrum contains frequencies $|\omega \pm n\Omega|$, where $n = 0, 1, 2, \ldots$, and that the corresponding pressure amplitudes are proportional to $J_n(ka \sin \theta)$, where $k = \omega/c$, θ is the angle between the direction of propagation and the x axis, and J_n is the nth-order Bessel function.

12 A rigid cylinder of diameter $d = 1$ cm is mounted between two opposite walls of a square duct perpendicular to the duct axis. The cross section of the duct is 30 by 30 sq cm. When air is passed through the duct, at what average flow speeds will stimulated Kármán-vortex shedding and the related sound emission take place?

13 For isentropic fluid motion, we have $dS/dt = \partial S/\partial t + \mathbf{V} \cdot \nabla S = 0$. If the unperturbed fluid is inhomogeneous, so that $S_0 = S_0(x,z,y)$, and moving with a velocity $\mathbf{V}_0$, which may be a function of the space and time coordinates, show that the linearized wave equation for the pressure is

$$\frac{1}{c^2}\frac{\partial^2 p}{\partial t^2} - \nabla^2 p = \frac{\gamma-1}{c^2}\frac{\partial H}{\partial t} - \nabla\cdot\mathbf{F} + \nabla\cdot(\rho\mathbf{V}_0\,\mathbf{V}_0)\cdot\nabla$$
$$+ \nabla\cdot(\rho\,\mathbf{u}\,\mathbf{V}_0)\cdot\nabla - \frac{\rho_0}{c_v}\frac{\partial}{\partial t}(\mathbf{u}\cdot\nabla S_0 + \mathbf{V}_0\cdot\nabla_0)$$

where H = rate of energy transfer per unit volume of fluid
         F = external force per unit volume
        $\mathbf{uV}_0$ = tensor with components $u_x V_{0x}$, $U_x V_{0y}$, etc.

Show that the source term can be written in a form that contains the gradient of the sound pressure. Also use the equation of state for an ideal gas $S = C_v \log P + C_p \log \rho + \text{const}$ to express the entropy source term as a function of the gradients of the unperturbed pressure and density. What external force is required to maintain the fluid in this unperturbed state? Consider as an example sound propagation in a moving isothermal atmosphere.

14 Derive Eq. (11.4.8).

15 With reference to Eq. (11.3.24), show that the Green's functions, expressing the sound pressure fields from an axial and a tangential force of frequency ω and unit amplitude at point (r_1,φ_1) in a cylindrical duct of radius r_0, are [see Eq. (9.2.11)]

$$G_z(r\varphi \mid r_1\varphi_1) = \sum_{m,n}\frac{1}{S\Lambda_{mn}}J_m(\kappa_{mn}r_1)J_m(\kappa_{mn}r)e^{im(\varphi-\varphi_1)}\,e^{ik_{mn}z-i\omega t}$$

$$= \sum_{m,n}\frac{m}{S\Lambda_{mn}k_{mn}}J_m(\kappa_{mn}r_1)J_m(\kappa_{mn}r)e^{im(\varphi-\varphi_1)}e^{ik_{mn}z-i\omega t}$$

$$k_{mn}{}^2 = (\omega/c)^2 - \kappa_{mn}{}^2$$

Use these functions to determine the sound pressure of Eq. (11.3.24) from the fan-force distributions f_z and f_φ in an inhomogeneous flow, as defined in the discussion preceding Eq. (11.3.24). Discuss the amplitude of the (m,n)th mode of the resulting sound as a function of the tip speed $V = \Omega r_0$, where Ω is the angular velocity of the fan. In particular, investigate the relative contributions of the axial and the tangential force components when $V \simeq c$. Show that the nth-order harmonic component in the Fourier expansion in φ of the flow perturbation gives rise to a *plane* wave in the duct if n is a multiple of the number of blades B in the fan, and that, under these conditions, the contribution from the tangential (drag) force is zero. Assume that α_s, β_q, f_{z0}, and $f_{\varphi 0}$ are independent of r and calculate the values of the coupling coefficients $A_{mn}{}^{sq}$ and $B_{mn}{}^{sq}$.

CHAPTER
12

PLASMA ACOUSTICS

12.1 WEAKLY IONIZED GAS

In the discussion of sound waves in a neutral gas we have had little occasion to consider effects of external forces acting in the bulk of the gas or fluid.   In this chapter the situation is different, since we shall now be concerned with the dynamics of an ionized gas or plasma.   External electric and magnetic fields produce forces in the gas and internal fields, and corresponding forces are produced by the gas itself as a result of its motion.   These fields and forces lead to many dynamical phenomena: we shall study here only those which may be considered to be acoustic in nature.   The dynamical effects of external or internal fields and forces depend to a great extent on the degree of ionization of the gas, i.e., on the ratio between the number densities of the charged particles and the neutrals.   For this reason we shall treat separately the weakly ionized and the highly ionized gas.

In the weakly ionized gas the neutral gas plays a dominant part, and in an ordinary sound wave the ions and the electrons are forced to follow the neutrals.   From this standpoint the presence of the charged particles is inconsequential.   However, the charged particles draw energy from external electric fields, and some of this energy is transferred to the neutral gas, which then can lead to sound generation and amplification in the neutral gas.

The energy transfer from the electrons to the neutrals leads not only to a heating of the neutral gas, but also to ionization.   As we shall see, this creation of charged particles, under certain conditions, will travel as a wave through the gas.   In the fully ionized gas we shall find, in addition to the ordinary sound wave, a wave mode in which there is charge separation. Such a particular kind of sound wave is often called a *plasma wave*, or *plasma oscillation*.   Finally, in the presence of an external magnetic field, the dynamics of the ionized gas is rather complicated; we shall consider in this chapter only some aspects of the problem which involve low-frequency phenomena, such as the magnetoacoustic waves in a conducting fluid.

Coupling to the electromagnetic wave

A weakly ionized gas can be regarded as a mixture of three different gas components, the neutral-gas atoms (or molecules), the ions, and the electrons. Such a gas mixture, often called a plasma, plays an important role in modern electronics. From an acoustic standpoint the basic difference between the plasma and a neutral-gas mixture is the fact that the plasma couples strongly to electric and magnetic fields, whereas for a neutral gas this coupling is very weak, being due to electrostriction only. Actually, a plasma not only responds to an electromagnetic field, but ordinarily depends on an electric field for its existence. The energy of the electrons is drawn from the external electric field that maintains the discharge, and in steady state an electron energy density is reached such that the energy-loss rate equals the rate of energy transfer to the electrons.

The energy losses of the electrons are due to several processes. In a weakly ionized gas, as encountered in an ordinary glow discharge, the main energy loss of the electrons is the energy transfer to the neutral-gas particles through elastic collisions. Other loss mechanisms correspond to diffusion of electrons to the walls of the chamber and to ionization and excitation of the neutrals. All these losses depend on the average speed or temperature of the electrons, and steady-state conditions are reached at a particular temperature of the electrons. For example, in a typical weakly ionized gas, such as in a glow discharge with a neutral-gas density of about 10^{16} cm^{-3} and an electron and ion density of 10^{10} to 10^{11} cm^{-3}, the steady-state electron temperature is of the order of 10^4 to 10^5 °K, whereas the neutral-gas temperature is about 500°K. In other words, in contrast to the situation in a mixture of neutral-gas components, the steady-state temperatures of the gas components in a plasma differ widely, the electron temperature being considerably higher than the ion and neutral-gas temperatures. Because of this temperature difference, there is a continuous energy flow from the electrons to the neutral-gas component, and if this transfer is made to vary with time, sound will be produced in the neutral gas.

The mechanism of the energy transfer from the electrons (or the ions) to the neutrals is roughly as follows: In the time interval between collisions, an electron is accelerated in the direction of the electric field and receives a certain amount of energy corresponding to the distance of travel between collisions. As a result of collisions with the neutral-gas particles, the motion of the electrons is randomized, and most of the energy received from the electric field between collisions goes into random motion. However, there remains a small net drift motion of the electron, and the drift velocity $\mathbf{V}_e$ is proportional to the electric field

$$\mathbf{V}_e = b_e \mathbf{E} \qquad (12.1.1)$$

where b_e is called the *electron mobility*. A similar relation holds for the ions, $\mathbf{V}_i = b_i \mathbf{E}$. The average rate of energy absorbed by the electrons per unit volume can then be expressed as

$$H_e = eN_eb_eE^2 \qquad (12.1.2)$$

with a similar expression for the ions (e, of course, is the electronic charge). Since we have $N_e \simeq N_i$, it follows that the energy contributions from the electrons and the ions are related as are their mobilities. Ordinarily, the mobility of the electrons is considerably larger than for the ions, and therefore the energy contribution from the ions can be neglected.

If the electric field or the electron density varies with time, it follows from Eq. (12.1.2) that there is a corresponding time dependence of the energy transfer to the neutrals. There is very little time delay in this process, between the time variation of H and of E^2 or N_e, the characteristic time being of the order of the electron collision time, typically of the order of 10^{-8} sec in a glow discharge. Thus, if the electric field is modulated at acoustic frequencies, we may neglect the collision time and treat (12.1.2) as true also when N_e and E are time-dependent.

In a similar manner we can express the rate of momentum transfer from the electrons and the ions to the neutrals. However, the contributions from the ions and the electrons tend to cancel each other, and the net effect is very small at the frequencies that are considered here.

Since the energy transfer from the electric field may be regarded as instantaneous, and the energy loss from the gas due to heat conduction is a comparatively slow process, with a much longer relaxation time than the acoustic periods of interest here, we may neglect the effect of heat conduction and regard Eq. (12.1.2) as the time-dependent *net* heat transfer to the gas. We have already seen in Eq. (7.1.19) that the corresponding acoustic source term in the equation for the sound pressure is given by $(\gamma - 1)c^{-2}(\partial H/\partial t)$, so that

$$\frac{1}{c^2}\frac{\partial^2 p}{\partial t^2} - \nabla^2 p = \frac{\gamma - 1}{c^2}\frac{\partial H}{\partial t} = \frac{\gamma - 1}{c^2} e\frac{\partial}{\partial t}(N_e b_e E^2) \qquad (12.1.3)$$

After having obtained the expression for the source term in the wave equation for the acoustic pressure in the neutral gas, we can now proceed to calculate the sound pressure field in terms of the space-time dependence of the electric field and the electron density if these quantities are known *a priori*. This will be assumed to be the case in the two examples, which follow, but later, in the discussion of wave amplification and spontaneous oscillations, the reaction of the sound wave back on the plasma will be accounted for.

Sound generation by ionization waves

As mentioned briefly in the introduction, and as will be discussed in more detail later, in a plasma traveling waves frequently exist in the electric field and in the electron density. For example, it is well known that an ordinary glow discharge often contains traveling "striations" that move in the direction from the anode to the cathode with a velocity of approximately 10^4 cm per sec. The frequency of oscillations typically is in the range 10^3 to 5×10^4 sec^{-1}.

Thus, with the static field $\mathbf{E}_0$ and the perturbation $\epsilon_1 \mathbf{E}_0$, we can express the total electric field as $E = E_0[1 + \epsilon_1 \cos \omega(t - x/V)]$, where V is the wave speed. Similarly, for the electron density, we have $N_e = N_{e0}[1 + \epsilon_2 \cos \omega(t - x/V)]$, where we have assumed that the two waves are in phase with each other; x is the coordinate along the axis of the tube, positive in the direction of the electric field. For simplicity we shall consider a somewhat idealized geometry of the plasma such that the unperturbed values N_{e0} and E_0 of the electron density and the electric field are both constant along the tube in the interval $-L < x < L$ and zero outside this region, where only the neutral-gas component is present. Also, we shall be interested only in plane sound waves along the axis of the tube; consequently, only the average values of N_{e0} and E_0 across the tube will contribute to the generation of these waves. In the present example we assume that N_{e0} and $\mathbf{E}_0$ stand for these averages.

If the perturbations are sufficiently small, we need include only first-order terms in ϵ_1 and ϵ_2 in the source term in Eq. (12.1.3). Furthermore, we shall neglect the perturbation in the mobility resulting from the perturbation in the electric field. The quantity H in Eq. (12.1.2) can then be written in the form

$$H = H_0 + h_0 \exp\left(\frac{i\omega x}{V} - i\omega t\right)$$

$$\equiv eN_{e0}b_e E_0{}^2\left[1 + (\epsilon_1 + 2\epsilon_2) \exp\left(\frac{i\omega x}{V} - i\omega t\right)\right] \quad (12.1.4)$$

in the region $-L < x < L$, and zero outside this region. This function is now used in the source term in the wave equation (12.1.3), and to obtain the solution to this equation, we conveniently use the Green's-function techniques, which we have done so frequently throughout the book. The one-dimensional Green's function $G(x \mid x_0)$, being a solution to $(1/c^2)\partial^2 p/\partial t^2 - \partial^2 p/\partial x^2 = \delta(x - x_0) \exp(-i\omega t)$, is

$$G(x \mid x_0) = \frac{i}{2k} e^{ik|x-x_0|}e^{-i\omega t}$$

where $k = \omega/c$. The pressure field in the region $|x| > L$ outside the plasma is then given by

$$p = -i\omega \frac{\gamma - 1}{c^2} h_0 e^{\pm ikx - i\omega t} - \frac{i}{2k} \int_{-L}^{L} e^{i\omega x_0/V} e^{\mp i\omega x_0/c} \, dx_0$$

or

$$p = -(\epsilon_1 + 2\epsilon_2)p_0 \frac{\sin [(\omega L/V) \mp (\omega L/c)]}{(\omega L/V) \mp (\omega L/c)} e^{\pm ikx - i\omega t} \quad (12.1.5$$

where

$$p_0 = \frac{(\gamma - 1)LH_0}{c} = \frac{(\gamma - 1)eN_{e0}b_e E_0{}^2 L}{c}$$

The minus and plus signs refer to the acoustic waves emitted in the positive and negative x directions, respectively. We note that the amplitude of the wave in the positive x direction, the direction of the source wave, reaches a maximum value p_0 when the speed V of the source wave is the same as the speed of sound c in the neutral gas. In the opposite direction the amplitude of the sound wave is always less than p_0.

It is interesting to rewrite the expression for the characteristic sound pressure p_0 somewhat differently in order to relate it to some characteristic pressure in the plasma. To do this, we express the rate of energy transfer to the neutrals in a different manner, making use of the energy transfer of an electron to a neutral particle in an elastic collision. Since the electron temperature of the electrons is considerably higher than the temperature of the neutrals, we can regard the neutrals as stationary in the collisions with the electrons. Then the average energy transfer in a single collision is $\dfrac{4m_e}{m_n}\dfrac{m_e v_e^2}{2}$, where v_e is an average value of the speed of the incident electrons. The number of collisions per second can be expressed as v_e/l_e, where l_e is an average mean-free-path of the electrons. The total rate of energy transfer from the N_e electrons per unit volume, then, is

$$H = \frac{4m_e}{m_n}\frac{m_e v_e^3}{2}\frac{N_e}{l_e} = \frac{4m_e}{m_n}\frac{v_e}{l_e}p_e \qquad (12.1.6)$$

where we have introduced $p_e = N_e m_e v_e^2/2$. The characteristic sound pressure in (12.1.5) thus can be expressed as

$$p_0 = 4(\gamma - 1)\frac{m_e}{m_n}\frac{v_e}{c}\frac{L}{l_e}p_e \qquad (12.1.7)$$

To get an idea of the order of magnitude of the sound-pressure amplitudes obtained, we consider a He plasma at a neutral-gas pressure $\simeq 1$ mm Hg, an electron density $N_{e0} \simeq 10^{10}$ cm^{-3}, and an electron temperature of about 1 eV. Assuming $\epsilon_1 = \epsilon_2 = 0.1$, and with $L = 10$ cm, we find a value of the sound-pressure amplitude $(\epsilon_1 + 2\epsilon_2)p_0$ of the order of about 1 dyne per sq cm.

Inside the plasma the sound field is a superposition of waves traveling in both directions. We leave it as one of the problems to determine this field.

Plasma afterglow

A second example of some importance in plasma acoustics is the generation of sound waves in the plasma afterglow. By plasma afterglow is meant the period of decay of the plasma after the electric field has been turned off. In the afterglow period the electron temperature decays quite quickly to the temperature of the neutral-gas component. For the purposes of the present discussion of sound generation, this decay can be regarded as instantaneous. The electron density, on the other hand, decays much more slowly and, as far as sound generation is concerned, is of minor

importance.   However, the sound produced at the instant of turnoff of the
electric field, due to the change in the heating of the neutral gas, frequently
produces acoustic oscillations in the discharge tube which modulate the
decaying electron density at the acoustic frequency.   Such a modulation
actually has been observed in several experiments.

To get an idea of the strength of these oscillations, we shall consider the
sound produced as a result of a sudden turnoff of the electric field E_0 that
maintains a discharge.   In other words, the time dependence of the heating
function H in (12.1.3) is now such that $H = eN_{e0}b_eE_0^2$ for $t < 0$, and $H = 0$
for $t > 0$.   We let the plasma be contained in a tube with rectangular cross
section, and for simplicity, only the modes perpendicular to the tube axis
and to the tube walls at $y = 0$ and $y = D$ will be treated.   The coupling
to these modes depends on the y dependence of the heating function H.
The electron density is approximately of the form $N_{e0} = N_0 \sin(\pi y/D)$,
and therefore the y dependence of the heating function also is given by
$\sin(\pi y/D)$.

Since the time dependence of H is described by a step function as discussed above, it follows that the wave equation describing the excitation of
sound in the afterglow is of the form

$$\frac{1}{c^2}\frac{\partial^2 p}{\partial t^2} - \frac{\partial^2 p}{\partial y^2} = H_0 \sin\left(\frac{\pi y}{D}\right)\delta(t) \tag{12.1.8}$$

where
$$H_0 = \frac{eNb_eE_0^2(\gamma - 1)}{c^2}$$

To obtain the solution, we conveniently make a Fourier transform in time
and express the y dependence of the pressure in terms of a Fourier series of
modes symmetrical with respect to the axis of the tube (since the source is
symmetrical).   Thus

$$p(t,y) = \sum_n \int_{-\infty}^{+\infty} A_n(\omega)\cos(k_n y)e^{-i\omega t}\,d\omega \qquad k_n = \frac{2\pi n}{D} \tag{12.1.9}$$

By expanding the right-hand side of Eq. (12.1.8) in a similar way, we obtain

$$A_n = \frac{2}{\pi^2}H_0\frac{c^2}{4n^2 - 1}\frac{1}{\omega^2 - k_n^2 c^2}$$

Finally, after we have carried out the integration in Eqs. (12.1.9), the pressure
field in the plasma afterglow is found to be

$$p(t,y) = \sum_{n=1}^{\infty}\frac{2p_0}{\pi^2 n(4n^2 - 1)}\sin\frac{2\pi nct}{D}\cos\frac{2\pi ny}{D} \tag{12.1.10}$$

where p_0 is the same as in (12.1.5) or (12.1.7), with L replaced by D.   We note
that the fundamental mode has an amplitude about ten times larger than the
amplitude of the second mode, and for large n the amplitude decreases with
n approximately as n^{-3}.

Acoustic wave amplification

The preceding examples have been concerned with an electric field, imposed from outside the system, generating a sound wave. Next we should study the cases where there is a static electric field, and an acoustic wave is imposed from outside, or else is generated spontaneously from the interaction between acoustic density variations and thermal energy from the electrons. In a plasma the free electrons represent a constant source of additional thermal energy supplied to the gas; when this source is coupled to the acoustic wave, a positive feedback can occur which can transform electronic energy into acoustic energy.

To investigate this problem in great detail would require the analysis of the equations of motion for the neutrals, electrons, and ions in the plasma, accounting for the interaction between them in terms of momentum and energy transfer. Such a general analysis would include not only the acoustic mode of motion, but other wave modes as well. We shall not go into such an analysis here, but merely point out that, as far as the acoustic mode of motion is concerned, the neutrals, ions, and electrons all move approximately in phase with each other and with the same amplitude of oscillation. The relative density fluctuations for all components are also the same, so that a perturbation n_n in the neutral-particle density produces a perturbation n_e in the electron density such that

$$n_e \simeq \frac{N_e}{N_n} n_n \qquad (12.1.11)$$

We have already seen in Eq. (12.1.3) that the source term in the wave equation for the sound pressure contains a term proportional to the rate of change of the total electron density. If this rate of change is produced as a result of a perturbation in the acoustic density, as in Eq. (12.1.11), and if the unperturbed electron density N_{e0}, as well as the electric field E_0, is time-independent, the corresponding contribution to the source term in Eq. (12.1.3) is $(\gamma - 1)c^{-2}eb_eE_0^2(\partial N_e/\partial t) = (\gamma - 1)c^{-2}eb_eE_0^2(N_{e0}/N_n)(\partial n_n/\partial t)$. Since the acoustic pressure p is related to the perturbation in the density through the relation $p = c^2 m_n n_n$, where m_n is the neutral particle mass, we can express $\partial n_n/\partial t$ in terms of $\partial p/\partial t$, and the wave equation (12.1.3) takes the form

$$\frac{1}{c^2}\frac{\partial^2 p}{\partial t^2} - \nabla^2 p = (\gamma - 1)\frac{H_0}{\rho_0 c^4}\frac{\partial p}{\partial t} \qquad (12.1.12)$$

where we have made use of Eq. (12.1.2).

It should be noted that the sign of the term on the right-hand side is opposite to that for a resistance in the simple-oscillator equation

$$M(d^2x/dt^2) + R(dx/dt) + Kx = 0$$

($-\nabla^2$ takes the place of K). Thus energy is being added, rather than lost.

This energy of course comes from the electrons, kept at high temperature by the constant electric field. If the wavelength $\lambda = 2\pi/k$ is specified, the frequency will have a positive-imaginary part, representing an exponential increase with time; if the frequency $\omega/2\pi$ is specified, the wavenumber k will have a negative exponential part, representing an exponential buildup of the wave as it travels along.

To study the behavior of such a plane wave, transmitted through the plasma at a frequency ω, we introduce $p \sim \exp(ikx - i\omega t)$ into Eq. (12.1.12), and find that the corresponding value for the propagation constant is given by the dispersion relation

$$k = \frac{\omega}{c}\sqrt{1 - \frac{i}{\omega\tau}} = \alpha - i\beta \qquad (12.1.13)$$

where

$$\tau = \frac{\rho_0 c^2}{(\gamma - 1)H_0}$$

This time can be described approximately as the time required to transfer from the electrons an amount of energy equal to the thermal energy of the neutral gas. Using the expression (12.1.6), we can express this time in terms of the electron collision time l_e/v_e and obtain

$$\tau = \frac{\gamma}{3(\gamma - 1)} \frac{m_n}{m_e} \frac{N_n}{N_e} \frac{T_n}{T_e} \left(\frac{l_e}{v_e}\right) \qquad (12.1.14)$$

where T_n, T_e are the gas and electron temperatures. Under normal glow-discharge conditions, this time is of the order of 0.1 sec, and therefore for the frequencies of interest here, $\omega\tau \gg 1$. Under special conditions, however, τ may be considerably shorter.

Returning to the dispersion relation (12.1.13), we obtain the following expressions for the real and imaginary parts of the propagation constant:

$$\alpha = \frac{\omega/c}{\sqrt{2}}[\sqrt{1 + (\omega\tau)^{-2}} + 1]^{\frac{1}{2}} = \begin{cases} \sqrt{\dfrac{\omega}{2c^2\tau}} & \omega\tau \ll 1 \\[2ex] \dfrac{\omega}{c}[1 + \tfrac{1}{2}(2\omega\tau)^{-2}] & \omega\tau \gg 1 \end{cases}$$

$$(12.1.15)$$

$$\beta = \frac{\omega/c}{\sqrt{2}}[\sqrt{1 + (\omega\tau)^{-2}} - 1]^{\frac{1}{2}} = \begin{cases} \sqrt{\dfrac{\omega}{2c^2\tau}} = \beta_1 & \omega\tau \ll 1 \\[2ex] \dfrac{1}{2c\tau} = \beta_2 & \omega\tau \gg 1 \end{cases}$$

Then, for $\omega \ll 1/\tau$, the growth rate increases with frequency as $\sqrt{\omega}$, and in the limit of high frequencies the growth rate is independent of frequency.

In obtaining these results, we have neglected the attenuation of the acoustic wave resulting from various loss mechanisms. When sufficiently small, the growth and the attenuation rates are additive, so that we can treat them separately. For a monatomic gas the acoustic losses are caused solely by viscosity and heat conduction. Normally, the plasma is contained in a tube, and we can make use of the results obtained in Chap. 6 to obtain the attenuation rate caused by viscous and heat-conduction losses. As we recall, it is convenient to consider separately the losses in the bulk of the gas outside the viscous and thermal boundary layers and the losses in the boundary layers themselves. These loss contributions lead to spatial attenuation rates for the fundamental mode [see Eq. (6.4.14) and page 519].

$$a_1 = \frac{\omega/c}{2d}\left[\sqrt{\frac{2\nu}{\omega}} + (\gamma - 1)\sqrt{\frac{2K}{\omega\rho C_p}}\right] \simeq \frac{1}{d}\sqrt{\frac{2\omega l_n}{c}} \quad \text{(boundary)}$$

$$\text{(12.1.16)}$$

$$a_2 = \frac{1}{2}\left(\frac{\omega}{c}\right)^2\left[\frac{4}{3}\frac{\nu}{c} + (\gamma - 1)\frac{K}{\rho C_p c}\right] \simeq 1.16\left(\frac{\omega}{c}\right)^2 l_n \quad \text{(bulk)}$$

where $l_n = \nu/c$ is approximately equal to the mean-free-path of the neutrals, K is the heat-conduction coefficient, and ν is the kinematic viscosity $\nu = \mu/\rho_0$. In obtaining the approximate expressions (12.1.16), we have set $\mu \simeq 0.67 \times K/c_p$, which is approximately valid for a monatomic gas.

When β and the a's are small enough, the growth and attenuation effects are additive and the sound-pressure amplitude will depend on x through the factor $\exp(\beta - a_1 - a_2)x$; amplification will occur if $\beta > a_1 + a_2$. At very low frequencies all the quantities are small; in fact a_2 is much smaller than a_1. Thus in the very low frequency range, it appears that amplification occurs if $4l_n c\tau < d^2$.

However, in the frequency range of major interest, β and the a's are not small and the effects are not additive. To obtain a more generally valid result, we would have to solve the coupled equations of motion for the three fluid components, electrons, ions, and neutrals. When this is done we find that the three fluids do not move exactly in phase, so that a sound wave produces charge separation and consequent electric fields. The situation is analogous to the coupled waves of Eq. (10.1.2). We shall not go into details but shall mention a few results of the more complete analysis.

We find that the effects are additive and the pressure depends on the factor $\exp(\beta - a_1 - a_2)x$ only for frequencies such that

$$\omega \ll \Omega_n \frac{N_n T_n}{N_e T_e}$$

where Ω_n is the mean frequency of collision per second of a neutral particle with charged particles and the N's and T's are the particle densities and absolute temperatures of neutrals and electrons, respectively.   The wave-number k of the acoustic-type wave is obtained from

$$k^2 = \left(\frac{\omega}{c}\right)^2\left[1 - \left(\frac{\Omega_n}{\omega} + \frac{1}{\omega\tau}\right)\frac{\delta^2}{1+\delta^2}\right.$$

$$\left. + i\left(\frac{1}{\omega\tau_n} + \frac{\Omega_n}{\omega}\frac{\delta^2}{1+\delta^2} - \frac{1}{\omega\tau}\frac{1}{1+\delta^2}\right)\right] \quad (12.1.17)$$

where

$$\delta = \frac{1}{\gamma}\frac{T_e N_e}{T_n N_n}\frac{\omega}{\Omega_n}$$

and

$$\frac{1}{\tau_n} = \left(\frac{\omega}{c}\right)^2\left[\tfrac{4}{3}\nu + (\gamma-1)\frac{K}{\rho C_p}\right] + \frac{\omega}{d}\left[\sqrt{\frac{2\nu}{\omega}} + (\gamma-1)\sqrt{\frac{2K}{\omega\rho C_p}}\right]$$

Quantity τ is defined in Eq. (12.1.14).   It is not difficult to show that when $\delta \ll 1$ *and* when both $\omega\tau$ and $\omega\tau_n$ are *larger* than unity, the imaginary part of k reduces to $-\beta + a_1 + a_2$ and the implications of Eqs. (12.1.13) to (12.1.16) have validity.

Spontaneous excitations

In addition to the study of the behavior of an externally produced wave transmitted into the plasma, it is interesting to investigate the possibility of spontaneous excitations of normal modes.   For simplicity, we consider the lateral modes between two rigid walls in a rectangular tube.   The pressure field is then of the form $\cos(\pi n y/B)\exp(-i\omega_n t)$, the reflecting walls being at $y = 0$ and $y = B$.   If we insert this into the wave equation (12.1.12), we obtain

$$\omega_n = \frac{\pi n c}{B}\sqrt{1 - \left(\frac{B}{2\pi n c\tau}\right)^2} + \frac{i}{2\tau} \quad (12.1.18)$$

where, as before, τ stands for the characteristic time defined in Eq. (12.1.14). Under the conditions of interest here, $B/2\pi n c\tau \ll 1$, and we note that the oscillations grow in time at a rate $\exp(t/2\tau)$.

To obtain a criterion for the onset of spontaneous oscillations, we must compare the growth rate with the decay caused by the acoustic losses in the gas.   We now conveniently apply the general results obtained in Chap. 6 to the calculations of the energy losses in the lateral modes of oscillation of interest here.

In Eq. (6.4.13), the loss rate per unit volume in a fluid is given by

$$L = D + \frac{K}{T}(\operatorname{grad}\theta)^2 + \operatorname{div}\left(\frac{K}{T}\operatorname{grad}T\right) \quad (12.1.19)$$

where D = viscous dissipation function defined in Eq. (6.4.10)
  K = thermal conductivity coefficient
  θ = temperature in the sound field

Since we are interested in the average of the loss rate over one period of oscillation, the last term in Eq. (12.1.19) does not contribute.   As in Chap. 6, to calculate L in the bulk of the fluid, we use the loss-free field distribution, and the temperature field then can be expressed in terms of the sound pressure (assuming adiabatic conditions), to give

$$\langle L_{\text{bulk}} \rangle = \langle D \rangle + \frac{(\gamma - 1)K}{\rho C_P \rho c^2} \langle \text{grad } p \rangle^2$$

where the brackets indicate time average.   For the one-dimensional motion of interest here we have $D = \left(\frac{4\mu}{3}\right)\left(\frac{\partial u_y}{\partial y}\right)^2$.   As was shown in Chap. 6, the evaluation of L in the thin boundary layers leads to an expression for the average loss rate per unit area of the surface,

$$\langle L_s \rangle = \sqrt{\frac{\mu\omega}{2\rho}} \frac{\rho |u_s|^2}{2} + (\gamma - 1)\sqrt{\frac{K\omega}{2\rho C_P}} \frac{|p|^2}{2\rho c^2}$$

where $|u_s|$ is the amplitude of the tangential particle velocity just outside the boundary layer (which is approximately the same as the tangential particle velocity at the boundary in the loss-free solution), and $|p|$ is the sound-pressure amplitude at the boundary (loss-free solution).

In the case of the lateral oscillations in a rectangular tube of width B, we have $p = p_0 \cos (\pi n y/B)$, and $u_y = (p_0/\rho c) \sin (\pi n y/B)$.   Since the motion is perpendicular to the walls $y = 0$ and $y = B$, only conduction losses will obtain, whereas at the remaining two walls we have both viscous and conduction losses.   The total energy in the sound wave per unit length of the tube is $W = (p_0^2/4\rho c^2)A$, where A is the cross-sectional area of the duct. The loss rate is proportional to W, and if we call the constant of proportionality $2/\tau_n$, we have $dW/dt = -(2/\tau_n)W$.   The time dependence of the pressure field is then given by $\exp (-t/\tau_n)$.   From the expressions for the losses $\langle L_{\text{bulk}} \rangle$ and $\langle L_s \rangle$, we find for a duct with square cross section that the decay time for the nth mode is given by

$$\frac{T_1}{\tau_n} = \frac{8\pi^2}{3} \frac{n^2 l_n}{B} + \sqrt{2\pi}\left[1 + 3(\gamma - 1)\sqrt{\frac{K}{\eta c_P}}\right]\sqrt{\frac{nl}{B}} \simeq 26\frac{n^2 l_n}{B} + 8\sqrt{\frac{nl_n}{B}}$$

$$(12.1.20)$$

where we have used $K/C_P \approx 1.5\mu$ approximately valid for a monatomic gas. T_1 is the period of the first mode corresponding to $n = 1$.   The two terms in (12.1.20) correspond to the bulk and surface losses, respectively.   The criterion for the onset of spontaneous oscillations is given by $1/2\tau > 1/\tau_1$, where the growth and attenuation rates $1/2\tau$ and $1/\tau_1$ are given in Eqs. (12.1.18) and (12.1.20).

Ionization waves

Although the ionization wave, or traveling striation, is not an acoustic wave in the ordinary sense of the word, we shall describe it briefly, since we used it earlier in this section as a particular type of sound source in one of the examples of sound generation in a plasma. The ionization wave actually is a self-sustained oscillation of the plasma, which may be described approximately as follows: The plasma is maintained by a constant electric field E_0, which under steady-state conditions produces as many electrons (and ions) through ionization as are lost by diffusion to the walls in the discharge tube.

A perturbation in the electric field in the plasma produces a perturbation in the ionization rate, and hence a perturbation in the charge density, the electron density and the ion density being initially perturbed in the same manner. However, the electrons are more mobile than the ions and seek to distribute themselves according to a Boltzmann distribution in the periodic potential that corresponds to the electric field perturbation. Thus the electrons move away from the regions of large electric field, leaving an excess of ions. This change, in turn, gives rise to an electric field which is in the same direction as E_0 on one side of the ions and in the opposite direction to E_0 on the other side. This perturbation in the electric field is displaced with respect to the initial perturbation in the direction of E_0, which is the direction toward the cathode. We have then closed the loop, and we realize that under certain conditions the process can be self-sustained, so that the perturbation in the electric field travels toward the cathode without decrease in amplitude.

We can put these observations in somewhat more quantitative terms through the following linearized equations: The first expresses the fact that the perturbation in the production rate of the ion density is proportional to the electric field perturbation E and the equilibrium density N. We assume that the characteristic time of change is much longer than the time required for the electrons to become diffused. Thus

$$\frac{\partial n_+}{\partial t} = \gamma E N$$

The second equation is simply the Poisson equation

$$\frac{\partial E}{\partial x} = 4\pi e n_+$$

where e is the unit charge.

A plane wave of the form $\exp(ikx - i\omega t)$ satisfies these equations of motion if the following dispersion relation obtains:

$$k = 4\pi e \gamma N \frac{1}{\omega}$$

This dispersion is unusual inasmuch as the wavelength is proportional to the frequency. As a result, the group velocity $d\omega/dk$ [Eq. (9.1.18)] is equal to the phase velocity but opposite in direction.

$$\frac{d\omega}{dk} = -\frac{\omega}{k} = -4\pi\frac{\gamma eN}{k^2}$$

From a more detailed analysis of this type of wave motion the conditions for wave growth can be established, but it is found that the relation between the group and phase velocity as given above is approximately correct under most conditions.

12.2 HIGHLY IONIZED GAS

In the weakly ionized gas considered in the preceding section, the number density of the charged particles was very small by comparison with the neutral particle density, say, one electron (and ion) for every 10^5 neutral particles. The dynamics of the gas is then much the same as for the neutral gas, and were it not for the energy transfer from the electrons, the charged particles could be regarded merely as "tracer" elements, which make possible the study of the motion of the gas by a study of the behavior of the charged particles. This is so because the charged particles are strongly coupled to the neutrals through collisions and, as was mentioned in the previous section, the relative changes in the density are all the same in the acoustic mode of motion. For this motion, charge separation and the creation of (fluctuating) electric field perturbations in the gas are of little or no consequence. The speed of sound in the gas is, for all practical purposes, the same as in the neutral gas. The contribution from the ions or even from the electrons to the total pressure in the gas is negligible, and the speed of sound is determined by the thermal speed in the neutral gas.

Sound waves

In the highly ionized gas the situation is different. The effect of the neutrals can be neglected, and we are left with the ions and the electrons, which have different thermal speeds and, in general, even different temperatures; the electron temperature usually is much greater than the ion temperature. Thus there is no longer one unique thermal speed that might be approximately identified with the sound speed, as in the case of the neutral gas. However, we can express the speed of sound as $\sqrt{\gamma P/\rho}$, where γ is an appropriate specific-heat ratio. The total pressure P in the plasma is the sum of the ion pressure and the electron pressure $P = P_i + P_e$, and the total density is $\rho = \rho_i + \rho_e = N(m_i + m_e) \simeq Nm_i$, where N is the particle density of the ions (and the electrons) and m_i and m_e are the ion mass and the electron mass. Since under ordinary conditions $P_e \gg P_i$, we expect that the speed of an ordinary sound wave in the highly ionized gas will be

$$c = \sqrt{\frac{P_e \gamma}{N m_i}} = \sqrt{\frac{\gamma k T_e}{m_i}} \qquad (12.2.1)$$

An analysis of the various wave modes in a plasma shows that there is indeed an acoustic wave mode that travels with a speed given by (12.2.1). We note that in this mode, often called an ion-acoustic wave, the ions provide the inertia, whereas the electrons provide the "restoring" force.

Plasma oscillations

In the acoustic mode of motion of the plasma there is no charge separation, and the wave motion represents a perturbation of the density and the corresponding (adiabatic) change in the temperature. We shall now consider a mode of motion which is longitudinal but in which charge separation and the related electric field are the essential characteristics.

It is very simple to get a qualitative idea about the nature of this mode of motion. Consider an initially unperturbed portion of the plasma with an equal number N of ions and electrons. Suppose the electrons in this region are displaced an amount ξ from their equilibrium position. This will leave an excess $N\xi$ of ions per unit area, and an electric field is established between the electrons and the ions, $E = 4\pi e N \xi$. The corresponding restoring force on one of the displaced electrons is then $4\pi e^2 N \xi$, proportional to the displacement. As a result, the electrons will perform an oscillatory motion with a frequency given by

$$\omega_p{}^2 = \frac{4\pi e^2 N}{m_e} \qquad (12.2.2)$$

According to Eq. (12.2.2), the frequency of oscillation does not depend on the wavelength of the oscillations, which means that the oscillation has zero group velocity, $d\omega/dk = 0$. In this qualitative analysis, however, the thermal motion of the electrons, clearly, is not accounted for, and the expression for the frequency obtained can only be approximately correct. Furthermore, this simple picture of the motion does not provide for any damping of the oscillation.

To explore the motion further, we shall carry out here a hydrodynamic analysis of plasma oscillations. Let the electron density be $\rho = N m_e$, with the perturbation $\delta = n m_e$. We assume the ions to be stationary to form a uniform positive background to keep the plasma neutral on the average. The electric field E in the plasma is then due solely to the perturbation in the electron density, and we have

$$\text{div } \mathbf{E} = -4\pi n e \qquad (12.2.3)$$

The equations of mass and momentum balance of the electrons are

$$\frac{\partial \delta}{\partial t} + \rho \,\mathrm{div}\, \mathbf{u} = 0 \tag{12.2.4}$$

$$\rho \frac{\partial \mathbf{u}}{\partial t} = -\mathrm{grad}\, p - N e \mathbf{E} \tag{12.2.5}$$

where p is the perturbation in the electron pressure. Introducing a characteristic thermal speed v, we may set $p = v^2 \delta = v^2 nm$. Eliminating $\mathbf{u}$ and $\mathbf{E}$ and using $\delta = nm$, these equations lead to the following wave equation for δ:

$$\frac{\partial^2 \delta}{\partial t^2} = v^2 \nabla^2 \delta - \frac{4\pi e^2 N}{m}\delta \tag{12.2.6}$$

To find the dispersion relation, we insert the plane wave $\delta = \delta_0 \exp(ikr - i\omega t)$ and obtain

$$\omega^2 = \omega_p{}^2 + k^2 v^2 \qquad k = \frac{\omega}{v}\sqrt{1 - \left(\frac{\omega_p}{\omega}\right)^2} \tag{12.2.7}$$

where

$$\omega_p = \sqrt{\frac{4\pi e^2 N}{m}}$$

is 2π times the plasma frequency. We note that k becomes imaginary when $\omega < \omega_p$, and the plasma wave then decays exponentially.

Comparing Eqs. (12.2.7) with the result (12.2.2) obtained from the qualitative observations, we see that by accounting for the thermal motion of the electrons (as we did here by introducing the electron pressure in the equations of motion), we get a dispersion relation that indicates that the oscillation is not stationary but that it propagates. The group velocity is $d\omega/dk = kv^2/\omega$, or $\frac{\omega}{k}\frac{d\omega}{dk} = v^2$; the product of the phase velocity V_p and the group velocity V_g is $V_p V_g = v^2$. The two velocities can be expressed as

$$V_p = \frac{\omega}{k} = \frac{v}{\sqrt{1 - (\omega/\omega_p)^2}}$$

$$V_g = v\sqrt{1 - (\omega/\omega_p)^2} \tag{12.2.8}$$

which shows that the phase velocity is always larger than v for a propagating wave and that the opposite holds true for the group velocity. We can also express the phase velocity as

$$V_p = \frac{\omega_p}{k}\sqrt{1 + \left(\frac{\lambda_d}{\lambda}\right)^2}$$

where we have introduced $\lambda_d = v(2\pi/\omega_p)$, the characteristic length traveled by an electron in thermal motion in a time equal to a period of plasma oscillation. This length is called the *Debye length*. For wavelengths considerably longer than the Debye length, the phase velocity is approximately equal to ω_p/k, that is, proportional to the wavelength. For wavelengths of the order of λ_d or smaller, a more detailed analysis of plasma waves will show that the waves are strongly damped. This will be explained under the next heading.

The plasma wave is a longitudinal wave, a kind of acoustic wave, that involves fluid motion in the direction of propagation, as well as a longitudinal electric field. Returning to the equations of motion (12.2.3) to (12.2.5), we can now determine the relationship between the various field quantities. We find, for example,

$$p = \rho V_g u \qquad \delta = \frac{u}{V_p}\rho$$

which in our acoustic terminology means that the plasma-wave "impedance" relating the electron pressure and the particle velocity in the wave is the density times the group velocity. Similarly, we find

$$\frac{E^2}{8\pi} = \left(\frac{\omega_p}{\omega}\right)^2 \frac{\rho u^2}{2} \tag{12.2.9}$$

which says that the electric field energy density in the propagating plasma wave ($\omega > \omega_p$) is always smaller than the kinetic energy density due to the collective motion of the electrons. For $\omega = \omega_p$ they are equal.

After having obtained Eq. (12.2.9), it is interesting to determine the total energy density in the plasma wave. It is the sum of three parts, the kinetic energy of the electrons U_K, the electrostatic field energy U_E, and the "potential" energy U_p due to the compression of the electron gas (resulting in a perturbation of thermal energy). The potential energy in the wave can be expressed as $\kappa p^2/2$, where κ is the compressibility,

$$\kappa = \frac{\delta}{p\rho} = \frac{1}{\rho V_p V_g} = \frac{1}{\rho v^2}$$

Hence the potential energy density is

$$U_P = \frac{\rho^2 V_g^2}{2\rho v^2} u^2 = \left[1 - \left(\frac{\omega_p}{\omega}\right)^2\right] \frac{\rho u^2}{2}$$

The total energy density in the wave then can be written

$$U = \frac{\rho u^2}{2} + \left[1 - \left(\frac{\omega_p}{\omega}\right)^2\right] \frac{\rho u^2}{2} + \frac{E^2}{8\pi} = \rho u^2 = \frac{E^2}{4\pi}\left(\frac{\omega}{\omega_p}\right)^2 \tag{12.2.10}$$

where we have made use of Eq. (12.2.9).

Landau damping

In our hydrodynamical analysis of plasma oscillations we have neglected viscous stresses as well as heat conduction in the electron fluid, and the resulting plasma waves were undamped.   However, we can account for the effect of viscosity and heat conduction in the same way as we did in the preceding discussion and in Chap. 6.

Actually, even in the absence of viscosity and heat conduction, a kinetic theory of plasma oscillations leads to a damping of the plasma waves. This damping, however, cannot be explained on the basis of the hydrodynamical model, since it is related to the trapping of electrons in the traveling periodic potential produced by the plasma wave and depends on the form of the velocity distribution function of the electrons.   The hydrodynamic analysis deals only with averages of the distribution function.   The electrons that will be trapped by the wave are those with a thermal velocity component v_x in the direction of wave propagation about equal to the phase velocity ω/k of the wave.   As we shall see, the degree of damping resulting from this trapping depends on the slope of the velocity distribution function in the neighborhood of $v_x = \omega/k$.

To study the trapping mechanism to obtain an expression for the Landau damping, let the unperturbed velocity distribution function of the electrons be $f(v_x,v_y,v_z)$ such that

$$N = \iiint f(v_x,v_y,v_z)\,dv_x\,dv_y\,dv_z$$

If propagation of the plasma wave is chosen to be in the x direction, we shall be interested mainly in the dependence of f on v_x.   It is convenient, therefore, to introduce

$$g(v_x) = \frac{1}{N} \iint f(v_y,v_z)\,dv_y\,dv_z \tag{12.2.11}$$

The plasma wave represents a periodic electric field $E = E_0 \sin kx$ and a corresponding   potential $V = (E_0/k) \cos kx$ that travels   with a   velocity $V_p = \omega/k$ in a direction which we choose to be the x direction.   Imagine that the wave is formed at $t = 0$.   The electrons which at this time have a velocity $(V_p - \Delta v) < v_x < (V_p + \Delta v)$ will be trapped in the potential well of the plasma wave if the velocity increment Δv is such that

$$\frac{m}{2}(\Delta v)^2 = \frac{eE_0}{k} \tag{12.2.12}$$

This value of Δv should be regarded as an average since for any particular electron the actual value of the velocity required to make the electron go over the potential barrier depends upon the location of the electron at $t = 0$ in the potential wave trough.   Electrons that have a velocity in the range $(V_p - \Delta v) < v_x < V_p$ will be speeded up and receive energy as a result of the trapping, whereas the opposite holds true for the electrons in the range $V_p < v_x < V_p + \Delta v$. The corresponding energy will be drawn from or transferred to the plasma wave.

An electron that has been trapped will perform an oscillatory motion in the potential wave trough with a frequency of oscillation which, at least for sufficiently small amplitudes, is

$$\omega_c = \sqrt{\frac{eE_0 k}{m}} \tag{12.2.13}$$

A trapped electron with the initial velocity v_x oscillates back and forth with a velocity amplitude $v_x - V_p$, and its average total energy of oscillation is $(m/2)(v_x - V_p)^2$. In addition, the trapped electron has the kinetic energy of translation $(m/2)V_p^2$ with respect to the laboratory system. Therefore the net amount of energy gained by the electrons, and hence lost by the plasma wave, is

$$\Delta W = \frac{m}{2} \int\limits_{V_p - \Delta v}^{V_p + \Delta v} g(v_x) \, dv_x \, [V_p^2 + (v_x - V_p)^2 - v_x^2] \tag{12.2.14}$$

Since we are interested in values of v_x in the neighborhood of V_p, we expand $g(v_x)$.

$$g(v_x) = g(V_p) + (v_x - V_p)g'(V_p) + \cdots$$

where $g'(V_p) = (dg/dv_x)_{V_p}$. If we insert into (12.2.14) this expansion of $g(v_x)$, keeping only the first two terms, we obtain

$$\Delta W = -\frac{2m}{3} V_p g'(V_p)(\Delta v)^3 \tag{12.2.15}$$

The trapping process, and hence the energy exchange between the electrons and the wave, takes place approximately in one period $T_c = 2\pi/\omega_c$ of the oscillations of the electrons in the potential well, and as an average rate of energy transfer we may use $\omega_c \Delta W$. Then, if the internal energy density of the plasma oscillations is denoted by U, we have

$$\frac{\partial U}{\partial t} = -\omega_c \Delta W \tag{12.2.16}$$

Now we can express Δv and ω_c in terms of the electric field E, using Eqs. (12.2.12) and (12.2.13), which yields ΔW in terms of E^2. From Eq. (12.2.10), E^2 can be expressed in terms of U to give

$$\frac{\partial U}{\partial t} = -\frac{4\sqrt{2}}{3} \frac{\omega}{k} \frac{\omega_p^2}{k} g'\left(\frac{\omega}{k}\right)\left(\frac{\omega_p}{\omega}\right)^2 U \tag{12.2.17}$$

The corresponding decay of the amplitude of the plasma oscillations is then

$\exp(-\beta t)$, where

$$\frac{\beta}{\omega} = \frac{2\sqrt{2}}{3}\left(\frac{\omega_p}{k}\right)^2\left(\frac{\omega_p}{\omega}\right)^2 g'\left(\frac{\omega}{k}\right) \qquad (12.2.18)$$

For a Maxwellian velocity distribution of the electrons, we have $g(v_x) = (1/c\sqrt{\pi})\exp(-v_x^2/c^2)$ and $g'(\omega/k) = -(2\omega/kc^3\sqrt{\pi})\exp(-\omega^2/k^2c^2)$. The corresponding expression for the decay constant becomes

$$\beta = \frac{4\sqrt{2}}{3\sqrt{\pi}}\,\omega_p\left(\frac{\omega_p}{kc}\right)^3\exp\left(-\frac{\omega^2}{k^2c^2}\right) \qquad (12.2.19)$$

where $c^2 = 2kT_e/m$, T_e being the electron temperature.

12.3 MAGNETOACOUSTIC WAVES

In the presence of a magnetic field, the plasma becomes nonisotropic, and the wave motion becomes considerably more complicated than in the absence of the field. A systematic study of this problem would carry us far beyond the scope of this chapter, and we shall limit the discussion to low-frequency phenomena in which we can neglect charge-separation effects.

To save 4π's in this section, we shall use a "normalized" set of electromagnetic units; in this section only, $\mathbf{B} = \mathbf{B}_c/\sqrt{4\pi}$; $\mathbf{D} = \mathbf{D}_c/\sqrt{4\pi}$; and $\mathbf{j} = \sqrt{4\pi}\,\mathbf{j}_c$; $\mathbf{B}_c$, $\mathbf{D}_c$, and $\mathbf{j}_c$ being in the usual cgs units.

Magnetic stress tensor

The electrons and ions then move together as in a single conducting fluid. Under these conditions the principal effect of a magnetic field $\mathbf{B}$ on the motion of the conducting fluid is to produce a force per unit volume of the fluid,

$$\mathbf{F} = \frac{1}{c}\,(\mathbf{j} \times \mathbf{B}) \qquad (12.3.1)$$

where $\mathbf{j}$ is the current density. The current density, in turn, is related to the magnetic field through the Maxwell equation

$$c\,\mathrm{curl}\,\mathbf{B} = \mathbf{j} + \frac{\partial\mathbf{D}}{\partial t} \simeq \mathbf{j} \qquad (12.3.2)$$

In the last step we have neglected the displacement current $\partial\mathbf{D}/\partial t$, which is justified in a fluid with high conductivity, at least for the low-frequency phenomena we wish to study here. To see this, we note that, in a periodic field with a frequency ω, $\partial\mathbf{D}/\partial t$ is of the order of $\omega\epsilon\mathbf{E}$ and $\mathbf{j} = \sigma\mathbf{E}$, where ϵ is the dielectric constant, σ the conductivity, and $\mathbf{E}$ the electric field. Thus the ratio between the magnitudes of the displacement current and the conduction current is then $(\omega\epsilon/\sigma)$, which is much less than unity for the problems to be considered here.

Now if we introduce $\mathbf{j} = c\,\text{curl}\,\mathbf{B}$ into (12.3.1) and make use of the well-known relation $\mathbf{B} \times \text{curl}\,\mathbf{B} = \text{grad}\,(B^2/2) - (\mathbf{B} \cdot \nabla)\mathbf{B}$, it follows that the ith component of the force in Eq. (12.3.1) can be written in the form, remembering that $\Sigma\,\partial B_j/\partial x_j = 0$,

$$F_i = -\sum_j \frac{\partial}{\partial x_j}\left(\frac{B^2}{2}\delta_{ij} - B_i B_j\right)$$

where $\delta_{ij} = 0$ for $i \neq j$ and $\delta_{ij} = 1$ for $i = j$. This expression for F_i makes it possible to write the equation of motion in the form

$$\frac{\partial(\rho u_i)}{\partial t} + \sum_j \frac{\partial}{\partial x_j}\left(P\delta_{ij} + \rho u_i u_j + \frac{B^2}{2}\delta_{ij} - B_i B_j\right) = 0 \qquad (12.3.3)$$

This way of writing the equation of motion demonstrates that the magnetic field is dynamically equivalent to a mechanical stress tensor $(B^2/2)\delta_{ij} - B_i B_j$. To see the physical meaning of this stress a little better, we express it with reference to a coordinate system, with the x_1 axis in the direction of the magnetic field. The components of the stress are then given by $(B^2/2)\delta_{ij} - B^2\delta_{i1}\delta_{j1}$. In other words, the effect of the magnetic field is equivalent to an isotropic magnetic pressure component $B^2/2$ and a "pull" B^2 per unit area in the direction of the field lines.

The magnetic Reynolds number

Before proceeding to study specific motions, it is interesting to mention an additional general property of motion which has to do with the coupling between the magnetic field and the fluid velocity field. We shall find that in the limit of infinite conductivity, the coupling is so strong that the magnetic field lines are convected with the fluid as if they were "trapped by" or "frozen into" the fluid. To prove this important aspect of the motion, we first seek an expression for the time rate of change of the magnetic field in a coordinate system moving with the fluid. We start from the expression for the current density,

$$\mathbf{j} = \sigma\left(\mathbf{E} + \frac{\mathbf{U}}{c} \times \mathbf{B}\right) \qquad (12.3.4)$$

together with Maxwell's equations,

$$c\,\text{curl}\,\mathbf{B} = \mathbf{j} + \frac{\partial \mathbf{E}}{\partial t} \qquad \text{div}\,\mathbf{B} = 0 \qquad (12.3.5)$$

$$c\,\text{curl}\,\mathbf{E} = -\frac{\partial \mathbf{B}}{\partial t} \qquad \text{div}\,\mathbf{E} = 0 \qquad (12.3.6)$$

where we have assumed zero net charge density in the fluid.

From these equations follows

$$\frac{1}{c}\frac{\partial \mathbf{B}}{\partial t} = -\text{curl}\left(\frac{\mathbf{j}}{\sigma} - \frac{\mathbf{U}}{c} \times \mathbf{B}\right) \qquad (12.3.7)$$

and if we neglect the displacement current $\partial \mathbf{E}/\partial t$ we get

$$\frac{\partial \mathbf{B}}{\partial t} - \text{curl} (\mathbf{U} \times \mathbf{B}) = -\frac{1}{\sigma} c^2 \, \text{curl} (\text{curl} \, \mathbf{B}) = \frac{c^2}{\sigma} \nabla^2 \mathbf{B} \qquad (12.3.8)$$

By making use of curl $(\mathbf{U} \times \mathbf{B}) = (\mathbf{B} \cdot \nabla)\mathbf{U} + (\mathbf{U} \cdot \nabla)\mathbf{B}$ and introducing $d/dt = \partial/\partial t + \mathbf{U} \cdot \nabla$, this equation can be expressed as

$$\frac{d\mathbf{B}}{dt} = (\mathbf{B} \cdot \nabla)\mathbf{U} + \frac{c^2}{\sigma} \nabla^2 \mathbf{B} \qquad (12.3.9)$$

Next consider two fluid particles located at the positions $\mathbf{r}$ and $\mathbf{r} + d\mathbf{r}$ on a particular magnetic field line at time t. The separation between the particles is specified by the vector $d\mathbf{r}$, which has the same direction as $\mathbf{B}$, so that we can set $d\mathbf{r} = \epsilon \mathbf{B}$, where ϵ is a small quantity. As the fluid moves, the separation between the particles will also change, in general, so that at time $t + dt$ the new vector is given by

$$d\mathbf{r}' = d\mathbf{r} + (d\mathbf{r} \cdot \nabla)\mathbf{U} \, dt = \epsilon[\mathbf{B} + (\mathbf{B} \cdot \nabla)\mathbf{U} \, dt]$$

However, for the time dependence of the $\mathbf{B}$ vector, we have already seen that, for $\sigma \to \infty$

$$\frac{d\mathbf{B}}{dt} = (\mathbf{B} \cdot \nabla)\mathbf{U} \qquad (12.3.10)$$

so that

$$d\mathbf{r}' = \epsilon(\mathbf{B} + d\mathbf{B}) = \epsilon \mathbf{B}'$$

where $\mathbf{B}'$ is the magnetic field vector at $t + dt$. In other words, $d\mathbf{r}$ and $\mathbf{B}$ remain lined up at all times. We can interpret this result pictorially as a comovement of the magnetic field lines with the fluid, just as if the lines were embedded in the fluid like rubber bands that stretch and bend to conform to the variations in the fluid flow field.

This interpretation, taken together with the result obtained earlier that the electromagnetic stress corresponds to a tension along the field lines equal to B^2 per unit area, suggests the existence of a transverse wave motion in the fluid in much the same way as on a stretched string. To see this, let the unperturbed magnetic field be a uniform field in the z direction of magnitude B. The fluid is initially at rest. Consider now perturbation of the fluid in the form of a local displacement perpendicular to the z axis. This leads to a local transverse displacement of the magnetic field lines. Since there is a force B^2 along the field lines per unit area, and since a fluid string of unit area has a mass per unit length equal to ρ, the density of the fluid, we expect, by analogy with wave motion on a string, that the transverse displacement of the fluid and the magnetic field "string" will propagate with a speed $\sqrt{B^2/\rho} = B/\sqrt{\rho}$. We shall show under the next heading that this observation is consistent with more quantitative detail.

If the conductivity of the fluid is finite, the coupling between the magnetic field and the fluid motion is not complete; the field lines will diffuse, or "slip" through the fluid. As a result, the wave motion described above will be damped. To discuss the degree of coupling on somewhat more quantitative terms, let us return to Eq. (12.3.9). If, for simplicity, we consider two-dimensional motion of the fluid in a magnetic field perpendicular to the flow velocity, the equation reduces to

$$\frac{d\mathbf{B}}{dt} = \left(\frac{c^2}{\sigma}\right) \nabla^2 \mathbf{B}$$

which describes the diffusion of the magnetic field in a coordinate system moving with the fluid. We rewrite this equation in dimensionless form, introducing a characteristic length L and a characteristic time T and a corresponding characteristic velocity $U = L/T$ of the fluid. Then, with $\tau = t/T$ and $\xi = x/L$, we get

$$\frac{d\mathbf{B}}{d\tau} = \frac{c^2 T}{\sigma L^2} \nabla_\xi{}^2 \mathbf{B} = \frac{1}{R_m} \nabla_\xi{}^2 \mathbf{B}$$

The dimensionless quantity

$$R_m = \frac{UL}{(1/\sigma)c^2} \tag{12.3.11}$$

is called the *magnetic Reynolds number*. It plays the same role in regard to diffusion of the magnetic field as does the ordinary Reynolds number $UL/(\mu/\rho)$ in regard to diffusion of vorticity. The characteristic time of diffusion of the magnetic field is given by

$$T_d = \frac{L^2 \sigma}{c^2}$$

The characteristic time available for the diffusion can be expressed as $T = L/U$, and a necessary condition that the magnetic field be trapped in the fluid is $T \ll T_d$, which means $U \gg c^2/L\sigma$.

This, obviously, is not a sufficient condition for good coupling, however. We must also have a magnetic field present, and for this magnetic field to have a pronounced effect on the motion, it is necessary that the magnetic-field energy density $B^2/2$ be at least of the same order of magnitude as the kinetic energy density $\rho U^2/2$ of the fluid, so that $B^2 > \rho U^2$. If we now use the condition for the velocity $U \gg c^2/L\sigma$ obtained above, we get a condition for the magnetic field,

$$B \gg \frac{c^2 \sqrt{\rho}}{L\sigma} \quad \text{or} \quad D_m = \frac{BL\sigma}{c^2 \sqrt{\rho}} \gg 1 \tag{12.3.12}$$

to assure good coupling between the magnetic field and the velocity. This condition is well satisfied in astrophysical situations, where D_m may range from, say, 10 for the ionosphere to 10^{10} for the solar corona and hot interstellar gases. On the laboratory scale D_m is considerably smaller, say, between 1 and 100 in experiments with liquid metals and hot plasmas.

Transverse waves (Alfvén waves)

The qualitative observations in the preceding discussion about transverse waves traveling along magnetic field lines will now be verified by a study of the linearized equations of motion. We shall choose a particularly simple situation in which the fluid performs one-dimensional motion perpendicular to a uniform external magnetic field. Thus, with an external magnetic field B_0 in the z direction and with a fluid motion in the y direction, we have $\partial/\partial x = \partial/\partial y = 0$. The induced current $\mathbf{j}$ in the fluid then has only an x component, and the corresponding induced perturbation in the magnetic field $\mathbf{b}$ has only a y component. Under these conditions, and assuming an incompressible fluid, Maxwell's equations (12.3.4) to (12.3.6) and the momentum equation (12.3.3) are considerably simplified, and the linearized forms of the equations are

$$j_x = -c\frac{\partial b_y}{\partial z} \qquad \frac{\partial b_y}{\partial t} = -c\frac{\partial E_x}{\partial z}$$

$$j_x = \sigma E_x + \frac{v_y B_0}{c} \qquad \rho\frac{\partial b_y}{\partial t} = -c^{-1}j_x B_0 \tag{12.3.13}$$

From these equations follows the "wave equation" for b_y:

$$\frac{\partial^2 b_y}{\partial t^2} - \frac{B_0{}^2}{\rho}\frac{\partial^2 b_y}{\partial z^2} - \frac{c^2}{\sigma}\frac{\partial^3 b_y}{\partial t\,\partial z^2} = 0 \tag{12.3.14}$$

Seeking a wavelike solution of the form $\exp(ikz - i\omega t)$, we have $\partial^2/\partial t^2 = -\omega^2$, $\partial^2/\partial z^2 = -k^2$, and $\partial^3/\partial t\,\partial z^2 = +ik^2\omega$. Equation (12.3.14) then yields

$$k^2 = \frac{\omega^2}{c_A{}^2}\frac{1}{1 - i(c^2\omega/\sigma c_A{}^2)} = \frac{\omega^2}{c_A{}^2}\frac{1}{1 - i\omega\tau} \tag{12.3.15}$$

where

$$c_A = \frac{B}{\sqrt{\rho}} \qquad \tau = \frac{c^2}{c_A{}^2}\frac{1}{\sigma}$$

The time $\tau = \dfrac{c^2}{c_A{}^2}\dfrac{1}{\sigma}$ can be regarded as a "relaxation time" of the magnetic field lines.

If the conductivity is infinite, we get $k = \omega/c_A$, which indicates that the wave speed is $c_A = B/\sqrt{\rho}$, in agreement with the preliminary observations in the previous subsection. If σ is large enough so that $\omega\tau \ll 1$,

$$k \simeq \frac{\omega}{c_A}\left(1 + i\frac{\omega\tau}{2}\right)$$

which means that in this limit the phase velocity is c_A and the wave amplitude is attenuated exponentially, corresponding to the factor $\exp(-\alpha x)$, where

$$\alpha = \frac{c^2}{2\sigma}\frac{\omega^2}{c_A{}^3} \qquad \text{or} \qquad \alpha\lambda = \pi(\omega\tau)$$

The general expressions for the phase velocity and the attenuation constants are

$$c = c_A \frac{\sqrt{2}\sqrt{1 + (\omega\tau)^2}}{\sqrt{1 + \sqrt{1 + (\omega\tau)^2}}} \rightarrow \begin{cases} c_A & \omega\tau \ll 1 \\ c_A\sqrt{2\omega\tau} & \omega\tau \gg 1 \end{cases}$$

$$(12.3.16)$$

$$\lambda\alpha = \frac{\sqrt{2}\,\pi\omega\tau}{\sqrt{1 + \sqrt{1 + (\omega\tau)^2}}} \rightarrow \begin{cases} \pi(\omega\tau) & \omega\tau \ll 1 \\ \pi\sqrt{2\omega\tau} & \omega\tau \gg 1 \end{cases}$$

where $\lambda = c_A(2\pi/\omega)$.

As an example, let us calculate the energy flow in an Alfvén wave. The wave is generated by a source which produces an oscillatory motion of the fluid, $v_y = v_0 \exp(-i\omega t)$ in the plane $z = 0$. The constant external magnetic field B_0 points in the z direction, and as we have seen before, the transverse motion generated at $z = 0$ will be propagated in the z direction with a speed $c_A = B_0/\sqrt{\rho}$. From Eq. (12.2.13) it follows that the perturbation in the magnetic field is proportional to v_y. In fact, we have $\rho\, \partial v_y/\partial t = B_0\, \partial b_y/\partial z$, and with $v_y = v_0 \exp[i(\omega z/c_A) - i\omega t]$, it follows that

$$B_0 b_y = -\rho c_A v_y \qquad \text{or} \qquad \frac{b_y{}^2}{2} = \frac{\rho v_y{}^2}{2} \qquad (12.3.17)$$

This shows that the magnetic and kinetic energy densities in the wave are the same.

The energy flow in the wave can be obtained in several different ways. For example, we can express it as c_A times the total energy density, $I_z = [(\rho v_y{}^2/2) + (b_y{}^2/2)]c_A = (\rho v_y{}^2)c_A$. It is instructive to derive this result in a different manner, using the magnetic stress tensor. As we recall, the magnetic stress tensor $(B^2/2)\delta_{ij} - B_i B_j$ in Eq. (12.3.3) plays the same role as the pressure tensor, and the components of the related energy flow vector $\mathbf{I}$ can be expressed as

$$I_i = \sum_j \left[\left(\frac{B^2}{2}\right)\delta_{ij} - B_i B_j \right] v_j$$

In this example the velocity $\mathbf{v}$ has only the component v_y, and the energy flow in the z direction again is found to be

$$I_z = -B_z B_y v_y = -B_0 b_y v_y = \rho c_A v_y{}^2$$

where we have used Eq. (12.3.17).

The force required in the source plane $z = 0$ to produce the wave must be such that $F_y v_y = \rho c_A v_y{}^2$, or

$$F_y = \rho c_A v_y$$

The ratio $F_y/v_y = \rho c_A$ is the wave impedance of the Alfvén wave.

Longitudinal waves

Next we wish to study the effect of a magnetic field on the propagation of a sound wave in a conducting fluid. It is clear that there will be no effect if the waves are propagated in the direction of the magnetic lines, and that the largest effect is to be expected when the direction of propagation is perpendicular to the magnetic field. We shall start by studying this particular case. As a preliminary observation we note that on the basis of the previous findings about the coupling between the fluid and the magnetic field, we may expect that the magnetic field will provide an additional "restoring" force for the particle motion in the fluid, and this should result in an increase in the speed of sound.

In the mathematical analysis of the problem let the uniform magnetic field be in the z direction and the sound wave in the x direction. The linearized equations of motion are then

$$\frac{\partial \delta}{\partial t} + \rho_0 \frac{\partial v_x}{\partial x} = 0$$

$$\rho_0 \frac{\partial v_x}{\partial t} = -\frac{\partial p}{\partial x} - c^{-1}j_y B_z$$

$$j_y = \sigma(E_y - c^{-1}v_x B_x) \qquad j_y = -c\frac{\partial b_z}{\partial x}$$

$$p = c_s^2 \delta \qquad \frac{\partial b_z}{\partial t} = -c\frac{\partial E_y}{\partial x}$$

(12.3.18)

A plane wave of the form $\exp(ikx - i\omega t)$ inserted in these equations leads to the following dispersion relation:

$$\left(\frac{\omega}{k}\right)^2 = c_s^2 + c_A^2 \frac{1}{1 + i\omega\tau_1(c_s k/\omega)^2}$$

where $\tau_1 = c^2/c_s^2\sigma$, $c_s^2 = 1/\kappa\rho_0$, and $c_A^2 = B^2/\rho$. As expected, the magnetic field increases the phase velocity of the sound wave. In fact, in the limiting case of infinite conductivity, so that $\omega\tau_1 = 0$, the phase velocity is $\omega/k = \sqrt{c_s^2 + c_A^2}$. In the other limiting case of zero conductivity, we get $\omega/k = c_s$, as expected. For intermediate values of the conductivity, the propagation constant k is complex for real values of ω, and the sound wave is attenuated through conduction losses. If $\omega\tau_1 \ll 1$, we can insert $c_s^2 + c_A^2$ for $(\omega/k)^2$ in the denominator, and with $\tau_2 = \tau_1[c_s^2/(c_A^2 + c_s^2)] = c^2/(c_A^2 + c_s^2)\sigma$, we obtain

$$k \simeq \frac{\omega}{\sqrt{c_s^2 + c_A^2}}\left(1 + \frac{1}{2}\frac{c_A^2}{c_A^2 + c_s^2} i\omega\tau_2\right)$$

$$\tau_2 = \frac{c^2}{c_A^2 + c_s^2}\frac{1}{\sigma}$$

from which it follows that for $\omega\tau_2 \ll 1$, the attenuation of the wave is proportional to ω^2. It is interesting to note, however, that in both limits, $\omega\tau_2 \to 0$ and $\omega\tau_2 \to \infty$, the attenuation is zero, and consequently attenuation will attain a maximum value for some value of $\omega\tau_2$ (of the order of unity). We leave it as a problem to study the dispersion relation in further detail.

Coupled waves

Under the last two headings we have considered two special cases of motion: first, the transverse (Alfvén) waves traveling along the magnetic field lines, and second, ordinary sound waves traveling perpendicular to the field lines. In the analysis of these waves an arbitrary value of the conductivity was used which enabled us to determine not only the phase velocity of the waves, but also the attenuation.

To carry through a similar general analysis of waves traveling at an arbitrary angle with respect to the magnetic field lines is quite involved, and we shall consider here only the case when the conductivity is infinite.

In analyzing this problem, we first have to deal with the questions concerning the definition of directions, etc. The two prominent directions involved in the problem defined by $\mathbf{k}$ and $\mathbf{B}$ establish a plane, and as we shall see, the nature of the wave depends markedly on whether the fluid velocity is parallel with or perpendicular to this plane. In the latter case we expect the wave motion to be, simply, an Alfvén wave having a speed equal to $B \cos\theta/\sqrt{\rho}$, where θ is the angle between the direction of propagation and magnetic field. As we recall, the perturbation $\mathbf{b}$ of the magnetic field in such a wave is in the same direction as the particle velocity in the wave motion, and if the plane of $\mathbf{k}$ and $\mathbf{B}$ is taken to be the xy plane, the only components of $\mathbf{u}$ and $\mathbf{b}$ then are u_z and b_z. If, on the other hand, the particle velocity is in the plane of $\mathbf{k}$ and $\mathbf{B}$, the wave mode is expected to contain both a transverse and a longitudinal motion, corresponding to the particle velocity components perpendicular to and parallel with $\mathbf{k}$, respectively. The perturbation in the magnetic field again should be perpendicular to $\mathbf{k}$. If the direction of $\mathbf{k}$ is the x axis, this magnetic field perturbation will have a b_y component only. These qualitative observations are easily verified from the linearized equations of motion of a conducting fluid in a magnetic field. In the case of infinite conductivity, and when the wave perturbation is of the form $\exp(i\mathbf{k}\cdot\mathbf{r} - i\omega t)$, these equations are

$$\omega\delta = \rho_0(\mathbf{k}\cdot\mathbf{u}) \qquad \omega\mathbf{b} = -[\mathbf{k} \times (\mathbf{u} \times \mathbf{B}_0)] \qquad (12.3.19)$$

$$\rho_0\omega\mathbf{u} = kc^2\delta + [\mathbf{B}_0 \times (\mathbf{k} \times \mathbf{b})]$$

where δ, $\mathbf{u}$, and $\mathbf{b}$ are the complex amplitudes of the perturbations in density, velocity, and magnetic field.

It follows from the first of these equations that a density fluctuation is produced only if there is a velocity component in the direction of propagation and the perturbation in the magnetic field is always perpendicular to $\mathbf{k}$.

Furthermore, if the plane defined by $\mathbf{k}$ and $\mathbf{B}$ is chosen to be the xy plane, with the x axis along $\mathbf{k}$, we find that the velocity component u_z belongs to a transverse wave motion (Alfvén wave) that travels in the x direction with a phase velocity

$$c_1 = \frac{B_0}{\sqrt{\rho_0}} \cos \theta = \frac{B_x}{\sqrt{\rho_0}}$$

where θ is the angle between $\mathbf{k}$ and $\mathbf{B}_0$, and $B_0/\sqrt{\rho_0} = c_A$ is the characteristic Alfvén speed in the fluid. This mode of motion represents one of three possible characteristic modes of motion that can exist in the fluid. The remaining two modes correspond to wave motion with the fluid velocity components in the xy plane. We can regard these modes as coupled waves involving both transverse and longitudinal motions. One of them is more like an Alfvén wave with a touch of sound in it, and in the other mode the opposite holds true. To find the dispersion relation for these waves, we insert a wave motion with the particle velocity in the $\mathbf{kB}$ plane. After some algebra we then find that two modes exist with the phase velocities

$$c_{2,3} = \frac{c_s}{2} (\sqrt{1 + \alpha^2 + 2\alpha \cos \theta} \pm \sqrt{1 + \alpha^2 - 2\alpha \cos \theta}) \quad (12.3.20)$$

where $\alpha = c_A/c_s$, and $c_s = $ sound speed $= \sqrt{1/\kappa\rho_0}$. For small values of the magnetic field, so that $\alpha < 1$, the two modes, apart from second-order terms in α, reduce to an ordinary sound wave with velocity c_s and an Alfvén wave with the velocity $c_A \cos \theta$. We note that if propagation is in the direction of the magnetic field so that $\theta = 0$, the two waves reduce to a pure sound wave and a pure Alfvén wave. For propagation perpendicular to the magnetic field, so that $\theta = \pi/2$, the Alfvén-wave component is absent, and we have only a sound wave with a speed $c_s\sqrt{1 + \alpha^2}$.

Problems

1 A uniform plasma is maintained in a region of length L in a long tube by means of an electric field $\mathbf{E}_0$. The electric field is modulated by a periodic electric field $E_1 \sin (\omega t)$ with $E_1 \ll E_0$. As a result of this modulation, sound is generated. Derive an expression for the sound field outside the plasma region as well as inside this region. At what frequencies of modulation is the sound pressure field outside the plasma a maximum (minimum)? The tube is so long that reflections can be neglected.

2 The intensity of a microwave beam that has been transmitted through a plasma depends on the electron density of the plasma, and the fractional change in this intensity is approximately equal to the fractional change in electron density. If such a modulation is to be produced by a traveling sound wave, what should be the pressure amplitude of the wave to produce a 1 percent modulation of the transmitted microwave intensity? The plasma parameters are helium gas, neutral-particle density $= 10^{16}$ cm^{-3}, electron density $= 10^{11}$ cm^{-3}, neutral-gas temperature $= 500°$K.

3 In Prob. 2 let the electron temperature be $10^{4\circ}$K, the diameter of the discharge tube 3 cm, and the electron-neutral elastic collision cross section 2×10^{-16} sq cm. Under these conditions is acoustic wave amplification possible? If so, determine the frequency range in which amplification is expected.

4 In the study of the amplification and spontaneous excitation of a sound wave we considered only the case of plane waves in a rectangular tube. Consider now the problems of spontaneous excitation of sound waves in a spherical discharge, and derive an expression for the growth rates of radial modes of oscillation. In this case the distribution of electron density and electron temperature is such that Eq. (12.1.12) is of the form

$$\frac{\partial^2 p}{\partial t^2} - c^2 \nabla^2 p = 2\beta j_0 \left(\frac{\pi r}{R}\right) \frac{\partial p}{\partial t}$$

where R is the radius of the spherical cavity, and β a constant that depends on the electron density and the electron temperature. Set $p = \Sigma P_n(t) j_0(\pi \alpha_n r/R)$, where j_0 is the zeroth-order spherical Bessel function, and $j_0'(\pi \alpha_n) = 0$. Show that the equation for P_n is of the form

$$\frac{d^2 P_n}{dt^2} + \omega_n{}^2 P_n = 2\beta \sum_m f_{n,m} \frac{dP_m}{dt}$$

where $\omega_n = \pi \alpha_n c/R$, and determine $f_{n,m}$ ($f_{1,1} \simeq 0.60, f_{1,2} \simeq 0.145$, etc.). Then set $P_n(t) = p_n \exp(-i\omega t)$ and derive, in determinant form, an equation for the eigenfrequencies ω. If, in the determinant, nondiagonal terms containing $f_{m,n}$ ($m \neq n$) are neglected, show that

$$\omega = i\beta f_{n,n} \pm \sqrt{\omega_n{}^2 - \beta^2 f_{n,n}} \simeq i\beta f_{n,n} \pm \omega_n$$

where $\beta f_{n,n}$ is the growth rate.

CHAPTER
13

ACOUSTOOPTICAL INTERACTION

13.1 LIGHT SCATTERING BY A SOUND BEAM

In the preceding chapter we studied some aspects of the acoustics of an ionized gas. We found that an important feature of the plasma is that the electron temperature is considerably higher than the neutral-gas temperature. Consequently, energy flows from the electrons to the neutrals, which may lead to the generation of sound. The problem of the interaction between light and sound can be viewed in much the same manner if we consider the light field as a photon gas. The photons have a considerably higher temperature than the medium carrying the sound wave, and as a result an energy flow into the fluid medium is to be expected.

The interaction between the electric field and the material medium is due to the response of the atomic electrons to the electric field, which results in a polarization of the material and a resulting index of refraction. This index depends on the density in the material; consequently, the density fluctuations in a sound wave produce a perturbation in the index of refraction. As a result an electromagnetic wave transmitted through the medium will be modulated by the sound wave, and scattering and refraction will occur. Conversely, a spatial variation of the electromagnetic field intensity and the corresponding electromagnetic stress lead to a volume force distribution in the medium (electrostriction), which, under certain conditions, leads to sound generation. This mutual interaction between light and sound may also lead to instabilities and wave amplification, just as in the case of electron-sound interaction discussed in the preceding chapter.

The coupling equation

To study the scattering and diffraction of an electromagnetic wave by sound, we start from the wave equation for the electromagnetic field $\mathbf{E}$.

$$\nabla^2 \mathbf{E} - \left(\frac{n}{c}\right)^2 \frac{\partial^2 \mathbf{E}}{\partial t^2} = 0$$

809

where n is the index of refraction, and c is the speed of light in vacuum. This equation is valid also for a time-dependent index of refraction as long as $\dfrac{1}{n}\dfrac{\partial n}{\partial t}T \ll 1$, where T is the period of the oscillation in the E field. We write the unperturbed value of the index of refraction as n_0. If the perturbation n_1 caused by the sound wave is small, $n_1/n_0 \ll 1$, we have $n^2 = n_0{}^2 + 2n_0 n_1$, and the wave equation can be expressed as

$$\nabla^2 \mathbf{E} - \frac{1}{c^2}\frac{\partial^2 \mathbf{E}}{\partial t^2} = \frac{2(n_1/n_0)}{c^2}\frac{\partial^2 \mathbf{E}}{\partial t^2} = \frac{\delta\epsilon/\epsilon_0}{c^2}\frac{\partial^2 \mathbf{E}}{\partial t^2} \qquad (13.1.1)$$

In the last term we have introduced the relative variation in the dielectric constant, which is related to the index of refraction as $n \sim \sqrt{\epsilon}$, so that $2\delta n/n = \delta\epsilon/\epsilon_0$.

Similarly, the effect of the electric field on the sound field can be represented as a source term in the acoustic-wave equation, the source term being the divergence of the force distribution produced by the electromagnetic stress. The sound pressure, and therefore the perturbation in the dielectric constant, depends on the electric field, and the wave equations for the electric and acoustic fields are thus coupled, as will be discussed further in Sec. 13.3. However, in the study of the scattering of light by sound, we can ordinarily neglect the influence of the electric field on the sound field, so that $\delta\epsilon$ in Eq. (13.1.1) can be regarded as a known function of space and time, independent of $\mathbf{E}$. The wave equation is of the same form as encountered in many of our acoustic-scattering problems.

As an example we shall treat here the problem of light diffraction by a sound beam (usually referred to as the Debye-Sears effect). If the width of the sound beam is not too large, we expect it to act in much the same way as an ordinary optical transmission grating with a grating constant equal to the wavelength λ_s of the sound wave. If, for example, a light wave is incident at right angles to the sound beam, we should expect that the angle for the first-order diffraction line from the "grating" would be given by $\sin\theta = \Lambda/\lambda_s$, where Λ is the wavelength of the incident light and θ is measured from the incident light beam. From the angle of diffraction θ and the wavelength Λ of light, the acoustic wavelength λ_s can be measured. However, the finite width of the ultrasonic grating, its time dependence, and motion make the acoustic grating somewhat more complex than the ordinary optical grating, and we shall devote our initial discussion to this problem.

The beam of sound is assumed to travel in the x direction and to fill the region between the two planes $z = L$ and $z = -L$, as indicated in Fig. 13.1. The incident light beam travels parallel to the xz plane at an angle θ with respect to the z axis. Inside the sound beam the index of refraction varies with the acoustic density fluctuations at the frequency $\omega_s/2\pi$ of the sound. This frequency is considerably lower than the frequency of the light $\Omega/2\pi$.

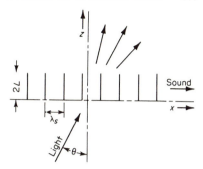

FIGURE 13.1
A sinusoidal wave of sound is a diffraction grating for light.

Actually, this fact justifies the omission of terms involving the rate of change $\partial n/\partial t$ of the index of refraction in the wave equation (13.1.1).

After a Fourier transform in time (and using the result of Prob. 23 of Chap. 1), this equation becomes

$$\nabla^2 \mathbf{E}(x,z,\omega) + k^2 \mathbf{E}(x,z,\omega) = -c^{-2}\int_{-\infty}^{\infty} (\omega')^2 \mathbf{E}(x,z,\omega') f(\omega - \omega')\, d\omega'$$

(13.1.2)

where $k = \omega/c$, and f is the Fourier transform of $\delta\epsilon/\epsilon_0$. This is a vector wave equation; its general solution would involve dyadic Green's functions. However, for the purposes of this chapter, we need only deal with polarized waves, such that scalar-wave solutions are sufficient.

The Born approximation

The space-time dependence of $\delta\epsilon$ and the corresponding Fourier transform $f(\omega)$ depend on the sound beam that forms the diffraction grating. We shall consider here a single harmonic collimated beam of sound, with the acoustic pressure in the region between $z = -L$ and $z = L$, where the sound pressure may be set equal to $p = p_0 \cos(k_s x - \omega_s t)$, where $2\pi/k_s = \lambda_s$ is the acoustic wavelength, and $\omega_s/2\pi$ is the frequency. The perturbation in the index of refraction will have the same form, and we shall set

$$\frac{\delta\epsilon}{\epsilon_0} = \eta \cos(k_s x - \omega_s t)$$

(13.1.3)

The Fourier transform f of $\delta\epsilon/\epsilon_0$ in Eq. (13.1.2) then is

$$f(\omega - \omega') = (\eta/2)[\delta(\omega - \omega' - \omega_s)\exp(ik_s x)$$
$$+ \delta(\omega - \omega' + \omega_s)\exp(-ik_s x)]$$

Furthermore, if the incident light is monochromatic with a frequency $\Omega/2\pi$ and with $K = \Omega/c$, we have for the incident light $E = E_0 \cos(\mathbf{K} \cdot \mathbf{r} - \Omega t)$, where $\mathbf{K} \cdot \mathbf{r} = Kx \sin\theta + Kz \cos\theta$ (Fig. 13.1).

If we use the transform of this wave in the integral of Eq. (13.1.2), we shall be using a variety of the Born approximation to calculate the diffracted light. The Fourier transform of the incident field is $E(x,z,\omega') = (E_0/2)[\delta(\omega' - \Omega) \exp(i\mathbf{K} \cdot \mathbf{r}) + \delta(\omega' + \Omega) \exp(-i\mathbf{K} \cdot \mathbf{r})]$.

If we use these expressions for $f(\omega - \omega')$ and $E(x,z,\omega')$ in the integral of Eq. (13.1.2), this equation becomes

$$\nabla^2 E + k^2 E = -\frac{\Omega^2 \eta E_0}{2} \{\delta(\omega - \Omega - \omega_s) \exp[i(\mathbf{K} \cdot \mathbf{r} + k_s x)]$$
$$+ \delta(\omega - \Omega + \omega_s) \exp[i(\mathbf{K} \cdot \mathbf{r} - k_s x)]$$
$$+ \delta(\omega + \Omega - \omega_s) \exp[-i(\mathbf{K} \cdot \mathbf{r} - k_s x)]$$
$$+ \delta(\omega + \Omega + \omega_s) \exp[-i(\mathbf{K} \cdot \mathbf{r} + k_s x)]\} \equiv S(x,z;\omega) \quad (13.1.4)$$

This equation expresses the fact that the sound wave modulates the incident light so that electric-field components with frequencies $\Omega \pm \omega_s$ are produced in the interaction region.

The corresponding scattered field, when E_0 is parallel with the y axis, is conveniently calculated with the help of the two-dimensional Green's function of Eq. (7.3.17).

$$G_\omega(x,z \mid x_0,z_0) = \frac{i}{4} H_0^{(1)}(k\sqrt{(x - x_0)^2 + (z - z_0)^2}) \quad (13.1.5)$$

which is a solution of the equation $\nabla^2 G + k^2 G = -\delta(\mathbf{r} - \mathbf{r}_0)$, representing the field produced by a line source of unit strength at (x_0,z_0). The electric field scattered from the sound wave can then be expressed as

$$E(x,z,\omega) = \int_{-\infty}^{+\infty} dx_0 \int_{-L}^{+L} dz_0\, S(x_0,z_0;\omega) G_\omega(x,z \mid x_0,z_0) \quad (13.1.6)$$

where the source term S is defined in Eq. (13.1.4).

Setting $\mathbf{K} \cdot \mathbf{r}_0 = x_0 K \sin\theta + z_0 K \cos\theta = K_x x_0 + K_z z_0$ in S, we note that in the integral over x_0 in (13.1.6), the source term contributes exponentials of the form $\exp[\pm i(K_x \pm k_s)x_0] \exp(\pm iK_s z_0)$. Thus we are computing the Fourier transform of the Hankel function $H_0^{(1)}(KR)$, where $R^2 = (x - x_0)^2 + (y - y_0)^2$ and $K = \omega/c$. The trick here is to use Eqs. (7.3.14), (7.3.17), and (1.2.15). We have the two forms for the Green's function in two dimensions,

$$\frac{i}{4} H_0^{(1)}(KR) = \frac{1}{4\pi^2} \iint \frac{\exp[iK_x(x - x_0) + iK_z(z - z_0)]}{K_x^2 + K_z^2 - K^2} dK_x\, dK_z$$
$$= \frac{i}{4\pi} \int dK_x \frac{\exp[iK_x(x - x_0) + i\sqrt{K^2 - K_x^2}(z - z_0)]}{\sqrt{K^2 - K_x^2}}$$

and the inverse transform of this is

$$\frac{i}{8\pi} \int_{-\infty}^{\infty} e^{i\mu x_0} H_0^{(1)}(KR)\, dx_0 = \frac{i}{4\pi} \frac{\exp[i\mu x + i\sqrt{K^2 - \mu^2}(z - z_0)]}{\sqrt{K^2 - \mu^2}}$$

For a typical term of S, μ would be $\pm(K_x \pm k_s)$; so the integral of the exponential times $H_0^{(1)}$ for this term would be

$$\frac{2}{(\omega/c)\cos\theta_s^{\pm}} \exp\left[\pm i\frac{\omega}{c}x\sin\theta_s^{\pm} + i\frac{\omega}{c}(z-z_0)\cos\theta_0^{\pm}\right] \quad (13.1.7)$$

where $(\omega/c)\sin\theta_s^{\pm} = K_x \pm K_s = K\sin\theta \pm k_s$. This expression represents a scattered plane wave traveling in a direction that makes an angle θ_s with the z axis.

It remains to carry out the integration over z_0 in Eq. (13.1.6).

$$\int_{-L}^{+L} dz_0 \exp\left[i\left(K_z - \frac{\omega}{c}\cos\theta_s\right)z_0\right] = 2L\frac{\sin[K_z - (\omega/c)\cos\theta_s]L}{[K_z - (\omega/c)\cos\theta_s]L} \quad (13.1.8)$$

Collecting the contributions to $E(x,z,\omega)$ from the four exponential terms in S [Eq. (13.1.4)] and inverting the Fourier transform to obtain

$$E(x,z,t) = \int_{-\infty}^{+\infty} E(x,z,\omega)e^{-i\omega t}\,d\omega$$

we find that the scattered electric field consists of two plane waves traveling in different directions, one with the frequency $\Omega + \omega_s$, the other with a frequency $\Omega - \omega_s$.

$$E(x,z,t) = E_+ \sin[K_+ \cdot r - (\Omega + \omega_s)t] + E_- \sin[K_- \cdot r - (\Omega - \omega_s)t]$$

where
$$(13.1.9)$$

$$E_{\pm} = \frac{\Omega^2 L\eta E_0}{c(\Omega \pm \omega_s)\cos\theta_s^{\pm}}\frac{\sin[(\Omega L/c)(\cos\theta\cos\theta_s^{\pm})]}{(\Omega L/c)(\cos\theta - \cos\theta_s^{\pm})}$$

$$K_{\pm} \cdot r = \frac{\Omega}{c}x\sin\theta_s^{\pm} + \frac{\Omega \pm \omega_s}{c}z\cos\theta_s^{\pm}$$

Here the amplitude of the sound wave is expressed in terms of the parameter η [Eq. (13.1.3)], which is the amplitude of the relative change $\delta\epsilon/\epsilon_0$ of the index of refraction. We can express η in terms of the sound-pressure amplitude p_0 as

$$\eta = \frac{1}{\epsilon_0}\frac{\partial\epsilon}{\partial\rho}\delta\rho = \frac{1}{\epsilon_0}\frac{\partial\epsilon}{\partial\rho}\frac{p_0}{c_s^2} \quad (13.1.10)$$

The diffracted waves

The relation between the angle of incidence θ and the angle θ_1 giving the direction of the diffracted wave is shown in Eq. (13.1.7).

$$\sin\theta_1^{\pm} = \frac{K\sin\theta}{(\Omega \pm \omega_s)/c} \pm \frac{k_s}{(\Omega \pm \omega_s)/c}$$

$$\simeq \sin\theta \pm \frac{k_s}{K} = \sin\theta \pm \frac{\Lambda}{\lambda_s} \quad (13.1.11)$$

where λ_s and Λ are the wavelengths of sound and light. If the incident light beam is perpendicular to the sound beam so that $\theta = 0$, we see that the diffraction lines corresponding to the frequencies $\Omega + \omega_s$ and $\Omega - \omega_s$ are found in the directions given by

$$\sin \theta_1 \simeq \pm \frac{\Lambda}{\lambda_s} \qquad (13.1.12)$$

as expected.

To calculate the amplitude of the scattered wave under these conditions, we have, for small values of θ,

$$\cos \theta_s = \sqrt{1 - \sin^2 \theta_s} \simeq \cos \theta + \frac{1}{2} \frac{k_s}{K} \tan \theta \qquad (13.1.13)$$

and from (13.1.9),

$$\mathbf{E}_\pm \simeq \frac{\Omega L}{c} \eta \mathbf{E}_0 \frac{\sin(\frac{1}{2} k_s L \theta)}{\frac{1}{2} k_s L \theta} \simeq \frac{\Omega L}{c} \eta \mathbf{E}_0 = 2\pi \frac{\eta L}{\lambda} \mathbf{E}_0$$

In other words, the scattered amplitude is proportional to $\eta L = \Delta L$, the perturbation in the optical path length caused by the sound wave. The validity of the Born approximation clearly implies that the corresponding phase shift $2\pi(\Delta L/\lambda)$ be small.

It should be emphasized that the *sinusoidal grating* corresponding to the traveling sound wave produces only a *first-order* diffraction pattern with a single line on either side of the incident-beam direction. If the sound wave is periodic but contains many Fourier components, such as encountered, for example, in a large-amplitude sound wave with nonlinear distortion (as will be discussed in the next chapter), an additional order of diffraction is obtained for each harmonic component of the sound wave. For the nth-harmonic acoustic component we then find a diffraction line at an angle

$$\theta_n = \theta \pm \frac{n\Lambda}{\lambda}$$

and the frequencies of the diffracted light in these directions are $\Omega + n\omega_s$ and $\Omega - n\omega_s$.

Partial-wave analysis

Returning to the expression for the scattered light, we note that the analysis based on the Born approximation is good only as long as the perturbation of the optical path length across the beams is small compared with the wavelength of light (i.e., as long as the width $2L$ in Fig. 13.1 is small). In order to obtain a solution valid over a wider range, we can of course improve upon the Born approximation by using a better approximation than the unperturbed incident field in the source term in Eq. (13.1.2). To illustrate another procedure, however, we return to the basic wave equation (13.1.1) and seek to satisfy it directly by series expansion of the electric field.

In this approach we wish to study how the light beam gradually changes from the initial plane monochromatic wave as it enters the sound beam at $z = -L$ to a superposition of partial waves of different frequencies and directions inside the sound beam. The amplitude of these partial waves will increase gradually with z from zero at the plane of entrance $z = -L$ of the sound beam to some final value at the exit plane $z = L$. To study this gradual deformation of the light beam, we seek a solution to the wave equation (13.1.1) subject to the boundary condition that all partial waves are zero at $z = -L$ except for the initial incident beam.

The acoustic frequency ω_s is always considerably smaller than the optical frequency Ω, and if we express the electric field as

$$E(x,z,t) = \psi(x,z,t) \exp\left(iKx \sin\theta + iKz\cos\theta - i\Omega t\right) + \text{complex conjugate}$$

$$(13.1.14)$$

the amplitude ψ will be a slowly varying function of time. Therefore, if we insert this expression in Eq. (13.1.1) and neglect derivatives of ψ with respect to t, we find that the equation for ψ is

$$\frac{\partial^2\psi}{\partial x^2} + \frac{\partial^2\psi}{\partial z^2} + 2iK\left(\sin\theta\,\frac{\partial\psi}{\partial x} + \cos\theta\,\frac{\partial\psi}{\partial z}\right) = -\frac{\delta\epsilon}{\epsilon_0}\left(\frac{\Omega}{c}\right)^2\psi \quad (13.1.15)$$

If the sound wave travels in the x direction and is periodic in x and t, the amplitude function ψ will have the same periodicity in x and t. The z dependence of ψ is of main interest here, and to obtain the equation describing this z dependence, it is convenient to make a Fourier transform of Eq. (13.1.15) with respect to x and t. Then, if the Fourier transforms of ψ and $\delta\epsilon/\epsilon_0$ are $\psi(z,k_x,\omega)$ and $f(z,k_x,\omega)$, respectively, we obtain from Eq. (13.1.15)

$$2iK\cos\theta\,\frac{\partial\psi}{\partial z} - k_x^2\psi - 2k_xK\sin\theta\,\psi$$

$$= -\left(\frac{\Omega}{c}\right)^2\int\limits_{-\infty}^{+\infty}\!\!\int d\omega'\,dk_x'\,\psi(z,k_x',\omega')f(\omega - \omega', k_x - k_x') \quad (13.1.16)$$

Since $\psi(z,k_x,\omega)$ varies slowly with z, we have neglected $\partial^2\psi/\partial z^2$.

As an example we consider, first, the case when we have a single periodic traveling sound wave so that

$$\frac{\delta\epsilon}{\epsilon_0} = \eta\sin\left(k_s x - \omega_s t\right) \quad (13.1.17)$$

with the corresponding Fourier transform

$$f(\omega - \omega', k_x - k_x') = \tfrac{1}{2}\eta[\delta(\omega - \omega' - \omega_s)\delta(k_x - k_x' - k_s)$$
$$- \delta(\omega - \omega' + \omega_s)\delta(k_x - k_x' + k_s)] \quad (13.1.18)$$

Under these conditions the amplitude function $\psi(x,z,t)$ will be periodic in x and t with the period of the sound wave, and the Fourier transform is then a sum of delta functions,

$$\psi(z,k_x',\omega') = \sum_{n=-\infty}^{+\infty} \zeta_n(z)\delta(k_x' - nk_s)\delta(\omega' - n\omega_s) \qquad (13.1.19)$$

Inserting Eqs. (13.1.18) and (13.1.19) into Eq. (13.1.16) and identifying terms of equal frequency, we find the following recursion formula for the wave amplitude ζ_n:

$$2iK\cos\theta \frac{d\zeta_n}{dz} - i\eta K^2(\zeta_{n-1} - \zeta_{n+1}) = [(nk_s)^2 + 2nk_s K\sin\theta]\zeta_n$$

$$(13.1.20)$$

$$k_s = \frac{\omega_s}{c_s} = \frac{2\pi}{\lambda_s} \qquad K = \frac{\Omega}{c} = \frac{2\pi}{\Lambda}$$

which can be solved to find the dependence on penetration distance of the various diffracted beams of light.

For example, if the light wave is moving at right angles to the sound wave ($\theta = 0$, the incident light moving in the positive z direction), we can neglect the right-hand side compared with the left-hand side, since $k_s \ll K$. In this case we can take advantage of the second of Eqs. (5.2.21), which may be written

$$\frac{d}{dz}J_n(\alpha z) = \frac{\alpha}{2}[J_{n-1}(\alpha z) - J_{n+1}(\alpha z)]$$

to show that ζ_n should be proportional to the Bessel function $J_n(\eta K z)$. In fact, if the lower bound of the sound beam is the plane $z = 0$, then the electric vector of the light wave will be, approximately,

$$\mathbf{E}_0 \exp(iKz - i\Omega t) \qquad\qquad z < 0$$

$$(13.1.21)$$

$$\mathbf{E}_0 \sum_{n=-\infty}^{\infty} J_n(\eta K z)\exp[ink_s x + iKz - i(\Omega + n\omega_s)t] \qquad z > 0$$

Thus, as the light penetrates the sound wave, the incident beam $n = 0$ diminishes in amplitude, slowly at first, then oscillates, proportional to $J_0(\eta K z)$. The first-order waves ($n = \pm 1$) rise linearly from zero at $z = 0$, reaching maximum value of about $0.6E_0$ at $z \simeq 0.3(\Lambda/\eta)$ (which is many wavelengths since the amplitude η of $\delta\epsilon/\epsilon_0$ is usually very small), and thereafter oscillate. The second-order waves ($n = \pm 2$) rise from zero at $z = 0$ proportional to z^2 and reach their first maximum amplitude (about $0.5E_0$) at $z \simeq 0.5(\Lambda/\eta)$, and so on, each higher order rising more slowly with z initially and reaching its first maximum for greater z.

I3.2   BRILLOUIN SCATTERING

The analysis in the preceding section is best applicable to comparatively narrow beams of sound, when the effect of the beam is essentially equivalent to that of a plane grating.   As we have seen, these are the conditions for constructive interference in the scattered field at all angles of incidence of the primary light beam.   In the other extreme case, when the width of the

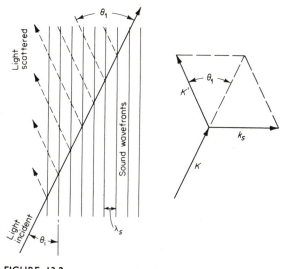

FIGURE I3.2
Brillouin scattering of light from sound waves.

beam is very large so that it can be regarded as a succession of a set of infinite partially reflecting planes, we expect the incident light to be scattered in a manner analogous to Bragg reflection from a crystal.   In order for such scattering to take place, however, the width of the sound beam must be considerably larger than $\lambda/\sin\theta$, where λ is the acoustic wavelength and θ is the angle of incidence measured from the direction perpendicular to the sound beam as shown in Fig. 13.2.   (In order to get interference, we must have at least two reflected rays for each incident ray.)

Bragg reflection

The condition for constructive interference of the light reflected from successive planes is the familiar Bragg condition for the first-order diffraction line,

$$\sin\theta_i = \frac{\Lambda}{2\lambda_s} \tag{I3.2.I}$$

where λ_s, the separation of the planes, is the wavelength of the sound wave, and Λ is the wavelength of the incident light.   Thus, for a given wavelength and direction of propagation of the sound, there is only *one* direction of the incident light which will give rise to a (first-order) diffraction line.   This is in contrast to the scattering characteristics observed in the case of a narrow beam of sound, discussed in the previous section.

The scattering angle θ_1, that is, the angle between the direction of propagation of the observed scattered light and the direction of the incident light beam, as indicated in Fig. 13.2, is twice the angle of incidence, $\theta_1 = 2\theta_i$. Thus, rewriting Eq. (13.2.1), the angle θ_1 that corresponds to the first-order diffraction line can be obtained from the formula

$$2 \sin \frac{\theta_1}{2} = \frac{\Lambda}{\lambda} = \frac{\omega}{\Omega} \frac{c}{n v_s}$$

or

$$k_s = 2K \sin \frac{\theta_1}{2} \qquad k = \frac{2\pi}{\lambda_s}, \; K = \frac{2\pi}{\Lambda}$$

where n = mean index of refraction of medium
 ω = acoustic frequency
 Ω = light frequency
 v_s = speed of sound

Since the reflecting planes in Fig. 13.2 are moving with the speed of sound, the Bragg scattered light will be Doppler-shifted, and this shift equals $2\Omega(n v_s/c) \sin (\theta_1/2)$.   This follows directly from the fact that the relative speed between the fixed point of observation and the (moving) image of the primary light source is $\pm 2 v_s \sin (\theta_1/2)$.   The plus and minus signs refer to motion of the sound toward and away from the observer.   Comparison with Eqs. (13.2.2) below shows that the frequency shift $\Delta\Omega$ equals the frequency of the sound wave, $\Delta\Omega = \omega$.   This frequency shift will make the magnitudes of $K' = \Omega'/c$ and $K = \Omega/c$ slightly different, but since $\omega \ll \Omega$, we may use K' instead of K in Eqs. (13.2.2).   Then, since the direction of propagation of the sound wave is the same as the direction of $\mathbf{K} - \mathbf{K}'$, it follows from Eqs. (13.2.2) that $\mathbf{k} = \mathbf{K} - \mathbf{K}'$, corresponding to the vector diagram in Fig. 13.2.   The kinematic relations that exist between the acoustic wave and the incident and scattered light beams then can be summarized as follows:

$$\mathbf{k} = \mathbf{K} - \mathbf{K}' \qquad \omega = \Omega - \Omega'$$

$$k = 2K \sin \frac{\theta_1}{2} \qquad \frac{\Lambda}{\lambda} = 2 \sin \frac{\theta_1}{2} \tag{13.2.2}$$

$$\Delta\Omega = \Omega - \Omega' = \omega = 2n \frac{v_s}{c} \Omega \sin \frac{\theta_1}{2}$$

the unprimed quantities belonging to the incident wave, the primed quantities to the scattered wave. In other words, from measurements of the frequency shift $\Omega - \Omega' = \omega$ (which determines the acoustic frequency ω) and the corresponding angle of scattering θ_1, we can determine the speed of sound v_s in terms of the speed of light c, the optical frequency Ω, and the index of refraction n of the medium.

If the conditions are such that the scattered wave is in the backward direction, $\theta_1 = \pi$, the frequency shift, and hence the frequency of the target sound wave, is $2n(v_s/c)$. With $v_c \simeq 10^5$ cm per sec, $c \simeq 3 \times 10^{10}$ cm per sec, and $\Omega \simeq 10^{15}$ cps, we see that in order for this type of scattering to take place, the frequency of the acoustic wave involved must be of the order of 10^{10} cps. For scattering in a direction close to the forward direction ($\theta_1 \simeq 0$), the acoustic frequency involved is of course smaller; but from a practical standpoint, measurements cannot be made too close to the forward direction, and the acoustic frequencies usually involved in measurements of this kind lie in the range 10^9 to 10^{10} cps. Thus, because of the high frequencies involved, this method of measurement lends itself to the study of the dispersion relation $v_s = v_s(\omega)$ for high-frequency sound waves, hypersonic waves that are excited in the thermal motion of matter.

Thermal sound waves travel in all directions and have a wide frequency spectrum. Therefore, when a light beam is scattered from these waves, there will be frequency-shifted scattered light at any direction of observation, corresponding to an arbitrary angle θ_1. For each direction, the light beam selects an acoustic target wave with the appropriate wavelength and direction of propagation so as to fulfill the conditions in Eqs. (13.2.2). This phenomenon is usually called *Brillouin scattering*. The scattered light contains two frequencies, $\Omega + \omega$ and $\Omega - \omega$; in a spectroscope they would show as a doublet, the Brillouin doublet. In addition, there is scattered light with unshifted frequency, called the Rayleigh scattered line.

Thermal motion of the medium

In this context the scattered-light components may be regarded as a result of the interaction of the light wave with collective modes of motion of the fluid, in thermal motion. There are several types of such collective modes, and if the fluid is in thermal equilibrium, they are all excited. In our general analysis of the (small-amplitude) dynamics of a real fluid in Chap. 6, we found three types of such collective motions. The first is the ordinary sound wave [the propagational mode of Eqs. (6.4.23)], traveling with the speed v_s and being attenuated at a rate given in Eq. (6.4.14). The second mode of motion is a rapidly decaying shear, or transverse, wave, obeying the second of Eqs. (6.4.17).

The third mode is the thermal wave of Eqs. (6.4.25), also with a comparatively short "lifetime." These two latter modes both contribute to the entropy fluctuations in the fluid, but the viscous wave contributes only to second order in the field variables, the rate of entropy production being proportional to $\mu(\text{curl } \mathbf{u})^2$. Thus it may be neglected in comparison with the entropy fluctuations resulting from the thermal-wave modes. For these thermal waves the space-time dependence is described by Eq. (6.4.18), $T\rho(\partial\sigma/\partial t) = K\nabla^2 T$, with K the thermal conductivity, and where, at constant pressure, $T(\partial\sigma/\partial t) = C_p(\partial T/\partial t)$. Thus a periodic temperature field with a spatial distribution $\delta T(q,t)\cos(qx)$ will decay in time according to the equation $\partial(\delta T)/\partial t = -(K/\rho C_p)q^2\,\delta T$, which has the solution

$$\delta T = \delta T(0)\exp(-\beta t) \qquad \text{where } \beta = \frac{K}{\rho C_p}q^2 \qquad (13.2.3)$$

A corresponding initial perturbation in pressure, $\delta p \cos(k_s x)$, on the other hand, will decay in time as

$$\tfrac{1}{2}\,\delta p\,e^{-\alpha t}[\cos(k_s x + \omega_s t) + \cos(k_s x - \omega_s t)]$$

representing two decaying disturbances that travel with the speed $v_s = \omega_s/k_s$ in opposing directions. A light wave interacting with these traveling, decaying waves will be scattered into two lines with frequencies $\Omega \pm \omega_s$, as was demonstrated previously. Because of the decay of the waves, these lines, the Brillouin lines, will be broadened by an amount of the order of α, the inverse of the lifetime of the propagational waves in the fluid.

In our previous discussion of the decay of acoustic waves (Sec. 4.1), we have considered mainly the spatial decay of a wave driven at a frequency $\omega/2\pi$ and in the corresponding dispersion relation $k = k(\omega)$ [Eq. (6.4.26)] ω has been regarded as real and k as a complex quantity. However, in the present problem of the thermal excitation of acoustic modes in a medium, we are concerned with the decay, in time, of an acoustic wave of given wavelength $2\pi/k$. To obtain this decay we have to return to Eq. (6.4.26) and express ω (now complex) in terms of k (real). Although this inversion of the dispersion relation can be made in principle, it is not always simple, particularly because of the frequency dependence of the bulk viscosity coefficient η. In the cases where a, the spatial decay constant, times the wavelength of the sound is very small compared to unity, the approximate relation $\alpha = v_s a$ is valid for the temporal decay constant α, where v_s is the speed of sound. In many cases of interest for Brillouin scattering this is not the case and a much more complicated formula, relating a and α, holds. We shall not pursue this matter further, but will simply remark that in monatomic fluids, where η is negligible and when the a of Eq. (6.4.14) times λ is quite small, the temporal decay constant has the simple form

$$\alpha = \left[\frac{4}{3}\frac{\mu}{\rho} + \frac{(\gamma - 1)K}{\rho C_p}\right]\frac{\omega^2}{2v_s^2} \qquad (13.2.4)$$

Since μ/ρ and $K/\rho C_p$ are both of the order of lv_s, where in a gas l is the mean-free-path, we see that $a \simeq \dfrac{l}{v_s}\dfrac{\omega^2}{v_s}$. The decay time of the sound wave is thus of the order of $1/a = 1/v_s a \simeq (\lambda_s/l)\omega_s^{-1}$, where λ_s is the wavelength of the sound wave.

For the thermal disturbance, on the other hand, the decay is aperiodic, and there are no traveling waves involved. The light scattered from such a disturbance will not be frequency-shifted, and the corresponding scattered line, the Rayleigh line, will have the same frequency as the incident line, except for a line broadening by an amount of the order of $\beta = (K/\rho C_p)q^2$.

The scattered intensity

The previous considerations, which have been mainly kinematic, do not say anything about the intensity of the scattered light or about the details of the line shapes. To study these aspects of the problem, we return to the wave equation (13.1.1).

Again we use the Born approximation, so that the electric field in the source term is replaced by the incident field $E_0 \exp i(K \cdot r - i\Omega t)$. We express the function $[\delta\epsilon(r,t)/\epsilon_0]$ by its spatial Fourier transform

$$\frac{\delta\epsilon(r,t)}{\epsilon_0} = \int \eta(k,t) \exp(ik \cdot r)\, dv_k \tag{13.2.5}$$

Using the free-field Green's function $(1/4\pi\,|r - r_0|)\,\delta(t_0 - t + |r - r_0|/c)$, we find, in a plane perpendicular to E_0, that the scattered field is

$$E' = \frac{E_0\Omega^2}{4\pi r c^2} \int dv_k \int \exp\left(iK \cdot r_0 - i\Omega\,\frac{t - |r - r_0|}{c}\right) \eta(k,t)e^{ik\cdot r_0}\, dv_0 \tag{13.2.6}$$

where we have assumed that $1/|r - r_0| \simeq 1/r$, which is true if r is large enough.

We introduce $|r - r_0| \simeq r - K' \cdot r_0/K'$ in the exponential, where K' is the propagation vector for the scattered light wave and $|K'| = \Omega/c$. The integration over r_0 gives $V\delta(k - K' + K)$, where V is the volume of the source region, and δ the delta function. Consequently, the scattered field is given by

$$E'(r,K',t) = \frac{E_0\Omega^2 V}{4\pi r c^2} \eta(k,t)e^{iK'r - i\Omega t} \qquad k = K - K' \tag{13.2.7}$$

To determine the frequency spectrum of the scattered-light intensity, we make use of the fact that the spectrum density of $E^2(r,t)$ is the Fourier transform of the autocorrelation function $\langle E(r, K', t - \tau)E^*(r,K',t)\rangle = \psi(\tau)$, where $\langle\ \rangle$ indicates the time average.

Fluctuations of the refractive index

In order to apply this equation to the problem of scattering from the thermally excited collective modes in a substance, as discussed above, we must first relate the fluctuations in the dielectric constant to these modes of motion. From a thermodynamic standpoint this is, in effect, accomplished when we consider ϵ as a function of the two independent thermodynamic variables, pressure and entropy. We have

$$\delta\epsilon = \left(\frac{\partial\epsilon}{\partial P}\right)_S \delta P + \left(\frac{\partial\epsilon}{\partial S}\right)_P \delta S \tag{13.2.8}$$

The fluctuations in ϵ, then, are a result of fluctuations in P and S, and these fluctuations, as we have seen from the previous qualitative considerations, correspond to two of the independent types of collective modes of the fluid, the propagational waves and the thermal waves (the shear waves contribute only to second order in the entropy fluctuations and will not be included in a first-order analysis).

After these observations regarding the fluctuation in ϵ, we can proceed to derive an expression for the scattered electric field and the frequency spectrum of the scattered intensity. We obtain, using Eqs. (13.2.7),

$$I(\Omega',r) = |E'(r,\Omega')|^2 = \frac{1}{2\pi}\int_{-\infty}^{\infty}\langle E'(\mathbf{r}, t+\tau)E^*(\mathbf{r},t)\rangle e^{i\Omega'\tau}\,d\tau$$

$$= E_0^2\left(\frac{\Omega}{c}\right)^4\frac{1}{(4\pi r)^2}\frac{1}{2\pi}\int\langle\eta(\mathbf{k}, t+\tau)\eta^*(\mathbf{k},t)\rangle e^{-i\Omega\tau}\,e^{i\Omega'\tau}\,d\tau \tag{13.2.9}$$

Since $\eta = \delta\epsilon/\epsilon_0$, we obtain from Eq. (13.2.4), realizing that the fluctuations in P and S are uncorrelated (they represent different collective modes of motion),

$$\Upsilon(\tau) = \langle\eta(\mathbf{k}, t+\tau)\eta^*(\mathbf{k},t)\rangle = A\,\Upsilon_P(\tau) + B\,\Upsilon_S(\tau) \tag{13.2.10}$$

where

$$A = \frac{1}{\epsilon_0}\left(\frac{\partial\epsilon}{\partial P}\right)_S \qquad B = \frac{1}{\epsilon_0}\left(\frac{\partial\epsilon}{\partial S}\right)_P$$

and

$$\Upsilon_P(\tau) = \langle\delta P(\mathbf{k}, t+\tau)\,\delta P^*(\mathbf{k},t)\rangle$$

$$\Upsilon_S(\tau) = \langle\delta S(\mathbf{k}, t+\tau)\,\delta S^*(\mathbf{k},t)\rangle$$

are the autocorrelation functions for the pressure and entropy fluctuations. Here $\delta P(\mathbf{k},t)$ is the amplitude of the spatial component of pressure, having the wavelength $2\pi/k$.

$$\delta P(\mathbf{k},t) = \left(\frac{1}{2\pi}\right)^3\iiint e^{-i\mathbf{k}\cdot\mathbf{r}}\,\delta P(\mathbf{r},t)\,dv_r \tag{13.2.11}$$

with a similar equation for the amplitude $\delta S(\mathbf{k},t)$. Note the parallel with the discussion following Eq. (8.1.26), for the scattering of sound from turbulence.

To obtain more specific formulas for the spectrum of the scattered light, we must be more explicit about the autocorrelation functions Υ_P and Υ_S for the pressure and entropy fluctuations.   As we have seen, in the preceding discussion of the basic properties of these modes of motion, a pressure mode of a certain wavelength $2\pi/k$, which happens to be excited at a certain time, will decay like a damped harmonic oscillator with a temporal decay rate α, given on page 820.   Similarly, a thermal wave will decay exponentially without oscillations with a decay rate β, given in Eq. (13.2.3).

In the course of time these modes are excited at random intervals, and the resulting time dependence of δP and δS is analogous to the time dependence of the damped harmonic oscillator driven by a random force, as discussed in Sec. 1.6.   There we determined not only the average spectrum density of the oscillator displacement, but also the corresponding autocorrelation function, which we expressed as the root-mean-square displacement of the oscillator multiplied by a function of the time delay variable τ, as shown in Eq. (2.3.15). In a completely analogous manner we can express the autocorrelation functions Υ_P and Υ_S for these randomly excited acoustic and thermal oscillators (the thermal mode corresponds to a damped oscillator with resonance frequency equal to zero) in terms of the mean-square wave amplitudes. Thus, with reference to Eq. (2.3.15), we find

$$\Upsilon_P = \langle |\delta P(\mathbf{k})|^2 \rangle \, e^{-\alpha|\tau|} \cos (\omega_s \tau)$$

$$\Upsilon_S = \langle |\delta S(\mathbf{k})|^2 \rangle \, e^{-\beta|\tau|}$$

(13.2.12)

The scattered lines

The corresponding spectrum density of the light intensity $I' \sim |E'|^2$ scattered from the acoustic oscillations in the medium is then obtained from Eq. (13.2.9).

$$I_P'(\mathbf{k},\Omega') = I_0 D \langle \delta P(\mathbf{k})|^2 \rangle \int_{-\infty}^{\infty} e^{-\alpha|\tau|} \cos (\omega_s \tau) e^{i(\Omega - \Omega')\tau} \, d\tau$$

$$= I_0 D \langle |\delta P(\mathbf{k})|^2 \rangle \left\{ \frac{\alpha}{\alpha^2 + [\Omega' - (\Omega - \omega_s)]^2} \right.$$

$$\left. + \frac{\alpha}{\alpha^2 + [\Omega' - (\Omega + \omega_s)]^2} \right\}^2 \quad (13.2.13)$$

where $D = \left(\dfrac{\Omega}{c} \right)^4 \left(\dfrac{1}{4\pi r} \right)^2 \dfrac{V}{\pi} \left(\dfrac{\partial P}{\partial \epsilon} \right)_S^2$, and I_0 is the intensity of the incident light.   These represent the two *Brillouin lines*, centered at the frequencies $\Omega' = \Omega - \omega_s$ and $\Omega' = \Omega + \omega_s$, broadened by an amount determined by the "lifetime" $1/\alpha$ of the frequency ω_s.   The line shape is what is commonly called *Lorentzian*.   Thus, from measurements of $\Omega' - \Omega = \omega_s$, the scattering angle θ_1 [Eq. (13.2.1)], and the line width, we can determine not only the speed of sound at the frequency ω_s, but also the decay rate.   In other words, the complete (complex) dispersion relation $\omega = \omega(k)$ can be measured in this way for sound waves in the kilomegacycle region, which are those involved in the Brillouin scattering.

In a completely similar manner we obtain the intensity contribution I_S from the entropy fluctuations of wavenumber k. In fact, the result is obtained directly from Eq. (13.2.13) by putting $\omega_s = 0$ and replacing α by the decay constant β given in Eq. (13.2.3).

$$I_S'(\mathbf{k},\Omega') = I_0 \left(\frac{\Omega}{c}\right)^4 \frac{1}{(4\pi r)^2} \frac{\pi}{1} \left(\frac{\partial \epsilon}{\partial S}\right)_P^2 V \langle|\delta S(\mathbf{k})|^2\rangle \frac{\beta}{\beta^2 + (\Omega' - \Omega)^2} \qquad (13.2.14)$$

This represents the *Rayleigh line*, centered at the incident-light frequency Ω and broadened by an amount corresponding to the lifetime $1/\beta$ of the entropy waves. As we note from Eq. (13.2.3), this lifetime decreases with k^2.

It remains to relate the mean-square amplitudes $\langle|\delta p(\mathbf{k})|^2\rangle$ and $\langle|\delta S(\mathbf{k})|^2\rangle$ in terms of the intensity of thermal agitation, as measured by the temperature of the medium. For sufficiently long wavelengths the various modes are excited with equal strength, and we may set $\langle|\delta P(\mathbf{k})|^2\rangle \simeq \langle|\delta P(0)|^2\rangle$. Then, from the inverse Fourier transform in Eq. (13.2.11), we get $\langle|\delta P(\mathbf{k},t)|^2\rangle \simeq (2\pi)^{-3}\langle|\delta P(\mathbf{r},t)|^2\rangle V^2$, where $\langle|\delta p|^2\rangle$ is the mean-square pressure fluctuation, and V the volume element under consideration [see also Eq. (1.3.17), for an alternative derivation]. Similarly, we have $\langle|\delta S(\mathbf{k},t)|^2\rangle \simeq \langle|\delta S(0,t)|^2\rangle \simeq (2\pi)^{-3}\langle|\delta S(\mathbf{r},t)|^2\rangle V^2$. Then, from well-known thermodynamic relations, we have $\langle|\delta P|^2\rangle = (k_b T)(V/\kappa_S) = k_b TV\rho v_s^2$, $\langle|\delta S|^2\rangle = k_b C_p\rho V^2$, where k_b is the Boltzmann constant, T is the temperature, $\kappa_S = \frac{1}{\rho}\left(\frac{\partial P}{\partial \rho}\right)_S = (\rho v_s^2)^{-1}$ is the adiabatic compressibility, and C_p is the specific heat at constant pressure.

We have now determined all the quantities in the expression for $I_P(\Omega',k)$ and $I_S(\Omega',k)$ to enable the numerical calculation of the spectral-intensity distribution in terms of the incident intensity and the thermodynamic properties of the medium. Actually, in such calculations it is convenient to express the power scattered into the solid angle $d\Omega$ at the point of observation a distance r from the region of scattering. This scattered power is $I'r^2 d\Omega$; from it we can define a scattering cross section $d\Sigma$ given by the equation

$$\frac{d\Sigma}{d\Omega} = \frac{r^2 I'}{I_0} = \left(\frac{\partial \Sigma}{\partial \Omega}\right)_P + \left(\frac{\partial \Sigma}{\partial \Omega}\right)_S \qquad (13.2.15)$$

and it follows directly from the sum of Eqs. (13.2.13) and (13.2.14).

We could study several other properties of the scattered light on the basis of the relations we have derived, but we shall stop here, and take up some of these questions in the problems.

13.3 STIMULATED BRILLOUIN SCATTERING

In the discussion of scattering in the two preceding sections, the perturbation in the dielectric constant was assumed to be a known function of space and time, independent of the electric field. However (as was mentioned in the introduction), the electric field does produce small mechanical effects on the medium.

Electrostrictive pressure

Although it is well known from classical electromagnetic theory, we shall demonstrate the presence of electrostrictive pressure in the following manner: The change of the electrostatic energy $\epsilon E^2/8\pi$ per unit volume element V, when a substance is compressed so that the density changes by $d\rho$, is $\dfrac{\partial \epsilon}{\partial \rho} \dfrac{E^2}{8\pi} d\rho$. To produce this change, a pressure P_E is required such that the work done by the pressure, $P_E(d\rho/\rho)$ per unit volume, equals the change in the electric energy. Thus

$$P_E = \rho \frac{\partial \epsilon}{\partial \rho} \frac{E^2}{8\pi} = \delta\epsilon \frac{E^2}{8\pi}$$

where $\rho(\partial\epsilon/\partial\rho)$ is of the order of unity for many liquids and solids. As far as the dynamics of the medium is concerned, this pressure has the same effect as the ordinary pressure P. Under more general conditions the effect of an electric field on a material medium is known to be the same as a mechanical stress tensor, $(\epsilon/4\pi)[E_i E_j - (E^2/2)\delta_{ij}]$.

Thus, in a more careful study of the interaction of light and sound waves, we should allow for the feedback from the light to sound and express it as a source term in the acoustic-wave equation. In the same way as the acoustic field enters in the source term of the electric-field equation (13.1.1) through the variation in ϵ, we also have a source term in the acoustic-wave equation that contains the electric field. From our previous discussion it follows that these two equations are

$$\nabla^2 \mathbf{E} - \frac{1}{c^2}\frac{\partial^2 \mathbf{E}}{\partial t^2} = \frac{\delta\epsilon/\epsilon_0}{c^2}\frac{\partial^2 \mathbf{E}}{\partial t^2} = \frac{1}{c^2}\frac{1}{\epsilon_0}\left(\frac{\partial \epsilon}{\partial P}\right)_S p\,\frac{\partial^2 \mathbf{E}}{\partial t^2} \tag{13.3.1}$$

for the electric field, and

$$\nabla^2 p - \frac{1}{v_s^2}\frac{\partial^2 p}{\partial t^2} = -\nabla^2\left(\frac{\epsilon}{8\pi}E^2\right) \tag{13.3.2}$$

for the sound field. We have neglected losses in both equations. To study the coupling between the electric field $\mathbf{E}$ and the sound pressure p in connection with Brillouin scattering, we make use of the previous analysis and consider the electric field to consist of three components: a primary wave $\mathbf{E}_0$ with frequency Ω, and two scattered waves $\mathbf{E}_+$ and $\mathbf{E}_-$, with frequencies $\omega_+ = \Omega + \omega_s$ and $\omega_- = \Omega - \omega_s$.

Studying the general character of the source terms in these coupled equations, we note that the source term in the equation for $\mathbf{E}$, because of the product of the primary field $\mathbf{E}_0$ and the sound field p, contains a term with frequency $\omega_- = \Omega - \omega_s$ (the term with the frequency $\omega_+ = \Omega + \omega_s$ is of no particular interest now). If p and $\mathbf{E}_0$ are traveling waves, this source term is also a traveling wave, which, under the conditions of Brillouin scattering, has a propagation constant $\mathbf{K}_- = \mathbf{K} - \mathbf{k}_s$, where $\mathbf{K}$ and $\mathbf{k}_s$ are the propagation constants for the primary light wave and the sound wave, respectively, as shown in Fig. 13.2.

In fact, if the sound wave and the primary electric wave are $p = p_0 \cos(\omega_s t - \mathbf{k}_s \cdot \mathbf{r})$ and $\mathbf{E} = \mathbf{E}_0 \cos(\Omega t - \mathbf{K} \cdot \mathbf{r})$, respectively, the wave equation for the electric field produced by the term with the frequency $\Omega - \omega_s = \Omega_-$ will be

$$\nabla^2 \mathbf{E}_- - \frac{1}{c^2} \frac{\partial^2 \mathbf{E}_-}{\partial t^2} = \left[-\frac{1}{2c^2} \frac{1}{\epsilon_0} \left(\frac{\partial \epsilon}{\partial P} \right)_S p_0 \Omega^2 \mathbf{E}_0 \right] \cos\left[(\Omega - \omega_s) t - (\mathbf{K} - \mathbf{k}_s) \cdot \mathbf{r} \right]$$

$$(13.3.3)$$

The particular solution to this equation, which satisfies the boundary condition $\mathbf{E}_- = 0$ at $\mathbf{K}' \cdot \mathbf{r} = 0$, is

$$\mathbf{E}_- = \frac{1}{4} \frac{1}{\epsilon_0} \left(\frac{\partial \epsilon}{\partial P} \right)_S p_0 \mathbf{E}_0 (\mathbf{K}' \cdot \mathbf{r}) \sin(\Omega_- t - \mathbf{K}_- \cdot \mathbf{r})$$

where, under conditions of Brillouin scattering, we have $\mathbf{K}' = \mathbf{K} - \mathbf{k}_s$. In other words, we see that the amplitude of the electric field $\mathbf{E}_-$ grows linearly with distance if p_0 and $\mathbf{E}_0$ are independent of $\mathbf{r}$.

Let us now see what effect this growing electric field has on the sound field. We note that the contribution to the source term in Eq. (13.3.2), arising from $(\epsilon_0/8\pi)\nabla^2(\mathbf{E}_0 \cdot \mathbf{E}_-)$, has a wave component with the frequency ω_s and propagation constant $\mathbf{k}_s$. Consequently, the corresponding sound pressure field is obtained from the acoustic-wave equation (13.3.2), with a source term that is a wave traveling with the speed of sound in the direction of $\mathbf{k}_s$. Just as in Eq. (13.3.3), we find that the generated sound also will grow with the distance of travel.

Thus the coupled electroacoustic wave system is unstable. In other words, the electric field $\mathbf{E}_-$ feeds energy into the sound field through Brillouin scattering, and the sound field in turn feeds energy into $\mathbf{E}_-$ (and $\mathbf{E}_0$). Through this feedback phenomenon a small initial sound field will build up, and with it the electric field component $\mathbf{E}_-$. However, this growth does not apply to the higher frequency component $\mathbf{E}_+$.

In this discussion we have neglected both acoustic and electric losses. If these were accounted for, we could then show that there exists a critical threshold for the primary electric field strength, which must be exceeded for the stimulated Brillouin scattering to occur. This threshold has been exceeded experimentally in high-intensity (pulsed) laser beams, and stimulated Brillouin scattered light has been produced both in solids and liquids.

Problems

I In the discussion of Debye-Sears scattering the sound-pressure amplitude in the sound beam was assumed to be uniform across the beam. (*a*) Generalize the results in Eqs. (13.1.9) and (13.1.13) to include a variation of the sound-pressure amplitude across the beam so that $p_0 = p_0(z)$ and $\eta = \eta(z)$ [Eq. (13.1.10)]. (*b*) If the light beam is incident at right angles to the sound beam, discuss the possibility of determining the sound-pressure-amplitude distribution $p_0(z)$ from measurements of the scattered light. (*c*) Suppose the amplitude of the sound wave is uniform across the beam but large enough so that distortion of wave-shape takes place (as discussed in Chap. 14).   Then the sound wave will be periodic, but not simple-harmonic.   What information about the waveshape can be obtained from studies of the scattered light?

2 In Brillouin scattering of a laser beam (frequency 5×10^{15} cps) from a liquid the scattered light in the direction perpendicular to the incident beam is observed. In this direction the frequency separation of the two Brillouin lines is found to be 6×10^9 cps, and the width of a line (at an intensity equal to half the maximum intensity) is 10^8 cps.   If the index of refraction of the liquid is $n = 1.4$, determine the frequency, phase velocity, and spatial attenuation of the thermally excited target sound waves involved.

3 Determine the total scattered power in the Brillouin and Rayleigh lines by integrating I_p' and I_s' in Eqs. (13.2.13) and (13.2.14) over the frequency Ω'. Obtain an expression for the ratio between these powers.

4 In our discussion of Brillouin scattering only first-order diffraction (corresponding to an optical-path difference of one wavelength between the light rays scattered from neighboring Bragg planes) takes place.   Under what conditions (if at all) can higher-order diffraction lines occur?

5 Brillouin scattering can be regarded as an elastic collision between phonons with momentum $\hbar k_s$ and energy $\hbar \omega_s$ and photons with momentum $\hbar K$ and energy $\hbar \Omega$.   Show that the scattering formulas (13.2.2) follow from conservation of momentum and energy in such a collision.

6 A circular light beam of radius r_0 is intensity-modulated so that the intensity is of the form $I = I_0 [1 + \epsilon \sin (\omega t)]$, where $\epsilon \ll 1$.   Derive an expression for the sound field produced by such a beam in a liquid as a result of electrostriction. Also discuss the case of a pulsed light beam, where the intensity is $I = I_0$ for $-\tau < t < \tau$ and $I = 0$ for $t > \tau$.

CHAPTER
14

NONLINEAR OSCILLATIONS AND WAVES

14.1 THE SIMPLE OSCILLATOR

Throughout the preceding chapters we have dealt with linear, or linearized, equations of motion. The unknown quantities in these equations, the displacements, velocities, pressures, densities, and temperatures, all enter the equations to the first power; higher-power terms were either neglected or, as in the case of turbulence, were considered as known source terms. As was mentioned many times, the reasons for this restriction are quite persuasive. In the first place, the usual acoustical motions are small enough so that the nonlinear terms are considerably smaller than the linear ones, and their neglect does not alter the nature of the solution in any important respect.

An equally weighty reason, as far as this book is concerned, is the fact that the techniques for solving linear equations are simpler and much more completely worked out than are the techniques for solving nonlinear equations. Even though there are many acoustical phenomena in which nonlinear effects cannot be neglected, a book dealing with the theory of sound and explaining the methods for computing acoustical behavior must devote most of its space to linear equations, simply because there is so much more that can be said about them. The whole gamut of general techniques, so useful in linear theory, such as the Green's function, Fourier transforms, indeed the whole procedure of adding solutions to produce yet other solutions, is for the most part useless in dealing with nonlinear systems. We cannot employ the useful complex factor $e^{-i\omega t}$, since real and imaginary parts do not separate in a nonlinear equation.

In this final chapter we shall present a representative sample of the methods of solution of those few nonlinear problems to which analytic solutions have been obtained to date. The first section will discuss the motion of a simple oscillator, a point mass moving in one direction, acted on by nonlinear forces, to bring out some of the general characteristics which also apply to more complex systems. The rest of the chapter will present a few examples of nonlinear wave motion. An increasing number of nonlinear problems are being solved numerically, for specific cases, by the use of electronic computers. Such numerical techniques, for either linear or nonlinear equations, applicable for individual cases but lacking the generality of analytic solutions, are not discussed in this book.

Examples of nonlinear forces

Even the usual examples of the simple oscillator, a mass on the end of a coiled spring, exhibits nonlinear aspects if the amplitude of motion is large. If the spring is stretched too far, it becomes a straight piece of wire, with much greater stiffness than that of the coiled spring. Likewise, if the coiled spring is compressed sufficiently, the "pitch" of the spiral eventually will become zero, and the coil spring will become like a solid tube, with a stiffness which again considerably exceeds that of the original spring.

We need not, however, go to such extremes in order to illustrate nonlinearity. In fact, some oscillators are nonlinear even for very small displacements, as we shall see shortly. Consider, for example, the oscillator shown schematically in Fig. 14.1. Here both ends of a coil spring are held fixed a constant distance L apart. A mass is attached to the center of the spring and is set in oscillation transverse to the length of the spring. If the relaxed length of the spring is L_0 and the spring constant is K, the potential energy of the spring is $\frac{1}{2}K[2\sqrt{(L/2)^2 + x^2} - L_0]^2$, and the restoring force on the mass is

$$F = -4Kx\left(1 - \frac{L_0}{\sqrt{L^2 + 4x^2}}\right)$$

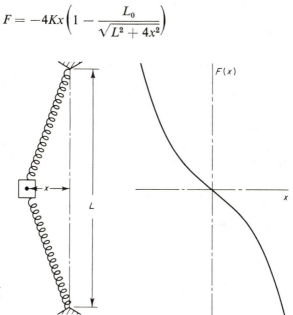

FIGURE 14.1
Example of a nonlinear oscillator with a "hard" restoring force.

For sufficiently small values of x, this can be expanded into the series

$$F = -4K(L - L_0)\frac{x}{L} - 8KL_0\left[\left(\frac{x}{L}\right)^3 - 3\left(\frac{x}{L}\right)^5 + \cdots\right] \qquad x < L$$

which contains all the odd powers of x.

For x very much smaller than L, we can usually neglect the higher terms and consider the force to be linear in x. But note that if the relaxed length L_0 of the spring is equal to the distance L between the supports, the linear term vanishes, and even for small values of x, the restoring force, $F \simeq -(8K/L^2)x^3$, is proportional to the cube of the displacement.

The oscillator in Fig. 14.1 is said to have a "hard" restoring force, which means that the slope of the F versus x curve increases with x. There are many other oscillators with this general property, as illustrated by some of the problems. An extreme but simple example is the "particle-in-a-box" oscillator, where the particle bounces elastically back and forth between parallel walls a distance L apart. The force on the particle in the region between the walls is zero; it rises suddenly to a very large value at the walls.

Another familiar example of a nonlinear oscillator is the simple pendulum. In terms of the angle of deflection, the equation of motion of the pendulum is $ML(d^2\theta/dt^2) + MgL \sin \theta = 0$, where M is the mass and L is the length of the pendulum. Thus, for sufficiently small angles of deflection, the restoring torque is $MgL\theta - MgL\theta^3/6$. In this case the nonlinearity has a "softening" effect on the restoring torque, in contrast to the effect of the nonlinearity in previous examples.

Deviations from linear oscillator

As a result of the nonlinearity, the oscillator characteristics will differ from the well-known properties of the linear oscillator. For example, the period of oscillation no longer is independent of the amplitude. In the special case of the particle-in-a-box oscillator, for example, the period clearly is proportional to the distance between the parallel walls, and hence proportional to the amplitude of oscillation. Also, it is inversely proportional to the velocity amplitude of the motion. In other cases it may not be so simple to find an explicit expression for the period, but we shall find that a "hard" nonlinearity generally serves to decrease the period with increasing amplitude and that the opposite is true for the "soft" nonlinearity.

So far we have implied a conservative oscillator in which the force depends only upon the position coordinates. However, there are many examples of oscillators in which velocity-dependent forces, such as friction, play an essential role. Since a friction force changes sign with the velocity, it follows that it will contain only odd powers of the velocity. We shall consider, for example, a force of the form $av + bv^3$. Although in most cases both a and b are positive, there are oscillators which correspond to a negative value of a. Under these conditions self-sustained oscillations will result; the energy in these oscillations often is produced by a time-inde-pendent force, such as in the bowing of a violin string.

When we go from the free to the forced oscillations of a nonlinear oscillator, we find further unfamiliar features in terms of our experience with linear oscillators. Physically, the most important of these features is, perhaps, the fact that the principle of linear superposition of motions produced by two or more driving forces no longer applies. For example, two harmonic forces with frequencies ω_1 and ω_2 will produce a displacement which contains not only these frequencies, but also "combination tones," with frequencies $n\omega_1 \pm m\omega_2$, and in some cases also "subharmonic" frequencies.

The mathematical analysis of a one-dimensional nonlinear oscillator thus involves the solution of an ordinary differential equation which contains nonlinear terms in the position coordinates and/or its derivatives. These can be solved numerically, but there is no general theory for differential equations of this kind; exact solutions are known only in very special cases, which rarely are of interest in problems of practical interest. It should be mentioned in this connection, however, that approximate equations of motion of an oscillator sometimes can be solved exactly. An important example is a conservative oscillator in which the restoring force can be described approxi-mately as $\alpha x + \beta x^3$, typical of the oscillator of Fig. 14.1, to the third order of approximation. This can be solved exactly in terms of elliptic functions which have been tabulated. When more general types of nonlinearities are considered, we have to resort to graphical or other approximate procedures in the analysis of the motion; the purpose of this section is to survey some of these methods and apply them to a few special examples. The survey is not intended to be exhaustive; we refer the reader to the specialized treatises mentioned in the bibliography for further details.

Conservative forces

Let us consider first the problem of free oscillations of a nonlinear one-dimensional oscillator in which the restoring force is conservative, dependent only upon the position coordinates of the oscillating mass. The equation of motion is then [Eqs. (1.3.1) to (1.3.4)]

$$M\frac{d^2x}{dt^2} = Mv\frac{dv}{dx} = F(x) \tag{14.1.1}$$

If we introduce the potential energy

$$V(x) = -\int F(x)\,dx$$

we obtain the energy relation

$$M\frac{v^2}{2} + V(x) = E \equiv \text{const} \tag{14.1.2}$$

which, with $v = dx/dt$, can also be written

$$t - t_0 = \int_{x_0}^{x} \frac{dx}{\sqrt{2(E - V)/M}} \tag{14.1.3}$$

Formally, this relation represents the solution to the equation of motion, and even if we cannot express the integral explicitly, in closed form, we can determine x as a function of time and the period of oscillation T_p of Eq. (1.3.4) to any degree of accuracy by numerical analysis.

In the particular case of a simple pendulum of length L, the position coordinate along the path of motion is $x = L\theta$, where θ is the angle of deflection. The potential energy is $MgL(1 - \cos \theta)$, and if the maximum angle of deflection in the oscillation is θ_0, we have $E = MgL(1 - \cos \theta_0)$. Then, setting $t_1 = 0$ when $x_1 = 0$, we obtain for the pendulum

$$t = \sqrt{\frac{L}{2g}} \int_0^\theta \frac{\pm \, d\theta}{\sqrt{\cos \theta - \cos \theta_0}} = \frac{1}{2}\sqrt{\frac{L}{g}} \int_0^\theta \frac{\pm \, d\theta}{\sqrt{\sin^2 (\theta_0/2) - \sin^2 (\theta/2)}}$$

If we introduce the new variable φ, given by $\sin (\theta/2) = k \sin \varphi$, where $k = \sin (\theta_0/2)$, this equation becomes

$$t = \sqrt{\frac{L}{g}} \int_0^\varphi \frac{d\varphi}{\sqrt{1 - k^2 \sin^2 \varphi}} \equiv \sqrt{\frac{L}{g}} \, F(k,\varphi) \tag{14.1.4}$$

By expanding $(1 - k^2 \sin^2 \varphi)^{-\frac{1}{2}}$ in a power series and integrating term by term, we obtain t as a function of φ. The function $F(k,\varphi)$, known as an elliptic integral, is available in tabulated form.

$$F(k,\varphi) = \varphi + \tfrac{1}{2}k^2 I_1(\varphi) + \frac{1 \cdot 3}{2 \cdot 4} k^4 I_2(\varphi) + \frac{1 \cdot 3 \cdot 5}{2 \cdot 4 \cdot 6} k^6 I_3(\varphi) + \cdots$$

where

$$I_n(\varphi) = \int_0^\varphi \sin^{2n} \varphi \, d\varphi$$

Using the first term only, we obtain the harmonic motion $\theta = \theta_0 \sin (\sqrt{g/L} t)$, familiar from the linear analysis. The period of oscillation can be expressed as

$$T = 4\sqrt{\frac{L}{g}} \, F\left(\frac{k, \pi}{2}\right) = 2\pi\sqrt{\frac{L}{g}} \left(1 + \frac{k^2}{2} + \frac{3}{8} k^4 + \cdots\right) \tag{14.1.5}$$

For an amplitude of oscillation corresponding to $\theta_0 = 90°$, we have $k = 1/\sqrt{2}$, and the period of oscillation is $T \simeq 1.4 T_0$, where T_0 is the small-amplitude period, $2\pi\sqrt{L/g}$.

Thus the motion of a simple pendulum is an example of a nonlinear oscillation that can be described exactly in terms of elliptic functions. These functions can also be used for an exact description of a conservative oscillation in which the restoring force is of the form $\alpha x + \beta x^3$, as will be discussed further in one of the problems.

The phase plane

Even if the integral of Eq. (14.1.3) cannot be evaluated explicitly in terms of known functions, it is always possible to describe the behavior of the oscillator qualitatively by plotting the momentum p or the velocity v as a function of the position coordinate x. This representation of the motion in the vx plane, the *phase plane*, follows directly from conservation of energy. In the case of a harmonic oscillator, we have

$$\tfrac{1}{2}Mv^2 + \tfrac{1}{2}Kx^2 = E$$

Thus, for each given energy of oscillation E, the relation between v and x is represented by an ellipse in the phase plane. It is often convenient to normalize the coordinates v and x. For example, if we introduce v/v_0 and x/x_0, where v_0 and x_0 are the velocity and displacement amplitudes in the harmonic motion, we obtain

$$\left(\frac{x}{x_0}\right)^2 + \left(\frac{v}{v_0}\right)^2 = 1 \qquad x_0{}^2 = \frac{2E}{K} \qquad v_0{}^2 = \frac{2E}{M} \tag{14.1.6}$$

Thus, in terms of the coordinates x/x_0 and v/v_0, the phase-space trajectory representing the motion is a circle of unit radius.

A deviation of the restoring force from linearity will produce a distortion of the circle. For example, in the extreme case of the particle-in-a-box oscillator, the circle is replaced by a square path. In this particular oscillator the force can be considered to be of the form $(x/x_0)^n$, where n is a very large odd number. For more moderate nonlinearities the deviation of the phase-space trajectory from the circle depends on the energy of oscillation, as will be demonstrated in the following discussion.

As pointed out in connection with Fig. 1.2, if the potential energy of the oscillator increases monotonically from a stable equilibrium point, as in the linear oscillator, the motion is always periodic and the phase-plane trajectories are all closed paths about the equilibrium point. If, on the other hand, the potential energy has a maximum corresponding to an unstable equilibrium point, we shall find that phase-plane trajectories intersect at that point. To illustrate these characteristics and the use of the phase plane in general, we consider the mass-spring oscillator mentioned in connection with Fig. 14.1.

If, as before, the relaxed length of the spring in Fig. 14.1 is L_0 and the distance between the supports is L, we obtain, for the potential energy corresponding to a transverse displacement x,

$$V(x) = 2Kx^2 + KLL_0\left(1 - \sqrt{1 + \frac{4x^2}{L^2}}\right) + \tfrac{1}{2}K(L - L_0)^2$$

or in dimensionless form, and omitting the useless constant,

$$\frac{V(x)}{2KL^2} = \eta^2 + \frac{L_0}{2L}(1 - \sqrt{1 + 4\eta^2})$$

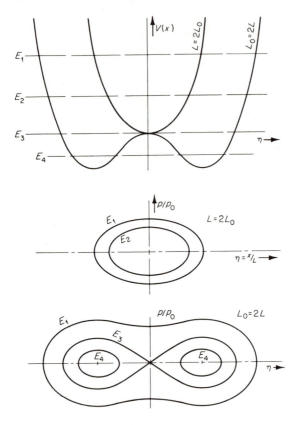

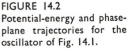

FIGURE 14.2
Potential-energy and phase-plane trajectories for the oscillator of Fig. 14.1.

where $\eta = x/L$. We have already seen that the behavior of the oscillator depends, essentially, upon the ratio (L_0/L) between the relaxed length of the spring and the distance between the fixed end points of the spring. If $L_0 < L$, the equilibrium position $x = 0$ is stable, and the potential-energy curve has a minimum. On the other hand, if $L_0 > L$, the spring is compressed when $x = 0$ and the potential-energy curve has a maximum as shown in Fig. 14.2, where the special cases $L_0/L = 0.5$ and $L_0 = 2$ are considered. Note that in the latter case there are two stable equilibrium points, at approximately $x = \pm L$, about which oscillatory motion can occur for sufficiently small energies.

In Fig. 14.2 are shown the integral curves in phase space for some different energies. For $L_0/L = \frac{1}{2}$, the energy curves are all ellipses centered about $p = 0$, $x = 0$. For $L_0/L = 2$, however, the curves are essentially different when $E > 0$ and when $E < 0$. In the former case the integral curves are closed paths around the origin. For $E = 0$, on the other hand, the integral curve runs through the origin, and the slope of the curve at the origin depends on the direction of motion; the particle comes momentarily to rest as it passes through the origin.   Finally, for $E < 0$, the integral curve splits up into two separate closed branches centered about the two stable equilibrium points mentioned earlier.

In the calculation of the period of oscillation, we shall consider only the case when $L > L_0$.   For sufficiently small values of the amplitude of oscillation, the restoring force becomes

$$F(x) \simeq \frac{4Kx(L - L_0)}{L} + \frac{8Kx^3}{L^2} = kx\left[1 + \left(\frac{x}{d}\right)^2\right]$$

where $k = 4K(L - L_0)/L$ is positive, and $d = L\sqrt{2L_0/(L - L_0)}$ is the displacement at which the linear and nonlinear parts of the restoring force are equal.   The potential energy is then (again omitting the constant term)

$$V(x) = \frac{kx^2}{2}\left[1 + \frac{1}{2}\left(\frac{x_0}{d}\right)^2\right] \tag{14.1.7}$$

If the amplitude of oscillation is x_0, the total mechanical energy of oscillation can be expressed as

$$E = \frac{kx_0^2}{2}\left[1 + \frac{1}{2}\left(\frac{x_0}{d}\right)^2\right]$$

and the period of oscillation, according to Eq. (14.1.3), becomes

$$T_p = 4\sqrt{\frac{M}{2}}\int_x^{x_0}(E - V)^{-\frac{1}{2}}\,dx = \frac{8d}{\sqrt{2}}\sqrt{\frac{M}{k}}\int_x^{x_0}\frac{dx}{\sqrt{(x_0^2 - x^2)(x_0^2 + 2d + x^2)}}$$

If we introduce $x = x_0\cos\psi$, this integral reduces to

$$T_p = T_0\left[\frac{2}{\pi}\int_0^{2/\pi}\frac{d\psi}{\sqrt{1 + (x_0^2/2d^2)(1 + \cos^2\psi)}}\right]$$

where $T_0 = 2\pi\sqrt{M/k}$.   Expanding the square root, we obtain

$$T = T_0\left[1 - \frac{3}{8}\left(\frac{x_0}{d}\right)^2 + \cdots\right] \tag{14.1.8}$$

and T_0 is shown to be the limiting period, when the amplitude of motion x_0 is much smaller than d, so that the motion is simple-harmonic.   We note that the period decreases with increasing amplitude, a characteristic feature of an oscillator with a "hard" spring.   For an oscillator with a "soft" nonlinearity, as demonstrated already in the case of a pendulum, the period increases with increasing amplitude.

Nonconservative forces

In many cases it is necessary to consider, in addition to the conservative force $F(x)$, a nonconservative force $G(x,v)$, which depends on the velocity v.

In the damped linear oscillator in Chap. 2 we had, for example, $G(x,v) = -Rv$. The force $G(x,v)$ clearly will change the total mechanical energy H of the oscillator at a rate given by

$$\frac{dH}{dt} = G(x,v)v$$

where
$$H = \frac{Mv^2}{2} + \int (-F)\,dx$$

This result follows, of course, from Eq. (1.3.2).

If we again use the linear damped oscillator as an example, we have $Gv = -Rv^2$; that is, energy is removed from the oscillation at a rate equal to Rv^2. As a result, the amplitude of oscillation decreases exponentially with time. In other oscillators, which will be discussed later, we shall find that energy is not always removed from the oscillation. Here the function Gv will be found to be positive, at least in some parts of the phase plane, and in these regions the total mechanical energy of the oscillator will increase. Under such conditions the relation $Gv = 0$ defines a curve in the phase plane which divides the plane into regions of negative and positive values of Gv, which we shall call *dissipation* and *activation* regions, respectively. For example, if $G(x,v) = Av - Bv^3$, it follows that $v > \sqrt{A/B}$ defines a dissipation region, and $v < \sqrt{A/B}$ an activation region, if A and B are positive.

When both dissipation and activation regions exist, so-called *self-sustained* oscillations can be produced. For such oscillations, the phase-plane trajectory of such a periodic oscillation must pass through both the dissipation and the activation region, so that the energy dissipated in one region is regained in the other. In the particular case referred to above, in which $G = Av - Bv^3$, it follows that the velocity amplitude of any self-sustained oscillation must have a velocity amplitude which is larger than $\sqrt{A/B}$; otherwise the phase-plane trajectory would not be able to enter the dissipation region. We shall return to these oscillations later, with special examples.

Graphical analysis

In the presence of a nonconservative force it is no longer possible to obtain an integral solution similar to that in Eq. (14.1.3); generally, we have to resort to graphical or other numerical methods to construct solutions to the equation of motion. In the graphical procedure it is convenient to write the equation of motion in the form

$$\frac{dv}{dx} = \frac{F(x) + G(x,v)}{Mv} \tag{14.1.9}$$

from which the slope of the phase-plane trajectory can be calculated directly from F and G. The trajectory can then be constructed when the initial conditions are given.

The graphical construction becomes particularly simple in the special case when the conservative force is linear and the nonconservative force depends only on v. Then, with $F(x) = -Kx$, we obtain $dv/dx = (-\omega_0^2 x + g)/v$, where $\omega_0^2 = K/M$ and $g = G/M$. Furthermore, if the new coordinate $x_1 = \omega_0 x$ is introduced, the equation reduces to

$$\frac{dv}{dx_1} = \frac{-x_1 + (g/\omega_0)}{v} = -\frac{x_1 - g_1(v)}{v} \qquad (14.1.10)$$

where $g_1 = g/\omega_0$. The slope dv/dx_1 can now be constructed at a certain

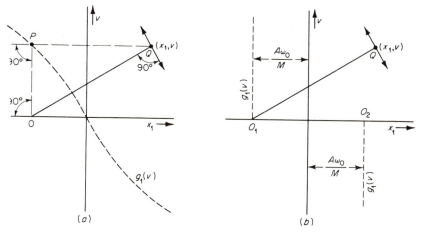

FIGURE 14.3
Graphical construction of slope dv/dx_1 at Q.

point x_1, v in the following manner: First plot the function $g_1(v)$ as indicated schematically in Fig. 14.3a.   The horizontal line through the point $Q(x_1,v)$ intersects g_1 in the point P.   The distance PQ is then equal to $x_1 - g_1(v)$. Draw the vertical through P and mark the intersection O with the x_1 axis. The required slope of the trajectory through x_1, v is then perpendicular to the line OQ.   It is noted that the trajectory must have infinite and zero slopes when it crosses the x_1 axis and the "guide" curve $g_1(v)$, respectively.

In the special case when $G(x_1,v)$ is a constant friction force of magnitude A, the solution is particularly simple.   The guide curve is represented by the straight lines $g_1(v) = -(A/\omega_0 M)$ for $v > 0$ and $g_1(v) = +(A/\omega_0 M)$ for $v < 0$.   We find that the amplitude of oscillation decreases along a spirallike phase-plane trajectory which consists of a succession of semicircles with centers alternating between the points O_1 and O_2 given by $x_1 = -(A/\omega_0 M)$

and $x_1 = +(A/\omega_0 M)$, as indicated in Fig. 14.3b.   At the points O_1 and O_2, the magnitudes of the spring force and the friction force are equal.   Between these points the friction force is larger than the spring force, and once the velocity has become zero in this region, the oscillations cannot be continued. Thus, in the motion illustrated in the figure, the amplitude of the oscillations decays until the mass comes to rest between O_1 and O_2, at which point the oscillator stops.

Slowly varying amplitudes

If the nonconservative and nonlinear forces on an oscillator are small compared with the linear one (which is often the case at sufficiently small amplitudes of oscillation), the motion can be analyzed formally by various approximate methods.   It is not our intention here to attempt a systematic survey of these various methods; we shall select one which is simple, yet typical of many situations encountered in practice.

We start by considering again a linear oscillator of mass m.   If we use the same coordinates in the phase plane as in Eq. (14.1.10), that is, v and $\omega_0 x$, the trajectory representing the linear oscillator will be the circle

$$v^2 + (\omega_0 x)^2 = \frac{2E}{m}$$

with a radius equal to the velocity amplitude $v_0 = \sqrt{2E/m}$.   Actually, if we wish the projections of the radius on the two axes to represent the displacement and the velocity, respectively, we should use as the coordinates $-v$ and $\omega_0 x$, since $\omega_0 x = v_0 \cos(\omega_0 t)$ and $-v = -dx/dt = v_0 \sin(\omega_0 t)$.

A nonlinearity in the oscillator produces a deviation from the circular trajectory and makes both the length of the radius vector A and its angular velocity in the phase plane time-dependent.   Instead of the constant velocity amplitude v_0 and the phase $\omega_0 t$, the time-dependent polar coordinates A and ψ of the point $(\omega_0 x, v)$ in the phase plane will be used.   In many cases of practical interest the fractional changes of A and $\psi - \omega_0 t$ during the course of a period $T_0 = 2\pi/\omega_0$ are both small, and under these conditions it is possible, as we shall see, to obtain simple approximate expressions for the rates of change of the amplitude and the angular velocity.   To see how these relations are obtained, we start by expressing the equations of motion in terms of the polar coordinates A and ψ, defined by $\omega_0 x = A \cos \psi$ and $-v = A \sin \psi$.

This appears, at first, to be substituting two independent variables, A and ψ, for one, x.   This is not so, since we have required that the velocity

$$v = \frac{dx}{dt} = \frac{1}{\omega_0} \frac{dA}{dt} \cos \psi - \frac{A}{\omega_0} \frac{d\psi}{dt} \sin \psi$$

be equal to $-A \sin \psi$.   The second equation determining A and ψ is, of course, the equation of motion

$$\frac{dv}{dt} = -\frac{dA}{dt}\sin\psi - A\frac{d\psi}{dt}\cos\psi = -\omega_0^2 x + g(x,v)$$

$$= -\omega_0 A\cos\psi + g(A,\psi)$$

As before $g = G(x,v)/m$, where $G(x,v)$ is the nonlinear or nonconservative part of the force on the oscillator. Solving these two equations simultaneously for dA/dt and $d\psi/dt$, we obtain

$$\frac{dA}{dt} = g(A,\psi)\sin\psi$$

$$A\left(\frac{d\psi}{dt} - \omega_0\right) = g(A,\psi)\cos\psi \qquad (14.1.11)$$

These equations reduce to the linear oscillator equations $A = $ const and $\psi = \omega_0 t$ when force $g(x,v) = 0$. Since $g(x,v) \equiv g(\omega_0^{-1}A\cos\psi, -A\sin\psi)$, it then follows that the right-hand members of the equations are both periodic functions of ψ, and thus can be expanded in Fourier series.

The first terms in these series, those independent of ψ, are, respectively,

$$a(A) = \frac{1}{2\pi}\int_0^{2\pi} g(A,\psi)\sin\psi\, d\psi$$

and $\qquad\qquad\qquad\qquad\qquad\qquad\qquad\qquad\qquad (14.1.12)$

$$b(A) = \frac{1}{2\pi}\int_0^{2\pi} g(A,\psi)\cos\psi\, d\psi$$

These terms represent the average, or *secular*, rate of change of A and $\psi - \omega_0 t$ over a period. If A and $\psi - \omega_0 t$ vary slowly, so that their fractional change is much less than unity in a period, these average rates should be good approximations for the determination of the time dependence of A and ψ.

Under these conditions we can approximate Eqs. (14.1.11) by

$$\frac{dA}{dt} = a(A) \qquad A\left(\frac{d\psi}{dt} - \omega_0\right) = b(A) \qquad (14.1.13)$$

The first of these equations can be integrated directly, yielding t as a function of A; then ψ can be determined from the second equation.

$$\psi = \omega_0 t + \int \frac{b(A)}{A}\, dt \qquad (14.1.14)$$

Before we apply these results to some specific cases, we observe that if the system is conservative, g depends only on $x = \omega_0^{-1}\cos\psi$, and not on v, and is consequently an even function of ψ. As a result, $\int_0^{2\pi} g(A,\psi)\sin\psi = 0$, and according to the first of Eqs. (14.1.12), the average rate of change of the amplitude A vanishes. Thus, to the approximation considered here, the amplitude of oscillation remains constant, $A = v_0$, and only the phase is affected by a conservative perturbation of the linear oscillator. The phase is then given by

$$\psi = \omega_0 t + \left[\frac{b(v_0)}{v_0}\right]t = \left[\omega_0 + \frac{1}{2\pi m v_0}\int_0^{2\pi} G(A,\psi)\cos\psi\,d\psi\right]t \quad (14.1.15)$$

The period of the oscillation, i.e., the time required to change the phase by an angle 2π, is then

$$T = \frac{2\pi}{\omega_0 + (b/v_0)} = \frac{T_0}{1 + (b/\omega_0 v_0)}$$

The quantity $(b/\omega_0 v_0)$ can be interpreted as the ratio between the mean value of the nonlinear force, averaged over the motion, and the maximum linear restoring force $m\omega_0 v_0$. If this average nonlinear force is negative (soft spring), the period is increased; if it is positive (hard spring), the period is decreased by the nonlinearity.

Frictional forces

Having considered the effect of a conservative perturbation, we next treat the case when the force g depends only on the velocity. We get $b = (2\pi)^{-1}\int_0^{2\pi} g(-A\sin\psi)\cos\psi\,d\psi$, and if g is an odd function of v, this is zero. Therefore, according to Eqs. (14.1.13), only the amplitude, and not the phase, is influenced by the perturbation force g.

It is instructive to use the present approximate method on the familiar example of a linear oscillator in which the damping force is $G = -Rv$. With $v = -A\sin\psi$ and $g = G/m$, we have $g = (2/\tau)A\sin\psi$, where $2/\tau = R/m$. The rates of change of the amplitude and the phase of the oscillator then reduce to $a(A) = (2\pi)^{-1}\int_0^{2\pi} g\sin\psi\,d\psi = (2\pi)^{-1}(2/\tau)\int_{v_0}^{2\pi}\sin^2\psi\,d\psi = A/\tau$ and $b(A) = 0$; that is,

$$\frac{dA}{dt} = -\frac{A}{\tau} \qquad A\left(\frac{d\psi}{dt} - \omega_0\right) = 0$$

The solutions are $A = v_0 e^{-t/\tau}$ and $\psi = \omega_0 t + \phi_0$. Thus the expression for the velocity amplitude agrees with the exact solution in Chap. 2, whereas the angular velocity $\psi = \omega_0$ falls short of the exact solution by a factor $1/\sqrt{1 - (\omega_0\tau)^{-2}}$. This deviation from the exact solution is negligible for practical purposes when the "lifetime" τ of the oscillation is long compared with the period $T_0 = 2\pi/\omega_0$ of the undamped oscillator, as we have assumed.

As another illustration of the method, we consider again the problem of the transverse mass-spring oscillator already shown in Fig. 14.1. In that case, from Eq. (14.1.7), the restoring force for sufficiently small displacements is of the form $kx[1 + (x/d)^2]$. The nonlinear part is then $G(x) = kx^3/d^2$, or $g(x) = G/m = \omega_0^2 x^3/d^2$, which, with $x = (A/\omega_0) \cos \psi$, becomes $g(A,\psi) = (A^3/\omega_0 d^2) \cos^3 \psi$. Since in this case $a(A) = 0$, the amplitude of the oscillation remains constant, $A = v_0$. The expression for the rate of change of the phase, according to Eqs. (14.1.12) and (14.1.13), is then

$$b(A) = (2\pi)^{-1} \frac{v_0^3}{\omega_0 d^2} \int_0^{2\pi} \cos^4 \psi \, d\psi = \frac{3}{8} \frac{v_0^3}{\omega_0 d^2}$$

and Eq. (14.1.15) gives the corresponding expression for the period of oscillation,

$$T = \frac{T_0}{1 + \frac{3}{8}(x_0/d)^2} = T_0\left[1 - \frac{3}{8}\left(\frac{x_0}{d}\right)^2 + \cdots\right]$$

where we have introduced $v_0 = \omega_0 x_0$, x_0 being the displacement amplitude. It is interesting to see that this result for the period is the same as that in Eq. (14.1.8), obtained from the exact solution of the equation of motion.

As a final example, let us consider an oscillator with a linear restoring force and with a friction force of the form $G = Rv + Cv^3 = Rv[1 + (v/\mu)^2]$. Here the quantity μ is the velocity at which the linear friction force is equal to the nonlinear one. Introducing the polar coordinates, we have $v = -A \sin \psi$ and $g = G/m = -(2/\tau) A \sin \psi + \mu^{-2} A^3 \sin^3 \psi$, where $\tau = 2m/R$. After having performed the required integrations for the calculation of $a(A)$, we obtain, according to Eqs. (14.1.11) and (14.1.12),

$$\frac{dA}{dt} = -\frac{1}{\tau} A\left[1 + \frac{3}{4}\left(\frac{A}{\mu}\right)^2\right] \tag{14.1.16}$$

It so happens that this equation can be solved exactly, in closed form; the solution is

$$A = v_0 e^{-t/\tau}\left[1 + \frac{3}{4}\left(\frac{v_0}{\mu}\right)^2 (1 - e^{-2t/\tau})\right]^{-\frac{1}{2}} \tag{14.1.17}$$

where v_0 is the initial velocity amplitude. Thus

$$x \simeq \left(\frac{1}{\omega_0}\right) A(t) \cos(\omega_0 t + \Phi)$$

neglecting higher terms in the Fourier series for x and v, which, as we saw, is allowable as long as $\tau \gg 1/\omega_0$, that is, as long as the Q of the oscillator, $\sqrt{Km/R^2} = \omega_0 m/R$ of Eq. (2.2.4), is a large number. The amplitude of motion diminishes with time, a little faster than exponentially at first, because of the nonlinear term.

Self-sustained oscillations

When the dissipation function $vG(x,v)$ is positive as well as negative here is the opportunity for the oscillator to gain energy, as well as to lose it.,

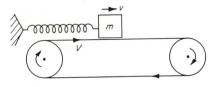

FIGURE 14.4
Example of a self-sustained oscillator.

And when the path in phase space traverses activation regions (where vG is positive) as well as dissipation regions (where vG is negative), steady-state motion is again possible if the amount of energy gained just equals the energy lost per cycle. A large number of mechanical and electronic systems display this characteristic. In some of these, self-sustained oscillations are produced by means of some "feedback" mechanism, the energy required to sustain the motion being derived from a time-independent energy source. Electronic tube oscillators belong to this category. In other systems the oscillations are driven more or less directly by some steady force; wind instruments and aircraft wings are driven into oscillation by an air blast, violin strings by the friction of the moving bow, and clocks by the steady force of the mainspring. The activation of the oscillator by these forces in most cases can be regarded conveniently as a negative resistance, and the conditions for self-sustained oscillations then can be expressed by the requirement that the total resistance in the oscillator should be zero (or negative).

As an illustration, let us consider the oscillator shown in Fig. 14.4. A mass m is attached to the end of a spring which provides a linear restoring force. The mass slides on a horizontal surface which is moving with a velocity V. We assume that air friction or some other damping mechanism produces a damping force $-Rv$ proportional to the velocity v of the mass. In addition, there is the friction force from the belt, a force $F(V - v)$, which depends on the relative velocity $V - v$ between the mass and the belt. The equation of motion of m is

$$m\frac{d^2x}{dt^2} = -Kx - Rv + F(V - v) \qquad v = \frac{dx}{dt}$$

We shall consider $v < V$ and expand, $F(V - v) = F(V) - F'(V)v + F''(V)(v^2/2) + \cdots$. The contact friction force decreases with increasing relative velocity (i.e., with *decrease* of v) so that $F'(V) < 0$. If we set $F'(V) = -R'$ and neglect higher-order terms in v, we have $F(V - v) \simeq F(V) + R'(dx/dt)$, and the resistive force in the equation of motion becomes $-(R - R')(dx/dt)$. In other words, the effect of the moving belt, for small velocities v, can be interpreted as a negative resistance R', and if $R' > R$, the total resistance in the oscillator becomes negative and the oscillations grow with time. These oscillations do not grow indefinitely because the higher-order terms eventually will introduce further damping, and the amplitude of oscillation will be limited, as we shall see in the following discussion. The mechanism illustrated by this example describes in principle the oscillations of a bowed violin string or the squeaking of a door or the brake on a car.

To demonstrate the amplitude-limiting effect of the higher-order terms in the expression for the contact friction, we shall consider only the third-order term. We are interested now in the case when the total linear resistance $R - R'$ in the oscillator is negative, and we set $R - R' = -R_1$. The total resistive force is then of the form

$$-R_1v + Cv^3 = -R_1v\left[1 - \left(\frac{v}{\mu}\right)^2\right]$$

where μ is the velocity at which the magnitudes of the linear and nonlinear terms are the same.

Except for the difference in sign, this nonconservative force is of the same form as in the example discussed earlier in this section [Eq. (14.1.16)]. If this force is small compared with the linear restoring force, the same approximate method of solution can be used here. Under these conditions the expression for the velocity amplitude of the oscillation can be written down directly from the result already given in Eq. (14.1.17). The only change required is to replace τ by $-\tau$ and μ^2 by $-\mu^2$, and the time dependence of the (velocity) amplitude of the oscillations becomes

$$A(t) = v_0 e^{t/\tau}\left[1 + \frac{3}{4}\left(\frac{v_0}{\mu}\right)^2(e^{2t/\tau} - 1)\right]^{-\frac{1}{2}} \qquad (14.1.18)$$

where $\tau = 2m/R_1$, and v_0 is the velocity amplitude at $t = 0$. The "relaxation time" τ must be much larger than $1/\omega_0$ for the approximation

$$x = (1/\omega_0)A(t)\cos(\omega_0 t + \Phi)$$

to be valid. We see that, no matter what its initial value v_0, the velocity amplitude A eventually becomes $\sqrt{\frac{4}{3}}\,\mu$; if v_0 is larger than this, the amplitude decreases; if v_0 is smaller, the oscillator picks up energy to reach the stable amplitude. This differs from the purely dissipative case of Eq. (14.1.17), for which the amplitude goes to zero, no matter what its initial value was.

This result is consistent with our previous findings. The nonconservative force in the present case is $G(v) = R_1v[1 - (v/\mu)^2]$, and the power transferred to the oscillation is $Gv^2[1 - (v/\mu)^2]$. Thus, when $v < \mu$, the energy transfer is positive, and when $v > \mu$, it is negative; the region $|v| < \mu$ in phase space is the activation region, and the region $|v| > \mu$ is the dissipation region. Therefore, when the oscillation is stationary, the phase-space trajectory must go through both these regions, and consequently the velocity amplitude must be larger than μ.

When the oscillation amplitude has reached the stationary value $\sqrt{\frac{4}{3}}\,\mu$, the rate of change of the amplitude is zero, as can be seen directly from

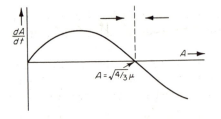

FIGURE 14.5
Plot of time rate of change of amplitude.

Eqs. (14.1.12) and (14.1.13), which in this case result in

$$\frac{dA}{dt} = a(A) = \frac{1}{\tau} A\left[1 - \frac{3}{4}\left(\frac{A}{\mu}\right)^2\right] \qquad (14.1.19)$$

Clearly, the rate of change dA/dt is zero also at $A = 0$. At this point, however, the motion is unstable, which is characterized by the fact that the curve dA/dt versus A has a positive slope at that point, as shown in Fig. 14.5. A small disturbance from the point $A = 0$ makes the rate of change of A increase further in the same direction. At $A = \sqrt{\frac{4}{3}}\,\mu$, on the other hand, the slope is negative, and an increase of A beyond this point will be reduced until the amplitude is brought back to the value $\sqrt{\frac{4}{3}}\,\mu$; likewise, a reduction of A will cause a countertrend to increase A again. In other words, regardless of the starting point in the phase plane, all phase-space trajectories end up in the same stable path $A = \sqrt{\frac{4}{3}}\,\mu$, circular to this order of approximation.

There are other types of force fields, producing self-sustained oscillations, for which this is not the case, forces resulting in a rate of change of the amplitude which depends on A, in the manner shown in Fig. 14.6. In this case, if the initial amplitude of oscillation is less than A_1, the oscillation will decay to the stable equilibrium point $A = 0$. If, on the other hand, the initial amplitude is larger than A_1, the oscillation will be pulled toward a stable motion in which the amplitude is A_2. A familiar example of an oscillation of this kind is an ordinary pendulum clock. To get the clock going, the pendulum must be started with an amplitude that exceeds a certain value A_1; otherwise the motion of the pendulum will decay.

Graphical solutions

When the nonconservative forces in the oscillator are not small compared with the restoring force, the approximate formulas of Eqs. (14.1.13) are no

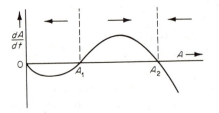

FIGURE 14.6
Oscillator with two distinct points of stable equilibrium, 0 and A_2.

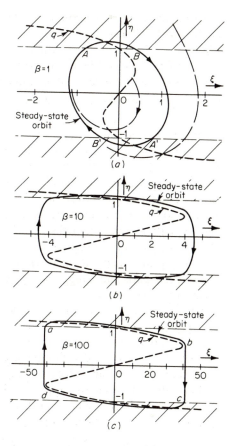

FIGURE 14.7
Phase-plane trajectories for systems with negative and strong positive resistive forces.

longer satisfactory. Rather than make use of improved versions of this method or other analytical techniques, we shall construct solutions graphically, using the method described in connection with Fig. 14.3. For this purpose it is convenient to rewrite the equation of motion $mv(dv/dx) = -Kx + Rv[1 - (v/\mu)^2]$ in terms of the normalized coordinates $\xi = \omega_0 x/\mu$ and $\eta = v/\mu$. In this way we reduce the number of parameters in the equation to one, and obtain

$$\frac{d\eta}{d\xi} = \frac{-\xi + \beta\eta(1 - \eta^2)}{\eta} \qquad (14.1.20)$$

where $\beta = 2/\omega_0\tau = R/\omega_0 m = 1/Q$ is the reciprocal of the Q of the oscillator. We are now assuming that Q is not very large; therefore β is not necessarily small. This equation is the same as Eq. (14.1.10), with η instead of v, ξ instead of x_1, and $q = \beta\eta(1 - \eta^2)$ instead of $g_1(v)$.

In the graphical solution of this equation the curve $q = \beta\eta(1 - \eta^2)$ is shown as the dotted line of Fig. 14.7a. Using it as a guide curve, we construct the slope of the trajectory following the procedure described earlier in Fig. 14.3. Trajectories thus obtained for the case $\beta = 1$ are shown in Fig. 14.7a. Two trajectories with different initial oscillation amplitudes are shown to be "pulled" in toward the same stable orbit, as expected. We recall from the analytical treatment that the phase-space trajectory for small values of β is approximately circular, with an amplitude $A = \sqrt{\frac{4}{3}}\,\mu$, that is, $\eta = \sqrt{\frac{4}{3}}$. The graphically constructed limiting orbit for $\beta = 1$ is not too far different from this circular path.

In Fig. 14.7a are also shown the lines $\eta = 1$ and $\eta = -1$ which divide the phase plane into the dissipation and activation regions, $|\eta| > 1$ and $|\eta| < 1$, respectively. In the dissipative region the amplitude of oscillation decreases, and in the activation region it increases. When the limit cycle is reached, the net energy exchange is zero, so that the motion becomes periodic; the dissipation along the paths $A - B$ and $A' - B'$ is then compensated by the energy gain in the regions $B' - A$ and $B - A'$.

As β is increased, the steady-state orbit will deviate more and more from the circular path and cling to the guide curve, as illustrated in Fig. 14.7b, where the steady-state orbit for $\beta = 10$ is shown. To study Eq. (14.1.20) for larger values of β, it is convenient to introduce $\zeta = \xi/\beta$ and obtain $\eta(d\eta/d\zeta) = \beta^2[\eta(1 - \eta^2) - \zeta]$. Then, if $\beta = \infty$, the slope $d\eta/d\zeta$ is infinite at all points in the phase plane, except on the guide curve $\zeta = \eta(1 - \eta^2)$. In view of this fact, and with reference to Fig. 14.7a and b, it is reasonable to assume (and it can indeed be shown rigorously) that as β approaches infinity, the limit cycle will approach the q curve, as indicated in Fig. 14.7c.

In the sections bc and da of this curve, the position coordinate remains constant (at $\zeta = \pm 2/3\sqrt{3}$). (The corresponding values of η are $\eta_b = 1/\sqrt{3}$, $\eta_d = -1/\sqrt{3}$, $\eta_a = 2/\sqrt{3}$, $\eta_c = -2/\sqrt{3}$.) Therefore, in the calculation of the period of oscillation, these sections do not contribute. Since the times between the points a and b and between c and d are the same, the period of oscillation for large values of β, say, $\beta > 10$, can be approximated by

$$T = 2 \int_a^b |dx/v|.$$ Introducing $\xi = \omega_0 x/\mu$ and $\eta = v/\mu$, we get

$$T = (2/\omega_0) \int d\xi/\eta$$

which, with $\xi = \beta\eta(1 - \eta^2)$, can be expressed as

$$T = \frac{2\beta}{\omega_0} \int_{\eta_a}^{\eta_b} |1 - 3\eta^2| \frac{d\eta}{\eta}$$

$$= (3 - 2\ln 2)\frac{\beta}{\omega_0} \simeq 1.61 \frac{\beta}{\omega_0} = \frac{T_0}{10.2\tau} T_0 \qquad (14.1.21)$$

where $T_0 = 2\pi\sqrt{m/K}$, $\tau = 2m/R_1$, and $q = \sqrt{mK/R_1^2} = 1/\beta$. As pointed out above, this expression is a good approximation only for sufficiently large values of β, say, $\beta > 10$, which means that the value of τ must be less than about $T_0/10\pi$ for the validity of Eq. (14.1.21). We can also express this limitation directly in terms of T. With $\beta > 10$ inserted in Eq. (14.1.21), we see that the period T should be larger than three times T_0.

14.2 FORCED MOTION

So far we have studied the effect of nonlinearities only upon the free and self-sustained motion of an oscillator. Of at least equal interest is the response of such an oscillator to an external harmonic driving force. The forced motion of an undamped linear oscillator, as we know, has infinite amplitude at resonance; therefore it is to be expected that, in practice, nonlinear effects should be of importance, at least in the neighborhood of the resonance frequency. The study of the nonlinear distortion of the amplitude-frequency response curve is of particular interest in this connection. Non-linearities are responsible also for other phenomena, such as the generation of motion containing harmonics, and under certain conditions even sub-harmonics, of the driving frequency. Furthermore, when the driving force contains two or more frequencies, the resulting motion will contain fre-quencies which are linear combinations of the driving frequencies.

A simple example

In the following discussion of these questions, we shall consider only a special example, an oscillator with a "hard" restoring force of the form $Kx[1 + (x/d)^2]$ and with a linear damping force $R(dx/dt)$. The equation of motion is then

$$m\frac{d^2x}{dt^2} + R\left(\frac{dx}{dt}\right) + Kx\left[1 + \left(\frac{x}{d}\right)^2\right] = F\cos(\omega t) \qquad (14.2.1)$$

where we must use the cosine, instead of the exponential $e^{-i\omega t}$, since the equation is nonlinear. It is convenient to rewrite this equation in dimen-sionless form. Two characteristic displacements are involved, the static displacement $x_{st} = F/K$ corresponding to the force amplitude F, and the displacement d at which the nonlinear restoring force becomes equal to the linear one. We pick the former for the normalization of the displacement, and introduce $\xi = x/x_{st}$. Furthermore, we set $\omega_0 = \sqrt{K/m}$, $\theta = \omega_0 t$, $\gamma = \omega/\omega_0$, and $\epsilon = (x_{st}/d)^2$. First let us consider the resistance to be negli-gible, in which case the equation of motion reduces to

$$\frac{d^2\xi}{d\theta^2} + \xi(1 + \epsilon\xi^2) = \cos(\gamma\theta) \qquad (14.2.2)$$

We seek a periodic solution with the fundamental frequency equal to the driving frequency. Since the right-hand side is symmetrical with respect to $\theta = 0$ and antisymmetrical with respect to $\gamma\theta = \pi/2$, the Fourier expansion of the left-hand side must have the same properties; that is, ξ must be of the form

$$\xi = \sum_n \xi_n \cos(n\gamma\theta)$$

where $n = 1, 3, 5, \ldots$.

The left-hand side of the equation will be of the form $A_n \cos(n\gamma\theta)$, where the A_n's are functions of ξ_n; to satisfy the equation we must have $A_1 = 1$ and $A_n = 0$ for all other values of n. This condition corresponds to an infinite set of equations from which, in principle, we can determine the amplitudes ξ_n. As a first approximation, we consider now only two terms in the expansion and set $\xi = \xi_1 \cos(\gamma\theta) + \xi_3 \cos(3\gamma\theta)$. By using the trigonometric relation $4\cos^3(\gamma\theta) = 3\cos(\gamma\theta) + \cos(3\gamma\theta)$, we find that the coefficient for the $\cos(\gamma\theta)$ term on the left-hand side of Eq. (14.2.2) is

$$A_1 = (1 - \gamma^2)\xi_1 + \tfrac{3}{4}\epsilon\xi_1(\xi_1^2 + \xi_1\xi_3 + 2\xi_3^2)$$

and similarly, the coefficient for the $\cos(3\gamma\theta)$ term is

$$A_3 = (1 - 9\gamma^2)\xi_3 + \tfrac{1}{4}\epsilon(\xi_1^3 + 6\xi_1^2\xi_3 + 3\xi_3^3)$$

In addition to these terms, we also get a series of terms containing the frequencies 5γ, 7γ, and 9γ, namely,

$$\frac{\epsilon\xi_3}{4}[3\xi_1(\xi_1 + \xi_3)\cos(5\gamma\theta) + 3\xi_1\xi_3\cos(7\gamma\theta) + \xi_3^2\cos(9\gamma\theta)]$$

These remaining terms are small compared with A_1 and A_3 if $\epsilon \ll 1$ an $|\xi_3| \ll |\xi_1|$, and only under these conditions will our approximation be goo The terms containing the product $\epsilon\xi_3$ in A_1 and A_3 also can be neglected, an

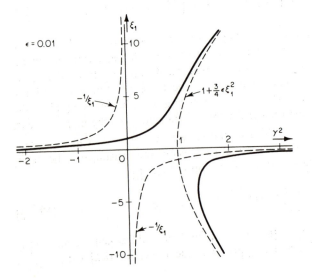

FIGURE 14.8
Construction of am
tude-frequency respo
curves. Solid curves
plots of $\gamma^2 = (1 + \tfrac{3}{4}\epsilon$
$- (1/\xi_1)$ versus ξ_1.

we obtain the following equations for the determination of the coefficients ξ_1 and ξ_3:

$$1 - \gamma^2 + \tfrac{3}{4}\epsilon\xi_1^2 = \frac{1}{\xi_1} \qquad (1 - 9\gamma^2)\xi_3 + \tfrac{1}{4}\epsilon\xi_1^3 = 0 \qquad (14.2.3)$$

The first of these equations represents the amplitude-frequency response for the fundamental frequency in terms of the nonlinearity parameter ϵ.

The amplitude of the third harmonic follows from Eqs. (14.2.3).

$$\xi_3 = - \frac{\epsilon\xi_1^3}{4(1 - 9\gamma^2)}$$

a valid approximation as long as $|\xi_3| < |\xi_1|$. If we wish to obtain improved

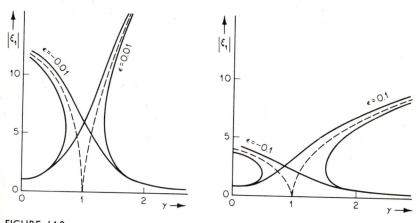

FIGURE 14.9
Amplitude-frequency response curves for nonlinear oscillator.
Amplitude ξ_1 of fundamental plotted against driving frequency
$\omega/2\pi = \gamma\omega_0/2\pi$.

expressions for ξ_1 and ξ_3, we start with three terms in the initial expansion and satisfy Eq. (14.2.2) for terms including the fifth harmonic. From the three equations then obtained, the amplitudes ξ_1, ξ_3, and ξ_5 can be determined. Higher approximations are obtained in a similar manner.

The dependence of the amplitude ξ_1 of the fundamental frequency in Eqs. (14.2.3) on the frequency ratio $\gamma = \omega/\omega_0$ is displayed in Fig. 14.8 as the sum of the parabola $1 + \tfrac{3}{4}\epsilon\xi_1^2$ and the hyperbola $-1/\xi_1$. As in the case of the linear oscillator, the response curve contains two branches; in one the displacement is in phase, in the other it is out of phase with the driving force. Both branches approach the curve $1 + \tfrac{3}{4}\epsilon\xi_1^2$ asymptotically. Response curves obtained in this manner for $\epsilon = 0.01$ and $\epsilon = 0.1$ are shown in Fig. 14.9.

The important difference between the linear and nonlinear oscillators is exhibited in the response curve; for the nonlinear oscillator it is multivalued over part of the frequency range. For a hard nonlinearity, $\epsilon > 0$, this effect occurs above, and for a soft nonlinearity, $\epsilon < 0$, below the linear resonance frequency, as can be seen in Fig. 14.9. This result suggests that in these frequency regions the motion is unstable; the oscillator amplitude can jump suddenly from one value to another.

The distortion of the frequency-response curve is consistent with the observation made earlier in the discussion of the free oscillations, that the characteristic frequency of the oscillator changes with the amplitude of oscillation. As the driving frequency increases from zero, the amplitude of oscillation increases, and if $\epsilon > 0$, the "resonance" frequency increases also. The maximum of the displacement, therefore, is not reached at the customary value $\gamma = 1$, but is pushed toward higher frequencies. In fact, when damping is absent, the maximum value of the curve is moved out to $\gamma = \infty$.

The effects of friction

Just as in the linear oscillator, damping will limit the amplitude of oscillation. It is again useful to determine the amplitude and the corresponding frequency in terms of the oscillator parameters. For this purpose we rewrite the equation of motion (14.2.1), containing a linear damping term, in terms of the dimensionless variables ξ and θ, as in the foregoing analysis, and obtain

$$\frac{d^2\xi}{d\theta^2} + \frac{1}{Q}\frac{d\xi}{d\theta} + \xi(1 + \epsilon\xi^2) = \cos(\gamma\theta) \tag{14.2.4}$$

where $\xi = Kx/F$

$\theta = \omega_0 t = t\sqrt{K/m}$

$\epsilon = (F/Kd)^2$

$Q = (1/R)\sqrt{Km} = m\omega_0/R$

The frequency dependence of the amplitude of the fundamental follows from this equation by inserting $\xi = \xi_1 \cos(\gamma\theta)$. The left-hand side of this equation then will be of the form $A \cos(\gamma\theta) + B \sin(\gamma\theta)$, where $A = \xi_1(1 - \gamma^2) + \frac{3}{4}\epsilon\xi_1^3$ and $B = -(\gamma/Q)\xi_1$. The expression for A, of course, is the same as in the previous analysis. The amplitude of the right-hand side is unity, and to satisfy the equation we must have $A^2 + B^2 = 1$. To satisfy the phase relations in the equation, a phase angle should be added in the terms in the Fourier series for ξ, but this does not alter the relation between the amplitudes, which is

$$\left(\frac{\gamma}{Q}\right)^2 \xi_1^2 + (1 - \gamma^2 + \tfrac{3}{4}\epsilon\xi_1^2)^2\xi_1^2 = 1 \tag{14.2.5}$$

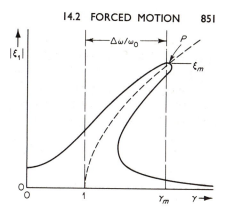

FIGURE 14.10
Response curve for oscillator with non-
linear force proportional to x^3 plus friction.
Resonance is at point P.

This response curve intersects the former asymptotic curve $1 - \gamma^2 + \frac{3}{4}\epsilon\xi_1^2 = 0$ at the point P (Fig. 14.10), with the coordinates $\gamma^2 = \frac{1}{2}(1 + \sqrt{1 + 3\epsilon Q^2})$ and $\xi_1^2 = Q^2\gamma^2$. In the case of a linear oscillator ($\epsilon = 0$), these coordinates are simply $\gamma = 1$ and $\xi_1 = Q$, and the Q value [see discussion of Eq. (2.3.2)] can be interpreted as the ratio between the amplitude of oscillation x_1 at resonance and the static displacement x_{st}, $Q = x_1/x_{st}$. We recall that the parameter ϵ is defined by $\epsilon = (x_{st}/d)^2$, where d is the displacement at which the linear and nonlinear restoring forces are equal. The quantity ϵQ^2 thus can be given the simple interpretation $(x_1/d)^2$; it is a measure of the degree of nonlinearity present in the region of maximum amplitude. If $x_1/d \ll 1$, the coordinates of the point P reduce to $\gamma^2 \simeq 1 + (3\epsilon Q^2/4)$ and $\xi_1 \simeq Q(1 + \frac{3}{8}\epsilon Q^2)$.

The maximum of ξ_1, which appears a little to the left of the point P in Fig. 14.10, is found to have the coordinates

$$\gamma_m{}^2 = \frac{1}{2}\left(1 - \frac{1}{Q^2}\right)\left(1 + \sqrt{1 + 3\epsilon Q^2 \frac{Q^2 + 1}{Q^2 - 1}}\right)$$

$$\xi_m{}^2 = (Q^2 + 1)\left(1 + \frac{3\epsilon Q^2}{4}\frac{Q^2 + 1}{Q^2 - 1}\right)$$

(14.2.6)

For small values of ϵQ^2 these expressions reduce to

$$\gamma_m{}^2 - 1 \simeq \tfrac{3}{4}\epsilon(Q^2 + 1) - \frac{1}{Q^2}$$

$$\xi_m \simeq (\sqrt{1 + Q^2})(1 + \tfrac{3}{8}\epsilon Q^2)$$

If we set $\gamma_m{}^2 - 1 \simeq 2(\gamma - 1) = 2\Delta\omega/\omega_0$, and if $Q \gg 1$, the frequency corresponding to the maximum of the response curve can be written $\omega_0 + \Delta\omega$, where

$$\frac{\Delta\omega}{\omega_0} = \frac{3}{8}\epsilon Q^2 = \frac{3}{8}\left(\frac{x_1}{d}\right)^2$$

(14.2.7)

To investigate the properties of the response curves further, we turn our attention to the points D_1 and D_2 in Fig. 14.11, where the slope of the curves is infinite. The section of the curve between D_1 and D_2, as a detailed analysis will show, represents an unstable motion of the oscillator, and as the frequency is increased continuously from zero, the oscillator amplitude will jump suddenly from the point D_1 to D_1' when the frequency γ_m is reached, as illustrated schematically in the figure. Similarly, if the frequency is decreased continuously, the amplitude of oscillation follows the lower branch of the response curve until the point D_2 is reached, at which the amplitude

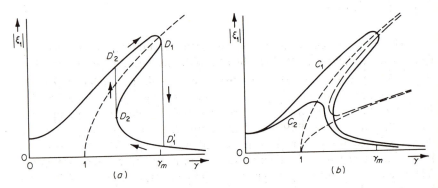

FIGURE 14.11
Illustrating the instabilities of the nonlinear oscillator near resonance. Dashed curve of b is locus of the points D_1 and D_2, where jumps occur.

jumps suddenly to the point D_2'. As the nonlinearity in the oscillator is decreased, the region of instability is decreased until, at a certain value of ϵ, the two points D_1 and D_2 coincide.

To determine this critical value of ϵ, we proceed to find the coordinates of the points D_1 and D_2. Differentiating Eq. (14.2.5) with respect to ξ, we obtain, for $d\gamma/d\xi = 0$,

$$(1 - \gamma^2 + \tfrac{9}{4}\epsilon\xi_1{}^2)(1 - \gamma^2 + \tfrac{3}{4}\epsilon\xi_1{}^2) + \left(\frac{\gamma}{Q}\right)^2 = 0 \qquad (14.2.8)$$

This relation is represented by the dashed curve in Fig. 14.11b. The points D_1 and D_2 are given by the intersections between the curves (14.2.5) and (14.2.8). These intersections merge at the point where the function of Eq. (14.2.8) has a vertical tangent. The response curve which passes through that point will not be multivalued, and the corresponding value of ϵ then gives the limit below which no jump phenomenon occurs. After differentiation of Eq. (14.2.8) with respect to ξ_1 and with $d\gamma/d\xi_1 = 0$, we find that $1 - \gamma^2 = -9\epsilon\xi_1{}^2/8$, and the merging point is found to be given by $1 - \gamma^2 = -(3/2Q^2) \times [1 \pm \sqrt{1 + (4Q^2/3)}]$ and $\epsilon\xi_1{}^2 = (4/3Q^2)[1 \pm \sqrt{1 + (4Q^2/3)}]$. (The minus sign corresponds to negative ϵ.) Finally, if these values are inserted into Eq. (14.2.5), the particular value of ϵ is found below which no jump phenomenon will occur. The complete expression is somewhat lengthy, but if $Q \gg 1$, we find that the amplitude-frequency-response curve is single-valued and jump-free if

$$|\epsilon| < \frac{16}{9} (\sqrt{4/3}) \frac{1}{Q^3} \simeq \frac{2}{Q^3}$$

Since $\epsilon = (x_{st}/d)^2$, we can also express this condition as $x_{st} < d(1.4/Q^{\frac{3}{2}})$, which means that the amplitude of the driving force must be smaller than $Kd(1.4Q^{-\frac{3}{2}})$.

Subharmonics

Which steady-state motions are actually produced depends on the initial conditions; relatively slight changes may result in quite different behavior. All we can do here is to mention some of the alternative steady-state solutions which may eventuate. It is not difficult to see that in the particular oscillator with a restoring force $Kx[1 + (x/d)^2]$, the equation of motion can be satisfied with a periodic motion with a fundamental frequency which is one-third of the driving frequency. Assuming the existence of such a periodic motion, we shall now investigate the conditions that must be satisfied. The calculations are analogous to those in the foregoing section, and as we set the amplitude of the subharmonic component equal to zero, we should, of course, obtain the result found earlier.

Starting with the equation of motion without damping,

$$\frac{d^2\xi}{d\theta^2} + \xi(1 + \epsilon\xi^2) = \cos(\gamma\theta)$$

as in Eq. (14.2.2), where the normalized coordinates are defined, we set $\xi = \xi_{\frac{1}{3}} \cos(\gamma\theta/3) + \xi_1 \cos(\gamma\theta) + \cdots$ and use in the first approximation only the first two terms in this expansion. The left-hand side of the equation then will be of the form $A_{\frac{1}{3}} \cos(\gamma\theta/3) + A_1 \cos \gamma\theta + \cdots$, where the coefficients are functions of $\xi_{\frac{1}{3}}$ and ξ_1. As a first approximation, we then get the equations $A_{\frac{1}{3}} = 0$ and $A_1 = 1$, to satisfy the equation of motion. These equations are

$$\xi_{\frac{1}{3}} \left[1 - \frac{\gamma^2}{9} + \tfrac{3}{4}\epsilon(\xi_{\frac{1}{3}}^2 + 2\xi_1^2 + \xi_{\frac{1}{3}}\xi_1) \right] = 0$$

$$(1 - \gamma^2)\xi_1 + \frac{\epsilon}{4} (\xi_{\frac{1}{3}}^3 + 6\xi_1^2\xi_1 + 3\xi_1^3) = 1$$

(14.2.9)

From these equations we can determine the amplitudes $\xi_{\frac{1}{3}}$ and ξ_1 as functions of the frequency parameter γ. From the first equation we obtain

$$\xi_{\frac{1}{3}} = -\tfrac{1}{2}\xi_1 \left[1 \pm \sqrt{\frac{16}{27\epsilon\xi_1^2} (\gamma^2 - 9) - 7} \right]$$

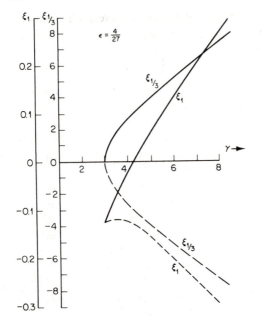

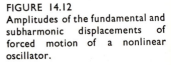

FIGURE 14.12
Amplitudes of the fundamental and subharmonic displacements of forced motion of a nonlinear oscillator.

In order that $\xi_{\frac{1}{3}}$ be real, the condition $\gamma^2 > 9 + (189\epsilon\xi_1^2/16)$ must be satisfied. For small values of ϵ this reduces to $\gamma > 3$; for example, the driving frequency must be larger than three times the (linear) resonance frequency $\omega_0/2\pi$.

If ϵ is small, the amplitude of the subharmonic is large, approximately equal to $\pm\sqrt{(4/27\epsilon)(\gamma^2 - 9)}$. To obtain a corresponding approximation for the amplitude ξ_1 of the fundamental, we must solve the second of Eqs. (14.2.9). The largest term on the left-hand side is $(\epsilon/4)\xi_{\frac{1}{3}}^3$, which is of order $1/\sqrt{\epsilon}$, but to compute ξ_1 we must also include the terms $(1 - \gamma^2)\xi_1$ and $(2\epsilon/3)\xi_{\frac{1}{3}}^2\xi_1$, which are independent of ϵ; terms in positive powers of ϵ can be neglected. To this approximation, then,

$$\xi_{\frac{1}{3}} \simeq \pm\sqrt{\frac{4}{27\epsilon}(\gamma^2 - 9)}$$

$$\epsilon \ll 1 \qquad (14.2.10)$$

$$\xi_1 \simeq \frac{\frac{3}{8}\epsilon\xi_{\frac{1}{3}}^3 - \frac{3}{2}}{1 + \frac{11}{9}\gamma^2}$$

Figure 14.12 shows a typical set of curves for the two amplitudes vs. the driving-frequency parameter $\gamma = \omega/\omega_0$. The solid curves are one solution, the dashed curves the other; which pair is used depends on the initial conditions (Note, however, that a combination of both solutions is *not* a solution the steady-state motion corresponds either to the solid curves of Fig. 14.1:

or to the dashed ones.) For the case of the solid curves, the fundamental amplitude goes to zero when

$$\gamma^2 \simeq 9 + \frac{1}{27}\left(\frac{4}{\epsilon}\right)^{\frac{1}{2}}$$

At this driving frequency, and for this mode of vibration, the driven motion is almost entirely at the subharmonic frequency.

Bias forces

In the study of the motion of a linear oscillator we are accustomed to disregard the influence of a constant force since it merely produces a change of the equilibrium position of the motion. For example, a mass-spring oscillator will have the same natural frequency when the motion is along a vertical line, under the influence of gravity, as when the motion is along a horizontal line. For a nonlinear oscillator, however, this is no longer the case; a constant force, which may be called a "bias" force, does influence the response of the oscillator to a periodic driving force.

As an illustration, let us again study the oscillator with a restoring force of the form $Kx[1 + (x/d)^2]$, where K is the spring constant for small displacements, and d is the characteristic displacement at which the linear and nonlinear contributions to the restoring force are the same. Let this oscillator be acted upon by a constant force F plus a periodic force f. If $f \ll F$, the equilibrium position X of the oscillator is determined solely by the time-independent force, according to the equation

$$\frac{K}{d^2}X^3 + KX = F$$

which has the solution

$$X = d[(a + b)^{\frac{1}{3}} + (a - b)^{\frac{1}{3}}]$$

where $a = F/Kd$, and $b = \sqrt{\frac{1}{27} + (F/2Kd)^2}$. A small periodic force produces an oscillation about this equilibrium, and as long as the corresponding displacement is sufficiently small, the response of the oscillator is the same as that of a linear oscillator, with a spring constant given by the slope of the force vs. displacement curve at the point X. The slope of the curve at X is

$$\frac{dF}{dX} = K\left[1 + 3\left(\frac{X}{d}\right)^2\right] = K_1$$

and the effective spring constant of the oscillator at that point is then $K_1 = K[1 + 3(X/d)^2]$. The frequency of small oscillations about the equilibrium point is then $\omega_1 = \omega_0\sqrt{1 + 3(X/d)^2}$, where ω_0 is the natural frequency of small oscillations without a bias force.

A bias force of some acoustical importance is produced by a moving airstream. This force depends on the velocity of the stream relative to the

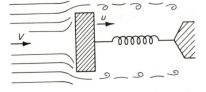

FIGURE 14.13
Airstream impinging on an oscillator produces
additional damping of the oscillator.

oscillator. It produces not merely a change in the effective stiffness of the oscillator (assuming, of course, that the restoring force is nonlinear), but also an additional damping in the system. To demonstrate this effect, consider as a simple example a mass-spring oscillator in an airstream moving along the line of oscillation of the oscillator, as shown in Fig. 14.13. If the velocity of the airstream is V, the force produced on the oscillator is of the form

$$F = AC_d(\rho V^2) \qquad (14.2.11)$$

where A is the area of the body (or rather, the projection of the area on a plane perpendicular to the airstream), and C_d is a constant, the so-called drag coefficient. This coefficient, which has to be determined experimentally, depends on the geometry of the body and, in general, also upon V. In some cases, however, C_d is essentially independent of velocity over a wide range of velocities; we shall assume this to be the case in these calculations. (Clearly, at very low velocities, C_d varies inversely as the velocity since, in the limit $V \to 0$, the force must become the viscous force, which is proportional to the velocity. We consider velocities large enough so that viscous effects are not important.)

If the oscillator has a velocity u, the relative velocity between the airstream and the oscillator is $V - u$, and the force in Eq. (14.2.11) becomes

$$F + \Delta F = AC_d\rho(V^2 + u^2) - 2\rho AC_d Vu$$

If $u \ll V$, we see that, in addition to the time-independent force F, we now have also the time-dependent force $\Delta F = -(2AC_d\rho V)u$, which is proportional to the velocity u and is equivalent to a frictional force. In the more general case, when C_d is velocity-dependent, the equivalent frictional force can be expressed as $(dF/dV)_X u$. This aerodynamically produced frictional force in the oscillator has a considerable influence on various acoustical resonators operating in relative motion with the surrounding medium. The free oscillations of cavity resonators, strings, and membranes, etc., have been found to decay more rapidly when in motion than when at rest. The reason for this behavior is much the same as in the present simple example.

14.3   LARGE-AMPLITUDE STRING WAVES

In keeping with the treatment in the rest of the book, in this chapter we have first dealt with the nonlinear behavior of a point mass; next we should take up nonlinear waves on a string, and then come to the nonlinear acoustics of a fluid. This section is concerned with the wave motion of a string when its amplitude of transverse displacement is large enough so that the simple linear wave equation does not adequately describe its motion.

Of course we shall not relax all the idealizations imposed in Chap. 4. We still assume that the string is uniform in mass density and is perfectly flexible, though corrections to the formulas to allow for stiffness can be applied analogous to those of Sec. 5.1. We also shall here continue to assume that the end supports, holding the string and applying the tension, are perfectly rigid, though the extension of the theory to allow for support motion is not difficult—indeed, is an obvious extension of the discussions on pages 112 and 172.

The nonlinearity discussed in this section comes about when the transverse displacement of the string is large enough to modify the tension by a material amount; the string is stretched more in some parts than in others. We thus must consider the longitudinal motion of the string as well as its transverse motion—in fact, the two are coupled—and we must introduce the relationship between longitudinal strain and stress resulting from the differences in stretch, the string's equation of state, so to speak. In contrast to the last two sections of this chapter, where the effects of nonlinearities of the equation of state of the fluid are discussed, in this section we shall assume that the linear relationship of Hooke's law holds for the equation of state of the stretched string.

General wave motion

In equilibrium we suppose the string to be stretched along the x axis and its longitudinal strain $\Delta l / l_0$ to be produced by being stretched between rigid supports, so that the equilibrium stress is tension $T_0 = QA(\Delta l / l_0)$, Q being the Young's modulus [Eq. (5.1.1)] of the string and A its cross-sectional area. In this stretched condition, at rest along the x axis, the mass of the string per unit length is ϵ_0. When in motion, the element of string which was, in equilibrium, at point $(x,0,0)$ will be at point $(x + \xi, \eta, \zeta)$, its displacement from equilibrium being given by the vector $\mathbf{r}$, so that the vector from the origin to its instantaneous position is

$$\mathbf{R}(x,t) = (x + \xi)\mathbf{e}_x + \eta\mathbf{e}_y + \zeta\mathbf{e}_z = x\mathbf{e}_x + \mathbf{r}(x,t) \qquad (14.3.1)$$

where $\mathbf{e}_x$, $\mathbf{e}_y$, $\mathbf{e}_z$ are unit vectors along the coordinate axes. Thus we are using the Lagrange description, shown in Fig. 6.2, to specify the motion.

The element of length dx in equilibrium will now have the length $|\partial \mathbf{R}/\partial x| \, dx$, where

$$\left|\frac{\partial \mathbf{R}}{\partial x}\right| = \sqrt{\left(1 + \frac{\partial \xi}{\partial x}\right)^2 + \left(\frac{\partial \eta}{\partial x}\right)^2 + \left(\frac{\partial \zeta}{\partial x}\right)^2} \qquad (14.3.2)$$

$$\epsilon_0 = \epsilon \left|\frac{\partial \mathbf{R}}{\partial x}\right|$$

Therefore the instantaneous mass density of the string, originally at $(x,0,0)$, is $\epsilon_0/|\partial \mathbf{R}/\partial x|$, and its tension $T(x,t)$ will be modified by the change in length. If Hooke's law holds, the change will be linear, so that

$$T(x,t) = T_0 + QA\left(\left|\frac{\partial \mathbf{R}}{\partial x}\right| - 1\right) \tag{14.3.3}$$

The force on the displaced element (Fig. 4.2) will arise, not only because the pull across the two ends may be in slightly different directions, but also because the magnitude of T differs slightly between the two ends. If $\mathbf{s}(x,t)$ is the unit vector pointing along the tangent to the string at the point labeled x,

$$\mathbf{s} = \frac{\partial \mathbf{R}/\partial x}{|\partial \mathbf{R}/\partial x|} = \frac{[1 + (\partial \xi/\partial x)]\mathbf{e}_x + (\partial \eta/\partial x)\mathbf{e}_y + (\partial \zeta/\partial x)\mathbf{e}_z}{\sqrt{[1 + (\partial \xi/\partial x)]^2 + (\partial \eta/\partial x)^2 + (\partial \zeta/\partial x)^2}}$$

then the equation of motion of the element of string which was of length dx and at point $(x,0,0)$ in equilibrium (and thus always has mass $\epsilon_0\, dx$) is

$$\epsilon_0 \frac{\partial^2 \mathbf{R}}{\partial t^2} = \frac{\partial(T\mathbf{s})}{\partial x} = \mathbf{s}\frac{\partial T}{\partial x} + T\frac{\partial \mathbf{s}}{\partial x} \tag{14.3.4}$$

In Chap. 4 we neglected $\partial T/\partial x$ and assumed that $\mathbf{s} = \mathbf{e}_x + (\partial \eta/\partial x)\mathbf{e}_y$. The exact equation includes longitudinal, as well as transverse, forces.

Inserting Eq. (14.3.3) into this equation of motion and using the formula

$$\frac{\partial \mathbf{s}}{\partial x}\left|\frac{\partial \mathbf{R}}{\partial x}\right| = \frac{\partial^2 \mathbf{R}}{dx^2} - \mathbf{s}\frac{\partial\,|\partial \mathbf{R}/\partial x|}{\partial x}$$

we obtain

$$\frac{\partial^2 \mathbf{R}}{\partial t^2} = c_l^2 \frac{\partial^2 \mathbf{R}}{\partial x^2} - (c_l^2 - c_t^2)\frac{\partial \mathbf{s}}{\partial x} \tag{14.3.5}$$

$$c_l^2 = \frac{AQ}{\epsilon_0} \qquad c_t^2 = \frac{T_0}{\epsilon_0}$$

When $\mathbf{s} \simeq \mathbf{e}_x + (\partial \eta/\partial x)\mathbf{e}_y \simeq \partial \mathbf{R}/\partial x$, this reduces to the linear form of Eq. (4.1.2). Constants c_l and c_t are the familiar longitudinal and transverse wave speeds for small-amplitude waves. It is interesting to note that the nonlinearity is all contained in the second term on the right-hand side of the equation of motion, and that this term is proportional to the difference in the squares of the longitudinal and transverse wave speeds. In other words, a string for which these speeds are equal would be linear; it would transmit large-amplitude waves in just the same way as small-amplitude waves. However, to obtain such a string, with $c_l = c_t$, the tension T_0 must be large enough to extend the string to twice its relaxed length. Except possibly for rubber strings or coil springs, this condition can hardly be satisfied for most strings used in practice; generally, c_l is considerably larger than c_t. The general wave equation (14.3.5) is written in vector form, but if we use the explicit expression for $\mathbf{R}$ and $\mathbf{s}$ as they were defined, we obtain three coupled differential equations in terms of the components ξ, η, ζ.   Before we discuss these equations, it is interesting to consider, first, some general properties of motion.

Constant tension

If the boundary conditions of the string are properly chosen, it seems possible that a wave can propagate on the string without altering the initial tension.   In order to investigate whether this is consistent with the equation of motion, we first observe from Eq. (14.3.4) that on the assumption of Hooke's law, a constant tension implies a constant value of $|\partial \mathbf{R}/\partial x|$ and also a constant value of ϵ.   In that case we get $\partial s/\partial x = \dfrac{\partial^2 \mathbf{R}}{\partial x^2} \Big/ \left|\dfrac{\partial \mathbf{R}}{\partial x}\right|$, and the wave equation (14.3.3) then can be expressed in the form

$$\epsilon \frac{\partial^2 \mathbf{R}}{\partial t^2} = T \frac{\partial^2 \mathbf{R}}{\partial x^2} \qquad \epsilon, T \text{ const} \qquad (14.3.6)$$

where we have introduced ϵ instead of ϵ_0 according to Eqs. (14.3.2).   This equation is just like the ordinary equation for small-amplitude motion; the only difference is that we use the actual values of ϵ and T of the deformed string rather than the unperturbed values ϵ_0 and T_0.   Equation (14.3.6) is valid for arbitrary amplitudes, but limited to constant values of T (and ϵ), and it follows that under these conditions all portions of a large-amplitude wave will travel with the same wave speed $\sqrt{T/\epsilon}$; no distortion of the waveform will take place.   It is easy to see, however, that this conclusion is consistent with the requirement of a constant value of the quantity $|\partial \mathbf{R}/\partial x|^2$ (which is equivalent to constant tension) only for a single traveling wave along the string.   For such a wave we can maintain a constant value of $|\partial \mathbf{R}/\partial x|$, once established, because, as we have seen, the wave will travel without change in its shape and therefore without change of $\partial r/\partial x$; the mere translation of the wave pattern that occurs in the traveling wave cannot alter the constant value of $|\partial \mathbf{R}/\partial x|$.   For waves traveling in the opposite direction, on the other hand, a constant value cannot be maintained; the wave pattern, and consequently the derivatives $\partial \xi/\partial x$, etc., will vary as different parts of the two waves interact.

A wave of this sort can be generated by the rather impractical method illustrated in Fig. 14.14.   The string is displaced from equilibrium along the portion from a to b by being fitted inside a frictionless tube, as shown, originally at rest.   This transverse displacement will of course stretch the string so that the tension throughout its length will equalize to a new value $T > T_0$, and its density will also take the uniform value $\epsilon < \epsilon_0$.   At point P of the tube, where the radius of curvature is ρ, the element ds of string will have acting on it a net force $T \, ds/\rho$ toward the center of curvature.

Now set the string into motion, maintaining tension, with a velocity v, when the element ds will also have acting a centrifugal force $\epsilon v^2 \, ds/\rho$ away

FIGURE 14.14
Large-amplitude constant-tension traveling wave in string.

from the center of curvature. If the speed happens to equal $c = \sqrt{T/\epsilon}$, these two forces will exactly balance, *no matter what value ρ has.* In other words, at this speed the tube could be removed, and the stationary wave would remain; at every point the forces balance, and there would be no change in waveform, and thus no change in string length or tension or density. Of course, the situation of the string moving with velocity c and the waveform standing still is no different dynamically from the case of the string standing still (the part outside of ab, that is) and the wave moving with velocity c. Thus the wave, started in this artificial manner, will travel with constant shape and velocity and will maintain constant tension throughout the string. The wave must go in just one direction, however; one could not start two inter-penetrating waves, going in opposite directions, by this method. Also, if two such waves, originally noninterpenetrating, going in opposite directions, should meet, the conditions would no longer hold, the wave would change form, and the tension would no longer be uniform.

Third-order equations

Apart from the situation just outlined, of constant tension, or when $c_l = c_t$, exact solutions of Eq. (14.3.5) are not known. With waves of modest amplitudes, however, an approximate form of the equation can be obtained from a power expansion of the nonlinear term $\partial s / \partial x$ in Eq. (14.3.5), in which only the first few terms are retained. The equation thus obtained is a convenient starting point for perturbation calculations of nonlinear behavior of these waves.

In obtaining the power series expansion, we start from the explicit form of $s = \dfrac{\partial \mathbf{R}/\partial x}{|\partial \mathbf{R}/\partial x|}$ in terms of the slopes $\partial \xi / \partial x$, etc., as given in Eq. (14.3.5). Thus, after having expanded the square root in this expression and performed the differentiation $\partial s / \partial x$, we obtain, after some algebra, the following three equations for the components ξ, η, and ζ:

$$\frac{\partial^2 \xi}{\partial t^2} - c_l^2 \frac{\partial^2 \xi}{\partial x^2} = \tfrac{1}{2}(c_l^2 - c_t^2) \frac{\partial}{\partial x}[(\eta_x^2 + \zeta_x^2)(1 - 2\xi_x)] \tag{14.3.7}$$

$$\frac{\partial^2 \eta}{\partial t^2} - c_l^2 \frac{\partial^2 \eta}{\partial x^2} = \tfrac{1}{2}(c_l^2 - c_t^2) \frac{\partial}{\partial x}[\eta_x(\eta_x^2 + \zeta_x^2) + 2\eta_x \xi_x(1 - \xi_x)]$$

$$\tag{14.3.8}$$

$$\frac{\partial^2 \zeta}{\partial t^2} - c_t^2 \frac{\partial^2 \zeta}{\partial x^2} = \tfrac{1}{2}(c_l^2 - c_t^2) \frac{\partial}{\partial x}[\zeta_x(\eta_x^2 + \zeta_x^2) + 2\zeta_x \xi_x(1 - \xi_x)]$$

$$\tag{14.3.9}$$

where we have here written $\partial\eta/\partial x$ as η_x, etc. In these equations the non-linear terms have been written as "source terms" in the wave equations for the three modes of motion, the longitudinal mode and the two transverse modes. These modes are all coupled with each other as a result of the nonlinearity.

Even these equations cannot be solved exactly, and we have to resort to approximate methods of analysis much in the same way as in the case of the nonlinear one-dimensional oscillator in Sec. 14.1. In many cases we can start from the first-order motions being solutions of the linear string equations. Using these first-order solutions for the evaluation of the nonlinear source terms in Eqs. (14.3.7) to (14.3.9), nonlinear corrections to the linear solutions can be obtained, including the longitudinal-transverse and transverse-transverse mode-coupling effects.

The generation of a longitudinal wave mode by a transverse wave is to be expected on account of the stretching of the string which is associated with the transverse motion. This effect is expressed by Eq. (14.3.7), which shows that longitudinal motion is induced by transverse motion through a "source strength" of second order in the transverse motion, proportional to $\partial(\eta_x^2 + \zeta_x^2)/\partial x$. Less obvious is the coupling between the two transverse modes η and ζ. This coupling consists of two contributions; the first is a "direct" coupling, which for the η motion corresponds to the source term $\partial(\zeta_x^3 + \zeta_x\eta_x^2)/\partial x$, and the second is a coupling carried via longitudinal motion and expressed by the source term proportional to $\partial(\zeta_x\xi_x)/\partial x$. This term is also of third order in the transverse field variables since, according to Eq. (14.3.7), ξ_x is of second order in η and ζ. Consequently, in any study of interactions in lateral motions of a string, the *longitudinal motion generally cannot be neglected.*

Transverse-longitudinal mode coupling

As an example, we shall determine the longitudinal motion induced by the (first-order) fundamental mode of a transverse motion of a string clamped at both ends. Thus, let the first-order transverse motion be $\eta_1 = \eta_0 \sin(\pi x/L) \sin(\omega_t t)$, where $\omega_t/2\pi = c_t/2L$ is the fundamental frequency. This transverse motion generates a longitudinal motion, which we find from Eq. (14.3.7). By using η_1 in the source term of this equation, we get for the second-order longitudinal motion ξ_2,

$$\frac{\partial^2 \xi_2}{\partial t^2} - c_l^2 \frac{\partial^2 \xi_2}{\partial x^2} = \tfrac{1}{2}(c_l^2 - c_t^2) \frac{\partial}{\partial x} \eta_x^2$$

$$= (c_l^2 - c_t^2)(k_t\eta_0^2) \frac{\partial}{\partial x} [\cos^2(k_t x) \sin^2(\omega_t t)]$$

where $k_t = \omega_t/c_t$. The steady-state solution to this equation is

$$\xi_2 = \frac{\pi}{16} \frac{\eta_0}{L} \eta_0 \left[\sin(2k_t x) \cos(2\omega_t t) - \left(1 - \frac{c_t^2}{c_l^2}\right) \sin(2k_t x) \right]$$

$$(14.3.10)$$

The time-dependent part of ξ_2 varies with twice the transverse frequency; the string stretches at every transverse excursion of the string, regardless of its direction. The second term represents the time-average displacement $\langle \xi_2 \rangle$. Its space average must be zero since, for fixed supports, the projection of the string on the x axis is not altered by the motion. The axial displacement $\langle \xi_2 \rangle$ is negative on the left half of the string, where each point of the string moves slightly to the left as the string makes its transverse displacement. The center point of the string moves vertically, so that ξ_2 is zero there, and the points on the right half of the string move slightly to the right, making $\langle \xi_2 \rangle$ positive. We leave it for one of the problems to calculate the corresponding changes in the mass per unit length and the tension.

It is interesting to find that the amplitude of this solution is independent of the relation between the "driving frequency" and the resonance frequency ω_l of the longitudinal motion; one may expect resonance excitation to occur if $\omega_t = \omega_l$. However, the corresponding resonant denominator is canceled by the fact that the nonlinearity of the string, as we have seen, is proportional to $c_l^2 - c_t^2$.

Lateral-lateral mode coupling

In much the same manner we can proceed to calculate the coupling between transverse modes, for example, the nonlinear distortion of a plane transverse motion, starting, say, from the fundamental η mode. Or we can determine the coupling between an η mode and a ζ mode. Although we shall not carry out these calculations here, we merely point out that even when we deal with lateral-lateral coupling problems, the associated excitation of longitudinal motion cannot be neglected as a coupling mechanism. Suppose, for example, we are interested in the excitation of higher order ζ motion by (first-order) η and ζ motion. The source term to be considered is then proportional to the x derivative of $\zeta_x[(\eta_x^2 + \zeta_x^2) + 2\xi_x]$, neglecting the term involving ξ_x^2, which turns out to be of higher order than the other terms. The longitudinal motion ξ_x excited by the first-order η and ζ motions can be determined from Eq. (14.3.7), which, after differentiation with respect to x, can be written

$$\frac{\partial^2 \xi_x}{\partial t^2} - c_l^2 \frac{\partial^2 \xi_x}{\partial x^2} = \frac{1}{2}(c_l^2 - c_t^2)\frac{\partial^2}{\partial x^2}(\eta_x^2 + \zeta_x^2) \qquad (14.3.11)$$

where we have neglected the terms containing ξ_x on the right-hand side, being of higher order. Suppose now that $\eta_x^2 + \zeta_x^2$ contains terms of the form $\sin(mk_t x)\sin(m\omega_t t)$ or $\sin(mk_t x)\cos(m\omega_t t)$ or some other combination of these trigonometric functions. We denote this part of $\eta_x^2 + \zeta_x^2$ by $(\eta_x^2 + \zeta_x^2)_{m,m}$. The longitudinal motion induced by these terms, as expressed by the solution ξ_x in Eq. (14.3.11), will have the same x, t dependence. In fact, the solution is

$$(\xi_x)_{m,m} = -\tfrac{1}{2}(\eta_x^2 + \zeta_x^2)_{m,m}$$

If we now introduce this expression for $(\xi_x)_{m,m}$ in the source term

$$\zeta_x[(\eta_x{}^2 + \zeta_x{}^2) + 2\xi_x]$$

we find the rather interesting result that the (m,m) component of $(\eta_x{}^2 + \zeta_x{}^2)$ is exactly canceled by the (m,m) component of $2\xi_x$. In other words, the fact that longitudinal motion necessarily is excited by transverse motion plays an important role also in regard to the lateral-lateral coupling problems.

14.4 THE SECOND-ORDER EQUATIONS OF SOUND

In previous chapters we have repeatedly pointed out that the nonlinear terms in the equations of motion have been neglected but that the resulting linearized equations adequately describe the acoustic wave motion one usually encounters. Although, in most of this linear analysis, we omitted viscosity and heat conduction, they were included in the study of wave attenuation in Chap. 6. As we pointed out there, the effects of viscosity and heat conduction can be regarded as small perturbations in the equations of motion. We now ask, under what conditions is it justified to omit nonlinear terms compared with the viscous and heat-conduction terms? The role of nonlinearity is not as simple to foresee as is the effect of viscosity; we shall find that its effect makes itself felt even at low amplitudes, in connection with quantities such as mass flow.

Relative magnitudes

To compare the relative importance of the viscous and nonlinear terms, it is sufficient to consider the one-dimensional form of the momentum equation

$$\frac{\partial u}{\partial t} + u\frac{\partial u}{\partial x} + \frac{1}{\rho}\frac{\partial P}{\partial x} = \frac{\mu'}{\rho}\frac{\partial^2 u}{\partial x^2} \qquad (14.4.1)$$

where $\mu' = \frac{4}{3}\mu + \eta$ [Eq. (6.4.15)]. Since we are interested in the relative order of magnitude of the terms in this equation in connection with the propagation of sound waves, we introduce for the characteristic time and length the period T and the wavelength λ of a harmonic sound wave, so that $\partial/\partial t \sim 1/T$ and $\partial/\partial x \sim 1/\lambda$. Then, to obtain the relative orders of magnitude of the different terms, we use the properties of plane waves discussed in Chap. 6, where we have seen that if the perturbation of the pressure in the wave is p, the corresponding perturbations in density, fluid velocity, and temperature are $\delta = p/c^2 = (u/c)\rho$, $u = p/\rho c$, and $\theta = (\gamma - 1)(u/c)T = (\gamma - 1)pT/\gamma P$.

The order of magnitude of the first and third terms in Eq. (14.4.1) is then u/T, and $p/\rho\lambda = u/T$; that is, they are of the same order of magnitude. The second term, on the other hand, which is nonlinear, is seen to be of the order of $\dfrac{u}{c}\dfrac{u}{T}$. Thus the relative orders of magnitude of the three terms on the left-hand side of Eq. (14.4.1) are 1, u/c, and 1.   Similarly, the order of magnitude of the viscous term, using the first term as reference, is

$$\frac{(\mu/\rho)(\partial^2 u/\partial x_2)}{\partial u/\partial t} \simeq \frac{(\mu/\rho)(u/\lambda^2)}{u/T} = \frac{\nu T}{\lambda^2} \simeq \frac{\nu\omega}{c^2}$$

where ν is the kinematic viscosity.   It follows, then, that the viscous and nonlinear terms are of the same order of magnitude when at a wave amplitude given by $u = \nu\omega/c$.   If we introduce the fluid particle displacement $\xi = u/\omega$ and the characteristic length $\nu/c = l$, which is of the order of the mean-free-path in the gas (page 272), we find that the nonlinear and viscous terms are of the same order of magnitude when the particle displacement in the wave is of the same order of magnitude as the mean-free-path.   Thus the omission of the nonlinear terms is justified in comparison with viscous terms only when $\xi \ll l$.   In air at atmospheric pressure the mean-free-path is of the order of 10^{-5} cm, and if we consider a sound wave with a frequency of 1,000 cps, the condition $\xi = l$ corresponds to a particle velocity of about 0.1 cm per sec and a sound pressure of about 4 dynes per cm^2 (85 db).   In other words, nonlinear effects as compared with viscous ones become significant at a rather modest sound-pressure level.   At higher levels, in an analysis of a problem such as the attenuation of a harmonic wave, an omission of nonlinear effects in comparison with viscosity may lead to noticeable errors, particularly at low frequencies.

The second-order equations

After these introductory remarks concerning the relative importance of nonlinear and viscous terms, we shall consider more quantitatively the influence of these effects on sound propagation, and we shall start by deriving the acoustic equations to second order in the field variables.   The meaning of this needs some explanation.   As we know, the basic equations of fluid motion are nonlinear; they contain terms that are products of the field variables and their derivatives.   Complete solutions of these equations are rare.   However, if we are interested only in small perturbations of the field variables, linear equations in the perturbations can be developed, as we have seen, and these equations form the basis of the linear theory of sound.   Then, if the unperturbed value of a field variable ψ is ψ_0, the linearized equations contain terms only of first order in the perturbation $\psi_1 = \psi - \psi_0$; terms of the order of $\psi_1{}^2$ and higher-order terms are omitted.   To develop the theory of sound one step further, second-order terms, of the order of $\psi_1{}^2$, will be included, in addition to the linear terms.   To do this in a systematic manner, each field variable ψ is written

$$\psi = \psi_0 + \psi_1 + \psi_2 + \cdots \qquad (14.4.2)$$

i.e., as a sum of the unperturbed value ψ_0, a first-order term ψ_1 (which is defined by the fact that it is a solution of the linearized equations), and a second-order term ψ_2, which is of the order of $\psi_1{}^2$.

If we expand each field variable in this way and insert the expansions into the general equations of fluid motion (Chap. 6) and require that the equations are to be satisfied separately to each order in ψ, we can extract sets of equations containing only zeroth-, first-, and second-order terms. It should be pointed out that a procedure of this kind is meaningful only if the expansion (14.4.2) is a succession of decreasing terms, so that $\psi_2 \ll \psi_1 \ll \psi_0$. Actually, it can be regarded as a Taylor expansion of ψ in which the expansion parameter is a measure of the strength of the wave. Thus, if the exact ψ is assumed to be of the form $\psi(x,t,s)$, where s is an amplitude parameter, we have $\psi_0 = \psi(x,t,0)$, $\psi_1 = s\left(\dfrac{\partial \psi}{\partial s}\right)_0$, $\psi_2 = \dfrac{s^2}{2}\left(\dfrac{\partial^2 \psi}{\partial s^2}\right)_0$, etc.

We start from the general equations in the form given in Chap. 6.

$$\frac{\partial \rho}{\partial t} + \mathbf{\nabla} \cdot (\rho \mathbf{V}) = 0$$

$$\frac{\partial \rho \, \mathbf{V}}{\partial t} + \mathbf{\nabla} \cdot (\mathfrak{J} + P\mathcal{I} + \mathfrak{D}) = 0 \tag{14.4.3}$$

where $\mathfrak{J} =$ momentum tensor of Eq. (6.2.10)
$P\mathcal{I} =$ pressure tensor
$\mathfrak{D} =$ viscous-stress tensor of Eq. (6.4.8)

$$\mathcal{I}_{ij} = \delta_{ij}$$

The equation of state is assumed to be expressible in the form

$$\rho = \rho(P) \tag{14.4.4}$$

so that the density is a function of pressure only.

In the present discussion we shall assume that the coordinate system is chosen so that the unperturbed fluid is at rest; that is, $\mathbf{V}_0 = 0$. Then, if we insert $\mathbf{V} = \mathbf{V}_1 + \mathbf{V}_2$, $\rho = \rho_0 + \rho_1 + \rho_2$, etc., into Eqs. (14.4.3) and (14.4.4) and extract the corresponding first-order equations, we get

$$\frac{\partial \rho_1}{\partial t} + \rho_0 \mathbf{\nabla} \cdot \mathbf{V}_1 = 0 \tag{14.4.5}$$

$$\rho_0 \frac{\partial \mathbf{V}_1}{\partial t} + \mathbf{\nabla} \cdot (P_1 \mathcal{I} + \mathfrak{D}_1) = 0 \tag{14.4.6}$$

$$\rho_1 = \left(\frac{\partial \rho}{\partial P}\right)_0 P_1 = \frac{1}{c^2} P_1 \tag{14.4.7}$$

If we use the expression for the viscous-stress tensor given in Chap. 6 and eliminate $\mathbf{V}_1$ from these equations by taking the time derivative of the first equation and the divergence of the second, the following wave equation for the first-order pressure results.

$$\nabla^2\left(P_1 + \frac{v'}{c^2}\frac{\partial P}{\partial t}\right) - \frac{1}{c^2}\frac{\partial^2 P_1}{\partial t^2} = 0 \qquad (14.4.8)$$

where $\rho v' = (4\mu/3) + \eta$, μ and η being the shear and bulk viscosities [Eq. (6.4.10)]. Similarly, for the second-order acoustic quantities we obtain the following equations:

$$\frac{\partial \rho_2}{\partial t} + \nabla \cdot \rho_0 \mathbf{V}_2 = -\nabla \cdot \rho_1 \mathbf{V}_1 \qquad (14.4.9)$$

$$\rho_0 \frac{\partial \mathbf{V}_2}{\partial t} + \nabla P_2 + \nabla \cdot \mathfrak{D}_2 = -\nabla \cdot (\mathfrak{J}_2) - \frac{\partial(\rho_1 \mathbf{V}_1)}{\partial t} \qquad (14.4.10)$$

$$\rho_2 = \left(\frac{\partial P}{\partial \rho}\right)_0 P_2 + \frac{1}{2}\left(\frac{\partial^2 \rho}{\partial P^2}\right)_0 P_1^2 = \frac{1}{c_0^2}P_2 + \frac{\Gamma}{\rho_0 c_0^4}P_1^2 \qquad (14.4.11)$$

where $\Gamma = \frac{1}{2}\rho_0 c^4(\partial^2\rho/\partial P^2)$, which, for an ideal gas, is $\Gamma = -(\gamma - 1)/2\gamma$. We have used here the notation in Eq. (6.4.10), so that in component form,

$$\nabla \cdot \mathfrak{D} = \sum_j \frac{\partial D_{ij}}{dx_j} \quad \text{and} \quad \nabla \cdot \mathfrak{D} \cdot \nabla = \sum_i \sum_j \frac{\partial^2 D_{ij}}{\partial x_i \, \partial x_j}$$

with similar expressions for the tensor $\mathfrak{J}$.

The corresponding wave equation for P_2 is

$$\nabla^2\left(P_2 + \frac{v'}{c^2}\frac{\partial P_2}{\partial t}\right) - \frac{1}{c^2}\frac{\partial^2 P_2}{\partial t}$$
$$= -\nabla \cdot \mathfrak{J}_2 \cdot \nabla - \frac{\Gamma}{\rho_0 c_0^4}\frac{\partial}{\partial t}\left(\frac{\partial P_1^2}{\partial t} + v'\nabla^2 P_1^2\right) \qquad (14.4.12)$$

Since in the present case we assume the unperturbed fluid to be homogeneous and at rest, the zeroth-order equations are identically satisfied. We note that the equations for the first- and second-order contributions to the field variables are of the same form except for the occurrence of *source terms* in the second-order equations. These source terms are products of first-order terms, and if the first-order solutions can be regarded as known, these equations are sufficient for the calculation of the second-order contribution to the sound field. For example, the solution to the wave equation (14.4.12) for P_2 can be obtained in a standard manner, using an appropriate Green's function.

Mass and energy flow

As a first application of Eqs. (14.4.9) to (14.4.12) we shall discuss the mass-flow paradox that was mentioned briefly in Chap. 6. On the basis of linear acoustics, one can be led to the erroneous conclusion that the time average of the mass flow in the sound field is $\langle \rho_1 \mathbf{V}_1 \rangle$, which for a plane harmonic wave is different from zero.   The error lies in the fact that $\langle \rho_1 V_1 \rangle$ is a quantity of second order and represents only one part of the correct expression for the mass flow, including terms up to second order, $\langle \rho V \rangle = \langle \rho_0 V_1 \rangle + \langle \rho_1 V_1 \rangle + \langle \rho_0 V_2 \rangle$.   For a harmonic time dependence the first term is zero.   The second term, on the other hand, is different from zero since, in a traveling plane wave, ρ_1 and V_1 are in phase, as we have already indicated. The third term, $\langle \rho_0 V_2 \rangle$, follows directly from the time average of Eq. (14.4.9), which becomes $\nabla \cdot \langle \rho_0 V_2 \rangle = -\nabla \cdot \langle \rho_1 V_1 \rangle$, since, for a finite value of a field variable ψ, we must have $\langle \partial \psi / \partial t \rangle = 0$.   Then, if the flow is irrotational, it follows that $\langle \rho_0 V_2 \rangle = -\langle \rho_1 V_1 \rangle$, and therefore the mass flow $\langle \rho_0 V_1 \rangle + \langle \rho_1 V_1 \rangle + \langle \rho_0 V_2 \rangle$, *in a sound wave*, is *zero* to this order.

Although the expression for the mass flow obtained on the basis of linear theory led to an erroneous result, we have nevertheless accepted without concern the expression $\langle P_1 V_1 \rangle$ for the acoustic intensity, and this has been used throughout the book.   This quantity, however, like the mass flow $\langle \rho_1 V_1 \rangle$, is of second order, and there is no guarantee that it is the true value of the intensity, correct to second order.   We are now in a position to reexamine this question, and to do this we start from the general expression for the energy-flow vector in a fluid, in the absence of heat conduction and viscosity;

$$\mathbf{I} = \left\langle \left(\rho \frac{V^2}{2} + \rho E \right) \mathbf{V} + P \mathbf{V} \right\rangle \tag{14.4.13}$$

The first term represents the convection of the energy density in the fluid, and the second term represents the work done by pressure. E is the internal energy of the fluid per unit mass.   Since we are restricting ourselves to a loss-free fluid, the first law of thermodynamics gives a simple relation between the variation in internal energy and the density $dE = P\, d\rho/\rho^2$. Since the unperturbed fluid is at rest, the velocity entering into Eq. (14.4.13) is due solely to the sound wave; that is, $V_0 = 0$.   Then, to collect all the terms up to second order that contribute to the acoustic intensity, we introduce the expansions $E = E_0 + E_1 + E_2 + \cdots$, $\mathbf{V} = \mathbf{V}_1 + \mathbf{V}_2 + \cdots$, $P = P_0 + P_1 + P_2$, $\rho = \rho_0 + \rho_1 + \rho_2$, into (14.4.13), and collect the terms of zero, first, and second order.   Since $V_0 = 0$, we have no zeroth-order contributions.   Since we assume harmonic time dependence of the first-order quantities, their time average is zero.   Those remaining are of second order, and we have

$$\mathbf{I} = E_0(\langle \rho_1 V_1 \rangle + \rho_0 \langle V_2 \rangle) + \langle E_1 V_1 \rangle \rho_0 + \langle P_1 V_1 \rangle + P_0 \langle V_2 \rangle \tag{14.4.14}$$

Since $\langle \rho_1 V_1 \rangle = -\rho_0 \langle V_2 \rangle$, as we have seen from Eq. (14.4.9), the term containing E_0 is zero.   Furthermore, from $dE = P\, d\rho/\rho^2$ follows $E_1 = (P_0/\rho_0^2)\rho_1$, and therefore $\langle E_1 V_1 \rangle \rho_0 = (P_0/\rho_0)\langle \rho_1 V_1 \rangle = -P_0 \langle V_2 \rangle$.   Consequently, the only surviving term in the expression for the acoustic intensity is

$$\mathbf{I} = \langle P_1 V_1 \rangle \tag{14.4.15}$$

This, then, is the proof that the expression for the acoustic intensity that we have used in the linear theory, indeed, is correct. It should be emphasized, however, that this result refers to a coordinate system with respect to which the unperturbed fluid velocity is zero. If this is not the case, the expression for the intensity is considerably more complicated.

The *momentum flux tensor* $\mathfrak{I}$ with components $\rho V_i V_j$ correct to second order is simply $\rho_0 V_1 V_1$ if the unperturbed fluid is at rest so that $V_0 = 0$. The corresponding rate of momentum transfer in the direction n per unit area has the components $\sum_j \rho_0 V_i V_j n_j$, and the time average of this is sometimes called the acoustic radiation pressure on a perfectly absorbing (no-reflection) surface. For one-dimensional motion the radiation pressure in the direction of propagation is $\langle \rho V_1^2 \rangle$, which is equal to the energy density in the wave. However, one should carefully distinguish between this quantity and the *total* time-average pressure, or rate of momentum transfer to a surface, which is $\langle \sum_j [(P - P_0)\delta_{ij} + \rho V_i V_j] n_j \rangle$, which, for one-dimensional motion, is $\langle P - P_0 \rangle + \langle \rho_0 V_1^2 \rangle$. To evaluate it, we must determine also $\langle P - P_0 \rangle$, and we shall make some comments on it under the following heading.

Other time averages

In addition to the time averages of mass, momentum, and energy flow discussed above, some properties of the time averages of the field variables themselves can be inferred from the second-order equations (14.4.9) to (14.4.11). For example, if we consider irrotational flow and set the viscous stress equal to zero, we find from Eq. (14.4.9) that the time average of the (Eulerian) velocity is given by $\langle V_2 \rangle = -\langle \rho_1 V_1 \rangle / \rho_0$. For a traveling harmonic sound wave with $V_1 = u \sin(\omega t - kx)$ and $\rho_1 = \delta \sin(\omega t - kx)$, where $\delta = p/\rho c = (u/c)\rho_0$, this time average becomes $\langle V_2 \rangle = -u^2/2c$. This is in agreement with the result obtained in Chap. 6 in connection with the discussion of the relationship between the Lagrangian and the Eulerian description of fluid motion.

Similarly, on the assumption of irrotational motion, we see from Eq. (14.4.10) that the time average of the pressure in one-dimensional motion is $\langle P_2 \rangle = -\rho_0 \langle V_1^2 \rangle$. This result suggests that in a traveling beam of sound the average pressure in the beam should be lower than the pressure in the unperturbed surrounding fluid. Such a pressure difference will induce secondary flow, however, and the assumption of irrotational flow and the omission of viscosity in the discussion are hardly justified. Under such conditions the question of whether a pressure decrease in the sound field will result or not will depend on boundary conditions. Actually, this question is of importance in the interpretation of experiments designed to measure acoustic radiation pressure.

In the special case, however, when we have a sound wave in a closed tube, so that there is no communication with the outside unperturbed fluid, our expression for the time-average pressure is expected to be a good approximation to the actual behavior in the tube. Thus, if the tube is terminated at $x = 0$ and $x = L$, a standing wave with the velocity field $u = u_0 \sin(\pi x/L) \sin(\omega t)$ will produce a time-average pressure field $\langle P \rangle = P_0 - \rho(u_0{}^2/2) \times \sin^2(\pi x/L) = P_0 - \rho_0(u_0/2)^2[1 - \cos(2\pi x/L)]$. In other words, the time-average pressure has minima at the velocity maxima, and the distance between these minima is half a wavelength. Such a pressure distribution, indeed, is found to exist in the tube, and an amusing way to demonstrate it is to let the standing wave act on a water surface (by partially filling the tube with water, the tube being horizontal). The water is pulled up in the region of the pressure minima and is maintained in this way by the sound field.

Vorticity

In our discussion of the linear theory of a viscous fluid in Chap. 6 we have already pointed out that the velocity field can be considered to be a superposition of two contributions. The first, representing the actual sound field, is related to the density fluctuations and is curl-free; that is, $\nabla \times \mathbf{V}_1 = 0$ and div $\mathbf{V}_1 = -(1/\rho_0)\,\partial\rho_1/\partial t$. The second part is a shear motion, and there are no density fluctuations associated with it, so that div $\mathbf{V}_1' = 0$. The vorticity, on the other hand, is different from zero; in fact, with $\boldsymbol{\Omega}_1 = \nabla \times \mathbf{V}_1'$, we have

$$\mu \nabla^2 \boldsymbol{\Omega}_1 - \frac{\partial \boldsymbol{\Omega}_1}{\partial t} = 0$$

This velocity field, as we have seen, is present close to boundaries, where it serves to bring the total tangential velocity in to zero at the boundary.

So far in our discussion of second-order effects in a sound field, we have been concerned, basically, with the curl-free velocity field related to the actual sound field. It remains to make some comments about second-order effects related to the second type of motion involving the vorticity. The equation for the second-order vorticity vector $\boldsymbol{\Omega}_2 = \nabla \times \mathbf{V}_2$ is obtained from the second-order momentum equation (14.4.10). The divergence of the viscous-stress tensor that occurs in this equation is

$$\nabla \cdot \mathbf{D}_2 = \mu \nabla^2 \mathbf{V}_2 + (\tfrac{1}{3}\mu + \eta)\,\mathrm{grad}\,(\mathrm{div}\,\mathbf{V}_2)$$

Thus, by taking the curl of Eq. (14.4.10), we obtain, for the second-order vorticity $\boldsymbol{\Omega}_2 = \mathrm{curl}\,\mathbf{V}_2$,

$$\mu \nabla^2 \boldsymbol{\Omega}_2 - \frac{\partial \boldsymbol{\Omega}_2}{\partial t} = \nabla \times \nabla \cdot \mathfrak{J} + \frac{\partial}{\partial t}[\nabla \times (\rho_1 \mathbf{V}_1)]$$

This diffusion equation, in contrast with the equation for $\boldsymbol{\Omega}_1$, contains a volume distribution of sources involving the first-order acoustic field. Thus, assuming that the first-order field for the viscous fluid has been determined from Eqs. (14.4.5) to (14.4.7), we can in principle solve the equation for Ω_2. It is interesting to note that, in general, Ω_2 has a nonzero time average, given by

$$\mu\nabla^2\langle\Omega_2\rangle = \nabla \times \nabla \cdot \langle\mathfrak{J}\rangle$$

Steady vortex patterns can readily be observed in the standing wave in a tube and about cylinders and other objects with which the sound wave interacts. A quantitative analysis of these vortex patterns requires, first, that a solution be found for the first-order sound field in a viscous fluid satisfying the boundary conditions of zero tangential velocity on the boundaries. After this field has been determined, the inhomogeneous vorticity equation for $\langle\Omega_2\rangle$ has to be solved. As with other second-order effects, the general nature of the phenomenon is of interest, but detailed solutions, for specific situations, do not usually merit the considerable additional computational effort.

Wave interaction

As another application of the second-order equations we shall discuss some solutions to the second-order wave equation (14.4.12) when viscosity is neglected. The source term in this equation contains the square of first-order acoustic field variables, and therefore a harmonic sound wave with a frequency ω will result in a source term which contains a term with frequency 2ω. Thus the harmonic sound wave will continuously generate a second-order wave field of frequency 2ω as it propagates through the fluid. As we shall see shortly, this second harmonic wave will grow with distance. This wave of frequency 2ω will interact with the primary wave of frequency ω to produce a source term containing the frequency 3ω, etc. In other words, the initially single harmonic wave will be distorted as it travels. Our quantitative analysis of this problem of wave (self) interaction, however, is limited only to comparatively small distortions, since we have developed the acoustic equations only to second order. Under the next heading we shall discuss the behavior of the other extreme case of a largely distorted wave.

In addition to its application in the problem of a sound wave interacting with itself, Eq. (14.4.12) also allows the calculation of the second-order field resulting from the interaction between two or more sound fields, P_a, P_b, etc., emitted from different sources. In that case, $P_1 = P_a + P_b + \cdots$, and it follows that if these waves are harmonic, with frequencies ω_a, ω_b, ..., the resulting wavefield will contain combination tones with frequencies $\omega_a \pm \omega_b$, etc. In the present analysis we shall, for simplicity, consider only two such waves, P_a and P_b, and assume the sources of these waves to be at the same positions.

In carrying out such calculations it is found convenient to introduce a change of variables which will in effect eliminate $\mathfrak{J}$ from the source term in Eq. (14.4.12). Thus, if we set

$$P_2 = P_2' - \frac{\rho_0 V_1^2}{2} - \frac{P_1^2}{2\rho_0 c_0^2} - \frac{1}{\rho_0 c_0^2} \frac{\partial P_1}{\partial t} \int P_1 \, dt$$

$$V_2 = V_2' - \frac{1}{\rho_0 c_0^2} \frac{\partial}{\partial t} \left(V_1 \int P_1 \, dt \right) \tag{14.4.16}$$

$$\rho_2 = \rho_2' - \frac{\rho_0 V_1^2}{2c_0^2} - (1 - \Gamma) \frac{P_1^2}{2\rho_0 c_0^4} - \frac{1}{\rho_0 c_0^2} \frac{\partial P_1}{\partial t} \int P_1 \, dt$$

the wave equation for P_2' is then, simply (for Γ see page 866),

$$\nabla^2 P_2' - \frac{1}{c_0^2} \frac{\partial^2 P_2'}{\partial t^2} = -(1 - \Gamma) \frac{1}{\rho_0 c_0^4} \frac{\partial^2 P_1^2}{\partial t^2} \tag{14.4.17}$$

The problem of determining the second-order sound field thus has been reduced to a problem, familiar from linear acoustics, of determining a field produced by a source which can be interpreted as a monopole distribution. As indicated earlier, we shall consider the second-order field produced when the primary first-order pressure wave P_1 consists of two harmonic components, P_a and P_b, with frequencies ω_a and ω_b. As has been the practice throughout most of this book, we assume that these first-order field components are expressed in complex form. For example, in the case of a plane wave, we have $P_a = A \exp(ik_a x - i\omega_a t)$.

The pressure P_1 in Eq. (14.4.17) then is

$$P_1 = \text{Re}\,(P_a + P_b) = \tfrac{1}{2}[(P_a + P_a^*) + (P_b + P_b^*)]$$

where P_a^* is the complex conjugate of P_a. The second-order component P_2' then will contain harmonic terms with frequencies $2\omega_a$, $2\omega_b$, and $\omega_a \pm \omega_b$. The first two of these correspond to the interaction of the wave components with themselves, whereas the mutual interaction results in the frequencies $\omega_+ = \omega_a + \omega_b$ and $\omega_- = \omega_a - \omega_b$. It is sufficient to consider here only the portion of the source term that contains these combination frequencies, since the result thus obtained reduces to the case of self-interaction if we put $\omega_a = \omega_b$ and $P_a = P_b$. The corresponding portion of P_1^2 that contains ω_+ is $\tfrac{1}{2}(P_a P_b + P_a^* P_b^*) = \text{Re}\,(P_a P_b)$, and similarly, the ω_- term is $\tfrac{1}{2}(P_a P_b^* + P_a^* P_b) = \text{Re}\,(P_a P_b^*)$.

Then, if we express the solution $P_{2\pm}'$ of Eq. (14.4.17) that corresponds to the frequency $\omega_\pm$ as $\text{Re}\,P_\pm$, the equation for $P_\pm$ is

$$(\nabla^2 + k_\pm^2)P_\pm = \frac{1 - \Gamma}{\rho_0 c^2} k_\pm^2 P_a P_{b\pm} \tag{14.4.18}$$

where $k_+ = \omega_+/c = (\omega_a + \omega_e)/c$, and $P_{b+} = P_b$, $P_{b-} = P_b^*$ (complex conjugate). To obtain the corresponding equation for P_-, we replace k_+ by k_- and P_b by P_b^*.

The interaction fields produced by various primary waves P_a and P_b

can now be determined by solving this equation. As illustrations we shall consider here the important cases when the primary waves are plane, cylindrical, and spherical. For two plane waves traveling in the same direction along the x axis, we have $P_a = A \exp(ik_a x - i\omega_a t)$ and $P_b = B \exp(ik_b x - i\omega_b t)$, where A and B are real; the equation for $P_\pm$ is

$$(\nabla^2 + k_\pm{}^2)P_\pm = (\rho_0 c_0{}^2)^{-1}(1 - \Gamma)k_\pm{}^2 AB \exp(ik_\pm x - i\omega_\pm t)$$
$$= C_\pm \exp(ik_\pm x - i\omega_\pm t) \quad (14.4.19)$$

This equation is of the same type that we encountered in the discussion of stimulated Brillouin scattering; the particular solution, for the plus sign, is

$$\begin{aligned} \cdot P'_{2+} = \operatorname{Re} P_+ &= \operatorname{Re} \frac{C_+ x}{2ik_+} \exp(ik_+ x - i\omega_+ t) \\ &= \frac{1 - \Gamma}{2\rho_0 c_0{}^2} AB(k_+ x) \sin(k_+ x - \omega_+ t) \qquad (14.4.20) \\ &= (1 - \Gamma)\frac{\rho_0}{2} v_A v_B(k_+ x) \sin(k_+ x - \omega_+ t) \end{aligned}$$

where v_A and v_B are the velocity amplitudes of the two primary waves. In other words, the interaction field amplitude increases with distance; there is a continuous energy transfer from the primary harmonic components to the combination tone.

The experimental arrangement that corresponds to this case of two interacting plane waves (which could very well be two harmonic components of one and the same wave) might be a plane piston moving with a velocity $V_x = V_a \exp(-i\omega_a t) + V_b \exp(-i\omega_b t)$. Similarly, we may consider the combination tones in the cylindrical and spherical wavefields as being produced by a pulsating cylinder and a sphere, respectively. Thus, consider a cylinder of radius a pulsating so that the radial surface velocity of the cylinder is

$$V_r = V_a \exp(-i\omega_a t) + V_b \exp(-i\omega_b t)$$

Referring to the discussion of the radiation from a cylinder in Chap. 7, we find that the corresponding sound pressure field is

$$P_{a,b} = -\rho_0 c_0 V_{a,b} \frac{iH_0^{(1)}(k_{a,b} r)}{H_1^{(1)}(k_{a,b} a)} \exp(-i\omega_{a,b} t) \qquad (14.4.21)$$

We now proceed in a manner completely analogous to the plane-wave case and determine the second-order pressures $P_\pm$ from Eq. (14.4.18), using the values for P_a and P_b given in Eq. (14.4.21). The following expressions for the combination tones $\omega_+ = \omega_a + \omega_b$ and $\omega_- = \omega_a - \omega_b$ are found to satisfy the wave equation (14.4.18):

$$P_\pm = \mp \frac{\rho_0 V_a V_b(1 - \Gamma)k_\pm r}{2} \frac{H_1^{(1)}(k_a r)H_{0\pm}(k_b r) \pm H_0^{(1)}(k_a r)H_{1\pm}(k_b r)}{H_1^{(1)}(k_a a)H_{1\pm}(k_b a)}$$
$$\times \exp(-i\omega_\pm t) \quad (14.4.22)$$

and the actual pressure is $P'_{2\pm} = \mathrm{Re}\, P_\pm$. Here $H_0^{(1)}$ and $H_1^{(1)}$ are Hankel functions of the first kind, of zeroth and first order. The plus sign and the minus sign on the Hankel functions indicate functions of first and second kind, $H_- = H^{(2)}$ being the complex conjugate of $H_+ = H^{(1)}$. This difference between the expressions for P_+ and P_- occurs because of the difference in the source terms as shown in Eq. (14.4.18).

It is interesting to investigate the asymptotic form of these waves when $k_\pm r \to \infty$. If the cylinder radius is small, $k_\pm a \ll 1$, this asymptotic form is

$$P'_{2\pm} \Rightarrow \mp \rho_0 V_a V_b \frac{\pi}{2}(1 - \Gamma)(k_a k_b)^{\frac{1}{2}} k_\pm a^2 \begin{cases} \cos(k_+ r - \omega_+ t) \\ \sin(k_- r - \omega_- t) \end{cases} \quad (14.4.23)$$

This pressure amplitude is independent of r, and the total energy flux in this wave grows with the distance from the source, just as in the case of plane waves. However, the solution ceases to be valid beyond a distance at which the amplitude is of the same order of magnitude as that of the primary wave pressures.

Finally, we consider a sphere of radius a pulsating with a radial velocity $V_r = V_a \exp(-i\omega_a t) + V_b \exp(-i\omega_b t)$. The corresponding pressure fields P_a and P_b, with frequencies ω_a and ω_b, are

$$P_{a,b} = -\frac{i\rho_0 c_0 k_{a,b} a^2}{1 - ik_{a,b} a} V_{a,b} \exp(ik_{a,b} r - i\omega_{a,b} t)$$

Again, using these expressions for P_a and P_b in Eq. (14.4.18), we find (by direct insertion into the equation) the following function:

$$P_\pm = \pm(1 - \Gamma)\rho_0 V_a V_b \frac{ik_\pm k_a k_b a^4}{(1 - ik_a a)(1 - ik_b a)}$$

$$\times \frac{\exp i(k_\pm r - \omega_\pm t)}{r}\left[\ln\frac{r}{a} - e^{-i\phi(r)}\right] \quad (14.4.24)$$

where $\phi(r) = 2k_\pm r \int_a^r r^{-1} \exp(2ik_\pm r)\, dr$. For a small radiating sphere, $k_\pm a \ll 1$, the asymptotic form of the corresponding pressure field $P'_{2\pm} = \mathrm{Re}\, P_\pm$ for large values of r is

$$P'_{2\pm} \to \pm(1 - \Gamma)\rho_0 V_a V_b (k_\pm k_a k_b a^4)\frac{\ln(k_\pm r)}{2r}\sin(k_\pm r - \omega_\pm t) \quad (14.4.25)$$

As in the cylindrical case, the amplitude of the combination tone does not decrease as rapidly with distance as do the primary-wave components, the radial dependence being of the form $r^{-1}\ln(k_\pm r)$ as compared with r^{-1} for the primary waves. The range of validity of the solution, of course, is subject to the same restriction as in the plane and cylindrical cases, that the second-order pressure field be small compared with the primary pressures.

14.5 SHOCKED SOUND WAVES

As has already been pointed out, the analysis in the preceding section is restricted to comparatively weak nonlinear effects and does not lend itself very well to the study of large distortions encountered, for example, in shock waves. Actually, a detailed analysis of the continuous large-amplitude distortion of a sound wave, and the formation of a shock front, is a problem not as yet fully clarified in all its details. However, a good deal of understanding of the process of wave distortion and the properties of shocks can be obtained from comparatively simple considerations; we shall devote this section to that end.

Wave steepening and shock formation

If we neglect the effects of viscosity and heat conduction, the distortion of a sound wave is a cumulative effect resulting from the fact that the sound wave produces a small perturbation of the medium so that the wave speed varies somewhat from one point to another in the wave. As a result, different portions of the wave will travel with different speeds, and wave distortion results.

To investigate this phenomenon somewhat more closely, we consider a traveling plane wave with a fluid velocity field $u = u[t - (x/c)]$. Associated with this wave, as we have seen in Chap. 6, is a pressure, a density, and a temperature wave. Of particular interest here is the temperature wave, which is

$$\theta = \frac{u}{c_0}(\gamma - 1)T_0 \qquad (14.5.1)$$

where T_0 is the unperturbed temperature of the fluid, and c_0 the corresponding speed of sound. In the plane sound wave these waves are all in phase.

Since the speed of sound is proportional to the square root of the temperature, it follows that at a point in the wave where the excess temperature produced by the sound field is θ, the speed of sound at that point is $c = c_0(1 + \theta/2T_0)$. At the same point (the velocity and temperature waves being in phase), the fluid velocity in the wave is $u = \dfrac{\theta}{T_0}\dfrac{c_0}{\gamma - 1}$. The total speed of propagation of the disturbance at that point is the sum of the local sound speed and the fluid velocity

$$c + u = c_0\left(1 + \frac{\theta}{2T_0}\right) + u = c_0 + \frac{\gamma + 1}{2}u \qquad (14.5.2)$$

Thus, if the plane wave is harmonic with a velocity amplitude u_0, the wave speed in the crest of the wave is $c_0 + \frac{1}{2}u_0(\gamma + 1)$ and in the trough it is $c_0 - \frac{1}{2}u_0(\gamma + 1)$. As a result, the wave will be distorted, the crest

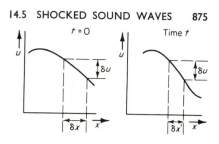

FIGURE 14.15
Increasing steepness of acoustic wavefront of finite size.

eventually will overtake the trough, and the velocity gradient in the wave will be infinite. This corresponds to shock formation.

To determine the distance of wave travel required for shock formation, we consider two points in the fluid with an initial difference in fluid velocity δu and separation δx, as shown in Fig. 14.15. The difference in wave speed between these points, according to Eq. (14.5.2), is $\frac{1}{2}(\gamma + 1)\,\delta u$. Thus, after a time t, the separation δx of the points characterized by the velocity difference δu has changed to

$$\delta x' = \delta x + (\gamma + 1)\frac{\delta u}{2}\,t$$

and the initial velocity gradient $\delta u/\delta x$ has changed to $\delta u/\delta x'$. The separation $\delta x'$ becomes zero, and the velocity gradient infinite, at a time $t = -2[(\gamma + 1)(\partial u/\partial x)]^{-1}$. In other words, the larger the initial velocity gradient, the shorter the time required for shock formation. If, initially, the largest negative gradient in the wave is $(-\partial u/\partial x)_m$, the time of travel to shock formation is

$$t_s = \frac{2}{(\gamma + 1)(-\partial u/\partial x)_m} \qquad (14.5.3)$$

As an example we consider an initially harmonic wave $u = u_0 \sin(kx - \omega t)$. We have $|\partial u/\partial x|_m = ku_0 = 2\pi u_0/\lambda$, and the corresponding distance of travel to the shocked state is

$$c_0 t_s = \frac{1}{\pi(\gamma + 1)}\frac{c}{u_0}\lambda = \frac{\gamma}{\gamma + 1}\frac{1}{\pi}\frac{P_0}{p}\lambda$$

where λ is the wavelength, $P_0 = \rho c^2/\gamma$ is the static pressure, and $p = \rho c u_0$ is the sound pressure. Thus, with a sound-pressure amplitude about one-tenth of the static pressure, a shock front will form after only a few wavelengths of wave travel.

Factors opposing shock formation

This phenomenon of nonlinear wave distortion is often referred to as a result of "convection," which implies that fluid velocity contributes to the signal speed. This is not the only effect; the change in the local temperature, and hence in the local sound speed, also contributes to the distortion.

When an infinite velocity gradient has been reached at a certain point, the properties of the fluid have a discontinuity at this point, not only in the flow velocity, but also in the state variables. Continued influence of the convective wave distortion would lead to a multivalued function, since the wave crest would overtop the wave trough. This, naturally, is not physically possible. Thus the wave appears to be unstable, and on the basis of this discussion, an *ideal compressible* fluid should be unstable for any (one-dimensional) arbitrarily small perturbation, since eventually a discontinuity and a multivalued function would result from it.

The wave-steepening convective effect is not, however, the only one involved. Other competing processes tend to relax the buildup of the gradients in the fluid. The most apparent is the diffusion of velocity and temperature resulting from viscosity and heat conduction. In addition, relaxation of internal degrees of freedom in the fluid resists rapid changes in the temperature, which again tends to limit the gradients.

To estimate the significance of this effect, consider the influence of viscosity and heat conduction on the waveform of a plane wave-pulse. We recall that a harmonic wave is attenuated by the factor $\exp(-\alpha x)$, where $\alpha \simeq \omega^2 l/c^2$, with l of the order of the mean-free-path. An arbitrary wave, which at $x = 0$ has a time dependence $u_0(t)$,

$$u_0(t) = \int_{-\infty}^{+\infty} F(\omega) e^{-i\omega t}\, d\omega$$

then will have the form

$$u(t,x) = \int_{-\infty}^{+\infty} F(\omega) \exp\left(-i\omega\, \frac{t-x}{c}\right) \exp\left(\frac{-\omega^2 l x}{c^2}\right) d\omega$$

at the position x. For simplicity, let us make $u_0(t)$ a delta-function pulse $u_0(t) = \delta(t)$. The spectral density $F(\omega)$ is then simply $1/2\pi$, and we obtain for $u(t,x)$

$$u(t,x) = \frac{1}{2\pi} \int_{-\infty}^{+\infty} \exp\left(-iA\omega\right) \exp\left(-B^2\omega^2\right) d\omega$$

$$= \frac{1}{2\pi} \int_{-\infty}^{+\infty} \exp\left[-B^2\left(\omega + \frac{iA}{2B^2}\right)^2 - \frac{A^2}{4B^2}\right] d\omega$$

Since $\int_{-\infty}^{+\infty} \exp(-ay^2)\, dy = \sqrt{\pi/a}$, we obtain, with $A = t - x/c$, $B = \sqrt{xl/c^2}$,

$$u(t,x) = \frac{c}{\sqrt{2\pi(2lx)}} \exp\left[-\frac{(t-x/c)^2 c^2}{4lx}\right]$$

In other words, the wave is a Gaussian distribution with a standard deviation, $\sqrt{lx}$, which increases as the square root of the travel distance. The fact that the gradient-relaxing effect by diffusion is proportional to $\sqrt{x}$

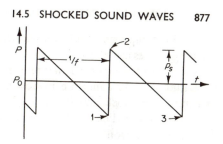

FIGURE 14.16
Pressure fluctuations in a shocked wave.

is significant. The wave distortion produced by the convective effect is proportional to x, as we have seen. Thus, for very small x, the diffusion effects predominate. If we start out with a discontinuity at $t = 0$ at $x = 0$, the convective effects, which tend to make the wave multivalued, are more than checked by the diffusive effects, and the slope of the wavefront initially is decreased. After this initial predominant influence of diffusion, its effect decreases, and the competing convective effects may be able to balance the diffusive effects and thus establish a stable waveform. From these considerations we expect that a sufficiently strong sound wave eventually will reach a *stable* sawtoothlike shape, as portrayed in Fig. 14.16. We shall call such a wave a shocked sound wave, or a repeated shock wave.

Shock relations

We shall now assume that a balance has been struck between the competing effects of nonlinearity and diffusion so that a stable shocked sound wave has been established. In each of the shock fronts in the wave the state of the fluid changes quite rapidly; a fluid element which is carried through the shock experiences a sudden change in velocity, density, pressure, and temperature and, as we shall see later, also in entropy. For the present we shall not be interested in the spatial dependence of this change of state. Instead, we ask what can be said about the relationship between the state of the fluid element before and after the passage through the shock.

The situation is much the same as in the study of collisions in particle dynamics, where, without the knowledge of the interaction, a good deal can be said about the relationship between the state of motion before and after the collision merely by applying conservation laws. Similar results can be obtained for the fluid. To show this, it is convenient to describe the fluid motion with respect to a coordinate system that moves with the shock wave. In this coordinate system the fluid moves into the shock on the front side and leaves it on the other side. If we integrate the equations of mass, momentum, and energy balance over a volume element that includes the shock, we find (replacing the volume integral of a divergence with a surface integral, by means of Green's theorem) that the components normal to the shock front of the mass, momentum, and energy fluxes are the same on both sides of the shock. This is an obvious result, since there are no sources of mass, momentum, or energy in the shock.

In component form the expressions for the mass, momentum, and energy fluxes are $j_i = \rho V_i$, $P\delta_{ij} + \rho V_i V_j$, and $(\rho E + \frac{1}{2}\rho V^2 + P)V_i$, where ρE is the internal energy of the fluid per unit volume. In order to save some writing, we shall indicate by square brackets the difference of a quantity behind the shock and in front of the shock. The conservation of mass, momentum, and energy flux then takes the form, if $\mathbf{n}$ is the direction normal to the shock front,

$$\sum_i [\rho V_i n_i] = 0 \quad \text{or} \quad \sum_i [j_i n_i] = 0$$

$$\sum_j [P\delta_{ij} + \rho V_i V_j]n_j = 0 \quad \text{or} \quad \sum_j [(P\delta_{ij} + j_i V_j)n_j] = 0 \quad (14.5.4)$$

$$\sum_i [(\rho E + \frac{1}{2}\rho V^2 + P)V_i n_i] = 0 \quad \text{or} \quad \sum_i [(H + \frac{1}{2}V^2)j_i n_i] = 0$$

where $H = E + (P/\rho)$ is the enthalpy. These five equations, usually called the Rankine-Hugeniot relations, together with an equation of state, are sufficient to determine the six quantities V_1, V_2, V_3, ρ, and P on one side of the shock in terms of the corresponding quantities on the other side. Although we shall be concerned with normal shock waves, in which the flow velocity is perpendicular to the shock front, Eqs. (14.5.4) apply to more general situations. To write the equations in a more explicit form, we choose the x_1 direction as the direction of propagation so that the components of n_i are $(1,0,0)$. Equations (14.5.4) then take the simpler form

$$[\rho V_1] = 0 \qquad (14.5.5)$$

$$[P + \rho V_1^2] = 0 \qquad [\rho V_1 V_2] = 0 \qquad [\rho V_1 V_3] = 0 \qquad (14.5.6)$$

$$[(H + \frac{1}{2}V^2)\rho V_1] = 0 \qquad (14.5.7)$$

If we have a velocity component V_1 different from zero, perpendicular to the shock, it follows from Eq. (14.5.5) that ρV_1 is continuous across the shock, and from Eqs. (14.5.6) that also the tangential components V_2 and V_3 are continuous. If V_2 and V_3 are zero, we have a *normal* shock, which is of interest here. Otherwise the directions of propagation and fluid motion are different, and we have an *oblique* shock.

Since we shall restrict ourselves here to the normal shock traveling in the x direction, we can drop the coordinate subscripts, and we shall use the subscript 0 to designate the state of the fluid in front of the shock. Then with $j = \rho V = \rho_0 V_0$, we obtain from Eqs. (14.5.6), with $\rho V^2 = j^2/\rho$,

$$j^2 = \frac{P - P_0}{\rho_0^{-1} - \rho^{-1}}$$

The velocity of the fluid in front of the shock, $V_0 = j/\rho_0$, follows directly from this equation, and since our frame of reference moves with the speed of the shock wave, it follows that V_0 is the speed of the shock with respec to the laboratory coordinate system, with respect to which the fluid in front of the shock is at rest.   In other words, the shock speed is

$$U = |V_0| = \sqrt{\frac{P - P_0}{\rho - \rho_0}\frac{\rho}{\rho_0}} = \frac{1}{\rho_0}\sqrt{\frac{P - P_0}{\rho_0^{-1} - \rho^{-1}}} \qquad (14.5.8)$$

For a very weak shock wave this expression reduces to $\sqrt{\partial P/\partial \rho}$, which, if the flow is isentropic, is the same as the speed of sound.

In order to express the shock speed in terms of the pressure ratio (P/P_0), we must obtain the relation between ρ and P.   To obtain this relation, we may first use Eq. (14.5.7) to obtain $H - H_0 = \frac{1}{2}(V_0^2 - V^2)$, which, with $V = j/\rho$, becomes

$$H - H_0 = \frac{j^2}{2}\left(\frac{1}{\rho_0^2} - \frac{1}{\rho^2}\right) = \frac{1}{2}(P - P_0)(\rho_0^{-1} + \rho^{-1}) \qquad (14.5.9)$$

This expression for $H - H_0$ does not contain the flow velocities, and relates the thermodynamic state variables on the two sides of the shock.   If the enthalpy function H is known in terms of the state variables P and ρ, we have the desired relation between ρ and P.   If we specialize to an ideal gas, we have

$$H = E + \frac{P}{\rho} = C_v T + (C_p - C_v)T = C_p T = \frac{\gamma P}{\rho(\gamma - 1)}$$

and Eq. (14.5.9) can be expressed as

$$\frac{\rho}{\rho_0} = \frac{P + P_0[(\gamma - 1)/(\gamma + 1)]}{P_0 + P[(\gamma - 1)/(\gamma + 1)]} \qquad (14.5.10)$$

We now return to Eq. (14.5.8) for the shock speed.   Using Eq. (14.5.10) for the density ratio, we can express the shock speed in terms of a single shock-strength parameter, which we choose to be $\eta = (P - P_0)/P_0$, so that $P/P_0 = 1 + \eta$.   The expression for the shock speed can then be written

$$U_0 = c_0\sqrt{1 + \frac{(\gamma + 1)\eta}{2\gamma}} \qquad (14.5.11)$$

where c_0 is the speed of sound in the unperturbed fluid in front of the shock. It is interesting to express also other properties of the shock in terms of this parameter.   For example, the flow velocity behind the shock, measured with respect to the laboratory coordinate system, is found to be

$$V_L = c_0\frac{\eta}{\sqrt{1 + [(\gamma + 1)\eta/2\gamma]}} \qquad (14.5.12)$$

and the temperature and the corresponding local speed of sound in the shock are obtained from

$$\frac{T}{T_0} = \frac{c^2}{c_0^2} = (1 + \eta)\frac{2\gamma + (\gamma + 1)\eta}{2\gamma + (\gamma - 1)\eta} \qquad (14.5.13)$$

The derivation and further study of these relations are left as one of the problems.

For given values of ρ_0 and P_0, Eq. (14.5.10) defines the relation between ρ and P in the shock. It is interesting to note that the density ratio has a maximum value $(\gamma + 1)/(\gamma - 1)$ which corresponds to $P/P_0 = \infty$. For a monatomic gas this maximum value is 4, and for a diatomic gas it is 6. Clearly, this relation is not isentropic; therefore, as a fluid particle passes through the shock, its entropy will increase. To calculate the entropy increase in the case of an ideal gas, we use the equation of state $S = S(\rho,P)$,

$$S - S_0 = C_v \log \left[\frac{P}{P_0} \left(\frac{\rho_0}{\rho} \right)^{\gamma} \right]$$

and use Eq. (14.5.10) to express ρ/ρ_0 in terms of P/P_0. Then, if we introduce $\eta = (P - P_0)/P_0$ as a measure of the shock strength, we can express $S - S_0$ in terms of η. In a power series expansion of $S - S_0$ in terms of η, the contributions from η and η^2 are found to be absent, and the first nonzero term is of third order in η. Thus, for $\eta < 1$, we obtain

$$S - S_0 \simeq \mathscr{R} \frac{\gamma + 1}{12\gamma^2} \eta^3 \qquad \eta = \frac{P - P_0}{P_0} \qquad (14.5.14)$$

where $\mathscr{R}$ is the gas constant, and $\gamma = C_p/C_v$. Physically, this increase in entropy must correspond to the diffusion processes caused by viscosity and heat conduction in the shock front, where large gradients of velocity and temperature are maintained. If these gradients were known, the heat generation and the corresponding entropy production could have been calculated explicitly. It is interesting to see, however, that even without this detailed knowledge of the structure of the shock, the conditions imposed by the conservation laws are strong enough to determine the entropy production in the shock front. This reminds one of the similar situation in an inelastic one-dimensional collision between two particles in classical mechanics. There the conservation of momentum enables one to determine the amount of kinetic energy transformed into heat without the detailed knowledge of the interaction forces.

Attenuation of a shocked sound wave

We shall now apply the result in Eqs. (14.5.14) to the calculation of the attenuation of a shocked sound wave. The basic idea in this calculation is to determine the rate of entropy production in the fluid per unit volume as a result of the passage of the shocked sound wave. The time dependence of the state of the fluid element is indicated schematically in Fig. 14.16. The particle goes through the shock at point 1, and the state changes to the value corresponding to point 2.

A fluid element passes through a shock front f times a second, where f is the frequency of the sound wave, and each time the entropy is increased by an amount that can be determined from Eqs. (14.5.14). The amplitude parameter η was defined in this equation, and in the present case we have $\eta = (P_2 - P_1)/P_1$. Thus, if the pressure amplitude of the shocked sound wave is p_s, so that $P_2 = P_0 + p_s$ and $P_1 = P_0 - p_s$, the amplitude parameter is $\eta = 2p_s/(P_0 - p_s) \simeq 2p_s/P_0$. The rate of heat production per unit volume in the fluid, $dQ/dt = \rho Tf\,\Delta S$, resulting from the heating of the fluid by the shock fronts, then can be expressed as

$$L_1 = \rho \frac{dQ}{dt} = \rho Tf\,\Delta S = \rho \mathscr{R} T \frac{\gamma + 1}{12\gamma^2}\,\eta^3 f = P_0 \frac{\gamma + 1}{12\gamma^2} \frac{8p_s^3}{P_0^3} f \quad (14.5.15)$$

In the region between successive shocks (between points 2 and 3 in Fig. 14.16) the gradient in the fluid is comparatively small, and to a first approximation the entropy here may be neglected. If we do wish, however, to account for the entropy production also in this region, we may use the general expression developed in Eqs. (6.4.18) and (6.4.21), which, for one-dimensional motion, becomes

$$\rho T \frac{dS}{dt} = \mu' \left(\frac{\partial u}{\partial x}\right)^2 + \frac{K}{T} \left(\frac{\partial T}{\partial x}\right)^2 \quad (14.5.16)$$

where $\mu' = (4\mu/3) + \eta[(6.4.14)]$, and K is the heat conductivity. In the present case the gradient is constant in the regions between successive shocks, as shown in Fig. 14.16, and therefore, if the particle velocity amplitude in the wave is u_0, the velocity gradient is $\partial u/\partial x = 2u_s/\lambda$, where λ is the wavelength. Similarly, the gradient in temperature is $2\theta_s/\lambda$, where θ_s is the temperature amplitude in the wave.

For modestly large amplitudes we may use the plane-wave acoustic relations between the field variables of Eqs. (6.4.23) and express both u_s and θ_s in terms of the pressure amplitude p_s of the shocked sound wave. The energy loss in the region between shocks, from Eq. (14.5.16), then can be written in the form

$$L_2 = \frac{\delta}{\rho c\lambda^2} \frac{p_s^2}{\rho_0 c} \quad (14.5.17)$$

where $\delta = 4[\mu' + (\gamma - 1)K/C_p]$. The total energy loss is $L_1 + L_2$, and in order to determine the attenuation caused by it, we have to calculate the average energy density in the wave. If p is the excess pressure in the linearly varying portion of the wave between the shock fronts, the potential wave energy is $p^2/2\rho_0 c^2$, and with an equal amount of kinetic energy, the total energy density is $p^2/\rho_0 c^2$.

Since the pressure varies linearly with position between the shocks from 0 to a maximum value p_s in a distance of $\lambda/2$, we have, at a position x from the zero point, $p = 2xp_s/\lambda$. The average energy density in the wave is then

$$W = \frac{1}{\lambda/2} \int_0^{\lambda/2} \frac{1}{\rho c^2} \frac{p_s^2}{(\lambda/2)^2} x^2\,dx = \frac{1}{3}\frac{p_s^2}{\rho c^2} \quad (14.5.18)$$

The spatial rate of change of W is related to the loss rate $L_1 + L_2$ through $c\, dW/dx = -(L_1 + L_2)$, and from Eqs. (14.5.16) to (14.5.18), the corresponding equation for p_s becomes

$$\frac{2}{3}\frac{1}{\rho c^2}\frac{dp_s}{dx} = -\frac{2}{3}\frac{\gamma+1}{\gamma^2\lambda}\frac{p_s^2}{P_0^2} - \frac{\delta}{\rho c\lambda^2}\frac{p_s}{\rho_0 c^2}$$

This equation is of the form

$$\frac{dp_s}{(Ap_s + B)p_s} = -dx$$

with the solution

$$\frac{p_s}{p_{s0}}\frac{p_{s0}+(B/A)}{p_s+(B/A)} = \exp\left[-B(x-x_0)\right] \qquad B = \frac{3}{2}\frac{l}{\lambda^2} \qquad (14.5.19)$$

where $l = \mu'/\rho c$ is of the order of the mean-free-path, and

$$A = \frac{\gamma+1}{\gamma\lambda P_0} \qquad \frac{B}{A} = \frac{3}{2}\frac{l}{\lambda^2}\frac{P_0\lambda}{\gamma+1} = \frac{3}{2}\frac{\gamma}{\gamma+1}\frac{l}{\lambda}P_0$$

In this equation p_{s0} is the value of p_s at $x = x_0$.

For very small values of B and B/A, we set $\exp(-Bx) = 1 - Bx$, and Eqs. (14.5.19) reduces to

$$\frac{1}{p_{s0}} - \frac{1}{p_s} = \frac{\gamma+1}{\gamma}\frac{1}{P_0}\frac{(x-x_0)}{\lambda} \qquad (14.5.20)$$

The dependence of p_s on x is not exponential in this limiting case, since here the loss rate is not proportional to the energy density. The pressure amplitude p_s decreases from its initial value p_{s0}, at $x = x_0$, to half this value a distance $(P_0/2p_{s0})[\gamma/(\gamma+1)]$ wavelengths further on. Eventually, p_s is inversely proportional to $x - x_0$.

In the other limit, on the other hand, when $p_{s0} \ll (l/\lambda)P_0$, or correspondingly, $\xi_0 < l$, where ξ_0 is the particle displacement in the wave, the decay is exponential, $p_s = p_{s0}\exp\left[-B(x-x_0)\right]$. However, in this limit, $\xi < l$, as we have seen in the introductory comparison between the relative importance of nonlinearity and diffusive effects, the nonlinear wave steepening is more than checked by diffusion, and the sound wave no longer can maintain its shocked state, but decays into an ordinary sound wave.

Problems

I   A simple oscillator of mass M has a conservative restoring force acting on it, $F(x) = -(2lM/\pi)\omega_0^2 \sin(\pi x/2l)$, where $\omega_0/2\pi$ is the frequency of small-amplitude oscillations. Plot the ratio (ω/ω_0) of natural frequency to small-amplitude frequency, as function of the amplitude of oscillation A, from $A = 0$ to $A = 2l$. What happens when $A = 2l$? Use tables of the elliptic integral F, if you wish.

2 A simple oscillator of mass M is under the influence of a conservative potential energy $V = Ae^{-2x/l} - 2Ae^{-x/l}$. Sketch V as a function of x/l. If the energy of the system is $E = -\gamma A$ $(0 < \gamma < 1)$, show that the displacement of the oscillator is given by the equation

$$\frac{x}{l} = \ln\left[\frac{1}{\gamma} + \frac{\sqrt{1-\gamma}}{\gamma}\sin\left(\sqrt{\frac{2\gamma A}{m}}\frac{t-t_0}{l}\right)\right]$$

(*Hint*: Set $x/l = -\ln y$ in the integral for t.) What is the period T of oscillation? Plot x/l against t/T for one-half cycle, for $\gamma = 0.8$. What is the equation relating x and t when E is a positive quantity? What is then the period of oscillation?

3 If we account for the nonlinear dissipative effects in the neck of a Helmholtz cavity resonator, an approximate equation for free oscillations of such a resonator is of the form

$$m\frac{d^2x}{dt^2} + R\frac{dx}{dt}\left[1 + \beta\left(\frac{dx}{dt}\right)^2\right] + Kx = 0$$

With $Q = \sqrt{Km}/R = 30$, and $\beta = 2 \times 10^{-5}$ (cm per sec)$^{-2}$, determine graphically the phase-space trajectory of free oscillations, using the method described in the text. The initial conditions are $dx/dt = 700$ cm per sec, and $x = 0$. From this construction show that the maximum excursion of the oscillator in this motion is approximately 0.15 cm.

4 A body A of mass m is attracted by a fixed body B by a force which is inversely proportional to the distance between the bodies. Body A is restrained by a spring (force constant K). Thus, if the distance between A and B is x_0 when the spring is relaxed, the force between A and B is of the form $A/(x_0 - x)$, where x is the displacement of A from the relaxed position of the spring. Plot phase-space trajectories for the motion of A for different values of the total energy of motion. For what energies is motion periodic? Show that no periodic motion is possible if $A > Kx_0^2/4$.

5 Solve Eq. (14.1.10) graphically, for the case

$$g_1(v) = \begin{cases} -v & v > 1 \\ -1 & 1 > v > 0 \\ +1 & 0 > v > -1 \\ -v & -1 > v \end{cases}$$

for the phase-plane trajectory starting upward from the initial point $v = 0$, $x_1 = -3$.

6 A string of length L, clamped at both ends, oscillates in its nth harmonic mode, with a displacement amplitude η_0, in the xy plane. Use the procedure indicated in the text to determine the longitudinal motion induced in the string by this transverse motion. What is the spatial perturbation in the mass density and tension? What is the change in resonant frequency produced by the interaction?

7 A transverse traveling wave $\eta = \eta_0 \sin(k_t x - \omega t)$ is generated by a transverse displacement $-\eta_0 \sin \omega t$ at the end $x = 0$. By taking the coupling between transverse and longitudinal motion into account to the third order, show that the wave speed is not the square root of T_0/ϵ_0, but equals the square root of $(T_0/\epsilon_0)\{1 + \frac{1}{4}[(AQ/T_0) - 1](k_t\eta_0)^2\}$.

8   The $g_1(v)$ of Eq. (14.1.10) for an idealized force having both activation and dissipation regions is

$$g_1(v) = \begin{cases} -\sqrt{3} & v > 1 \\ +\sqrt{3} & 1 > v > 0 \\ -\sqrt{3} & 0 > v > -1 \\ +\sqrt{3} & -1 > v \end{cases}$$

By piecing the phase-plane trajectory out of parts of circles, show that the steady-state self-sustaining oscillations have displacement amplitude $(x_1)_{max} = 2\sqrt{2}$. What is the velocity amplitude?

9   Suppose the frictional force $F(V - v)$ of the discussion of Fig. 14.4 is

$$F(u) = \begin{cases} \Delta - R'u & u > 0 \\ -\Delta - R'u & u < 0 \end{cases} \qquad u = V - v$$

Use Eqs. (14.1.19) to show that the velocity amplitude of steady-state oscillations of the mass is

$$A \simeq V\left[1 + \frac{1}{2}\left(\frac{\pi R'V}{4\Delta}\right)^2\right] \qquad \text{if } \frac{\pi R'V}{4\Delta} \ll 1$$

What is the frequency of oscillation, to first order, in $\pi R'V/4\Delta$?

10   Work out the solution of Eq. (14.2.2), to first approximation in ϵ, including both the third harmonic and the $\frac{1}{3}$ subharmonic; i.e., set

$$\xi = \xi_{\frac{1}{3}} \cos \frac{\gamma\theta}{3} + \xi_1 \cos (\gamma\theta) + \xi_3 \cos (3\gamma\theta)$$

and solve for all three coefficients to first approximation.

11   In the Lagrangian description of fluid motion as discussed in Chap. 6, $\phi_L(t,x)$ designates the value of the field variable ψ at time t for a fluid element which, at $t = 0$, was at the position x. If in time t the displacement of this fluid element from this position is ξ, the position of the element at time t is $x + \xi$. The value of ϕ at this point can be expressed in the Eulerian description as $\phi(x + \xi, t)$, and it follows that the relation between the Lagrangian and the Eulerian description can be expressed as

$$\phi_L(t,x) = \phi(x + \xi, t)$$

Suppose φ is a perturbation of the field variable ϕ so that $\phi = \phi_0 + \varphi$. Expand ϕ in a Taylor series, and show that, to second order in the field variable φ, the relation between the Lagrangian and Eulerian values ϕ_L and ϕ is

$$\phi_L = \phi_0 + \sum_i (\xi_{1i} + \xi_{2i}) \frac{\partial\phi_0}{\partial x_i} + \sum_i \xi_{1i} \frac{\partial\varphi_1}{\partial x_i} + \sum_{ij} \frac{\xi_{1i}\xi_{1j}}{2} \frac{\partial^2\phi_0}{\partial x_i \partial x_j} + \cdots$$

As in the text, ξ_1 and φ_1 represent the solutions to the linearized equations of motion, whereas ξ_2, φ_2 are of second order in ξ_1 and φ_1.

As a specific example, let φ_1 be the velocity component in a plane sound wave $\varphi_1 = u_0 \sin (\omega t - kx)$. From the result above show that

$$\langle u \rangle - \langle u_L \rangle = -\frac{u_0^2}{2c}$$

consistent with the result obtained in Chap. 6.

12 Consider the equations for mass and momentum conservation for isentropic motion of an ideal gas. Introduce the local speed of sound c as a dependent variable, and show that these equations can be written

(a)
$$\frac{1}{\gamma - 1} \frac{\partial c}{\partial t} + \frac{V}{\gamma - 1} \frac{\partial c}{\partial x} + \frac{c}{2} \frac{\partial V}{\partial x} = 0$$

(b)
$$\frac{1}{2} \frac{\partial V}{\partial t} + \frac{V}{2} \frac{\partial V}{\partial x} + \frac{c}{\gamma - 1} \frac{\partial c}{\partial x} = 0$$

Next introduce the "characteristic" variables $R_+ = (V/2) + [c/(\gamma - 1)]$ and $R_- = -(V/2) + [c/(\gamma - 1)]$, and show that the equations for R_+ and R_- are

$$\frac{\partial R_+}{\partial t} + (V + c) \frac{\partial R_+}{\partial x} = 0$$

$$\frac{\partial R_-}{\partial t} + (V - c) \frac{\partial R_-}{\partial x} = 0$$

(R_+ and R_- are the Riemann invariants.) From $dR_\pm = (\partial R_\pm / \partial x) dx + (\partial R_\pm / \partial t) dt$, show that $R_\pm$ is constant along the "characteristics" in the xt plane, defined by the differential equation

$$\frac{dx}{dt} = V \pm c$$

13 Show that the speed of a shock wave has a value that lies between the local sound speeds c_0 and c_1 in the unperturbed fluid in front of the shock and in the fluid behind the shock. In particular, for weak shock, show that the shock speed is $(c_0 + c_1)/2$.

14 With reference to Eq. (14.5.10), sometimes called the "shock adiabatic," make a rough graph of P as a function of $1/\rho$, starting the graph from the unperturbed state ρ_0, P_0 to the point ρP, representing the state of the fluid in the shock. Draw a straight line between these points, and show that the slope of this line is a direct measure of the speed of the shock wave. Also, show that the ordinary adiabatic curve through $\rho_0 P_0$ is tangent to the shock adiabatic and that the slope of this curve represents the speed of sound in the unperturbed fluid.

15 With reference to the calculation of the distance to shock formation in a plane wave in the text, show that the corresponding distance in a spherical wave, which at $x = x_0$ has a maximum "slope" $\delta u_0 / \delta x_0$, is given by

$$\log \frac{x_s}{x_0} = \frac{2}{\gamma + 1} \frac{c}{x_0} \frac{1}{\partial u_0 / \partial x_0}$$

Also in analogy with the calculation carried out for the plane wave, determine the attenuation of the spherical shocked sound wave.

16 In the text, both in the linear and nonlinear analyses of sound waves, we have for the most part used the Eulerian description of motion. In the study of one-dimensional nonlinear wave motion, however, it is sometimes convenient to use the Lagrangian description. Show that under isentropic conditions the equations for mass and momentum balance, the equation of state, and the resulting "wave equation" are

$$\rho \left(1 + \frac{\partial \xi}{\partial x}\right) = \rho_0 \qquad \rho_0 \frac{\partial^2 \xi}{\partial t^2} = -\frac{\partial p}{\partial x} \qquad p \left(1 + \frac{\partial \xi}{\partial x}\right)^\gamma = P_0$$

$$c_0^2 \frac{\partial^2 \xi}{\partial x^2} = \left(1 + \frac{\partial \xi}{\partial x}\right)^{\gamma + 1} \frac{\partial^2 \xi}{\partial t^2}$$

BIBLIOGRAPHY

The following list is not presented as an exhaustive summary all of publications on theoretical acoustics; it is simply a list of the books and articles which the authors have found particularly useful as collateral reading or as references where certain problems are discussed in more detail than can be given in this volume.

BERGMANN, L.: "Der Ultraschall," 6th ed., S. Hirzel Verlag KG, Stuttgart, 1954.

BLOKHINTZEV, D.: "The Acoustics of an Inhomogeneous Moving Medium" (transl. by R. T. Beyer and D. Mintzer), Brown University, Providence, R.I., 1952.

BORGNIS, F. E.: Acoustic Radiation Pressure of Plane Compressional Waves, *Rev. Mod. Phys.*, **25**, 653 (1953).

BRILLOUIN, L.: "Wave Propagation in Periodic Structures," McGraw-Hill Book Company, New York, 1946.

CHERNOV, L. A.: "Wave Propagation in a Random Medium" (transl. from the Russian by R. A. Silverman), McGraw-Hill Book Company, New York, 1960.

CRANDALL, S. H. (ed.): "Random Vibration," vols. 1 and 2, M.I.T. Press, Cambridge, Mass., 1959–1963.

CREMER, L.: "Die wissenschaftlichen Grundlagen der Raumakustik," vols. 1–3, S. Hirzel Verlag KG, Stuttgart, 1948–1961.

HERZFELD, K. F., and T. A. LITOVITZ: "Absorption and Dispersion of Ultrasonic Waves," Academic Press Inc., New York, 1959.

KAUDERER, H.: "Nichtlineare Mechanik," Springer–Verlag OHG, Berlin, 1958.

LAMB, H.: "Dynamical Theory of Sound," 2d ed., Edward Arnold (Publishers) Ltd., London, 1931.

———: "Hydrodynamics," 6th ed., Dover Publications, Inc., New York, 1945.

LANDAU, L. D., and E. M. LIFSHITZ: "Fluid Mechanics "(transl. by J. B. Sykes and W. H. Reid), Addison–Wesley Publishing Company, Inc., Reading, Mass., 1959.

LIGHTHILL, M. J.: Viscosity Effects in Sound Waves of Finite Amplitude, in G. K. Batchelor and R. M. Davies (eds.), "Surveys in Mechanics," Cambridge University Press, New York, 1956.

LINDSAY, R. B.: "Mechanical Radiation," McGraw-Hill Book Company, New York, 1960.

MARKHAM, J. J., R. T. BEYER, and R. B. LINDSAY: Absorption of Sound in Fluids, *Rev. Mod. Phys.*, **23**, 353 (1951).

MASON, W. P. (ed.): "Physical Acoustics, Principles and Methods," vols. 1–3, Academic Press Inc., New York, 1964-1965.

MINORSKY, N.: "Nonlinear Oscillations," D. Van Nostrand Company, Inc., Princeton, N.J., 1962.

MORSE, P. M.: "Vibration and Sound," 2d ed., McGraw-Hill Book Company, New York, 1948.

——— and R. H. BOLT: Sound Waves in Rooms, *Rev. Mod. Phys.*, **16**, 69 (1944).

——— and H. FESHBACH: "Methods of Theoretical Physics," McGraw-Hill Book Company, New York, 1953.

——— and K. U. INGARD: Linear Acoustic Theory, vol. 11/1 of S. Flügge (ed.), "Handbuch der Physik," Springer–Verlag OHG, Berlin, 1961.

RAYLEIGH, LORD: "Theory of Sound," 2d ed., Dover Publications, Inc., New York, 1945.

RICHARDSON, E. G.: "Technical Aspects of Sound," vols. 1 and 2, and E. G. RICHARDSON and E. MEYER, "Technical Aspects of Sound," vol. 3, Elsevier Publishing Company, Amsterdam, 1953–1962.

SCHOCH, A.: Schallreflexion, Schallbrechung und Schallbeugung, *Ergeb. Exakt. Naturw.* **23**: 127–234 (1950).

SKUDRZYK, E.: "Die Grundlagen der Akustik," Springer–Verlag OHG, Berlin, 1954.

SOMMERFELD, A.: "Mechanics of Deformable Bodies" (Lectures on Theoretical Physics, vol. 2, transl. by G. Kuerti), Academic Press Inc., New York, 1950.

STEPHENS, R. W. B., and A. E. BATE: "Acoustics and Vibrational Physics," 2d ed., Edward Arnold and Co., London, 1966.

STEWART, G. W., and R. B. LINDSAY: "Acoustics," D. Van Nostrand Company, Inc., Princeton, N.J., 1930.

STOKER, J. J.: "Nonlinear Vibrations," Interscience Publishers, Inc., New York, 1950.

TATARSKII, V. I.: "Wave Propagation in a Turbulent Medium" (transl. from the Russian by R. A. Silverman), McGraw-Hill Book Company, New York, 1961.

WOOD, A.: "Acoustics," Interscience Publishers, Inc., New York, 1941.

GLOSSARY OF SYMBOLS

In general, italic, roman, and Greek symbols are used for scalar quantities (real or complex), boldface roman and Greek for vectors, and German capitals for vector operators (dyadics, tensors). The symbol $|\ |$, in general, means "the magnitude of"; $|\mathbf{A}| = A$ is equal to the square root of the sum of the squares of the components of the vector $\mathbf{A}$; $|f|$ is the square root of the sum of the squares of the real and imaginary parts of the complex number $f = u + iv$. The asterisk indicates the complex conjugate, $f^* = u - iv$. The symbol $\simeq$ means "approximately equal to"; the symbol $\rightarrow$ means "approaches, in the limit"; the symbol $\equiv$ means "equals, by definition." For a property f of a fluid, the expression df/dt represents the time rate of change of f, moving with the fluid; the expression $\partial f/\partial t$ represents the time rate of f at a fixed point in space [Eq. (6.1.10)].

The list of symbols given here includes only those used in more than a single section. The commonly used meaning or meanings are given, followed by the equation or page number where the symbol is defined (if a definition is not obvious).

a Radius of cylinder or disk or sphere

A Amplitude of vibration

b_e Electron mobility (12.1.1)

$B_m = |dh_m(\zeta)/d\zeta|$ Radiation amplitude for a sphere (7.2.15), Table VIII

$\mathbf{B}$ Magnetic induction (12.3.1)

c Wave velocity (4.1.3), (6.1.8)

c_φ Phase velocity (9.1.17)

c_g Group velocity (9.1.18)

$c(\infty)$ Signal velocity (see page 479)

$\cos z$ Cosine function (1.2.3), Tables I and II

$\cosh z$ Hyperbolic cosine (1.2.10), Tables I and II

C Amplitude (1.2.8)

$C = 0.8905$ (7.3.2), (10.1.6)

C_a Analogous capacitance (9.1.11), (9.1.21)

C_p, C_v Heat capacities at constant pressure or volume (6.1.3), (6.4.3)

C_r Reflection coefficient (6.3.5), (6.3.14)

d_v, d_h Thermal and viscous boundary thickness (6.4.31)

D Viscous loss rate (6.4.10)

D_i Components of **D** (7.1.7)

D_{ij} Components of $\mathfrak{D}$ (6.4.8)

D   Dipole strength (7.1.7), (7.1.13)

$\mathfrak{D}$ Shear-stress tensor (6.4.8)

e Electronic charge (12.1.2)

$e = 2.71828 \cdots$ Base of natural logarithms

E Energy of system (1.3.2), internal energy (see page 296)

$E_m = |dH_m(u)/du|$ Radiation amplitude for a cylinder (7.3.7), Table V

E   Electric intensity (12.1.1)

f General function, sometimes a force

$f(x,t)$ Force density (4.4.3)

F or **F**   Mechanical force (1.1.1), (6.2.11)

$F(\omega)$ Fourier transform of $f(t)$ (1.3.16)

$F_L(s)$ Laplace transform of $f(t)$ (1.3.23), Table 2.1

F_0 Magnitude of F

$g(x \mid x_0)$ Green's function for infinite string or plate (4.4.11), (5.3.17)

$g_\omega(\mathbf{r} \mid \mathbf{r}_0) = e^{ikR}/4\pi R$ Green's function for infinite space (7.2.31), (7.3.14)

G General Green's function (4.4.7), (7.1.15)

G_ω Green's function for frequency $\omega/2\pi$

h Half-thickness of a plate (5.3.1), (10.1.21)

$h_n(x)$ Spherical Hankel function (7.2.9), Table VII

H Hamiltonian function (1.3.2), (4.1.10)

$H_m^{(1)}(z)$ Cylindrical Hankel function (10.1.37), Table III

i An integer

$i = \sqrt{-1} = -j$ (see page 12)

$I_m(x)$ Hyperbolic Bessel function (5.3.3), Table IV

$\operatorname{Im} f$ Imaginary part of f

I   Intensity (6.2.15)

$\mathscr{I}$ Identity tensor, $(\mathscr{I})_{ij} = \delta_{ij}$

j An integer

$j = -i$ (see page 12)

j   Current density (12.3.1)

$J_m(x)$ Bessel function (1.2.5), (5.2.20), Table III

$j_n(z)$ Spherical Bessel function (7.2.12), Table VII

J   Flux vector (6.1.15)

$\mathfrak{J}$ Momentum flux tensor (6.2.10)
k Damping constant (2.2.1), (3.2.8)
$k = \omega/c = 2\pi/\lambda$ Wavenumber (4.3.4), (7.1.5)
$\mathbf{k}$ Wavenumber vector (8.1.14)
K Stiffness constant of spring (2.1.1), (3.1.1)
K Thermal conductivity (6.4.1)
$\mathbf{K}$ Wavenumber vector in spatial Fourier transform (7.1.31), (8.1.15)
l Length of string (4.3.9)
$l_h = k/\rho c C_p$ Characteristic thermal length (6.4.20)
$l_v = \mu/\rho c$ Characteristic viscous length (6.4.20)
$\ln z$ Natural logarithm of z
L Lagrange density (4.1.8), (6.2.13)
L_a Analogous inductance (9.1.10), (9.1.22)
L_{bh}, L_{bv} Power loss at surface (6.4.37)
m An integer
m Mass (1.1.1)
M mass
M Molecular weight (6.1.4)
$M = V/c$ Mach number (11.1.1)
$\mathbf{M}$ Wave momentum (6.2.16)
n An integer
$n_m(z)$ Spherical Neumann function (7.2.12), Table VII
N Number of particles per unit volume (12.1.2)
$N_m(x)$ Neumann function (5.3.15), (7.3.2), Table III
p Acoustic pressure (6.1.2)
P Acoustic-pressure amplitude
P Equilibrium pressure (6.1.2)
$P_n(\eta)$ Legendre function (7.2.4), Table VI
$\mathfrak{P}$ Stress tensor (6.2.10), (6.4.9)
q Rate of production of fluid (7.1.19)
$q = \mu - i\nu$ Standing-wave constant for rectangular duct (9.2.14), Plate IV
$q = \sqrt{k^2 - K^2}$ (10.1.11), (10.1.23)
Q Young's modulus (5.1.1)
$Q = \omega_0 m/R$ The Q of the system (2.2.4)
Q_{ij} Components of $\mathfrak{Q}$ (7.1.9)
$\mathfrak{Q}$ Quadrupole strength tensor (7.1.10)
r Radial distance from origin (7.1.1); occasionally, radial distance in cylindrical coordinates, when there is no danger of confusion with spherical coordinates
$\mathbf{r}$ Radius vector from origin
R Distance between source and observer (7.1.6)
R Mechanical resistance (2.2.1), (2.3.2)
R Reynolds number (11.3.28), (12.3.11)
R_a Analogous resistance (9.1.12), (9.1.23)

Re f Real part of f

$\mathbf{R} = \mathbf{r} - \mathbf{r}_0$ Radius vector from source to observer (7.1.6)

s Monopole source-strength density (7.1.22)

s Poisson's ratio (5.3.1)

$\sin z$ Sine function (1.2.3), Tables I and II

$\sinh z$ Hyperbolic sine (1.2.10), Tables I and II

S Cross-sectional area (5.1.1), (9.1.10)

S Equilibrium value of entropy (6.4.12)

S Strength of monopole source (7.1.4)

t Time

T Equilibrium temperature (6.1.3)

T Tension (4.1.3), (5.2.1)

T_p Period of oscillation (1.3.4), (14.1.8)

$\mathfrak{T} = \mathfrak{J} + \mathfrak{D}$ (6.2.11), (7.1.20)

u Acoustic velocity (6.1.6)

$u(x)$ Unit step function (1.3.22)

U Kinetic energy (4.3.3)

U Velocity amplitude

v Velocity

dv Volume element

V Potential energy (1.1.4)

V Volume

$V(u)$ Variational expression (4.5.22)

$\mathbf{V}$ Flow velocity

w Acoustic energy density (6.2.15), (9.5.1)

w Radial distance from cylindrical axis, when there is likelihood of confusion with spherical radius r (7.1.1)

$w = v + i\mu$ Standing-wave constant for cylindrical duct (9.2.25), Plate V

x, y, z Rectangular coordinates

X_m Mechanical reactance (2.3.2)

y Displacement of string (4.1.1)

$Y_{mn}{}^\sigma(\vartheta,\varphi)$ Spherical harmonic (7.2.7)

$z = p/u = \rho c\zeta = \rho c/\beta$ Acoustic impedance (6.3.4)

$Z_m = F/u$ Mechanical impedance (2.3.2), (4.4.17)

Z_a Analogous impedance (9.1.10), (9.1.29)

Z_s Analogous impedance of simple source (7.1.11), (9.2.12)

$\alpha = c/c_s$ (10.1.5); $= c/c_p$ (10.2.1)

α_m, β_m Characteristic numbers for Bessel functions (5.2.23), (9.2.23), Table X

β Coefficient of thermal expansion (6.1.1)

$\beta = \rho c/z = \xi - i\sigma$ Specific acoustic admittance (6.3.4), (7.4.9)

γ Wavenumber parameter for plate (5.3.9), (10.1.21)

γ Ratio of specific heats (6.1.3)

γ_m Phase shift for cylindrical scattering (7.3.7), Table V

δ Acoustic density change (6.1.2)

$\delta(x)$ Dirac delta function (1.3.24)

δ_m Phase shift for spherical scattering (7.2.15), Table VIII

$\delta_{mn} = 1$ when $m = n$; $= 0$ when $m \neq n$;   Kronecker delta symbol (1.3.9)

∂ Partial-derivative symbol

ϵ Dielectric constant (13.1.1)

ϵ Linear density (3.3.2), (4.1.3)

$\epsilon_m = 1$ when $m = 0$; $= 2$ when $m \neq 0$;   Neumann symbol (7.2.7)

$\zeta = z/\rho c = \theta - i\chi$ Specific acoustic impedance (6.3.4)

η Coefficient of bulk viscosity (6.4.8), (6.4.47)

η Displacement of membrane or plate (5.2.1), (10.1.21)

ϑ Angle of incidence (6.3.5)

ϑ Azimuthal angle in spherical coordinates (7.1.1)

θ Specific acoustic resistance (6.3.4)

κ Compressibility, usually adiabatic, κ_s (6.1.3)

κ_T Isothermal compressibility (6.1.1)

$\varkappa$ Eigenvalue for transverse mode in duct (9.2.6)

$\lambda = 2\pi/k = 2\pi c/\omega$ Wavelength (4.3.4)

Λ Wavelength of light (13.1.11)

Λ_n Mean-square amplitude of cross-duct mode (9.2.8)

μ Coefficient of viscosity (6.4.4)

μ Coupling constant (3.1.2)

μ_n Wave-distribution parameter (9.2.19), Plates IV and V

ν Kinematic viscosity (page 750), (12.1.16)

$\nu = \omega/2\pi$ Frequency (page 6), (2.1.3)

ν_n Wave-distribution parameter (9.2.9), Plates IV and V

ξ Specific acoustic conductance (6.3.4)

$\pi = 3.14159 \cdots$ Ratio of circumference to diameter of circle

Π Total radiated power (7.1.5)

ρ Mean density of medium (6.1.2)

ρ Volume density (5.1.5)

σ Acoustic entropy change (6.4.18)

σ Areal density (5.2.1)

σ Specific acoustic susceptance (6.3.4)

$\sum$ Summation symbol (page 4)

Σ_s, Σ_a Absorption and scattering cross sections (8.2.16), (8.2.17)

τ Acoustic temperature change (6.1.3)

τ Time shift in autocorrelation function (1.3.13)

Υ Autocorrelation function (1.3.13), (7.1.32)

Υ Power transmitted (4.5.9)

φ Axial angle in spherical coordinates (7.1.1)

ϕ Axial angle for cylindrical coordinates (7.1.1)

ϕ_n Characteristic function for cross-duct mode (9.3.24)

Φ Flow resistance (6.2.23)

Φ Phase angle (1.2.8), (9.1.3)

$\Phi_s(\vartheta)$ Angle-distribution factor (8.1.14), (8.2.15)

χ Specific acoustic reactance (6.3.4)

χ Variational trial function (4.5.25), (9.3.13)

ψ_n Characteristic function (5.2.16)

Ψ Velocity potential (6.2.9)

Ψ_n Characteristic function for cross-duct mode (9.2.8)

$\omega = 2\pi\nu$ Angular velocity (1.2.1), (page 6)

Ω Porosity (6.2.22)

$\Omega = 2\pi$ times frequency of light (13.1.4)

$\nabla^2 = \text{div grad}$ Laplacian operator (5.2.3), (6.2.8), (7.1.3)

$\nabla \equiv \text{grad}$ Gradient operator (1.1.5), (7.1.1)

$\nabla \cdot \equiv \text{div}$ Divergence operator (6.1.15), (7.1.2)

$\nabla \times \equiv \text{curl}$ Curl operator (6.4.6)

$\mathbf{u} \cdot \nabla \equiv \mathbf{u} \cdot \text{grad}$ Directional derivative (6.1.14)

$\nabla \cdot \mathfrak{D} = \sum_i \partial D_{ij}/\partial x_i; \ \mathfrak{D} \cdot \nabla = \sum_j \partial D_{ij}/\partial x_j$

TABLE I

Trigonometric and hyperbolic functions [see Eqs. (1.2.3) and (1.2.10)]

x	$\sin(x)$	$\cos(x)$	$\tan(x)$	$\sinh(x)$	$\cosh(x)$	$\tanh(x)$	e^x	e^{-x}
0.0	0.0000	1.0000	0.0000	0.0000	1.0000	0.0000	1.0000	1.0000
0.2	0.1987	0.9801	0.2127	0.2013	1.0201	0.1974	1.2214	0.8187
0.4	0.3894	0.9211	0.4228	0.4018	1.0811	0.3799	1.4918	0.6703
0.6	0.5646	0.8253	0.6841	0.6367	1.1855	0.5370	1.8221	0.5488
0.8	0.7174	0.6967	1.0296	0.8881	1.3374	0.6640	2.2255	0.4493
1.0	0.8415	0.5403	1.5574	1.1752	1.5431	0.7616	2.7183	0.3679
1.2	0.9320	0.3624	2.5722	1.5095	1.8106	0.8337	3.3201	0.3012
1.4	0.9854	+0.1700	+5.7979	1.9043	2.1509	0.8854	4.0552	0.2466
1.6	0.9996	−0.0292	−34.233	2.3756	2.5775	0.9217	4.9530	0.2019
1.8	0.9738	−0.2272	−4.2863	2.9422	3.1075	0.9468	6.0496	0.1553
2.0	0.9093	−0.4161	−2.1850	3.6269	3.7622	0.9640	7.3891	0.1353
2.2	0.8085	−0.5885	−1.3738	4.4571	4.5679	0.9757	9.0250	0.1108
2.4	0.6755	−0.7374	−0.9160	5.4662	5.5569	0.9837	11.023	0.0907
2.6	0.5155	−0.8569	−0.6016	6.6947	6.7690	0.9890	13.464	0.0742
2.8	0.3350	−0.9422	−0.3555	8.1919	8.2527	0.9926	16.445	0.0608
3.0	+0.1411	−0.9900	−0.1425	10.018	10.068	0.9951	20.086	0.0498
3.2	−0.0584	−0.9983	+0.0585	12.246	12.287	0.9967	24.533	0.0407
3.4	−0.2555	−0.9668	0.2643	14.965	14.999	0.9978	29.964	0.0333
3.6	−0.4425	−0.8968	0.4935	18.285	18.313	0.9985	36.598	0.0273
3.8	−0.6119	−0.7910	0.7736	22.339	22.362	0.9990	44.701	0.0223
4.0	−0.7568	−0.6536	1.1578	27.290	27.308	0.9993	54.598	0.0183
4.2	−0.8716	−0.4903	1.7778	33.335	33.351	0.9996	66.686	0.0150
4.4	−0.9516	−0.3073	3.0963	40.719	40.732	0.9997	81.451	0.0123
4.6	−0.9937	−0.1122	+8.8602	49.737	49.747	0.9998	99.484	0.0100
4.8	−0.9962	+0.0875	−11.385	60.751	60.759	0.9999	121.51	0.0082
5.0	−0.9589	0.2837	−3.3805	74.203	74.210	0.9999	148.41	0.0067
5.2	−0.8835	0.4685	−1.8856	90.633	90.639	0.9999	181.27	0.0055
5.4	−0.7728	0.6347	−1.2175	110.70	110.71	1.0000	221.41	0.0045
5.6	−0.6313	0.7756	−0.8139	135.21	135.22	1.0000	270.43	0.0037
5.8	−0.4646	0.8855	−0.5247	165.15	165.15	1.0000	330.30	0.0030
6.0	−0.2794	0.9602	−0.2910	201.71	201.71	1.0000	403.43	0.0025
6.2	−0.0831	0.9965	−0.0834	246.37	246.37	1.0000	492.75	0.0020
6.4	+0.1165	0.9932	+0.1173	300.92	300.92	1.0000	601.85	0.0016
6.6	0.3115	0.9502	0.3279	367.55	367.55	1.0000	735.10	0.0013
6.8	0.4941	0.8694	0.5683	448.92	448.92	1.0000	897.85	0.0011
7.0	0.6570	0.7539	0.8714	548.32	548.32	1.0000	1096.6	0.0009
7.2	0.7937	0.6084	1.3046	669.72	669.72	1.0000	1339.4	0.0007
7.4	0.8987	0.4385	2.0493	817.99	817.99	1.0000	1636.0	0.0006
7.6	0.9679	0.2513	3.8523	999.10	999.10	1.0000	1998.2	0.0005
7.8	0.9985	+0.0540	+18.507	1220.3	1220.3	1.0000	2440.6	0.0004
8.0	0.9894	−0.1455	−6.7997	1490.5	1490.5	1.0000	2981.0	0.0003

TABLE II
Trigonometric and hyperbolic functions [see Eqs. (1.2.3) and (1.2.10)]

x	$\sin(\pi x)$	$\cos(\pi x)$	$\tan(\pi x)$	$\sinh(\pi x)$	$\cosh(\pi x)$	$\tanh(\pi x)$	$e^{\pi x}$	$e^{-\pi x}$
0.00	0.0000	1.0000	0.0000	0.0000	1.0000	0.0000	1.0000	1.0000
0.05	0.1564	0.9877	0.1584	0.1577	1.0124	0.1558	1.1701	0.8546
0.10	0.3090	0.9511	0.3249	0.3194	1.0498	0.3042	1.3691	0.7304
0.15	0.4540	0.8910	0.5095	0.4889	1.1131	0.4392	1.6019	0.6242
0.20	0.5878	0.8090	0.7265	0.6705	1.2040	0.5569	1.8745	0.5335
0.25	0.7071	0.7071	1.0000	0.8687	1.3246	0.6558	2.1933	0.4559
0.30	0.8090	0.5878	1.3764	1.0883	1.4780	0.7363	2.5663	0.3897
0.35	0.8910	0.4540	1.9626	1.3349	1.6679	0.8003	3.0028	0.3330
0.40	0.9511	0.3090	3.0777	1.6145	1.8991	0.8502	3.5136	0.2846
0.45	0.9877	+0.1564	+6.3137	1.9340	2.1772	0.8883	4.1111	0.2432
0.50	1.0000	0.0000	∞	2.3013	2.5092	0.9171	4.8105	0.2079
0.55	0.9877	−0.1564	−6.3137	2.7255	2.9032	0.9388	5.6287	0.1777
0.60	0.9511	−0.3090	−3.0777	3.2171	3.3689	0.9549	6.5861	0.1518
0.65	0.8910	−0.4540	−1.9626	3.7883	3.9180	0.9669	7.7062	0.1298
0.70	0.8090	−0.5878	−1.3764	4.4531	4.5640	0.9757	9.0170	0.1109
0.75	0.7071	−0.7071	−1.0000	5.2280	5.3228	0.9822	10.551	0.09478
0.80	0.5878	−0.8090	−0.7265	6.1321	6.2131	0.9870	12.345	0.08100
0.85	0.4540	−0.8910	−0.5095	7.1879	7.2572	0.9905	14.437	0.06922
0.90	0.3090	−0.9511	−0.3249	8.4214	8.4806	0.9930	16.902	0.05916
0.95	+0.1564	−0.9877	−0.1584	9.8632	9.9137	0.9949	19.777	0.05056
1.00	0.0000	−1.0000	0.0000	11.549	11.592	0.9962	23.141	0.04321
1.05	−0.1564	−0.9877	+0.1584	13.520	13.557	0.9973	27.077	0.03693
1.10	−0.3090	−0.9511	0.3249	15.825	15.857	0.9980	31.682	0.03156
1.15	−0.4540	−0.8910	0.5095	18.522	18.549	0.9985	37.070	0.02697
1.20	−0.5878	−0.8090	0.7265	21.677	21.700	0.9989	43.376	0.02305
1.25	−0.7071	−0.7071	1.0000	25.367	25.387	0.9992	50.753	0.01970
1.30	−0.8090	−0.5878	1.3764	29.685	29.702	0.9994	59.387	0.01683
1.35	−0.8910	−0.4540	1.9626	34.737	34.751	0.9996	69.484	0.01438
1.40	−0.9511	−0.3090	3.0777	40.647	40.660	0.9997	81.307	0.01230
1.45	−0.9877	−0.1564	+6.3137	47.563	47.573	0.9998	95.137	0.01051
1.50	−1.0000	0.0000	∞	55.654	55.663	0.9998	111.32	0.00898
1.55	−0.9877	+0.1564	−6.3137	65.122	65.130	0.9999	130.25	0.00767
1.60	−0.9511	0.3090	−3.0777	76.200	76.206	0.9999	152.41	0.00656
1.65	−0.8910	0.4540	−1.9626	89.161	89.167	0.9999	178.33	0.00561
1.70	−0.8090	0.5878	−1.3764	104.32	104.33	1.0000	208.66	0.00479
1.75	−0.7071	0.7071	−1.0000	122.07	122.08	1.0000	244.15	0.00409
1.80	−0.5878	0.8090	−0.7265	142.84	142.84	1.0000	285.68	0.00350
1.85	−0.4540	0.8910	−0.5095	167.13	167.13	1.0000	334.27	0.00299
1.90	−0.3090	0.9511	−0.3249	195.56	195.56	1.0000	391.12	0.00256
1.95	−0.1564	0.9877	−0.1584	228.82	228.82	1.0000	457.65	0.00219
2.00	0.0000	1.0000	0.0000	267.75	267.75	1.0000	535.49	0.00187

TABLE III
Bessel functions for cylindrical coordinates
$J_n(x)$ and $N_n(x)$
[see Eqs. (1.2.5), (5.2.20), (5.3.15), and (7.3.2)]

x	$J_0(x)$	$N_0(x)$	$J_1(x)$	$N_1(x)$	$J_2(x)$	$N_2(x)$
0.0	1.0000	$-\infty$	0.0000	$-\infty$	0.0000	$-\infty$
0.1	0.9975	-1.5342	0.0499	-6.4590	0.0012	-127.64
0.2	0.9900	-1.0811	0.0995	-3.3238	0.0050	-32.157
0.4	0.9604	-0.6060	0.1960	-1.7809	0.0197	-8.2983
0.6	0.9120	-0.3085	0.2867	-1.2604	0.0437	-3.8928
0.8	0.8463	-0.0868	0.3688	-0.9781	0.0758	-2.3586
1.0	0.7652	$+0.0883$	0.4401	-0.7812	0.1149	-1.6507
1.2	0.6711	0.2281	0.4983	-0.6211	0.1593	-1.2633
1.4	0.5669	0.3379	0.5419	-0.4791	0.2074	-1.0224
1.6	0.4554	0.4204	0.5699	-0.3476	0.2570	-0.8549
1.8	0.3400	0.4774	0.5815	-0.2237	0.3061	-0.7259
2.0	0.2239	0.5104	0.5767	-0.1070	0.3528	-0.6174
2.2	0.1104	0.5208	0.5560	$+0.0015$	0.3951	-0.5194
2.4	$+0.0025$	0.5104	0.5202	0.1005	0.4310	-0.4267
2.6	-0.0968	0.4813	0.4708	0.1884	0.4590	-0.3364
2.8	-0.1850	0.4359	0.4097	0.2635	0.4777	-0.2477
3.0	-0.2601	0.3768	0.3391	0.3247	0.4861	-0.1604
3.2	-0.3202	0.3071	0.2613	0.3707	0.4835	-0.0754
3.4	-0.3643	0.2296	0.1792	0.4010	0.4697	$+0.0063$
3.6	-0.3918	0.1477	0.0955	0.4154	0.4448	0.0831
3.8	-0.4026	$+0.0645$	$+0.0128$	0.4141	0.4093	0.1535
4.0	-0.3971	-0.0169	-0.0660	0.3979	0.3641	0.2159
4.2	-0.3766	-0.0938	-0.1386	0.3680	0.3105	0.2690
4.4	-0.3423	-0.1633	-0.2028	0.3260	0.2501	0.3115
4.6	-0.2961	-0.2235	-0.2566	0.2737	0.1846	0.3425
4.8	-0.2404	-0.2723	-0.2985	0.2136	0.1161	0.3613
5.0	-0.1776	-0.3085	-0.3276	0.1479	$+0.0466$	0.3677
5.2	-0.1103	-0.3312	-0.3432	0.0792	-0.0217	0.3617
5.4	-0.0412	-0.3402	-0.3453	$+0.0101$	-0.0867	0.3429
5.6	$+0.0270$	-0.3354	-0.3343	-0.0568	-0.1464	0.3152
5.8	0.0917	-0.3177	-0.3110	-0.1192	-0.1989	0.2766
6.0	0.1507	-0.2882	-0.2767	-0.1750	-0.2429	0.2299
6.2	0.2017	-0.2483	-0.2329	-0.2223	-0.2769	0.1766
6.4	0.2433	-0.2000	-0.1816	-0.2596	-0.3001	0.1188
6.6	0.2740	-0.1452	-0.1250	-0.2858	-0.3119	$+0.0586$
6.8	0.2931	-0.0864	-0.0652	-0.3002	-0.3123	-0.0019
7.0	0.3001	-0.0259	-0.0047	-0.3027	-0.3014	-0.0605
7.2	0.2951	$+0.0339$	$+0.0543$	-0.2934	-0.2800	-0.1154
7.4	0.2786	0.0907	0.1096	-0.2731	-0.2487	-0.1652
7.6	0.2516	0.1424	0.1592	-0.2428	-0.2097	-0.2063
7.8	0.2154	0.1872	0.2014	-0.2039	-0.1638	-0.2395
8.0	0.1716	0.2235	0.2346	-0.1581	-0.1130	-0.2630

TABLE IV
Hyperbolic Bessel functions
$I_m(z) = i^{-m}J_m(iz)$
[see Eq. (5.3.3)]

z	$I_0(z)$	$I_1(z)$	$I_2(z)$
0.0	1.0000	0.0000	0.0000
0.1	1.0025	0.0501	0.0012
0.2	1.0100	0.1005	0.0050
0.4	1.0404	0.2040	0.0203
0.6	1.0921	0.3137	0.0464
0.8	1.1665	0.4329	0.0843
1.0	1.2661	0.5652	0.1358
1.2	1.3937	0.7147	0.2026
1.4	1.5534	0.8861	0.2876
1.6	1.7500	1.0848	0.3940
1.8	1.9895	1.3172	0.5260
2.0	2.2796	1.5906	0.6890
2.2	2.6292	1.9141	0.8891
2.4	3.0492	2.2981	1.1111
2.6	3.5532	2.7554	1.4338
2.8	4.1574	3.3011	1.7994
3.0	4.8808	3.9534	2.2452
3.2	5.7472	4.7343	2.7884
3.4	6.7848	5.6701	3.4495
3.6	8.0278	6.7926	4.2538
3.8	9.5169	8.1405	5.2323
4.0	11.302	9.7594	6.4224
4.2	13.443	11.705	7.8683
4.4	16.010	14.046	9.6259
4.6	19.097	16.863	11.761
4.8	22.794	20.253	14.355
5.0	27.240	24.335	17.505
5.2	32.584	29.254	21.332
5.4	39.010	35.181	25.980
5.6	46.738	42.327	31.621
5.8	56.039	50.945	38.472
6.0	67.235	61.341	46.788
6.2	80.717	73.888	56.882
6.4	96.963	89.025	69.143
6.6	116.54	107.31	84.021
6.8	140.14	129.38	102.08
7.0	168.59	156.04	124.01
7.2	202.92	188.25	150.63
7.4	244.34	227.17	182.94
7.6	294.33	274.22	222.17
7.8	354.68	331.10	269.79
8.0	427.57	399.87	327.60

TABLE V
Phase angles and amplitudes for radiation and scattering from a cylinder
$ka = 2\pi a/\lambda = \omega a/c$
[see Eqs. (7.3.7)]

ka	E_0	γ_0	E_1	γ_1	E_2	γ_2	E_3	γ_3	E_4	γ_4
0.0	∞	0.00°	∞	0.00°	∞	0.00°	∞	0.00°	∞	0.00°
0.1	12.92	0.44	63.06	$-$ 0.45	2546	0.00	—	0.00	—	0.00
0.2	6.651	1.71	15.55	$-$ 1.82	318.2	$-$ 0.01	9565	0.00	—	0.00
0.4	3.583	6.28	3.875	$-$ 6.97	39.71	$-$ 0.14	600.7	0.00	—	0.00
0.6	2.585	12.82	1.844	-13.62	11.72	$-$ 0.69	119.6	$-$ 0.01	1595	0.00
0.8	2.091	20.66	1.199	-18.73	4.922	$-$ 2.09	38.20	$-$ 0.06	382.9	0.00
1.0	1.793	29.39	0.9283	-20.50	2.529	$-$ 4.77	15.81	$-$ 0.20	127.3	0.00
1.2	1.593	38.74	0.7884	-18.94	1.503	$-$ 8.91	7.712	$-$ 0.57	52.03	$-$ 0.02
1.4	1.447	48.52	0.7035	-14.80	1.012	-14.06	4.212	$-$ 1.35	24.54	$-$ 0.06
1.6	1.335	58.62	0.6453	$-$ 8.84	0.7627	-19.03	2.504	$-$ 2.77	12.85	$-$ 0.16
1.8	1.246	68.96	0.6019	$-$ 1.61	0.6309	-22.49	1.596	$-$ 5.08	7.290	$-$ 0.37
2.0	1.173	79.49	0.5676	$+$ 6.52	0.5573	-23.69	1.086	$-$ 8.44	4.405	$-$ 0.79
2.2	1.112	90.15	0.5392	15.31	0.5130	-22.56	0.7898	-12.71	2.801	$-$ 1.55
2.4	1.060	100.93	0.5152	24.57	0.4836	-19.45	0.6158	-17.32	1.861	$-$ 2.80
2.6	1.014	111.81	0.4944	34.20	0.4624	-14.75	0.5136	-21.41	1.287	$-$ 4.73
2.8	0.9743	122.75	0.4760	44.11	0.4457	$-$ 8.84	0.4535	-24.15	0.9265	$-$ 7.46
3.0	0.9389	133.76	0.4597	54.24	0.4319	$-$ 1.99	0.4175	-25.09	0.6965	-11.01
3.2	0.9071	144.82	0.4450	64.55	0.4198	$+$ 5.59	0.3952	-24.19	0.5496	-15.13
3.4	0.8785	155.92	0.4317	75.01	0.4090	13.73	0.3804	-21.64	0.4566	-19.29
3.6	0.8524	167.06	0.4195	85.58	0.3992	2.33	0.3698	-17.71	0.3987	-22.81
3.8	0.8286	178.23	0.4084	96.25	0.3901	31.29	0.3617	-12.66	0.3631	-25.11
4.0	0.8067	189.42	0.3980	107.01	0.3816	40.55	0.3549	$-$ 6.72	0.3412	-25.90
4.2	0.7865	200.64	0.3885	117.83	0.3737	50.06	0.3489	$-$ 0.04	0.3275	-25.14
4.4	0.7678	211.88	0.3796	128.72	0.3662	59.77	0.3434	$+$ 7.22	0.3187	-22.95
4.6	0.7503	223.14	0.3713	139.65	0.3592	69.66	0.3383	14.97	0.3126	-19.54
4.8	0.7341	234.42	0.3635	150.64	0.3525	79.70	0.3334	23.13	0.3081	-15.10
5.0	0.7188	245.71	0.3562	161.66	0.3462	89.87	0.3287	31.62	0.3044	$-$ 9.81

ka	E_5	γ_5	E_6	γ_6	E_7	γ_7	E_8	γ_8	E_9	γ_9
2.0	22.07	$-$ 0.04°	130.8	0.00°	903.5	0.00°	7144	0.00	—	0.00
2.2	12.82	$-$ 0.10	68.99	0.00	432.1	0.00	3099	0.00	—	0.00
2.4	7.834	$-$ 0.22	38.65	$-$ 0.01	221.4	0.00	1452	0.00	—	0.00
2.6	4.999	$-$ 0.45	22.78	$-$ 0.03	120.2	0.00	725.6	0.00	4941	0.00
2.8	3.309	$-$ 0.86	14.03	$-$ 0.06	68.58	0.00	383.4	0.00	2418	0.00
3.0	2.261	$-$ 1.53	8.967	$-$ 0.13	40.86	$-$ 0.01	212.6	0.00	1248	0.00
3.2	1.590	$-$ 2.59	5.922	$-$ 0.25	25.27	$-$ 0.02	122.9	0.00	674.6	0.00
3.4	1.149	$-$ 4.18	4.025	$-$ 0.47	16.16	$-$ 0.04	73.80	0.00	380.0	0.00
3.6	0.8534	$-$ 6.41	2.805	$-$ 0.83	10.65	$-$ 0.07	45.80	0.00	222.1	0.00
3.8	0.6539	$-$ 9.35	2.000	$-$ 1.41	7.200	$-$ 0.14	29.29	$-$ 0.01	134.1	0.00
4.0	0.5190	-12.92	1.456	$-$ 2.30	4.985	$-$ 0.26	19.24	$-$ 0.02	83.43	0.00
4.2	0.4287	-16.83	1.082	$-$ 3.60	3.526	$-$ 0.46	12.95	$-$ 0.04	53.32	0.00
4.4	0.3693	-20.62	0.8211	$-$ 5.42	2.542	$-$ 0.77	8.907	$-$ 0.08	34.92	$-$ 0.01
4.6	0.3312	-23.74	0.6374	$-$ 7.85	1.865	$-$ 1.26	6.252	$-$ 0.14	23.40	$-$ 0.01
4.8	0.3071	-25.76	0.5081	-10.88	1.391	$-$ 2.00	4.471	$-$ 0.25	16.00	$-$ 0.02
5.0	0.2921	-26.44	0.4177	-14.40	1.054	$-$ 3.06	3.251	$-$ 0.42	11.15	$-$ 0.04

TABLE VI
Legendre functions for spherical coordinates [see Eqs. (7.2.4)]

ϑ	$P_{-1} = P_0$	$P_1(\cos\vartheta)$	$P_2(\cos\vartheta)$	$P_3(\cos\vartheta)$	$P_4(\cos\vartheta)$
0°	1.0000	1.0000	1.0000	1.0000	1.0000
5	1.0000	0.9962	0.9886	0.9773	0.9623
10	1.0000	0.9848	0.9548	0.9106	0.8352
15	1.0000	0.9659	0.8995	0.8042	0.6847
20	1.0000	0.9397	0.8245	0.6649	0.4750
25	1.0000	0.9063	0.7321	0.5016	0.2465
30	1.0000	0.8660	0.6250	0.3248	+0.0234
35	1.0000	0.8192	0.5065	+0.1454	−0.1714
40	1.0000	0.7660	0.3802	−0.0252	−0.3190
45	1.0000	0.7071	0.2500	−0.1768	−0.4063
50	1.0000	0.6428	+0.1198	−0.3002	−0.4275
55	1.0000	0.5736	−0.0065	−0.3886	−0.3852
60	1.0000	0.5000	−0.1250	−0.4375	−0.2891
65	1.0000	0.4226	−0.2321	−0.4452	−0.1552
70	1.0000	0.3420	−0.3245	−0.4130	−0.0038
75	1.0000	0.2588	−0.3995	−0.3449	+0.1434
80	1.0000	0.1736	−0.4548	−0.2474	0.2659
85	1.0000	0.0872	−0.4886	−0.1291	0.3468
90	1.0000	0.0000	−0.5000	0.0000	0.3750

ϑ	$P_5(\cos\vartheta)$	$P_6(\cos\vartheta)$	$P_7(\cos\vartheta)$	$P_8(\cos\vartheta)$	$P_9(\cos\vartheta)$
0°	1.0000	1.0000	1.0000	1.0000	1.0000
5	0.9437	0.9216	0.8962	0.8675	0.8358
10	0.7840	0.7045	0.6164	0.5218	+0.4228
15	0.5471	0.3983	+0.2455	+0.0962	−0.0428
20	0.2715	+0.0719	−0.1072	−0.2518	−0.3517
25	+0.0009	−0.2040	−0.3441	−0.4062	−0.3896
30	−0.2233	−0.3740	−0.4102	−0.3388	−0.1896
35	−0.3691	−0.4114	−0.3096	−0.1154	+0.0965
40	−0.4197	−0.3236	−0.1006	+0.1386	0.2900
45	−0.3757	−0.1484	+0.1271	0.2983	0.2855
50	−0.2545	+0.0564	0.2854	0.2947	+0.1041
55	−0.0868	0.2297	0.3191	+0.1422	−0.1296
60	+0.0898	0.3232	0.2231	−0.0763	−0.2679
65	0.2381	0.3138	+0.0422	−0.2411	−0.2300
70	0.3281	0.2089	−0.1485	−0.2780	−0.0476
75	0.3427	+0.0431	−0.2731	−0.1702	+0.1595
80	0.2810	−0.1321	−0.2835	+0.0233	0.2596
85	0.1577	−0.2638	−0.1778	0.2017	0.1913
90	0.0000	−0.3125	0.0000	0.2734	0.0000

TABLE VII

Bessel functions for spherical coordinates

$$j_n(x) = \sqrt{\pi/2x}\, J_{n+\frac{1}{2}}(x), \quad n_n(x) = \sqrt{\pi/2x}\, N_{n+\frac{1}{2}}(x)$$

[see Eqs. (7.2.11)]

x	$j_0(x)$	$n_0(x)$	$j_1(x)$	$n_1(x)$	$j_2(x)$	$n_2(x)$
0.0	1.0000	$-\infty$	0.0000	$-\infty$	0.0000	$-\infty$
0.1	0.9983	-9.9500	0.0333	-100.50	0.0007	-3005.0
0.2	0.9933	-4.9003	0.0664	-25.495	0.0027	-377.52
0.4	0.9735	-2.3027	0.1312	-6.7302	0.0105	-48.174
0.6	0.9411	-1.3756	0.1929	-3.2337	0.0234	-14.793
0.8	0.8967	-0.8709	0.2500	-1.9853	0.0408	-6.5740
1.0	0.8415	-0.5403	0.3012	-1.3818	0.0620	-3.6050
1.2	0.7767	-0.3020	0.3453	-1.0283	0.0865	-2.2689
1.4	0.7039	-0.1214	0.3814	-0.7906	0.1133	-1.5728
1.6	0.6247	$+0.0183$	0.4087	-0.6133	0.1416	-1.1682
1.8	0.5410	0.1262	0.4268	-0.4709	0.1703	-0.9111
2.0	0.4546	0.2081	0.4354	-0.3506	0.1985	-0.7340
2.2	0.3675	0.2675	0.4346	-0.2459	0.2251	-0.6028
2.4	0.2814	0.3072	0.4245	-0.1534	0.2494	-0.4990
2.6	0.1983	0.3296	0.4058	-0.0715	0.2700	-0.4121
2.8	0.1196	0.3365	0.3792	$+0.0005$	0.2867	-0.3359
3.0	$+0.0470$	0.3300	0.3457	0.0630	0.2986	-0.2670
3.2	-0.0182	0.3120	0.3063	0.1157	0.3084	-0.2035
3.4	-0.0752	0.2844	0.2623	0.1588	0.3066	-0.1442
3.6	-0.1229	0.2491	0.2150	0.1921	0.3021	-0.0890
3.8	-0.1610	0.2082	0.1658	0.2158	0.2919	-0.0378
4.0	-0.1892	0.1634	0.1161	0.2300	0.2763	$+0.0091$
4.2	-0.2075	0.1167	0.0673	0.2353	0.2556	0.0514
4.4	-0.2163	0.0699	$+0.0207$	0.2321	0.2304	0.0884
4.6	-0.2160	$+0.0244$	-0.0226	0.2213	0.2013	0.1200
4.8	-0.2075	-0.0182	-0.0615	0.2037	0.1691	0.1456
5.0	-0.1918	-0.0567	-0.0951	0.1804	0.1347	0.1650
5.2	-0.1699	-0.0901	-0.1228	0.1526	0.0991	0.1871
5.4	-0.1431	-0.1175	-0.1440	0.1213	0.0631	0.1850
5.6	-0.1127	-0.1385	-0.1586	0.0880	$+0.0278$	0.1856
5.8	-0.0801	-0.1527	-0.1665	0.0538	-0.0060	0.1805
6.0	-0.0466	-0.1600	-0.1678	$+0.0199$	-0.0373	0.1700
6.2	-0.0134	-0.1607	-0.1629	-0.0124	-0.0654	0.1547
6.4	$+0.0182$	-0.1552	-0.1523	-0.0425	-0.0896	0.1353
6.6	0.0472	-0.1440	-0.1368	-0.0690	-0.1094	0.1126
6.8	0.0727	-0.1278	-0.1172	-0.0915	-0.1243	0.0875
7.0	0.0939	-0.1077	-0.0943	-0.1029	-0.1343	0.0609
7.2	0.1102	-0.0845	-0.0692	-0.1220	-0.1391	0.0337
7.4	0.1215	-0.0593	-0.0429	-0.1294	-0.1388	$+0.0068$
7.6	0.1274	-0.0331	-0.0163	-0.1317	-0.1338	-0.0189
7.8	0.1280	-0.0069	$+0.0095$	-0.1289	-0.1244	-0.0427
8.0	0.1237	$+0.0182$	0.0336	-0.1214	-0.1111	-0.0637

TABLE VIII
Phase angles and amplitudes for radiation and scattering from a sphere
$ka = 2\pi a/\lambda = \omega a/c$
[see Eq. (7.2.15)]

ka	B_0	δ_0	B_1	δ_1	B_2	δ_2	B_3	δ_3	B_4	δ_4
0.0	∞	0.00°	∞	0.00°	∞	0.00°	∞	0.00°	∞	0.00°
0.1	100.5	0.02	2000	−0.01	—	0.00	—	0.00	—	0.00
0.2	25.50	0.15	250.1	−0.08	5637	0.00	—	0.00	—	0.00
0.4	6.731	1.12	31.35	−0.58	354.6	− 0.01	5906	0.00	—	0.00
0.6	3.239	3.41	9.408	−1.82	70.73	− 0.06	785.5	0.00	—	0.00
0.8	2.001	7.18	4.101	−3.80	22.67	− 0.25	188.9	− 0.01	2058	0.00
1.0	1.414	12.30	2.236	−6.14	9.434	− 0.70	62.97	− 0.02	547.8	0.00
1.2	1.085	18.56	1.426	−8.11	4.646	− 1.59	25.82	− 0.08	186.9	0.00
1.4	0.8778	25.75	1.021	−8.97	2.583	− 3.07	12.22	− 0.22	75.74	− 0.01
1.6	0.7370	33.68	0.7931	−8.25	1.584	− 5.19	6.426	− 0.51	34.84	− 0.02
1.8	0.6355	42.19	0.6529	−5.87	1.057	− 7.77	3.667	− 1.05	17.66	− 0.06
2.0	0.5590	51.16	0.5590	−1.97	0.7629	−10.40	2.236	− 1.97	9.669	− 0.15
2.2	0.4993	60.49	0.4918	+3.21	0.5901	−12.49	1.444	− 3.38	5.635	− 0.32
2.4	0.4514	70.13	0.4411	9.44	0.4837	−13.51	0.9823	− 5.34	3.459	− 0.64
2.6	0.4121	80.01	0.4011	16.50	0.4148	−13.14	0.7036	− 7.80	2.220	− 1.18
2.8	0.3792	90.08	0.3686	24.23	0.3676	−11.31	0.5308	−10.54	1.481	− 2.02
3.0	0.3514	100.32	0.3415	32.49	0.3333	− 8.11	0.4214	−13.16	1.024	− 3.27
3.2	0.3274	110.70	0.3184	41.18	0.3071	− 3.73	0.3508	−15.17	0.7334	− 5.00
3.4	0.3066	121.20	0.2985	50.23	0.2862	+ 1.65	0.3042	−16.17	0.5443	− 7.23
3.6	0.2883	131.79	0.2811	59.57	0.2688	7.86	0.2723	−15.94	0.4195	− 9.83
3.8	0.2721	142.47	0.2657	69.15	0.2540	14.75	0.2496	−14.41	0.3364	−12.58
4.0	0.2577	153.22	0.2519	78.92	0.2411	22.20	0.2326	−11.67	0.2807	−15.10
4.2	0.2448	164.03	0.2396	88.88	0.2296	30.12	0.2193	− 7.84	0.2432	−17.00
4.4	0.2331	174.91	0.2285	98.97	0.2194	38.44	0.2084	− 3.08	0.2174	−17.95
4.6	0.2225	185.83	0.2184	109.20	0.2101	47.08	0.1992	+ 2.47	0.1994	−17.79
4.8	0.2128	196.79	0.2091	119.55	0.2016	56.00	0.1912	8.70	0.1863	−16.47
5.0	0.2040	207.79	0.2006	129.98	0.1939	65.16	0.1840	15.48	0.1764	−14.04

ka	B_5	δ_5	B_6	δ_6	B_7	δ_7	B_8	δ_8	B_9	δ_9
2.0	51.31	− 0.01°	323.7	0.00°	2370	0.00°	—	0.00°	—	0.00°
2.2	27.14	− 0.02	155.2	0.00	1030	0.00	7790	0.00	—	0.00
2.4	15.25	− 0.04	79.69	0.00	483.5	0.00	3343	0.00	—	0.00
2.6	9.021	− 0.10	43.38	−0.01	242.2	0.00	1541	0.00	—	0.00
2.8	5.573	− 0.20	24.83	−0.01	128.2	0.00	755.6	0.00	5002	0.00
3.0	3.576	− 0.37	14.83	−0.03	71.28	0.00	390.7	0.00	2407	0.00
3.2	2.371	− 0.68	9.206	−0.06	41.34	0.00	211.7	0.00	1219	0.00
3.4	1.620	− 1.16	5.907	−0.11	24.88	−0.01	119.5	0.00	645.8	0.00
3.6	1.137	− 1.91	3.904	−0.21	15.49	−0.02	70.01	0.00	356.1	0.00
3.8	0.8183	− 2.99	2.649	−0.38	9.933	−0.03	42.39	0.00	203.6	0.00
4.0	0.6043	− 4.48	1.841	−0.66	6.545	−0.06	26.44	0.00	120.2	0.00
4.2	0.4583	− 6.43	1.308	−1.09	4.418	−0.12	16.94	−0.01	73.09	0.00
4.4	0.3577	− 8.79	0.9486	−1.73	3.050	−0.22	11.13	−0.02	45.66	0.00
4.6	0.2881	−11.45	0.7015	−2.65	2.148	−0.37	7.479	−0.04	29.24	0.00
4.8	0.2399	−14.15	0.5290	−3.92	1.542	−0.61	5.130	−0.07	19.16	0.00
5.0	0.2065	−16.56	0.4072	−5.58	1.126	−0.98	3.587	−0.12	12.82	−0.01

TABLE IX
Impedance functions for piston in infinite plane wall
$\theta_0 - i\chi_0 = 1 - (2/w)J_1(w) - iM(w) = \tanh\left[\pi(\alpha_p - i\beta_p)\right]$
$w = 4\pi a/\lambda$
[see Eqs. (7.4.31)]

w	θ_0	χ_0	α_p	β_p
0.0	0.0000	0.0000	0.0000	0.0000
0.5	0.0309	0.2087	0.0094	0.0655
1.0	0.1199	0.3969	0.0330	0.1216
1.5	0.2561	0.5471	0.0628	0.1663
2.0	0.4233	0.6468	0.0939	0.2020
2.5	0.6023	0.6905	0.1247	0.2316
3.0	0.7740	0.6801	0.1552	0.2572
3.5	0.9215	0.6238	0.1858	0.2800
4.0	1.0330	0.5349	0.2175	0.3008
4.5	1.1027	0.4293	0.2517	0.3194
5.0	1.1310	0.3231	0.2899	0.3353
5.5	1.1242	0.2300	0.3344	0.3460
6.0	1.0922	0.1594	0.3868	0.3456
6.5	1.0473	0.1159	0.4450	0.3207
7.0	1.0013	0.0989	0.4788	0.2600
7.5	0.9639	0.1036	0.4594	0.2050
8.0	0.9413	0.1220	0.4241	0.1887
8.5	0.9357	0.1456	0.3980	0.1958
9.0	0.9454	0.1663	0.3839	0.2132
9.5	0.9661	0.1782	0.3799	0.2344
10.0	0.9913	0.1784	0.3845	0.2565
10.5	1.0150	0.1668	0.3964	0.2774
11.0	1.0321	0.1464	0.4153	0.2958
11.5	1.0397	0.1216	0.4410	0.3097
12.0	1.0372	0.0973	0.4734	0.3158
12.5	1.0265	0.0779	0.5101	0.3083
13.0	1.0108	0.0662	0.5421	0.2810
13.5	0.9944	0.0631	0.5490	0.2409
14.0	0.9809	0.0676	0.5316	0.2117
14.5	0.9733	0.0770	0.5073	0.2032
15.0	0.9727	0.0881	0.4877	0.2092
15.5	0.9784	0.0973	0.4758	0.2231
16.0	0.9887	0.1021	0.4718	0.2406
16.5	1.0007	0.1013	0.4750	0.2591
17.0	1.0115	0.0948	0.4852	0.2767
17.5	1.0187	0.0843	0.5017	0.2914
18.0	1.0209	0.0719	0.5247	0.3007
18.5	1.0180	0.0602	0.5522	0.3010
19.0	1.0111	0.0515	0.5798	0.2879
19.5	1.0021	0.0470	0.5968	0.2610
20.0	0.9933	0.0473	0.5940	0.2314

TABLE X
Zeros and associated values of Bessel functions
$J_m(j_{mn}) = 0$; $j_{mn} = \pi\beta_{mn}$; $J_m'(j_{mn}') = 0$; $j_{mn}' = \pi\alpha_{mn}$

n	j_{0n}	$J_0'(j_{0n})$	j_{1n}	$J_1'(j_{1n})$	j_{2n}	$J_2'(j_{2n})$	j_{3n}	$J_3'(j_{3n})$
0			0	0.5000	0	0	0	0
1	2.4048	−0.5191	3.8317	−0.4028	5.1356	−0.3397	6.3802	−0.2983
2	5.5201	+0.3403	7.0156	+0.3001	8.4172	+0.2714	9.7610	+0.2494
3	8.6537	−0.2715	10.1735	−0.2497	11.6198	−0.2324	13.0152	−0.2183
4	11.7915	+0.2325	13.3237	+0.2184	14.7960	+0.2065	16.2235	+0.1964

$$j_{mn} \to \pi(n + \tfrac{1}{2}m - \tfrac{1}{4}) - \frac{4m^2 - 1}{8\pi(n + \tfrac{1}{2}m - \tfrac{1}{4})} \qquad n \gg m$$

n	j_{0n}'	$J_0(j_{0n}')$	J_{1n}'	$J_1(j_{1n}')$	j_{2n}'	$J_2(j_{2n}')$	j_{3n}'	$J_3(j_{3n}')$
0	0	1.0000	1.8412	+0.5819	3.0542	+0.4865	4.2012	+0.4344
1	3.8317	−0.4028	5.3314	−0.3461	6.7061	−0.3135	8.0152	−0.2912
2	7.0156	+0.3001	8.5363	+0.2733	9.9695	+0.2547	11.3459	+0.2407
3	10.1735	−0.2497	11.7060	−0.2333	13.1704	−0.2209	14.5859	−0.2110
4	13.3237	+0.2184	14.8636	+0.2070	16.3475	+0.1979	17.7888	+0.1904

$$j_{mn}' \to \pi(n + \tfrac{1}{2}m - \tfrac{3}{4}) - \frac{4m^2 + 3}{8\pi(n + \tfrac{1}{2}m - \tfrac{3}{4})} \qquad n \gg m$$

TABLE XI
Angle-distribution functions for transmission through a membrane
values of $F_n{}^m(x)$
[see Eq. (10.2.14)]

x	$F_1{}^0(x)$	$F_2{}^0(x)$	$F_3{}^0(x)$	$F_1{}^1(x)$	$F_2{}^1(x)$	$F_3{}^1(x)$	$F_1{}^2(x)$	$F_2{}^2(x)$	$F_3{}^2(x)$
0.0	1.6020	−1.0648	0.8514	0.0000	0.0000	0.0000	0.0000	0.0000	0.0000
0.5	1.5714	−1.0076	0.8017	+0.3194	−0.2313	+0.1912	+0.0354	−0.0269	+0.0227
1.0	1.4821	−0.8424	0.6603	0.6120	−0.4267	0.3498	0.1369	−0.1020	0.0857
1.5	1.3421	−0.5884	0.4493	0.8539	−0.5554	0.4490	0.2909	−0.2099	0.1747
2.0	1.1633	−0.2744	+0.2014	1.0273	−0.5963	0.4723	0.4768	−0.3274	0.2692
2.5	0.9600	+0.0648	−0.0449	1.1217	−0.5409	0.4165	0.6702	−0.4283	0.3463
3.0	0.7489	0.3930	−0.2517	1.1354	−0.3941	0.2924	0.8460	−0.4875	0.3857
3.5	0.5446	0.6769	−0.3869	1.0749	−0.1737	+0.1227	0.9819	−0.4855	0.3735
4.0	0.3601	0.8904	−0.4300	0.9533	+0.0929	−0.0615	1.0614	−0.4118	0.3059
4.5	0.2053	1.0175	−0.3741	0.7896	0.3729	−0.2262	1.0756	−0.2671	0.1898
5.0	0.0856	1.0532	−0.2270	0.6041	0.6324	−0.3400	1.0254	−0.0630	+0.0423
5.5	+0.0026	1.0025	−0.0098	0.4173	0.8415	−0.3798	0.9154	+0.1793	−0.1120
6.0	−0.0462	0.8841	+0.2470	0.2649	0.9786	−0.3340	0.7630	0.4323	−0.2452
6.5	−0.0661	0.7165	0.5081	0.1062	1.0321	−0.2047	0.5856	0.6668	−0.3313
7.0	−0.0643	0.5255	0.7390	+0.0026	1.0015	−0.0070	0.4028	0.8558	−0.3503
7.5	−0.0489	0.3352	0.9112	−0.0619	0.8993	+0.2332	0.2331	0.9784	−0.2923
8.0	−0.0273	0.1661	1.0053	−0.0905	0.7421	0.0840	+0.0908	1.0233	−0.1591
8.5	−0.0058	+0.0326	1.0142	−0.0903	0.5544	0.7122	−0.0147	0.9887	+0.0356
9.0	+0.0111	−0.0580	0.9421	−0.0704	0.3609	0.8884	−0.0802	0.8852	0.2681
9.5	0.0213	−0.1053	0.8048	−0.0406	0.1837	0.9917	−0.1079	0.7286	0.5089
10.0	0.0242	−0.1148	0.6244	−0.0097	0.0400	1.0121	−0.1046	0.5419	0.7270

values of $\theta_n^m(x)$ and of $\chi_n^m(x)$ [see Eq. (10.2.14)]

x	θ_{11}^m	χ_{11}^m	θ_{22}^m	χ_{22}^m	θ_{33}^m	χ_{33}^m	θ_{12}^m	χ_{12}^m	θ_{13}^m	χ_{13}^m	θ_{23}^m	χ_{23}^m
						$m = 0$						
0	$2.566x$	5.541	$1.134x$	4.516	$0.724x$	3.704	$-1.706x$	−1.803	$1.363x$	1.481	$-0.906x$	−1.170
1	2.310	4.699	0.847	4.021	0.527	3.582	−1.399	−1.008	1.105	0.838	−0.668	−0.723
2	3.421	2.805	0.645	3.555	0.389	3.456	−1.409	+0.464	1.082	+0.283	−0.524	−0.231
3	3.204	1.099	0.526	4.301	0.285	3.700	−0.360	1.108	+0.323	−0.697	−0.387	−0.444
4	2.398	0.233	1.822	5.138	0.595	3.892	+0.477	0.664	−0.148	−0.384	−1.024	−0.440
5	1.722	0.018	3.867	4.451	0.578	3.986	0.501	+0.014	−0.098	−0.113	−1.385	+0.564
6	1.340	0.017	4.758	2.428	0.461	4.646	+0.166	−0.196	−0.026	−0.137	−0.490	1.499
7	1.116	0.014	4.147	0.711	1.625	5.910	−0.013	−0.098	−0.127	−0.138	+0.749	1.155
8	0.978	0.004	3.053	0.088	4.088	5.796	−0.014	−0.018	−0.173	−0.033	1.001	+0.101
9	0.845	0.001	2.401	0.060	5.770	3.647	−0.001	−0.013	−0.082	+0.091	0.428	−0.420
						$m = 1$						
0	$0.280x^3$	4.072	$0.151x^3$	3.779	$0.103x^3$	3.582	$-0.205x^3$	−0.710	$0.170x^3$	0.615	$-0.125x^3$	−0.545
1	0.246	4.494	0.118	3.932	0.078	3.712	−0.170	−0.925	0.140	0.791	−0.097	−0.658
2	1.559	4.991	0.588	3.976	0.374	3.731	−0.953	−0.885	0.767	0.748	−0.471	−0.593
3	3.308	4.255	0.712	3.700	0.430	3.516	−1.505	+0.083	1.157	+0.000	−0.553	−0.222
4	4.144	2.468	0.434	4.050	0.261	3.638	−0.893	1.173	+0.687	−0.742	−0.338	−0.214
5	3.717	0.848	1.047	5.202	0.413	3.948	+0.239	1.214	−0.036	−0.700	−0.654	−0.480
6	2.810	0.137	3.087	5.562	0.630	4.007	0.759	+0.386	−0.218	−0.267	−1.352	+0.046
7	2.131	0.026	4.989	4.054	0.442	4.276	0.512	−0.242	−0.064	−0.149	−1.107	1.204
8	1.753	0.039	5.197	1.752	0.841	5.613	+0.042	−0.261	−0.108	−0.235	+0.244	1.649
9	1.539	0.020	4.146	0.358	2.932	6.555	−0.032	−0.079	−0.271	−0.126	1.194	0.723
						$m = 2$						
0	$0.010x^5$	3.918	$0.006x^5$	3.633	$0.006x^5$	3.451	$-0.009x^5$	−0.465	$0.007x^5$	0.417	$-0.005x^5$	−0.378
1	0.009	4.108	0.005	3.696	0.004	3.585	−0.007	−0.533	0.006	0.472	−0.004	−0.416
2	0.257	4.752	0.116	3.939	0.090	3.702	−0.179	−0.771	0.148	0.664	−0.105	−0.538
3	1.316	5.405	0.492	4.000	0.315	3.794	−0.803	−0.845	0.647	0.702	−0.393	−0.536
4	3.144	5.204	0.720	3.902	0.429	3.687	−1.481	−0.147	1.137	+0.160	−0.554	−0.250
5	4.607	3.664	0.491	4.047	0.289	3.663	−1.254	1.046	0.935	−0.650	−0.374	−0.117
6	4.683	1.677	0.667	5.153	0.315	4.069	−0.061	1.536	+0.171	−0.901	−0.457	−0.393
7	3.755	0.398	2.348	6.137	0.589	4.284	+0.917	0.852	−0.273	−0.486	−1.148	−0.263
8	2.840	0.051	4.740	5.374	0.539	4.358	0.861	−0.100	−0.158	−0.175	−1.426	+0.848
9	2.321	0.058	5.905	3.020	0.497	5.375	0.279	−0.403	−0.079	−0.253	−0.369	1.807

Entries for $x = 0$ give limiting values for $x \to 0$.

TABLE XIII
Transmission factor for transmission through a membrane
values of $H_m(uvw) = H_m(vuw) = H_m(wvu)$
[see Eq. (10.2.22)]

uvw	$H_m(uvw)$	uvw	$H_m(uvw)$	uvw	$H_m(uvw)$	uvw	$H_m(uvw)$
			$m = 0$				
000	0.062500	044	0.004531	228	−0.000534	448	−0.000162
002	0.044104	046	0.001040	244	0.002575	466	−0.001249
004	0.011379	048	−0.000184	246	0.000088	468	−0.000267
006	−0.003373	066	0.001570	248	−0.000150	488	−0.000297
008	−0.000886	068	0.000440	266	0.000427	666	−0.001224
022	0.031701	088	0.000582	268	0.000124	668	−0.000195
024	0.009236	222	0.022937	288	0.000147	688	0.000103
026	−0.001680	224	0.006954	444	−0.000625	888	0.000582
028	−0.000715	226	−0.001020	446	−0.001290		
			$m = 1$				
000	$0.001302uvw$	044	$0.002516u$	228	−0.000606	448	−0.000356
002	$0.002015uv$	046	$0.000728u$	244	0.003986	466	0.000076
004	$0.001680uv$	048	$-0.000228u$	246	0.001087	468	−0.000267
006	$0.000199uv$	066	$0.000630u$	248	−0.000345	488	−0.000023
008	$-0.000284vu$	068	$0.000211u$	266	0.000845	666	−0.000857
022	$0.003133u$	088	$0.000317u$	268	0.000282	668	−0.000683
024	$0.001965u$	222	0.004415	288	0.000403	688	−0.000545
026	$-0.000710u$	224	0.004249	444	0.003693	888	−0.000348
028	$0.000116u$	226	0.000733	446	0.000948		
			$m = 2$				
000	$0.0000068u^2v^2w^2$	044	$0.000324u^2$	228	−0.000030	448	0.000086
002	$0.0000221u^2v^2$	046	$0.000661u^2$	244	0.001069	466	0.001232
004	$0.0000458u^2v^2$	048	$-0.000023u^2$	246	0.000694	468	0.000253
006	$0.0000259u^2v^2$	066	$0.000174u^2$	248	0.000000	488	0.000165
008	$-0.0000042u^2v^2$	068	$0.000264u^2$	266	0.000573	666	0.000794
022	$0.000072u^2$	088	$0.000043u^2$	268	0.000153	668	−0.000016
024	$0.001154u^2$	222	0.000237	288	0.000128	688	−0.000133
026	$-0.000210u^2$	224	0.000497	444	0.002334	888	−0.000422
028	$-0.000089u^2$	224	0.000295	446	0.001586		

Entry for 024, for example, gives limiting value for u small. Second-order interpolation of $(1 + u^2 + v^2 + w^2)H_0$, of $(1 + u^3 + v^3 + w^3)H_1/uvw$, or of $(1 + u^4 + v^4 + w^4)H_2/u^2v^2w^2$ will produce fairly accurate intermediate values.

TABLE XIV

A. Density, sound velocity, viscosity, and thermal properties of a few gases and liquids, at 0°C and I atm, in cgs units

MATERIAL	ρ_0	c_0	μ	γ	C_p	K
Air	0.00129	33,100	171×10^{-6}	1.40	0.240	0.055×10^{-3}
Argon	0.00178	31,900	210×10^{-6}	1.67	0.125	0.039×10^{-3}
CO_2	0.00198	25,800	138×10^{-6}	1.30	0.199	0.034×10^{-3}
Helium	0.00018	97,000	186×10^{-6}	1.66	1.250	0.344×10^{-3}
H_2	0.00009	127,000	84×10^{-6}	1.41	3.389	0.416×10^{-3}
Neon	0.00090	43,500	311×10^{-6}	1.64		0.110×10^{-3}
O_2	0.00143	31,700	189×10^{-6}	1.40	0.218	0.057×10^{-3}

	(4°C)					
Pure water	1.000	143,000	179×10^{-4}		1.009	1.20×10^{-3}
Sea water	1.023	144,700				

At pressure of P atm, temperature $T°K$, for the gases, $\rho = \rho_0 P(273°/T)$ and $c = c_0 \sqrt{T/273°}$ for moderate ranges of T and P. For sea water (32 g salt per 1,000 g water) $c = c_0 + 317(T - 273°)$

B. Density, Young's modulus Q, and values of $\gamma^2 h/\omega = [3\rho(1 - s^2)/Q]^{\frac{1}{4}}$, where γ is the propagational parameter for a plate of Eq. (5.3.2)

MATERIAL	ρ	Q	$\gamma^2 h/\omega$
Aluminum	2.70	7×10^{11}	3.21×10^{-6}
Brass	8.44	9×10^{11}	5.00×10^{-6}
Copper	8.89	12×10^{11}	4.45×10^{-6}
Iron	7.2	9×10^{11}	4.62×10^{-6}
Nickel	8.7	20×10^{11}	3.41×10^{-6}
Silver	10.5	8×10^{11}	5.92×10^{-6}
Steel	7.8	20×10^{11}	3.22×10^{-6}

C. Comparison of mks and cgs units

MKS UNIT		CGS UNIT
1 meter	$= 100$	centimeters
1 kilogram	$= 1,000$	grams
1 newton	$= 10^5$	dynes
1 joule	$= 10^7$	ergs
1 kg/m^3	$= 10^{-3}$	g/cm^3
1 newton/m^2	$= 10$	dynes/cm^2
1 kg/(m^2)(sec)	$= 1/10$	g/(cm^2)(sec) (ρc)
1 amp/m^2	$= 3 \times 10^5$	statamp/cm^2
1 volt	$= 1/300$	statvolt
1 weber/m^2	$= 10^4$	B in emu
1 newton-sec/m^2	$= 10$	poises (viscosity)
1 joule/(sec)(m)(deg)	$\simeq 1/420$	cal/(sec)(cm)(deg) (thermal conductivity)
1 joule/(mole)(deg)	$\simeq 1/4200M$	cal/g (deg) (specific heat) (M = molecular weight)
1 atm	$\simeq 10^5$ newtons/m$^2 = 10^6$ dynes/cm^2	

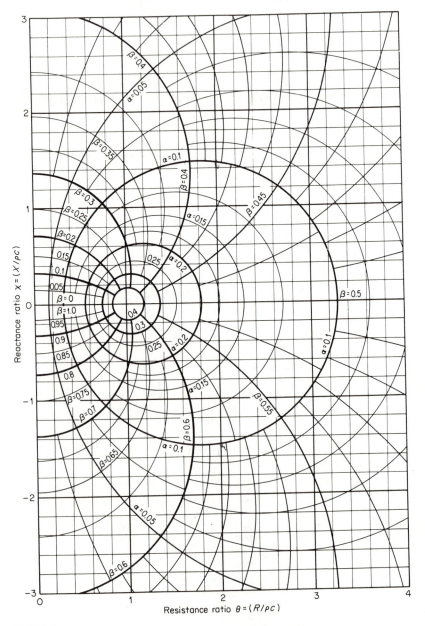

PLATE I
Conformal transformation from specific acoustic impedance
$\zeta = \theta - i\chi = z/\rho c = \tanh [\pi(\alpha - i\beta)]$ of duct end, to wave-
reflection parameter $(\Phi/\pi) + \frac{1}{2}i = \alpha - i\beta$. See Eqs. (9.1.4)
and (9.3.1).

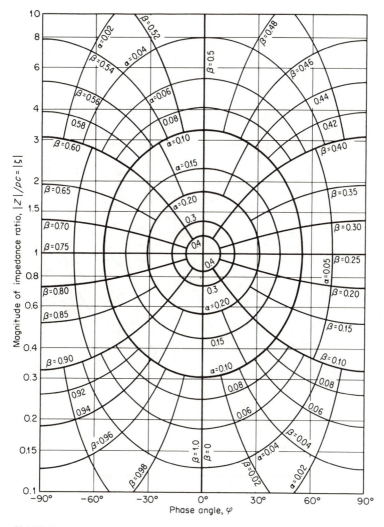

Magnitude of impedance ratio, $|Z|/\rho c = |\zeta|$

Phase angle, φ

PLATE II
Conformal transformation from impedance magnitude and phase parameter $\ln \zeta = \ln |\zeta| - i\varphi$ to wave-reflection parameter $(\Phi/\pi) + \frac{1}{2}i = \alpha - i\beta$. See Eqs. (9.1.4) and (9.3.1).

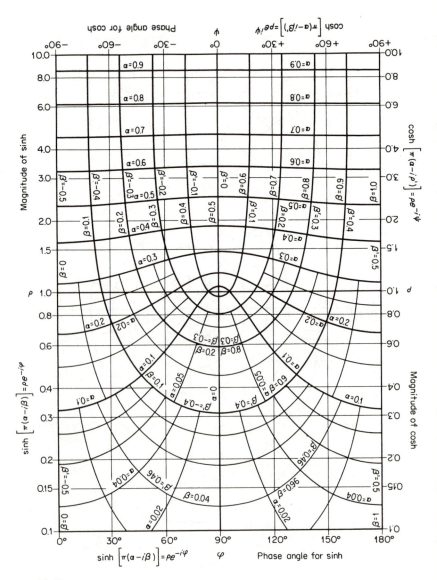

PLATE III

Conformal transformation from $\ln \begin{bmatrix} \sinh \\ \cosh \end{bmatrix} \pi(\alpha - i\beta) = \ln \rho - i\varphi$ to argument $\alpha - i\beta$. See Eq. (9.1.5).

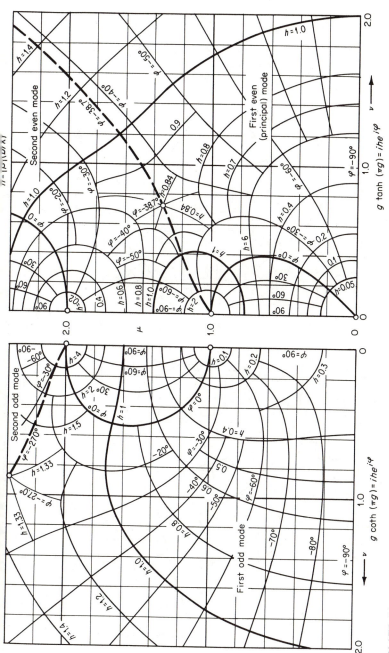

PLATE IV

Conformal transformation between admittance magnitude-and-phase parameter $\ln(\beta b/\lambda)$ and wavenumber parameter $iq_m = 2g = v + i\mu$, for a rectangular duct of distance b between two walls, both having specific acoustic admittance $\beta = \rho c/z = |\beta| e^{i\varphi}$. If only one wall has admittance β, the other being rigid, use the right-hand chart only, and set $h = |2\beta b/\lambda|$; $q_m = \mu - iv$ then has half the values given on the chart. See Eqs. (9.2.19) and (9.2.20).

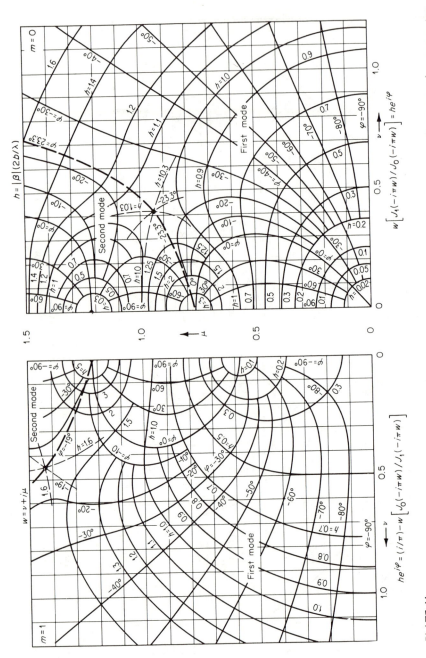

PLATE V
Conformal transformation between admittance magnitude-and-phase parameter $\ln{(\beta b/\lambda)} = \ln h + i\varphi$ and wavenumber parameter $iq_m = w = v + i\mu$, for a circular duct of radius b, the wall of the duct having specific acoustic admittance $\beta = \rho c/z = \beta e^{i\varphi}$. For modes with angle factor $\frac{\cos}{\sin} \phi$ ($m = 1$), use the left-hand chart; for modes symmetric about the duct axis ($m = 0$), use the right-hand chart.

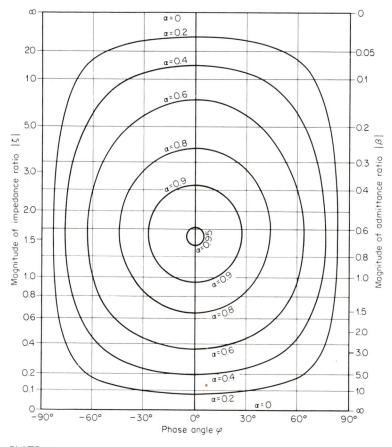

PLATE VI
Contour plot for mean absorption coefficient α for various
values of magnitude and phase angle of specific acoustic
impedance $z/\rho c = |\zeta|\,e^{-i\varphi} = 1/\beta$ of wall. Valid only when
geometrical acoustics is applicable. See Eq. (9.5.8).